Spon's Architects' and Builders' Price Book

Spon's Architects' and Builders' Price Book

Edited by
DAVIS LANGDON & EVEREST
Chartered Quantity Surveyors

1991

One hundred and sixteenth Edition

E. & F.N. SPON
An imprint of Chapman and Hall
London · New York · Tokyo · Melbourne · Madras

UK	Chapman and Hall, 2–6 Boundary Row, London SE1 8HN
USA	Van Nostrand Reinhold, 115 5th Avenue, New York NY10003
Japan	Chapman and Hall Japan, Thomson Publishing Japan, Hirakawacho Nemoto Building, 7F, 1–7–11 Hirakawa-cho, Chiyoda-ku, Tokyo 102
Australia	Chapman and Hall Australia, Thomas Nelson Australia, 480 La Trobe Street, PO Box 4725, Melbourne 3000
India	Chapman and Hall, India, R. Seshadri, 32 Second Main Road, CIT East, Madras 600 035

First published 1873
One hundered and sixteenth edition 1990

Printed in Great Britain by
Richard Clay Ltd, Bungay, Suffolk

ISBN 0 419 16790 0 0 442 31297 0 (USA)
ISSN 0306 3046

British Library Cataloguing in Publication Data

Spon's Architects' and builders' price book.
–116th ed. (1991)
I. Buildings – Estimates – Great Britain – Periodicals
692′.5 TH435
ISBN 0–419–16790–0

Preface

A year ago high interest rates seemed likely simply to dampen demand at a time when tender prices were moving well ahead of general inflation, but 12 months on they are now acting as a real disincentive to investment funds that are more profitably held in bank deposits.

The lack of funds in the market place has heightened fears amongst the trader developers of takeovers and even bankruptcy for those that are highly geared. The residential market is still some way from recovery and poor performance in the high street by many of the major retailers leaves the construction industry in a state of uncertainty as to the source of future work load.

In the first quarter of 1990 tender prices as measured by the Davis Langdon & Everest Tender Price Index actually fell by 4.5% and there is little prospect that there will be a significant rise before the second half of 1991. But bad news for Contractors is good news for Clients; with general inflation continuing to rise, 1990 and the early part of 1991 will be a good time to go out to tender.

The reduction in tender prices has occurred as a result of keener sub-Contractors' quotations, plus lower overheads and profit additions. Furthermore site establishment costs have been cut.

Against this background the Tender Price Index for the A & B has been reduced to 330. The key adjustments in setting this price level are as follows:

(i) A reduction in the level of bonus to site operatives from 55% to 25%.

(ii) A reduction in overheads and profit from 9% in the Major Works to 5% and from 12.5% in the Minor Works to 7.5%.

(iii) A reduction in Preliminaries from 14% to 12.5%.

The net effect on the price level for the items in the Measured Work section is a reduction of 10% relative to that in the 1990 edition. The current edition includes the annual NJC wage award agreed on 25 June 1990 and the latest material prices.

The Tender Price Index on the front cover reflects our view as to the price level for competitive tenders let in outer London in the fourth quarter of 1990. Adjustments to other locations can be made by references to the Regional Variation factors shown in Part IV - Approximate Estimating.

A major advantage of the A & B is that, unlike other Price Books it is geared to the market conditions that affect building tender prices. It is possible for the reader to make such adjustments as may be necessary during the year to reflect trends resulting from changing workload. This is achieved by comparing the Tender Price Index for the current A & B with subsequent revisions to that Index as circulated in Spons Price Book Update. The Update is circulated free every 3 months to those readers who have registered with the publishers; the colour card bound in with this volume should be completed and returned in order to register.

In 1988, the Measured Works Sections of Spons A & B 1989 were restructured in accordance with SMM7, which became operative in July 1988. However, the use of SMM7 has not become as widespread as might have been anticipated, so we have retained a number of SMM6 labours/items to help estimators build up additional rates or to break composite SMM7 rates back to their SMM6 equivalents. We have also retained our SMM6/SMM7 index to assist readers during the transition period.

Additional items in this year's book include:

- Access floors
- Concrete shot blasting
- Fire-packed expansion joints
- Granite claddings
- 'Hardrow' roof slates
- 'Metsec' lattice beams
- Protective boards
- Sand lime bricks
- Shot-firing
- 'Tico' anti-vibration pads

and during the production of this edition the Authors have particularly revised and expanded items for:

Asphalt roads, Cement and sand renderings, Expansion joints, Floor tiles and sheeting, Flue linings, Fire-resisting doors, Fire-resisting glass, Formwork and reinforcement items, Insulation, Painting, Plywood linings, Rectangular block pavings, Rooflights, Steel cladding, Vapour barriers and Wall tiles.

The reader's attention is drawn to the fact that since 1989 all non-domestic building, including refurbishment and alterations, is now subject to VAT, currently at 15%.

Prices included within this edition do not include Value Added Tax, which must be added if appropriate.

For the benefit of new readers, a brief guide to the book follows:

Part I: Fees and Daywork
This section contains Fees for Professional Services, Daywork and Prime Cost.

Part II: Rates of Wages
This section includes authorized wage agreements currently applicable to the Building and Associated Industries.

Part III: Prices for Measured Work
This section contains Prices for Measured Work - Major Works, and Prices for Measured Work - Minor Works (on coloured paper).

Part IV: Approximate Estimating
This section contains the Building Cost and Tender Price Index, information on regional price variations, prices per square metre for various types of buildings, approximate estimates, cost limits and allowances and a procedure for valuing property for insurance purposes.

Part V: Tables and Memoranda
This section contains general formulae, weights and quantities of materials, other design criteria and useful memoranda associated with each trade and a list of useful Trade Associations.

While every effort is made to ensure the accuracy of the information given in this publication, neither the Editors nor Publishers in any way accept liability for loss of any kind resulting from the use made by any person of such information.

DAVIS LANGDON & EVEREST
Chartered Quantity Surveyors
Princes House
39 Kingsway
London WC2B 6TP

Contents

PART IV
APPROXIMATE ESTIMATING

PART V
TABLES AND MEMORANDA

SMM6/SMM7 Index

For the benefit of readers, during a year of transition, we have included a brief index of comparative trade/work sections between SMM6 and SMM7, used in producing this years edition of 'Spons'.

	SMM6		SMM7
A	GENERAL RULES		
B	PRELIMINARIES	A	PRELIMINARIES/GENERAL CONDITIONS
	Preliminary Particulars	A10	Project particulars
		A11	Drawings
		A12	The site/Existing buildings
		A13	Description of the work
	Contract	A20	The Contract/Sub-contract
		A30	Employer's requirements: Tendering/Sub-letting/Supply
		A31	Employer's requirements: Provision, content and use of documents
		A32	Employer's requirements: Management of the Works
		A33	Employer's requirements: Quality standards/control
		A34	Employer's requirements: Security/Safety/Protection
		A35	Employer's requirements: Specific limitations on method/ sequence/timing
		A36	Employer's requirements: Facilities/Temporary works/Services
		A37	Employer's requirements: Operation/Maintenance of the finished building
		A50	Work/Materials by the Employer
	Works by Nominated Sub Contractors	A51	Nominated sub-contractors
	Goods and Materials from Nominated	A52	Nominated suppliers
	Suppliers and works by Public Bodies	A53	Work by statutory authorities
	General facilities and obligations	A40	Contractor's general cost items: Management and staff
		A41	Contractor's general cost items: Site accommodation
		A42	Contractor's general cost items: Services and facilities

	SMM6		SMM7
B	PRELIMINARIES - cont'd		
		A43	Contractor's general cost items: Mechanical plant
		A44	Contractor's general cost items: Temporary works
	Contingencies	A54	Provisional work
		A55	Dayworks
C	DEMOLITION	C	DEMOLITION/ALTERATION/RENOVATION
	Generally	C10	Demolishing structures
		C20	Alterations - spot items
		C30	Shoring
		C40	Repairing/Renovating concrete/brick/block/stone
		C41	Chemical dpcs to existing walls
		C50	Repairing/Renovating metal
		C51	Repairing/Renovating timber
		C52	Fungus/Beetle eradication
	Protection	A42	Contractor's services and facilities
D	EXCAVATION AND EARTHWORK	D	GROUNDWORK
	Site preparation	D10	Ground investigation
		D11	Soil stabilization
		D12	Site dewatering
	Excavation	D20	Excavation and filling
	for services etc.	P30	Trenches/Pipeways/Pits for buried engineering services
	Earthwork support		
	Disposal of water		
	Disposal of excavated material		
	Filling		
	Surface treatments		
	Protection	A42	Contractor's services and facilities
E	PILING AND DIAPHRAGM WALLING		
	Piling	D30	Cast in place concrete piling
		D31	Preformed concrete piling
		D32	Steel piling
	Diaphragm walls	D40	Diaphragm walling
	Protection	A42	Contractor's services and facilities
F	CONCRETE WORK	E	IN SITU CONCRETE/LARGE PRECAST CONCRETE
	In situ concrete	E10	In situ concrete
		E11	Gun applied concrete

	SMM6		SMM7
	Labours/sundries	E40	Designed joints in in situ concrete
		E41	Worked finishes/Cutting to in situ concrete
		E42	Accessories cast into in situ concrete
		P22	Sealant joints
		J31	Liquid applied waterproof roof coatings
		J32	Sprayed vapour barriers
		J33	In situ glass reinforced plastics
	damp proof membranes	J30	Liquid applied tanking/damp proof membranes
		J40	Flexible sheet tanking/damp proof membranes
	surface sealers	M60	Painting/Clear finishing
	holes and chases for services	P31	Holes/Chases/Covers/Supports for services
	Reinforcement	E30	Reinforcement for in situ concrete
		E31	Post tensioned reinforcement for in situ concrete
	Formwork	E20	Formwork for in situ concrete
	Precast concrete		
	small units	F31	Precast concrete sills/lintels/copings/features
		H14	Concrete rooflights/pavement lights
	large units	E50	Precast concrete large units
		H40	Glass reinforced cement cladding/features
		H50	Precast concrete slab cladding/features
	Composite construction	E60	Precast/Composite concrete decking
	Hollow block suspended construction		-
	Prestressed concrete work		-
	Contractor-designed construction		-
	Protection	A42	Contractor's services and facilities
G	BRICKWORK AND BLOCKWORK	F	MASONRY
	Brickwork	F10	Brick/Block walling
	Brick facework	F10	Brick/Block walling

	SMM6	SMM7
G	BRICKWORK AND BLOCKWORK - cont'd	
	Brickwork in connection with boilers	-
	Blockwork	F10 Brick/Block walling
	Glass blockwork	F11 Glass block walling
	Damp-proof courses	F30 Accessories/Sundry items for brick/block/stone walling
	Sundries	
	generally	F30 Accessories/Sundry items for brick/block/stone walling
	bedding and pointing frames	L10-12 Windows L20-22 Doors etc.
		F30 Accessories/Sundry items for brick/block/stone walling
	cavity insulation	P11 Foamed/Fibre/Bead cavity wall insulation
	holes	- deemed included - except P31 Holes/Chases/Covers/Supports for services
	centering	- deemed included
	Protection	A42 Contractor's services and facilities
H	UNDERPINNING	
	Work in all trades	D50 Underpinning
	Protection	A42 Contractor's services and facilities
J	RUBBLE WALLING	
	Stone rubble work	F20 Natural stone rubble walling
	Sundries	F30 Accessories/Sundry items
	Centering	F20 Natural stone rubble walling
	Protection	A42 Contractor's services and facilities
K	MASONRY	
	Natural stonework	F21 Natural stone/ashlar walling/dressing
		H51 Natural stone slab cladding/features
	Cast stonework	F22 Cast stone walling/dressings
		H52 Cast stone slab cladding/features
	Clayware work	
	Sundries	F30 Accessories/Sundry items
	holes and chases for services	P31 Holes/Chases/Covers/Supports for services

	SMM6		SMM7
	other holes, cramps etc.	F21	Natural stone/ashlar walling/dressings
		F22	Cast stone walling/dressings
	Centering	F21	Natural stone/ashlar walling/dressings
		F22	Cast stone walling/dressings
	Protection	A42	Contractor's services and facilities
L	ASPHALT WORK	J	WATERPROOFING
	Mastic asphalt	J20	Mastic asphalt tanking/damp proof membranes
		J21	Mastic asphalt roofing/insulation/finishes
		M11	Mastic asphalt flooring
	Asphalt tiling		-
	Protection	A42	Contractor's services and facilities
M	ROOFING	H	CLADDING/COVERING
	Slate or tile roofing	H60	Clay/Concrete roof tiling
		H61	Fibre cement slating
		H62	Natural slating
		H63	Reconstructed stone slating/tiling
		H64	Timber shingling
		H76	Fibre bitumen thermoplastic sheet coverings/flashings
	Corrugated or troughed sheet roofing or cladding	H30	Fibre cement profiled sheet cladding/covering/siding
		H31	Metal profiled/flat sheet cladding/covering/siding
		H32	Plastics profiled sheet cladding/covering/siding
		H33	Bitumen and fibre profiled sheet cladding/covering
		H41	Glass reinforced plastics cladding/features
		K12	Under purlin/Inside rail panel linings
	Roof decking	G30	Metal profiled sheet decking
		G31	Prefabricated timber unit decking
		G32	Edge supported/Reinforced woodwool slab decking
		J22	Proprietary roof decking with asphalt finish
		J43	Proprietary roof decking with felt finish
		K11	Rigid sheetflooring/sheathing/linings casings

	SMM6		SMM7
M	ROOFING - cont'd		
	Bitumen-felt roofing	J41	Built up felt roof coverings
		J42	Single layer plastics roof coverings
	Sheet metal roofing	H70	Malleable metal sheet prebonded coverings/cladding
	Sheet metal flashings and gutters	H71	Lead sheet coverings/flashings
		H72	Aluminium sheet coverings/flashings
		H73	Copper sheet coverings/flashings
		H74	Zinc sheet coverings/flashings
		H75	Stainless steel sheet coverings/ flashings
	Protection	A42	Contractor's services and facilities
N	WOODWORK	G	STRUCTURAL/CARCASSING METAL/TIMBER
	Carcassing	G20	Carpentry/Timber framing/First fixing
	First fixings boardings, grounds, framework etc.	G20	Carpentry/Timber framing/First fixing
	external weatherboarding	H21	Timber weatherboarding
	flooring	K20	Timber board flooring/sheathing/ linings/casings
		K21	Timber narrow strip flooring/linings
	Second fixings skirtings, architraves etc.	P20	Unframed isolated trims/skirtings/ sundry items
	sheet linings and casings	K11	Rigid sheet flooring/sheathing/ linings/casings
		K13	Rigid sheet fine linings/panelling
		L42	Infill panels/sheets
	Composite items trussed rafters etc.	G20	Carpentry/Timber framing/First fixing
		L	WINDOWS/DOORS/STAIRS
	windows, window frames etc.	L10	Timber windows/rooflights/screens/ louvres
	doors, door frames etc.	L20	Timber doors/shutters/hatches
	staircases	L30	Timber stairs/walkways/balustrades

SMM6		SMM7	
	fittings	N	FURNITURE/EQUIPMENT
		N10	General fixtures/furnishings/ equipment
		N11	Domestic kitchen fittings
		N20,	N21, N22, N23 Special purpose fixtures/furnishings/equipment
	Sundries	P	BUILDING FABRIC SUNDRIES
	plugging	-	deemed included with fixed items (however rates retained under G20)
	holes in timber	-	deemed included - except P31 Holes/ Chases/Covers/Supports for services
	insulating materials	P10	Sundry insulation/proofing work/fire stops
	metalwork	G20	Carpentry/Timber framing/First fixing
	Ironmongery	N15	Signs/Notices
		P21	Ironmongery
	Protection	A42	Contractor's services and facilities
P	STRUCTURAL STEELWORK		
	Steelwork	G10	Structural steel framing
		G11	Structural aluminium framing
		G12	Isolated structural metal members
	Protection	A42	Contractor's services and facilities
Q	METALWORK		
	Composite items		
	curtain walling	H11	Curtain walling
		H12	Plastics glazed vaulting/walling
		H13	Structural glass assemblies
	windows	L11	Metal windows/rooflights/screens/ louvres
		L12	Plastics windows/rooflights/screens/ louvres
	doors	L21	Metal doors/shutters/hatches
		L22	Plastics/Rubber doors/Shutters/ hatches
	rooflights	L11	Metal windows/rooflights/screens/ louvres
		L12	Plastics windows/rooflights/screens/ louvres
	balustrades and staircases	L31	Metal stairs/walkways/balustrades

		SMM6	SMM7
Q	METALWORK - cont'd		
	sundries		
	duct covers etc.	P31	Holes/Covers/Chases/Supports for services
	gates/shutters/hatches	L21	Metal doors/shutters/hatches
	cloakroom fittings etc.	N10	General fixtures/furnishings/ equipment
	steel lintels	F30	Accessories/Sundry items for brick/ block/stone walling
	Plates, bars etc.		
	floor plates	L31	Metal stairs/walkways/balustrades
	matwells	N10	General fixtures/furnishings/ equipment
	Sheet metal, wiremesh and expanded metal	L21	Metal doors/shutters/hatches
	Holts, bolts, screws and rivets	E42	Accessories cast into in situ concrete
		G20	Carpentry/Timber framing/First fixing
		L10-L12	Windows
		L20-L22	Doors
		L31	Metal stairs/walkways/balustrades
	Protection	A42	Contractor's services and facilities
R	PLUMBING AND MECHANICAL ENGINEERING INSTALLATIONS	R	DISPOSAL SYSTEMS
		S	PIPED SUPPLY SYSTEMS
		T	MECHANICAL HEATING/COOLING REFRIGERATION SYSTEMS
		U	VENTILATION/AIR CONDITIONING SYSTEMS
		X	TRANSPORT SYSTEMS
		Y	MECHANICAL AND ELECTRICAL SERVICES MEASUREMENT
	Classification of work		
	a. rainwater installation	R10	Rainwater pipework/gutters
	b. sanitary installation (including traps)	R11	Foul drainage above ground

SMM6	SMM7
c. cold water installation	S10 Cold water
	S13 Pressurised water
	S14 Irrigation
	S15 Fountains/Water features
	S20 Treated/Deionised/Distilled water
	S21 Swimming pool water treatment
d. firefighting installation	S60 Fire hose reels
	S61 Dry risers
	S62 Wet risers
	S63 Sprinklers
	S64 Deluge
	S65 Fire hydrants
	S70 Gas fire fighting
	S71 Foam fire fighting
e. heated, hot water installations etc.	S11 Hot water
	S12 Hot and cold water (small scale)
	S51 Steam
	T20 Primary heat distribution
	T30 Medium temperature hot water heating
	T31 Low temperature hot water heating
	T32 Low temperature hot water heating (small scale)
	T33 Steam heating
f. fuel oil installation	S40 Petrol/Oil-lubrication
	S41 Fuel oil storage/distribution
g. fuel gas installation	S32 Natural gas
	S33 Liquid petroleum gas
h. refrigeration installation	T61 Primary/Secondary cooling distribution
	T70 Local cooling units
	T71 Cold rooms
	T72 Ice pads
j. compressed air installation	S30 Compressed air
	S31 Instrument air
k. hydraulic installation	-
l. chemical installation	-
m. special gas installation	S34 Medical/Laboratory gas
n. medical suction installation	S50 Vaccuum
p. pneumatic tube installation	S50 Vaccuum
q. vaccuum installation	R30 Centralised vacuum cleaning
	S50 Vacuum

SMM6		SMM7
R PLUMBING AND MECHANICAL ENGINEERING INSTALLATIONS - cont'd		
Classification of work - cont'd		
r. refuse disposal installation	R14	Laboratory/Industrial waste drainage
	R20	Sewage pumping
	R21	Sewage treatment/sterilisation
	R31	Refuse chutes
	R32	Compactors/Macerators
	R33	Incineration plant
s. air handling installation	T40	Warm air heating
	T41	Warm air heating (small scale)
	T42	Local heating units
	T50	Heat recovery
	U10	General supply/extract
	U11	Toilet extract
	U12	Kitchen extract
	U13	Car parking extract
	U14	Smoke extract/Smoke control
	U15	Safety cabinet/Fume cupboard extract
	U16	Fume extract
	U17	Anaesthetic gas extract
	U20	Dust collection
	U30	Low velocity air conditioning
	U31	VAV air conditioning
	U32	Dual-duct air conditioning
	U33	Multi-zone air conditioning
	U40	Induction air conditioning
	U41	Fan-coil air conditioning
	U42	Terminal re-heat air conditioning
	U43	Terminal heat pump air conditioning
	U50	Hybrid system air conditioning
	U60	Free standing air conditioning units
	U61	Window/Wall air conditioning units
	U70	Air curtains
	-	Trace heating (*Y24)
	-	Cleaning & chemical treatment (*Y25)
t. automatic control installation	-	Control components-mechanical (*Y53)
u. special equipment eg kitchen equipment	N12	Catering equipment
v. other specialist installations	X10	Lifts
	X11	Escalators
	X12	Moving pavements
	X20	Hoists
	X21	Cranes
	X22	Travelling cradles

* section ref. for measurement rules

SMM6		SMM7
	X23	Goods distribution/Mechanised warehousing
	X30	Mechanical document conveying
	X31	Pneumatic document conveying
	X32	Automatic document filing and retrieval
Gutterwork	R10	Rainwater pipework/gutters
Pipework (including fittings)	S10/S11/S12/T30	Pipelines (*Y10)
Ductwork	-	Air ductlines (*Y30)
	-	Air ductline ancillaries (*Y31)
	-	Air handling units (*Y40)
	-	Fans (*Y41)
	-	Air filtration (*Y42)
	-	Heating/Cooling coils (*Y43)
	-	Humidifiers (*Y44)
	-	Silencers/Acoustic treatment (*Y45)
	-	Grilles/Diffusers/Louvres (*Y46)
Equipment and ancillaries	N13	Sanitary appliances/fittings
	T10	Gas/Oil fired boilers
	T11	Coal fired boilers
	T12	Electrode/Direct electric boilers
	T13	Packaged steam generators
	T14	Heat pumps
	T15	Solar collectors
	T16	Alternative fuel boilers
	T60	Central refrigeration plant
	S10/S11/S12/T30	Pipeline ancillaries (*Y11)
	-	Pumps (*Y20)
	S10/S11/S12	Water tanks/cisterns (*Y21)
	-	Heat exchangers (*Y22)
	S10/S11/S12	Storage cylinders/calorifiers (*Y23)
	T30	Vibration isolation mountings (*Y52)
	-	Identification-mechanical (*Y54)
Insulation	S10/S11/S12	Thermal insulation (*Y50)
Sundries	-	Testing and commissioning of mechanical services (*Y51)
	-	Sundry common mechanical items (*Y59)
Builder's work	P31	Holes/Chases/Covers/Supports for services
Protection	A42	Contractor's services and facilities

* section ref. for measurement rules

SMM6			SMM7	
S	ELECTRICAL INSTALLATIONS		V	ELECTRICAL SUPPLY/POWER LIGHTING SYSTEMS
			W	COMMUNICATIONS/SECURITY/CONTROL SYSTEMS
			Y	MECHANICAL AND ELECTRICAL MEASUREMENT SERVICES
	Classification of work			
	a.	incoming services	V11	HV supply/distribution/public utility supply
			V12	LV supply/public utility supply
	b.	standby equipment	V32	Uninterrupted power supply
			V40	Emergency lighting
	c.	mains installation excluding final sub circuits	V20	LV distribution
	d.	power installation	V22	General LV power
	e.	lighting installation	V21	General lighting
			V41	Street/Area/Flood lighting
			V42	Studio/Auditorium/Arena lighting
			V90	General lighting and power (small scale)
			W21	Projection
			W22	Advertising display
	f.	electric heating installation	V50	Electric underfloor heating
			V51	Local electric heating units
	g.	electrical appliances		-
	h.	electrical work associated with plumbing and mechanical engineering installations	-	Control components - mechanical (*Y53)
	j.	telephone installations	W10	Telecommunications
	k.	clock installation	W23	Clocks
	l.	sound distribution installation	W11	Staff paging/location
			W12	Public address/Sound amplification
			W13	Centralized dictation
	m.	alarm system installation	W41	Security detection and alarm
			W50	Fire detection and alarm

* section ref. for measurement rules

SMM6		SMM7	
n. earthing system installation		W51	Earthing and bonding
p. lightning protection installation		W52	Lightning protection
q. special services		V30	Extra low voltage supply
		V31	DC supply
		W20	Radio/TV/CCTV
		W30	Data transmission
		W40	Access control
		W53	Electromagnetic screening
		W60	Monitoring
		W61	Central control
		W62	Building automation
Equipment and control gear		V10	Electricity generation plant
		-	HV switchgear (*Y70)
		-	LV switchgear and distribution boards (*Y71)
		-	Contactors and starters (*Y72)
		-	Motor drives - electric (*Y92)
Fittings and accessories		-	Luminaires and lamps (*Y73)
		-	Accessories for electrical services (*Y74)
Conduit, trunking and cable trays		-	Conduit and cable trunking (*Y60)
		-	Busbar trunking (*Y62)
Cables/Final sub-circuits		-	HV/LV cables and wiring (*Y61)
Earthing		-	Earthing and bonding components (*Y80)
Ancillaries		-	Identification - electrical (*Y82)
		-	Sundry common electrical items (*Y89)
Sundries		-	Testing and commissioning of electrical services (*Y81)
Builders work		P31	Holes/Chases/Covers/Supports for services
Protection		A42	Contractor's services and facilities
T	FLOOR WALL AND CEILING FINISHINGS	M	SURFACE FINISHES
	In situ finishings/Lathing and base boarding/Beds and backings	J10	Specialist waterproof rendering (including accessories)
		J33	In situ glass reinforced plastics

*section ref. for measurement rules

	SMM6		SMM7
T	FLOOR WALL AND CEILING FINISHINGS - cont'd	M10	Sand cement/Concrete/Granolithic screeds/flooring
		M12	Trowelled bitumen/resin/rubber-latex flooring
		M20	Plastered/Rendered/Roughcast coatings including backings
		M21	Insulation with rendered finish
		M22	Sprayed mineral fibre coatings
		M23	Resin bound mineral coatings
		M30	Metal mesh lathing/Anchored reinforcement for plastered coatings
		M41	Terrazzo tiling/In situ terrazzo
	surface sealers	M60	Painting/Clear finishing
	Tile, slab and block finishings/Mosaic work	M40	Stone/Concrete/Quarry/Ceramic tiling/ Mosaic
		M41	Terrazzo tiling/In situ terrazzo
		M42	Wood block/Composition block/Parquet flooring
	Flexible sheet finishings	M50	Rubber/Plastics/Cork/Lino/Carpet tiling/sheeting
	Dry linings and partitions	K	LININGS/SHEATHING DRY PARTITIONING
		K10	Plasterboard dry lining
		K30	Demountable partitions
		K31	Plasterboard fixed partitions/inner walls/linings
		K32	Framed panel cubicle partitions
		K33	Concrete/Terrazzo partitions
	Raised floors	K41	Raised access floors
	Suspended ceilings, linings, and support work	K40	Suspended ceilings
	Fibrous plaster	M31	Fibrous plaster
	Fitted carpeting	M51	Edge fixed carpeting
	Protection	A42	Contractor's services and facilities
U	GLAZING		
	Glass in openings	L40	General glazing
	Leaded lights and copper lights in openings	L41	Lead light glazing
	Mirrors	N10	General fixtures/furnishings/ equipment
		N20,	N21, N22, N23 Special purpose furnishings/equipment
	Patent glazing	H10	Patent glazing

	SMM6		SMM7
	Domelights	L12	Plastics windows/rooflights/screens/ louvres
	Protection	A42	Contractor's services and facilities
V	PAINTING AND DECORATING		
	Painting, polishing and similar work	M60	Painting/Clear finishing
	Signwriting	N15	Signs/Notices
	Decorative paper, sheet plastic or fabric backing and lining	M52	Decorative papers/fabrics
	Protection	A42	Contractor's services and facilities
W	DRAINAGE	R	DISPOSAL SYSTEMS
	Pipe trenches Manholes, soakaways, cesspits and septic tanks Connections to sewers Testing drains	R12 R13	Drainage below ground Land drainage
	Protection	A42	Contractor's services and facilities
X	FENCING	Q40	Fencing
	Open type fencing Close type fencing Gates Sundries Protection		
	EXTERNAL WORKS	D	GROUNDWORK
		F	MASONRY
		Q	PAVING/PLANTING/FENCING/SITE FURNITURE
		Q10	Stone/Concrete/Brick kerbs/edgings/ channels
		Q20	Hardcore/Granular/Cement bound bases/sub-bases to roads/pavings

SMM6	SMM7
EXTERNAL WORKS - cont'd	
	Q21 In situ concrete roads/pavings/bases
	Q22 Coated macadam/Asphalt roads/pavings
	Q23 Gravel/Hoggin roads/pavings
	Q24 Interlocking brick/block roads/ pavings
	Q25 Slab/Brick/Sett/Cobble pavings
	Q26 Special surfacings/pavings for sport
	Q30 Seeding/Turfing
	Q31 Planting
	Q50 Site/Street furniture/equipment

Acknowledgements

The Editors wish to record their appreciation of the assistance given by many individuals and organisations in the compilation of this edition.

Material Suppliers and Sub-Contractors who have contributed this year include:-

Aggregates	
Redland Aggregates	0992-586600
lightweight	
Boral Lytag	03752-77181
Aluminium pipes and fittings	
Alumasc	0536-722121
Anchors, ties etc	
Abbey Building Supplies	021-550-7674
Harris & Edgar	081-686-4891
Asbestos pipes and fittings	
Eternit Building Prods	0763-60421
Asphalt work	
Prater Asphalt	0737-772331
Beams,	
laminated	
Moelven (UK)	0703-454944
lattice	
Metal Sections	021-552-1541
Bearing pads	
Tico	0483-757757
Blockwork	
Aerated Concrete Ltd	0375-673344
ARC Concrete	0235-848808
Boral Edenhall	0708-862881
Celcon Ltd	071-242-9766
Forticrete	0533-320277
Tarmac	0442-54321
Thermalite	0675-62081
Brickwork	
Armitage	0532-822141
Ibstock Bricks	071-402-1227
London Brick	0525-405858
Redland Bricks	0293-786688
Ryarsh Brick	0732-870100
Building admixtures etc.	
Ardex U.K.	0440-63939
Aston Building Products	0785-57265
Beton Construct Mats	0256-53146
Cementone Beaver	0280-823823
FEB	061-794-7411
Kerner-Greenwood	0553-772293
RIW Protective Products	0734-792566
Sealocrete	0703-777331
Sika	0707-329241
Vandex (UK) Ltd	081-394-2766
W Hawley & Son	0332-840294
Washington Mills Electro Minerals	0793-28131
Building papers	
British Sisalcraft	0634-290505
Davidson Packaging	0602-844022
Cement	
Blue Circle Cement	071-731-7762
Chequer plate flooring	
Westwood	0633-614555
Cisterns/tanks/cylinders etc.	
Harvey Fabrication	081-981-7811
IMI Range	061-338-3353
Cladding,	
aluminium	
British Alcan	0905-754030
asbestos etc.	
Eternit Building Prods	0763-60421
steel	
Briggs Amasco	0306-885933
British steel	071-735-7654
Plannja	0628-37313
transluscent	
BIP Chemicals	021-353-0814
Column guards	
Huntley & Sparks	0460-72222

- Concrete,
  - paving
    - BDC Concrete Products 0375-673921
    - Marley 0708-852201
    - Marshalls Mono 0332-792301
    - Redland Aggregates 0509-812601
  - pipes/manholes etc.
    - Drainage Systems 071-286-5151
    - Hume Pipe 0420-80086
    - Milton pipes 0795-25191
    - RBS Brooklyns 09295-6656
  - precast floors etc.
    - Bison 0753-652909
  - ready-mixed
    - Greenham Concrete 071-736-6592
- Copper,
  - fittings
    - IMI York Imp Fittings 0532-701104
  - tubes
    - IMI Yorkshire Copper Tube 051-546-2700

- Damp proof courses, closers etc.
  - Cavity Trays 0935-74769
  - Colas Building products 0268-728811
  - IMI Rolled Metal 021-356-3344
  - Marley Waterproofing 0732-741400
  - Products 0268-728811
  - Westbrick Plastics 0722-331933
- Door sets
  - Swedoors 0602-725231
- Doors
  - Crosby Doors 0793-729555
  - Leaderflush Doors 0773-530500
  - Sarek Joinery 0787-60808
- Doors, garage
  - Catnic Garador 0222-885955
- Drainage, stoneware
  - Hepworth Iron Co 0226-763561
  - Sandell Perkins 071-258-0257
- Duct covers
  - Glynwed Brickhouse 0952-641414

- Expansion joint fillers
  - Expandite 081-965-8877
  - Servicised 0753-692929
- Fencing & gates
  - Binns Fencing 081-802-5211
- Finishes, textured
  - Artex 0273-513100
- Fire,
  - protection
    - Morceau-Aaronite 0773-812505
  - resisting glass
    - Solaglass 071-928-8010
- Firecheck boards/channels/strips
  - Cape Boards & Panels 0895-37111
  - Nullifire 0203-470022
- Fixings etc
  - BAT 0952-680193
  - Halfen 0296-20141
  - Hilti 061-872-5010
- Flashings
  - Evode 0785-57755
- Flooring,
  - access
    - H. H. Robertson 051-355-3622
    - Phoenix Floors 0708-851441
  - accessories
    - Carpet & Flooring 021-550-9131
    - Roberts Smoothedge 0403-40721
  - hardwood
    - Hewetsons 04827-81701
    - Junckers 0376-517512
    - Viger Floors 037882-3035
  - tiles, sheet flooring etc.
    - Altro 0454-412992
    - Armstrong 0642-763224
    - Daniel Platt 0782-577187
    - Forbo Nairn 0923-52323
    - James Halstead 061-766-3781
    - Marley Floors 0622-858877
    - Ruabon, Dennis 0978-843276
    - Wicanders 0293-27700
- Flooring Sub-Contractor
  - GC Flooring & Furnishings 081-991-1000
- Flue linings & blocks
  - True Flue 0242-862551
- Formers/linings etc.
  - Cordek 0403-783383
  - Dufaylite Dev 0480-215000
  - Exxon Chemical Geopolymers 04955-57722
  - Richard Lees 0335-60601
- Fosalsil bricks, aggregates etc
  - Molar products 0206-73191
- French polishing/staining etc
  - Mosford Joinery 081-459-6241

- Galvanised steel gutters etc.
  - WP Metals 0922-743111
- Glass & glazing
  - Solaglass 071-928-8010
- Granite linings
  - Banmoor Masonry 0527-510515
  - Whiteheads 071-498-3111
- Granolithic Sub-Contractor
  - Malcolm MacLeod 081-520-1147
- Granular fillings etc.
  - Yeoman Aggregates 081-993-6411

Hardcore, Cart away	
Western Foundations	081-684-7700
Hardwood	
see Joinery	
Insulation	
Celotex	081-579-0811
Coolag Purlboard	0524-55611
DOW Chemicals	021-705-6363
Erisco Bauder	0473-57671
Gyroc Insulation	0928-712627
Isocrete Group Sales	081-906-1077
Pilkington Insulation	0533-717202
Pittsburgh Corning	0734-500655
Plaschem	061-766-9711
Vencel Resil	0322-27299
Intumescent strips	
Sealmaster	0223-832851
Iron pipes & fittings	
Stanton Pipeline Services	081-459-7801
Ironmongery, etc	
Comyn Ching	071-253-8414
Jewson	071-736-5511
Joinery, purpose-made	
Llewellyns	0323-21300
Joinery, standard hardwood	
Sarek Joinery	0787-60808
Joinery, standard softwood	
Boulton & Paul	0603-660133
Kerbs, channels etc.	
ECC Quarries Ltd	0335-70600
F.R.Dangerfield	071-435-8044
Lathing, expanded metal	
Expanded Metal Co	0429-266633
Lathing, waterproof	
J Newton	071-629-5752
Lifts, escalators	
Express Lifts	0604-51221
Lightning protection	
R.C.Cutting	081-348-0052
Lime	
Blue Circle Industries	0482-633381
Lintols, steel	
Catnic Components	0222-885955
Manhole covers	
Glynwed Foundries	0952-641414
Marble linings	
Banmoor Masonry	0527-510515
Whiteheads	071-498-3111
Matwell frames	
Nuway Manufacturing	0952-680400
Metal,	
framing	
Unistrut Midland	021-784-4178
staircases	
Crescent of Cambridge	0480-301522
Mortars, ready-mixed	
Tilcon (Midland & South)	0732-453633
Paints, preservatives etc.	
Akzo Coatings	0235-815141
Cuprinol	0373-65151
Crown Paints	0254-704951
J. P. MacDougall	081-749-0111
Sadolin	0480-496868
Solignium	0322-526966
Tretol	0753-24164
Patent glazing	
Mellowes PPG	021-553-4011
Pavement lights, glass block walling	
Luxcrete	081-965-7292
Paving slabs	
Redland Aggregates	050-981-2601
Piling, diaphragm walling	
Cementation	0923-776666
Plant Hire	
Agent Plant Hire Ltd	0322-22221
LPH Equipment	0494-21481
Plaster products	
Mineralite Products	05435-71312
Tilcon Special Products	0423-862841
Plaster/plasterboard	
British Gypsum	0602-844844
Plastering Sub-Contractor	
Jonathon James	04027-56921
Polythene sheeting	
McArthur Steel and Metal	0272-656242
Preservatives etc.	
See Paints etc.	
PVC pipes and fittings	
Caradon Terrain	0622-77811
Hunter Building Products	081-855-9851
Uponur	0532-701160
Rainwater outlets	
Harmer Holdings	07072-73481
Rawlplugs, Rawlbolts etc.	
Rawlplug Co.	081-546-2191
Reconstructed stone walling/roof tiles	
ECC Quarries	0793-28131
Reinforcement bars	
Barfab	081-878-7771

Reinforcement mesh
- BRC 0785-57777

Roof decking
- Briggs Amasco 0306-885933
- Plannja 0628-37313

Roofing/cladding fasteners
- Sela Fasteners 0532-430541

Rooflights/windows
- Coxdome 044-282-4222
- Velux 0438-312570

Roofing products
- Nuralite Roofing Systems 0474-82-3451
- Permanite 0992-550511
- Ruberoid Building Products 081-805-3434

Roofing slates, natural
- Burlington Slate 022-989-661
- Greaves Welsh Slates 0766-830522
- Hardrow 0283-64389

Roofing tiles, clay
- Goxhill Tileries 0427-872696
- Hinton Perry Davenhills 0384-77405
- Keymer Brick 04446-2931
- William Blyth 0652-32175

concrete
- Marley roof tiles 0732-460055
- Redland Roof Tiles 07372-42488

Roofing Sub-Contractors
- Ruberoid Contracts 0256-461431
- Standard Flat Roofing 081-981-2422

Rubble walling
- ARC Southern 0285-712471

Sand
- Tilcon (Midland & South) 0732-452325

Sanitary fittings
- Ideal Standard 0482-499425
- Stelrad Bathroom Products 0782-49191
- W & G Sissons 0433-30791

Screed Sub Contractor
- Alan Milne 081-998-9961

Shelving, adjustable
- Spur Systems 0923-26071

Sliding door gear
- Hillaldam Coburn 081-397-5151

Slots, ties etc.
- Abbey Building Supplies 021-550-7674
- Harris & Edgar 081-686-4891

Softwood, panel products etc.
- C.F.Anderson 071-226-1212
- James Latham 0454-315421

Stainless Steel tube and fittings
- Lancashire Fittings 0423-522355

Stairs
- Boulton & Paul (Joinery) 0603-660133

Steel arch frames
- Truline Building Products 0245-450450

Steel tubes and fittings
- British Tube Stockholdings 0633-290290

Steelwork
- British Steel 071-735-7654
- Graham Wood 071-586-6094

Stone Cladding Sub-Contractor
- Bath & Portland Stone 0305-820331

Stonework
- Bath & Portland Stone 0225-810456
- Gregory Quarries 0623-23092

Suspended ceilings
- Thermal and Acoustic installations 0992-38311

Tarmacadam Sub-Contractor
- Constable Hart 0483-224522

Terrazzo Sub-Contractor
- Marriot & Price 081-521-2821

Timber preservation etc
- Peter Cox Preservation 0895-443788
- Rentokil 0342-833022

Trussed rafters
- Montague L. Meyer 081-594-7111

Wall tiles
- Langley London 071-407-4444
- Pilkington tiles 061-794-2024

Waterbars etc.
- Servicised 0753-692929

Windows uPVC
- Anglian Windows 0603-619471

Windows, aluminium and metal
- Crittall Windows 0376-24106

Windows, hardwood
- Sarek Joinery 0787-60808

Windows, softwood
- Boulton & Paul 0603-660133

Woodwool slabs
- Torvale Building Products 05447-262

SPON'S 1991 PRICE BOOKS ORDER FORM

Please send me:

____ copy/ies of **Spon's ARCHITECTS' AND BUILDERS' Price Book 1991** 116th Edition @ £49.50

____ copy/ies of **Spon's MECHANICAL AND ELECTRICAL SERVICES Price Book 1991** 22nd Edition @ £52.50

____ copy/ies of **Spon's CIVIL ENGINEERING AND HIGHWAY WORKS Price Book 1991** 5th Edition @ £55.00

____ copy/ies of **Spon's LANDSCAPE AND EXTERNAL WORKS Price Book 1991** 11th Edition @ £42.50

____ copy/ies of **Spon's PLANT AND EQUIPMENT Price Guide** @ £130.00

Please tick as appropriate:

☐ I enclose a cheque/PO for £__________payable to E & F N Spon

☐ Please debit my Access/Visa/Diners/AmEx card

Card Number______________________________

Expiry Date______________________________

☐ I enclose an official order, please send invoice with books

Name____

Address____

_______________________________________Postcode__________________

Signature________________________________Date____________________

Return to: *The Promotion Dept., E & F N Spon, 2-6 Boundary Row, London SE1 8HN*

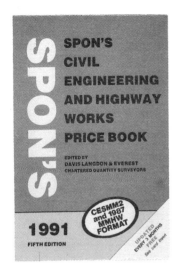

PART I

Fees and Daywork

This part of the book contains the following sections:

Fees for Professional Services

Extracts from the scales of fees for architects, quantity surveyors and consulting engineers are given together with extracts from the Town and Country Planning Regulations 1989 and the Building (Prescribed Fees etc.) Regulations 1986. These extracts are reproduced by kind permission of the bodies concerned, in the case of Building Regulation Fees by kind permission of the Controller HM Stationary Office. Attention is drawn to the fact that the full scales are not reproduced here and that the extracts are given for guidance only. The full authorized scales should be studied before concluding any agreement and the reader should ensure that the fees quoted here are still current at the time of reference

ARCHITECTS' FEES

EFFECTIVE 1st JULY 1982
including amendments up to November 1988

INTRODUCTION

The RIBA requires of its members that before making an engagement for professional services they shall define the terms of the engagement including the scope of the service, the allocation of resposibilities and any limitation of liability, the method of calculation of remuneration and the provision for termination.

Architect's Appointment consists of four related parts:

Part 1 Architect's Services

Preliminary and Basic services normally provided by the architect. These services progress through work stages based on the RIBA Plan of Work (RIBA Publications Limited). The sequence of work stages may be varied or two or more work stages may be combined to suit the particular circumstances.

Preliminary Services

Work stage A: Inception

Work stage B: Feasibility

Basic Services

Work stage C: Outline proposals

Work stage D: Scheme design

Work stage E: Detail design

Work stage F: Production informatation

Work stage G: Bills of Quantities

Work stage H: Tender action

Work stage J: Project planning

Work stage K: Operation on site

Work stage L: Completion

Preliminary Services are normally charged on a time basis and Basic Services on a percentage basis, as described in Part 4.

Part 2 Other Services

Services which may augment the Preliminary and Basic Services or which may be the subject of a separate appointment. Fees for these services are normally charged on a time or lump sum basis as described in Part 4.

Part 3 Conditions of Appointment

Conditions which apply to an architect's appointment.

Part 4 Recommended Fees and Expenses

Recommended and not mandatory methods of calculating the architect's fees and expenses and of apportioning fees between work stages.
A sample Memorandum of Agreement and Schedule of Services and Fees are included with the document for information. They are published separately.

ARCHITECTS' FEES

The client and the architect should discuss the architect's appointment and agree in writing the services, conditions and fee basis. These should be stated in the Schedule of Services and Fees and referred to in the Memorandum of Agreement between client and architect; alternatively, they should be stated in a letter of appointment.

PART 1 ARCHITECT'S SERVICES

This part describes Preliminary and Basic Services which an architect will normally provide.

Preliminary Services

Word stage A: Inception

1.1. Discuss the client's requirements including timescale and any financial limits; assess these and give general advice on how to proceed; agree the architect's services.
1.2. Obtain from the client information on ownership and any lessors and lessees of the site, any existing buildings on the site, boundary fences and other enclosures, and any known easement, encroachments, underground services, rights of way, rights of support and other relevant matters.
1.3. Visit the site and carry out an initial appraisal.
1.4. Advise on the need for other consultants' services and on the scope of these services.
1.5. Advise on the need for specialist contractors, sub-contractors and suppliers to design and execute part of the works to comply with the architect's requirements.
1.6. Advise on the need for site staff.
1.7. Prepare where required an outline timetable and fee basis for further services for the client's approval.

Work stage B: Feasibility

1.8. Carry out such studies as may be necessary to determine the feasibility of the client's requirements; review with the client alternative design and construction approaches and cost implications; advise on the need to obtain planning permissions, approvals under building acts or regulations, and other similar statutory requirements.

BASIC SERVICES

Work stage C: Outline proposals

1.9. With other consultants where appointed, analyse the client's requirements; prepare outline proposals and an approximation of the construction cost for the client's preliminary approval.

Work stages D: Scheme design

1.10. With other consultants where appointed, develop a scheme design from the outline proposals taking into account amendments requested by the client; prepare a cost estimate; where applicable give an indication of possible start and completion dates for the building contract. The scheme design will illustrate the size and character of the project in sufficient detail to enable the client to agree the spatial arrangements, materials and appearance.

ARCHITECTS' FEES

1.11. With other consultants where appointed, advise the client of the implications of any subsequent changes on the cost of the project and on the overall programme.
1.12. Make where required application for planning permission. The permission itself is beyond the architect's control and no guarantee that it will be granted can be given.

Work stage E: Detail Design

1.13. With other consultants where appointed, develop the scheme design; obtain the client's approval of the type of construction, quality of materials and standard of workmanship; co-ordinate any design work done by consultants, specialist contractors, sub-contractors and suppliers; obtain quotations and other information in connection with specialist work.
1.14. With other consultants where appointed, carry out cost checks as necessary; advise the client of the consequences of any subsequent change on the cost and programme.
1.15. Make and negotiate where required applications for approvals under building acts, regulations or other statutory requirements.

Work stage F and G: Production information and bills of quantities.

1.16. With other consultants where appointed, prepare production information including drawings, schedules and specification of material and workmanship; provide information for bills of quantities, if any, to be prepared: all information complete in sufficient detail to enable a contractor to prepare a tender.

Work stage H: Tender action

1.17. Arrange, where relevant, for other contracts to be let prior to the contractor commencing work.
1.18. Advise on and obtain the client's approval to a list of tenderers.
1.19. Invite tenders from approved contractors; appraise and advise on tenders submitted. Alternatively, arrange for a price to be negotiated with a contractor.

Work stage J: Project planning

1.20. Advise the client on the appointment of contractor and on the responsibilities of the client, contractor and architect under the terms of the building contract; where required prepare the building contract and arrange for it to be signed by the client and the contractor; provide production information as required by the building contract.

Work stage K: Operations on site

1.21. Administer the terms of the building contract during operations on site.
1.22. Visit the site as appropriate to inspect generally the progress and quantity of the work.
1.23. With other consultants where appointed, make where required periodic financial reports to the client including the effect of any variations on the construction cost.

ARCHITECTS' FEES

Work stage L: Completion

1.24. Administer the terms of the building contract relating to the completion of the work.
1.25. Give general guidance on maintenance.
1.26. Provide the client with a set of drawings showing the building and the main lines drainage; arrange for drawings of the services installations to be provided.

PART 2 OTHER SERVICES

This part describes services which may be provided by the architect to augment the Preliminary and Basic Service described in Part 1 or which may be subject of a separate appointment. The list of services so described is not exhaustive.

Surveys and investigations

2.1. Advise on the selection and suitability of sites; conduct negotiations concerned with sites and buildings.
2.2. Make measured surveys, take levels and prepare plans of sites and buildings
2.3. Provide services in connection with soil and other similar investigations.
2.4. Make inspections, prepare reports or give general advise on the condition of premises.
2.5. Prepare schedules of dilapidations; negotiate them on behalf of landlords or tenants.
2.6. Make structural surveys to ascertain whether there are defects in the walls, roof, floors, drains or other parts of a building which may materially affect its saftey, life and value.
2.7. Investigate building failures; arrange and supervise exploratory work by contractors or specialists.
2.8. Take particulars on site; prepare specifications and/or schedules for repairs and restoration work, and inspect their execution.
2.9. Investigate and advise on problems in existing buildings such as fire protection, floor loadings, sound insulation, or change of use.
2.10. Advise on the efficient use of energy in new and existing buildings.
2.11. Carry out life cycle analyses of buildings to determine their cost in use.
2.12. Make an inspection and valuation for morgage or other purposes.

Development services

2.13. Prepare special drawings, models or technical information for the use of the client or for applications under planning, building act, building regulation or other statutory requirements, or for negotiations with ground landlords, ajoining owners, public authorities, licensing authorities, mortgagors and others; prepare plans for conveyancing, land registry and other legal purposes.
2.14. Prepare development plans for a large building or complex of buildings; prepare a layout only, or prepare a layout for a greater area than that which is to be developed immediately.
2.15. Prepare layouts for housing, industrial or other estates showing the siting of buildings and other works such as roads and sewers.
2.16. Prepare drawings and specification of materials and workmanship for the construction of housing, industrial or other estate roads and sewers.

ARCHITECTS' FEES

2.17. Provide services in connection with demolitions works.
2.18. Provide services in connection with environmental studies.

Design services

2.19. Design or advise on the section of furniture and fittings; inspect the making up of such furnishings.
2.20. Advise on and prepare detailed designs for works of special quality such as shop-fitting or exhibition design, either independently or within the shell of an existing building.
2.21. Advise on the commissioning or selection of works of art; supervise their installation.
2.22. Carry out specialist acoustical investigations.
2.23. Carry out special constructional research in connection with a scheme design, including the design, construction or testing of prototype buildings or models.
2.24. Develop a building system or mass-produced building components; examine and advise on existing building systems; monitor the testing of prototype buildings and models.

Cost estimating and financial advisory services

2.25. Carry out planning for a building project, including the cost of associated design services, site development, landscaping, furnature and equiptment; advise on cash flow requirements for design cost, construction cost, and cost in use.
2.26. Prepare schedules of rates or schedules of quantities for tendering purposes; value work excuted where no quantity suveyor is appointed. Fees for this work are recommended to be in accordance with the Professional Charges of the Royal Institution of Chartered Surveyors.
2.27. Carry out inspections and surveys; prepare estimates for the replacement and reinstatement of buildings and plant; submit and negotiate claims following damage by fire or other causes.
2.28. Provide information; make applications for and conduct negotiations in connection with local authority, government or other grants.

Negotiations

2.29. Conduct exceptional negotiations with a planning authority.
2.30. Prepare and submit an appeal under planning acts; advise on other works in connection with planning appeals.
2.31. Conduct exceptional negotiations for approvals under building acts or regulations; negotiate waivers or relaxations.
2.32. Make submissions to the Royal Fine Art Commission and other non-statutory bodies.
2.33. Submit plans of proposed building works for approval of landlords, mortgagors, Freeholders or others.
2.34. Advise on the rights and responsibilities of owners or lessees including rights of light, rights of support, and rights of way; provide information; undertake any negotiations.
2.35. Provide services in connection with party wall negotiations.
2.36. Prepare and give evidence; settle proofs; confer with solicitors and counsel; attend court and arbitrations; appear before other tribunals; act as arbitrator.

ARCHITECTS' FEES

Administration and management of building projects

2.37. Provide site staff for frequent or constant inspection of the works.
2.38. Provide management from inception to completion: prepare briefs; appoint and co-ordinate consultants, construction managers, agents and contractors; monitor time, cost and agreed targets; monitor progress of the works; hand over the building on completion; equip, commission and set up any operational organizations.
2.39. Provide services to the client, whether employer or contractor, in carrying out duties under a design and build contract.
2.40. Private services in connection with separate trades contracts; agree a programme of work; act as co-ordinator for the duration of the contract.
2.41. Provide services in connection with labour employed directly by the client; agree a programme of work; co-ordinate the supply of labour and materials; provide general supervision; agree the final account.
2.42. Provide specially prepared drawings of a building 'as built'.
2.43. Compile maintainance and operational manuals; incorporate information prepared by other consultants, specialist contractors, sub-contractors and suppliers.
2.44. Prepare a programme for the maintenance of a building; arrange maintenance contracts.

Services normally provided by consultants

2.45. Provide such services as:

a Quantity surveying
b Structural engineering
c Mechanical engineering
d Electrical engineering
e Landscape and garden design
f Civil engineering
g Town planning
h Furniture design
j Graphic design
k Industrial design
l Interior design

Where consultants' services are provides from within the architect's own office or by consultants in association with the architect it is recommended that fees be in accordance with the scales of charges of the relevant professional body.

Consultancy services

2.46. Provide services as a consultant architect on a regular or intermittent basis.

Note:

In January 1990 the RIBA issued a supplement to Part 2 of the 'Architects Appointment' entitled 'Community Architectural Services' which describes the additional services which may be provided by the architect for a community architecture project to augment the services described in Parts 1 and 2.

Fees for these services will normally be charged on a time or lump sum basis.

ARCHITECTS' FEES

PART 3 CONDITIONS OF APPOINTMENT

This part describes the conditions which normally apply to an architect's appointement. If different or additional conditions are to apply, they should be set out in the Schedule of Services and Fees or letter of appointment.

3.1. The architect will exercise reasonable skill and care in conformity with with the normal standards of the architect's profession.

Architect's authority

3.2. The architect will act on behalf of the client in the matters set out or implied in the architect's appointment; the architect will obtain the authority of the client before initiating any service or work stage.

3.3. The architect shall not make any material alteration, additional to or omission from the approved design without the knowledge and consent of the client, except if found necessary during construction for constructional reasons in which case the architect shall inform the client without delay.

3.4. The architect will inform the client if the total authorised expenditure or the building contract period is likely to be materially valued.

Consultants

3.5. Consultants may be nominated by either the client or the architect, subject to acceptance by each party.

3.6. Where the client employes the consultants, either directly or through the agency of the architect, the client will hold each consultant, and not the architect, responsible for the competence, general inspection and performance of the work entrusted to that consultant; provided that in relation to the execution of such work under the contract between the client and the contractor nothing in this clause shall affect any responsibility of the architect for issuing instructions or for other functions ascribed to the architect under that contract.

3.7. The architect will have the authority to co-ordinate and integrate into the overall design the services provided by any consultant, however employed.

Contractors, sub-contractors and suppliers

3.8. A specialist contractor, sub-contractor or supplier who is to be employed by the client to design any part of the works may be nominated by either the architect or the client, subject to acceptance by each party. The client will hold such contractor, sub-contractor or supplier, and not the architect, responsible for the competence, proper excution and performance of the work thereby entrusted to that contractor, sub-contractor or supplier. The architect will have the authority to co-ordinate and integrate such work into the overall design.

3.9. The client will employ a contractor under a separate agreement to undertake construction or other works. The client will hold the contractor, and not the architect, responsible for the contractor's operational methods and for the proper execution of the works.

ARCHITECTS' FEES

Site inspection

3.10. The architect will visit the site at intervals appropriate to the stage of construction to inspect the progess and quality of the works and to determine that they are being executed generally in accordance with the contract documents. The architect will not be required to make frequent or constant inspections.

3.11. Where frequent or constant inspection is required a clerk or clerk of works will be employed. They may be employed either by the client or by the architect and will be either event be under the architect's direction and control.

3.12. Where frequent or constant inspection by the architect is agreed to be necessary a resident architect may be appointed by the architect on a part or full time basis.

Client's instructions

3.13. The client will provide the architect with such information and make such decisions as are necessary for the proper performance of the agreed service.

3.14. The client, if a firm or other body of persons, will, when requested by the architect, nominate a responsible representative through whom all instructions will be given.

Copyright

3.15. Copyright in all documents and drawings prepared by the architect and in any works executed from those documents and drawings shall, otherwise agreed, remain the property of the architect.

3.16. The client, unless otherwise agreed, will be entitled to reproduce the architect's design by proceeding to execute the project provided that:

- the entitlement applies only to the site or part of the site to which the design relates; and
- the architect has completed work stages D or has provided detail design and production information in work stages E, F and G; and
- any fees due to the architect have been paid or tendered.

This entitlement will also apply to the maintainence, repair and renewal of the works.

3.17. Where an architect has not completed work stage D, or where the client and the architect have agreed that clause 3.16 shall not apply, the client may not reproduce the design by proceeding to execute the project without the consent of the architect and payment of any additional fee that may be agreed in exchange for the architect's consent.

3.18. The architect shall not unreasonably withhold his consent under clause 3.17 but where his services are limited to making and negotiating planning applications he may withhold his consent unless otherwise determined by an arbitrator appointed in accordance with clause 3.26.

ARCHITECTS' FEES

Assignment

3.19. Neither the architect nor the client may assign the whole or any part of his duties without the other's written consent.

Suspension and termination

3.20. The architect will give immediate notice in writing to the client of any siutation arising from force majeure which makes it impraticable to carry out any of the agreed services, and agree with the client a suitable course of action.

3.21. The client may suspend the performance of any or all of the agrees services by giving reasonable notice in writing to the architect.

3.22. If the architect has not been given instructions to resume any suspended service within six months from the date of suspension the architect will make written request for such instructions which must be given in writing. If these have not been received within 30 days of the date of such a request will have the right to treat the appointment as terminated upon the expiry of the 30 days.

3.23. The architect's appointment may be terminated by either party on the expiry of reasonable notice given in writing.

3.24. Should the architect through death or incapacity be unable to provide the agreed services, the appointment will thereby be terminated. In such an event the client may, on payment or tender of all outstanding fees and expenses, make full use of reports, drawings or other documents prepared by the architect in accordance with and for use under the agreement, but only for the purpose for which they were prepared.

Settlement of disputes

3.25. A difference or dispute arising on the application of the Architect's Appointment to fees charged by a member may, by agreement between the parties, be referred to the RIBA, RIAS or RSUA for an opinion provided that:

- The member's appointment is based on this document and has been agreed and confirmed in writing; and
- the opinion is sought on a joint statement of undisputed facts; and
- the parties undertake to accept the opinion as final and binding upon them.

3.26. Any other difference or dispute arising out of the appointment and any difference or dispute arising on the fees charged which cannot be resolved in accordance with clause 3.25 shall be referred to the arbitration by a person to be agreed between the parties or, failing agreement within 14 days after either party has given to the other a written request to concur in the appointment of an arbitrator, a person to be nominated at the request of either party by the President of the Chartered Institute of Arbitrators, provided that in a difference or dispute arising out of provisions relating to copyright, cause 3.15 to 3.18 above, the arbitrator shall, unless otherwise agreed be an architect.

ARCHITECTS' FEES

3.26S In Scotland, any difference or dispute arising out of the appointment which cannot be resolved in accordance with clause 3.25 shall be referred to arbitration by a person to be agreed between the parties or, failing agreement within 14 days after either party has given to the other a written request to concur in the appointment of an arbiter, a person has to be nominated at the request of either party by the Dean of the Faculty of Advocates, provided that in a difference or dispute arising out of provisions relating to copyright, clauses 3.15 to 3.18 above, the arbiter shall, unless otherwise agreed, be an architect.

3.27. Nothing herein shall prevent the parties agreeing to settle any difference or dispute arising out of the appointment without recourse to arbitration.

Governing laws

3.28. The application of these conditions shall be governed by the laws of England and Wales.

or

3.28S The application of these conditions shall be governed by the laws of Scotland

or

3.28NI The application of these conditions shall be governed by the laws of Northern Ireland.

PART 4 RECOMMENDED FEES AND EXPENSES

This part describes the recommended and not mandatory methods of calculating fees for the architect's services and expenses. Fees may be based on a percentage of the total construction cost or on time expended, or may be a lump sum. This part should be read in conjunction with Parts 1, 2 and 3.

Percentage fees

4.1. The percentage fee scales shown in Figure 1 (graph) are for use where the architect's appointment is for the Basic Services described in Part 1 for new works having a total construction cost between £20,000 and £5,000,000. Where the total construction is less than £20,000 or more than £5,000,000 client and architect should agree an appropiate fee basis at the time of appointment.

4.2. Percentage fees are based on the total construction cost of the works; on the issue of the final certificate fees should be recalculated on the actual total construction cost.

4.3. Total construction cost is defined as the cost, as certified by the architect, of all works including site works executed under the architect's direction, subject to the following:

a The total construction cost includes the cost of all work designed by consultants and co-ordinated by the architect, irrespective of whether such work is carried out under separate building contracts for which the architect may not be responsible. The architect will be informed of the cost of any such separate contracts.

b The total construction cost does not include specialist sub-contractors' design fees for work on which consultants would otherwise have been employed. Where such fees are not known, the architect will estimate a reduction from the total construction cost.

ARCHITECTS' FEES

c For the purpose of calculating the appropriate fee, the total construction cost - includes the actual or estimated cost of any work executed which is excluded from the contract but otherwise designed by the Architect; - is not subject to any deductions made in respect of work not in accordance with the building contract.

d The total construction cost includes the cost of built-in furniture and equipment. Where the the cost of any special equipment is excluded from the total construction cost, the architect may charge additionally for work in connection with such items.

e Where any material, labour or carriage is supplied by a client who is not the contractor, the cost will be estimated by the architect as if it were supplied by the contractor, and included in the total cost.

f Where the client is the contractor, a statement of the acertained gross cost of the works may be used in caluclating the total construction cost of the works. In the absence of such a statement, the architect's own estimate will be used. In both a statement of ascertained gross cost and an architect's estimate there will be included an allowance for the contractor's profit and overheads.

4.4. Buildings are divided into five classes for fee calculation purposes. For guidance only the building types most likely to fall into each class are shown in Figure 3.

Repetition

4.5. The classification of buildings in Figure 3 takes account of reduced design work arising from the nature of the building.

4.6. Where a building is repeated for the same client the recommended fee for the superstructure may be reduced on all except the first three of any houses of the same design and on all except the first of all other building types of the same design.

4.7. Where a single building incorporates a number of identical compartments such as floors or complete structural bays the recommended fee may be reduce on all identical compartments in excess of ten.

4.8. Reductions should be made by waiving the fee for work stages E, F and G where a complete design can be re-used without modification other than the handing of a plan.

Time charge fees

4.9. Time charges are based on hourly rates for principals and other techical staff. In assessing the hourly rate all relevant factors should be considered, including the complexity of the work, the qualifications, experience and responsibility of the architect, and the character of any negotiations. Hourly rates for principals shall be agreed. The hourly rate for technical staff should be not less than 18 pence per £100 of gross annual income.

4.10. Technical staff are defined as architectural and other professional and technical staff, where the architect is responsible for deducting PAYE and National Insurance contributions from those persons' salaries on behalf of the Inland Revenue.

4.11. Gross annual income includes bonus payments plus the employer's share of contributions towards National Insurance, pension and private medical schemes and other emoluments such as car and accomodation allowances.

ARCHITECTS' FEES

4.12. Where the staff are provided by an agency hourly rates shall be agreed.
4.13. Where site staff are employed by the architect hourly rates shall be agreed.
4.14. Unless otherwise agreed no separate time charges will be made for secretarial staff or staff engaged on general accountancy or administrative duties.
4.15. The architect will maintain records of time spent on services performed on a time basis. The architect will make such records available to the client on reasonable request.

Lump sum fees

4.16. The architect may agree with the client to charge a lump sum fee for any of the services described in Parts 1 and 2 in appropriate circumstances, for example where:

- the client's requirements are provided in a form such that the architect is not obliged to undertake any additional preparatory work;
- the full extent of the service can be determined when the architect is appointed; and
- the architect's service can be completed within an agreed period.

Works to existing building

4.17. The percentage fee scales shown in Figure 2 (graph) are for use where the architect's appointment is for the Basic Services described in Part 1 for alterations or extensions to an existing building having a total construction cost of between £20 000 and £5 000 000. Where the total construction cost is less than £20 000 or more than £5 000 000 client and architect should agree an appropriate fee basis at the time of the appointment.
4.18. Where extensions to existing buildings are substantially independent, percentage fees should be as Figure 1 for new works, but the fee for those sections of the works which marry existing buildings to the new should be charged separately as Figure 2 applicable to an independent commission of similar value.
4.19. Where the architect's appointment is for repair and restoration work fees should be on a time basis: alternatively a percentage fee may be agreed.
4.20. Where the architect's appointment is in connection with works to a building of architectural or historical interest, or to a building in a conservation area, higher fees may be charged.

Compounding of fees

4.21. By agreement the percentage or lump sum fee may be compounded to cover all or any part of the architect's services and expenses.

Interim payments

4.22. Fees and expenses should be paid in instalments either at regular intervals or on completion of work stages of the Basic services (Part 1).

ARCHITECTS' FEES

4.23. Where interim payment of percentage or lump sum fees is related to completion of work stages of the Basic Services the recommended apportionment is as follows:

Work stage	Proportion of fee	Cumulative total
C	15%	15%
D	20%	35%
E	20%	55%
F G	20%	75%
H J K L	25%	100%

Fees in respect of work stages E to L should be paid in instalments proportionate to the work completed or the values of the works certified from time to time. Interim payment should be based on the current estimated cost of the works.

The apportionment of fees is a means of assessing interim payments and does not necessarily reflect the amount of work completed in any work stage. By agreement an adjustment in the apportionment may be made.

Interest on "Any sums remaining un-paid at the expiry of 30 days from the date of submission of the fee account shall bear intrest thereafter, such interest to accrue from day to day at the rate of 2% per annum above the current base rate of the architect's principal bank".

Partial services

4.24. The architect may be required to provide part only of the Basic Services (Part 1). In such cases the architect will be entitled to a commensurate fee.

4.25. Where work is to be done by or on behalf of the client, resulting in the omission of part of work stages C to L, or a sponsored constructional method is to be used, a commensurate reduction in the recommended percentage fee may be agreed. In accessing the reduction, due account should be taken for the need for the architect to become thoroughly familiar with the work done by others, and a familiarization fee will be charged for this work.

4.26. All percentage fees for partial services should be based on the architect's current estimate of the total construction cost of the works. Such estimates may be based on an accepted tender or, subject to the following, on the lowest of unaccepted tenders. Where partial services are provided in respect of works for which the executed cost is not known and no tender has been accepted, percentage fees should be based either on the architect's estimated total construction cost or on the most recent cost limit agreed with the client, whichever is the lower.

4.27. Fees for partial services may alternatively be on a time or lump sum basis.

Suspension, resumption and termination

4.28. On suspension, or termination of the architect's appointment the architect will be entitled to fees for all work complete at that time. Fees will be charged on a partial service basis.

4.29. During such period of suspension the architect will be reimbursed by the client for all expenses and disbursement necessarily incurred under the appointment.

ARCHITECTS' FEES

4.30. On the resumption of a suspended service within six months, previous payments will be regarded solely as payment on account towards the total fee.

4.31. Where the architect's appointment is terminated by the client the architect will be reimbursed by the client for all expenses and disbursements necessarily incurred in connection with work then in progress and arising as a result of the termination.

Expenses and disbursements

4.32. In addition to the fees charged the architect will be reimbursed for all expenses and disbursements properly incurred in connection with the appointment, including the following:

a Printing, reproduction or purchase costs of all documents, drawings, maps, models, photographs, and other records, including all those used in communication between architect, client, consultants and contractors, and for enquiries to contractors, sub-contractors and suppliers, notwithstanding any obligation on the part of the architect to supply such documents to those concerned, except that contractors will pay for any prints additional to those to which they are entitled under the contract.

b Hotel and travelling expenses, including milage allowance for cars at rates stated in the Schedule of Services and Fees and other similar disbursements.

c All payments made on behalf of the client, such as expenses incurred in advertising for tenders and resident site staff including the time and expense of interviewers and reasonable expenses for interviewees.

d Fees and other charges for specialist professional advise, including legal advise, which have been incurred by the architect with the specific authority of the client.

e The cost of postage, telephone charges, telex messages, telegrams, cables, facsimilies, air-freight and courier services.

f Rental and hire charges for specialised equipment, including computers, where required and agreed by the client.

g Where work charged on a percentage fee is at such distance that an exceptional amount of time is spent travelling, additional charges may be made.

4.33. The architect will maintain records of all such expenses and disbursement and will make these records available to the client on reasonable request.

4.34. Expenses and disbursements may by agreement be estimated or standardised in whole or in part, or compounded for an increase in the percentage or lump sum fee.

4.35. The client will pay all fees in respect of applications under planning and building acts and other statutory requirements.

ARCHITECTS' FEES

Variations

4.36. Where the scope of the architect's services is varied fees may be adjusted accordingly.

4.37. Where the architect is involved in extra work and expense for reasons beyond the architect's control and for which the architect would not otherwise be renumerated additional fees are due. Any of the following is likely to involve the architect in extra work and expense:

- a The need to revise reports, drawings, specifications or other documents due to changes in interpretation or enactment or revisions of the laws, statutory or other regulations.
- b Changes in the client's instructions, or delay by the client in providing information.
- c Consideration of notices, applications or claims by the contractor under a building contract; delays resulting from defects or deficiencies in the work of the contractor, sub-contractors or suppliers; default, bankruptcy or liquidation of the contractor, sub-contractors or suppliers.
- d Any other cause beyond the architect's control.

Value Added Tax

4.38. The amount of any Value Added Tax on the services and expenses of the architect arising under the Finance Act 1972 will be chargeable to the client in addition to the architect's fees and expenses.

ARCHITECTS' FEES

Figure 1: Recommended percentage fee scales: New works

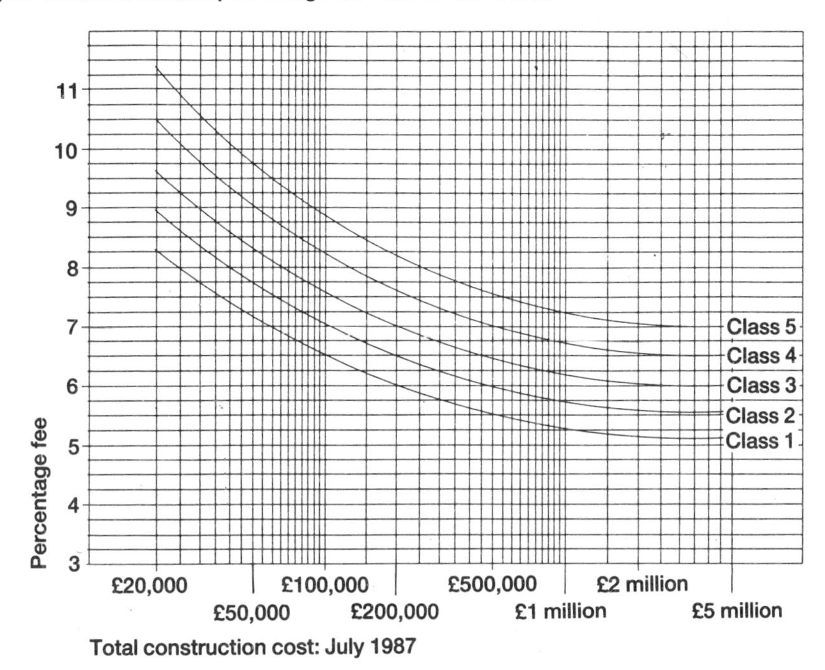

Figure 2: Recommended percentage fee scales: Works to existing buildings

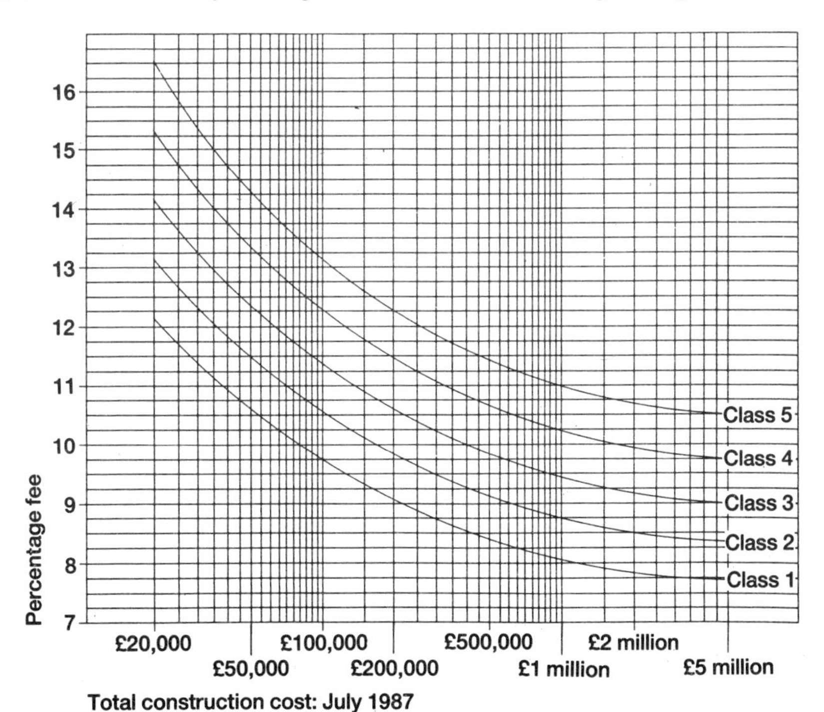

ARCHITECTS' FEES

Figure 3: Classification of buildings, for guidance

Type	Class 1	Class 2	Class 3	Class 4	Class 5
Industrial	Storage sheds	Speculative factories and warehouses Assembly and machine workshops Transport garages	Purpose-built factories and warehouses Motor garages/ showrooms		
Agricultural	Barns and sheds Stables	Animal breeding units			
Commercial	Single-storey car parks	Mutli-storey car parks	Supermarkets Banks Shops Retall warehouses Offices	Department stores Shopping centres Food processing units Breweries Telecommunications and computer accommodation Restaurants Public houses	High risk research and production buildings Research and development laboratories Radio, TV and recording studios
Community		Communal halls	Community centres Branch libraries Ambulance and fire stations Bus stations Police stations Prisons Postal accommodation	Civic centres Churches and crematoria Concert halls Specialist libraries Museums Art galleries Magistrates/ county/ sheriff courts	Theatres Opera houses Crown/high courts
Residential		Dormitory hostels	Estate housing and flats Sheltered housing Housing for single people	Parsonages manses Hotels Accommodation for the handicapped Accommodation for the frail elderly	Houses and flats for individual clients
Education			Primary/nursery/ first schools	Other schools including middle and secondary University complexes	University laboratories
Recreation			Sports halls Squash courts	Swimming pools Leisure complexes	Leisure pools Specialised complexes
Medical social services			Clinics	Health centres General hospital complexes Nursing homes Surgeries	Teaching hospitals Hospital laboratories Dental surgeries

QUANTITY SURVEYORS' FEES

Scale 36 inclusive of professional charges for quantity surveying services for building works issued by The Royal Institution of Chartered Surveyors. The scale is recommended and not mandatory.

EFFECTIVE FROM JULY 1988

1.0. GENERALLY

1.1. This scale is for use when an inclusive scale of professional charges is considered to be appropriate by mutual agreement between the employer and the quantity surveyor.

1.2. This scale does not apply to civil engineering works, housing schemes financed by local authorities and the Housing Corporation and housing improvement work for which separate scales of fees have been published.

1.3. The fees cover quantity surveying services as may be required in connection with a building project irrespective of the type of contract from initial appointment to final certification of the contractor's account such as:

(a) Budget estimating; cost planning and advice on tendering procedures and contract arrangements.

(b) Preparing tendering documents for main contract and specialist sub-contracts; examining tenders received and reporting thereon or negotiating tenders and pricing with a selected contractor and or sub-contractors.

(c) Preparing recommendations for interim payments on account to the contractor; preparing periodic assessments of anticipated final cost and reporting thereon; measuring work and adjusting variations in accordance with the terms of the contract and preparing final account, pricing same and agreeing total with the contractor.

(d) Providing a reasonable number of copies of bills of quantities and other documents; normal travelling and other expenses. Additional copies of documents, abnormal travelling and other expenses (e.g. in remote areas or overseas) and the provision of checkers on site shall be charged in addition by prior arrangement with the employer

1.4. If any of the materials used in the works are supplied by the employer or charged at a preferential rate, then the actual or estimated market value thereof shall be included in the amounts upon which fee are to be calculated.

1.5. If the quantity surveyor incurs additional costs due to exceptional delays in building operations or any other cause beyond the control of the quantity surveyor then the fees may be adjusted by agreement between the employer and the quantity surveyor to cover the reimbursement of these additional costs.

1.6. The fees and charges are in all cases exclusive of value added tax which will be applied in accordance with legislation.

1.7. Copyright in bills of quantities and other documents prepared by the quantity surveyor is reserved to the quantity surveyor.

QUANTITY SURVEYORS' FEES

2.0. INCLUSIVE SCALE

2.1. The fees for the services outlined in para.1.3, subject to the provision of para. 2.2, shall be as follows:

(a) **Category A:** Relatively complex works and/or works with little or no repetition.

Examples:
Ambulance and fire stations; banks; cinemas; clubs; computer buildings; council offices; crematoria; houses; fitting out of existing buildings; homes for the elderly; hospitals and nursing homes; laboratories; law courts; libraries; 'one off' houses; petrol stations; places of religious worship; police stations; public houses, licensed premises; restaurants; sheltered housing; sports pavillions; theatres;town halls; universities, polytechnics and colleges of further education (other than halls of residence and hostels); and the like.

Value of work £	Category A fee £	£
Up to 150,000	380 + 6.0%	(Minimum fee £3,380)
150,000 - 300,000	9,380 + 5.0%	on balance over 150,000
300,000 - 600,000	16,880 + 4.3%	on balance over 300,000
600,000 - 1,500,000	29,780 + 3.4%	on balance over 600,000
1,500,000 - 3,000,000	60,380 + 3.0%	on balance over 1,500,000
3,000,000 - 6,000,000	105,380 + 2.8%	on balance over 3,000,000
Over 6,000,000	189,380 + 2.4%	on balance over 6,000,000

(b) **Category B:** Less complex works and/or works with some element of repetition.

Examples:
Adult education facilities; canteens; church halls; community centres; departmental stores; enclosed sports stadia and swimming baths; halls of residence; hospitals; hostels; motels; offices other than those included in Categories A and C; railway stations; recreation and leisure centres; residential hotels; schools; self-contained flats and maisonettes; shops and shopping centres; supermarkets and hypermarkets; telephone exchanges; and the like.

Value of Work £	Category B fee £	£
Up to 150,000	360 + 5.8%	(Minimum fee £3,260)
150,000 - 300,000	9,060 + 4.7%	on balance over 150,000
300,000 - 600,000	16,110 + 3.9%	on balance over 300,000
600,000 - 1,500,000	27,810 + 2.8%	on balance over 600,000
1,500,000 - 3,000,000	53,010 + 2.6%	on balance over 1,500,000
3,000,000 - 6,000,000	92,010 + 2.4%	on balance over 3,000,000
Over 6,000,000	164,010 + 2.0%	on balance over 6,000,000

QUANTITY SURVEYORS' FEES

(c) **Category C:** Simple works and/or works with a substantial element of repetition.

Examples:
Factories; garages; multi-storey car parks; open-air sports stadia; structural shell offices not fitted out; warehouses; workshops; and the like,

Value of work £	Category C fee £	£
Up to 150,000	300 + 4.9% (Minimum fee £2,750)	
150,000 - 300,000	7,650 + 4.1% on balance over	150,000
300,000 - 600,000	13,800 + 3.3% on balance over	300,000
600,000 - 1,500,000	23,700 + 2.5% on balance over	600,000
1,500,000 - 3,000,000	46,200 + 2.2% on balance over	1,500,000
3,000,000 - 6,000,000	79,200 + 2.0% on balance over	3,000,000
Over 6,000,000	139,200 + 1.6% on balance over	6,000,000

(d) Fees shall be calculated upon the total of the final account for the whole of the work including all nominated sub-contractors' and nominated supplier's accounts. When work normally included in a building contract is the subject of a separate contract for which the quantity surveyor has not been paid fees under any other clause hereof, the value of such work shall be included in the amount upon which fees are charged.

(e) When a contract comprises buildings which fall into more than one category, the fee shall be calculated as follows:

(i) The amount upon which fees are chargeable shall be allocated to the categories of work applicable and the amounts so allocated expressed as percentages of the total amount upon which fees are chargeable.

(ii) Fees shall then be calculated for each category on the total amount upon which fees are chargeable.

(iii) The fee chargeable shall then be calculated by applying the percentages of work in each category to the appropriate total fee and adding the resultant amounts.

(iv) A consolidated percentage fee applicable to the total value of the work may be charged by prior agreement between the employer and the quantity surveyor. Such a percentage shall be based on this scale and on the estimated cost of the various categories of work and calculated in accordance with the principles stated above.

(f) When a project is subject to a number of contracts then, for the purpose of calculating fees, the values of such contracts shall not be aggregated but each contract shall be taken separately and the scale of charges (paras. 2.1 (a) to (e)) applied as appropriate.

QUANTITY SURVEYORS' FEES

2.2. Air conditioning, heating, ventilating and electrical services

(a) When the services outlined in para. 1.3 are provided by the quantity surveyor for the air conditioning, heating, ventilating and electrical services there shall be a fee for these services in addition to the fee calculated in accordance with para. 2.1 as follows:

Value of work £	Additional fee £	£
Up to 120,000	5.0%	
120,000 - 240,000	6,000 + 4.7% on balance over	120,000
240,000 - 480,000	11,640 + 4.0% on balance over	240,000
480,000 - 750,000	21,240 + 3.6% on balance over	480,000
750,000 -1,000,000	30,960 + 3.0% on balance over	750,000
1,000,000 -4,000,000	38,460 + 2.7% on balance over	1,000,000
Over 4,000,000	119,460 + 2.4% on balance over	4,000,000

(b) The value of such services, whether the subject of separate tenders or not, shall be aggregated and the total value of work so obtained used for the purpose of calculating the additional fee chargeable in accordance with para. (a). (Except that when more than one firm of consulting engineers is engaged on the design of these services, the separate values for which each such firm is responsible shall be aggregated and the additional fees charged shall be calculated independently on each such total value so obtained.)

(c) Fees shall be calculated upon the basis of the account for the whole of the air conditioning, heating, ventilating and electrical services for which bills of quantities and final accounts have been prepared by the quantity surveyor.

2.3. Works of alteration

On works of alteration or repair, or on those sections of the work which are mainly works of alteration or repair, there shall be a fee of 1.0% in addition to the fee calculated in accordance with paras. 2.1 and 2.2.

2.4. On works of redecoration and associated minor repairs

On works of redecoration and associated minor repairs, there shall be a fee of 1.5% in addition to the fee calculated in accordance with paras. 2.1 and 2.2.

QUANTITY SURVEYORS' FEES

2.5. **Generally**

If the works are substantially varied at any stage or if the quantity surveyor is involved in an excessive amount of abortive work, then the fees shall be adjusted by agreement between the employer and the quantity surveyor.

3.0. ADDITIONAL SERVICES

3.1. For additional services not normally necessary, such as those arising as a result of the termination of a contract before completion, liquidation, fire damage to the buildings, services in connection with arbitration, litigation and investigation of the validity of contractors' claims, services in connection with taxation matters and all similar services where the employer specifically instructs the quantity surveyor, the charges shall be in accordance with para. 4.0 below.

4.0. TIME CHARGES

4.1. (a) For consultancy and other services performed by a principal, a fee by arrangement according to the circumstances including the professional status and qualifications of the quantity surveyor.

(b) When a principal does work which would normally be done by a member of staff, the charge shall be calculated as para. 4.2 below.

4.2. (a) For services by a member of staff, the charges for which are to be based on the time involved, such charges shall be calculated on the hourly cost of the individual involved plus 145%.

(b) A member of staff shall include a principal doing work normally done by an employee (as para. 4.1 (b) above), technical and supporting staff, but shall exclude secretarial staff or staff engaged upon general administration.

(c) For the purpose of para. 4.2 (b) above, a principal's time shall be taken at the rate applicable to a senior assistant in the firm.

(d) The supervisory duties of a principal shall be deemed to be included in the addition of 145% as para. 4.2 (a) above and shall not be charged separately.

(e) The hourly cost to the employer shall be calculated by taking the sum of the annual cost of the member of staff of:

(i) Salary and bonus but excluding expenses;

(ii) Employer's contributions payable under any Pension and Life Assurance Schemes;

(iii) Employer's contributions made under the National Insurance A Acts, the Redundancy Payments Act and any other payments made in respect of the employee by virtue of any statutory requirements; and

(iv) Any other payments or benefits made or granted by the employer in pursuance of the terms of employment of the member of staff;

and dividing by 1,650.

QUANTITY SURVEYORS' FEES

5.0. INSTALMENT PAYMENTS

5.1. In the absence of agreement to the contrary, fees shall be paid by instalments as follows:

(a) Upon acceptance by the employer of a tender for the works, one half of the fee calculated on the amount of the accepted tender.

(b) The balance by instalments at intervals to be agreed between the date of the first certificate and one month after final certification of the contractor's account.

5.2. (a) In the event of no tender being accepted, one half of the fee shall be paid within three months of completion of the tender documents. The fee shall be calculated upon the basis of the lowest original bona fide tender received. In the event of no tender being received, the fee shall be calculated upon a reasonable valuation of the works based upon the tender documents.

(b) In the event of the project being abandoned at any stage other than those covered by the foregoing, the proportion of fee payable shall be by agreement between the employer and the quantity surveyor.

NOTE: In the foregoing context 'bona fide tender' shall be deemed to mean a tender submitted in good faith without major errors of computation and not subsequently withdrawn by the tenderer.

QUANTITY SURVEYORS' FEES

Scale 37 itemized scale of professional charges for quantity surveying services for building work issued by the Royal Institution of Chartered Surveyors. The scale is recommended and not mandatory.

EFFECTIVE FROM JULY 1988

1.0. GENERALLY

1.1. The fees are in all cases exclusive of travelling and other expenses (for which the actual disbursement is recoverable unless there is some prior arrangement for such charges) and of the cost of reproduction of bills of quantities and other documents, which are chargeable in addition at net cost.
1.2. The fees are in all cases exclusive of services in connection with the allocation of the cost of the works for purposes of calculating value added tax for which there shall be an additional fee based on the time involved (see paras. 19.1 and 19.2).
1.3. If any of the materials used in the works are supplied by the employer or charged at a preferential rate, then the actual or estimated market value thereof shall be included in the amounts upon which fees are to be calculated.
1.4. The fees are in all cases exclusive of preparing a specification of the materials to be used and the works to be done, but the fees for preparing bills of quantities and similar documents do include for incorporating preamble clauses describing the materials and workmanship (from instructions given by the architect and/or consulting engineer).
1.5. If the quantity surveyor incurs additional costs due to exceptional delays in building operations or any other cause beyond the control of the quantity surveyor then the fees may be adjusted by agreement between the employer and the quantity surveyor to cover the reimbursement of these additional costs.
1.6. The fees and charges are in all cases exclusive of value added tax which will be applied in accordance with legislation.
1.7. Copyright in bills of quantities and other documents prepared by the quantity surveyor is reserved to the quantity surveyor.

CONTRACTS BASED ON BILLS OF QUANTITIES: PRE-CONTRACT SERVICES

2.0. BILLS OF QUANTITIES

2.1. **Basic scale**

For preparing bills of quantities and examining tenders received and reporting thereon.

(a) **Category A**: Relatively complex works and/or works with little or no repetition.

QUANTITY SURVEYORS' FEES

Examples:
Ambulance and fire stations; banks; cinemas; clubs; computer buildings; council offices; crematoria; fitting out of existing buildings; homes for the elderly; hospitals and nursing homes; laboratories; law courts; libraries; 'one off' houses; petrol stations; places of religious worship; police stations; public houses, licensed premises; restaurants; sheltered housing; sports pavilions; theatres; town halls; universities, polytechnics and colleges of further education (other than halls of residence and hostels); and the like.

Value of work £	Category A fee £	£
Up to 150,000	230 + 3.0% (Minimum fee £1,730)	
150,000 - 300,000	4,730 + 2.3%on balance over	150,000
300,000 - 600,000	8,180 + 1.8% on balance over	300,000
600,000 -1,500,000	13,580 + 1.5% on balance over	600,000
1,500,000 -3,000,000	27,080 + 1.2% on balance over	1,500,000
3,000,000 -6,000,000	45,080 + 1.1% on balance over	3,000,000
Over 6,000,000	78,080 + 1.0% on balance over	6,000,000

(b) **Category B:** Less complex works and/or works with some element of repetition.

Examples:
Adult education facilities; canteens; church halls; community centres; departmental stores; enclosed sports stadia and swimming baths; halls of residence; hospitals; hostels; motels; offices other than those included in Categories in A and C; railway stations; recreation and leisure centres; residential hotels; schools; self-contained flats and maisonettes; shops and shopping centres; supermarkets and hypermarkets; telephone exchanges; and the like.

Value of work £	Category B fee £	£
Up to 150,000	210 + 2.8% (Minimum fee £1,610)	
150,000 - 300,000	4,410 + 2.0% on balance over	150,000
300,000 - 600,000	7,410 + 1.5% on balance over	300,000
600,000 - 1,500,000	11,910 + 1.1% on balance over	600,000
1,500,000 - 3,000,000	21,810 + 1.0% on balance over	1,500,000
3,000,000 - 6,000,000	36,810 + 0.9% on balance over	3,000,000
Over 6,000,000	63,810 + 0.8% on balance over	6,000,000

(c) **Category C:** Simple works and/or works with a substantial element of repetition

Examples:
Factories; garages; multi-storey car parks; open-air sports stadia; structural shell offices not fitted out; warehouses; workshops and the like.

QUANTITY SURVEYORS' FEES

Value of Work £		Category C fee £		£
Up to	150,000	180 + 2.5%	(Minimum fee £1,430)	
150,000 -	300,000	3,930 + 1.8%	on balance over	150,000
300,000 -	600,000	6,630 + 1.2%	on balance over	300,000
600,000 -	1,500,000	10,230 + 0.9%	on balance over	600,000
1,500,000 -	3,000,000	18,330 + 0.8%	on balance over	1,500,000
3,000,000 -	6,000,000	30,330 + 0.7%	on balance over	3,000,000
Over	6,000,000	51,330 + 0.6%	on balance over	6,000,000

(d) The scales of fees for preparing bills of quantities (paras. 2.1 (a) to (c).) are overall scales based upon the inclusion of all provisional and prime cost items, subject to the provision of para. 2.1 (g). When work normally included in a building contract is the subject of a separate contract for which the quantity surveyor has not been paid fees under any other clause hereof, the value of such work shall be included in the amount upon which fees are charged.

(e) Fees shall be calculated upon the accepted tender for the whole of the work subject to the provisions of para. 2.6. In the event of no tender being accepted, fees shall be calculated upon the basis of the lowest original bona fide tender received. In the event of no such tender being received, the fees shall be calculated upon a reasonable valuation of the works based upon the original bills of quantities.

NOTE: In the foregoing context 'bona fide tender' shall be deemed to mean a tender submitted in good faith without major errors of computation and not subsequently withdrawn by the tenderer.

(f) In calculating the amount upon which fees are charged the total of any credits and the totals of any alternative bills shall be aggregated and added to the amount described above. The value of any omission or addition forming part of an alternative bill shall not be added unless measurement or abstraction from the original dimension sheets was necessary.

(g) Where the value of the air conditioning, heating, ventilating and electrical services included in the tender documents together exceeds 25% of the amount calculated as described in paras. 2.1 (d) and (e), then, subject to the provisions of para. 2.2, no fee is chargeable on the amount by which the value of these services exceeds the said 25%. In this context the term 'value' excludes general contractor's profit, attendance, builder's work in connection with the services, preliminaries and any similar additions.

(h) When a contract comprises buildings which fall into more than one category, the fee shall be calculated as follows:

(i) The amount upon which fees are chargeable shall be allocated to the categories of work applicable and the amounts so allocated expressed as percentages of the total amount upon which fees are chargeable.

(ii) Fees shall then be calculated for each category on the total amount upon which fees are chargeable.

(iii) The fee chargeable shall then be calculated by applying the percentages of work in each category to the appropriate total fee and adding the resultant amounts.

QUANTITY SURVEYORS' FEES

(j) When a project is the subject of a number of contracts then, for the purpose of calculating fees, the values of such contracts shall not be aggregated but each contract shall be taken seperately and the scale of charges (paras. 2.1 (a) to (h)) applied as appropriate.

(k) Where the quantity surveyor is specifically instructed to provide cost planning services the fee calculated in accordance with paras. 2.1 (a) to (j) shall be increased by a sum calculated in accordance with the following table and based upon the same value of work as that upon which the aforementioned fee has been calculated:

Categories A & B: (as defined in paras. 2.1 (a) and (b)).

Value of work £	Fee £	£
Up to 600,000	0.7%	
600,000 - 3,000,000	4,200 + 0.4% on balance over	600,000
3,000,000 - 6,000,000	13,800 + 0.35% on balance over	3,000,000
Over 6,000,000	24,300 + 0.3% on balance over	6,000,000

Category C: (as defined in paras. 2.1 (c))

Value of work £	Fee £	£
Up to 600,000	0.5%	
600,000 -3,000,000	3,000 + 0.3% on balance over	600,000
3,000,000 -6,000,000	10,200 + 0.25% on balance over	3,000,000
Over 6,000,000	17,700 + 0.2% on balance over	6,000,000

2.2. **Air conditioning, heating, ventilating and electrical services**

(a) Where bills of quantities are prepared by the quantity surveyor for the air conditioning, heating, ventilating and electrical services there shall be a fee for these services (which shall include examining tenders received and reported thereon), in addition to the fee calculated in accordance with para. 2.1, as follows:

Value of work £	Additional fee £	£
Up to 120,000	2.5%	
120,000 - 240,000	3,000 + 2.25% on balance over	120,000
240,000 - 480,000	5,700 + 2.0% on balance over	240,000
480,000 - 750,000	10,500 + 1.75% on balance over	480,000
750,000 - 1,000,000	15,225 + 1.25% on balance over	750,000
Over - 1,000,000	18,350 + 1.15% on balance over	1,000,000

(b) The values of such services, whether the subject of separate tenders or not, shall be aggregated and the total value of work so obtained used for the purpose of calculating the additional fee chargeable in accordance with para. (a).
(Except that when more than one firm of consulting engineers is engaged on the design of these services, the separate values for which each such firm is responsible shall be aggregated and the additional fees charged shall be calculated independently on each such total value so obtained.)

QUANTITY SURVEYORS' FEES

(c) Fees shall be calculated upon the accepted tender for the whole of the air conditioning, heating, ventilating and electrical services for which bills of quantities have been prepared by the quantity surveyor. In the event of no tender being accepted, fees shall be calculated upon the basis of the lowest original bona fide tender received. In the event of no such tender being received, the fees shall be calculated upon a reasonable valuation of the services based upon the original bills of quantities.

NOTE: In the foregoing context 'bona fide tender' shall be deemed to mean a tender submitted in good faith without major errors of computation and not subsequently withdrawn by the tenderer.

(d) When cost planning services are provided by the quantity surveyor for air conditioning, heating, ventilating and electrical services (or for any part of such services) there shall be an additional fee based on the time involved (see paras. 19.1 and 19.2). Alternatively the fee may be on a lump sum or percentage basis agreed between the employer and the quantity surveyor.

NOTE: The incorporation of figures for air conditioning, heating, ventilating and electrical services provided by the consulting engineer is deemed to be included in the quantity surveyor's services under para. 2.1.

2.3. **Works on alteration**

On works of alteration or repair, or on those sections of the works hich are mainly works of alteration or repair, there shall be a fee of 1.0% in addition to the fee calculated in accordance with paras. 2.1 and 2.2.

2.4. **Works of redecoration and associated minor repairs,**

On works of redecoration and associated minor repairs, there shall be a fee of 1.5% in addition to the fee calculated in accordance with paras. 2.1 and 2.2.

2.5. **Bills of quantities prepared in special forms**

Fees calculated in accordance with paras. 2.1, 2.2, 2.3, and 2.4 include for the preparation of bills of quantities on a normal trade basis. If the employer requires additional information to be provided in the bills of quantities or the bills to be prepared in an elemental, operational or similar form, then the fee may be adjusted by agreement between the employer and the quantity surveyor.

2.6. **Reduction of tenders**

(a) When cost planning services have been provided by the quantity surveyor and a tender, when received, is reduced before acceptance and if the reduction are not necessitated by amended instructions of the employer or by the inclusion in the bills of quantities of items which the quantity surveyor has indicated could not be contained within the approved estimate, then in such a case no charge shall be made by the quantity surveyor for the preparation of bills of reductions and the fee for the preparation of the bills of quantities shall be based on the amount of the reduced tender.

QUANTITY SURVEYORS' FEES

(b) When cost planning services have not been provided by the quantity surveyor and if a tender, when received, is reduced before acceptance, fees are to be calculated upon the amount of the unreduced tender. When the preparation of bills of reductions is required, a fee is chargeable for preparing such bills of reductions as follows:

(i) 2.0% upon the gross amount of all omissions requiring measurement or abstraction from original dimensional sheets.
(ii) 3.0% upon the gross amount of all additions requiring measurement.
(iii) 0.5% upon the gross amount of all remaining additions.

NOTE: The above scale for the preparation of bills of reductions applies to work in all categories.

2.7. **Generally**

If the works are substantially varied at any stage or if the quantity surveyor is involved in an excessive amount of abortive work, then the fees shall be adjusted by agreement between the employer and the quantity surveyor.

3.0. NEGOTIATING TENDERS

3.1. (a) For negotiating and agreeing prices with a contractor:

Value of work £		Fee £		£
Up to	150,000	0.5%		
150,000 -	600,000	750 + 0.3%	on balance over	150,000
600,000 -	1,200,000	2,100 + 0.2%	on balance over	600,000
Over	1,200,000	3,300 + 0.1%	on balance over	1,200,000

(b) The fee shall be calculated on the total value of the works as defined in paras. 2.1 (d), (e), (f), (g) and (j).
(c) For negotiating and agreeing prices with a contractor for air conditioning, heating, ventilating and electrical services there shall be an additional fee as para. 3.1 (a) calculated on the total value of such services as defined in para. 2.2 (b).

4.0. CONSULTATIVE SERVICES AND PRICING BILLS OF QUANTITIES

4.1. **Consultative services**

Where the quantity surveyor is appointed to prepare approximate estimates, feasibility studies or submissions for the approval of financial grants or similar services, then the fee shall be based on the time involved (see paras. 19.1 and 19.2) or alternatively, on a lump sum or percentage basis agreed between the employer and the quantity surveyor.

QUANTITY SURVEYORS' FEES

4.2. **Pricing bills of quantities**

(a) For pricing bills of quantities, if instructed, to provide an estimate comparable with tenders, the fee shall be one-third (33.33%) of the fee for negotiating and agreeing prices with a contractor, calculated in accordance with paras. 3.1 (a) and (b).
(b) For pricing bills of quantities, if instructed, to provide an estimate comparable with tenders for air conditioning, heating, ventilating and electrical services the fee shall be one-third (33.33%) of the fee calculated in accordance with para. 3.1. (c).

CONTRACTS BASED ON BILLS OF QUANTITIES: POST-CONTRACT SERVICES

Alternative scales (I and II) for post-contract services are set out below to be used at the quantity surveyor's discretion by prior agreement with the employer.

5.0. ALTERNATIVE I: OVERALL SCALE OF CHARGES FOR POST-CONTRACT SERVICES

5.1. If the quantity surveyor appointed to carry out the post-contract services did not prepare the bills of quantities then the fees in paras. 5.2 and 5.3 shall be increased to cover the additional services undertaken by the quantity suveyor.

5.2. **Basic scale**

For taking particulars and reporting valuations for interim certificates for payments on account to the contractor, preparing periodic assessments of anticipated final cost and reporting thereon, measuring and making up bills of variations including pricing and agreeing totals with the contractor, and adjusting fluctuations in the cost of labour and materials if required by the contract.

(a) **Category A:** Relatively complex works and/or works with little or no repetition.

Examples:
Ambulance and fire stations; banks; cinemas; clubs; computer buildings; council offices; crematoria; fitting out existing buildings; homes for the elderly; hospitals and nursing homes; laboratories; law courts; libraries; 'one-off' houses; petrol stations; places of religious worship; police stations; public houses, licensed premises; restaurants; sheltered housing; sports pavilions; theatres; town halls; universities, polytechnics and colleges of further education (other than halls of residence and hostels); and the like.

Value of work £		Category A fee £		£
Up to	150,000	150 +	2.0% (Minimum fee £1,150)	
150,000 -	300,000	3,150 +	1.7% on balance over	150,000
300,000 -	600,000	5,700 +	1.6% on balance over	300,000
600,000 -	1,500,000	10,500 +	1.3% on balance over	600,000
1,500,000 -	3,000,000	22,200 +	1.2% on balance over	1,500,000
3,000,000 -	6,000,000	40,200 +	1.1% on balance over	3,000,000
Over	6,000,000	73,200 +	1.0% on balance over	6,000,000

QUANTITY SURVEYORS' FEES

(b) **Category B**: Less complex works and/or works with some element of repetition.

Examples:
Adult education facilities; canteens; church halls; community centres; departmental stores; enclosed sports stadia and swimming baths; halls of residence; hostels; motels; offices other than those included in Categories A and C; railway stations; recreation and leisure centres; residential hotels; schools; self-contained flats and maisonettes; shops and shopping centres; supermarkets and hypermarkets; telephone exchanges; and the like.

Value of work £		Category B fee £	£
Up to	150,000	150 + 2.0% (Minimum fee £1,150)	
150,000 -	300,000	3,150 + 1.7% on balance over	150,000
300,000 -	600,000	5,700 + 1.5% on balance over	300,000
600,000 -	1,500,000	10,200 + 1.1% on balance over	600,000
1,500,000 -	3,000,000	20,100 + 1.0% on balance over	1,500,000
3,000,000 -	6,000,000	35,100 + 0.9% on balance over	3,000,000
Over	6,000,000	62,100 + 0.8% on balance over	6,000,000

(c) **Category C**: Simple works and/or works with a substantial element of repetition.

Examples:
Factories; garages; multi-storey car parks; open-air sports stadia; structural shell offices not fitted out; warehouses; workshops; and the like.

Value of work £		Category C fee £	£
Up to	150,000	120 + 1.6% (Minimum fee £920)	
150,000 -	300,000	2,520 + 1.5% on balance over	150,000
300,000 -	600,000	4,770 + 1.4% on balance over	300,000
600,000 -	1,500,000	8,970 + 1.1% on balance over	600,000
1,500,000 -	3,000,000	18,870 + 0.9% on balance over	1,500,000
3,000,000 -	6,000,000	32,370 + 0.8% on balance over	3,000,000
Over	6,000,000	56,370 + 0.7% on balance over	6,000,000

(d) The scales of fees for post-contract services (paras. 5.2 (a) to (c)) are overall scales based upon the inclusion of all nominated sub-contractors' and nominated suppliers' accounts, subject to the provision of para. 5.2 (g). When work normally included in a building contract is the subject of a separate contract for which the quantity surveyor has not been paid fees under any other clause hereof, the value of such work shall be included in the amount on which fees are charges.

(e) Fees shall be calculated upon the basis of the account for the whole of the work, subject to the provisions of para. 5.3.

(f) In calculating the amount on which fees are charged the total of any credits is to be added to the amount described above.

(g) Where the value of air conditioning, heating, ventilating and electrical services included in the tender documents together exceeds 25% of the amount calculated as described in paras. 5.2. (d) and (e) above, then, subject to provisions of para. 5.3, no fee is chargeable on the amount by which the value of these services exceeds the said 25%. In this context the term 'value' excludes general contractors' profit, attendance, builders work in connection with the services, preliminaries and other similar additions.

(h) When a contract comprises buildings which fall into more than one category, the fee shall be calculated as follows:

(i) The amount upon which fees are chargeable shall be allocated to the categories of work applicable and the amounts so allocated expressed as percentages of the old total amount upon which fees are chargeable.

(ii) Fees shall then be calculated for each category on the total amount upon which fees are chargeable.

(iii) The fee chargeable shall then be calculated by applying the percentages of work in each category to the appropriate total fee and adding the resultant amounts.

(j) When a project is the subject of a number of contracts then, for the purposes of calculating fees, the values of such contracts shall not be aggregated but each contract shall be taken separately and the scale of charges (paras. 5.2 (a) to (h)), applied as appropriate.

(k) When the quantity surveyor is required to prepare valuations of materials or goods off site, an additional fee shall be charged based on the time involved (see paras. 19.1 and 19.2).

(l) The basic scale for post-contract services includes for a simple routine of periodically estimating final costs. When the employer specifically requests a cost monitoring service which involves the quantity surveyor in additional or abortive measurement an additional fee shall be charged based on the time involved (see paras. 19.1 and 19.2), or alternatively on a lump sum or percentage basis agreed between the employer and the quantity surveyor.

(m) The above overall scales of charges for post-contract services assume normal conditions when the bills of quantities are based on drawing accurately depicting the building work the employer requires. If the works are materially varied to the extent that substantial remeasurements is necessary then the fee for post-contract services shall be adjusted by agreement between the employer and the quantity surveyor.

5.3. **Air conditioning, heating, ventilating and electrical services**

(a) Where final accounts are prepared by the quantity surveyor for the air conditioning, heating, ventilating and electrical services there shall be a fee for these services, in addition to the fee calculated in accordance with para. 5.2, as follows:

QUANTITY SURVEYORS' FEES

Value of Work £		Additional Fee £		£
Up to	120,000	2.0%		
120,000 -	240,000	2,400 + 1.6%	on balance over	120,000
240,000 -	1,000,000	4,320 + 1.25%	on balance over	240,000
1,000,000 -	4,000,000	13,820 + 1.0%	on balance over	1,000,000
Over	4,000,000	43,820 + 0.9%	on balance over	4,000,000

(b) The values of such services, whether the subject of separate tenders or not, shall be aggregated and the total value of work so obtained used for the purpose of calculating the addtional fee chargeable in accordance with para. (a).
(Except that when more than one firm of consulting engineers is engaged on the design of these services the separate values of which each such firm is responsible shall be aggregated and the additional fee charged shall be calculated independently on each such total value obtained.)

(c) The scope of the scale of the services to be provided by the quantity surveyor under para. (a) above shall be deemed to be equivalent to those described for the basic scale for post-contract services.

(d) When the quantity surveyor is required to prepare periodic valuations of materials or goods off site, an additional fee shall be charged based on the time involved (see paras. 19.1 and 19.2).

(e) The basic scale for post-contract services included for a simple routine of periodically estimating final costs. When the employer specifically requests a cost monitoring services which involves the quantity surveyor in addtional or abortive measurement an additional fee shall be based on the time involved (see paras. 19.1 and 19.2), or alternatively on a lump sum or percentage basis agreed between the employer and the quantity surveyor.

(f) Fees shall be calculated upon the basis of the account for the whole of the air conditioning, heating, ventilating and electrical services for which final accounts have been prepared by the quantity surveyor.

6.0. ALTERNATIVE II: SCALE OF CHARGES FOR SEPARATE STAGES OF POST-CONTRACT SERVICES

6.1. If the quantity surveyor appointed to carry out the post-contract services did not prepare the bills of quantities then the fees in paras. 6.2 and 6.3 shall be increased to cover the additional services undertaken by the quantity surveyor.

NOTE: The scales of fees in paras. 6.2 and 6.3 apply to work in all categories (including air conditioning, heating, ventilating and electrical services).

6.2. **Valuations for interim certificates**

(a) For taking particulars and reporting valuations for interim certificates for payments on account to the contractor.

QUANTITY SURVEYORS' FEES

Total of valuations		Fee		
£		£		£
Up to	300,000	0.5%		
300,000	- 1,000,000	1,500 + 0.4%	on balance over	300,000
1,000,000	- 6,000,000	4,300 + 0.3%	on balance over	1,000,000
Over	6,000,000	19,300 + 0.2%	on balance over	6,000,000

NOTES:

1 Subject to note 2 below, the fees are to be calculated on the total of all interim valuations (i.e. the amount of the final acount less only the net amount of the final valuation).

2 When consulting engineers are engaged in supervising the installation of air conditioning, heating, ventilating and electrical services and their duties include reporting valuations for inclusion in interim certificates for payments on account in respect of such services, then valuations so reported shall be excluded from any total amount of valuations used for calculating fees.

(b) When the quantity surveyor is required to prepare valuations of materials or goods off site, an additional fee shall be charged based on the time involved (see paras. 19.1 and 19.2).

6.3. **Preparing accounts of variation upon contracts**

For measuring and making up bills of variations including pricing and agreeing totals with the contractor:

(a) An initial lump sum of £600 shall be payable on each contract.

(b) 2.0% upon the gross amount of omissions requiring measurement or abstraction from the original dimension sheets.

(c) 3.0% upon the gross amount of additions requiring measurement and upon dayworks.

(d) 0.5% upon the gross amount of remaining additions which shall be deemed to include all nominated sub-contractors' and nominated suppliers' accounts which do not involve measurement or checking of quantities but only checking against lump sum estimates.

(e) 3.0% upon the aggregate of the amounts of the increases and/or decreases in the cost of labour and materials in accordance with any fluctuations clause in the conditions of contract, except where a price adjustment formula applies.

(f) On contracts where fluctuations are calculated by the use of a price adjustment formula method the following scale shall be applied to the account for the whole of the work:

Value of work		Fee		
£		£		£
Up to	300,000	300 + 0.5%		
300,000	-1,000,000	1,800 + 0.3%	on balance over	300,000
Over	1,000,000	3,900 + 0.1%	on balance over	1,000,000

(g) When consulting engineers are engaged in supervising the installation of air conditioning, heating, ventilating and electrical services and their duties include for the adjustment of accounts and pricing and agreeing totals with the sub-contractors for inclusion in the measured account, then any totals so agreed shall be excluded from any amounts used for calculating fees.

6.4. **Cost monitoring services**

The fee for providing all approximate estimates of final cost and/or a cost monitoring service shall be based on the time involved (see paras. 19.1 and 19.2), or alternatively on a lump sum or percentage basis agreed between the employer and the quantity surveyor.

7.0. BILLS OF APPROXIMATE QUANTITIES, INTERIM CERTIFICATES AND FINAL ACCOUNTS

7.1. **Basic scale**

For preparing bills of approximate quantities suitable for obtaining competitive tenders which will provide a schedule of prices and a reasonably close forecast of the cost of the works, but subject to complete remeasurement, examining tenders and reporting thereon, taking particulars and reporting valuations for interim certificates for payments on account to the contractor, preparing periodic assessments of anticipated final cost and reporting thereon, measuring and preparing final account, including pricing and agreeing totals with the contractor and adjusting fluctations in the cost of labour and materials if required by the contract:

(a) **Category A**: Relatively complex works and/or works with little or no repetition.

Examples: Ambulance and fire stations; banks; cimemas; clubs; computer buildings; council offices; crematoria; fitting out existing buildings; homes for the elderly; hospitals and nursing homes; laboratories; law courts; libraries; 'one-off' houses; petrol stations; places of religious worship; police stations; public houses; licensed premises; resturants; sheltered housing; sports pavilions; theatres; town halls; universities, polytechnics and colleges of further education (other than halls of residence and hostels); and the like.

Value of work £	Category A fee £	£
Up to 150,000	380 + 5.0%	(Minimum fee £2,880)
150,000 - 300,000	7,880 + 4.0%	on balance over 150,000
300,000 - 600,000	13,880 + 3.4%	on balance over 300,000
600,000 -1,500,000	24,080 + 2.8%	on balance over 600,000
1,500,000 -3,000,000	49,280 + 2.4%	on balance over 1,500,000
3,000,000 -6,000,000	85,280 + 2.2%	on balance over 3,000,000
Over 6,000,000	151,280 + 2.0%	on balance over 6,000,000

(b) **Category B**: Less complex works and/or works with some element of repetition

Examples: Adult education facilities; canteens; church halls; comunity centres; departmental stores; enclosed sports stadia and swimming baths; halls of residence; hostels; motels; offices other than those included in Categories A and C; railway stations; recreation and leisure centres; residential hotels; schools; self-contained flats and maisonettes shops and shopping centres; supermarkets and hypermarkets; telephone exchanges; and the like.

Value of work £	Category B fee £		£
Up to 150,000	360 + 4.8%	(Minimum fee £2,760)	
150,000 - 300,000	7,560 + 3.7%	on balance over	150,000
300,000 - 600,000	13,110 + 3.0%	on balance over	300,000
600,000 -1,500,000	22,110 + 2.2%	on balance over	600,000
1,500,000 -3,000,000	41,910 + 2.0%	on balance over	1,500,000
3,000,000 -6,000,000	71,910 + 1.8%	on balance over	3,000,000
Over 6,000,000	125,910 + 1.6%	on balance over	6,000,000

(c) **Category C**: Simple works and/or works with a substantial element of repetition.
Examples: Factories; garages; multi-storey car parks; open air sports stadia; structural shell offices not fitted out; warehouses; workshops; and the like.

Value of work £	Category C fee £		£
Up to 150,000	300 + 4.1%	(Minimum fee £2,350)	
150,000 - 300,000	6,450 + 3.3%	on balance over	150,000
300,000 - 600,000	11,400 + 2.6%	on balance over	300,000
600,000 - 1,500,000	19,200 + 2.0%	on balance over	600,000
1,500,000 - 3,000,000	37,200 + 1.7%	on balance over	1,500,000
3,000,000 - 6,000,000	62,700 + 1.5%	on balance over	3,000,000
Over 6,000,000	107,700 + 1.3%	on balance over	6,000,000

(d) The scales of fees for pre-contract and post-contract services (paras. 7.1 (a) to (c)) are overall scales based upon the inclusion of all nominated sub-contractors' and nominated suppliers' accounts, subject to the provision of para. 7.1. (g). When work normally included in a building contract is the subject of a separate contract for which the quantity surveyor has not been paid fees under any other clause hereof, the value of such work shall be included in the amount on which fees are charged.

(e) Fees shall be calculated upon the basis of the account for the whole of the work, subject to the provisions of para. 7.2.

(f) In calculating the amount on which fees are charged the total of any credits is to be added to the amount described above.

(g) Where the value of air conditioning, heating, ventilating and electrical services included in tender documents together exceeds 25% of the amount calculated as described in paras. 7.1. (d) and (e), then, subject to the provisions of para. 7.2 no fee is chargeable on the amount by which the value of theses services exceeds the said 25%. In this context the term 'value' excludes general contractors' profit, attendance, builders' work in connection with the services, preliminaries and any other similar additions.

QUANTITY SURVEYORS' FEES

(h) When a contract comprises buildings which fall into more than one category, the fee shall be calculated as follows.

(i) The amount upon which fees are chargeable shall be allocated to the categories of work applicable and the amount so allocated expressed as percentages of the total amount upon which fees are chargeable.
(ii) Fees shall then be calculated for each category on the total amount upon which fees are chargeable.
(iii) The fee chargeable shall then be calculated by applying the percentages of work in each category to the approp- riate total fee adding the resultant amounts.

(j) When a project is the subject of a number of contracts then, for the purpose of calculating fees, the values of such contracts shall not be aggregated but each contract shall be taken separately and the scale of charges (paras. 7.1 (a) to (h)) applied as approp- riate.

(k) Where the quantity surveyor is specifically instructed to provide cost planning services, the fee calculated in accordance with paras. 7.1 (a) to (j) shall be increased by a sum calculated in accordance with the following table and based upon the same value of work as that upon which the aforementioned fee has been calculated:

Categories A & B: (as defined in paras. 7.1 (a) and (b))

Value of work		Fee		
£		£		£
Up to	600,000	0.7%		
600,000	-3,000,000	4,200 + 0.4%	on balance over	600,000
3,000,000	-6,000,000	13,800 + 0.35%	on balance over	3,000,000
Over	6,000,000	24,300 + 0.3%	on balance over	6,000,000

Category C: (as defined in para. 7.1 (c))

Value of work		Fee		
£		£		£
Up to	600,000	0.5%		
600,000	-3,000,000	1,500 + 0.3%	on balance over	600,00
3,000,000	-6,000,000	10,200 + 0.25%	on balance over	3,000,000
Over	6,000,000	17,700 + 0.2%	on balance over	6,000,000

(l) When the quantity surveyor is required to prepare valuations of materials or goods off site, an additional fee shall be charged based on the time involved (see paras. 19.1 and 19.2).
(m) The basic scale for post-contract services includes for a simple routine of periodically estimating final costs. When the employer specifically requests a cost monitering service which involves the quantity surveyor in additional or abortive measurement an additional fee shall be charged based on the time involved (see paras. 19.1 and 19.2), or alternatively on a lump sum or percentage basis agreed between the employer and the quantity surveyor.

QUANTITY SURVEYORS' FEES

7.2. Air conditioning, heating, ventilating and electrical services

(a) Where bills of approximate quantities and final accounts are prepared by the quantity surveyor for the air conditioning, heating, ventilating and electrical services there shall be a fee for these services in addition to the fee calculated in accordance with para. 7.1 as follows:

Value of work £	Additional fee £		£
Up to 120,000	4.5%		
120,000 - 240,000	5,400 + 3.85%	on balance over	120,000
240,000 - 480,000	10,020 + 3.25%	on balance over	240,000
480,000 - 750,000	17,820 + 3.0%	on balance over	480,000
750,000 -1,000,000	25,920 + 2.5%	on balance over	750,000
1,000,000 -4,000,000	32,170 + 2.15%	on balance over	1,000,000
Over 4,000,000	96,670 + 2.05%	on balance over	4,000,000

(b) The value of such services, whether the subject of separate tenders or not, shall be aggregated and the value of work so obtained used for the purpose of calculating the additional fee chargeable in accordance with para. (a).
(Except that when more than one firm of consulting engineers is engaged on the design of these services, the separate values for which each such firm is responsible shall be aggregated and the additional fees charged shall be calculated independently on each such total value so obtained.)

(c) The scope of the services to be provided by the quantity surveyor under para. (a) above shall be deemed to be equivalent to those described for the basic scale for pre-contract and post-contract services.

(d) When the quantity surveyor is required to prepare valuations of materials or goods off site, an additional fee shall be charged based on the time involved (see paras. 19.1 and 19.2).

(e) The basic scale for post-contract services includes for a simple routine of periodically estimating final costs. When the employer specifically requests a cost monitoring service, which involves the quantity surveyor in additional or abortive measurement, an additional fee shall be charged based on the time involved (see paras. 19.1 and 19.2), or alternatively on a lump sum or percentage basis agreed between the employer and the quantity surveyor.

(f) Fees shall be calculated upon the basis of the account for the whole of the air conditioning, heating, ventilating and electrical services for which final accounts have been prepared by the quantity surveyor.

(g) When cost planning services are provided by the quantity surveyor for air conditioning, heating, ventilating and electrical services (or for any part of such services) there shall be an additional fee based on the time involved (see paras. 19.1 and 19.2) or alternatively on a lump sum or percentage basis agreed between the employer and quantity surveyor.

NOTE: The incorperation of figures for air conditioning, heating, ventilating and electrical services provided by the consulting engineer is deemed to be included in the quantity surveyor's services under para 7.1.

QUANTITY SURVEYORS' FEES

7.3. **Works of alteration**

On works of alteration or repair, or on those sections of the work which are mainly works of alteration or repair, there shall be a fee of 1.0% in addition to the fee calculated in accordance with paras. 7.1 and 7.2

7.4. **Works of redecoration and associated minor repairs**

On works of redecoration and associated minor repairs, there shall be a fee of 1.5% in addition to the fee calculated in accordance with paras. 7.1 and 7.2.

7.5. **Bills of quantities and/or final accounts prepared in special forms**

Fees calculated in accordance with paras. 7.1, 7.2, 7.3 and 7.4 include for the preparation of bills of quantities and/or final accounts on a normal trade basis. If the employer requires additional information to be provided in the bills of quantities and/or final accounts or the bills and/or final accounts to be prepared in an elemental, operational or similar form, then the fee may be adjusted by agreement between the employer and the quantity surveyor.

7.6. **Reduction of tenders**

(a) When cost planning services have been provided by the quantity surveyor and a tender, when received, is reduced before acceptance and if the reductions are not necessitated by amended instructions of the employer or by the inclusion in the bills of approximate quantities of items which the quantity surveyor has indicated could not be contained within the approved estimate, then in such a case no charge shall be made by the quantity surveyor for the preparation of bills of reductions and the fee for the preparation of bills of approximate quantities shall be based on the amount of the reduced tender.

(b) When cost planning services have not been provided by the quantity surveyor and if a tender, when received, is reduced before acceptance, fees are to be calculated upon the amount of the unreduced tender. When the preparation of bills of reductions is required, a fee is chargeable for preparing such bills of reductions as follows:

(i) 2.0% upon the gross amount of all omissions requiring measurement or abstraction from original dimension sheets.
(ii) 3.0% upon the gross amount of all additions requiring measurement.
(iii) 0.5% upon the gross amount of all remaining additions.

NOTE: The above scale for the preparation of bills of reductions applies to work in all categories.

7.7. **Generally**

If the works are substantially varied at any stage or if the quantity surveyor is involved in an excessive amount of abortive work, then the fees shall be adjusted by agreement between the employer and the quantity surveyor.

QUANTITY SURVEYORS' FEES

8.0. NEGOTIATING TENDERS

8.1. (a) For negotiating and agreeing prices with a contractor:

Value of Work		Fee	
£		£	£
Up to	150,000	0.5%	
150,000 -	600,000	750 + 0.3% on balance over	150,000
600,000 -	1,200,000	2,100 + 0.2% on balance over	600,000
Over	1,200,000	3,300 + 0.1% on balance over	1,200,000

(b) The fee shall be calculated on the total value of the works as defined in paras. (d), (e), (f), (g) and (j).

(c) For negotiating and agreeing prices with a contractor for air conditioning, heating, ventilating and electrical services there shall be an additional fee as para. 8.1 (a) calculated on the total value of such services as defined in para. 7.2 (b).

9.0. CONSULTATIVE SERVICES AND PRICING BILLS OF APPROXIMATE QUANTITIES

9.1. **Consultative services**

Where the quantity surveyor is appointed to prepare approximate, feasibility studies or submissions for the approval of financial grants or similar services, then the fee shall be based on the time involved (see paras. 19.1 and 19.2) or alternatively, on a lump sum or percentage basis agreed between the employer and the quantity surveyor.

9.2. **Pricing bills of approximate quantities**

For pricing bills of approximate quantities, if instructed, to provide an estimate comparable with tenders, the fees shall be the same as for the corresponding services in paras. 4.2 (a) and (b).

10.0. INSTALMENT PAYMENTS

10.1. For the purpose of instalment payments the fee for preparation of bills of approximate quantities only shall be the equivalent of forty per cent (40%) of the fees calculated in accordance with the appropriate sections of paras. 7.1 to 7.5 and the fee for providing cost planning services shall be in accordance with the appropriate section of para. 7.1 (k); both fees shall be based on the total value of the bills of approximate quantities ascertained in accordance with the provisions of para. 2.1 (e).

10.2. In the absence of agreement to the contary, fees shall be paid by installments as follows:

(a) Upon acceptance by the employer of a tender for the works the above defined fees for the preparation of bills of approximate quantities and for providing cost planning services.

(b) In the event of no tender being accepted, the aforementioned fees shall be paid within three months of completion of the bills of approximate quantities.

(c) The balance by instalments at intervals to be agreed between the date of the first certificate and one month after certification of the contractor's account.

QUANTITY SURVEYORS' FEES

10.3. In the event of the project being abandoned at any stage other than those covered by the foregoing, the proportion of fee payable shall be by agreement between the employer and the quantity surveyor.

11.0. SCHEDULES OF PRICES

11.1. The fee for preparing, pricing and agreeing schedules of prices shall be based on the time involved (see paras. 19.1 and 19.2). Alternatively, the fee may be on a lump sum or percentage basis agreed between the employer and the quantity surveyor.

12.0. COST PLANNING AND APPROXIMATE ESTIMATES

12.1 The fee for providing cost planning services or for preparing approximate estimates shall be based on the time involved (see paras. 19.1 and 19.2). Alternatively, the fee may be on a lump sum basis agreed between the employer and the quantity surveyor.

CONTRACTS BASED ON SCHEDULES OF PRICES: POST-CONTRACT SERVICES

13.0. FINAL ACCOUNTS

13.1. **Basic Scale**

(a) For taking particulars and reporting valuations for interim certificates for payments on account to the contractor, preparing periodic assessments of anticipated final cost and reporting thereon, measuring and preparing final account includiong pricing and agreeing totals with the contractor, and adjusting fluctuations in the cost of labour and materials if required by the contract, the fee shall be equivalent to sixty per cent (60%) of the fee calculated in accordance with paras. 7.1 (a) to (j).

(b) When the quantity surveyor is required to prepare valuations of materials or goods off site, an additional fee shall be charged on the basis of the time involved (see paras. 19.1 and 19.2).

(c) The basic scale for post-contract services includes for a simple routine of periodically estimating final costs. When the employer specifically requests a cost monitoring service, which involves the quantity surveyor in additional or abortive measurement, an additional fee shall be charged based on the time involved (see paras. 19.1 and 19.2), or alternatively on a lump sum or percentage basis agreed between the employer and the quantity surveyor.

13.2. **Air conditioning, heating, ventilating and electrical services**

Where final accounts are prepared by the quantity surveyor for the air conditioning, heating, ventilating services there shall be a fee for these services, in addition to the fee calculated in accordance with para. 13.1, equivalent to sixty per cent (60%) of the fee calculated in accordance with paras. 7.2 (a) to (f).

13.3. **Works of alterations**

On works of alteration or repair, or on those sections of the work which are mainly works of alteration or repair, there shall be a fee of 1.0% in addition to the fee calculated in accordance with paras. 13.1 and 13.2.

QUANTITY SURVEYORS' FEES

13.4. **Works of redecoration and associated minor repairs**

On works of redecoration and associated minor repairs, there shall be a fee of 1.5% in addition to the fee calculated in accordance with paras. 13.1 and 13.2.

13.5. **Final accounts prepared in special forms**

Fees calculated in accordance with paras. 13.1,13.2,13.2 and 13.4 include for the preparation of final accounts on a normal trade basis. If the employer requires additional information to be provided in the final accounts or the accounts to be prepared in an elemental, operational or similar form, then the fee may be adjusted by agreement between the employer and the quantity surveyor.

14.0. COST PLANNING

14.1. The fee for providing a cost planning service shall be based on the time involved (see paras. 19.1 and 19.2). Alternatively, the fee may be on a lump sum or percentage basis agreed between the employer and the quantity surveyor.

15.0. ESTIMATES OF COST

15.1.(a) For preparing an approximate estimate, calculated by measurement, of the cost of work, and, if required under the terms of the contract, negotiating, adjusting and agreeing the estimate:

Value of work	Fee	
£	£	£
Up to 30,000	1.25%	
30,000 - 150,000	375 + 1.0% on balance over	30,000
150,000 - 600,000	1,575 + 0.75%on balance over	150,000
Over 600,000	4,950 + 0.5% on balance over	600,000

(b) The fee shall be calculated upon the total of the approved estimates.

16.0. FINAL ACCOUNTS

16.1.(a) For checking prime costs, reporting for interim certificates for payments on account to the contractor and preparing final accounts:

Value of work	Fee	
£	£	£
Up to 30,000	2.25%	
30,000 - 150,000	375 + 2.0% on balance over	30,000
150,000 - 600,000	3,150 + 1.5% on balance over	150,000
Over 600,000	9,900 + 1.25%on balance over	600,000

(b) The fee shall be calculated upon the total of the final account with the addition of the value of credits received for old materials removed and less the value of any work charged for in accordance with para. 16.1 (c).

QUANTITY SURVEYORS' FEES

(c) On the value of any work to be paid for on a measured basis, the fee shall be 3%.

(d) When the quantity surveyor is required to prepare valuations of material or goods off site, an additional fee shall be charged based on the time involved (see paras. 19.1 and 19.2).

(e) The above charges do not include the provision of checkers on the site. If the quantity surveyor is required to provide such checkers an additional charge shall be made by arrangement.

17.0. COST REPORTING AND MONITORING SERVICES

17.1. The fee for providing cost reporting and/or monitoring services (e.g. preparing periodic assessments of anticipated final costs and reporting thereon) shall be based on the time involved (see paras. 19.1 and 19.2) or alternatively, on a lump sum or percentage basis agreed between the employer and the quantity surveyor.

18.0. ADDITIONAL SERVICES

18.1. For additional services not normally necessary, such as those arising as a result of the termination of a contract before completion, liquidation, fire damage to the buildings, services in connection with arbitration, litigation and investigation of the validity of contractors' claims, services in connection with taxation matters and all similar services where the employer specifically instructs the quantity surveyor, the charges shall be in accordance with paras. 19.1 and 19.2.

19.0. TIME CHARGES

19.1.(a) For consultancy and other services performed by a principle, a fee by arrangment according to the circumstances including the professional status and qualifications of the quantity surveyor.

(b) When a principal does work which would normally be done by a member of staff, the charge shall be calculated as para. 19.2 below.

19.2.(a) For services by a member of staff, the charges for which are to be based on the time involved, such charges shall be calculated on the hourly cost of the individual involved plus 145%.

(b) A member of staff shall include a principal doing work normally done by an employee (as para. 19.1 (b) above), technical and supporting staff, but shall exclude secretarial staff or staff engaged upon general administration.

(c) For the purpose of para. 19.2 (b) above, a principal's time shall be taken at the rate applicable to a senior assistant in the firm.

(d) The supervisory duties of a principal shall be deemed to be included in the addition of 145% as para. 19.2 (a) above and shall not be charged seperately.

(e) the hourly cost to the employer shall be calculated by taking the sum of the annual cost of the member of staff of:

(i) Salary and bonus but excluding expenses;

(ii) Empolyer's contributions payable under any Pension and Life Assurance Schemes;

QUANTITY SURVEYORS' FEES

(iii) Employer's contributions made under the National Insurance Acts, the Redundancy Payments Act and any other payments made in respect of the employee by virtue of any statutory requirements; and

(iv) Any other payments or benifits made or granted by the employer in pursuance of the terms of employment of the member of staff;

and dividing by 1,650.

19.3. The foregoing Time Charges under paras. 19.1 and 19.2 are intended for use where other paras. of the Scale (not related to Time Charges) form a significant proportion of the overall fee. In all other cases an increased time charge may be agreed.

20.0. INSTALMENT PAYMENTS

20.1. In the absence of agreement to the contrary, payments to the quantity surveyor shall be made by instalments by arrangement between the employer and the quantity surveyor.

QUANTITY SURVEYORS' FEES

Scale 40 professional charges for quantity surveying services in connection with Housing schemes for Local Authorities

EFFECTIVE FROM FEBRUARY 1983

1.0 GENERALLY

1.1 The scale is applicable to housing schemes of self-contained dwellings regardless of type (e.g. houses, maisonettes, bungalows or flats) and irrespective of the amount of repitition of identical types or blocks within an individual housing scheme and shall also apply to all external works forming part of the contract for the housing scheme. This scale does not apply to improvement to existing dwellings.

1.2 The fees set out below cover the following quantity surveying services as may be required:

(a) Preparing bills of quantities or other tender documents; checking tenders received or negotiating tenders and pricing with a selected contractor; reporting thereon.

(b) Preparing recommendations for interim payments on account to the contractor; measuring work and adjusting variations in accordance with the terms of the contract and preparing the final account; pricing same and agreeing totals with the contractor; adjusting fluctuations in the cost of labour and materials if required by the contract.

(c) Preparing periodic financial statements showing the anticipated final cost by means of a simple routine of estimating final costs and reporting thereon, but excluding cost monitoring (see para. 1.4).

1.3 Where the quantity surveyor is appointed to prepare approximate estimates to establish and substantiate the economic viability of the scheme and to obtain the necessary approvals and consents, or to enable the scheme to be designed and constructed within approved cost criteria an additional fee shall be charged based on the time involved (see para. 7.0) or, alternatively, on a lump sum or percentage basis agreed between the employer and the quantity surveyor. (Cost planning services, see para. 3.0).

1.4 When the employer specifically requests a post-contract cost monitoring service which involves the quantity surveyor in additional or abortive work an additional fee shall be charged based on the time involved (see para. 7.0) or, alternatively, on a lump sum or percentage basis agreed between the employer and the quantity surveyor.

1.5 The fees are in all cases exclusive og travelling and other expenses (for which the actual disbursement is recoverable unless there is some prior arrangement for such charges) and of the cost of reproduction of bills of quantities and other documents, which are chargeable in addition at net cost.

1.6 The fees are in all cases exclusive of services in connection with the allocation of the cost of the works for purposes of calculating value added tax for which there shall be an additional fee based on the time involved (see para. 7.0).

QUANTITY SURVEYORS' FEES

1.7 When work normally included in a building contract is the subject of a separate contract for which the quantity surveyor has not beem paid fees under any other clause thereof, the value of such work shall be included in the amount upon which fees are charged.

1.8 If any of the materials used in the works are supplied by the employer or charged at a preferential rate, then the estimated or actual value thereof shall be included in the amount upon which fees are to be calculated.

1.9 The fees are in all cases exclusive of preparing a specification of the materials to be used and the works to be done, but the fees for preparing bills of quantities and similar documents do include for incorporating preamble claused describing the materials and workmanship (from inform-ation given by the architect and/or consulting engineer).

1.10 If the quantity surveyir incurs additional costs due to exceptional delays in building operations or any other casue beyond the control of the quantity surveyor, then the fees shall be adjusted by agreement between the employer and the quantity surveyor to cover the reimbursement of these additional costs.

1.11 When a project is the subject of a number of contracts then for the purposes of calculating fees, the values of such contracts should not be aggregated but each contract shall be taken separately and the scale of charges applied as appropriate.

1.12 The fees and charges are in all cases exclusive of value added tax which will be applied in accordance with legislation.

1.13 Copyright in bills of quantities and other documents prepared by the quantity surveyor is reserved to the quantity surveyor.

2.0 BASIC SCALE

2.1 The basic fee for the services outlined in para. 1.2 shall be as follows:-

Value of Work £	Fee £	£
Up to 75,000	250 + 4.6%	
75,000 - 150,000	3,700 + 3.6% on balance over	75,000
150,000 - 750,000	6,400 + 2.3% on balance over	150,000
750,000 - 1,500,000	20,200 + 1.7% on balance over	750,000
Over 1,500,000	32,950 + 1.5% on balance over	1,500,000

2.2 Fees shall be calculated upon the total of the final account for the whole of the work including all nominated sub-contractors' and nominated suppliers' accounts.

2.3 For services in connection with accommodation designed for the elderly or the disabled or other special category occupants for whom special facilities are required an addition of 10% shall be made to the fee calculated in accordance with para. 2.1.

QUANTITY SURVEYORS' FEES

2.4 When additional fees under para. 2.3 are chargeable on a part or parts of a scheme, the value of basic fee to which the additional percentages shall be applied shall be determined by the proportion that the values of the various types of accommodation bear to the total of those values.

2.5 When the quantity surveyor is required to prepare an interim valuation of materials or goods off site, an additional fee shall be charged based on the time involved (see para. 7.0).

2.6 If the works are substantially varied at any stage and if the quantity surveyor is involved in an excessive amount of abortive work, then the fee shall be adjusted by agreement between the employer and the quantity surveyor.

2.7 The fees payable under paras. 2.1 and 2.3 include for the preparation of bills of quantities or other tender documents on a normal trade basis. If the employer requires additional information to be provided in bills of quantities, or bills of quantities to be prepared in an elemental, operational or similar form, then the fee may be adjusted by agreement between the employer and the quantity surveyor.

3.0 COST PLANNING

3.1 When the quantity surveyor is specifically instructed to provide cost planning services, the fee shall calculated in accordance with paras. 2.1 and 2.3 shall be increased by a sum calculated in accordance with the following table and based upon the amount of the accepted tender.

Value of Work		Fee		
	£	£		£
Up to	150,000		0.45%	
150,000 -	750,000	675 +	0.35% on balance over	150,000
Over	750,000	2,775 +	0.25% on balance over	750,000

3.2 Cost planning is defined as the process of ascertaining a cost limit, where necessary, within the guidelines set by any appropriate Authority, and thereafter checking the cost of the project within that limit throughout the design process. It includes the preparation of a cost plan (based upon elemental analysis or other suitable criterion) checking and revising it where required and effecting the necessary liaison with other consultants employed.

3.3 (a) When cost planning services have been provided by the quantity surveyor and bills of reductions are required, then no charge shall be made by the quantity surveyor for the bills of reductions unless the reductions are necessitated by amended instructions of the employer or by the inclusion in the bills of quantities of items which the quantity surveyor has indicated could not be contained within the approved estimate.

(b) When cost planning services have not been provided by the quantity surveyor and bills of reductions are required, a fee is chargeable for preparing such bills of reductions as follows:-

(i) 2.0% upon the gross amount of all omissions requiring measurement or abstraction from original dimension sheets.

(ii) 3.0% upon the gross amount of all additions requiring measurement.

(iii) 0.5% upon the gross amount of all remaining additions.

QUANTITY SURVEYORS' FEES

4.0 HEATING, VENTILATING AND ELECTRICAL SERVICES

(a) When bills of quantities and the final account are prepared by the quantity surveyor for the heating, ventilating and electrical services, there shall be a fee for these services in addition to the fee calculated in accordance with paras. 2.1 and 2.3 as follows:-

Value of Work £	Fee £	£
Up to 60,000	4.5%	
60,000 - 120,000	2,700 + 3.85% on balance over	60,000
120,000 - 240,000	5,010 + 3.25% on balance over	120,000
240,000 - 375,000	8,910 + 3.00% on balance over	240,000
375,000 - 500,000	12,960 + 2.50% on balance over	375,000
Over 500,000	16,085 + 2.15% on balance over	500,000

(b) The value of such services, whether the subject of separate tenders or not shall be aggregated and the total value of work so obtained used for the purpose of calculating the additional fee chargeable in accordance with para. (a). (Except that when more than one firm of consulting engineers is engaged on the design of these services, the separate values for which each such firm is responsible shall be aggregated and the additional fee charged shall be calculated indepe-ndently on each such total value so obtained).

(c) The scope of the services to be provided by the quantity surveyor under para. (a) above shall be deemed to be equivalent to those outlined in para. 1.2.

(d) Fee shall be calculated upon the basis of the account for the whole of the heating, ventilating and electrical services for which final accounts have been prepared by the quantity surveyor.

5.0 INSTALMENT PAYMENTS

5.1 In the absence of agreement to the contrary, fees shall be paid by instalments as follows:-

(a) Upon receipt by the employer of a tender for the works sixty per cent (60%) of the fees calculated in accordance with paras. 2.0 and 4.0 in the amount of the accepted tender plus the appropriate recoverable expenses and the full amount of the fee for cost planning services if such services have been instructed by the employer.

(b) The balance of fees and expenses by instalments at intervals to be agreed between the date of the first certificate and one month after final certification of the contractor's account.

5.2 In the event of no tender being accepted, sixty per cent (60%) of the fees, plus the appropriate recoverable expenses, and the full amount of the fee for cost planning services if such services have been instructed by the employer, shall be paid within three months of the completion of the tender documents. The fee shall be calculated on the amount of the lowest original bona fide tender received. In the event of no tender being received, the fee shall be calculated on a reasonable valuation of the work based upon the tender documents.

NOTE: In the foregoing context "bona fide tender" shall be deemed to mean a tender submitted in good faith without major errors of computation and not subsequently withdrawn by the tenderer.

5.3 In the event of the project being abandoned at any stage other than those covered by the foregoing, the proportion of fee payable shall be by agreement between the employer and the quantity surveyor.

5.4 When the quantity surveyor is appointed to carry out post-contract services only and has not prepared the bills of quantities then the fees shall be agreed between the employer and the quantity surveyor as a proportion of the scale set out in paras. 2.0 and 4.0 with an allowance for the necessary familiarisation and any additional services undertaken by the quantity surveyor. The percentages stated in paras. 5.1 and 5.2 are not intended to be used as a means of calculating the fees payable for post-contract services only.

6.0 ADDITIONAL SERVICES

6.1 For additional services not normally necessary such as those arising as a result of the termination of a contract before completion, liquidation, fire damage to the buildings, services in connection with arbitration, litigation and investigation of the validity of contractors' claims, services where the employer specially instructs the quantity surveyor, the charge shall be in accordance with para. 7.0.

7.0 TIME CHARGES

7.1 (a) For consultancy and other services performed by a principal, a fee by arrangement according to the circumstances, including the professional status and qualifications of the quantity surveyor.

(b) When a principal does work which would normally be done by a member of staff, the charge shall be calculated as para. 7.2.

7.2 (a) For services by a member of staff, the charges for which are to be based on the time involved, such hourly charges shall be calculated on the basis of annual salary (including bonus and any other payments or benefits previously sgreed with the employer) multiplied by a factor of 2.5, plus reimbursement of payroll costs, all divided by 1600. Payroll costs shall include inter alia employer's contributions payable under any Pension and Life Assurance Schemes, employer's contributions made under the National Insurance Acts, the Redundancy Payments Act and any other payments made in respect of the employee by virtue of any statutory requirements. In this connection it would not be unreasonable in individual cases to take account of the cost of providing a car as part of the "salary" of staff engaged on time charge work when considering whether the salaries paid to staff engaged on such work are reasonable.

QUANTITY SURVEYORS' FEES

(b) A member of staff shall include a principal doing work normally done by an employee (as para. 7.1 (b) above), technical and supporting staff, but shall exclude secretarial staff or staff engaged upon general administration.

(c) For the purpose of para. 7.2 (b) above a principal's time shall be taken at the rate applicable to a senior assistant in the firm.

(d) The supervisory duties of a principal shall be deemed to be included in the multiplication factor as para. 7.2 (a) above and shall not be charged separately.

7.3 The foregoing Time Charges under paras. 7.1 and 7.2 are intended for use where other paras. of the scale (not related to Time Charges) form a significant proportion of the overall fee in all other cases an increased Time Charge may be agreed.

QUANTITY SURVEYORS' FEES

Scale 44 professional charges for quantity surveying services in connection with improvements to existing housing and enviromental improvement works

EFFECTIVE FROM FEBRUARY 1987

1. This scale of charges is applicable to all works of improvement to existing housing for local authorities, development corporations, housing associations and the like and to environmental improvement works associated therewith or of a similar nature.

2. The fees set out below cover such quantity surveying services as may be required in connection with an improvement project irrespective of the type of contract or contract documentation from initial appointment to final certification of the contractor's account such as:-

   (a) Preliminary cost exercises and advice on tendering procedures and contract arrangements.
   (b) Providing cost advice to assist the design and construction of the project within approved cost limits.
   (c) Preliminary inspection of a typical dwelling of each type.
   (d) Preparation of tender documents; checking tenders received and reporting thereon or negotiating tenders and agreeing prices with a selected contractor.
   (e) Making recommendations for and, where necessary, preparing bills of reductions except in cases where the reductions are necessitated by amended instructions of the employer or by the inclusion in the bills of quantities of items which the quantity surveyor has indicated could not be contained within the approved estimate.
   (f) Analysing tenders and preparing details for submission to a Ministry or Government Department and attending upon the employer in any negotiations with such Ministry or Government Department.
   (g) Recording the extent of work required to every dwelling before work commences.
   (h) Preparing recommendations for interim payments on account to the contractor; preparing periodic assessments of the anticipated final cost of the works and reporting thereon
   (j) Measurement of work and adjustment of variations and fluctuations in the cost of labour and materials in accordance with the terms of the contract and preparing final account, pricing same and agreeing totals with the contractor.

3. The services listed in para. 2 do not include the carrying out of structural surveys.

4. The fees set out below have been calculated on the basis of experience that all of the services described above will not normally be required and in consequence these scales shall not be abated if, by agreement, any of the services are not required to be provided by the quantity surveyor.

IMPROVEMENT WORKS TO HOUSING

5. The fee for quantity services in connection with improvement works to existing housing and external works in connection therewith shall be calculated from a sliding scale based upon the total number of houses or flats in a project divided by the total number of types substantially the same in design and plan as follows:-

QUANTITY SURVEYORS' FEES

Total number of houses or flats divided by total number of types substantially the same in design and plan	Fee
not exceeding 1	see note below
exceeding 1 but not exceeding 2	7.0%
exceeding 2 but not exceeding 3	5.0%
exceeding 3 but not exceeding 4	4.5%
exceeding 4 but not exceeding 20	4.0%
exceeding 20 but not exceeding 50	3.6%
exceeding 50 but not exceeding 100	3.2%
exceeding 100	3.0%

and to the result of the computation shall be added 12.5%

Note: For schemes of only one house or flat per type an appropriate fee is to be agreed between the employer and the quantity surveyor on a percentage, lump sum or time basis.

ENVIRONMENTAL IMPROVEMENT WORKS

6. The fee for quantity surveying services in connection with environmental improvement works associated with improvements to existing housing or enivronmental improvement works of a similar nature shall be as follows:-

Value of Work £	Fee £	£
Up to 50,000	4.5%	
50,000 - 200,000	2,250 + 3.0% on balance over	50,000
200,000 - 500,000	6,750 + 2.1% on balance over	200,000
Over 500,000	13,050 + 2.0% on balance over	500,000

and to the result of that computation shall be added 12.5%

GENERALLY

7. When tender documents prepared by a quantity surveyor for an earlier scheme are re-used without amendment by the quantity surveyor for a subsequent scheme or part thereof for the same employer, the percentage fee in respect of such subsequent scheme or the part covered by such re-used documents shall be reduced by 20%.

8. The foregoing fees shall be calculated upon the separate totals of the final account for improvement works to housing and environmental Goverment works respectively including all nominated sub-contractors' and nominated suppliers' accounts abd (subject to para. 5 above) regardless of the amount of repetition within the scheme. When environmental improvement works are the subject of a number of contracts then for the purpose of calculating fees, the values of such contracts shall not be aggregated but each contract shall be taken separately and the scale of charges in para. 6 above applied as appropriate.

9. In cases where any of the materials used in the works are supplied by the employer, the estimated or actual value thereof is to be included in the total on which the fee is calculated.

QUANTITY SURVEYORS' FEES

10. In the absence of agreement to the contrary, fees shall be paid by instalments as follows:-

    (a) Upon acceptence by the employer of a tender for the works, one half of the fee calculated on the amount of the accepted tender.
    (b) The balance by instalments at intervals to be agreed between the date of the first certificate and one month after final certification of the contractor's account.

11. (a) In the event of no tender being accepted, one half of the fee shall be paid within three months of completion of the tender documents. The fee shall be calculated on the amount of the lowest original bona fide tender received. If no such tender has been received, the fee shall be calculated upon a reasonable valuation of the work based upon the tender documents.
    (b) In the event of the project being abandoned at any stage other than those covered by the foregoing, the proportion of fee payable shall be by agreement between the employer and the quantity surveyor.

12. If the works are substantially varied at any stage or if the quantity surveyor is involved in an excessive amount of abortive work, then the fee shall be adjusted by agreement between the employer and the quantity surveyor.

13. When the quantity surveyor is required to perform additional services in connection with the allocation of the costs of the works for purposes of calculating value added tax there shall be an additional fee based on the time involved.

14. For additional services not normally necessary such as those arising as a result of the termination of the contract before completion, liquidation, fire damage to the buildings, services in connection with arbitration, litigation and clains on which the employer and the quantity surveyor.

15. Copyright in the bills of quantities and other documents prepared by the quantity surveyor is reserved to the quantity surveyor.

16. The foregoing fees are in all cases exclusive of travelling expenses and lithography or other charges for copies of documents, the net amount of such expenses and charges to be paid for in addition. Subsistence expenses, if any, to be charged by arrangement with the employer.

17. The foregoing fees and charges are in all cases exclusive of value added tax which shall be applied in accordance with legislation current at the time the account is rendered.

QUANTITY SURVEYORS' FEES

Scale 45 professional charges for quantity surveying services in connection with housing schemes financed by the Housing Corporation

EFFECTIVE FROM JANUARY 1982

1. (a) This scale of charges has been agreed between The Royal Institution of Chartered Surveyors and the Housing Corporation and shall apply to housing schemes of self-contained dwellings financed by the Housing Corporation regardless of type (e.g. houses, maisonettes, bungalows or flats) and irrespective of the amount of repetition of identical types or blocks within an individual housing scheme.
   (b) This scale does not apply to services in connection with improvements to existing dwellings.

2. The fees set out below cover the following quantity surveying services as may be required in connection with the particular project:-

   (a) Preparing such estimates if cost as are required by the employer to establish and substantiate the economic viability of the scheme and to obtain the necessary approvals and consents from the Housing Corporation but excluding cost planning services (see para. 10)
   (b) Providing pre-contract cost advice (e.g. approximate estimates on a floor area or similar basis) to enable the scheme to be designed and constructed within the approved cost criteria but excluding cost planning services (see para. 10).
   (c) Preparing bills of quantities or other tender documents; checking tenders received or negotiating tenders and pricing with a selected contractor; reporting thereon.
   (d) Preparing an elemental analysis of the accepted tender (RICS/BCIS Detailed Form of Cost Analysis excluding the specification notes or equivalent).
   (e) Preparing recommendations for interim payments on account to the contractor; measuring the work and adjusting variations in accordance with the terms of the contract and preparing the final account, pricing same and agreeing totals with the contractor; adjusting fluctuations in the cost of labour and materials if required by the contract.
   (f) Preparing periodic post-contract assessments of the anticipated final cost by means of a simple routine of periodically estimating final costs and reporting thereon, but excluding a cost monitoring service specifically required by the employer.

3. The fees set out below are exclusive of travelling and of other expenses (for which the actual disbursement is recoverable unless there is some special prior arrangement for such charges) and the cost of reproduction of bills of quantities and other documents, which are chargeable in addition at net cost.

4. Copyright in the bills of quantities and other documents prepared by the quantity surveyor is reserved to the quantity surveyor.

QUANTITY SURVEYORS' FEES

5. (a) The basic fee for the services outlined in para. 2 (regardless of the extent of services described in para. 2) shall be as follows:-

Value of Work £	Fee £	£
Up to 75,000	210 + 3.8%	
75,000 - 150,000	3,060 + 3.0% on balance over	75,000
150,000 - 750,000	5,310 + 2.0% on balance over	150,000
750,000 - 1,500,000	17,310 + 1.5% on balance over	750,000
Over 1,500,000	28,560 + 1.3% on balance over	1,500,000

(b) (i) For services in connection with Categories 1 and 2 Accommodation designed for Old People in accordance with the standards described in Ministry of Housing and Local Government Circulars 82/69 and 27/70 (Welsh Office Circulars 84/69 & 30/70), there shall be a fee in addition to that in accordance with para. 5 (a), calculated as follows:-

Category 1 An addition of five per cent (5%) to the basic fee calculated in accordance with para. 5 (a)

Category 2 An addition of twelve and a half per cent (12.1/2%) to the basic fee calculated in accordance with para. 5 (a).

(ii) For services in connection with Accommodation designed for the Elderly in Scotland in accordance with the standards described in Scottish Housing Handbook Part 5, Housing for the Elderly, the fee shall be calculated as follows:-

Mainstream and Amenity Housing Basic fee in accordance with para. 5 (a)

Basic sheltered Housing (i.e. Amenity Housing plus Warden's accommodation and alarm system)	An addition of five per cent (5%) to the basic fee calculated in accordance with para. 5 (a)
Sheltered Housing, including optional additional facilities	An addition of twelve and a half per cent (12.1/2%) to the basic fee calculated in accordance with para. 5 (a)

(c) (i) For services in connection with Accommodation designed for Disabled People in accordance with the standards described in Department of Environment Circular 92/75 (Welsh Office Circular 163/75), there shall be an addition of fifteen per cent (15%) to the fee calculated in accordance with paragraph 5 (a).

(ii) For services in connection with Accommodation designed for the Disabled in Scotland on accordance with the standards described in Scottish Housing Handbook Part 6, Housing for the Disabled, there shall be an addition of fifteen per cent (15%) to the fee calculated in accordance with para. 5 (a).

QUANTITY SURVEYORS' FEES

(d) For services in connection with Accommodation designed for Disabled Old People, the fee shall be calculated in accordance with para. 5 (c).

(e) For services in connection with Subsidised Fair Rent New Build Housing, there shall be a fee, in addition to that in accordance with paras. 5 (a) to (d), calculated as follows:-

Value of Work £	Fee £	£
Up to 75,000	20 + 0.40%	
75,000 - 150,000	320 + 0.20% on balance over	75,000
150,000 - 500,000	470 + 0.07% on balance over	150,000
Over 500,000	715	

6\. (a) Where additional fees under paras. 5 (b) to (d) are chargeable on a part or parts of a scheme, the value of basic fee to which the additional percentages shall be applied shall be determined by the proportion that the values of the various types of accommodation bear to the total of those values.

(b) Fees shall be calculated upon the total of the final account for the whole of the work including all nominated sub-contractors' and nominated suppliers' accounts.

(c) If any of the materials used in the works are supplied free of charge to the contractor, the estimated or actual value thereof shall be included in the amount upon which fees are to be calculated.

(d) When a project is the subject of a number of contracts then, for the purpose of calculating fees, the values of such contracts shall not be aggregated but each contract shall be taken separately and the scale of charges applied as appropriate.

7\. If bills of quantities and final accounts are prepared by the quantity surveyor for the heating, ventilating or electrical services, there shall be an additional fee by agreement between the employer and the quantity surveyor subject to the approval of the Housing Corporation.

8\. In the absence of agreement to the contrary, fees shall be paid by instalments as follows:-

(a) Upon receipt by the employer of a tender for the works, or when the employer certifies to the Housing Corporation that the tender documents have been completed, a sum on account representing ninety per cent (90%) if the anticipated sum under para. 8 (b) below.

(b) Upon aceptance by the employer of a tender for the works, sixty per cent (60%) of the fee calculated on the amount of the accepted tender, plus the appropriate recoverable expenses.

(c) The balance of fees and expenses by instalments at intervals to be agreed between the date of the first certificate and one month after final certification of the contractor's account.

QUANTITY SURVEYORS' FEES

9. (a) In the event of no tender being accepted, sixty per cent (60%) of the fee and the appropriate recoverable expenses shall be paid within six months of completion of the tender documents. The fee shall be calculated on the amount of the lowest original bona fide tender received. In the event of no tender being received, the fee shall be calculated upon a reasonable valuation of the work based upon the tender documents.

NOTE: In the foregoing context "bona fide tender" shall be deemed to mean a tender submitted in good faith without major errors of computation and not subsequently withdrawn by the tenderer.

(b) In the event of part of the project being postponed or abandoned after the preparation of the bills of quantities or other tender documents, sixty per cent (60%) of the fee in this part shall be paid within three months of the date of postponement or abandonment.

(c) In the event of the project being postponed or abandoned at any stage other than those covered by the foregoing, the proportion of fee payable shall be by agreement between the employer and the quantity surveyor.

10. (a) Where with the approval of the Housing Corporation the employer instructs the quantity surveyor to carry out cost planning services there shall be a fee additional to that charged under para. 5 as follows:-

Value of Work		Fee	
£		£	£
Up to	150,000	0.45%	
150,000 -	750,000	675 + 0.35% on balance over	150,000
Over	750,000	2 775 + 0.25% on balance over	750,000

(b) Cost planning is defined as the process of ascertaining a cost limit where necessary, within guidelines set by any appropriate Authority, and thereafter checking the cost of the project within that limit throughout the design process. It includes the preparation of a cost plan (based upon elemental analysis or other suitable criterion) checking and revising it where required and effecting the necessary liaison with the other consultants employed.

11. If the quantity surveyor incurs additional costs due to exceptional delays in building operations or any other cause beyond the control of the quantity surveyor, then the fees shall be adjusted by agreement between the employer and the quantity surveyor to cover the reimbursement of these additional costs.

12. When the quantity surveyor is required to prepare an interim valuation of materials or goods off site, an additional fee shall be charged based on the time involved (see paras. 15 and 16) in respect of each such valuation.

QUANTITY SURVEYORS' FEES

13. If the Works are materially varied to the extent that substantial re-measurement is necessary, then the fee may be adjusted by agreement between the employer and the quantity surveyor.

14. For additional services not normally necessary, such as those arising as a result of the termination of a contract before completion, fire damage to the buildings, cost monitoring (see para. 2 (f), services in connection with arbitration, litigation and investigation of the validity of contractors' claims, services where the employer specifically instructs the quantity surveyor, the charges shall be in accordance with paras. 15 and 16.

15. (a) For consultancy and other services performed by a principal, a fee by arrangement according to the circumstances, including the professional status and qualifications of the quantity surveyor.
(b) When a principal does work which would normally be done by a member of staff, the charge shall be calculated as para. 16.

16. (a) For services by a member of staff, the charges for which are to be based on the time involved, such hourly charges shall be calculated on the basis of annual salary (including bonus and any other payments or benefits previously agreed with the employer) multiplied by a factor of 2.5, plus reimbursement of payroll costs, all divided by 1600. Payroll costs shall include inter alia employer's contributions payable under any Pension and Life Assurance Schemes, employer's contributions made under the National Insurance Acts, the Redundancy Payments Act and any other payments made in respect of the employee by virtue of any statutory requirements in this connection it would not be unreasonable in individual cases to take account of the cost of providing a car as part of the "salary" of staff engaged on time charge work when considering whether the salaries paid to staff engaged on such work are reasonable.
(b) A member of staff shall include a principle doing work normally done by an employee (as para. 15 (b) above), technical and supporting staff, but shall exclude secretarial staff or staff engaged upon general administration.
(c) For the purpose of para. 16 (b) above a principal's time shall be taken at the rate applicable to a senior assistant in the firm.
(d) The supervisory duties of a principal shall be deemed to be included in the multiplication factor as para. 16 (a) above and shall not be charged separately.

17. The foregoing Time Charges under paras. 15 and 16 are intended for use where other paras. of the scale (not related to Time Charges) form a significant proportion of the overall fee. In all other cases an increased time charge may be agreed.

18. (a) In the event of the employment of the contractor being determined due to bankruptcy or liquidation, the fee for the services outlined in para. 2, and for the additional services required, shall be recalculated to the aggregate of the following:-

QUANTITY SURVEYORS' FEES

(i) Fifty per cent (50%) of the fee in accordance with paragraphs 5 and 6 calculated upon the total of the Notional Final Account in accordance with the terms of the original contract;

(ii) Fifty per cent (50%) of the fee in accordance with paragraphs 5 and 6 calculated upon the aggregate of the total value (which may differ from the total of interim valuations) of work up to the date of determination in accordance with the terms of the original contract plus the total of the final account for the completion contract;

(iii) A charge based upon time involved (in accordance with paragraphs 15 and 16) in respect of dealing with those matters specifically generated by the liquidation (other than normal post-contract services related to the completion contract), which may include (inter alia):-

Site inspection and (where required) security (initial and until the replacement contractor takes possession);
Taking instructions from and/or advising the employer;
Representing the employer at meeting(s) of creditors;
Making arrangements for the continued employment of sub-contractors and similar related matters;
Preparing bills of quantities or other appropriate documents for the completion contract, obtaining tenders, checking and reporting thereon;
The additional cost (over and above the preparation of the final account for the completion contract) of preparing the Notional Final Account; pricing the same;
Negotiations with the liquidator (trustee or receiver).

(b) In calculating fees under para. 18 (a) (iii) above, regard shall be taken of any services carried out by the quantity surveyor for which a fee will ultimately be chargeable under para. 18 (a) (i) and (ii) above in respect of which a suitable abatement shall be made from the fee charged (e.g. measurement of variations for purposes of the completion contract where such would contribute towards the preparation of the contract final account).

(c) Any interim instalments of fees paid under para. 8 in respect of services outlined in para. 2 shall be deducted from the overall fee computed as outlined herein.

(d) In the absence of agreement to the contrary fees and expenses in respect of those services outlined in para. 18 (a) (iii) above up to acceptance of a completion tender shall be paid upon such acceptence; the balance of fees and expenses shall be paid in accordance with para. 8 (c)

(e) For the purpose of this Scale the term "Notional Final Account" shall be deemed to mean an account indicating that which would have been payable to the original contractor had he completed the whole of the works and before deduction of interim payments to him.

19. The fees and charges are in all cases exclusive of Value Added Tax which will be applied in accordance with legislation.

QUANTITY SURVEYORS' FEES

Scale 46 professional charges for quantity surveying services in connection with loss assessment of damage to buildings from fire, etc issued by The Royal Institution of Chartered Surveyors. The scale is recommended and not mandatory.

EFFECTIVE FROM JULY 1988

1. This scale of professional charges is for use in assessing loss resulting from damage to buildings by fire etc., under the "building" section of an insurance policy and is applicable to all categories of buildings.
2. The fees set out below cover the following quantity surveying services as may be required in connection with the particular loss assessment:-

    (a) Examining the insurance policy.
    (b) Visiting the building and taking all necessary site notes.
    (c) Measuring at site and/or from drawings and preparing itemised statement of claim and pricing same.
    (d) Negotiating and agreeing claim with the loss adjuster.

3. The fees set out below are exclusive of the following:-

    (a) Travelling and other expenses (for which the actual disbursement is recoverable unless there is some special prior arrangement for such charge.)
    (b) Cost of reproduction of all documents, which are chargeable in addition at net cost.

4. Copyright in all documents prepared by the quantity surveyor is reserved

5. (a) The fees for the services outlined in paragragh 2 shall be as follows:-

Agreed Amount of Damage		Fee	
	£	£	£
Up to	60,000	See note 5(c)below	
60,000 -	180,000	2.5%	
180,000 -	360,000	4,500 + 2.3% on balance over	180,000
360,000 -	720,000	8,640 + 2.0% on balance over	360,000
Over	720,000	15,840 + 1.5% on balance over	720,000

and to the result of that computation shall be added 12.5%

(b) The sum on which the fees above shall be calculated shall be arrived at after having given effect to the following:-

(i) The sum shall be based on the amount of damage, including such amounts in respect of architects', surveyors and other consultants' fees for reinstatement, as admitted by the loss adjuster.

(ii) When a policy is subject to an average clause, the sum shall be the agreed amount before the adjustment for"average".

(iii) When, in order to apply the average clause, the reinstatement value of the whole subject is calculated and negotiated an additional fee shall be charged commensurate with the work involved.

QUANTITY SURVEYORS' FEES

(c) Subject to 5 (b) above, when the amount of the sum on which fees shall be calculated is under £60,000 the fee shall be based on time involved as defined in Scale 37 (July 1988) para. 19 or on a lump sum or percentage basis agreed between the building owner and the quantity surveyor.

6. The foregoing scale of charges is exclusive of any services in connection with litigation and arbitration.

7. The fees and charges are in all cases exclusive of value added tax which shall be applied in accordance with legislation.

QUANTITY SURVEYORS' FEES

Scale 47 professional charges for the assessment of replacement costs of buildings for insurance, current cost accounting and other purposes issued by The Royal Institution of Chartered Surveyors. The scale is recommended and not mandatory.

EFFECTIVE FROM JULY 1988

1.0. GENERALLY

1.1. The fees are in all cases exclusive of travelling and other expenses (for which the actual disbursement is recoverable unless there is some prior arrangement for such charges).

1.2. The fees and charges are in all cases exclusive of value added tax which will be applied in accordance with legislation.

2.0. ASSESSMENT OF REPLACEMENT COSTS OF BUILDINGS FOR INSURANCE PURPOSES

2.1. Assessing the current replacement cost of buildings where adequate drawings for the purpose are available.

Assessed current cost		Fee		
	£	£		£
Up to	140,000	0.2%		
140,000 -	700,000	280 + 0.075%	on balance over	140,000
700,000 -	4,200,000	700 + 0.025%	on balance over	700,000
Over	4,200,000	1,575 + 0.01%	on balance over	4,200,000

2.2. Fees are to be calculated on the assessed current cost, i.e. base value, for replacement purposes including allowances for demolition and site clearance but excluding inflation allowances and professional fees.

2.3. Where drawings adequate for the assessed of cost are not available or where other circumstances require that measurements of the whole or part of the buildings are taken, an additional fee shall be charged based on the time involved or alternatively on a lump sum basis agreed between the employer and the surveyor.

2.4. When the assessment is for buildings of different character or on more than one site, the costs shall not be aggregated for the purpose of calculating fees.

2.5. For current cost accounting purposes this scale refers only to the assessment of replacement costs of buildings.

2.6. The scale is appropriate for initial assessments but for annual review or a regular reassessment the fee should be by arangement having regard to the scale and to the amount of work involved and the time taken.

2.7. The fees are exclusive of services in connection with negotiations with brokers, accountants or insurance companies for which there shall be an additional fee based upon the time involved.

CONSULTING ENGINEERS' FEES

CONDITIONS OF ENGAGEMENT - 1981

INTRODUCTION

The following paragraphs describe the scope of professional services provided by the Consulting Engineer and give general advise about his appointment.

The Association of Consulting Engineers has drawn up standard Conditions of Engagement to form the basis of the agreement between the Client and the Consulting Engineer, for five different types of appointment as hereafter described. Each of the standard Conditions of Engagement is accompanied by a recommended Memorandum of Agreement. These two documents, taken together, constitute the recommended form of Agreement in each case.

When the standard Conditions of Engagement are not so used, it is in the interests of both the Client and the Consulting Engineer that there should be an exchange of letters defining the duties which the Consulting Engineer is to perform and the terms of payment.

REPORT AND ADVISORY WORK

For reports, and for advisory work, the services required from the Consulting Engineer will usually comprise one or more of the following:

(a) investigating and advising on a project and submitting a report thereon. The Consulting Engineer may be asked to examine alternatives; review all technical aspects; make an economic appraisal of costs and benifits; draw conclusions and make recommendations.
(b) inspecting existing works (e.g. a reservoir or a building or an installation) and reporting thereon. If the Client requires continuing advise on maintenance or operation of an existing project, the Consulting Engineer may be appointed to make periodic visits.
(c) making a special investigation of an engineering problem and reporting thereon.
(d) making valuations of plant and undertakings.

Payment for services provied under Items (a) to (d) above should normally be on a time basis or, where the duration and extent of the services can be defined clearly, by an agreed lump sum. It is recommended that the conditions governing an appointment for these services should be based on:

Agreement 1 - Conditions of Engagement for Reporting and Advisory Work.

DESIGN AND SUPERVISION OF CONSTRUCTION

When the Client has decided to proceed with the construction of engineering works or the installation of plant, it is normal practice for the Consulting Engineer who prepared or assisted in the preparation of any report to be appointed for the subsequent stages of the work or the relevent part thereof. The Consulting Engineeer will assist the Client in obtaining the requisite approvals, then prepare the designs and tender documents to enable competitive tenders to be obtained or orders to be placed, and will be responsible for the technical control and administration of the construction of he Works.

It is recommended that the agreement for this type of appointment should be based on the most appropriate of the following:

Agreement 2 - Conditions of Engagement for Civil, Mechanical and Electrical Work and for Structural Engineering Work where an Architect is not appointed by the Client.

CONSULTING ENGINEERS' FEES

Agreement 3 - Conditions of Engagement for Structural Engineering Work where an Architect is appointed by the Client.

Agreement 4A - Conditions of Engagement for Engineering Services in relation to Sub-contract Works.

Agreement 4B - Conditions of Engagement for Engineering Services in relation to Direct Contract Works.

TERMS OF PAYMENT

The Association considers that, in normal circumstances, the level of renumeration represented by the scales of percentage fees and hourly charging rates set out in the ACE Conditions of Engagement is such as to ensure the provision by a Consulting Engineer to his Client of a full, competent and reliable standard of service.

These scales and rates are not mandatory but solely guidelines, and may by negotiation be adjusted upwards or downwards to take account of the abnormal complexity or simplicity of design, increase or diminution of the extent of services to be provided, long-standing client relationships or other circumstances.

However the Association strongly advises clients to satisfy themselves that the level and quality of services they want will be covered if they make appointments based on charges which are appreciably lower than those shown.

DIRECT LABOUR WORKS

The A.C.E. Conditions of Engagement require to be modified and supplemented when the Client intends to have the Works constructed wholly or partly by direct labour under the control of he Consulting Engineer.

The additional services and substantially greater responsibilities undertaken by the Consulting Engineer in connection with work carried out by direct labour usually entitle him to a higher level of remuneration than that payable in respect of Works carried out by contract.

PARTIAL SERVICES

When the Client wishes to appoint the Consulting Engineer for partial services only, it is important that both parties recognize the limitation which such an appointment places upon the responsibiliy of the Consulting Engineer who cannot be held liable for matters that are outside his control.

The terms of reference for the appointment should be carefully drawn up and the relevant A.C.E. Conditions of Engagement should be adapted to suit the scope of services required.

Professional charges for partial services are usually best calculated on a time basis, but may, in suitable cases, be a commensurate part of the percentage fee for normal services shown in the standard Conditions of Engagement.

INSPECTION SERVICES

The inspection of materials and plant during manufacture or on site is usually required during the construction stage of a project. Consequently, the arrangements for this service are described in the standard Conditions of Engagement. If, however the Client wishes to engage the Consulting Engineer to provide only inspection services, the charges of the Consulting Engineer may be either on a time basis or a percentage of the cost of the materials to be inspected.

CONSULTING ENGINEERS' FEES

ACTING AS ARBITRATOR, UMPIRE OR EXPERT WITNESS

A Consulting Engineer may be appointed to act as Arbitrator or Umpire, or be required to attend as an Expert Witness at Parliamentary Committees, Courts of Law, Arbitrations or Official Inquiries. Payment for any of these services should be on a basis of a lump sum retainer plus time charges, not less than three hours per day being chargeable for attendance, however short, either before or after a mid-day adjournment.

PAYMENT ON A TIME BASIS

When it is not possible to estimate in advance the duration and extent of the Consulting Engineer's services, neither a lump sum payment alone nor a percentage of the estimated construction cost would normally be a fair basis of remuneration. The most satisfactory and equitable method of payment in these cases makes allowance for the actual time occupied in providing the services required, and comprises the following elements, as applicable:

(1) A charge in the form of hourly rate(s) for the services of a self-employed Principal or a Consultant of the firm. The hourly rate will depend upon his standing, the nature of the work and any special circumstances. Alternatively a lump sum fee may be charged instead of the said hourly rate(s).

(2) A charge which covers salaried Principals, Directors and technical and supporting staff salary and other payroll costs actually incurred by the Consulting Engineer, together with a fair proportion of his overhead costs, plus an element of profit. This charge is most conveniently calculated by applying a multiplier to the salary cost and then adding the net amount of other pay roll costs. The major part of the multiplier is attributable to the Consulting Engineer's overheads which may include, inter alia, the following indirect costs and expenses:

    (a) rent, rates and other expenses of up-keep of his office, its furnishings, equipment and supplies;
    (b) insurance premiums other than those recovered in the payroll cost;
    (c) administrative, accounting, secretarial and financing costs;
    (d) the expence of keeping abreast of advances in engineering;
    (e) the expense of preliminary arrangements for new or prospective projects;
    (f) loss of productive time of technical staff between assignments.

(3) A charge for use of a computer or other special equipment.

When calculating amounts chargeable on a time basis, a Consulting Engineer is entitled to include time spent by Partners, Directors,Consultants and technical and supporting staff in travelling in connection with the performance of the services. The time spent by secretarial staff or by staff engaged on general accountancy or administration duties in the Consulting Engineer's office is not chargeable unless otherwise agreed.

DISBURSEMENTS AND OUT OF POCKET EXPENSES

In addition to the charges and percentage fees referred to in the preceeding Sections, the Consulting Engineer is entitled to recover from the Client all disbursements and out of pocket expenses incurred in performing his services.

CONSULTING ENGINEERS' FEES

APPOINTMENTS OUTSIDE THE UNITED KINGDOM

The standard A.C.E. Conditions of Engagement are suitable for appointments in the United Kingdom. For work overseas, it is impracticable to make definite recommendations as the conditions vary widely from country to country. There are added complications in documentation relating to import customs, conditions of payment, insurance, freight, etc. Furthermore, it is necessary to arrange for site visits to be undertaken by Partners or senior staff whose absence abroad during such periods represents a serious reduction of their earning power.

The additional duties, responsiblities and non-recoverable costs involved, and the extra work on general co-ordination, therefore justify higher fees in such cases. Special arrangements are also necessary to cover travelling and other out of pocket expenses in excess of those normally incurred on similar work in the United Kingdom - including such matters as local cost-of-living allowances and the cost of providing home-leave facilities to expatriate staff.

AGREEMENT 1 CONDITIONS OF ENGAGEMENT

1. DEFINITIONS

In construing this Agreement the following expressions shall have the meanings hereby assigned to them except where the context otherwise requires:

'The Consulting Engineer' means the person or firm named in the Memorandum of Agreement and shall include any other person or persons taken into partnership by such person or firm during the currency of this Agreement and the surviving member or members of any such partnership.

'The Task' means the work described in the Memorandum of Agreement in respect of which the Client has engaged the Consulting Engineer to provide professional services.

'Contractor' means any person or persons firm or company under contract to the Client to perform work and/or supply goods in connection with the Task.

'Salary Cost' means the total annual taxable remuneration paid by the Consulting Engineer to any person employed by him, divided by 1600 (being deemed to be the average annual total of effective working hours of an employee) and multiplied by the number of working hours spent by such person in performing any of the services in respect of which payment under this Agreement is to be made to the Consulting Engineer upon the basis of Salary Cost. For the purposes of this definition the annnual remuneration of a person employed by the Consulting Engineer for a period less than a full year shall be calculated pro rata to such person's remuneration for such lesser period.

'Other Payroll Cost' means the annual amount of all contributions and payments made by the Consulting Engineer on behalf of or in respect of a person employed by him for staff pension and life assurance schemes, and also for National Insurance Contributions and for any other tax, charge, levy, impost or payment of any kind whatsoever which the Consulting Engineer at any time during the performance of this Agreement is obliged by law to make on behalf of or in respect of such person, divide by 1600 (being deemed to be the average annual total of effective working hours of an employee) and multiplied by the number of working hours spent by such person in performing any of the services in respect of which payment under this Agreement is to be made to the Consulting Engineer upon the basis of Other Payroll Cost.

CONSULTING ENGINEERS' FEES

For the purposes of this definition the annual amount of all contribution and payments made by the Consulting Engineer on behalf of or in respect of a person employed by him for a period less than a full year shall be calculated pro rata to the amount of such contributions and payments for such lesser period.

Words importing the singular include the plural and vice versa where the context requires.

2. DURATION OF ENGAGEMENT

2.1. The appointment of the Consulting Engineer shall commence from the date stated in the Memorandum of Agreement or from the time when the Consulting Engineer shall have begun to perform for the Client any of the services specified in Clauses 6 and 7 of this agreement, whichever is the earlier.

2.2. The Consulting Engineer shall not, without the consent of the Client, assign the benefit or in any way transfer the obligations of this Agreement or any part thereof.

2.3. If at any time the Client decides to postpone or abandon the Task, he may there-upon by notice in writing to the Consulting Engineer terminate the Consulting Engineer's appointment under this Agreement provided that, in any case in which the Consulting Engineer is paid for his services under Clause 6 in accordance with Clause 9.1 or Clause 9.2, the client may, when the Task is postponed, in lieu of so terminating the Consulting Engineer's appointment require the Consulting Engineer in writing to suspend the carrying out of his services under this Agreement for the time being.

2.4. If the Client shall not have required the Consulting Engineer to resume the performance of services in respect of the Task within a period of 12 months from the date of the Client's requirement to the Consulting Engineer to suspend the carrying out of his services, the Task shall be considered to have been abandoned and this Agreement shall terminate.

2.5. In the event of the failure of the Client to comply with any of his obligations under this Agreement, or upon the occurence of any circumstances beyond the control of the Consulting Engineer which are such as to delay for a period of more than 12 months or prevent or unreasonably impede the carrying out by the Consulting Engineer of his services under this Agreement, the Consulting Engineer may upon not less than 60 days' notice in writing to the Client terminate his appointment under this Agreement, provided that, in lieu of so terminating his appointment, the Consulting Engineer may:

(a) forthwith upon any such failure or the occurence of any such circumstances suspend the carrying out of his services hereunder for a period of 60 days (provided that he shall as soon as practicable inform the Client in writing of such suspension and the reasons therefor), and

(b) at the expiry of such period of suspension either continue with the carrying out of his services under this Agreement or else, if any of the reasons for the suspension then remain, forthwith in writing to the Client terminate his appointment under this Agreement.

CONSULTING ENGINEERS' FEES

2.6. The Consulting Engineer shall, upon receipt of any notice or requirement in writing in accordance with Clause 2.3 or the termination by him of his appointment in persuance of Clause 2.5, proceed in an orderly manner but with all reasonable speed and economy to take such steps as are necessary to bring to an end his services under this Agreement.

2.7. Any termination of the Consulting Engineer's appointment under this Agreement shall not prejudice or affect the accrued rights or claims of either party to this Agreement.

3. OWNERSHIP OF DOCUMENTS AND COPYRIGHT

3.1. The copyright in all drawings, reports, calculations and other documents provided by the Consulting Engineer in connection with the Task shall remain vested in the Consulting Engineer, but the Client shall have a licence to use such drawings and other documents for any purpose related to the Task. Save as aforesaid, the Client shall not make copies of such drawings or other documents nor shall he use the same in connection with the making or improvement of any works other than those to which the Task relates without the prior written approval of the Consulting Engineer and upon such terms as may be agreed between the Client and the Consulting Engineer.

3.2. The Consulting Engineer may with the consent of the Client, which consent shall not be unreasonably withheld, publish alone or in conjunction with any other person any articles, photographs or other illustrations relating to the Task.

4. SETTLEMENT OF DISPUTES

Any dispute or difference arising out of this Agreement shall be refered to the arbitration of a person to be agreed upon between the Client and the Consulting Engineer or, failing agreement, nominated by the President for the time being of the Chartered Insitute of Arbitrators.

OBLIGATIONS OF THE CONSULTING ENGINEER

5. CARE AND DILIGENCE

5.1. The Consulting Engineer shall exercise all reasonable skill, care and diligence in the discharge of the services agreed to be performed by him. If in the performance of his services the Consulting Engineer has a descretion exercisable as between the Client and the Contractor, the Consulting Engineer shall exercise his discretion fairly.

5.2. Where any person or persons are engaged, whether by the Client or by the Consulting Engineer on the Client's behalf under Clause 7.4, the Consulting Engineer shall not be liable for acts of negligence, default or omission by such person or persons.

6. NORMAL SERVICES

The services to be provided by the Consulting Engineer shall comprise:

(a) all or any of the services stated in the Appendix to the Memorandum of Agreement and

CONSULTING ENGINEERS' FEES

(b) advising the Client as to the need for the Client to be provided with additional services in accordance with Clause 7.

7. ADDITIONAL SERVICES NOT INCLUDED IN NORMAL SERVICES

7.1. As services additional to those specified in Clause 6, the Consulting Engineer shall, if so requested by the Client, provide any of the services specified in Clause 7.2 and provide or take all reasonable steps to arrange for the provision of any of the services specified in Clause 7.3.

7.2. (a) Carrying out works consequent upon a decision by the Client to seek parlimentary powers.
(b) Carrying out work in connection with any application by the Client for any order, sanction, licence, permit or other consent, approval or authorization necessary to enable the Task to proceed.
(c) Carrying out work arising from the failure of the Client to award a contract in due time.
(d) Carrying out work consequent upon any assignment of a contract by the Contractor or upon the failure of the Contractor properly to perform any contract or upon delay by the Client in fulfilling his obligations under Clause 8 or in taking any other step necessary for the due performance of the Task.
(e) Advising the Client upon and carrying out work following the taking of any step in or towards and litigation or arbitration relating to the Task.
(f) Carrying out work in conjunction with others employed to any of the services specified in Clause 7.3.

7.3. (a) Specialist technical advise on any abnormal aspects of the Task.
(b) Architectual, legal, financial and other professional services.
(c) Services in connection with the valuation, purchase, sale or leasing of lands and the obtaining of wayleaves.
(d) The carrying out of marine, air and land surveys, and the making of model tests or special investigations.

7.4. The Consulting Engineer shall obtain the prior agreement of the Client to the arrangements which he proposes to make on the Client's behalf for the provision of any services specified in Clause 7.3. The Client shall be responsible to any person or persons providing such services for the cost thereof.

OBLIGATIONS OF THE CLIENT

8. INFORMATION TO BE SUPPLIED TO THE CONSULTING ENGINEER

8.1. The Client shall supply to the Consulting Engineer without charge and within a reasonable time all necessary and relevant data and information in the possession of the Client and shall give such assistance as shall reasonably be required by the Consulting Engineer in the performance of the Task.

CONSULTING ENGINEERS' FEES

8.2. The Client shall give his decision on all sketches, drawings, reports, recommendations, tender documents and other matters properly referred to him for decision by the Consulting Engineer in such reasonable time as not to delay or disrupt the performance by the Consulting Engineer of the Task.

9. PAYMENT FOR SERVICES - EXPLANATORY NOTE:

Three different methods of payment for services carried out under Clause 6 are detailed in the three succeeding Clauses 9.1, 9.2 and 9.3. The method of payment which the Client and the Consulting Engineer agree to adopt must be specified in the Memorandum of Agreement.

9.1. **Payment at hourly rates**

In respect of services provided by the Consulting Engineer under Clauses 6 and 7 the Client shall pay the Consulting Engineer:

(a) For time spent by Principals and Consultants of the firm, including time spent in travelling in connection with the Task, at the hourly rate or rates specified in Article 3 (a) of the Memorandum of Agreement.

(b) For Directors, salaried Principals and technical and supporting staff working in or based on the Consulting Engineer's Office: Salary Cost times the multiplier stated in Article 3 (b) of the Memorandum of Agreement, plus Other Payroll Cost. Time spent in travelling in connection with the Task shall be chargeable.

(c) For Directors, salaried Principals and technical and supporting staff, working as field staff, in or based on any field office established in pursuance of Clause 10: Salary Cost times the appropriate multiplier specified in Article 3 (c) of the Memorandum of Agreement, plus Other Payroll Cost.

(d) A reasonable charge for the use of a computer or other special equipment which charge shall be agreed between the Client and the Consulting Engineer before the work is put in hand.

(e) For time spent by Directors, salaried Principals and technical and supporting staff in connection with the use of a computer or other special equipment, including the development and writing of programmes and the operation of the computer in trial and final runs in accordance with Clause 9.1 (b) above. Unless otherwise agreed between the Client and the Consulting Engineer, the Consulting Engineer shall not be entitled to any payment in respect of time spent by secretarial staff or by staff engaged on general accountancy or administration duties in the Consulting Engineer's office.

9.2. **Payment at hourly rates plus a fixed fee**

(a) In respect of services provided by the Consulting Engineer under Clause 6, the Client shall pay the Consulting Engineer in accordance with Clause 9.1, except that in lieu of charging self-employed Principals and Consultants at hourly rates the Client shall pay the Consulting Engineer the fee stated in Article 3 (d) of the Memorandum of Agreement.

(b) In respect of all other services provided by the Consulting Engineer, the Client shall pay the Consulting Engineer on the basis specified in Clause 9.1.

CONSULTING ENGINEERS' FEES

9.3. **Payment of a fixed sum**

(a) The sum payable by the Client to the Consulting Engineer for his services under Clause 6 shall be the sum stated in Article 3 (e) of the Memorandum of Agreement.
(b) In respect of all other services provided by the Consulting Engineer, the Client shall pay the Consulting Engineer on the basis specified in Clause 9.1.

10. PAYMENT FOR FIELD STAFF FACILITIES

The Client shall be responsible for the cost of providing such field office accommodation, furniture, telephones, equipment and transport as shall be necessary for the use of field staff, and for the reasonable running costs of such necessary field office accommodation and other facilities, including those of stationery, telephone calls, telex, telegrams and postage. Unless otherwise agreed between the Client and the Consulting Engineer, the Consulting Engineer shall arrange for the provision of field office accommodation and facilities for the use of field staff.

11. DISBURSEMENTS

The Client shall reimburse the Consulting Engineer in respect of all the Consulting Engineer's disbursements properly made in connection with:

(a) Printing, reproduction and purchase of all documents, drawings, maps, records and photographs.
(b) Telegrams, telex and telephone calls (other than local telephone calls).
(c) Postage and similar delivery charges except in the case of items weighting less than 250 grams sent by ordinary inland post.
(d) Travelling, hotel expenses and other similar disbursements.
(e) Advertising for tenders and for field staff.
(f) The provision of additional services to the Client pursuant to Clause 7.4.
(g) Professional Indemnity Insurance taken out by the Consulting Engineer to accord with the wishes of the Client and as set out in the Memorandum of Agreement.

The Client, by agreement with the Consulting Engineer and in satisfaction of his liability to the Consulting Engineer in respect of these disbursements, may take to the Consulting Engineer a lump sum payment as specified in Article 3 (f) of the Memorandum of Agreement.

12. PAYMENT FOLLOWING TERMINATION OR SUSPENSION BY THE CLIENT

12.1. Upon a termination or suspension by the Client in pursuance of Clause 2.3, the Client shall pay the Consulting Engineer the sums specified in (a), (b) and (c) of this Clause less the amount of payments previously made to the Consulting Engineer under the terms of this Agreement.

CONSULTING ENGINEERS' FEES

(a) All amounts due to the Consulting Engineer at hourly rates in accordance with Clause 9 in respect of services rendered up to the date of termination or suspension together with a sum calculated in accordance with Clause 9 in respect of time worked by the Consulting Engineer's staff in complying with Clause 2.6.

(b) A fair and reasonable proportion of any lump sum specified in Articles 3 (d), 3 (e) and 3 (f) of the Memorandum of Agreement. In the assessment of such proportion, the services carried out by the Consulting Engineer up to the date of termination or suspension and in pursuance of Clause 2.6 shall be compared with a reasonable assessment of the services which the Consulting Engineer would have carried out but for the termination or suspension.

(c) Amounts due to the Consulting Engineer under any other clauses of this Agreement.

12.2. In any case in which the Client has required the Consulting Engineer to suspend the carrying out of the Consulting Engineer's services in pursuance of the power conferred by Clause 2.3, the Client, may at any time within a period of 12 months from the date of his requirement in writing to the Consulting Engineer to suspend the carrying out of the Consulting Engineer's services require the Consulting Engineer in writing to resume the performances of such services. In such event the Consulting Engineer shall within a reasonable time of receipt by him of Client's said requirement in writing resume the performance of his services in accordance with this Agreement. Upon such a resumption, the amount of any payment made to the Consulting Engineer under Clause 12.1 (b) shall rank as a payment made on account of the total sum payable to the Consulting Engineer under this Agreement, but no adjustment shall be made of any sum paid or payable to the Consulting Engineer upon suspension.

12.3. If the Consulting Engineer shall need to perform any additional services in connection with the resumption of his services in accordance with Clause 12.1, th client shall pay the Consulting Engineer in respect of the performance of such additional services in accordance with Clause 9.1 and any appropriate reimbursements in accordance with Clause 11.

13. PAYMENT FOLLOWING TERMINATION BY THE CONSULTING ENGINEER

Upon a termination by the Consulting Engineer in pursuance of Clause 2.5, the Client shall pay to the Consulting Engineer the sums specified in Clause 12.1 (a), (b) and (c) less the amount of payments perviously made to the Consulting Engineer under the terms of this Agreement. Upon payment of such sums, the Consulting Engineer shall deliver to the Client such completed drawings and other similar documents relevant to the Task as are in his possession. The Consulting Engineer shall be permitted to retain copies of any documents so delivered to the Client.

The provisions of this Clause are without prejudice to any other rights and remedies which the Consulting Engineer may possess.

CONSULTING ENGINEERS' FEES

14. PAYMENT OF ACCOUNTS

14.1. Unless otherwise agreed between the Client and the Consulting Engineer from time to time

(a) The fee referred to in Clause 9.2 (a) and the sum payable to the Consulting Engineer under Clause 9.3 (a) shall be paid by the Client in the instalments and at the intervals stated in Article 3 (g) of the Memorandum of Agreement. Any lump sum payable under Clause 11 shall be paid as stated in Article 3 (f).

(b) All sums due to the Consulting Engineer, other than those referred to in Clause 14.1 (a), shall be paid by the Client on accounts rendered monthly by the Consulting Engineer.

14.2. The Consulting Engineer shall submit to the Client at the time of submission of the monthly accounts such supporting data as may be agreed between the Client and the Consulting Engineer.

14.3. All sums due from the Client to the Consulting Engineer in accordance with the items of this Agreement shall be paid within 40 days of the submission by the Consulting Engineer of his accounts therefore to the Client, and any sums remaining unpaid at the expiry of such period of 40 days shall bear interest thereafter, such interest to accrue from day to day at the rate of 2% per annum above the Base Rate of the Engineer's Principal Bank as stated in Article 3 (h) of the Memorandum of Agreement.

14.4. If any item or part of an item of an account rendered by the Consulting Engineer is disputed or subject to question by the Client, the payment by the Client of the remainder of that account shall not be withheld on those grounds and the provisions of Clause 14.3 shall apply to such remainder and also to the disputed of questioned item, to the extent that it shall subsequently be agreed or determined to have been due to the Consulting Engineer.

14.5 All fees set out in this Agreement are exclusive of Value Added Tax, the amount of which, at the rate and in the manner prescribed by law, shall be paid by the Client to the Consulting Engineer.

CONSULTING ENGINEERS' FEES

MEMORANDUM OF AGREEMENT

BETWEEN CLIENT AND CONSULTING ENGINEER
FOR REPORT AND ADVISORY WORK

MEMORANDUM OF AGREEMENT made the ....

day of .... 19......

BETWEEN

....

....(hereinafter called 'the Client')

of the one part and...

....

(hereinafter called 'the Consulting Engineer') of the other part.

WHEREAS the Client has requested the Consulting Engineer to provide professional services as described in the Appendix hereto in connection with

....

....

....

..................................(referred to in this Agreement as 'the Task')

NOW IT IS HEREBY AGREED as follows:

1. The Client agrees to engage the Consulting Engineer subject to and in accordance with the Conditions of Engagement attached hereto and the Consulting Engineer agrees to provide professional services subject to and in accordance with the said Conditions of Engagement.
2. This memorandum of Agreement and the said Conditions of Engagement shall together constitute the Agreement between the Client and the Consulting Engineer.
3. In the said Conditions of Engagement:

   (a) the rate or rates referred to in Clause 9.1 (a) shall be.............

   ....

   ....

   (b) the multiplier referred to in Clause 9.1 (b) shall be................

   (c) the multiplier referred to in Clause 9.1 (c) shall be:
   for field staff who are permanent employees of
   the Consulting Engineer...
   for field staff who are recruited specifically
   for the Task..

CONSULTING ENGINEERS' FEES

(d) the fee referred to in Clause 9.2 (a) shall be......................

(e) the sum referred to in Clause 9.3 (a) shall be......................

(f) the lump sum referred to in Clause 11 shall be................payable in ....equal monthly instalments.

(g) the intervals for the payment of instalments under Clause 14.1 (a) and the instalments referred to in the said sub-clause shall be:

....

....

(h) the Engineer's Principal Bank referred to in Clause 14.3 shall be

....

4. The method of payment for services under Clause 6 of the said Conditions of Engagement shall be that described in Clause 9.1*, 9.2*, 9.3*, thereof.

5. PII

The amount of professional idemnity insurance referred to in Clause ii........+ of the said Conditions of Engagement shall be................ ....pounds(£) for any one occurrence or series of occurrences arising out of this engagement.

This professional indemnity insurance shall be maintained for a period of .... years from the date of this Memorandum of Agreement, unless such insurance cover ceases to be available in which event the Consulting Engineer will notify the Client immediately.

The sum payable by the Client to the Consulting Engineer as a contribution to the additional cost of the professional indemnity insurance thus provided shall be ....pounds(£) and such amount shall become due to the Consulting Engineer immediately upon acceptance by or on behalf of the Client of any tender in respect of the Works or any part thereof/immediately upon submission by the Consulting Engineer to the Client of his report in relation to the Task.*

6. LIMITATION OF LIABILITY

Notwithstanding anything to the contary contained elsewhere in this Agreement, the total liability of the Consulting Engineer under or in connection with this Agreement, whether in contract, in tort, for breach of statutory duty or otherwise, shall not exceed.............pounds(£). The Client shall indemnify and keep indemnified the Consulting Engineer from and against all claims, demands, proceedings, damages, costs, charges and expenses arising out of or in connection with this Agreement, the Works and/or the Project in excess thereof.

AS WITNESS the hands of the parties the day and year first above written.

CONSULTING ENGINEERS' FEES

Duly Authorised Representative of the Client	Consulting Engineer ....
Witness	Witness ....

* Delete as appropriate.

APPENDIX TO THE MEMORANDUM OF AGREEMENT

The services to be provided by the Consulting Engineer shall be as follows:

BASIS FOR COMPLETING MEMORANDUM OF AGREEMENT- EXPLANATORY NOTE:

1. The Appendix to the Memorandum of Agreement should be a concise but comprehensive description of the services to be provided by the Consulting Engineer including, for example, reference to

   Objective
   Scope
   Timing
   Surveys
   Geotechnical Investigation
   Field Staff
   Particpation by Client or other parties
   Budget
   Cost Control
   Style of report to suit Client's probable funding agency
   Number of copies of report or method of submitting data to the Client.

2. Articles 3 (a) and 3 (b) must be completed even where payment is to be made for normal services under Clause 9.2 or 9.3. In Article 3 (b) the normal multiplier is 2.6.
3. In Article 3 (c) the normal multiplier for field staff who are permanent employees of the Consulting Engineer is 2.6. The normal multiplier for staff recruited specifically for the Task is 1.3.
4. If disbursements are to be reimbursed at cost then Article 3 (f) should be deleted in entirety.
5. In Article 3 (g) the intervals for payment of instalments should be monthly or quarterly. The amount and number of instalments need to be stated.
6. The amount inserted (in words, figures and time) as P11 shall be based on what cover the Client decides appropriate. It should normally be available on an each and every occurrence basis. Exceptionally it can be on a special basis. It should not be less than the Consulting Engineer provides as a general rule.

The sum payable by the Client shall be subject to negotiation. It will have to be based on the notional premiums which the Consulting Engineer expects to be having to pay over the whole period involved. Such a figure will need to be assessed in consultation with the Consulting Engineer's insurance broker allowing for inflation and based on the anticipated fees to be received during the whole project.

CONSULTING ENGINEERS' FEES

Account may also need to be taken of the level of insurance taken within the Consulting Engineer's firm itself in the form of an excess or self-insurance. The proportion of this assessed sum which the Client should be invited to contribute should take into account whether greater than normal cover is to be provided and whether the fees being paid for the project are to be below the guideline level.

The limit of the Consulting Engineer's liability should be the amount of P11 as completed in the Memorandum of Agreement.

AGREEMENT 3 CONDITIONS OF ENGAGEMENT

1. DEFINITIONS

In construing this Agreement the following expressions shall have the meanings hereby assigned to them except where the context otherwise requires:

'The Consulting Engineer' means the person or firm named in the Memorandum of Agreement and shall include any other person or persons taken into partnership by such person or firm during the currency of this Agreement and the surviving member or members of any such partnership.

'The Architect' means any Architect appointed by the Client to act as the Architect of the project.

'The Project' means the project with which the Client is proceeding and of which the Works form a part.

'The Works' means the Work in connection with which the Client has engaged the Consulting Engineer to perform professional services.

'Contractor' means any person or persons firm or company under contract to the Client to perform work and/or supply goods in connection with the Works.

'Salary Cost' means the total annual taxable remuneration paid by the Consulting Engineer to any person employed by him, divided by 1600 (being deemed to be the average annual total of effective working hours of an employee) and multiplied by the number of working hours spent by such person in performing any of the services in respect of which payment under this Agreement is to be made to the Consulting Engineer upon the basis of Salary Cost. For the purposes of this definition the annual remuneration of a person employed by the Consulting Engineer for a period less than a full year shall be calculated pro rata to such person's remuneration for such lesser period.

'Other Payroll Cost' means the annual amount of all contributions and payment made by the Consulting Engineer on behalf of or in respect of a person employed by him for staff pension and life assurance schemes, and also for National Insurance Contributions and for any other tax, charge, levy, impost or payment of any kind whatsoever which the Consulting Engineer at any time during the performance of this Agreement is obliged by law to make on behalf of or in respect of such person, divided by 1600 (being deemed to be the average annual total of effective working hours of an employee) and multiplied by the number of working hours spent by such person in performing any of the services in respect of which payment under this Agreement is to be made to the Consulting Engineer upon the basis of Other Payroll Cost. For the purpose of this definition the annual amount of all contributions and payments made by the Consulting Engineer on behalf of in respect of a person employed by him for a period less than a full year shall be calculated pro rata to the amount of such contributions and payments for such lesser period.

CONSULTING ENGINEERS' FEES

Words importing the singular include the plural and vice versa where the context requires.

2. DURATION OF ENGAGEMENT

2.1. The appointment of the Consulting Engineer shall commence from the date stated in the Memorandum of Agreement or from the time when the Consulting Engineer shall have begun to perform for the Client any of the services in Clauses 6 and 7 hereof, whichever is the earlier.

2.2. The Consulting Engineer shall not, without the consent of the Client, assign the benefit or in any way transfer the obligations of this Agreement or any part thereof.

2.3. If at any time the Client decides to postpone or abandon the Works, he may thereupon by notice in writing to the Consulting Engineer forthwith terminate the Consulting Engineer's appointment under this Agreement, provided that the Client may, when the Works or any part thereof are postponed, in lieu of so terminating the Consulting Engineer's appointment require the Consulting Engineer in writing to suspend the carrying out of his services under this Agreement for the time being.

2.4. If at any time the Client decides to postpone or abandon any part of the Works, he may thereupon by notice in writing to the Consulting Engineer seek to vary this Agreement either by excluding the services to be performed by the Consulting Engineer in relation to such part of the Works, or by suspending performance of the same and in such notice the Client shall specify the services affected. The Consulting Engineer shall forthwith comply with the Client's notice and the Client shall pay to the Consulting Engineer a sum caluclated in accordance with the provisions of Clause 18.2 in respect of such compliance.

2.5. If the Client shall not have required the Consulting Engineer to resume the performance of services in respect of the whole or any part of the Works suspended under Clause 2.3 or Clause 2.4 hereof within a period of 12 months from the date of the Client's notice:

(i) In the case of a suspension under Clause 2.3 this Agreement shall forthwith automatically terminate; or

(ii) In the case of a suspension under Clause 2.4 the suspended services shall be deemed to have been excluded from the services to be performed by the Consulting Engineer under this Agreement.

2.6. In the event of the failure of the Client to comply with any of his obligations under this Agreement, or upon the occurrence of any circumstances beyond the control of the Consulting Engineer which are such as to delay for a period of more than 12 months or prevent or unreasonably impede the carrying out by the Consulting Engineer of his services under this Agreement, the Consulting Engineer may upon not less than 60 days' notice in writing to the Client terminate his appointment under this Agreement, provided that, in lieu of so terminating his appointment, the Consulting Engineer may

(a) forthwith upon any such failure or the occurrence of any such circumstances suspend the carrying out of his services hereunder for a period of 60 days (provided that he shall as soon as practicable inform the Client in writing of such suspension and the reasons therefore), and

(b) at the expiry of such period of suspension either continue with the carrying out of his services under this Agreement or else, if any of the reasons for the suspension then remain, forthwith in writing to the Client terminated his appointment under this Agreement.

2.7. The Consulting Engineer shall, upon receipt of any notice in accordance with Clause 2.3 or in the event of the termination by him of his appointment in pursuance of Clause 2.6, proceed in an orderly manner but with all reasonable speed and economy to take such steps as are necessary to bring to an end his services under this Agreement.

2.8. Unless terminated under this Clause the Consulting Engineer's appointment under this Agreement shall terminate when the certificate authorising the final payment to the Contractor is issued.

2.9. Any termination of the Consulting Engineer's appointment under this Agreement shall not prejudice or affect the accrued rights or claims of either party to this Agreement.

3. OWNERSHIP OF DOCUMENTS AND COPYRIGHT

3.1. The copyright in all drawing, reports, specifications, bills of quantities, calculations and other documents provided by the Consulting Engineer in connection with the Works shall remain vested in the Consulting Engineer, but the Client shall have a licence to use such drawings and other documents for any purpose related to the Works. Save as aforesaid, the Client shall not make copies of such drawings or other documents nor shall he use the same in connection with the making or improvement of any works other than those to which the Works relate without the prior wrtten approval of the Consulting Engineer and upon such terms as may be agreed between the Client and the Consulting Engineer.

3.2. The Consulting Engineer may with the consent of the Client, which consent shall not be unreasonable withheld,publish alone or in conjunction with any other person any articles, photographs or other illustrations relating to the Works.

4. SETTLEMENT OF DISPUTES

Any dispute or difference arising out of this Agreement shall be referred to the arbitration of a person to be agreed upon between the Client and the Consulting Engineer or, failing agreement, nominated by the President for the time being of the Chartered Institute of Arbitrators.

OBLIGATIONS OF THE CONSULTING ENGINEER

5. CARE AND DILIGENCE

5.1. The Consulting Engineer shall exercise all reasonable skill, care and diligence in the discharge of the services agreed to be performed by him. If in the performance of his services the Consulting Engineer has a discretion exercisable as between the Client and the Contractor, the Consulting Engineer shall exercise his discretion fairly.

CONSULTING ENGINEERS' FEES

5.2. The Consulting Engineer may recommend that specialist suppliers and/or contractors should design and execute certain part or parts of the Works in which circumstances the Consulting Engineer shall co-ordinate and intergrate the design of such part or parts with the overall design of the Works but he shall be relieved of all responsibility for the design, manufacture, installation and performance of any such part or parts of the works. Where any persons are engaged in accordance with Clause 7.4, the Consulting Engineer shall be under no liability for any negligence, default or omission of such persons.

6. NORMAL SERVICES

6.1. **Preliminary or Sketch Plan Stage.**

The services to be provided by the Consulting Engineer at this stage shall comprise all or any of the following as may be necessary in the particular case:

(a) Investigating data and information relating to the Project and relevant to the Works which are reasonably accessible to the Consulting Engineer and considering any reports relating to the Works which have either been prepared by the Consulting Engineer or else prepared by others and made available to the Consulting Engineer by the Client.
(b) Advising the Client on the need to carry out any geotechnical investigations which may be necessary to supplement the geotechnical information already available to the Consulting Engineer, arranging for such investigations when authorised by the Client, certifying the amount of any payments to be made by the Client to the persons or firm carrying out such investigations under the Consulting Engineer's direction, and advising the Client on the results of such investigations.
(c) Advising the client on the need for arrangements to be made, in accordance with Clause 7, for the carrying out of special surveys, special investigations or model tests, and advising the Client of the results of any such surveys, investigations or tests carried out.
(d) Consulting any local or other authorities on matters of principle in connection with the structural design of the Works.
(e) Providing sufficient structural information to enable the Architect to produce his sketch plans.

6.2. **Tender Stage**

The services to be provided by the Consulting Engineer at this stage shall include all or any of the following as may be necessary in the particular case, for the purpose of enabling tenders to be obtained:

(a) Developing the design of the Works in collaboration with the Architect and preparing calculations, drawings and specifications of the Works to enable Bills of Quantities to be prepared.
(b) Advising on conditions of contract relevant to the Works and forms of tender and invitations to tender as they relate to the Works.

(c) Consulting any local or other authorites in connection with the structural design of the Works, and preparing typical details and typical calculations.
(d) Advising the Client as to the suitability for carrying out the Works of persons and firms tendering.

6.3. **Working Drawing Stage**

The services to be provided by the Consulting Engineer at this stage shall include all or any of the following as may be necessary in the particular case:

(a) Advising the Client as to the relative merits of tenders, prices and estimates received for carrying out the Works.
(b) Preparing such calculations and details relating to the Works as may be required for submission to any appropriate authority.
(c) Preparing any further designs, specifications and drawings, including bar bending schedules, necessary for the information of the Contractor to enable him to carry out the Works, but excepting the preparation of any shop details relating to the Works or any part thereof.
(d) Examining shop details for general dimensions and adequancy of members and connections.

6.4 **Construction Stage**

The Consulting Engineer shall not accept any tender in respect of the Works unless the Client gives him instructions in writing to do so, and any acceptance so made by the Consulting Engineer on the instructions of the Client shall be on behalf of the Client. The services to be provided by the Consulting Engineer at this stage shall include all or any of the following as may be necessary in the particular case:

(a) Advising on the preparation of formal contract documents relating to accepted tenders for carrying out the Works or any part thereof.
(b) Advising the Client on the need for special inspections or tests.
(c) Advising the Client and the Architect on the appointment of site staff in accordance with Clause 8.
(d) Examining the Contractor's proposals.
(e) Making such visits to site as the Consulting Engineer shall consider necessary to satisfy himself as to the performance of any site staff appointed pursuant to Clause 8, and to satisfy himself that the Works are executed generally according to the contract and otherwise in accordance with good engineering practice.
(f) With the prior agreement of the Architect, giving all necessary instructions to the contractor, provided that the Consulting Engineer shall not without prior approval of the Client give any instructions which in the opinion of the Consulting Engineer are likely substantially to increase the cost of the Works unless it is not in the circumstances practicable for the Consulting Engineer to obtain such prior approval.
(g) Advising on certificates for payment to the Contractor.

(h) Performing any service which the Consulting Engineer may be required to carry out under any contract for the execution of the Works including where appropriate the supervision of any specified tests, provided that the Consulting Engineer may decline to perform any services specified in a contract the terms of which have not initially been approved by the Consulting Engineer.

(j) Delivering to the Client on the completion of the Works such records as are reasonably necessary to enable the Client to operate and maintain the Works.

(k) Assisting in settling any dispute or difference relating to the Works which may arise between the Client and the Contractor provided that such assistance shall not relate to the detailed examination of any financial claim and shall not extend to advising the Client following the taking of any steps in or towards any arbitration or litigation in connection with the Works.

6.5. **General**

Without prejudice to the preceding provisions of this clause, the Consulting Engineer shall from time to time as may be necessary advise the Client as to the need for the Client to be provided with additional services in accordance with Clause 7.

7. ADDITIONAL SERVICES NOT INCLUDED IN NORMAL SERVICES

7.1. As services additional to those specified in Clause 6, the Consulting Engineer shall, if so requested by the Client, provide any of the services specified in Clause 7.2 and provide or take all reasonable steps to arrange for the provision by others of any of the services specified in Clause 7.3.

7.2. (a) Preparing any report or additional contract documents required for consideration of proposals for the carrying out of alternative works.

(b) Carrying out work consequent upon a decision by the Client to seek parliamentary powers.

(c) Carrying out work in connection with any application by the Client for any order, sanction, licence, permit or other consent, approval or authorisation necessary to enable the Works to proceed.

(d) Carrying out work arising from the failure of the Client to award a contract in due time.

(e) Preparing preliminary estimates.

(f) Preparing details for shop fabrication of ductwork, metal or plastic frameworks.

(g) Checking and advising upon any part of the Project not designed by the Consulting Engineer.

(h) Preparing intrim or other reports or detailed valuations, including estimates or cost analyses based on measurement or forming an element of a cost planning service.

(j) Carrying out work consequent upon any assignment of a contract by the Contractor or upon the failure of the Contractor to properly perform any contract on or upon delay by the Client in fulfilling his obligations under Clause 9 or in taking any other step necessary for the due performance of the Works.

(k) Advising the Client with regard to any dispute or difference which involves matters excluded from Clause 6.4 (k) above.

(l) Carrying out work in conjunction with others employed to provide any of the services specified in Clause 7.3.

(m) Carrying out such other additional services, if any, as are specified in Article 5 of the Memorandum of Agreement.

7.3. (a) Specialist technical advice on any abnormal aspects of the Works.

(b) Legal, financial and other professional services.

(c) Services in connection with the valuation, purchase, sale or leasing of lands and the obtaining of wayleaves.

(d) The surveying of sites or existing works.

(e) Investigation of the nature and strength works and the making of models tests or special investigations.

(f) The carrying out of special inspections or tests advised by the Consulting Engineer under Clause 6.4 (b).

7.4. The Consulting Engineer shall obtain the prior agreemnet of the Client to the arrangements which he proposes to make on the Client's behalf for the provision of any of the services specified in Clause 7.3. The Client shall be responsible to any person or persons providing such services for the cost thereof.

8. SUPERVISION ON SITE

8.1. If in the opinion of the Consulting Engineer the nature of the Works, including the carrying out of any geotechnical investigation pursuant to Clause 6.1, warrents full-time or part-time engineering supervision on site, the Client shall agree to the appointment of suitably qualified technical and clerical site staff as the Consulting Engineer shall consider reasonably necessary to enable such supervision to be carried out.

8.2. Persons appointed pursuant to Clause 8.1 shall be employed either by the Consulting Engineer or, if the Client and the Consulting Engineer shall so agree, by the Client directly, provided that the Client shall not employ any person as a member of the site staff unless the Consulting Engineer has first selected or approved such person as suitable for employment.

8.3. The terms of service of all site staff to be employed by the Consulting Engineer shall be subject to the approval of the Client, which approval shall not be unreasonably withheld.

8.4. The Client shall procure that the contracts of employment of the site staff employed by the Client shall stipulate that the staff so employed shall in no circumstances take or act upon instructions other than those of the Consulting Engineer.

8.5. Where a clerk of works nominated by the Architect is charged with the supervision of the Works on site, his selection and appointment shall be subject to the approval of the Consulting Engineer. In respect of the Works, the Client shall ensure that such clerk of works shall take instructions solely from the Consulting Engineer, who shall inform the Architect of all such instructions.

8.6. Where any supervision on site is performed by staff employed other than by the Consulting Engineer, the Consulting Engineer shall not be responsible for any failure on the part of such staff properly to comply with any instructions given by the Consulting Engineer.

CONSULTING ENGINEERS' FEES

OBLIGATIONS OF THE CLIENT

9. INFORMATION TO BE SUPPLIED TO THE CONSULTING ENGINEER

9.1. The Client shall supply to the Consulting Engineer, without charge and in such reasonable time as not to delay or disrupt the performance by the Consulting Engineer of his services under this Agreement, all necessary and relevant data and information in the possession of the Client or the Contractors or their respective servants, agents or sub-contractors and the Client shall give and shall procure that such persons give such assistance as shall reasonably be required by the Consulting Engineer in the performance of his services under this Agreement. The information to be provided by the Client to the Consulting Engineer shall include:

(a) All such drawings as may be necessary to make the Client's or the Architect's requirements clear, including plans and sections of all buildings (to a scale of not less than 1 to 100) and essential details (to a scale of not less 1 to 25) together with site plans (to a scale of not less than 1 to 1,250) and levels.
(b) Copies of all contract documents, including priced bills of quanities, relating to those parts of the Project which are relevant to the Works.
(c) Copies of all variation orders and supporting documents relating to the Project.

9.2. The Client shall give his decision on all sketches, drawings, reports, recommendation, tender documents and other matters properly referred to him for decision by the Consulting Engineer in such reasonable time as not to delay or disrupt the performance by the Consulting Engineer of his services under this Agreement.

10. PAYMENT FOR NORMAL SERVICES - EXPLANATORY NOTE:

Alternative methods of payment for services carried out under Clause 6 are detailed in the two succeeding Clauses 10.1 and 10.2. The method of payment which the Client and the Consulting Engineer agree to adopt must be specified in the Memorandum of Agreement. It should be noted that the method of payment specified in Clause 10.2 may also be used for services other than those carried out under Clause 6.

10.1. **Payment depending upon the actual cost of the Works.**

10.1.1. The sum payable to the Client to the Consulting Engineer for his services under Clause 6 shall be:

(a) On the total cost of the Works including reinforced concrete and brickwork or blockwork, but excluding unreinforced brickwork or blockwork, a fee calculated in accordance with Article 3(a) of the Memorandum of Agreement.
(b) A further fee on the cost of reinforced concrete work and/or reinforced brickwork or blockwork calculated in accordance with the percentage stated in Article 3(b) of the Memorandum of Agreement.
(c) A fee on the cost of unreinforced load-bearing brickwork or blockwork calculated in accordance with the percentage stated in Article 3(c) of the Memorandum of Agreement.

10.1.2. The total cost of the Works shall be calculated in accordance with Clause 20.

10.1.3. If the Works are to be constructed in more than one phase and as a consequence the services which it may be necessary for the Consulting Engineer to perform under Clause 6 have to be undertake by the Consulting Engineer separately in respect of each phase, then the fee entered in Article 3(a) of the Memorandum of Agreement shall be adjusted for each phase.

10.2. **Payment on a Time Basis**

10.2.1. The Client shall pay the Consulting Engineer in accordance with the Scale of Charges set out in Clause 10.2.2.

10.2.2. Scales of Charges:

(a) Self-employed Principals and Consultants: At the hourly rate or rates specified in Article 3 (d) of the Memorandum of Agreement.

(b) Directors, salaried Principals and Technical and supporting staff: Salary Cost times the multiplier specified in Article 3 (e) of the Memorandum of Agreement, plus Other Payroll Cost.

(c) Time spent by Principals, Directors,Consultants, technical and supporting staff in travelling in connection with the Works shall be chargeable on the above basis.

(d) Unless otherwise agreed between the Client and the Consulting Engineer, the Consulting Engineer shall not be entitled to any payment in respect of time spent by secretarial staff or by staff engaged on general accountancy or administration duties in the Consulting Engineer's office.

11. PAYMENT FOR ADDITIONAL SERVICES

In respect of additional services provided by the Consulting Engineer under Clause 7, the Client shall pay the Consulting Engineer in accordance with the Scales of Charges set out in Clause 10.2.2.

12. PAYMENT FOR QUANTITY SURVEYING SERVICES

If the Client requires the Consulting Engineers to carry out quantity surveying services in respect of the Works designed by him, the Client shall pay the Consulting Engineer in respect of such services in accordance with the appropriate scale of professional charges for quantity surveying services published by The Royal Institution of Chartered Surveyors current when such services are carried out.

CONSULTING ENGINEERS' FEES

13. PAYMENT FOR USE OF COMPUTER OR OTHER SPECIAL EQUIPMENT

Where the Consulting Engineer decides to use a computer or other special equipment in carrying out any additional services in accordance with Clause 7 or is expressly required by the Client to use a computer or other special equipment in carrying out of his services under Clause 6, the Client shall,unless otherwise agreed between the Client and Consulting Engineer, pay the Consulting Engineer:

(a) for the time spent in connection with the use of a computer or other special equipment including the development and writing of programmes and the operation of the computer in trial and final runs, in accordance with the Scale of charges set out in Clause 10.2.2 and
(b) a reasonable charge for the use of the Computer or other special equipment which charge shall be agreed between the Client and the Consulting Engineer before work is put in hand.

14. PAYMENT FOR SITE SUPERVISION

14.1 In addition to any other payment to be made by the Client to the Consulting Engineer under the Agreement, the Client shall

(a) reimburse the Consulting Engineer in respect of all salary and wage payments made by the Consulting Engineer to site staff employed by the Consulting Engineer pursuant to Clause 8 and in respect of all other expenditure incurred by the Consulting Engineer in connection with the selection, engagement and employment of site staff, and
(b) pay to the Consulting Engineer an increment on the amount due under sub-clause 14.1 (a), calculated at the percentage specified in Article 3 (f) of the Memorandum of Agreement in respect of head office overhead costs incurred on site staff administration.

provided that in lieu of payments under (a) and (b) above the Client and the Consulting Engineer may agree upon inclusive monthly or other rates to be paid by the Client to the Consulting Engineer for each member of site staff employed by the Consulting Engineer.

14.2 The Client shall also be responsible for the cost of providing such local office accommodation, furniture, telephones, equipment and transport as shall be reasonably necessary for the use of site staff appointed pursuant to Clause 8, and for the reasonable running costs of such necessary local office accommodation and other facilities, including those of stationery, telephone calls, telex, telegrams and postage. Unless otherwise agreed between the Client and the Consulting Engineer, the Consulting Engineer shall arrange, whether through the Contractor or otherwise, for the provision of local office accommodation and facilities for the use of site staff.

14.3 In cases where the Consulting Engineer has thought it proper that site staff should not be appointed, or where the necessary site staff is not available at site due to sickness or any other cause, the Consulting Engineer shall be paid in accordance with the Scale of Charges set out in Clause 10.2.2 for site visits which would have been unnecessary but for the absence or non-availability of site staff.

CONSULTING ENGINEERS' FEES

15. DISBURSEMENTS

15.1 The Client shall in all cases reimburse the Consulting Engineer in respect of all the Consulting Engineer's disbursements properly made in connection with:

(a) Printing, reproduction and purchase of all documents, drawings, maps, records and photographs.
(b) Telegrams, telex and telephone calls (other than local telephone calls).
(c) Postage and similar delivery charges except in the case of items weighing less than 250 grams sent by ordinary inland post.
(d) Travelling, hotel expenses and other similar disbursements.
(e) Advertising for tenders and for site staff.
(f) The provision of additional services to the Client pursuance to Clause 7.4.
(g) Professional Indemnity Insurance taken out by the Consulting Engineer to accord with the wishes of the Client as set out in the Memorandum of Agreement.

15.2. The Client by agreement with the Consulting Engineer, and in full satisfaction of his liability to the Consulting Engineer in respect of these disbursements, may make to the Consulting Engineer a lump sum payment or sum calculated as a percentage of the fees and charges falling due under Clauses 10, 11, 12 and 13, as specified in Article 3 (g) of the Memorandum of Agreement.

16. PAYMENT FOR ALTERATION OR MODIFICATION TO DESIGN

16.1 Subject to Clause 16.2 if at any time after the commencement of the Consulting Engineer's appointment under Clause 2.1, any design whether completed or in progress or any specification, drawing or other document prepared in whole or in part by the Consulting Engineer shall require to be modified or revised by reason of instructions received by the Consulting Engineer from or on behalf of the Client, or by reason of circumstances which could not reasonably have been foreseen by the Consulting Engineer, then the Client shall make additional payment to the Consulting Engineer for making necessary modifications or revisions and for any consequential reproduction of documents. Unless otherwise agreed between the Client and the Consulting Engineer, the additional sum to be paid to the Consulting Engineer shall be calculated in accordance with the Scale of Charges set out in Clause 10.2, and shall also include any appropriate reimbursements in accordance with Clause 15 provided always that the Consulting Engineer shall give the Client written notice as soon as it is reasonably apparent that additional payments will arise under this Clause.

16.2 Where in the Consulting Engineer's opinion the Client's instruction necessitates a fundamental redesign of of the part or parts of the Works affected by the instruction such that designs, specifications, drawings and other documents prepared by the Consulting Engineer cannot be modified or revised to take account thereof or where the modification or revision instructed by the Client results in a reduction in the cost of the part or parts of the Works affected thereby as contained in the Consulting Engineer's most recent estimate under Clause 21.1 (a) by 10% or more, then such part or parts of the Works shall be deemed to have been abandoned and the Consulting Engineer shall be paid therefor in accordance with Clause 18.2.

The Consulting Engineer shall carry out such further work and shall produce such further designs, specifications, drawings and other such documents as may be necessary to comply with the Client's instructions and the Consulting Engineer shall be paid therefor in accordance with the provisions of this Agreement.

17. PAYMENT WHEN WORKS ARE DAMAGED OR DESTROYED

If at any time before completion of the Works any part of the Works or any materials, plant or equipment whether incorporated in the Works or not shall be damaged or destroyed, the Client shall make additional payment to the Consulting Engineer in respect of any expenses incurred or additional work required to be carried out by the Consulting Engineer as a result of such damage or destruction. The amount of such additional payment shall be calculated in accordance with the Scale of Charges set out in Clause 10.2 and shall also include any appropriate reimbursements in accordance with Clause 15.

18. PAYMENT FOLLOWING TERMINATION OR SUSPENSION BY THE CLIENT

18.1. Upon a termination or suspension by the Client in pursuance of Clause 2.3, the Client shall pay to the Consulting Engineer the sums specified in (a), (b) and (c) below less the amount of payments previously made to the Consulting Engineer under the terms of this Agreement.

(a) A fair and reasonable proportion of the sum which would have been payable to the Consulting Engineer under Clauses 10 and 12 if no such termination or suspension had taken place. In the assessment of such proportion, the services carried out by the Consulting Engineer up to the date of termination or suspension and in pursuance of Clause 2.7 shall be compared with a reasonable assessment of the services which the Consulting Engineer would have carried out but for the termination or suspension. In any case in which it is necessary to assess the payment to be made to the Consulting Engineer in accordance with this sub-clause by reference to the cost of the Works, then to the extent that such cost is not known the assessment shall be made upon the basis of the Consulting Engineer's best estimate of the cost.
(b) Amounts due to the Consulting Engineer under any other clauses of this Agreement.
(c) A disruption charge equal to one-sixth of the difference between the sum which would have been payable to the Consulting Engineer under Clauses 10 and 12 but for the termination or suspension, and the sum payable under (a).

18.2. Upon a termination or suspension by the Client in pursuance of Clause 2.4, the Client shall pay to the Consulting Engineer the sums specified in (a) and (b) below.

(a) A fair and reasonable proportion of the sum which would have been payable to the Consulting Engineer under Clauses 10 and 12 in respect of the services affected if no such termination or suspension had taken place. The proportion shall be calculated in accordance with the provisions of Clause 18.1 (a); and
(b) A disruption charge, to be calculated in accordance with the provisions of Clause 18.1 (c)

18.3. In any case in which the Client has required the Consulting Engineer to suspend the carrying out of the Consulting Engineer's services in pursuance of Clauses 2.3 or 2.4 hereof, the Client may, at any time within the period of 12 months from the date of the Client's notice, require the Consulting Engineer in writing to resume the performance of such services. In such event:

(a) the Consulting Engineer shall within a reasonable time of receipt by him of the Client's said requirement in writing resume the performance of his service in accordance with this Agreement, the payment made under Clause 18.1 (a) or 18.2. (a) as the case may be ranking as payment on account towards the total sum payable to the Consulting Engineer under Clauses 10 and 12 but,

(b) not withstanding such resumption, the Consulting Engineer shall be entitled to retain or receive as an additional payment due in accordance with this Agreement the disruption charge referred to in Clause 18.1 (c) or 18.2 (b) as the case may be.

18.4. If the Consulting Engineer shall need to perform any additional services in connection with the resumption of his services in accordance with Clause 18.3 the Client shall pay the Consulting Engineer in respect of the performance of such additional services in accordance with the Scale of Charges set out in Clause 10.2 and any appropriate reimbursements in accordance with Clause 15.

19. PAYMENT FOLLOWING TERMINATION BY THE CONSULTING ENGINEER

Upon a termination by the Consulting Engineer in pursuance of Clause 2.6 the Client shall pay to the Consulting Engineer the sums specified in Clauses 18.1 (a) and (b) less the amount of payments previously made to the Consulting Engineer under the terms of this Agreement. Upon payment of such sums, the Consulting Engineer shall deliver to the Client such completed drawings, specifications and other similar documents relevant to the Works as are in his possession. The Consulting Engineer shall be permitted to retain copies of any document so delivered to the Client. The provisions of this Clause are without prejudice to any other rights and remedies which the Consulting Engineer may possess.

20. COST OF THE WORKS

20.1. The cost of the Works or any part thereof shall be deemed to include:

(a) The cost to the Client of the Works however incurred, without deduction of any liquidated damages or penalties payable by the Contractor to the Client but including any payments made by the Client to the Contractor by way of bonus, incentive or ex-gratia payments, or in settlement of claims, including but not limited to:

(i) all excavations necessary to enable the Works to be carried out and supports thereof, filling, shoring pumping and other operations for the control of water

(ii) concrete, reinforcement, prestressing tendons and anchorages, formwork and inserts

(iii) load-bearing brickwork or other masonry, including facings and damp-proof courses, which has to resist forces from vertical loads, wind or other loading or upon which the stability of the structure depends.

(iv) all labours, sundries and materials associated with the Works and the cost of the preliminaries and general items in the proportion that the cost of the Works bears to the total cost of the Project.

(b) A fair valuation of any labour, materials, manufactured goods, machinery or other facilities provided by the Client, and of the full benifit accruing to the Contractor from the use of construction plant and equipment belonging to the Client which the Client has required to be used in the execution of the Works.

(c) The market value, as if purchased new, of any second-hand materials, manufactured goods and machinery incorporated in the Works.

(d) The cost of geotechnical investigation (Clause 6.1 (b)).

20.2.The cost of the Works shall not include:

(a)Administration expenses incurred by the Client.

(b) Costs incurred by the Client under this Agreement.

(c) Interest on capital during construction, and the cost of raising moneys required for carrying out the construction of the Works.

(d) Cost of land and wayleaves.

21. PAYMENT OF ACCOUNTS

21.1. Unless otherwise agreed between the Client and the Consulting Engineer from time to time

(a) The sum payable to the Consulting Engineer under Clause 10.1 shall, until the cost of the Works is known, be paid by the Client to the Consulting Engineer in instalments and shall be calculated by reference to the Consulting Engineer's most recent estimate of the cost of the Works taking into account any acceptable tender or tenders when available and any fluctuations clauses contained therein. Such instalments shall be paid during each of the several stages of the Consulting Engineer's services at the intervals specified in Article 4 of the Memorandum of Agreement so that by the end of each stage the cumulative total of all instalments then paid shall amount to the relevant proportion, as specified in Article 4 of the Memorandum of Agreement, of the estimate total sum payable to the Consulting Engineer under this Agreement.

(b) Any lump sum payable to the Consulting Engineer under Clause 15.2 shall be paid by the Client in the instalments and at the intervals stated in Article 3 (g) (2) of the Memorandum of Agreement.

(c) All other sums due to the Consulting Engineer shall be paid by the Client on accounts rendered monthly by the Consulting Engineer.

21.2. Instalments paid by the Client to the Consulting Engineer in accordance with Clause 21.1 (a) shall constitute no more than payments on account. A statement of the total sum due to the Consulting Engineer shall be prepared when the cost of the Works is fully known. Such statement, after giving credit to the Client for all instalments previously paid, shall state the balance (if any) due from the Client to the Consulting Engineer or from the Consulting Engineer to the Client as the case may be which balance shall be paid to or by the Consulting Engineer as the case may require.

21.3. All sums due from the Client to the Consulting Engineer in accordance with the terms of this agreement shall be paid within 40 days of the submission by the Consulting Engineer of his accounts therefor to the Client, and any sums remaining unpaid at the expiry of such period of 40 days shall bear interest thereafter, such interest to accrue from day to day at the rate of 2% per annum above the Base Rate of the Engineers' Principal Bank as stated in Article 3 (h) of the Memorandum of Agreement.

21.4. If any item or part of an item of an account rendered by the Consulting Engineer is disputed or subject to question by the Client, the payment by the Client of the remainder of that account shall not be withheld on those grounds and the provisions of Clause 21.3 shall apply to such remainder and also to the disputed or questioned item to the extent that it shall subsequently be agreed or determined to have been due to the Consulting Engineer.

21.5. All fees set out in this Agreement are exclusive of Value Added Tax, the amount of which, at the rate and in the manner prescribed by law, shall be paid by the Client to the Consulting Engineer.

MEMORANDUM OF AGREEMENT

BETWEEN CLIENT AND CONSULTING ENGINEER FOR STRUCTURAL ENGINEERING WORK WHERE AN ARCHITECT IS APPOINTED BY THE CLIENT

MEMORANDUM OF AGREEMENT made the ....

day of........... 19.........

BETWEEN ....

....(hereinafter called 'the Client')

of the one part and ....

....

(hereinafter called 'the Consulting Engineer') of the other part.

WHEREAS the Client has appointed or proposes to appoint ....

....

CONSULTING ENGINEERS' FEES

to be the Architect for the Project and intends to proceed with ....

....

....

....

and has requested the Consulting Engineer to provide professional services in

connection with ....

....

....
(referred to in this Agreement as 'the Works').

NOW IT IS HEREBY AGREED as follows:

1. The Client agrees to engage the Consulting Engineer subject to and in accordance with the Conditions of Engagement attached hereto and the Consulting Engineer agrees to provide professional services subject to and in accordance with the said Conditions of Engagement.
2. This Memorandum of Agreement and the said Conditions of Engagement shall together constitute the Agreement between the Client and the Consulting Engineer.
3. The method of payment for services under Clause 6 of the said Conditions of Engagement shall be that described in Clause 10.1*/Clause 10.2* thereof.

   (a) The fee referred to in Clause 10.1.1 (a) shall be ......% of the cost of the Works.

   (b) The percentage referred to in Clause 10.1.1 (b) shall be ....%

   (c) The percentage referred to in Clause 10.1.1 (c) shall be ....%

   (d) The rate or rates referred to in Clause 10.2.2 (a) shall be ....

   ....

   (e) The multiplier referred to in Clause 10.2.2 (b) shall be ....

   ....

   (f) The percentage to be be applied to the cost of site staff referred to in Clause 14.1 (b) shall be ....% for specifically recruited staff and ....% for seconded staff.

   (g) The payment for Disbursements referred to in Clause 15 shall be:

   (1) reimbursed at cost*

   (2) a lump sum of ....

   ....(£) payable in ....equal monthly instalments*

CONSULTING ENGINEERS' FEES

(3) ....% of the total fees and charges payable in .... equal monthly instalments*.

(h) The Engineer's Principal Bank referred to in Clause 21.3 shall be

....

4. The intervals for the payment of instalment under Clause 21.1 (a) of the said Conditions of Engagement shall be equal monthly/quarterly* intervals reckoned from the commencement of the Consulting Engineer's appointment, and the cumulative proportions referred to in the said sub-clause shall be as follows:

On completion of Preliminary or Sketch Plan Stage.......% of the total fee
On completion of Tender Stage ....% of the total fee
On completion of Working Drawing Stage ....% of the total fee
On completion of Construction Stage 100 % of the total fee

5. The additional services to be carried out in accordance with Clause 7.2 (m) of the said Conditions of Engagement shall be ....

....

....

....

....

6. PII

The amount of Professional Indemnity Insurance referred to in Clause 15.1 (g) of the said Conditions of Engagement shall be ....
....pounds(£)
for any one occurence or series of occurences arising out of this engagement.

This Professional Indemnity Insurance shall be maintained for a period of....... years from the date of this Memorandum of Agreement, unless such insurance cover ceases to be available in which event the Consulting Engineer will notify the Client immediately.

The sum payable by the Client to the Consulting Engineer as a contribution to the additional cost of the Professional Indemnity Insurance thus provided shall be...pounds(£)
and such amount shall become due to the Consulting Engineer immediately upon acceptance by or on behalf of the Client of any tender in respect of the Works or any part thereof/immediately upon submission by the Consulting Engineer to the Client of his report in relation to the Works.*

7. LIMITATION OF LIABILITY

Notwithstanding anything to the contrary contained elsewhere in this Agreement, the total liability of the Consulting Engineer under or in connection with this Agreement, whether in connection, in tort, for breach of statutory duty or otherwise, shall not exceed...................................pounds(£).

CONSULTING ENGINEERS' FEES

The Client shall indemnify and keep indemnified the Consulting Engineer from and against all claims, demands, proceedings, damages, costs, charges and expenses arising out of or in connection with this Agreement, the Works and/or the Project in excess thereof.

AS WITNESS the hands of the parties the and the year first above written

Duly Authorised
Representative
of the Client ....

Witness ....

Consulting
Engineer ....

Witness ....

*Delete as appropriate.

CONSULTING ENGINEERS' FEES

BASIS FOR COMPLETING MEMORANDUM OF AGREEMENT - EXPLANATORY NOTES:

1. In Article 3 (a) the percentages for fee calculations are fixed for the duration of the Agreement.
2. For works of average complexity the percentage for fee calculation may be determined from the graph and notes on page 75. The estimated cost of works shall be agreed with the Client and the costs shall be divided by the current Output Price Index (1975 = 100) published by the Department of the Enviroment.
   The current index is 327 (provisional - Q4 1989). From the figures so obtained the percentage fee appropriate for the works can be determined from the graph.

   e.g. If the cost of works is estimated to be £400,000 and the Output Price Index is 200, the percentage would be 6.1.

3. The normal percentage to be entered in Article 3 (b) is 3%.
4. The normal percentage to be entered in Article 3 (c) is 3.5%.
5. Articles 3 (d) and 3 (e) must be completed in all cases.
6. In Article 3 (e) the normal multiplier is 2.6.
7. In Article 3 (f) the percentage to be added to the direct costs of providing site supervisory staff is to cover all overheads and head office charges incurred in administering such staff and should include for administration of payroll, finance costs and an allowance for potential redundancy payments. The normal figure is 15% for staff recruited on a contract basis. For seconded staff the terms are to be negotiated between the Client and the Consulting Engineer.
8. In Article 4, the following table, provided for guidance only, is typical of the proportionate amount of the total services performed at the end of each stage:

On completion of Preliminary or Sketch Plan Stage	15%
On completion of Tender Stage	60%
On completion of Working Drawing Stage	85%
On completion of Construction Stage	100%

9. The amount inserted in Article 6 (in words, figures and time) as P11 shall be based on what cover the Client decides appropriate. It should normally be available on an each and every occurrence basis. Exceptionally it can be on a special basis. It should not be less than the Consulting Engineer provides as a general rule.

   The sum payable by the Client shall be subject to negotiation. It will have to be based on the notional premiums which the Consulting Engineer expects to be having to pay over the whole period involved. Such a figure will need to be assessed in consultation with the Consulting Engineer's insurance broker allowing for inflation and based on the the anticipated fees to be received during the whole of the project. Account may also need to be taken of the level of insurance taken within the Consulting Engineer's firm itself in the form of an excess or self-insurance. The proportion of this assessed sum which the Client should be invited to contribute should take into account whether greater than normal cover is to be provided and whether the fees being paid for the project are to be below the guideline level. As a guide the Client should contribute say half of the P11 normal insurance premiums and all of the premium for additional cover.

CONSULTING ENGINEERS' FEES

10. The limit of the Consulting Engineer's liability should be the amount of P11 as completed in the Memorandum of Agreement.

Graph showing relationship of percentage fee with Cost of Works (excluding unreinforced load bearing brickwork or blockwork)

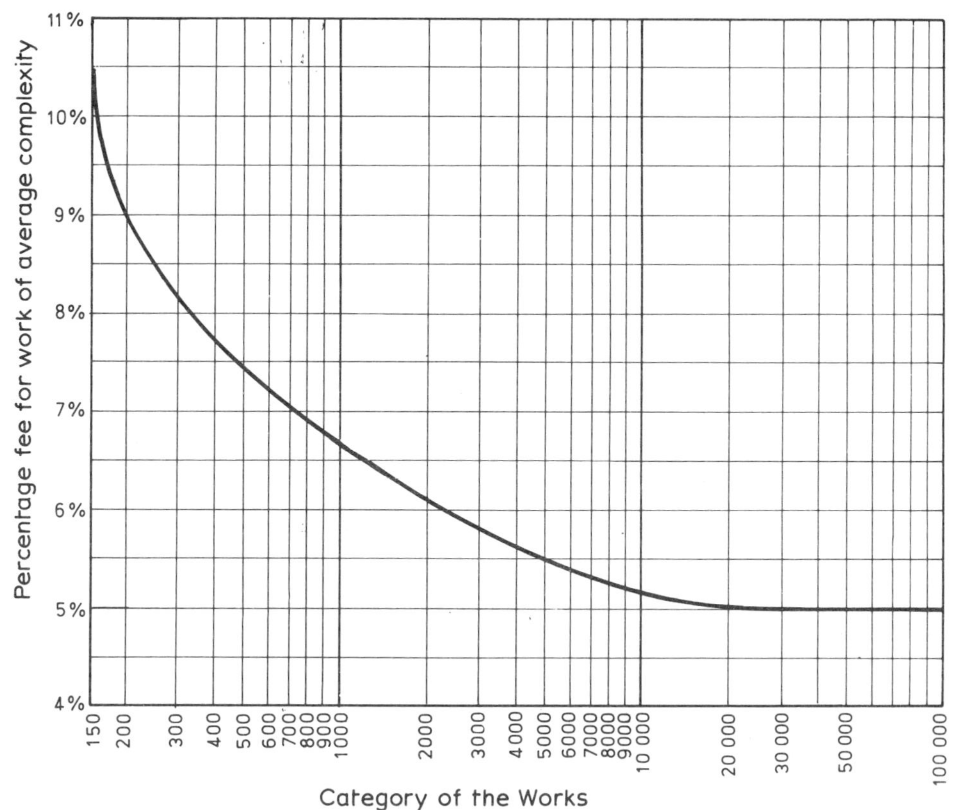

(i) For small projects, i.e. those where the Cost of the Works (as defined in Clause 20, cost of unreinforced brickwork or blockwork) divided by the Output Price Index is less than 150, time charges should be used.

(ii) The minimum fee should not be less than 5% of the Cost of the Works (Clause 20, as above).

CONSULTING ENGINEERS' FEES

AGREEMENT 3 (1984) CONDITIONS OF ENGAGEMENT
HARMONISED WITH THE 'ARCHITECTS' APPOINTMENT

1. DEFINITIONS

In construing this Agreement the following expressions shall have the meanings hereby assigned to them except where the context otherwise requires:

'The Consulting Engineer' means the person, partnership or company named in the Memorandum of Agreement.

'The Architect' means the Architect appointed by the Client to act as the Architect for the project.

'The Project' means the project with which the Client is proceeding and of which the Works form a part.

'The Works' means that part of the Project in connection with which the Client has engaged the Consulting Engineer to perform professional services.

'Contractor' means any person or persons firm or company under contract to the Client to perform work and/or supply goods in connection with the Works.

'The Quantity Surveyor' means any person, partnership or company appointed to provide Quantity Surveying or cost control services in connection with the Project and/or the Works.

'Gross Annual Remuneration' means the annual salary and other emoluments provided by the Consulting Engineer to any person employed by him. It shall include the annual amount of all contributions and payments made by the Consulting Engineer on behalf of and in respect of such person for staff pension and life assurance schemes, National Insurance and private medical schemes; together with any other emoluments such as car and accommodation allowances; and any other tax, charge, levy, impost or payment of any kind whatsoever which the Consulting Engineer at any time during the performance of this agreement is obliged by law to make on behalf of or in respect of such person. For the purposes of this definition the annual amount of all contributions and payments made by the Consulting Engineers on behalf of or in respect of a person employed by him for a period of less than a full year shall be calculated pro rata to the amount of such contribution and payment for a lesser period.

Words importing the singular include the plural and vice versa where the context requires.

2. DURATION OF ENGAGEMENT

2.1. The appointment of the Consulting Engineer shall commence from the date stated in the Memorandum of Agreement or from the time when the Consulting Engineer shall have begun to perform for the Client any of the services in Clauses 6 and 7 whichever is the earlier.

2.2. The Consulting Engineer shall not, without the consent of the Client, assign the benefit or in any way transfer the obligations of this Agreement or any part of it.

CONSULTING ENGINEERS' FEES

2.3. If at any time the Client decides to postpone or abandon the Project, he may thereupon by notice in writing to the Consulting Engineer forthwith terminate the Consulting Engineer's appointment under this Agreement, provided that the Client may, when the Project is postponed, in lieu of so terminating the Consulting Engineer's appointment require the Consulting Engineer in writing to suspend the carrying out of his services under this Agreement for the time being. Upon such a termination the Client shall pay to the Consulting Engineer a sum calculated in accordance with the provisions of Clause 16.1.

2.4. If at any time the Client decides to postpone or abandon any part of the Works, he may thereupon by notice in writing to the Consulting Engineer seek to vary this Agreement either by excluding the services to be performed by the Consulting Engineer in relation to such part of the Works, or by suspending performance of the same and in such notice the Client shall specify the services affected. The Consulting Engineer shall forthwith comply with the Client's notice and the Client shall pay to the Consulting Engineer a sum calculated in accordance with the provisions of Clause 16.2 in respect of such compliance.

2.5. If the Client shall not have required the Consulting Engineer to resume the performance of services in respect of the whole or any part of the Works suspended under Clause 2.3 or Clause 2.4 within a period of 6 months from the date of the Client's notice:-

(i) In the case of a suspension under Clause 2.3 this Agreement shall forthwith automatically terminate; or
(ii) In the case of a suspension under Clause 2.4 the suspended services shall be deemed to have been excluded from the services to be performed by the Consulting Engineer under this Agreement.

2.6. In the event of the failure of the Client to comply with any of his obligations under this Agreement, or upon the occurrence of any circumstances beyond the control of the Consulting Engineer which are such as to delay for a period of more than 6 months or prevent or unreasonably impede the carrying out by the Consulting Engineer of his services under this Agreement, the Consulting Engineer may upon not less than 60 days' notice in writing to the Client terminate his appointment under this Agreement, provided that, in lieu of so terminating his appointment, the Consulting Engineer may:-

(a) forthwith upon any such failure or the occurrence of any such circumstances suspend the carrying out of his services hereunder for a period of 60 days (provided that he shall as soon as practicable inform the Client in writing of such suspension and the reasons therefore), and
(b) at the expiry of such period of suspension either continue with the carrying out of his services under this Agreement or else, if any of the reasons for the suspension then remain, forthwith in writing to the Client terminate his appointment under this Agreement. Upon such a termination the Client shall pay the Consulting Engineer a sum calculated in accordance with the provisions of Clause 17.

CONSULTING ENGINEERS' FEES

2.7. The Consulting Engineer shall, upon receipt of any notice in accordance with Clause 2.3 or in the event of the termination by him of his appointment in pursuance of Clause 2.6, proceed in an orderly manner but with all reasonable speed and economy to take such steps as are necessary to bring to an end his services under this Agreement.

2.8. Any termination of the Consulting Engineer's appointment under this Agreement shall not prejudice or affect the accrued rights or claims of either party to this Agreement.

3. OWNERSHIP OF DOCUMENTS AND COPYRIGHT

3.1. The copyright in all drawings, reports, specifications, bills of quantities, calculations and other documents provided by the Consulting Engineer in connection with the Works shall remain vested in the Consulting Engineer, but the Client shall have a licence to copy and use such drawings and other documents for any purpose related to the construction, letting or maintenance of the Works. Save as aforesaid, the Client shall not make copies of such drawings or other documents nor shall he use the same in connection with any other works without the prior written approval of the Consulting Engineer and upon such terms as may be agreed between the Client and the Consulting Engineer.

3.2. The Consulting Engineer may with the consent of the Client, which consent shall not be unreasonably withheld, publish alone or in conjunction with any other person any articles, photographs or other illustrations relating to the Works.

4. SETTLEMENT OF DISPUTES

Any dispute or difference arising out of this Agreement shall be referred to the arbitration of a person to be agreed upon between the Client and the Consulting Engineer or, failing agreement, nominated by the President for the time being of the Chartered Institute of Arbitrators.

OBLIGATIONS OF THE CONSULTING ENGINEER

5. CARE AND DILIGENCE

5.1. The Consulting Engineer shall exercise all reasonable skill, care and diligence in the discharge of the services agreed to be performed by him. If in the performance of his services the Consulting Engineer has a discretion exercisable as between the Client and the Contractor, the Consulting Engineer shall exercise his discretion fairly.

5.2. The Consulting Engineer may recommend that specialist suppliers and/or contractors should design and execute certain part or parts of the Works in which circumstances the Consulting Engineer shall co-ordinate the design of such part or parts with the overall design of the Works but he shall be relieved of all responsibility for the design, manufacture, installation and performance of any such part or parts of the Works. Where any persons are engaged in accordance with Clauses 6.1.(b) and 6.1.(c) or 7.4, the Consulting Engineer shall not be liable for any negligence, default or omission by such person or persons.

CONSULTING ENGINEERS' FEES

6. NORMAL SERVICES

The services to be provided by the Consulting Engineer at each stage shall comprise all or any of the following as may be necessary in the particular case:-

6.1. **Outline Proposal Stage (broadly falling within the RIBA Plan of Work Stage C).**

(a) Visiting the site and studying data and information relating to the Project and relevant to the Works which are reasonably accessible to the Consulting Engineer, and considering reports relating to the Works which have either been prepared by the Consulting Engineer or else prepared by others and made available to the Consulting Engineer by the Client.
(b) Advising the Client or the Architect on the need for arrangements to be made for the carrying out of geotechnical investigations of the site, arranging as agent for the Client when authorised by him for such investigations, certifying the amount of any payments to be made by the Client to the persons or firms carrying out such investigations under the Consulting Engineer's direction and advising the Client on the results of such investigations.
(c) Advising the client or the Architect on the need for arrangements to be made for the carrying out of topographical and dimensional surveys of the site, surveys to obtain details of construction in existence on or adjacent to the site, special investigations or model tests, arranging as agent for the Client for such works when authorised by him, certifying the amount of any payments to be made by the Client to the persons or firms carrying out such investigations under the Consulting Engineer's direction, and advising the Client on the results of such works.
(d) Consulting any local or other authorities on matter of principle in connection with the structural design of the Works.
(e) Providing sufficient preliminary structural information by means of advice, sketches, reports or outline specifications to enable the Architect to produce his outline proposals and assisting the Quantity Surveyor to finalise his outline cost plan.

6.2. **Scheme Design Stage (broadly falling within the RIBA Plan of Work Stage D).**

(a) Liaising with the Architect and other design team members as necessary to agree a programme for the completion of subsequent design and production information stages subject to a scheme design being approved by the client.
(b) Developing the design of the outline proposal for the Works in collaboration with the Architect and other members of the design team.
(c) Preparing such representative sketches, drawings, specifications, and/or calculations as are necessary to enable the Quantity Surveyor to finalise his cost plan.
(d) Consulting any local or other authorites in connection with the structural design of the Works.

CONSULTING ENGINEERS' FEES

6.3. Detail Design and Tender Documentation Stage (broadly falling within the RIBA Plan of Work Stage E but including part of Stages F & G)

(a) Developing the detail design of the approved scheme design of the Works in collaboration with the Architect and other members of the design team.
(b) Preparing sufficient calculations, drawings, estimates of reinforcement, and final specifications of the Works to enable the Quantity Surveyor to prepare a Bill of Quantities and/or other tender documents.
(c) Assisting the Architect in advising the Client as to the technical suitability for carrying out the Works of persons and firms tendering for the main contract and for any specialist sub-contract involving the supply and/or installation of parts of the Works.
(d) Advising on the need for any special conditions of contract relevant to the Works and forms of tender and invitations to tender as they relate to the Works.

6.4 Production Information Stage (broadly falling within the RIBA Plan of Work Stage F)

(a) Assisting the Architect and Quantity Surveyor in advising the Client as to the relative merits of tenders, prices and estimates received for carrying out the Works.
(b) Preparing such final calculations and details relating to the Works as may be required for submission to any appropriate authority including the co-ordination of structural information supplied by specialist suppliers and/or contractors which need to be included in such submissions.
(c) Submission of general arrangement drawings for the Works to enable the Architect to co-ordinate dimensional and detailed requirements including builders work, provisions for building services,and any special provisions for the fixing of non-structural cladding and other components.
(d) Preparing any further detail drawings including reinforcement bending schedules where appropriate, necessary for the information of the Contractor to enable him to carry out the Works, but excluding drawings and designs for formwork and other temporary works and excluding shop fabrication details relating to the Works or any part thereof.
(e) Examining shop fabrication drawings prepared by the Contractor, his Sub-contractor or Suppliers for the Works, or a part thereof, in respect of general dimensions and structural adequacy of members and connections.

6.5. Construction Stage (broadly falling within the RIBA Plan of Work Stages J to L)

(a) Advising on the preparation of formal contract documents relating to accepted tenders for carrying out the Works or any part of them. The Consulting Engineer shall not accept any tender in respect of the Works.
(b) Advising the Client and the Architect in relation to the need for special inspections or tests arising during the construction of the project.

(c) Advising the Client and the Architect on the appointment of site staff in accordance with Clause 8.
(d) Assisting the Architect in examining the Contractor's proposals where required by the building contract documents.
(e) Attending relevant site meetings and making other periodic visits to site in order to assist the Architect to monitor that the Works are being executed generally in accordance with the contract and advising the Architect on the need for instructions to the Contractor. The frequency of site visits allowed for under Clause 10 is fortnightly throughout the construction of the Works unless agreed otherwise at the appointment stage. The presence of the Consulting Engineer, his employees, agents or any site staff appointed pursuant to Clause 8, on site shall not relieve the Contractor of his responsibility for the correctness of the materials and methods used by the Contractor, nor for the safety of the Works or any temporary works during the course of the construction.
(f) Advising the Architect on Certificates for Payment to the Contractor.
(g) Performing any services which the Consulting Engineer may be required to carry out under any contract for the execution of the Works including where appropriate the witnessing of any specified tests, provided that the Consulting Engineer may decline to perform any services specified in a contract, the terms of which have not initially been approved by the Consulting Engineer.
(h) Delivering upon request to the Client on the completion of the Works, copies of the drawings supplied to the Contractor for the purpose of constructing the Works.
(i) Assisting the Architect in settling any dispute or difference relating to the Works which may arise between the Client and the Contractor provided that such assistance shall not relate to the detailed examination of any financial claim and shall not extend to advising the Client following the taking of any step in or towards any arbitration or litigation in connection with the Works.

6.6 **General**

Without prejudice to the preceding provisions of this clause, the Consulting Engineer shall from time to time as may be necessary advise the Client as to the need for the Client to be provided with additional services in accordance with Clause 7.

7. ADDITIONAL SERVICES NOT INCLUDED IN NORMAL SERVICES

7.1. As services additional to those specified in Clause 6, the Consulting Engineer shall, if agreed with the Client, provide any of the services specified in Clause 7.2 and provide or take all reasonable steps to arrange for the provision by others of any of the services specified in Clause 7.3.

CONSULTING ENGINEERS' FEES

7.2. (a) Preparing any report or additional contract documents required for consideration of proposals for the carrying out of alternative Works.
(b) Carrying out work consequent upon a decision by the Client to seek parliamentary powers.
(c) Carrying out work in connection with any application by the Client for any order, sanction, licence, permit or other consent, approval or authorisation necessary to enable the Works to proceed.
(d) Carrying out work arising from the failure of the Client to award a contract in due time.
(e) Providing quantity surveying services.
(f) Preparing details for shop fabrication of structural steelwork, ductwork, metal, timber, or plastic frameworks or other specialist supplied components.
(g) Checking and/or advising upon any part of the Project not designed by the Consulting Engineer.
(h) Providing services in connection with the funding, sale or letting of the Project including all necessary liaison with legal and financial advisors and any checking consultants or surveyors appointed on behalf of funding organisations and/or potential purchasers or lessees.
(i) Making visits on site in addition to those allowed for in Clause 6.5 (e).
(j) Carrying out work including additional site visits consequent upon any assignment of a contract by the Contractor or upon the failure of the Contractor or any of his sub-contractors properly to perform any contractual obligation or upon delay by the Client in fulfilling his obligations under Clause 9 or in taking any other step necessary for the due performance of the Works.
(k) Advising the Client with regard to any dispute or difference which involves matters excluded from Clause 6.5 (i) above.
(l) Carrying out work in conjunction with others employed to provide any of the services specified in Clause 7.3.
(m) Carrying out such other additional services, if any, as are specified in Article 5 of the Memorandum of Agreement.
(n) Providing 'as-built' drawings of the Works.

7.3. (a) Specialist technical advice on any aspects of the Works.
(b) Legal, financial and other professional services.
(c) Services in connection with the valuation, purchase, sale or leasing of lands and the obtaining of wayleaves.
(d) The surveying of sites or existing works.
(e) Investigation of the nature and strength of existing works and the making of model tests or special investigations.
(f) The carrying out of special inspections or tests advised by the Consulting Engineer under Clause 6.5 (b).

7.4. The Consulting Engineer shall obtain the prior agreement of the Client to the arrangements which he proposes to make as agent for the Client for the provision of any of the services specified in Clause 7.3. The Client shall be responsible to any person or persons providing such services for the cost of them.

CONSULTING ENGINEERS' FEES

8. SITE STAFF

8.1. If in the opinion of the Consulting Engineer, the construction of the Works, including the carrying out of any geotechnical investigations pursuant to Clause 6.1, warrants full-time or part-time engineering staff to be deployed on site at any stage, the Client shall agree to the appointment of such suitably qualified technical and clerical site staff as the Consulting Engineer shall reasonably consider necessary.

8.2. Persons appointed pursuant to Clause 8.1 shall be employed either by the Consulting Engineer or, if the Client and the Consulting Engineer shall so agree, by the Client directly, provided that the Client shall not employ any person as a member of the site staff unless the Consulting Engineer has first approved such person as suitable for employment.

8.3. The terms of service of all site staff to be employed by the Consulting Engineer shall be subject to the approval of the Client, which approval shall not be unreasonably withheld.

8.4. The Client shall procure that the contracts of employment of site staff employed by the Client empower the Consulting Engineer to issue instructions in relation to the works to such staff and shall stipulate that the staff so employed shall in no circumstances take or act upon instructions other than those of the Consulting Engineer.

8.5. Where site duties are performed by staff employed other than by the Consulting Engineer, the Consulting Engineer shall not be responsible for any failure on the part of such staff properly to comply with any instructions given by the Consulting Engineer.

OBLIGATIONS OF THE CLIENT

9. INFORMATION TO BE SUPPLIED TO THE CONSULTING ENGINEER

9.1. The Client shall supply to the Consulting Engineer, without charge and in such reasonable time as not to delay or disrupt the performance by the Consulting Engineer of his services under this Agreement, all necessary and relevant data and information in possession of the Client or the Contractors or their respective servants, agents or sub-contractors and the Client shall give and shall procure that such persons give such assistance as shall reasonably be required by the Consulting Engineer in the performance of his services under this Agreement. The information to be provided by the Client to the Consulting Engineer shall include:-

(a) All such documents as may be necessary to make the Client's or the Architect's requirements clear, including plans and sections of all buildings (to a scale of not less than 1 to 100) and essential details (to a scale of not less 1 to 25) together with site plans (to a scale of not less than 1 to 1,250) and levels.

(b) Copies of all contract documents, including a priced bill of quanities, relating to those parts of the Project which are relevant to the Works.

(c) Copies of all variation orders and supporting documents relating to the Project.

CONSULTING ENGINEERS' FEES

9.2. The Client shall give his decision on all sketches, drawings, reports, recommendations, tender documents and other matters properly referred to him for decision by the Consulting Engineer in such reasonable time as not to delay or disrupt the performance by the Consulting Engineer of his services under this Agreement.

10.PAYMENT FOR NORMAL SERVICES

10.1. **Alternative methods of payment**

10.1.1. The payment for services carried out under Clause 6 shall be agreed upon by the Client and the Consulting Engineer according to one or more of the following methods as specified in the Memorandum of Agreement:
(a) Calculated upon the cost of the Project.
(b) Calculated upon the cost of the Works.
(c) As a Lump Sum.
(d) On a Time Basis.

10.2. **Payment calculated upon the cost of the Project**

10.2.1. On the total cost of the Project a fee calculated in accordance with the percentage stated in Article 3(a) of the Memorandum of Agreement.
10.2.2. The total cost of the Project shall be calculated in accordance with Clause 18.

10.3 **Payment calculated upon the cost of the Works**

10.3.1. The sum payable by the Client to the Consulting Engineer shall be:-
(a)On the total cost of the Works including reinforced concrete and reinforced brickwork or blockwork a fee calculated in accordance with the percentage stated in Article 3(b)(1) of the Memorandum of Agreement.
(b)A further fee on the cost of the reinforced concrete work and reinforced brickwork or blockwork (including the general and preliminary items thereon) calculated in accordance with the percentage stated in Article 3(b)(2) of the Memorandum of Agreement.
10.3.2 The total cost of the Works shall be calculated in accordance with Clause 18.

10.4 **Payment as a Lump Sum**

10.4.1 The sum payable by the Client to the Consulting Engineer shall be the Lump Sum fee stated in Article 3(c) of the Memorandum of Agreement.

10.5 **Payment on a Time Basis**

10.5.1 The client shall pay the Consulting Engineer in accordance with the Scale of Charges set out in Clause 10.5.2.

CONSULTING ENGINEERS' FEES

10.5.2 Scale of Charges:

(a) Self-employed Principles and Consultants: At the hourly rate or rates specified in Article 3(d) of the Memorandum of Agreement.

(b) Directors, salaried Principles and technical and supporting staff: At the rate of 15.6 pence per hour per £100 of Gross Annual Remuneration.

(c) Time spent by Principles, Directors, Consultants and technical and supporting staff in travelling in connection with the Works shall be chargeable on the above basis.

(d) Unless otherwise agreed between the Client and the Consulting Engineer, the Consulting Engineer shall not be entitled to any payment in respect of time spent by secretarial staff or by staff engaged on general accountancy or administration duties in the Consulting Engineer's office.

(e) Charges for the use of computers or other special equipment, which charges shall be agreed between Client and Consulting Engineer before the work commences.

10.6 When the Works are to be constructed in more than one phase then the percentage fees stated in Article 3(a) or Article 3(b) as applicable and the provisions of this Agreement shall apply separately to each phase as if the expression 'the Works' meant, in the case of each phase, the work comprised in that phase.

11. PAYMENT FOR ADDITIONAL SERVICES

In respect of additional services provided by the Consulting Engineer under Clause 7, the Client shall pay the Consulting Engineer in accordance with the Scale of Charges set out in Clause 10.5 or as otherwise agreed between the Client and Consulting Engineer.

12. PAYMENT FOR SITE STAFF

12.1 In addition to any other payment to be made by the Client to the Consulting Engineer under this Agreement the Client shall pay the Consulting Engineer for all site staff employed by the Consulting Engineer pursuant to Clause 8 at the rate of 10 pence per hour per £100 of Gross Annual Remuneration. Where site staff are employed on a part-time basis or full-time for a period of less than nine weeks duration, the Consulting Engineer shall be paid in accordance with the Scale of Charges set out in Clause 10.5

12.2 The Client shall also be responsible for the cost of providing such local office accommodation, furniture, telephones, equipment and transport as shall be reasonably necessary for the use of site staff appointed pursuant to Clause 8, and for the reasonable running costs of such necessary local office accommodation and other facilities including those of stationery, telephone calls, telex, and postage. Unless otherwise agreed between the Client and the Consulting Engineer, the Consulting Engineer shall arrange, whether through the Contractor or otherwise, for the provision of local office accommodation and facilities for the use of site staff. Where site staff are employed on a part-time basis or full-time for a period of less than nine weeks' duration, the Consulting Engineer shall be paid in accordance with the Scale of Charges set out in Clause 10.5.

CONSULTING ENGINEERS' FEES

12.3 In cases where the Consulting Engineer has thought it proper that site staff should not be appointed, or where the necessary site staff is not available at site due to sickness or any other cause, the Consulting Engineer shall be paid in accordance with the Scale of Charges set out in Clause 10.5 for site visits which would have been unnecessary but for the absence or non-availability of site staff.

13. DISBURSEMENTS

The Client shall in all cases reimburse the Consulting Engineer in respect of all the Consulting Engineer's disbursements properly made in connection with:-

13.1 (a) Printing, reproduction and purchase of all documents, drawings, maps, records and photographs.

(b) Telex and telephone calls (other than local telephone calls).

(c) Postage and similar delivery charges.

(d) Travelling, hotel expenses and other similar disbursements.

(e) Advertising for tenders and site staff and any Agency fees in connection therewith.

(f) The provision of additional services to the Client pursuant to Clause 7.4.

(g) Professional Indemnity Insurance taken out by the Consulting Engineer to accord with the wishes of the Client and as set out in the Memorandum of Agreement.

13.2 The Client by agreement with the Consulting Engineer, and in full satisfaction of his liability to the Consulting Engineer in respect of these disbursements, may make to the Consulting Engineer a lump sum payment or a sum calculated as a percentage of the fees and charges falling due under Clause 10 as specified in Article 3(e) of the Memorandum of Agreement.

14. PAYMENT FOR ALTERATION OR MODIFICATION TO DESIGN

14.1 Subject to Clause 14.2 if at any time after the commencement of the Consulting Engineer's appointment under Clause 2.1, any design whether completed or in progress or any specification, drawing or other document prepared in whole or in part by the Consulting Engineer shall require to be modified or revised by reason of instructions received by the Consulting Engineer from or on behalf of the Client, or by reason of circumstances which could not reasonably have been foreseen by the Consulting Engineer then the Client shall make additional payment to the Consulting Engineer for making any necessary modifications or revisions and for any consequential reproduction of documents. Unless otherwise agreed between the Client and the Consulting Engineer, the additional sum to be paid to the Consulting Engineer shall be calculated in accordance with the Scale of Charges set out in Clause 10.5 and shall also include any appropriate reimbursements in accordance with Clause 13 provided always that the Consulting Engineer shall give the Client written notice as soon as it is reasonably apparent that additional payments will arise under this Clause.

CONSULTING ENGINEERS' FEES

14.2 Where in the Consulting Engineer's opinion the Client's instruction necessitates a fundamental redesign of the part or parts of the Works affected by the instruction such that designs, specifications, drawings and other documents prepared by the Consulting Engineer cannot be modified or revised to take account thereof or where the modification or revision instructed by the Client results in a reduction in the cost of the part or parts of the Works affected thereby as contained in the most recent estimate under Clause 19.1 (a) by 10% or more, then such part or parts of the Works shall be deemed to have been abandoned and the Consulting Engineer shall be paid therefore in accordance with Clause 16.2. The Consulting Engineer shall carry out such further work and shall produce such further designs, specifications, drawings and other documents as may be necessary to comply with the Client's instructions and the Consulting Engineer shall be paid therefore in accordance with the provisions of this Agreement.

15. PAYMENT WHEN WORKS ARE DAMAGED OR DESTROYED

If at any time before completion of the Works any part of the Works or any materials, plant or equipment whether incorporated in the Works or not shall be damaged or destroyed, the Client shall make additional payment to the Consulting Engineer in respect of any expenses incurred or additional work required to be carried out by the Consulting Engineer as a result of such damage or destruction. The amount of such additional payment shall be calculated in accordance with the Scale of Charges set out in Clause 10.5 and shall also include any appropriate reimbursements in accordance with Clause 13.

16. PAYMENT FOLLOWING TERMINATION OR SUSPENSION BY THE CLIENT

16.1 Upon a termination or suspension by the Client in pursuance of Clause 2.3, the Client shall pay to the Consulting Engineer the sums specified in (a), (b) and (c) below less the amount of payments previously made to the Consulting Engineer under the terms of this Agreement.

(a) A fair and reasonable proportion of the sum which would have been payable to the Consulting Engineer under Clause 10 if no such termination or suspension had taken place. In the assessment of such proportion, the services carried out by the Consulting Engineer up to the date of termination or suspension and in pursuance of Clause 2.7 shall be compared with a reasonable assessment of the services which the Consulting Engineer would have carried out but for the termination or suspension. In any case in which it is necessary to assess the payment to be made to the Consulting Engineer in accordance with this Sub-clause by reference to the cost of the Project or the Works, then to the extent that such cost is not known the assessment shall be made upon the basis of the Consulting Engineer's best estimates of cost.

(b) Amounts due to the Consulting Engineer under any other Clauses of this Agreement.

(c) A disruption charge equal to one-sixth of the difference between the sum which would have been payable to the Consulting Engineer under Clause 10 but for the termination or suspension, and the sum payable under (a).

CONSULTING ENGINEERS' FEES

16.2 Upon a termination or suspension by the Client in pursuance of Clause 2.4, the Client shall pay to the Consulting Engineer the sums specified in (a), (b) and (c) below.

(a) A fair and reasonable proportion of the sum which would have been payable to the Consulting Engineer under Clause 10 in respect of the services affected if no such termination or suspension had taken place. The proportion shall be calculated in accordance with the provisions of Clause 16.1(a).
(b) Amounts due to the Consulting Engineer under any other clauses of this Agreement.
(c) A disruption charge, to be calculated in accordance with the provisions of Clause 16.1(c).

16.3 In any case in which the Client has required the Consulting Engineer to suspend the carrying out of the Consulting Engineer's services in pursuance of Clauses 2.3 or 2.4 hereof, the Client may, at any time within the period of 6 months from the date of the Client's notice, require the Consulting Engineer in writing to resume the performance of such services. In the event:

(a) The Consulting Engineer shall within a reasonable time of receipt by him of the Client's said requirements in writing resume the performance of his services in accordance with this Agreement, the payment made under Clause 16.1(a) or 16.2(a) as the case may be ranking as payment on account towards the total sum payable to the Consulting Engineer under Clause 10.

(b) Notwithstanding such resumption, the Consulting Engineer shall be entitled to retain or receive as an additional payment due in accordance with this Agreement the disruption charge refered to in Clause 16.1(c) or 16.2(c).

16.4 If the Consulting Engineer shall need to perform any additional services in connection with the resumption of his services in accordance with Clause 16.3 the Client shall pay the Consulting Engineer in respect of the performance of such additional services in accordance with the Scale of Charges set out in Clause 10.5 and any appropriate reimbursements in accordance with Clause 13.

17. PAYMENT FOLLOWING TERMINATION BY THE CONSULTING ENGINEER

Upon a termination by the Consulting Engineer in pursuance of Clause 2.6 the Client shall pay to the Consulting Engineer the sum specified in Clause 16.1(a) and (b) less the amounts of payments previously made to the Consulting Engineer under the terms of this Agreement. Upon payment of such sums, the Consulting Engineer shall deliver to the Client such completed drawings, specifications and other similar documents relevant to the Works as are in his possession. The Consulting Engineer shall be permitted to retain copies of any documents so delivered to the Client. The provisions of this Clause are without prejudice to any other rights and remedies which the Consulting Engineer may possess.

CONSULTING ENGINEERS' FEES

18. COSTS OF THE PROJECT AND THE WORKS

18.1. The cost of the Project or Works or any part thereof shall include:-

(a) The cost to the Client of the Project or Works however incurred, without deduction of any liquidated damages or penalties payable by the Contractor to the Client but including any payments made by the Client to the Contractor by way of bonus, incentive or ex-gratia payments, or in settlement of claims.
(b) A fair valuation of any labour, materials, manufactured goods, machinery or other facilities provided by the Client, and of the full benefit accruing to the Contractor from the use of construction plant and equipment belonging to the Client which the Client has required to be used in the execution of the Project or Works.
(c) The market value, as if purchased new, of any second-hand materials, manufactured goods and machinery incorporated in the Project or Works.
(d) The cost of geotechnical investigations.

18.2. The cost of the Works or any part therof shall include without limitation:-

(a) All excavations necessary to enable the Works to be carried out and supports thereof, filling, disposal of excavated material, shoring, pumping and other operations for the control of water.
(b) Concrete, reinforcement, prestressing tendons and anchorages, formwork, inserts and fixings. All types of concrete piling shall be classified as reinforced concrete work.
(c) Structural steelwork, structural timberwork and any other structural elements including associated connections, fixings, bearings and bracings.
(d) Load-bearing brickwork or other masonry, including reinforcement, facings and damp-proof courses, which has to resist forces from vertical loads, wind or other loadings or upon which the stability of the structure depends.
(e) All labours, sundries, materials and equipment costs associated with the Works and the cost of preliminaries and other general items in the proportion that the cost of the Works bears to the total cost of the Project.

18.3. The cost of the Project or Works shall not include:

(a) Administration expenses incurred by the Client.
(b) Costs incurred by the Client under this Agreement.
(c) Interest on capital during construction, and the cost of raising moneys required for carrying out the construction of the Project or Works.
(d) Cost of land and wayleaves.

CONSULTING ENGINEERS' FEES

19. PAYMENT OF ACCOUNTS

19.1 Unless otherwise agreed between the Client and the Consulting Engineer from time to time

(a) The sum payable to the Consulting Engineer under Clauses 10.2 or 10.3 as applicable shall, until the cost of the Project or Works is known, be paid by the Client to the Consulting Engineer in instalments and shall be calculated in the case of 10.2 and 10.3 by reference to the Consulting Engineer's most recent estimate of the cost of the Project or Works taking into account any accepted tender or tenders when available and any fluctuation clauses contained therein where relevant. Such instalments shall be paid during each of the several stages of the Consulting Engineer's services at the intervals specified in Article 4 of the Memorandum of Agreement so that by the end of each stage the cumulative total of all instalments then paid shall amount to the relevant proportion, as specified in Article 4 of the Memorandum of Agreement, of the estimated total sum payable to the Consulting Engineer under this Agreement. Any sum payable to the Consulting Engineer under Clause 10.4 shall be paid in instalments as defined in the Memorandum of Agreement.

(b) Any sum payable to the Consulting Engineer under Clause 13.1 or 13.2 shall be paid by the Client in the instalments and at the intervals stated in Article 3(e) of the Memorandum of Agreement.

(c) All other sums due to the Consulting Engineer shall be paid by the Client on accounts rendered monthly by the Consulting Engineer.

19.2 Instalments paid by the Client to the Consulting Engineer in accordance with Clause 19.1(a) shall consitute no more than payments on account. A statement of the total sum due to the Consulting Engineer shall be prepared when the cost of the Project or the Works is fully known. Such statement, after giving credit to the Client for all relevant instalments previously paid, shall state the balance (if any) due from the Client to the Consulting Engineer or from the Consulting Engineer to the Client as the case may require.

19.3 All sums due from the Client to the Consulting Engineer in accordance with the terms of this Agreement shall be paid within 30 days of the submission by the Consulting Engineer of his accounts therefor to the Client, and any sums remaining unpaid at the expiry of such period of 30 days shall bear interest thereafter, such interest to accrue from day to day at the rate of 2% per annum above the current base rate of the Consulting Engineer's principal bank, as specified in Article 3(f) of the Memorandum of Agreement.

19.4 If any item or part of an item of an account rendered by the Consulting Engineer is disputed or subject to question by the Client, the payment by the Client of the remainder of that account shall not be withheld on those grounds and the provisions of Clause 19.3 shall apply to such remainder and also to the disputed or questioned item to the extent that it shall subsequently be agreed or determined to have been due to the Consulting Engineer.

19.5 All fees set out in this Agreement are exclusive of Value Added Tax, the amount of which at the rate and in the manner prescribed by law, shall be paid by the Client to the Consulting Engineer.

CONSULTING ENGINEERS' FEES

MEMORANDUM OF AGREEMENT

BETWEEN CLIENT AND CONSULTING
ENGINEER FOR STRUCTURAL ENGINEERING
WORK WHERE AN ARCHITECT IS APPOINTED
BY THE CLIENT

MEMORANDUM OF AGREEMENT made the ....

day of........... 19........

BETWEEN ....

....

....(hereinafter called 'the Client')

of the one part and ....

....

(hereinafter called 'the Consulting Engineer') of the other part.

WHEREAS the Client has appointed or proposes to appoint ....

....

to be the Architect for the Project and intends to proceed with ....

....

....

and has requested the Consulting Engineer to provide professional services in

connection with ....

....
referred to in this Agreement as 'the Works').

NOW IT IS HEREBY AGREED as follows:

1. The Client agrees to engage the Consulting Engineer subject to and in accordance with the Conditions of Engagement attached hereto and the Consulting Engineer agrees to provide professional services subject to and in accordance with the said Conditions of Engagement.

2. The Memorandum of Agreement and the said Conditions of Engagement shall together constitute the Agreement between the Client and the Consulting Engineer.

3. The method of payment for services under Clause 6 of the said Conditions of Engagement shall be determined as described in Clause 10 thereof in accordance with the method of Clause 10.2/10.3/10.4/10.5.*

CONSULTING ENGINEERS' FEES

*(a) The fee referred to in Clause 10.2.1 shall be ......% of the cost of the Project

*(b) (1) The percentage referred to in Clause 10.3.1 (a) shall be ......% of the cost of the Works

(2) The fee referred to in Clause 10.3.1(b) shall be......% of the cost of the reinforced concrete work and reinforced brickwork or blockwork

*(c) The lump sum referrred to in Clause 10.4.1 shall be..................

*(d) The rate or rates referred to in Clause 10.5.2(a) shall be...........

(e) The payment for Disbursements referred to in Clause 13 shall be

* (1) reimbursement of costs against monthly/quarterly* statements

* (2) a lump sum of..................payable in...................... equal monthly/quarterly* instalments.

* (3) ....% of the total fees and charges payable * in .... equal monthly/quarterly* instalments * pro rata with fees

(f) The Engineer's Principal Bank referred to in Clause 19.3 shall be..... ....

4. The intervals for the payment of instalment under Clause 19.1(a) of the said Conditions of Engagement shall be equal monthly/quarterly/.......* intervals reckoned from the commencement of the Consulting Engineer's appointment, and the cumulative proportions referred to in the said sub-clauses shall be as follows:

On Completion of Outline Proposal Stage..............% of the fee
On Completion of Scheme Design Stage.................% of the fee
On Completion of Detail Design and Tender
Documentation Stage..................................% of the fee
On Completion of Production Information Stage........% of the fee
On Completion of the Works.......................100% of the fee

5. The additional services to be carried out in accordance with Clause 7.2 of the said Conditions of Engagement shall be ....

....

....

....

6.P11

The amount of Professional Indemnity Insurance referred to in Clause 13 of the said Conditions of Engagement shall be.............................. ....pounds(£) for any one occurence or series of occurences arising out of this engagement.

* Delete as appropriate

CONSULTING ENGINEERS' FEES

This Professional Indemnity Insurance shall be maintained for a period of .... years from the date of this Memorandum of Agreement, unless such insurance cover ceases to be available in which event the Consulting Engineer will notify the Client immediately.

The sum payable by the Client to the Consulting Engineer as a contribution to the additional cost of the Professional Indemnity Insurance thus provided shall be..pounds(£) and such amount shall become due to the Consulting Engineer immediately upon acceptance by or on behalf of the Client of any tender in respect of the Works or any part thereof/immediately upon submission by the Consulting Engineer to the Client of his report in relation to the Task.*

7. LIMITATION OF LIABILITY

Notwithstanding anything to the contrary contained elsewhere in this Agreement, the total liability of the Consulting Engineer under or in connection with this Agreement, whether in contract, in tort, for breach of statutory duty or otherwise, shall not exceed.......................... ....pounds(£). The Client shall indemnify and keep indemnified the Consulting Engineer from and against all claims, demands, proceedings, damages, costs, charges and expenses arising out of or in connection with this Agreement, the Works and/or the Project in excess thereof.

AS WITNESS the hands of the parties the day and the year first above written

Duly Authorised Representative of the Client ....

Consulting Engineer ....

Witness ....

Witness ....

*Delete as appropriate.

CONSULTING ENGINEERS' FEES

BASIS FOR COMPLETING MEMORANDUM OF AGREEMENT - EXPLANATORY NOTES:

1. In Article 3 (a) the percentages for fee calculations are fixed for the duration of the Agreement.
2. For works of average complexity the percentage for fee calculation may be determined from the graph and notes on page 75. The estimated cost of works shall be agreed with the Client and the costs shall be divided by the current Output Price Index (1975 = 100) published by the Department of the Enviroment.
   The current index is 327 (provisional - Q4 1989). From the figures so obtained the percentage fee appropriate for the works can be determined from the graph.

   e.g. If the cost of works is estimated to be £400,000 and the Output Price Index is 200, the percentage would be 6.1.

3. The normal percentage to be entered in Article 3 (b) is 3%.
4. The normal percentage to be entered in Article 3 (c) is 3.5%.
5. Articles 3 (d) and 3 (e) must be completed in all cases.
6. In Article 3 (e) the normal multiplier is 2.6.
7. In Article 3 (f) the percentage to be added to the direct costs of providing site supervisory staff is to cover all overheads and head office charges incurred in administering such staff and should include for administration of payroll, finance costs and an allowance for potential redundancy payments. The normal figure is 15% for staff recruited on a contract basis. For seconded staff the terms are to be negotiated between the Client and the Consulting Engineer.
8. In Article 4, the following table, provided for guidance only, is typical of the proportionate amount of the total services performed at the end of each stage:

On completion of Preliminary or Sketch Plan Stage	15%
On completion of Tender Stage	60%
On completion of Working Drawing Stage	85%
On completion of Construction Stage	100%

9. The amount inserted in Article 6 (in words, figures and time) as PII shall be based on what cover the Client decides appropriate. It should normally be available on an each and every occurrence basis. Exceptionally it can be on a special basis. It should not be less than the Consulting Engineer provides as a general rule.

   The sum payable by the Client shall be subject to negotiation. It will have to be based on the notional premiums which the Consulting Engineer expects to be having to pay over the whole period involved. Such a figure will need to be assessed in consultation with the Consulting Engineer's insurance broker allowing for inflation and based on the the anticipated fees to be received during the whole of the project. Account may also need to be taken of the level of insurance taken within the Consulting Engineer's firm itself in the form of an excess or self-insurance. The proportion of this assessed sum which the Client should be invited to contribute should take into account whether greater than normal cover is to be provided and whether the fees being paid for the project are to be below the guideline level. As a guide the Client should contribute say half of the PII normal insurance premiums and all of the premium for additional cover.

CONSULTING ENGINEERS' FEES

10. The limit of the Consulting Engineer's liability should be the amount of P11 as completed in the Memorandum of Agreement.

Graph showing relationship of percentage fee with Cost of Works (excluding unreinforced load bearing brickwork or blockwork)

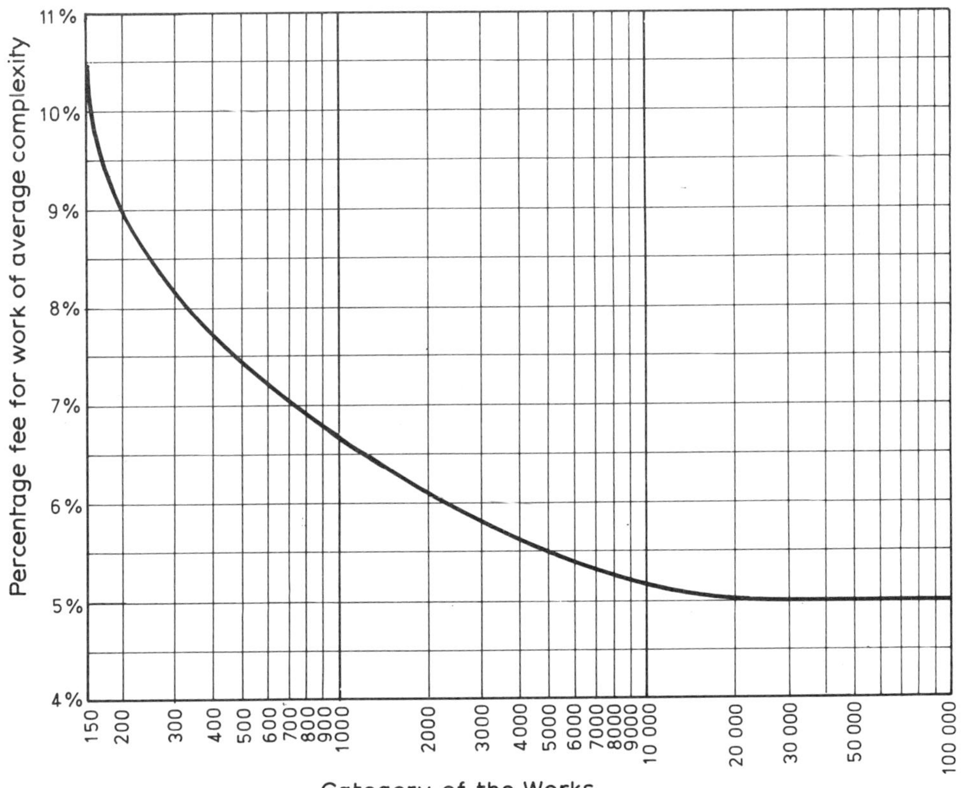

(i) For small projects, i.e. those where the Cost of the Works (as defined in Clause 20, cost of unreinforced brickwork or blockwork) divided by the Output Price Index is less than 150, time charges should be used.

(ii) The minimum fee should not be less than 5% of the Cost of the Works (Clause 20, as above).

THE TOWN AND COUNTRY PLANNING (FEES FOR APPLICATIONS AND DEEMED APPLICATIONS) (AMENDMENTS) REGULATIONS 1989

- operative from 14th March 1989

The following extracts from the Town and Country Planning Fees Regulations, available from HMSO, relate only to those applications which meet the "deemed to qualify clauses" laid down in regulations 1 to 11 of S.I. No. 1989/193.

Further advice on the interpretation of these regulations can be found in the 1971, 1981 and 1984 Town and County Planning Acts and DOE circulars S.I. 1981/369, 1983/1674, 1987/764 and 1988/1813.

SCHEDULE 1

PART 1: GENERAL PROVISIONS

1. (1) Subject to paragraphs 3 to 11, the fee payable under regulation 3 or regulation 10 shall be calculated in accordance with the table set out in Part II of this Schedule and paragraphs 2 and 12 to 16.
   (2) In the case of an applicationn for approval of reserved matters, references in this Schedule to the category of development to which an application relates shall be construed as references to the category of development authorised by the relevant outline planning permission.

2. Where an application or deemed application relates to the retention of buildings or works or to the continuance of a use of land, the fee payable shall be calculated as if the application or deemed application, were one for planning permission to construct or carry out those buildings or works or to institute the use.

3. Where an application or deemed application is made or deemed to be made by or on behalf of a parish council or by or on behalf of a community council, the fee payable shall be one-half of the amount as would otherwise be payable.

4. (1) Where and application or deemed application for planning permission is made or deemed to be made by or on behalf of a club, society or other organisation(including any persons administering a trust) which is not established or conducted for profit and whose objects are the provision of facilities for sport or recreation, and the conditions specified in subparagragh (2) below satisfied, the amount of the fee payable in respect of the application or deemed application shall be £76.

   (2) The conditions referred to in subparagraph (1) above are:

   (a) that the application or deemed application relates to:-
      (i) the making of a material change in the use of land to use as a playing field; or
      (ii) the carrying out of operations (other than the erection of a building containing floor space) for purposes ancillary to the use of land as a playing field, and to no other development: and
   (b) that the local planning authority with whom the application is lodged, or (in the case of a deemed application) the Secretary of State, is satisfied that the development is to be carried out on land which is, or is intended to be, occupied by the club, society or organisation and used wholly or mainly for the carrying out of its objects.

THE TOWN AND COUNTRY PLANNING (FEES FOR APPLICATIONS AND DEEMED APPLICATIONS) (AMENDMENTS) REGULATIONS 1989

5. (1) Where an application for planning permission or an application for approval of reserved matters is made not more than 28 days after the lodging with the local planning authority of an application for planning permission or, as the case may be, an application for approval of reserved matters:

(a) made by or on behalf of the same applicant;
(b) relating to the same site; and
(c) relating to the same development or, in the case of an application for approval of reserved matters, relating to the same reserved matters in respect of the same building or buildings authorised by the relevant outline planning permission,

and a fee of the full amount payable in respect of the category or categories of development to which the applications relate has been paid in respect of the earlier application, the fee payable in respect of the later application shall, subject sub-paragraph (2) below, be one-quarter of the full amount paid in respect of the earlier application.

(2) Sub-paragraph (1) apply only in respect of one application made by or on behalf of the same applicant in relation to the same development or in relation to the same reserved matters (as the case may be).

6. (1) This paragraph applies where:

(a) an application is made for approval of one or more reserved matters ("the current application"); and
(b) the applicant has previously applied for such approval under the same outline planning permission and paid fees in relation to one or more such applications; and
(c) no application has been made under that permission other than by or on behalf of the applicant.

(2) Where the amount as mentioned in sub-paragraph (1) (b) is not less than the amount which would be payable if the applicant were by his current application seeking approval of all the matters reserved by the outline permission (and in relation to the whole of the development authorised by the permission), the amount of the fee payable in respect of the current application shall be £76.

(3) Where:
(a) a fee has been paid as mentioned in sub-paragraph (1) (b) at a rate lower than that prevailing at the date of the current application; and
(b) subparagraph (2) would apply if that fee had been paid at the rate applying at that date, the fee in respect of the current application shall be the amount specified in sub-paragraph (2).

7. Where application is made pursuant to section 31A of the 1971 Act the fee payable in respect of the application shall be £38.

THE TOWN AND COUNTRY PLANNING (FEES FOR APPLICATIONS AND DEEMED APPLICATIONS) (AMENDMENTS) REGULATIONS 1989

8. (1) This paragraph applies where applications are made for planning permission or for the approval of reserved matters in respect of the development of land lying in the areas of:
(a) two or more local planning authorities in a Metropolitan County or in Greater London; or
(b) two or more district planning authorities in a Non-metropolitan County; or
(c) one or more such local planning authorities and one or more such district planning authorities.

(2) A fee shall be payable only to the local planning authority or district planning authority in whose area the largest part of the relevant land is situated: and the amount payable shall not exceed:
(a) where the applications relate wholly or partly to a county matter within the meaning of paragraph 32 of Schedule 16 to the Local Government Act 1972 (b), and all the land is situated in a single non-metropolitan County, the amount which would have been payable if application had fallen to be made to one authority in relation to the whole development;
(b) in any other case, one and a half times the amount which would have been payable if application had fallen to be made to a single authority.

9. (1) This paragraph applies where application for planning permission is deemed to have been made by virtue of section 88B (3) of the 1971 Act in respect of such land as is mentioned in paragraph 8 (1).

(2) The fee payable to the Secretary of State shall be a fee of the amount which would be payable by virtue of paragraph 8 (2) if application for the like permission had been made to the relevant local or district planning authority on the date on which notice of appeal was given in accordance with section 88 (3) of the 1971 Act.

10. (1) Where:

(a) application for planning permission is made in respect of two or more alternative proposals for the development of the same land; or
(b) application for approval of reserved matters is made, in respect of two or more alternative proposals for the carrying out of the development authorised by an outline planning permission, and application is made in respect of all of the alternative proposals on the same date and by or on behalf of the same applicant, a single fee shall be payable in respect of all such alternative proposals, calculated as provided in sub-paragraph (2).

(2) Calculations shall be made, in accordance with this Schedule, of the fee appropriate to each of the alternative proposals and the single fee payable in respect of all the alternative proposals shall be the sum of:
(i) an amount equal to the highest of the amounts calculated in respect of each of the alternative proposals; and
(ii) an amount calculated by adding together the amounts appropriate to all of the alternative proposals, other than the amount referred to in subparagraph (i) above, and dividing that total by the figure of 2.

THE TOWN AND COUNTRY PLANNING (FEES FOR APPLICATIONS AND DEEMED APPLICATIONS) (AMENDMENTS) REGULATIONS 1989

11. In the case of an application for planning permission which is deemed to have been made by virtue of section 95 (6) of the 1971 Act, the fee payable shall be the sum of £76.

12. Where, in respect of any category of development specified in the table set out in Part II of this Schedule, the amount of the fee is to be calculated by reference to the site area:-

(a) that area shall be taken as consisting of the area of land to which the application relates or, in the case of an application for planning permission which is deemed to have been made by virtue of section 88B (3) of the 1971 Act, the area of land to which the relevant enforcement notice relates; and
(b) where the area referred to in sub-paragraph (a) above is not an exact multiple of the unit of measurement specified in respect of the relevant category of development, the fraction of a unit remaining after division of the total area by the unit of measurement shall be treated as a complete unit.

13. (1) In relation to development within any of the categories 2 to 4 specified in the table in Part II of this Schedule, the area of gross floor space to be created by the development shall be ascertained by external measurement of the floor space, whether or not it is to be bounded (wholly or partly) by external walls of a building.

(2) In relation to development within category 2 specified in the said table, where the area of gross floor space to be created by the development exceeds 75 sq metres and is not an exact multiple of 75 sq metres, the area remaining after division of the total number of square metres of gross floor space by the figure of 75 shall be treated as being 75 sq metres.

(3) In relation to development within category 3 specified in the said table, where the area of gross floor space exceeds 540 sq metres and the amount of the excess is not an exact multiple of 75 sq metres, the area remaining after division of the number of square metres of that excess area of gross floor space by the figure of 75 shall be treated as being 75 sq metres.

14. (1) Where an application (other than an outline application) or a deemed application relates to development which is within category 1 in the table set out in Part II of this Schedule and in part within category 2,3, or 4, the following sub-paragraphs shall apply for the purpose of calculating the fee payable in respect of the application or deemed application.

(2) An assessment shall be made of the total amount of gross floor space which is to be created by that part of the development which is within category 2,3 or 4 ("the non-residential floor space"), and the sum payable in respect of the non-residential floor space to be created by the development shall be added to the sum payable in respect of that part of the development which is within category 1, and subject to sub-paragraph (4), the sum so calculated shall be the fee payable in respect of the application or deemed application.

THE TOWN AND COUNTRY PLANNING (FEES FOR APPLICATIONS AND DEEMED APPLICATIONS) (AMENDMENTS) REGULATIONS 1989

(3) For the purpose of calculating the fee under sub-paragraph (2)-

(a) Where any of the buildings is to contain floor space which it is proposed to use for the purposes of providing common access or common services or facilities for persons occupying or using that building for residential purposes and for persons occupying or using it for non-residential purposes ("common floor space"), the amount of non-residential floor space shall be assessed, in relation to that building, as including such proportion of the common floor space as the amount of non-residential floor space in the building bears to the total amount of gross floor space in the building to be created by the development;
(b) where the development falls within more than one of categories 2,3, and 4 an amount shall be calculated in accordance with each such category highest amount so calculated shall be taken as the sum payable in respect of all of the non-residental floor space.

(4) Where an application or deemed application to which this paragraph applies relates to development which is also within one one or more than one of the categories 5 to 13 in the table set out in Part II of this Schedule, an amount shall be calculated in accordance with the with each such category and if any of the amounts so calculated exceeds the amount calculated in accordance with sub-paragraph (2) that higher amount shall be the fee payable in respect of all of the development to which the application or deemed application relates.

15. (1) Subject to paragraph 14, and sub-paragraph (2), where an application or deemed application relates to development which is within more than one of the categories specified in the table set out in Part II of this Schedule-

(a) an amount shall be calculated in accordance with each such category; and
(b) the highest amount so calculated shall be the fee payable in respect of the application or deemed application.

(2) Where an application is for outline planning permission and relates to development which is within more than one of the categories specified in the said table, the fee payable in respect of the application shall be £76 for each 0.1 hectacres of the site area, subject to a maximum of £1900.

16. In the case of an application for planning permission which is deemed to have been made by virtue of section 88B (3) of the 1971 Act, references in this Schedule to the development to which an application relates shall be construed as references to the use of land or the operations (as the case may be) to which the relevant enforcement notice relates; references to the development shall be construed as references to the amount of floor space or the number of dwellinghouses to be created by the development shall be construed as references to the amount of floor space or the number of dwellinghouses to which that enforcement notice relates; and references to the purposes for which it is proposed that floor space be used shall be construed as references to the purposes for which floor space was stated to be used in the enforcement notice.

THE TOWN AND COUNTRY PLANNING (FEES FOR APPLICATIONS AND DEEMED APPLICATIONS) (AMENDMENTS) REGULATIONS 1989

I. **Operations**

PART II: SCALE OF FEES

Category of development	Fee payable
1. The erection of dwellinghouses (other than development within category 6 below).	(a) Where the application is for outline planning permission £76 for each 0.1 hectare of the site area, subject to a maximum of £1,900, (b) in other cases, £76 for each dwellinghouse to be created by the development, subject to a maximum of £3,800.
2. The erection of buildings (other than buildings coming within categories 1,3, 4, 5 or 7.	(a) Where the application is for outline planning permission £76 for each 0.1 hectare of the site area, subject to a maximum of £1,900; (b) in other cases: (i) where no floor space is to be created by the development, £38; (ii) where the area of gross floor space to be created by the development does not exceed 40 sq metres, £38; (iii) where the area of gross floor space to be created by the development exceeds 40 sq metres but does not exceed 75 sq metres, £76; and (iv) where the area of gross floor space to be created by the development exceeds 75 sq metres, £76 for each 75 sq metres, subject to a maximum of £3,800.
3. The erection, on land used for the purposes of agriculture, of buildings to be used for agricultural purposes (other than buildings coming within category 4).	(a) Where the application is for outline planning permission £76 for each 0.1 hectare of the site area, subject to a maximum of £,1900;

THE TOWN AND COUNTRY PLANNING (FEES FOR APPLICATIONS AND DEEMED APPLICATIONS) (AMENDMENTS) REGULATIONS 1989

Category of development	Fee payable
	(b) in other cases: (i) where the area of gross floor space to be created by the development does not exceed 465 sq metres nil; (ii) where the area of gross floor space to be created by the development exceeds 465 sq metres but does not exceed 540 sq metres, £76; (iii) where the area of gross floor space to be created by development exceeds 540 sq metres, £76 for the first 540 sq metres and £76 for each 75 sq metres in excess of that figure, subject to a maximum of £3,800.
4. The erection of glasshouses on land used for the purposes of agriculture.	(a) Where the area of gross floor space to be created by the development does not exceed 465 sq metres, nil; (b) where the area of gross floor space to be created by the development exceeds 465 sq metres, £450.
5. The erection, alteration or replacement of plant or machinery.	£76 for each 0.1 hectare of the area, subject to a maximum of £3,800.
6. The enlargement, improvement or other alteration of existing dwellinghouses.	(a) Where the applications relates to one dwellinghouse, £38; (b) where the application relates to 2 or more dwellinghouses, £76
7. (a) The carrying out of operations (including the erection of a building) within the curtilage of an existing dwellinghouse, for purposes ancillary to the enjoyment of the dwellinghouse as such, or the erection or construction of gates, fences, walls or other means of enclosure along a boundary of the curtilage of an existing dwellinghouse; or	£38

THE TOWN AND COUNTRY PLANNING (FEES FOR APPLICATIONS AND DEEMED APPLICATIONS) (AMENDMENTS) REGULATIONS 1989

Category of development	Fee payable
(b) the construction of car parks, service roads and other means of access on land used for the purposes of a single undertaking, where the development is required for a purpose incidental to the existing use of the land.	
8. The carrying out of any operations connected with exploratory drilling for oil or natural gas.	£76 for each 0.1 hectare of the site area, subject to a maximum of £5,700.
9. The carrying out of any operations not coming within any of the above categories.	£38 for each 0.1 hectare of the site area, subject to a maximum of: (a) in the case of operations for the winning and working of minerals, £5,700; (b) in other case, £380.
II. Uses of Land	
10. The change of use of a building to use as one or more separate dwellinghouses.	(a) Where the change is from a previous use as a single dwellinghouse to use as a two or more single dwellinghouses, £76 for each additional dwellinghouse to be created by the development subject to a maximum of £3,800. (b) in other cases, £76 for each dwellinghouse to be created by the development, subject to a maximum of £3,800.
11. (a) The use of land for the disposal of refuse or waste materials or for the deposit of material remaining after minerals have been extracted from land; or (b) the use of land for the storage of minerals in the open.	£38 for each 0.1 hectare of the site area, subject to a maximum of £5,700.
12. The making of a material change in the use of a building or land (other than a material change of use coming within any of the above categories).	£76.

THE TOWN AND COUNTRY PLANNING (FEES FOR APPLICATIONS AND DEEMED APPLICATIONS) (AMENDMENTS) REGULATIONS 1989

	Fee payable
13. The continuance of a use of land, or the retention of buildings or works on land, without compliance with a condition subject to which a previous planning permission has been granted (including a condition requiring the discontinuance of the use or the removal of the building or works at the end of a specified period).	£38.

SCHEDULE 2

SCALE OF FEES IN RESPECT OF APPLICATIONS FOR CONSENT TO DISPLAY ADVERTISEMENTS

Category of advertisments

1. Advertisements displayed on business premises, on the forecourt of business premises or on other land within the curtilage of business premises, wholly with reference to all or any of the following matters:- — £21

   (a) the nature of the business or other activity carried on on the premises;
   (b) the goods sold or the services provided on the premises; or
   (c) the name and qualifications of the person carrying on such business or activity or supplying such goods or services.

2. Avertisemnents for the purpose of directing members of the public to, or otherwise drawing attention to the existence of, business premises which are in the same locality as the site on which the advertisement is to be displayed but which are not visible from that site. — £21

3. All other advertisements. — £76

THE BUILDING (PRESCRIBED FEES ETC.) REGULATIONS 1989

- operative from 1st September 1989

The following extracts are from the Building Regulations 1985, supplemented by DoE circular S.I. 1989/1118, which are available from HMSO.

PART I: GENERAL

Interpretation

2 (1)
'cost" does not include any professional fees paid to an architect, quantity surveyor or any other person;

(2)
(a) the total floor area of a dwelling or extension is the total of the floor areas of all the storeys in it; and
(b) the floor area of -
(i) any storey of a dwelling or extension, or
(ii) a garage or carport,
is the total floor area calculated by reference to the finished internal faces of the wall enclosing the area, or if at any point there is no enclosing wall, by reference to the outermost edge of the floor.

PART II: FEES CHARGED BY LOCAL AUTHORITIES

Prescribed functions

3. The prescribed functions in relation to which local authorities are authorised to charge fees are the following -

(a) the passing or rejection by the local authority, in accordance with section 16 of the Act, of plans of proposed work deposited with them (including plans of work proposed to be carried out by or on behalf of the authority);
(b) the inspection in connection with the principal regulations of work for which such plans have been deposited;
(c) the inspection in connection with the principal regulations of work for which a building notice has been given to the local authority; and
(d) the consideration of plans of work reverting to local authority control, and the inspection of that work.

PART III: FEES FOR DETERMINATION OF QUESTIONS BY THE SECRETARY OF STATE

18. (1) Where in accordance with section 16 (10) of the Act (determination of questions arising under that section) a person has referred a question to the Secretary of State for his determination, that application shall be accompanied by a fee calculated in accordance with paragraph 1 or 2 of Schedule 4.
(2) Where in accordance with section 50 (2) of the Act (determination of questions arising between approved inspector and developer) the person proposing to carry out work has referred a question to the Secretary of State for his determination, his application shall be accompanied by a fee calculated in accordance with paragraph 3 of Schedule 4.

THE BUILDING (PRESCRIBED FEES ETC.)
REGULATIONS 1989

SCHEDULE 1
SMALL DOMESTIC BUILDINGS

Fees for single small domestic buildings

Plan fee

1. Where a plan fee is chargeable in respect of -

   (a) plans for the erection of a single small domestic building, or
   (b) plans for the execution of works of drainage in connection with the erection of such a building deposited before plans for the erection of the building are deposited,

the plan fee for that building is the amount specified in column 2 of Table 1 below for the number of dwellings in that building.

Inspection fee

2. In relation to the erection of a small domestic building and any work in connection with the erection of such a building, the inspection fee is the aggregate of -

   (a) £88 multiplied by the number of dwellings in that building; and
   (b) where any of those dwellings has a total floor area (excluding the floor area of any garage comprised in the building) exceeding 64 m2, the amount specified in column (2) of Table 2 below for the number of such dwellings in that building.

Building notice fee

3. In relation to the erection of a single small domestic building and any work in connection with the erection of such a building, the building notice fee is the aggregate of the plan fee and the inspection fee which would have been payable had plans been deposited for the purpose of section 16 of the Act.

Fees for small domestic buildings in multiple work scheme

Plan fee for more than one small domestic building

4. (1) Where a plan fee is chargeble in respect of -

   (a) plans for the erection of a small domestic building, or
   (b) plans for the execution of works of drainage in connection with the erection of such buildings deposited before the plans for the erection of those buildings are deposited, and

the works forms part of a multiple work scheme which consists of or includes -

   (i) the erection of two or more small domestic buildings, or
   (ii) the execution of works of drainage in connection with such buildings,

THE BUILDING (PRESCRIBED FEES ETC.)
REGULATIONS 1989

the plan fee for each small domestic building is determined in accordance with sub-paragraph (2).

(2) In a case falling within sub-paragraph (1), the plan fee is determined by the formula -

$$\frac{A}{B} \times C$$

where -

A is the amount specified in column 2 of Table 1 below for the number of dwellings comprises in all the small domestic buildings in the multiple work scheme,

B is the number of dwellings comprised in all the small domestic buildings in the multiple work scheme, and

C is the number of dwellings in the small domestic building for which the fee is being determined.

Inspection fee for more than one small domestic building

5. In relation to -

(a) the erection of small domestic buildings, and
(b) any work in connection with the erection of small domestic buildings.

Where that work forms part of a multiple work scheme which consists of or includes the erection of two or more small domestic buildings, the inspection fee for each small domestic building is determined by the formula -

$$\frac{D + E}{B} \times C$$

Where -

D is the amount which results from multiplying £88 by the number of dwellings in all the small domestic buildings in the multiple work scheme;

E is the amount specified in column 2 of Table 2 below for the number of dwellings (if any) comprised in the small domestic buildings in the multiple work scheme with a total floor area (excluding the floor area of any garage) exceeding 64m2;

B is the number of dwellings in the small domestic buildings in the multiple work scheme; and

C is the number of dwellings in the small domestic building for which the fee is being determined.

THE BUILDING (PRESCRIBED FEES ETC.)
REGULATIONS 1989

TABLE 1

Plan fee for erection of small domestic buildings

(1) No. of dwellings	(2) Amount of or for determining plan fee	(1) No. of dwellings	(2) Amount of or for determining plan fee
	£		£
1	53	11	459
2	106	12	476
3	159	13	494
4	212	14	511
5	265	15	529
6	300	16	547
7	335	17	564
8	370	18	582
9	405	19	600
10	440	20 or more	617

Building notice fee for more than one small domestic building

6. In relation to the erection of small domestic buildings and any work in connection with the erection of such buildings, where that work forms part of a multiple work scheme which consists of or includes the erection of two or more small domestic buildings, the building notice fee for each small domestic building is the aggregate of the plan fee and the inspection fee which would have been payable had plans been deposited for the purposes of section 16 of the Act.

TABLE 2

Inspection fee for erection of small domestic building

(1) Number of dwellings each having a floor area exceeding 64m2	(2) Amount for determining inspection fee
	£
1	53
2	106
3	159
4	212
5	265
6	283
7	300
8	318
9	335
10 or more	353

THE BUILDING (PRESCRIBED FEES ETC.)
REGULATIONS 1989

SCHEDULE 2

SMALL GARAGES, CARPORTS AND CERTAIN ALTERATIONS AND EXTENSIONS

Fees for the erection of small garages and carports

1) In relation to the erection of a detached building which -

(a) consists of a garage or carport or both,
(b) has a total floor area not exceeding 40 m2, and
(c) is intended to be used in common with an existing building, and which is not an exempt building,

(i) the plan fee is £14;
(ii) the inspection fee of £43; and
(iii) the building notice fee is £57.

Fees for domestic extensions and alteration

2. (1) This paragraph applies to work to or in connection with -

(a) a small domestic building; or
(b) a building, other than a small domestic building, which consists of flats or maisonettes or both; or
(c) a building consisting of a garage or carport or both which is occupied in common with a building of the kind described in paragraphs (a) and (b),

(2) Where work shown in deposited plan to be carried out to or in connection with a buiilding to which this paragraph applies is described in column 1 of the Table in Schedule 2 of the Regulation, the plan fee is the amount shown in column 2 for that description of work.

(3) In relation to work described in column 1 of the Table in Schedule 2 of the Regulation to be carried out to or in connection with a building to which this paragraph applies, the inspection fee is the amount shown in column 3 for that description of work.

(4) In relation to work described in column (1) of the Table in Schedule 2 of the Regulation to be carried out to or in connection with a building to which this paragraph applies, the building notice fee is the aggregate of the amounts shown in column 2 and 3 for that description of work.

THE BUILDING (PRESCRIBED FEES ETC.)
REGULATIONS 1989

TABLE 3

Fees for certain domestic extensions and alterations

(1) Type of work	(2) Amount of plan fee	(3) Amount of inspection fee
1. An extension or alteration consisting of the provision of one or more rooms in roof space, including means of access.	£28	£86
2. Any extension (not falling within paragraph 1 above) the total floor area of which does not exceed 20 m2, including means of access	£14	£43
3. Any extension (not falling within paragraph 1 above) the total floor area of which exceeds 20 m2 but does not exceed 40 m2, including means of access.	£28	£86

References in this table to square metres relate to the total area of the work in question as it is shown on the plans

SCHEDULE 3

WORK OTHER THAN THAT DESCRIBED IN SCHEDULE 1 OR SCHEDULE 2

Interpretation

1. In this Schedule the appropriate percentage figure is 70% of the estimate of cost for the building in question supplier in accordance with regulation 13 (2) (a).

Plan fee for single building

2. Where a plan fee is chargeable in a case described in regulation 5 (3), the plan fee is, unless paragraph 3 applies, a fee of the amount shown in column 2 of the Table in Schedule 3 of the Regulation for the appropriate percentage figure in Column 1.

Plan fee for multiple work scheme

3. Where a plan fee is chargeable in a case described in regulation 5 (3) in relation to work which forms part of a multiple work scheme, and the scheme includes other such proposed work to or in connection with at least one other building, the plan fee is the amount determined in accordance with the formula -

$$\frac{E}{T} \times A$$

where -

E is the appropriate percentage figure;

THE BUILDING (PRESCRIBED FEES ETC.)
REGULATIONS - 1989

T is 70% of the aggregate figure supplied in accordance with regulation 13 (2) (b); and
A is the amount shown in column 2 of the Table below for the amount of T.

Inspection fee

4. In relation to work described in regulation 5 (3), the inspection fee is a fee of the amount shown in column 3 of the Table in Schedule 3 of the Regulation for the appropriate percentage figure in column 1.

Building notice fee

5. In relation to work described in regulation 5 (3), the building notice fee is the aggregate of the plan fee and the inspection fee which would have been payable had plans been deposited for the purposes of section 16 of the Act.

SCHEDULE 4

FEES FOR DETERMINATION OF QUESTIONS BY THE SECRETARY OF STATE

Questions about comformity of plans with building regulations

1. Where an application is made for the determination under section 16 (10) (a) of the Act of the question whether plans of proposed work are in conformity with building regulations, the fee payable to the Secretary of State is half the plan fee which is payable in relation to the work shown in the plans, subject to a minimum fee of £25 and a maximum of £250.

Questions about rejection of certificated plans

2. Where an application is made for the detemination under section 16 (10) (b) of the Act of the question whether the local authority are prohibited by virtue of section 16 (9) of the Act (certificates that plans comply with certain regulations) from rejecting plans of proposed work, the fee payable to the Secretary of State is £25.

Question arising between approved inspectors and developers

3. Where an application is made for the determination under section 50 (2) of the Act of the question whether plans of proposed work are in conformity with building regulations, the fee payable to the Secretary of State is half the plan fee which would have been payable in relating to the work shown in the plans had plans been deposited in accordance with section 16 of the Act, subject to a minimum fee of £25 and a maximum of £250.

THE BUILDING (PRESCRIBED FEES ETC.)
REGULATIONS 1989

TABLE 4

Fees for other work

(1) 70% of estimated cost	(2) Amount of plan fee or for determining plan fee	(3) Amount of inspection fee
£	£	£
Under 1,000	6	18
1,000 and under 2,000	10	30
2,000 and under 3,000	12	36
3,000 and under 4,000	16	48
4,000 and under 5,000	20	60
5,000 and under 6,000	24	72
6,000 and under 7,000	28	84
7,000 and under 8,000	32	96
8,000 and under 9,000	34	102
9,000 and under 10,000	36	108
10,000 and under 12,000	40	120
12,000 and under 14,000	46	138
14,000 and under 16,000	52	156
16,000 and under 18,000	58	174
18,000 and under 20,000	64	192
20,000 and under 25,000	75	225
25,000 and under 30,000	85	255
30,000 and under 35,000	95	285
35,000 and under 40,000	110	330
40,000 and under 45,000	120	360
45,000 and under 50,000	130	390
50,000 and under 60,000	145	435
60,000 and under 70,000	170	510
70,000 and under 80,000	195	585
80,000 and under 90,000	210	630
90,000 and under 100,000	230	690
100,000 and under 140,000	255	765
140,000 and under 180,000	330	990
180,000 and under 240,000	410	1230
240,000 and under 300,000	510	1530
300,000 and under 400,000	610	1830
400,000 and under 500,000	775	2325
500,000 and under 700,000	910	2730
700,000 up to and including 1,000,000	1185	3555
Thereafter for each additional 100,000 and part thereof	200	600

Daywork and Prime Cost

When work is carried out which cannot be valued in any other way it is customary to assess the value on a cost basis with an allowance to cover overheads and profit. The basis of costing is a matter for agreement between the parties concerned, but definitions of prime cost for the building industry have been prepared and published jointly by the Royal Institution of Chartered Surveyors and the National Federation of Building Trades Employers (now the Building Employers Confederation) for the convenience of those who wish to use them. These documents are reproduced on the following pages by kind permission of the publishers.

The daywork schedule published by the Federation of Civil Engineering Contractors has been omitted from this section, as it is included in the A&B's companion title, 'Spons Civil Engineering and Highway Works Price Book'.

For larger Prime Cost contracts the reader is referred to the form of contract issued by the Royal Institute of British Architects.

BUILDING INDUSTRY

DEFINITION OF PRIME COST OF DAYWORK CARRIED OUT UNDER A BUILDING CONTRACT (DECEMBER 1975 EDITION)

This definition of Prime Cost is published by the Royal Institution of Chartered Surveyors and the National Federation of Building Trades Employers, for convenience and for use by people who choose to use it. Members of the National Federation of Building Trades Employers are not in any way debarred from defining Prime Cost and rendering their accounts for work carried out on that basis in any way they choose. Building owners are advised to reach agreement with contractors on the Definition of Prime Cost to be used prior to issuing instructions.

SECTION 1 - APPLICATION

1.1. This definition provides a basis for the valuation of daywork executed under such building contracts as provide for its use (e.g. contracts embodying the Standard Forms issued by the Joint Contracts Tribunal).

1.2. It is not applicable in any other circumstances, such as jobbing or other work carried out as a separate or main contract nor in the case of daywork executed during the Defects Liability Period of contracts embodying the above mentioned Standard Forms.

SECTION 2 - COMPOSITION OF TOTAL CHARGES

2.1. The prime cost of daywork comprises the sum of the following costs:

(a) Labour as defined in Section 3.
(b) Material and goods as defined in Section 4.
(c) Plant as defined in Section 5.

BUILDING INDUSTRY

2.2. Incidental costs, overheads and profit as defined in Section 6, as provided in the building contract and expressed therein as percentage adjustments are applicable to each of 2.1 (a)-(c).

SECTION 3 - LABOUR

3.1. The standard wage rates, emoluments and expenses referred to below and the standard working hours referred to in 3.2 are those laid down for the time being in the rules or decisions of the National Joint Council for the Building Industry and the terms of the Building and Civil Enginnering Annual and Public Holiday Agreements applicable to the works, or the rules or decisions or agreements of such body, other than the National Joint Council for the Building Industry, as may be applicable relating to the class of labour concerned at the time when and in the area where the daywork is executed.

3.2. Hourly base rates for labour are computed by dividing the annual prime cost of labour, based upon standard working hours and as defined in 3.4 (a)-(i), by the number of standard working hours per annum.

3.3. The hourly rates computed in accordance with 3.2 shall be applied in respect of the time spent by operatives directly engaged on daywork, including those operating mechanical plant and transport and erecting and dismantling other plant (unless otherwise expressly provided in the building contract).

3.4. The annual prime cost of labour comprises the following:

(a) Guaranteed minimum weekly earnings (e.g. Standard Basic Rate of Wages, Joint Board Supplement and Guaranteed Minimum Bonus Payment in the case of NJCBI rules).
(b) All other guaranteed minimum payments (unless included in Section 6).
(c) Differentials or extra payments in respect of skill, responsibility, discomfort, inconvenience or risk (excluding those in respect of supervisory responsibility - see 3.5).
(d) Payments in respect of public holidays.
(e) Any amounts which may become payable by the Contractor to or in respect of operatives arising from the operation of the rules referred to in 3.1 which are not provided for in 3.4 (a)-(d) or in Section 6.
(f) Employer's National Insurance contributions applicable to 3.4 (a)-(e).
(g) Employer's contributions to annual holiday credits.
(h) Employer's contributions to death benefit scheme.
(i) Any contribution, levy or tax imposed by statute, payable by the contractor in his capacity as an employer.

3.5. **Note:**

Differentials or extra payments in respect of supervisory responsibility are excluded from the annual prime cost (see Section 6). The time of principals, foremen, gangers, leading hands and similar categories, when working manually, is admissible under this Section at the appropriate rates for the trades concerned.

SECTION 4 - MATERIALS AND GOODS

4.1. The prime cost of materials and goods obtained from stockists or manufacturers is the invoice cost after deduction of all trade discounts but including cash discounts not exceeding 5 per cent and includes the cost of delivery to site.

BUILDING INDUSTRY

4.2. The prime cost of materials and goods supplied from the Contractor's stock is based upon the current market prices plus any appropriate handling charges.

4.3. Any Value Added Tax which is treated, or is capable of being treated, as input tax (as defined in the Finance Act, 1972) by the Contractor is excluded.

SECTION 5 - PLANT

5.1. The rates for plant shall be as provided in the building contract.

5.2. The costs included in this Section comprise the following:

(a) Use of mechanical plant and transport for the time employed on daywork.
(b) Use of non-mechanical plant (excluding non-mechanical hand tools) for the time employed on daywork.

5.3. **Note:**

The use of non-mechanical hand tools and of erected scaffolding, staging, trestles or the like is excluded (see Section 6).

SECTION 6 - INCIDENTAL COSTS, OVERHEADS AND PROFIT

6.1. The percentage adjustments provided in the building contract, which are applicable to each of the totals of Sections 3, 4 and 5, comprise the following:

(a) Head Office charges.
(b) Site staff, including site supervision.
(c) The additional cost of overtime (other than that referred to in 6.2).
(d) Time lost due to inclement weather.
(e) The additional cost of bonuses and all other incentive payments in excess of any guaranteed minimum included in 3.4. (a).
(f) Apprentices study time.
(g) Subsistence and periodic allowances.
(h) Fares and travelling allowances.
(i) Sick pay or insurance in respect thereof.
(j) Third-party and employers' liability insurance.
(k) Liability in respect of redundancy payments to employees.
(l) Employers' National Insurance contributions not included in Section 3.4.
(m) Tool allowances.
(n) Use, repair and sharpening of non-mechanical hand tools.
(o) Use of erected scaffolding, staging, trestles or the like.
(p) Use of tarpaulins, protective clothing, artificial lighting, safety and welfare facilities, storage and the like that may be available on the site.
(q) Any variation to basic rates required by the Contractor in cases where the building contract provides for the use of a specified schedule of basic plant charges (to the extent that no other provision is made for such variation).
(r) All other liabilities and obligations whatsoever not specifically referred to in this Section nor chargeable under any other Section.
(s) Profit.

6.2. **Note:**

The additional cost of overtime, where specifically ordered by the Architect/Supervising Officer shall only be chargeable in the terms of prior written agreement between the parties to the building contract.

BUILDING INDUSTRY

Example of calculation of typical standard hourly base rate (as defined in Section 3) for NJCBI building craftsman and labourer in Grade A areas at 1st July, 1975.

		Rate £	Craftsman £	Rate £	Labourer £
Guaranteed minimum weekly earnings					
Standard Basic Rate	49 wks	37.00	1813.00	31.40	1538.60
Joint Board Supplement	49 wks	5.00	245.00	4.20	205.80
Guaranteed Minimum Bonus	49 wks	4.00	196.00	3.60	176.40
			2254.00		1920.80
Employer's National Insurance Contribution at 8.5%			191.59		163.27
			2445.59		2084.07
Employer's Contribution to:					
CITB annual levy			15.00		3.00
Annual holiday credits	49 wks	2.80	137.20	2.80	137.20
Public holidays (included in guaranteed minimum weekly earnings above)					
Death benefit scheme	49 wks	0.10	4.90	0.10	4.90
Annual labour cost as defined in Section 3			£2602.69		£2229.17
Hourly rate of labour as defined in Section 3, Clause 3.2		£2602.69 / 1904 =	£1.37	£2229.17 / 1904 =	£1.17

Note:

1. Standard working hours per annum calculated as follows:

52 weeks @ 40 hours	=	2080
Less		
3 weeks holiday @ 40 hours	= 120	
7 days public holidays @ 8 hours	= 56	176
		1904

2. It should be noted that all labour costs incurred by the Contractor in his capacity as an employer other than those contained in the hourly rate, are to be taken into account under Section 6.

3. The above example is for the convenience of users only and does not form part of the Definition; all basic costs are subject to re-examination according to the time when and in the area where the daywork is executed.

BUILDING INDUSTRY

NOTE: For the convenience of readers the example which appears on the previous page has been updated by the Editors for London and Liverpool rates as at 30th July 1990.

		Rate £	Craft operative £	Rate £	Labourer £
Guaranteed minimum weekly earnings					
Standard Basic Rate	47.80 wks	132.02	6310.56	112.52	5378.46
Guaranteed Minimum Bonus	47.80 wks	16.38	782.96	14.04	671.11
			7093.52		6049.57
Employer's National Insurance Contribution 9%			638.42		544.46
		£	7731.94	£	6594.03
Employer's Contributions to:					
CITB annual levy			75.00		18.00
Annual holiday credits	47 wks	16.15	759.05	16.15	759.05
Public holidays (included in guaranteed minimum weekly earnings above)					
Retirement and death benefit scheme	47 wks	1.55	72.85	1.55	72.85
Annual labour cost as defined in Section 3			8638.84		7443.93
Hourly rate of labour as defined in Section 3, Clause 3.02		£8638.84 / 1801.80 =	£4.79	£7443.93 / 1801.80 =	£4.13

Note:

1. Standard working hours per annum calculated as follows:

52 weeks @ 39 hours	=	2028.0
Less		
4.20 weeks holiday @ 39 hours	= 163.8	
8 days public holidays @ 7.8 hours	= 62.4	226.2
		1801.8

2. It should be noted that all labour costs incurred by the Contractor in his capacity as an employer other than those contained in the hourly rate, are to be taken into account under Section 6.

3. The above example is for the convenience of users only and does not form part of the Definition; all the basic costs are subject to re-examination according to the time when and in the area where the daywork is executed.

4. The rates for annual holiday credits and retirement and death benefits are those which came into force in 30th July 1990 and do not include the discount available.

BUILDING INDUSTRY

DEFINITION OF PRIME COST OF BUILDING WORKS OF A JOBBING OR MAINTENANCE CHARACTER (1980 EDITION)

This Definition of Prime Cost is published by the Royal Institution of Chartered Surveyors and the National Federation of Building Trades Employers for convenience and for use by people who choose to use it. Members of the National Federation of Building Trades Employers are not in any way debarred from defining Prime Cost and rendering their accounts for work carried out on that basis in any way they choose.

Building owners are advised to reach agreement with contractors on the Definition of Prime Cost to be used prior to issuing instructions.

SECTION 1 - APPLICATION

1.1. This definition provides a basis for the valuation of work of a jobbing or maintenance character executed under such building contracts as provide for its use.

1.2. It is not appplicable in any other circumstances, such as daywork executed under or incidental to a building contract.

SECTION 2 - COMPOSITION OF TOTAL CHARGES

2.1. The prime cost of jobbing work comprises the sum of the following costs:

(a) Labour as defined in Section 3.
(b) Materials and goods as defined in Section 4.
(c) Plant, consumable stores and services as defined in Section 5.
(d) Sub-contracts as defined in Section 6.

2.2. Incidental costs, overhead and profit as defined in Section 7 and expressed as percentage adjustments are applicable to each of 2.1 (a)-(d).

SECTION 3 - LABOUR

3.1. Labour costs comprise all payments made to or in respect of all persons directly engaged upon the work, whether on or off the site, except those included in Section 7.

3.2. Such payments are based upon the standard wage rates, emoluments and expenses as laid down for the time being in the rules or decisions of the National Joint Council for the Building Industry and the terms of the Building and Civil Engineering Annual and Public Holiday Agreements applying to the works, or the rules of decisions or agreements of such other body as may relate to the class of labour concerned, at the time when and in the area where the work is executed, together with the Contractor's statutory obligations, including:

(a) Guaranteed minimum weekly earnings (e.g. Standard Basic Rate of Wages and Guaranteed Minimum Bonus Payment in the case of NJCBI rules).
(b) All other guaranteed minimum payments (unless included in Section 7).
(c) Payments in respect of incentive schemes or productivity agreements applicable to the works.
(d) Payments in respect of overtime normally worked; or necessitated by the particular circumstances of the work; or as otherwise agreed between the parties.
(e) Differential or extra payments in respect of skill, responsibility, discomfort, inconvenience or risk.
(f) Tool allowance.

(g) Subsistence and periodic allowances.
(h) Fares, travelling and lodging allowances.
(j) Employer's contributions to annual holiday credits.
(k) Employer's contributions to death benefit schemes.
(l) Any amounts which may become payable by the Contractor to or in respect of operatives arising from the operation of the rules referred to in 3.2 which are not provided for in 3.2 (a)-(k) or in Section 7.
(m) Employer's National Insurance contributions and any contribution, levy or tax imposed by statute, payable by the Contractor in his capacity as employer.

Note:

Any payments normally made by the Contractor which are of a similar character to those described in 3.2 (a)-(c) but which are not within the terms of the rules and decisions referred to above are applicable subject to the prior agreement of the parties, as an alternative to 3.2 (a)-(c).

3.3. The wages or salaries of supervisory staff, timekeepers, storekeepers, and the like, employed on or regularly visiting site, where the standard wage rates etc. are not applicable, are those normally paid by the Contractor together with any incidental payments of a similar character to 3.2 (c)-(k).

3.4. Where principals are working manually their time is chargeable, in respect of the trades practised, in accordance with 3.2.

SECTION 4 - MATERIALS AND GOODS

4.1. The prime cost of materials and goods obtained by the Contractor from stockists or manufacturers is the invoice cost after deduction of all trade discounts but including cash discounts not exceeding 5 per cent, and includes the cost of delivery to site.

4.2. The prime cost of materials and goods supplied from the Contractor's stock is based upon the current market prices plus any appropriate handling charges.

4.3. The prime cost under 4.1 and 4.2 also includes any costs of:

(a) non-returnable crates or other packaging.
(b) returning crates and other packaging less any credit obtainable.

4.4. Any Value Added Tax which is treated, or is capable of being treated, as input tax (as defined in the Finance Act, 1972 or any re-enactment thereof) by the Contractor is excluded.

SECTION 5 - PLANT, CONSUMABLE STORES AND SERVICES

5.1. The prime cost of plant and consumable stores as listed below is the cost at hire rates agreed between the parties or in the absence of prior agreement at rates not exceeding those normally applied in the locality at the time when the works are carried out, or on a use and waste basis where applicable:

(a) Machinery in workshops.
(b) Mechanical plant and power-operated tools.
(c) Scaffolding and scaffold boards.
(d) Non-mechanical plant excluding hand tools.
(e) Transport including collection and disposal of rubbish.

(f) Tarpaulins and dust sheets.
(g) Temporary roadways, shoring, planking and strutting, hoarding, centering, formwork, temporary fans, partitions or the like.
(h) Fuel and consumable stores for plant and power-operated tools unless included in 5.1 (a), (b), (d) or (e) above.
(j) Fuel and equipment for drying out the works and fuel for testing mechanical services.

5.2. The prime cost also includes the net cost incurred by the Contractor of the following services, excluding any such cost included under Sections 3, 4 or 7:

(a) Charges for temporary water supply including the use of temporary plumbing and storage.
(b) Charges for temporary electricity or other power and lighting including the use of temporary installations.
(c) Charges arising from work carried out by local authorities or public undertakings.
(d) Fees, royalties and similar charges.
(e) Testing of materials.
(f) The use of temporary buildings including rates and telephone and including heating and lighting not charged under (b) above.
(g) The use of canteens, sanitary accommodation, protective clothing and other provision for the welfare of persons engaged in the work in accordance with the current Working Rule Agreement and any Act of Parliament, statutory instrument, rule, order, regulation or bye-law.
(h) The provision of safety measures necessary to comply with any Act of Parliament.
(j) Premiums or charges for any performance bonds or insurances which are required by the Building Owner and which are not referred to elsewhere in this Definition.

SECTION 6 - SUB-CONTRACTS

6.1. The prime cost of work executed by sub-contractors, whether nominated by the Building Owner or appointed by the Contractor, is the amount which is due from the Contractor to the sub-contractors in accordance with the terms of the sub-contracts after deduction of all discounts except any cash discount offered by any sub-contractor to the Contractor not exceeding 2.5%.

SECTION 7 - INCIDENTAL COSTS, OVERHEADS AND PROFIT

7.1. The percentage adjustments provided in the building contract, which are applicable to each of the totals of Sections 3-6, provide for the following:
(a) Head Office charges.
(b) Off-site staff including supervisory and other administrative staff in the Contractor's workshops and yard.
(c) Payments in respect of public holidays.
(d) Payments in respect of apprentices' study time.
(e) Sick pay or insurance in respect thereof.
(f) Third party employer's liability insurance.
(g) Liability in respect of redundancy payments made to employees.
(h) Use, repair and sharpening of non-mechanical hand tools.
(j) Any variations to basic rates required by the Contractor in cases where the building contract provides for the use of a specified schedule of basic plant charges (to the extent that no other provision is made for such variation).
(k) All other liabilities and obligations whatsoever not specifically referred to in this Section nor chargeable under any other section.
(l) Profit.

BUILDING INDUSTRY

SPECIMEN ACCOUNT FORMAT

If this Definition of Prime Cost is followed the Contractor's account could be in the following format:

	£
Labour (as defined in Section 3)	
Add ... % (see Section 7)	
Materials and goods (as defined in Section 4)	
Add ... % (see Section 7)	
Plant, consumable stores and services (as defined in Section 5)	
Add ... % (see Section 7)	
Sub-contracts (as defined in Section 6)	
Add ... % (see Section 7)	

	£

VAT to be added if applicable.

SCHEDULE OF BASIC PLANT CHARGES (JANUARY 1990 ISSUE)

This Schedule is published by the Royal Institution of Chartered Surveyors and is for use in connection with Dayworks under a Building Contract.

EXPLANATORY NOTES

1. The rates in the Schedule are intended to apply solely to daywork carried out under and incidental to a Building Contract. They are NOT intended to apply to:
   (i) Jobbing or any other work carried out as a main or separate contract; or
   (ii) Work carried out after the date of commencement of the Defects Liability Period.
2. The rates in the Schedule are basic and may be subject to an overall adjustment to be quoted by the Contractor prior to the placing of the Contract.
3. The rates apply to plant and machinery already on site, whether hired or owned by the Contractor.
4. The rates, unless otherwise stated, include the cost of fuel and power of every description, lubricating oils, grease, maintenance, sharpening of tools, replacement of spare parts, all consumable stores and for licences and insurances applicable to items of plant. They do not include the costs of drivers and attendants (unless otherwise stated).
5. The rates should be applied to the time during which the plant is actually engaged in daywork.
6. Whether or not plant is chargeable on daywork depends on the daywork agreement in use and the inclusion of an item of plant in this schedule does not necessarily indicate that that item is chargeable.
7. Rates for plant not included in the Schedule or which is not on site and is specifically hired for daywork shall be settled at prices which are reasonably related to the rates in the Schedule having regard to any overall adjustment quoted by the Contractor in the Conditions of Contract.

BUILDING INDUSTRY

MECHANICAL PLANT AND TOOLS

Item of plant	Description		Unit	Rate per hour £
BAR-BENDING AND SHEARING MACHINES				
Bar bending machine	Power driven - up to 2 in (51 mm) dia rods		each	2.01
Bar shearing machine	1.5 in (38 mm) dia rods		each	2.00
	2 in (51 mm) dia rods		each	2.00
Bar cropper machine	2 in (51 mm) dia rods		each	1.51
BLOCK AND STONE SPLITTER	Hydraulic		each	0.83
BRICK SAWS				
Brick saw (use of abrasive disc to be charged net and credited)	Power driven (bench type clipper or similar or portable)		each	1.83
COMPRESSORS				
Portable compressor (machine only)	Nominal delivery of free air per min at 100 lb/sq in (7 kg/sq.m) pressure.			
	(cfm)	(m^3/min)		
	80/85	2.41	each	1.30
	125-140	3.50-3.92	each	1.58
	160-175	4.50-4.95	each	2.37
	250	7.08	each	3.25
	380	10.75	each	4.80
	600-630	16.98-17.84	each	6.00
LORRY MOUNTED COMPRESSORS				
(machine plus lorry only)	Nominal delivery of free air per min at 100 PSI (7 kg/cm^2)			
	101-150	(2.86-4.24)	each	4.12
TRACTOR MOUNTED COMPRESSORS				
(Machine plus rubber tyred tractors only	101-120	(2.86-3.40)	each	3.63
COMPRESSED AIR EQUIPMENT				
(with and including up to 50 ft (15.24 m) of air hose)				
Breakers (with six steels)	Light		each	0.47
	Medium (65 lbs)		each	0.47
	Heavy (85 lbs)		each	0.47
Light pneumatic pick (with six steels)			each	0.37
Pneumatic clay spade (one blade)			each	0.22
Chipping hammer (plus six steels)			each	0.65
Hand-held rock drill with rod (bits to be paid for at net cost and credited)	7 kg (15 lbs) class (light)		each	0.33
	16 kg (35 lbs) class (medium)		each	0.37
	20 kg (45 lbs) class (medium)		each	0.37
	25 kg (55 lbs) class (heavy)		each	0.39
Rotary drill (bits to be paid for at net cost and credited)	Up to 0.75 in (19 mm)		each	0.37
	1.25 in (32 mm)		each	0.73
Sander/Grinder			each	0.39
Scabbler (heads extra)	Single head		each	0.48
	Triple head		each	0.65

BUILDING INDUSTRY

Item of plant	Description		Unit	Rate per hour
				£
COMPRESSOR EQUIPMENT ACCESSORIES				
Additional hoses	Per 50 ft (15 m) length		each	0.09
Muffler, tool silencer			each	0.08
CONCRETE BREAKER				
Concrete breaker, portable hydraulic complete with power pack			each	1.33
CONCRETE/MORTAR MIXERS				
Concrete mixer	Diesel, electric or petrol			
	cu ft	m^3		
Open drum without hopper	3/2	0.09/0.06	each	0.40
	4/3	0.12/0.09	each	0.44
	5/3.5	0.15/0.10	each	0.54
Open drum with hopper	7/5	0.20/0.15	each	0.67
Closed drum	8/6.5	0.25/0.18	each	0.90
Reversing drum with hopper	10/7	0.28/0.20	each	2.44
weigher and feed shovel	21/14	0.60/0.40	each	4.17
CONCRETE PUMP				
Lorry mounted concrete pump (metreage charge to be added net)			each	13.70
CONCRETE EQUIPMENT				
Vibrator, poker type	Petrol, diesel or electric			
	Up to 3 in dia		each	0.87
	Air, excluding compressor and hose Up to 3 in dia		each	0.65
	Extra heads		each	0.50
Vibrator, tamper	With tamping board		each	0.87
	Double beam screeder		each	1.19
Power float	29 in - 36 in		each	1.13
CONVEYOR BELTS	Power operated up to 25 ft (7.62 m) long, 16 in (400 mm) wide		each	3.15
CRANES				
Mobile rubber tyred	Maximum capacity up to			
	15 cwt (762 kg)		each	5.15
Lorry mounted, telescopic jib, 2 wheel (All rates inclusive of driver)	6 tons (tonnes)		each	10.47
	7 tons (tonnes)		each	11.22
	8 tons (tonnes)		each	13.67
	10 tons (tonnes)		each	14.99
	12 tons (tonnes)		each	16.91
	15 tons (tonnes)		each	19.21
	18 tons (tonnes)		each	20.34
	20 tons (tonnes)		each	22.04
	25 tons (tonnes)		each	24.86

BUILDING INDUSTRY

MECHANICAL PLANT AND TOOLS

Item of plant	Description		Unit	Rate per hour £
Lorry mounted, telescopic jib, 4 wheel (All rates inclusive of driver)	10 tons (tonnes)		each	15.30
	12 tons (tonnes)		each	17.26
	15 tons (tonnes)		each	19.59
	20 tons (tonnes)		each	22.48
	25 tons (tonnes)		each	25.36
	30 tons (tonnes)		each	28.82
	45 tons (tonnes)		each	31.12
	50 tons (tonnes)		each	33.64
Track-mounted tower crane (electric) (Capacity = max lift in tons X max radius at which can be lifted) (All rates inclusive of driver)	Capacity (metre/ tonnes) up	Height under hook above ground (m)		
	to 10	17	each	7.40
	15	17	each	7.95
	20	18	each	8.50
	25	20	each	10.70
	30	22	each	12.76
	40	22	each	16.75
	50	22	each	21.70
	60	22	each	22.52
	70	22	each	23.07
	80	22	each	23.99
	110	22	each	24.49
	125	30	each	27.20
	150	30	each	29.95
Static tower	To be charged at a percentage of the above rates		each	90% of the above

CRANE EQUIPMENT

Item of plant	Description	Unit	Rate per hour £
Tipping bucket	Circular		
	0.19 m^3 (0.25 cu.yd)	each	0.25
	0.57 m^3 (0.75 cu.yd)	each	0.28
Skip, muck	Up to		
	0.38 m^3 (.5 cu.yd)	each	0.25
	0.57 m^3 (0.75 cu.yd)	each	0.25
	0.76 m^3 (1.0 cu.yd)	each	0.30
Skip, concrete	Up to		
	0.38 m^3 (.5 cu.yd)	each	0.44
	0.57 m^3 (0.75 cu.yd)	each	0.53
	0.76 m^3 (1 cu.yd)	each	0.76
	1.15 m^3 (1.5 cu.yd)	each	0.61
Skip, concrete lay down or roll over	0.38 m^3 (.5 cu yd)	each	0.44

DEHUMIDIFIERS

Item of plant	Description	Unit	Rate per hour £
110/240v Water extraction per 24 hours	68 litres (15 gallons)	each	0.84
	90 litres (20 gallons)	each	0.97

DIAMOND DRILLING AND CHASING

Item of plant	Description	Unit	Rate per hour £
Chasing machine drilling (Diamond core)	6 in (152 mm)	each	1.35
	Mini rig up to 35 mm	each	3.16
	3 in - 8 in (76mm - 203 mm) electric 110v	each	3.33
	3 in - 8 in (76mm - 203 mm) Air	each	4.23

BUILDING INDUSTRY

Item of plant	Description	Unit	Rate per hour £
DUMPERS			
Dumper (site use only excluding Tax, Insurance and cost of DERV etc., when operating on highway)	Makers capacity		
2 wheel drive			
Gravity tip	15 cwt (762 kg)	each	1.04
Hydraulic tip	20 cwt (1016 kg)	each	1.25
Hydraulic tip	23 cwt (1168 kg)	each	1.31
4 wheel drive			
Gravity tip	15 cwt (762 kg)	each	1.42
Hydraulic tip	25 cwt (1270 kg)	each	1.61
Hydraulic tip	30 cwt (1524 kg)	each	1.83
Hydraulic tip	35 cwt (1778 kg)	each	2.08
Hydraulic tip	40 cwt (2032 kg)	each	2.33
Hydraulic tip	50 cwt (2540 kg)	each	2.96
Hydraulic tip	60 cwt (3048 kg)	each	3.68
Hydraulic tip	80 cwt (4064 kg)	each	4.72
Hydraulic tip	100 cwt (5080 kg)	each	6.38
Hydraulic tip	120 cwt (6096 kg)	each	7.54
ELECTRIC HAND TOOLS			
Breakers	Heavy: Kango 1800	each	0.87
	Kango 2500	each	0.90
	Medium: Hilti TP800	each	0.73
Rotary hammers	Heavy: Kango 950	each	0.55
	Medium: Kango 627/637	each	0.47
	Light: Hilti TE12/TE17	each	0.39
Pipe drilling tackle	Ordinary type	set	0.51
	Under pressure type	set	0.83
EXCAVATORS			
Hydraulic full circle slew	Crawler mounted, backactor		
	5/8 cu.yd (0.50 m^3)	each	5.01
	3/4 cu.yd (0.60 m^3)	each	5.92
	7/8 cu.yd (0.70 m^3)	each	8.24
	1 cu.yd (0.75 m^3)	each	13.25
Wheeled tractor type	Hydraulic excavator, JCB type 3C or similar	each	9.23
MINI EXCAVATORS			
Kubota mini excavators	360 tracked - 1 tonne	each	5.38
	360 tracked - 3.5 tonnes	each	7.23

FORKLIFTS	**Payload**	**Max. lift**		
2 wheel drive "Rough terrain"	20 cwt (1016 kg)	21 ft 4 in (6.50 m)	each	4.38
	20 cwt (1016 kg)	26 ft 0 in (7.92 m)	each	4.38
	30 cwt (1524 kg)	20 ft 0 in (6.09 m)	each	4.50
	36 cwt (1829 kg)	18 ft 0 in (5.48 m)	each	4.50
	50 cwt (2540 kg)	12 ft 0 in (3.66 m)	each	4.73
4 wheel drive "Rough terrain"	30 cwt (1524 kg)	20 ft 0 in (6.09 m)	each	5.60
	40 cwt (2032 kg)	20 ft 0 in (6.09 m)	each	5.78
	50 cwt (2540 kg)	12 ft 0 in (3.66 m)	each	7.79

BUILDING INDUSTRY

MECHANICAL PLANT AND TOOLS

Item of plant	Description	Unit	Rate per hour £
HAMMERS			
Cartridge (excluding cartridges and studs)	Hammer DX450	each	0.40
	Hammer DX600	each	0.48
	Hammer spit TS	each	0.48
HEATERS, SPACE	Paraffin/electric Btu/hr		
	50 000- 75 000	each	0.73
	80 000-100 000	each	0.77
	150 000	each	0.88
	320 000	each	1.20
HOISTS			
Scaffold	Up to 5 cwt (254 kg)	each	0.90
Mobile (goods only)	Up to 10 cwt (508 kg)	each	1.43
Static (goods only)	10 to 15 cwt (508-762 kg)	each	1.67
Rack and pinion (goods only)	16 cwt (813 kg)	each	2.75
	20 cwt (1016 kg)	each	3.00
	25 cwt (1270 kg)	each	3.30
Rack and pinion (goods and and passenger)	8 person, 1433 lbs (650 kg)	each	4.50
	12 person, 2205 lbs (1000 kg)	each	4.95
LORRIES	Plated gross vehicle Weight, (ton/tonnes)		
Fixed body	Up to 5.50	each	9.60
	Up to 7.50	each	10.20
Tipper	Up to 16.00	each	11.73
	Up to 24.00	each	14.98
	Up to 30.00	each	17.73
PIPE WORK EQUIPMENT			
Pipe bender	Power driven, 50-150 mm dia	each	0.75
	Hydraulic, 2 in capacity	each	0.96
Pipe cutter	Hydraulic	each	1.02
Pipe defrosting equipment	Electrical	set	1.83
Pipe testing equipment	Compressed air	set	1.04
	Hydraulic	set	0.68
Pipe threading equipment	Up to 4 in	set	1.65
PUMPS			
Including 20′ (6 m) length of suction/delivery hose, couplings, valves and strainers			
Diaphragm:			
'Simplite' 2 in (50 mm)	Petrol or electric	each	0.79
'Wickham' 3 in (76 mm)	Diesel	each	1.22
Submersible:			
2 in (50 mm)	Electric	each	0.85
Induced flow:			
'Spate' 3 in (76 mm)	Diesel	each	1.42
'Spate' 4 in (102 mm)	Diesel	each	1.88

BUILDING INDUSTRY

Item of plant	Description		Unit	Rate per hour £
Centrifugal, self priming:				
'Univac' 2 in (50 mm)	Diesel		each	1.87
'Univac' 4 in (102 mm)	Diesel		each	2.40
'Univac' 6 in (152 mm)	Diesel		each	3.69
PUMPING EQUIPMENT	in diameter	mm		
Pump hoses (per 20 ft (6 m)	2	51	each	0.15
flexible suction/delivery	3	76	each	0.17
including coupling valve and	4	102	each	0.19
strainer	6	152	each	0.32
RAMMERS AND COMPACTORS				
Power rammer, Pegson or similar			each	0.88
Soil compactor, plate type				
12 x 13 in (305 x 330 mm)	172 lb (78 kg)		each	0.78
20 x 18 in (508 x 457 mm)	264 lb (120 kg)		each	0.93
ROLLERS	cwt	kg		
Vibrating rollers	7.25-8.25	368-420	each	1.00
Single roller	10.5	533	each	1.53
Twin roller	13.75	698	each	1.75
	16.75	851	each	2.01
Twin roller with seat and steering	21	1067	each	2.75
wheel	27.5	1397	each	2.88
	ton/tonne			
Pavement rollers dead weight	3-4		each	2.89
	over 4- 6		each	3.75
	over 6-10		each	4.40
SAWS, MECHANICAL	in	m		
Chain saw	21	0.53	each	0.78
	30	0.76	each	1.12
Bench saw	Up to 20	0.51 blade	each	0.80
	Up to 24	0.61 blade	each	1.20
SCREED PUMP	Maximum delivery			
Working volume 7 cu ft (200 litres)	Vertical 300 ft (91 m)			
	Horizontal 600 ft (182 m)		each	8.33
Screed pump hose	50/65 mm dia			
	13.3 m long		each	1.83
SCREWING MACHINES	13- 50 mm dia		each	0.43
	25-100 mm dia		each	0.87
TRACTORS	cu yd	cu m		
	Up to			
Shovel, tractor (crawler), any	0.75	0.57	each	4.70
type of bucket	1	0.76	each	5.70
	1.25	0.96	each	6.62
	1.5	1.15	each	8.15
	1.75	1.34	each	8.98
	2	1.53	each	10.69
	2.75 - 3.5	2.10 - 2.70	each	17.27

BUILDING INDUSTRY

Item of plant	Description		Unit	Rate per hour
	cu.yd	cu. m		£
	Up to			
Shovel, tractor (wheeled)	0.5	0.38	each	3.36
	0.75	0.57	each	4.14
	1.0	0.76	each	5.02
	2.5	1.91	each	7.30
Tractor (crawler) with dozer	Maker's rated flywheel horsepower			
	75		each	9.31
	140		each	12.52
Tractor-wheeled (rubber tyred)	Light 48 hp		each	4.23
Agricultural type	Heavy 65 hp		each	4.68
TRAFFIC LIGHTS	Mains/generator 2-way		set	2.23
	Mains/generator 3-way		set	4.40
	Mains/generator 4-way		set	5.45
	Trailer mounted 2-way		set	2.21
WELDING AND CUTTING AND BURNING SETS				
Welding and cutting set (including oxygen and acetylene, excluding underwater equipment and thermic boring)			each	4.33
Welding set, diesel (excluding electrodes)	300 amp single operator		each	1.84
	480 amp double operator		each	1.89
	600 amp double operator		each	2.35
NON-MECHANICAL PLANT				
BAR BENDING AND SHEARING MACHINES				
Bar bending machine, manual	Up to 1 in (25 mm) dia rods		each	0.26
Shearing machine, manual	Up to 0.625 (16 mm) dia rods		each	0.15
BROTHER OR SLING CHAINS	Not exceeding 2 ton/tonne		set	0.28
	Exceeding 2 ton/tonne, not exceeding 5 ton/tonne		set	0.40
	Exceeding 5 ton/tonne, not exceeding 10 ton/tonne		set	0.48
DRAIN TESTING EQUIPMENT			set	0.42
LIFTING AND JACKING GEAR	ton/tonne			
Pipe winch - including shear legs	0.5		set	0.35
Pipe winch - including gantry	2		set	0.72
	3		set	0.84
Chain blocks up to 20 ft (6.10 m) lift	1		each	0.35
	2		each	0.44
	3		each	0.51
	4		each	0.66
Pull lift (Tirfor type)	0.75		each	0.35
	1.5		each	0.47
	3		each	0.61
PIPE BENDERS	13- 75 mm dia		each	0.32
	50-100 mm dia		each	0.52

BUILDING INDUSTRY

NON-MECHANICAL PLANT

Item of plant	Description	Unit	Rate per hour £
PLUMBER'S FURNACE	Calor gas or similar	each	1.20
ROAD WORKS - EQUIPMENT			
Barrier trestles or similar		10	0.50
Crossing plates (steel sheets)		each	0.32
Danger lamp, including oil		each	0.03
Warning sign		each	0.07
Road cone		10	0.11
Flasher unit (battery to be charged at cost)		each	0.06
Flashing bollard (battery to be charged at cost)		each	0.10
RUBBISH CHUTES			
Length of sections 154 cm giving working length of 130 cm			
Standard section		each	0.15
Brand section		each	0.20
Hopper		each	0.16
Fixing frame		each	0.20
Winch Type A	20 m for erection	each	0.29
Winch Type B	40 m for erection	each	0.67
SCAFFOLDING			
Boards		100 ft (30.48m)	0.06
Castor wheels	Steel or rubber tyred	100	0.87
Fall ropes	Up to 200 ft (61 m)	each	0.44
Fittings (including couplers, base plates	Steel or alloy	100	0.07
Ladders, pole	20 rung	each	0.11
	30 rung	each	0.15
	40 rung	each	0.26
Ladders, extension	Extended length ft (m)		
	20 (6.10)	each	0.31
	26 (7.92)	each	0.43
	35 (10.67)	each	0.64
Putlogs	Steel or alloy	100	0.06
Splithead	small	10	0.12
	medium	10	0.15
	large	10	0.16
Staging, lightweight		100ft (30.48 m)	1.00
Tube	Steel	100 ft	0.02
	Alloy	(30.48 m)	0.04
Wheeled tower	Working platform up to 20 ft		
7 ft x 7 ft (2.13 x 2.13 m)	(6 m high. Including castors	each	0.58
10ft x 10 ft (3.05 x 3.05 m)	and boards	each	0.59
TARPAULINS		10 m2	0.04

BUILDING INDUSTRY

NON-MECHANICAL PLANT

Item of plant	Description	Unit	Rate per hour £
TRENCH STRUTS AND SHEETS			
Adjustable steel trench strut	All sizes from 1 ft (305 mm) to 5 ft 6 in (1.8 m) extended	10	0.06
Steel trench sheet	5-14 ft (1.52-4.27 m) lengths	100 ft (30.48 m)	0.16

PART II

Rates of Wages

BUILDING INDUSTRY - ENGLAND, WALES AND SCOTLAND

Authorized rates of wages, etc., in the building industry in England, Wales and Scotland agreed by the National Joint Council for the Building Industry

AND EFFECTIVE FROM 25th JUNE 1990

Subject to the conditions prescribed in the Working Rule Agreement guaranteed minimum weekly earnings shall be as follows:

	Craft operatives £	Labourers £
LONDON AND LIVERPOOL DISTRICT	148.395	126.555
GRADE A AND SCOTLAND	148.200	126.360

These guaranteed minimum weekly earnings shall be made as follows:

	Craft operatives £	Labourers £
LONDON AND LIVERPOOL DISTRICT		
Standard basic rates of wages	132.015	112.513
Guaranteed minimum bonus payment	16.380	14.040
Guaranteed minimum weekly earning	148.395	126.555

	Craft operatives £	Labourers £
GRADE A AND SCOTLAND		
Standard basic rates of wages	131.820	112.320
Guaranteed minimum bonus payment	16.380	14.040
Guaranteed minimum weekly earnings	148.200	126.360

NOTE

The guaranteed minimum bonus payment shall be set off against existing bonus payments of all kinds or against existing extra payments other than those prescribed in the Working Rule Agreement. The entitlement to the guaranteed minimum bonus shall be pro rata to normal working hours for which the operative is available for work.

BUILDING INDUSTRY - ENGLAND, WALES AND SCOTLAND

Young Labourers

The rates of wages for young labourers shall be the following proportions of the labourers' rates:

At 16 years of age 50%
At 17 years of age 70%
At 18 years of age 100%

Watchmen

The weekly pay for watchmen shall be equivalent to the labourers' guaranteed minimum weekly earnings provided that not less than five shifts (day or night) are worked.

Apprentices/Trainees under 19

Six-month period	**Rate per week** £	**GMB** £
First	62.010	-
Second	83.460	-
Third	90.870	11.310
*Thereafter, until skills test is passed	108.810	13.550
On passing skills test and until completion of training period	122.850	15.405

*The NJCBI has agreed to the introduction of skills testing under the National Joint Training Scheme. The first practical skills tests commenced in May 1986.

Apprentices/Trainees over 19

Six-month period	**Rate per week** £	**GMB** £
First and Second	105.885	13.260
Third and Fourth	112.125	14.040
Fifth and Sixth	118.170	14.820

BUILDING AND ALLIED TRADES
JOINT INDUSTRIAL COUNCIL

Authorized rates of wages in the building industry in England and Wales agreed by the Building & Allied Trades Joint Industrial Council

AND EFFECTIVE FROM 25th JUNE 1990

Subject to the conditions prescribed in the Working Rule Agreement the standard weekly rates of wages shall be as follows:

	£
Craft operatives	154.05
Adult general operatives	132.60

ROAD HAULAGE WORKERS EMPLOYED IN THE
BUILDING INDUSTRY

Authorized rates of pay for road haulage workers in the building industry recommended by the Builders Employers Confederation.

AND EFFECTIVE FROM 25th JUNE 1990

Employers
The Building Employers
Confederation,
82 New Cavendish Street,
London W1M 8AD
Telephone: 071-580 5588

Operatives
The Transport and General
Workers Union,
Transport House,
Smith Square,
Westminster,
London SW1P 3JB
Telephone: 071-828 7788

DEFINITIONS OF GRADES
London
London Region as defined by the National Joint Council for the Building Industry.

Grade 1
The Grade A districts as defined by the National Joint Council for the Building Industry, including Liverpool and District (i.e., the area covered by the Liverpool Regional Federation of Building Trades Employers).

	LONDON per week £	**GRADE 1** per week £	**GMB** per week £
Drivers of vehicles of gross vehicle weight up to 3.5 tonnes	132.015	131.820	16.380
over 3.5 tonnes and up to and including 7.5 tonnes	132.405	132.210	16.380
over 7.5 tonnes and up to and including 10 tonnes	132.990	132.795	16.380
over 10 tonnes and up to and including 16 tonnes	133.380	133.185	16.380
over 16 tonnes and up to and including 24 tonnes	134.745	134.550	16.380
over 24 tonnes	135.525	135.330	16.380

ROAD HAULAGE WORKERS EMPLOYED IN THE BUILDING INDUSTRY

Subject to certain conditions workers covered by this agreement are entitled to additonal payments corresponding to the NJCBI's Guaranteed Minimum Bonus.
The gross vehicle weight in every case is to include, where applicable, the unladen weight of a trailer designed to carry goods.

BUILDING INDUSTRY - ISLE OF MAN

Authorized rates of wages in the building industry in the Isle of Man agreed by the Isle of Man Joint Industrial Council for the Building Industry

AND EFFECTIVE FROM 25th JUNE 1990

SECRETARIES

Employers
Mr Parnell,
Herondene,
Crossag Road,
Ballasalla,
Isle of Man
Telephone: 0624 824454

Operatives
Mr Moffat,
Transport House,
Prospect Hill,
Douglas,
Isle of Man
Telephone: 0624 21156

	Standard basic Wage rate per 39-hour week	Guaranteed minimum bonus payment	Guaranteed minimum weekly earnings
	£	£	£
Craft operatives	131.820	16.380	148.20
Labourers	112.320	14.040	126.36

PLUMBING MECHANICAL ENGINEERING SERVICES INDUSTRY

Authorized rates of wages agreed by the Joint Industry Board for the Plumbing Mechanical Engineering Services Industry in England and Wales

The Joint Industry Board for Plumbing Mechanical
Engineering Services in England and Wales,
Brook House, Brook Street,
St Neots,
Huntingdon, Cambs PE19 2HW
Telephone: 0480 76925-8

Rates of pay effective from 2nd APRIL 1990

	Per hour
	£
Technical plumber	5.29
Advanced plumber	4.66
Trained plumber	4.22
Apprentices	
1st year of training	1.22
2nd year of training	1.82
3rd year of training	2.25
4th year of training	2.90

PART III

Prices for Measured Work

This part of the book contains the following sections:

Prices for Measured Work – Major Works

INTRODUCTION

The 'Prices for Measured Work - Major Works' are intended to apply to a project costing (excluding Preliminaries) about £1 650 000 in the outer London area and assume that reasonable quantities of all types of work are required. Similarly it has been necessary to assume that the size of the job warrants the sub-letting of all types of work normally sub-let.

The distinction between builders' work and work normally sub-let is stressed because prices for work which can be sub-let may well be quite inadequate for the contractor who is called upon to carry out relatively small quantities of such work himself

As explained in more detail later the prices are generally based on wage rates which came into force in June 1990, material costs as stated and include an allowance for overheads and profit. They do not allow for preliminary items which are dealt with under a separate heading (see page 167) or for any Value Added Tax which will be payable on non-domestic buildings after 1st April 1989.

The format of this section is so arranged that, in the case of work normally undertaken by the Main Contractor, the constituent parts of the total rate are shown enabling the reader to make such adjustments as may be required in particular circumstances.

Similar details have also been given for work normally sub-let although it has not been possible to provide this in all instances.

As explained in the Preface, there is now a facility available to readers which enables a comparison to be made between the level of prices in this section and current tenders by means of a tender index.

The tender index for this Major Works section of Spon's is 330 (as shown on the front cover) which coincides with our forecast tender price level index for the outer London region applicable to the fourth quarter of 1990

To adjust prices for other regions/times, the reader is recommended to refer to the explanations and examples on how to apply these tender indexes, given on pages 769-771.

There follow explanations and definitions of the basis of costs in the 'Prices for Measured Work' section under the following headings:

Overhead charges and profit
Labour hours and Labour £ column
Material £ column
Material/Plant £ columns
Total rate £ column

OVERHEAD CHARGES AND PROFIT

For those items where detailed breakdowns have been given an allowance of 5% has been added to labour, plant and material costs for overhead charges and profit.

In other cases 2.5% has been included for the General Contractor's attendance, overhead charges and profit.

LABOUR HOURS AND LABOUR £ COLUMNS

'Labour rates are based upon typical gang costs divided by the number of primary working operatives for the trade concerned, and for general building work include an allowance for trade supervision (see below). 'Labour hours' multiplied by 'Labour rate' with the appropriate addition for overhead charges and profit gives 'Labour £'. In some instances, due to variations in gangs used, 'Labour rate' figures have not been indicated, but can be calculated by dividing 'Labour £' by 'Labour hours'. For building operatives the rates used are as set out below and allow for a bonus of plus 25% (in lieu of Guaranteed Minimum Bonus), which is considered to be representative of current labour costs in the outer London area.

Alternative labour rates are also given based upon bonuses of other values.

Building craft operatives and labourers

From 25th June 1990 guaranteed minimum weekly earnings in the London area for craft operatives and labourers are £148.395 and £126.555 respectively; to these rates have been added allowances for the items below in accordance with the recommended procedure of the Institute of Building in its 'Code of Estimating Practice'. The resultant hourly rates on which the 'Prices for Measured Work' have generally been based are £5.97 and £5.12 for craft operatives and labourers, respectively.

The items referred to above for which allowances have been made are:

25% bonus (in lieu of guaranteed minimum bonus)
Lost time
Non-productive overtime
Extra payments under National Working Rule 3, 17 and 18
Construction Industry Training Board Levy
Holidays with pay
Pension and death benefit scheme
Sick pay
National Insurance
Severance pay and sundry costs
Employers' liability and third party insurance

NOTE: For travelling allowances and site supervision see 'Preliminaries' section.

The table which follows shows how the 'all-in' hourly rates referred to above have been calculated. Productive time has been based on a total of 2063 hours worked per year.

		Craft operatives		Labourers	
		£	£	£	£
Wages at standard basic rate					
productive time	52.88 wks	132.02	6981.22	112.52	5950.06
Lost time allowance	1.06 wks	132.02	139.94	112.52	119.27
Non-productive overtime	4.43 wks	132.02	584.85	112.52	498.46
			-------		-------
			7706.01		6567.79
Allowance for bonus	Plus	25%	1926.50	25%	1641.95
Extra payments under					
National Working Rule 3	46.80 wks	1.82	85.18	3.43	160.52
Sick pay	1 wk	-	-	-	-
CITB levy	1 year	-	75.00	-	18.00
Public holiday pay	1.60 wks	148.40	237.43	126.56	202.50
			-------		-------
		C/F	10030.12		8590.76

		Craft operatives		Labourers	
		£	£	£	£
		B/F	10030.12		8590.76
Employer's contribution to:					
annual holiday pay scheme	47 wks	16.15	759.05	16.15	759.05
pension and death benefit scheme	47 wks	1.55	72.85	1.55	72.85
National Insurance	47.80 wks	21.76	1040.13	16.14	771.49
			11902.15		10194.15
Severance pay and sundry costs	Plus	1.5%	178.53	1.5%	152.91
			12080.68		10347.06
Employer's liability and third-party insurance	Plus	2%	241.61	2%	206.94
Total cost per annum			12322.29		10554.00
Total cost per hour			5.97		5.12

NOTES:

1. Absence due to sickness has been assumed to be for individual days when no payment would be due.
2. The annual holiday pay pension and death benefit scheme contributions are those which came into force on 30th July 1990.
3. Employers' contribution to the holiday stamp assumes that the Contractor makes use of the special credit arrangements on buying the stamps and does not therefore receive the discount from the Management Company.

From these rates have been calculated the 'labour rate' applicable to each trade which generally include an allowance for supervision by a foreman or ganger and are based on a bonus rate plus 25%.

Gang	Gang rate £ (bonus + 25%)	Labour(man)rate £/hour (bonus + 25%)	Alternative labour rates £/hour GMB	+20%	+30%	+40%
Groundwork gang						
1 Ganger	1 x 5.62 = 5.62					
6 Labourers	6 x 5.12 = 30.72					
	36.34 / 6.5	= 5.59	5.03	5.38	5.79	6.24
Concreting gang						
1 Ganger	1 x 5.62 = 5.62					
4 Labourers	4 x 5.12 = 20.48					
	26.10 / 4.5	= 5.80	5.21	5.58	6.00	6.47
Steelfixing gang						
1 Foreman	1 x 6.42 = 6.42					
4 Steelfixers	4 x 5.97 = 23.88					
	30.30 / 4.5	= 6.73	6.00	6.49	6.98	7.46
Formwork gang						
1 Foreman	1 x 6.42 = 6.42					
10 Carpenters	10 x 5.97 = 59.70					
1 Labourer	1 x 5.12 = 5.12					
	71.24 / 10.5	= 6.78	6.05	6.54	7.03	7.52

Gang	Gang rate £ (bonus + 25%)	Labour(man)rate £/hour (bonus + 25%)	Alternative labour rates £/hour GMB	+20%	+30%	+40%
Bricklaying/Ltwt blockwork gang						
1 Foreman	1 x 6.42 = 6.42					
6 Bricklayers	6 x 5.97 = 35.82					
4 Labourers	4 x 5.12 = 20.48					
	62.72	/ 6.5 = 9.65	8.64	9.31	10.00	10.72
Dense blockwork gang						
1 Foreman	1 x 6.42 = 6.42					
6 Bricklayers	6 x 5.97 = 35.82					
6 Labourers	6 x 5.12 = 30.72					
	72.96	/ 6.5 = 11.22	10.06	10.83	11.63	12.48
Carpentry/joinery gang						
1 Foreman	1 x 6.42 = 6.42					
5 Carpenters	5 x 5.97 = 29.85					
1 Labourer	1 x 5.12 = 5.12					
	41.39	/ 5.5 = 7.53	6.72	7.26	7.80	8.34
Craft Operatives only						
1 Craft operative	1 x 5.97 = 5.97	/ 1 = 5.97	5.32	5.76	6.19	6.61
Craft operatives/Labourer pair						
1 Craft operative	1 x 5.97 = 5.97					
1 Labourer	1 x 5.12 = 5.12					
	11.09	/ 1 = 11.09	9.94	10.70	11.49	12.33
Small labouring gang (making good)						
1 Ganger	1 x 5.62 = 5.62					
4 Labourers	4 x 5.12 = 20.48					
	26.10	/ 4.5 = 5.80	5.21	5.58	6.00	6.47
2 Craft Operatives/1 Labourer						
2 Craft Operatives	2 x 5.97 = 11.94					
1 Labourer	1 x 5.12 = 5.12					
	17.06	/ 2 = 8.53	7.63	8.23	8.84	9.47
Drain laying gang/clayware						
2 Labourers	2 x 5.23 = 10.46	/ 2 = 5.23	4.73	5.05	5.41	5.83

Sub-Contractor's operatives

Similar labour rates are shown in respect of sub-let trades where applicable.

Plumbing operatives

From 2nd April 1990 the hourly earnings for technical and trained plumbers are £5.29 and £4.22, respectively; to these rates have been added allowances similar to those added for building operatives (see below). The resultant average hourly rate on which the 'Prices for Measured Work' have been based is £8.38

The items referred to above for which allowance has been made are:

25% Bonus
Non-productive overtime
Tool allowance
Plumbers' welding supplement

Construction Industry Training Board Levy
Holidays with pay
Pension and welfare stamp
National Insurance 'contracted out' contributions
Severance pay and sundry costs
Employer's liability and third party insurance

No allowance has been made for supervision as we have assumed the use of a team of technical or trained plumbers who are able to undertake such relatively straightforward plumbing works, e.g. on housing schemes, without further supervision.

The table which follows shows how the average hourly rate referred to above has been calculated. Productive time has been based on a total of 2034 hours worked per year.

		Technical	plumber	Trained	plumber
		£	£	£	£
Wages at standard basic rate					
productive time	2034 hrs	5.29	10 759.86	4.22	8 583.48
Non-productive overtime	113 hrs	5.29	597.77	4.22	476.86
Plumbers' welding supplement	2034 hrs	0.40	813.60	-	-
			12 171.23		9 060.34
Allowance for bonus	Plus	25%	3 042.81	25%	2 265.09
Tool allowance	45.20 wks	2.00	90.40	2.00	90.40
CITB levy	1 year	-	110.00	-	110.00
Public holiday pay	60 hrs	5.29	317.40	4.22	253.20
Employer's contributions to holiday credit/welfare stamps	47 wks	20.10	944.70	16.48	774.56
pension (6.50% of earnings)	47.8 wks	21.24	1 015.27	15.87	758.59
National Insurance	47.8 wks	22.60	1 080.28	17.98	859.44
			18 772.09		14 171.62
Severance pay and sundry costs	Plus	1.5%	281.58	1.5%	212.57
			19 053.67		14 384.19
Employers' liability and third-party insurance	Plus	2%	381.07	2%	287.68
Total cost per annum			19 434.74		14 671.87
Total cost per hour			9.55		7.21
Average all-in rate per hour			£8.38 per hour		

MATERIAL £ COLUMN

Many items have reference to a 'PC' value. This indicates the prime cost of the major material delivered to site in the outer London area for large quantities, assuming maximum discounts.

When obtaining material prices from other sources, it is important to identify any discount that may apply. Some manufacturers only offer 5 to 10% discount for the largest of orders; or 'firm' orders (as distinct from quotations).

For other materials, discounts of 30% to 40% may be obtainable, depending on value of order and any preferential position of the purchaser.

The 'Material £' column indicates the total cost of the material, based upon the PC, including delivery, waste, sundry materials and an allowance for overhead charges and profit for the unit of work concerned. Alternative material prices are given at the beginning of many sections, by means of which alternative 'Total rate £' prices may be calculated.

All material prices quoted are exclusive of Value Added Tax.

MATERIAL PLANT £ COLUMN

Plant costs have been based on current weekly hire charges and estimated weekly cost of oil, grease, ropes (where necessary), site servicing and cartage charges.

The total amount is divided by 30 (assuming 25% idle time) to arrive at a cost per working hour of plant. To this hourly rate is added one hour fuel consumption and one hour for each of the following operators; the rate to be calculated in accordance with the principles set out earlier in this section, i.e. with an allowance for plus rates, etc.

For convenience the all-in rates per hour used in the calculations of 'Prices for Measured Work' are shown below and where included in Material/Plant £ column will include the appropriate addition allowance of 5% for overhead charges and profit for Major Works and 7.5% for overhead charges and profit for Minor Works.

Plant	Labour	'All-in' rate Per hour £
Excavator (4 wheeled - 0.76 m3 shovel, 0.24 m3 bucket)	Driver	20.00
Excavator (JCB 3C - 0.24 m3 bucket)	Driver	17.00
Excavator (JCB 3C off centre - 0.24 m3 bucket)	Driver	19.00
Excavator (Hitachi EX120 - 0.53 m3 bucket)	Driver	22.00
Dumper (2.30 m3)	Driver	13.50
Two tool portable compressor (125 cfm)		
per breaking tool	*	2.10
per 'jumping jack' rammer	*	2.50
Roller		
0.75 tonnes	Driver	8.60
6-8 tonnes	Driver	12.00
Concrete mixer 10/7	Operator	8.75
Kango heavy duty hammer	-	1.05
Power float	-	2.35
Light percussion drill	-	0.45

* Operation of compressor by tool operator

TOTAL RATE £ COLUMN

'Total rate £' column is the sum of 'Labour £' and 'Material £' columns and therefore includes the allowances for overhead charges and profit previously described.

This column excludes any allowance for 'Preliminaries' which must be taken into account if one is concerned with the total cost of work.

The example of 'Preliminaries' in the following section indicates that in the absence of detailed calculations at least 12.5% must be added to all prices for measured work to arrive at total cost.

A PRELIMINARIES

The number of items priced in the 'Preliminaries' section of Bills of Quantities and the manner in which they are priced vary considerably between Contractors. Some Contractors, by modifying their percentage factor for overheads and profit, attempt to cover the costs of 'Preliminary' items in their 'Prices for Measured Work'. However, the cost of 'Preliminaries' will vary widely according to job size and complexity, site location, accessibility, degree of mechanization practicable, position of the Contractor's head office and relationships with local labour/domestic Sub-Contractors. It is therefore usually far safer to price 'Preliminary' items separately on their merits according to the job.

In amending the Preliminaries/General Conditions section for SMM7, the Joint Committee stressed that the preliminaries section of a bill should contain two types of cost significant item:

1. Items which are not specific to work sections but which have an identifiable cost which is useful to consider separately in tendering e.g. contractual requirements for insurances, site facilities for the employer's representative and payments to the local authority.
2. Items for fixed and time-related costs which derive from the contractor's expected method of carrying out the work, e.g. bringing plant to and from site, providing temporary works and supervision.

A fixed charge is for work the cost of which is to be considered as independent of duration. A time related charge is for work the cost of which is to be considered as dependent on duration.

The fixed and time-related subdivision given for a number of preliminaries items will enable tenderers to price the elements separately should they so desire. Tenderers also have the facility at their discretion to extend the list of fixed and time-related cost items to suit their particular methods of construction.

The opportunity for Tenderers to price fixed and time-related items in A30-A37, A40-A44 and A51-A52 have been noted against the following appropriate items although we have not always provided guidance as costs can only be assessed in the light of circumstances of a particular job.

Works of a temporary nature are deemed to include rates, fees and charges related thereto in Sections A36, A41, A42, and A44, all of which will probably be dealt with as fixed charges.

In addition to the cost significant items required by the method, other preliminaries items which are important from other points of view, e.g. quality control requirements, administrative procedures, may need to be included to complete the Preliminaries/General conditions as a comprehensive statement of the employer's requirements.

Typical clause descriptions from a 'Preliminaries/General Conditions' section are given below together with details of those items that are usually priced in detail in tenders. As SMM7 only becomes operative on 1 July 1988, it may be that Tenderers may adopt a different approach to pricing these sections; possibly necessitating further revisions to this example in the light of their response.

An example in pricing 'Preliminaries' follows, and this assumes the form of contract used is the Standard Form of Building Contract 1980 Edition and the value, excluding 'Preliminaries', is £1 650 000. The contract is estimated to take 80 weeks to complete and the value is built up as follows:

	£
Labour value	610 000
Material value	500 000
Provisional sums and all Sub-Contractors	540 000
	£1 650 000

At the end of the section the examples are summarized to give a total value of 'Preliminaries' for the project.

A PRELIMINARIES/GENERAL CONDITIONS

Preliminary particulars

A10 Project particulars - Not priced

A11 Drawings - Not priced

A12 The site/Existing buildings - Generally not priced

The reference to existing buildings relates only to those buildings which could have an influence on cost. This could arise from their close proximity making access difficult, their heights relative to the possible use of tower cranes or the fragility of, for example, an historic building, necessitating special care.

A13 Description of the work - Generally not priced

A20 The Contract/Sub-contract

(The Standard Form of Building Contract 1980 Edition is assumed)

Clause no

1. **Interpretation, definitions, etc.** - Not priced
2. **Contractor's obligations** - Not priced
3. **Contract Sum - adjustment - Interim certificates** - Not priced
4. **Architect's Instructions** - Not priced
5. **Contract documents - other documents - issue of certificates** The contract conditions may require a master programme to be prepared. This will normally form part of head office overheads and therefore is 'Not priced' separately here.
6. **Statutory obligations, notices, fees and charges** - Not priced. Unless the Contractor is specifically instructed to allow for these items.
7. **Level and setting out of the works** - Not priced
8. **Materials, goods and workmanship to conform to description, testing and inspection** - Not priced
9. **Royalties and patent rights** - Not priced

NOTE: The term 'Not priced' or 'Generally not priced' where used throughout this section means either that the cost implication is negligible or that is is usually included elsewhere in the tender.

A PRELIMINARIES

10. **Person-in-charge**

Under this heading are usually priced any staff that will be required on site. The staff required will vary considerably according to the size, layout and complexity of the scheme, from one foreman-in-charge to a site agent, general foreman, assistants, checkers and storemen etc.

The costs included for such people should include not only their wages, but their total cost to the site including statutory payments, pension, expenses, holiday relief, overtime, etc.

Part of the foreman's time, together with that of an assistant will be spent on setting out the site. Allow say £2. 00 per day for levels, staff, pegs and strings, plus the assistant's time if not part of the general management team. Most sites usually include for one operative to clean up generally and do odd jobs around the site.

Cost of other staff, such as buyers, and quantity surveyors, are usually part of head office overhead costs, but alternatively, may now be covered under A40 Management and staff.

A typical build-up of a foreman's costs might be:

	£
Annual salary	16 500.00
Expenses	1 500.00
Bonus - say	4 500.00
Employer's National Insurance contribution on salary and bonus (10.75% on £21 000.00) say	2 250.00
Training levy	75.00
Pension scheme (say 6% of salary and bonus)	1 260.00
Sundries, including Employer's Liability and Third Party (say 3.5% of salary and bonus)	735.00
	£26 820.00
Divide by 47 to allow for holidays:	
Per week	570.63
Say	£570.00

Corresponding costs for other site staff should be calculated in a similar manner.

Example

Site administration

	£
General foreman 80 weeks @ £570	45 600.00
Holiday relief 4 weeks @ £570	2 280.00
Assistant foreman 60 weeks @ £320	19 200.00
Storeman/checker 60 weeks @ £230	13 800.00
	80 880.00
Add 5% for overheads and profit - say	4 020.00
	£84 900.00

A PRELIMINARIES

A20 The Contract/Sub-contract - cont'd.

11. **Access for Architect to the Works** - Not priced
12. **Clerk of Works** - Not priced
13. **Variations and provisional sums** - Not priced
14. **Contract Sum** -Not priced
15. **Value Added Tax - supplemental provisions**
Major changes to the VAT status of supplies of goods and services by Contractors came into effect on 1 April 1989. It is clear that on and from that date the majority of work supplied under contracts on JCT Forms will be chargeable on the Contractor at the standard rate of tax. In April 1989, the JCT issued Amendment 8 and a guidance note dealing with the amendment to VAT provisions. This involves a revision to clause 15 and the Supplemental Provisions (the VAT agreement). Although the standard rating of most supplies should reduce the amount of VAT analysing previously undertaken by contractors, he should still allow for any incidental costs and expenses which may be incurred.
16. **Materials and goods unfixed or off-site** - Not priced
17. **Practical completion and Defects Liability**
Inevitably some defects will arise after practical completion and an allowance will often be made to cover this. An allowance of say 0.25 to 0.5% should be sufficient.

Example

Defects after completion	£
Based on £0.25% of the contract sum	
£1 650 000 @ £0.25% - say	4 125.00
Add 5% for overheads and profit - say	205.00

	£4 330.00
	--------

18. **Partial possession by Employer** - Not priced
19. **Assignment and Sub-Contracts** - Not priced
19A. **Fair wages** - Not priced
20. **Injury to persons and property and Employers indemnity**
(See Clause no 21)
21. **Insurance against injury to persons and property**
The Contractor's Employer's Liability and Public Liability policies (which would both be involved under this heading) are often in the region of 0.5 to 0.6% on the value of his own contract work (excluding provisional sums and work by Sub-Contractors whose prices should allow for these insurances). However, this allowance can be included in the all-in hourly rate used in the calculation of 'Prices for Measured Work' (see page 146).
Under Clause 21.2 no requirement is made upon the Contractor to insure as stated by the clause unless a provisional sum is allowed in the Contract Bills.
22. **Insurance of the works against Clause 22 Perils**
If at the Contractor's risk the insurance cover must be sufficient to include the full cost of reinstatement, all increases in cost, professional fees and any consequential costs such as demolition. The average provision for fire risk is £0.10% of the value of the work after adding for increased costs and professional fees.

A PRELIMINARIES

Example	£
Contractor's Liability - Insurance of works against fire, etc.	
Contract value (including 'Preliminaries'), say	1 855 000.00
Estimated increased costs during contract period, say 4%	74 200.00
	1 929 200.00
Estimated increased costs incurred during period of reinstatement, say an average of 4%	77 200.00
Fees @ 16%- say	321 000.00
	£2 327 400.00
Allow £0.10% on say £2 327 400.00 plus 5% for overheads and profit - say	£2 431.00

NOTE: Insurance premiums are liable to considerable variation, depending on the Contractor, the nature of the work and the market in which the insurance is placed.

23. **Date of possession, completion and postponement** - Not priced
24. **Damages for non-completion** - Not priced
25. **Extension of time** - Not priced
26. **Loss and expense caused by matters materially affecting regular progress of the works** - Not priced
27. **Determination by Employer** - Not priced
28. **Determination by Contractor** -Not priced
29. **Works by Employer or persons employed or engaged by Employer** - Not priced
30. **Certificates and payments** - Not priced
31. **Finance (No.2) Act 1975 - Statutory tax deduction scheme** - Not priced
32. **Outbreak of hostilities** -Not priced
33. **War damage** - Not priced
34. **Antiquities** - Not priced
35. **Nominated Sub-Contractors**
    Not priced here. An amount should be added to the relevant PC sums, if required, for profit and a further sum for special attendance.
36. **Nominated Suppliers**
    Not priced here. An amount should be added to the relevant PC sums, if required, for profit.
37. **Choice of fluctuation provisions - entry in Appendix**
    The amount which the Contractor may recover under the fluctuations clauses (Clause nos 38, 39 and 40) will vary depending on whether the Contract is 'firm', i.e. Clause no 38 is included, or 'fluctuating', whether the traditional method of assessment is used, i.e. Clause no 39 or the formula method, i.e. Clause no 40.
    An allowance should be made for any shortfall in reimbursement under fluctuating contracts.
38. **Contribution Levy and Tax Fluctuations** (see Clause no 37)
39. **Labour and Materials Cost and Tax Fluctuations** (see Clause no 37)
40. **Use of Price Adjustment Formulae** (see Clause no 37)

A PRELIMINARIES

Details should include special conditions or amendments to standard conditions, the Appendix insertions and the Employer's insurance responsibilities.

Additional obligations may include the provision of a performance bond. If the Contractor is required to provide sureties for the fulfilment of the work the usual method of providing this is by a bond provided by one or more insurance companies. The cost of a performance bond depends largely on the financial standing of the applying Contractor. Figures tend to range from £0.25 to £0.50% of the contract sum.

A30-A37 EMPLOYERS' REQUIREMENTS

These include the following items but costs can only be assessed in the light of circumstances on a particular job.

Details should be given for each item and the opportunity for the Tenderer to separately price items related to fixed charges and time related charges.

A30 Tendering/Sub-letting/Supply

A31 Provision, content and use of documents

A32 Management of the works

A33 Quality standards/control

A34 Security/Safety/Protection

This includes noise and pollution control, maintaining adjoining buildings, public and private roads, live services, security and the protection of work in all sections.

(i) Control of noise, pollution and other obligations

The Local Authority, Landlord or Management Company may impose restrictions on the timing of certain operations, particularly noisy of dust-producing operations, which may necessitate the carrying out of these works outside normal working hours or using special tools and equipment.

The situation is most likely to occur in built-up areas such as city centres, shopping malls etc., where the site is likely to be in close proximity to offices, commercial or residential property.

(ii) Maintenance of public and private roads

Some additional value or allowance may be required against this item to insure/protect against damage to entrance gates, kerbs or bridges caused by extraordinary traffic in the execution of the works.

A PRELIMINARIES

A35 Specific limitations on method/sequence/timing

This includes design constraints, method and sequence of work, access, possession and use of the site, use or disposal of materials found, start of work, working hours, employment of labour and sectional possession or partial possession etc.

A36 Facilities/Temporary work/Services

This includes offices, sanitary accommodation, temporary fences, hoardings, screens and roofs, name boards, technical and surveying equipment, temperature and humidity, telephone/facsimile installation and rental/maintenance, special lighting and other general requirements etc.

The attainment and maintenance of suitable levels necessary for satisfactory completion of the work including the installation of joinery, suspended ceilings, lift machinery etc. is the responsibility of the contractor.

The following is an example how to price a mobile office for a Clerk of Works

		£
(a)	Fixed charge	
	Haulage to and from site - say	60.00
	Add 5% for overheads and profit - say	3.00

		£63.00

(b)	Time related charge	
	Hire charge - 15 m2 x 76 weeks @ £2.32 m2	2 650.00
	Lighting, heating and attendance on office, say 76 weeks @ £22.50	1 710.00
	Rates on temporary building based on £10.00/m2 per annum	230.00
		--------
		4 590.00
Add 5% for overheads and profit - say		227.00

		£4 817.00

(c)	Combined charge	£4 880.00

The installation of telephones or facsimiles for the use of the Employer, and all related charges therewith, shall be given as a provisional sum.

A37 Operation/Maintenance of the finished building

A40-A44 CONTRACTORS GENERAL COST ITEMS

For items A41-A44 it shall be clearly indicated whether such items are to be 'Provided by the Contractor' or 'Made available (in any part) by the Employer'.

A PRELIMINARIES

A40 Management and staff (Provided by the Contractor)

NOTE: The cost of site administrative staff has previously been included against Clause no 10 of the Conditions of Contract, where Readers will find an example of management and staff costs.

When required allow for the provision of a watchman or inspection by a security organization.

Other general administrative staff costs, e.g. Engineering, Programming and production and Quantity Surveying could be priced as either fixed or time related charges, under this section.

For the purpose of this example - allow say 1% of contract value for other administrative staff costs.

		£
(a)	Time related charge	
	Based on 1% of £1 650 000 - say	16 500.00
	Add 5% for overheads and profit	825.00
		£17 325.00

A41 Site accommodation (Provided by the Contractor or made available by the Employer)

This includes all temporary offices laboratories, cabins, stores, compounds, canteens, sanitary facilities and the like for the Contractor's and his domestic sub-contractors' use (temporary office for a Clerk of Works is covered under obligations and restrictions imposed by the Employer).

Typical costs for mobile offices are as follows, based upon a twelve months minimum hire period they exclude furniture which could add a further £12.50 - £15.00 per week.

Size	Rate per week £
12ft x 7ft 6in (8.36 m2)	26.40 (£3.15 m2)
16ft x 7ft 6in (11.15 m2)	30.00 (£2.70 m2)
22ft x 7ft 6in (15.33 m2)	35.60 (£2.32 m2)
32ft x 9ft (26.75 m2)	47.80 (£1.78 m2)

Typical rates for timber huts are as follows:

Size	Rate per week £
6ft x 12ft (6.69 m2)	11.04 (£1.65 m2)
18ft x 12ft (20.07 m2)	19.32 (£0.96 m2)
24ft x 12ft (26.75 m2)	25.56 (£0.96 m2)
30ft x 12ft (33.44 m2)	31.80 (£0.95 m2)

A PRELIMINARIES

The following example is for one Foreman's office and two storage sheds.

		£
(a)	Fixed charge	
	Foreman's office - haulage to and from site - say	60.00
	Storage sheds - haulage to and from site - erection and dismantling - say	340.00
		400.00
	Add 5% for overheads and profit - say	20.00
		£420.00
(b)	Time related charge	
	Hire charge - Foreman's office - 15m2 x 76 weeks @ £2.32 m2	2 650.00
	Hire charge - 2 No. storage sheds - 60 m2 x 70 weeks @ £0.95 m2	3 990.00
		6 640.00
	Lighting, heating and attendance on office say, 76 weeks @ £22.50	1 710.00
	Rates on temporary buildings based on £10.00/m2 per annum	1 100.00
		9 450.00
	Add 5% for overheads and profit - say	470.00
		£ 9 920.00
(c)	Combined charge	£10 340.00

A42 Services and facilities (Provided by the Contractor or made available by the Employer)

This generally includes the provision of all of the Contractor's own services, power, lighting, fuels, water, telephone and administration, safety, health and welfare, storage of materials, rubbish disposal, cleaning, drying out, protection of work in all sections, security, maintaining public and private roads, small plant and tools and general attendance on nominated sub-contractors.

However, this section does not cover fuel for testing and commissioning permanent installations which would be measured under Sections Y51 and Y81.

Examples of build-ups/allowances for some of the major items are provided below:

(i) Lighting and power for the works

The Contractor is usually responsible for providing all temporary lighting and power for the works and all charges involved. On large sites this could be expensive and involve sub-stations and the like, but on smaller sites it is often limited to general lighting (depending upon time of year), power for power operated tools, a small diesel generator and some transformers.

Typical costs are:

Low voltage diesel generator	£52.00 - £85.00 per week
1.5 to 10 kVA transformer	£4.00 - £9.00 per week

A PRELIMINARIES

A42 Services and facilities - cont'd

A typical allowance, including charges, installation and fitting costs could be 1% of contract value.

Example
Lighting and power for the works
(c) Combined charge
The fixed charge would normally represent a proportion of the following allowance applicable to connection and supply charges. The residue would be allocated to time related charges.

	£
Dependant on the nature of the work, time of year and incidence of power operated tools, allow say, 1% on £1 650 000	16 500.00
Add 5% for overheads and profit	825.00
	£17 325.00

(ii) Water for the works
Charges should properly be ascertained from the local Water Authority. If these are not readily available, an allowance of £0.33% of the value of the contract is probably adequate, providing water can be obtained directly from the mains. Failing this, each case must be dealt with on its merits. In all cases an allowance should also be made for temporary plumbing including site storage of water if required.

Useful rates for temporary plumbing include:

Piping	£6.00 per metre
Connection	£150.00
Standpipe	£60.00

Plus an allowance for barrels and hoses.

Example
Water for the works
(c) Combined charge
The fixed charge would normally represent a proportion of the following allowance applicable to connection and supply charges. The residue would be allocated to time related charges.

£0.33% on £1 650 000	5 450.00
Temporary plumbing - say	450.00
	5 900.00
Add 5% for overheads and profit - say	300.00
	£6 200.00

A PRELIMINARIES

(iii) Temporary telephones for the use of the Contractor

Against this item should be included the cost of installation, rental and an assessment of the cost of calls made during the contract.

Installation costs	£126.74 (for more than 12 months rental)
Rental	£25.83 per quarter
Cost of calls	For sites with one telephone allow about £20.00 per week.

Example

Temporary telephones	£
(a) Fixed charge	
Connection charges	
for telephone	126.74
for outside bell	10.00

	136.74
Add 5% for overheads and profit - say	6.86

	£143.60

(b) Time related charge	
Rental	
for telephone 6 quarters @ £24.83	150.00
for outside bell 6 quarters @ £1.00	6.00
Calls - 76 weeks @ £20.00	1 520.00
	--------
	1 676.00
Add 5% for overheads and profit - say	84.40

	£1 760.40

(c) Combined charge	£1 904.00

(iv) Safety, health and welfare of workpeople

The Contractor is required to comply with the Code of Welfare Conditions for the Building Industry which sets out welfare requirements as follows:

1. Shelter from inclement weather
2. Accommodation for clothing
3. Accommodation and provision for meals
4. Provision of drinking water
5. Sanitary conveniences
6. Washing facilities
7. First aid
8. Site conditions

A PRELIMINARIES

(iv) Safety, health and welfare of workpeople - cont'd

A variety of self-contained mobile or jack-type units are available for hire and a selection of rates is given below:

	£
Kitchen with cooker, fridge, sink unit, water heater and basin 32ft x 9ft	£54.80 per week
Mess room with water heater, wash basin and seating	
16ft x 7ft 6in	£32.00 per week
16ft x 9ft	£36.00 per week
Welfare unit with drying rack, lockers, tables, seating, cooker, heater, sink and basin	
22ft x 7ft 6in	£46.00 per week
Toilets (mains type)	
Single unit	£14.15 per week
Two unit	£28.35 per week
Four unit	£43.60 per week
Add for wheels or jack mounting	£ 7.00 per week

Allowance must be made in addition for transport costs to and from site, setting up costs, connection to mains, fuel supplies and attendance

Site first aid kit £3.50 per week

A general provision to comply with the above code is often £0.50 to £0.75% of the contract value.

The costs of safety supervisors (required for firms employing more than 20 people) are usually part of head office overhead costs.

Example

Safety, health and welfare

(c) Combined charge

The fixed charge would normally represent a proportion of the following allowance, with the majority allocated to time related charges.

	£
Based on £0.66% of £1 650 000 - say	10 890.00
Add 5% for overheads and profit - say	545.00
	£11 435.00

(v) Removing rubbish, protective casings and coverings and cleaning the works on completion

This includes removing surplus materials and final cleaning of the site prior to handover.

Allow for sufficient 'bins' for the site throughout the contract duration and for some operatives time at the end of the contract for final clearing and cleaning ready for handover.

Cost of 'bins' - approx. £25.00 each

A general allowance of £0.20% of contract value is probably sufficient.

A PRELIMINARIES

Example
Removing rubbish, etc., and cleaning
(c) Combined charge
The fixed charge would normally represent an allowance for final clearing of the works on completion with the residue for cleaning
throughout the contract period

	£
Say £0.2% of contract value of £1 650 000	3 300.00
Add 5% for overheads and profit - say	165.00
	£3 465.00

(vi) Drying the works
Use or otherwise of an installed heating system will probably determine the value to be placed against this item.
Dependant upon the time of year, say allow 0.1% to 0.2% of the contract value to cover this item.

Example
Drying of the works £

(c) Combined charge
Generally this cost is likely to be related to a fixed charge only

Say 0.2% of contract value of £1 650 000	3 300.00
Add 5% for overheads and profit - say	165.00
	£3 465.00

(vii) Protecting the works from inclement weather
In areas likely to suffer particularly inclement weather, some nominal allowance should be included for tarpaulins, polythene sheeting, battening, etc., and the effect of any delays in concreting or brickwork by such weather.

(viii) Small plant and tools
Small plant and hand tools are usually assessed as between 0.5% and 1.5% of total labour value.

(c) Combined charge

	£
Say 1% of labour value of £610 000	6 100.00
Add 5% for overheads and profit	305.00
	£6 405.00

A PRELIMINARIES

(ix) General attendance on nominated sub-contractors

In the past this item was located after each PC sum. Under SMM7 it is intended that two composite items (one for fixed charges and the other for time related charges) should be provided in the preliminaries bill for general attendance on all nominated sub-contractors.

A43 Mechanical plant

This includes for cranes, hoists, personnel transport, transport, earthmoving plant, concrete plant, piling plant, paving and surfacing plant etc.

SMM6 required that items for protection or for plant be given in each section, whereas SMM7 provides for these items to be covered under A34, A42 and A43, as appropriate.

(i) Plant

Quite often, the Contractors own plant and plant employed by sub-contractors are included in with measured rates e.g. for earthmoving, concrete or piling plant, and the Editors have adopted this method of pricing where they believe it to be 'appropriate'.

As for other items of plant e.g. cranes, hoists, these tend to be used by a variety of trades. An example of such an item might be:

Example

Tower crane

- static 30 m radius - 4/5 tonne max. load

		£
(a)	Fixed charge	
	Haulage of crane to and from the site	
	erection, testing, commisioning & dismantling - say	5 000.00
	Add 5% for overheads and profit - say	250.00
		£5 250.00
(b)	Time related charge	
	Crane, on hire say 30 weeks @ 650.00 per week	19 500.00
	Electricity, fuel and oil - say	4 500.00
	Operator 30 weeks @ 40 hours @ £11.00	13 200.00
		37 200.00
	Add 5% for overheads and profit - say	1 850.00
		£39 050.00
(c)	Combined charge	£44 300.00

For the purpose of this example pricing of Preliminaries, the Editors have assumed that the costs of the above tower crane are included in with 'Measured Rates' items.

(ii) Personnel transport

The labour rates per hour on which 'Prices for Measured Work' have been based do not cover travel and lodging allowances which must be assessed according to the appropriate working rule agreement.

A PRELIMINARIES

Example
Personnel transport

Assuming all labour can be found within the London region, the labour value of £610 000 represents approximately 2200 man weeks. Assume each man receives an allowance of £1.40 per day or £7.00 per week of five days.

		£
(c)	Combined/time related charge	
	2200 man weeks at £7.00	15 400.00
	Add 5% for overheads and profit - say	770.00
		£16 170.00

A44 Temporary works (Provided by the Contractor or made available by the Employer)

This includes for temporary roads, temporary walkways, access scaffolding, support scaffolding and propping, hoardings, fans, fencing etc., hardstanding and traffic regulations etc.

The Contractor should include maintaining any temporary works in connection with the items, adapting, clearing away and making good, and all notices and fees to Local Authorities and public undertakings. On fluctuating contracts, i.e. where Clause no 39 or no 40 is incorporated there is no allowance for fluctuations in respect of plant and temporary works and in such instances allowances must be made for any increases likely to occur over the contract period.

Examples of build-ups/allowances for some items are provided below:

(i) Temporary roads, hardstandings, crossings and similar items

Quite often consolidated bases of eventual site roads are used throughout a contract to facilitate movement of materials around the site. However, during the initial setting up of a site, with drainage works outstanding, this is not always possible and occasionally temporary roadways have to be formed and ground levels later reinstated.

Typical costs are:

Removal of topsoil and provision of 225 mm stone/hardcore base blinded with ashes as a temporary roadway 3.50 m wide and subsequent reinstatement:

on level ground	£24.00 per metre
on sloping ground including 1 m of cut or fill	£30.00 per metre

Removal of topsoil and provision of 225 mm stone/hardcore base blinded with ashes as a temporary hardstanding £9.00 per m2

NOTE: Any allowance for special hardcore hardstandings for piling Sub-Contractors is usually priced against the 'special attendance' clause after the relevant Prime Cost sum.

A PRELIMINARIES

(ii) Scaffolding

The General Contractor's standing scaffolding is usually undertaken by specialist Sub-Contractors who will submit quotations based on the specific requirements of the works. It is not possible to give rates here for the various types of scaffolding that may be required but for the purposes of this section is is assumed that the cost of supplying, erecting, maintaining and subsequently dismantling the scaffolding required would amount to £11 000.00 inclusive of overheads and profit.

(iii) Temporary fencing, hoarding, screens, fans, planked footways, guardrails, gantries, and similar items.

This item must be considered in some detail as it is dependant on site perimeter, phasing of the work, work within existing buildings, etc.

Useful rates include:

Hoarding 2.3 m high of 18 mm plywood with 50 x 100 mm sawn softwood studding, rails and posts including later dismantling

undecorated	£30.00/m (£13.05/m^2)
decorated one side	£34.00/m (£14.87/m^2)
Pair of gates for hoarding	extra £150.00 per pair
Cleft Chestnut fencing 1.2 m high including dismantling	£7.00 m

Example

Temporary hoarding	£
(c) Combined/fixed charge	
Plywood decorated hoarding	
100 metres @ £34.00	3 400.00
Extra for one pair of gates	150.00
	3 550.00
Add 5% for overheads and profit - say	175.00
	£3 725.00

(iv) Traffic regulations

Waiting and unloading restrictions can occasionally add considerably to costs, resulting in forced overtime or additional weekend working. Any such restrictions must be carefully assessed for the job in hand.

A50 Work/Materials by the Employer

A description shall be given of works by others directly engaged by the Employer and any attendance that is required shall be priced in the same way as works by nominated sub-contractors.

A PRELIMINARIES

A51 Nominated sub-contractors

This section governs how nominated sub-contractors should be covered in the bills of quantities for main contracts. Bills of quantities used for inviting tenders from potential nominated sub-contractors should be drawn up in accordance with SMM7 as a whole as if the work was main contractor's work. This means, for example, that bills issued to potential nominated sub-contractors should include preliminaries and be accompanied by the drawings.

As much information as possible should be given in respect of nominated sub-contractors' work in order that tenderers can make due allowance when assessing the overall programme and establishing the contract period if not already laid down. A simple list of the component elements of the work might not be sufficient, but a list describing in addition the extent and possible value of each element would be more helpful. The location of the main plant e.g. whether in the basement or on the roof would clearly have a bearing on tenderers' programmes. It would be good practice to seek programme information when obtaining estimates from sub-contractors so that this can be incorporated in the bills of quantities, for the benefit of tenderers.

A percentage should be added for the main contractor's profit together with items for fixed and time related charges for special attendances required by nominated sub-contractors.

Special attendances to include scaffolding (additional to the Contractor's standing scaffolding), access roads, hardstandings, positioning (including unloading, distributing, hoisting or placing in position items of significant weight or size), storage, power, temperature and humidity etc.

A52 Nominated suppliers

Goods and materials which are required to be obtained from a nominated supplier shall be given as a prime cost sum to which should be added, if required, a percentage for profit.

A53 Work by statutory authorities

Works which are to be carried out by a Local Authority or statutory undertakings shall be given as provisional sums.

A54 Provisional work

One of the more significant revisions in SMM7 is the introduction of two types of provisional sum (defined and undefined work).

The new rules require that each sum for defined work should be accompanied in the bills of quantities by a description of the work sufficiently detailed for the tenderer to make allowance for its effect in the pricing of relevant preliminaries. The information should also enable the length of time required for execution of the work to be estimated and its position in the sequence of construction to be determined and incorporated into the programme.

Where Provisional Sums are given for undefined work the Contractor will be deemed not to have made any allowance in programming, planning and pricing preliminaries.

Any provision for Contingencies shall be given as an undefined provisional sum.

A PRELIMINARIES

A55 Dayworks

To include provisional sums for:

Labour
Materials and goods
Plant

SAMPLE SUMMARY

Item		£
A20.10	Site administration	84 900.00
A20.17	Defects after completion	4 330.00
A20.22	Insurance of the works against fire, etc.	2 431.00
A36	Clerk of Work's Office	4 880.00
A40	Additional management and staff	17 325.00
A41	Contractor's accommodation	10 340.00
A42(i)	Lighting and power for the works	17 325.00
A42(ii)	Water for the works	6 200.00
A42(iii)	Temporary telephones	1 904.00
A42(iv)	Safety, health and welfare	11 435.00
A42(v)	Removing rubbish, etc., and cleaning	3 465.00
A42(vi)	Drying the works	3 465.00
A42(viii)	Small plant and tools	6 405.00
A43(ii)	Personnel transport	16 170.00
A44(ii)	Scaffolding	11 000.00
A44(iii)	Temporary hoarding	3 725.00
	TOTAL	£205 300.00

It is emphasized that the above is an example only of the way in which 'Preliminaries' may be priced and it is essential that for any particular contract or project the items set out in 'Preliminaries' should be assessed on their respective values.

'Preliminaries' as a percentage of a total contract will vary considerably according to each scheme and each Contractors' estimating practice. A recent national study undertaken by the authors indicates that the current trend for preliminaries to be priced at, is approximately 12 to 15%.

The value of the 'Preliminaires' in the above example is about a 12.5% addition to the value of work; this fact should not be forgotten when using the rates given in the trades following this section.

D GROUNDWORK
Including overheads and profit at 5.00%

Prices are applicable to excavation in firm soil. Multiplying factors for other soils are as follows:-

	Mechanical	Hand
Clay	x 2.00	x 1.20
Compact gravel	x 3.00	x 1.50
Soft chalk	x 4.00	x 2.00
Hard rock	x 5.00	x 6.00
Running sand or silt	x 6.00	x 2.00

	Labour hours	Labour £	Material Plant £	Unit	Total rate £
D20 EXCAVATION AND FILLING					
Site preparation					
Removing trees					
600 mm - 1.50 m girth	20.00	117.39	-	nr	**117.39**
1.50 - 3.00 m girth	35.00	205.43	-	nr	**205.43**
over 3.00 m girth	50.00	293.48	-	nr	**293.48**
Removing tree stumps					
600 mm - 1.50 m girth	1.00	5.87	28.35	nr	**34.22**
1.50 - 3.00 m girth	1.50	8.80	42.53	nr	**51.33**
over 3.00 m girth	2.00	11.74	56.70	nr	**68.44**
Clearing site vegetation	0.04	0.23	-	m2	**0.23**
Lifting turf for preservation; stacking	0.35	2.05	-	m2	**2.05**
Mechanical excavation using a tracked hydraulic excavator with a 0.76 m3 shovel					
Excavating topsoil to be preserved					
average 150 mm deep	0.03	0.18	0.63	m2	**0.81**
Add or deduct for each 25 mm variation in depth	0.01	0.03	0.11	m2	**0.14**
Excavating to reduce levels; not exceeding					
0.25 m deep	0.09	0.53	1.89	m3	**2.42**
1 m deep	0.07	0.41	1.47	m3	**1.88**
2 m deep	0.09	0.53	1.89	m3	**2.42**
4 m deep	0.11	0.65	2.31	m3	**2.96**
Mechanical excavation using a tracked hydraulic excavator with a 0.53 m3 bucket					
Excavating basements and the like; starting from reduced level; not exceeding					
1 m deep	0.09	0.53	2.08	m3	**2.61**
2 m deep	0.07	0.41	1.62	m3	**2.03**
4 m deep	0.09	0.53	2.08	m3	**2.61**
6 m deep	0.10	0.59	2.31	m3	**2.90**
8 m deep	0.12	0.70	2.77	m3	**3.47**
Excavating pits to receive bases; not exceeding					
0.25 m deep	0.26	1.53	6.01	m3	**7.54**
1 m deep	0.23	1.35	5.31	m3	**6.66**
2 m deep	0.26	1.53	6.01	m3	**7.54**
4 m deep	0.28	1.64	6.47	m3	**8.11**
6 m deep	0.30	1.76	6.93	m3	**8.69**
Extra over pit excavation for commencing below ground or reduced level					
1 m below	0.03	0.18	0.69	m3	**0.87**
2 m below	0.04	0.23	0.92	m3	**1.15**
3 m below	0.05	0.29	1.16	m3	**1.45**
4 m below	0.07	0.41	1.62	m3	**2.03**

D GROUNDWORK Including overheads and profit at 5.00%	Labour hours	Labour £	Material Plant £	Unit	Total rate £
D20 EXCAVATION AND FILLING - cont'd					
Mechanical excavation using a tracked hydraulic excavator with a 0.53 m3 bucket - cont'd					
Excavating trenches; not exceeding 0.30 m wide; not exceeding					
0.25 m deep	0.22	1.29	5.08	m3	**6.37**
1 m deep	0.19	1.12	4.39	m3	**5.51**
2 m deep	0.22	1.29	5.08	m3	**6.37**
4 m deep	0.26	1.53	6.01	m3	**7.54**
6 m deep	0.30	1.76	6.93	m3	**8.69**
Excavating trenches; over 0.30 m wide; not exceeding					
0.25 m deep	0.20	1.17	4.62	m3	**5.79**
1 m deep	0.17	1.00	3.93	m3	**4.93**
2 m deep	0.20	1.17	4.62	m3	**5.79**
4 m deep	0.23	1.35	5.31	m3	**6.66**
6 m deep	0.28	1.64	6.47	m3	**8.11**
Extra over trench excavation for commencing below ground or reduced level					
1 m below	0.03	0.18	0.69	m3	**0.87**
2 m below	0.04	0.23	0.92	m3	**1.15**
3 m below	0.05	0.29	1.16	m3	**1.45**
4 m below	0.07	0.41	1.62	m3	**2.03**
Extra over trench excavation for curved trench	0.01	0.06	0.23	m3	**0.29**
Excavating for pile caps and ground beams between piles; not exceeding					
0.25 m deep	0.33	1.94	7.62	m3	**9.56**
1 m deep	0.30	1.76	6.93	m3	**8.69**
2 m deep	0.33	1.94	7.62	m3	**9.56**
Excavating to bench sloping ground to receive filling; not exceeding					
0.25 m deep	0.09	0.53	2.08	m3	**2.61**
1 m deep	0.07	0.41	1.62	m3	**2.03**
2 m deep	0.09	0.53	2.08	m3	**2.61**
Extra over any type of excavation at any depth for excavating					
alongside services	1.95	10.29		m3	**10.29**
around services crossing excavation	4.50	22.72		m3	**22.72**
below ground water level	0.10	0.59	2.31	m3	**2.90**
Extra over excavation for breaking out existing materials					
rock	2.30	13.50	10.71	m3	**24.21**
concrete	1.95	11.45	8.06	m3	**19.51**
reinforced concrete	2.80	16.43	11.81	m3	**28.24**
brickwork; blockwork or stonework	1.40	8.22	5.91	m3	**14.13**
Extra over excavation for breaking out existing hard pavings					
tarmacadam 75 mm thick	0.15	0.88	0.50	m2	**1.38**
tarmacadam and hardcore 150 mm thick	0.20	1.17	0.54	m2	**1.71**
concrete 150 mm thick	0.30	1.76	1.18	m2	**2.94**
reinforced concrete 150 mm thick	0.45	2.64	1.72	m2	**4.36**
Working space allowance to excavations					
reduced levels; basements and the like	0.08	0.47	1.85	m2	**2.32**
pits	0.17	1.00	3.93	m2	**4.93**
trenches	0.16	0.94	3.70	m2	**4.64**
pile caps and ground beams between piles	0.18	1.06	4.16	m2	**5.22**
Extra over excavation for working space for filling in with					
hardcore	0.10	0.59	5.20	m2	**5.79**
coarse ashes	0.10	0.59	6.30	m2	**6.89**
sand	0.10	0.59	11.85	m2	**12.44**

D GROUNDWORK Including overheads and profit at 5.00%	Labour hours	Labour £	Material Plant £	Unit	Total rate £
40 - 20 mm gravel	0.10	0.59	14.88	m2	**15.47**
in situ concrete - 7.50 N/mm2 - 40 mm aggregate (1:8)	0.72	4.38	26.28	m2	**30.66**
Hand excavation					
Excavating topsoil to be preserved					
average 150 mm deep	0.25	1.47	-	m2	**1.47**
Add or deduct for each 25 mm variation in depth	0.04	0.23	-	m2	**0.23**
Excavating to reduce levels; not exceeding					
0.25 m deep	1.80	10.57	-	m3	**10.57**
1 m deep	2.00	11.74	-	m3	**11.74**
2 m deep	2.20	12.91	-	m3	**12.91**
4 m deep	2.40	14.09	-	m3	**14.09**
Excavating basements and the like; starting from reduced level; not exceeding					
1 m deep	2.40	14.09	-	m3	**14.09**
2 m deep	2.70	15.85	-	m3	**15.85**
4 m deep	3.50	20.54	-	m3	**20.54**
6 m deep	4.30	25.24	-	m3	**25.24**
8 m deep	5.20	30.52	-	m3	**30.52**
Excavating pits to receive bases; not exceeding					
0.25 m deep	2.80	16.43	-	m3	**16.43**
1 m deep	3.00	17.61	-	m3	**17.61**
2 m deep	3.50	20.54	-	m3	**20.54**
4 m deep	4.50	26.41	-	m3	**26.41**
6 m deep	5.70	33.46	-	m3	**33.46**
Extra over pit excavation for commencing below ground or reduced level					
1 m below	0.50	2.93	-	m3	**2.93**
2 m below	1.00	5.87	-	m3	**5.87**
3 m below	1.50	8.80	-	m3	**8.80**
4 m below	2.00	11.74	-	m3	**11.74**
Excavating trenches; not exceeding 0.30 m wide; not exceeding					
0.25 m deep	2.50	14.67	-	m3	**14.67**
1 m deep	2.75	16.14	-	m3	**16.14**
2 m deep	3.20	18.78	-	m3	**18.78**
4 m deep	4.05	23.77	-	m3	**23.77**
6 m deep	5.15	30.23	-	m3	**30.23**
Excavating trenches; over 0.30 m wide; not exceeding					
0.25 m deep	2.30	13.50	-	m3	**13.50**
1 m deep	2.50	14.67	-	m3	**14.67**
2 m deep	2.90	17.02	-	m3	**17.02**
4 m deep	3.70	21.72	-	m3	**21.72**
6 m deep	4.70	27.59	-	m3	**27.59**
Extra over trench excavation for commencing below ground or reduced level					
1 m below	0.50	2.93	-	m3	**2.93**
2 m below	1.00	5.87	-	m3	**5.87**
3 m below	1.50	8.80	-	m3	**8.80**
4 m below	2.00	11.74	-	m3	**11.74**
Extra over trench excavation for curved trench	0.50	2.93	-	m3	**2.93**
Excavating for pile caps and ground beams between piles; not exceeding					
0.25 m deep	3.30	19.37	-	m3	**19.37**
1 m deep	3.60	21.13	-	m3	**21.13**
2 m deep	4.20	24.65	-	m3	**24.65**

D GROUNDWORK Including overheads and profit at 5.00%	Labour hours	Labour £	Material Plant £	Unit	Total rate £
D20 EXCAVATION AND FILLING - cont'd					
Hand excavation - cont'd					
Excavating to bench sloping ground to receive filling; not exceeding					
0.25 m deep	1.50	8.80	-	m3	**8.80**
1 m deep	1.80	10.57	-	m3	**10.57**
2 m deep	2.00	11.74	-	m3	**11.74**
Extra over any type of excavation at any depth for excavating					
alongside services	1.00	5.87	-	m3	**5.87**
around services crossing excavation	2.00	11.74	-	m3	**11.74**
below ground water level	0.35	2.05	-	m3	**2.05**
Extra over excavation for breaking out existing materials					
rock	5.00	29.35	6.62	m3	**35.97**
concrete	4.50	26.41	5.51	m3	**31.92**
reinforced concrete	6.00	35.22	7.72	m3	**42.94**
brickwork; blockwork or stonework	3.00	17.61	3.31	m3	**20.92**
Extra over excavation for breaking out existing hard pavings					
tarmacadam 75 mm thick	0.40	2.35	0.44	m2	**2.79**
tarmacadam and hardcore 150 mm thick	0.50	2.93	0.55	m2	**3.48**
concrete 150 mm thick	0.70	4.11	0.77	m2	**4.88**
reinforced concrete 150 mm thick	0.90	5.28	1.10	m2	**6.38**
Extra for taking up precast concrete paving slabs	0.30	1.76	-	m2	**1.76**
Working space allowance to excavations					
reduced levels; basements and the like	2.60	15.26	-	m2	**15.26**
pits	2.70	15.85	-	m2	**15.85**
trenches	2.35	13.79	-	m2	**13.79**
pile caps and ground beams between piles	2.75	16.14	-	m2	**16.14**
Extra over excavation for working space for filling in with					
hardcore	0.80	4.70	3.94	m2	**8.64**
coarse ashes	0.72	4.23	5.04	m2	**9.27**
sand	0.80	4.70	10.59	m2	**15.29**
40 - 20 mm gravel	0.80	4.70	13.62	m2	**18.32**
in situ concrete - 7.50 N/mm2 - 40 mm aggregate (1:8)	1.10	6.70	25.02	m2	**31.72**
Earthwork support (average 'risk' prices)					
Not exceeding 2 m between opposing faces; not exceeding					
1 m deep	0.10	0.59	0.31	m2	**0.90**
2 m deep	0.13	0.76	0.35	m2	**1.11**
4 m deep	0.16	0.94	0.41	m2	**1.35**
6 m deep	0.19	1.12	0.48	m2	**1.60**
2 - 4 m between opposing faces; not exceeding					
1 m deep	0.11	0.65	0.35	m2	**1.00**
2 m deep	0.14	0.82	0.41	m2	**1.23**
4 m deep	0.17	1.00	0.48	m2	**1.48**
6 m deep	0.21	1.23	0.58	m2	**1.81**
Over 4 m between opposing faces; not exceeding					
1 m deep	0.12	0.70	0.41	m2	**1.11**
2 m deep	0.15	0.88	0.48	m2	**1.36**
4 m deep	0.19	1.12	0.58	m2	**1.70**
6 m deep	0.24	1.41	0.69	m2	**2.10**

D GROUNDWORK Including overheads and profit at 5.00%	Labour hours	Labour £	Material Plant £	Unit	Total rate £
Earthwork support (open boarded)					
Not exceeding 2 m between opposing faces; not exceeding					
1 m deep	0.30	1.76	0.96	m2	**2.72**
2 m deep	0.38	2.23	1.16	m2	**3.39**
4 m deep	0.48	2.82	1.45	m2	**4.27**
6 m deep	0.60	3.52	1.66	m2	**5.18**
2 - 4 m between opposing faces; not exceeding					
1 m deep	0.34	2.00	1.16	m2	**3.16**
2 m deep	0.42	2.47	1.47	m2	**3.94**
4 m deep	0.54	3.17	1.74	m2	**4.91**
6 m deep	0.66	3.87	2.05	m2	**5.92**
Over 4 m between opposing faces; not exceeding					
1 m deep	0.38	2.23	1.35	m2	**3.58**
2 m deep	0.48	2.82	1.70	m2	**4.52**
4 m deep	0.60	3.52	2.05	m2	**5.57**
6 m deep	0.76	4.46	2.35	m2	**6.81**
Earthwork support (close boarded)					
Not exceeding 2 m between opposing faces; not exceeding					
1 m deep	0.80	4.70	1.74	m2	**6.44**
2 m deep	1.00	5.87	2.12	m2	**7.99**
4 m deep	1.25	7.34	2.60	m2	**9.94**
6 m deep	1.56	9.16	3.09	m2	**12.25**
2 - 4 m between opposing faces; not exceeding					
1 m deep	0.88	5.17	1.93	m2	**7.10**
2 m deep	1.10	6.46	2.41	m2	**8.87**
4 m deep	1.25	7.34	2.89	m2	**10.23**
6 m deep	1.65	9.68	3.38	m2	**13.06**
Over 4 m between opposing faces; not exceeding					
1 m deep	0.97	5.69	2.12	m2	**7.81**
2 m deep	1.20	7.04	2.70	m2	**9.74**
4 m deep	1.40	8.22	3.18	m2	**11.40**
6 m deep	1.90	11.15	3.67	m2	**14.82**
Extra over earthwork support for					
extending into running silt or the like	0.50	2.93	0.58	m2	**3.51**
extending below ground water level	0.30	1.76	0.29	m2	**2.05**
support next to roadways	0.40	2.35	0.48	m2	**2.83**
curved support	0.03	0.18	0.33	m2	**0.51**
support left in	0.65	3.82	11.58	m2	**15.40**
Mechanical disposal of excavated materials					
Removing from site to tip not exceeding 13 km (using lorries)	-	-	-	m3	**7.87**
Depositing on site in spoil heaps					
average 25 m distant	-	-	-	m3	**0.63**
average 50 m distant	-	-	-	m3	**1.05**
average 100 m distant	-	-	-	m3	**1.68**
average 200 m distant	-	-	-	m3	**2.52**
Spreading on site					
average 25 m distant	0.20	1.17	0.63	m3	**1.80**
average 50 m distant	0.20	1.17	1.05	m3	**2.22**
average 100 m distant	0.20	1.17	1.68	m3	**2.85**
average 200 m distant	0.20	1.17	2.52	m3	**3.69**

D GROUNDWORK Including overheads and profit at 5.00%	Labour hours	Labour £	Material Plant £	Unit	Total rate £
D20 EXCAVATION AND FILLING - cont'd					
Hand disposal of excavated materials					
Removing from site to tip not exceeding 13 km (using lorries)	1.00	5.87	9.84	m3	**15.71**
Depositing on site in spoil heaps					
average 25 m distant	1.00	5.87	-	m3	**5.87**
average 50 m distant	1.30	7.63	-	m3	**7.63**
average 100 m distant	1.90	11.15	-	m3	**11.15**
average 200 m distant	2.80	16.43	-	m3	**16.43**
Spreading on site					
average 25 m distant	1.30	7.63	-	m3	**7.63**
average 50 m distant	1.60	9.39	-	m3	**9.39**
average 100 m distant	2.20	12.91	-	m3	**12.91**
average 200 m distant	3.10	18.20	-	m3	**18.20**
Mechanical filling with excavated material					
Filling to excavations	0.15	0.88	1.68	m3	**2.56**
Filling to make up levels over 250 mm thick; depositing; compacting	0.22	1.29	1.26	m3	**2.55**
Filling to make up levels not exceeding 250 mm thick; compacting	0.26	1.53	1.68	m3	**3.21**
Mechanical filling with imported soil					
Filling to make up levels over 250 mm thick; depositing; compacting in layers	0.22	1.29	11.76	m3	**13.05**
Filling to make up levels not exceeding 250 mm thick; compacting	0.26	1.53	15.12	m3	**16.65**
Mechanical filling with hardcore; PC £5.00/m3					
Filling to excavations	0.17	1.00	8.66	m3	**9.66**
Filling to make up levels over 250 mm thick; depositing; compacting	0.26	1.53	8.24	m3	**9.77**
Filling to make up levels not exceeding 250 mm thick; compacting	0.30	1.76	10.50	m3	**12.26**
Mechanical filling with granular fill type 1; PC £9.15/t (PC £15.55/m3)					
Filling to excavations	0.17	1.00	22.51	m3	**23.51**
Filling to make up levels over 250 mm thick; depositing; compacting	0.26	1.53	22.09	m3	**23.62**
Filling to make up levels not exceeding 250 mm thick; compacting	0.30	1.76	28.22	m3	**29.98**
Mechanical filling with granular fill type 2; PC £8.75/t (PC £15.00/m3)					
Filling to excavations	0.17	1.00	21.79	m3	**22.79**
Filling to make up levels over 250 mm thick; depositing; compacting	0.26	1.53	21.37	m3	**22.90**
Filling to make up levels not exceeding 250 mm thick; compacting	0.30	1.76	27.30	m3	**29.06**
Mechanical filling with coarse ashes; PC £6.00/m3					
Filling to make up levels over 250 mm thick; depositing; compacting	0.24	1.41	9.85	m3	**11.26**
Filling to make up levels not exceeding 250 mm thick; compacting	0.28	1.64	12.60	m3	**14.24**
Mechanical filling with sand; PC £8.40/t (PC £13.45/m3)					
Filling to make up levels over 250 mm thick; depositing; compacting	0.26	1.53	19.33	m3	**20.86**
Filling to make up levels not exceeding 250 mm thick; compacting	0.30	1.76	24.70	m3	**26.46**

D GROUNDWORK Including overheads and profit at 5.00%	Labour hours	Labour £	Material Plant £	Unit	Total rate £
Hand filling with excavated material					
Filling to excavations	1.00	5.87	-	m3	**5.87**
Filling to make up levels over 250 mm thick; depositing; compacting	1.10	6.46	-	m3	**6.46**
Filling to make up levels exceeding 250 mm thick; multiple handling via spoil heap average - 25 m distant	2.40	14.09	-	m3	**14.09**
Filling to make up levels not exceeding 250 mm thick; compacting	1.35	7.92	-	m3	**7.92**
Hand filling with imported soil					
Filling to make up levels over 250 mm thick; depositing; compacting in layers	1.10	6.46	10.50	m3	**16.96**
Filling to make up levels not exceeding 250 mm thick; compacting	1.35	7.92	13.44	m3	**21.36**
Hand filling with hardcore; PC £5.00/m3					
Filling to excavations	1.30	7.63	6.56	m3	**14.19**
Filling to make up levels over 250 mm thick; depositing; compacting	0.55	3.23	11.53	m3	**14.76**
Filling to make up levels not exceeding 250 mm thick; compacting	0.66	3.87	14.36	m3	**18.23**
Hand filling with granular fill type 1; PC £9.15/t (PC £15.55/m3)					
Filling to excavations	1.30	7.63	20.41	m3	**28.04**
Filling to make up levels over 250 mm thick; depositing; compacting	0.55	3.23	25.38	m3	**28.61**
Filling to make up levels not exceeding 250 mm thick; compacting	0.66	3.87	32.08	m3	**35.95**
Hand filling with granular fill type 2; PC £8.75/t (PC £15.00/m3)					
Filling to excavations	1.30	7.63	19.69	m3	**27.32**
Filling to make up levels over 250 mm thick; depositing; compacting	0.55	3.23	24.65	m3	**27.88**
Filling to make up levels not exceeding 250 mm thick; compacting	0.66	3.87	31.16	m3	**35.03**
Hand filling with coarse ashes; PC £6.00/m3					
Filling to make up levels over 250 mm thick; depositing; compacting	0.50	2.93	12.89	m3	**15.82**
Filling to make up levels not exceeding 250 mm thick; compacting	0.60	3.52	16.13	m3	**19.65**
Hand filling with sand; PC £8.40/t (PC £13.45/m3)					
Filling to excavations	0.50	2.93	22.17	m3	**25.10**
Filling to make up levels over 250 mm thick; depositing; compacting	0.65	3.82	23.52	m3	**27.34**
Filling to make up levels not exceeding 250 mm thick; compacting	0.77	4.52	29.55	m3	**34.07**
Surface packing to filling					
To vertical or battered faces	0.18	1.06	0.47	m2	**1.53**
Surface treatments					
Compacting					
surfaces of ashes	0.04	0.23	-	m2	**0.23**
filling; blinding with ashes	0.08	0.47	0.46	m2	**0.93**
filling; blinding with sand	0.08	0.47	1.07	m2	**1.54**
bottoms of excavations	0.05	0.29	-	m2	**0.29**

D GROUNDWORK Including overheads and profit at 5.00%	Labour hours	Labour £	Material Plant £	Unit	Total rate £
D20 EXCAVATION AND FILLING - cont'd					
Surface treatments - cont'd					
Trimming					
sloping surfaces	0.18	1.06	-	m2	**1.06**
sloping surfaces; in rock	1.00	5.38	2.21	m2	**7.59**
Filter membrane; one layer; laid on earth to receive granular material					
'Terram 500'	0.05	0.29	0.39	m2	**0.68**
'Terram 700'	0.05	0.29	0.39	m2	**0.68**
'Terram 1000'	0.06	0.35	0.53	m2	**0.88**

D30 CAST IN PLACE PILING

The following approximate prices for the quantities of piling quoted, are for work on clear open sites with reasonable access. They are based on 500 mm nominal diameter piles normal concrete mix 20 N/mm2 reinforced for loading up to 40 000 kg depending on ground conditions and include up to 0.16 m of projecting reinforcement at top of pile. The prices do not allow for removal of spoil.

	Labour hours	Labour £	Material Plant £	Unit	Total rate £
Tripod bored cast-in-place concrete piles					
Provision of all plant; including bringing to and removing from site; maintenance, erection and dismantling at each pile position for 100 nr piles	-	-	-	item	**8000.00**
500 mm diameter piles; reinforced; 10 m long	-	-	-	nr	**550.00**
Add for additional pile length up to 15 m	-	-	-	m	**55.00**
Deduct for reduction in pile length	-	-	-	m	**20.00**
Cutting off heads of piles*	1.30	11.65	-	nr	**11.65**
Blind boring; 500 mm diameter	-	-	-	m	**20.00**
Plant; provisional standing time; per rig unit	-	-	-	hour	**70.00**
Extra for boring through artificial obstructions; per rig unit	-	-	-	hour	**95.00**
Testing piles; working to 600 kN/t; using tension piles as reaction					
first pile	-	-	-	nr	**3500.00**
subsequent piles	-	-	-	nr	**3250.00**
Rotary bored cast-in-place concrete piles					
Provision of all plant; including bringing to and removing from site; maintenance, erection and dismantling at each pile position for 100 nr piles	-	-	-	item	**8000.00**
500 mm diameter piles; reinforced; 10 m long	-	-	-	nr	**275.00**
Add for additional pile length up to 15 m	-	-	-	m	**27.50**
Deduct for reduction in pile length	-	-	-	m	**15.00**
Cutting off heads of piles*	1.30	11.65	-	nr	**11.65**
Blind boring; 500 mm diameter	-	-	-	m	**15.00**
Plant; provisional standing time; per rig unit	-	-	-	hour	**140.00**
Extra for boring through artificial obstructions; per rig unit	-	-	-	hour	**160.00**
Testing piles; working to 600 kN/t; using tension piles as reaction					
first pile	-	-	-	nr	**3500.00**
subsequent piles	-	-	-	nr	**3250.00**

*Work normally carried out by the Main Contractor

D GROUNDWORK Including overheads and profit at 5.00%	Labour hours	Labour £	Material Plant £	Unit	Total rate £
D33 STEEL PILING					
'Frodingham' steel sheet piling; BS 4360; grade 43A; pitched and driven					
Provision of all plant; including bringing to and removing from site; maintenance, erection and dismantling; assuming one rig for 1500 m2 of piling	-	-	-	item	**4800.00**
Type 1N; 99.1 kg/m2	2.60	23.29	42.71	m2	**66.00**
Type 2N; 112.3 kg/m2	2.75	24.63	48.37	m2	**73.00**
Type 3N; 137.1 kg/m2	3.00	26.87	59.13	m2	**86.00**
Type 4N; 170.8 kg/m2	3.35	30.00	73.80	m2	**103.80**
Type 5N; 236.9 kg/m2	4.00	35.83	102.27	m2	**138.10**
Burn off tops of piles level	1.00	8.96	-	m	**8.96**
'Frodingham' steel sheet piling; extract only					
Provision of all plant; including bringing to and removing from site; maintenance, erection and dismantling; assuming one rig as before	-	-	-	item	**3750.00**
Type 1N; 99.1 kg/m2	1.25	11.20	-	m2	**11.20**
Type 2N; 112.3 kg/m2	1.25	11.20	-	m2	**11.20**
Type 3N; 137.1 kg/m2	1.25	11.20	-	m2	**11.20**
Type 4N; 170.8 kg/m2	1.25	11.20	-	m2	**11.20**
Type 5N; 236.9 kg/m2	1.40	12.54	-	m2	**12.54**
D40 DIAPHRAGM WALLING					
Provision of all plant; including bringing to and removing from site; maintenance, erection and dismantling; assuming one rig for 1000 m2 of walling	-	-	-	item	**50000.00**
Excavation for diaphragm wall; excavated material removed from site; Bentonite slurry supplied and disposed of					
600 mm thick walls	-	-	-	m3	**184.54**
1000 mm thick walls	-	-	-	m3	**159.60**
Ready mixed reinforced in situ concrete; normal Portland cement; mix 30.00 N/mm2-10mm aggregate in walls					
over 300 mm thick	-	-	-	m3	**89.25**
Reinforcement bars; BS 4461 cold rolled deformed square high yield steel bars; straight or bent					
25 - 40 mm	-	-	-	t	**682.50**
20 mm	-	-	-	t	**682.50**
16 mm	-	-	-	t	**735.00**
Formwork 75 mm thick to form chases	-	-	-	m2	**50.00**
Construct twin guide walls in reinforced concrete; together with reinforcement and formwork along the axis of the diaphragm wall*	-	-	-	m	**200.00**
Plant; provisional standing time; per rig unit	-	-	-	hour	**250.00**
Extra for delay through obstructions; per rig unit	-	-	-	hour	**275.00**

*Work normally carried out by the Main Contractor

D GROUNDWORK Including overheads and profit at 5.00%	Labour hours	Labour £	Material Plant £	Unit	**Total rate £**
D50 UNDERPINNING					
Mechanical excavation using a wheeled hydraulic off-centre excavator with a 0.24 m3 bucket					
Excavating preliminary trenches; not exceeding					
1 m deep	0.25	1.47	4.99	m3	**6.46**
2 m deep	0.30	1.76	5.99	m3	**7.75**
4 m deep	0.35	2.05	6.98	m3	**9.03**
Extra for breaking up					
concrete 150 mm thick	0.70	4.11	0.77	m2	**4.88**
Hand excavation					
Excavating preliminary trenches; not exceeding					
1 m deep	2.90	17.02	-	m3	**17.02**
2 m deep	3.30	19.37	-	m3	**19.37**
4 m deep	4.25	24.95	-	m3	**24.95**
Extra for breaking up					
concrete 150 mm thick	0.30	1.76	1.55	m2	**3.31**
Excavating underpinning pits starting from 1 m below ground level; not exceeding					
0.25 m deep	4.40	25.83	-	m3	**25.83**
1 m deep	4.80	28.17	-	m3	**28.17**
2 m deep	5.75	33.75	-	m3	**33.75**
Excavating underpinning pits starting from 2 m below ground level; not exceeding					
0.25 m deep	5.40	31.70	-	m3	**31.70**
1 m deep	5.80	34.04	-	m3	**34.04**
2 m deep	6.75	39.62	-	m3	**39.62**
Excavating underpinning pits starting from 4 m below ground level; not exceeding					
0.25 m deep	6.40	37.56	-	m3	**37.56**
1 m deep	6.80	39.91	-	m3	**39.91**
2 m deep	7.75	45.49	-	m3	**45.49**
Extra over any type of excavation at any depth for excavating below ground water level	0.35	2.05	-	m3	**2.05**
Earthwork support (open boarded) in 3 m lengths					
To preliminary trenches; not exceeding 2 m between opposing faces; not exceeding					
1 m deep	0.40	2.35	1.25	m2	**3.60**
2 m deep	0.50	2.93	1.54	m2	**4.47**
4 m deep	0.64	3.76	1.93	m2	**5.69**
To underpinning pits; not exceeding 2 m between opposing faces; not exceeding					
1 m deep	0.44	2.58	1.35	m2	**3.93**
2 m deep	0.55	3.23	1.74	m2	**4.97**
4 m deep	0.70	4.11	2.12	m2	**6.23**
Earthwork support (closed boarded) in 3 m lengths					
To preliminary trenches; not exceeding 2 m between opposing faces; not exceeding					
1 m deep	1.00	5.87	2.12	m2	**7.99**
2 m deep	1.25	7.34	2.70	m2	**10.04**
4 m deep	1.55	9.10	3.28	m2	**12.38**

D GROUNDWORK Including overheads and profit at 5.00%	Labour hours	Labour £	Material Plant £	Unit	Total rate £
To underpinning pits; not exceeding 2 m between opposing faces; not exceeding					
1 m deep	1.10	6.46	2.32	m2	**8.78**
2 m deep	1.38	8.10	2.89	m2	**10.99**
4 m deep	1.70	9.98	3.67	m2	**13.65**
Extra over all types of earthwork support for earthwork support left in	0.75	4.40	11.58	m2	**15.98**
Preparation/cutting away					
Cutting away projecting plain concrete foundations					
150 x 150 mm	0.16	0.94	0.35	m	**1.29**
150 x 225 mm	0.24	1.41	0.53	m	**1.94**
150 x 300 mm	0.32	1.88	0.71	m	**2.59**
300 x 300 mm	0.63	3.70	1.39	m	**5.09**
Cutting away masonry					
one course high	0.05	0.29	0.11	m	**0.40**
two courses high	0.14	0.82	0.31	m	**1.13**
three courses high	0.27	1.58	0.60	m	**2.18**
four courses high	0.45	2.64	0.99	m	**3.63**
Preparing underside of existing work to receive new underpinning					
380 mm wide	0.60	3.52	-	m	**3.52**
600 mm wide	0.80	4.70	-	m	**4.70**
900 mm wide	1.00	5.87	-	m	**5.87**
1200 mm wide	1.20	7.04	-	m	**7.04**
Hand disposal of excavated materials					
Removing from site to tip not exceeding 13 km (using lorries)	1.00	5.87	9.84	m3	**15.71**
Hand filling with excavated material					
Filling to excavations	1.00	5.87	-	m3	**5.87**
Surface treatments					
Compacting bottoms of excavations	0.05	0.29	-	m2	**0.29**
Plain insitu ready mixed concrete; 11.50 N/mm2 - 40 mm aggregate (1:3:6); poured against faces of excavation					
Underpinning					
over 450 mm thick	2.70	16.44	46.75	m3	**63.19**
150 - 450 mm thick	3.10	18.88	46.75	m3	**65.63**
not exceeding 150 mm thick	3.70	22.53	46.75	m3	**69.28**
Plain insitu ready mixed concrete; 21.00 N/mm2 - 20 mm aggregate (1:2:4); poured against faces of excavation					
Underpinning					
over 450 mm thick	2.70	16.44	49.78	m3	**66.22**
150 - 450 mm thick	3.10	18.88	49.78	m3	**68.66**
not exceeding 150 mm thick	3.70	22.53	49.78	m3	**72.31**
Extra for working around reinforcement	0.30	1.83	-	m3	**1.83**
Sawn formwork					
Sides of foundations in underpinning					
over 1 m high	1.60	11.39	4.63	m2	**16.02**
not exceeding 250 mm high	0.55	3.92	1.30	m	**5.22**
250 - 500 mm high	0.85	6.05	2.45	m	**8.50**
500 mm - 1 m high	1.30	9.25	4.63	m	**13.88**

D GROUNDWORK Including overheads and profit at 5.00%		Labour hours	Labour £	Material £	Unit	Total rate £
D50 UNDERPINNING - cont'd						
Reinforcement						
Reinforcement bars; BS 4449; hot rolled plain round mild steel bars; bent						
20 mm	PC £337.05	17.00	120.13	404.82	t	**524.95**
16 mm	PC £342.45	20.00	141.33	413.96	t	**555.29**
12 mm	PC £369.45	23.00	162.53	446.93	t	**609.46**
10 mm	PC £391.05	27.00	190.80	473.94	t	**664.74**
8 mm	PC £418.95	31.00	219.06	507.90	t	**726.96**
6 mm	PC £458.55	36.00	254.39	554.76	t	**809.15**
8 mm; links or the like	PC £418.95	42.00	296.79	517.49	t	**814.28**
6 mm; links or the like	PC £458.55	49.00	346.26	567.55	t	**913.81**
Reinforcement bars; BS 4461; cold worked deformed square high yield steel bars; bent						
20 mm	PC £340.65	17.00	120.13	408.79	t	**528.92**
16 mm	PC £346.05	20.00	141.33	417.93	t	**559.26**
12 mm	PC £373.05	23.00	162.53	450.90	t	**613.43**
10 mm	PC £394.65	27.00	190.80	477.91	t	**668.71**
8 mm	PC £422.55	31.00	219.06	511.86	t	**730.92**
6 mm	PC £462.15	36.00	254.39	558.73	t	**813.12**
Common bricks; PC £120.00/1000; in cement mortar (1:3)						
Walls in underpinning						
one brick thick		2.40	24.32	18.62	m2	**42.94**
one and a half brick thick		3.30	33.44	28.00	m2	**61.44**
two brick thick		4.10	41.54	37.24	m2	**78.78**
Add or deduct for variation of £10.00/1000 in PC of common bricks						
one brick thick		-	-	-	m2	**1.32**
one and a half brick thick		-	-	-	m2	**1.98**
two brick thick		-	-	-	m2	**2.65**
Class A engineering bricks; PC £340.00/1000; in cement mortar (1:3)						
Walls in underpinning						
one brick thick		2.60	26.34	55.01	m2	**81.35**
one and a half brick thick		3.55	35.97	82.57	m2	**118.54**
two brick thick		4.40	44.58	110.01	m2	**154.59**
Add or deduct for variation of £10.00/1000 in PC of common bricks						
one brick thick		-	-	-	m2	**1.32**
one and a half brick thick		-	-	-	m2	**1.98**
two brick thick		-	-	-	m2	**2.65**
Class B engineering bricks; PC £175.00/1000; in cement mortar (1:3)						
Walls in underpinning						
one brick thick		2.60	26.34	25.90	m2	**52.24**
one and a half brick thick		3.55	35.97	38.92	m2	**74.89**
two brick thick		4.40	44.58	51.79	m2	**96.37**
Add or deduct for variation of £10.00/1000 in PC of common bricks						
one brick thick		-	-	-	m2	**1.32**
one and a half brick thick		-	-	-	m2	**1.98**
two brick thick		-	-	-	m2	**2.65**
'Pluvex' (hessian based) damp proof course or similar; PC £3.62/m2; 200 mm laps; in cement mortar (1:3)						
Horizontal						
over 225 mm wide		0.25	2.53	4.08	m2	**6.61**
not exceeding 225 mm wide		0.50	5.07	4.18	m2	**9.25**

D GROUNDWORK Including overheads and profit at 5.00%	Labour hours	Labour £	Material £	Unit	Total rate £
'Hyload' (pitch polymer) damp proof course or similar; PC £4.06/m2; 150 mm laps; in cement mortar (1:3)					
Horizontal					
over 225 mm wide	0.25	2.53	4.58	m2	**7.11**
not exceeding 225 mm wide	0.50	5.07	4.69	m2	**9.76**
'Ledkore' grade A (bitumen based lead cored) damp proof course or similar; PC £11.34/m2; 200 mm laps; in cement mortar (1:3)					
Horizontal					
over 225 mm wide	0.33	3.34	12.80	m2	**16.14**
not exceeding 225 mm wide	0.66	6.69	13.09	m2	**19.78**
Two courses of slates in cement mortar (1:3)					
Horizontal					
over 225 mm wide	1.50	15.20	22.39	m2	**37.59**
not exceeding 225 mm wide	2.50	25.33	22.79	m2	**48.12**
Wedging and pinning					
To underside of existing construction with slates in cement mortar (1:3)					
102 mm wall	1.10	11.15	4.23	m	**15.38**
215 mm wall	1.30	13.17	6.40	m	**19.57**
317 mm wall	1.50	15.20	8.50	m	**23.70**

E IN SITU CONCRETE/LARGE PRECAST CONCRETE
Including overheads and profit at 5.00%

BASIC CONCRETE PRICES

	£		£		£
Concrete aggregates (£/tonne)					
40 mm all-in	9.70	40 mm shingle	9.70	10 mm shingle	9.80
20 mm all-in	9.80	20 mm shingle	9.80	sharp sand	8.40

	£		£
Formwork items			
plywood (£/m2)	6.75	timber (£/m3)	183.75
Lightweight aggregates			
'Lytag' (£/m3)			
'fines'	20.21	6-12 mm granular	19.23
Portland cement (£/tonne)			
normal - in bags	64.01	normal - in bulk to silos	57.23
sulphate - resisting - plus	9.63	rapid - hardening - plus	3.74

Tying wire for reinforcement - £0.47/kg

MIXED CONCRETE PRICES (£/m3)

The following prices are for ready or site mixed concrete ready for placing including 2.5% for waste and 5% for overheads and profit	Mix 7.50 N/mm2 - 40mm aggregate (1:8) £	Mix 11.50 N/mm2 - 40mm aggregate (1:3:6) £	Mix 15.00 N/mm2 -40mm aggregate £	Mix 21.00 N/mm2 - 20mm aggregate (1:2:4) £	Mix 26.00 N/mm2 - 20mm aggregate (1:1.5:3) £	Mix 31.00 N/mm2 - 20mm aggregate £	Mix 40.00 N/mm2 - 20mm aggregate £
Ready mixed concrete							
Normal Portland cement	41.70	42.50	43.75	45.25	47.70	50.00	53.00
Sulphate - resistant cement	43.40	44.20	45.80	47.60	50.50	53.00	56.50
Normal Portland cement with water-repellent additive	43.80	44.60	45.85	47.35	49.80	52.10	55.10
Normal Portland cement; air-entrained	43.80	44.60	45.85	47.35	49.80	52.10	55.10
Lightweight concrete using Lytag medium and natural sand	-	-	60.00	59.60	61.60	63.45	-
Site mixed concrete							
Normal Portland cement	-	44.00	46.00	48.00	51.00	55.00	60.00
Sulphate - resistant cement	-	46.00	48.50	51.00	54.50	59.00	65.00

	Labour hours	Labour £	Material £	Unit	Total rate £
E10 IN SITU CONCRETE					
Plain in situ ready mixed concrete; 11.50 N/mm2 - 40 mm aggregate (1:3:6)					
Foundations	1.60	9.74	42.50	m3	**52.24**
Isolated foundations	1.90	11.57	42.50	m3	**54.07**

E IN SITU CONCRETE/LARGE PRECAST CONCRETE Including overheads and profit at 5.00%	Labour hours	Labour £	Material £	Unit	Total rate £
Beds					
over 450 mm thick	1.25	7.61	42.50	m3	**50.11**
150 - 450 mm thick	1.65	10.05	42.50	m3	**52.55**
not exceeding 150 mm thick	2.40	14.62	42.50	m3	**57.12**
Filling to hollow walls					
not exceeding 150 mm thick	4.25	25.88	42.50	m3	**68.38**
Plain in situ ready mixed concrete; 11.50 N/mm2 - 40 mm aggregate (1:3:6); poured on or against earth or unblinded hardcore					
Foundations	1.70	10.35	44.63	m3	**54.98**
Isolated foundations	2.00	12.18	44.63	m3	**56.81**
Beds					
over 450 mm thick	1.30	7.92	44.63	m3	**52.55**
150 - 450 mm thick	1.75	10.66	44.63	m3	**55.29**
not exceeding 150 mm thick	2.50	15.23	44.63	m3	**59.86**
Plain in situ ready mixed concrete; 21.00 N/mm2 - 20 mm aggregate (1:2:4)					
Foundations	1.60	9.74	45.26	m3	**55.00**
Isolated foundations	1.90	11.57	45.26	m3	**56.83**
Beds					
over 450 mm thick	1.25	7.61	45.26	m3	**52.87**
150 - 450 mm thick	1.65	10.05	45.26	m3	**55.31**
not exceeding 150 mm thick	2.40	14.62	45.26	m3	**59.88**
Filling to hollow walls					
not exceeding 150 mm thick	4.25	25.88	45.26	m3	**71.14**
Plain in situ ready mixed concrete; 21.00 N/mm2 - 20 mm aggregate (1:2:4); poured on or against earth or unblinded hardcore					
Foundations	1.70	10.35	47.52	m3	**57.87**
Isolated foundations	2.00	12.18	47.52	m3	**59.70**
Beds					
over 450 mm thick	1.30	7.92	47.52	m3	**55.44**
150 - 450 mm thick	1.75	10.66	47.52	m3	**58.18**
not exceeding 150 mm thick	2.50	15.23	47.52	m3	**62.75**
Reinforced in situ ready mixed concrete; 21.00 N/mm2 - 20 mm aggregate (1:2:4)					
Foundations	2.00	12.18	45.26	m3	**57.44**
Ground beams	3.50	21.32	45.26	m3	**66.58**
Isolated foundations	2.30	14.01	45.26	m3	**59.27**
Beds					
over 450 mm thick	1.60	9.74	45.26	m3	**55.00**
150 - 450 mm thick	2.00	12.18	45.26	m3	**57.44**
not exceeding 150 mm thick	2.75	16.75	45.26	m3	**62.01**
Slabs					
over 450 mm thick	3.10	18.88	45.26	m3	**64.14**
150 - 450 mm thick	3.50	21.32	45.26	m3	**66.58**
not exceeding 150 mm thick	4.40	26.80	45.26	m3	**72.06**
Coffered or troughed slabs					
over 450 mm thick	3.50	21.32	45.26	m3	**66.58**
150 - 450 mm thick	4.00	24.36	45.26	m3	**69.62**
Extra over for laying to slopes					
not exceeding 15 degrees	0.30	1.83	-	m3	**1.83**
over 15 degrees	0.60	3.65	-	m3	**3.65**
Walls					
over 450 mm thick	3.25	19.79	45.26	m3	**65.05**
150 - 450 mm thick	3.70	22.53	45.26	m3	**67.79**
not exceeding 150 mm thick	4.60	28.01	45.26	m3	**73.27**

E IN SITU CONCRETE/LARGE PRECAST CONCRETE Including overheads and profit at 5.00%	Labour hours	Labour £	Material £	Unit	Total rate £
E10 IN SITU CONCRETE - cont'd					
Reinforced in situ ready mixed concrete; 21.00 N/mm2 - 20 mm aggregate (1:2:4) - cont'd					
Isolated beams	5.00	30.45	45.26	m3	**75.71**
Isolated deep beams	5.50	33.50	45.26	m3	**78.76**
Attached deep beams	5.00	30.45	45.26	m3	**75.71**
Isolated beam casings	5.50	33.50	45.26	m3	**78.76**
Isolated deep beam casings	6.00	36.54	45.26	m3	**81.80**
Attached deep beam casings	5.50	33.50	45.26	m3	**78.76**
Columns	6.00	36.54	45.26	m3	**81.80**
Column casings	6.60	40.19	45.26	m3	**85.45**
Staircases	7.50	45.68	45.26	m3	**90.94**
Upstands	4.80	29.23	45.26	m3	**74.49**
Reinforced in situ ready mixed concrete; 26.00 N/mm2 - 20 mm aggregate (1:1.5:3)					
Foundations	2.00	12.18	47.70	m3	**59.88**
Ground beams	3.50	21.32	47.70	m3	**69.02**
Isolated foundations	2.30	14.01	47.70	m3	**61.71**
Beds					
over 450 mm thick	1.60	9.74	47.70	m3	**57.44**
150 - 450 mm thick	2.00	12.18	47.70	m3	**59.88**
not exceeding 150 mm thick	2.75	16.75	47.70	m3	**64.45**
Slabs					
over 450 mm thick	3.10	18.88	47.70	m3	**66.58**
150 - 450 mm thick	3.50	21.32	47.70	m3	**69.02**
not exceeding 150 mm thick	4.40	26.80	47.70	m3	**74.50**
Coffered or troughed slabs					
over 450 mm thick	3.50	21.32	47.70	m3	**69.02**
150 - 450 mm thick	4.00	24.36	47.70	m3	**72.06**
Extra over for laying to slopes					
not exceeding 15 degrees	0.30	1.83	-	m3	**1.83**
over 15 degrees	0.60	3.65	-	m3	**3.65**
Walls					
over 450 mm thick	3.25	19.79	47.70	m3	**67.49**
150 - 450 mm thick	3.70	22.53	47.70	m3	**70.23**
not exceeding 150 mm thick	4.60	28.01	47.70	m3	**75.71**
Isolated beams	5.00	30.45	47.70	m3	**78.15**
Isolated deep beams	5.50	33.50	47.70	m3	**81.20**
Attached deep beams	5.00	30.45	47.70	m3	**78.15**
Isolated beam casings	5.50	33.50	47.70	m3	**81.20**
Isolated deep beam casings	6.00	36.54	47.70	m3	**84.24**
Attached deep beam casings	5.50	33.50	47.70	m3	**81.20**
Columns	6.00	36.54	47.70	m3	**84.24**
Column casings	6.60	40.19	47.70	m3	**87.89**
Staircases	7.50	45.68	47.70	m3	**93.38**
Upstands	4.80	29.23	47.70	m3	**76.93**
Reinforced in situ ready mixed concrete; 31.00 N/mm2 - 20 mm aggregate (1:1:2)					
Beds					
over 450 mm thick	1.60	9.74	50.00	m3	**59.74**
150 - 450 mm thick	2.00	12.18	50.00	m3	**62.18**
not exceeding 150 mm thick	2.75	16.75	50.00	m3	**66.75**
Slabs					
over 450 mm thick	3.10	18.88	50.00	m3	**68.88**
150 - 450 mm thick	3.50	21.32	50.00	m3	**71.32**
not exceeding 150 mm thick	4.40	26.80	50.00	m3	**76.80**
Coffered or troughed slabs					
over 450 mm thick	3.50	21.32	50.00	m3	**71.32**
150 - 450 mm thick	4.00	24.36	50.00	m3	**74.36**

E IN SITU CONCRETE/LARGE PRECAST CONCRETE Including overheads and profit at 5.00%	Labour hours	Labour £	Material £	Unit	Total rate £
Extra over for laying to slopes					
not exceeding 15 degrees	0.30	1.83	-	m3	**1.83**
over 15 degrees	0.60	3.65	-	m3	**3.65**
Walls					
over 450 mm thick	3.25	19.79	50.00	m3	**69.79**
150 - 450 mm thick	3.70	22.53	50.00	m3	**72.53**
not exceeding 150 mm thick	4.60	28.01	50.00	m3	**78.01**
Isolated beams	5.00	30.45	50.00	m3	**80.45**
Isolated deep beams	5.50	33.50	50.00	m3	**83.50**
Attached deep beams	5.00	30.45	50.00	m3	**80.45**
Isolated beam casings	5.50	33.50	50.00	m3	**83.50**
Isolated deep beam casings	6.00	36.54	50.00	m3	**86.54**
Attached deep beam casings	5.50	33.50	50.00	m3	**83.50**
Columns	6.00	36.54	50.00	m3	**86.54**
Column casings	6.60	40.19	50.00	m3	**90.19**
Staircases	7.50	45.68	50.00	m3	**95.68**
Upstands	4.80	29.23	50.00	m3	**79.23**
Extra over vibrated concrete for					
reinforcement content over 5%	0.55	3.35	-	m3	**3.35**
Grouting with cement mortar (1:1)					
Stanchion bases					
10 mm thick	1.00	6.09	0.38	nr	**6.47**
25 mm thick	1.25	7.61	0.91	nr	**8.52**
Grouting with epoxy resin					
Stanchion bases					
10 mm thick	1.25	7.61	0.27	nr	**7.88**
25 mm thick	1.50	9.13	0.65	nr	**9.78**
Filling; plain in situ concrete;					
21.00 N/mm2 - 20 mm aggregate (1:2:4)					
Mortices	0.10	0.61	0.25	nr	**0.86**
Holes	0.25	1.52	1.71	m3	**3.23**
Chases					
over 0.01 m2	0.20	1.22	1.00	m3	**2.22**
not exceeding 0.01 m2	0.15	0.91	0.50	m	**1.41**
Sheeting to prevent moisture loss					
Building paper; lapped joints					
subsoil grade; horizontal on foundations	0.03	0.18	0.44	m2	**0.62**
standard grade; horizontal on slabs	0.05	0.30	0.70	m2	**1.00**
Polyethylene sheeting; lapped joints;					
horizontal on slabs					
250 microns; 0.25 mm thick	0.05	0.30	0.25	m2	**0.55**
'Visqueen' sheeting; lapped joints;					
horizontal on slabs					
1000 Grade; 0.25 mm thick	0.05	0.30	0.26	m2	**0.56**
1200 Super; 0.30 mm thick	0.06	0.37	0.33	m2	**0.70**
E20 FORMWORK FOR IN SITU CONCRETE					
Note: Generally all formwork based on four					
uses unless otherwise stated					
Sides of foundations					
over 1 m high	1.40	9.97	3.61	m2	**13.58**
not exceeding 250 mm high	0.45	3.20	1.14	m	**4.34**
250 - 500 mm high	0.75	5.34	1.94	m	**7.28**
500 mm - 1 m high	1.15	8.19	3.61	m	**11.80**

E IN SITU CONCRETE/LARGE PRECAST CONCRETE Including overheads and profit at 5.00%	Labour hours	Labour £	Material £	Unit	Total rate £
E20 FORMWORK FOR IN SITU CONCRETE - cont'd					
Sides of foundations; left in					
over 1 m high	1.40	9.97	12.31	m2	**22.28**
not exceeding 250 mm high	0.45	3.20	3.18	m	**6.38**
250 - 500 mm high	0.75	5.34	6.23	m	**11.57**
500 mm - 1 m high	1.15	8.19	12.31	m	**20.50**
Sides of ground beams and edges of beds					
over 1 m high	1.65	11.75	5.01	m2	**16.76**
not exceeding 250 mm high	0.50	3.56	1.40	m	**4.96**
250 - 500 mm high	0.90	6.41	2.58	m	**8.99**
500 mm - 1 m high	1.25	8.90	5.01	m	**13.91**
Edges of suspended slabs					
not exceeding 250 mm high	0.75	5.34	1.67	m	**7.01**
250 - 500 mm high	1.10	7.83	3.31	m	**11.14**
500 mm - 1 m high	1.75	12.46	6.34	m	**18.80**
Sides of upstands					
over 1 m high	2.00	14.24	6.66	m2	**20.90**
not exceeding 250 mm high	0.63	4.48	1.74	m	**6.22**
250 - 500 mm high	1.00	7.12	3.47	m	**10.59**
500 mm - 1 m high	1.75	12.46	6.66	m	**19.12**
Steps in top surfaces					
not exceeding 250 mm high	0.50	3.56	1.81	m	**5.37**
250 - 500 mm high	0.80	5.70	3.75	m	**9.45**
Steps in soffits					
not exceeding 250 mm high	0.55	3.92	1.81	m	**5.73**
250 - 500 mm high	0.88	6.26	3.75	m	**10.01**
Machine bases and plinths					
over 1 m high	1.60	11.39	5.01	m2	**16.40**
not exceeding 250 mm high	0.50	3.56	1.40	m	**4.96**
250 - 500 mm high	0.85	6.05	2.58	m	**8.63**
500 mm - 1 m high	1.25	8.90	5.01	m	**13.91**
Soffits of slabs; 1.5 - 3 m height to soffit					
not exceeding 200 mm thick	1.70	12.10	4.90	m2	**17.00**
not exceeding 200 mm thick (5 uses)	1.65	11.75	4.48	m2	**16.23**
not exceeding 200 mm thick (6 uses)	1.60	11.39	4.17	m2	**15.56**
200 - 300 mm thick	1.80	12.81	6.63	m2	**19.44**
300 - 400 mm thick	1.85	13.17	7.05	m2	**20.22**
400 - 500 mm thick	1.95	13.88	7.47	m2	**21.35**
500 - 600 mm thick	2.10	14.95	7.87	m2	**22.82**
Soffits of slabs; not exceeding 200 mm thick					
not exceeding 1.5 m height to soffit	1.80	12.81	6.14	m2	**18.95**
3 - 4.5 m height to soffit	1.70	12.10	5.99	m2	**18.09**
4.5 - 6 m height to soffit	1.80	12.81	7.03	m2	**19.84**
Soffits of landings; 1.5 - 3m height to soffit					
not exceeding 200 mm thick	1.80	12.81	5.09	m2	**17.90**
200 - 300 mm thick	1.90	13.53	6.89	m2	**20.42**
300 - 400 mm thick	1.95	13.88	7.29	m2	**21.17**
400 - 500 mm thick	2.05	14.59	7.71	m2	**22.30**
500 - 600 mm thick	2.20	15.66	8.13	m2	**23.79**
Extra over for sloping					
not exceeding 15 degrees	0.20	1.42	-	m2	**1.42**
over 15 degrees	0.40	2.85	-	m2	**2.85**
Soffits of coffered or troughed slabs; including 'Cordek' troughed forms; 300 mm deep; ribs at 600 mm centres and cross ribs at centres of bay; 300 - 400 mm thick					
1.5 - 3 m height to soffit	2.50	17.80	7.39	m2	**25.19**
3 - 4.5 m height to soffit	2.60	18.51	8.12	m2	**26.63**
4.5 - 6 m height to soffit	2.70	19.22	9.16	m2	**28.38**

E IN SITU CONCRETE/LARGE PRECAST CONCRETE Including overheads and profit at 5.00%	Labour hours	Labour £	Material £	Unit	Total rate £
Soffits of bands and margins to troughed slabs					
horizontal; 300 - 400 mm thick	2.00	14.24	5.58	m2	**19.82**
Top formwork	1.50	10.68	3.77	m2	**14.45**
Walls					
vertical	2.00	14.24	5.83	m2	**20.07**
vertical; interrupted	2.10	14.95	6.10	m2	**21.05**
vertical; exceeding 3 m high; inside stairwells	2.20	15.66	6.16	m2	**21.82**
vertical; exceeding 3 m high; inside lift shaft	2.40	17.09	6.66	m2	**23.75**
battered	2.80	19.93	6.90	m2	**26.83**
Beams attached to insitu slabs					
square or rectangular; 1.5 - 3 m height to soffit	2.20	15.66	7.38	m2	**23.04**
square or rectangular; 3 - 4.5 m height to soffit	2.30	16.37	8.44	m2	**24.81**
square or rectangular; 4.5 - 6 m height to soffit	2.40	17.09	9.50	m2	**26.59**
Beams attached to walls					
square or rectangular; 1.5 - 3 m height to soffit	2.30	16.37	7.38	m2	**23.75**
Isolated beams					
square or rectangular; 1.5 - 3 m height to soffit	2.40	17.09	7.38	m2	**24.47**
square or rectangular; 3 - 4.5 m height to soffit	2.50	17.80	8.44	m2	**26.24**
square or rectangular; 4.5 - 6 m height to soffit	2.60	18.51	9.50	m2	**28.01**
Beam casings attached to insitu slabs					
square or rectangular; 1.5 - 3 m height to soffit	2.30	16.37	7.38	m2	**23.75**
square or rectangular; 3 - 4.5 m height to soffit	2.40	17.09	8.44	m2	**25.53**
Beam casings attached to walls					
square or rectangular; 1.5 - 3 m height to soffit	2.40	17.09	7.38	m2	**24.47**
Isolated beam casings					
square or rectangular; 1.5 - 3 m height to soffit	2.50	17.80	7.38	m2	**25.18**
square or rectangular; 3 - 4.5 m height to soffit	2.60	18.51	8.44	m2	**26.95**
Extra over for sloping					
not exceeding 15 degrees	0.30	2.14	0.71	m2	**2.85**
over 15 degrees	0.60	4.27	1.42	m2	**5.69**
Columns attached to walls					
square or rectangular	2.20	15.66	5.83	m2	**21.49**
Isolated columns					
square or rectangular	2.30	16.37	5.83	m2	**22.20**
Column casings attached to walls					
square or rectangular	2.30	16.37	5.83	m2	**22.20**
Isolated column casings					
square or rectangular	2.40	17.09	5.83	m2	**22.92**
Extra over for					
Throat	0.05	0.36	0.16	m	**0.52**
Chamfer					
30 mm wide	0.06	0.43	0.22	m	**0.65**
60 mm wide	0.07	0.50	0.36	m	**0.86**
90 mm wide	0.08	0.57	0.75	m	**1.32**
Rebate or horizontal recess					
12 x 12 mm	0.07	0.50	0.09	m	**0.59**
25 x 25 mm	0.07	0.50	0.15	m	**0.65**
25 x 50 mm	0.07	0.50	0.15	m	**0.65**
50 x 50 mm	0.07	0.50	0.53	m	**1.03**

E IN SITU CONCRETE/LARGE PRECAST CONCRETE Including overheads and profit at 5.00%	Labour hours	Labour £	Material £	Unit	Total rate £
E20 FORMWORK FOR IN SITU CONCRETE - cont'd					
Nibs					
50 x 50 mm	0.55	3.92	1.14	m	**5.06**
100 x 100 mm	0.78	5.55	2.22	m	**7.77**
100 x 200 mm	1.04	7.40	2.96	m	**10.36**
Extra over basic formwork for rubbing down, filling and leaving face of concrete smooth					
general surfaces	0.33	2.35	0.10	m2	**2.45**
edges	0.50	3.56	0.11	m2	**3.67**
Add to prices for basic formwork for					
for curved radius 6 m	27.5%				
for curved radius 2 m	50%				
coating with retardant agent	0.02	0.14	0.48	m2	**0.62**
Wall kickers to both sides					
150 mm high	0.50	3.56	1.15	m	**4.71**
225 mm high	0.65	4.63	1.50	m	**6.13**
150 mm high; one side suspended	0.63	4.48	1.84	m	**6.32**
Wall ends, soffits and steps in walls					
over 1 m wide	1.90	13.53	6.12	m	**19.65**
not exceeding 250 mm wide	0.60	4.27	1.74	m	**6.01**
250 - 500 mm wide	0.95	6.76	3.36	m	**10.12**
500 mm - 1 m wide	1.50	10.68	6.12	m	**16.80**
Openings in walls					
over 1 m wide	2.10	14.95	7.37	m	**22.32**
not exceeding 250 mm wide	0.65	4.63	1.86	m	**6.49**
250 - 500 mm wide	1.10	7.83	3.73	m	**11.56**
500 mm - 1 m wide	1.70	12.10	7.37	m	**19.47**
Stair flights					
1 m wide; 150 mm waist; 150 mm undercut risers	5.00	35.60	18.71	m	**54.31**
2 m wide; 200 mm waist; 150 mm vertical risers	9.00	64.07	30.17	m	**94.24**
Mortices; not exceeding 250 mm deep					
not exceeding 500 mm girth	0.15	1.07	0.89	nr	**1.96**
Holes; not exceeding 250 mm deep					
not exceeding 500 mm girth	0.20	1.42	0.92	nr	**2.34**
500 mm - 1 m girth	0.25	1.78	1.77	nr	**3.55**
1 - 2 m girth	0.45	3.20	3.54	nr	**6.74**
2 - 3 m girth	0.60	4.27	5.32	nr	**9.59**
Holes; 250 - 500 mm deep					
not exceeding 500 mm girth	0.30	2.14	1.77	nr	**3.91**
500 mm - 1 m girth	0.38	2.71	3.54	nr	**6.25**
1 - 2 m girth	0.67	4.77	7.09	nr	**11.86**
2 - 3 m girth	0.90	6.41	10.63	nr	**17.04**
Permanent shuttering; left in					
Dufaylite 'Clayboard' shuttering; type KN30; horizontal; under concrete beds; left in					
50 mm thick	0.15	1.07	6.76	m2	**7.83**
75 mm thick	0.16	1.14	7.24	m2	**8.38**
100 mm thick	0.17	1.21	7.84	m2	**9.05**
150 mm thick	0.20	1.42	8.39	m2	**9.81**
Dufaylite 'Clayboard' shuttering; type KN30; horizontal or vertical (including temporary supports); beneath or to sides of foundations; left in					
50 mm thick; vertical	0.25	1.78	7.34	m2	**9.12**
400 x 50 mm thick; horizontal	0.09	0.64	2.76	m	**3.40**
600 x 50 mm thick; horizontal	0.12	0.85	4.16	m	**5.01**
800 x 50 mm thick; horizontal	0.15	1.07	5.54	m	**6.61**
75 mm thick; vertical	0.27	1.92	7.82	m2	**9.74**
400 x 75 mm thick; horizontal	0.10	0.71	2.97	m	**3.68**

E IN SITU CONCRETE/LARGE PRECAST CONCRETE Including overheads and profit at 5.00%		Labour hours	Labour £	Material £	Unit	Total rate £
600 x 75 mm thick; horizontal		0.13	0.93	4.45	m	**5.38**
800 x 75 mm thick; horizontal		0.16	1.14	5.92	m	**7.06**
100 mm thick; vertical		0.30	2.14	8.44	m2	**10.58**
400 x 100 mm thick; horizontal		0.11	0.78	3.21	m	**3.99**
600 x 100 mm thick; horizontal		0.14	1.00	4.82	m	**5.82**
800 x 100 mm thick; horizontal		0.17	1.21	6.42	m	**7.63**
Hyrib permanent shuttering and reinforcement ref 2411 to soffits of slabs; left in						
horizontal		1.50	10.68	12.29	m2	**22.97**
0.9 mm 'Super Holorib' steel deck permanent shuttering; to soffit of slabs; left in						
1.5 - 3 m height to soffit		0.85	6.05	13.87	m2	**19.92**
3 - 4.5 m height to soffit		1.00	7.12	14.59	m2	**21.71**
4.5 - 6 m height to soffit		1.15	8.19	15.63	m2	**23.82**
1.2 mm 'Super Holorib' steel deck permanent shuttering; to soffit of slabs; left in						
1.5 - 3 m height to soffit		1.00	7.12	17.68	m2	**24.80**
3 - 4.5 m height to soffit		1.15	8.19	18.82	m2	**27.01**
3 - 4.5 m height to soffit; deck stud welded to steelwork		1.50	13.43	21.08	m2	**34.51**
4.5 - 6 m height to soffit		1.30	9.25	20.37	m2	**29.62**
E30 REINFORCEMENT FOR IN SITU CONCRETE						
Reinforcement bars; BS 4449; hot rolled plain round mild steel bars; straight						
40 mm	PC £328.95	11.00	77.73	386.30	t	**464.03**
32 mm	PC £317.25	11.50	81.26	376.59	t	**457.85**
25 mm	PC £310.95	13.50	95.40	372.84	t	**468.24**
20 mm	PC £309.15	15.50	109.53	374.05	t	**483.58**
16 mm	PC £310.05	18.00	127.20	378.24	t	**505.44**
12 mm	PC £328.05	21.00	148.40	401.29	t	**549.69**
10 mm	PC £336.15	24.00	169.60	413.42	t	**583.02**
8 mm	PC £346.05	27.00	190.80	427.53	t	**618.33**
6 mm	PC £381.15	32.00	226.13	469.42	t	**695.55**
Reinforcement bars; BS 4449; hot rolled plain round mild steel bars; bent						
40 mm	PC £351.45	11.00	77.73	411.10	t	**488.83**
32 mm	PC £340.65	13.00	91.86	402.39	t	**494.25**
25 mm	PC £338.85	15.00	106.00	403.60	t	**509.60**
20 mm	PC £337.05	17.00	120.13	404.82	t	**524.95**
16 mm	PC £342.45	20.00	141.33	413.96	t	**555.29**
12 mm	PC £369.45	23.00	162.53	446.93	t	**609.46**
10 mm	PC £391.05	27.00	190.80	473.94	t	**664.74**
8 mm	PC £418.95	31.00	219.06	507.90	t	**726.96**
6 mm	PC £458.55	36.00	254.39	554.76	t	**809.15**
8 mm; links or the like	PC £418.95	42.00	296.79	517.49	t	**814.28**
6 mm; links or the like	PC £458.55	49.00	346.26	567.55	t	**913.81**
Reinforcement bars; BS 4461; cold worked deformed square high steel bars; straight						
40 mm	PC £332.55	10.00	70.67	390.26	t	**460.93**
32 mm	PC £321.75	11.50	81.26	381.55	t	**462.81**
25 mm	PC £315.45	13.50	95.40	377.80	t	**473.20**
20 mm	PC £313.65	15.50	109.53	379.02	t	**488.55**
16 mm	PC £314.55	18.00	127.20	383.21	t	**510.41**
12 mm	PC £332.55	21.00	148.40	406.25	t	**554.65**
10 mm	PC £340.65	24.00	169.60	418.37	t	**587.97**
8 mm	PC £350.55	27.00	190.80	432.48	t	**623.28**
6 mm	PC £385.65	32.00	226.13	474.38	t	**700.51**

E IN SITU CONCRETE/LARGE PRECAST CONCRETE Including overheads and profit at 5.00%		Labour hours	Labour £	Material £	Unit	Total rate £
E30 REINFORCEMENT FOR IN SITU CONCRETE - cont'd						
Reinforcement bars; BS 4461; cold worked deformed square high steel bars; bent						
40 mm	PC £355.05	11.00	77.73	415.07	t	**492.80**
32 mm	PC £344.25	13.00	91.86	406.36	t	**498.22**
25 mm	PC £342.45	15.00	106.00	407.57	t	**513.57**
20 mm	PC £340.65	17.00	120.13	408.79	t	**528.92**
16 mm	PC £346.05	20.00	141.33	417.93	t	**559.26**
12 mm	PC £373.05	23.00	162.53	450.90	t	**613.43**
10 mm	PC £394.65	27.00	190.80	477.91	t	**668.71**
8 mm	PC £422.55	31.00	219.06	511.86	t	**730.92**
6 mm	PC £462.15	36.00	254.39	558.73	t	**813.12**
Reinforcement fabric; BS 4483; lapped; in beds or suspended slabs						
Ref A98 (1.54 kg/m2)	PC £0.55	0.12	0.85	0.69	m2	**1.54**
Ref A142 (2.22 kg/m2)	PC £0.69	0.12	0.85	0.87	m2	**1.72**
Ref A193 (3.02 kg/m2)	PC £0.94	0.12	0.85	1.18	m2	**2.03**
Ref A252 (3.95 kg/m2)	PC £1.21	0.13	0.92	1.52	m2	**2.44**
Ref A393 (6.16 kg/m2)	PC £1.92	0.15	1.06	2.42	m2	**3.48**
Ref B196 (3.05 kg/m2)	PC £1.05	0.12	0.85	1.32	m2	**2.17**
Ref B283 (3.73 kg/m2)	PC £1.23	0.12	0.85	1.55	m2	**2.40**
Ref B385 (4.53 kg/m2)	PC £1.47	0.13	0.92	1.84	m2	**2.76**
Ref B503 (5.93 kg/m2)	PC £1.80	0.15	1.06	2.27	m2	**3.33**
Ref B785 (8.14 kg/m2)	PC £2.56	0.17	1.20	3.22	m2	**4.42**
Ref B1131 (10.90 kg/m2)	PC £3.47	0.19	1.34	4.37	m2	**5.71**
Ref C283 (2.61 kg/m2)	PC £0.89	0.12	0.85	1.13	m2	**1.98**
Ref C385 (3.41 kg/m2)	PC £1.14	0.12	0.85	1.44	m2	**2.29**
Ref C503 (4.34 kg/m2)	PC £1.38	0.13	0.92	1.75	m2	**2.67**
Ref C636 (5.55 kg/m2)	PC £1.78	0.14	0.99	2.25	m2	**3.24**
Ref C785 (6.72 kg/m2)	PC £2.15	0.15	1.06	2.71	m2	**3.77**
Reinforcement fabric; BS 4483; lapped; in casings to steel columns or beams						
Ref D49 (0.77 kg/m2)	PC £0.55	0.25	1.77	0.69	m2	**2.46**
Ref D98 (1.54 kg/m2)	PC £0.55	0.25	1.77	0.69	m2	**2.46**
E40 DESIGNED JOINTS IN IN SITU CONCRETE						
Expandite 'Flexcell' impregnated fibreboard joint filler; or similar						
Formed joint; 10 mm thick						
not exceeding 150 mm wide		0.15	1.07	0.91	m	**1.98**
150 - 300 mm wide		0.20	1.42	1.57	m	**2.99**
300 - 450 mm wide		0.25	1.78	2.06	m	**3.84**
Formed joint; 12.5 mm thick						
not exceeding 150 mm wide		0.15	1.07	0.96	m	**2.03**
150 - 300 mm wide		0.20	1.42	1.13	m	**2.55**
300 - 450 mm wide		0.25	1.78	2.20	m	**3.98**
Formed joint; 20 mm thick						
not exceeding 150 mm wide		0.20	1.42	1.34	m	**2.76**
150 - 300 mm wide		0.25	1.78	1.75	m	**3.53**
300 - 450 mm wide		0.30	2.14	3.08	m	**5.22**
Formed joint; 25 mm thick						
not exceeding 150 mm wide		0.20	1.42	1.52	m	**2.94**
150 - 300 mm wide		0.25	1.78	2.68	m	**4.46**
300 - 450 mm wide		0.30	2.14	3.52	m	**5.66**

E IN SITU CONCRETE/LARGE PRECAST CONCRETE Including overheads and profit at 5.00%		Labour hours	Labour £	Material £	Unit	Total rate £
Sealing top of joint with Expandite 'Pliastic' hot poured rubberized bituminous compound						
10 x 25 mm		0.18	1.28	0.37	m	**1.65**
12.5 x 25 mm		0.19	1.35	0.45	m	**1.80**
20 x 25 mm		0.20	1.42	0.62	m	**2.04**
25 x 25 mm		0.21	1.49	0.87	m	**2.36**
Sealing top of joint with Expandite 'Thioflex 600' cold poured polysulphide rubberized compound						
10 x 25 mm		0.06	0.43	2.48	m	**2.91**
12.5 x 25 mm		0.07	0.50	2.99	m	**3.49**
20 x 25 mm		0.08	0.57	4.14	m	**4.71**
25 x 25 mm		0.09	0.64	5.96	m	**6.60**
Servicised 'Kork-pak' waterproof bonded cork joint filler board; or similar						
Formed joint; 10 mm thick						
not exceeding 150 mm wide		0.15	1.07	2.39	m	**3.46**
150 - 300 mm wide		0.20	1.42	3.70	m	**5.12**
300 - 450 mm wide		0.25	1.78	5.89	m	**7.67**
Formed joint; 13 mm thick						
not exceeding 150 mm wide		0.15	1.07	2.74	m	**3.81**
150 - 300 mm wide		0.20	1.42	3.81	m	**5.23**
300 - 450 mm wide		0.25	1.78	5.57	m	**7.35**
Formed joint; 19 mm thick						
not exceeding 150 mm wide		0.20	1.42	2.89	m	**4.31**
150 - 300 mm wide		0.25	1.78	4.92	m	**6.70**
300 - 450 mm wide		0.30	2.14	7.82	m	**9.96**
Formed joint; 25 mm thick						
not exceeding 150 mm wide		0.20	1.42	3.40	m	**4.82**
150 - 300 mm wide		0.25	1.78	5.81	m	**7.59**
300 - 450 mm wide		0.30	2.14	9.46	m	**11.60**
Sealing top of joint with 'Paraseal' polysulphide sealing compound						
10 x 25 mm		0.06	0.43	1.96	m	**2.39**
13 x 25 mm		0.07	0.50	2.25	m	**2.75**
19 x 25 mm		0.08	0.57	3.33	m	**3.90**
25 x 25 mm		0.09	0.64	4.03	m	**4.67**
Servicised water stops or similar						
Formed joint; PVC water stop; flat dumbell type; heat welded joints						
100 mm wide	PC £39.96/15m	0.22	1.57	2.95	m	**4.52**
Flat angle	PC £2.54	0.27	1.92	2.80	nr	**4.72**
Vertical angle	PC £4.42	0.27	1.92	4.87	nr	**6.79**
Flat three way intersection	PC £5.00	0.37	2.63	5.52	nr	**8.15**
Vertical three way intersection	PC £5.38	0.37	2.63	5.93	nr	**8.56**
Four way intersection	PC £6.28	0.47	3.35	6.92	nr	**10.27**
170 mm wide	PC £56.30/15m	0.25	1.78	4.16	m	**5.94**
Flat angle	PC £2.57	0.30	2.14	2.84	nr	**4.98**
Vertical angle	PC £4.42	0.30	2.14	4.87	nr	**7.01**
Flat three way intersection	PC £5.05	0.40	2.85	5.56	nr	**8.41**
Vertical three way intersection	PC £6.26	0.40	2.85	6.90	nr	**9.75**
Four way intersection	PC £6.85	0.50	3.56	7.55	nr	**11.11**
210 mm wide	PC £66.56/15m	0.28	1.99	4.91	m	**6.90**
Flat angle	PC £4.29	0.32	2.28	4.73	nr	**7.01**
Vertical angle	PC £4.90	0.32	2.28	5.41	nr	**7.69**
Flat three way intersection	PC £6.25	0.42	2.99	6.89	nr	**9.88**
Vertical three way intersection	PC £7.74	0.42	2.99	8.54	nr	**11.53**
Four way intersection	PC £7.85	0.52	3.70	8.65	nr	**12.35**

E IN SITU CONCRETE/LARGE PRECAST CONCRETE Including overheads and profit at 5.00%		Labour hours	Labour £	Material £	Unit	Total rate £
E40 DESIGNED JOINTS IN IN SITU CONCRETE - cont'd						
Servicised water stops or similar - cont'd						
Formed joint; PVC water stop; flat dumbell type; heat welded joints						
250 mm wide	PC £97.34/15m	0.30	2.14	7.19	m	**9.33**
Flat angle	PC £5.22	0.34	2.42	5.75	nr	**8.17**
Vertical angle	PC £5.52	0.34	2.42	6.09	nr	**8.51**
Flat three way intersection	PC £7.01	0.44	3.13	7.73	nr	**10.86**
Vertical three way intersection	PC £8.72	0.44	3.13	9.62	nr	**12.75**
Four way intersection	PC £9.03	0.54	3.84	9.95	nr	**13.79**
Formed joint; PVC water stop; centre bulb type; heat welded joints						
160 mm wide	PC £54.00/15m	0.25	1.78	3.99	m	**5.77**
Flat angle	PC £3.37	0.30	2.14	3.71	nr	**5.85**
Vertical angle	PC £5.54	0.30	2.14	6.12	nr	**8.26**
Flat three way intersection	PC £7.10	0.40	2.85	7.83	nr	**10.68**
Vertical three way intersection	PC £8.78	0.40	2.85	9.69	nr	**12.54**
Four way intersection	PC £8.46	0.50	3.56	9.32	nr	**12.88**
210 mm wide	PC £77.89/15m	0.28	1.99	5.75	m	**7.74**
Flat angle	PC £5.00	0.32	2.28	5.52	nr	**7.80**
Vertical angle	PC £6.59	0.32	2.28	7.26	nr	**9.54**
Flat three way intersection	PC £8.05	0.42	2.99	8.87	nr	**11.86**
Vertical three way intersection	PC £9.97	0.42	2.99	11.00	nr	**13.99**
Four way intersection	PC £9.66	0.52	3.70	10.64	nr	**14.34**
260 mm wide	PC £110.70/15m	0.30	2.14	8.18	m	**10.32**
Flat angle	PC £6.53	0.34	2.42	7.21	nr	**9.63**
Vertical angle	PC £7.17	0.34	2.42	7.91	nr	**10.33**
Flat three way intersection	PC £9.81	0.44	3.13	10.82	nr	**13.95**
Vertical three way intersection	PC £12.16	0.44	3.13	13.41	nr	**16.54**
Four way intersection	PC £11.75	0.54	3.84	12.96	nr	**16.80**
325 mm wide	PC £186.03/15m	0.33	2.35	13.74	m	**16.09**
Flat angle	PC £9.47	0.36	2.56	10.43	nr	**12.99**
Vertical angle	PC £9.77	0.36	2.56	10.78	nr	**13.34**
Flat three way intersection	PC £13.91	0.46	3.27	15.34	nr	**18.61**
Vertical three way intersection	PC £14.21	0.46	3.27	15.67	nr	**18.94**
Four way intersection	PC £15.53	0.56	3.99	17.13	nr	**21.12**
Formed joint; rubber water stop; flat dumbell type; sleeved joints						
150 mm wide	PC £97.08/9m	0.20	1.42	13.08	m	**14.50**
Flat angle	PC £27.63	0.20	1.42	30.46	nr	**31.88**
Vertical angle	PC £27.63	0.20	1.42	30.46	nr	**31.88**
Flat three way intersection	PC £30.48	0.25	1.78	33.60	nr	**35.38**
Vertical three way intersection	PC £30.48	0.25	1.78	33.60	nr	**35.38**
Four way intersection	PC £33.68	0.30	2.14	37.13	nr	**39.27**
230 mm wide	PC £146.86/9m	0.25	1.78	19.49	m	**21.27**
Flat angle	PC £33.43	0.22	1.57	36.86	nr	**38.43**
Vertical angle	PC £33.43	0.22	1.57	36.86	nr	**38.43**
Flat three way intersection	PC £36.42	0.27	1.92	40.16	nr	**42.08**
Vertical three way intersection	PC £36.42	0.27	1.92	40.16	nr	**42.08**
Four way intersection	PC £39.24	0.33	2.35	43.26	nr	**45.61**
Formed joint; rubber water stop; centre bulb type; sleeved joints						
150 mm wide	PC £111.71/9m	0.20	1.42	14.85	m	**16.27**
Flat angle	PC £30.27	0.20	1.42	33.37	nr	**34.79**
Vertical angle	PC £30.27	0.20	1.42	33.37	nr	**34.79**
Flat three way intersection	PC £33.40	0.25	1.78	36.82	nr	**38.60**
Vertical three way intersection	PC £33.40	0.25	1.78	36.82	nr	**38.60**
Four way intersection	PC £36.54	0.30	2.14	40.29	nr	**42.43**

E IN SITU CONCRETE/LARGE PRECAST CONCRETE Including overheads and profit at 5.00%		Labour hours	Labour £	Material £	Unit	Total rate £
230 mm wide	PC £166.91/9m	0.25	1.78	22.05	m	23.83
Flat angle	PC £35.55	0.22	1.57	39.20	nr	40.77
Vertical angle	PC £35.55	0.22	1.57	39.20	nr	40.77
Flat three way intersection	PC £37.19	0.27	1.92	41.00	nr	42.92
Vertical three way intersection	PC £37.19	0.27	1.92	41.00	nr	42.92
Four way intersection	PC £43.10	0.33	2.35	47.52	nr	49.87
305 mm wide	PC £275.25/9m	0.30	2.14	36.08	m	38.22
Flat angle	PC £55.86	0.24	1.71	61.59	nr	63.30
Vertical angle	PC £55.86	0.24	1.71	61.59	nr	63.30
Flat three way intersection	PC £67.54	0.30	2.14	74.46	nr	76.60
Vertical three way intersection	PC £67.54	0.30	2.14	74.46	nr	76.60
Four way intersection	PC £84.57	0.36	2.56	93.24	nr	95.80

E41 WORKED FINISHES/CUTTING ON IN SITU CONCRETE

	Labour hours	Labour £	Material £	Unit	Total rate £
Tamping by mechanical means	0.03	0.18	0.12	m2	0.30
Power floating	0.17	1.04	0.42	m2	1.46
Trowelling	0.33	2.01	-	m2	2.01
Lightly shot blast surface of concrete to receive finishes	0.40	2.44	-	m2	2.44
Hacking					
by mechanical means	0.33	2.01	0.36	m2	2.37
by hand	0.70	4.26	-	m2	4.26
Wood float finish	0.13	0.79	-	m2	0.79
Tamped finish	0.05	0.30	-	m2	0.30
to falls	0.07	0.43	-	m2	0.43
to crossfalls	0.10	0.61	-	m2	0.61
Spade finish	0.15	0.91	-	m2	0.91
Cutting chases					
not exceeding 50 mm deep; 10 mm wide	0.33	2.01	0.16	m	2.17
not exceeding 50 mm deep; 50 mm wide	0.50	3.05	0.24	m	3.29
not exceeding 50 mm deep; 75 mm wide	0.66	4.02	0.31	m	4.33
50 - 100 mm deep; 75 mm wide	0.90	5.48	0.43	m	5.91
50 - 100 mm deep; 100 mm wide	1.00	6.09	0.47	m	6.56
100 - 150 mm deep; 100 mm wide	1.30	7.92	0.61	m	8.53
100 - 150 mm deep; 150 mm wide	1.60	9.74	0.76	m	10.50
Cutting chases in reinforced concrete					
50 - 100 mm deep; 100 mm wide	1.50	9.13	0.71	m	9.84
100 - 150 mm deep; 100 mm wide	2.00	12.18	0.94	m	13.12
100 - 150 mm deep; 150 mm wide	2.40	14.62	1.13	m	15.75
Cutting rebates					
not exceeding 50 mm deep; 50 mm wide	0.50	3.05	0.24	m	3.29
50 - 100 mm deep; 100 mm wide	1.00	6.09	0.47	m	6.56
Cutting mortices; not exceeding 100 mm deep; making good					
20 mm dia	0.15	0.91	0.06	nr	0.97
50 mm dia	0.17	1.04	0.08	nr	1.12
150 x 150 mm	0.35	2.13	0.18	nr	2.31
300 x 300 mm	0.70	4.26	0.38	nr	4.64
Cutting mortices in reinforced concrete; not exceeding 100 mm deep; making good					
150 x 150 mm	0.55	3.35	0.25	nr	3.60
300 x 300 mm	1.05	6.39	0.52	nr	6.91
Cutting holes; not exceeding 100 mm deep					
50 mm dia	0.35	2.13	0.39	nr	2.52
100 mm dia	0.40	2.44	0.44	nr	2.88
150 x 150 mm	0.45	2.74	0.50	nr	3.24
300 x 300 mm	0.55	3.35	0.61	nr	3.96
Cutting holes; 100 - 200 mm deep					
50 mm dia	0.50	3.05	0.55	nr	3.60
100 mm dia	0.60	3.65	0.66	nr	4.31
150 x 150 mm	0.75	4.57	0.83	nr	5.40
300 x 300 mm	0.95	5.79	1.05	nr	6.84

E IN SITU CONCRETE/LARGE PRECAST CONCRETE Including overheads and profit at 5.00%	Labour hours	Labour £	Material £	Unit	Total rate £
E41 WORKED FINISHES/CUTTING ON IN SITU CONCRETE - cont'd					
Cutting holes; 200 - 300 mm deep					
50 mm dia	0.75	4.57	0.83	nr	5.40
100 mm dia	0.90	5.48	0.99	nr	6.47
150 x 150 mm	1.10	6.70	1.21	nr	7.91
300 x 300 mm	1.40	8.53	1.54	nr	10.07
Add for making good fair finish one side					
50 mm dia	0.05	0.30	0.02	nr	0.32
100 mm dia	0.12	0.73	0.02	nr	0.75
150 x 150 mm	0.20	1.22	0.04	nr	1.26
300 x 300 mm	0.40	2.44	0.08	nr	2.52
Add for fixing only sleeve					
50 mm dia	0.10	0.61	-	nr	0.61
100 mm dia	0.22	1.34	-	nr	1.34
150 x 150 mm	0.33	2.01	-	nr	2.01
300 x 300 mm	0.60	3.65	-	nr	3.65
Cutting holes in reinforced concrete; not exceeding 100 mm deep					
50 mm dia	0.55	3.35	0.61	nr	3.96
100 mm dia	0.60	3.65	0.66	nr	4.31
150 x 150 mm dia	0.70	4.26	0.77	nr	5.03
300 x 300 mm dia	0.85	5.18	0.94	nr	6.12
Cutting holes in reinforced concrete; 100 - 200 mm deep					
50 mm dia	0.75	4.57	0.83	nr	5.40
100 mm dia	0.90	5.48	0.99	nr	6.47
150 x 150 mm dia	1.15	7.00	1.27	nr	8.27
300 x 300 mm dia	1.45	8.83	1.60	nr	10.43
Cutting holes in reinforced concrete; 200 - 300 mm deep					
50 mm dia	1.15	7.00	1.27	nr	8.27
100 mm dia	1.35	8.22	1.49	nr	9.71
150 x 150 mm dia	1.65	10.05	1.82	nr	11.87
300 x 300 mm dia	2.10	12.79	2.32	nr	15.11
E42 ACCESSORIES CAST INTO IN SITU CONCRETE					
Temporary plywood foundation bolt boxes					
75 x 75 x 150 mm	0.45	2.74	0.35	nr	3.09
75 x 75 x 250 mm	0.50	3.05	0.57	nr	3.62
'Expamet' cylindrical expanded steel foundation boxes					
76 mm dia x 152 mm high	0.30	1.83	2.83	nr	4.66
76 mm dia x 305 mm high	0.20	1.22	0.91	nr	2.13
102 mm dia x 457 mm high	0.25	1.52	1.63	nr	3.15
10 mm dia x 100 mm long	0.25	1.52	1.07	nr	2.59
12 mm dia x 120 mm long	0.25	1.52	1.22	nr	2.74
16 mm dia x 160 mm long	0.30	1.83	2.80	nr	4.63
20 mm dia x 200 mm long	0.30	1.83	2.83	nr	4.66
'Abbey' galvanized steel masonry slots; 18 G (1.22 mm)					
3.048 m lengths	0.35	2.13	0.95	m	3.08
76 mm long	0.08	0.49	0.13	nr	0.62
102 mm long	0.08	0.49	0.15	nr	0.64
152 mm long	0.09	0.55	0.19	nr	0.74
229 mm long	0.10	0.61	0.27	nr	0.88

E IN SITU CONCRETE/LARGE PRECAST CONCRETE Including overheads and profit at 5.00%	Labour hours	Labour £	Material £	Unit	Total rate £
'Unistrut' galvanized steel slotted metal inserts; 2.5 mm thick; end caps and foam filling					
41 x 41 mm; ref P3270	0.40	2.44	4.24	m	6.68
41 x 41 x 75 mm; ref P3249	0.10	0.61	1.59	nr	2.20
41 x 41 x 100 mm; ref P3250	0.10	0.61	1.70	nr	2.31
41 x 41 x 150 mm; ref P3251	0.10	0.61	1.92	nr	2.53
Butterfly type wall ties; casting one end into concrete; other end built into joint of brickwork					
galvanized steel	0.10	0.61	0.07	nr	0.68
stainless steel	0.10	0.61	0.09	nr	0.70
Mild steel fixing cramp; once bent; one end shot fired into concrete; other end fanged and built into joint of brickwork					
200 mm girth	0.15	0.91	0.51	nr	1.42
Sherardized steel floor clips; pinned to surface of concrete					
50 mm wide; standard type	0.08	0.49	0.10	nr	0.59
50 mm wide; direct fix acoustic type	0.10	0.61	0.65	nr	1.26
Hardwood dovetailed fillets					
50 x 50/40 x 1000 mm	0.10	0.61	1.09	nr	1.70
50 x 50/40 x 100 mm	0.08	0.49	0.16	nr	0.65
50 x 50/40 x 200 mm	0.08	0.49	0.22	nr	0.71
'Rigifix' galvanized steel plate column guard; 1 m long					
75 mm x 75 mm x 3 mm	0.60	3.65	7.81	nr	11.46
75 mm x 75 mm x 4.5 mm	0.60	3.65	10.44	nr	14.09
'Rigifix' white nylon coated steel plate corner guard; plugged and screwed to concrete with chromium plated domed headed screws					
75 x 75 x 1.5 mm x 1 m long	0.80	4.87	11.50	nr	16.37
E60 PRECAST/COMPOSITE CONCRETE DECKING					
Prestressed precast flooring planks; Bison 'Drycast' or similar; cement and sand (1:3) grout between planks and on prepared bearings					
100 mm thick suspended slabs; horizontal					
400 mm wide planks	-	-	-	m2	28.35
1200 mm wide planks	-	-	-	m2	26.25
150 mm thick suspended slabs; horizontal					
400 mm wide planks	-	-	-	m2	28.88
1200 mm wide planks	-	-	-	m2	26.78
Prestressed precast concrete beam and block floor; Bison 'Housefloor' or similar; in situ concrete 30 N/mm2 - 10 mm aggregate in filling at wall abutments; cement and sand (1:6) grout brushed in between beams and blocks					
155 mm thick suspended slab at ground level; 440 x 215 x 100 mm blocks; horizontal					
beams at 520 mm centres; up to 3.30 m span with a superimposed load of 5 kN/m2	-	-	-	m2	16.80
beams at 295 mm centres; up to 4.35 m span with a superimposed load of 5 kN/m2	-	-	-	m2	19.42

E IN SITU CONCRETE/LARGE PRECAST CONCRETE Including overheads and profit at 5.00%	Labour hours	Labour £	Material £	Unit	Total rate £
E60 PRECAST/COMPOSITE CONCRETE DECKING - cont'd					
Composite floor comprising reinforced in situ ready-mixed concrete 31.00 N/mm2; on and including 1.2 mm 'Super Holorib' steel deck permanent shuttering; complete with reinforcment to support imposed loading and A142 anti-crack reinforcement					
150 mm thick suspended slab; 5 kN/m2 loading					
1.5 - 3 m height to soffit	1.75	11.77	25.54	m2	**37.31**
3 - 4.5 m height to soffit	1.90	12.84	26.25	m2	**39.09**
4.5 - 6 m height to soffit	2.05	13.91	27.29	m2	**41.20**
200 mm thick suspended slab; 7.5 kN/m2 loading					
1.5 - 3 m height to soffit	1.79	12.02	29.36	m2	**41.38**
3 - 4.5 m height to soffit	1.94	13.09	30.50	m2	**43.59**
4.5 - 6 m height to soffit	2.09	14.15	32.06	m2	**46.21**

F MASONRY
Including overheads and profit at 5.00%

BASIC MORTAR PRICES

	£		£		£		£
Coloured mortar materials (£/tonne); (excluding cement)							
light	26.72	medium	28.22	dark	30.97	extra dark	30.97
Mortar materials (£/tonne)							
cement	64.01	lime	100.72	sand	8.86	white cement	70.90

Mortar plasticizer - £1.67/Litre

	Labour hours	Labour £	Material £	Unit	Total rate £
F10 BRICK/BLOCK WALLING					
Common bricks; PC £120.00/1000; in cement mortar (1:3)					
Walls					
half brick thick	1.25	12.67	8.97	m2	**21.64**
one brick thick	2.10	21.28	18.62	m2	**39.90**
one and a half brick thick	2.85	28.88	28.00	m2	**56.88**
two brick thick	3.50	35.46	37.24	m2	**72.70**
Walls; facework one side					
half brick thick	1.40	14.19	8.97	m2	**23.16**
one brick thick	2.25	22.80	18.62	m2	**41.42**
one and a half brick thick	3.00	30.40	28.00	m2	**58.40**
two brick thick	3.65	36.98	37.24	m2	**74.22**
Walls; facework both sides					
half brick thick	1.50	15.20	8.97	m2	**24.17**
one brick thick	2.35	23.81	18.62	m2	**42.43**
one and a half brick thick	3.10	31.41	28.00	m2	**59.41**
two brick thick	3.75	38.00	37.24	m2	**75.24**
Walls; built curved mean radius 6 m					
half brick thick	1.65	16.72	9.66	m2	**26.38**
one brick thick	2.75	27.86	19.98	m2	**47.84**
Walls; built curved mean radius 1.50 m					
half brick thick	2.10	21.28	10.15	m2	**31.43**
one brick thick	3.45	34.96	20.98	m2	**55.94**
Walls; built overhand					
half brick thick	1.55	15.71	8.97	m2	**24.68**
Walls; building up against concrete including flushing up at back					
half brick thick	1.35	13.68	10.10	m2	**23.78**
Walls; backing to masonry; cutting and bonding					
one brick thick	2.50	25.33	19.00	m2	**44.33**
one and a half brick thick	3.35	33.94	28.56	m2	**62.50**
Honeycomb walls					
half brick thick	1.00	10.13	6.33	m2	**16.46**
Dwarf support wall					
half brick thick	1.55	15.71	8.97	m2	**24.68**
one brick thick	2.50	25.33	18.62	m2	**43.95**
Battering walls					
one and a half brick thick	3.30	33.44	28.56	m2	**62.00**
two brick thick	4.10	41.54	37.99	m2	**79.53**
Walls; tapering one side; average					
337 mm thick	3.65	36.98	29.13	m2	**66.11**
450 mm thick	4.70	47.62	38.75	m2	**86.37**

F MASONRY Including overheads and profit at 5.00%	Labour hours	Labour £	Material £	Unit	Total rate £
F10 BRICK/BLOCK WALLING - cont'd					
Common bricks; PC £120.00/1000;					
in cement mortar (1:3) - cont'd					
Walls; tapering both sides; average					
337 mm thick	4.20	42.56	29.13	m2	**71.69**
450 mm thick	5.25	53.20	38.75	m2	**91.95**
Isolated piers					
one brick thick	3.20	32.42	19.00	m2	**51.42**
two brick thick	5.00	50.66	37.99	m2	**88.65**
three brick thick	6.30	63.83	56.99	m2	**120.82**
Isolated casings to steel columns					
half brick thick	1.60	16.21	9.16	m2	**25.37**
one brick thick	2.75	27.86	19.00	m2	**46.86**
Chimney stacks					
one brick thick	3.20	32.42	19.00	m2	**51.42**
two brick thick	5.00	50.66	37.99	m2	**88.65**
three brick thick	6.30	63.83	56.99	m2	**120.82**
Projections; vertical					
225 x 112 mm	0.40	4.05	2.13	m	**6.18**
225 x 225 mm	0.75	7.60	4.10	m	**11.70**
337 x 225 mm	1.10	11.15	6.90	m	**18.05**
440 x 225 mm	1.25	12.67	8.08	m	**20.75**
Bonding ends to existing					
half brick thick	0.40	4.05	0.65	m	**4.70**
one brick thick	0.55	5.57	1.30	m	**6.87**
one and a half brick thick	0.85	8.61	1.96	m	**10.57**
two brick thick	1.20	12.16	2.62	m	**14.78**
ADD or DEDUCT to walls for variation of £10.00/1000 in PC of common bricks					
half brick thick	-	-	-	m2	**0.68**
one brick thick	-	-	-	m2	**1.35**
one and a half brick thick	-	-	-	m2	**2.03**
two brick thick	-	-	-	m2	**2.65**
Extra over walls for sulphate-resisting cement mortar (1:3) in lieu of cement mortar (1:3)					
half brick thick	-	-	-	m2	**0.11**
one brick thick	-	-	-	m2	**0.29**
one and a half brick thick	-	-	-	m2	**0.44**
two brick thick	-	-	-	m2	**0.58**
Common bricks; PC £120.00/1000;					
in gauged mortar (1:1:6)					
Walls					
half brick thick	1.25	12.67	8.91	m2	**21.58**
one brick thick	2.10	21.28	18.46	m2	**39.74**
one and a half brick thick	2.85	28.88	27.75	m2	**56.63**
two brick thick	3.50	35.46	36.91	m2	**72.37**
Walls; facework one side					
half brick thick	1.40	14.19	8.91	m2	**23.10**
one brick thick	2.25	22.80	18.46	m2	**41.26**
one and a half brick thick	3.00	30.40	27.75	m2	**58.15**
two brick thick	3.65	36.98	36.91	m2	**73.89**
Walls; facework both sides					
half brick thick	1.50	15.20	8.91	m2	**24.11**
one brick thick	2.35	23.81	18.46	m2	**42.27**
one and a half brick thick	3.10	31.41	27.75	m2	**59.16**
two brick thick	3.75	38.00	36.91	m2	**74.91**
Walls; built curved mean radius 6 m					
half brick thick	1.65	16.72	9.59	m2	**26.31**
one brick thick	2.75	27.86	19.81	m2	**47.67**

F MASONRY Including overheads and profit at 5.00%	Labour hours	Labour £	Material £	Unit	Total rate £
Walls; built curved mean radius 1.50 m					
half brick thick	2.10	21.28	10.08	m2	31.36
one brick thick	3.45	34.96	20.80	m2	55.76
Walls; built overhand					
half brick thick	1.55	15.71	8.91	m2	24.62
Walls; built up against concrete including flushing up at back					
half brick thick	1.35	13.68	9.97	m2	23.65
Walls; backing to masonry; cutting and bonding					
one brick thick	2.50	25.33	18.83	m2	44.16
one and a half brick thick	3.35	33.94	28.32	m2	62.26
Honeycomb walls					
half brick thick	1.00	10.13	6.27	m2	16.40
Dwarf support wall					
half brick thick	1.55	15.71	8.91	m2	24.62
one brick thick	2.50	25.33	18.46	m2	43.79
Battering walls					
one and a half brick thick	3.30	33.44	28.32	m2	61.76
two brick thick	4.10	41.54	37.67	m2	79.21
Walls; tapering one side; average					
337 mm thick	3.65	36.98	28.88	m2	65.86
450 mm thick	4.70	47.62	38.43	m2	86.05
Walls; tapering both sides; average					
337 mm thick	4.20	42.56	28.88	m2	71.44
450 mm thick	5.25	53.20	38.43	m2	91.63
Isolated piers					
one brick thick	3.20	32.42	18.83	m2	51.25
two brick thick	5.00	50.66	37.67	m2	88.33
three brick thick	6.30	63.83	56.50	m2	120.33
Isolated casings to steel columns					
half brick thick	1.60	16.21	9.10	m2	25.31
one brick thick	2.75	27.86	18.83	m2	46.69
Chimney stacks					
one brick thick	3.20	32.42	18.83	m2	51.25
two brick thick	5.00	50.66	37.67	m2	88.33
three brick thick	6.30	63.83	56.50	m2	120.33
Projections; vertical					
225 x 112 mm	0.40	4.05	2.11	m	6.16
225 x 225 mm	0.75	7.60	4.08	m	11.68
337 x 225 mm	1.10	11.15	6.86	m	18.01
440 x 225 mm	1.25	12.67	8.03	m	20.70
Bonding ends to existing					
half brick thick	0.40	4.05	0.64	m	4.69
one brick thick	0.55	5.57	1.28	m	6.85
one and a half brick thick	0.85	8.61	1.93	m	10.54
two brick thick	1.20	12.16	2.58	m	14.74
ADD or DEDUCT to walls for variation of £10.00/1000 in PC of common bricks					
half brick thick	-	-	-	m2	0.66
one brick thick	-	-	-	m2	1.32
one and a half brick thick	-	-	-	m2	1.98
two brick thick	-	-	-	m2	2.65
Segmental arches; one ring, 102 mm high on face					
102 mm wide on exposed soffit	2.20	19.58	2.52	m	22.10
215 mm wide on exposed soffit	2.65	22.78	5.24	m	28.02
Segmental arches; two ring, 215 mm high on face					
102 mm wide on exposed soffit	2.80	25.66	3.88	m	29.54
215 mm wide on exposed soffit	3.25	28.86	7.95	m	36.81

F MASONRY Including overheads and profit at 5.00%	Labour hours	Labour £	Material £	Unit	Total rate £
F10 BRICK/BLOCK WALLING - cont'd					
Common bricks; PC £120.00/1000; in gauged mortar (1:1:6) - cont'd					
Semi-circular arches; one ring, 102 mm high on face					
102 mm wide on exposed soffit	2.80	24.75	3.07	m	**27.82**
215 mm wide on exposed soffit	3.20	27.60	6.12	m	**33.72**
Semi-circular arches; two ring, 215 mm high on face					
102 mm wide on exposed soffit	3.60	32.86	4.69	m	**37.55**
215 mm wide on exposed soffit	4.00	35.71	8.51	m	**44.22**
Labours on brick fairface					
Fair returns					
half brick wide	0.03	0.30	-	m	**0.30**
one brick wide	0.05	0.51	-	m	**0.51**
Class A engineering bricks; PC £340.00/1000; in cement mortar (1:3)					
Walls					
half brick thick	1.35	13.68	22.99	m2	**36.67**
one brick thick	2.25	22.80	46.65	m2	**69.45**
one and a half brick thick	3.00	30.40	70.05	m2	**100.45**
two brick thick	3.75	38.00	93.31	m2	**131.31**
Walls; facework one side					
half brick thick	1.50	15.20	23.53	m2	**38.73**
one brick thick	2.40	24.32	47.72	m2	**72.04**
one and a half brick thick	3.15	31.92	71.65	m2	**103.57**
two brick thick	3.90	39.52	95.45	m2	**134.97**
Walls; facework both sides					
half brick thick	1.60	16.21	23.53	m2	**39.74**
one brick thick	2.50	25.33	47.72	m2	**73.05**
one and a half brick thick	3.25	32.93	71.65	m2	**104.58**
two brick thick	4.00	40.53	95.45	m2	**135.98**
Walls; built curved mean radius 6 m					
one brick thick	3.00	30.40	50.15	m2	**80.55**
Walls; backing to masonry; cutting and bonding					
one brick thick	2.70	27.36	48.80	m2	**76.16**
one and a half brick thick	3.60	36.48	73.26	m2	**109.74**
Walls; tapering one side; average					
337 mm thick	3.90	39.52	73.26	m2	**112.78**
450 mm thick	5.00	50.66	97.59	m2	**148.25**
Walls; tapering both sides; average					
337 mm thick	4.50	45.60	74.87	m2	**120.47**
450 mm thick	5.70	57.76	99.73	m2	**157.49**
Isolated piers					
one brick thick	3.50	35.46	47.72	m2	**83.18**
two brick thick	5.50	55.73	95.45	m2	**151.18**
three brick thick	6.75	68.39	143.26	m2	**211.65**
Isolated casings to steel columns					
half brick thick	1.75	17.73	24.06	m2	**41.79**
one brick thick	3.00	30.40	48.80	m2	**79.20**
Projections; vertical					
225 x 112 mm	0.45	4.56	5.60	m	**10.16**
225 x 225 mm	0.80	8.11	10.81	m	**18.92**
337 x 225 mm	1.20	12.16	18.32	m	**30.48**
440 x 225 mm	1.35	13.68	21.24	m	**34.92**
Bonding ends to existing					
half brick thick	0.45	4.56	1.49	m	**6.05**
one brick thick	0.60	6.08	2.98	m	**9.06**
one and a half brick thick	0.90	9.12	4.48	m	**13.60**
two brick thick	1.30	13.17	5.97	m	**19.14**

F MASONRY Including overheads and profit at 5.00%	Labour hours	Labour £	Material £	Unit	Total rate £
ADD or DEDUCT to walls for variation of £10.00/1000 in PC of engineering bricks					
half brick thick	-	-	-	m2	**0.66**
one brick thick	-	-	-	m2	**1.32**
one and a half brick thick	-	-	-	m2	**1.98**
two brick thick	-	-	-	m2	**2.65**
Class B engineering bricks; PC £175.00/1000; in cement mortar (1:3)					
Walls					
half brick thick	1.35	13.68	12.33	m2	**26.01**
one brick thick	2.25	22.80	25.35	m2	**48.15**
one and a half brick thick	3.00	30.40	38.09	m2	**68.49**
two brick thick	3.75	38.00	50.69	m2	**88.69**
Walls; facework one side					
half brick thick	1.50	15.20	12.62	m2	**27.82**
one brick thick	2.40	24.32	25.90	m2	**50.22**
one and a half brick thick	3.15	31.92	38.92	m2	**70.84**
two brick thick	3.90	39.52	51.79	m2	**91.31**
Walls; facework both sides					
half brick thick	1.60	16.21	12.62	m2	**28.83**
one brick thick	2.50	25.33	25.90	m2	**51.23**
one and a half brick thick	3.25	32.93	38.92	m2	**71.85**
two brick thick	4.00	40.53	51.79	m2	**92.32**
Walls; built curved mean radius 6 m					
one brick thick	3.00	30.40	27.23	m2	**57.63**
Walls; backing to masonry; cutting and bonding					
one brick thick	2.70	27.36	26.45	m2	**53.81**
one and a half brick thick	3.60	36.48	39.73	m2	**76.21**
Walls; tapering one side; average					
337 mm thick	3.90	39.52	39.73	m2	**79.25**
450 mm thick	5.00	50.66	52.89	m2	**103.55**
Walls; tapering both sides; average					
337 mm thick	4.50	45.60	40.56	m2	**86.16**
450 mm thick	5.70	57.76	54.00	m2	**111.76**
Isolated piers					
one brick thick	3.50	35.46	25.90	m2	**61.36**
two brick thick	5.50	55.73	51.79	m2	**107.52**
three brick thick	6.75	68.39	77.78	m2	**146.17**
Isolated casings to steel columns					
half brick thick	1.75	17.73	12.89	m2	**30.62**
one brick thick	3.00	30.40	26.45	m2	**56.85**
Projections					
225 x 112 mm	0.45	4.56	2.99	m	**7.55**
225 x 225 mm	0.80	8.11	5.78	m	**13.89**
337 x 225 mm	1.20	12.16	9.76	m	**21.92**
440 x 225 mm	1.35	13.68	11.37	m	**25.05**
Bonding ends to existing					
half brick thick	0.45	4.56	0.85	m	**5.41**
one brick thick	0.60	6.08	1.71	m	**7.79**
one and a half brick thick	0.90	9.12	2.56	m	**11.68**
two brick thick	1.30	13.17	3.42	m	**16.59**
ADD or DEDUCT to walls for variation of £10.00/1000 in PC of engineering bricks					
half brick thick	-	-	-	m2	**0.66**
one brick thick	-	-	-	m2	**1.32**
one and a half brick thick	-	-	-	m2	**1.98**
two brick thick	-	-	-	m2	**2.65**

F MASONRY Including overheads and profit at 5.00%	Labour hours	Labour £	Material £	Unit	Total rate £
F10 BRICK/BLOCK WALLING - cont'd					
Refractory bricks; PC £531.00/1000; stretcher bond lining to flue; in fireclay cement mortar (1:4); built 50 mm clear of flues; one header per m2					
Walls; vertical; facework one side					
half brick thick	1.70	17.23	34.63	m2	**51.86**

ALTERNATIVE FACING BRICK PRICES (£/1000)

	£		£
Ibstock facing bricks; 215 x 102.5 x 65 mm			
Aldridge brown blend	302.00	Leicester Anglican Red Rustic	294.00
Cattybrook Gloucester Golden	300.00	Leicester Red Stock	326.00
Himley Dark Brown Rustic	345.00	Roughdales Red Multi Rustic	330.00
Himley Mixed Russet	297.00	Roughdales Trafford Buff Multi	340.00
London Brick Company facing bricks; 215 x 102.5 x 65 mm			
Brecken Grey	121.26	Orton Multi Buff	132.86
Chiltern	133.58	Regency	122.70
Claydon Red Multi	123.94	Sandfaced	133.53
Delph Autumn	121.60	Saxon Gold	128.89
Edwardian	127.26	Tudor	136.04
Georgian Red Multi	124.55	Victorian	134.50
Heather	131.72	Wansford Multi	121.74
Ironstone	123.66	Windsor	124.13
Milton Buff	127.77		
Redland facing bricks; 215 x 102.5 x 65 mm			
Arun	409.00	Southwater class B	280.00
Beare Green restoration red	572.00	Sheppy 'matured' yellow	420.00
Chailey yellow multicoloured	413.00	Stourbridge Sherbourne range	317.00
Cottage mixed multicoloured	303.00	Stourbridge Henley range	297.00
Crowborough multicoloured	378.00	Stourbridge Pennine range	286.00
Dorking	314.00	Stourbridge Stratford range	241.00
Funton yellow London	300.00	Surrey bronze multicoloured	326.00
Hamsey multicoloured	397.00	Tonbridge handmade	518.00
Holbrook Sherbourne textured	311.00	Tonbridge handmade (50 mm deep)	518.00
Holbrook smooth red	355.00	Tudor (53 mm deep)	657.00
Nutbourne sandfaced	288.00	Wealden	409.00
Pevensey red multicoloured	381.00	Wealdmade	450.00
Pluckley multicoloured	396.00		

	Labour hours	Labour £	Material £	Unit	Total rate £
Facing bricks; sand faced; PC £130.00/1000 (unless otherwise stated); in gauged mortar (1:1:6)					
Extra over common bricks; PC £120.00/1000; for facing bricks in					
stretcher bond	0.40	4.05	0.68	m2	**4.73**
flemish bond with snapped headers	0.50	5.07	3.61	m2	**8.68**
english bond with snapped headers	0.50	5.07	3.53	m2	**8.60**
ADD or DEDUCT for variation of £10.00/1000 in PC of facing bricks	-	-	-	m2	**0.90**

F MASONRY Including overheads and profit at 5.00%	Labour hours	Labour £	Material £	Unit	Total rate £
Half brick thick; stretcher bond; facework one side					
walls	1.65	16.72	9.78	m2	26.50
walls; building curved mean radius 6 m	2.40	24.32	10.52	m2	34.84
walls; building curved mean radius 1.50 m	3.00	30.40	11.05	m2	41.45
walls; building overhand	2.00	20.27	9.78	m2	30.05
walls; building up against concrete including flushing up at back	1.75	17.73	11.03	m2	28.76
walls; as formwork; temporary strutting	2.40	24.32	11.90	m2	36.22
walls; panels and aprons; not exceeding 1 m2	2.10	21.28	9.81	m2	31.09
isolated casings to steel columns	2.50	25.33	9.78	m2	35.11
bonding ends to existing	0.65	6.59	1.46	m	8.05
projections; vertical					
225 x 112 mm	0.40	4.05	2.27	m	6.32
337 x 112 mm	0.75	7.60	3.50	m	11.10
440 x 112 mm	1.10	11.15	4.74	m	15.89
Half brick thick; flemish bond with snapped headers; facework one side					
walls	1.90	19.25	10.73	m2	29.98
walls; building curved mean radius 6 m	2.70	27.36	11.26	m2	38.62
walls; building curved mean radius 1.50 m	3.50	35.46	11.80	m2	47.26
walls; building overhand	2.25	22.80	10.73	m2	33.53
walls; building up against concrete including flushing up at back	2.00	20.27	11.98	m2	32.25
walls; as formwork; temporary strutting	2.65	26.85	12.85	m2	39.70
walls; panels and aprons; not exceeding 1 m2	2.35	23.81	10.73	m2	34.54
isolated casings to steel columns	2.75	27.86	9.78	m2	37.64
bonding ends to existing	0.65	6.59	1.46	m	8.05
projections; vertical					
225 x 112 mm	0.50	5.07	2.56	m	7.63
337 x 112 mm	0.85	8.61	3.79	m	12.40
440 x 112 mm	1.20	12.16	5.18	m	17.34
One brick thick; two stretcher skins tied together; facework both sides					
walls	2.80	28.37	20.37	m2	48.74
walls; building curved mean radius 6 m	3.90	39.52	21.84	m2	61.36
walls; building curved mean radius 1.50 m	4.80	48.64	25.05	m2	73.69
isolated piers	3.30	33.44	22.12	m2	55.56
bonding ends to existing	0.85	8.61	2.92	m	11.53
One brick thick; flemish bond; facework both sides					
walls	2.90	29.38	20.19	m2	49.57
walls; building curved mean radius 6 m	4.00	40.53	21.67	m2	62.20
walls; building curved mean radius 1.50 m	5.00	50.66	24.88	m2	75.54
isolated piers	3.40	34.45	21.95	m2	56.40
bonding ends to existing	0.85	8.61	2.92	m	11.53
projections; vertical					
225 x 225 mm	0.80	8.11	2.99	m	11.10
337 x 225 mm	1.50	15.20	4.34	m	19.54
440 x 225 mm	2.20	22.29	5.82	m	28.11
ADD or DEDUCT for variation of £10.00/1000 in PC of facing bricks; in stretcher bond					
half brick thick	-	-	-	m2	0.68
one brick thick	-	-	-	m2	1.37
ADD or DEDUCT for variation of £10.00/1000 in PC of facing bricks; in flemish bond					
half brick thick	-	-	-	m2	0.85
one brick thick	-	-	-	m2	1.70
Extra over facing bricks for					
recessed joints	0.03	0.30	-	m2	0.30
raking out joints and pointing in black mortar	0.50	5.07	0.16	m2	5.23
bedding and pointing half brick wall in black mortar	-	-	-	m2	0.78

F MASONRY Including overheads and profit at 5.00%	Labour hours	Labour £	Material £	Unit	Total rate £
F10 BRICK/BLOCK WALLING - cont'd					
Facing bricks; sand faced; PC £130.00/1000 (unless otherwise stated); in gauged mortar (1:1:6) - cont'd					
Extra over facing bricks for					
bedding and pointing one brick wall in black mortar	-	-	-	m2	2.07
flush plain bands; 225 mm wide stretcher bond; horizontal; bricks; PC £150.00/1000	0.25	2.53	0.31	m	2.84
flush quoins; average 320 mm girth; block bond vertical; facing bricks; PC £150.00/1000	0.40	4.05	0.29	m	4.34
Flat arches; 215 mm high on face					
102 mm wide exposed soffit	1.15	10.45	1.46	m	11.91
215 mm wide exposed soffit	1.72	15.62	2.93	m	18.55
Flat arches; 215 mm high on face; bullnosed specials; PC £46.08/100					
102 mm wide exposed soffit	1.20	10.95	8.15	m	19.10
215 mm wide exposed soffit	1.80	16.43	16.32	m	32.75
Segmental arches; one ring; 215 mm high on face					
102 mm wide exposed soffit	2.10	18.26	2.05	m	20.31
215 mm wide exposed soffit	3.15	27.40	4.30	m	31.70
Segmental arches; two ring; 215 mm high on face					
102 mm wide exposed soffit	2.70	24.34	2.05	m	26.39
215 mm wide exposed soffit	4.05	36.52	4.30	m	40.82
Segmental arches; 215 mm high on face; cut voussoirs; PC £63.99/100					
102 mm wide exposed soffit	2.20	19.28	11.28	m	30.56
215 mm wide exposed soffit	3.30	28.92	22.75	m	51.67
Segmental arches; one and a half ring; 320 mm high on face; cut voussoirs; PC £63.99/100					
102 mm wide exposed soffit	3.00	27.38	21.39	m2	48.77
215 mm wide exposed soffit	4.50	41.08	42.98	m2	84.06
Semi circular arches; one ring; 215 mm high on face					
102 mm wide exposed soffit	3.60	32.86	4.97	m	37.83
215 mm wide exposed soffit	5.20	47.87	9.94	m	57.81
Semi circular arches; two ring; 215 mm high on face					
102 mm wide exposed soffit	4.80	45.02	4.97	m	49.99
215 mm wide exposed soffit	5.20	47.87	9.94	m	57.81
Semi circular arches; one ring; 215 mm high on face; cut voussoirs PC £63.99/100					
102 mm wide exposed soffit	2.70	23.74	11.54	m	35.28
215 mm wide exposed soffit	3.85	34.19	36.98	m	71.17
Bullseye window 600 mm dia; two rings; 215 mm high on face					
102 mm wide exposed soffit	6.00	57.78	3.12	nr	60.90
215 mm wide exposed soffit	9.00	86.67	6.03	nr	92.70
Bullseye window 1200 mm dia; two rings; 215 mm high on face					
102 mm wide exposed soffit	10.50	100.36	4.74	nr	105.10
215 mm wide exposed soffit	15.75	150.55	9.46	nr	160.01
Bullseye window 600 mm dia; one ring; 215 mm high on face; cut voussoirs PC £63.99/100					
102 mm wide exposed soffit	5.00	47.65	26.64	nr	74.29
215 mm wide exposed soffit	7.50	71.47	53.08	nr	124.55

F MASONRY Including overheads and profit at 5.00%	Labour hours	Labour £	Material £	Unit	Total rate £
Bullseye window 1200 mm dia; one ring; 215 mm high on face; cut voussoirs					
102 mm wide exposed soffit	9.00	85.17	45.71	nr	130.88
215 mm wide exposed soffit	13.50	127.75	91.43	nr	219.18
ADD or DEDUCT for variation of £10.00/1000 in PC of facing bricks	-	-	-	m	0.28
Sills; horizontal; headers on edge; pointing top and one side; set weathering					
150 x 102 mm	0.70	7.09	2.32	m	9.41
150 x 102 mm; cant headers; PC £59.49/100	0.75	7.60	9.65	m	17.25
Sills; horizontal; headers on flat; pointing top and one side					
150 x 102 mm; bullnosed specials; PC £46.08/100	0.65	6.59	5.38	m	11.97
Coping; horizontal; headers on edge; pointing top and both sides					
215 x 102 mm	0.56	5.67	2.27	m	7.94
260 x 102 mm	0.90	9.12	3.50	m	12.62
215 x 102 mm; double bullnosed specials; PC £47.07/100	0.60	6.08	7.65	m	13.73
260 x 120 mm; single bullnosed specials; PC £46.08/100	0.90	9.12	12.25	m	21.37
ADD or DEDUCT for variation of £10.00/1000 in PC of facing bricks	-	-	-	m	0.28
Facing bricks; white sandlime; PC £134.00/1000 in gauged mortar (1:1:6)					
Extra over common bricks; PC £120.00/1000; for facing bricks in					
stretcher bond	0.40	4.05	1.19	m2	5.24
flemish bond with snapped headers	0.50	5.07	4.28	m2	9.35
ADD or DEDUCT for variation of £10.00/1000 in PC of facing bricks	-	-	-	m2	0.90
Half brick thick; stretcher bond; facework one side					
walls	1.65	16.72	10.29	m2	27.01
One brick thick; flemish bond; facework both sides					
walls	2.90	29.38	21.21	m2	50.59
ADD or DEDUCT for variation of £10.00/1000 in PC of facing bricks; in flemish bond					
half brick thick	-	-	-	m2	0.85
one brick thick	-	-	-	m2	1.70
Facing bricks; machine made facings; PC £320.00/1000 (unless otherwise stated; in gauged mortar (1:1:6)					
Extra over common bricks; PC £120.00/1000; for facing bricks in					
stretcher bond	0.40	4.05	13.55	m2	17.60
flemish bond with snapped headers	0.50	5.07	16.94	m2	22.01
english bond with snapped headers	0.50	5.07	19.18	m2	24.25
ADD or DEDUCT for variation of £10.00/1000 in PC of facing bricks	-	-	-	m2	0.85
Half brick thick; stretcher bond; facework one side					
walls	1.65	16.72	22.65	m2	39.37
walls; building curved mean radius 6 m	2.40	24.32	24.35	m2	48.67
walls; building curved mean radius 1.50 m	3.00	30.40	25.53	m2	55.93
walls; building overhand	2.00	20.27	22.65	m2	42.92
walls; building up against concrete including flushing up at back	1.75	17.73	23.90	m2	41.63

F MASONRY Including overheads and profit at 5.00%	Labour hours	Labour £	Material £	Unit	Total rate £
F10 BRICK/BLOCK WALLING - cont'd					
Facing bricks; machine made facings; PC £320.00/1000 (unless otherwise stated; in gauged mortar (1:1:6) - cont'd					
Half brick thick; stretcher bond; facework one side					
walls; as formwork; temporary strutting	2.40	24.32	24.77	m2	49.09
walls; panels and aprons; not exceeding 1 m2	2.10	21.28	23.15	m2	44.43
isolated casings to steel columns	2.50	25.33	22.65	m2	47.98
bonding ends to existing	0.65	6.59	1.46	m	8.05
projections; vertical					
225 x 112 mm	0.40	4.05	5.27	m	9.32
337 x 112 mm	0.75	7.60	8.00	m	15.60
440 x 112 mm	1.10	11.15	10.75	m	21.90
Half brick thick; flemish bond with snapped headers; facework one side					
walls	1.90	19.25	25.00	m2	44.25
walls; building curved mean radius 6 m	2.70	27.36	26.19	m2	53.55
walls; building curved mean radius 1.50 m	3.50	35.46	27.38	m2	62.84
walls; building overhand	2.25	22.80	25.00	m2	47.80
walls; building up against concrete including flushing up at back	2.00	20.27	27.12	m2	47.39
walls; as formwork; temporary strutting	2.65	26.85	26.25	m2	53.10
walls; panels and aprons; not exceeding 1 m2	2.35	23.81	25.00	m2	48.81
isolated casings to steel columns	2.75	27.86	22.65	m2	50.51
bonding ends to existing	0.65	6.59	1.46	m	8.05
projections; vertical					
225 x 112 mm	0.50	5.07	5.99	m	11.06
337 x 112 mm	0.85	8.61	8.73	m	17.34
440 x 112 mm	1.20	12.16	11.83	m	23.99
One brick thick; two stretcher skins tied together; facework both sides					
walls	2.80	28.37	46.11	m2	74.48
walls; building curved mean radius 6 m	3.90	39.52	49.32	m2	88.84
walls; building curved mean radius 1.50 m	4.80	48.64	51.87	m2	100.51
isolated piers	3.30	33.44	46.10	m2	79.54
bonding ends to existing	0.85	8.61	2.92	m	11.53
One brick thick; flemish bond; facework both sides					
walls	2.90	29.38	45.92	m2	75.30
walls; building curved mean radius 6 m	4.00	40.53	49.32	m2	89.85
walls; building curved mean radius 1.50 m	5.00	50.66	51.70	m2	102.36
isolated piers	3.40	34.45	46.11	m2	80.56
bonding ends to existing	0.85	8.61	2.92	m	11.53
projections; vertical					
225 x 225 mm	0.80	8.11	10.18	m	18.29
337 x 225 mm	1.50	15.20	17.25	m	32.45
440 x 225 mm	2.20	22.29	19.99	m	42.28
ADD or DEDUCT for variation of £10.00/1000 in PC of facing bricks; in stretcher bond					
half brick thick	-	-	-	m2	0.68
one brick thick	-	-	-	m2	1.37
ADD or DEDUCT for variation of £10.00/1000 in PC of facing bricks; in flemish bond					
half brick thick	-	-	-	m2	0.85
one brick thick	-	-	-	m2	1.70
Extra over facing bricks for					
recessed joints	0.03	0.30	-	m2	0.30
raking out joints and pointing in black mortar	0.50	5.07	0.16	m2	5.23
bedding and pointing half brick wall in black mortar	-	-	-	m2	0.78

All four Spon Price Books – *Architects' and Builders', Civil Engineering and Highway Works, Landscape and External Works* and *Mechanical and Electrical Services* – are supported by an updating service. Updates are issued each November, February and May and give details of changes in prices of materials, wages rates and other significant items, with regional price level adjustments for Northern Ireland, Scotland and Wales and regions of England.

As a purchaser of this edition you are entitled to this updating service – *free of charge*. Simply complete this registration card and return it to us. Your completion of the reader survey data section – whilst not obligatory – would help ensure that we continue to provide the ideal reference book to facilitate your work.

REGISTRATION CARD for Spon's Price Book Update 1991

Please print your details clearly

Name..

(Please indicate membership of any professional body eg RICS, RIBA, CIOB etc.)

Address ....

....

.... Postcode ....

Signature .... Date ....

READERSHIP SURVEY – *Your assistance here would be appreciated*

How many other persons access your copy of the Price Book?

☐ myself alone ☐ 2 ☐ 3-5 ☐ more than 5

Profession:

☐ Architect ☐ Building Contractor ☐ Building Surveyor
☐ Civil Engineer ☐ Chemical Engineer ☐ Civil Engineer
☐ Cost Engineer ☐ Electrical Engineer ☐ Landscape Contractor
☐ Landscape Architect ☐ Mechanical Engineer ☐ Quantity Surveyor

☐ Other – please specify...

Would your employer be best described as:

☐ Commercial concern ☐ Local Authority
☐ Professional practice ☐ Public Utility

Thank you very much for your assistance.

AFFIX
STAMP
HERE

F MASONRY Including overheads and profit at 5.00%	Labour hours	Labour £	Material £	Unit	Total rate £
bedding and pointing one brick wall in black mortar	-	-	-	m2	**2.07**
flush plain bands; 225 mm wide stretcher bond; horizontal; bricks PC £350.00/1000	0.25	2.53	0.47	m	**3.00**
flush quoins; average 320 mm girth; black bond vertical; bricks PC £350.00/1000	0.40	4.05	0.44	m	**4.49**
Flat arches; 215 mm high on face					
102 mm wide exposed soffit	1.15	10.45	2.32	m	**12.77**
215 mm wide exposed soffit	1.72	15.62	4.65	m	**20.27**
Flat arches; 215 mm high on face; bullnosed specials; PC £113.00/100					
102 mm wide exposed soffit	1.20	10.95	18.74	m	**29.69**
215 mm wide exposed soffit	1.80	16.43	37.46	m	**53.89**
Segmental arches; one ring; 215 mm high on face					
102 mm wide exposed soffit	2.10	18.26	3.34	m	**21.60**
215 mm wide exposed soffit	3.15	27.40	6.89	m	**34.29**
Segmental arches; two ring; 215 mm high on face					
102 mm wide exposed soffit	2.70	24.34	3.34	m	**27.68**
215 mm wide exposed soffit	4.05	36.52	6.89	m	**43.41**
Segmental arches; 215 mm high on face; cut voussoirs; PC £245.00/100					
102 mm wide exposed soffit	2.20	19.28	39.88	m	**59.16**
215 mm wide exposed soffit	3.30	28.92	79.96	m	**108.88**
Segmental arches; one and a half ring; 320 mm high on face; cut voussoirs PC £245.00/100					
102 mm wide exposed soffit	3.00	27.38	78.60	m	**105.98**
215 mm wide exposed soffit	4.50	41.08	157.40	m	**198.48**
Semi circular arches; one ring; 215 mm high on face					
102 mm wide exposed soffit	3.60	32.86	10.14	m	**43.00**
215 mm wide exposed soffit	5.20	47.87	20.28	m	**68.15**
Semi circular arches; two ring; 215 mm high on face					
102 mm wide exposed soffit	4.80	45.02	10.14	m	**55.16**
215 mm wide exposed soffit	5.20	47.87	20.28	m	**68.15**
Semi circular arches; one ring; 215 mm high on face; cut voussoirs PC £245.00/100					
102 mm wide exposed soffit	2.70	23.74	40.15	m	**63.89**
215 mm wide exposed soffit	3.85	34.19	94.18	m	**128.37**
Bullseye window 600 mm dia; two rings; 215 mm high on face					
102 mm wide exposed soffit	6.00	57.78	44.90	nr	**102.68**
215 mm wide exposed soffit	9.00	86.67	89.60	nr	**176.27**
Bullseye window 1200 mm dia; two rings; 215 mm high on face					
102 mm wide exposed soffit	10.50	100.36	8.18	nr	**108.54**
215 mm wide exposed soffit	15.75	150.55	16.36	nr	**166.91**
Bullseye window 600 mm dia; one ring; 215 mm high on face; cut voussoirs PC £245.00/100					
102 mm wide exposed soffit	5.00	47.65	98.14	nr	**145.79**
215 mm wide exposed soffit	7.50	71.47	196.10	nr	**267.57**
Bullseye window 1200 mm dia; one ring; 215 mm high on face; cut voussoirs PC £245.00/100					
102 mm wide exposed soffit	9.00	85.17	169.07	nr	**254.24**
215 mm wide exposed soffit	13.50	127.75	338.15	nr	**465.90**
ADD or DEDUCT for variation of £10.00/1000 in PC of facing bricks	-	-	-	m2	**0.28**

F MASONRY Including overheads and profit at 5.00%	Labour hours	Labour £	Material £	Unit	Total rate £
F10 BRICK/BLOCK WALLING - cont'd					
Facing bricks; machine made facings; PC £320.00/1000 (unless otherwise stated; in gauged mortar (1:1:6) - cont'd					
Sills; horizontal; headers on edge; pointing top and one side; set weathering					
150 x 102 mm;	0.70	7.09	5.31	m	**12.40**
150 x 102 mm; cant headers; PC £113.00/100	0.75	7.60	18.11	m	**25.71**
Sills; horizontal; headers on flat; pointing top and one side					
150 x 102 mm; bullnosed specials; PC £113.00/100	0.65	6.59	12.98	m	**19.57**
Coping; horizontal; headers on edge; pointing top and both sides					
215 x 102 mm	0.56	5.67	11.83	m	**17.50**
260 x 102 mm	0.90	9.12	8.00	m	**17.12**
215 x 102 mm; double bullnosed specials; PC £113.00/100	0.60	6.08	18.07	m	**24.15**
260 x 120 mm; single bullnosed specials; PC £113.00/100	0.90	9.12	29.63	m	**38.75**
ADD or DEDUCT for variation of £10.00/1000 in PC of facing bricks	-	-	-	m	**0.28**
Facing bricks; hand made; PC £525.00/1000 (unless otherwise stated); in gauged mortar (1:1:6)					
Extra over common bricks; PC £120.00/1000; for facing bricks in					
stretcher bond	0.40	4.05	27.43	m2	**31.48**
flemish bond with snapped headers	0.50	5.07	34.29	m2	**39.36**
english bond with snapped headers	0.50	5.07	38.86	m2	**43.93**
ADD or DEDUCT for variation of £10.00/1000 in PC of facing bricks	-	-	-	m2	**0.85**
Half brick thick; stretcher bond; facework one side					
walls	1.65	16.72	36.53	m2	**53.25**
walls; building curved mean radius 6 m	2.40	24.32	39.26	m2	**63.58**
walls; building curved mean radius 1.50 m	3.00	30.40	41.17	m2	**71.57**
walls; building overhand	2.00	20.27	36.53	m2	**56.80**
walls; building up against concrete including flushing up at back	1.75	17.73	37.78	m2	**55.51**
walls; as formwork; temporary strutting	2.40	24.32	38.65	m2	**62.97**
walls; panels and aprons; not exceeding 1 m2	2.10	21.28	36.53	m2	**57.81**
isolated casings to steel columns	2.50	25.33	36.53	m2	**61.86**
bonding ends to existing	0.65	6.59	1.46	m	**8.05**
projections; vertical					
225 x 112 mm	0.40	4.05	8.51	m	**12.56**
337 x 112 mm	0.75	7.60	12.87	m	**20.47**
440 x 112 mm	1.10	11.15	17.22	m	**28.37**
Half brick thick; flemish bond with snapped headers; facework one side					
walls	1.90	19.25	40.38	m2	**59.63**
walls; building curved mean radius 6 m	2.70	27.36	42.29	m2	**69.65**
walls; building curved mean radius 1.50 m	3.50	35.46	44.20	m2	**79.66**
walls; building overhand	2.25	22.80	40.38	m2	**63.18**
walls; building up against concrete including flushing up at back	2.00	20.27	41.64	m2	**61.91**
walls; as formwork; temporary strutting	2.65	26.85	42.51	m2	**69.36**
walls; panels and aprons; not exceeding 1 m2	2.35	23.81	40.38	m2	**64.19**
isolated casings to steel columns	2.75	27.86	36.53	m2	**64.39**

F MASONRY Including overheads and profit at 5.00%	Labour hours	Labour £	Material £	Unit	Total rate £
bonding ends to existing	0.65	6.59	1.46	m	**8.05**
projections; vertical					
225 x 112 mm	0.50	5.07	9.69	m	**14.76**
337 x 112 mm	0.85	8.61	14.06	m	**22.67**
440 x 112 mm	1.20	12.16	19.00	m	**31.16**
One brick thick; two stretcher skins tied together; facework both sides					
walls	2.80	28.37	74.20	m2	**102.57**
walls; building curved mean radius 6 m	3.90	39.52	77.95	m2	**117.47**
walls; building curved mean radius 1.50 m	4.80	48.64	82.96	m2	**131.60**
isolated piers	3.30	33.44	74.19	m2	**107.63**
bonding ends to existing	0.85	8.61	2.92	m	**11.53**
One brick thick; flemish bond; facework both sides					
walls	2.90	29.38	73.69	m2	**103.07**
walls; building curved mean radius 6 m	4.00	40.53	79.16	m2	**119.69**
walls; building curved mean radius 1.50 m	5.00	50.66	82.96	m2	**133.62**
isolated piers	3.40	34.45	73.69	m2	**108.14**
bonding ends to existing	0.85	8.61	2.92	m	**11.53**
projections; vertical					
225 x 225 mm	0.80	8.11	16.42	m	**24.53**
337 x 225 mm	1.50	15.20	27.89	m	**43.09**
440 x 225 mm	2.20	22.29	32.26	m	**54.55**
ADD or DEDUCT for variation of £10.00/1000 in PC of facing bricks; in stretcher bond					
half brick thick	-	-	-	m2	**0.68**
one brick thick	-	-	-	m2	**1.37**
ADD or DEDUCT for variation of £10.00/1000 in PC of facing bricks; in flemish bond					
half brick thick	-	-	-	m2	**0.85**
one brick thick	-	-	-	m2	**1.70**
Extra over facing bricks for					
recessed joints	0.03	0.30	-	m2	**0.30**
raking out joints and pointing in black mortar	0.50	5.07	0.16	m2	**5.23**
bedding and pointing half brick wall in black mortar	-	-	-	m2	**0.78**
bedding and pointing one brick wall in black mortar	-	-	-	m2	**2.07**
flush plain bands; 225 mm wide stretcher bond; horizontal; bricks PC £565.00/1000	0.25	2.53	0.63	m	**3.16**
flush quoins; average 320 mm girth; block bond vertical; bricks PC £565.00/1000	0.40	4.05	0.59	m	**4.64**
Flat arches; 215 mm high on face					
102 mm wide exposed soffit	1.15	10.45	3.26	m	**13.71**
215 mm wide exposed soffit	1.72	15.62	6.52	m	**22.14**
Flat arches; 215 mm high on face; bullnosed specials; PC £113.00/100					
102 mm wide exposed soffit	1.20	10.95	18.74	m	**29.69**
215 mm wide exposed soffit	1.80	16.43	37.46	m	**53.89**
Segmental arches; one ring; 215 mm high on face					
102 mm wide exposed soffit	2.10	18.26	4.74	m	**23.00**
215 mm wide exposed soffit	3.15	27.40	9.67	m	**37.07**
Segmental arches; two ring; 215 mm high on face					
102 mm wide exposed soffit	2.70	24.34	4.74	m	**29.08**
215 mm wide exposed soffit	4.05	36.52	9.67	m	**46.19**
Segmental arches; 215 mm high on face; cut voussoirs; PC £245.00/100					
102 mm wide exposed soffit	2.20	19.28	39.88	m	**59.16**
215 mm wide exposed soffit	3.30	28.92	79.96	m	**108.88**

F MASONRY Including overheads and profit at 5.00%	Labour hours	Labour £	Material £	Unit	Total rate £
F10 BRICK/BLOCK WALLING - cont'd					
Facing bricks; hand made; PC £525.00/1000 (unless otherwise stated); in gauged mortar (1:1:6) - cont'd					
Segmental arches; one and a half ring; 320 mm high on face; cut voussoirs PC £245.00/100					
102 mm wide exposed soffit	3.00	27.38	78.60	m	**105.98**
215 mm wide exposed soffit	4.50	41.08	157.40	m	**198.48**
Semi circular arches; one ring; 215 mm high on face					
102 mm wide exposed soffit	3.60	32.86	15.72	m	**48.58**
215 mm wide exposed soffit	5.20	47.87	31.44	m	**79.31**
Semi circular arches; two ring; 215 mm high on face					
102 mm wide exposed soffit	4.80	45.02	15.72	m	**60.74**
215 mm wide exposed soffit	5.20	47.87	31.44	m	**79.31**
Semi circular arches; one ring; 215 mm high on face; cut voussoirs PC £245.00/100					
102 mm wide exposed soffit	2.70	23.74	40.15	m	**63.89**
215 mm wide exposed soffit	3.85	34.19	94.18	m	**128.37**
Bullseye window 600 mm dia; two rings; 215 mm high on face					
102 mm wide exposed soffit	6.00	57.78	72.80	nr	**130.58**
215 mm wide exposed soffit	9.00	86.67	145.40	nr	**232.07**
Bullseye window 1200 mm dia; two rings; 215 mm high on face					
102 mm wide exposed soffit	10.50	100.36	11.90	nr	**112.26**
215 mm wide exposed soffit	15.75	150.55	23.80	nr	**174.35**
Bullseye window 600 mm dia; one ring; 215 mm high on face; cut voussoirs; PC £245.00/100					
102 mm wide exposed soffit	5.00	47.65	98.08	nr	**145.73**
215 mm wide exposed soffit	7.50	71.47	196.10	nr	**267.57**
Bullseye window 1200 mm dia; one ring; 215 mm high on face; cut voussoirs; PC £245.00/100					
102 mm wide exposed soffit	9.00	85.17	169.07	nr	**254.24**
215 mm wide exposed soffit	13.50	127.75	338.15	nr	**465.90**
ADD or DEDUCT for variation of £10.00/1000 in PC of facing bricks	-	-	-	m	**0.28**
Sills; horizontal; headers on edge; pointing top and one side; set weathering					
150 x 102 mm;	0.70	7.09	8.55	m	**15.64**
150 x 102 mm; cant headers; PC £115.00/100	0.75	7.60	18.43	m	**26.03**
Sills; horizontal; headers on flat; pointing top and one side					
150 x 102 mm; bullnosed specials; PC £113.00/100	0.65	6.59	12.98	m	**19.57**
Coping; horizontal; headers on edge; pointing top and both sides					
215 x 102 mm	0.56	5.67	15.06	m	**20.73**
260 x 102 mm	0.90	9.12	12.87	m	**21.99**
215 x 102 mm; double bullnosed specials; PC £113.00/100	0.60	6.08	18.07	m	**24.15**
260 x 120 mm; single bullnosed specials; PC £113.00/100	0.90	9.12	29.63	m	**38.75**
ADD or DEDUCT for variation of £10.00/1000 in PC of facing bricks	-	-	-	m2	**0.28**

F MASONRY Including overheads and profit at 5.00%	Labour hours	Labour £	Material £	Unit	Total rate £
50 mm facing bricks slips; PC £100.00/100; in gauged mortar (1:1:6) built up against concrete including flushing up at back (ties measured elsewhere)					
Walls	2.50	25.33	69.45	m2	**94.78**
Edges of suspended slabs 200 mm wide	0.75	7.60	13.98	m	**21.58**
Columns 400 mm wide	1.50	15.20	27.87	m	**43.07**
Engineering bricks; PC £340.00/1000; and specials at PC £113.00/100; in cement mortar (1:3)					
Steps; all headers-on-edge; edges set with					
215 x 102 mm; horizontal; set weathering	0.70	7.09	17.67	m	**24.76**
Returned ends pointed	0.20	2.03	2.99	nr	**5.02**
430 x 102 mm; horizontal; set weathering	1.00	10.13	23.27	m	**33.40**
Returned ends pointed	0.25	2.53	5.48	nr	**8.01**
Labours on brick facework					
Fair cutting					
to curve	0.20	2.03	1.68	m	**3.71**
Fair returns					
half brick wide	0.05	0.51	-	m	**0.51**
one brick wide	0.07	0.71	-	m	**0.71**
one and a half brick wide	0.09	0.91	-	m	**0.91**
Fair angles formed by cutting					
squint	0.80	8.11	-	m	**8.11**
birdsmouth	0.70	7.09	-	m	**7.09**
external; chamfered 25 mm wide	1.00	10.13	-	m	**10.13**
external; rounded 100 mm radius	1.20	12.16	-	m	**12.16**
Fair chases					
100 x 50 mm; horizontal	1.50	15.20	-	m	**15.20**
100 x 50 mm; vertical	1.50	15.20	-	m	**15.20**
300 x 50 mm; vertical	2.70	27.36	-	m	**27.36**
Bonding ends to existing					
half brick thick	0.65	6.59	1.46	m	**8.05**
one brick thick	0.85	8.61	2.92	m	**11.53**
Centering to brickwork soffits; (prices included within arches rates)					
Flat soffits; not exceeding 2 m span					
over 0.3 m wide	2.00	14.24	5.36	m2	**19.60**
102 mm wide	0.40	2.85	0.96	m	**3.81**
215 mm wide	0.60	4.27	1.93	m	**6.20**
Segmental soffits; 1500 mm span 200 mm rise					
102 mm wide	1.33	9.47	1.99	nr	**11.46**
215 mm wide	2.00	14.24	3.97	nr	**18.21**
Semicircular soffits; 1500 mm span					
102 mm wide	1.80	12.81	2.35	nr	**15.16**
215 mm wide	2.40	17.09	4.72	nr	**21.81**
Bullseye window; 265 mm wide					
600 mm dia	1.50	10.68	2.77	nr	**13.45**
1200 mm dia	3.00	21.36	5.22	nr	**26.58**

ALTERNATIVE BLOCK PRICES (£/m2)

	£		£		£		£
Aerated Concrete Durox 'Supablocs'; 630 x 225 mm							
75 mm	4.19	125 mm	7.87	175 mm	11.02	225 mm	14.17
90 mm	5.39	130 mm	8.19	190 mm	11.38	250 mm	15.75
100 mm	5.99	140 mm	8.82	200 mm	11.98	280 mm	17.64
115 mm	7.24	150 mm	9.45	215 mm	12.87		

F MASONRY
Including overheads and profit at 5.00%

F10 BRICK/BLOCK WALLING - cont'd

ALTERNATIVE BLOCK PRICES (£/m2) - cont'd

	£		£		£
ARC Conbloc blocks; 450 x 225 mm					
Cream fair faced					
75 mm solid	5.65	190 mm solid	14.11	190 mm hollow	11.49
100 mm solid	6.76	100 mm hollow	6.54	215 mm hollow	12.89
140 mm solid	9.70	140 mm hollow	8.89		
Fenlite					
90 mm solid	3.90	100 mm solid	4.35	140 mm solid	6.25
Leca					
75 mm solid	6.90	100 mm solid	13.30	140 mm solid	9.10
Standard facing					
100 mm solid	5.95	190 mm solid	12.20	190 mm hollow	10.05
140 mm solid	8.80	140 mm hollow	8.20	215 mm hollow	11.45

	£		£		£		£
Celcon 'Standard' blocks; 450 x 225 mm							
75 mm	4.56	125 mm	7.60	190 mm	11.55	230 mm	14.00
90 mm	5.47	140 mm	8.52	200 mm	12.16	250 mm	15.20
100 mm	6.08	150 mm	9.12	215 mm	13.07	300 mm	18.24

'Solar' blocks are also available at the same price, in a limited range

	£		£		£
Forticrete painting quality blocks; 450 x 225 mm					
100 mm solid	6.55	215 mm solid	14.11	190 mm hollow	10.65
140 mm solid	9.74	100 mm hollow	5.64	215 mm hollow	10.96
190 mm solid	13.02	140 mm hollow	7.84		
Lytag blocks; 450 x 225 mm					
3.5 N/mm2 Insulating blocks					
75 mm solid	3.33	140 mm solid	6.25	100 mm cellular	6.87
90 mm solid	6.50	190 mm solid	8.47	215 mm cellular	8.84
100 mm solid	4.43				
7.0 N/mm2 Insulating extra strength blocks					
100 mm solid	4.59	140 mm solid	6.49	190 mm solid	8.79
10.5 N/mm2 High strength blocks					
100 mm solid	5.07	140 mm solid	7.21	190 mm solid	9.74
3.5 N/mm2 and 7.0 N/mm2 Close textured blocks					
100 mm solid (3.5)	5.71	140 mm solid (3.5)	8.03	190 mm solid (3.5)	10.85
100 mm solid (7.0)	5.93	140 mm solid (7.0)	8.35	190 mm solid (7.0)	11.28
Tarmac 'Topblocks'; 450 x 225 mm					
3.5 N/mm2 'Hemelite' blocks					
70/75 mm solid	4.15	100 mm solid	4.00	215 mm solid	9.75
90 mm solid	6.80	190 mm solid	8.40		
7.0 N/mm2 'Hemelite' blocks					
90 mm solid	6.40	140 mm solid	7.27	215 mm solid	10.56
100 mm solid	4.71	190 mm solid	9.08		

	£		£		£		£
'Toplite' standard blocks							
75 mm	4.09	100 mm	5.45	150 mm	8.18	215 mm	11.72
90 mm	4.91	140 mm	7.63	200 mm	10.90		
'Toplite' GTI (thermal) blocks							
115 mm	6.44	130 mm	7.28	150 mm	8.40	215 mm	12.04
125 mm	7.00	140 mm	7.84	200 mm	11.20		

Discounts of 0 - 7.5% available depending on quantity/status

F MASONRY Including overheads and profit at 5.00%		Labour hours	Labour £	Material £	Unit	Total rate £
Lightweight aerated concrete blocks; Thermalite 'Shield'/'Turbo' blocks or similar; in gauged mortar (1:2:9)						
Walls or partitions or skins of hollow walls						
75 mm thick	PC £4.65	0.60	6.08	5.81	m2	**11.89**
90 mm thick	PC £5.58	0.66	6.69	6.95	m2	**13.64**
100 mm thick	PC £6.20	0.70	7.09	7.75	m2	**14.84**
115 mm thick	PC £7.50	0.74	7.50	9.33	m2	**16.83**
125 mm thick	PC £8.15	0.77	7.80	10.19	m2	**17.99**
130 mm thick	PC £8.48	0.80	8.11	10.58	m2	**18.69**
140 mm thick	PC £8.68	0.82	8.31	10.85	m2	**19.16**
150 mm thick	PC £9.31	0.85	8.61	11.57	m2	**20.18**
190 mm thick	PC £11.78	1.00	10.13	14.65	m2	**24.78**
200 mm thick	PC £12.40	1.05	10.64	15.37	m2	**26.01**
215 mm thick	PC £13.33	1.10	11.15	16.55	m2	**27.70**
255 mm thick	PC £15.81	1.20	12.16	19.57	m2	**31.73**
Isolated piers or chimney stacks						
190 mm thick		1.40	14.19	14.65	m2	**28.84**
215 mm thick		1.50	15.20	16.45	m2	**31.65**
255 mm thick		1.70	17.23	19.57	m2	**36.80**
Isolated casings						
75 mm thick		0.70	7.09	5.81	m2	**12.90**
90 mm thick		0.76	7.70	7.00	m2	**14.70**
100 mm thick		0.80	8.11	7.75	m2	**15.86**
115 mm thick		0.84	8.51	9.33	m2	**17.84**
125 mm thick		0.87	8.82	10.19	m2	**19.01**
140 mm thick		0.92	9.32	10.85	m2	**20.17**
Extra over for fair face; flush pointing						
walls; one side		0.10	1.01	-	m2	**1.01**
walls; both sides		0.17	1.72	-	m2	**1.72**
Bonding ends to common brickwork						
75 mm blockwork		0.15	1.52	0.47	m	**1.99**
90 mm blockwork		0.15	1.52	0.55	m	**2.07**
100 mm blockwork		0.20	2.03	0.61	m	**2.64**
115 mm blockwork		0.20	2.03	0.73	m	**2.76**
125 mm blockwork		0.23	2.33	0.84	m	**3.17**
130 mm blockwork		0.23	2.33	0.87	m	**3.20**
140 mm blockwork		0.25	2.53	0.88	m	**3.41**
150 mm blockwork		0.25	2.53	0.94	m	**3.47**
190 mm blockwork		0.30	3.04	1.21	m	**4.25**
200 mm blockwork		0.33	3.34	1.27	m	**4.61**
215 mm blockwork		0.36	3.65	1.36	m	**5.01**
255 mm blockwork		0.40	4.05	1.63	m	**5.68**
Lightweight smooth face aerated concrete blocks; Thermalite 'Smooth Face' blocks or similar; in gauged mortar (1:2:9); flush pointing one side						
Walls or partitions or skins of hollow walls						
100 mm thick	PC £8.77	0.88	8.59	10.46	m2	**19.05**
140 mm thick	PC £12.28	1.02	9.93	14.67	m2	**24.60**
150 mm thick	PC £12.54	1.05	10.23	14.97	m2	**25.20**
190 mm thick	PC £16.67	1.22	11.88	19.90	m2	**31.78**
200 mm thick	PC £17.54	1.27	12.38	20.90	m2	**33.28**
215 mm thick	PC £18.86	1.34	13.01	22.48	m2	**35.49**
Isolated piers or chimney stacks						
190 mm thick		1.62	15.93	19.90	m2	**35.83**
200 mm thick		1.72	16.94	20.90	m2	**37.84**
215 mm thick		1.94	19.09	22.48	m2	**41.57**
Isolated casings						
100 mm thick		0.98	9.61	10.46	m2	**20.07**
140 mm thick		1.12	10.94	14.67	m2	**25.61**

F MASONRY Including overheads and profit at 5.00%		Labour hours	Labour £	Material £	Unit	Total rate £
F10 BRICK/BLOCK WALLING - cont'd						
Lightweight smooth face aerated concrete blocks; Thermalite 'Smooth Face' blocks or similar; in gauged mortar (1:2:9); flush pointing one side - cont'd						
Extra over for flush pointing						
walls; both sides		0.07	0.71	-	m2	0.71
Bonding ends to common brickwork						
100 mm blockwork		0.25	2.53	0.85	m	3.38
140 mm blockwork		0.28	2.84	1.22	m	4.06
150 mm blockwork		0.30	3.04	1.24	m	4.28
190 mm blockwork		0.35	3.55	1.66	m	5.21
200 mm blockwork		0.38	3.85	1.74	m	5.59
215 mm blockwork		0.40	4.05	1.86	m	5.91
Lightweight aerated high strength concrete blocks (7 N/mm2); Thermalite 'High Strength' blocks or similar; in cement mortar (1:3)						
Walls or partitions or skins of hollow walls						
100 mm thick	PC £8.00	0.74	7.34	9.56	m2	16.90
140 mm thick	PC £11.20	0.85	8.41	13.42	m2	21.83
150 mm thick	PC £12.01	0.90	8.92	14.36	m2	23.28
190 mm thick	PC £15.20	1.06	10.50	18.20	m2	28.70
200 mm thick	PC £16.00	1.11	11.00	18.48	m2	29.48
215 mm thick	PC £17.20	1.17	11.57	20.55	m2	32.12
Isolated piers or chimney stacks						
190 mm thick		1.41	14.04	16.61	m2	30.65
200 mm thick		1.51	15.06	19.13	m2	34.19
215 mm thick		1.72	17.14	20.51	m2	37.65
Isolated casings						
100 mm thick		0.84	8.35	9.56	m2	17.91
140 mm thick		0.95	9.42	13.42	m2	22.84
150 mm thick		1.00	9.93	14.36	m2	24.29
190 mm thick		1.16	11.51	18.20	m2	29.71
200 mm thick		1.26	12.52	19.13	m2	31.65
215 mm thick		1.32	13.09	20.55	m2	33.64
Extra over for fair face; flush pointing						
walls; one side		0.10	1.01	-	m2	1.01
walls; both sides		0.17	1.72	-	m2	1.72
Bonding ends to common brickwork						
100 mm blockwork		0.25	2.53	0.78	m	3.31
140 mm blockwork		0.28	2.84	1.12	m	3.96
150 mm blockwork		0.30	3.04	1.19	m	4.23
190 mm blockwork		0.35	3.55	1.52	m	5.07
200 mm blockwork		0.38	3.85	1.60	m	5.45
215 mm blockwork		0.40	4.05	1.71	m	5.76
Dense aggregate concrete blocks; 'ARC Conbloc' or similar; in gauged mortar (1:2:9)						
Walls or partitions or skins of hollow walls						
75 mm thick; solid	PC £3.77	0.70	8.25	4.60	m2	12.85
100 mm thick; solid	PC £3.77	0.80	9.42	4.68	m2	14.10
140 mm thick; hollow	PC £6.45	0.90	10.60	7.89	m2	18.49
140 mm thick; solid	PC £6.45	1.00	11.78	7.89	m2	19.67
190 mm thick; hollow	PC £8.15	1.15	13.55	10.06	m2	23.61
215 mm thick; hollow	PC £8.35	1.25	14.73	10.34	m2	25.07
Isolated piers or chimney stacks						
140 mm thick; hollow		1.25	14.73	7.89	m2	22.62
190 mm thick; hollow		1.60	18.85	10.06	m2	28.91
215 mm thick; hollow		1.80	21.21	10.34	m2	31.55

F MASONRY Including overheads and profit at 5.00%		Labour hours	Labour £	Material £	Unit	Total rate £
Isolated casings						
75 mm thick; solid		0.80	9.42	4.60	m2	**14.02**
100 mm thick; solid		0.90	10.60	4.68	m2	**15.28**
140 mm thick; solid		1.10	12.96	7.89	m2	**20.85**
Extra over for fair face; flush pointing						
walls; one side		0.10	1.18	-	m2	**1.18**
walls; both sides		0.17	2.00	-	m2	**2.00**
Bonding ends to common brickwork						
75 mm blockwork		0.20	2.36	0.39	m	**2.75**
100 mm blockwork		0.26	3.06	0.39	m	**3.45**
140 mm blockwork		0.32	3.77	0.68	m	**4.45**
190 mm blockwork		0.38	4.48	0.88	m	**5.36**
215 mm blockwork		0.44	5.18	0.90	m	**6.08**
Dense aggregate concrete blocks; (7 N/mm2) Forticrete 'Leicester Common' blocks or similar; in cement mortar (1:3)						
Walls or partitions or skins of hollow walls						
75 mm thick; solid	PC £4.63	0.70	8.25	5.61	m2	**13.86**
100 mm thick; hollow	PC £5.12	0.80	9.42	6.27	m2	**15.69**
100 mm thick; solid	PC £5.94	0.80	9.42	7.22	m2	**16.64**
140 mm thick; hollow	PC £7.09	0.90	10.60	8.68	m2	**19.28**
140 mm thick; solid	PC £8.86	1.00	11.78	10.73	m2	**22.51**
190 mm thick; hollow	PC £9.61	1.15	13.55	11.82	m2	**25.37**
190 mm thick; solid	PC £11.86	1.25	14.73	14.42	m2	**29.15**
215 mm thick; hollow	PC £9.95	1.25	14.73	12.25	m2	**26.98**
215 mm thick; solid	PC £12.78	1.35	15.90	15.53	m2	**31.43**
Dwarf support wall						
140 mm thick; solid		1.40	16.49	10.73	m2	**27.22**
190 mm thick; solid		1.60	18.85	14.42	m2	**33.27**
215 mm thick; solid		1.80	21.21	15.53	m2	**36.74**
Isolated piers or chimney stacks						
140 mm thick; hollow		1.25	14.73	8.68	m2	**23.41**
190 mm thick; hollow		1.60	18.85	11.82	m2	**30.67**
215 mm thick; hollow		1.80	21.21	12.25	m2	**33.46**
Isolated casings						
75 mm thick; solid		0.80	9.42	5.61	m2	**15.03**
100 mm thick; solid		0.90	10.60	7.22	m2	**17.82**
140 mm thick; solid		1.10	12.96	10.73	m2	**23.69**
Extra over for fair face; flush pointing						
walls; one side		0.10	1.18	-	m2	**1.18**
walls; both sides		0.17	2.00	-	m2	**2.00**
Bonding ends to common brickwork						
75 mm blockwork		0.20	2.36	0.48	m	**2.84**
100 mm blockwork		0.26	3.06	0.60	m	**3.66**
140 mm blockwork		0.32	3.77	0.91	m	**4.68**
190 mm blockwork		0.38	4.48	1.23	m	**5.71**
215 mm blockwork		0.44	5.18	1.31	m	**6.49**
Dense aggregate coloured concrete blocks; Forticrete 'Leicester Bathstone'; in coloured gauged mortar (1:1:6); flush pointing one side						
Walls or partitions or skins of hollow walls						
100 mm thick; hollow	PC £11.46	0.90	10.60	13.81	m2	**24.41**
100 mm thick; solid	PC £13.76	0.90	10.60	16.47	m2	**27.07**
140 mm thick; hollow	PC £15.18	1.00	11.78	18.32	m2	**30.10**
140 mm thick; solid	PC £20.35	1.10	12.96	24.30	m2	**37.26**
190 mm thick; hollow	PC £18.95	1.25	14.73	23.04	m2	**37.77**
190 mm thick; solid	PC £27.72	1.35	15.90	33.16	m2	**49.06**
215 mm thick; hollow	PC £20.46	1.35	15.90	24.85	m2	**40.75**
215 mm thick; solid	PC £29.58	1.45	17.08	35.38	m2	**52.46**

F MASONRY Including overheads and profit at 5.00%	Labour hours	Labour £	Material £	Unit	Total rate £
F10 BRICK/BLOCK WALLING - cont'd					
Dense aggregate coloured concrete blocks; Forticrete 'Leicester Bathstone'; in coloured gauged mortar (1:1:6); flush pointing one side - cont'd					
Isolated piers or chimney stacks					
140 mm thick; solid	1.50	17.67	24.30	m2	**41.97**
190 mm thick; solid	1.70	20.03	33.16	m2	**53.19**
215 mm thick; solid	1.90	22.38	35.53	m2	**57.91**
Extra over for flush pointing					
walls; both sides	0.07	0.82	-	m2	**0.82**
Extra over blocks for					
100 mm thick lintol blocks; ref D14	0.25	2.95	7.84	m	**10.79**
140 mm thick lintol blocks; ref H14	0.30	3.53	8.63	m	**12.16**
140 mm thick quoin blocks; ref H16	0.40	4.71	18.36	m	**23.07**
140 mm thick cavity closer blocks; ref H17	0.40	4.71	21.16	m	**25.87**
140 mm thick cill blocks; ref H21	0.30	3.53	12.80	m	**16.33**
190 mm thick lintol blocks; ref A14	0.40	4.71	9.97	m	**14.68**
190 mm thick cill blocks; ref A21	0.35	4.12	13.71	m	**17.83**
F11 GLASS BLOCK WALLING					
NOTE: The following specialist prices for glass block walling assume standard blocks; panels of 50 m2; no fire rating; work in straight walls at ground floor level; and all necessary ancillary fixing; strengthening; easy access; pointing and expansion materials etc.					
Hollow glass block walling; Luxcrete sealed 'Luxblocks' or similar; in cement mortar 'Luxfix' joints; reinforced with 6 mm dia. stainless steel rods; with 'Luxfibre' at head and jambs; pointed both sides with 'Luxseal' mastic					
Panel walls; facework both sides					
115 x 115 x 80 mm flemish blocks	-	-	-	m2	**547.63**
190 x 190 x 80 mm flemish; cross reeded or clear blocks	-	-	-	m2	**265.34**
240 x 240 x 80 mm flemish; cross reeded or clear blocks	-	-	-	m2	**227.43**
240 x 115 x 80 mm flemish or clear blocks	-	-	-	m2	**342.14**
F20 NATURAL STONE RUBBLE WALLING					
Cotswold Guiting limestone; laid dry					
Uncoursed random rubble walling					
275 mm thick	1.80	20.96	28.35	m2	**49.31**
350 mm thick	2.10	24.45	35.86	m2	**60.31**
425 mm thick	2.35	27.36	43.85	m2	**71.21**
500 mm thick	2.60	30.28	51.37	m2	**81.65**
Cotswold Guiting limestone; bedded; jointed and pointed in cement - lime mortar (1:2:9)					
Uncoursed random rubble walling; faced and pointed; both sides					
275 mm thick	1.70	19.80	29.97	m2	**49.77**
350 mm thick	1.80	20.96	37.88	m2	**58.84**
425 mm thick	1.90	22.12	46.28	m2	**68.40**
500 mm thick	2.00	23.29	54.61	m2	**77.90**

F MASONRY Including overheads and profit at 5.00%	Labour hours	Labour £	Material £	Unit	Total rate £
Coursed random rubble walling; rough dressed; faced and pointed one side					
114 mm thick	1.48	15.00	42.55	m2	**57.55**
150 mm thick	1.74	17.63	48.47	m2	**66.10**
Fair returns on walling					
114 mm wide	0.03	0.30	-	m	**0.30**
150 mm wide	0.04	0.41	-	m	**0.41**
275 mm wide	0.07	0.71	-	m	**0.71**
350 mm wide	0.09	0.91	-	m	**0.91**
425 mm wide	0.11	1.11	-	m	**1.11**
500 mm wide	0.13	1.32	-	m	**1.32**
Fair raking cutting on walling					
114 mm thick	0.20	2.03	6.00	m	**8.03**
150 mm thick	0.24	2.43	6.84	m	**9.27**
Level uncoursed rubble walling for damp proof courses and the like					
275 mm wide	0.21	2.45	1.82	m	**4.27**
350 mm wide	0.22	2.56	2.06	m	**4.62**
425 mm wide	0.23	2.68	2.29	m	**4.97**
500 mm wide	0.24	2.79	2.77	m	**5.56**
Copings formed of rough stones; faced and pointed all round					
275 x 200 mm (average) high	0.51	5.94	6.39	m	**12.33**
350 x 250 mm (average) high	0.67	7.80	9.64	m	**17.44**
425 x 300 mm (average) high	0.85	9.90	13.45	m	**23.35**
500 x 350 mm (average) high	1.05	12.23	18.38	m	**30.61**
F22 CAST STONE WALLING/DRESSINGS					
Reconstructed limestone walling; 'Bradstone' 100 mm bed weathered Cotswold or North Cerney masonry blocks or similar; laid to pattern or course recommended; bedded, jointed and pointed in approved coloured cement - lime mortar (1:2:9)					
Walls; facing and pointing one side					
masonry blocks; random uncoursed	1.12	11.35	19.36	m2	**30.71**
Extra for					
Return ends	0.40	4.05	1.12	m	**5.17**
Plain 'L' shaped quoins	0.13	1.32	6.01	m	**7.33**
traditional walling; coursed squared	1.40	14.19	18.61	m2	**32.80**
squared random rubble	1.40	14.19	19.25	m2	**33.44**
squared coursed rubble (large module)	1.30	13.17	19.11	m2	**32.28**
squared coursed rubble (small module)	1.35	13.68	19.40	m2	**33.08**
squared and pitched rock faced walling; coursed	1.45	14.69	18.61	m2	**33.30**
rough hewn rockfaced walling; random	1.50	15.20	18.47	m2	**33.67**
Extra for return ends	0.16	1.62	-	m	**1.62**
Isolated piers or chimney stacks; facing and pointing one side					
masonry blocks; random uncoursed	1.55	15.71	20.29	m2	**36.00**
traditional walling; coursed squared	1.95	19.76	19.50	m2	**39.26**
squared random rubble	1.95	19.76	20.17	m2	**39.93**
squared coursed rubble (large module)	1.80	18.24	20.01	m2	**38.25**
squared coursed rubble (small module)	1.90	19.25	20.33	m2	**39.58**
squared and pitched rock faced walling; coursed	2.05	20.77	19.50	m2	**40.27**
rough hewn rockfaced walling; random	2.10	21.28	19.34	m2	**40.62**

F MASONRY Including overheads and profit at 5.00%	Labour hours	Labour £	Material £	Unit	Total rate £
F22 CAST STONE WALLING/DRESSINGS - cont'd					
Reconstructed limestone walling; 'Bradstone' 100 mm bed weathered Cotswold or North Cerney masonry blocks or similar; laid to pattern or course recommended; bedded, jointed and pointed in approved coloured cement - lime mortar (1:2:9) - cont'd					
Isolated casings; facing and pointing one side					
masonry blocks; random uncoursed	1.35	13.68	20.29	m2	**33.97**
traditional walling; coursed squared	1.70	17.23	19.50	m2	**36.73**
squared random rubble	1.70	17.23	20.17	m2	**37.40**
squared coursed rubble (large module)	1.55	15.71	20.01	m2	**35.72**
squared coursed rubble (small module)	1.65	16.72	20.33	m2	**37.05**
squared and pitched rock faced walling; coursed	1.75	17.73	19.50	m2	**37.23**
rough hewn rockfaced walling; random	1.80	18.24	19.34	m2	**37.58**
Fair returns 100 mm wide					
masonry blocks; random uncoursed	0.12	1.22	-	m	**1.22**
traditional walling; coursed squared	0.15	1.52	-	m	**1.52**
squared random rubble	0.15	1.52	-	m	**1.52**
squared coursed rubble (large module)	0.14	1.42	-	m	**1.42**
squared coursed rubble (small module)	0.14	1.42	-	m	**1.42**
squared and pitched rock faced walling; coursed	0.15	1.52	-	m	**1.52**
rough hewn rockfaced walling; random	0.16	1.62	-	m	**1.62**
Fair raking cutting on masonry blocks					
100 mm thick	0.18	1.82	-	m	**1.82**
Reconstructed limestone dressings; 'Bradstone Architectural' dressings in weathered Cotswold or North Cerney shades or similar; bedded, jointed and pointed in approved coloured cement - lime mortar (1:2:9)					
Copings; twice weathered and throated					
152 x 76 mm; type A	0.33	3.34	8.02	m	**11.36**
178 x 64 mm; type B	0.33	3.34	8.62	m	**11.96**
305 x 76 mm; type A	0.40	4.05	15.31	m	**19.36**
Extra for					
Fair end	-	-	-	nr	**2.98**
Returned mitred fair end	-	-	-	nr	**2.98**
Copings; once weathered and throated					
191 x 76 mm	0.33	3.34	9.33	m	**12.67**
305 x 76 mm	0.40	4.05	15.04	m	**19.09**
365 x 76 mm	0.40	4.05	16.18	m	**20.23**
Extra for					
Fair end	-	-	-	nr	**2.98**
Returned mitred fair end	-	-	-	nr	**2.98**
Chimney caps; four times weathered and throated; once holed					
553 x 533 x 76 mm	0.40	4.05	20.71	nr	**24.76**
686 x 686 x 76 mm	0.40	4.05	34.04	nr	**38.09**
Pier caps; four times weathered and throated					
305 x 305 mm	0.25	2.53	6.87	nr	**9.40**
381 x 381 mm	0.25	2.53	9.66	nr	**12.19**
457 x 457 mm	0.30	3.04	13.42	nr	**16.46**
533 x 533 mm	0.30	3.04	18.63	nr	**21.67**
Splayed corbels					
457 x 102 x 229 mm	0.15	1.52	8.65	nr	**10.17**
686 x 102 x 229 mm	0.20	2.03	13.79	nr	**15.82**
Air bricks					
229 x 142 x 76 mm	0.08	0.81	3.38	nr	**4.19**

F MASONRY Including overheads and profit at 5.00%		Labour hours	Labour £	Material £	Unit	Total rate £
102 x 152 mm lintels; rectangular; reinforced with mild steel bars						
not exceeding 1.22 m long		0.24	2.43	14.06	m	**16.49**
1.37 - 1.67 m long		0.26	2.63	19.33	m	**16.96**
1.83 - 1.98 m long		0.28	2.84	15.68	m	**18.52**
102 x 229 mm lintels; rectangular; reinforced with mild steel bars						
not exceeding 1.67 m long		0.26	2.63	17.55	m	**20.18**
1.83 - 1.98 m long		0.28	2.84	17.97	m	**20.81**
2.13 - 2.44 m long		0.30	3.04	18.09	m	**21.13**
2.59 x 2.90 m long		0.32	3.24	19.73	m	**22.97**
197 x 67 mm sills to suit standard softwood windows; stooled at ends						
0.56 - 2.50 m long		0.30	3.04	22.11	m	**25.15**
Window surround; traditional with label moulding; for single light; sill 146 x 133 mm; jambs 146 x 146 mm; head 146 x 105 mm; including all dowels and anchors						
window size 508 x 1479 mm	PC £102.43	0.90	9.12	107.99	nr	**117.11**
Window surround; traditional with label moulding; three light; for windows 508 x 1219 mm; sill 146 x 133 mm; jambs 146 x 146 mm; head 146 x 103 mm; mullions 146 x 108 mm; including all dowels and anchors						
overall size 1975 x 1479 mm	PC £223.10	2.35	23.81	234.97	nr	**258.78**
Door surround; moulded continuous jambs and head with label moulding; including all dowels and anchors						
door 839 x 1981 mm in 102 x 64 mm frame		1.65	16.72	285.28	nr	**302.00**
F30 ACCESSORIES/SUNDRY ITEMS FOR BRICK/BLOCK/ STONE WALLING						
Sundries - brick/block walling						
Forming cavities in hollow walls; three wall ties per m2						
50 mm cavity; polypropylene ties		0.06	0.61	0.15	m2	**0.76**
50 mm cavity; galvanized steel butterfly wall ties		0.06	0.61	0.17	m2	**0.78**
50 mm cavity; galvanized steel twisted wall ties		0.06	0.61	0.23	m2	**0.84**
50 mm cavity; stainless steel butterfly wall ties		0.06	0.61	0.22	m2	**0.83**
50 mm cavity; stainless steel twisted wall ties		0.06	0.61	0.45	m2	**1.06**
75 mm cavity; polypropylene ties		0.06	0.61	0.15	m2	**0.76**
75 mm cavity; galvanised steel butterfly wall ties		0.06	0.61	0.17	m2	**0.78**
75 mm cavity; galvanised steel twisted wall ties		0.06	0.61	0.23	m2	**0.84**
75 mm cavity; stainless steel butterfly wall ties		0.06	0.61	0.22	m2	**0.83**
75 mm cavity; stainless steel twisted wall ties		0.06	0.61	0.45	m2	**1.06**
Closing at jambs with common brickwork half brick thick						
50 mm cavity		0.30	3.04	0.89	m	**3.93**
50 mm cavity; including damp proof course		0.40	4.05	1.39	m	**5.44**
75 mm cavity		0.30	3.04	1.17	m	**4.21**
75 mm cavity; including damp proof course		0.40	4.05	1.67	m	**5.72**

F MASONRY Including overheads and profit at 5.00%	Labour hours	Labour £	Material £	Unit	Total rate £
F30 ACCESSORIES/SUNDRY ITEMS FOR BRICK/BLOCK/ STONE WALLING - cont'd					
Sundries - brick/block walling - cont'd					
Closing at jambs with blockwork 100 mm thick					
50 mm cavity	0.25	2.53	0.44	m	**2.97**
50 mm cavity; including damp proof course	0.30	3.04	0.95	m	**3.99**
75 mm cavity	0.25	2.53	0.62	m	**3.15**
75 mm cavity; including damp proof course	0.30	3.04	1.13	m	**4.17**
Closing at sill with one course common brickwork					
50 mm cavity	0.30	3.04	1.99	m	**5.03**
50 mm cavity; including damp proof course	0.35	3.55	2.48	m	**6.03**
75 mm cavity	0.30	3.04	1.99	m	**5.03**
75 mm cavity; including damp proof course	0.35	3.55	2.48	m	**6.03**
Closing at top with					
single course of slates	0.15	1.52	2.21	m	**3.73**
'Westbrick' cavity closer	0.15	1.52	2.86	m	**4.38**
Cavity wall insulation; 'Dritherm'					
filling 50 mm cavity	0.15	1.52	1.59	m2	**3.11**
filling 75 mm cavity	0.17	1.72	2.10	m2	**3.82**
Cavity wall insulation; fixing with insulation retaining ties					
30 mm Celotex RR cavity insulation	0.20	2.03	3.56	m2	**5.59**
30 mm cavity wall batts	0.20	2.03	2.13	m2	**4.16**
25 mm Plaschem 'Aerobuild' foil faced polyurethene cavity slabs	0.20	2.03	3.36	m2	**5.39**
25 mm Styrofoam cavity wall insulation	0.20	2.03	3.12	m2	**5.15**

ALTERNATIVE DAMP PROOF COURSE PRICES (£/m2)

	£		£		£
Asbestos based					
'Astos'	3.97	'Barchester'	3.97		
Fibre based					
'Challenge'	2.31	'Stormax'	2.42		
Hessian based					
'Callendrite'	3.23	'Nubit'	3.59	'Permaseal'	3.59
Leadcored					
'Astos'	9.66	'Ledumite' grade 2	13.01	'Permalead'	10.76
'Ledkore' grade B	12.86	'Ledumite' grade 3	16.37	'Pluvex'	8.61
'Ledumite' grade 1	10.52	'Nuled'	8.53	'Trindos'	9.59
Pitch polymer					
'Aquagard'	4.06	'Permaflex'	4.19		

Further discounts may be available depending on quantity/status

	Labour hours	Labour £	Material £	Unit	Total rate £
'Pluvex' (hessian based) damp proof course or similar; PC £3.62/m2; 200 mm laps; in gauged mortar (1:1:6)					
Horizontal					
over 225 mm wide	0.25	2.53	4.10	m2	**6.63**
over 225 mm wide; forming cavity gutters in hollow walls	0.40	4.05	4.10	m2	**8.15**
not exceeding 225 mm wide	0.50	5.07	4.10	m2	**9.17**
Vertical					
not exceeding 225 mm wide	0.75	7.60	4.10	m2	**11.70**

F MASONRY Including overheads and profit at 5.00%	Labour hours	Labour £	Material £	Unit	Total rate £
'Pluvex' (fibre based) damp proof course or similar; PC £2.42/m2; 200 mm laps; in gauged mortar (1:1:6)					
Horizontal					
over 225 mm wide	0.25	2.53	2.74	m2	**5.27**
over 225 mm wide; forming cavity gutters in hollow walls	0.40	4.05	2.73	m2	**6.78**
not exceeding 225 mm wide	0.50	5.07	2.73	m2	**7.80**
Vertical					
not exceeding 225 mm wide	0.75	7.60	2.73	m2	**10.33**
'Asbex' (asbestos based) damp proof course or similar; PC £3.97/m2; 200 mm laps; in gauged mortar (1:1:6)					
Horizontal					
over 225 mm wide	0.25	2.53	4.48	m2	**7.01**
over 225 mm wide; forming cavity gutters in hollow walls	0.40	4.05	4.48	m2	**8.53**
not exceeding 225 mm wide	0.50	5.07	4.48	m2	**9.55**
Vertical					
not exceeding 225 mm wide	0.75	7.60	4.48	m2	**12.08**
Polyethelene damp proof course; PC £0.81/m2; 200 mm laps; in gauged mortar (1:1:6)					
Horizontal					
over 225 mm wide	0.25	2.53	0.92	m2	**3.45**
over 225 mm wide; forming cavity gutters in hollow walls	0.40	4.05	0.92	m2	**4.97**
not exceeding 225 mm wide	0.50	5.07	0.92	m2	**5.99**
Vertical					
not exceeding 225 mm wide	0.75	7.60	0.92	m2	**8.52**
'Permabit' bitumen polymer damp proof course or similar; PC £4.19/m2; 150 mm laps; in gauged mortar (1:1:6)					
Horizontal					
over 225 mm wide	0.25	2.53	4.76	m2	**7.29**
over 225 mm wide; forming cavity gutters in hollow walls	0.40	4.05	4.76	m2	**8.81**
not exceeding 225 mm wide	0.50	5.07	4.76	m2	**9.83**
Vertical					
not exceeding 225 mm wide	0.75	7.60	4.73	m2	**12.33**
'Hyload' (pitch polymer) damp proof course or similar; PC £4.06/m2; 150 mm laps; in gauged mortar (1:1:6)					
Horizontal					
over 225 mm wide	0.25	2.53	4.60	m2	**7.13**
over 225 mm wide; forming cavity gutters in hollow walls	0.40	4.05	4.60	m2	**8.65**
not exceeding 225 mm wide	0.50	5.07	4.60	m2	**9.67**
Vertical					
not exceeding 225 mm wide	0.75	7.60	4.60	m2	**12.20**
'Ledkore' grade A (bitumen based lead cored); damp proof course or similar; PC £11.34/m2; 200 mm laps; in gauged mortar (1:1:6)					
Horizontal					
over 225 mm wide	0.33	3.34	12.86	m2	**16.20**
over 225 mm wide; forming cavity gutters in hollow walls	0.53	5.37	12.86	m2	**18.23**
not exceeding 225 mm wide	0.50	5.07	12.86	m2	**17.93**

F MASONRY Including overheads and profit at 5.00%	Labour hours	Labour £	Material £	Unit	Total rate £
F30 ACCESSORIES/SUNDRY ITEMS FOR BRICK/BLOCK/ STONE WALLING - cont'd					
'Ledkore' grade A (bitumen based lead cored); damp proof course or similar; PC £11.34/m2; 200 mm laps; in gauged mortar (1:1:6) - cont'd					
Vertical					
not exceeding 225 mm wide	0.75	7.60	12.80	m2	**20.40**
Milled lead damp proof course; PC £23.73/m2; BS 1178; 1.80 mm (code 4), 175 mm laps; in cement-lime mortar (1:2:9)					
Horizontal					
over 225 mm wide	2.00	20.27	26.91	m2	**47.18**
not exceeding 225 mm wide	3.00	30.40	26.91	m2	**57.31**
Two courses slates in cement mortar (1:3)					
Horizontal					
over 225 mm wide	1.50	15.20	17.14	m2	**32.34**
Vertical					
over 225 mm wide	2.25	22.80	17.55	m2	**40.35**
'Synthaprufe' damp proof membrane; PC £28.37/25L; three coats brushed on					
Vertical					
not exceeding 150 mm wide	0.08	0.82	0.47	m2	**1.29**
150 - 225 mm wide	0.10	1.01	0.72	m2	**1.73**
225 - 300 mm wide	0.13	1.32	0.94	m2	**2.26**
over 300 mm wide	0.28	2.84	3.13	m2	**5.97**
'Type X' polypropylene abutment cavity tray by Cavity Trays Ltd; built into facing brickwork as the work proceeds; complete with Code 4 lead flashing					
Intermediate tray with short leads (requiring soakers); to suit roof of					
17 - 20 degrees pitch	0.06	0.61	4.29	nr	**4.90**
21 - 25 degrees pitch	0.06	0.61	3.98	nr	**4.59**
26 - 45 degrees pitch	0.06	0.61	3.35	nr	**3.95**
Intermediate tray with long leads (suitable only for corrugated roof tiles); to suit roof of					
17 - 20 degrees pitch	0.06	0.61	5.08	nr	**5.69**
21 - 25 degrees pitch	0.06	0.61	4.67	nr	**5.28**
26 - 45 degrees pitch	0.06	0.61	4.31	nr	**4.92**
Extra for					
ridge tray	0.06	0.61	6.59	nr	**7.20**
catchment end tray	0.06	0.61	8.95	nr	**9.56**
Servicised 'Bitu-thene' self-adhesive cavity flashing; type 'CA'; well lapped at joints; in gauged mortar (1:1:6)					
Horizontal					
over 225 mm wide	0.85	8.61	0.52	m2	**9.13**
'Brickforce' galvanised steel joint reinforcement					
In walls					
40 mm wide; ref. GBF 40	0.02	0.20	0.17	m	**0.37**
60 mm wide; ref. GBF 60	0.03	0.30	0.17	m	**0.47**
100 mm wide; ref. GBF 100	0.04	0.41	0.23	m	**0.64**
160 mm wide; ref. GBF 160	0.05	0.51	0.32	m	**0.83**

F MASONRY Including overheads and profit at 5.00%	Labour hours	Labour £	Material £	Unit	Total rate £
'Brickforce' stainless steel joint reinforcement					
In walls					
40 mm wide; ref. SBF 40	0.02	0.20	1.67	m	**1.87**
60 mm wide; ref. SBF 60	0.03	0.30	1.83	m	**2.13**
100 mm wide; ref. SBF 100	0.04	0.41	1.89	m	**2.30**
160 mm wide; ref. SBF 160	0.05	0.51	1.96	m	**2.47**
Fillets/pointing/wedging and pinning etc.					
Weather or angle fillets in cement mortar (1:3)					
50 mm face width	0.12	1.22	0.07	m	**1.29**
100 mm face width	0.20	2.03	0.27	m	**2.30**
Pointing wood frames or sills with mastic					
one side	0.07	0.71	0.38	m	**1.09**
each side	0.14	1.42	0.77	m	**2.19**
Pointing wood frames or sills with polysulphide sealant					
one side	0.07	0.71	0.81	m	**1.52**
each side	0.14	1.42	1.61	m	**3.03**
Bedding wood plates in cement mortar (1:3)					
100 mm wide	0.06	0.61	0.04	m	**0.65**
Bedding wood frame in cement mortar (1:3) and point					
one side	0.08	0.81	0.04	m	**0.85**
each side	0.10	1.01	0.06	m	**1.07**
one side in mortar; other side in mastic	0.15	1.52	0.43	m	**1.95**
Wedging and pinning up to underside of existing construction with slates in cement mortar (1:3)					
102 mm wall	0.80	8.11	2.30	m	**10.41**
215 mm wall	1.00	10.13	4.24	m	**14.37**
327 mm wall	1.20	12.16	6.16	m	**18.32**
Raking out joint in brickwork or blockwork for turned-in edge of flashing					
horizontal	0.15	1.52	0.04	m	**1.56**
stepped	0.20	2.03	0.04	m	**2.07**
Raking out and enlarging joint in brickwork or blockwork for nib of asphalt					
horizontal	0.20	2.03	-	m	**2.03**
Cutting grooves in brickwork or blockwork					
for water bars and the like	0.30	3.04	-	m	**3.04**
for nib of asphalt; horizontal	0.30	3.04	-	m	**3.04**
Preparing to receive new walls					
top existing 215 mm brick wall	0.20	2.03	-	m	**2.03**
Expansion joints					
Cleaning and priming both faces; filling with pre-formed closed cell joint filler and pointing one side with polysulphide sealant; 12 mm deep					
12 mm joint	0.25	2.53	2.11	m	**4.64**
19 mm joint	0.27	2.74	2.67	m	**5.41**
25 mm joint	0.30	3.04	3.21	m	**6.25**
Fire resisting horizontal expansion joint; filling with 'Nullifire System J' joint filler; fixed with high temparature slip adhesive; between top of wall and soffit					
10 mm joint with 30 mm deep ref. J60/10 filler (one hour fire seal)					
Wall 100 mm wide	0.25	2.53	5.00	m	**7.53**
Wall 150 mm wide	0.25	2.53	5.00	m	**7.53**
Wall 200 mm wide	0.25	2.53	5.00	m	**7.53**

F MASONRY Including overheads and profit at 5.00%	Labour hours	Labour £	Material £	Unit	Total rate £
F30 ACCESSORIES/SUNDRY ITEMS FOR BRICK/BLOCK/ STONE WALLING - cont'd					
Fire resisting horizontal expansion joint; filling with 'Nullifire System J' joint filler; fixed with high temparature slip adhesive; between top of wall and soffit - cont'd					
10 mm joint with 30 mm deep ref. J120/10 filler (two hour fire seal)					
Wall 100 mm wide	0.25	2.53	5.00	m	**7.53**
Wall 150 mm wide	0.25	2.53	5.00	m	**7.53**
Wall 200 mm wide	0.25	2.53	5.00	m	**7.53**
20 mm joint with 45 mm deep ref. J120/20 filler (two hour fire seal)					
Wall 100 mm wide	0.30	3.04	8.18	m	**11.22**
Wall 150 mm wide	0.30	3.04	8.18	m	**11.22**
Wall 200 mm wide	0.30	3.04	8.18	m	**11.22**
30 mm joint with 75 mm deep ref. J180/30 filler (three hour fire seal)					
Wall 100 mm wide	0.35	3.55	25.63	m	**29.18**
Wall 150 mm wide	0.35	3.55	25.63	m	**29.18**
Wall 200 mm wide	0.35	3.55	25.63	m	**29.18**
Fire resisting vertical expansion joint; filling with 'Nullifire System J' joint filler; fixed with high temparature slip adhesive; with polysulphide sealant one side; between end of wall and concrete					
20 mm joint with 45 mm deep ref. J120/20 filler (two hour fire seal)					
Wall 100 mm wide	0.35	3.55	10.89	m	**14.44**
Wall 150 mm wide	0.35	3.55	10.89	m	**14.44**
Wall 200 mm wide	0.35	3.55	10.89	m	**14.44**
Sills and tile creasings					
Sills; two courses of machine made plain roofing tiles; set weathering; bedded and pointed	0.60	6.08	2.77	m	**8.85**
Extra over brickwork for two courses of machine made tile creasing; bedded and pointed; projecting 25 mm each side; horizontal					
215 mm wide copings	0.50	5.07	4.18	m	**9.25**
260 mm wide copings	0.70	7.09	5.39	m	**12.48**
Galvanised steel coping cramp; built in	0.10	1.01	0.42	nr	**1.43**
Flue linings etc					
Flue linings; True Flue 200 mm refractory concrete square flue linings; rebated joints in refractory mortar (1:2.5)					
linings	0.30	3.04	8.92	m	**11.96**
bottom svivel unit; ref 2u	0.10	1.01	6.39	nr	**7.40**
45 deg. bend; ref 4u	0.10	1.01	6.39	nr	**7.40**
offset unit; ref 5u	0.10	1.01	6.52	nr	**7.53**
offset unit; ref 6u	0.10	1.01	5.85	nr	**6.86**
off set unit; ref 7u	0.10	1.01	5.48	nr	**6.49**
pot; ref 8u	0.10	1.01	3.65	nr	**4.66**
single cap unit; ref 10u	0.25	2.53	14.70	nr	**17.23**
double cap unit; ref 11u	0.30	3.04	13.30	nr	**16.34**
U-type lintol with U1 attachment	0.30	3.04	15.52	nr	**18.56**

F MASONRY Including overheads and profit at 5.00%	Labour hours	Labour £	Material £	Unit	Total rate £
Gas flue system; True Flue 'Typex HP'; concrete blocks built in; in refractory mortar (1:2.5); cutting brickwork or blockwork around					
recess; ref HP.1	0.10	1.01	2.55	nr	**3.56**
cover ref HP.2	0.10	1.01	3.61	nr	**4.62**
222 mm standard block; ref HP.3	0.10	1.01	2.38	nr	**3.39**
112 mm standard block; ref HP.3	0.10	1.01	2.37	nr	**3.38**
72 mm standard block; ref HP.3	0.10	1.01	2.38	nr	**3.39**
vent block; ref HP.3/H	0.10	1.01	4.82	nr	**5.83**
222 mm standard block; ref HP.4	0.10	1.01	2.30	nr	**3.31**
112 mm standard block; ref HP.4	0.10	1.01	2.30	nr	**3.31**
72 mm standard block; ref HP.4	0.10	1.01	2.30	nr	**3.31**
120 mm side offset; ref HP.5	0.10	1.01	2.84	nr	**3.85**
70 mm back offset; ref HP.6	0.10	1.01	6.98	nr	**7.99**
vertical exit; ref HP.7	0.10	1.01	4.69	nr	**5.70**
angled entry/exit; ref HP.8	0.10	1.01	4.69	nr	**5.70**
reverse rebate; ref HP.9	0.10	1.01	3.76	nr	**4.77**
corbel block; ref HP.10	0.10	1.01	4.45	nr	**5.46**
lintol block; ref HP.11	0.10	1.01	4.22	nr	**5.23**
Parging and coring flues with refractory mortar (1:2.5); sectional area not exceeding 0.25 m2					
900 x 900 mm	0.60	6.08	4.96	m	**11.04**
Ancillaries					
Forming openings; one ring arch over 225 x 225 mm; one brick facing brickwork making good facings both sides	1.50	15.20	-	nr	**15.20**
Forming opening through hollow wall; slate lintel over; sealing 50 cavity with slates in cement mortar (1:3); making good fair face or facings one side					
225 x 75 mm	0.20	2.03	1.60	nr	**3.63**
225 x 150 mm	0.25	2.53	2.35	nr	**4.88**
225 x 225 mm	0.30	3.04	3.11	nr	**6.15**
Air bricks; red terracotta; building into prepared openings					
215 x 65 mm	0.08	0.81	0.91	nr	**1.72**
215 x 140 mm	0.08	0.81	1.49	nr	**2.30**
215 x 215 mm	0.08	0.81	3.16	nr	**3.97**
Air bricks; cast iron; building into prepared openings					
215 x 65 mm	0.08	0.81	2.67	nr	**3.48**
215 x 140 mm	0.08	0.81	4.87	nr	**5.68**
215 x 215 mm	0.08	0.81	6.43	nr	**7.24**
Bearers; mild steel					
51 x 10 mm flat bar	0.25	2.24	3.33	m	**5.57**
64 x 13 mm flat bar	0.25	2.24	5.33	m	**7.57**
50 x 50 x 6 mm angle section	0.35	3.13	2.72	m	**5.85**
Ends fanged for building in	-	-	-		**2.10**
Proprietary items					
Ties in walls; 150 mm long butterfly type; building into joints of brickwork or blockwork					
galvanized steel or polypropylene	0.03	0.30	0.07	nr	**0.37**
stainless steel	0.03	0.30	0.09	nr	**0.39**
copper	0.03	0.30	0.13	nr	**0.43**

F MASONRY Including overheads and profit at 5.00%		Labour hours	Labour £	Material £	Unit	Total rate £
F30 ACCESSORIES/SUNDRY ITEMS FOR BRICK/BLOCK/ STONE WALLING - cont'd						
Proprietary items - cont'd						
Ties in walls; 20 x 3 x 150 mm long twisted wall type; building into joints of brickwork or blockwork						
galvanized steel		0.04	0.41	0.10	nr	**0.51**
stainless steel		0.04	0.41	0.19	nr	**0.60**
copper		0.04	0.41	0.33	nr	**0.74**
Anchors in walls; 25 x 3 x 100 mm long; one end dovetailed; other end building into joints of brickwork or blockwork						
galvanized steel		0.06	0.61	0.21	nr	**0.82**
stainless steel		0.06	0.61	0.34	nr	**0.95**
copper		0.06	0.61	0.52	nr	**1.13**
Fixing cramp 25 x 3 x 250 mm long; once bent; fixed to back of frame; other end building into joints of brickwork or blockwork						
galvanized steel		0.06	0.61	0.10	nr	**0.71**
Chimney pots; red terracotta; plain or cannon-head; setting and flaunching in cement mortar (1:3)						
185 mm dia x 300 mm long	PC £11.48	1.80	18.24	12.80	nr	**31.04**
185 mm dia x 600 mm long	PC £19.80	2.00	20.27	21.75	nr	**42.02**
185 mm dia x 900 mm long	PC £26.33	2.20	22.29	28.78	nr	**51.07**
Galvanized steel lintels; Catnic or similar; built into brickwork or blockwork						
'CN7' combined lintel; 143 mm high; for standard cavity walls						
750 mm long	PC £10.82	0.25	2.24	11.93	nr	**14.17**
900 mm long	PC £13.02	0.30	2.69	14.36	nr	**17.05**
1200 mm long	PC £17.31	0.35	3.13	19.08	nr	**22.21**
1500 mm long	PC £22.48	0.40	3.58	24.78	nr	**28.36**
1800 mm long	PC £27.53	0.45	4.03	30.36	nr	**34.39**
2100 mm long	PC £32.10	0.50	4.48	35.40	nr	**39.88**
'CN8' combined lintel; 219 mm high; for standard cavity walls						
2400 mm long	PC £43.59	0.60	5.37	48.06	nr	**53.43**
2700 mm long	PC £49.18	0.70	6.27	54.22	nr	**60.49**
3000 mm long	PC £59.98	0.80	7.17	66.12	nr	**73.29**
3300 mm long	PC £66.76	0.90	8.06	73.60	nr	**81.66**
3600 mm long	PC £73.11	1.00	8.96	80.61	nr	**89.57**
3900 mm long	PC £97.36	1.10	9.85	107.34	nr	**117.19**
4200 mm long	PC £104.81	1.20	10.75	115.56	nr	**126.31**
'CN92' single lintel; for 75 mm internal walls						
900 mm long	PC £1.94	0.30	2.69	2.14	nr	**4.83**
1050 mm long	PC £2.28	0.30	2.69	2.51	nr	**5.20**
1200 mm long	PC £2.56	0.35	3.13	2.82	nr	**5.95**
'CN102' single lintel; for 100 mm internal walls						
900 mm long	PC £2.39	0.30	2.69	2.63	nr	**5.32**
1050 mm long	PC £2.78	0.30	2.69	3.07	nr	**5.76**
1200 mm long		0.35	3.13	3.34	nr	**6.47**

F MASONRY Including overheads and profit at 5.00%		Labour hours	Labour £	Material £	Unit	Total rate £
F31 PRECAST CONCRETE SILLS/LINTELS/COPINGS/ FEATURES						
Mix 21.00 N/mm2 - 20 mm aggregate (1:2:4)						
Lintels; plate; prestressed; bedded						
100 x 65 x 750 mm	PC £2.25	0.40	4.05	2.81	nr	**6.86**
100 x 65 x 900 mm	PC £2.47	0.40	4.05	3.05	nr	**7.10**
100 x 65 x 1050 mm	PC £2.93	0.40	4.05	3.52	nr	**7.57**
100 x 65 x 1200 mm	PC £3.33	0.40	4.05	3.95	nr	**8.00**
150 x 65 x 900 mm	PC £3.42	0.50	5.07	4.04	nr	**9.11**
150 x 65 x 1050 mm	PC £4.23	0.50	5.07	4.89	nr	**9.96**
150 x 65 x 1200 mm	PC £4.59	0.50	5.07	5.27	nr	**10.34**
220 x 65 x 900 mm	PC £5.31	0.60	6.08	6.03	nr	**12.11**
220 x 65 x 1200 mm	PC £7.02	0.60	6.08	7.82	nr	**13.90**
220 x 65 x 1500 mm	PC £8.73	0.70	7.09	9.62	nr	**16.71**
265 x 65 x 900 mm	PC £6.52	0.60	6.08	7.30	nr	**13.38**
265 x 65 x 1200 mm	PC £8.91	0.60	6.08	9.81	nr	**15.89**
265 x 65 x 1500 mm	PC £11.12	0.70	7.09	12.12	nr	**19.21**
265 x 65 x 1800 mm	PC £12.38	0.80	8.11	13.44	nr	**21.55**
Lintels; rectangular; reinforced with mild steel bars; bedded						
100 x 150 x 900 mm	PC £7.20	0.60	6.08	8.01	nr	**14.09**
100 x 150 x 1050 mm	PC £8.10	0.60	6.08	8.95	nr	**15.03**
100 x 150 x 1200 mm	PC £13.38	0.60	6.08	14.49	nr	**20.57**
225 x 150 x 1200 mm	PC £16.88	0.80	8.11	18.17	nr	**26.28**
225 x 225 x 1800 mm	PC £31.25	1.50	15.20	33.26	nr	**48.46**
Lintels; boot; reinforced with mild steel bars; bedded						
250 x 225 x 1200 mm	PC £25.00	1.20	12.16	26.70	nr	**38.86**
275 x 225 x 1800 mm	PC £33.75	1.80	18.24	36.34	nr	**54.58**
Padstones						
300 x 100 x 75 mm	PC £2.50	0.30	3.04	3.07	nr	**6.11**
225 x 225 x 150 mm	PC £3.75	0.40	4.05	4.39	nr	**8.44**
450 x 450 x 150 mm	PC £10.00	0.60	6.08	10.95	nr	**17.03**
Mix 31.00 N/mm2 - 20 mm aggregate (1:1:2)						
Copings; once weathered; once throated bedded and pointed						
152 x 76 mm	PC £8.28	0.70	7.09	9.15	m	**16.24**
178 x 64 mm	PC £8.28	0.70	7.09	9.15	m	**16.24**
305 x 76 mm	PC £10.74	0.80	8.11	11.72	m	**19.83**
Extra for						
fair end	PC £3.03	-	-	3.18	nr	**3.18**
angles	PC £6.60	-	-	6.93	nr	**6.93**
Copings; twice weathered; twice throated; bedded and pointed						
152 x 76 mm	PC £7.66	0.70	7.09	8.49	m	**15.58**
178 x 64 mm	PC £7.66	0.70	7.09	8.49	m	**15.58**
305 x 76 mm	PC £10.11	0.80	8.11	11.06	m	**19.17**
Extra for						
fair end	PC £3.03	-	-	3.18	nr	**3.18**
angles		-	-	-	nr	**6.93**

G STRUCTURAL/CARCASSING METAL/TIMBER
Including overheads and profit at 5.00%

BASIC STEEL PRICES

NOTE: The following basic prices are based on **Basis** quantities (over 20 tonnes, of one quality, one section, for delivery to one destination) plus the addition of transport charges to the London area.

Prices are based on grade 43A steel. The additional cost of grade 50C steel is £40.00 per tonne.

Figures in brackets are available weights in kg per metre, at the prices indicated.

	£		£
Universal beams (£/tonne)			
914x419 mm (343,388)	427.40	356x171 mm (57,67)	367.55
914x305 mm (201,224,253,289)	422.40	356x127 mm (33,39)	357.55
838x292 mm (176,194,226)	417.40	305x165 mm (40)	347.55
762x267 mm (147,173,197)	417.40	305x165 mm (46,54)	357.55
686x254 mm (125,140,152,170)	417.40	305x127 mm (37)	352.55
610x305 mm (149,179,238)	397.40	305x127 mm (42,48)	352.55
610x229 mm (101,113,125,140)	397.40	305x102 mm (25,28,33)	352.55
533x210 mm (82,92,101)	382.40	254x146 mm (31,37)	337.55
533x210 mm (109,122)	392.40	254x146 mm (43)	337.55
457x191 mm (67,74)	352.40	254x102 mm (22,25)	352.55
457x191 mm (82,89,98)	357.40	254x102 mm (28)	352.55
457x152 mm (52,60,67,74,82)	367.40	203x133 mm (25,30)	322.55
406x178 mm (54)	362.40	203x102 mm (33)	317.55
406x178 mm (60,67,74)	362.40	178x102 mm (19)	312.55
406x140 mm (39,46)	357.55	152x 89 mm (16)	352.55
356x171 mm (45,51)	357.55	127x 76 mm (13)	373.20
Universal columns (£/tonne)			
356x406 mm (235,287,340)	427.40	254x254 mm (107)	357.40
356x406 mm (393,467,551,634)	432.40	254x254 mm (167)	387.40
356x368 mm (129,153,177,202)	427.40	203x203 mm (46)	353.80
305x305 mm (97)	397.40	203x203 mm (52)	358.80
305x305 mm (118)	397.40	203x203 mm (60)	373.80
305x305 mm (137,158)	397.40	203x203 mm (71)	363.80
305x305 mm (240,283)	402.40	203x203 mm (86)	383.80
305x305 mm (198)	387.40	152x152 mm (23)	353.80
254x254 mm (73)	357.40	152x152 mm (30,37)	353.80
254x254 mm (89,132)	367.40		
Joists (£/tonne)			
254x203 mm (81.85)	377.40	114x114 mm (26.79)	357.40
203x152 mm (52.09)	367.40	102x102 mm (23.06)	357.40
152x127 mm (37.20)	367.40	89x89 mm (19.35)	357.40
127x114 mm (26.79,29.76)	357.40	76x76 mm (12.65)	407.40
Channels (£/tonne)			
432x102 mm (65.54)	407.40	203x89 mm (29.78)	373.20
381x102 mm (55.10)	403.20	203x76 mm (23.82)	318.20
305x102 mm (46.18)	373.20	178x89 mm (26.81)	363.20
305x89 mm (41.69)	373.20	178x76 mm (20.84)	318.20
254x89 mm (35.74)	373.20	152x89 mm (23.84)	363.20
254x76 mm (28.29)	373.20	152x76 mm (17.88)	318.20
229x89 mm (32.76)	373.20	127x64 mm (14.90)	308.20
229x76 mm (26.06)	373.20		

G STRUCTURAL/CARCASSING METAL/TIMBER
Including overheads and profit at 5.00%

Equal angles (£/tonne)

	£		£		£
200x200x16 mm	327.40	150x150x18 mm	338.20	100x100x12 mm	288.20
200x200x18 mm	332.40	120x120x8 mm	308.20	90x90x6 mm	268.20
200x200x20 mm	332.40	120x120x10 mm	298.20	90x90x8 mm	268.20
200x200x24 mm	332.40	120x120x12 mm	293.20	90x90x10 mm	268.20
150x150x10 mm	323.20	120x120x15 mm	298.20	90x90x12 mm	268.20
150x150x12 mm	323.20	100x100x8 mm	263.20		
150x150x15 mm	328.20	100x100x10 mm	263.20		
150x150x18 mm	338.20	100x100x12 mm	263.20		

Unequal angles (£/tonne)

	£		£		£
200x150x12 mm	342.40	150x90x15 mm	332.40	125x75x15 mm	322.40
200x150x15 mm	347.40	150x75x10 mm	322.40	100x75x8 mm	278.20
200x150x18 mm	347.40	150x75x12 mm	322.40	100x75x10 mm	278.20
200x100x10 mm	337.40	150x75x15 mm	322.40	100x75x12 mm	278.20
200x100x12 mm	337.40	125x75x8 mm	287.40	100x65x7 mm	282.40
200x100x15 mm	357.40	125x75x10 mm	287.40	100x65x8 mm	282.40
150x90x10 mm	332.40	125x75x12 mm	287.40	100x65x10 mm	282.40
150x90x12 mm	332.40				

Structural hollow sections to BS 4848: Part 2 (£/100 m)	BS 4360 grades			BS 4360 grades	
	43C	50C		43C	50C
	£	£		£	£
Circular			Rectangular and square		
21.3x3.2 mm (1.43)	57.91	-	20x20x2.5 mm (1.35)	52.56	-
26.9x3.2 mm (1.87)	75.73	-	50x50x3.2 mm (4.66)	177.97	195.77
42.4x3.2 mm (3.09)	124.53	-	60x60x4.0 mm (6.97)	281.94	310.14
48.3x4.0 mm (4.37)	177.81	195.59	80x80x5.0 mm (11.70)	501.77	551.95
60.3x4.0 mm (5.55)	234.64	258.10	100x50x4.0 mm (8.86)	368.83	405.72
76.1x5.0 mm (8.77)	370.77	407.85	100x100x6.3 mm (18.40)	789.10	868.03
114.3x6.3 mm (16.80)	710.25	781.28	100x100x10.0 mm (27.90)	1293.00	1422.29
139.7x6.3 mm (20.70)	914.05	1005.46	120x80x6.3 mm (18.40)	789.10	868.02
168.3x6.3 mm (25.20)	1112.76	1224.04	150x150x6.3 mm (28.30)	1279.87	1407.87
219.1x10.0 mm (51.60)	2855.44	3283.77	200x100x8.0 mm (35.40)	1600.96	1761.08
273x10.0 mm (64.90)	3591.44	4130.17	200x200x10.0 mm (59.30)	3361.54	3865.77
323.9x12.5 mm (96.00)	5607.84	6448.99	300x200x12.5 mm (92.60)	5541.00	6372.18

NOTE: PC prices indicated below relate to an average price for the sections used, assuming a clean and level site with easy access.		Labour hours	Labour £	Material £	Total Unit rate £
G10 STRUCTURAL STEEL FRAMING					
Fabricated steelwork; BS 4360; grade 50					
Single beams; universal beams, joists or channels	PC £425.00	-	-	-	t **1020.00**
Single beams; castellated universal beams		-	-	-	t **2465.00**
Single beams; rectangular hollow sections		-	-	-	t **1240.00**
Single cranked beams; rectangular hollow sections		-	-	-	t **1460.00**
Built-up beams; universal beams, joists or channels and plates	PC £425.00	-	-	-	t **1475.00**
Latticed beams; angle sections	PC £310.00	-	-	-	t **1175.00**
Latticed beams; circular hollow sections		-	-	-	t **1320.00**
Single columns; universal beams, joists or channels	PC £425.00	-	-	-	t **905.00**
Single columns; circular hollow sections	PC £1062.60	-	-	-	t **1745.00**
Single columns; rectangular and square hollow sections	PC £945.00	-	-	-	t **1622.50**

G STRUCTURAL/CARCASSING METAL/TIMBER Including overheads and profit at 5.00%	Labour hours	Labour £	Material £	Unit	Total rate £
G10 STRUCTURAL STEEL FRAMING - cont'd					
Fabricated steelwork; BS 4360; grade 50 - cont'd					
Built-up columns; universal beams, joists or channels and plates PC £425.00	-	-	-	t	**1525.00**
Roof trusses; angle sections PC £310.00	-	-	-	t	**1430.00**
Roof trusses; circular hollow sections PC £520.00	-	-	-	t	**1585.00**
Bolted fittings; other than in connections; consisting of cleats, brackets etc.	-	-	-	t	**1615.00**
Welded fittings; other than in connections; consisting of cleats, brackets, etc.	-	-	-	t	**1460.00**
Erection of fabricated steelwork					
Erection of steelwork on site	-	-	-	t	**157.50**
Wedging					
stanchion bases	-	-	-	nr	**7.30**
Holes for other trades; made on site					
16 mm dia; 6 mm thick metal	-	-	-	nr	**3.32**
16 mm dia; 10 mm thick metal	-	-	-	nr	**4.65**
16 mm dia; 20 mm thick metal	-	-	-	nr	**6.52**
Anchorage for stanchion base; including plate; holding down bolts; nuts and washers; for stanchion size					
152 x 152 mm	-	-	-	nr	**9.46**
254 x 254 mm	-	-	-	nr	**14.13**
356 x 406 mm	-	-	-	nr	**21.15**
Surface treatments off site					
On steelwork; general surfaces					
galvanizing	-	-	-	t	**462.00**
shot blasting	-	-	-	m2	**0.74**
grit blast and one coat zinc chromate primer	-	-	-	m2	**1.97**
touch up primer and one coat of two pack epoxy zinc phosphate primer	-	-	-	m2	**2.14**
G12 ISOLATED STRUCTURAL METAL MEMBERS					
Unfabricated steelwork; BS 4360; grade 43A					
Single beams; joists or channels PC £425.00	-	-	-	t	**810.00**
Erection of steelwork on site	-	-	-	t	**168.00**
Metsec open web steel lattice beams; in single members; raised 3.50 m above ground; ends built in					
Beams; one coat zinc phosphate primer at works					
20 cm dp; to span 5m (7.69kg/m); ref NA20	0.20	1.79	17.14	m	**18.93**
25 cm dp; to span 6m (7.64kg/m); ref NA25	0.20	1.79	17.16	m	**18.95**
30 cm dp; to span 7m (10.26kg/m); ref NBs30	0.25	2.24	19.97	m	**22.21**
35 cm dp; to span 8.5m (10.6kg/m); ref NBs35	0.25	2.24	20.10	m	**22.34**
35 cm dp; to span 10m (12.76kg/m); ref NCs35	0.30	2.69	22.00	m	**24.69**
40 cm dp; to span 11.5m (17.08kg/m); ref E40	0.35	3.13	32.04	m	**35.17**
45 cm dp; to span 13m (25.44kg/m); ref G45	0.50	4.48	39.84	m	**44.32**
Beams; galvanised					
20 cm dp; to span 5m (7.69kg/m); ref NA20	0.20	1.79	18.79	m	**20.58**
25 cm dp; to span 6m (7.64kg/m); ref NA25	0.20	1.79	18.82	m	**20.61**
30 cm dp; to span 7m (10.26kg/m); ref NBs30	0.25	2.24	22.26	m	**24.50**
35 cm dp; to span 8.5m (10.6kg/m); ref NBs35	0.25	2.24	22.33	m	**24.57**
35 cm dp; to span 10m (12.76kg/m); ref NCs35	0.30	2.69	24.74	m	**27.43**
40 cm dp; to span 11.5m (17.08kg/m); ref E40	0.35	3.13	35.94	m	**39.07**
45 cm dp; to span 13m (25.44kg/m); ref G45	0.50	4.48	45.32	m	**49.80**

G STRUCTURAL/CARCASSING METAL/TIMBER
Including overheads and profit at 5.00%

BASIC TIMBER PRICES

	£		£		£
Hardwood; fair average (£/m3 ex-wharf)					
African Walnut	575.00	European Oak	1200.00	Sapele	520.00
Afrormosia	710.00	Iroko	500.00	Teak	1500.00
Agba	575.00	Maple	700.00	Utile	685.00
Beech	500.00	Obeche	340.00	W.A.Mahogany	470.00
Brazil. Mahogany	635.00				
Softwood					
Carcassing quality (£/m3)					
2-4.8 m lengths	190.00	G.S.grade	13.00		
4.8-6 m lengths	180.50	S.S.grade	25.00		
6-9 m lenths	199.50				
Joinery quality - £310.00/m3					
'Treatment'(£/m3)					
Pre-treatment of timber by vacuum/pressure impregnation, excluding transport costs and any subsequent seasoning:-					
interior work; min. salt ret. 4 kg/m3			30.00		
exterior work; min. salt ret. 5.30 kg/m3			35.00		
Pre-treatment of timber including flame proofing					
all purposes; min. salt ret. 36 kg/m3			104.50		

	Labour hours	Labour £	Material £	Unit	Total rate £
G20 CARPENTRY/TIMBER FRAMING/FIRST FIXING					
Sawn softwood; untreated					
Floor members					
38 x 75 mm	0.12	0.95	0.71	m	**1.66**
38 x 125 mm	0.12	0.95	0.91	m	**1.86**
38 x 125 mm	0.13	1.03	1.12	m	**2.15**
38 x 150 mm	0.14	1.11	1.32	m	**2.43**
50 x 75 mm	0.12	0.95	0.79	m	**1.74**
50 x 100 mm	0.14	1.11	1.05	m	**2.16**
50 x 125 mm	0.14	1.11	1.30	m	**2.41**
50 x 150 mm	0.15	1.19	1.56	m	**2.75**
50 x 175 mm	0.15	1.19	1.81	m	**3.00**
50 x 200 mm	0.16	1.27	2.15	m	**3.42**
50 x 225 mm	0.16	1.27	2.48	m	**3.75**
50 x 250 mm	0.17	1.34	2.92	m	**4.26**
75 x 125 mm	0.16	1.27	2.05	m	**3.32**
75 x 150 mm	0.16	1.27	2.42	m	**3.69**
75 x 175 mm	0.16	1.27	2.90	m	**4.17**
75 x 200 mm	0.17	1.34	3.38	m	**4.72**
75 x 225 mm	0.17	1.34	3.86	m	**5.20**
75 x 250 mm	0.18	1.42	4.37	m	**5.79**
100 x 150 mm	0.22	1.74	3.71	m	**5.45**
100 x 200 mm	0.23	1.82	4.94	m	**6.76**
100 x 250 mm	0.25	1.98	6.17	m	**8.15**
100 x 300 mm	0.27	2.13	6.70	m	**8.83**
Wall or partition members					
25 x 25 mm	0.07	0.55	0.19	m	**0.74**
25 x 38 mm	0.07	0.55	0.27	m	**0.82**
25 x 75 mm	0.09	0.71	0.49	m	**1.20**
38 x 38 mm	0.09	0.71	0.41	m	**1.12**

G STRUCTURAL/CARCASSING METAL/TIMBER Including overheads and profit at 5.00%	Labour hours	Labour £	Material £	Unit	Total rate £
G20 CARPENTRY/TIMBER FRAMING/FIRST FIXING - cont'd					
Sawn softwood; untreated - cont'd					
Wall or partition members					
38 x 50 mm	0.09	0.71	0.48	m	1.19
38 x 75 mm	0.12	0.95	0.71	m	1.66
38 x 100 mm	0.15	1.19	0.91	m	2.10
50 x 50 mm	0.12	0.95	0.60	m	1.55
50 x 75 mm	0.15	1.19	0.80	m	1.99
50 x 100 mm	0.18	1.42	1.06	m	2.48
50 x 125 mm	0.19	1.50	1.32	m	2.82
75 x 75 mm	0.18	1.42	1.31	m	2.73
75 x 100 mm	0.21	1.66	1.79	m	3.45
100 x 100 mm	0.21	1.66	2.47	m	4.13
Flat roof members					
38 x 75 mm	0.14	1.11	0.72	m	1.83
38 x 100 mm	0.14	1.11	0.91	m	2.02
38 x 125 mm	0.14	1.11	1.12	m	2.23
38 x 150 mm	0.14	1.11	1.32	m	2.43
50 x 100 mm	0.14	1.11	1.05	m	2.16
50 x 125 mm	0.14	1.11	1.30	m	2.41
50 x 150 mm	0.15	1.19	1.56	m	2.75
50 x 175 mm	0.15	1.19	1.81	m	3.00
50 x 200 mm	0.16	1.27	2.15	m	3.42
50 x 225 mm	0.16	1.27	2.48	m	3.75
50 x 250 mm	0.17	1.34	2.92	m	4.26
75 x 150 mm	0.16	1.27	2.40	m	3.67
75 x 175 mm	0.16	1.27	2.89	m	4.16
75 x 200 mm	0.17	1.34	3.38	m	4.72
75 x 225 mm	0.17	1.34	3.86	m	5.20
75 x 250 mm	0.18	1.42	4.37	m	5.79
Pitched roof members					
25 x 100 mm	0.12	0.95	0.60	m	1.55
25 x 125 mm	0.12	0.95	0.73	m	1.68
25 x 150 mm	0.15	1.19	0.87	m	2.06
25 x 150 mm; notching over trussed rafters	0.30	2.37	0.87	m	3.24
25 x 175 mm	0.15	1.19	1.05	m	2.24
25 x 175 mm; notching over trussed rafters	0.30	2.37	1.05	m	3.42
25 x 200 mm	0.18	1.42	1.22	m	2.64
32 x 150 mm; notching over trussed rafters	0.33	2.61	1.23	m	3.84
32 x 175 mm; notching over trussed rafters	0.33	2.61	1.48	m	4.09
32 x 200 mm; notching over trussed rafters	0.33	2.61	1.72	m	4.33
38 x 100 mm	0.15	1.19	0.91	m	2.10
38 x 125 mm	0.15	1.19	1.12	m	2.31
38 x 150 mm	0.15	1.19	1.32	m	2.51
50 x 50 mm	0.12	0.95	0.59	m	1.54
50 x 75 mm	0.15	1.19	0.79	m	1.98
50 x 100 mm	0.18	1.42	1.05	m	2.47
50 x 125 mm	0.18	1.42	1.30	m	2.72
50 x 150 mm	0.21	1.66	1.56	m	3.22
50 x 175 mm	0.21	1.66	1.81	m	3.47
50 x 200 mm	0.21	1.66	2.15	m	3.81
50 x 225 mm	0.21	1.66	2.48	m	4.14
75 x 100 mm	0.25	1.98	1.78	m	3.76
75 x 125 mm	0.25	1.98	2.05	m	4.03
75 x 150 mm	0.25	1.98	2.42	m	4.40
100 x 150 mm	0.30	2.37	3.73	m	6.10
100 x 175 mm	0.30	2.37	4.33	m	6.70
100 x 200 mm	0.30	2.37	4.95	m	7.32
100 x 225 mm	0.33	2.61	5.55	m	8.16
100 x 250 mm	0.33	2.61	6.17	m	8.78

G STRUCTURAL/CARCASSING METAL/TIMBER Including overheads and profit at 5.00%	Labour hours	Labour £	Material £	Unit	Total rate £
Kerbs, bearers and the like					
19 x 100 mm	0.04	0.32	0.48	m	**0.80**
19 x 125 mm	0.04	0.32	0.60	m	**0.92**
19 x 150 mm	0.04	0.32	0.71	m	**1.03**
25 x 75 mm	0.05	0.40	0.49	m	**0.89**
25 x 100 mm	0.05	0.40	0.60	m	**1.00**
38 x 75 mm	0.06	0.47	0.71	m	**1.18**
38 x 100 mm	0.06	0.47	0.91	m	**1.38**
50 x 75 mm	0.06	0.47	0.79	m	**1.26**
50 x 100 mm	0.07	0.55	1.05	m	**1.60**
75 x 100 mm	0.07	0.55	1.76	m	**2.31**
75 x 125 mm	0.08	0.63	2.04	m	**2.67**
75 x 150 mm	0.08	0.63	2.40	m	**3.03**
75 x 150 mm; splayed and rounded	0.10	0.79	4.21	m	**5.00**
Kerbs, bearers and the like; fixing by bolting					
19 x 100 mm	0.08	0.63	0.48	m	**1.11**
19 x 125 mm	0.08	0.63	0.60	m	**1.23**
19 x 150 mm	0.08	0.63	0.71	m	**1.34**
25 x 75 mm	0.10	0.79	0.49	m	**1.28**
25 x 100 mm	0.10	0.79	0.60	m	**1.39**
38 x 75 mm	0.12	0.95	0.71	m	**1.66**
38 x 100 mm	0.12	0.95	0.91	m	**1.86**
50 x 75 mm	0.12	0.95	0.79	m	**1.74**
50 x 100 mm	0.14	1.11	1.05	m	**2.16**
75 x 100 mm	0.14	1.11	1.78	m	**2.89**
75 x 125 mm	0.16	1.27	2.05	m	**3.32**
75 x 150 mm	0.16	1.27	2.40	m	**3.67**
Herringbone strutting 50 x 50 mm					
to 150 mm deep joists	0.50	3.95	1.37	m	**5.32**
to 175 mm deep joists	0.50	3.95	1.40	m	**5.35**
to 200 mm deep joists	0.50	3.95	1.43	m	**5.38**
to 225 mm deep joists	0.50	3.95	1.46	m	**5.41**
to 250 mm deep joists	0.50	3.95	1.49	m	**5.44**
Solid strutting to joists					
50 x 150 mm	0.30	2.37	1.77	m	**4.14**
50 x 175 mm	0.30	2.37	2.04	m	**4.41**
50 x 200 mm	0.30	2.37	2.38	m	**4.75**
50 x 225 mm	0.30	2.37	2.73	m	**5.10**
50 x 250 mm	0.30	2.37	3.18	m	**5.55**
Cleats					
225 x 100 x 75 mm	0.20	1.58	0.62	nr	**2.20**
Sprockets					
50 x 50 x 200 mm	0.15	1.19	0.62	nr	**1.81**
Extra for stress grading to above timbers					
general structural (GS) grade	-	-	-	m3	**13.65**
special structural (SS) grade	-	-	-	m3	**26.25**
Extra for protecting and flameproofing timber with 'Celcure F' protection					
small sections	-	-	-	m3	**115.50**
large sections	-	-	-	m3	**109.73**
Wrought faces					
generally	0.30	2.37	-	m2	**2.37**
50 mm wide	0.04	0.32	-	m	**0.32**
75 mm wide	0.05	0.40	-	m	**0.40**
100 mm wide	0.06	0.47	-	m	**0.47**
Raking cutting					
50 mm thick	0.20	1.58	0.48	m	**2.06**
75 mm thick	0.25	1.98	0.73	m	**2.71**
100 mm thick	0.30	2.37	0.96	m	**3.33**
Curved cutting					
50 mm thick	0.25	1.98	0.68	m	**2.66**

G STRUCTURAL/CARCASSING METAL/TIMBER Including overheads and profit at 5.00%	Labour hours	Labour £	Material £	Unit	Total rate £
G20 CARPENTRY/TIMBER FRAMING/FIRST FIXING - cont'd					
Sawn softwood; untreated - cont'd					
Scribing					
50 mm thick	0.30	2.37	-	m	**2.37**
Notching and fitting ends to metal	0.12	0.95	-	nr	**0.95**
Trimming to openings					
760 x 760 mm; joists 38 x 150 mm	2.20	17.39	0.77	nr	**18.16**
760 x 760 mm; joists 50 x 175 mm	2.30	18.18	1.35	nr	**19.53**
1500 x 500 mm; joists 38 x 200 mm	2.50	19.77	1.54	nr	**21.31**
1500 x 500 mm; joists 50 x 200 mm	2.50	19.77	1.93	nr	**21.70**
3000 x 1000 mm; joists 50 x 225 mm	4.50	35.58	4.24	nr	**39.82**
3000 x 1000 mm; joists 75 x 225 mm	5.00	39.53	6.37	nr	**45.90**
Sawn softwood; 'Tanalised'					
Floor members					
38 x 75 mm	0.12	0.95	0.88	m	**1.83**
38 x 100 mm	0.12	0.95	1.13	m	**2.08**
38 x 125 mm	0.13	1.03	1.33	m	**2.36**
38 x 150 mm	0.14	1.11	1.57	m	**2.68**
50 x 75 mm	0.12	0.95	0.94	m	**1.89**
50 x 100 mm	0.14	1.11	1.30	m	**2.41**
50 x 125 mm	0.14	1.11	1.62	m	**2.73**
50 x 150 mm	0.15	1.19	1.94	m	**3.13**
50 x 175 mm	0.15	1.19	2.26	m	**3.45**
50 x 200 mm	0.16	1.27	2.67	m	**3.94**
50 x 225 mm	0.16	1.27	3.10	m	**4.37**
50 x 250 mm	0.17	1.34	3.65	m	**4.99**
75 x 125 mm	0.16	1.27	2.56	m	**3.83**
75 x 150 mm	0.16	1.27	3.00	m	**4.27**
75 x 175 mm	0.16	1.27	3.60	m	**4.87**
75 x 200 mm	0.17	1.34	4.22	m	**5.56**
75 x 225 mm	0.17	1.34	4.81	m	**6.15**
75 x 250 mm	0.18	1.42	5.46	m	**6.88**
100 x 150 mm	0.22	1.74	4.63	m	**6.37**
100 x 200 mm	0.23	1.82	6.16	m	**7.98**
100 x 250 mm	0.25	1.98	7.68	m	**9.66**
100 x 300 mm	0.27	2.13	9.24	m	**11.37**
Wall or partition members					
25 x 25 mm	0.07	0.55	0.24	m	**0.79**
25 x 38 mm	0.07	0.55	0.32	m	**0.87**
25 x 75 mm	0.09	0.71	0.61	m	**1.32**
38 x 38 mm	0.09	0.71	0.50	m	**1.21**
38 x 50 mm	0.09	0.71	0.60	m	**1.31**
38 x 75 mm	0.12	0.95	0.88	m	**1.83**
38 x 100 mm	0.15	1.19	1.13	m	**2.32**
50 x 50 mm	0.12	0.95	0.70	m	**1.65**
50 x 75 mm	0.15	1.19	0.95	m	**2.14**
50 x 100 mm	0.18	1.42	1.32	m	**2.74**
50 x 125 mm	0.19	1.50	1.64	m	**3.14**
75 x 75 mm	0.18	1.42	1.66	m	**3.08**
75 x 100 mm	0.21	1.66	2.21	m	**3.87**
100 x 100 mm	0.21	1.66	3.12	m	**4.78**
Flat roof members					
38 x 75 mm	0.14	1.11	0.88	m	**1.99**
38 x 100 mm	0.14	1.11	1.13	m	**2.24**
38 x 125 mm	0.14	1.11	1.33	m	**2.44**
38 x 150 mm	0.14	1.11	1.57	m	**2.68**
50 x 100 mm	0.14	1.11	1.30	m	**2.41**
50 x 125 mm	0.14	1.11	1.62	m	**2.73**
50 x 150 mm	0.15	1.19	1.94	m	**3.13**
50 x 175 mm	0.15	1.19	2.26	m	**3.45**
50 x 200 mm	0.16	1.27	2.67	m	**3.94**
50 x 225 mm	0.16	1.27	3.10	m	**4.37**

G STRUCTURAL/CARCASSING METAL/TIMBER Including overheads and profit at 5.00%	Labour hours	Labour £	Material £	Unit	Total rate £
50 x 250 mm	0.17	1.34	3.65	m	**4.99**
75 x 150 mm	0.16	1.27	3.00	m	**4.27**
75 x 175 mm	0.16	1.27	3.59	m	**4.86**
75 x 200 mm	0.17	1.34	4.22	m	**5.56**
75 x 225 mm	0.17	1.34	4.81	m	**6.15**
75 x 250 mm	0.18	1.42	5.46	m	**6.88**
Pitched roof members					
25 x 100 mm	0.12	0.95	0.73	m	**1.68**
25 x 125 mm	0.12	0.95	0.91	m	**1.86**
25 x 150 mm	0.15	1.19	1.08	m	**2.27**
25 x 150 mm; notching over trussed rafters	0.30	2.37	1.08	m	**3.45**
25 x 175 mm	0.15	1.19	1.30	m	**2.49**
25 x 175 mm; notching over trussed rafters	0.30	2.37	1.30	m	**3.67**
25 x 200 mm	0.18	1.42	1.53	m	**2.95**
32 x 150 mm; notching over trussed rafters	0.33	2.61	1.53	m	**4.14**
32 x 175 mm; notching over trussed rafters	0.33	2.61	1.83	m	**4.44**
32 x 200 mm; notching over trussed rafters	0.33	2.61	2.15	m	**4.76**
38 x 100 mm	0.15	1.19	1.13	m	**2.32**
38 x 125 mm	0.15	1.19	1.33	m	**2.52**
38 x 150 mm	0.15	1.19	1.57	m	**2.76**
50 x 50 mm	0.12	0.95	0.69	m	**1.64**
50 x 75 mm	0.15	1.19	0.94	m	**2.13**
50 x 100 mm	0.18	1.42	1.30	m	**2.72**
50 x 125 mm	0.18	1.42	1.62	m	**3.04**
50 x 150 mm	0.21	1.66	1.94	m	**3.60**
50 x 175 mm	0.21	1.66	2.26	m	**3.92**
50 x 200 mm	0.21	1.66	2.67	m	**4.33**
50 x 225 mm	0.21	1.66	3.10	m	**4.76**
75 x 100 mm	0.25	1.98	2.20	m	**4.18**
75 x 125 mm	0.25	1.98	2.56	m	**4.54**
75 x 150 mm	0.25	1.98	3.00	m	**4.98**
100 x 150 mm	0.30	2.37	4.64	m	**7.01**
100 x 175 mm	0.30	2.37	5.40	m	**7.77**
100 x 200 mm	0.30	2.37	6.17	m	**8.54**
100 x 225 mm	0.33	2.61	6.93	m	**9.54**
100 x 250 mm	0.33	2.61	7.70	m	**10.31**
Kerbs, bearers and the like					
19 x 100 mm	0.04	0.32	0.60	m	**0.92**
19 x 125 mm	0.04	0.32	0.73	m	**1.05**
19 x 150 mm	0.04	0.32	0.88	m	**1.20**
25 x 75 mm	0.05	0.40	0.61	m	**1.01**
25 x 100 mm	0.05	0.40	0.73	m	**1.13**
38 x 75 mm	0.06	0.47	0.88	m	**1.35**
38 x 100 mm	0.06	0.47	1.13	m	**1.60**
50 x 75 mm	0.06	0.47	0.94	m	**1.41**
50 x 100 mm	0.07	0.55	1.30	m	**1.85**
75 x 100 mm	0.07	0.55	2.18	m	**2.73**
75 x 125 mm	0.08	0.63	2.55	m	**3.18**
75 x 150 mm	0.08	0.63	2.99	m	**3.62**
75 x 150 mm; splayed and rounded	0.10	0.79	5.25	m	**6.04**
Kerbs, bearers and the like; fixing by bolting					
19 x 100 mm	0.08	0.63	0.60	m	**1.23**
19 x 125 mm	0.08	0.63	0.73	m	**1.36**
19 x 150 mm	0.08	0.63	0.88	m	**1.51**
25 x 75 mm	0.10	0.79	0.62	m	**1.41**
25 x 100 mm	0.10	0.79	0.73	m	**1.52**
38 x 75 mm	0.12	0.95	0.88	m	**1.83**
38 x 100 mm	0.12	0.95	1.13	m	**2.08**
50 x 75 mm	0.12	0.95	0.94	m	**1.89**
50 x 100 mm	0.14	1.11	1.30	m	**2.41**
75 x 100 mm	0.14	1.11	2.20	m	**3.31**
75 x 125 mm	0.16	1.27	2.56	m	**3.83**
75 x 150 mm	0.16	1.27	2.99	m	**4.26**

G STRUCTURAL/CARCASSING METAL/TIMBER Including overheads and profit at 5.00%	Labour hours	Labour £	Material £	Unit	Total rate £
G20 CARPENTRY/TIMBER FRAMING/FIRST FIXING - cont'd					
Sawn softwood; 'Tanalised' - cont'd					
Herringbone strutting 50 x 50 mm					
to 150 mm deep joists	0.50	3.95	1.59	m	5.54
to 175 mm deep joists	0.50	3.95	1.62	m	5.57
to 200 mm deep joists	0.50	3.95	1.66	m	5.61
to 225 mm deep joists	0.50	3.95	1.69	m	5.64
to 250 mm deep joists	0.50	3.95	1.72	m	5.67
Solid strutting to joists					
50 x 150 mm	0.30	2.37	2.16	m	4.53
50 x 175 mm	0.30	2.37	2.49	m	4.86
50 x 200 mm	0.30	2.37	2.90	m	5.27
50 x 225 mm	0.30	2.37	3.34	m	5.71
50 x 250 mm	0.30	2.37	3.92	m	6.29
Cleats					
225 x 100 x 75 mm	0.20	1.58	0.74	nr	2.32
Sprockets					
50 x 50 x 200 mm	0.15	1.19	0.74	nr	1.93
Extra for stress grading to above timbers					
general structural (GS) grade	-	-	-	m3	13.65
special structural (SS) grade	-	-	-	m3	26.25
Extra for protecting and flameproofing timber with 'Celcure F' protection					
small sections	-	-	-	m3	115.50
large sections	-	-	-	m3	109.73
Wrought faces					
generally	0.30	2.37	-	m2	2.37
50 mm wide	0.04	0.32	-	m	0.32
75 mm wide	0.05	0.40	-	m	0.40
100 mm wide	0.06	0.47	-	m	0.47
Raking cutting					
50 mm thick	0.20	1.58	0.58	m	2.16
75 mm thick	0.25	1.98	0.88	m	2.86
100 mm thick	0.30	2.37	1.16	m	3.53
Curved cutting					
50 mm thick	0.25	1.98	0.81	m	2.79
Scribing					
50 mm thick	0.30	2.37	-	m	2.37
Notching and fitting ends to metal	0.12	0.95	-	nr	0.95
Trimming to openings					
760 x 760 mm; joists 38 x 150 mm	2.20	17.39	0.92	nr	18.31
760 x 760 mm; joists 50 x 175 mm	2.30	18.18	1.62	nr	19.80
1500 x 500 mm; joists 38 x 200 mm	2.50	19.77	1.85	nr	21.62
1500 x 500 mm; joists 50 x 200 mm	2.50	19.77	2.31	nr	22.08
3000 x 1000 mm; joists 50 x 225 mm	4.50	35.58	5.08	nr	40.66
3000 x 1000 mm; joists 75 x 225 mm	5.00	39.53	7.62	nr	47.15
Trussed rafters, stress graded sawn softwood pressure impregnated; raised through two storeys and fixed in position					
'W' type truss (Fink); 22.5 degree pitch; 450 mm eaves overhang					
5.00 m span	1.60	12.65	16.95	nr	29.60
7.60 m span	1.75	13.84	25.55	nr	39.39
10.00 m span	2.00	15.81	40.49	nr	56.30
'W' type truss (Fink); 30 degree pitch; 450 mm eaves overhang					
5.00 m span	1.60	12.65	17.90	nr	30.55
7.60 m span	1.75	13.84	26.56	nr	40.40
10.00 m span	2.00	15.81	42.82	nr	58.63

G STRUCTURAL/CARCASSING METAL/TIMBER Including overheads and profit at 5.00%	Labour hours	Labour £	Material £	Unit	Total rate £
'W' type truss (Fink); 45 degree pitch; 450 mm eaves overhang					
4.60 m span	1.60	12.65	20.19	nr	**32.84**
7.00 m span	1.75	13.84	30.73	nr	**44.57**
'Mono' type truss; 17.5 degree pitch; 450 mm eaves overhang					
3.30 m span	1.40	11.07	14.73	nr	**25.80**
5.60 m span	1.60	12.65	23.80	nr	**36.45**
7.00 m span	1.85	14.63	31.76	nr	**46.39**
'Mono' type truss; 30 degree pitch; 450 mm eaves overhang					
3.30 m span	1.40	11.07	16.56	nr	**27.63**
5.60 m span	1.60	12.65	27.02	nr	**39.67**
7.00 m span	1.85	14.63	38.09	nr	**52.72**
Attic type truss; 45 degree pitch; 450 mm eaves overhang					
5.00 m span	3.15	24.91	37.51	nr	**62.42**
7.60 m span	3.30	26.09	50.21	nr	**76.30**
9.00 m span	3.50	27.67	61.39	nr	**89.06**
Standard 'Toreboda' glulam timber beams; Moelven (UK) Ltd.; LB grade whitewood; pressure impregnated; phenol resorcinal adhesive; clean planed finish; fixed					
Laminated roof beams					
56 x 255 mm	0.55	4.35	14.08	m	**18.43**
66 x 315 mm	0.70	5.53	18.91	m	**24.44**
90 x 315 mm	0.90	7.12	23.71	m	**30.83**
90 x 405 mm	1.15	9.09	29.68	m	**38.77**
115 x 405 mm	1.45	11.46	35.76	m	**47.22**
115 x 495 mm	1.80	14.23	47.45	m	**61.68**
115 x 630 mm	2.20	17.39	59.35	m	**76.74**

ALTERNATIVE FIRST FIXING MATERIAL PRICES

	£		£		£
Chipboard roofing (£/10 m2)					
12 mm	27.64	18 mm	38.31	25 mm	53.22
Non-asbestos boards (£/10 m2)					
'Masterboard'					
6 mm	40.77	9 mm	74.07	12 mm	97.65
'Masterclad'; sanded finish					
4.5 mm	33.21	6 mm	42.84	9 mm	71.64
Plywood (£/10 m2)					
External quality					
12 mm	67.18	15 mm	82.45	25 mm	120.55
Marine quality					
12 mm	55.74	15 mm	70.70	25 mm	113.07

Discounts of 0 - 10% available depending on quantity/status

G STRUCTURAL/CARCASSING METAL/TIMBER Including overheads and profit at 5.00%	Labour hours	Labour £	Material £	Unit	Total rate £
G20 CARPENTRY/TIMBER FRAMING/FIRST FIXING - cont'd					
'Masterboard'; 6 mm thick; PC £40.77/10m2					
Boarding to eaves, verges, fascias and the like					
over 300 mm wide	0.70	5.53	4.88	m2	**10.41**
75 mm wide	0.21	1.66	0.44	m	**2.10**
150 mm wide	0.24	1.90	0.82	m	**2.72**
200 mm wide	0.27	2.13	1.07	m	**3.20**
225 mm wide	0.28	2.21	1.21	m	**3.42**
250 mm wide	0.29	2.29	1.36	m	**3.65**
Raking cutting	0.05	0.40	0.24	m	**0.64**
Plywood; external quality; 15 mm thick; PC £83.40/10m2					
Boarding to eaves, verges, fascias and the like					
over 300 mm wide	0.82	6.48	9.69	m2	**16.17**
75 mm wide	0.25	1.98	0.88	m	**2.86**
150 mm wide	0.29	2.29	1.64	m	**3.93**
225 mm wide	0.33	2.61	2.43	m	**5.04**
Plywood; external quality; 18 mm thick; PC £97.96/10m2					
Boarding to eaves, verges, fascias and the like					
over 300 mm wide	0.82	6.48	11.32	m2	**17.80**
75 mm wide	0.25	1.98	1.03	m	**3.01**
150 mm wide	0.29	2.29	1.93	m	**4.22**
225 mm wide	0.33	2.61	2.84	m	**5.45**
Plywood; marine quality; 18 mm thick; PC £81.41/10m2					
Boarding to gutter bottoms or sides; butt joints					
over 300 mm wide	0.93	7.35	9.46	m2	**16.81**
150 mm wide	0.33	2.61	1.61	m	**4.22**
225 mm wide	0.37	2.93	2.46	m	**5.39**
300 mm wide	0.42	3.32	3.12	m	**6.44**
Boarding to eaves, verges, fascias and the like					
over 300 mm wide	0.82	6.48	9.46	m2	**15.94**
75 mm wide	0.25	1.98	0.85	m	**2.83**
150 mm wide	0.29	2.29	1.61	m	**3.90**
225 mm wide	0.31	2.45	2.07	m	**4.52**
Plywood; marine quality; 25 mm thick; PC £121.75/10m2					
Boarding to gutter bottoms or sides; butt joints					
150 mm wide	1.00	7.91	14.01	m	**21.92**
150 mm wide	0.35	2.77	2.39	m	**5.16**
225 mm wide	0.40	3.16	3.65	m	**6.81**
300 mm wide	0.45	3.56	4.64	m	**8.20**
Boarding to eaves, verges, fascias and the like					
over 300 mm wide	0.88	6.96	14.01	m2	**20.97**
75 mm wide	0.26	2.06	1.26	m	**3.32**
150 mm wide	0.31	2.45	2.39	m	**4.84**
225 mm wide	0.35	2.77	3.08	m	**5.85**
Sawn softwood; untreated					
Boarding to gutter bottoms or sides; butt joints					
19 mm thick; sloping	1.25	9.88	5.05	m2	**14.93**
19 mm thick x 75 mm wide	0.35	2.77	0.40	m	**3.17**
19 mm thick x 150 mm wide	0.40	3.16	0.76	m	**3.92**
19 mm thick x 225 mm wide	0.45	3.56	1.18	m	**4.74**

G STRUCTURAL/CARCASSING METAL/TIMBER Including overheads and profit at 5.00%	Labour hours	Labour £	Material £	Unit	Total rate £
25 mm thick; sloping	1.25	9.88	6.17	m2	16.05
25 mm thick x 75 mm wide	0.35	2.77	0.51	m	3.28
25 mm thick x 150 mm wide	0.40	3.16	0.92	m	4.08
25 mm thick x 225 mm wide	0.45	3.56	1.46	m	5.02
Cesspools with 25 mm thick sides and bottom					
225 x 225 x 150 mm	1.20	9.49	2.92	nr	12.41
300 x 300 x 150 mm	1.40	11.07	4.15	nr	15.22
Firrings					
50 mm wide x 36 mm average depth	0.15	1.19	0.62	m	1.81
50 mm wide x 50 mm average depth	0.15	1.19	0.74	m	1.93
50 mm wide x 75 mm average depth	0.15	1.19	1.10	m	2.29
Bearers					
25 x 50 mm	0.10	0.79	0.37	m	1.16
38 x 50 mm	0.10	0.79	0.52	m	1.31
50 x 50 mm	0.10	0.79	0.63	m	1.42
50 x 75 mm	0.10	0.79	0.83	m	1.62
Angle fillets					
38 x 38 mm	0.10	0.79	0.30	m	1.09
50 x 50 mm	0.10	0.79	0.45	m	1.24
75 x 75 mm	0.12	0.95	0.86	m	1.81
Tilting fillets					
19 x 38 mm	0.10	0.79	0.17	m	0.96
25 x 50 mm	0.10	0.79	0.25	m	1.04
38 x 75 mm	0.10	0.79	0.48	m	1.27
50 x 75 mm	0.10	0.79	0.63	m	1.42
75 x 100 mm	0.15	1.19	1.00	m	2.19
Grounds or battens					
13 x 19 mm	0.05	0.40	0.13	m	0.53
13 x 32 mm	0.05	0.40	0.16	m	0.56
25 x 50 mm	0.05	0.40	0.35	m	0.75
Grounds or battens; plugged and screwed					
13 x 19 mm	0.15	1.19	0.17	m	1.36
13 x 32 mm	0.15	1.19	0.21	m	1.40
25 x 50 mm	0.15	1.19	0.39	m	1.58
Open-spaced grounds or battens; at 300 mm centres one way					
25 x 50 mm	0.15	1.19	1.15	m2	2.34
25 x 50 mm; plugged and screwed	0.45	3.56	1.31	m2	4.87
Framework to walls; at 300 mm centres one way and 600 mm centres the other					
25 x 50 mm	0.75	5.93	1.91	m2	7.84
38 x 50 mm	0.75	5.93	2.69	m2	8.62
50 x 50 mm	0.75	5.93	3.28	m2	9.21
50 x 75 mm	0.75	5.93	4.46	m2	10.39
75 x 75 mm	0.75	5.93	7.29	m2	13.22
Framework to walls; at 300 mm centres one way and 600 mm centres the other way; plugged and screwed					
25 x 50 mm	1.25	9.88	2.14	m2	12.02
38 x 50 mm	1.25	9.88	2.93	m2	12.81
50 x 50 mm	1.25	9.88	3.52	m2	13.40
50 x 75 mm	1.25	9.88	4.71	m2	14.59
75 x 75 mm	1.25	9.88	7.54	m2	17.42
Framework to bath panel; at 500 mm centres both ways					
25 x 50 mm	0.90	7.12	2.08	m2	9.20
Framework as bracketing and cradling around steelwork					
25 x 50 mm	1.40	11.07	2.48	m2	13.55
50 x 50 mm	1.50	11.86	4.26	m2	16.12
50 x 75 mm	1.60	12.65	5.80	m2	18.45

G STRUCTURAL/CARCASSING METAL/TIMBER Including overheads and profit at 5.00%	Labour hours	Labour £	Material £	Unit	Total rate £
G20 CARPENTRY/TIMBER FRAMING/FIRST FIXING - cont'd					
Sawn softwood; untreated - cont'd					
Blockings wedged between flanges of steelwork					
50 x 50 x 150 mm	0.12	0.95	0.29	nr	1.24
50 x 75 x 225 mm	0.13	1.03	0.43	nr	1.46
50 x 100 x 300 mm	0.14	1.11	0.57	nr	1.68
Sawn softwood; 'Tanalised'					
Boarding to gutter bottoms or sides; butt joints					
19 mm thick; sloping	1.25	9.88	6.21	m2	16.09
19 mm thick x 75 mm wide	0.35	2.77	0.49	m	3.26
19 mm thick x 150 mm wide	0.40	3.16	0.93	m	4.09
19 mm thick x 225 mm wide	0.45	3.56	1.40	m	4.96
25 mm thick; sloping	1.25	9.88	7.66	m2	17.54
25 mm thick x 75 mm wide	0.35	2.77	0.63	m	3.40
25 mm thick x 150 mm wide	0.40	3.16	1.13	m	4.29
25 mm thick x 225 mm wide	0.45	3.56	1.74	m	5.30
Firrings					
50 mm wide x 36 mm average depth	0.15	1.19	0.74	m	1.93
50 mm wide x 50 mm average depth	0.15	1.19	0.90	m	2.09
50 mm wide x 75 mm average depth	0.15	1.19	1.34	m	2.53
Bearers					
25 x 50 mm	0.10	0.79	0.46	m	1.25
38 x 50 mm	0.10	0.79	0.64	m	1.43
50 x 50 mm	0.10	0.79	0.74	m	1.53
50 x 75 mm	0.10	0.79	0.99	m	1.78
Cesspools with 25 mm thick sides and bottom					
225 x 225 x 150 mm	1.20	9.49	3.64	nr	13.13
300 x 300 x 150 mm	1.40	11.07	5.18	nr	16.25
Angle fillets					
38 x 38 mm	0.10	0.79	0.35	m	1.14
50 x 50 mm	0.10	0.79	0.54	m	1.33
75 x 75 mm	0.12	0.95	1.06	m	2.01
Tilting fillets					
19 x 38 mm	0.10	0.79	0.21	m	1.00
25 x 50 mm	0.10	0.79	0.31	m	1.10
38 x 75 mm	0.10	0.79	0.59	m	1.38
50 x 75 mm	0.10	0.79	0.77	m	1.56
75 x 100 mm	0.15	1.19	1.24	m	2.43
Grounds or battens					
13 x 19 mm	0.05	0.40	0.15	m	0.55
13 x 32 mm	0.05	0.40	0.19	m	0.59
25 x 50 mm	0.05	0.40	0.43	m	0.83
Grounds or battens; plugged and screwed					
13 x 19 mm	0.15	1.19	0.17	m	1.36
13 x 32 mm	0.15	1.19	0.24	m	1.43
25 x 50 mm	0.15	1.19	0.48	m	1.67
Open-spaced grounds or battens; at 300 mm centres one way					
25 x 50 mm	0.15	1.19	1.42	m2	2.61
25 x 50 mm; plugged and screwed	0.45	3.56	1.59	m2	5.15
Framework to walls; at 300 mm centres one way and 600 mm centres the other					
25 x 50 mm	0.75	5.93	2.36	m2	8.29
38 x 50 mm	0.75	5.93	3.34	m2	9.27
50 x 50 mm	0.75	5.93	3.88	m2	9.81
50 x 75 mm	0.75	5.93	5.33	m2	11.26
75 x 75 mm	0.75	5.93	9.33	m2	15.26

G STRUCTURAL/CARCASSING METAL/TIMBER Including overheads and profit at 5.00%	Labour hours	Labour £	Material £	Unit	Total rate £
Framework to walls; at 300 mm centres one way and 600 mm centres the other way; plugged and screwed					
25 x 50 mm	1.25	9.88	2.60	m2	12.48
38 x 50 mm	1.25	9.88	3.59	m2	13.47
50 x 50 mm	1.25	9.88	4.13	m2	14.01
50 x 75 mm	1.25	9.88	5.57	m2	15.45
75 x 75 mm	1.25	9.88	9.57	m2	19.45
Framework to bath panel; at 500 mm centres both ways					
25 x 50 mm	0.90	7.12	2.58	m2	9.70
Framework as bracketing and cradling around steelwork					
25 x 50 mm	1.40	11.07	3.08	m2	14.15
50 x 50 mm	1.50	11.86	5.04	m2	16.90
50 x 75 mm	1.60	12.65	6.90	m2	19.55
Blockings wedged between flanges of steelwork					
50 x 50 x 150 mm	0.12	0.95	0.35	nr	1.30
50 x 75 x 225 mm	0.13	1.03	0.54	nr	1.57
50 x 100 x 300 mm	0.14	1.11	0.71	nr	1.82
Floor fillets set in or on concrete					
38 x 50 mm	0.12	0.95	0.58	m	1.53
50 x 50 mm	0.12	0.95	0.71	m	1.66
Floor fillets fixed to floor clips (priced elsewhere)					
38 x 50 mm	0.10	0.79	0.61	m	1.40
50 x 50 mm	0.10	0.79	0.74	m	1.53
Wrought softwood					
Boarding to gutter bottoms or sides; tongued and grooved joints					
19 mm thick; sloping	1.50	11.86	6.86	m2	18.72
19 mm thick x 75 mm wide	0.40	3.16	0.67	m	3.83
19 mm thick x 150 mm wide	0.45	3.56	1.03	m	4.59
19 mm thick x 225 mm wide	0.50	3.95	1.63	m	5.58
25 mm thick; sloping	1.50	11.86	8.89	m2	20.75
25 mm thick x 75 mm wide	0.40	3.16	0.69	m	3.85
25 mm thick x 150 mm wide	0.45	3.56	1.30	m	4.86
25 mm thick x 225 mm wide	0.50	3.95	2.10	m	6.05
Boarding to eaves, verges, fascias and the like					
19 mm thick x over 300 mm wide	1.24	9.80	6.71	m2	16.51
19 mm thick x 150 mm wide; once grooved	0.20	1.58	2.41	m	3.99
25 mm thick x 150 mm wide; once grooved	0.20	1.58	2.92	m	4.50
25 mm thick x 175 mm wide; once grooved	0.22	1.74	3.30	m	5.04
32 mm thick x 225 mm wide; moulded	0.25	1.98	4.99	m	6.97
Mitred angles	0.10	0.79	-	nr	0.79
Rolls					
32 x 44 mm	0.12	0.95	0.56	m	1.51
50 x 50 mm	0.12	0.95	0.90	m	1.85
50 x 75 mm	0.13	1.03	1.41	m	2.44
75 x 75 mm	0.14	1.11	2.08	m	3.19
Wrought softwood; 'Tanalised'					
Boarding to gutter bottoms or sides; tongued and grooved joints					
19 mm thick; sloping	1.50	11.86	8.46	m2	20.32
19 mm thick x 75 mm wide	0.40	3.16	0.83	m	3.99
19 mm thick x 150 mm wide	0.45	3.56	1.27	m	4.83
19 mm thick x 225 mm wide	0.50	3.95	2.03	m	5.98
25 mm thick; sloping	1.50	11.86	10.97	m2	22.83
25 mm thick x 75 mm wide	0.40	3.16	0.84	m	4.00
25 mm thick x 150 mm wide	0.45	3.56	1.62	m	5.18
25 mm thick x 225 mm wide	0.50	3.95	2.62	m	6.57

G STRUCTURAL/CARCASSING METAL/TIMBER Including overheads and profit at 5.00%	Labour hours	Labour £	Material £	Unit	Total rate £
G20 CARPENTRY/TIMBER FRAMING/FIRST FIXING - cont'd					
Wrought softwood; 'Tanalised' - cont'd					
Boarding to eaves, verges, fascias and the like					
19 mm thick x over 300 mm wide	1.24	9.80	8.31	m2	**18.11**
19 mm thick x 150 mm wide; once grooved	0.20	1.58	2.98	m	**4.56**
25 mm thick x 150 mm wide; once grooved	0.20	1.58	3.64	m	**5.22**
25 mm thick x 175 mm wide; once grooved	0.22	1.74	4.13	m	**5.87**
32 mm thick x 225 mm wide; moulded	0.25	1.98	6.22	m	**8.20**
Mitred angles	0.10	0.79	-	nr	**0.79**
Rolls					
32 x 44 mm	0.12	0.95	0.73	m	**1.68**
50 x 50 mm	0.12	0.95	1.17	m	**2.12**
50 x 75 mm	0.13	1.03	1.84	m	**2.87**
75 x 75 mm	0.14	1.11	2.72	m	**3.83**
Labours on softwood boarding					
Raking cutting					
12 mm thick	0.06	0.47	0.31	m	**0.78**
19 mm thick	0.08	0.63	0.27	m	**0.90**
25 mm thick	0.10	0.79	0.38	m	**1.17**
Boundary cutting					
19 mm thick	0.09	0.71	0.27	m	**0.98**
Curved cutting					
19 mm thick	0.12	0.95	0.38	m	**1.33**
Tongued edges and mitred angle					
19 mm thick	0.14	1.11	0.38	m	**1.49**
Plugging					
Plugging blockwork					
300 mm centres; both ways	0.12	0.95	0.04	m2	**0.99**
300 mm centres; one way	0.06	0.47	0.02	m	**0.49**
isolated	0.03	0.24	0.01	nr	**0.25**
Plugging brickwork					
300 mm centres; both ways	0.20	1.58	0.04	m2	**1.62**
300 mm centres; one way	0.10	0.79	0.02	m	**0.81**
isolated	0.05	0.40	0.01	nr	**0.41**
Plugging concrete					
300 mm centres; both ways	0.36	2.85	0.04	m2	**2.89**
300 mm centres; one way	0.18	1.42	0.02	m	**1.44**
isolated	0.09	0.71	0.01	nr	**0.72**
Shot-firing					
Concrete					
1 m centres; one way	0.08	0.63	0.23	m	**0.86**
isolated	0.10	0.79	0.23	nr	**1.02**
Steel					
1 m centres; one way	0.08	0.63	0.23	m	**0.86**
isolated	0.10	0.79	0.23	nr	**1.02**

BASIC BOLT PRICES

	£		£		£		£
Black bolts, nuts and washers (£/100)							
Mild steel hex. hdd. bolts/nuts							
M6x50 mm	5.78	M10x50 mm	13.93	M12x100 mm	28.59	M16x140 mm	76.95
M6x80 mm	9.68	M10x80 mm	19.88	M12x140 mm	43.33	M16x180 mm	112.55
M6x100 mm	10.52	M10x100 mm	24.50	M12x180 mm	81.90	M20x80 mm	79.26
M8x50 mm	9.16	M10x140 mm	33.69	M16x50 mm	34.94	M20x100 mm	92.48
M8x80 mm	12.70	M12x50 mm	19.80	M16x80 mm	44.74	M20x140 mm	125.01
M8x100 mm	15.84	M12x80 mm	25.09	M16x100 mm	49.54	M20x180 mm	170.37

G STRUCTURAL/CARCASSING METAL/TIMBER
Including overheads and profit at 5.00%

Mild steel cup. hdd. bolts/nuts

M6x50 mm	5.32	M8x75 mm	9.63	M10x100 mm	22.04	M12x100 mm	31.13
M6x75 mm	6.51	M8x100 mm	15.84	M10x150 mm	31.62	M12x150 mm	43.20
M6x100 mm	10.14	M8x150 mm	21.65	M10x200 mm	66.54	M12x200 mm	105.87
M6x150 mm	15.44	M10x50 mm	13.03	M12x50 mm	20.37		
M8x50 mm	8.16	M10x75 mm	14.76	M12x75 mm	25.55		
Mild steel washers; round							
M6	1.50	M10	1.85	M16	3.94	M20	4.56
M8	1.30	M12	2.86				
Mild steel washers; square							
38x38 mm	8.54	50x50 mm	5.91				

		Labour hours	Labour £	Material £	Unit	Total rate £
Metalwork; mild steel; galvanized						
Fix only bolts; 50-200 mm long						
6 mm dia		0.04	0.32	-	nr	**0.32**
8 mm dia		0.04	0.32	-	nr	**0.32**
10 mm dia		0.05	0.40	-	nr	**0.40**
12 mm dia		0.05	0.40	-	nr	**0.40**
16 mm dia		0.06	0.47	-	nr	**0.47**
20 mm dia		0.06	0.47	-	nr	**0.47**
Straps; standard twisted vertical restraint; fixing to softwood and brick or blockwork						
30 x 2.5 x 400 mm girth		0.25	1.98	0.64	nr	**2.62**
30 x 2.5 x 600 mm girth		0.26	2.06	0.81	nr	**2.87**
30 x 2.5 x 800 mm girth		0.27	2.13	1.06	nr	**3.19**
30 x 2.5 x 1000 mm girth		0.30	2.37	1.29	nr	**3.66**
30 x 2.5 x 1200 mm girth		0.31	2.45	1.51	nr	**3.96**
Timber connectors; round toothed plate; for 10 mm or 12 mm dia bolts						
38 mm dia; single sided		0.02	0.16	0.15	nr	**0.31**
38 mm dia; double sided		0.02	0.16	0.17	nr	**0.33**
50 mm dia; single sided		0.02	0.16	0.17	nr	**0.33**
50 mm dia; double sided		0.02	0.16	0.19	nr	**0.35**
63 mm dia; single sided		0.02	0.16	0.24	nr	**0.40**
63 mm dia; double sided		0.02	0.16	0.27	nr	**0.43**
75 mm dia; single sided		0.02	0.16	0.35	nr	**0.51**
75 mm dia; double sided		0.02	0.16	0.39	nr	**0.55**
Framing anchor		0.15	1.19	0.43	nr	**1.62**
Joist hangers 1.0 mm thick; for fixing to softwood; joint sizes						
50 x 100 mm	PC £0.61	0.12	0.95	0.78	nr	**1.73**
50 x 125 mm	PC £0.61	0.12	0.95	0.78	nr	**1.73**
50 x 150 mm	PC £0.61	0.13	1.03	0.78	nr	**1.81**
50 x 175 mm	PC £0.61	0.13	1.03	0.78	nr	**1.81**
50 x 200 mm	PC £0.61	0.14	1.11	0.78	nr	**1.89**
50 x 225 mm	PC £0.61	0.14	1.11	0.78	nr	**1.89**
50 x 250 mm	PC £0.61	0.15	1.19	0.78	nr	**1.97**
75 x 150 mm	PC £0.65	0.13	1.03	0.82	nr	**1.85**
75 x 175 mm	PC £0.65	0.13	1.03	0.82	nr	**1.85**
75 x 200 mm	PC £0.65	0.14	1.11	0.82	nr	**1.93**
75 x 225 mm	PC £0.65	0.14	1.11	0.82	nr	**1.93**
75 x 250 mm	PC £0.65	0.15	1.19	0.82	nr	**2.01**
100 x 200 mm	PC £0.70	0.15	1.19	0.88	nr	**2.07**

G STRUCTURAL/CARCASSING METAL/TIMBER Including overheads and profit at 5.00%		Labour hours	Labour £	Material £	Unit	Total rate £
G20 CARPENTRY/TIMBER FRAMING/FIRST FIXING - cont'd						
Metalwork; mild steel; galvanized - cont'd						
Joist hangers 2.7 mm thick; for building in; joist sizes						
50 x 100 mm	PC £1.05	0.08	0.63	1.20	nr	**1.83**
50 x 125 mm	PC £1.05	0.08	0.63	1.21	nr	**1.84**
50 x 150 mm	PC £1.08	0.09	0.71	1.24	nr	**1.95**
50 x 175 mm	PC £1.11	0.09	0.71	1.28	nr	**1.99**
50 x 200 mm	PC £1.22	0.10	0.79	1.39	nr	**2.18**
50 x 225 mm	PC £1.30	0.10	0.79	1.49	nr	**2.28**
50 x 250 mm	PC £1.70	0.11	0.87	1.93	nr	**2.80**
75 x 150 mm	PC £1.72	0.09	0.71	1.95	nr	**2.66**
75 x 175 mm	PC £1.58	0.09	0.71	1.80	nr	**2.51**
75 x 200 mm	PC £1.74	0.10	0.79	1.97	nr	**2.76**
75 x 225 mm	PC £1.81	0.10	0.79	2.05	nr	**2.84**
75 x 250 mm	PC £2.01	0.11	0.87	2.26	nr	**3.13**
100 x 200 mm		0.10	0.79	2.27	nr	**3.06**
Herringbone joist struts; to suit joists at						
400 mm centres	PC £20.58/100	0.30	2.37	1.08	m	**3.45**
450 mm centres	PC £23.25/100	0.27	2.13	1.07	m	**3.20**
600 mm centres	PC £25.93/100	0.24	1.90	1.09	m	**2.99**
Expanding bolts; 'Rawlbolt' projecting type; plated; one nut; one washer						
6 mm dia; ref M6 10P		0.08	0.63	0.40	nr	**1.03**
6 mm dia; ref M6 25P		0.08	0.63	0.45	nr	**1.08**
6 mm dia; ref M6 60P		0.08	0.63	0.47	nr	**1.10**
8 mm dia; ref M8 25P		0.08	0.63	0.54	nr	**1.17**
8 mm dia; ref M8 60P		0.08	0.63	0.57	nr	**1.20**
10 mm dia; ref M10 15P		0.10	0.79	0.70	nr	**1.49**
10 mm dia; ref M10 30P		0.10	0.79	0.73	nr	**1.52**
10 mm dia; ref M10 60P		0.10	0.79	0.76	nr	**1.55**
12 mm dia; ref M12 15P		0.10	0.79	1.11	nr	**1.90**
12 mm dia; ref M12 30P		0.10	0.79	1.19	nr	**1.98**
12 mm dia; ref M12 75P		0.10	0.79	1.48	nr	**2.27**
16 mm dia; ref M16 35P		0.12	0.95	2.74	nr	**3.69**
16 mm dia; ref M16 75P		0.12	0.95	2.87	nr	**3.82**
Expanding bolts; 'Rawlbolt' loose bolt type; plated; one bolt; one washer						
6 mm dia; ref M6 10L		0.10	0.79	0.40	nr	**1.19**
6 mm dia; ref M6 25L		0.10	0.79	0.43	nr	**1.22**
6 mm dia; ref M6 40L		0.10	0.79	0.43	nr	**1.22**
8 mm dia; ref M8 25L		0.10	0.79	0.53	nr	**1.32**
8 mm dia; ref M8 40L		0.10	0.79	0.56	nr	**1.35**
10 mm dia; ref M10 10L		0.10	0.79	0.68	nr	**1.47**
10 mm dia; ref M10 25L		0.10	0.79	0.70	nr	**1.49**
10 mm dia; ref M10 50L		0.10	0.79	0.74	nr	**1.53**
10 mm dia; ref M10 75L		0.10	0.79	0.76	nr	**1.55**
12 mm dia; ref M12 10L		0.10	0.79	1.00	nr	**1.79**
12 mm dia; ref M12 25L		0.10	0.79	1.11	nr	**1.90**
12 mm dia; ref M12 40L		0.10	0.79	1.16	nr	**1.95**
12 mm dia; ref M12 60L		0.10	0.79	1.22	nr	**2.01**
16 mm dia; ref M16 30L		0.12	0.95	2.60	nr	**3.55**
16 mm dia; ref M16 60L		0.12	0.95	2.81	nr	**3.76**
Metalwork; mild steel; galvanised						
Ragbolts; mild steel; one nut; one washer						
M10 x 120 mm long		0.10	0.90	0.94	nr	**1.84**
M12 x 160 mm long		0.13	1.16	1.16	nr	**2.32**
M20 x 200 mm long		0.15	1.34	2.54	nr	**3.88**

G STRUCTURAL/CARCASSING METAL/TIMBER Including overheads and profit at 5.00%		Labour hours	Labour £	Material £	Unit	Total rate £
G32 EDGE SUPPORTED/REINFORCED WOODWOOL SLAB DECKING						
Woodwool interlocking reinforced slabs; Torvale 'Woodcelip' or similar; natural finish; fixing to timber or steel with galvanized nails or clips; flat or sloping						
50 mm slabs; type 503; max. span 2100 mm						
1800 - 2100 mm lengths	PC £11.33	0.50	3.95	12.72	m2	**16.67**
2400 mm lengths	PC £11.89	0.50	3.95	13.33	m2	**17.28**
2700 - 3000 mm lengths	PC £12.08	0.50	3.95	13.66	m2	**17.61**
75 mm slabs; type 751; max. span 2100 mm						
1800 - 2400 mm lengths	PC £17.00	0.55	4.35	19.00	m2	**23.35**
2700 - 3000 mm lengths	PC £17.08	0.55	4.35	19.09	m2	**23.44**
75 mm slabs; type 752; max. span 2100 mm						
1800 - 2400 mm lengths	PC £16.94	0.55	4.35	18.94	m2	**23.29**
2700 - 3000 mm lengths	PC £16.99	0.55	4.35	19.14	m2	**23.49**
75 mm slabs; type 753; max. span 3600 mm						
2400 mm lengths	PC £16.44	0.55	4.35	18.38	m2	**22.73**
2700 - 3000 mm lengths	PC £17.15	0.55	4.35	19.17	m2	**23.52**
3300 - 3900 mm lengths		0.55	4.35	22.62	m2	**26.97**
Raking cutting; including additional trim		0.22	1.74	4.05	m	**5.79**
Holes for pipes and the like		0.12	0.95	-	nr	**0.95**
100 mm slabs; type 1001; max. span 3600 mm						
3000 mm lengths	PC £22.21	0.60	4.74	24.79	m2	**29.53**
3300 - 3600 mm lengths	PC £24.26	0.60	4.74	27.05	m2	**31.79**
100 mm slabs; type 1002; max. span 3600 mm						
3000 mm lengths	PC £21.69	0.60	4.74	24.21	m2	**28.95**
3300 - 3600 mm lengths	PC £23.02	0.60	4.74	25.85	m2	**30.59**
100 mm slabs; type 1003; max. span 4000 mm						
3000 - 3600 mm lengths	PC £20.61	0.60	4.74	23.02	m2	**27.76**
3900 - 4000 mm lengths	PC £20.61	0.60	4.74	23.02	m2	**27.76**
125 mm slabs; type 1252; max. span 3000 mm						
2400 - 3000 mm lengths	PC £23.43	0.60	4.74	26.13	m2	**30.87**
Extra over slabs for						
pre-screeded deck		-	-	-	m2	**0.88**
pre-screeded soffit		-	-	-	m2	**2.01**
pre-screeded deck and soffit		-	-	-	m2	**2.55**
pre-screeded and proofed deck		-	-	-	m2	**1.62**
pre-screeded and proofed deck plus pre-screeded soffit		-	-	-	m2	**3.80**
pre-felted deck (glass fibre)		-	-	-	m2	**2.14**
pre-felted deck plus pre-screeded soffit		-	-	-	m2	**4.09**
'Weatherdeck'		-	-	-	m2	**1.38**

H CLADDING/COVERING Including overheads and profit at 5.00%	Labour hours	Labour £	Material £	Unit	Total rate £
H10 PATENT GLAZING					
Patent glazing; aluminium alloy bars 2.44 mm long at 622 mm centres; fixed to supports					
Roof cladding; glazing with 7 mm thick Georgian wired cast glass	-	-	-	m2	**56.46**
Extra for associated code 4 lead flashings					
top flashing; 210 mm girth	-	-	-	m	**17.56**
bottom flashing; 240 mm girth	-	-	-	m	**22.83**
end flashing; 300 mm girth	-	-	-	m	**32.49**
Wall cladding; glazing with 7 mm thick Georgian wired cast glass	-	-	-	m2	**62.18**
Wall cladding; glazing with 6 mm thick plate glass	-	-	-	m2	**87.28**
Extra for aluminium alloy members					
38 x 38 x 3 mm angle jamb	-	-	-	m	**16.31**
extruded cill member	-	-	-	m	**18.42**
extruded channel head and PVC came	-	-	-	m	**11.01**
H14 CONCRETE ROOFLIGHTS/PAVEMENT LIGHTS					
NOTE: The following specialist prices for rooflights/pavement lights assume panels of 50 m2; pavement lights at ground level and rooflights at 2nd floor level; easy access; and all necessary ancillary fixing; strengthening; pointing and expansion materials etc.					
Reinforced concrete rooflights/pavement lights; 'Luxcrete' or similar; with glass lenses; supplied and fixed complete					
Rooflights					
2.5 KN/m2 loading; ref. R254/125	-	-	-	m2	**335.48**
2.5 KN/m2 loading; ref. R200/90 or R254/B191	-	-	-	m2	**352.49**
2.5 KN/m2 loading; home office; double glazed	-	-	-	m2	**435.65**
Pavement lights					
20 KN/m2; pedestrian traffic; ref. P150/100	-	-	-	m2	**292.01**
60 KN; vehicular traffic; ref. P165/165	-	-	-	m2	**299.57**
Brass terrabond	-	-	-	m2	**6.24**
150 x 75 mm identification plates	-	-	-	m2	**14.18**
Escape hatch	-	-	-	m2	**2311.47**
H20 RIGID SHEET CLADDING					
Eternit 2000 'Glasal' sheet; Eternit TAC Ltd; flexible neoprene gasket joints; fixing with stainless steel screws and coloured caps					
7.5 mm thick cladding to walls					
over 300 mm wide	2.10	24.45	35.72	m2	**60.17**
not exceeding 300 mm wide	0.70	8.15	17.23	m	**25.38**
External angle	0.10	1.16	6.33	m	**7.49**
7.5 mm thick cladding to eaves; verges fascias or the like					
100 mm wide	0.50	5.82	9.88	m	**15.70**
150 mm wide	0.55	6.40	11.45	m	**17.85**
200 mm wide	0.60	6.99	13.02	m	**20.01**
250 mm wide	0.65	7.57	14.59	m	**22.16**
300 mm wide	0.70	8.15	16.69	m	**24.84**

H CLADDING/COVERING Including overheads and profit at 5.00%		Labour hours	Labour £	Material £	Unit	Total rate £
H30 FIBRE CEMENT PROFILED SHEET CLADDING/ COVERING/SIDING						
Asbestos-free corrugated sheets; Eternit '2000' or similar						
Roof cladding; sloping not exceeding 50 degrees; fixing to timber purlins with drive screws						
'Profile 3'; natural	PC £5.76	0.20	2.33	9.05	m2	**11.38**
'Profile 3'; coloured	PC £6.34	0.20	2.33	9.66	m2	**11.99**
'Profile 6'; natural	PC £5.90	0.25	2.91	9.00	m2	**11.91**
'Profile 6'; coloured		0.25	2.91	9.59	m2	**12.50**
'Profile 6'; natural; insulated; 60 mm glass fibre infill; lining panel		0.45	5.24	21.15	m2	**26.39**
Roof cladding; sloping not exceeding 50 degrees; fixing to steel purlins with hook bolts						
'Profile 3'; natural		0.25	2.91	9.48	m2	**12.39**
'Profile 3'; coloured		0.25	2.91	10.08	m2	**12.99**
'Profile 6'; natural		0.30	3.49	9.38	m2	**12.87**
'Profile 6'; coloured		0.30	3.49	9.96	m2	**13.45**
'Profile 6'; natural; insulated; 60 mm glass fibre infill; lining panel		0.50	5.82	19.64	m2	**25.46**
Wall cladding; vertical; fixing to steel rails with hook bolts						
'Profile 3'; natural		0.30	3.49	9.48	m2	**12.97**
'Profile 3'; coloured		0.30	3.49	10.25	m2	**13.74**
'Profile 6'; natural		0.35	4.08	9.38	m2	**13.46**
'Profile 6'; coloured		0.35	4.08	9.96	m2	**14.04**
'Profile 6'; natural; insulated; 60 mm glass fibre infill; lining panel		0.55	6.40	19.64	m2	**26.04**
Raking cutting		0.15	1.75	1.39	m	**3.14**
Holes for pipes and the like		0.15	1.75	-	nr	**1.75**
Accessories; to 'Profile 3' cladding; natural						
eaves filler		0.10	1.16	6.19	m	**7.35**
vertical corrugation closure		0.12	1.40	6.19	m	**7.59**
apron flashing		0.12	1.40	6.97	m	**8.37**
underglazing flashing		0.12	1.40	6.69	m	**8.09**
plain wing or close fitting two-piece adjustable capping to ridge		0.17	1.98	15.60	m	**17.58**
ventilating two-piece adjustable capping to ridge		0.17	1.98	15.60	m	**17.58**
Accessories; to 'Profile 3' cladding; coloured						
eaves filler		0.10	1.16	6.69	m	**7.85**
vertical corrugation closure		0.12	1.40	7.44	m	**8.84**
apron flashing		0.12	1.40	9.44	m	**10.84**
underglazing flashing		0.12	1.40	8.04	m	**9.44**
plain wing or close fitting two-piece adjustable capping to ridge		0.17	1.98	16.80	m	**18.78**
ventilating two-piece adjustable capping to ridge		0.17	1.98	18.62	m	**20.60**
Accessories; to 'Profile 6' cladding; natural						
eaves filler		0.10	1.16	4.73	m	**5.89**
vertical corrugation closure		0.12	1.40	5.26	m	**6.66**
apron flashing		0.12	1.40	4.07	m	**5.47**
plain cranked crown to ridge		0.17	1.98	10.48	m	**12.46**
plain wing, close fitting or north light two-piece adjustable capping to ridge		0.17	1.98	9.12	m	**11.10**
ventilating two-piece adjustable capping to ridge		0.17	1.98	13.67	m	**15.65**

H CLADDING/COVERING Including overheads and profit at 5.00%	Labour hours	Labour £	Material £	Unit	Total rate £
H30 FIBRE CEMENT PROFILED SHEET CLADDING/ COVERING/SIDING - cont'd					
Asbestos-free corrugated sheets; Eternit '2000' or similar - cont'd					
Accessories; to 'Profile 6' cladding; coloured					
eaves filler	0.10	1.16	6.31	m	**7.47**
vertical corrugation closure	0.12	1.40	6.31	m	**7.71**
apron flashing	0.12	1.40	4.88	m	**6.28**
plain cranked crown to ridge	0.17	1.98	12.50	m	**14.48**
plain wing, close fitting or north light two-piece adjustable capping to ridge	0.17	1.98	12.00	m	**13.98**
ventilating two-piece adjustable capping to ridge	0.17	1.98	16.32	m	**18.30**
H31 METAL PROFILED/FLAT SHEET CLADDING/ COVERING/SIDING					
Galvanised steel strip troughed sheets; BSC Strip Mill Products; colorcoat 'Plastisol' finish					
Roof cladding; sloping not exceeding 50 degrees; fixing to steel purlins with plastic headed self-tapping screws					
0.7 mm type 12.5/3 in. corrugated PC £8.91/m	0.35	4.08	15.19	m2	**19.27**
0.7 mm Long Rib 1000; 35 mm deep PC £8.91/m	0.40	4.66	15.30	m2	**19.96**
Wall cladding; vertical; fixing to steel rails with plastic headed self-tapping screws					
0.7 mm type 12.5/3 in. corrugated	0.40	4.66	15.19	m2	**19.85**
0.7 mm Scan Rib 1000; 19 mm deep	0.45	5.24	14.00	m2	**19.24**
Raking cutting	0.22	2.56	2.24	m	**4.80**
Holes for pipes and the like	0.40	4.66	-	nr	**4.66**
Accessories; colorcoat silicone polyester finish 0.9 mm standard flashings; bent to profile					
250 mm girth	0.20	2.33	3.85	m	**6.18**
375 mm girth	0.22	2.56	5.18	m	**7.74**
500 mm girth	0.24	2.79	6.45	m	**9.24**
625 mm girth	0.30	3.49	7.78	m	**11.27**
Galvanised steel profile sheet cladding; Briggs Amasco Ltd. 'Bitumetal'					
Roof cladding; sloping not exceeding 50 degrees; fixing to steel purlins with plastic headed self-tapping screws					
0.7 mm; type ST.D60	-	-	-	m2	**10.67**
0.7 mm; type ST.D60; PVF2 coated soffit finish	-	-	-	m2	**15.30**
Galvanised steel profile sheet cladding; Plannja Ltd.; PVF2 coated finish					
Roof cladding; sloping not exceeding 50 degrees; fixing to steel purlins with plastic headed self-tapping screws					
0.72 mm 'profile 20B'	-	-	-	m2	**16.40**
0.72 mm 'profile TOP 40'	-	-	-	m2	**15.77**
0.72 mm 'profile 45'	-	-	-	m2	**18.37**
Extra for 80 mm insulation and 0.4 mm coated inner lining sheet	-	-	-	m2	**10.88**

H CLADDING/COVERING Including overheads and profit at 5.00%		Labour hours	Labour £	Material £	Unit	Total rate £
Wall cladding; vertical; fixing to steel rails with plastic headed self-tapping screws						
0.60 mm 'profile 20B'; corrugations vertical		-	-	-	m2	**17.65**
0.60 mm 'profile 30'; corrugations vertical		-	-	-	m2	**17.65**
0.60 mm 'profile TOP 40'; corrugations vertical		-	-	-	m2	**16.47**
0.6 mm 'profile 60B'; corrugations vertical		-	-	-	m2	**21.29**
0.60 mm 'profile 30'; corrugations horizontal		-	-	-	m	**18.11**
0.60 mm 'profile 60B'; corrugations horizontal		-	-	-	nr	**21.75**
Extra for 80 mm insulation and 0.4 mm coated inner lining sheet		-	-	-	m	**10.93**
Accessories for roof/vertical cladding; PVF2 coated finish; 0.60 mm thick flashings; once bent						
250 mm girth		-	-	-	m	**9.68**
375 mm girth		-	-	-	m	**11.71**
500 mm girth		-	-	-	m	**13.29**
625 mm girth		-	-	-	m	**14.93**
Extra bends - each		-	-	-	m	**0.14**
Profile fillers		-	-	-	m	**0.51**
Aluminium troughed sheets; British Aluminium 'Rigidal' range; pre-painted finish						
Roof cladding; sloping not exceeding 50 degrees; fixing to steel purlins with plastic headed self-tapping screws						
0.7 mm type WA6	PC £8.59/m	0.45	5.24	14.60	m2	**19.84**
0.9 mm type A7	PC £10.86/m	0.45	5.24	18.31	m2	**23.55**
Wall cladding; vertical; fixing to steel rails with plastic headed self-tapping screws						
0.7 mm type WA6	PC £8.59/m	0.50	5.82	14.60	m2	**20.42**
0.9 mm type A7	PC £10.86/m	0.50	5.82	18.31	m2	**24.13**
0.9 mm type MM10	PC £10.86/m	0.50	5.82	16.58	m2	**22.40**
Raking cutting		0.22	2.56	2.74	m	**5.30**
Holes for pipes and the like		0.40	4.66	-	nr	**4.66**
Accessories; 0.9 mm pre-painted standard flashings; bent to profile						
200 mm girth		0.25	2.91	3.16	m	**6.07**
312 mm girth		0.27	3.14	4.24	m	**7.38**
380 mm girth		0.30	3.49	5.18	m	**8.67**
500 mm girth		0.35	4.08	7.15	m	**11.23**
H41 GLASS REINFORCED PLASTICS CLADDING/ FEATURES						
Glass fibre translucent sheeting grade AB class 3						
Roof cladding; sloping not exceeding 50 degrees; fixing to timber purlins with drive screws; to suit						
'Profile 3'	PC £4.99	0.20	2.33	7.79	m2	**10.12**
'Profile 6'	PC £6.94	0.25	2.91	10.38	m2	**13.29**
Roof cladding; sloping not exceeding 50 degrees; fixing to steel purlins with hook bolts; to suit						
'Profile 3'	PC £4.99	0.25	2.91	8.23	m2	**11.14**
'Profile 6'	PC £6.94	0.30	3.49	10.81	m2	**14.30**
'Longrib 1000'	PC £7.22	0.30	3.49	11.17	m2	**14.66**

H CLADDING/COVERING
Including overheads and profit at 5.00%

BASIC NATURAL STONE BLOCK PRICES

	£			£
Block prices (£/m3)				
Ancaster Hardwhite	375.00	Westwood		275.00
Ancaster Weatherbed	375.00	Portland Whitbed	200.00 -	350.00
Doulting	267.00	Westmorland green slate		375.00
Monks Park	197.00			

	Labour hours	Labour £	Material £	Unit	Total rate £
H51 NATURAL STONE SLAB CLADDING FEATURES					
Portland Whitbed limestone bedded and jointed in cement - lime - mortar (1:2:9); slurrying with weak lime and stonedust mortar; flush pointing and cleaning on completion (cramps etc measured separately)					
Facework; one face plain and rubbed; bedded against backing					
50 mm thick stones	-	-	-	m2	**182.00**
63 mm thick stones	-	-	-	m2	**202.00**
75 mm thick stones	-	-	-	m2	**225.00**
100 mm thick stones	-	-	-	m2	**255.00**
Fair returns on facework					
50 mm wide	-	-	-	m	**0.53**
63 mm wide	-	-	-	m	**0.53**
75 mm wide	-	-	-	m	**0.53**
100 mm wide	-	-	-	m	**0.64**
Fair raking cutting on facework					
50 mm thick	-	-	-	m	**5.23**
63 mm thick	-	-	-	m	**5.62**
75 mm thick	-	-	-	m	**6.16**
100 mm thick	-	-	-	m	**6.39**
Copings; once weathered; and throated; rubbed; set horizontal or raking					
250 x 50 mm	-	-	-	m	**50.50**
300 x 50 mm	-	-	-	m	**60.60**
350 x 75 mm	-	-	-	m	**81.28**
400 x 100 mm	-	-	-	m	**107.56**
450 x 100 mm	-	-	-	m	**118.01**
500 x 125 mm	-	-	-	m	**134.10**
Extra for angles on copings					
250 x 50 mm	-	-	-	nr	**84.31**
300 x 50 mm	-	-	-	nr	**90.65**
375 x 75 mm	-	-	-	nr	**98.52**
400 x 100 mm	-	-	-	nr	**102.56**
450 x 100 mm	-	-	-	nr	**106.17**
500 x 125 mm	-	-	-	nr	**111.50**
Band courses; plain; rubbed; horizontal					
225 x 112 mm	-	-	-	m	**80.77**
300 x 112 mm	-	-	-	m	**90.86**
Extra for					
Ends	-	-	-	nr	**22.93**
Angles	-	-	-	nr	**27.61**

H CLADDING/COVERING Including overheads and profit at 5.00%	Labour hours	Labour £	Material £	Unit	Total rate £
Band courses; moulded 100 mm girth on face; rubbed; horizontal					
125 x 75 mm	-	-	-	m	117.27
150 x 75 mm	-	-	-	m	122.18
200 x 100 mm	-	-	-	m	132.46
250 x 150 mm	-	-	-	m	143.28
300 x 250 mm	-	-	-	m	157.06
Extra for ends on band courses					
125 x 75 mm	-	-	-	nr	33.93
150 x 75 mm	-	-	-	nr	45.24
200 x 100 mm	-	-	-	nr	56.55
250 x 150 mm	-	-	-	nr	67.86
300 x 250 mm	-	-	-	nr	90.47
Extra for angles on band courses					
125 x 75 mm	-	-	-	nr	22.61
150 x 75 mm	-	-	-	nr	35.21
200 x 100 mm	-	-	-	nr	45.40
250 x 150 mm	-	-	-	nr	58.11
300 x 250 mm	-	-	-	nr	68.02
Coping apex block; two sunk faces; rubbed					
650 x 450 x 225 mm	-	-	-	nr	167.23
Coping kneeler block; three sunk faces; rubbed					
350 x 350 x 375 mm	-	-	-	nr	121.59
450 x 450 x 375 mm	-	-	-	nr	144.47
Corbel; turned and moulded; rubbed					
225 x 225 x 375 mm	-	-	-	nr	231.44
Slab surrounds to openings; one face splayed; rubbed					
200 x 75 mm	-	-	-	m	69.28
75 x 100 mm	-	-	-	m	63.81
100 x 100 mm	-	-	-	m	69.38
125 x 100 mm	-	-	-	m	76.73
125 x 150 mm	-	-	-	m	84.58
175 x 175 mm	-	-	-	m	91.79
225 x 175 mm	-	-	-	m	101.05
300 x 175 mm	-	-	-	m	108.65
300 x 225 mm	-	-	-	m	118.03
Slab surrounds to openings; one face sunk splayed; rubbed					
200 x 75 mm	-	-	-	m	70.25
75 x 100 mm	-	-	-	m	78.64
100 x 100 mm	-	-	-	m	82.91
125 x 100 mm	-	-	-	m	88.02
125 x 150 mm	-	-	-	m	94.76
175 x 175 mm	-	-	-	m	103.25
225 x 175 mm	-	-	-	m	111.11
300 x 175 mm	-	-	-	m	119.74
300 x 225 mm	-	-	-	m	127.50
Extra for					
Throating	-	-	-	m	6.60
Rebates and grooves	-	-	-	m	7.92
Stooling	-	-	-	nr	37.18
Sundries - stone walling					
Coating backs of stones with brush applied cold bitumen solution; two coats					
limestone facework	0.20	1.22	1.68	m2	2.90
Cutting grooves in limestone masonry for water bars or the like	-	-	-	m	5.32
Mortices in limestone masonry for					
metal dowel	-	-	-	nr	0.39
metal cramp	-	-	-	nr	0.92

H CLADDING/COVERING Including overheads and profit at 5.00%		Labour hours	Labour £	Material £	Unit	Total rate £
H51 NATURAL STONE SLAB CLADDING FEATURES - cont'd						
Cramps and dowels; Harris and Edgar's 'Delta' range or similar; one end built into brickwork or set in slot in concrete; Stainless Steel						
Dowel						
8 mm dia x 75 mm long	PC £13.82/100	0.05	0.51	0.15	nr	**0.66**
10 mm dia x 150 mm long	PC £38.29/100	0.05	0.51	0.40	nr	**0.91**
Pattern 'J' tie						
25 x 3 x 100 mm	PC £28.16/100	0.07	0.71	0.30	nr	**1.01**
Pattern 'S' cramp; with two 20 mm turndowns (190 mm girth)						
25 x 3 x 150 mm	PC £46.30/100	0.07	0.71	0.49	nr	**1.20**
Pattern 'B' anchor; with 8 x 75 mm loose dowel						
25 x 3 x 150 mm	PC £46.30/100	0.10	1.01	0.49	nr	**1.50**
Pattern 'Q' tie						
25 x 3 x 200 mm	PC £47.44/100	0.07	0.71	0.50	nr	**1.21**
38 x 3 x 250 mm (special)	PC £54.45/100	0.07	0.71	0.57	nr	**1.28**
Pattern 'H1' halftwist tie						
25 x 3 x 200 mm	PC £57.56/100	0.07	0.71	0.60	nr	**1.31**
38 x 3 x 250 mm (special)	PC £64.57/100	0.07	0.71	0.68	nr	**1.39**

ALTERNATIVE TILE PRICES (£/1000)

	£		£		£
Clay tiles; plain, interlocking and pantile					
'Langleys' 'Sterreberg' pantiles					
anthracite	1078.00	natural red	762.00	rustic	762.00
deep brown	1203.00				
Sandtoft pantiles					
Bold roll 'Roman'	795.40	'Gaelic'	522.83	'County' i'locking	480.15
William Blyth pantiles					
'Barco' bold roll	416.10	'Celtic' (French)	463.60		
Concrete tiles, plain and interlocking					
Marley roof tiles					
'Anglia'	382.85	plain	215.65	'Roman'	537.70
'Ludlow +'	308.75				
Redland roof tiles					
'49'-granule	299.25	'Grovebury'	592.80	'Stonewold Mk 1	803.70
'50 Roman'	532.95				

Discounts of 2.5 - 15% available depending on quantity/status

NOTE: The following items of tile roofing unless otherwise described, include for conventional fixing assuming 'normal exposure' with appropriate nails and/or rivets or clips to pressure impregnated softwood battens fixed with galvanized nails; Prices also include for all bedding and pointing at verges; beneath ridge tiles, etc.

H CLADDING/COVERING Including overheads and profit at 5.00%	Labour hours	Labour £	Material £	Unit	Total rate £
H60 CLAY/CONCRETE ROOF TILING					
Clay interlocking pantiles; Sandtoft Goxhill 'Tudor' red sand faced; PC £824.50/1000; 470 x 285 mm; to 100 mm lap; on 25 x 38 mm battens and type 1F reinforced underlay					
Roof coverings	0.40	4.66	12.92	m2	**17.58**
Extra over coverings for					
fixing every tile	0.02	0.23	0.28	m2	**0.51**
eaves course with plastic filler	0.30	3.49	4.69	m	**8.18**
verges; extra single undercloak course of plain tiles	0.30	3.49	2.06	m	**5.55**
open valleys; cutting both sides	0.18	2.10	4.67	m	**6.77**
ridge tiles	0.60	6.99	9.49	m	**16.48**
hip tiles; cutting both sides	0.75	8.73	14.16	m	**22.89**
Holes for pipes and the like	0.20	2.33	-	nr	**2.33**
Clay interlocking pantiles; Langley's 'Sterreberg' black glazed; or similar; PC £1203.00/1000; 355 x 240 mm: to 75 mm lap; on 25 x 38 mm battens and type 1F reinforced underlay					
Roof coverings	0.45	5.24	25.69	m2	**30.93**
Extra over coverings for					
double course at eaves	0.33	3.84	4.99	m	**8.83**
verges; extra single undercloak course of plain tiles	0.30	3.49	5.78	m	**9.27**
open valleys; cutting both sides	0.18	2.10	9.60	m	**11.70**
saddleback ridge tiles	0.60	6.99	16.55	m	**23.54**
saddleback hip tiles; cutting both sides	0.75	8.73	26.15	m	**34.88**
Holes for pipes and the like	0.20	2.33	-	nr	**2.33**
Clay pantiles; Sandtoft Goxhill 'Old English' red sand faced; PC £475.30/1000 342 x 241 mm; to 75 mm lap; on 25 x 38 mm battens and type 1F reinforced underlay					
Roof coverings	0.45	5.24	12.87	m2	**18.11**
Extra over coverings for					
fixing every tile	0.03	0.35	0.47	m2	**0.82**
other colours	-	-	-	m2	**0.75**
double course at eaves	0.33	3.84	3.24	m	**7.08**
verges; extra single undercloak course of plain tiles	0.30	3.49	2.06	m	**5.55**
open valleys; cutting both sides	0.18	2.10	3.79	m	**5.89**
ridge tiles; tile slips	0.60	6.99	10.19	m	**17.18**
hip tiles; tile slips; cutting both sides	0.75	8.73	13.98	m	**22.71**
Holes for pipes and the like	0.20	2.33	-	nr	**2.33**
Clay pantiles; William Blyth's 'Lincoln' natural; 343 x 280 mm; to 75 mm lap; PC £566.25/1000; on 19 x 38 mm battens and type 1F reinforced underlay					
Roof coverings	0.45	5.24	14.60	m2	**19.84**
Extra over coverings for					
fixing every tile	0.03	0.35	0.47	m2	**0.82**
other colours	-	-	-	m2	**0.55**
double course at eaves	0.33	3.84	3.05	m	**6.89**
verges; extra single undercloak course of plain tiles	0.30	3.49	5.38	m	**8.87**
open valleys; cutting both sides	0.18	2.10	4.76	m	**6.86**
ridge tiles; tile slips	0.60	6.99	10.29	m	**17.28**
hip tiles; tile slips; cutting both sides	0.75	8.73	15.04	m	**23.77**
Holes for pipes and the like	0.20	2.33	-	nr	**2.33**

H CLADDING/COVERING Including overheads and profit at 5.00%	Labour hours	Labour £	Material £	Unit	Total rate £
H60 CLAY/CONCRETE ROOF TILING - cont'd					
Clay plain tiles; Hinton, Perry and Davenhill 'Dreadnought' smooth red machine-made; PC £262.20/1000; 265 x 165 mm; on 19 x 38 mm battens and type 1F reinforced underlay					
Roof coverings; to 64 mm lap	1.05	12.23	21.18	m2	33.41
Wall coverings; to 38 mm lap	1.25	14.56	18.92	m2	33.48
Extra over coverings for					
25 x 38 mm battens in lieu	-	-	-	m2	0.33
other colours	-	-	-	m2	3.59
ornamental tiles in lieu	-	-	-	m2	7.60
double course at eaves	0.25	2.91	1.86	m	4.77
verges; extra single undercloak course	0.33	3.84	2.69	m	6.53
valley tiles; cutting both sides	0.65	7.57	29.13	m	36.70
bonnet hip tiles; cutting both sides	0.80	9.32	30.17	m	39.49
external vertical angle tiles; supplementary nail fixings	0.40	4.66	21.70	m	26.36
half round ridge tiles	0.50	5.82	10.07	m	15.89
Holes for pipes and the like	0.20	2.33	-	nr	2.33
Clay plain tiles; Keymer best hand-made sand-faced tiles; PC £507.30/1000; 265 x 165 mm; on 19 x 38 mm battens and type 1F reinforced underlay					
Roof coverings; to 64 mm lap	1.05	12.23	37.40	m2	49.63
Wall coverings; to 38 mm lap	1.25	14.56	33.24	m2	47.80
Extra over coverings for					
25 x 38 mm battens in lieu	-	-	-	m2	0.33
ornamental tiles in lieu	-	-	-	m2	3.59
double course at eaves	0.25	2.91	3.40	m	6.31
verges; extra single undercloak course	0.33	3.84	5.01	m	8.85
valley tiles; cutting both sides	0.65	7.57	20.18	m	27.75
bonnet hip tiles; cutting both sides	0.80	9.32	29.74	m	39.06
external vertical angle tiles; supplementary nail fixings	0.40	4.66	17.55	m	22.21
half round ridge tiles	0.50	5.82	8.89	m	14.71
Holes for pipes and the like	0.20	2.33	-	nr	2.33
Concrete interlocking tiles; Marley 'Bold Roll' granule finish tiles or similar; PC £595.65/1000; 419 x 330 mm; to 75 mm lap; on 22 x 38 mm battens and type 1F reinforced underlay					
Roof coverings	0.35	4.08	8.62	m2	12.70
Extra over coverings for					
fixing every tile	0.03	0.35	0.60	m2	0.95
25 x 38 mm battens in lieu	-	-	-	m2	0.03
eaves; eave filler	0.05	0.58	0.73	m	1.31
verges; 150 mm asbestos cement strip undercloak	0.23	2.68	1.59	m	4.27
valley trough tiles; cutting both sides	0.55	6.40	12.59	m	18.99
segmental ridge tiles; tile slips	0.55	6.40	9.62	m	16.02
segmental ridge tiles; tile slips; cutting both sides	0.70	8.15	12.12	m	20.27
dry ridge tiles; segmental including batten sections; unions and filler pieces	0.30	3.49	9.51	m	13.00
segmental monoridge tiles	0.55	6.40	11.06	m	17.46
gas ridge terminal	0.50	5.82	38.85	nr	44.67
Holes for pipes and the like	0.20	2.33	-	nr	2.33

H CLADDING/COVERING Including overheads and profit at 5.00%	Labour hours	Labour £	Material £	Unit	Total rate £
Concrete interlocking tiles; Marley 'Ludlow Major' granule finish tiles or similar; PC £532.00/1000; 413 x 330 mm; to 75 mm lap; on 22 x 38 mm battens					
Roof coverings	0.35	4.08	8.14	m2	**12.22**
Extra over coverings for					
fixing every tile	0.03	0.35	0.36	m2	**0.71**
25 x 38 mm battens in lieu	-	-	-	m2	**0.03**
verges; 150 mm asbestos cement strip undercloak	0.23	2.68	1.59	m	**4.27**
dry verge system; extruded white pvc	0.15	1.75	5.46	m	**7.21**
Segmental ridge cap	0.03	0.35	1.68	nr	**2.03**
valley trough tiles; cutting both sides	0.55	6.40	14.20	m	**20.60**
segmental ridge tiles	0.50	5.82	4.88	m	**10.70**
segmental hip tiles; cutting both sides	0.65	7.57	7.22	m	**14.79**
dry ridge tiles; segmental including batten sections; unions and filler pieces	0.30	3.49	9.51	m	**13.00**
segmental monoridge tiles	0.50	5.82	9.77	m	**15.59**
gas ridge terminal	0.50	5.82	38.85	nr	**44.67**
Holes for pipes and the like	0.20	2.33	-	nr	**2.33**
Concrete interlocking tiles; Marley 'Mendip' granule finish double pantiles or similar; PC £595.65/1000; 413 x 330 mm; to 75 mm lap; on 22 x 38 mm battens and type 1F reinforced underlay					
Roof coverings	0.35	4.08	8.85	m2	**12.93**
Extra over coverings for					
fixing every tile	0.03	0.35	0.36	m2	**0.71**
25 x 38 mm battens in lieu	-	-	-	m2	**0.03**
eaves; eave filler	0.03	0.35	0.06	m	**0.41**
verges; 150 mm asbestos cement strip undercloak	0.23	2.68	1.59	m	**4.27**
dry verge system; extruded white pvc	0.15	1.75	5.46	m	**7.21**
valley trough tiles; cutting both sides	0.55	6.40	14.48	m	**20.88**
segmental ridge tiles	0.55	6.40	7.81	m	**14.21**
segmental hip tiles; cutting both sides	0.70	8.15	10.43	m	**18.58**
dry ridge tiles; segmental including batten sections; unions and filler pieces	0.30	3.49	9.51	m	**13.00**
segmental monoridge tiles	0.50	5.82	9.77	m	**15.59**
gas ridge terminal	0.50	5.82	38.85	nr	**44.67**
Holes for pipes and the like	0.20	2.33	-	nr	**2.33**
Concrete interlocking tiles; Marley 'Modern' smooth finish tiles or similar; PC £610.85/1000; 413 x 330 mm; to 75 mm lap; on 22 x 38 mm battens and type 1F reinforced underlay					
Roof coverings	0.35	4.08	9.11	m2	**13.19**
Extra over coverings for					
fixing every tile	0.04	0.47	0.31	m2	**0.78**
25 x 38 mm battens in lieu	-	-	-	m2	**0.03**
verges; 150 mm asbestos cement strip undercloak	0.28	3.26	2.55	m	**5.81**
dry verge system, extruded white pvc	0.20	2.33	6.59	m	**8.92**
'Modern' ridge cap	0.03	0.35	1.68	nr	**2.03**
valley trough tiles; cutting both sides	0.55	6.40	14.55	m	**20.95**
'Modern' ridge tiles	0.50	5.82	5.12	m	**10.94**
'Modern' hip tiles; cutting both sides	0.65	7.57	7.81	m	**15.38**
dry ridge tiles; 'Modern'; including batten sections, unions and filler pieces	0.30	3.49	8.89	m	**12.38**
monoridge tiles	0.50	5.82	9.77	m	**15.59**
gas ridge terminal	0.50	5.82	38.85	nr	**44.67**
Holes for pipes and the like	0.20	2.33	-	nr	**2.33**

H CLADDING/COVERING Including overheads and profit at 5.00%	Labour hours	Labour £	Material £	Unit	Total rate £
H60 CLAY/CONCRETE ROOF TILING - cont'd					
Concrete interlocking tiles; Marley 'Wessex' smooth finish tiles or similar; PC £688.75/1000; 413 x 330 mm; to 75 mm lap; on 22 x 38 mm battens and type 1F reinforced underlay					
Roof coverings	0.35	4.08	9.99	m2	14.07
Extra over coverings for					
fixing every tile	0.04	0.47	0.31	m2	0.78
25 x 38 mm battens in lieu	-	-	-	m2	0.03
verges; 150 mm asbestos cement strip undercloak	0.23	2.68	1.59	m	4.27
dry verge system, extruded white pvc	0.15	1.75	5.46	m	7.21
'Modern' ridge cap	0.03	0.35	1.68	nr	2.03
valley trough tiles; cutting both sides	0.55	6.40	14.89	m	21.29
'Modern' ridge tiles	0.55	6.40	6.56	m	12.96
'Modern' hip tiles; cutting both sides	0.70	8.15	9.59	m	17.74
dry ridge tiles; 'Modern'; including batten sections, unions and filler pieces	0.30	3.49	9.75	m	13.24
monoridge tiles	0.50	5.82	9.77	m	15.59
gas ridge terminal	0.50	5.82	38.85	nr	44.67
Holes for pipes and the like	0.20	2.33	-	nr	2.33
Concrete interlocking tiles; Redland 'Delta' smooth finish tiles or similar; PC £884.45/1000; 430 x 380 mm; to 75 mm lap; on 22 x 38 mm battens and type 1F reinforced underlay					
Roof coverings	0.35	4.08	10.10	m2	14.18
Extra over coverings for					
fixing every tile	0.03	0.35	0.33	m2	0.68
25 x 38 mm battens in lieu	-	-	-	m2	0.03
eaves; eave filler	0.03	0.35	0.10	m	0.45
verges; extra single undercloak course of plain tiles	0.25	2.91	3.41	m	6.32
dry verge system; extruded white pvc	0.20	2.33	8.41	m	10.74
Ridge end unit	0.03	0.35	2.00	nr	2.35
valley trough tiles; cutting both sides	0.55	6.40	15.63	m	22.03
universal 'Delta' ridge tiles	0.50	5.82	4.97	m	10.79
universal 'Delta' hip tiles; cutting both sides	0.65	7.57	8.31	m	15.88
universal 'Delta' mono-pitch ridge tiles	0.50	5.82	8.85	m	14.67
gas flue terminal; 'Delta' type	0.50	5.82	38.85	nr	44.67
Holes for pipes and the like	0.20	2.33	-	nr	2.33
Concrete interlocking tiles; Redland 'Norfolk' smooth finish pantiles or similar; PC £387.60/1000; 381 x 229 mm; to 75 mm lap; on 22 x 38 mm battens and type 1F reinforced underlay					
Roof coverings	0.45	5.24	9.15	m2	14.39
Extra over coverings for					
fixing every tile	0.05	0.58	0.30	m2	0.88
25 x 38 mm battens in lieu	-	-	-	m2	0.04
eaves; eave filler	0.05	0.58	0.58	m	1.16
verges; extra single undercloak course of plain tiles	0.30	3.49	1.05	m	4.54
valley trough tiles; cutting both sides	0.60	6.99	15.38	m	22.37
segmental ridge tiles	0.60	6.99	7.41	m	14.40
segmental hip tiles; cutting both sides	0.75	8.73	10.50	m	19.23
Holes for pipes and the like	0.20	2.33	-	nr	2.33

H CLADDING/COVERING Including overheads and profit at 5.00%	Labour hours	Labour £	Material £	Unit	Total rate £
Concrete interlocking tiles; Redland 'Regent' granule finish bold roll tiles or similar; PC £592.80/1000; 418 x 332 mm; to 75 mm lap; on 22 x 38 mm battens and type 1F reinforced underlay					
Roof coverings	0.35	4.08	8.63	m2	**12.71**
Extra over coverings for					
fixing every tile	0.04	0.47	0.50	m2	**0.97**
25 x 38 mm battens in lieu	-	-	-	m2	**0.03**
eaves; eave filler	0.05	0.58	0.72	m	**1.30**
verges; extra single undercloak course of plain tiles	0.25	2.91	2.07	m	**4.98**
dry verge system; extruded white pvc	0.15	1.75	6.76	m	**8.51**
Ridge end unit	0.03	0.35	2.00	nr	**2.35**
cloaked verge system	0.15	1.75	3.39	m	**5.14**
Blocked end ridge unit	0.03	0.35	3.59	nr	**3.94**
valley trough tiles; cutting both sides	0.55	6.40	14.78	m	**21.18**
segmental ridge tiles; tile slips	0.55	6.40	7.41	m	**13.81**
segmental hip tiles; tile slips; cutting both sides	0.70	8.15	9.90	m	**18.05**
dry ridge system; segmental ridge tiles; including fixing straps; 'Nuralite' fillets and seals	0.25	2.91	12.40	m	**15.31**
half round mono-pitch ridge tiles	0.55	6.40	12.89	m	**19.29**
gas flue terminal; half round type	0.50	5.82	36.81	nr	**42.63**
Holes for pipes and the like	0.20	2.33	-	nr	**2.33**
Concrete interlocking tiles; Redland 'Renown' granule finish tiles or similar; PC £532.95/1000; 418 x 330 mm; to 75 mm lap; on 22 x 38 mm battens and type 1F reinforced underlay					
Roof coverings	0.35	4.08	7.83	m2	**11.91**
Extra over coverings for					
fixing every tile	0.03	0.35	0.42	m2	**0.77**
25 x 38 mm battens in lieu	-	-	-	m2	**0.03**
verges; extra single undercloak course of plain tiles	0.25	2.91	1.05	m	**3.96**
dry verge system; extruded white pvc	0.15	1.75	6.76	m	**8.51**
Ridge end unit	0.03	0.35	2.00	nr	**2.35**
cloaked verge system	0.15	1.75	3.59	m	**5.34**
Blocked end ridge unit	0.03	0.35	3.59	nr	**3.94**
valley trough tiles; cutting both sides	0.55	6.40	14.53	m	**20.93**
segmental ridge tiles	0.50	5.82	4.68	m	**10.50**
segmental hip tiles; cutting both sides	0.65	7.57	6.92	m	**14.49**
dry ridge system; segmental ridge tiles; including fixing straps; 'Nuralite' fillets and seals	0.25	2.91	12.40	m	**15.31**
half round mono-pitch ridge tiles	0.50	5.82	8.49	m	**14.31**
gas flue terminal; half round type	0.50	5.82	36.81	nr	**42.63**
Holes for pipes and the like	0.20	2.33	-	nr	**2.33**
Concrete interlocking tiles; Redland 'Stonewold' smooth finish tiles or similar; PC £737.20/1000; 430 x 380 mm; to 75 mm lap; on 22 x 38 mm battens and type 1F reinforced underlay					
Roof coverings	0.35	4.08	8.71	m2	**12.79**
Extra over coverings for					
fixing every tile	0.03	0.35	0.33	m2	**0.68**
25 x 38 mm battens in lieu	-	-	-	m2	**0.03**
verges; extra single undercloak course of plain tiles	0.30	3.49	3.18	m	**6.67**

H CLADDING/COVERING Including overheads and profit at 5.00%	Labour hours	Labour £	Material £	Unit	Total rate £
H60 CLAY/CONCRETE ROOF TILING - cont'd					
Concrete interlocking tiles; Redland 'Stonewold' smooth finish tiles or similar; PC £737.20/1000; 430 x 380 mm; to 75 mm lap; on 22 x 38 mm battens and type 1F reinforced underlay - cont'd					
Extra over coverings for					
dry verge system; extruded white pvc	0.20	2.33	8.16	m	10.49
Ridge end unit	0.03	0.35	2.00	nr	2.35
valley trough tiles; cutting both sides	0.55	6.40	15.08	m	21.48
universal 'Stonewold' ridge tiles	0.50	5.82	4.97	m	10.79
universal 'Stonewold' hip tiles; cutting both sides	0.65	7.57	7.75	m	15.32
dry ridge system; universal 'Stonewold' ridge tiles; including fixing straps; 'Nuralite' fillets and seals	0.25	2.91	12.89	m	15.80
universal 'Stonewold' mono-pitch ridge tiles	0.50	5.82	8.85	m	14.67
gas flue terminal; 'Stonewold' type	0.50	5.82	38.85	nr	44.67
Holes for pipes and the like	0.20	2.33	-	nr	2.33
Concrete plain tiles; BS 473 and 550 group A; PC £209.95/1000; 267 x 165 mm; on 19 x 38 mm battens and type 1F reinforced underlay					
Roof coverings; to 64 mm lap	1.05	12.23	17.72	m2	29.95
Wall coverings; to 38 mm lap	1.25	14.56	15.87	m2	30.43
Extra over coverings for					
25 x 38 mm battens in lieu	-	-	-	m2	0.33
ornamental tiles in lieu	-	-	-	m2	6.94
double course at eaves	0.25	2.91	1.53	m	4.44
verges; extra single undercloak course	0.33	3.84	2.20	m	6.04
valley tiles; cutting both sides	0.65	7.57	16.23	m	23.80
bonnet hip tiles; cutting both sides	0.80	9.32	17.26	m	26.58
external vertical angle tiles; supplementary nail fixings	0.40	4.66	10.33	m	14.99
segmental ridge tiles	0.50	5.82	6.82	m	12.64
segmental hip tiles; cutting both sides	0.75	8.73	7.43	m	16.16
Holes for pipes and the like	0.20	2.33	-	nr	2.33
Sundries					
Hip irons					
galvanized mild steel; fixing with screws	0.10	1.16	1.75	nr	2.91
Fixing					
lead soakers (supply included elsewhere)	0.08	0.93	-	nr	0.93
Pressure impregnated softwood counter battens; 25 x 50 mm					
450 mm centres	0.07	0.82	0.74	m2	1.56
600 mm centres	0.05	0.58	0.56	m2	1.14
Underlay; BS 747 type 1B; bitumen felt weighing 14 kg/10 m2; PC £14.67/20m2; 75 mm laps					
To sloping or vertical surfaces	0.03	0.35	0.83	m2	1.18
Underlay; BS 747 type 1F; reinforced bitumen felt; weighing 22.5 kg/15 m2; PC £16.15/15m2; 75 mm laps					
To sloping or vertical surfaces	0.03	0.35	1.21	m2	1.56

H CLADDING/COVERING Including overheads and profit at 5.00%	Labour hours	Labour £	Material £	Unit	Total rate £
H61 FIBRE CEMENT SLATING					
Asbestos-cement slates; Eternit or similar; to 75 mm lap; on 19 x 50 mm battens and type 1F reinforced underlay					
Coverings; 600 x 300 mm 'blue/black' slates					
roof coverings	0.50	5.82	13.49	m2	**19.31**
wall coverings	0.65	7.57	13.49	m2	**21.06**
Extra over slate coverings for					
double course at eaves	0.25	2.91	2.22	m	**5.13**
verges; extra single undercloak course	0.33	3.84	1.75	m	**5.59**
open valleys; cutting both sides	0.20	2.33	4.57	m	**6.90**
valley gutters; cutting both sides	0.55	6.40	16.25	m	**22.65**
half round ridge tiles	0.50	5.82	11.95	m	**17.77**
Stop end	0.10	1.16	4.13	nr	**5.29**
roll top ridge tiles	0.50	5.82	14.94	m	**20.76**
Stop end	0.10	1.16	6.18	nr	**7.34**
mono-pitch ridges	0.50	5.82	17.22	m	**23.04**
Stop end	0.10	1.16	18.75	nr	**19.91**
duo-pitch ridges	0.50	5.82	14.72	m	**20.54**
Stop end	0.10	1.16	13.76	nr	**14.92**
mitred hips; cutting both sides	0.20	2.33	4.57	m	**6.90**
half round hip tiles; cutting both sides	0.65	7.57	16.52	m	**24.09**
Holes for pipes and the like	0.20	2.33	-	nr	**2.33**
Asbestos-free artificial slates; Eternit '2000' or similar; to 75 mm lap; on 19 x 50 mm battens and type 1F reinforced underlay					
Coverings; 400 x 200 mm 'blue/black' slates					
roof coverings	0.80	9.32	17.14	m2	**26.46**
wall coverings	1.05	12.23	17.14	m2	**29.37**
Coverings; 500 x 250 mm 'blue/black' slates					
roof coverings	0.65	7.57	15.35	m2	**22.92**
wall coverings	0.85	9.90	15.35	m2	**25.25**
Coverings; 600 x 300 mm 'blue/black' slates					
roof coverings	0.50	5.82	14.63	m2	**20.45**
wall coverings	0.65	7.57	14.63	m2	**22.20**
Coverings; 600 x 300 mm 'brown' or 'rose nuit' slates					
roof coverings	0.50	5.82	14.63	m2	**20.45**
wall coverings	0.65	7.57	14.63	m2	**22.20**
Extra over slate coverings for					
double course at eaves	0.25	2.91	2.43	m	**5.34**
verges; extra single undercloak course	0.33	3.84	1.92	m	**5.76**
open valleys; cutting both sides	0.20	2.33	5.07	m	**7.40**
valley gutters; cutting both sides	0.55	6.40	16.75	m	**23.15**
half round ridge tiles	0.50	5.82	11.95	m	**17.77**
Stop end	0.10	1.16	4.13	nr	**5.29**
roll top ridge tiles	0.60	6.99	14.94	m	**21.93**
Stop end	0.10	1.16	6.18	nr	**7.34**
mono-pitch ridges	0.50	5.82	14.08	m	**19.90**
Stop end	0.10	1.16	18.75	nr	**19.91**
duo-pitch ridges	0.50	5.82	14.08	m	**19.90**
Stop end	0.10	1.16	13.76	nr	**14.92**
mitred hips; cutting both sides	0.20	2.33	5.07	m	**7.40**
half round hip tiles; cutting both sides	0.65	7.57	17.03	m	**24.60**
Holes for pipes and the like	0.20	2.33	-	nr	**2.33**

H CLADDING/COVERING
Including overheads and profit at 5.00%

ALTERNATIVE SLATE PRICES (£/1000)

	£		£		£		£
Natural slates							
Greaves Portmadoc Welsh blue-grey							
Mediums (Class 1)							
305x255 mm	503.50	405x255 mm	802.75	510x255 mm	1292.00	610x305 mm	2204.00
355x255 mm	636.50	460x255 mm	1026.00	560x305 mm	1700.50	610x355 mm	2351.25
Strongs (Class 2)							
305x255 mm	489.25	405x255 mm	750.50	510x255 mm	1235.00	610x305 mm	2061.50
355x255 mm	612.75	460x255 mm	964.25	560x305 mm	1648.25	610x355 mm	2242.00

Discounts of 2.5 - 15% available depending on quantity/status

NOTE: The following items of slate roofing unless otherwise described, include for conventional fixing assuming 'normal exposure' with appropriate nails and/or rivets or clips to pressure impregnated softwood battens fixed with galvanized nails; Prices also include for all bedding and pointing at verges; beneath ridge tiles, etc.	Labour hours	Labour £	Material £	Unit	Total rate £
H62 NATURAL SLATING					
Natural slates; BS 680 Part 2; Welsh blue; uniform size; to 75 mm lap; on 25 x 50 mm battens and type 1F reinforced underlay					
Coverings; 405 x 255 mm slates					
roof coverings	0.90	10.48	25.01	m2	**35.49**
wall coverings	1.15	13.39	25.01	m2	**38.40**
Coverings; 510 x 255 mm slates					
roof coverings	0.75	8.73	29.24	m2	**37.97**
wall coverings	0.90	10.48	29.24	m2	**39.72**
Coverings; 610 x 305 mm slates					
roof coverings	0.60	6.99	32.84	m2	**39.83**
wall coverings	0.75	8.73	32.84	m2	**41.57**
Extra over coverings for					
double course at eaves	0.30	3.49	5.92	m	**9.41**
verges; extra single undercloak course	0.42	4.89	4.85	m	**9.74**
open valleys; cutting both sides	0.22	2.56	12.96	m	**15.52**
blue/black glazed ware 152 mm half round ridge tiles	0.50	5.82	9.23	m	**15.05**
blue/black glazed ware 125 x 125 mm plain angle ridge tiles	0.50	5.82	12.33	m	**18.15**
mitred hips; cutting both sides	0.22	2.56	12.96	m	**15.52**
blue/black glazed ware 152 mm half round hip tiles; cutting both sides	0.70	8.15	22.19	m	**30.34**
blue/black glazed ware 125 x 125 mm plain angle hip tiles; cutting both sides	0.70	8.15	25.29	m	**33.44**
Holes for pipes and the like	0.20	2.33	-	nr	**2.33**
Natural slates; Westmorland green; PC £1235.00/t; random lengths; 457 - 229 mm proportionate widths to 75 mm lap; in diminishing courses; on 25 x 50 mm battens and type 1F underlay					
Roof coverings	1.15	13.39	86.31	m2	**99.70**
Wall coverings	1.45	16.88	86.31	m2	**103.19**

H CLADDING/COVERING Including overheads and profit at 5.00%	Labour hours	Labour £	Material £	Unit	Total rate £
Extra over coverings for					
double course at eaves	0.66	7.69	14.81	m	22.50
verges; extra single undercloak course					
slates 152 mm wide	0.75	8.73	11.32	m	20.05
Holes for pipes and the like	0.30	3.49	-	nr	3.49
H63 RECONSTRUCTED STONE SLATING/TILING					
Reconstructed stone slates; 'Hardrow Slates' standard colours; or similar; PC £553.35/1000 75 mm lap; on 25 x 50 mm battens and type 1F reinforced underlay					
Coverings; 457 x 305 mm slates					
roof coverings	0.80	9.32	13.30	m2	22.62
wall coverings	1.00	11.64	13.30	m2	24.94
Coverings; 457 x 457 mm slates					
roof coverings	0.65	7.57	13.31	m2	20.88
wall coverings	0.85	9.90	13.31	m2	23.21
Extra over coverings for					
double course at eaves	0.30	3.49	2.44	m	5.93
verges; extra single undercloak course	0.42	4.89	2.49	m	7.38
open valleys; cutting both sides	0.22	2.56	4.89	m	7.45
ridge tiles	0.50	5.82	9.41	m	15.23
hip tiles	0.70	8.15	12.82	m	20.97
Holes for pipes and the like	0.20	2.33	-	nr	2.33
Reconstructed stone slates; 'Hardrow Slates' green/'oldstone' colours or similar; PC £598.37/1000 75 mm lap; on 25 x 50 mm battens and type 1F reinforced underlay					
Coverings; 457 x 305 mm slates					
roof coverings	0.80	9.32	14.09	m2	23.41
wall coverings	1.00	11.64	14.09	m2	25.73
Coverings; 457 x 457 mm slates					
roof coverings	0.65	7.57	14.12	m2	21.69
wall coverings	0.85	9.90	14.12	m2	24.02
Extra over coverings for					
double course at eaves	0.30	3.49	2.62	m	6.11
verges; extra single undercloak course	0.42	4.89	2.68	m	7.57
open valleys; cutting both sides	0.22	2.56	5.28	m	7.84
ridge tiles	0.50	5.82	10.13	m	15.95
hip tiles	0.70	8.15	13.41	m	21.56
Holes for pipes and the like	0.20	2.33	-	nr	2.33
Reconstructed stone slates; Bradstone 'Cotswold' style or similar; PC £17.60/m2; random lengths 550 - 300 mm; proportional widths; to 80 mm lap; in diminishing courses; on 25 x 50 mm battens and type 1F reinforced underlay					
Roof coverings	1.05	12.23	24.30	m2	36.53
Wall coverings	1.35	15.72	25.68	m2	41.40
Extra over coverings for					
double course at eaves	0.50	5.82	3.98	m	9.80
verges; extra single undercloak course	0.66	7.69	3.22	m	10.91
open valleys; cutting both sides	0.45	5.24	8.32	m	13.56
ridge tile	0.66	7.69	9.71	m	17.40
mitred hips; cutting both sides	0.45	5.24	8.32	m	13.56
hip tile; cutting both sides	1.05	12.23	17.81	m	30.04
Holes for pipes and the like	0.30	3.49	-	nr	3.49

H CLADDING/COVERING Including overheads and profit at 5.00%	Labour hours	Labour £	Material £	Unit	Total rate £
H63 RECONSTRUCTED STONE SLATING/TILING - cont'd					
Reconstructed stone slates; Bradstone 'Moordale' style or similar; PC £17.65/m2; random lengths 550 - 450 mm; proportional widths; to 80 mm lap; in diminishing courses; on 25 x 50 mm battens and type 1F reinforced underlay					
Roof coverings	0.95	11.06	23.93	m2	**34.99**
Wall coverings	1.25	14.56	25.53	m2	**40.09**
Extra over coverings for					
double course at eaves	0.50	5.82	3.99	m	**9.81**
verges; extra single undercloak course	0.66	7.69	3.23	m	**10.92**
ridge tile	0.66	7.69	9.71	m	**17.40**
mitred hips; cutting both sides	0.45	5.24	8.34	m	**13.58**
Holes for pipes and the like	0.30	3.49	-	nr	**3.49**
H64 TIMBER SHINGLING					
Red cedar sawn shingles preservative treated; PC £28.73 per bundle (2.11 m2 cover); uniform length 450 mm; varying widths; to 125 mm lap; on 25 x 100 mm battens and type 1F reinforced underlay					
Roof coverings	1.05	12.23	21.77	m2	**34.00**
Wall coverings	1.35	15.72	21.77	m2	**37.49**
Extra over coverings for					
double course at eaves; three rows of battens	0.30	3.49	4.28	m	**7.77**
verges; extra single undercloak course	0.50	5.82	4.47	m	**10.29**
open valleys; cutting both sides	0.20	2.33	3.32	m	**5.65**
selected shingles to form hip capping	1.20	13.97	8.39	m	**22.36**
Double starter course on last	0.20	2.33	1.82	nr	**4.15**
Holes for pipes and the like	0.15	1.75	-	nr	**1.75**
H71 LEAD SHEET COVERINGS/FLASHINGS					
Milled lead; BS 1178; PC £1217.43/t					
1.25 mm (code 3) roof coverings					
flat	2.70	23.76	16.89	m2	**40.65**
sloping 10 - 50 degrees	3.00	26.40	16.89	m2	**43.29**
vertical or sloping over 50 degrees	3.30	29.04	16.89	m2	**45.93**
1.80 mm (code 4) roof coverings					
flat	2.90	25.52	24.30	m2	**49.82**
sloping 10 - 50 degrees	3.20	28.16	24.30	m2	**52.46**
vertical or sloping over 50 degrees	3.50	30.80	24.30	m2	**55.10**
1.80 mm (code 4) dormer coverings					
flat	3.40	29.92	24.88	m2	**54.80**
sloping 10 - 50 degrees	3.90	34.32	24.88	m2	**59.20**
vertical or sloping over 50 degrees	4.20	36.96	24.88	m2	**61.84**
2.24 mm (code 5) roof coverings					
flat	3.10	27.28	30.24	m2	**57.52**
sloping 10 - 50 degrees	3.40	29.92	30.24	m2	**60.16**
vertical or sloping over 50 degrees	3.70	32.56	30.24	m2	**62.80**
2.24 mm (code 5) dormer coverings					
flat	3.70	32.56	30.97	m2	**63.53**
sloping 10 - 50 degrees	4.10	36.08	30.97	m2	**67.05**
vertical or sloping over 50 degrees	4.50	39.60	30.97	m2	**70.57**
2.50 mm (code 6) roof coverings					
flat	3.30	29.04	33.77	m2	**62.81**
sloping 10 - 50 degrees	3.60	31.68	33.77	m2	**65.45**
vertical or sloping over 50 degrees	3.90	34.32	33.77	m2	**68.09**

H CLADDING/COVERING Including overheads and profit at 5.00%	Labour hours	Labour £	Material £	Unit	Total rate £
2.50 mm (code 6) dormer coverings					
flat	4.00	35.20	34.58	m2	**69.78**
sloping 10 - 50 degrees	4.30	37.84	34.58	m2	**72.42**
vertical or sloping over 50 degrees	4.70	41.36	34.58	m2	**75.94**
Dressing over glazing bars and glass	0.33	2.90	-	m	**2.90**
Soldered dot	1.25	11.00	2.14	nr	**13.14**
Copper nailing 75 mm spacing	0.20	1.76	0.28	m	**2.04**
1.80 mm (code 4) lead flashings, etc.					
Flashings; wedging into grooves					
150 mm girth	0.80	7.04	3.86	m	**10.90**
240 mm girth	0.90	7.92	6.22	m	**14.14**
Stepped flashings; wedging into grooves					
180 mm girth	0.90	7.92	4.64	m	**12.56**
270 mm girth	1.00	8.80	6.95	m	**15.75**
Linings to sloping gutters					
390 mm girth	1.20	10.56	10.08	m	**20.64**
450 mm girth	1.30	11.44	11.58	m	**23.02**
750 mm girth	1.60	14.08	18.32	m	**32.40**
Cappings to hips or ridges					
450 mm girth	1.50	13.20	11.58	m	**24.78**
600 mm girth	1.60	14.08	15.48	m	**29.56**
Soakers					
200 x 200 mm	0.15	1.32	1.11	nr	**2.43**
300 x 300 mm	0.20	1.76	2.48	nr	**4.24**
Saddle flashings; at intersections of hips and ridges; dressing and bossing					
450 x 600 mm	1.80	15.84	7.43	nr	**23.27**
Slates; with 150 mm high collar					
450 x 450 mm; to suit 50 mm pipe	1.70	14.96	10.81	nr	**25.77**
450 x 450 mm; to suit 100 mm pipe	2.00	17.60	15.55	nr	**33.15**
2.24 mm (code 5) lead flashings, etc.					
Flashings; wedging into grooves					
150 mm girth	0.80	7.04	4.82	m	**11.86**
240 mm girth	0.90	7.92	7.67	m	**15.59**
Stepped flashings; wedging into grooves					
180 mm girth	0.90	7.92	5.78	m	**13.70**
270 mm girth	1.00	8.80	8.66	m	**17.46**
Linings to sloping gutters					
390 mm girth	1.20	10.56	12.52	m	**23.08**
450 mm girth	1.30	11.44	14.42	m	**25.86**
750 mm girth	1.60	14.08	22.89	m	**36.97**
Cappings to hips or ridges					
450 mm girth	1.50	13.20	14.42	m	**27.62**
600 mm girth	1.60	14.08	19.27	m	**33.35**
Soakers					
200 x 200 mm	0.15	1.32	1.38	nr	**2.70**
300 x 300 mm	0.20	1.76	3.03	nr	**4.79**
Saddle flashings; at intersections of hips and ridges; dressing and bossing					
450 x 600 mm	1.80	15.84	10.32	nr	**26.16**
Slates; with 150 mm high collar					
450 x 450 mm; to suit 50 mm pipe	1.70	14.96	12.18	nr	**27.14**
450 x 450 mm; to suit 100 mm pipe	2.00	17.60	16.91	nr	**34.51**
H72 ALUMINIUM SHEET COVERINGS/FLASHINGS					
Aluminium roofing; commercial grade; PC £1850.00/t					
0.90 mm roof coverings					
flat	3.00	26.40	4.79	m2	**31.19**
sloping 10 - 50 degrees	3.30	29.04	4.79	m2	**33.83**
vertical or sloping over 50 degrees	3.60	31.68	4.79	m2	**36.47**

H CLADDING/COVERING Including overheads and profit at 5.00%	Labour hours	Labour £	Material £	Unit	Total rate £
H72 ALUMINIUM SHEET COVERINGS/FLASHINGS - cont'd					
Aluminium roofing; commercial grade; PC £1850.00/t - cont'd					
0.90 mm dormer coverings					
flat	3.60	31.68	4.90	m2	**36.58**
sloping 10 - 50 degrees	4.00	35.20	4.90	m2	**40.10**
vertical or sloping over 50 degrees	4.40	38.72	4.90	m2	**43.62**
Aluminium nailing; 75 mm spacing	0.20	1.76	0.21	m	**1.97**
0.90 mm commercial grade aluminium flashings, etc.					
Flashings; wedging into grooves					
150 mm girth	0.80	7.04	0.74	m	**7.78**
240 mm girth	0.90	7.92	1.20	m	**9.12**
300 mm girth	1.05	9.24	1.50	m	**10.74**
Stepped flashings; wedging into grooves					
180 mm girth	0.90	7.92	0.90	m	**8.82**
270 mm girth	1.00	8.80	1.37	m	**10.17**
H73 COPPER SHEET COVERINGS/FLASHINGS					
Copper roofing; BS 2870					
0.56 mm (24 swg) roof coverings					
flat PC £3200.00/t	3.20	28.16	17.18	m2	**45.34**
sloping 10 - 50 degrees	3.50	30.80	17.18	m2	**47.98**
vertical or sloping over 50 degrees	3.80	33.44	17.18	m2	**50.62**
0.56 mm (24 swg) dormer coverings					
flat	3.80	33.44	17.59	m2	**51.03**
sloping 10 - 50 degrees	4.20	36.96	17.59	m2	**54.55**
vertical or sloping over 50 degrees	4.60	40.48	17.59	m2	**58.07**
0.61 mm (23 swg) roof coverings					
flat PC £3100.00/t	3.20	28.16	19.82	m2	**47.98**
sloping 10 - 50 degrees	3.50	30.80	19.82	m2	**50.62**
vertical or sloping over 50 degrees	3.80	33.44	19.82	m2	**53.26**
0.61 mm (23 swg) dormer coverings					
flat	3.80	33.44	20.30	m2	**53.74**
sloping 10 - 50 degrees	4.20	36.96	20.30	m2	**57.26**
vertical or sloping over 50 degrees	4.60	40.48	20.30	m2	**60.78**
Copper nailing; 75 mm spacing	0.20	1.76	0.28	m	**2.04**
0.56 mm copper flashings, etc.					
Flashings; wedging into grooves					
150 mm girth	0.80	7.04	2.69	m	**9.73**
240 mm girth	0.90	7.92	4.32	m	**12.24**
300 mm girth	1.05	9.24	5.40	m	**14.64**
Stepped flashings; wedging into grooves					
180 mm girth	0.90	7.92	3.25	m	**11.17**
270 mm girth	1.00	8.80	4.88	m	**13.68**
0.61 mm copper flashings, etc.					
Flashings; wedging into grooves					
150 mm girth	0.80	7.04	3.12	m	**10.16**
240 mm girth	0.90	7.92	5.01	m	**12.93**
300 mm girth	1.05	9.24	6.23	m	**15.47**
Stepped flashings; wedging into grooves					
180 mm girth	0.90	7.92	3.76	m	**11.68**
270 mm girth	1.00	8.80	5.62	m	**14.42**

H CLADDING/COVERING Including overheads and profit at 5.00%	Labour hours	Labour £	Material £	Unit	Total rate £
H74 ZINC SHEET COVERINGS/FLASHINGS					
Zinc BS 849; PC £2000.00/t					
0.81 mm roof coverings					
flat	3.20	28.16	12.55	m2	**40.71**
sloping 10 - 50 degrees	3.50	30.80	12.55	m2	**43.35**
vertical or sloping over 50 degrees	3.80	33.44	12.55	m2	**45.99**
0.81 mm dormer coverings					
flat	3.80	33.44	12.84	m2	**46.28**
sloping 10 - 50 degrees	4.20	36.96	12.84	m2	**49.80**
vertical or sloping over 50 degrees	4.60	40.48	12.84	m2	**53.32**
0.81 mm zinc flashings, etc.					
Flashings; wedging into grooves					
150 mm girth	0.80	7.04	1.96	m	**9.00**
240 mm girth	0.90	7.92	3.16	m	**11.08**
300 mm girth	1.05	9.24	3.95	m	**13.19**
Stepped flashings; wedging into grooves					
180 mm girth	0.90	7.92	2.38	m	**10.30**
270 mm girth	1.00	8.80	3.56	m	**12.36**
H75 STAINLESS STEEL SHEET COVERINGS/ FLASHINGS					
Terne coated stainless steel roofing					
0.38 mm roof coverings					
flat	3.20	28.16	26.46	m2	**54.62**
sloping 10 - 50 degrees	3.50	30.80	26.46	m2	**57.26**
vertical or sloping over 50 degrees	3.80	33.44	26.46	m2	**59.90**
Flashings; wedging into grooves					
150 mm girth	1.00	8.80	5.54	m	**14.34**
240 mm girth	1.15	10.12	8.04	m	**18.16**
300 mm girth	1.30	11.44	9.70	m	**21.14**
Stepped flashings; wedging into grooves					
180 mm girth	1.15	10.12	6.37	m	**16.49**
270 mm girth	1.25	11.00	8.87	m	**19.87**
H76 FIBRE BITUMEN THERMOPLASTIC SHEET COVERINGS/FLASHINGS					
Glass fibre reinforced bitumen strip slates; 'Langhome 1000' or similar; PC £19.11/2 m2; strip pack; 900 x 300 mm mineral finish; fixed to external plywood boarding (measured separately)					
Roof coverings	0.25	2.91	8.81	m2	**11.72**
Wall coverings	0.40	4.66	8.81	m2	**13.47**
Extra over coverings for					
double course at eaves; felt soaker	0.20	2.33	1.71	m	**4.04**
verges; felt soaker	0.25	2.91	1.62	m	**4.53**
valley slate; cut to shape; felt soaker both sides; cutting both sides	0.45	5.24	4.17	m	**9.41**
ridge slate; cut to shape	0.30	3.49	2.15	m	**5.64**
hip slate; cut to shape; cutting both sides	0.45	5.24	4.13	m	**9.37**
Holes for pipes and the like	0.05	0.58	-	nr	**0.58**

H CLADDING/COVERING Including overheads and profit at 5.00%	Labour hours	Labour £	Material £	Unit	Total rate £
H76 FIBRE BITUMEN THERMOPLASTIC SHEET COVERINGS/FLASHINGS - cont'd					
'Evode Flashband' sealing strips and flashings; special grey finish					
Flashings; wedging at top if required; pressure bonded; flashband primer before application; to walls					
100 mm girth	0.25	2.20	0.99	m	3.19
150 mm girth	0.33	2.90	1.48	m	4.38
225 mm girth	0.40	3.52	2.25	m	5.77
300 mm girth	0.45	3.96	2.92	m	6.88
450 mm girth	0.60	5.28	5.67	m	10.95
'Nuralite' semi-rigid bitumen membrane roofing DC12 jointing strip system PC £21.41/3m2					
Roof coverings					
flat	0.50	4.40	10.97	m2	15.37
sloping 10 - 50 degrees	0.55	4.84	10.97	m2	15.81
vertical or sloping over 50 degrees	0.70	6.16	10.97	m2	17.13
'Nuralite' flashings, etc.					
Flashings; wedging into grooves					
150 mm girth	0.50	4.40	1.39	m	5.79
200 mm girth	0.60	5.28	1.93	m	7.21
250 mm girth	0.60	5.28	2.18	m	7.46
Linings to sloping gutters					
450 mm girth	0.85	7.48	4.74	m	12.22
490 mm girth	0.95	8.36	5.25	m	13.61
Cappings to hips or ridges					
450 mm girth	0.90	7.92	7.00	m	14.92
600 mm girth	1.00	8.80	10.51	m	19.31
Undersoakers					
150 mm girth	0.12	1.06	0.54	nr	1.60
250 mm girth	0.18	1.58	0.75	nr	2.33
Oversoakers					
200 mm girth	0.15	1.32	7.21	nr	8.53
Cavity tray without apron					
450 mm long	0.45	3.96	1.35	nr	5.31
250 mm long	0.25	2.20	1.05	nr	3.25
350 mm long	0.35	3.08	1.25	nr	4.33

J WATERPROOFING Including overheads and profit at 5.00%	Labour hours	Labour £	Material £	Unit	Total rate £
J10 SPECIALIST WATERPROOF RENDERING					
'Sika' waterproof rendering; steel trowelled					
20 mm work to walls; three coat; to concrete base					
over 300 mm wide	-	-	-	m2	29.14
not exceeding 300 mm wide	-	-	-	m2	45.65
25 mm work to walls; three coat; to concrete base					
over 300 mm wide	-	-	-	m2	33.02
not exceeding 300 mm wide	-	-	-	m2	52.45
40 mm work to walls; four coat; to concrete b ase					
over 300 mm wide	-	-	-	m2	50.51
not exceeding 300 mm wide	-	-	-	m2	77.70
J20 MASTIC ASPHALT TANKING/DAMP PROOF MEMBRANES					
Mastic asphalt to BS 1097					
13 mm one coat coverings to concrete base; flat; subsequently covered					
over 300 mm wide	-	-	-	m2	6.24
225 - 300 mm wide	-	-	-	m2	11.45
150 - 225 mm wide	-	-	-	m2	13.07
not exceeding 150 mm wide	-	-	-	m2	13.71
20 mm two coat coverings to concrete base; flat; subsequently covered					
over 300 mm wide	-	-	-	m2	8.44
225 - 300 mm wide	-	-	-	m2	17.61
150 - 300 mm wide	-	-	-	m2	20.11
not exceeding 150 mm wide	-	-	-	m2	21.09
30 mm three coat coverings to concrete base; flat; subsequently covered					
over 300 mm wide	-	-	-	m2	12.10
225 - 300 mm wide	-	-	-	m2	26.43
150 - 225 mm wide	-	-	-	m2	30.19
not exceeding 150 mm wide	-	-	-	m2	31.68
13 mm two coat coverings to brickwork base; vertical; subsequently covered					
over 300 mm wide	-	-	-	m2	25.67
225 - 300 mm wide	-	-	-	m2	30.56
150 - 225 mm wide	-	-	-	m2	36.26
not exceeding 150 mm wide	-	-	-	m2	47.67
20 mm three coat coverings to brickwork base; vertical; subsequently covered					
over 300 mm wide	-	-	-	m2	36.63
225 - 300 mm wide	-	-	-	m2	42.77
150 - 225 mm wide	-	-	-	m2	50.31
not exceeding 150 mm wide	-	-	-	m2	65.38
Turning 20 mm into groove	-	-	-	m	0.63
Internal angle fillets; subsequently covered	-	-	-	m	3.23
Mastic asphalt to BS 6577					
13 mm one coat coverings to concrete base; flat; subsequently covered					
over 300 mm wide	-	-	-	m2	8.62
225 - 300 mm wide	-	-	-	m2	13.55
150 - 225 mm wide	-	-	-	m2	15.46
not exceeding 150 mm wide	-	-	-	m2	16.19

J WATERPROOFING Including overheads and profit at 5.00%	Labour hours	Labour £	Material £	Unit	Total rate £
J20 MASTIC ASPHALT TANKING/DAMP PROOF MEMBRANES - cont'd					
Mastic asphalt to BS 6577 - cont'd					
20 mm two coat coverings to concrete base; flat; subsequently covered					
over 300 mm wide	-	-	-	m2	12.50
225 - 300 mm wide	-	-	-	m2	20.84
150 - 225 mm wide	-	-	-	m2	23.78
not exceeding 150 mm wide	-	-	-	m2	24.92
30 mm three coat coverings to concrete base; flat; subsequently covered					
over 300 mm wide	-	-	-	m2	18.19
225 - 300 mm wide	-	-	-	m2	31.28
150 - 225 mm wide	-	-	-	m2	35.70
not exceeding 150 mm wide	-	-	-	m2	37.41
13 mm two coat coverings to brickwork base; vertical; subsequently covered					
over 300 mm wide	-	-	-	m2	28.31
225 - 300 mm wide	-	-	-	m2	34.54
150 - 225 mm wide	-	-	-	m2	40.74
not exceeding 150 mm wide	-	-	-	m2	53.13
20 mm three coat coverings to brickwork base; vertical; subsequently covered					
over 300 mm wide	-	-	-	m2	39.14
225 - 300 mm wide	-	-	-	m2	48.55
150 - 225 mm wide	-	-	-	m2	56.79
not exceeding 150 mm wide	-	-	-	m2	73.22
Turning 20 mm into groove	-	-	-	m	0.63
Internal angle fillets; subsequently covered	-	-	-	m	3.66
J21 MASTIC ASPHALT ROOFING/INSULATION/ FINISHES					
Mastic asphalt to BS 988					
20 mm two coat coverings; felt isolating membrane; to concrete (or timber) base; flat or to falls or slopes not exceeding 10 degrees from horizontal					
over 300 mm wide	-	-	-	m2	8.76
225 - 300 mm wide	-	-	-	m2	16.77
150 - 225 mm wide	-	-	-	m2	17.87
not exceeding 150 mm wide	-	-	-	m2	20.09
Add to the above for covering with:					
10 mm limestone chippings in hot bitumen	-	-	-	m2	1.76
coverings with solar reflective paint	-	-	-	m2	1.58
300 x 300 x 8 mm g.r.p. tiles in hot bitumen	-	-	-	m2	23.63
Cutting to line; jointing to old asphalt	-	-	-	m	3.99
13 mm two coat skirtings to brickwork base					
not exceeding 150 mm girth	-	-	-	m	7.15
150 - 225 mm girth	-	-	-	m	8.16
225 - 300 mm girth	-	-	-	m	9.17
13 mm three coat skirtings; expanded metal lathing reinforcement nailed to timber base					
not exceeding 150 mm girth	-	-	-	m	11.24
150 - 225 mm girth	-	-	-	m	13.35
225 - 300 mm girth	-	-	-	m	15.46
13 mm two coat fascias to concrete base					
not exceeding 150 mm girth	-	-	-	m	7.15
150 - 225 mm girth	-	-	-	m	8.16

J WATERPROOFING Including overheads and profit at 5.00%		Labour hours	Labour £	Material £	Unit	Total rate £
20 mm two coat linings to channels to concrete base						
not exceeding 150 mm girth		-	-	-	m	17.31
150 - 225 mm girth		-	-	-	m	17.31
225 - 300 mm girth		-	-	-	m	17.31
20 mm two coat lining to cesspools						
250 x 150 x 150 mm deep		-	-	-	nr	16.88
Collars around pipes, standards and like members		-	-	-	nr	8.14
Mastic asphalt to BS 6577						
20 mm two coat coverings; felt isolating membrane; to concrete (or timber) base; flat or to falls or slopes not exceeding 10 degrees from horizontal						
over 300 mm wide		-	-	-	m2	12.37
225 - 300 mm wide		-	-	-	m2	19.84
150 - 225 mm wide		-	-	-	m2	21.14
not exceeding 150 mm wide		-	-	-	m2	23.73
Add to the above for covering with:						
10 mm limestone chippings in hot bitumen		-	-	-	m2	1.76
solar reflective paint		-	-	-	m2	1.58
300 x 300 x 8 mm g.r.p. tiles in hot bitumen		-	-	-	m2	23.63
Cutting to line; jointing to old asphalt		-	-	-	m	3.99
13 mm two coat skirtings to brickwork base						
not exceeding 150 mm girth		-	-	-	m	7.97
150 - 225 mm girth		-	-	-	m	9.17
225 - 300 mm girth		-	-	-	m	10.36
13 mm three coat skirtings; expanded metal lathing reinforcement nailed to timber base						
not exceeding 150 mm girth		-	-	-	m	12.24
150 - 225 mm girth		-	-	-	m	14.64
225 - 300 mm girth		-	-	-	m	17.03
13 mm two coat fascias to concrete base						
not exceeding 150 mm girth		-	-	-	m	7.97
150 - 225 mm girth		-	-	-	m	9.17
20 mm two coat linings to channels to concrete base						
not exceeding 150 mm girth		-	-	-	m	19.50
150 - 225 mm girth		-	-	-	m	19.50
225 - 300 mm girth		-	-	-	m	19.50
20 mm two coat lining to cesspools						
250 x 150 x 150 mm deep		-	-	-	nr	17.47
Collars around pipes, standards and like members		-	-	-	nr	8.14
Accessories						
Eaves trim; extruded aluminium alloy; working asphalt into trim						
'Alutrim'; type A roof edging	PC £8.26/2.5m	0.40	4.66	3.71	m	8.37
Angle	PC £2.85	0.12	1.40	3.14	nr	4.54
Roof screed ventilator - aluminium alloy						
Extr-aqua-vent; set on screed over and including dished sinking; working collar around ventilator	PC £3.18	1.10	12.81	3.51	nr	16.32
J30 LIQUID APPLIED TANKING/DAMP PROOF MEMBRANES						
'Synthaprufe'; blinding with sand; horizontal on slabs						
two coats		0.20	1.22	1.82	m2	3.04
three coats		0.28	1.71	2.66	m2	4.37

J WATERPROOFING Including overheads and profit at 5.00%	Labour hours	Labour £	Material £	Unit	Total rate £
J30 LIQUID APPLIED TANKING/DAMP PROOF MEMBRANES - cont'd					
'Tretolastex 202T'; on vertical surfaces of concrete					
two coats	0.20	1.22	1.64	m2	2.86
three coats	0.28	1.71	2.46	m2	4.17
One coat Vandex 'Super' 0.75/m2 slurry; one consolidating coat of Vandex 'Premix' 1kg/m2 slurry; horizontal on beds					
over 225 mm wide	-	-	-	m2	5.17
'Ventrot' hot applied damp proof membrane; one coat; horizontal on slabs					
over 225 mm wide	-	-	-	m2	3.65
J40 FLEXIBLE SHEET TANKING/DAMP PROOF MEMBRANES					
'Bituthene' sheeting; lapped joints; horizontal on slabs					
standard 500 grade	0.10	0.61	4.24	m2	4.85
1000 grade	0.11	0.67	4.72	m2	5.39
1200 grade	0.12	0.73	6.45	m2	7.18
heavy duty grade	0.13	0.79	5.35	m2	6.14
'Bituthene' sheeting; lapped joints; dressed up vertical face of concrete					
1000 grade	0.18	1.10	5.22	m2	6.32
'Servi-pak' protection board; butt jointed; to horizontal surfaces; including 'Bituthene' standard sheeting; fixed with adhesive dabs					
3 mm thick	0.40	2.44	7.35	m2	9.79
6 mm thick	0.45	2.74	8.78	m2	11.52
12 mm thick	0.50	3.05	12.12	m2	15.17
'Servi-pak' protection board; butt jointed; to vertical surfaces; including 'Bituthene' standard sheeting; fixed with adhesive dabs					
3 mm thick	0.50	3.05	7.35	m2	10.40
6 mm thick	0.55	3.35	8.78	m2	12.13
12 mm thick	0.60	3.65	12.12	m2	15.77
'Bituthene' fillet					
40 x 40 mm	0.10	0.61	3.73	m	4.34
'Bituthene' reinforcing strip; 300 mm wide					
1000 grade	0.10	0.61	1.62	m	2.23
Expandite 'Famflex' waterproof tanking; 150 mm laps					
horizontal; over 300 mm wide	0.40	2.44	6.05	m2	8.49
vertical; over 300 mm wide	0.65	3.96	6.05	m2	10.01
J41 BUILT UP FELT ROOF COVERINGS					
NOTE: The following items of felt roofing, unless otherwise described, include for conventional lapping, laying and bonding between layers and to base; and laying flat or to falls to cross-falls or to slopes not exceeding 10 degrees - but exclude any insulation etc. (measured separately)					
Felt roofing; BS 747; suitable for flat roofs					
Three layer coverings type 1B (two 18 kg/10 m2 and one 25 kg/10 m2) bitumen fibre based felts	-	-	-	m2	12.11
Extra over top layer type 1B for mineral surfaced layer type 1E	-	-	-	m2	1.60

J WATERPROOFING Including overheads and profit at 5.00%	Labour hours	Labour £	Material £	Unit	Total rate £
Three layer coverings type 2B bitumen asbestos based felts	-	-	-	m2	**15.97**
Extra over top layer type 2B for mineral surfaced layer type 2E	-	-	-	m2	**1.98**
Three layer coverings first layer type 3G; subsequent layers type 3B bitumen glass fibre based felt	-	-	-	m2	**12.77**
Extra over top layer type 3B for mineral surfaced layer type 3E	-	-	-	m2	**1.80**
Extra over felt for covering with and bedding in hot bitumen					
13 mm granite chippings	-	-	-	m2	**5.04**
300 x 300 x 8 mm asbestos tiles	-	-	-	m2	**23.63**
Working into outlet pipes and the like	-	-	-	nr	**5.04**
Skirtings; three layer; top layer mineral surfaced; dressed over tilting fillet; turned into groove					
not exceeding 200 mm girth	-	-	-	m	**8.01**
200 - 400 mm girth	-	-	-	m	**10.23**
Coverings to kerbs; three layer					
400 - 600 mm girth	-	-	-	m	**12.01**
Linings to gutters; three layer					
400 - 600 mm (average) girth	-	-	-	m	**20.52**
Collars around pipes and the like; three layer mineral surfaced; 150 mm high					
not exceeding 55 mm nominal size	-	-	-	nr	**5.93**
55 - 110 mm nominal size	-	-	-	nr	**7.26**
Felt roofing; BS 747; suitable for pitched timber roofs; sloping not exceeding 50 degrees					
Two layer coverings; first layer type 2B; second layer 2E; mineral surfaced asbestos based felts	-	-	-	m2	**16.94**
Three layer coverings; first two layers type 2B; top layer type 2E; mineral surfaced asbestos based felts	-	-	-	m2	**23.86**
'Andersons' high performance polyester-based roofing system					
Two layer coverings; first layer HT 125 underlay; second layer HT 350; fully bonded to wood; fibre or cork base	0.30	3.49	9.53	m2	**13.02**
Extra over for					
Top layer mineral surfaced	-	-	-	m2	**1.53**
13 mm granite chippings	-	-	-	m2	**5.04**
Third layer of type 3B as underlay for concrete or screeded base	0.15	1.75	1.61	m2	**3.36**
Working into outlet pipes and the like	0.50	5.82	-	nr	**5.82**
Skirtings; two layer; top layer mineral surfaced; dressed over tilting fillet; turned into groove					
not exceeding 200 mm girth	0.15	1.75	2.54	m	**4.29**
200 - 400 mm girth	0.20	2.33	4.81	m	**7.14**
Coverings to kerbs; two layer					
400 - 600 mm girth	0.25	2.91	6.59	m	**9.50**
Linings to gutters; three layer					
400 - 600 mm (average) girth	0.60	6.99	8.19	m	**15.18**
Collars around pipes and the like; two layer; 150 mm high					
not exceeding 55 mm nominal size	0.33	3.84	0.40	nr	**4.24**
55 - 110 mm nominal size	0.40	4.66	0.66	nr	**5.32**

J WATERPROOFING Including overheads and profit at 5.00%	Labour hours	Labour £	Material £	Unit	Total rate £
J41 BUILT UP FELT ROOF COVERINGS - cont'd					
'Ruberglas 120 GP' high performance roofing					
Two layer coverings; first and second layers 'Ruberglas 120 GP'; fully bonded to wood; fibre or cork base	-	-	-	m2	9.30
Extra over for					
Top layer mineral surfaced	-	-	-	m2	1.63
13 mm granite chippings	-	-	-	m2	5.04
Third layer of 'Rubervent 3G' as underlay for concrete or screeded base	-	-	-	m2	3.97
Working into outlet pipes and the like	-	-	-	nr	5.04
Skirtings; two layer; top layer mineral surfaced; dressed over tilting fillet; turned into groove					
not exceeding 200 mm girth	-	-	-	m	5.11
200 - 400 mm girth	-	-	-	m	6.71
Coverings to kerbs; two layer					
400 - 600 mm girth	-	-	-	m	8.42
Linings to gutters; three layer					
400 - 600 mm (average) girth	-	-	-	m	20.65
Collars around pipes and the like; two layer; 150 mm high					
not exceeding 55 mm nominal size	-	-	-	nr	8.67
55 - 110 mm nominal size	-	-	-	nr	10.08
'Ruberfort HP 350' high performance roofing					
Two layer coverings; first layer 'Ruberfort HP 180'; second layer 'Ruberfort HP 350'; fully bonded; to wood; fibre or cork base	-	-	-	m2	14.56
Extra over for					
Top layer mineral surfaced	-	-	-	m2	2.06
13 mm granite chippings	-	-	-	m2	5.04
Third layer of 'Rubervent 3G' as underlay for concrete or screeded base	-	-	-	m2	3.97
Working into outlet pipes and the like	-	-	-	nr	5.25
Skirtings; two layer; top layer mineral surfaced; dressed over tilting fillet; turned into groove					
not exceeding 200 mm girth	-	-	-	m	6.44
200 - 400 mm girth	-	-	-	m	9.12
Coverings to kerbs; two layer					
400 - 600 mm girth	-	-	-	m	12.05
Linings to gutters; three layer					
400 - 600 mm (average) girth	-	-	-	m	25.40
Collars around pipes and the like; two layer; 150 mm high					
not exceeding 55 mm nominal size	-	-	-	nr	8.67
55 - 110 mm nominal size	-	-	-	nr	10.08
'Polybit 350' elastomeric roofing					
Two layer coverings; first layer 'Polybit 180'; second layer 'Polybit 350'; fully bonded to wood; fibre or cork base	-	-	-	m2	18.33
Extra over for					
Top layer mineral surfaced	-	-	-	m2	2.14
13 mm granite chippings	-	-	-	m2	5.04
Third layer of 'Rubervent 3G' as underlay for concrete or screeded base	-	-	-	m2	3.97
Working into outlet pipes and the like	-	-	-	nr	5.73

J WATERPROOFING Including overheads and profit at 5.00%	Labour hours	Labour £	Material £	Unit	Total rate £
Skirtings; two layer; top layer mineral surfaced; dressed over tilting fillet; turned into groove					
not exceeding 200 mm girth	-	-	-	m	6.52
200 - 400 mm girth	-	-	-	m	9.28
Coverings to kerbs; two layer					
400 - 600 mm girth	-	-	-	m	12.82
Linings to gutters; three layer					
400 - 600 mm (average) girth	-	-	-	m	26.73
Collars around pipes and the like; two layer; 150 mm high					
not exceeding 55 mm nominal size	-	-	-	nr	10.08
55 - 110 mm nominal size	-	-	-	nr	11.29
'Hyload 150 E' elastomeric roofing					
Two layer coverings; first layer 'Ruberglas 120 GP'; second layer 'Hyload 150 E' fully bonded to wood; fibre or cork base	-	-	-	m2	14.77
Extra over for					
13 mm granite chippings	-	-	-	m2	5.04
Third layer of 'Rubervent 3G' as underlay for concrete or screeded base	-	-	-	m2	3.97
Three layer coverings; finished with 13 mm granite chippings					
Working into outlet pipes and the like	-	-	-	nr	6.77
Skirtings; two layer; dressed over tilting fillet; turned into groove					
not exceeding 200 mm girth	-	-	-	m	8.38
200 - 400 mm girth	-	-	-	m	12.76
Coverings to kerbs; two layer					
400 - 600 mm girth	-	-	-	m	18.42
Linings to gutters; three layer					
400 - 600 mm (average) girth	-	-	-	m	28.65
Collars around pipes and the like; two layer; 150 mm high					
not exceeding 55 mm nominal size	-	-	-	nr	8.67
55 - 110 mm nominal size	-	-	-	nr	10.08
Felt; 'Paradiene' elastomeric bitumen roofing; perforated crepe paper isolating membrane; first layer 'Paradiene 20'; second layer 'Paradiene 30' pre-finished surface					
Two layer coverings	-	-	-	m2	16.46
Working into outlet pipes and the like	-	-	-	nr	6.48
Skirtings; three layer; dressed over tilting fillet; turned into groove					
not exceeding 200 mm girth	-	-	-	m	5.20
200 - 400 mm girth	-	-	-	m	9.13
Coverings to kerbs; two layer					
400 - 600 mm girth	-	-	-	m	11.13
Linings to gutters; three layer					
400 - 600 mm (average) girth	-	-	-	m	18.17
Collars around pipes and the like; three layer; 150 mm high					
not exceeding 55 mm nominal size	-	-	-	nr	4.73
55 - 110 mm nominal size	-	-	-	nr	5.25

J WATERPROOFING Including overheads and profit at 5.00%		Labour hours	Labour £	Material £	Unit	Total rate £
J41 BUILT UP FELT ROOF COVERINGS - cont'd						
Metal faced 'Veral' glass cloth reinforced bitumen roofing; first layer 'Veralvent' perforated underlay; second layer 'Veralglas'; third layer 'Veral' natural slate aluminium surfaced						
Three layer coverings		-	-	-	m2	**23.56**
Working into outlet pipes and the like		-	-	-	nr	**6.83**
Skirtings; three layer; dressed over tilting fillet; turned into groove						
not exceeding 200 mm girth		-	-	-	m	**4.74**
200 - 400 mm girth		-	-	-	m	**8.14**
Coverings to kerbs; three layer						
400 - 600 mm girth		-	-	-	m	**11.83**
Linings to gutters; three layer						
400 - 600 mm (average) girth		-	-	-	m	**15.97**
Collars around pipes and the like; three layer; 150 mm high						
not exceeding 55 mm nominal size		-	-	-	nr	**4.73**
55 - 110 mm nominal size		-	-	-	nr	**5.25**
Accessories						
Eaves trim; extruded aluminium alloy; working felt into trim						
'Alutrim' type F roof edging	PC £7.77/2.5m	0.25	2.91	3.50	m	**6.41**
Angle	PC £2.85	0.12	1.40	3.14	nr	**4.54**
Roof screed ventilator - aluminium alloy 'Extr-aqua-vent'; set on screed over and including dished sinking; working collar around ventilator	PC £3.18	0.60	6.99	3.51	nr	**10.50**
Insulation board underlays						
Vapour barrier						
reinforced; metal lined		0.03	0.35	3.87	m2	**4.22**
Cork boards; density 112 - 125 kg/m3						
60 mm thick		0.30	3.49	5.05	m2	**8.54**
Foamed glass boards; density 125 - 135 kg/m2						
60 mm thick		0.30	3.49	11.73	m2	**15.22**
Glass fibre boards; density 120 - 130 kg/m2						
60 mm thick		0.30	3.49	8.39	m2	**11.88**
Perlite boards; density 170 - 180 kg/m3						
60 mm thick		0.30	3.49	8.43	m2	**11.92**
Polyurethene boards; density 32 kg/m3						
30 mm thick		0.20	2.33	4.07	m2	**6.40**
35 mm thick		0.20	2.33	4.27	m2	**6.60**
50 mm thick		0.30	3.49	5.47	m2	**8.96**
Wood fibre boards; impregnated; density 220 - 350 kg/m3						
12.7 mm thick		0.20	2.33	1.82	m2	**4.15**
Insulation board overlays						
Dow 'Roofmate SL' extruded polystyrene foam boards						
50 mm thick		0.30	3.49	7.79	m2	**11.28**
75 mm thick		0.30	3.49	11.17	m2	**14.66**
Dow 'Roofmate LG' extruded polystyrene foam boards						
50 mm thick		0.30	3.49	14.77	m2	**18.26**
75 mm thick		0.30	3.49	16.24	m2	**19.73**
100 mm thick		0.30	3.49	19.22	m2	**22.71**

J WATERPROOFING Including overheads and profit at 5.00%	Labour hours	Labour £	Material £	Unit	Total rate £
J42 SINGLE LAYER PLASTICS ROOF COVERINGS					
Felt; 'Derbigum' special polyester 4 mm roofing; first layer 'Ventilag' (partial bond) underlay; second layer 'Derbigum SF'; glass reinforced weathering surface					
Two layer coverings	-	-	-	m2	22.51
Skirtings; two layer; dressed over tilting fillet; turned into groove					
not exceeding 200 mm girth	-	-	-	m	11.00
200 - 400 mm girth	-	-	-	m	15.15
Coverings to kerbs; two layer					
400 - 600 mm girth	-	-	-	m	23.10
Collars around pipes and the like; two layer; 150 mm high					
not exceeding 55 mm nominal size	-	-	-	nr	7.26
55 - 110 mm nominal size	-	-	-	nr	10.82
J43 PROPRIETARY ROOF DECKING WITH FELT FINISH					
'Bitumetal' flat roof construction fixing to timber, steel or concrete; flat or sloping; vapour check; 32 mm polyurethane insulation; 3G perforated felt underlay; two layers of glass fibre base felt roofing; stone chipping finish					
0.7 mm galvanized steel					
35 mm profiled decking; 2.38 m span	-	-	-	m2	25.96
46 mm profiled decking; 2.96 m span	-	-	-	m2	26.03
60 mm profiled decking; 3.74 m span	-	-	-	m2	27.15
100 mm profiled decking; 5.13 m span	-	-	-	m2	32.51
0.9 mm aluminium; mill finish					
35 mm profiled decking; 1.79 m span	-	-	-	m2	30.08
60 mm profiled decking; 2.34 m span	-	-	-	m2	31.89
'Bitumetal' flat roof construction fixing to timber, steel or concrete; flat or sloping; vapour check; 32 mm polyurethane insulation; 3G perforated felt underlay; two layers of polyester based roofing; stone chipping finish					
0.7 mm galvanized steel					
35 mm profiled decking; 2.38 m span	-	-	-	m2	28.92
46 mm profiled decking; 2.96 m span	-	-	-	m2	28.99
60 mm profiled decking; 3.74 m span	-	-	-	m2	30.11
100 mm profiled decking; 5.13 m span	-	-	-	m2	35.47
0.9 mm aluminium; mill finish					
35 mm profiled decking; 1.79 m span	-	-	-	m2	33.04
60 mm profiled decking; 2.34 m span	-	-	-	m2	34.85

J WATERPROOFING Including overheads and profit at 5.00%	Labour hours	Labour £	Material £	Unit	Total rate £
J43 PROPRIETARY ROOF DECKING WITH FELT FINISH - cont'd					
'Plannja' flat roof construction; fixing to timber; steel or concrete; flat or sloping; 3B vapour check; 32 mm polyurethene insulation; 3G perforated felt underlay; two layers of glass fibre bitumen felt roofing type 3B; stone chipping finish					
0.72 mm galvanised steel					
45 mm profiles decking; 3.12 m span	-	-	-	m2	32.96
70 mm profiled decking; 4.40 m span	-	-	-	m2	34.50
'Plannja' flat roof construction; fixing to timber; steel or concrete; flat or sloping; 3B vapour check; 32 mm polyurethene insulation; 3G perforated underlay; one layer of Anderson's HT 125 underlay and HT 350 sanded; stone chipping finish					
0.72 mm galvanised steel					
45 mm profiles decking; 3.12 m span	-	-	-	m2	37.16
70 mm profiled decking; 4.40 m span	-	-	-	m2	38.70

K LININGS/SHEATHING/DRY PARTITIONING
Including overheads and profit at 5.00%

ALTERNATIVE SHEET LINING MATERIAL PRICES

	£		£		£		£
Asbestos cement flat sheets (£/10 m2)							
Fully compressed							
4.5 mm	53.53	6 mm	66.42	9 mm	100.49	12 mm	125.10
Semi compressed							
4.5 mm	27.22	6 mm	37.03	9 mm	55.11	12 mm	91.81
Blockboard							
Gaboon faced (£/10 m2)							
16 mm	86.56	18 mm	90.94	22 mm	102.94	25 mm	115.57

	£		£		£
18 mm Decorative faced (£/10 m2)					
Ash	112.50	Mahogany	84.37	Teak	106.85
Beech	95.62	Oak	100.00		
Edgings; self adhesive (£/25 m roll)					
19 mm Mahogany	2.47	19 mm Oak	3.46	19 mm Ash	3.46
25 mm Mahogany	3.63	25 mm Oak	4.12	19 mm Teak	3.46

	£		£		£		£
Chipboard (£/10 m2)							
Standard grade							
3.2 mm	7.94	9 mm	16.90	16 mm	23.09	22 mm	31.69
4 mm	9.88	12 mm	18.50	18 mm	25.20	25 mm	35.97
6 mm	14.58						
Melamine faced							
12 mm	32.88	18 mm	39.88				
Laminboard; Birch faced (£/10 m2)							
18 mm	118.12	22 mm	138.40	25 mm	153.88		
Medium density fibreboard (£/10 m2)							
6.5 mm	20.28	12 mm	32.40	17.5 mm	41.90	25 mm	59.32
9 mm	24.61	16 mm	38.90	19 mm	45.82		

	£		£		£		£
Plasterboard (£/100m2)							
Wallboard plank							
9.5 mm	104.91	12.5 mm	125.12	15 mm	176.94	19 mm	220.45
Lath (Thistle baseboard)							
9.5 mm	109.94	12.5	131.45				

	£		£		£
Industrial board				**Fireline Industrial board**	
9.5 mm	211.76	12.5 mm	238.28	12.5 mm	308.68
Fireline board					
12.5 mm	188.43	15 mm	218.87		

	£		£		£
Plywood (£/10 m2)					
Decorative					
6 mm Ash	44.55	9 mm Ash	56.82	12 mm Ash	70.12
6 mm Oak	47.00	9 mm Oak	59.57	12 mm Oak	72.07
6 mm Sapele	39.50	9 mm Sapele	53.25	12 mm Sapele	64.61
6 mm Teak	43.41	9 mm Teak	56.85	12 mm Teak	69.85

Discounts of 0 - 10% available depending on quantity/status

K LININGS/SHEATHING/DRY PARTITIONING Including overheads and profit at 5.00%	Labour hours	Labour £	Material £	Unit	Total rate £
K10 PLASTERBOARD DRY LINING					
Gypsum plasterboard; BS 1230; fixing with nails; joints left open to receive 'Artex'; to softwood base					
Plain grade tapered edge wallboard					
9.5 mm board to ceilings					
over 300 mm wide	0.28	3.26	1.12	m2	**4.38**
9.5 mm board to beams					
over 300 mm wide	0.35	4.08	0.70	m	**4.78**
12.5 mm board to ceilings					
total girth not exceeding 600 mm	0.45	5.24	1.34	m	**6.58**
total girth 600 - 1200 mm	0.30	3.49	1.55	m2	**5.04**
12.5 mm board to beams					
total girth not exceeding 600 mm	0.36	4.19	0.97	m	**5.16**
total girth 600 - 1200 mm	0.48	5.59	1.85	m	**7.44**
Gypsum plasterboard to BS 1230; fixing with nails; joints filled with joint filler and joint tape to receive direct decoration; to softwood base					
Plain grade tapered edge wallboard					
9.5 mm board to walls					
wall height 2.40 - 2.70	1.00	11.64	3.87	m	**15.51**
wall height 2.70 - 3.00	1.15	13.39	4.32	m	**17.71**
wall height 3.00 - 3.30	1.30	15.14	4.76	m	**19.90**
wall height 3.30 - 3.60	1.50	17.47	5.20	m	**22.67**
9.5 mm board to reveals and soffits of openings and recesses					
not exceeding 300 mm wide	0.20	2.33	0.46	m	**2.79**
width 300 - 600 mm	0.40	4.66	0.92	m	**5.58**
9.5 mm board to faces of columns					
total girth not exceeding 600 mm	0.50	5.82	0.89	m	**6.71**
total girth 600 - 1200 mm	1.00	11.64	1.85	m	**13.49**
total girth 1200 - 1800 mm	1.30	15.14	2.72	m	**17.86**
9.5 mm board to ceilings					
over 300 mm wide	0.42	4.89	1.44	m2	**6.33**
9.5 mm board to faces of beams					
total girth not exceeding 600 mm	0.53	6.17	0.92	m	**7.09**
total girth 600 - 1200 mm	1.06	12.34	1.85	m	**14.19**
total girth 1200 - 1800 mm	1.38	16.07	2.72	m	**18.79**
Add for 'Duplex' insulating grade	-	-	-	m2	**0.43**
12.5 mm board to walls					
wall height 2.40 - 2.70	1.05	12.23	4.54	m	**16.77**
wall height 2.70 - 3.00	1.20	13.97	5.03	m	**19.00**
wall height 3.00 - 3.30	1.35	15.72	5.56	m	**21.28**
wall height 3.30 - 3.60	1.60	18.63	6.09	m	**24.72**
12.5 mm board to reveals and soffits of openings and recesses					
not exceeding 300 mm wide	0.21	2.45	0.54	m	**2.99**
width 300 - 600 mm	0.42	4.89	1.08	m	**5.97**
12.5 mm board to faces of columns					
total girth not exceeding 600 mm	0.52	6.06	1.08	m	**7.14**
total girth 600 - 1200 mm	1.04	12.11	2.16	m	**14.27**
total girth 1200 - 1800 mm	1.35	15.72	3.17	m	**18.89**
12.5 mm board to ceilings					
over 300 mm wide	0.44	5.12	1.70	m2	**6.82**
12.5 mm board to faces of beams					
total girth not exceeding 600 mm	0.56	6.52	1.08	m	**7.60**
total girth 600 - 1200 mm	1.12	13.04	2.16	m	**15.20**
total girth 1200 - 1800 mm	1.45	16.88	3.17	m	**20.05**

K LININGS/SHEATHING/DRY PARTITIONING Including overheads and profit at 5.00%	Labour hours	Labour £	Material £	Unit	Total rate £
External angle; with joint tape bedded					
in joint filler; covered with joint finish	0.12	1.40	0.42	m	**1.82**
Add for 'Duplex' insulating grade	-	-	-	m2	**0.43**
Tapered edge plank					
19 mm plank to walls					
wall height 2.40 - 2.70	1.10	12.81	7.47	m	**20.28**
wall height 2.70 - 3.00	1.25	14.56	8.29	m	**22.85**
wall height 3.00 - 3.30	1.40	16.30	9.12	m	**25.42**
wall height 3.30 - 3.60	1.70	19.80	9.51	m	**29.31**
19 mm plank to reveals and soffits of openings and recesses					
not exceeding 300 mm wide	0.22	2.56	0.89	m	**3.45**
width 300 - 600 mm	0.44	5.12	1.76	m	**6.88**
19 mm plank to faces of columns					
total girth not exceeding 600 mm	0.54	6.29	1.76	m	**8.05**
total girth 600 - 1200 mm	1.08	12.58	3.53	m	**16.11**
total girth 1200 - 1800 mm	1.40	16.30	5.14	m	**21.44**
19 mm plank to ceilings					
over 300 mm wide	0.46	5.36	2.76	m2	**8.12**
19 mm plank to faces of beams					
total girth not exceeding 600 mm	0.58	6.75	1.72	m	**8.47**
total girth 600 - 1200 mm	1.16	13.51	3.53	m	**17.04**
total girth 1200 - 1800 mm	1.52	17.70	5.14	m	**22.84**
Thermal board					
25 mm board to walls					
wall height 2.40 - 2.70	1.15	13.39	9.09	m	**22.48**
wall height 2.70 - 3.00	1.30	15.14	10.07	m	**25.21**
wall height 3.00 - 3.30	1.45	16.88	11.10	m	**27.98**
wall height 3.30 - 3.60	1.75	20.38	12.13	m	**32.51**
25 mm board to reveals and soffits of openings and recesses					
not exceeding 300 mm wide	0.23	2.68	1.07	m	**3.75**
width 300 - 600 mm	0.46	5.36	2.14	m	**7.50**
25 mm board to faces of columns					
total girth not exceeding 600 mm	0.56	6.52	2.14	m	**8.66**
total girth 600 - 1200 mm	1.12	13.04	4.28	m	**17.32**
total girth 1200 - 1800 mm	1.45	16.88	6.27	m	**23.15**
25 mm board to ceilings					
over 300 mm wide	0.50	5.82	3.35	m2	**9.17**
25 mm board to faces of beams					
total girth not exceeding 600 mm	0.60	6.99	2.14	m	**9.13**
total girth 600 - 1200 mm	1.20	13.97	4.28	m	**18.25**
total girth 1200 - 1800 mm	1.60	18.63	6.27	m	**24.90**
50 mm board to walls					
wall height 2.40 - 2.70	1.25	14.56	15.53	m	**30.09**
wall height 2.70 - 3.00	1.40	16.30	17.28	m	**33.58**
wall height 3.00 - 3.30	1.55	18.05	18.99	m	**37.04**
wall height 3.30 - 3.60	1.85	21.54	20.73	m	**42.27**
50 mm board to reveals and soffits of openings and recesses					
not exceeding 300 mm wide	0.25	2.91	1.81	m	**4.72**
width 300 - 600 mm	0.50	5.82	3.63	m	**9.45**
50 mm board to faces of columns					
total girth not exceeding 600 mm	0.60	6.99	3.63	m	**10.62**
total girth 600 - 1200 mm	1.20	13.97	7.26	m	**21.23**
total girth 1200 - 1800 mm	1.55	18.05	10.67	m	**28.72**
50 mm board to ceilings					
over 300 mm wide	0.53	6.17	5.75	m2	**11.92**
50 mm board to faces of beams					
total girth not exceeding 600 mm	0.63	7.34	3.63	m	**10.97**
total girth 600 - 1200 mm	1.27	14.79	7.26	m	**22.05**
total girth 1200 - 1800 mm	1.70	19.80	10.67	m	**30.47**

K LININGS/SHEATHING/DRY PARTITIONING Including overheads and profit at 5.00%	Labour hours	Labour £	Material £	Unit	Total rate £
K10 PLASTERBOARD DRY LINING - cont'd					
White plastic faced gypsum plasterboard to BS 1230; fixing with screws; butt joints; to softwood base					
Insulating grade square edge wallboard					
9.5 mm board to walls					
wall height 2.40 - 2.70	1.00	11.64	7.51	m	19.15
wall height 2.70 - 3.00	1.15	13.39	8.34	m	21.73
wall height 3.00 - 3.30	1.30	15.14	9.18	m	24.32
wall height 3.30 - 3.60	1.50	17.47	10.01	m	27.48
9.5 mm board to reveals and soffits of openings and recesses					
not exceeding 300 mm wide	0.20	2.33	0.87	m	3.20
width 300 - 600 mm	0.40	4.66	1.74	m	6.40
9.5 mm board to faces of columns					
total girth not exceeding 600 mm	0.50	5.82	1.74	m	7.56
total girth 600 - 1200 mm	1.00	11.64	3.44	m	15.08
total girth 1200 - 1800 mm	1.30	15.14	5.08	m	20.22
9.5 mm board to ceilings					
over 300 mm wide	0.42	4.89	2.78	m2	7.67
9.5 mm board to faces of beams					
total girth not exceeding 600 mm	0.53	6.17	1.74	m	7.91
total girth 600 - 1200 mm	1.06	12.34	3.44	m	15.78
total girth 1200 - 1800 mm	1.38	16.07	5.08	m	21.15
12.5 mm board to walls					
wall height 2.40 - 2.70	1.05	12.23	7.51	m	19.74
wall height 2.70 - 3.00	1.20	13.97	8.34	m	22.31
wall height 3.00 - 3.30	1.35	15.72	9.18	m	24.90
wall height 3.30 - 3.60	1.60	18.63	10.01	m	28.64
12.5 mm board to reveals and soffits of openings and recesses					
not exceeding 300 mm wide	0.21	2.45	0.87	m	3.32
width 300 - 600 mm	0.42	4.89	1.74	m	6.63
12.5 mm board to faces of columns					
total girth not exceeding 600 mm	0.52	6.06	1.74	m	7.80
total girth 600 - 1200 mm	1.04	12.11	3.44	m	15.55
total girth 1200 - 1800 mm	1.35	15.72	5.08	m	20.80
12.5 mm board to ceilings					
over 300 mm wide	0.44	5.12	3.07	m2	8.19
12.5 mm board to faces of beams					
total girth not exceeding 600 mm	0.56	6.52	1.92	m	8.44
total girth 600 - 1200 mm	1.12	13.04	3.81	m	16.85
total girth 1200 - 1800 mm	1.45	16.88	5.61	m	22.49
Two layers of gypsum plasterboard to BS 1230; fixing with nails; joints filled					
19 mm two layer board to walls					
wall height 2.40 - 2.70	1.60	18.63	7.36	m	25.99
wall height 2.70 - 3.00	1.80	20.96	8.17	m	29.13
wall height 3.00 - 3.30	2.00	23.29	9.00	m	32.29
wall height 3.30 - 3.60	2.40	27.95	9.37	m	37.32
19 mm two layer board to reveals and soffits of openings and recesses					
not exceeding 300 mm wide	0.30	3.49	0.87	m	4.36
width 300 - 600 mm	0.60	6.99	1.74	m	8.73
19 mm two layer board to faces of columns					
total girth not exceeding 600 mm	0.80	9.32	1.74	m	11.06
total girth 600 - 1200 mm	1.50	17.47	3.48	m	20.95
total girth 1200 - 1800 mm	2.00	23.29	5.05	m	28.34
19 mm two layer board to ceilings					
over 300 mm wide	0.66	7.69	2.72	m2	10.41
19 mm two layer board to faces of beams					
total girth not exceeding 600 mm	0.85	9.90	1.70	m	11.60

K LININGS/SHEATHING/DRY PARTITIONING Including overheads and profit at 5.00%	Labour hours	Labour £	Material £	Unit	Total rate £
total girth 600 - 1200 mm	1.60	18.63	3.48	m	**22.11**
total girth 1200 - 1800 mm	2.15	25.04	5.05	m	**30.09**
25 mm two layer board to walls					
wall height 2.40 - 2.70	1.70	19.80	8.80	m	**28.60**
wall height 2.70 - 3.00	1.90	22.12	9.77	m	**31.89**
wall height 3.00 - 3.30	2.20	25.62	10.79	m	**36.41**
wall height 3.30 - 3.60	2.60	30.28	11.82	m	**42.10**
25 mm two layer board to reveals and soffits of openings and recesses					
not exceeding 300 mm wide	0.33	3.84	1.04	m	**4.88**
width 300 - 600 mm	0.65	7.57	2.06	m	**9.63**
25 mm two layer board to faces of columns					
total girth not exceeding 600 mm	0.86	10.01	2.06	m	**12.07**
total girth 600 - 1200 mm	1.60	18.63	4.13	m	**22.76**
total girth 1200 - 1800 mm	2.15	25.04	6.08	m	**31.12**
25 mm two layer board to ceilings					
over 300 mm wide	0.72	8.38	3.25	m2	**11.63**
25 mm two layer board to faces of beams					
total girth not exceeding 600 mm	0.92	10.71	2.06	m	**12.77**
total girth 600 - 1200 mm	1.75	20.38	4.13	m	**24.51**
total girth 1200 - 1800 mm	2.33	27.13	6.08	m	**33.21**
Gypsum plasterboard to BS 1230; 3 mm joints; fixed by the 'Thistleboard' system of dry linings; joints; filled with joint filler and joint tape; to receive direct decoration					
Plain grade tapered edge wallboard					
9.5 mm board to walls					
wall height 2.40 - 2.70	1.05	12.23	3.81	m	**16.04**
wall height 2.70 - 3.00	1.20	13.97	4.24	m	**18.21**
wall height 3.00 - 3.30	1.35	15.72	4.67	m	**20.39**
wall height 3.30 - 3.60	1.60	18.63	5.11	m	**23.74**
9.5 mm board to reveals and soffits of openings and recesses					
not exceeding 300 mm wide	0.21	2.45	0.44	m	**2.89**
width 300 - 600 mm	0.42	4.89	0.88	m2	**5.77**
9.5 mm board to faces of columns					
total girth not exceeding 600 mm	0.52	6.06	0.88	m	**6.94**
total girth 600 - 1200 mm	1.04	12.11	1.77	m	**13.88**
total girth 1200 - 1800 mm	1.35	15.72	2.60	m	**18.32**
Angle; with joint tape bedded in joint filler; covered with joint finish					
internal	0.06	0.70	0.17	m	**0.87**
external	0.12	1.40	0.41	m	**1.81**
Vermiculite gypsum cladding; 'Vicuclad' board on and including shaped noggins; fixed with nails and adhesive; joints pointed in adhesive					
25 mm column casings; 2 hour fire protection rating					
over 1 m girth	1.00	11.64	12.94	m2	**24.58**
150 mm girth	0.30	3.49	3.92	m	**7.41**
300 mm girth	0.45	5.24	5.87	m	**11.11**
600 mm girth	0.75	8.73	9.57	m	**18.30**
900 mm girth	0.95	11.06	12.30	m	**23.36**
30 mm beam casings; 2 hour fire protection rating					
over 1 m girth	1.10	12.81	15.89	m2	**28.70**
150 mm girth	0.33	3.84	4.81	m	**8.65**
300 mm girth	0.50	5.82	7.20	m	**13.02**
600 mm girth	0.83	9.66	11.78	m	**21.44**
900 mm girth	1.05	12.23	15.10	m	**27.33**

K LININGS/SHEATHING/DRY PARTITIONING Including overheads and profit at 5.00%		Labour hours	Labour £	Material £	Unit	Total rate £
K10 PLASTERBOARD DRY LINING - cont'd						
Vermiculite gypsum cladding; 'Vicuclad' board on and including shaped noggins; fixed with nails and adhesive; joints pointed in adhesive - cont'd						
55 mm column casings; 4 hour fire protection rating						
over 1 m girth		1.20	13.97	32.64	m2	**46.61**
150 mm girth		0.36	4.19	9.82	m	**14.01**
300 mm girth		0.55	6.40	14.74	m	**21.14**
600 mm girth		0.90	10.48	24.34	m	**34.82**
900 mm girth		1.15	13.39	31.01	m	**44.40**
60 mm beam casings; 4 hour fire protection rating						
over 1 m girth		1.30	15.14	34.41	m2	**49.55**
150 mm girth		0.39	4.54	10.36	m	**14.90**
300 mm girth		0.60	6.99	15.53	m	**22.52**
600 mm girth		0.97	11.30	25.68	m	**36.98**
900 mm girth		1.25	14.56	32.70	m	**47.26**
Add to the above for						
Plus 3% for work 3.5 - 5 m high						
Plus 6% for work 5 - 6.5 m high						
Plus 12% for work 6.5 - 8 m high						
Plus 18% for work over 8 m high						
Cutting and fitting around steel joists, angles, trunking, ducting, ventilators, pipes, tubes, etc						
over 2 m girth		0.30	3.49	-	m	**3.49**
not exceeding 0.30 m girth		0.20	2.33	-	nr	**2.33**
0.30 - 1 m girth		0.25	2.91	-	nr	**2.91**
1 - 2 m girth		0.35	4.08	-	nr	**4.08**
K11 RIGID SHEET FLOORING/SHEATHING/LININGS/ CASINGS						
Blockboard (Birch faced)						
12 mm lining to walls						
over 300 mm wide	PC £71.05/10m2	0.47	3.72	8.31	m2	**12.03**
not exceeding 300 mm wide		0.31	2.45	2.73	m	**5.18**
Raking cutting		0.08	0.63	0.30	m	**0.93**
Holes for pipes and the like		0.04	0.32	2.98	nr	**3.30**
18 mm lining to walls						
over 300 mm wide	PC £86.36/10m2	0.50	3.95	10.08	m2	**14.03**
not exceeding 300 mm wide		0.32	2.53	3.31	m	**5.84**
Raking cutting		0.10	0.79	0.45	m	**1.24**
Holes for pipes and the like		0.05	0.40	-	nr	**0.40**
Two-sided 18 mm thick pipe casing; 50 x 50 mm softwood framing; two members plugged to wall						
300 mm girth		1.25	9.88	5.27	m	**15.15**
450 mm girth		1.35	10.67	6.95	m	**17.62**
600 mm girth		1.45	11.46	8.62	m	**20.08**
750 mm girth		1.55	12.26	10.30	m	**22.56**
Three-sided 18 mm thick pipe casing; 50 x 50 mm softwood framing; two members plugged to wall						
450 mm girth		1.70	13.44	7.60	m	**21.04**
600 mm girth		1.80	14.23	9.27	m	**23.50**
750 mm girth		1.90	15.02	10.95	m	**25.97**
900 mm girth		2.00	15.81	12.63	m	**28.44**
1050 mm girth		2.10	16.60	14.31	m	**30.91**

K LININGS/SHEATHING/DRY PARTITIONING Including overheads and profit at 5.00%		Labour hours	Labour £	Material £	Unit	Total rate £
Extra for 400 x 400 mm removable access panel; brass cups and screws; additional framing		1.00	7.91	4.54	nr	12.45
25 mm lining to walls						
over 300 mm wide	PC £112.58/10m2	0.54	4.27	13.13	m2	17.40
not exceeding 300 mm wide		0.35	2.77	4.31	m	7.08
Raking cutting		0.12	0.95	0.59	m	1.54
Holes for pipes and the like		0.06	0.47	-	nr	0.47
Chipboard (plain)						
12 mm lining to walls						
over 300 mm wide	PC £18.50/10m2	0.38	3.00	2.25	m2	5.25
not exceeding 300 mm wide		0.22	1.74	0.74	m	2.48
Raking cutting		0.06	0.47	0.10	m	0.57
Holes for pipes and the like		0.03	0.24	-	nr	0.24
15 mm lining to walls						
over 300 mm wide	PC £21.55/10m2	0.40	3.16	2.60	m2	5.76
not exceeding 300 mm wide		0.24	1.90	0.86	m	2.76
Raking cutting		0.08	0.63	0.11	m	0.74
Holes for pipes and the like		0.04	0.32	-	nr	0.32
Two-sided 15 mm thick pipe casing; 50 x 50 mm softwood framing; two members plugged to wall						
300 mm girth		1.00	7.91	3.02	m	10.93
450 mm girth		1.08	8.54	3.57	m	12.11
600 mm girth		1.16	9.17	4.13	m	13.30
750 mm girth		1.24	9.80	4.69	m	14.49
Three-sided 15 mm thick pipe casing; 50 x 50 mm softwood framing; two members plugged to wall						
450 mm girth		1.49	11.78	4.22	m	16.00
600 mm girth		1.58	12.49	4.78	m	17.27
750 mm girth		1.70	13.44	5.34	m	18.78
900 mm girth		1.79	14.15	5.90	m	20.05
1050 mm girth		2.01	15.89	6.44	m	22.33
Extra for 400 x 400 mm removable access panel; brass cups and screws; additional framing		1.00	7.91	4.47	nr	12.38
18 mm lining to walls						
over 300 mm wide	PC £25.20/10m2	0.42	3.32	3.02	m2	6.34
not exceeding 300 mm wide		0.27	2.13	0.99	m	3.12
Raking cutting		0.10	0.79	0.13	m	0.92
Holes for pipes and the like		0.05	0.40	-	nr	0.40
Chipboard (Melamine faced white matt finish) 15 mm thick; PC £35.27/10m2; laminated masking strips						
Lining to walls						
over 300 mm wide		1.05	8.30	6.16	m2	14.46
not exceeding 300 mm wide		0.68	5.38	2.81	m	8.19
Raking cutting		0.13	1.03	0.19	m	1.22
Holes for pipes and the like		0.07	0.55	-	nr	0.55
Chipboard boarding and flooring						
Boarding to floors; butt joints						
18 mm thick	PC £26.96/10m2	0.30	2.37	3.10	m2	5.47
22 mm thick	PC £31.45/10m2	0.33	2.61	3.61	m2	6.22
Boarding to floors; tongued and grooved joints						
18 mm thick	PC £29.24/10m2	0.32	2.53	3.36	m2	5.89
22 mm thick	PC £38.44/10m2	0.35	2.77	4.37	m2	7.14
Boarding to roofs; butt joints						
18 mm thick; pre-felted	PC £25.20/10m2	0.33	2.61	2.92	m2	5.53
Raking cutting on 18 mm thick chipboard		0.08	0.63	1.42	m	2.05

K LININGS/SHEATHING/DRY PARTITIONING Including overheads and profit at 5.00%		Labour hours	Labour £	Material £	Unit	Total rate £
K11 RIGID SHEET FLOORING/SHEATHING/LININGS/ CASINGS - cont'd						
H.H.Robertson (U.K.) Ltd. raised flooring systems; laid on or fixed to concrete floor						
'Europlan 215' system; 80 mm overall height						
chipboard finish		-	-	-	m2	**19.95**
Extra for						
300 mm wide periphery duct		-	-	-	m	**29.40**
60 mm wide spine duct		-	-	-	m	**36.75**
600 mm wide segregated feeder duct		-	-	-	m	**49.35**
Three compartment service box		-	-	-	nr	**63.00**
'Cavity floor' full access system; 150 mm overall height; pedestal supports						
light grade; ref CMT/30L; chipboard finish		-	-	-	m2	**29.40**
medium grade; ref CMT/30M; steel finish		-	-	-	m2	**34.65**
heavy grade; ref CMT/40H; steel finish		-	-	-	m2	**43.05**
Extra for						
Factory applied needlepunch carpet		-	-	-	m2	**9.97**
Plywood flooring						
Boarding to floors; tongued and grooved joints						
15 mm thick	PC £111.10/10m2	0.40	3.16	12.39	m2	**15.55**
18 mm thick	PC £129.28/10m2	0.44	3.48	14.39	m2	**17.87**
Plywood; external quality; 18 mm thick; PC £97.96/10m2						
Boarding to roofs; butt joints						
flat to falls		0.40	3.16	10.94	m2	**14.10**
sloping		0.43	3.40	10.94	m2	**14.34**
vertical		0.57	4.51	10.94	m2	**15.45**
Hardboard to BS 1142						
3.2 mm lining to walls						
over 300 mm wide	PC £9.39/10m2	0.30	2.37	1.12	m2	**3.49**
not exceeding 300 mm wide		0.18	1.42	0.37	m	**1.79**
Raking cutting		0.02	0.16	0.05	m	**0.21**
Holes for pipes and the like		0.01	0.08	-	nr	**0.08**
6.4 mm lining to walls						
over 300 mm wide	PC £18.65/10m2	0.33	2.61	2.20	m2	**4.81**
not exceeding 300 mm wide		0.21	1.66	0.72	m	**2.38**
Raking cutting		0.03	0.24	0.10	m	**0.34**
Holes for pipes and the like		0.02	0.16	-	nr	**0.16**
Glazed hardboard to BS 1142; PC £14.10/10m2; on and including 38 x 38 mm wrought softwood framing						
3.2 mm thick panel						
to side of bath		1.80	14.23	4.35	nr	**18.58**
to end of bath		0.70	5.53	1.62	nr	**7.15**
Insulation board to BS 1142						
12.7 mm lining to walls						
over 300 mm wide	PC £16.40/10m2	0.24	1.90	2.00	m2	**3.90**
not exceeding 300 mm wide		0.14	1.11	0.66	m	**1.77**
Raking cutting		0.03	0.24	0.09	m	**0.33**
Holes for pipes and the like		0.01	0.08	-	nr	**0.08**
19 mm lining to walls						
over 300 mm wide	PC £26.75/10m2	0.26	2.06	3.20	m2	**5.26**
not exceeding 300 mm wide		0.16	1.27	1.05	m	**2.32**
Raking cutting		0.05	0.40	0.14	m	**0.54**
Holes for pipes and the like		0.02	0.16	-	nr	**0.16**

K LININGS/SHEATHING/DRY PARTITIONING Including overheads and profit at 5.00%		Labour hours	Labour £	Material £	Unit	Total rate £
25 mm lining to walls						
over 300 mm wide	PC £34.65/10m2	0.32	2.53	4.13	m2	**6.66**
not exceeding 300 mm wide		0.19	1.50	1.37	m	**2.87**
Raking cutting		0.06	0.47	0.18	m	**0.65**
Holes for pipes and the like		0.03	0.24	-	nr	**0.24**
Laminboard (Birch Faced); 18 mm thick PC £118.12/10m2						
Lining to walls						
over 300 mm wide		0.53	4.19	13.75	m2	**17.94**
not exceeding 300 mm wide		0.34	2.69	4.51	m	**7.20**
Raking cutting		0.11	0.87	0.62	m	**1.49**
Holes for pipes and the like		0.06	0.47	-	nr	**0.47**
Non-asbestos board; 'Masterboard'; sanded finish						
6 mm lining to walls						
over 300 mm wide	PC £40.77/10m2	0.33	2.61	4.79	m2	**7.40**
not exceeding 300 mm wide		0.20	1.58	1.56	m	**3.14**
6 mm lining to ceilings						
over 300 mm wide		0.44	3.48	4.79	m2	**8.27**
not exceeding 300 mm wide		0.27	2.13	1.56	m	**3.69**
Raking cutting		0.05	0.40	0.21	m	**0.61**
Holes for pipes and the like		0.03	0.24	-	nr	**0.24**
9 mm lining to walls						
over 300 mm wide	PC £74.07/10m2	0.36	2.85	8.65	m2	**11.50**
not exceeding 300 mm wide		0.22	1.74	2.84	m	**4.58**
9 mm lining to ceilings						
over 300 mm wide		0.45	3.56	8.65	m2	**12.21**
not exceeding 300 mm wide		0.29	2.29	2.84	m	**5.13**
Raking cutting		0.06	0.47	0.39	m	**0.86**
Holes for pipes and the like		0.04	0.32	-	nr	**0.32**
12 mm lining to walls						
over 300 mm wide	PC £104.31/10m2	0.40	3.16	12.15	m2	**15.31**
not exceeding 300 mm wide		0.24	1.90	3.99	m	**5.89**
12 mm lining to ceilings						
over 300 mm wide		0.53	4.19	12.15	m2	**16.34**
not exceeding 300 mm wide		0.32	2.53	3.99	m	**6.52**
Raking cutting		0.07	0.55	0.55	m	**1.10**
Holes for pipes and the like		0.05	0.40	-	nr	**0.40**
Non-asbestos board; 'Supalux'; sanded finish						
6 mm lining to walls						
over 300 mm wide	PC £51.66/10m2	0.33	2.61	6.05	m2	**8.66**
not exceeding 300 mm wide		0.20	1.58	1.98	m	**3.56**
6 mm lining to ceilings						
over 300 mm wide		0.44	3.48	6.05	m2	**9.53**
not exceeding 300 mm wide		0.27	2.13	1.98	m	**4.11**
Raking cutting		0.05	0.40	0.27	m	**0.67**
Holes for pipes and the like		0.03	0.24	-	nr	**0.24**
9 mm lining to walls						
over 300 mm wide	PC £78.84/10m2	0.36	2.85	9.20	m2	**12.05**
not exceeding 300 mm wide		0.22	1.74	3.02	m	**4.76**
9 mm lining to ceilings						
over 300 mm wide		0.45	3.56	9.20	m2	**12.76**
not exceeding 300 mm wide		0.29	2.29	3.02	m	**5.31**
Raking cutting		0.06	0.47	0.41	m	**0.88**
Holes for pipes and the like		0.04	0.32	-	nr	**0.32**
12 mm lining to walls						
over 300 mm wide	PC £104.31/10m2	0.40	3.16	12.15	m2	**15.31**
not exceeding 300 mm wide		0.24	1.90	3.99	m	**5.89**
12 mm lining to ceilings						
over 300 mm wide		0.53	4.19	12.15	m2	**16.34**
not exceeding 300 mm wide		0.32	2.53	3.99	m	**6.52**

K LININGS/SHEATHING/DRY PARTITIONING Including overheads and profit at 5.00%		Labour hours	Labour £	Material £	Unit	Total rate £
K11 RIGID SHEET FLOORING/SHEATHING/LININGS/ CASINGS - cont'd						
Non-asbestos board; 'Supalux'; sanded finish - cont'd						
12 mm lining to ceilings						
Raking cutting		0.07	0.55	0.55	m	1.10
Holes for pipes and the like		0.05	0.40	-	nr	0.40
Plywood (Russian Birch); internal quality						
4 mm lining to walls						
over 300 mm wide	PC £22.61/10m2	0.37	2.93	2.72	m2	5.65
not exceeding 300 mm wide		0.24	1.90	0.90	m	2.80
4 mm lining to ceilings						
over 300 mm wide		0.50	3.95	2.72	m2	6.67
not exceeding 300 mm wide		0.32	2.53	0.90	m	3.43
Raking cutting		0.05	0.40	0.12	m	0.52
Holes for pipes and the like		0.03	0.24	-	nr	0.24
6 mm lining to walls						
over 300 mm wide	PC £31.94/10m2	0.40	3.16	3.80	m2	6.96
not exceeding 300 mm wide		0.26	2.06	1.25	m	3.31
6 mm lining to ceilings						
over 300 mm wide		0.53	4.19	3.80	m2	7.99
not exceeding 300 mm wide		0.35	2.77	1.25	m	4.02
Raking cutting		0.05	0.40	0.17	m	0.57
Holes for pipes and the like		0.03	0.24	-	nr	0.24
Two-sided 6 mm thick pipe casings; 50 x 50 mm softwood framing; two members plugged to wall						
300 mm girth		1.20	9.49	3.38	m	12.87
450 mm girth		1.30	10.28	4.12	m	14.40
600 mm girth		1.40	11.07	4.85	m	15.92
750 mm girth		1.50	11.86	5.59	m	17.45
Three-sided 6 mm thick pipe casing; 50 x 50 mm softwood framing; to members plugged to wall						
450 mm girth		1.65	13.05	4.77	m	17.82
600 mm girth		1.75	13.84	5.50	m	19.34
750 mm girth		1.85	14.63	6.24	m	20.87
900 mm girth		1.95	15.42	6.97	m	22.39
1050 mm girth		2.00	15.81	7.71	m	23.52
9 mm lining to walls						
over 300 mm wide	PC £44.81/10m2	0.43	3.40	5.29	m2	8.69
not exceeding 300 mm wide		0.28	2.21	1.74	m	3.95
9 mm lining to ceilings						
over 300 mm wide		0.57	4.51	5.29	m2	9.80
not exceeding 300 mm wide		0.37	2.93	1.74	m	4.67
Raking cutting		0.06	0.47	0.24	m	0.71
Holes for pipes and the like		0.04	0.32	-	nr	0.32
12 mm lining to walls						
over 300 mm wide	PC £58.76/10m2	0.46	3.64	6.90	m2	10.54
not exceeding 300 mm wide		0.30	2.37	2.26	m	4.63
12 mm lining to ceilings						
over 300 mm wide		0.61	4.82	6.90	m2	11.72
not exceeding 300 mm wide		0.40	3.16	2.26	m	5.42
Raking cutting		0.06	0.47	0.31	m	0.78
Holes for pipes and the like		0.04	0.32	-	nr	0.32
Plywood (Finnish Birch); external quality						
4 mm lining to walls						
over 300 mm wide	PC £27.15/10m2	0.37	2.93	3.24	m2	6.17
not exceeding 300 mm wide		0.24	1.90	1.06	m	2.96
4 mm lining to ceilings						
over 300 mm wide		0.50	3.95	3.24	m2	7.19
not exceeding 300 mm wide		0.32	2.53	1.06	m	3.59

K LININGS/SHEATHING/DRY PARTITIONING Including overheads and profit at 5.00%		Labour hours	Labour £	Material £	Unit	Total rate £
Raking cutting		0.05	0.40	0.14	m	0.54
Holes for pipes and the like		0.03	0.24	-	nr	0.24
6 mm lining to walls						
over 300 mm wide	PC £40.80/10m2	0.40	3.16	4.82	m2	7.98
not exceeding 300 mm wide		0.26	2.06	1.58	m	3.64
6 mm lining to ceilings						
over 300 mm wide		0.53	4.19	4.82	m2	9.01
not exceeding 300 mm wide		0.35	2.77	1.58	m	4.35
Raking cutting		0.05	0.40	0.21	m	0.61
Holes for pipes and the like		0.03	0.24	-	nr	0.24
Two-sided 6 mm thick pipe casings; 50 x 50 mm softwood framing; two members plugged to wall						
300 mm girth		1.20	9.49	3.68	m	13.17
450 mm girth		1.30	10.28	4.58	m	14.86
600 mm girth		1.40	11.07	5.47	m	16.54
750 mm girth		1.50	11.86	6.35	m	18.21
Three-sided 6 mm thick pipe casing; 50 x 50 mm softwood framing; to members plugged to wall						
450 mm girth		1.65	13.05	5.23	m	18.28
600 mm girth		1.75	13.84	6.12	m	19.96
750 mm girth		1.85	14.63	7.00	m	21.63
900 mm girth		1.95	15.42	7.90	m	23.32
1050 mm girth		2.00	15.81	8.79	m	24.60
9 mm lining to walls						
over 300 mm wide	PC £52.67/10m2	0.43	3.40	6.19	m2	9.59
not exceeding 300 mm wide		0.28	2.21	2.04	m	4.25
9 mm lining to ceilings						
over 300 mm wide		0.57	4.51	6.19	m2	10.70
not exceeding 300 mm wide		0.37	2.93	2.04	m	4.97
Raking cutting		0.07	0.55	0.28	m	0.83
Holes for pipes and the like		0.05	0.40	-	nr	0.40
12 mm lining to walls						
over 300 mm wide	PC £67.18/10m2	0.46	3.64	7.87	m2	11.51
not exceeding 300 mm wide		0.30	2.37	2.59	m	4.96
12 mm lining to ceilings						
over 300 mm wide		0.61	4.82	7.88	m2	12.70
not exceeding 300 mm wide		0.40	3.16	2.59	m	5.75
Raking cutting		0.10	0.79	0.35	m	1.14
Holes for pipes and the like		0.08	0.63	-	nr	0.63
Extra over wall linings fixed with nails; for screwing		0.15	1.19	0.15	m2	1.34
Woodwool unreinforced slabs; Torvale 'Woodcemair' or similar; BS 1105 type SB; natural finish; fixing to timber or steel with galvanized nails or clips; flat or sloping						
50 mm slabs; type 500; max. span 600 mm						
1800 - 2400 mm lengths	PC £4.80	0.40	3.16	5.51	m2	8.67
2700 - 3000 mm lengths	PC £4.91	0.40	3.16	5.64	m2	8.80
75 mm slabs; type 750; max. span 900 mm						
2100 mm lengths	PC £6.55	0.45	3.56	7.48	m2	11.04
2400 - 2700 mm lengths	PC £6.64	0.45	3.56	7.62	m2	11.18
3000 mm lengths		0.45	3.56	7.62	m2	11.18
Raking cutting		0.14	1.11	1.58	m	2.69
Holes for pipes and the like		0.12	0.95	-	nr	0.95
100 mm slabs; type 1000; max. span 1200 mm						
3000 - 3600 mm lengths	PC £9.06	0.50	3.95	10.21	m2	14.16

K LININGS/SHEATHING/DRY PARTITIONING Including overheads and profit at 5.00%	Labour hours	Labour £	Material £	Unit	Total rate £
Internal quality West African Mahogany veneered plywood; 6 mm thick; PC £35.60/10m2; WAM cover strips					
Lining to walls					
over 300 mm wide	0.70	5.53	5.07	m2	**10.60**
not exceeding 300 mm wide	0.45	3.56	2.00	m	**5.56**
K13 RIGID SHEET FINE LININGS/PANELLING					
Formica faced chipboard; 17 mm thick; white matt finish; balancer; PC £187.35/10m2; aluminium cover strips and countersunk screws					
Lining to walls					
over 300 mm wide	1.90	15.02	23.30	m2	**38.32**
not exceeding 300 mm wide	1.24	9.80	8.24	m	**18.04**
Lining to isolated columns or the like					
over 300 mm wide	2.85	22.53	25.70	m2	**48.23**
not exceeding 300 mm wide	1.65	13.05	8.83	m	**21.88**
K20 TIMBER BOARD FLOORING/SHEATHING/LININGS/ CASINGS					
Sawn softwood; untreated					
Boarding to roofs; 150 mm wide boards; butt joints					
19 mm thick; flat; over 300 mm wide PC £60.67/100m	0.45	3.56	4.85	m2	**8.41**
19 mm thick; flat; not exceeding 300 mm wide	0.30	2.37	1.49	m	**3.86**
19 mm thick; sloping; over 300 mm wide	0.50	3.95	4.85	m2	**8.80**
19 mm thick; sloping; not exceeding 300 mm wide	0.33	2.61	1.49	m	**4.10**
19 mm thick; sloping; laid diagonally; over 300 mm wide	0.63	4.98	4.97	m2	**9.95**
19 mm thick; sloping; laid diagonally; not exceeding 300 mm wide	0.40	3.16	1.53	m	**4.69**
25 mm thick; flat; over 300 mm wide; PC £77.11/100m	0.45	3.56	6.10	m2	**9.66**
25 mm thick; flat; not exceeding 300 mm	0.30	2.37	1.87	m	**4.24**
25 mm thick; sloping; over 300 mm wide	0.50	3.95	6.10	m2	**10.05**
25 mm thick; sloping; not exceeding 300 mm wide	0.33	2.61	1.87	m	**4.48**
Boarding to tops or cheeks of dormers; 150 mm wide boards; butt joints					
19 mm thick; diagonally; over 300 mm wide	0.80	6.33	4.97	m2	**11.30**
19 mm thick; diagonally; not exceeding 300 mm wide	0.50	3.95	1.49	m	**5.44**
19 mm thick; diagonally; area not exceeding 1.00 m2; irrespective of width	1.00	7.91	5.07	nr	**12.98**
Sawn softwood; 'Tanalised'					
Boarding to roofs; 150 mm wide boards; butt joints					
19 mm thick; flat; over 300 mm wide PC £75.83/100m	0.45	3.56	6.00	m2	**9.56**
19 mm thick; flat; not exceeding 300 mm wide	0.30	2.37	1.84	m	**4.21**
19 mm thick; sloping; over 300 mm wide	0.50	3.95	6.00	m2	**9.95**
19 mm thick; sloping; not exceeding 300 mm wide	0.33	2.61	1.84	m	**4.45**
19 mm thick; sloping; laid diagonally; over 300 mm wide	0.63	4.98	6.14	m2	**11.12**
19 mm thick; sloping; laid diagonally; not exceeding 300 mm wide	0.40	3.16	1.89	m	**5.05**
25 mm thick; flat; over 300 mm wide; PC £96.39/100m	0.45	3.56	7.55	m2	**11.11**
25 mm thick; flat; not exceeding 300 mm	0.30	2.37	2.32	m	**4.69**

K LININGS/SHEATHING/DRY PARTITIONING Including overheads and profit at 5.00%	Labour hours	Labour £	Material £	Unit	Total rate £
25 mm thick; sloping; over 300 mm wide	0.50	3.95	7.55	m2	11.50
25 mm thick; sloping; not exceeding 300 mm wide	0.33	2.61	2.32	m	4.93
Boarding to tops or cheeks of dormers; 150 mm wide boards; butt joints					
19 mm thick; diagonally; over 300 mm wide	0.80	6.33	6.14	m2	12.47
19 mm thick; diagonally; not exceeding 300 mm wide	0.50	3.95	1.89	m	5.84
19 mm thick; diagonally; area not exceeding 1.00 m2; irrespective of width	1.00	7.91	6.14	nr	14.05
Wrought softwood					
Boarding to floors; butt joints					
19 mm thick x 75 mm wide boards PC £54.57/100m	0.60	4.74	8.67	m2	13.41
19 mm thick x 125 mm wide boards PC £64.92/100m	0.50	3.95	6.19	m2	10.14
22 mm thick x 150 mm wide boards PC £76.68/100m	0.45	3.56	6.07	m2	9.63
25 mm thick x 100 mm wide boards PC £75.13/100m	0.55	4.35	8.85	m2	13.20
25 mm thick x 150 mm wide boards PC £109.57/100m	0.45	3.56	8.56	m2	12.12
Boarding to floors; tongued and grooved joints					
19 mm thick x 75 mm wide boards PC £54.57/100m	0.70	5.53	9.18	m2	14.71
19 mm thick x 125 mm wide boards PC £64.92/100m	0.60	4.74	6.54	m2	11.28
22 mm thick x 150 mm wide boards PC £76.68/100m	0.55	4.35	6.42	m2	10.77
25 mm thick x 100 mm wide boards PC £75.13/100m	0.65	5.14	9.37	m2	14.51
25 mm thick x 150 mm wide boards PC £109.57/100m	0.55	4.35	9.06	m2	13.41
Nosings; tongued to edge of flooring					
19 x 75 mm; once rounded	0.25	1.98	1.40	m	3.38
25 x 75 mm; once rounded	0.25	1.98	1.64	m	3.62
Boarding to internal walls; tongued and grooved and V-jointed					
12 mm thick x 100 mm wide boards PC £58.32/100m	0.80	6.33	7.36	m2	13.69
16 mm thick x 100 mm wide boards PC £68.80/100m	0.80	6.33	8.61	m2	14.94
19 mm thick x 100 mm wide boards PC £81.68/100m	0.80	6.33	10.15	m2	16.48
19 mm thick x 125 mm wide boards PC £100.98/100m	0.75	5.93	9.99	m2	15.92
19 mm thick x 125 mm wide boards; chevron pattern	1.20	9.49	10.51	m2	20.00
25 mm thick x 125 mm wide boards PC £131.26/100m	0.75	5.93	12.89	m2	18.82
12 mm thick x 100 mm wide Knotty Pine boards PC £35.30/100m	0.80	6.33	4.60	m2	10.93
Boarding to internal ceilings; tongued and grooved and V-jointed					
12 mm thick x 100 mm wide boards	1.00	7.91	7.36	m2	15.27
16 mm thick x 100 mm wide boards	1.00	7.91	8.61	m2	16.52
19 mm thick x 100 mm wide boards	1.00	7.91	10.15	m2	18.06
19 mm thick x 125 mm wide boards	0.95	7.51	9.99	m2	17.50
19 mm thick x 125 mm wide boards; chevron pattern	1.40	11.07	10.51	m2	21.58
25 mm thick x 125 mm wide boards	0.95	7.51	12.89	m2	20.40

K LININGS/SHEATHING/DRY PARTITIONING Including overheads and profit at 5.00%		Labour hours	Labour £	Material £	Unit	Total rate £
K20 TIMBER BOARD FLOORING/SHEATHING/LININGS/ CASINGS - cont'd						
Wrought softwood - cont'd						
Boarding to internal ceilings;						
12 mm thick x 100 mm wide						
Knotty Pine boards		1.00	7.91	4.60	m2	12.51
Boarding to roofs; tongued and grooved joints						
19 mm thick; flat to falls	PC £84.48/100m	0.55	4.35	7.04	m2	11.39
19 mm thick; sloping		0.60	4.74	7.04	m2	11.78
19 mm thick; sloping; laid diagonally		0.75	5.93	7.17	m2	13.10
25 mm thick; flat to falls	PC £109.57/100m	0.55	4.35	9.06	m2	13.41
25 mm thick; sloping		0.60	4.74	9.06	m2	13.80
Boarding to tops or cheeks of dormers; tongued and grooved joints						
19 mm thick; laid diagonally		1.00	7.91	7.17	m2	15.08
Wrought softwood; 'Tanalised'						
Boarding to roofs; tongued and grooved joints						
19 mm thick; flat to falls	PC £105.61/100m	0.55	4.35	8.74	m2	13.09
19 mm thick; sloping		0.60	4.74	8.74	m2	13.48
19 mm thick; sloping; laid diagonally		0.75	5.93	8.89	m2	14.82
25 mm thick; flat to falls	PC £136.96/100m	0.55	4.35	11.25	m2	15.60
25 mm thick; sloping		0.60	4.74	11.25	m2	15.99
Boarding to tops or cheeks of dormers; tongued and grooved joints						
19 mm thick; laid diagonally		1.00	7.91	8.89	m2	16.80
Wood strip; 22 mm thick; 'Junckers' pre-treated or similar; tongued and grooved joints; pre-finished boards; level fixing to resilient battens; to cement and sand base						
Strip flooring; over 300 mm wide						
Beech; prime		-	-	-	m2	45.05
Beech; standard		-	-	-	m2	43.00
Beech; sylvia squash		-	-	-	m2	45.05
Oak; quality A		-	-	-	m2	50.16
Wrought hardwood						
Strip flooring to floors; 25 mm thick x 75 mm wide; tongued and grooved joints; secret fixing; surface sanded after laying						
American Oak		-	-	-	m2	31.74
Canadian Maple		-	-	-	m2	30.71
Gurjun		-	-	-	m2	26.62
Iroko		-	-	-	m2	30.71
Raking cutting		-	-	-	m	3.58
Curved cutting		-	-	-	m	6.65
K29 TIMBER FRAMED AND PANELLED PARTITIONS						
Purpose made screen components; wrought softwood						
Panelled partitions						
32 mm thick frame; 9 mm thick plywood panels; over 300 mm wide	PC £30.50	0.90	7.12	32.82	m2	39.94
38 mm thick frame; 9 mm thick plywood panels over 300 mm wide	PC £32.51	0.95	7.51	34.99	m2	42.50
50 mm thick frame; 12 mm thick plywood panels; over 300 mm wide	PC £39.56	1.00	7.91	42.58	m2	50.49

K LININGS/SHEATHING/DRY PARTITIONING Including overheads and profit at 5.00%		Labour hours	Labour £	Material £	Unit	Total rate £
Panelled partitions; mouldings worked on solid both sides						
32 mm thick frame; 9 mm thick plywood panels; over 300 mm wide	PC £35.23	0.90	7.12	37.92	m2	45.04
38 mm thick frame; 9 mm thick plywood panels over 300 mm wide	PC £36.86	0.95	7.51	39.67	m2	47.18
50 mm thick frame; 12 mm thick plywood panels; over 300 mm wide	PC £44.73	1.00	7.91	48.15	m2	56.06
Panelled partitions; mouldings planted on both sides						
32 mm thick frame; 9 mm thick plywood panels; over 300 mm wide	PC £35.66	0.90	7.12	38.38	m2	45.50
38 mm thick frame; 9 mm thick plywood panels over 300 mm wide	PC £37.19	0.95	7.51	40.02	m2	47.53
50 mm thick frame; 12 mm thick plywood panels; over 300 mm wide	PC £44.73	1.00	7.91	48.15	m2	56.06
Panelled partitions; diminishing stiles over 300 mm wide; upper portion open panels for glass; in medium panes						
32 mm thick frame; 9 mm thick plywood panels; over 300 mm wide	PC £50.65	0.90	7.12	54.52	m2	61.64
38 mm thick frame; 9 mm thick plywood panels over 300 mm wide	PC £56.64	0.95	7.51	60.96	m2	68.47
50 mm thick frame; 12 mm thick plywood panels; over 300 mm wide	PC £70.50	1.00	7.91	75.88	m2	83.79
Purpose made screen components; selected West African Mahogany; PC £470.00/m3						
Panelled partitions						
32 mm thick frame; 9 mm thick plywood panels; over 300 mm wide	PC £54.35	1.20	9.49	58.50	m2	67.99
38 mm thick frame; 9 mm thick plywood panels over 300 mm wide	PC £58.87	1.25	9.88	63.35	m2	73.23
50 mm thick frame; 12 mm thick plywood panels; over 300 mm wide	PC £74.11	1.35	10.67	79.76	m2	90.43
Panelled partitions; mouldings worked on solid both sides						
32 mm thick frame; 9 mm thick plywood panels; over 300 mm wide	PC £62.09	1.20	9.49	66.82	m2	76.31
38 mm thick frame; 9 mm thick plywood panels over 300 mm wide	PC £69.37	1.25	9.88	74.65	m2	84.53
50 mm thick frame; 12 mm thick plywood panels; over 300 mm wide	PC £76.97	1.35	10.67	82.83	m2	93.50
Panelled partitions; mouldings planted on both sides						
32 mm thick frame; 9 mm thick plywood panels; over 300 mm wide	PC £78.50	1.20	9.49	84.49	m2	93.98
38 mm thick frame; 9 mm thick plywood panels over 300 mm wide	PC £79.19	1.25	9.88	85.23	m2	95.11
50 mm thick frame; 12 mm thick plywood panels; over 300 mm wide	PC £86.23	1.35	10.67	92.81	m2	103.48
Panelled partitions; diminishing stiles over 300 mm wide; upper portion open panels for glass; in medium panes						
32 mm thick frame; 9 mm thick plywood panels; over 300 mm wide	PC £101.94	1.20	9.49	109.71	m2	119.20
38 mm thick frame; 9 mm thick plywood panels over 300 mm wide	PC £109.52	1.25	9.88	117.87	m2	127.75
50 mm thick frame; 12 mm thick plywood panels; over 300 mm wide	PC £116.24	1.35	10.67	125.11	m2	135.78

K LININGS/SHEATHING/DRY PARTITIONING Including overheads and profit at 5.00%		Labour hours	Labour £	Material £	Unit	Total rate £
K29 TIMBER FRAMED AND PANELLED PARTITIONS - cont'd						
Purpose made screen components; Afrormosia PC £710.00/m3						
Panelled partitions						
32 mm thick frame; 9 mm thick plywood panels; over 300 mm wide	PC £68.46	1.20	9.49	73.68	m2	**83.17**
38 mm thick frame; 9 mm thick plywood panels over 300 mm wide	PC £74.24	1.25	9.88	79.91	m2	**89.79**
50 mm thick frame; 12 mm thick plywood panels; over 300 mm wide	PC £93.57	1.35	10.67	100.71	m2	**111.38**
Panelled partitions; mouldings worked on solid both sides						
32 mm thick frame; 9 mm thick plywood panels; over 300 mm wide	PC £78.27	1.20	9.49	84.25	m2	**93.74**
38 mm thick frame; 9 mm thick plywood panels over 300 mm wide	PC £87.60	1.25	9.88	94.28	m2	**104.16**
50 mm thick frame; 12 mm thick plywood panels; over 300 mm wide	PC £96.98	1.35	10.67	104.37	m2	**115.04**
Panelled partitions; mouldings planted on both sides						
32 mm thick frame; 9 mm thick plywood panels; over 300 mm wide	PC £105.14	1.20	9.49	113.15	m2	**122.64**
38 mm thick frame; 9 mm thick plywood panels over 300 mm wide	PC £105.98	1.25	9.88	114.06	m2	**123.94**
50 mm thick frame; 12 mm thick plywood panels; over 300 mm wide	PC £108.79	1.35	10.67	117.08	m2	**127.75**
Panelled partitions; diminishing stiles over 300 mm wide; upper portion open panels for glass; in medium panes						
32 mm thick frame; 9 mm thick plywood panels; over 300 mm wide	PC £128.51	1.20	9.49	138.31	m2	**147.80**
38 mm thick frame; 9 mm thick plywood panels over 300 mm wide	PC £137.85	1.25	9.88	148.37	m2	**158.25**
50 mm thick frame; 12 mm thick plywood panels; over 300 mm wide	PC £146.74	1.35	10.67	157.93	m2	**168.60**
K31 PLASTERBOARD FIXED PARTITIONS/ INNER WALLS/ LININGS						
'Gyproc' laminated partition; comprising two skins of gypsum plasterboard bonded to a centre core of plasterboard square edge plank 19 mm thick; fixing with nails to softwood studwork (measured separately); joints filled with joint filler and joint tape; to receive direct decoration; (perimeter studwork measured separately)						
50 mm partition; two outer skins of 12.7 mm tapered edge wallboard						
height 2.10 - 2.40 m		2.15	25.04	15.82	m	**40.86**
height 2.40 - 2.70 m		2.40	27.95	17.86	m	**45.81**
height 2.70 - 3.00 m		2.70	31.44	19.93	m	**51.37**
height 3.00 - 3.30 m		3.00	34.93	22.00	m	**56.93**
height 3.30 - 3.60 m		3.40	39.59	24.08	m	**63.67**
65 mm partition; two outer skins of 19 mm tapered edge plank						
height 2.10 - 2.40 m		2.35	27.36	20.82	m	**48.18**
height 2.40 - 2.70 m		2.60	30.28	23.46	m	**53.74**
height 2.70 - 3.00 m		3.00	34.93	26.14	m	**61.07**
height 3.00 - 3.30 m		3.35	39.01	28.81	m	**67.82**
height 3.30 - 3.60 m		3.75	43.67	31.49	m	**75.16**

K LININGS/SHEATHING/DRY PARTITIONING Including overheads and profit at 5.00%	Labour hours	Labour £	Material £	Unit	Total rate £
Labours and associated additional wrought softwood studwork					
Floor, wall or ceiling battens					
25 x 38 mm	0.12	1.40	0.43	m	**1.83**
Forming openings					
25 x 38 mm framing	0.30	3.49	0.68	m	**4.17**
Fair ends	0.20	2.33	0.43	m	**2.76**
Angle	0.30	3.49	0.43	m	**3.92**
Intersection	0.20	2.33	-	m	**2.33**
Cutting and fitting around steel joists, angles, trunking, ducting, ventilators, pipes, tubes, etc					
over 2 m girth	0.09	1.05	-	m	**1.05**
not exceeding 0.30 m girth	0.05	0.58	-	nr	**0.58**
0.30 - 1 m girth	0.07	0.82	-	nr	**0.82**
1 - 2 m girth	0.11	1.28	-	nr	**1.28**
'Paramount' dry partition comprising Paramount panels with and including wrought softwood battens at vertical joints; fixing with nails to softwood studwork; (perimeter studwork measured separately)					
Square edge panels; joints filled with plaster and jute scrim cloth; to receive plaster (measured elsewhere);					
57 mm partition					
height 2.10 - 2.40 m	1.45	16.88	13.18	m	**30.06**
height 2.40 - 2.70 m	1.60	18.63	14.83	m	**33.46**
height 2.70 - 3.00 m	1.80	20.96	16.48	m	**37.44**
height 3.00 - 3.30 m	2.05	23.87	16.43	m	**40.30**
height 3.30 - 3.60 m	2.35	27.36	19.86	m	**47.22**
63 mm partition					
height 2.10 - 2.40 m	1.60	18.63	15.16	m	**33.79**
height 2.40 - 2.70 m	1.75	20.38	17.07	m	**37.45**
height 2.70 - 3.00 m	1.95	22.71	18.96	m	**41.67**
height 3.00 - 3.30 m	2.20	25.62	20.90	m	**46.52**
height 3.30 - 3.60 m	2.50	29.11	22.85	m	**51.96**
Tapered edge panels; joints filled with joint filler and joint tape to receive direct decoration;					
57 mm partition					
height 2.10 - 2.40 m	2.05	23.87	14.20	m	**38.07**
height 2.40 - 2.70 m	2.20	25.62	16.00	m	**41.62**
height 2.70 - 3.00 m	2.40	27.95	17.80	m	**45.75**
height 3.00 - 3.30 m	2.65	30.86	19.64	m	**50.50**
height 3.30 - 3.60 m	3.00	34.93	21.48	m	**56.41**
63 mm partition					
height 2.10 - 2.40 m	2.20	25.62	16.22	m	**41.84**
height 2.40 - 2.70 m	2.35	27.36	18.27	m	**45.63**
height 2.70 - 3.00 m	2.55	29.69	20.31	m	**50.00**
height 3.00 - 3.30 m	2.80	32.60	22.41	m	**55.01**
height 3.30 - 3.60 m	3.15	36.68	24.50	m	**61.18**
Labours and associated additional wrought softwood studwork on 57 mm partition					
Wall or ceiling battens					
19 x 37 mm	0.12	1.40	0.40	m	**1.80**
Sole plates; with 19 x 37 x 150 mm long battens spiked on at 600 mm centres					
19 x 57 mm	0.25	2.91	0.74	m	**3.65**
50 x 57 mm	0.30	3.49	1.17	m	**4.66**
Forming openings					
37 x 37 mm framing	0.30	3.49	0.40	m	**3.89**

K LININGS/SHEATHING/DRY PARTITIONING Including overheads and profit at 5.00%	Labour hours	Labour £	Material £	Unit	Total rate £
K31 PLASTERBOARD FIXED PARTITIONS/ INNER WALLS/ LININGS - cont'd					
Labours and associated additional wrought softwood studwork on 57 mm partition - cont'd					
Fair ends	0.20	2.33	0.40	m	2.73
Angle	0.30	3.49	0.85	m	4.34
Intersection	0.25	2.91	0.57	m	3.48
Cutting and fitting around steel joists, angles, trunking, ducting, ventilators, pipes, tubes, etc					
over 2 m girth	0.10	1.16	-	m	1.16
not exceeding 0.30 m girth	0.06	0.70	-	nr	0.70
0.30 - 1 m girth	0.08	0.93	-	nr	0.93
1 - 2 m girth	0.12	1.40	-	nr	1.40
Plugging; 300 mm centres; one way					
brickwork	0.10	1.16	0.03	m	1.19
concrete	0.18	2.10	0.03	m	2.13
'Gyroc' metal stud partition; comprising 146 mm metal stud frame; with floor channel plugged and screwed to concrete through 38 x 148 mm tanalised softwood sole plate					
Tapered edge panels; joints filled with joint filler and joint tape to receive direct decoration					
171 mm partition; one hour; one layer of 12.5 mm Fireline board each side					
height 2.10 - 2.40 m	4.20	48.91	19.93	m	68.84
height 2.40 - 2.70 m	4.85	56.48	22.00	m	78.48
height 2.70 - 3.00 m	5.40	62.88	24.21	m	87.09
height 3.00 - 3.30 m	6.25	72.78	26.52	m	99.30
height 3.30 - 3.60 m	6.85	79.76	28.36	m	108.12
height 3.60 - 3.90 m	8.20	95.48	31.01	m	126.49
height 3.90 - 4.20 m	8.80	102.47	33.22	m	135.69
Angles	0.20	2.33	1.15	m	3.48
T-junctions	0.20	2.33	1.08	m	3.41
Fair end	0.30	3.49	1.77	m	5.26
Tapered edge panels; joints filled with joint filler and joint tape to receive direct decoration					
196 mm partition; two hour; two layers of 12.5 mm Fireline board both sides					
height 2.10 - 2.40 m	6.00	69.87	31.32	m	101.19
height 2.40 - 2.70 m	6.75	78.60	34.84	m	113.44
height 2.70 - 3.00 m	7.50	87.33	38.45	m	125.78
height 3.00 - 3.30 m	8.75	101.89	41.96	m	143.85
height 3.30 - 3.60 m	9.60	111.79	45.46	m	157.25
height 3.60 - 3.90 m	11.70	136.24	49.53	m	185.77
height 3.90 - 4.20 m	12.60	146.72	53.17	m	199.89
Angles	0.20	2.33	1.15	m	3.48
T-junctions	0.20	2.33	1.08	m	3.41
Fair end	0.35	4.08	2.25	m	6.33

K LININGS/SHEATHING/DRY PARTITIONING Including overheads and profit at 5.00%	Labour hours	Labour £	Material £	Unit	Total rate £
K33 CONCRETE/TERRAZZO PARTITIONS					
Terrazzo faced partitions; cement and white Sicilian marble aggregate; polished					
Pre-cast reinforced terrazzo faced WC partitions					
38 mm thick; over 300 mm wide	6.75	95.68	9.24	m2	**104.92**
50 mm thick; over 300 mm wide	7.00	99.23	10.40	m2	**109.63**
Wall post; once rebated					
64 x 102 mm	3.75	53.16	4.04	m	**57.20**
64 x 152 mm	4.00	56.70	4.62	m	**61.32**
Centre post; twice rebated					
64 x 152 mm	4.75	67.33	5.20	m	**72.53**
64 x 203 mm	5.20	73.71	5.78	m	**79.49**
Lintel; once rebated					
64 x 102 mm	3.80	53.87	4.04	m	**57.91**
Brass topped plates or sockets cast into posts for fixings (measured elsewhere)	0.66	9.36	11.55	pr	**20.91**
K40 SUSPENDED CEILINGS					
Suspended ceilings, Gyproc M/F suspended ceiling system; hangers plugged and screwed to concrete soffite, 900 x 1800 x 12.7 mm tapered edge wallboard infill PC £1.80/m2; joints filled with joint filler and taped to receive direct decoration					
Lining to ceilings; hangers av. 400 mm long					
over 300 mm wide	1.10	12.81	6.60	m2	**19.41**
not exceeding 300 mm wide in isolated strips	0.55	6.40	2.34	m	**8.74**
300 - 600 mm wide in isolated strips	0.85	9.90	4.69	m	**14.59**
Vertical bulkhead; including additional hangers					
over 300 mm wide	1.40	16.30	8.80	m2	**25.10**
not exceeding 300 mm wide in isolated strips	0.70	8.15	2.93	m	**11.08**
300 - 600 mm wide in isolated strips	1.10	12.81	5.85	m	**18.66**
Suspended ceilings; 'Slimline' exposed suspended ceiling system; hangers plugged and screwed to concrete soffits, 600 x 600 x 19 mm Treetex 'Glacier' mineral tile infill; PC £8.29/m2					
Lining to ceilings; hangers av. 400 mm long					
over 300 mm wide	0.50	5.82	17.60	m2	**23.42**
not exceeding 300 mm wide in isolated strips	0.25	2.91	5.59	m	**8.50**
300 - 600 mm wide in isolated strips	0.40	4.66	11.18	m	**15.84**
Extra for					
suspension 800 mm high	0.05	0.58	0.08	m2	**0.66**
suspension 1200 mm high	0.07	0.82	0.11	m2	**0.93**
Extra for cutting and fitting around modular downlighting	0.20	2.33	-	nr	**2.33**
Vertical bulkhead; including additional hangers					
over 300 mm wide	0.60	6.99	22.10	m2	**29.09**
not exceeding 300 mm wide in isolated strips	0.30	3.49	7.01	m	**10.50**
300 - 600 mm wide in isolated strips	0.45	5.24	14.03	m	**19.27**
Edge detail 24 x 19 mm white finished angle edge trim	0.06	0.70	0.72	m	**1.42**
Cutting and fitting around pipes; not exceeding 0.30 m girth	0.10	1.16	-	nr	**1.16**

K LININGS/SHEATHING/DRY PARTITIONING Including overheads and profit at 5.00%	Labour hours	Labour £	Material £	Unit	Total rate £
K40 SUSPENDED CEILINGS - cont'd					
Suspended ceilings;'Z' demountable suspended ceiling system; hangers plugged and screwed to concrete soffite, 600 x 600 x 19 mm 'Echostop' glass reinforced fibrous plaster lightweight plain bevelled edge tiles; PC £10.43/m2					
Lining to ceilings; hangers av. 400 mm long					
over 300 mm wide	0.65	5.82	18.98	m2	24.80
not exceeding 300 mm wide in isolated strips	0.35	4.08	6.19	m	10.27
Suspended ceilings; concealed galvanised steel suspension system; hangers plugged and screwed to concrete soffite, Burgess white stove enamelled perforated mild steel tiles 600 x 600 mm; PC £11.21/m2					
Lining to ceilings; hangers av. 400 mm long					
over 300 mm wide	0.55	6.40	20.13	m2	26.53
not exceeding 300 mm wide in isolated strips	0.27	3.14	6.45	m	9.59
Suspended ceilings; concealed galvanised steel 'Trulok' suspension system; hangers plugged and screwed to concrete; Armstrong 'Travertone', 'Sanserra' or 'Highspire' 300 x 300 x 18 mm mineral ceiling tiles; PC £10.26/m2					
Linings to ceilings; hangers av. 700 mm long					
over 300 mm wide	0.65	7.57	18.03	m2	25.60
over 300 mm wide; 3.5 - 5 m high	0.85	9.90	17.96	m2	27.86
over 300 mm wide; in staircase areas or plant rooms	1.30	15.14	19.05	m2	34.19
not exceeding 300 mm wide in isolated strips	0.35	4.08	5.72	m	9.80
300 - 600 mm wide in isolated strips	0.70	6.27	11.43	m	17.70
Extra for cutting and fitting around modular downlighting	0.25	2.91	-	nr	2.91
Extra for					
suspension 800 mm high	0.05	0.58	0.08	m2	0.66
suspension 1200 mm high	0.07	0.82	0.11	m2	0.93
Vertical bulkhead; including additional hangers					
over 300 mm wide	0.80	9.32	22.89	m2	32.21
not exceeding 300 mm wide in isolated strips	0.40	4.66	7.23	m	11.89
300 - 600 mm wide in isolated strips	0.60	6.99	14.47	m	21.46
Shadow gap edge detail 24 x 19 mm white finished angle edge trim	0.09	1.05	0.86	m	1.91
Cutting and fitting around pipes; not exceeding 0.30 m girth	0.10	1.16	-	nr	1.16
Suspended ceilings; galvanised steel suspension system; hangers plugged and screwed to concrete soffite, 'Luxalon' stove enamelled aluminium linear panel ceiling, type 80B, complete with mineral insulation; PC £13.19/m2					
Linings to ceilings; hangers av. 700 mm long					
over 300 mm wide	0.65	7.57	22.20	m2	29.77
not exceeding 300 mm wide in isolated strips	0.33	3.84	7.10	m	10.94

L WINDOWS/DOORS/STAIRS
Including overheads and profit at 5.00%

ALTERNATIVE TIMBER WINDOW PRICES (£/each)

	£		£
Softwood shallow circular bay windows			
3 lightx1200 mm; ref CSB312CV	150.66	4 lightx1350 mm; ref CSB413CV	212.22
3 lightx1350 mm; ref CSB313CV	157.86	4 lightx1500 mm; ref CSB415T	240.54
3 light glass fibre roof unit	66.48	4 light glass fibre roof unit	71.64

		Labour hours	Labour £	Material £	Unit	Total rate £
L10 TIMBER WINDOWS/ROOFLIGHTS/SCREENS/LOUVRES						
Standard windows; 'treated' wrought softwood (refs. refer to Boulton & Paul cat. nos.)						
Side hung casement windows without glazing bars; with 140 mm wide softwood sills; opening casements and ventilators hung on rustproof hinges; fitted with aluminized laquered finish casement stays and fasteners; knotting and priming by manufacturer before delivery						
500 x 750 mm; ref N07V	PC £22.23	0.70	5.53	23.34	nr	**28.87**
500 x 900 mm; ref N09V	PC £26.01	0.80	6.33	27.31	nr	**33.64**
600 x 750 mm; ref 107V	PC £28.23	0.80	6.33	29.64	nr	**35.97**
600 x 750 mm; ref 107C	PC £25.80	0.80	6.33	27.09	nr	**33.42**
600 x 900 mm; ref 109V	PC £25.68	0.90	7.12	26.96	nr	**34.08**
600 x 900 mm; ref 109C	PC £30.72	0.80	6.33	32.26	nr	**38.59**
600 x 1050 mm; ref 110V	PC £30.06	0.80	6.33	31.56	nr	**37.89**
600 x 1050 mm; ref 110C	PC £32.10	1.00	7.91	33.71	nr	**41.62**
900 x 900 mm; ref 2N09W	PC £33.12	1.10	8.70	34.78	nr	**43.48**
900 x 1050 mm; ref 2N10W	PC £33.72	1.15	9.09	35.41	nr	**44.50**
900 x 1200 mm; ref 2N12W	PC £34.71	1.20	9.49	36.45	nr	**45.94**
900 x 1350 mm; ref 2N13W	PC £35.55	1.35	10.67	37.33	nr	**48.00**
900 x 1500 mm; ref 2N15W	PC £36.18	1.40	11.07	37.99	nr	**49.06**
1200 x 750 mm; ref 207C	PC £34.47	1.15	9.09	36.19	nr	**45.28**
1200 x 750 mm; ref 207CV	PC £45.09	1.15	9.09	47.34	nr	**56.43**
1200 x 900 mm; ref 209C	PC £36.42	1.20	9.49	38.24	nr	**47.73**
1200 x 900 mm; ref 209W	PC £38.43	1.20	9.49	40.35	nr	**49.84**
1200 x 900 mm; ref 209CV	PC £47.04	1.20	9.49	49.39	nr	**58.88**
1200 x 1050 mm; ref 210C	PC £37.86	1.35	10.67	39.75	nr	**50.42**
1200 x 1050 mm; ref 210W	PC £39.24	1.35	10.67	41.20	nr	**51.87**
1200 x 1050 mm; ref 210T	PC £47.04	1.35	10.67	49.39	nr	**60.06**
1200 x 1050 mm; ref 210CV	PC £48.63	1.35	10.67	51.06	nr	**61.73**
1200 x 1200 mm; ref 212C	PC £39.66	1.45	11.46	41.64	nr	**53.10**
1200 x 1200 mm; ref 212W	PC £40.14	1.45	11.46	42.15	nr	**53.61**
1200 x 1200 mm; ref 212T	PC £48.66	1.45	11.46	51.09	nr	**62.55**
1200 x 1200 mm; ref 212CV	PC £50.49	1.45	11.46	53.01	nr	**64.47**
1200 x 1350 mm; ref 213W	PC £41.10	1.55	12.26	43.16	nr	**55.42**
1200 x 1350 mm; ref 213CV	PC £53.34	1.55	12.26	56.01	nr	**68.27**
1200 x 1500 mm; ref 215W	PC £42.24	1.70	13.44	44.35	nr	**57.79**
1770 x 750 mm; ref 307CC	PC £54.54	1.40	11.07	57.27	nr	**68.34**
1770 x 900 mm; ref 309CC	PC £57.36	1.70	13.44	60.23	nr	**73.67**
1770 x 1050 mm; ref 310C	PC £45.87	1.75	13.84	48.16	nr	**62.00**
1770 x 1050 mm; ref 310T	PC £56.16	1.70	13.44	58.97	nr	**72.41**
1770 x 1050 mm; ref 310CC	PC £59.85	1.40	11.07	62.84	nr	**73.91**
1770 x 1050 mm; ref 310WW	PC £66.15	1.40	11.07	69.46	nr	**80.53**
1770 x 1200 mm; ref 312C	PC £47.43	1.80	14.23	49.80	nr	**64.03**

L WINDOWS/DOORS/STAIRS Including overheads and profit at 5.00%		Labour hours	Labour £	Material £	Unit	Total rate £
L10 TIMBER WINDOWS/ROOFLIGHTS/SCREENS/LOUVRES - cont'd						
Standard windows; 'treated' wrought softwood (refs. refer to Boulton & Paul cat. nos.) - cont'd						
Side hung casement windows without glazing bars; with 140 mm wide softwood sills; opening casements and ventilators hung on rustproof hinges; fitted with aluminized laquered finish casement stays and fasteners; knotting and priming by manufacturer before delivery						
1770 x 1200 mm; ref 312T	PC £57.93	1.80	14.23	60.83	nr	**75.06**
1770 x 1200 mm; ref 312CC	PC £62.37	1.80	14.23	65.49	nr	**79.72**
1770 x 1200 mm; ref 312WW	PC £68.01	1.80	14.23	71.41	nr	**85.64**
1770 x 1200 mm; ref 312CVC	PC £73.20	1.80	14.23	76.86	nr	**91.09**
1770 x 1350 mm; ref 313CC	PC £67.11	1.90	15.02	70.47	nr	**85.49**
1770 x 1350 mm; ref 312CC	PC £69.75	1.90	15.02	73.24	nr	**88.26**
1770 x 1350 mm; ref 313CVC	PC £77.97	1.90	15.02	81.87	nr	**96.89**
1770 x 1500 mm; ref 315T	PC £61.08	2.00	15.81	64.13	nr	**79.94**
2340 x 1050 mm; ref 410CWC	PC £83.88	1.95	15.42	88.07	nr	**103.49**
2340 x 1200 mm; ref 412CWC	PC £86.82	2.05	16.21	91.16	nr	**107.37**
2340 x 1350 mm; ref 413CWC	PC £92.01	2.20	17.39	96.61	nr	**114.00**
Top hung casement windows; with 140 mm wide softwood sills; opening casements and ventilators hung on rustproof hinges; fitted with aluminized laquered finish casement stays; knotting and priming by manufacturer before delivery						
600 x 750 mm; ref 107A	PC £28.20	0.80	6.33	29.61	nr	**35.94**
600 x 900 mm; ref 109A	PC £29.34	0.90	7.12	30.81	nr	**37.93**
600 x 1050 mm; ref 110A	PC £30.81	1.00	7.91	32.35	nr	**40.26**
900 x 750 mm; ref 2N07A	PC £35.19	1.05	8.30	36.95	nr	**45.25**
900 x 900 mm; ref 2N09A	PC £38.25	1.10	8.70	40.16	nr	**48.86**
900 x 1050 mm; ref 2N10A	PC £39.90	1.15	9.09	41.90	nr	**50.99**
900 x 1350 mm; ref 2N13AS	PC £46.17	1.35	10.67	48.48	nr	**59.15**
900 x 1500 mm; ref 2N15AS	PC £47.67	1.40	11.07	50.05	nr	**61.12**
1200 x 750 mm; ref 207A	PC £40.47	1.15	9.09	42.49	nr	**51.58**
1200 x 900 mm; ref 209A	PC £43.26	1.20	9.49	45.42	nr	**54.91**
1200 x 1050 mm; ref 210A	PC £44.91	1.35	10.67	47.16	nr	**57.83**
1200 x 1050 mm; ref 210AT	PC £60.21	1.35	10.67	63.22	nr	**73.89**
1200 x 1200 mm; ref 212A	PC £46.47	1.45	11.46	48.79	nr	**60.25**
1200 x 1200 mm; ref 212AT	PC £63.54	1.45	11.46	66.72	nr	**78.18**
1200 x 1350 mm; ref 213AS	PC £51.48	1.55	12.26	54.05	nr	**66.31**
1200 x 1500 mm; ref 215AS	PC £53.07	1.70	13.44	55.72	nr	**69.16**
1770 x 1050 mm; ref 310A	PC £54.27	1.70	13.44	56.98	nr	**70.42**
1770 x 1050 mm; ref 310AV	PC £66.57	1.70	13.44	69.90	nr	**83.34**
1770 x 1200 mm; ref 312A	PC £56.13	1.80	14.23	58.94	nr	**73.17**
1770 x 1220 mm; ref 312AV	PC £68.46	1.80	14.23	71.88	nr	**86.11**
2340 x 1200 mm; ref 412A	PC £63.06	2.05	16.21	66.21	nr	**82.42**
2340 x 1200 mm; ref 412AW	PC £79.83	2.05	16.21	83.82	nr	**100.03**
2340 x 1500 mm; ref 415AWS	PC £84.99	2.35	18.58	89.24	nr	**107.82**
High performance top hung reversible windows; with 140 mm wide softwood sills; adjustable ventilators weather stripping; opening sashes and fanlights hung on rustproof hinges; fitted with aluminized laquered espagnolette bolts; knotting and priming by manufacturer before delivery						
600 x 900 mm; ref R0609	PC £70.86	0.90	7.12	74.40	nr	**81.52**
900 x 900 mm; ref R0909	PC £77.49	1.10	8.70	81.36	nr	**90.06**
900 x 1050 mm; ref R0910	PC £79.02	1.15	9.09	82.97	nr	**92.06**
900 x 1200 mm; ref R0912	PC £81.21	1.25	9.88	85.27	nr	**95.15**
900 x 1500 mm; ref R0915	PC £92.01	1.40	11.07	96.61	nr	**107.68**

L WINDOWS/DOORS/STAIRS Including overheads and profit at 5.00%		Labour hours	Labour £	Material £	Unit	Total rate £
1200 x 1050 mm; ref R1210	PC £86.19	1.35	10.67	90.50	nr	**101.17**
1200 x 1200 mm; ref R1212	PC £88.50	1.45	11.46	92.93	nr	**104.39**
1200 x 1500 mm; ref R1215	PC £100.29	1.70	13.44	105.30	nr	**118.74**
1500 x 1200 mm; ref R1512	PC £99.33	1.70	13.44	104.30	nr	**117.74**
1800 x 1050 mm; ref R1810	PC £103.74	1.70	13.44	108.93	nr	**122.37**
1800 x 1200 mm; ref R1812	PC £105.60	1.80	14.23	110.88	nr	**125.11**
1800 x 1500 mm; ref R1815	PC £113.37	1.90	15.02	119.04	nr	**134.06**
2400 x 1500 mm; ref R2415	PC £125.79	2.35	18.58	132.08	nr	**150.66**
High performance double hung sash windows with glazing bars; solid frames; 63 x 175 mm softwood sills; standard flush external linings; spiral spring balances and sash catch; knotting and priming by manufacturer before delivery						
635 x 1050 mm; ref DH0610B	PC £101.46	2.00	15.81	106.53	nr	**122.34**
635 x 1350 mm; ref DH0613B	PC £110.37	2.20	17.39	115.89	nr	**133.28**
635 x 1650 mm; ref DH0616B	PC £118.29	2.45	19.37	124.20	nr	**143.57**
860 x 1050 mm; ref DH0810B	PC £111.21	2.30	18.18	116.77	nr	**134.95**
860 x 1350 mm; ref DH0813B	PC £120.81	2.60	20.56	126.85	nr	**147.41**
860 x 1650 mm; ref DH0816B	PC £130.41	3.00	23.72	136.93	nr	**160.65**
1085 x 1050 mm; ref DH1010B	PC £121.44	2.60	20.56	127.51	nr	**148.07**
1085 x 1350 mm; ref DH1013B	PC £133.74	3.00	23.72	140.43	nr	**164.15**
1085 x 1650 mm; ref DH1016B	PC £144.96	3.70	29.25	152.21	nr	**181.46**
1699 x 1050 mm; ref DH1710B	PC £221.37	3.70	29.25	232.44	nr	**261.69**
1699 x 1350 mm; ref DH1713B	PC £242.25	4.60	36.37	254.36	nr	**290.73**
1699 x 1650 mm; ref DH1716B	PC £260.76	4.70	37.16	273.80	nr	**310.96**
Purpose made window casements; 'treated' wrought softwood						
Casements; rebated; moulded						
38 mm thick	PC £17.52	-	-	18.40	m2	**18.40**
50 mm thick	PC £20.81	-	-	21.85	m2	**21.85**
Casements; rebated; moulded; in medium panes						
38 mm thick	PC £25.59	-	-	26.86	m2	**26.86**
50 mm thick	PC £31.65	-	-	33.24	m2	**33.24**
Casements; rebated; moulded; with semi-circular head						
38 mm thick	PC £31.97	-	-	33.57	m2	**33.57**
50 mm thick	PC £39.59	-	-	41.57	m2	**41.57**
Casements; rebated; moulded; to bullseye window						
38 mm thick x 600 mm dia	PC £57.48/nr	-	-	60.35	nr	**60.35**
38 mm thick x 900 mm dia	PC £85.27/nr	-	-	89.53	nr	**89.53**
50 mm thick x 600 mm dia	PC £66.12/nr	-	-	69.42	nr	**69.42**
50 mm thick x 900 mm dia	PC £98.03/nr	-	-	102.93	nr	**102.93**
Fitting and hanging casements						
square or rectangular		0.50	3.95	-	nr	**3.95**
semicircular		1.25	9.88	-	nr	**9.88**
bullseye		2.00	15.81	-	nr	**15.81**
Purpose made window frames; 'treated' wrought softwood						
Frames; rounded; rebated check grooved						
25 x 120 mm		0.14	1.11	5.86	m	**6.97**
50 x 75 mm		0.14	1.11	6.64	m	**7.75**
50 x 100 mm		0.16	1.27	8.18	m	**9.45**
50 x 125 mm		0.16	1.27	9.95	m	**11.22**
63 x 100 mm		0.16	1.27	10.18	m	**11.45**
75 x 150 mm		0.18	1.42	17.14	m	**18.56**
90 x 140 mm		0.18	1.42	22.19	m	**23.61**

L WINDOWS/DOORS/STAIRS Including overheads and profit at 5.00%		Labour hours	Labour £	Material £	Unit	Total rate £
L10 TIMBER WINDOWS/ROOFLIGHTS/SCREENS/LOUVRES - cont'd						
Purpose made window frames; 'treated' wrought softwood - cont'd						
Mullions and transoms; twice rounded, rebated and check grooved						
50 x 75 mm		0.10	0.79	8.06	m	**8.85**
50 x 100 mm		0.12	0.95	9.57	m	**10.52**
63 x 100 mm		0.12	0.95	11.60	m	**12.55**
75 x 150 mm		0.14	1.11	18.55	m	**19.66**
Sill; sunk weathered, rebated and grooved						
75 x 100 mm		0.20	1.58	16.77	m	**18.35**
75 x 150 mm		0.20	1.58	22.10	m	**23.68**
Add +5% to the above 'Material £ prices' for 'selected' softwood for staining						
Purpose made double hung sash windows; 'treated' wrought softwood						
Cased frames of 100 x 25 mm grooved inner linings; 114 x 25 mm grooved outer linings; 125 x 38 mm twice rebated head linings; 125 x 32 mm twice rebated grooved pulley stiles; 150 x 13 mm linings; 50 x 19 mm parting slips; 25 x 13 mm parting beads; 25 x 19 mm inside beads; 150 x 75 mm Oak twice sunk weathered throated sill; 50 mm thick rebated and moulded sashes; moulded horns; over 1.25 m2 each; both sashes in medium panes; including spiral spring balances	PC £129.04	2.25	17.79	194.08	m2	**211.87**
As above but with cased mullions	PC £136.16	2.50	19.77	201.56	m2	**221.33**
'Velux' pre-glazed roof windows; 'treated' Nordic Red Pine and aluminium trimmed 'Velux' windows, or equivalent; including type U flashings and soakers (for tiles and pantiles), and sealed double glazing unit; trimming opening measured separately						
550 x 700 mm; ref GGL-9	PC £90.45	9.00	71.16	101.26	nr	**172.42**
550 x 980 mm; ref GGL-6	PC £101.65	9.00	71.16	113.04	nr	**184.20**
700 x 1180 mm; ref GGL-5	PC £120.70	10.00	79.07	133.03	nr	**212.10**
780 x 980 mm; ref GGL-1	PC £116.20	9.50	75.11	128.30	nr	**203.41**
780 x 1400 mm; ref GGL-2	PC £136.81	10.50	83.02	149.95	nr	**232.97**
940 x 1600 mm; ref GGL-3	PC £161.08	10.50	83.02	175.44	nr	**258.46**
1140 x 1180 mm; ref GGL-4	PC £153.87	11.00	86.97	167.87	nr	**254.84**
1340 x 980 mm; ref GGL-7	PC £156.98	11.00	86.97	171.13	nr	**258.10**
1340 x 1400 mm; ref GGL-8	PC £181.61	11.50	90.92	196.99	nr	**287.91**
Standard windows; selected Philippine Mahogany; preservative stain finish						
Side hung casement windows; with 45 x 140 mm hardwood sills; weather stripping; opening sashes on canopy hinges; fitted with fasteners; aluminized lacquered finish ironmongery						
600 x 600 mm; ref SS0606/L	PC £54.30	0.95	7.51	57.02	nr	**64.53**
600 x 900 mm; ref SS0609/L	PC £60.18	1.20	9.49	63.19	nr	**72.68**
600 x 900 mm; ref SV0609/O	PC £81.90	0.95	7.51	86.00	nr	**93.51**
600 x 1050 mm; ref SS0610/L	PC £74.34	1.30	10.28	78.06	nr	**88.34**
600 x 1050 mm; ref SV0610/O	PC £84.35	1.30	10.28	88.57	nr	**98.85**
900 x 900 mm; ref SV0909/O	PC £109.13	1.50	11.86	114.59	nr	**126.45**
900 x 1050 mm; ref SV0910/O	PC £109.90	1.60	12.65	115.40	nr	**128.05**
900 x 1200 mm; ref SV0912/O	PC £112.42	1.70	13.44	118.04	nr	**131.48**
900 x 1350 mm; ref SV0913/O	PC £114.66	1.80	14.23	120.39	nr	**134.62**
900 x 1500 mm; ref SV0915/O	PC £117.04	1.90	15.02	122.89	nr	**137.91**

L WINDOWS/DOORS/STAIRS Including overheads and profit at 5.00%		Labour hours	Labour £	Material £	Unit	Total rate £
1200 x 900 mm; ref SS1209/L	PC £114.94	1.70	13.44	120.69	nr	134.13
1200 x 900 mm; ref SV1209/O	PC £125.30	1.70	13.44	131.57	nr	145.01
1200 x 1050 mm; ref SS1210/L	PC £104.28	1.80	14.23	109.49	nr	123.72
1200 x 1050 mm; ref SV1210/O	PC £127.75	1.80	14.23	134.14	nr	148.37
1200 x 1200 mm; ref SS1212/L	PC £109.86	1.95	15.42	115.35	nr	130.77
1200 x 1200 mm; ref SV1212/O	PC £130.06	1.95	15.42	136.56	nr	151.98
1200 x 1350 mm; ref SV1213/O	PC £132.16	2.10	16.60	138.77	nr	155.37
1200 x 1500 mm; ref SV1215/O	PC £134.75	2.20	17.39	141.49	nr	158.88
1800 x 900 mm; ref SS189D/O	PC £123.84	2.25	17.79	130.03	nr	147.82
1800 x 1050 mm; ref SS1810/L	PC £123.84	2.25	17.79	130.03	nr	147.82
1800 x 1050 mm; ref SS180D/O	PC £188.65	2.25	17.79	198.08	nr	215.87
1800 x 1200 mm; ref SS1812/L	PC £151.13	2.40	18.98	158.69	nr	177.67
1800 x 1200mm; ref SS182D/O	PC £180.42	2.40	18.98	189.44	nr	208.42
2400 x 1200 mm; ref SS242D/O	PC £200.04	2.80	22.14	210.04	nr	232.18
Top hung casement windows; with 45 x 140 mm hardwood sills; weather stripping; opening sashes on canopy hinges; fitted with fasteners; aluminized lacquered finish ironmongery						
600 x 900 mm; ref ST0609/O	PC £71.68	0.95	7.51	75.26	nr	82.77
600 x 1050 mm; ref ST0610/O	PC £77.14	1.30	10.28	81.00	nr	91.28
900 x 900 mm; ref ST0909/O	PC £91.42	1.50	11.86	95.99	nr	107.85
900 x 1050 mm; ref ST0910/O	PC £94.50	1.60	12.65	99.23	nr	111.88
900 x 1200 mm; ref ST0912/O	PC £104.44	1.70	13.44	109.66	nr	123.10
900 x 1350 mm; ref ST0913/O	PC £110.18	1.80	14.23	115.69	nr	129.92
900 x 1500 mm; ref ST0915/O	PC £119.21	1.90	15.02	125.17	nr	140.19
1200 x 900 mm; ref ST1209/O	PC £106.40	1.90	15.02	111.72	nr	126.74
1200 x 1050 mm; ref ST1210/O	PC £99.12	1.80	14.23	104.08	nr	118.31
1200 x 1200 mm; ref ST1212/O	PC £107.70	1.95	15.42	113.09	nr	128.51
1200 x 1350 mm; ref ST1213/O	PC £133.21	2.10	16.60	139.87	nr	156.47
1200 x 1500 mm; ref ST1215/O	PC £115.50	2.15	17.00	121.28	nr	138.28
1500 x 1050 mm; ref ST1510/L	PC £141.61	2.10	16.60	148.69	nr	165.29
1500 x 1200 mm; ref ST1512/L	PC £154.21	2.20	17.39	161.92	nr	179.31
1500 x 1350 mm; ref ST1513/L	PC £163.45	2.30	18.18	171.62	nr	189.80
1500 x 1500 mm; ref ST1515/L	PC £175.00	2.50	19.77	183.75	nr	203.52
1800 x 1050 mm; ref ST1810/L	PC £131.34	2.25	17.79	137.91	nr	155.70
1800 x 1200 mm; ref ST1812/L	PC £142.08	2.40	18.98	149.18	nr	168.16
1800 x 1350 mm; ref ST1813/L	PC £149.88	2.50	19.77	157.37	nr	177.14
1800 x 1500 mm; ref ST1815/L	PC £186.55	2.70	21.35	195.88	nr	217.23
Purpose made window casements; selected West African Mahogany; PC £470.00/m3						
Casements; rebated; moulded						
38 mm thick	PC £25.62	-	-	26.90	m2	26.90
50 mm thick	PC £33.39	-	-	35.06	m2	35.06
Casements; rebated; moulded; in medium panes						
38 mm thick	PC £31.09	-	-	32.64	m2	32.64
50 mm thick	PC £41.61	-	-	43.69	m2	43.69
Casements; rebated; moulded; with semi-circular head						
38 mm thick	PC £38.87	-	-	40.81	m2	40.81
50 mm thick	PC £52.02	-	-	54.62	m2	54.62
Casements; rebated; moulded; to bullseye window						
38 mm thick x 600 mm dia	PC £84.12/nr	-	-	88.33	nr	88.33
38 mm thick x 900 mm dia	PC £125.89/nr	-	-	132.19	nr	132.19
50 mm thick x 600 mm dia	PC £96.88/nr	-	-	101.73	nr	101.73
50 mm thick x 900 mm dia	PC £144.24/nr	-	-	151.46	nr	151.46
Fitting and hanging casements						
square or rectangular		0.70	5.53	-	nr	5.53
semicircular		1.70	13.44	-	nr	13.44
bullseye		2.70	21.35	-	nr	21.35

L WINDOWS/DOORS/STAIRS Including overheads and profit at 5.00%		Labour hours	Labour £	Material £	Unit	Total rate £
L10 TIMBER WINDOWS/ROOFLIGHTS/SCREENS/LOUVRES - cont'd						
Purpose made window frames; selected West African Mahogany; PC £470.00/m3						
Frames; rounded; rebated check grooved						
25 x 120 mm	PC £7.91	0.18	1.42	8.51	m	**9.93**
50 x 75 mm	PC £8.96	0.18	1.42	9.64	m	**11.06**
50 x 100 mm	PC £11.02	0.21	1.66	11.86	m	**13.52**
50 x 125 mm	PC £13.39	0.21	1.66	14.41	m	**16.07**
63 x 100 mm	PC £13.72	0.21	1.66	14.76	m	**16.42**
75 x 150 mm	PC £23.08	0.24	1.90	24.84	m	**26.74**
90 x 140 mm	PC £29.84	0.24	1.90	32.12	m	**34.02**
Mullions and transoms; twice rounded, rebated and check grooved						
50 x 75 mm	PC £10.87	0.14	1.11	11.70	m	**12.81**
50 x 100 mm	PC £12.94	0.16	1.27	13.92	m	**15.19**
63 x 100 mm	PC £15.63	0.16	1.27	16.82	m	**18.09**
75 x 150 mm	PC £24.98	0.18	1.42	26.88	m	**28.30**
Sill; sunk weathered, rebated and grooved						
75 x 100 mm	PC £22.65	0.27	2.13	24.38	m	**26.51**
75 x 150 mm	PC £29.85	0.27	2.13	32.13	m	**34.26**
Purpose made double hung sash windows; selected West African Mahogany; PC £470.00/m3						
Cased frames of 100 x 25 mm grooved inner linings; 114 x 25 mm grooved outer linings; 125 x 38 mm twice rebated head linings; 125 x 32 mm twice rebated grooved pulley stiles; 150 x 13 mm linings; 50 x 19 mm parting slips; 25 x 13 mm parting beads; 25 x 19 mm inside beads; 150 x 75 mm Oak twice sunk weathered throated sill; 50 mm thick rebated and moulded sashes; moulded horns; over 1.25 m2 each; both sashes in medium panes; including spiral spring balances	PC £176.47	3.00	23.72	243.89	m2	**267.61**
As above but with cased mullions	PC £186.17	3.33	26.33	254.06	m2	**280.39**
Purpose made window casements; Afrormosia; PC £710.00/m3						
Casements; rebated; moulded						
38 mm thick	PC £31.51	-	-	33.08	m2	**33.08**
50 mm thick	PC £38.71	-	-	40.65	m2	**40.65**
Casements; rebated; moulded; in medium panes						
38 mm thick	PC £40.84	-	-	42.88	m2	**42.88**
50 mm thick	PC £49.07	-	-	51.52	m2	**51.52**
Casements; rebated; moulded; with semi-circular head						
38 mm thick	PC £51.06	-	-	53.61	m2	**53.61**
50 mm thick	PC £61.36	-	-	64.42	m2	**64.42**
Casements; rebated; moulded; to bullseye window						
38 mm thick x 600 mm dia	PC £103.50/nr	-	-	108.67	nr	**108.67**
38 mm thick x 900 mm dia	PC £154.64/nr	-	-	162.37	nr	**162.37**
50 mm thick x 600 mm dia	PC £119.03/nr	-	-	124.98	nr	**124.98**
50 mm thick x 900 mm dia	PC £177.98/nr	-	-	186.88	nr	**186.88**
Fitting and hanging casements						
square or rectangular		0.70	5.53	-	nr	**5.53**
semicircular		1.70	13.44	-	nr	**13.44**
bullseye		2.70	21.35	-	nr	**21.35**

L WINDOWS/DOORS/STAIRS Including overheads and profit at 5.00%		Labour hours	Labour £	Material £	Unit	Total rate £
Purpose made window frames; Afrormosia; PC £710.00/m3						
Frames; rounded; rebated check grooved						
25 x 120 mm	PC £9.23	0.18	1.42	9.93	m	**11.35**
50 x 75 mm	PC £10.55	0.18	1.42	11.35	m	**12.77**
50 x 100 mm	PC £13.12	0.21	1.66	14.12	m	**15.78**
50 x 125 mm	PC £16.06	0.21	1.66	17.28	m	**18.94**
63 x 100 mm	PC £16.37	0.21	1.66	17.62	m	**19.28**
75 x 150 mm	PC £27.79	0.24	1.90	29.90	m	**31.80**
90 x 140 mm	PC £36.14	0.24	1.90	38.89	m	**40.79**
Mullions and transoms; twice rounded, rebated and check grooved						
50 x 75 mm	PC £12.44	0.14	1.11	13.39	m	**14.50**
50 x 100 mm	PC £15.02	0.16	1.27	16.17	m	**17.44**
63 x 100 mm	PC £18.27	0.16	1.27	19.67	m	**20.94**
75 x 150 mm	PC £29.69	0.18	1.42	31.95	m	**33.37**
Sill; sunk weathered, rebated and grooved						
75 x 100 mm	PC £25.80	0.27	2.13	27.76	m	**29.89**
75 x 150 mm	PC £34.60	0.27	2.13	37.25	m	**39.38**
Purpose made double hung sash windows; Afrormosia PC £710.00/m3						
Cased frames of 100 x 25 mm grooved inner linings; 114 x 25 mm grooved outer linings; 125 x 38 mm twice rebated head linings; 125 x 32 mm twice rebated grooved pulley stiles; 150 x 13 mm linings; 50 x 19 mm parting slips; 25 x 13 mm parting beads; 25 x 19 mm inside beads; 150 x 75 mm Oak twice sunk weathered throated sill; 50 mm thick rebated and moulded sashes; moulded horns; over 1.25 m2 each; both sashes in medium panes; including spiral spring balances	PC £199.99	3.00	23.72	243.89	m2	**267.61**
As above but with cased mullions	PC £210.98	3.33	26.33	254.06	m2	**280.39**
L11 METAL WINDOWS/ROOFLIGHTS/SCREENS/LOUVRES						
Aluminium fixed and fanlight windows; Crittall 'Luminaire' stock sliders or similar; white acrylic finish; fixed in position; including lugs plugged and screwed to brickwork or blockwork; or screwed to wooden sub-frame (measured elsewhere)						
Fixed lights; factory glazed with 3, 4 or 5 mm OQ clear float glass						
600 x 900 mm; ref 6FL9A	PC £39.75	2.20	19.70	42.43	nr	**62.13**
900 x 1500 mm; ref 9FL15A	PC £56.25	3.15	28.21	59.76	nr	**87.97**
1200 x 900 mm; ref 12FL9A	PC £60.75	3.15	28.21	64.48	nr	**92.69**
1200 x 1200 mm; ref 12FL12A	PC £62.25	3.15	28.21	66.06	nr	**94.27**
1500 x 1200 mm; ref 15FL12A	PC £69.00	3.15	28.21	73.14	nr	**101.35**
1500 x 1500 mm; ref 15FL15A	PC £140.25	4.00	35.83	147.96	nr	**183.79**
Fixed lights; site double glazed with 11 mm clear float glass						
600 x 900 mm; ref 6FL9A	PC £53.25	3.00	26.87	56.61	nr	**83.48**
900 x 1500 mm; ref 9FL15A	PC £73.50	4.75	42.54	77.87	nr	**120.41**
1200 x 900 mm; ref 12FL9A	PC £85.50	5.25	47.02	90.47	nr	**137.49**
1200 x 1200 mm; ref 12FL12A	PC £84.75	5.45	48.81	89.68	nr	**138.49**
Fixed lights with fanlight over; factory glazed with 3, 4 or 5 mm OQ clear float glass						
600 x 900 mm; ref 6FV9A	PC £105.75	2.20	19.70	111.73	nr	**131.43**
600 x 1500 mm; ref 6FV15A	PC £117.00	2.20	19.70	123.54	nr	**143.24**
900 x 1200 mm; ref 9FV12A	PC £129.00	3.15	28.21	136.14	nr	**164.35**
900 x 1500 mm; ref 9FV15A	PC £138.00	3.15	28.21	145.59	nr	**173.80**

L WINDOWS/DOORS/STAIRS Including overheads and profit at 5.00%		Labour hours	Labour £	Material £	Unit	Total rate £
L11 METAL WINDOWS/ROOFLIGHTS/SCREENS/LOUVRES - cont'd						
Aluminium fixed and fanlight windows; Crittall 'Luminaire' stock sliders or similar; white acrylic finish; fixed in position; including lugs plugged and screwed to brickwork or blockwork; or screwed to wooden sub-frame (measured elsewhere) - cont'd						
Fixed lights with fanlight over; site double glazed with 11 mm clear float glass						
600 x 900 mm; ref 6FV9A	PC £126.75	3.00	26.87	133.78	nr	**160.65**
600 x 1500 mm; ref 6FV15A	PC £151.50	3.55	31.80	159.77	nr	**191.57**
900 x 1200 mm; ref 9FV12A	PC £167.25	4.33	38.78	176.31	nr	**215.09**
900 x 1500 mm; ref 9FV15A	PC £180.75	5.25	47.02	190.48	nr	**237.50**
Vertical slider; factory glazed with 3; 4 or 5 mm OQ clear float glass						
600 x 900 mm; ref 6VV9A	PC £90.75	2.70	24.18	95.98	nr	**120.16**
900 x 1100 mm; ref 9VV11A	PC £107.25	2.70	24.18	113.31	nr	**137.49**
1200 x 1500 mm; ref 12VV15A	PC £130.50	3.75	33.59	137.72	nr	**171.31**
Vertical slider; factory double glazed with 11 mm clear float glass						
600 x 900 mm; ref 6VV9A	PC £147.75	3.40	30.45	155.83	nr	**186.28**
900 x 1100 mm; ref 9VV11A	PC £174.00	3.40	30.45	183.39	nr	**213.84**
1200 x 1500 mm; ref 12VV15A	PC £243.75	4.75	42.54	256.63	nr	**299.17**
Horizontal slider; factory glazed with 3, 4 or 5 mm OQ clear float glass						
1200 x 900 mm; ref 12HH9A	PC £116.25	3.75	33.59	122.76	nr	**156.35**
1200 x 1200 mm; ref 12HH12A	PC £126.75	3.75	33.59	133.78	nr	**167.37**
1500 x 900 mm; ref 15HH9A	PC £126.75	3.75	33.59	133.78	nr	**167.37**
1500 x 1300 mm; ref 15HH13A	PC £145.50	3.75	33.59	153.47	nr	**187.06**
1800 x 1200 mm; ref 18HH12A	PC £153.75	5.25	47.02	162.13	nr	**209.15**
Horizontal slider; factory double glazed with 11 mm clear float glass						
1200 x 900 mm; ref 12HH9A	PC £158.25	4.75	42.54	166.86	nr	**209.40**
1200 x 1200 mm; ref 12HH12A	PC £174.00	4.75	42.54	183.39	nr	**225.93**
1500 x 900 mm; ref 15HH9A	PC £173.25	4.75	42.54	182.61	nr	**225.15**
1500 x 1300 mm; ref 15HH13A	PC £204.75	4.75	42.54	215.68	nr	**258.22**
1800 x 1200 mm; ref 18HH12A	PC £216.75	5.75	51.50	228.28	nr	**279.78**
Galvanized steel fixed light; casement and fanlight windows; Crittall; 'Homelight' range or similar; site glazing measured elsewhere; fixed in position; including lugs cut and pinned to brickwork or blockwork						
Basic fixed lights; including easy-glaze beads						
628 x 292 mm; ref ZNG5	PC £10.31	1.20	10.75	11.12	nr	**21.87**
628 x 923 mm; ref ZNC5	PC £14.70	1.20	10.75	15.72	nr	**26.47**
628 x 1513 mm; ref ZNDV5	PC £20.91	1.20	10.75	22.24	nr	**32.99**
1237 x 292 mm; ref ZNG13	PC £15.75	1.20	10.75	16.83	nr	**27.58**
1237 x 923 mm; ref ZNC13	PC £20.38	1.75	15.67	21.69	nr	**37.36**
1237 x 1218 mm; ref ZND13	PC £23.99	1.75	15.67	25.48	nr	**41.15**
1237 x 1513 mm; ref ZNDV13	PC £26.28	1.75	15.67	27.88	nr	**43.55**
1846 x 292 mm; ref ZNG14	PC £20.42	1.20	10.75	21.72	nr	**32.47**
1846 x 923 mm; ref ZNC14	PC £25.16	1.75	15.67	26.71	nr	**42.38**
1846 x 1513 mm; ref ZNDV14	PC £30.48	2.20	19.70	32.29	nr	**51.99**
Basic opening lights; including easy-glaze beads and weatherstripping						
628 x 292 mm; ref ZNG1	PC £23.42	1.20	10.75	24.88	nr	**35.63**
1237 x 292 mm; ref ZNG13G	PC £34.76	1.20	10.75	36.79	nr	**47.54**
1846 x 292 mm; ref ZNG4	PC £59.45	1.20	10.75	62.71	nr	**73.46**

L WINDOWS/DOORS/STAIRS Including overheads and profit at 5.00%		Labour hours	Labour £	Material £	Unit	Total rate £
One-piece composites; including easy-glaze beads and weatherstripping						
628 x 923 mm; ref ZNC5F	PC £32.75	1.20	10.75	34.68	nr	**45.43**
628 x 1513 mm; ref ZNDV5F	PC £39.20	1.20	10.75	41.44	nr	**52.19**
1237 x 923 mm; ref ZNC2F	PC £64.64	1.75	15.67	68.16	nr	**83.83**
1237 x 1218 mm; ref ZND2F	PC £76.05	1.75	15.67	80.14	nr	**95.81**
1237 x 1513 mm; ref ZNDV2V	PC £91.29	1.75	15.67	96.14	nr	**111.81**
1846 x 923 mm; ref NC4F	PC £99.59	1.75	15.67	104.86	nr	**120.53**
1846 x 1218 mm; ref ZND10F	PC £96.50	2.20	19.70	101.61	nr	**121.31**
Reversible windows; including easy-glaze beads and weatherstripping						
997 x 923 mm; ref NC13R	PC £87.09	1.55	13.88	91.73	nr	**105.61**
997 x 1067 mm; ref NC013R	PC £92.34	1.55	13.88	97.25	nr	**111.13**
1237 x 923 mm; ref ZNC13R	PC £96.20	2.30	20.60	101.30	nr	**121.90**
1237 x 1218 mm; ref ZND13R	PC £105.15	2.30	20.60	110.70	nr	**131.30**
1237 x 1513 mm; ref ZNDV13RS	PC £118.23	2.30	20.60	124.43	nr	**145.03**
Pressed steel sills; to suit above window widths						
628 mm	PC £4.97	0.35	3.13	5.91	nr	**9.04**
997 mm	PC £6.98	0.45	4.03	8.02	nr	**12.05**
1237 mm	PC £7.88	0.55	4.93	8.96	nr	**13.89**
1486 mm	PC £9.14	0.65	5.82	10.29	nr	**16.11**
1846 mm	PC £10.38	0.75	6.72	11.59	nr	**18.31**
Factory finished steel fixed light; casement and fanlight windows; Crittall polyester powder coated 'Homelight' range or similar; site glazing measured elsewhere; fixed in position; including lugs cut and pinned to brickwork or blockwork						
Basic fixed lights; including easy-glaze beads						
628 x 292 mm; ref ZNG5	PC £12.89	1.20	10.75	13.82	nr	**24.57**
628 x 923 mm; ref ZNC5	PC £18.37	1.20	10.75	19.57	nr	**30.32**
628 x 1513 mm; ref ZNDV5	PC £26.13	1.20	10.75	27.73	nr	**38.48**
1237 x 292 mm; ref ZNG13	PC £19.67	1.20	10.75	20.94	nr	**31.69**
1237 x 923 mm; ref ZNC13	PC £25.48	1.75	15.67	27.04	nr	**42.71**
1237 x 1218 mm; ref ZND13	PC £29.99	1.75	15.67	31.78	nr	**47.45**
1237 x 1513 mm; ref ZNDV13	PC £32.84	1.75	15.67	34.77	nr	**50.44**
1846 x 292 mm; ref ZNG14	PC £25.52	1.20	10.75	27.08	nr	**37.83**
1846 x 923 mm; ref ZNC14	PC £31.45	1.75	15.67	33.31	nr	**48.98**
1846 x 1513 mm; ref ZNDV14	PC £38.10	2.20	19.70	40.29	nr	**59.99**
Basic opening lights; including easy-glaze beads and weatherstripping						
628 x 292 mm; ref ZNG1	PC £28.49	1.20	10.75	30.20	nr	**40.95**
1237 x 292 mm; ref ZNG13G	PC £42.36	1.20	10.75	44.77	nr	**55.52**
1846 x 292 mm; ref ZNG4	PC £72.44	1.20	10.75	76.35	nr	**87.10**
One-piece composites; including easy-glaze beads and weatherstripping						
628 x 923 mm; ref ZNC5F	PC £40.10	1.20	10.75	42.40	nr	**53.15**
628 x 1513 mm; ref ZNDV5F	PC £48.16	1.20	10.75	50.85	nr	**61.60**
1237 x 923 mm; ref ZNC2F	PC £78.49	1.75	15.67	82.70	nr	**98.37**
1237 x 1218 mm; ref ZND2F	PC £92.45	1.75	15.67	97.36	nr	**113.03**
1237 x 1513 mm; ref ZNDV2V	PC £110.79	1.75	15.67	116.62	nr	**132.29**
1846 x 923 mm; ref NC4F	PC £120.67	1.75	15.67	126.99	nr	**142.66**
1846 x 1218 mm; ref ZND10F	PC £117.52	2.20	19.70	123.68	nr	**143.38**
Reversible windows; including easy-glaze beads and weatherstripping						
997 x 923 mm; ref NC13R	PC £113.21	1.55	13.88	119.16	nr	**133.04**
997 x 1067 mm; ref NC013R	PC £120.05	1.55	13.88	126.34	nr	**140.22**
1237 x 923 mm; ref ZNC13R	PC £125.06	2.30	20.60	131.60	nr	**152.20**
1237 x 1218 mm; ref ZND13R	PC £136.69	2.30	20.60	143.81	nr	**164.41**
1237 x 1513 mm; ref ZNDV13RS	PC £153.70	2.30	20.60	161.67	nr	**182.27**

L WINDOWS/DOORS/STAIRS Including overheads and profit at 5.00%		Labour hours	Labour £	Material £	Unit	Total rate £
L11 METAL WINDOWS/ROOFLIGHTS/SCREENS/LOUVRES - cont'd						
Factory finished steel fixed light; casement and fanlight windows; Crittall polyester powder coated 'Homelight' range or similar; site glazing measured elsewhere; fixed in position; including lugs cut and pinned to brickwork or blockwork - cont'd						
Pressed steel sills; to suit above window widths						
628 mm	PC £6.21	0.35	3.13	7.21	nr	**10.34**
997 mm	PC £8.72	0.45	4.03	9.85	nr	**13.88**
1237 mm	PC £9.84	0.55	4.93	11.03	nr	**15.96**
1486 mm	PC £11.43	0.65	5.82	12.69	nr	**18.51**
1846 mm	PC £12.97	0.75	6.72	14.31	nr	**21.03**
L12 PLASTICS WINDOWS/ROOFLIGHTS/SCREENS/ LOUVRES						
uPVC windows to BS 2782; 'Anglian' or similar; reinforced where appropriate with aluminium alloy; including standard ironmongery; cills and glazing; fixed in position; including lugs plugged and screwed to brickwork or blockwork						
Fixed light; including e.p.d.m. glazing gaskets and weather seals						
600 x 900 mm; single glazed	PC £87.87	3.50	31.35	92.55	nr	**123.90**
600 x 900 mm; double glazed	PC £98.98	3.50	31.35	104.22	nr	**135.57**
Casement/fixed light; including e.p.d.m. glazing gaskets and weather seals						
600 x 1200 mm; single glazed	PC £126.25	3.75	33.59	132.85	nr	**166.44**
600 x 1200 mm; double glazed	PC £137.36	3.75	33.59	144.52	nr	**178.11**
1200 x 1200 mm; single glazed	PC £226.24	4.50	40.30	237.84	nr	**278.14**
1200 x 1200 mm; double glazed	PC £248.46	4.50	40.30	261.17	nr	**301.47**
1800 x 1200 mm; single glazed	PC £323.20	5.00	44.78	339.65	nr	**384.43**
1800 x 1200 mm; double glazed	PC £355.52	5.00	44.78	373.58	nr	**418.36**
'Tilt & Turn' light; including e.p.d.m. glazing gaskets and weather seals						
1200 x 1200 mm; single glazed	PC £181.80	4.50	40.30	191.18	nr	**231.48**
1200 x 1200 mm; double glazed	PC £203.01	4.50	40.30	213.45	nr	**253.75**
uPVC rooflights; plugged and screwed to concrete; or screwed to timber						
Rooflight; 'Coxdome Mark 2'; UPVC double skin; dome or pyramid						
619 x 619 mm	PC £98.87	1.50	9.40	109.00	nr	**118.40**
924 x 619 mm	PC £99.54	1.65	10.34	109.75	nr	**120.09**
924 x 924 mm	PC £138.02	1.80	11.28	152.16	nr	**163.44**
1229 x 924 mm	PC £217.35	1.95	12.22	239.63	nr	**251.85**
1229 x 1229 mm	PC £334.35	2.10	13.16	368.62	nr	**381.78**
1686 x 1076 mm		2.20	13.79	254.37	nr	**268.16**
Rooflight; 'Coxdome Mark 4'; UPVC double skin; GRP splayed upstand; hit and miss vent						
600 x 600 mm	PC £146.66	3.00	18.81	161.68	nr	**180.49**
900 x 600 mm	PC £230.63	3.30	20.69	254.26	nr	**274.95**
900 x 900 mm	PC £244.27	3.60	22.57	269.29	nr	**291.86**
1200 x 900 mm	PC £294.66	3.90	24.45	324.86	nr	**349.31**
1200 x 1200 mm	PC £308.74	4.20	26.33	340.39	nr	**366.72**
1800 x 1200 mm	PC £463.46	5.00	31.34	510.96	nr	**542.30**

L WINDOWS/DOORS/STAIRS
Including overheads and profit at 5.00%

ALTERNATIVE TIMBER DOOR PRICES (£/each)

	£		£		£
Hardwood doors					
Brazilian Mahogany period doors (£/each); 838 x 1981 x 44 mm					
4 panel	154.00	'Carolina'	178.00	'Gothic'	148.50
6 panel	158.40	'Elizabethan	184.80	'Kentucky'	169.20
Red Meranti period doors					
4 panel					
762x1981x44 mm	125.00	838x1981x44 mm	125.00	807x2000x44 mm	125.00
6 panel					
762x1981x44 mm	124.95	838x1981x44 mm	124.95	807x2000x44 mm	124.95
Half bow					
762x1981x44 mm	130.00	838x1981x44 mm	130.00	807x2000x44 mm	130.00
'Kentucky'					
762x1981x44 mm	126.15	838x1981x44 mm	126.15	807x2000x44 mm	126.15
Softwood doors					
Casement doors					
2 XG; two panel; beaded					
762x1981x44 mm	31.80	838x1981x44 mm	33.85	807x2000x44 mm	33.85
813x2032x44 mm	33.85	726x2040x44 mm	33.85	826x2040x44 mm	33.85
2 XGG; two panel; beaded					
762x1981x44 mm	30.75	838x1981x44 mm	32.80	807x2000x44 mm	32.80
813x2032x44 mm	32.80	726x2040x44 mm	32.80	826x2040x44 mm	32.80
10; one panel; beaded					
762x1981x44 mm	25.65	838x1981x44 mm	27.70	807x2000x44 mm	27.70
SA; 15 panel; beaded					
762x1981x44 mm	41.00	838x1981x44 mm	43.05	807x2000x44 mm	43.05
22; pair of two panel; beaded					
914x1981x44 mm	69.70	1168x1981x44 mm	69.70		
2 SA; pair of 10 panel; beaded					
914x1981x44 mm	92.25	1168x1981x44 mm	92.25		
Louvre doors					
533x1524x28 mm	12.65	610x1524x28 mm	13.86	686x1981x28 mm	18.80
533x1676x28 mm	13.70	610x1676x28 mm	15.17	762x1981x28 mm	20.84
533x1829x28 mm	14.59	610x1829x28 mm	16.12	737x1067x28 mm	
533x1981x28 mm	15.54	610x1981x28 mm	17.27	- pair	15.33
Period doors					
4 panel					
762x1981x44 mm	99.75	838x1981x44 mm	99.75	807x2000x44 mm	99.75
6 panel					
762x1981x44 mm	105.00	838x1981x44 mm	105.00	807x2000x44 mm	105.00
Half bow					
762x1981x44 mm	99.75	838x1981x44 mm	99.75	807x2000x44 mm	99.75
'Kentucky'					
762x1981x44 mm	96.75	838x1981x44 mm	96.75	807x2000x44 mm	96.75

L WINDOWS/DOORS/STAIRS Including overheads and profit at 5.00%		Labour hours	Labour £	Material £	Unit	Total rate £
L20 TIMBER DOORS/SHUTTERS/HATCHES						
Standard matchboarded doors; wrought softwood						
Matchboarded, ledged and braced doors; 25 mm ledges and braces; 19 mm tongued, grooved and V-jointed; one side vertical boarding						
762 x 1981 mm	PC £38.23	1.50	11.86	40.15	nr	**52.01**
838 x 1981 mm	PC £40.51	1.50	11.86	42.53	nr	**54.39**
Matchboarded, framed, ledged and braced doors; 44 mm framing; 25 mm intermediate and bottom rails; 19 mm tongued, grooved and V-jointed; one side vertical boarding						
762 x 1981 x 44 mm	PC £49.74	1.80	14.23	52.23	nr	**66.46**
838 x 1981 x 44 mm	PC £51.57	1.80	14.23	54.15	nr	**68.38**
Standard flush doors; softwood composition						
Flush door; internal quality; skeleton or cellular core; hardboard faced both sides;						
457 x 1981 x 35 mm	PC £10.55	1.25	9.88	11.08	nr	**20.96**
533 x 1981 x 35 mm	PC £10.55	1.25	9.88	11.08	nr	**20.96**
610 x 1981 x 35 mm	PC £12.00	1.25	9.88	12.92	nr	**22.80**
686 x 1981 x 35 mm	PC £10.55	1.25	9.88	11.08	nr	**20.96**
762 x 1981 x 35 mm	PC £10.55	1.25	9.88	11.08	nr	**20.96**
838 x 1981 x 35 mm	PC £11.55	1.25	9.88	12.13	nr	**22.01**
526 x 2040 x 40 mm	PC £12.00	1.25	9.88	12.60	nr	**22.48**
626 x 2040 x 40 mm	PC £12.00	1.25	9.88	12.60	nr	**22.48**
726 x 2040 x 40 mm	PC £12.30	1.25	9.88	12.92	nr	**22.80**
826 x 2040 x 40 mm	PC £12.30	1.25	9.88	12.92	nr	**22.80**
Flush door; internal quality; skeleton or cellular core; plywood faced both sides; lipped on two long edges						
457 x 1981 x 35 mm	PC £13.60	1.25	9.88	14.28	nr	**24.16**
533 x 1981 x 35 mm	PC £13.60	1.25	9.88	14.28	nr	**24.16**
610 x 1981 x 35 mm	PC £13.60	1.25	9.88	15.65	nr	**25.53**
686 x 1981 x 35 mm	PC £13.60	1.25	9.88	14.28	nr	**24.16**
762 x 1981 x 35 mm	PC £13.60	1.25	9.88	14.28	nr	**24.16**
838 x 1981 x 35 mm	PC £14.65	1.25	9.88	15.38	nr	**25.26**
526 x 2040 x 40 mm	PC £14.90	1.25	9.88	15.65	nr	**25.53**
626 x 2040 x 40 mm	PC £14.90	1.25	9.88	15.65	nr	**25.53**
726 x 2040 x 40 mm	PC £14.90	1.25	9.88	15.65	nr	**25.53**
826 x 2040 x 40 mm	PC £14.90	1.25	9.88	15.65	nr	**25.53**
Flush door; internal quality; skeleton or cellular core; Sapele faced both sides; lipped on all four edges						
457 x 1981 x 35 mm	PC £17.00	1.35	10.67	17.85	nr	**28.52**
533 x 1981 x 35 mm	PC £17.00	1.35	10.67	17.85	nr	**28.52**
610 x 1981 x 35 mm	PC £17.00	1.35	10.67	17.85	nr	**28.52**
686 x 1981 x 35 mm	PC £17.05	1.35	10.67	17.90	nr	**28.57**
762 x 1981 x 35 mm	PC £17.10	1.35	10.67	17.96	nr	**28.63**
838 x 1981 x 35 mm	PC £18.35	1.35	10.67	19.27	nr	**29.94**
Flush door; internal quality; skeleton or cellular core; Teak faced both sides; lipped on all four edges						
457 x 1981 x 35 mm	PC £65.00	1.35	10.67	68.25	nr	**78.92**
533 x 1981 x 35 mm	PC £57.15	1.35	10.67	60.01	nr	**70.68**
610 x 1981 x 35 mm	PC £57.15	1.35	10.67	60.01	nr	**70.68**
686 x 1981 x 35 mm	PC £57.15	1.35	10.67	60.01	nr	**70.68**
762 x 1981 x 35 mm	PC £68.00	1.35	10.67	71.40	nr	**82.07**
838 x 1981 x 35 mm	PC £59.00	1.35	10.67	61.95	nr	**72.62**
526 x 2040 x 40 mm	PC £58.30	1.35	10.67	61.22	nr	**71.89**
626 x 2040 x 40 mm	PC £58.30	1.35	10.67	61.22	nr	**71.89**
726 x 2040 x 40 mm	PC £58.32	1.35	10.67	61.24	nr	**71.91**
826 x 2040 x 40 mm		1.35	10.67	61.95	nr	**72.62**

L WINDOWS/DOORS/STAIRS Including overheads and profit at 5.00%		Labour hours	Labour £	Material £	Unit	Total rate £
Flush door; half-hour fire check (30/20); hardboard faced both sides;						
457 x 1981 x 44 mm	PC £28.10	1.75	13.84	29.50	nr	**43.34**
533 x 1981 x 44 mm	PC £28.10	1.75	13.84	29.50	nr	**43.34**
610 x 1981 x 44 mm	PC £28.10	1.75	13.84	29.50	nr	**43.34**
686 x 1981 x 44 mm	PC £28.10	1.75	13.84	29.50	nr	**43.34**
762 x 1981 x 44 mm	PC £28.10	1.75	13.84	29.50	nr	**43.34**
838 x 1981 x 44 mm	PC £29.55	1.75	13.84	31.03	nr	**44.87**
526 x 2040 x 44 mm	PC £29.18	1.75	13.84	30.64	nr	**44.48**
626 x 2040 x 44 mm	PC £29.18	1.75	13.84	30.64	nr	**44.48**
726 x 2040 x 44 mm	PC £29.18	1.75	13.84	30.64	nr	**44.48**
826 x 2040 x 44 mm	PC £30.66	1.75	13.84	32.19	nr	**46.03**
Flush door; half-hour fire check (30/20); plywood faced both sides; lipped on all four edges						
457 x 1981 x 44 mm	PC £26.96	1.75	13.84	28.31	nr	**42.15**
533 x 1981 x 44 mm	PC £26.96	1.75	13.84	28.31	nr	**42.15**
610 x 1981 x 44 mm	PC £26.96	1.75	13.84	28.31	nr	**42.15**
686 x 1981 x 44 mm	PC £26.96	1.75	13.84	28.31	nr	**42.15**
762 x 1981 x 44 mm	PC £26.96	1.75	13.84	28.31	nr	**42.15**
838 x 1981 x 44 mm	PC £28.38	1.75	13.84	29.80	nr	**43.64**
526 x 2040 x 44 mm	PC £27.71	1.75	13.84	29.09	nr	**42.93**
626 x 2040 x 44 mm	PC £27.71	1.75	13.84	29.09	nr	**42.93**
726 x 2040 x 44 mm	PC £27.71	1.75	13.84	29.09	nr	**42.93**
826 x 2040 x 44 mm	PC £29.11	1.75	13.84	30.57	nr	**44.41**
Flush door; half hour fire resisting (type B30); Red Meranti veneer for painting; hardwood lipping two long edges						
526 x 2040 x 44 mm	PC £62.90	1.85	14.63	66.05	nr	**80.68**
626 x 2040 x 44 mm	PC £64.80	1.85	14.63	68.04	nr	**82.67**
726 x 2040 x 44 mm	PC £66.70	1.85	14.63	70.04	nr	**84.67**
826 x 2040 x 44 mm	PC £68.60	1.85	14.63	72.03	nr	**86.66**
Flush door; half-hour fire check (30/20); Sapele faced both sides; lipped on all four edges						
457 x 1981 x 44 mm	PC £39.91	1.85	14.63	41.91	nr	**56.54**
533 x 1981 x 44 mm	PC £39.91	1.85	14.63	41.91	nr	**56.54**
610 x 1981 x 44 mm	PC £39.91	1.85	14.63	41.91	nr	**56.54**
686 x 1981 x 44 mm	PC £39.91	1.85	14.63	41.91	nr	**56.54**
762 x 1981 x 44 mm	PC £39.91	1.85	14.63	41.91	nr	**56.54**
526 x 2040 x 44 mm	PC £40.95	1.85	14.63	43.00	nr	**57.63**
626 x 2040 x 44 mm	PC £40.95	1.85	14.63	43.00	nr	**57.63**
726 x 2040 x 44 mm	PC £40.95	1.85	14.63	43.00	nr	**57.63**
826 x 2040 x 44 mm	PC £42.49	1.85	14.63	44.61	nr	**59.24**
Flush door; half-hour fire resisting (30/30) Sapele faced both sides; lipped on all four edges						
457 x 1981 x 44 mm	PC £31.74	1.85	14.63	33.33	nr	**47.96**
533 x 1981 x 44 mm	PC £32.22	1.85	14.63	33.83	nr	**48.46**
610 x 1981 x 44 mm	PC £32.58	1.85	14.63	34.21	nr	**48.84**
686 x 1981 x 44 mm	PC £37.10	1.85	14.63	38.96	nr	**53.59**
762 x 1981 x 44 mm	PC £38.04	1.85	14.63	39.95	nr	**54.58**
838 x 1981 x 44 mm	PC £39.69	1.85	14.63	41.67	nr	**56.30**
526 x 2040 x 44 mm	PC £38.08	1.85	14.63	39.98	nr	**54.61**
626 x 2040 x 44 mm	PC £38.68	1.85	14.63	40.61	nr	**55.24**
726 x 2040 x 44 mm	PC £38.47	1.85	14.63	40.39	nr	**55.02**
826 x 2040 x 44 mm	PC £40.39	1.85	14.63	42.41	nr	**57.04**
Flush door; half hour fire resisting (type B30); American light oak veneer; hardwood lipping all edges						
526 x 2040 x 44 mm	PC £100.15	1.85	14.63	105.16	nr	**119.79**
626 x 2040 x 44 mm	PC £102.05	1.85	14.63	107.15	nr	**121.78**
726 x 2040 x 44 mm	PC £103.95	1.85	14.63	109.15	nr	**123.78**
826 x 2040 x 44 mm	PC £105.85	1.85	14.63	111.14	nr	**125.77**

L WINDOWS/DOORS/STAIRS Including overheads and profit at 5.00%		Labour hours	Labour £	Material £	Unit	Total rate £
L20 TIMBER DOORS/SHUTTERS/HATCHES - cont'd						
Standard flush doors; softwood composition - cont'd						
Flush door; half hour fire-resisting (30/30); 'Melador'; laminate faced both sides; hardwood lipped on all edges						
610 x 1981 x 47 mm	PC £184.56	2.25	17.79	193.78	nr	**211.57**
686 x 1981 x 47 mm	PC £184.56	2.25	17.79	193.78	nr	**211.57**
762 x 1981 x 47 mm	PC £184.56	2.25	17.79	193.78	nr	**211.57**
838 x 1981 x 47 mm	PC £184.56	2.25	17.79	193.78	nr	**211.57**
526 x 2040 x 47 mm	PC £184.56	2.25	17.79	193.78	nr	**211.57**
626 x 2040 x 47 mm	PC £184.56	2.25	17.79	193.78	nr	**211.57**
726 x 2040 x 47 mm	PC £184.56	2.25	17.79	193.78	nr	**211.57**
826 x 2040 x 47 mm	PC £184.56	2.25	17.79	193.78	nr	**211.57**
Flush door; one hour fire check (60/45); plywood faced both sides; lipped on all four edges						
610 x 1981 x 54 mm	PC £143.77	2.00	15.81	150.95	nr	**166.76**
686 x 1981 x 54 mm	PC £143.77	2.00	15.81	150.95	nr	**166.76**
762 x 1981 x 54 mm	PC £93.21	2.00	15.81	97.87	nr	**113.68**
838 x 1981 x 54 mm	PC £100.11	2.00	15.81	105.12	nr	**120.93**
526 x 2040 x 54 mm	PC £143.77	2.00	15.81	150.95	nr	**166.76**
626 x 2040 x 54 mm	PC £143.77	2.00	15.81	150.95	nr	**166.76**
726 x 2040 x 54 mm	PC £93.21	2.00	15.81	97.87	nr	**113.68**
826 x 2040 x 54 mm	PC £100.11	2.00	15.81	105.12	nr	**120.93**
Flush door; one hour fire check (60/45); Sapele faced both sides; lipped on all four edges						
610 x 1981 x 54 mm	PC £150.36	2.10	16.60	157.88	nr	**174.48**
686 x 1981 x 54 mm	PC £150.36	2.10	16.60	157.88	nr	**174.48**
762 x 2040 x 54 mm	PC £150.36	2.10	16.60	157.88	nr	**174.48**
838 x 1981 x 54 mm	PC £150.36	2.10	16.60	157.88	nr	**174.48**
526 x 2040 x 54 mm	PC £150.36	2.10	16.60	157.88	nr	**174.48**
626 x 2040 x 54 mm	PC £150.36	2.10	16.60	157.88	nr	**174.48**
762 x 2040 x 54 mm	PC £150.36	2.10	16.60	157.88	nr	**174.48**
826 x 2040 x 54 mm	PC £150.36	2.10	16.60	157.88	nr	**174.48**
Flush door; one hour fire resisting (60/60); Sapele faced both sides; lipped on all four edges						
610 x 1981 x 54 mm	PC £169.06	2.10	16.60	177.52	nr	**194.12**
686 x 1981 x 54 mm	PC £169.06	2.10	16.60	177.52	nr	**194.12**
762 x 1981 x 54 mm	PC £169.06	2.10	16.60	177.52	nr	**194.12**
838 x 1981 x 54 mm	PC £169.06	2.10	16.60	177.52	nr	**194.12**
526 x 2040 x 54 mm	PC £169.06	2.10	16.60	177.52	nr	**194.12**
626 x 2040 x 54 mm	PC £169.06	2.10	16.60	177.52	nr	**194.12**
726 x 2040 x 54 mm	PC £169.06	2.10	16.60	177.52	nr	**194.12**
826 x 2040 x 54 mm	PC £169.06	2.10	16.60	177.52	nr	**194.12**
Flush door; one hour fire resisting (type B60); Afrormosia veneer; hardwood lipping all edges including groove and 'Leaderseal' intumescent strip						
457 x 1981 x 54 mm	PC £142.70	2.10	16.60	149.84	nr	**166.44**
533 x 1981 x 54 mm	PC £144.60	2.10	16.60	151.83	nr	**168.43**
610 x 1981 x 54 mm	PC £144.60	2.10	16.60	151.83	nr	**168.43**
686 x 1981 x 54 mm	PC £146.50	2.10	16.60	153.83	nr	**170.43**
762 x 1981 x 54 mm	PC £148.40	2.10	16.60	155.82	nr	**172.42**
838 x 1981 x 54 mm	PC £150.30	2.10	16.60	157.82	nr	**174.42**
526 x 2040 x 54 mm	PC £142.70	2.10	16.60	149.84	nr	**166.44**
626 x 2040 x 54 mm	PC £144.60	2.10	16.60	151.83	nr	**168.43**
726 x 2040 x 54 mm	PC £146.50	2.10	16.60	153.83	nr	**170.43**
826 x 2040 x 54 mm	PC £148.40	2.10	16.60	155.82	nr	**172.42**

L WINDOWS/DOORS/STAIRS Including overheads and profit at 5.00%		Labour hours	Labour £	Material £	Unit	Total rate £
Flush door; one hour fire-resisting (60/60); 'Melador'; laminate faced both sides; hardwood lipped on all edges						
610 x 1981 x 57 mm	PC £255.88	2.75	21.74	268.67	nr	**290.41**
686 x 1981 x 57 mm	PC £255.88	2.75	21.74	268.67	nr	**290.41**
762 x 1981 x 57 mm	PC £255.88	2.75	21.74	268.67	nr	**290.41**
838 x 1981 x 57 mm	PC £255.88	2.75	21.74	268.67	nr	**290.41**
526 x 2040 x 57 mm	PC £255.88	2.75	21.74	268.67	nr	**290.41**
626 x 2040 x 57 mm	PC £255.88	2.75	21.74	268.67	nr	**290.41**
726 x 2040 x 57 mm	PC £255.88	2.75	21.74	268.67	nr	**290.41**
826 x 2040 x 57 mm	PC £255.88	2.75	21.74	268.67	nr	**290.41**
Flush door; external quality; skeleton or cellular core; plywood faced both sides; lipped on all four edges						
762 x 1981 x 44 mm	PC £24.09	1.50	11.86	25.29	nr	**37.15**
838 x 1981 x 44 mm	PC £25.17	1.50	11.86	26.43	nr	**38.29**
Flush door; external quality with standard glass opening; skeleton or cellular core; plywood faced both sides; lipped on all four edges; including glazing beads						
762 x 1981 x 44 mm	PC £29.97	1.75	13.84	31.47	nr	**45.31**
838 x 1981 x 44 mm	PC £31.08	1.75	13.84	32.63	nr	**46.47**
Purpose made panelled doors; wrought softwood						
Panelled doors; one open panel for glass; including glazing beads						
686 x 1981 x 44 mm	PC £42.92	1.75	13.84	45.07	nr	**58.91**
762 x 1981 x 44 mm	PC £44.26	1.75	13.84	46.47	nr	**60.31**
838 x 1981 x 44 mm	PC £45.59	1.75	13.84	47.87	nr	**61.71**
Panelled doors; two open panel for glass; including glazing beads						
686 x 1981 x 44 mm	PC £59.11	1.75	13.84	62.06	nr	**75.90**
762 x 1981 x 44 mm	PC £62.29	1.75	13.84	65.40	nr	**79.24**
838 x 1981 x 44 mm	PC £65.41	1.75	13.84	68.68	nr	**82.52**
Panelled doors; four 19 mm thick plywood panel; mouldings worked on solid both sides						
686 x 1981 x 44 mm	PC £82.70	1.75	13.84	86.83	nr	**100.67**
762 x 1981 x 44 mm	PC £86.15	1.75	13.84	90.46	nr	**104.30**
838 x 1981 x 44 mm	PC £90.48	1.75	13.84	95.01	nr	**108.85**
Panelled doors; six 19 mm thick panels raised and fielded; mouldings worked on solid both sides						
686 x 1981 x 50 mm	PC £180.99	2.10	16.60	190.03	nr	**206.63**
762 x 1981 x 50 mm	PC £188.52	2.10	16.60	197.94	nr	**214.54**
838 x 1981 x 50 mm	PC £197.94	2.10	16.60	207.84	nr	**224.44**
Rebated edges beaded		-	-	-	m	**0.96**
Rounded edges or heels		-	-	-	m	**0.49**
Weatherboard; fixed to bottom rail		0.25	1.98	2.04	m	**4.02**
Stopped groove for weatherboard		-	-	-	m	**0.33**
Purpose made panelled doors; selected West African Mahogany; PC £470.00/m3						
Panelled doors; one open panel for glass; mouldings worked on the solid one side; 19 x 13 mm beads one side; fixing with brass screws and cups						
686 x 1981 x 50 mm	PC £74.26	2.50	19.77	77.97	nr	**97.74**
762 x 1981 x 50 mm	PC £76.66	2.50	19.77	80.50	nr	**100.27**
838 x 1981 x 50 mm	PC £79.07	2.50	19.77	83.02	nr	**102.79**
686 x 1981 x 63 mm	PC £90.08	2.75	21.74	94.58	nr	**116.32**
762 x 1981 x 63 mm	PC £93.10	2.75	21.74	97.76	nr	**119.50**
838 x 1981 x 63 mm	PC £96.09	2.75	21.74	100.89	nr	**122.63**

L WINDOWS/DOORS/STAIRS Including overheads and profit at 5.00%		Labour hours	Labour £	Material £	Unit	Total rate £
L20 TIMBER DOORS/SHUTTERS/HATCHES - cont'd						
Purpose made panelled doors; selected West African Mahogany; PC £470.00/m3 - cont'd						
Panelled doors; 250 mm wide cross-tongued intermediate rail; two open panels for glass mouldings worked on the solid one side; 19 x 13 mm beads one side; fixing with brass screws and cups						
686 x 1981 x 50 mm	PC £101.62	2.50	19.77	106.70	nr	**126.47**
762 x 1981 x 50 mm	PC £106.57	2.50	19.77	111.90	nr	**131.67**
838 x 1981 x 50 mm	PC £111.47	2.50	19.77	117.05	nr	**136.82**
686 x 1981 x 63 mm	PC £121.70	2.75	21.74	127.79	nr	**149.53**
762 x 1981 x 63 mm	PC £127.68	2.75	21.74	134.06	nr	**155.80**
838 x 1981 x 63 mm	PC £133.63	2.75	21.74	140.32	nr	**162.06**
Panelled doors; four panels; (19 mm for 50 mm thick doors and 25 mm for 63 mm thick doors); mouldings worked on solid both sides						
686 x 1981 x 50 mm	PC £128.86	2.50	19.77	135.30	nr	**155.07**
762 x 1981 x 50 mm	PC £134.23	2.50	19.77	140.94	nr	**160.71**
838 x 1981 x 50 mm	PC £140.93	2.50	19.77	147.98	nr	**167.75**
686 x 1981 x 63 mm	PC £151.62	2.75	21.74	159.20	nr	**180.94**
762 x 1981 x 63 mm	PC £157.96	2.75	21.74	165.86	nr	**187.60**
838 x 1981 x 63 mm	PC £165.85	2.75	21.74	174.15	nr	**195.89**
Panelled doors; 150 mm wide stiles in one width; 430 mm wide cross-tongued bottom rail; six panels raised and fielded one side (19 mm thick for 50 mm thick doors and 25 mm thick for 63 mm thick doors); mouldings worked on the solid both sides						
686 x 1981 x 50 mm	PC £277.19	2.50	19.77	291.04	nr	**310.81**
762 x 1981 x 50 mm	PC £288.73	2.50	19.77	303.17	nr	**322.94**
838 x 1981 x 50 mm	PC £303.16	2.50	19.77	318.32	nr	**338.09**
686 x 1981 x 63 mm	PC £340.36	2.75	21.74	357.38	nr	**379.12**
762 x 1981 x 63 mm	PC £354.61	2.75	21.74	372.34	nr	**394.08**
838 x 1981 x 63 mm	PC £372.28	2.75	21.74	390.90	nr	**412.64**
Rebated edges beaded		-	-	-	m	**1.46**
Rounded edges or heels		-	-	-	m	**0.72**
Weatherboard; fixed to bottom rail		0.33	2.61	3.89	m	**6.50**
Stopped groove for weatherboard		-	-	-	m	**0.63**
Purpose made panelled doors; Afrormosia; PC £710.00/m3						
Panelled doors; one open panel for glass; mouldings worked on the solid one side; 19 x 13 mm beads one side; fixing with brass screws and cups						
686 x 1981 x 50 mm	PC £87.44	2.50	19.77	91.81	nr	**111.58**
762 x 1981 x 50 mm	PC £90.35	2.50	19.77	94.87	nr	**114.64**
838 x 1981 x 50 mm	PC £93.27	2.50	19.77	97.93	nr	**117.70**
686 x 1981 x 63 mm	PC £106.52	2.75	21.74	111.85	nr	**133.59**
762 x 1981 x 63 mm	PC £110.20	2.75	21.74	115.71	nr	**137.45**
838 x 1981 x 63 mm	PC £113.82	2.75	21.74	119.51	nr	**141.25**
Panelled doors; 250 mm wide cross-tongued intermediate rail; two open panels for glass mouldings worked on the solid one side; 19 x 13 mm beads one side; fixing with brass screws and cups						
686 x 1981 x 50 mm	PC £119.02	2.50	19.77	124.97	nr	**144.74**
762 x 1981 x 50 mm	PC £124.96	2.50	19.77	131.21	nr	**150.98**
838 x 1981 x 50 mm	PC £130.82	2.50	19.77	137.36	nr	**157.13**
686 x 1981 x 63 mm	PC £143.43	2.75	21.74	150.60	nr	**172.34**
762 x 1981 x 63 mm	PC £150.61	2.75	21.74	158.14	nr	**179.88**
838 x 1981 x 63 mm	PC £157.76	2.75	21.74	165.65	nr	**187.39**

L WINDOWS/DOORS/STAIRS Including overheads and profit at 5.00%		Labour hours	Labour £	Material £	Unit	Total rate £
Panelled doors; four panels; (19 mm for 50 mm thick doors and 25 mm for 63 mm thick doors); mouldings worked on solid both sides						
686 x 1981 x 50 mm	PC £151.13	2.50	19.77	158.68	nr	**178.45**
762 x 1981 x 50 mm	PC £157.41	2.50	19.77	165.28	nr	**185.05**
838 x 1981 x 50 mm	PC £165.28	2.50	19.77	173.54	nr	**193.31**
686 x 1981 x 63 mm	PC £179.32	2.75	21.74	188.29	nr	**210.03**
762 x 1981 x 63 mm	PC £186.80	2.75	21.74	196.14	nr	**217.88**
838 x 1981 x 63 mm	PC £196.16	2.75	21.74	205.96	nr	**227.70**
Panelled doors; 150 mm wide stiles in one width; 430 mm wide cross-tongued bottom rail; six panels raised and fielded one side (19 mm thick for 50 mm thick doors and 25 mm thick for 63 mm thick doors); mouldings worked on the solid both sides						
686 x 1981 x 50 mm	PC £311.94	2.50	19.77	327.53	nr	**347.30**
762 x 1981 x 50 mm	PC £324.94	2.50	19.77	341.19	nr	**360.96**
838 x 1981 x 50 mm	PC £341.18	2.50	19.77	358.24	nr	**378.01**
686 x 1981 x 63 mm	PC £384.37	2.75	21.74	403.58	nr	**425.32**
762 x 1981 x 63 mm	PC £400.39	2.75	21.74	420.41	nr	**442.15**
838 x 1981 x 63 mm	PC £420.41	2.75	21.74	441.43	nr	**463.17**
Rebated edges beaded		-	-	-	m	**1.46**
Rounded edges or heels		-	-	-	m	**0.72**
Weatherboard; fixed to bottom rail		0.33	2.61	4.45	m	**7.06**
Stopped groove for weatherboard		-	-	-	m	**0.63**
Standard joinery sets; wrought softwood						
Internal door frame or lining set for 686 x 1981 mm door; all with loose stops unless rebated; 'finished sizes'						
27 x 94 mm lining	PC £13.95	0.80	6.33	14.65	nr	**20.98**
27 x 107 mm lining	PC £14.67	0.80	6.33	15.40	nr	**21.73**
35 x 107 mm rebated lining	PC £15.21	0.80	6.33	15.97	nr	**22.30**
27 x 121 mm lining	PC £15.84	0.80	6.33	16.63	nr	**22.96**
27 x 121 mm lining with fanlight over	PC £22.38	0.95	7.51	23.50	nr	**31.01**
27 x 133 mm lining	PC £17.07	0.80	6.33	17.92	nr	**24.25**
35 x 133 mm rebated linings	PC £17.52	0.80	6.33	18.40	nr	**24.73**
27 x 133 mm lining with fanlight over	PC £24.00	0.95	7.51	25.20	nr	**32.71**
33 x 57 mm frame	PC £11.37	0.80	6.33	11.94	nr	**18.27**
33 x 57 mm storey height frame	PC £13.95	0.85	6.72	14.65	nr	**21.37**
33 x 57 mm frame with fanlight over	PC £17.10	0.95	7.51	17.96	nr	**25.47**
33 x 64 mm frame	PC £12.18	0.80	6.33	12.79	nr	**19.12**
33 x 64 mm storey height frame	PC £14.91	0.85	6.72	15.66	nr	**22.38**
33 x 64 mm frame with fanlight over	PC £17.94	0.95	7.51	18.84	nr	**26.35**
44 x 94 mm frame	PC £18.66	0.92	7.27	19.59	nr	**26.86**
44 x 94 mm storey height frame	PC £22.23	1.00	7.91	23.34	nr	**31.25**
44 x 94 mm frame with fanlight over	PC £26.07	1.10	8.70	27.37	nr	**36.07**
44 x 107 mm frame	PC £21.45	0.92	7.27	22.52	nr	**29.79**
44 x 107 mm storey height frame	PC £25.08	1.00	7.91	26.33	nr	**34.24**
44 x 107 mm frame with fanlight over	PC £28.77	1.10	8.70	30.21	nr	**38.91**
Internal door frame or lining set for 762 x 1981 mm door; all with loose stops unless rebated; 'finished sizes'						
27 x 94 mm lining	PC £13.95	0.80	6.33	14.65	nr	**20.98**
27 x 107 mm lining	PC £14.67	0.80	6.33	15.40	nr	**21.73**
35 x 107 mm rebated lining	PC £15.21	0.80	6.33	15.97	nr	**22.30**
27 x 121 mm lining	PC £15.84	0.80	6.33	16.63	nr	**22.96**
27 x 121 mm lining with fanlight over	PC £22.38	0.95	7.51	23.50	nr	**31.01**
27 x 133 mm lining	PC £17.07	0.80	6.33	17.92	nr	**24.25**

L WINDOWS/DOORS/STAIRS Including overheads and profit at 5.00%		Labour hours	Labour £	Material £	Unit	Total rate £
L20 TIMBER DOORS/SHUTTERS/HATCHES - cont'd						
Standard joinery sets; wrought softwood - cont'd						
Internal door frame or lining set for 762 x 1981 mm door; all with loose stops unless rebated; 'finished sizes'						
35 x 133 mm rebated linings	PC £17.52	0.80	6.33	18.40	nr	**24.73**
27 x 133 mm lining with fanlight over	PC £24.00	0.95	7.51	25.20	nr	**32.71**
33 x 57 mm frame	PC £11.37	0.80	6.33	11.94	nr	**18.27**
33 x 57 mm storey height frame	PC £13.95	0.85	6.72	14.65	nr	**21.37**
33 x 57 mm frame with fanlight over	PC £17.10	0.95	7.51	17.96	nr	**25.47**
33 x 64 mm frame	PC £12.18	0.80	6.33	12.79	nr	**19.12**
33 x 64 mm storey height frame	PC £14.91	0.85	6.72	15.66	nr	**22.38**
33 x 64 mm frame with fanlight over	PC £17.94	0.95	7.51	18.84	nr	**26.35**
44 x 94 mm frame	PC £18.66	0.92	7.27	19.59	nr	**26.86**
44 x 94 mm storey height frame	PC £22.23	1.00	7.91	23.34	nr	**31.25**
44 x 94 mm frame with fanlight over	PC £26.07	1.10	8.70	27.37	nr	**36.07**
44 x 107 mm frame	PC £21.45	0.92	7.27	22.52	nr	**29.79**
44 x 107 mm storey height frame	PC £25.08	1.00	7.91	26.33	nr	**34.24**
44 x 107 mm frame with fanlight over	PC £28.77	1.10	8.70	30.21	nr	**38.91**
Internal door frame or lining set for 726 x 2040 mm door; with loose stops						
30 x 94 mm lining	PC £16.20	0.80	6.33	17.01	nr	**23.34**
30 x 94 mm lining with fanlight over	PC £23.22	0.95	7.51	24.38	nr	**31.89**
30 x 107 mm lining	PC £19.08	0.80	6.33	20.03	nr	**26.36**
30 x 107 mm lining with fanlight over	PC £26.19	0.95	7.51	27.50	nr	**35.01**
30 x 133 mm lining	PC £21.36	0.80	6.33	22.43	nr	**28.76**
30 x 133 mm lining with fanlight over	PC £28.23	0.95	7.51	29.64	nr	**37.15**
Internal door frame or lining set for 826 x 2040 mm door; with loose stops						
30 x 94 mm lining	PC £16.20	0.80	6.33	17.01	nr	**23.34**
30 x 94 mm lining with fanlight over	PC £23.22	0.95	7.51	24.38	nr	**31.89**
30 x 107 mm lining	PC £19.08	0.80	6.33	20.03	nr	**26.36**
30 x 107 mm lining with fanlight over	PC £26.19	0.95	7.51	27.50	nr	**35.01**
30 x 133 mm lining	PC £21.36	0.80	6.33	22.43	nr	**28.76**
30 x 133 mm lining with fanlight over		0.95	7.51	29.64	nr	**37.15**
Trap door set; 9 mm plywood in 35 x 107 mm (fin) rebated lining						
762 x 762 mm	PC £18.57	0.80	6.33	19.77	nr	**26.10**
Serving hatch set; plywood faced doors; rebated meeting stiles; hung on nylon hinges; in 35 x 140 mm (fin) rebated lining						
648 x 533 mm	PC £35.37	1.30	10.28	37.45	nr	**47.73**
Purpose made door frames and lining sets; wrought softwood						
Jambs and heads; as linings						
32 x 63 mm		0.17	1.34	3.28	m	**4.62**
32 x 100 mm		0.17	1.34	4.29	m	**5.63**
32 x 140 mm		0.17	1.34	5.57	m	**6.91**
Jambs and heads; as frames; rebated, rounded and grooved						
38 x 75 mm		0.17	1.34	4.24	m	**5.58**
38 x 100 mm		0.17	1.34	5.07	m	**6.41**
38 x 115 mm		0.17	1.34	5.92	m	**7.26**
38 x 140 mm		0.20	1.58	6.89	m	**8.47**

L WINDOWS/DOORS/STAIRS Including overheads and profit at 5.00%		Labour hours	Labour £	Material £	Unit	Total rate £
50 x 100 mm		0.20	1.58	6.37	m	**7.95**
50 x 125 mm		0.20	1.58	7.61	m	**9.19**
63 x 88 mm		0.20	1.58	7.44	m	**9.02**
63 x 100 mm		0.20	1.58	7.43	m	**9.01**
63 x 125 mm		0.20	1.58	9.40	m	**10.98**
75 x 100 mm		0.20	1.58	9.01	m	**10.59**
75 x 125 mm		0.22	1.74	11.18	m	**12.92**
75 x 150 mm		0.22	1.74	13.07	m	**14.81**
100 x 100 mm		0.22	1.74	11.76	m	**13.50**
100 x 150 mm		0.22	1.74	16.98	m	**18.72**
Mullions and transoms; in linings						
32 x 63 mm		0.12	0.95	5.31	m	**6.26**
32 x 100 mm		0.12	0.95	6.35	m	**7.30**
32 x 140 mm		0.12	0.95	7.59	m	**8.54**
Mullions and transoms; in frames; twice rebated, rounded and grooved						
38 x 75 mm		0.12	0.95	6.38	m	**7.33**
38 x 100 mm		0.12	0.95	7.22	m	**8.17**
38 x 115 mm		0.12	0.95	7.93	m	**8.88**
38 x 140 mm		0.14	1.11	8.92	m	**10.03**
50 x 100 mm		0.14	1.11	8.39	m	**9.50**
50 x 125 mm		0.14	1.11	9.77	m	**10.88**
63 x 88 mm		0.14	1.11	9.46	m	**10.57**
63 x 100 mm		0.14	1.11	9.89	m	**11.00**
75 x 100 mm		0.14	1.11	11.15	m	**12.26**
Extra for additional labours						
one		0.02	0.16	-	m	**0.16**
two		0.03	0.24	-	m	**0.24**
three		0.04	0.32	-	m	**0.32**
Add +5% to the above 'Material £ prices' for 'selected' softwood for staining						
Purpose made door frames and lining sets; selected West African Mahogany; PC £470.00/m3						
Jambs and heads; as linings						
32 x 63 mm	PC £4.74	0.23	1.82	5.10	m	**6.92**
32 x 100 mm	PC £6.10	0.23	1.82	6.56	m	**8.38**
32 x 140 mm	PC £7.77	0.23	1.82	8.36	m	**10.18**
Jambs and heads; as frames; rebated, rounded and grooved						
38 x 75 mm	PC £5.97	0.23	1.82	6.42	m	**8.24**
38 x 100 mm	PC £7.18	0.23	1.82	7.73	m	**9.55**
38 x 115 mm	PC £8.21	0.23	1.82	8.85	m	**10.67**
38 x 140 mm	PC £9.46	0.27	2.13	10.19	m	**12.32**
50 x 100 mm	PC £9.00	0.27	2.13	9.69	m	**11.82**
50 x 125 mm	PC £10.81	0.27	2.13	11.64	m	**13.77**
63 x 88 mm	PC £10.06	0.27	2.13	10.83	m	**12.96**
63 x 100 mm	PC £10.80	0.27	2.13	11.63	m	**13.76**
63 x 125 mm	PC £13.34	0.27	2.13	14.35	m	**16.48**
75 x 100 mm	PC £12.79	0.27	2.13	13.77	m	**15.90**
75 x 125 mm	PC £15.89	0.30	2.37	17.11	m	**19.48**
75 x 150 mm	PC £18.61	0.30	2.37	20.04	m	**22.41**
100 x 100 mm	PC £16.71	0.30	2.37	17.98	m	**20.35**
100 x 150 mm	PC £24.14	0.30	2.37	25.98	m	**28.35**
Mullions and transoms; in linings						
32 x 63 mm	PC £7.51	0.16	1.27	8.09	m	**9.36**
32 x 100 mm	PC £8.87	0.16	1.27	9.55	m	**10.82**
32 x 140 mm	PC £10.43	0.16	1.27	11.23	m	**12.50**

L WINDOWS/DOORS/STAIRS Including overheads and profit at 5.00%		Labour hours	Labour £	Material £	Unit	Total rate £
L20 TIMBER DOORS/SHUTTERS/HATCHES - cont'd						
Purpose made door frames and lining sets; selected West African Mahogany; PC £470.00/m3 - cont'd						
Mullions and transoms; in frames; twice rebated, rounded and grooved						
38 x 75 mm	PC £8.88	0.16	1.27	9.56	m	**10.83**
38 x 100 mm	PC £10.09	0.16	1.27	10.86	m	**12.13**
38 x 115 mm	PC £10.97	0.16	1.27	11.80	m	**13.07**
38 x 140 mm	PC £12.24	0.18	1.42	13.18	m	**14.60**
50 x 100 mm	PC £11.78	0.18	1.42	12.67	m	**14.09**
50 x 125 mm	PC £13.73	0.18	1.42	14.77	m	**16.19**
63 x 88 mm	PC £12.82	0.18	1.42	13.80	m	**15.22**
63 x 100 mm	PC £13.92	0.18	1.42	14.98	m	**16.40**
75 x 100 mm	PC £15.71	0.18	1.42	16.90	m	**18.32**
Sills; once sunk weathered; once rebated, three times grooved						
63 x 175 mm	PC £25.70	0.33	2.61	27.65	m	**30.26**
75 x 125 mm	PC £23.38	0.33	2.61	25.16	m	**27.77**
75 x 150 mm	PC £26.15	0.33	2.61	28.14	m	**30.75**
Extra for additional labours						
one		0.03	0.24	-	m	**0.24**
two		0.05	0.40	-	m	**0.40**
three		0.07	0.55	-	m	**0.55**
Purpose made door frames and lining sets; Afrormosia; PC £710.00/m3						
Jambs and heads; as linings						
32 x 63 mm	PC £5.69	0.23	1.82	6.12	m	**7.94**
32 x 100 mm	PC £7.44	0.23	1.82	8.01	m	**9.83**
32 x 140 mm	PC £9.56	0.23	1.82	10.29	m	**12.11**
Jambs and heads; as frames; rebated, rounded and grooved						
38 x 75 mm	PC £7.16	0.23	1.82	7.71	m	**9.53**
38 x 100 mm	PC £8.78	0.23	1.82	9.45	m	**11.27**
38 x 115 mm	PC £10.13	0.23	1.82	10.90	m	**12.72**
38 x 140 mm	PC £11.68	0.27	2.13	12.57	m	**14.70**
50 x 100 mm	PC £9.00	0.27	2.13	11.95	m	**14.08**
50 x 125 mm	PC £13.43	0.27	2.13	14.46	m	**16.59**
63 x 88 mm	PC £12.44	0.27	2.13	13.39	m	**15.52**
63 x 100 mm	PC £13.14	0.27	2.13	14.15	m	**16.28**
63 x 125 mm	PC £16.65	0.27	2.13	17.92	m	**20.05**
75 x 100 mm	PC £15.94	0.27	2.13	17.15	m	**19.28**
75 x 125 mm	PC £19.82	0.30	2.37	21.34	m	**23.71**
75 x 150 mm	PC £23.32	0.30	2.37	25.09	m	**27.46**
100 x 100 mm	PC £20.91	0.30	2.37	22.50	m	**24.87**
100 x 150 mm	PC £30.44	0.30	2.37	32.76	m	**35.13**
Mullions and transoms; in linings						
32 x 63 mm	PC £8.45	0.16	1.27	9.09	m	**10.36**
32 x 100 mm	PC £10.20	0.16	1.27	10.99	m	**12.26**
32 x 140 mm	PC £12.33	0.16	1.27	13.27	m	**14.54**
Mullions and transoms; in frames; twice rebated, rounded and grooved						
38 x 75 mm	PC £10.07	0.16	1.27	10.84	m	**12.11**
38 x 100 mm	PC £11.68	0.16	1.27	12.57	m	**13.84**
38 x 115 mm	PC £12.89	0.16	1.27	13.88	m	**15.15**
38 x 140 mm	PC £12.89	0.18	1.42	13.88	m	**15.30**
50 x 100 mm	PC £13.87	0.18	1.42	14.93	m	**16.35**
50 x 125 mm	PC £16.34	0.18	1.42	17.59	m	**19.01**
63 x 88 mm	PC £15.23	0.18	1.42	16.39	m	**17.81**
63 x 100 mm	PC £16.57	0.18	1.42	17.83	m	**19.25**
75 x 100 mm	PC £18.85	0.18	1.42	20.29	m	**21.71**

L WINDOWS/DOORS/STAIRS Including overheads and profit at 5.00%		Labour hours	Labour £	Material £	Unit	Total rate £
Sills; once sunk weathered; once rebated, three times grooved						
63 x 175 mm	PC £30.30	0.33	2.61	32.62	m	**35.23**
75 x 125 mm	PC £27.32	0.33	2.61	29.40	m	**32.01**
75 x 150 mm	PC £31.07	0.33	2.61	33.44	m	**36.05**
Extra for additional labours						
one		0.03	0.24	-	m	**0.24**
two		0.05	0.40	-	m	**0.40**
three		0.07	0.55	-	m	**0.55**
Door sills; European Oak PC £1200.00/m3						
Sills; once sunk weathered; once rebated, three times grooved						
63 x 175 mm	PC £50.70	0.33	2.61	54.57	m	**57.18**
75 x 125 mm	PC £45.19	0.33	2.61	48.63	m	**51.24**
75 x 150 mm	PC £51.60	0.33	2.61	55.53	m	**58.14**
Extra for additional labours						
one		0.03	0.24	-	m	**0.24**
two		0.06	0.47	-	m	**0.47**
three		0.09	0.71	-	m	**0.71**
Bedding and pointing frames						
Pointing wood frames or sills with mastic						
one side		0.07	0.71	0.38	m	**1.09**
each side		0.14	1.42	0.77	m	**2.19**
Pointing wood frames or sills with polysulphide sealant						
one side		0.07	0.71	0.81	m	**1.52**
each side		0.14	1.42	1.61	m	**3.03**
Bedding wood frame in cement mortar (1:3) and point						
one side		0.08	0.81	0.04	m	**0.85**
each side		0.10	1.01	0.06	m	**1.07**
one side in mortar; other side in mastic		0.15	1.52	0.43	m	**1.95**
L21 METAL DOORS/SHUTTERS/HATCHES						
Aluminium double glazed sliding patio doors; Crittall 'Luminaire' or similar; white acrylic finish; with and including 18 mm annealed double glazing; fixed in position; including lugs plugged and screwed to brickwork or blockwork; or screwed to wooden sub-frame (measured elsewhere)						
Patio doors						
1800 x 2100 mm; ref D18HDL21	PC £405.00	7.50	67.17	427.79	nr	**494.96**
2400 x 2100 mm; ref D24HDL21	PC £479.63	9.00	80.61	506.15	nr	**586.76**
3000 x 2100 mm; ref D30HDF21	PC £670.50	10.50	94.04	706.57	nr	**800.61**
Galvanized steel 'up and over' type garage doors; Catnic 'Garador' or similar; spring counterbalanced; fixed to timber frame (measured elsewhere)						
Garage door						
2135 x 1980 mm; ref MK 3C	PC £111.00	5.00	44.78	117.94	nr	**162.72**
2135 x 2135 mm; ref MK 3C	PC £122.25	5.25	47.02	129.75	nr	**176.77**
2400 x 2125 mm; ref MK 3C	PC £144.75	6.00	53.74	153.37	nr	**207.11**
4270 x 2135 mm; ref 'Carlton'	PC £484.88	9.00	80.61	510.50	nr	**591.11**

L WINDOWS/DOORS/STAIRS Including overheads and profit at 5.00%	Labour hours	Labour £	Material £	Unit	Total rate £
L30 TIMBER STAIRS/WALKWAYS/BALUSTRADES					
Standard staircases; wrought softwood					
Stairs; 25 mm treads with rounded nosings; 12 mm plywood risers; 32 mm once rounded strings; bullnose bottom tread; 50 x 75 mm hardwood handrail; two 32 x 140 mm balustrade knee rails; 32 x 50 mm stiffeners and 100 x 100 mm newel posts with hardwood newel caps on top					
straight flight; 838 mm wide; 2688 mm going; 2600 mm rise; with two newel posts	13.50	106.74	293.89	nr	**400.63**
straight flight; 838 mm wide; 2600 mm rise; with two newel posts and three top treads winding	18.00	142.32	364.04	nr	**506.36**
dogleg staircase; 838 mm wide; 2600 mm rise; quarter space landing third riser from top; with three newel posts	19.00	150.22	372.65	nr	**522.87**
as last but with half space landing; one 100 x 200 mm newel post; and two 100 x 100 mm newel posts	20.00	158.13	443.98	nr	**602.11**
Stairs; 25 mm treads with rounded nosings; 12 mm plywood risers; 32 mm once rounded strings; with string cappings; bullnose bottom tread; 50 x 75 mm hardwood handrail; two 32 x 32 mm balusters per tread and 100 x 100 mm newel post with hardwood newel caps on top					
straight flight; 838 mm wide 2688 going 2600 mm rise with two newel posts	15.00	118.60	288.61	nr	**407.21**
Standard balustrades; wrought softwood					
Landing balustrade; 50 x 75 mm hardwood handrail; three 32 x 140 mm balustrades knee rails; two 32 x 50 mm stiffeners; one end jointed to newel post; other end built into wall (newel post and mortices both measured separately)					
3 m long	3.00	23.72	108.25	nr	**131.97**
Landing balustrade; 50 x 75 mm hardwood handrail; 32 x 32 mm balusters; one end of handrail jointed to newel post; other end built into wall; balusters housed in at bottom (newel post and mortices both measured separately)					
3 m long	4.50	35.58	88.23	nr	**123.81**
Purpose made staircase components; wrought softwood					
Board landings; cross-tongued joints; 100 x 50 mm sawn softwood bearers					
25 mm thick	1.00	7.91	39.53	m2	**47.44**
32 mm thick	1.00	7.91	47.12	m2	**55.03**
Treads cross-tongued joints and risers; rounded nosings; tongued, grooved, glued and blocked together; one 175 x 50 mm sawn softwood carriage					
25 mm treads; 19 mm risers	1.50	11.86	51.20	m2	**63.06**
Ends; quadrant	-	-	-	nr	**24.82**
32 mm treads; 25 mm risers	1.50	11.86	60.05	m2	**71.91**
Ends; quadrant	-	-	-	nr	**29.24**
Ends; housed to hardwood	-	-	-	nr	**0.44**

L WINDOWS/DOORS/STAIRS Including overheads and profit at 5.00%	Labour hours	Labour £	Material £	Unit	Total rate £
Winders; cross-tongued joints and risers in one width; rounded nosings; tongued, grooved glued and blocked together; one 175 x 50 mm sawn softwood carriage					
25 mm treads; 19 mm risers	2.50	19.77	55.51	m2	**75.28**
32 mm treads; 25 mm risers	2.50	19.77	65.26	m2	**85.03**
Wide ends; housed to hardwood	-	-	-	nr	**0.88**
Narrow ends; housed to hardwood	-	-	-	nr	**0.67**
Closed strings; in one width; 230 mm wide; rounded twice					
32 mm thick	0.50	3.95	9.42	m	**13.37**
38 mm thick	0.50	3.95	11.41	m	**15.36**
50 mm thick	0.50	3.95	14.99	m	**18.94**
Closed strings; cross-tongued joints; 280 mm wide; once rounded; fixing with screws; plugging 450 mm centres					
32 mm thick	0.60	4.74	17.24	m	**21.98**
Extra for short ramp	0.12	0.95	8.62	nr	**9.57**
38 mm thick	0.60	4.74	19.54	m	**24.28**
Extra for short ramp	0.12	0.95	9.76	nr	**10.71**
50 mm thick	0.60	4.74	24.06	m	**28.80**
Ends; fitted	0.10	0.79	0.28	nr	**1.07**
Ends; framed	0.12	0.95	2.82	nr	**3.77**
Extra for tongued heading joint	0.12	0.95	1.46	nr	**2.41**
Extra for short ramp	0.12	0.95	12.02	nr	**12.97**
Closed strings; ramped; crossed tongued joints 280 mm wide; once rounded; fixing with screws; plugging 450 mm centres					
32 mm thick	0.60	4.74	18.99	m	**23.73**
38 mm thick	0.60	4.74	21.48	m	**26.22**
50 mm thick	0.60	4.74	26.46	m	**31.20**
Apron linings; in one width 230 mm wide					
19 mm thick	0.33	2.61	3.14	m	**5.75**
25 mm thick	0.33	2.61	2.86	m	**5.47**
Handrails; rounded					
44 x 50 mm	0.25	1.98	2.79	m	**4.77**
50 x 75 mm	0.27	2.13	3.51	m	**5.64**
63 x 87 mm	0.30	2.37	4.89	m	**7.26**
75 x 100 mm	0.35	2.77	5.56	m	**8.33**
Handrails; moulded					
44 x 50 mm	0.25	1.98	3.07	m	**5.05**
50 x 75 mm	0.27	2.13	3.79	m	**5.92**
63 x 87 mm	0.30	2.37	5.17	m	**7.54**
75 x 100 mm	0.35	2.77	5.82	m	**8.59**
Handrails; rounded; ramped					
44 x 50 mm	0.33	2.61	5.58	m	**8.19**
50 x 75 mm	0.36	2.85	7.04	m	**9.89**
63 x 87 mm	0.40	3.16	9.79	m	**12.95**
75 x 100 mm	0.45	3.56	11.09	m	**14.65**
Handrails; moulded; ramped					
44 x 50 mm	0.33	2.61	6.13	m	**8.74**
50 x 75 mm	0.36	2.85	7.59	m	**10.44**
63 x 87 mm	0.40	3.16	10.35	m	**13.51**
75 x 100 mm	0.45	3.56	11.65	m	**15.21**
Add to above for					
grooved once	-	-	-	m	**0.26**
ends; framed	0.10	0.79	2.24	nr	**3.03**
ends; framed on rake	0.15	1.19	2.82	nr	**4.01**
Heading joints on rake; handrail screws					
44 x 50 mm	0.15	1.19	14.29	nr	**15.48**
50 x 75 mm	0.15	1.19	14.29	nr	**15.48**
63 x 87 mm	0.15	1.19	18.14	nr	**19.33**
75 x 100 mm	0.15	1.19	18.14	nr	**19.33**

L WINDOWS/DOORS/STAIRS Including overheads and profit at 5.00%		Labour hours	Labour £	Material £	Unit	Total rate £
L30 TIMBER STAIRS/WALKWAYS/BALUSTRADES - cont'd						
Purpose made staircase components; wrought softwood - cont'd						
Mitres; handrail screws						
44 x 50 mm		0.20	1.58	14.29	nr	**15.87**
50 x 75 mm		0.20	1.58	14.29	nr	**15.87**
63 x 87 mm		0.20	1.58	18.14	nr	**19.72**
75 x 100 mm		0.20	1.58	18.14	nr	**19.72**
Balusters; stiffeners						
25 x 25 mm		0.08	0.63	1.15	m	**1.78**
32 x 32 mm		0.08	0.63	1.33	m	**1.96**
32 x 50 mm		0.08	0.63	1.66	m	**2.29**
Ends; housed		0.03	0.24	0.54	nr	**0.78**
Sub rails						
32 x 63 mm		0.33	2.61	2.27	m	**4.88**
Ends; housed to newel		0.10	0.79	2.24	nr	**3.03**
Knee rails						
32 x 140 mm		0.40	3.16	3.74	m	**6.90**
Ends; housed to newel		0.10	0.79	2.24	nr	**3.03**
Newel posts						
50 x 100 mm; half		0.40	3.16	3.84	m	**7.00**
75 x 75 mm		0.40	3.16	4.16	m	**7.32**
100 x 100 mm		0.50	3.95	6.49	m	**10.44**
Newel caps; splayed on four sides						
62.5 x 125 x 50 mm; half		0.15	1.19	2.65	nr	**3.84**
100 x 100 x 50 mm		0.15	1.19	2.65	nr	**3.84**
125 x 125 x 50 mm		0.15	1.19	2.85	nr	**4.04**
Purpose made staircase components; selected West African Mahogany; PC £470.00/m3						
Board landings; cross-tongued joints; 100 x 50 mm sawn softwood bearers						
25 mm thick	PC £53.09	1.50	11.86	56.76	m2	**68.62**
32 mm thick	PC £64.94	1.50	11.86	69.20	m2	**81.06**
Treads cross-tongued joints and risers; rounded nosings; tongued, grooved, glued and blocked together; one 175 x 50 mm sawn softwood carriage						
25 mm treads; 19 mm risers	PC £67.09	2.00	15.81	72.04	m2	**87.85**
Ends; quadrant	PC £39.31	-	-	41.28	nr	**41.28**
32 mm treads; 25 mm risers	PC £80.25	2.00	15.81	85.86	m2	**101.67**
Ends; quadrant	PC £56.18	-	-	58.98	nr	**58.98**
Ends; housed to hardwood		-	-	-	nr	**0.67**
Winders; cross-tongued joints and risers in one width; rounded nosings; tongued, grooved glued and blocked together; one 175 x 50 mm sawn softwood carriage						
25 mm treads; 19 mm risers	PC £73.79	3.35	26.49	78.42	m2	**104.91**
32 mm treads; 25 mm risers	PC £88.28	3.35	26.49	93.64	m2	**120.13**
Wide ends; housed to hardwood		-	-	-	nr	**1.32**
Narrow ends; housed to hardwood		-	-	-	nr	**0.99**
Closed strings; in one width; 230 mm wide; rounded twice						
32 mm thick	PC £14.04/m	0.67	5.30	14.81	m2	**20.11**
38 mm thick	PC £16.87/m	0.67	5.30	17.78	m2	**23.08**
50 mm thick	PC £22.27/m	0.67	5.30	23.46	m2	**28.76**
Closed strings; cross-tongued joints; 280 mm wide; once rounded; fixing with screws; plugging 450 mm centres						
32 mm thick	PC £25.02	0.80	6.33	26.34	m	**32.67**
Extra for short ramp	PC £12.51	0.18	1.42	13.13	nr	**14.55**
38 mm thick	PC £28.32	0.80	6.33	29.81	m	**36.14**
Extra for short ramp	PC £14.18	0.18	1.42	14.89	nr	**16.31**

L WINDOWS/DOORS/STAIRS Including overheads and profit at 5.00%		Labour hours	Labour £	Material £	Unit	Total rate £
50 mm thick	PC £34.94	0.80	6.33	36.75	m	**43.08**
Ends; fitted	PC £0.79	0.15	1.19	0.83	nr	**2.02**
Ends; framed	PC £4.73	0.18	1.42	4.96	nr	**6.38**
Extra for tongued heading joint	PC £2.33	0.18	1.42	2.45	nr	**3.87**
Extra for short ramp	PC £17.49	0.18	1.42	18.37	nr	**19.79**
Closed strings; ramped; crossed tongued joints 280 mm wide; once rounded; fixing with screws; plugging 450 mm centres						
32 mm thick	PC £27.55	0.80	6.33	28.99	m	**35.32**
38 mm thick	PC £31.18	0.80	6.33	32.81	m	**39.14**
50 mm thick	PC £38.43	0.80	6.33	40.42	m	**46.75**
Apron linings; in one width 230 mm wide						
19 mm thick	PC £5.40	0.44	3.48	5.74	m	**9.22**
25 mm thick	PC £6.48	0.44	3.48	6.87	m	**10.35**
Handrails; rounded						
44 x 50 mm	PC £4.69	0.33	2.61	5.17	m	**7.78**
50 x 75 mm	PC £5.76	0.36	2.85	6.36	m	**9.21**
63 x 87 mm	PC £7.32	0.40	3.16	8.07	m	**11.23**
75 x 100 mm	PC £8.80	0.45	3.56	9.70	m	**13.26**
Handrails; moulded						
44 x 50 mm	PC £5.06	0.33	2.61	5.58	m	**8.19**
50 x 75 mm	PC £6.14	0.36	2.85	6.77	m	**9.62**
63 x 87 mm	PC £7.70	0.40	3.16	8.48	m	**11.64**
75 x 100 mm	PC £9.17	0.45	3.56	10.11	m	**13.67**
Handrails; rounded; ramped						
44 x 50 mm	PC £9.38	0.44	3.48	10.34	m	**13.82**
50 x 75 mm	PC £11.52	0.48	3.80	12.70	m	**16.50**
63 x 87 mm	PC £14.64	0.53	4.19	16.14	m	**20.33**
75 x 100 mm	PC £17.58	0.60	4.74	19.39	m	**24.13**
Handrails; moulded; ramped						
44 x 50 mm	PC £10.13	0.44	3.48	11.17	m	**14.65**
50 x 75 mm	PC £12.27	0.48	3.80	13.53	m	**17.33**
63 x 87 mm	PC £15.40	0.53	4.19	16.98	m	**21.17**
75 x 100 mm	PC £18.33	0.60	4.74	20.22	m	**24.96**
Add to above for						
grooved once		-	-	-	m	**0.44**
ends; framed	PC £3.57	0.15	1.19	3.75	nr	**4.94**
ends; framed on rake	PC £4.39	0.22	1.74	4.61	nr	**6.35**
Heading joints on rake; handrail screws						
44 x 50 mm	PC £16.73	0.22	1.74	17.57	nr	**19.31**
50 x 75 mm	PC £16.73	0.22	1.74	17.57	nr	**19.31**
63 x 87 mm	PC £21.44	0.22	1.74	22.51	nr	**24.25**
75 x 100 mm	PC £21.44	0.22	1.74	22.51	nr	**24.25**
Mitres; handrail screws						
44 x 50 mm	PC £16.73	0.27	2.13	17.57	nr	**19.70**
50 x 75 mm	PC £16.73	0.27	2.13	17.57	nr	**19.70**
63 x 87 mm	PC £21.44	0.27	2.13	22.51	nr	**24.64**
75 x 100 mm	PC £21.44	0.27	2.13	22.51	nr	**24.64**
Balusters; stiffeners						
25 x 25 mm	PC £2.37	0.10	0.79	2.49	m	**3.28**
32 x 32 mm	PC £2.69	0.10	0.79	2.83	m	**3.62**
32 x 50 mm	PC £3.18	0.10	0.79	3.34	m	**4.13**
Ends; housed	PC £0.79	0.05	0.40	0.83	nr	**1.23**
Sub rails						
32 x 63 mm	PC £4.08	0.44	3.48	4.28	m	**7.76**
Ends; housed to newel	PC £3.21	0.15	1.19	3.37	nr	**4.56**
Knee rails						
32 x 140 mm	PC £6.15	0.53	4.19	6.46	m	**10.65**
Ends; housed to newel	PC £3.21	0.15	1.19	3.37	nr	**4.56**
Newel posts						
50 x 100 mm; half	PC £6.55	0.53	4.19	6.88	m	**11.07**
75 x 75 mm	PC £7.05	0.53	4.19	7.40	m	**11.59**
100 x 100 mm	PC £10.66	0.67	5.30	11.19	m	**16.49**

L WINDOWS/DOORS/STAIRS Including overheads and profit at 5.00%		Labour hours	Labour £	Material £	Unit	Total rate £
L30 TIMBER STAIRS/WALKWAYS/BALUSTRADES - cont'd						
Purpose made staircase components; selected West African Mahogany; PC £470.00/m3 - cont'd						
Newel caps; splayed on four sides						
62.5 x 125 x 50 mm; half	PC £4.53	0.20	1.58	4.76	nr	**6.34**
100 x 100 x 50 mm	PC £4.53	0.20	1.58	4.76	nr	**6.34**
125 x 125 x 50 mm	PC £4.86	0.20	1.58	5.10	nr	**6.68**
Purpose made staircase components; Oak; PC £1200.00/m3						
Board landings; cross-tongued joints; 100 x 50 mm sawn softwood bearers						
25 mm thick	PC £112.80	1.50	11.86	119.47	m2	**131.33**
32 mm thick	PC £138.83	1.50	11.86	146.80	m2	**158.66**
Treads cross-tongued joints and risers; rounded nosings; tongued, grooved, glued and blocked together; one 175 x 50 mm sawn softwood carriage						
25 mm treads; 19 mm risers	PC £124.38	2.00	15.81	132.19	m2	**148.00**
Ends; quadrant	PC £62.20	-	-	65.31	nr	**65.31**
32 mm treads; 25 mm risers	PC £152.71	2.00	15.81	161.94	m2	**177.75**
Ends; quadrant	PC £76.37	-	-	80.19	nr	**80.19**
Ends; housed to hardwood		-	-	-	nr	**0.88**
Winders; cross-tongued joints and risers in one width; rounded nosings; tongued, grooved glued and blocked together; one 175 x 50 mm sawn softwood carriage						
25 mm treads; 19 mm risers	PC £136.83	3.35	26.49	144.61	m2	**171.10**
32 mm treads; 25 mm risers	PC £167.99	3.35	26.49	177.33	m2	**203.82**
Wide ends; housed to hardwood		-	-	-	nr	**1.76**
Narrow ends; housed to hardwood		-	-	-	nr	**1.32**
Closed strings; in one width; 230 mm wide; rounded twice						
32 mm thick	PC £29.76	0.67	5.30	31.31	m	**36.61**
38 mm thick	PC £35.47	0.67	5.30	37.31	m	**42.61**
50 mm thick	PC £46.85	0.67	5.30	49.26	m	**54.56**
Closed strings; cross-tongued joints; 280 mm wide; once rounded; fixing with screws; plugging 450 mm centres						
32 mm thick	PC £48.11	0.80	6.33	50.58	m	**56.91**
Extra for short ramp	PC £24.05	0.18	1.42	25.25	nr	**26.67**
38 mm thick	PC £55.07	0.80	6.33	57.90	m	**64.23**
Extra for short ramp	PC £27.53	0.18	1.42	28.90	nr	**30.32**
50 mm thick	PC £69.12	0.80	6.33	72.64	m	**78.97**
Ends; fitted	PC £0.91	0.15	1.19	0.95	nr	**2.14**
Ends; framed	PC £6.08	0.18	1.42	6.38	nr	**7.80**
Extra for tongued heading joint	PC £3.03	0.18	1.42	3.19	nr	**4.61**
Extra for short ramp	PC £34.57	0.18	1.42	36.30	nr	**37.72**
Closed strings; ramped; crossed tongued joints 280 mm wide; once rounded; fixing with screws; plugging 450 mm centres						
32 mm thick	PC £52.91	0.80	6.33	55.62	m	**61.95**
38 mm thick	PC £60.60	0.80	6.33	63.69	m	**70.02**
50 mm thick	PC £76.03	0.80	6.33	79.89	m	**86.22**
Apron linings; in one width 230 mm wide						
19 mm thick	PC £13.39	0.44	3.48	14.13	m	**17.61**
25 mm thick	PC £16.76	0.44	3.48	17.67	m	**21.15**
Handrails; rounded						
44 x 50 mm	PC £9.72	0.33	2.61	10.72	m	**13.33**
50 x 75 mm	PC £13.12	0.36	2.85	14.47	m	**17.32**
63 x 87 mm	PC £18.04	0.40	3.16	19.88	m	**23.04**
75 x 100 mm	PC £22.69	0.45	3.56	25.01	m	**28.57**

L WINDOWS/DOORS/STAIRS Including overheads and profit at 5.00%		Labour hours	Labour £	Material £	Unit	Total rate £
Handrails; moulded						
44 x 50 mm	PC £10.23	0.33	2.61	11.28	m	13.89
50 x 75 mm	PC £13.63	0.36	2.85	15.02	m	17.87
63 x 87 mm	PC £18.55	0.40	3.16	20.45	m	23.61
75 x 100 mm	PC £23.19	0.45	3.56	25.57	m	29.13
Handrails; rounded; ramped						
44 x 50 mm	PC £19.46	0.44	3.48	21.45	m	24.93
50 x 75 mm	PC £26.22	0.48	3.80	28.91	m	32.71
63 x 87 mm	PC £36.08	0.53	4.19	39.77	m	43.96
75 x 100 mm	PC £45.38	0.60	4.74	50.03	m	54.77
Handrails; moulded; ramped						
44 x 50 mm	PC £20.46	0.44	3.48	22.56	m	26.04
50 x 75 mm	PC £27.23	0.48	3.80	30.02	m	33.82
63 x 87 mm	PC £37.09	0.53	4.19	40.89	m	45.08
75 x 100 mm	PC £46.37	0.60	4.74	51.12	m	55.86
Add to above for						
grooved once		-	-	-	m	0.51
ends; framed	PC £4.63	0.15	1.19	4.87	nr	6.06
ends; framed on rake	PC £5.73	0.22	1.74	6.01	nr	7.75
Heading joints on rake; handrail screws						
44 x 50 mm	PC £25.81	0.22	1.74	27.10	nr	28.84
50 x 75 mm	PC £25.81	0.22	1.74	27.10	nr	28.84
63 x 87 mm	PC £31.57	0.22	1.74	33.15	nr	34.89
75 x 100 mm	PC £31.57	0.22	1.74	33.15	nr	34.89
Mitres; handrail screws						
44 x 50 mm	PC £25.81	0.27	2.13	27.10	nr	29.23
50 x 75 mm	PC £25.81	0.27	2.13	27.10	nr	29.23
63 x 87 mm	PC £31.57	0.27	2.13	33.15	nr	35.28
75 x 100 mm	PC £31.57	0.27	2.13	33.15	nr	35.28
Balusters; stiffeners						
25 x 25 mm	PC £3.82	0.10	0.79	4.01	m	4.80
32 x 32 mm	PC £4.85	0.10	0.79	5.09	m	5.88
32 x 50 mm	PC £6.38	0.10	0.79	6.70	m	7.49
Ends; housed	PC £1.05	0.05	0.40	1.10	nr	1.50
Sub rails						
32 x 63 mm	PC £8.20	0.44	3.48	8.61	m	12.09
Ends; housed to newel	PC £4.27	0.15	1.19	4.49	nr	5.68
Knee rails						
32 x 140 mm	PC £14.66	0.53	4.19	15.39	m	19.58
Ends; housed to newel	PC £4.27	0.15	1.19	4.49	nr	5.68
Newel posts						
50 x 100 mm; half	PC £15.97	0.53	4.19	16.76	m	20.95
75 x 75 mm	PC £17.51	0.53	4.19	18.39	m	22.58
100 x 100 mm	PC £28.91	0.67	5.30	30.35	m	35.65
Newel caps; splayed on four sides						
62.5 x 125 x 50 mm; half	PC £6.53	0.20	1.58	6.86	nr	8.44
100 x 100 x 50 mm	PC £6.53	0.20	1.58	6.86	nr	8.44
125 x 125 x 50 mm	PC £7.55	0.20	1.58	7.93	nr	9.51
L31 METAL STAIRS/WALKWAYS/BALUSTRADES						
Cat ladders, balustrades and handrails, etc.; mild steel; BS 4360						
Cat ladders; welded construction; 64 x 13 mm bar strings; 19 mm rungs at 250 mm centres; fixing by bolting; 0.46 m wide; 3.05 m high		3.00	26.87	126.00	nr	152.87
Extra for						
ends of strings; bent once; holed once for 10 mm dia bolt		-	-	-	nr	2.63
ends of strings; fanged		-	-	-	nr	1.58
ends of strings; bent in plane		-	-	-	nr	7.35

L WINDOWS/DOORS/STAIRS Including overheads and profit at 5.00%	Labour hours	Labour £	Material £	Unit	Total rate £
L31 METAL STAIRS/WALKWAYS/BALUSTRADES - cont'd					
Cat ladders, balustrades and handrails, etc.; mild steel; BS 4360 - cont'd					
Chequer plate flooring; over 300 mm wide; bolted to steel supports					
6 mm thick	6.00	53.74	20.04	m2	**73.78**
8 mm thick	7.00	62.70	26.04	m2	**88.74**
Balustrades; welded construction; galvanized after manufacture; 1070 mm high; 50 x 50 x 3.2 mm r.h.s. top rail; 38 x 13 mm bottom rail, 50 x 50 x 3.2 mm r.h.s. standards at 1830 mm centres with base plate drilled and bolted to concrete; 13 x 13 mm balusters at 102 mm centres	1.50	13.43	52.50	m	**65.93**
Balusters; isolated; one end ragged and cemented in; one 76 x 25 x 6 mm flange plate welded on; ground to a smooth finish; countersunk drilled and tap screwed to underside of handrail					
19 x 19 x 914 mm square bar	-	-	-	nr	**9.45**
Core-rails; joints prepared, welded and ground to a smooth finish; fixing on brackets (measured elsewhere)					
38 x 10 mm flat bar	-	-	-	m	**12.60**
50 x 8 mm flat bar	-	-	-	m	**12.08**
Extra for					
ends fanged	-	-	-	nr	**2.10**
ends scrolled	-	-	-	nr	**3.15**
ramps in thickness	-	-	-	nr	**4.20**
wreaths	-	-	-	nr	**5.25**
Handrails; joints prepared, welded and ground to a smooth finish; fixing on brackets (measured elsewhere)					
38 x 12 mm half oval bar	-	-	-	m	**16.80**
44 x 13 mm half oval bar	-	-	-	m	**18.11**
Extra for					
ends fanged	-	-	-	nr	**2.10**
ends scrolled	-	-	-	nr	**3.15**
ramps in thickness	-	-	-	nr	**4.20**
wreaths	-	-	-	nr	**5.25**
Handrail bracket; comprising 40 x 5 mm plate with mitred and welded angle; one end welded to 100 mm dia x 5 mm backplate; three times holed and plugged and screwed to brickwork; other end scribed and welded to underside of handrail					
140 mm girth	-	-	-	nr	**9.97**
Holes					
Holes; countersunk; for screws or bolts					
6 mm dia wood screw; 3 mm thick	0.05	0.45	-	nr	**0.45**
6 mm dia wood screw; 6 mm thick	0.07	0.63	-	nr	**0.63**
8 mm dia bolt; 6 mm thick	0.07	0.63	-	nr	**0.63**
10 mm dia bolt; 6 mm thick	0.08	0.72	-	nr	**0.72**
12 mm dia bolt; 8 mm thick	0.10	0.90	-	nr	**0.90**

L WINDOWS/DOORS/STAIRS
Including overheads and profit at 5.00%

BASIC GLASS PRICES (£/m2)

	£		£		£		£
Trade cut prices - all to limiting sizes							
Ordinary transparent glass							
Float; GG quality							
2 mm	17.09	5 mm	35.48	12 mm	113.41	19 mm	181.31
3 mm	21.57	6 mm	39.66	15 mm	133.06	25 mm	256.73
4 mm	24.45	10 mm	84.28				
Ordinary transluscent/patterned glass							
Obscured ground sheet glass - extra on sheet glass prices £7.29/m2							
Patterned							
4 mm tint.	33.47	6 mm tint.	41.48	4 mm white	18.55	6 mm white	32.39
Rough cast							
6 mm	28.32	10 mm	40.95				
Ordinary Georgian wired							
7 mm cast	29.90	6 mm polish.	65.20				
Polycarbonate standard sheets							
2 mm	25.67	4 mm	49.31	6 mm	73.03	10 mm	114.55
3 mm	37.44	5 mm	61.17	8 mm	96.66	12 mm	144.30
Special glasses							
'Antisun' float; bronze or grey							
4 mm	47.90	6 mm	69.14	10 mm	152.28	12 mm	192.06
6 mm float	112.14	10 mm float	123.65				
'Cetuff' toughened							
float							
4 mm	33.21	6 mm	44.79	10 mm	88.89	12 mm	135.36
5 mm	41.19						
patterned							
6 mm rough	50.00	10 mm rough	75.00	4 mm white	33.68	6 mm white	47.58
solar control; bronze or grey							
4 mm	60.37	6 mm	77.82	10 mm	146.80	12 mm	191.04
Clear laminated							
'security'							
7.5 mm	102.39	9.5 mm	97.01	11.5 mm	116.32		
'safety'							
4.4 mm	50.81	6.4 mm	55.39	8.8 mm	86.03	10.8 mm	102.73
5.4 mm	51.99	6.8 mm	68.30				
'Permawal'							
'reflective'							
6 mm	50.00	10 mm	75.00				
'Permasol'							
6 mm	70.00						
'Silvered'							
2 mm	33.93	3 mm	34.32	4 mm	39.85	6 mm	51.65
'Silvered tinted'							
4 mm bronze	61.27	4 mm grey	61.27	6 mm bronze	84.02	6 mm grey	84.02
'Venetian striped'							
4 mm	112.14	6 mm	123.65				

Trade discounts of 20 - 30% are usually available off the above prices. The following 'measured rates' are provided by a glazing Sub-Contractor and assume quantities in excess of 500 m2, within 20 miles of the suppliers branch. Therefore deduction of 'trade prices for cut sizes' from 'measured rates' is not a reliable basis for identifying fixing costs.

L WINDOWS/DOORS/STAIRS Including overheads and profit at 5.00%	Labour hours	Labour £	Material £	Unit	Total rate £
EXTRAS:					
Panes under 0.10 m2 are charged as 0.10 m2 plus per pane:-		£			
putty or bradded beads		0.27			
bradded beads and butyl compound		0.31			
screwed beads		0.38			
screwed beads and butyl compound		0.46			
The following extras are charged per m2					
external nailed or bradded beads		1.16			
to preservative stained wood		2.12			
to contracts - 100 - 200 m2		2.19			
- 50 - 100 m2		3.14			
L40 GENERAL GLAZING					
Standard plain glass; BS 952; clear float; panes area 0.15 - 4.00 m2					
3 mm thick; glazed with					
putty or bradded beads	-	-	-	m2	13.52
bradded beads and butyl compound	-	-	-	m2	16.93
screwed beads	-	-	-	m2	18.42
screwed beads and butyl compound	-	-	-	m2	21.41
4 mm thick; glazed with					
putty or bradded beads	-	-	-	m2	16.18
bradded beads and butyl compound	-	-	-	m2	20.03
screwed beads	-	-	-	m2	21.59
screwed beads and butyl compound	-	-	-	m2	25.00
5 mm thick; glazed with					
putty or bradded beads	-	-	-	m2	21.37
bradded beads and butyl compound	-	-	-	m2	23.88
screwed beads	-	-	-	m2	25.44
screwed beads and butyl compound	-	-	-	m2	28.85
6 mm thick; glazed with					
putty or bradded beads	-	-	-	m2	22.50
bradded beads and butyl compound	-	-	-	m2	25.02
screwed beads	-	-	-	m2	26.57
screwed beads and butyl compound	-	-	-	m2	29.98
Standard plain glass; BS 952; white patterned; panes area 0.15 - 4.00 m2					
4 mm thick; glazed with					
putty or bradded beads	-	-	-	m2	15.33
bradded beads and butyl compound	-	-	-	m2	18.74
screwed beads	-	-	-	m2	20.24
screwed beads and butyl compound	-	-	-	m2	23.21
6 mm thick; glazed with					
putty or bradded beads	-	-	-	m2	22.15
bradded beads and butyl compound	-	-	-	m2	25.63
screwed beads	-	-	-	m2	27.02
screwed beads and butyl compound	-	-	-	m2	30.09
Standard plain glass; BS 952; rough cast; panes area 0.15 - 4.00 m2					
6 mm thick; glazed with					
putty or bradded beads	-	-	-	m2	18.40
bradded beads and butyl compound	-	-	-	m2	21.87
screwed beads	-	-	-	m2	23.27
screwed beads and butyl compound	-	-	-	m2	26.34

L WINDOWS/DOORS/STAIRS Including overheads and profit at 5.00%	Labour hours	Labour £	Material £	Unit	Total rate £
Standard plain glass; BS 952; Georgian wired cast; panes area 0.15 - 4.00 m2					
7 mm thick; glazed with					
putty or bradded beads	-	-	-	m2	19.75
bradded beads and butyl compound	-	-	-	m2	23.21
screwed beads	-	-	-	m2	24.62
screwed beads and butyl compound	-	-	-	m2	27.69
Extra for lining up wired glass	-	-	-	m2	2.08
Standard plain glass; BS 952; Georgian wired polished; panes area 0.15 - 4.00 m2					
6 mm thick; glazed with					
putty or bradded beads	-	-	-	m2	42.60
bradded beads and butyl compound	-	-	-	m2	44.87
screwed beads	-	-	-	m2	46.27
screwed beads and butyl compound	-	-	-	m2	49.34
Extra for lining up wired glass	-	-	-	m2	2.33
Special glass; BS 952; toughened clear float; panes area 0.15 - 4.00 m2					
4 mm thick; glazed with					
putty or bradded beads	-	-	-	m2	40.89
bradded beads and butyl compound	-	-	-	m2	43.14
screwed beads	-	-	-	m2	44.55
screwed beads and butyl compound	-	-	-	m2	47.62
5 mm thick; glazed with					
putty or bradded beads	-	-	-	m2	45.29
bradded beads and butyl compound	-	-	-	m2	47.54
screwed beads	-	-	-	m2	48.95
screwed beads and butyl compound	-	-	-	m2	52.02
6 mm thick; glazed with					
putty or bradded beads	-	-	-	m2	48.56
bradded beads and butyl compound	-	-	-	m2	50.81
screwed beads	-	-	-	m2	52.22
screwed beads and butyl compound	-	-	-	m2	55.29
10 mm thick; glazed with					
putty or bradded beads	-	-	-	m2	88.74
bradded beads and butyl compound	-	-	-	m2	90.99
screwed beads	-	-	-	m2	92.40
screwed beads and butyl compound	-	-	-	m2	95.47
Special glass; BS 952; clear laminated safety glass; panes area 0.15 - 4.00 m2					
4.4 mm thick; glazed with					
putty or bradded beads	-	-	-	m2	54.69
bradded beads and butyl compound	-	-	-	m2	56.95
screwed beads	-	-	-	m2	58.35
screwed beads and butyl compound	-	-	-	m2	61.43
5.4 mm thick; glazed with					
putty or bradded beads	-	-	-	m2	55.78
bradded beads and butyl compound	-	-	-	m2	58.03
screwed beads	-	-	-	m2	59.44
screwed beads and butyl compound	-	-	-	m2	62.51
6.4 mm thick; glazed with					
putty or bradded beads	-	-	-	m2	58.92
bradded beads and butyl compound	-	-	-	m2	61.18
screwed beads	-	-	-	m2	61.91
screwed beads and butyl compound	-	-	-	m2	65.66

L WINDOWS/DOORS/STAIRS Including overheads and profit at 5.00%	Labour hours	Labour £	Material £	Unit	Total rate £
L40 GENERAL GLAZING - cont'd					
Special glass; BS 952; 'Antisun' solar control float glass infill panels; panes area 0.15 - 4.00 m2					
4 mm thick; glazed with non-hardening compound to metal	-	-	-	m2	**58.12**
6 mm thick; glazed with non-hardening compound to metal	-	-	-	m2	**77.48**
10 mm thick; glazed with non-hardening compound to metal	-	-	-	m2	**153.22**
12 mm thick; glazed with non-hardening compound to metal	-	-	-	m2	**189.46**
Special glass; BS 952; 'Pyran' half-hour fire resisting glass					
6.5 mm thick rectangular panes; glazed with screwed hardwood beads and Interdens intumescent strip					
300 x 400 mm pane	1.40	16.30	28.55	nr	**44.85**
400 x 800 mm pane	2.40	27.95	69.14	nr	**97.09**
500 x 1400 mm pane	3.80	44.25	143.60	nr	**187.85**
600 x 1800 mm pane	4.80	55.89	215.97	nr	**271.86**
6.5 mm thick irregular panes; glazed with screwed hardwood beads and Intergens intumescent strip					
300 x 400 mm pane	1.75	20.38	43.31	nr	**63.69**
400 x 800 mm pane	3.00	34.93	108.50	nr	**143.43**
500 x 1400 mm pane	4.75	55.31	229.73	nr	**285.04**
600 x 1800 mm pane	6.00	69.87	348.86	nr	**418.73**
Special glass; BS 952; 'Pyrostop' one-hour fire resisting glass					
11 mm thick regular panes; glazed with screwed hardwood beads and Intergens intumescent strip					
300 x 400 mm pane	1.60	18.63	49.75	nr	**68.38**
400 x 800 mm pane	2.70	31.44	125.65	nr	**157.09**
500 x 1400 mm pane	4.20	48.91	267.24	nr	**316.15**
600 x 1800 mm pane	5.30	61.72	406.73	nr	**468.45**
Special glass; BS 952; clear laminated security glass					
7.5 mm thick regular panes; glazed with screwed hardwood beads and Intergens intumescent strip					
300 x 400 mm pane	1.40	16.30	27.11	nr	**43.41**
400 x 800 mm pane	2.40	27.95	42.72	nr	**70.67**
500 x 1400 mm pane	3.80	44.25	85.82	nr	**130.07**
600 x 1800 mm pane	4.80	55.89	126.83	nr	**182.72**
Glass louvres; BS 952; with long edges ground or smooth					
5 mm or 6 mm thick float					
100 mm wide	-	-	-	m	**5.31**
150 mm wide	-	-	-	m	**5.91**
7 mm thick Georgian wired cast					
100 mm wide	-	-	-	m	**11.18**
150 mm wide	-	-	-	m	**11.93**
6 mm thick Georgian polished wired					
100 mm wide	-	-	-	m	**14.96**
150 mm wide	-	-	-	m	**17.20**

L WINDOWS/DOORS/STAIRS Including overheads and profit at 5.00%	Labour hours	Labour £	Material £	Unit	Total rate £
Labours on glass/sundries					
Curved cutting on panes					
4 mm thick	-	-	-	m	1.85
6 mm thick	-	-	-	m	1.85
6 mm thick; wired	-	-	-	m	2.79
Imitation washleather or black velvet strip; as bedding to edge of glass	-	-	-	m	0.76
Intumescent paste to fire doors per side of glass	-	-	-	m	4.29
Drill hole exceeding 6 mm and not exceeding 15 mm dia. through panes					
not exceeding 6 mm thick	-	-	-	nr	2.04
not exceeding 10 mm thick	-	-	-	nr	2.64
not exceeding 12 mm thick	-	-	-	nr	3.29
not exceeding 19 mm thick	-	-	-	nr	4.09
not exceeding 25 mm thick	-	-	-	nr	5.14
Drill hole exceeding 16 mm and not exceeding 38 mm dia. through panes					
not exceeding 6 mm thick	-	-	-	nr	2.89
not exceeding 10 mm thick	-	-	-	nr	3.89
not exceeding 12 mm thick	-	-	-	nr	4.64
not exceeding 19 mm thick	-	-	-	nr	5.84
not exceeding 25 mm thick	-	-	-	nr	7.28
Drill hole over 38 mm dia. through panes					
not exceeding 6 mm thick	-	-	-	nr	5.84
not exceeding 10 mm thick	-	-	-	nr	7.08
not exceeding 12 mm thick	-	-	-	nr	8.38
not exceeding 19 mm thick	-	-	-	nr	10.32
not exceeding 25 mm thick	-	-	-	nr	12.87
Add to the above					
Plus 33% for countersunk holes					
Plus 50% for wired or laminated glass					
Factory made double hermetically sealed units; Pilkington's 'Insulight' or similar; to wood or metal with butyl non-setting compound and screwed or clipped beads					
Two panes; BS 952; clear float glass; GG; 3 mm or 4 mm thick; 13 mm air space					
2 - 4 m2	-	-	-	m2	53.73
1 - 2 m2	-	-	-	m2	51.88
0.75 - 1 m2	-	-	-	m2	57.61
0.5 - 0.75 m2	-	-	-	m2	65.59
0.35 - 0.5 m2	-	-	-	m2	72.74
0.25 - 0.35 m2	-	-	-	m2	89.72
not exceeding 0.25 m2	-	-	-	m2	89.72
Two panes; BS 952; clear float glass; GG; 5 mm or 6 mm thick; 13 mm air space					
2 - 4 m2	-	-	-	m2	69.21
1 - 2 m2	-	-	-	m2	65.93
0.75 - 1 m2	-	-	-	m2	72.48
0.5 - 0.75 m2	-	-	-	m2	81.07
0.35 - 0.5 m2	-	-	-	m2	89.66
0.25 - 0.35 m2	-	-	-	m2	104.28
not exceeding 0.25 m2	-	-	-	m2	104.28

M SURFACE FINISHES Including overheads and profit at 5.00%	Labour hours	Labour £	Material £	Unit	Total rate £
M10 SAND CEMENT/CONCRETE/GRANOLITHIC SCREEDS/ FLOORING					
Cement and sand (1:3); steel trowelled					
Work to floors; one coat; level and to falls not exceeding 15 degrees from horizontal; to concrete base; over 300 mm wide					
32 mm	0.40	4.66	1.77	m2	**6.43**
40 mm	0.42	4.89	2.11	m2	**7.00**
48 mm	0.45	5.24	2.50	m2	**7.74**
50 mm	0.46	5.36	2.56	m2	**7.92**
60 mm	0.49	5.71	2.94	m2	**8.65**
65 mm	0.51	5.94	3.19	m2	**9.13**
70 mm	0.53	6.17	3.43	m2	**9.60**
75 mm	0.55	6.40	3.67	m2	**10.07**
Finishing around pipes; not exceeding 0.30 m girth	0.01	0.09	-	nr	**0.09**
Add to the above for work					
to falls and crossfalls and to slopes not exceeding 15 degrees from horizontal	0.03	0.27	-	m2	**0.27**
to slopes over 15 degrees from horizontal	0.10	0.90	-	m2	**0.90**
three coats of surface hardener brushed on	0.08	0.72	1.33	m2	**2.05**
water-repellent additive incorporated in the mix	-	-	-	m2	**0.77**
oil-repellent additive incorporated in the mix	-	-	-	m2	**2.98**
Cement and sand (1:3) beds and backings					
Work to floors; one coat; level; to concrete base; screeded; over 300 mm wide					
25 mm thick	-	-	-	m2	**6.18**
50 mm thick	-	-	-	m2	**7.76**
75 mm thick	-	-	-	m2	**10.98**
100 mm thick	-	-	-	m2	**14.58**
Work to floors; one coat; level; to concrete base; steel trowelled; over 300 mm wide					
25 mm thick	-	-	-	m2	**6.18**
50 mm thick	-	-	-	m2	**7.76**
75 mm thick	-	-	-	m2	**10.98**
100 mm thick	-	-	-	m2	**14.58**
Granolithic paving; cement and granite chippings 5 mm down (1:2.5); steel trowelled					
Work to floors; one coat; level; laid on concrete while green; over 300 mm wide					
20 mm	0.30	4.25	3.58	m2	**7.83**
25 mm	0.35	4.96	4.33	m2	**9.29**
Work to floors; two coat; laid on hacked concrete with slurry; over 300 mm wide					
38 mm	0.60	8.51	5.77	m2	**14.28**
50 mm	0.70	9.92	7.39	m2	**17.31**
75 mm	0.90	12.76	10.89	m2	**23.65**
Work to landings; one coat; level; laid on concrete while green; over 300 mm wide					
20 mm	0.45	6.38	3.58	m2	**9.96**
25 mm	0.45	6.38	4.33	m2	**10.71**
Work to landings; two coat; laid on hacked concrete with slurry; over 300 mm wide					
38 mm	0.75	10.63	5.77	m2	**16.40**
50 mm	0.85	12.05	7.39	m2	**19.44**
75 mm	1.00	14.18	10.89	m2	**25.07**
Finishing around pipes; not exceeding 0.30 m girth	0.05	0.71	-	nr	**0.71**

M SURFACE FINISHES Including overheads and profit at 5.00%	Labour hours	Labour £	Material £	Unit	Total rate £
Add to the above over 300 mm wide for					
1.35 kg/m2 carborundum grains trowelled in	0.06	0.85	0.94	m2	**1.79**
three coats of surface hardener brushed on	0.05	0.71	1.33	m2	**2.04**
liquid hardening additive incorporated in the mix	-	-	-	m2	**1.39**
oil-repellent additive incorporated in the mix	-	-	-	m2	**2.98**
25 mm work to treads; one coat; to concrete base					
225 mm wide	0.90	12.76	4.72	m	**17.48**
275 mm wide	0.90	12.76	4.80	m	**17.56**
Return end	0.18	2.55	-	nr	**2.55**
Extra for 38 mm 'Ferodo' nosing	0.20	2.84	6.37	m	**9.21**
13 mm skirtings; rounded top edge and coved bottom junction; to brickwork or blockwork base					
75 mm wide on face	0.55	7.80	0.68	m	**8.48**
150 mm wide on face	0.75	10.63	1.02	m	**11.65**
Ends; fair	0.05	0.71	-	nr	**0.71**
Angles	0.07	0.99	-	nr	**0.99**
13 mm outer margin to stairs; to follow profile of and with rounded nosing to treads and risers; fair edge and arris at bottom; to concrete base					
75 mm wide	0.90	12.76	1.36	m	**14.12**
Angles	0.07	0.99	-	nr	**0.99**
13 mm wall string to stairs; fair edge and arris on top; coved bottom junction with treads and risers; to brickwork or blockwork base					
275 mm (extreme) wide	0.80	11.34	1.36	m	**12.70**
Ends	0.05	0.71	-	nr	**0.71**
Angles	0.07	0.99	-	nr	**0.99**
Ramps	0.08	1.13	-	nr	**1.13**
Ramped and wreathed corners	0.10	1.42	-	nr	**1.42**
13 mm outer string to stairs; rounded nosing on top at junction with treads and risers; fair edge and arris at bottom; to concrete base					
300 mm (extreme) wide	0.80	11.34	2.26	m	**13.60**
Ends	0.05	0.71	-	nr	**0.71**
Angles	-	-	-	nr	**0.99**
Ramps	0.08	1.13	-	nr	**1.13**
Ramped and wreathed corners	0.10	1.42	-	nr	**1.42**
19 mm skirtings; rounded top edge and coved bottom junction; to brickwork or blockwork base					
75 mm wide on face	0.55	7.80	1.13	m	**8.93**
150 mm wide on face	0.75	10.63	1.70	m	**12.33**
Ends; fair	0.05	0.71	-	nr	**0.71**
Angles	0.07	0.99	-	nr	**0.99**
19 mm riser; one rounded nosing; to concrete base					
150 mm high; plain	0.90	12.76	3.83	m	**16.59**
150 mm high; undercut	0.90	12.76	3.83	m	**16.59**
180 mm high; plain	0.90	12.76	3.90	m	**16.66**
180 mm high; undercut	0.90	12.76	3.90	m	**16.66**

M SURFACE FINISHES Including overheads and profit at 5.00%	Labour hours	Labour £	Material £	Unit	Total rate £
M11 MASTIC ASPHALT FLOORING					
Mastic asphalt paving to BS 1076; black					
20 mm one coat coverings; felt isolating membrane; to concrete base; flat					
over 300 mm wide	-	-	-	m2	8.51
225 - 300 mm wide	-	-	-	m2	10.54
150 - 225 mm wide	-	-	-	m2	16.36
not exceeding 150 mm wide	-	-	-	m2	15.91
25 mm one coat coverings; felt isolating membrane; to concrete base; flat					
over 300 mm wide	-	-	-	m2	10.12
225 - 300 mm wide	-	-	-	m2	12.55
150 - 225 mm wide	-	-	-	m2	19.48
not exceeding 150 mm wide	-	-	-	m2	18.95
20 mm three coat skirtings to brickwork base					
not exceeding 150 mm girth	-	-	-	m	9.81
150 - 225 mm girth	-	-	-	m	11.32
225 - 300 mm girth	-	-	-	m	12.83
Mastic asphalt paving; acid-resisting; black					
20 mm one coat coverings; felt isolating membrane; to concrete base; flat					
over 300 mm wide	-	-	-	m2	10.03
225 - 300 mm wide	-	-	-	m2	12.43
150 - 225 mm wide	-	-	-	m2	17.88
not exceeding 150 mm wide	-	-	-	m2	18.74
25 mm one coat coverings; felt isolating membrane; to concrete base; flat					
over 300 mm wide	-	-	-	m2	11.66
225 - 300 mm wide	-	-	-	m2	14.45
150 - 225 mm wide	-	-	-	m2	22.41
not exceeding 150 mm wide	-	-	-	m2	21.79
20 mm three coat skirtings to brickwork base					
not exceeding 150 mm girth	-	-	-	m	11.24
150 - 225 mm girth	-	-	-	m	13.13
225 - 300 mm girth	-	-	-	m	14.96
Mastic asphalt paving to BS 6577; black					
20 mm one coat coverings; felt isolating membrane; to concrete base; flat					
over 300 mm wide	-	-	-	m2	13.55
225 - 300 mm wide	-	-	-	m2	16.80
150 - 225 mm wide	-	-	-	m2	23.34
not exceeding 150 mm wide	-	-	-	m2	25.33
25 mm one coat coverings; felt isolating membrane; to concrete base; flat					
over 300 mm wide	-	-	-	m2	16.42
225 - 300 mm wide	-	-	-	m2	20.37
150 - 225 mm wide	-	-	-	m2	28.35
not exceeding 150 mm wide	-	-	-	m2	30.71
20 mm three coat skirtings to brickwork base					
not exceeding 150 mm girth	-	-	-	m	10.98
150 - 225 mm girth	-	-	-	m	12.60
225 - 300 mm girth	-	-	-	m	14.55
Mastic asphalt paving to BS 1451; red					
20 mm one coat coverings; felt isolating membrane; to concrete base; flat					
over 300 mm wide	-	-	-	m2	15.58
225 - 300 mm wide	-	-	-	m2	32.57
150 - 225 mm wide	-	-	-	m2	37.09
not exceeding 150 mm wide	-	-	-	m2	38.96

M SURFACE FINISHES Including overheads and profit at 5.00%	Labour hours	Labour £	Material £	Unit	Total rate £
20 mm three coat skirtings to brickwork base					
not exceeding 150 mm girth	-	-	-	m	13.22
150 - 225 mm girth	-	-	-	m	14.79
M12 TROWELLED BITUMEN/RESIN/RUBBER LATEX FLOORING					
Latex cement floor screeds; steel trowelled					
Work to floors; level; to concrete base; over 300 mm wide					
3 mm thick; one coat	0.17	1.52	1.60	m2	3.12
5 mm thick; two coats	0.22	1.97	2.49	m2	4.46
Isocrete K screeds; steel trowelled					
Work to floors; level; to concrete base; over 300 mm wide					
35 mm thick; plus polymer bonder coat	-	-	-	m2	9.62
40 mm thick	-	-	-	m2	10.14
45 mm thick	-	-	-	m2	10.65
50 mm thick	-	-	-	m2	11.26
Work to floors; to falls or cross-falls; to concrete base; over 300 mm wide					
55 mm (average) thick	-	-	-	m2	7.37
60 mm (average) thick	-	-	-	m2	7.99
65 mm (average) thick	-	-	-	m2	8.40
75 mm (average) thick	-	-	-	m2	9.50
90 mm (average) thick	-	-	-	m2	11.25
'Synthanite' floor screeds; steel trowelled					
Work to floors; one coat; level; paper felt underlay; to concrete base; over 300 mm wide					
25 mm thick	0.50	4.48	10.63	m2	15.11
50 mm thick	0.60	5.37	16.88	m2	22.25
75 mm thick	0.70	6.27	20.31	m2	26.58
Bituminous lightweight insulating roof screeds					
'Bit-Ag' or similar roof screed; to falls or cross-falls; bitumen felt vapour barrier; over 300 mm wide					
75 mm (average) thick	-	-	-	m2	12.05
100 mm (average) thick	-	-	-	m2	14.41

BASIC PLASTER PRICES

	£		£
Plaster prices (£/tonne)			
BS 1191 Part 1; class A			
CB stucco	49.82		
BS 1191 Part 1; class B			
'Thistle' board	56.04	'Thistle' finish	57.28
BS 1191 Part 1; class C			
'Limelite' - backing	135.24	'Limelite' - renovating	179.17
- finishing	107.12		
Pre-mixed lightweight; BS 1191 Part 2			
'Carlite' - bonding	93.73	'Carlite' - finishing	74.81
- browning	91.75	- metal lathing	94.76
- browning HSB	100.31		
Projection 86.64			
Tyrolean 'Cullamix'	163.20		

M SURFACE FINISHES
Including overheads and profit at 5.00%

Plastering sand £8.86/tonne

Sundries

ceramic tile - adhesive (£/5L)	5.00	plasterboard nails (£/25kg)	34.31
- grout (£/kg)	0.40	scrim (£/100 m roll)	3.36
impact adhesive (£/5L)	17.10		

	Labour hours	Labour £	Material £	Unit	Total rate £
M20 PLASTERED/RENDERED/ROUGHCAST COATINGS					
Cement and sand (1:3) beds and backings					
10 mm work to walls; one coat; to brickwork or blockwork base					
over 300 mm wide	-	-	-	m2	**5.73**
not exceeding 300 mm wide	-	-	-	m	**3.45**
12 mm work to walls; one coat; to brickwork or blockwork base					
over 300 mm wide	-	-	-	m2	**6.26**
not exceeding 300 mm wide	-	-	-	m	**3.79**
13 mm work to walls; one coat; to brickwork or blockwork base; screeded					
over 300 mm wide	-	-	-	m2	**6.90**
not exceeding 300 mm wide	-	-	-	m	**3.45**
15 mm work to walls; one coat; to brickwork or blockwork base					
over 300 mm wide	-	-	-	m2	**7.38**
not exceeding 300 mm wide	-	-	-	m	**4.42**
Cement and sand (1:3); steel trowelled					
13 mm work to walls; two coats; to brickwork or blockwork base					
over 300 mm wide	-	-	-	m2	**8.21**
not exceeding 300 mm wide	-	-	-	m	**4.86**
16 mm work to walls; two coats; to brickwork or blockwork base					
over 300 mm wide	-	-	-	m2	**8.94**
not exceeding 300 mm wide	-	-	-	m	**5.34**
19 mm work to walls; two coats; to brickwork or blockwork base					
over 300 mm wide	-	-	-	m2	**9.76**
not exceeding 300 mm wide	-	-	-	m	**4.88**
Add to the above over 300 mm wide for first coat in water-repellent cement	-	-	-	m2	**1.74**
finishing coat in coloured cement	-	-	-	m2	**4.22**
Cement-lime-sand (1:2:9); steel trowelled					
19 mm work to walls; two coats; to brickwork or blockwork base					
over 300 mm wide	-	-	-	m2	**9.23**
not exceeding 300 mm wide	-	-	-	m	**5.54**
Cement-lime-sand (1:1:6); steel trowelled					
13 mm work to walls; two coats; to brickwork or blockwork base					
over 300 mm wide	-	-	-	m2	**8.68**
not exceeding 300 mm wide	-	-	-	m	**5.21**
Add to the above over 300 mm wide for waterproof additive	-	-	-	m2	**1.10**

M SURFACE FINISHES Including overheads and profit at 5.00%	Labour hours	Labour £	Material £	Unit	Total rate £
19 mm work to ceilings; three coats; to metal lathing base					
over 300 mm wide	-	-	-	m2	12.76
not exceeding 300 mm wide	-	-	-	m	6.38
Plaster; first and finishing coats of 'Carlite' pre-mixed lightweight plaster; steel trowelled					
13 mm work to walls; two coats; to brickwork or blockwork base (or 10 mm work to concrete base)					
over 300 mm wide	-	-	-	m2	6.49
over 300 mm wide; in staircase areas or plant rooms	-	-	-	m2	8.59
not exceeding 300 mm wide	-	-	-	m	3.24
13 mm work to isolated piers or columns; two coats over 300 mm wide	-	-	-	m2	11.32
not exceeding 300 mm wide	-	-	-	m	5.66
10 mm work to ceilings; two coats; to concrete base					
over 300 mm wide	-	-	-	m2	6.78
over 300 mm wide; 3.5 - 5 m high	-	-	-	m2	7.31
over 300 mm wide; in staircase areas or plant rooms	-	-	-	m2	8.88
not exceeding 300 mm wide	-	-	-	m	4.44
10 mm work to isolated beams; two coats; to concrete base					
over 300 mm wide	-	-	-	m2	9.78
over 300 mm wide; 3.5 - 5 m high	-	-	-	m2	10.30
not exceeding 300 mm wide	-	-	-	m	4.89
Plaster; one coat 'Snowplast' plaster; steel trowelled					
13 mm work to walls; one coat; to brickwork or blockwork base					
over 300 mm wide	-	-	-	m2	7.09
over 300 mm wide; in staircase areas or plant rooms	-	-	-	m2	9.37
not exceeding 300 mm wide	-	-	-	m	3.55
13 mm work to isolated columns; one coat					
over 300 mm wide	-	-	-	m2	10.08
not exceeding 300 mm wide	-	-	-	m	5.04
Plaster; first coat of cement and sand (1:3); finishing coat of 'Thistle' class B plaster; steel trowelled					
13 mm work to walls; two coats; to brickwork or blockwork base					
over 300 mm wide	-	-	-	m2	8.58
over 300 mm wide; in staircase areas or plant rooms	-	-	-	m2	10.68
not exceeding 300 mm wide	-	-	-	m	4.29
13 mm work to isolated columns; two coats					
over 300 mm wide	-	-	-	m2	11.57
not exceeding 300 mm wide	-	-	-	m	5.79
Plaster; first coat of cement-lime-sand (1:1:6); finishing coat of 'Sirapite' class B plaster; steel trowelled					
13 mm work to walls; two coats; to brickwork or blockwork base					
over 300 mm wide	-	-	-	m2	8.58
not exceeding 300 mm wide	-	-	-	m	4.29

M SURFACE FINISHES Including overheads and profit at 5.00%	Labour hours	Labour £	Material £	Unit	Total rate £
M20 PLASTERED/RENDERED/ROUGHCAST COATINGS - cont'd					
Plaster; first coat of cement-lime-sand (1:1:6); finishing coat of 'Sirapite' class B plaster; steel trowelled - cont'd					
13 mm work to walls; two coats; to brickwork or blockwork base					
over 300 mm wide; in staircase areas or plant rooms	-	-	-	m2	**10.68**
not exceeding 300 mm wide	-	-	-	m	**4.29**
13 mm work to isolated columns; two coats					
over 300 mm wide	-	-	-	m2	**12.60**
not exceeding 300 mm wide	-	-	-	m	**7.56**
Plaster; first coat of 'Limelite' renovating plaster; finishing coat of 'Limelite' finishing plaster; steel trowelled					
13 mm work to walls; two coats; to brickwork or blockwork base					
over 300 mm wide	-	-	-	m2	**9.43**
over 300 mm wide; in staricase areas or compartments under 4 m2	-	-	-	m2	**11.31**
not exceeding 300 mm wide	-	-	-	m	**5.66**
Dubbing out existing walls with undercoat plaster; average 6 mm thick					
over 300 mm wide	-	-	-	m2	**3.74**
not exceeding 300 mm wide	-	-	-	m	**2.23**
Dubbing out existing walls with undercoat plaster; average 12 mm thick					
over 300 mm wide	-	-	-	m2	**5.05**
not exceeding 300 mm wide	-	-	-	m	**3.02**
Plaster; one coat 'Thistle' projection plaster; steel trowelled					
13 mm work to walls; one coat; to brickwork or blockwork base					
over 300 mm wide	-	-	-	m2	**9.18**
over 300 mm wide; in staircase areas or plant rooms	-	-	-	m2	**11.28**
not exceeding 300 mm wide	-	-	-	m	**4.59**
6 mm work to isolated columns; one coat					
over 300 mm wide	-	-	-	m2	**11.80**
not exceeding 300 mm wide	-	-	-	m	**5.90**
Plaster; first, second and finishing coats of 'Carlite' pre-mixed lightweight plaster; steel trowelled					
13 mm work to ceilings; three coats to metal lathing base					
over 300 mm wide	-	-	-	m2	**10.08**
over 300 mm wide; in staircase areas or plant rooms	-	-	-	m2	**12.18**
not exceeding 300 mm wide	-	-	-	m	**5.04**
13 mm work to swept soffit of metal lathing arch former					
not exceeding 300 mm wide	-	-	-	m	**14.96**
300 - 400 mm wide	-	-	-	m	**16.01**
13 mm work to vertical face of metal lathing arch former					
not exceeding 0.5 m2 per side	-	-	-	nr	**28.88**
0.5 m2 - 1 m2 per side	-	-	-	nr	**34.13**

M SURFACE FINISHES Including overheads and profit at 5.00%	Labour hours	Labour £	Material £	Unit	Total rate £
Tyrolean decorative rendering; 13 mm first coat of cement-lime-sand (1:1:6); finishing three coats of 'Cullamix' applied with approved hand operated machine external					
To walls; four coats; to brickwork or blockwork base					
over 300 mm wide	-	-	-	m2	**21.12**
not exceeding 300 mm wide	-	-	-	m	**10.56**
'Mineralite' decorative rendering; first coat of cement-lime-sand (1:0.5:4.5); finishing coat of 'Mineralite'; applied by Specialist Subcontractor					
17 mm work to walls; to brickwork or blockwork base					
over 300 mm wide	-	-	-	m2	**42.53**
not exceeding 300 mm wide	-	-	-	m	**21.26**
form 9 x 6 mm expansion joint and point in one part polysulphide mastic	-	-	-	m	**7.28**
Plaster; one coat 'Thistle' board finish; steel trowelled (prices included within plasterboard rates)					
3 mm work to walls or ceilings; one coat; to plasterboard base					
over 300 mm wide	-	-	-	m2	**4.84**
over 300 mm wide; in staircase areas or plant rooms	-	-	-	m2	**6.94**
not exceeding 300 mm wide	-	-	-	m	**2.43**
Plaster; one coat 'Thistle' board finish; steel trowelled 3 mm work to walls or ceilings; one coat on and including gypsum plasterboard; BS 1230; fixing with nails; 3 mm joints filled with plaster and jute scrim cloth; to softwood base; plain grade baseboard or lath with rounded edges					
9.5 mm board to walls					
over 300 mm wide	0.50	5.82	2.35	m2	**8.17**
not exceeding 300 mm wide	0.25	2.91	0.74	m	**3.65**
9.5 mm board to walls; in staircase areas or plant rooms					
over 300 mm wide	0.60	6.99	2.35	m2	**9.34**
not exceeding 300 mm wide	0.30	3.49	0.74	m	**4.23**
9.5 mm board to isolated columns					
over 300 mm wide	0.66	7.69	2.37	m2	**10.06**
not exceeding 300 mm wide	0.33	3.84	0.75	m	**4.59**
9.5 mm board to ceilings					
over 300 mm wide	0.56	6.52	2.35	m2	**8.87**
over 300 mm wide; 3.5 - 5 m high	0.66	7.69	2.35	m2	**10.04**
not exceeding 300 mm wide	0.28	3.26	0.74	m	**4.00**
9.5 mm board to ceilings; in staircase areas or plant rooms					
over 300 mm wide	0.66	7.69	2.35	m2	**10.04**
not exceeding 300 mm wide	0.33	3.84	0.74	m	**4.58**
9.5 mm board to isolated beams					
over 300 mm wide	0.70	8.15	2.37	m2	**10.52**
not exceeding 300 mm wide	0.35	4.08	0.72	m	**4.80**
Add for 'Duplex' insulating grade	-	-	-	m2	**0.43**
12.5 mm board to walls					
over 300 mm wide	0.56	6.52	2.79	m2	**9.31**
not exceeding 300 mm wide	0.28	3.26	0.88	m	**4.14**

M SURFACE FINISHES Including overheads and profit at 5.00%	Labour hours	Labour £	Material £	Unit	Total rate £
M20 PLASTERED/RENDERED/ROUGHCAST COATINGS - cont'd					
Plaster; one coat 'Thistle' board finish; steel trowelled 3 mm work to walls or ceilings; one coat on and including gypsum plasterboard; BS 1230; fixing with nails; 3 mm joints filled with plaster and jute scrim cloth; to softwood base; plain grade baseboard or lath with rounded edges - cont'd					
12.5 mm board to walls; in staircase areas or plant rooms					
over 300 mm wide	0.66	7.69	2.79	m2	**10.48**
not exceeding 300 mm wide	0.33	3.84	0.88	m	**4.72**
12.5 mm board to isolated columns					
over 300 mm wide	0.70	8.15	2.80	m2	**10.95**
not exceeding 300 mm wide	0.35	4.08	0.89	m	**4.97**
12.5 mm board to ceilings					
over 300 mm wide	0.60	6.99	2.79	m2	**9.78**
over 300 mm wide; 3.5 - 5 m high	0.70	8.15	2.79	m2	**10.94**
not exceeding 300 mm wide	0.30	3.49	0.88	m	**4.37**
12.5 mm board to ceilings; in staircase areas or plant rooms					
over 300 mm wide	0.70	8.15	2.79	m2	**10.94**
not exceeding 300 mm wide	0.35	4.08	0.88	m	**4.96**
12.5 mm board to isolated beams					
over 300 mm wide	0.76	8.85	2.80	m2	**11.65**
not exceeding 300 mm wide	0.38	4.42	0.89	m	**5.31**
Add for 'Duplex' insulating grade	-	-	-	m2	**0.43**
Accessories					
'Expamet' render beads; white PVC nosings; to brickwork or blockwork base					
external stop bead; ref 1222	0.06	0.70	1.54	m	**2.24**
'Expamet' plaster beads; to brickwork or blockwork base					
angle bead; ref 550	0.07	0.82	0.43	m	**1.25**
architrave bead; ref 579	0.09	1.05	0.67	m	**1.72**
stop bead; ref 563	0.06	0.70	0.50	m	**1.20**
M22 SPRAYED MINERAL FIBRE COATINGS					
Prepare and apply by spray 'Mandolite P20' fire protection on structural steel/metalwork					
16 mm (one hour) fire protection					
to walls and columns	-	-	-	m2	**7.88**
to ceilings and beams	-	-	-	m2	**8.66**
to isolated metalwork	-	-	-	m2	**19.69**
22 mm (one and a half hour) fire protection					
to walls and columns	-	-	-	m2	**9.97**
to ceilings and beams	-	-	-	m2	**10.97**
to isolated metalwork	-	-	-	m2	**22.94**
28 mm (two hour) fire protection					
to walls and columns	-	-	-	m2	**11.03**
to ceilings and beams	-	-	-	m2	**12.13**
to isolated metalwork	-	-	-	m2	**24.26**
52 mm (four hour) fire protection					
to walls and columns	-	-	-	m2	**22.58**
to ceilings and beams	-	-	-	m2	**24.83**
to isolated metalwork	-	-	-	m2	**45.15**

M SURFACE FINISHES Including overheads and profit at 5.00%		Labour hours	Labour £	Material £	Unit	Total rate £
M30 METAL MESH LATHING/ANCHORED REINFORCEMENT FOR PLASTERED COATING						
Accessories						
Pre-formed galvanised expanded steel arch-frames; 'Truline' or similar; semi-circular; to suit walls up to 230 mm thick						
375 mm radius; for 800 mm opening; ref SC 750	PC £14.78	0.25	2.91	15.52	nr	**18.43**
425 mm radius; for 850 mm opening; ref SC 850	PC £15.68	0.25	2.91	16.46	nr	**19.37**
450 mm radius; for 900 mm opening; ref SC 900	PC £18.20	0.25	2.91	19.11	nr	**22.02**
600 mm radius; for 1200 mm opening; ref SC 1200	PC £22.65	0.25	2.91	23.78	nr	**26.69**
Lathing; Expamet 'BB' expanded metal lathing or similar; BS 1369; 50 mm laps						
6 mm mesh linings to ceilings; fixing with staples; to softwood base; over 300 mm wide						
ref BB263; 0.500 mm thick	PC £2.24	0.30	3.49	2.47	m2	**5.96**
ref BB264; 0.675 mm thick	PC £2.61	0.30	3.49	2.88	m2	**6.37**
6 mm mesh linings to ceilings; fixing with wire; to steelwork; over 300 mm wide						
ref BB263; 0.500 mm thick		0.32	3.73	2.54	m2	**6.27**
ref BB264; 0.675 mm thick		0.32	3.73	2.95	m2	**6.68**
6 mm mesh linings to ceilings; fixings with wire; to steelwork; not exceeding 300 mm wide						
ref BB263; 0.500 mm thick		0.20	2.33	0.80	m	**3.13**
ref BB264; 0.675 mm thick		0.20	2.33	0.93	m	**3.26**
Raking cutting		0.20	2.33	0.81	m	**3.14**
Cutting and fitting around pipes; not exceeding 0.30 m girth		0.20	2.33	0.94	nr	**3.27**
Lathing; Expamet 'Riblath' or 'Spraylath' stiffened expanded metal lathing or similar; 50 mm laps						
10 mm mesh lining to walls; fixing with nails; to softwood base; over 300 mm wide						
Riblath ref 269; 0.30 mm thick	PC £2.82	0.25	2.91	3.11	m2	**6.02**
Riblath ref 271; 0.50 mm thick	PC £3.25	0.25	2.91	3.59	m2	**6.50**
Spraylath ref 273; 0.50 mm thick		0.25	2.91	4.36	m2	**7.27**
10 mm mesh lining to walls; fixing with nails; to softwood base; not exceeding 300 mm wide						
Riblath ref 269; 0.30 mm thick		0.15	1.75	0.97	m	**2.72**
Riblath ref 271; 0.50 mm thick		0.15	1.75	1.13	m	**2.88**
Spraylath ref 273; 0.50 mm thick		0.15	1.75	1.37	m	**3.12**
10 mm mesh lining to walls; fixing to brick or blockwork; over 300 mm wide						
Red-rib ref 274; 0.50 mm thick	PC £3.57	0.20	2.33	5.38	m2	**7.71**
Stainless steel Riblath ref 267; 0.30 mm thick	PC £7.37	0.20	2.33	9.57	m2	**11.90**
10 mm mesh lining to ceilings; fixing with wire; to steelwork; over 300 mm wide						
Riblath ref 269; 0.30 mm thick		0.32	3.73	3.20	m2	**6.93**
Riblath ref 271; 0.50 mm thick		0.32	3.73	3.68	m2	**7.41**
Spraylath ref 273; 0.50 mm thick		0.32	3.73	4.45	m2	**8.18**
Raking cutting		0.15	1.75	0.62	m	**2.37**
Cutting and fitting around pipes; not exceeding 0.30 m girth		0.05	0.58	-	nr	**0.58**

M SURFACE FINISHES Including overheads and profit at 5.00%	Labour hours	Labour £	Material £	Unit	Total rate £
M31 FIBROUS PLASTER					
Fibrous plaster; fixing with screws; plugging; countersinking; stopping; filling and pointing joints with plaster					
16 mm plain slab coverings to ceilings					
over 300 mm wide	-	-	-	m2	**87.57**
not exceeding 300 mm wide	-	-	-	m	**28.22**
Coves; not exceeding 150 mm girth					
per 25 mm girth	-	-	-	m	**4.09**
Coves; 150 - 300 mm girth					
per 25 mm girth	-	-	-	m	**5.04**
Cornices					
per 25 mm girth	-	-	-	m	**5.04**
Cornice enrichments					
per 25 mm girth; depending on degree of enrichments	-	-	-	m	**6.17**
Fibrous plaster; fixing with plaster wadding filling and pointing joints with plaster; to steel base					
16 mm plain slab coverings to ceilings					
over 300 mm wide	-	-	-	m2	**87.57**
not exceeding 300 mm wide	-	-	-	m	**28.22**
16 mm plain casings to stanchions					
per 25 mm girth	-	-	-	m	**2.46**
16 mm plain casings to beams					
per 25 mm girth	-	-	-	m	**2.46**
Gyproc cove; fixing with adhesive; filling and pointing joints with plaster					
Cove					
125 mm girth	0.20	1.79	0.68	m	**2.47**
Angles	0.04	0.36	0.21	nr	**0.57**

ALTERNATIVE TILE MATERIALS

	£		£
Dennis Ruabon clay floor quarries (£/1000)			
Heather brown			
150x72x12.5 mm	159.40	194x94x12.5 mm	226.00
150x150x12.5 mm	225.30	194x194x12.5 mm	437.10
Red			
150x150x12.5 mm; square	149.70	150x150x12.5 mm; hexagonal	219.30
Rustic			
150x150x12.5 mm	220.30		
Daniel Platt heavy duty floor tiles (£/m2)			
'Ferrolite' flat; 150x150x9 mm			
black	10.13	steel	9.46
cream	9.46	red	8.09
'Ferrolite' flat; 150x150x12 mm			
black	13.43	chocolate	13.43
cream	11.34	red	11.09
'Ferrundum' anti-slip; 150x150x12 mm			
black	21.28	red	19.32
chocolate	21.28		

M SURFACE FINISHES
Including overheads and profit at 5.00%

Langley wall and floor tiles (£/m2)

'Buchtal' fine grain ceramic wall and floor tiles			
240x115x11 mm - series 1	22.54	194x144x11 mm - series 1	21.85
- series 2	24.75	- series 2	23.99
'Buchtal' rustic glazed wall and floor tiles			
240x115x11 mm - series 1	22.54	194x94x11 mm - series 1	22.00
- series 2	24.12	- series 2	23.57
- series 3	21.25	- series 3	20.87
'Sinzag' glazed ceramic wall facings			
240x115x10 mm - series 1	20.89	240x52x10 mm - series 1	23.79
- series 2	28.68		
- series 3	58.38		
'Sinzag' vitrified ceramic floor tiles; 150x150x11/12 mm			
plain - grey/white speckled	23.18	textured - series 2	24.81

Marley floor tiles (£/m2); 300x300 mm

'Econoflex' - series 1/2	1.76	anti-static	19.85
- series 4	2.09	'Travertine' - 2.5 mm	3.40
'Marleyflex' - 2.0 mm	2.27	'Vylon'	2.38
- 2.5 mm	2.66		

		Labour hours	Labour £	Material £	Unit	Total rate £
M40 STONE/CONCRETE/QUARRY/CERAMIC TILING/ MOSAIC						
Clay floor quarries; BS 6431; class 1; Daniel Platt 'Crown' tiles or similar; level bedding 10 mm thick and jointing in cement and sand (1:3); butt joints straight both ways; flush pointing with grout; to cement and sand base						
Work to floors; over 300 mm wide						
150 x 150 x 12.5 mm; red	PC £202.23/1000	0.80	7.17	10.73	m2	**17.90**
150 x 150 x 12.5 mm; brown	PC £317.66/1000	0.80	7.17	16.34	m2	**23.51**
200 x 200 x 19 mm; brown	PC £774.79/1000	0.65	5.82	22.24	m2	**28.06**
Works to floors; in staircase areas or plant rooms						
150 x 150 x 12.5 mm; red		0.90	8.06	10.73	m2	**18.79**
150 x 150 x 12.5 mm; brown		0.90	8.06	16.34	m2	**24.40**
200 x 200 x 19 mm; brown		0.75	6.72	22.24	m2	**28.96**
Work to floors; not exceeding 300 mm wide						
150 x 150 x 12.5 mm; red		0.40	3.58	3.86	m	**7.44**
150 x 150 x 12.5 mm; brown		0.40	3.58	5.87	m	**9.45**
200 x 200 x 19 mm; brown		0.33	2.96	8.40	m	**11.36**
Fair square cutting against flush edges of existing finishings		0.10	0.90	0.50	m	**1.40**
Raking cutting		0.20	1.79	0.73	m	**2.52**
Cutting around pipes; not exceeding 0.30 m girth		0.15	1.34	-	nr	**1.34**
Extra for cutting and fitting into recessed manhole covers 600 x 600 mm; lining up with adjoining work		1.00	8.96	-	nr	**8.96**
Work to sills; 150 mm wide; rounded edge tiles						
150 x 150 x 12.5 mm; red	PC £378.10/1000	0.33	2.96	3.09	m	**6.05**
150 x 150 x 12.5 mm; brown	PC £407.56/1000	0.33	2.96	3.32	m	**6.28**
Fitted end		0.15	1.34	-	m	**1.34**

M SURFACE FINISHES Including overheads and profit at 5.00%		Labour hours	Labour £	Material £	Unit	Total rate £
M40 STONE/CONCRETE/QUARRY/CERAMIC TILING/ MOSAIC - cont'd						
Clay floor quarries; BS 6431; class 1; Daniel Platt 'Crown' tiles or similar; level bedding 10 mm thick and jointing in cement and sand (1:3); butt joints straight both ways; flush pointing with grout; to cement and sand base - cont'd						
Coved skirtings; 150 mm high; rounded top edge						
150 x 150 x 12.5 mm; red	PC £378.10/1000	0.25	2.24	3.09	m	**5.33**
150 x 150 x 12.5 mm; brown	PC £407.56/1000	0.25	2.24	3.32	m	**5.56**
Ends		0.05	0.45	-	nr	**0.45**
Angles		0.15	1.34	1.39	nr	**2.73**
Glazed ceramic wall tiles; BS 6431; fixing with adhesive; butt joints straight both ways; flush pointing with white grout; to plaster base						
Work to walls; over 300 mm wide						
152 x 152 x 5.5 mm; white	PC £8.50	0.60	6.99	10.82	m2	**17.81**
152 x 152 x 5.5 mm; light colours	PC £10.63	0.60	6.99	13.15	m2	**20.14**
152 x 152 x 5.5 mm; dark colours	PC £13.15	0.60	6.99	15.93	m2	**22.92**
Extra for RE or REX tile		-	-	-	nr	**0.07**
200 x 100 x 6.5 mm; white and light colours	PC £8.80	0.60	6.99	11.14	m2	**18.13**
250 x 200 x 7 mm; light colours	PC £14.03	0.50	5.82	16.91	m2	**22.73**
Work to walls; in staircase areas or plant rooms						
152 x 152 x 5.5 mm; white		0.67	7.80	10.82	m2	**18.62**
Work to walls; not exceeding 300 mm wide						
152 x 152 x 5.5 mm; white		0.30	3.49	3.38	m	**6.87**
152 x 152 x 5.5 mm; light colours		0.30	3.49	4.11	m	**7.60**
152 x 152 x 5.5 mm; dark colours		0.30	3.49	4.98	m	**8.47**
200 x 100 x 6.5 mm; white and light colours		0.30	3.49	3.48	m	**6.97**
250 x 200 x 7 mm; light colours		0.25	2.91	5.29	m	**8.20**
Cutting around pipes; not exceeding 0.30 m girth		0.10	1.16	-	nr	**1.16**
Work to sills; 150 mm wide; rounded edge tiles						
152 x 152 x 5.5 mm; white		0.25	2.91	1.72	m	**4.63**
Fitted end		0.10	1.16	-	nr	**1.16**
198 x 64.5 x 6 mm Langley 'Coloursound' wall tiles; fixing with adhesive; butt joints straight both ways; flush pointing with white grout; to plaster base						
Work to walls						
over 300 mm wide	PC £19.95	1.80	20.96	23.44	m2	**44.40**
not exceeding 300 mm wide		0.70	8.15	7.03	m	**15.18**
Glazed ceramic floor tiles; level; bedding 10 mm thick and jointing in cement and sand (1:3); butt joints straight both ways; flush pointing with white grout; to cement and sand base						
Work to floors; over 300 mm wide						
200 x 200 x 8 mm	PC £18.65	0.60	6.99	21.49	m2	**28.48**
Work to floors; not exceeding 300 mm wide						
200 x 200 x 8 mm		0.30	3.49	6.82	m	**10.31**

M SURFACE FINISHES Including overheads and profit at 5.00%	Labour hours	Labour £	Material £	Unit	Total rate £
Dakota mahogany granite cladding; polished finish; jointed and pointed in coloured mortar (1:2:8)					
20 mm Work to floors; level; to cement and sand base					
over 300 mm wide	-	-	-	m2	270.38
20 x 300 mm treads; plain nosing	-	-	-	m	100.80
Raking cutting	-	-	-	m	12.60
Polished edges	-	-	-	m	12.60
Birdsmouth	-	-	-	m	25.20
20 mm work to walls; to cement and sand base					
over 300 mm wide	-	-	-	m2	291.38
not exceeding 300 mm wide	-	-	-	m	114.98
40 mm work to walls; to cement and sand base					
over 300 mm wide	-	-	-	m2	371.70
not exceeding 300 mm wide	-	-	-	m	147.00
Roman Travertine marble cladding; polished finish; jointed and pointed in coloured mortar (1:2:8)					
20 mm Work to floors; level; to cement and sand base					
over 300 mm wide	-	-	-	m2	199.50
20 x 300 mm treads; plain nosing	-	-	-	m	84.00
Raking cutting	-	-	-	m	11.55
Polished edges	-	-	-	m	11.55
Birdsmouth	-	-	-	m	24.15
20 mm work to walls; to cement and sand base					
over 300 mm wide	-	-	-	m2	222.08
not exceeding 300 mm wide	-	-	-	m	90.30
40 mm work to walls; to cement and sand base					
over 300 mm wide	-	-	-	m2	278.25
not exceeding 300 mm wide	-	-	-	m	113.14
M41 TERRAZZO TILING/IN SITU TERRAZZO					
Terrazzo tiles; cement and white Sicilian marble aggregate; level; bedding 13 mm thick and jointing in cement and sand (1:3); butt joints straight both ways; pointing with white grout; margins laid in situ; polished; to concrete base					
Work to floors; over 300 mm wide					
305 x 305 x 25 mm	2.30	32.60	19.64	m2	52.24
Work to floors; not exceeding 300 mm wide					
305 x 305 x 25 mm	1.00	14.18	5.89	m	20.07
Add to the above for					
white or coloured cement	-	-	-	m2	2.43
coloured marble aggregate	-	-	-	m2	2.89
Terrazzo paving; cement and white Sicilian marble aggregate; polished					
16 mm work to floors; one coat; level; to cement and sand base					
over 300 mm wide	1.80	25.52	8.09	m2	33.61
not exceeding 300 mm wide	0.90	12.76	2.43	m	15.19
Finishing around pipes; not exceeding 0.30 m girth	0.25	3.54	-	nr	3.54
16 mm work to landings; one coat; level; to cement and sand base					
over 300 mm wide	2.40	34.02	8.09	m2	42.11
not exceeding 300 mm wide	1.00	14.18	2.43	m	16.61

M SURFACE FINISHES Including overheads and profit at 5.00%	Labour hours	Labour £	Material £	Unit	Total rate £
M41 TERRAZZO TILING/IN SITU TERRAZZO - cont'd					
Terrazzo paving; cement and white Sicilian marble aggregate; polished - cont'd					
Add to the above for					
35 mm cement and sand (1:3) screed	0.80	11.34	4.97	m2	**16.31**
carborundum grains 0.27 kg/m2	0.50	7.09	0.94	m2	**8.03**
white or coloured cement	-	-	-	m2	**3.47**
coloured marble aggregate	-	-	-	m2	**2.89**
Division strips; set in paving to form panelling					
25 x 3 mm ebonite strip	0.20	2.84	0.58	m	**3.42**
25 x 3 mm brass strip; including anchor wires	0.25	3.54	4.62	m	**8.16**
16 mm work to treads; one coat; including 16 mm cement and sand (1:3) screed; to concrete base					
225 mm wide	2.50	35.44	2.43	m	**37.87**
275 mm wide	2.70	38.27	2.66	m	**40.93**
Return end	0.60	8.51	-	m	**8.51**
Extra for					
white or coloured cement	-	-	-	m	**0.87**
38 mm 'Ferodo' nosing	0.25	3.54	5.20	m	**8.74**
projecting nosing	0.66	9.36	0.69	m	**10.05**
6 mm skirtings; rounded top edge; coved bottom junction; including 13 mm cement and sand screed; to brickwork or blockwork base					
75 mm wide on face	1.30	18.43	0.52	m	**18.95**
150 mm wide on face	1.50	21.26	1.16	m	**22.42**
Ends; fair	0.20	2.84	-	nr	**2.84**
Angles	0.20	2.84	-	nr	**2.84**
6 mm outer margin to stairs; to follow profile of and with rounded nosing to treads and risers; fair edge and arris at bottom; to concrete base					
75 mm wide	2.50	35.44	0.92	m	**36.36**
Angles	0.10	1.42	-	nr	**1.42**
6 mm riser; one rounded nosing; to concrete base					
150 mm high; plain	2.00	28.35	1.04	m	**29.39**
150 mm high; undercut	2.00	28.35	1.04	m	**29.39**
180 mm high; plain	2.10	29.77	1.39	m	**31.16**
180 mm high; undercut	2.10	29.77	1.39	m	**31.16**
6 mm wall string to stairs; fair edge and arris at top; coved bottom junction with treads and risers; including 13 mm cement and sand screed; to concrete base					
275 mm (extreme) wide	3.75	53.16	2.66	m	**55.82**
Ends	0.20	2.84	-	nr	**2.84**
Angles	0.25	3.54	-	nr	**3.54**
Ramps	0.33	4.68	-	nr	**4.68**
Ramped and wreathed corners	0.50	7.09	-	nr	**7.09**
6 mm outer string to stairs; rounded nosing on top at junction with treads and risers; fair edge and arris at bottom; including 13 mm cement and sand screed; to concrete base					
300 mm (extreme) wide	5.00	70.88	4.04	m	**74.92**
Ends	0.25	3.54	-	nr	**3.54**
Angles	0.40	5.67	-	nr	**5.67**
Ramps	0.50	7.09	-	nr	**7.09**
Ramped and wreathed corners	0.90	12.76	-	nr	**12.76**

M SURFACE FINISHES Including overheads and profit at 5.00%	Labour hours	Labour £	Material £	Unit	Total rate £
Terrazzo wall lining; cement and white Sicilian marble aggregate; polished					
6 mm work to walls; one coat; to normal dado height; on cement and sand base					
over 300 mm wide	6.00	85.05	5.78	m2	**90.83**
M42 WOOD BLOCK/COMPOSITION BLOCK/PARQUET FLOORING					
Wood block; Vigers, Stevens & Adams 'Feltwood'; 7.5 mm thick; level; fixing with adhesive; to cement and sand base					
Work to floors; over 300 mm wide					
Iroko	0.50	4.48	10.93	m2	**15.41**
Wood blocks 25 mm thick; tongued and grooved joints; herringbone pattern; level; fixing with adhesive; to cement and sand base					
Work to floors; over 300 mm wide					
Iroko	-	-	-	m2	**37.59**
Merbau	-	-	-	m2	**35.28**
French Oak	-	-	-	m2	**44.42**
American Oak	-	-	-	m2	**36.22**
Fair square cutting against flush edges of existing finishings	-	-	-	m	**2.10**
Extra for cutting and fitting into recessed duct covers 450 mm wide; lining up with adjoining work	-	-	-	m	**8.61**
Cutting around pipes; not exceeding 0.30 m girth	-	-	-	nr	**1.37**
Extra for cutting and fitting into recessed manhole covers 600 x 600 mm; lining up with adjoining work	-	-	-	nr	**5.88**
Add to wood block flooring over 300 mm wide for					
sanding; one coat sealer; one coat wax polish	-	-	-	m2	**3.53**
sanding; two coats sealer; buffing with steel wool	-	-	-	m2	**2.35**
sanding; three coats polyurethane lacquer; buffing down between coats	-	-	-	m2	**2.35**

ALTERNATIVE FLEXIBLE SHEET MATERIALS (£/m2)

	£		£
Forbo-Nairn flooring			
'Armourfloor' - 3.2 mm	7.59	'Amourflex' - 2.0 mm	5.70
'Armourflex' - 3.2 mm	6.89		
Marley sheet flooring			
'HD Acoustic'	4.59	'HD Safetread'	7.11
'Format +'	5.73	'Vynatred'	3.28

M SURFACE FINISHES Including overheads and profit at 5.00%	Labour hours	Labour £	Material £	Unit	Total rate £
M50 RUBBER/PLASTICS/CORK/LINO/CARPET TILING/ SHEETING					
Linoleum sheet; BS 810; Forbo-Nairn or similar; level; fixing with adhesive; butt joints; to cement and sand base					
Work to floors; over 300 mm wide					
3.2 mm; plain	0.40	3.58	7.18	m2	**10.76**
3.2 mm; marbled	0.40	3.58	9.11	m2	**12.69**
Vinyl sheet; Altro 'Safety' range or similar with welded seams; level; fixing with adhesive; to cement and sand base					
Work to floors; over 300 mm wide					
2 mm thick; Marine T20	0.60	5.37	9.66	m2	**15.03**
2.5 mm thick; Classic D25	0.70	6.27	12.71	m2	**18.98**
3.5 mm thick; stronghold	0.80	7.17	16.35	m2	**23.52**
Vinyl sheet; heavy duty; Marley 'HD' or similar; level; with welded seams; fixing with adhesive; level; to cement and sand base					
Work to floors; over 300 mm wide					
2 mm thick	0.45	4.03	5.26	m2	**9.29**
2.5 mm thick	0.50	4.48	5.91	m2	**10.39**
2 mm skirtings					
75 mm high	0.10	0.90	0.63	m	**1.53**
100 mm high	0.12	1.07	0.75	m	**1.82**
150 mm high	0.15	1.34	1.10	m	**2.44**
Vinyl sheet; 'Gerflex' standard sheet; 'Classic' range; level; with welded seams; fixing with adhesive; to cement and sand base					
Work to floors; over 300 mm wide					
2 mm thick	0.50	4.48	4.99	m2	**9.47**
Vinyl sheet; 'Armstrong Rhino Contract'; level; with welded seams; fixing with adhesive; to cement and sand base					
Work to floors; over 300 mm wide					
2.5 mm thick	0.50	4.48	9.34	m2	**13.82**
Vinyl sheet; Marmoleum'; level; with welded seams; fixing with adhesive; to cement and sand base					
Work to floors; over 300 mm wide					
2.5 mm thick	0.50	4.48	7.57	m2	**12.05**
Vinyl tiles; 'Accoflex'; level; fixing with adhesive; butt joints; straight both ways; to cement and sand base					
Work to floors; over 300 mm wide					
300 x 300 x 2.0 mm	0.25	2.24	3.75	m2	**5.99**

M SURFACE FINISHES Including overheads and profit at 5.00%	Labour hours	Labour £	Material £	Unit	Total rate £
Vinyl semi-flexible tiles; 'Arlon' or similar; level; fixing with adhesive; butt joints; straight both ways; to cement and sand base					
Work to floors; over 300 mm wide					
250 x 250 x 2.0 mm	0.25	2.24	4.08	m2	**6.32**
Vinyl semi-flexible tiles; Marley 'Marleyflex' or similar; level; fixing with adhesive; butt joints straight both ways; to cement and sand base					
Work to floors; over 300 mm wide					
250 x 250 x 2.0 mm	0.25	2.24	3.24	m2	**5.48**
250 x 250 x 2.5 mm	0.25	2.24	3.69	m2	**5.93**
Vinyl semi-flexible tiles; 'Vylon' or similar; level; fixing with adhesive; butt joints; straight both ways; to cement and sand base					
Work to floors; over 300 mm wide					
250 x 250 x 2.0 mm	0.25	2.24	3.24	m2	**5.48**
Vinyl tiles; anti-static; level; fixing with adhesive; butt joints straight both ways; to cement and sand base					
Work to floors; over 300 mm wide					
457 x 457 x 2 mm	0.45	4.03	23.96	m2	**27.99**
Vinyl tiles; 'Polyflex'; level; fixing with adhesive; butt joints; straight both ways;to cement and sand base					
Work to floors; over 300 mm wide					
300 x 300 x 1.5 mm	0.25	2.24	2.88	m2	**5.12**
300 x 300 x 2.0 mm	0.25	2.24	3.13	m2	**5.37**
Vinyl tiles; 'Polyflor XL'; level; fixing with adhesive; butt joints; straight both ways; to cement and sand base					
Work to floors; over 300 mm wide					
300 x 300 x 2.0 mm	0.25	2.24	5.21	m2	**7.45**
Vinyl tiles; 'Marley 'HD''; level; fixing with adhesive; butt joints; straight both ways; to cement and sand base					
Work to floors; over 300 mm wide					
300 x 300 x 2.0 mm	0.25	2.24	5.69	m2	**7.93**
Thermoplastic tiles; Marley 'Econoflex' or similar; level; fixing with adhesive; butt joints straight both ways; to cement and sand base					
Work to floors; over 300 mm wide					
250 x 250 x 2.0 mm; (series 2)	0.23	2.06	2.66	m2	**4.72**
250 x 250 x 2.0 mm; (series 4)	0.23	2.06	3.03	m2	**5.09**
Linoleum tiles; BS 810; Nairn Floors or similar; level; fixing with adhesive; butt joints straight both ways; to cement and sand base					
Work to floors; over 300 mm wide					
3.2 mm thick (marbled patterns)	0.30	2.69	9.30	m2	**11.99**

M SURFACE FINISHES Including overheads and profit at 5.00%		Labour hours	Labour £	Material £	Unit	Total rate £
M50 RUBBER/PLASTICS/CORK/LINO/CARPET TILING/ SHEETING - cont'd						
Cork tiles; Wicanders 'Corktile' or similar; level fixing with adhesive; butt joints straight both ways; to cement and sand base						
Work to floors; over 300 mm wide						
300 x 300 x 3.2 mm		0.40	3.58	13.78	m2	**17.36**
Rubber studded tiles; Altro 'Mondopave' or similar; level; fixing with adhesive; butt joints; straight; to cement and sand base						
Work to floors; over 300 mm wide						
500 x 500 x 2.5 mm; type GS; black		0.60	5.37	14.28	m2	**19.65**
500 x 500 x 4 mm; type BT; black		0.60	5.37	18.13	m2	**23.50**
Work to landings; over 300 mm wide						
500 x 500 x 4 mm; type BT; black		0.80	7.17	18.13	m2	**25.30**
4 mm thick to tread						
275 mm wide		0.50	4.48	5.87	m	**10.35**
4 mm thick to riser						
185 mm wide		0.60	5.37	4.06	m	**9.43**
Sundry floor sheeting underlays						
For floor finishings; over 300 mm wide						
building paper to BS 1521; class A;						
75 mm lap		0.06	0.54	0.62	m2	**1.16**
3.2 mm thick hardboard		0.20	1.79	0.99	m2	**2.78**
6.0 mm thick plywood		0.30	2.69	3.35	m2	**6.04**
Stair nosings						
'Ferodo' or equivalent; light duty hard aluminium alloy stair tread nosings; plugged and screwed to concrete						
type SD1	PC £5.06	0.25	2.24	5.83	m	**8.07**
type SD2	PC £6.98	0.30	2.69	7.95	m	**10.64**
'Ferodo' or equivalent; heavy duty aluminium alloy stair tread nosings; plugged and screwed to concrete						
type HD1	PC £5.98	0.30	2.69	6.85	m	**9.54**
type HD2	PC £8.33	0.35	3.13	9.44	m	**12.57**
Nylon needlepunch tiles; 'Marleyflex' or similar level; fixing with adhesive; to cement and sand base						
Work to floors						
over 300 mm wide		0.22	1.97	6.59	m2	**8.56**
Heavy duty carpet tiles; 'Heuga 581 Olympic' or similar; PC £11.90/m2; to cement and sand base						
Work to floors						
over 300 mm wide		0.25	2.24	13.74	m2	**15.98**
Nylon needlepunch carpet; 'Marleyflex' or similar; fixing; with adhesive; level; to cement						
Work to floors						
over 300 mm wide		0.25	2.24	6.16	m2	**8.40**

M SURFACE FINISHES Including overheads and profit at 5.00%	Labour hours	Labour £	Material £	Unit	Total rate £
M51 EDGE FIXED CARPETING					
Fitted carpeting; Wilton wool/nylon; 80/20 velvet pile; heavy domestic plain; PC £30.00/m2					
Work to floors					
over 300 mm wide	0.40	3.58	34.50	m2	**38.08**
Work to treads and risers					
over 300 mm wide	0.80	7.17	35.29	m2	**42.46**
Raking cutting	0.08	0.72	2.43	m2	**3.15**
Underlay to carpeting; PC £2.75/m2					
Work to floors					
over 300 mm wide	0.08	0.72	3.03	m2	**3.75**
Sundries					
Carpet gripper fixed to floor; standard edging	0.05	0.45	0.15	m	**0.60**
M52 DECORATIVE PAPERS/FABRICS					
Lining paper; PC £0.84/roll; roll; and hanging					
Plaster walls or columns					
over 300 mm girth	0.20	1.25	0.27	m2	**1.52**
Plaster ceilings or beams					
over 300 mm girth	0.25	1.57	0.27	m2	**1.84**
Decorative vinyl wallpaper; PC £4.80/roll; roll; and hanging					
Plaster walls or columns					
over 300 mm girth	0.25	1.57	1.18	m2	**2.75**

BASIC PAINT PRICES (£/5 Litre tin)

	£		£		£
Bituminous	8.79				
Emulsion					
matt	8.23	silk	8.68		
Knotting solution	23.00				
Oil					
gloss	10.06	undercoat	9.62		
Primer/undercoats					
acrylic	8.67	plaster	12.90	wood	9.58
aluminium	17.55	red lead	28.43	zinc phosphate	10.82
calcium plumbate	22.31	red oxide	9.59		
Road marking	18.57				

	£		£
Special paints			
aluminium	17.36	fire retardant	
anti-condensation	16.63	- undercoat	25.20
heat resisting	14.19	- top coat	33.84

M SURFACE FINISHES
Including overheads and profit at 5.00%

Stains and preservatives

creosote	2.80	Solignum - 'Architectural'	20.60
Cuprinol 'clear'	10.02	- brown	7.08
linseed oil	14.60	- cedar	10.43
Sadolin - 'Holdex'	26.89	- clear	10.01
- 'Classic'	22.76	- green	9.54
- 'Sadovac 35'	14.18	'Multiplus'	11.48

Varnishes

polyurethene	14.34	yacht	15.30

NOTE: The following prices include for preparing surfaces. Painting woodwork also includes for knotting prior to applying the priming coat and for all stopping of nail holes etc.	Labour hours	Labour £	Material £	Unit	Total rate £
M60 PAINTING/CLEAR FINISHING					
One coat primer; PC £9.58/5L; on wood surfaces before fixing					
General surfaces					
over 300 mm girth	0.15	0.94	0.20	m2	**1.14**
isolated surfaces not exceeding 300 mm girth	0.06	0.38	0.06	m	**0.44**
isolated surfaces not exceeding 0.50 m2	0.11	0.69	0.10	nr	**0.79**
One coat polyurethene sealer; PC £14.34/5L; on wood surfaces before fixing					
General surfaces					
over 300 mm girth	0.18	1.13	0.24	m2	**1.37**
isolated surfaces not exceeding 300 mm girth	0.07	0.44	0.08	m	**0.52**
isolated surfaces not exceeding 0.50 m2	0.13	0.81	0.12	nr	**0.93**
One coat clear wood preservative; PC £10.02/5L; on wood surfaces before fixing					
General surfaces					
over 300 mm girth	0.13	0.81	0.23	m2	**1.04**
isolated surfaces not exceeding 300 mm girth	0.05	0.31	0.07	m	**0.38**
isolated surfaces not exceeding 0.50 m2	0.10	0.63	0.12	nr	**0.75**
Two coats emulsion paint; PC £8.23/5L;					
Brick or block walls					
over 300 mm girth	0.20	1.25	0.35	m2	**1.60**
Cement render or concrete					
over 300 mm girth	0.18	1.13	0.26	m2	**1.39**
isolated surfaces not exceeding 300 mm girth	0.08	0.50	0.08	m	**0.58**
Plaster walls or plaster/plasterboard ceilings					
over 300 mm girth	0.17	1.07	0.26	m2	**1.33**
over 300 mm girth; in multi-colours	0.27	1.69	0.30	m2	**1.99**
over 300 mm girth; in staircase areas	0.20	1.25	0.26	m2	**1.51**
cutting in edges on flush surfaces	0.08	0.50	-	m	**0.50**
Plaster/plasterboard ceilings					
over 300 mm girth; 3.5 - 5.0 m high	0.19	1.19	0.26	m2	**1.45**

M SURFACE FINISHES Including overheads and profit at 5.00%	Labour hours	Labour £	Material £	Unit	Total rate £
One mist and two coats emulsion paint					
Brick or block walls					
over 300 mm girth	0.23	1.44	0.48	m2	**1.92**
Cement render or concrete					
over 300 mm girth	0.20	1.25	0.38	m2	**1.63**
isolated surfaces not exceeding 300 mm girth	0.09	0.56	0.10	m	**0.66**
Plaster walls or plaster/plasterboard ceilings					
over 300 mm girth	0.20	1.25	0.35	m2	**1.60**
over 300 mm girth; in multi-colours	0.30	1.88	0.39	m2	**2.27**
over 300 mm girth; in staircase areas	0.23	1.44	0.35	m2	**1.79**
cutting in edges on flush surfaces	0.11	0.69	-	m	**0.69**
Plaster/plasterboard ceilings					
over 300 mm girth; 3.5 - 5.0 m high	0.22	1.38	0.35	m2	**1.73**
Textured plastic; 'Artex' finish; PC £0.00/15L					
Plasterboard ceilings					
over 300 mm girth	0.25	1.57	1.09	m2	**2.66**
Concrete walls or ceilings					
over 300 mm girth	0.27	1.69	1.22	m2	**2.91**
Touch up primer; one undercoat and one finishing coat of gloss oil paint; PC £10.06/5L; on wood surfaces					
General surfaces					
over 300 mm girth	0.32	2.01	0.35	m2	**2.36**
isolated surfaces not exceeding 300 mm girth	0.13	0.81	0.11	m	**0.92**
isolated surfaces not exceeding 0.50 m2	0.24	1.50	0.17	nr	**1.67**
Windows and the like					
panes over 1.00 m2	0.32	2.01	0.13	m2	**2.14**
panes 0.50 - 1.00 m2	0.37	2.32	0.17	m2	**2.49**
panes 0.10 - 0.50 m2	0.43	2.70	0.22	m2	**2.92**
panes not exceeding 0.10 m2	0.53	3.32	0.26	m2	**3.58**
Touch up primer; two undercoats and one finishing coat of gloss oil paint; on wood surfaces					
General surfaces					
over 300 mm girth	0.45	2.82	0.51	m2	**3.33**
isolated surfaces not exceeding 300 mm girth	0.18	1.13	0.16	m	**1.29**
isolated surfaces not exceeding 0.50 m2	0.33	2.07	0.25	nr	**2.32**
Windows and the like					
panes over 1.00 m2	0.45	2.82	0.19	m2	**3.01**
panes 0.50 - 1.00 m2	0.52	3.26	0.25	m2	**3.51**
panes 0.10 - 0.50 m2	0.60	3.76	0.32	m2	**4.08**
panes not exceeding 0.10 m2	0.75	4.70	0.38	m2	**5.08**
Knot; one coat primer; stop; one undercoat and one finishing coat of gloss oil paint; on wood surfaces					
General surfaces					
over 300 mm girth	0.47	2.95	0.60	m2	**3.55**
isolated surfaces not exceeding 300 mm girth	0.19	1.19	0.19	m	**1.38**
isolated surfaces not exceeding 0.50 m2	0.35	2.19	0.30	nr	**2.49**
Windows and the like					
panes over 1.00 m2	0.47	2.95	0.22	m2	**3.17**
panes 0.50 - 1.00 m2	0.55	3.45	0.30	m2	**3.75**
panes 0.10 - 0.50 m2	0.65	4.07	0.37	m2	**4.44**
panes not exceeding 0.10 m2	0.80	5.01	0.45	m2	**5.46**

M SURFACE FINISHES Including overheads and profit at 5.00%	Labour hours	Labour £	Material £	Unit	Total rate £
M60 PAINTING/CLEAR FINISHING - cont'd					
Knot; one coat primer; stop; two undercoats and one finishing coat of gloss oil paint; on wood surfaces					
General surfaces					
over 300 mm girth	0.60	3.76	0.76	m2	4.52
isolated surfaces not exceeding 300 mm girth	0.24	1.50	0.24	m	1.74
isolated surfaces not exceeding 0.50 m2	0.45	2.82	0.38	nr	3.20
Windows and the like					
panes over 1.00 m2	0.60	3.76	0.28	m2	4.04
panes 0.50 - 1.00 m2	0.70	4.39	0.38	m2	4.77
panes 0.10 - 0.50 m2	0.80	5.01	0.47	m2	5.48
panes not exceeding 0.10 m2	1.00	6.27	0.57	m2	6.84
One coat primer; one undercoat and one finishing coat of gloss oil paint					
Plaster surfaces					
over 300 mm girth	0.42	2.63	0.72	m2	3.35
One coat primer; two undercoats and one finishing coat of gloss oil paint					
Plaster surfaces					
over 300 mm girth	0.55	3.45	0.88	m2	4.33
Touch up primer; one undercoat and one finishing coat of gloss paint; PC £10.06/5L; on iron or steel surfaces					
General surfaces					
over 300 mm girth	0.32	2.01	0.33	m2	2.34
isolated surfaces not exceeding 300 mm girth	0.13	0.81	0.10	m	0.91
isolated surfaces not exceeding 0.50 m2	0.24	1.50	0.17	nr	1.67
Windows and the like					
panes over 1.00 m2	0.32	2.01	0.12	m2	2.13
panes 0.50 - 1.00 m2	0.37	2.32	0.17	m2	2.49
panes 0.10 - 0.50 m2	0.43	2.70	0.21	m2	2.91
panes not exceeding 0.10 m2	0.53	3.32	0.25	m2	3.57
Structural steelwork					
over 300 mm girth	0.36	2.26	0.33	m2	2.59
Members of roof trusses					
over 300 mm girth	0.48	3.01	0.35	m2	3.36
Ornamental railings and the like; each side measured overall					
over 300 mm girth	0.55	3.45	0.37	m2	3.82
Iron or steel radiators					
over 300 mm girth	0.32	2.01	0.35	m2	2.36
Pipes or conduits					
over 300 mm girth	0.48	3.01	0.35	m2	3.36
not exceeding 300 mm girth	0.19	1.19	0.10	m	1.29
Touch up primer; two undercoats and one finishing coat of gloss oil paint; on iron or steel surfaces					
General surfaces					
over 300 mm girth	0.45	2.82	0.49	m2	3.31
isolated surfaces not exceeding 300 mm girth	0.18	1.13	0.15	m	1.28
isolated surfaces not exceeding 0.50 m2	0.33	2.07	0.25	nr	2.32

M SURFACE FINISHES Including overheads and profit at 5.00%	Labour hours	Labour £	Material £	Unit	Total rate £
Windows and the like					
panes over 1.00 m2	0.45	2.82	0.18	m2	**3.00**
panes 0.50 - 1.00 m2	0.52	3.26	0.25	m2	**3.51**
panes 0.10 - 0.50 m2	0.60	3.76	0.31	m2	**4.07**
panes not exceeding 0.10 m2	0.75	4.70	0.36	m2	**5.06**
Structural steelwork					
over 300 mm girth	0.51	3.20	0.49	m2	**3.69**
Members of roof trusses					
over 300 mm girth	0.68	4.26	0.52	m2	**4.78**
Ornamental railings and the like; each side measured overall					
over 300 mm girth	0.77	4.83	0.55	m2	**5.38**
Iron or steel radiators					
over 300 mm girth	0.45	2.82	0.52	m2	**3.34**
Pipes or conduits					
over 300 mm girth	0.68	4.26	0.52	m2	**4.78**
not exceeding 300 mm girth	0.27	1.69	0.15	m	**1.84**
One coat primer; one undercoat and one finishing coat of gloss oil paint; on iron or steel surfaces					
General surfaces					
over 300 mm girth	0.42	2.63	0.33	m2	**2.96**
isolated surfaces not exceeding 300 mm girth	0.17	1.07	0.10	m	**1.17**
isolated surfaces not exceeding 0.50 m2	0.32	2.01	0.17	nr	**2.18**
Windows and the like					
panes over 1.00 m2	0.42	2.63	0.12	m2	**2.75**
panes 0.50 - 1.00 m2	0.48	3.01	0.17	m2	**3.18**
panes 0.10 - 0.50 m2	0.56	3.51	0.21	m2	**3.72**
panes not exceeding 0.10 m2	0.70	4.39	0.25	m2	**4.64**
Structural steelwork					
over 300 mm girth	0.47	2.95	0.33	m2	**3.28**
Members of roof trusses					
over 300 mm girth	0.63	3.95	0.35	m2	**4.30**
Ornamental railings and the like; each side measured overall					
over 300 mm girth	0.72	4.51	0.37	m2	**4.88**
Iron or steel radiators					
over 300 mm girth	0.42	2.63	0.35	m2	**2.98**
Pipes or conduits					
over 300 mm girth	0.63	3.95	0.35	m2	**4.30**
not exceeding 300 mm girth	0.25	1.57	0.10	m	**1.67**
One coat primer; two undercoats and one finishing coat of gloss oil paint; on iron or steel surfaces					
General surfaces					
over 300 mm girth	0.55	3.45	0.49	m2	**3.94**
isolated surfaces not exceeding 300 mm girth	0.22	1.38	0.15	m	**1.53**
isolated surfaces not exceeding 0.50 m2	0.40	2.51	0.25	nr	**2.76**
Windows and the like					
panes over 1.00 m2	0.55	3.45	0.18	m2	**3.63**
panes 0.50 - 1.00 m2	0.65	4.07	0.25	m2	**4.32**
panes 0.10 - 0.50 m2	0.75	4.70	0.31	m2	**5.01**
panes not exceeding 0.10 m2	0.90	5.64	0.37	m2	**6.01**
Structural steelwork					
over 300 mm girth	0.62	3.89	0.49	m2	**4.38**
Members of roof trusses					
over 300 mm girth	0.82	5.14	0.52	m2	**5.66**

M SURFACE FINISHES Including overheads and profit at 5.00%	Labour hours	Labour £	Material £	Unit	Total rate £
M60 PAINTING/CLEAR FINISHING - cont'd					
One coat primer; two undercoats and one finishing coat of gloss oil paint; on iron or steel surfaces - cont'd					
Ornamental railings and the like; each side measured overall					
over 300 mm girth	0.93	5.83	0.55	m2	**6.38**
Iron or steel radiators					
over 300 mm girth	0.55	3.45	0.52	m2	**3.97**
Pipes or conduits					
over 300 mm girth	0.83	5.20	0.52	m2	**5.72**
not exceeding 300 mm girth	0.33	2.07	0.15	m	**2.22**
Two coats of bituminous paint; PC £8.79/5L; on iron or steel surfaces					
General surfaces					
over 300 mm girth	0.40	2.51	0.30	m2	**2.81**
Inside of galvanized steel cistern					
over 300 mm girth	0.60	3.76	0.30	m2	**4.06**
Two coats of boiled linseed oil; PC £14.60/5; on hardwood surfaces					
General surfaces					
over 300 mm girth	0.30	1.88	1.07	m2	**2.95**
isolated surfaces not exceeding 300 mm girth	0.12	0.75	0.32	m	**1.07**
isolated surfaces not exceeding 0.50 m2	0.23	1.44	0.54	nr	**1.98**
Two coats polyurethane; PC £14.34/5L; on wood surfaces					
General surfaces					
over 300 mm girth	0.30	1.88	0.48	m2	**2.36**
isolated surfaces not exceeding 300 mm girth	0.12	0.75	0.15	m	**0.90**
isolated surfaces not exceeding 0.50 m2	0.22	1.38	0.24	nr	**1.62**
Three coats polyurethane; on wood surfaces					
General surfaces					
over 300 mm girth	0.45	2.82	0.72	m2	**3.54**
isolated surfaces not exceeding 300 mm girth	0.18	1.13	0.23	m	**1.36**
isolated surfaces not exceeding 0.50 m2	0.34	2.13	0.36	nr	**2.49**
One undercoat; PC £25.20/5L; and one finishing coat; PC £33.84/5L; of 'Albi' clear flame retardent surface coating; on wood surfaces					
General surfaces					
over 300 mm girth	0.54	3.38	1.23	m2	**4.61**
isolated surfaces not exceeding 300 mm girth	0.22	1.38	0.39	m	**1.77**
isolated surfaces not exceeding 0.50 m2	0.42	2.63	0.60	nr	**3.23**
Two undercoats; PC £25.20/5L; and one finishing coat; PC £33.84/5L; of 'Albi' clear flame retardent surface coating; on wood surfaces					
General surfaces					
over 300 mm girth	0.54	3.38	1.89	m2	**5.27**
isolated surfaces not exceeding 300 mm girth	0.22	1.38	0.60	m	**1.98**
isolated surfaces not exceeding 0.50 m2	0.42	2.63	0.95	nr	**3.58**

M SURFACE FINISHES Including overheads and profit at 5.00%	Labour hours	Labour £	Material £	Unit	Total rate £
Seal and wax polish; dull gloss finish on wood surfaces					
General surfaces					
over 300 mm girth	-	-	-	m2	**5.48**
isolated surfaces not exceeding 300 mm girth	-	-	-	m	**2.46**
isolated surfaces not exceeding 0.50 m2	-	-	-	nr	**3.84**
One coat of 'Sadolins Classic'; PC £22.76/5L; clear or pigmented; one further coat of 'Holdex' clear interior silk matt laquer; PC £26.89/5L					
General surfaces					
over 300 mm girth	0.40	2.51	0.98	m2	**3.49**
isolated surfaces not exceeding 300 mm girth	0.16	1.00	0.31	m	**1.31**
isolated surfaces not exceeding 0.50 m2	0.30	1.88	0.49	nr	**2.37**
Windows and the like					
panes over 1.00 m2	0.40	2.51	0.36	m2	**2.87**
panes 0.50 - 1.00 m2	0.46	2.88	0.49	m2	**3.37**
panes 0.10 - 0.50 m2	0.53	3.32	0.62	m2	**3.94**
panes not exceeding 0.10 m2	0.66	4.14	0.72	m2	**4.86**
Two coats of 'Sadolins Classic'; PC £22.76/5L; clear or pigmented; two further coats of 'Holdex' clear interior silk matt laquer; PC £26.89/5L					
General surfaces					
over 300 mm girth	0.62	3.89	1.96	m2	**5.85**
isolated surfaces not exceeding 300 mm girth	0.25	1.57	0.62	m	**2.19**
isolated surfaces not exceeding 0.50 m2	0.47	2.95	0.98	nr	**3.93**
Windows and the like					
panes over 1.00 m2	0.62	3.89	0.72	m2	**4.61**
panes 0.50 - 1.00 m2	0.72	4.51	0.98	m2	**5.49**
panes 0.10 - 0.50 m2	0.83	5.20	1.23	m2	**6.43**
panes not exceeding 0.10 m2	1.04	6.52	1.47	m2	**7.99**
Body in and wax polish; dull gloss finish; on hardwood surfaces					
General surfaces					
over 300 mm girth	-	-	-	m2	**7.78**
isolated surfaces not exceeding 300 mm girth	-	-	-	m	**3.48**
isolated surfaces not exceeding 0.50 m2	-	-	-	nr	**5.45**
Stain; body in and wax polish; dull gloss finish; on hardwood surface					
General surfaces					
over 300 mm girth	-	-	-	m2	**9.73**
isolated surfaces not exceeding 300 mm girth	-	-	-	m	**4.35**
isolated surfaces not exceeding 0.50 m2	-	-	-	nr	**6.81**
Seal; two coats of synthetic resin lacquer; decorative flatted finish; wire down, wax and burnish; on wood surfaces					
General surfaces					
over 300 mm girth	-	-	-	m2	**12.29**
isolated surfaces not exceeding 300 mm girth	-	-	-	m	**5.53**
isolated surfaces not exceeding 0.50 m2	-	-	-	nr	**8.60**
Stain; body in and fully French polish; full gloss finish; on hardwood surfaces					
General surfaces					
over 300 mm girth	-	-	-	m2	**14.74**
isolated surfaces not exceeding 300 mm girth	-	-	-	m	**6.65**
isolated surfaces not exceeding 0.50 m2	-	-	-	nr	**10.32**

M SURFACE FINISHES Including overheads and profit at 5.00%	Labour hours	Labour £	Material £	Unit	Total rate £
M60 PAINTING/CLEAR FINISHING - cont'd					
Stain; fill grain and fully French polish; full gloss finish; on hardwood surfaces					
General surfaces					
over 300 mm girth	-	-	-	m2	**22.52**
isolated surfaces not exceeding 300 mm girth	-	-	-	m	**10.14**
isolated surfaces not exceeding 0.50 m2	-	-	-	nr	**15.77**
Stain black; body in and fully French polish; ebonized finish; on hardwood surfaces					
General surfaces					
over 300 mm girth	-	-	-	m2	**27.64**
isolated surfaces not exceeding 300 mm girth	-	-	-	m	**12.44**
isolated surfaces not exceeding 0.50 m2	-	-	-	nr	**19.35**
M60 PAINTING/CLEAR FINISHING - EXTERNALLY					
Two coats of cement paint, 'Sandtex Matt' or similar; PC £11.20/5L;					
Brick or block walls					
over 300 mm girth	0.45	2.82	1.41	m2	**4.23**
Cement render or concrete walls					
over 300 mm girth	0.40	2.51	0.94	m2	**3.45**
Roughcast walls					
over 300 mm girth	0.50	3.13	1.88	m2	**5.01**
One coat sealer and two coats of external grade emulsion paint, Dulux 'Weathershield' or similar					
Brick or block walls					
over 300 mm girth	0.60	3.76	2.01	m2	**5.77**
Cement render or concrete walls					
over 300 mm girth	0.50	3.13	1.34	m2	**4.47**
Concrete soffits					
over 300 mm girth	0.55	3.45	1.34	m2	**4.79**
One coat sealer (applied by brush) and two coats of external grade emulsion paint, Dulux 'Weathershield' or similar (spray applied)					
Roughcast					
over 300 mm girth	0.60	2.77	2.68	m2	**5.45**
Touch up primer; one undercoat and one finishing coat of gloss oil paint; PC £10.06/5L; on wood surfaces					
General surfaces					
over 300 mm girth	0.36	2.26	0.35	m2	**2.61**
isolated surfaces not exceeding 300 mm girth	0.15	0.94	0.11	m	**1.05**
isolated surfaces not exceeding 0.50 m2	0.27	1.69	0.17	nr	**1.86**
Windows and the like					
panes over 1.00 m2	0.36	2.26	0.13	m2	**2.39**
panes 0.50 - 1.00 m2	0.42	2.63	0.17	m2	**2.80**
panes 0.10 - 0.50 m2	0.48	3.01	0.22	m2	**3.23**
panes not exceeding 0.10 m2	0.60	3.76	0.26	m2	**4.02**
Windows and the like; casements in colours differing from frames					
panes over 1.00 m2	0.40	2.51	0.15	m2	**2.66**
panes 0.50 - 1.00 m2	0.45	2.82	0.19	m2	**3.01**
panes 0.10 - 0.50 m2	0.53	3.32	0.24	m2	**3.56**
panes not exceeding 0.10 m2	0.66	4.14	0.28	m2	**4.42**

M SURFACE FINISHES Including overheads and profit at 5.00%	Labour hours	Labour £	Material £	Unit	Total rate £
Touch up primer; two undercoats and one finishing coat of gloss oil paint; on wood surfaces					
General surfaces					
over 300 mm girth	0.50	3.13	0.51	m2	**3.64**
isolated surfaces not exceeding 300 mm girth	0.20	1.25	0.16	m2	**1.41**
isolated surfaces not exceeding 0.50 m2	0.38	2.38	0.25	m2	**2.63**
Windows and the like					
panes over 1.00 m2	0.50	3.13	0.19	m2	**3.32**
panes 0.50 - 1.00 m2	0.58	3.64	0.25	m2	**3.89**
panes 0.10 - 0.50 m2	0.67	4.20	0.32	m2	**4.52**
panes not exceeding 0.10 m2	0.83	5.20	0.38	m2	**5.58**
Windows and the like; casements in colours differing from frames					
panes over 1.00 m2	0.58	3.64	0.22	m2	**3.86**
panes 0.50 - 1.00 m2	0.67	4.20	0.28	m2	**4.48**
panes 0.10 - 0.50 m2	0.77	4.83	0.35	m2	**5.18**
panes not exceeding 0.10 m2	0.95	5.96	0.41	m2	**6.37**
Knot; one coat primer; one undercoat and one finishing coat of gloss oil paint; on wood surfaces					
General surfaces					
over 300 mm girth	0.53	3.32	0.60	m2	**3.92**
isolated surfaces not exceeding 300 mm girth	0.22	1.38	0.19	m	**1.57**
isolated surfaces not exceeding 0.50 m2	0.40	2.51	0.30	nr	**2.81**
Windows and the like					
panes over 1.00 m2	0.53	3.32	0.22	m2	**3.54**
panes 0.50 - 1.00 m2	0.62	3.89	0.30	m2	**4.19**
panes 0.10 - 0.50 m2	0.70	4.39	0.37	m2	**4.76**
panes not exceeding 0.10 m2	0.88	5.52	0.45	m2	**5.97**
Windows and the like; casements in colours differing from frames					
panes over 1.00 m2	0.58	3.64	0.25	m2	**3.89**
panes 0.50 - 1.00 m2	0.68	4.26	0.33	m2	**4.59**
panes 0.10 - 0.50 m2	0.77	4.83	0.58	m2	**5.41**
panes not exceeding 0.10 m2	0.96	6.02	0.48	m2	**6.50**
Knot; one coat primer; two undercoats and one finishing coat of gloss oil paint on wood surfaces					
General surfaces					
over 300 mm girth	0.66	4.14	0.76	m2	**4.90**
isolated surfaces not exceeding 300 mm girth	0.27	1.69	0.24	m	**1.93**
isolated surfaces not exceeding 0.50 m2	0.50	3.13	0.38	nr	**3.51**
Windows and the like					
panes over 1.00 m2	0.66	4.14	0.28	m2	**4.42**
panes 0.50 - 1.00 m2	0.77	4.83	0.38	m2	**5.21**
panes 0.10 - 0.50 m2	0.88	5.52	0.47	m2	**5.99**
panes not exceeding 0.10 m2	1.10	6.90	0.57	m2	**7.47**
Windows and the like; casements in colours differing from frames					
panes over 1.00 m2	0.76	4.76	0.32	m2	**5.08**
panes 0.50 - 1.00 m2	0.90	5.64	0.42	m2	**6.06**
panes 0.10 - 0.50 m2	1.02	6.39	0.69	m2	**7.08**
panes not exceeding 0.10 m2	1.25	7.84	0.61	m2	**8.45**
Touch up primer; one undercoat and one finishing coat of gloss oil paint; PC £10.06/5L; on iron or steel surfaces					
General surfaces					
over 300 mm girth	0.36	2.26	0.33	m2	**2.59**
isolated surfaces not exceeding 300 mm girth	0.15	0.94	0.10	m	**1.04**
isolated surfaces not exceeding 0.50 m2	0.27	1.69	0.17	nr	**1.86**

M SURFACE FINISHES Including overheads and profit at 5.00%	Labour hours	Labour £	Material £	Unit	Total rate £
M60 PAINTING/CLEAR FINISHING - EXTERNALLY - cont'd					
Touch up primer; one undercoat and one finishing coat of gloss oil paint; PC £10.06/5L; on iron or steel surfaces - cont'd					
Windows and the like					
panes over 1.00 m2	0.36	2.26	0.12	m2	**2.38**
panes 0.50 - 1.00 m2	0.42	2.63	0.17	m2	**2.80**
panes 0.10 - 0.50 m2	0.48	3.01	0.21	m2	**3.22**
panes not exceeding 0.10 m2	0.60	3.76	0.25	m2	**4.01**
Structural steelwork					
over 300 mm girth	0.40	2.51	0.33	m2	**2.84**
Members of roof trusses					
over 300 mm girth	0.53	3.32	0.35	m2	**3.67**
Ornamental railings and the like; each side measured overall					
over 300 mm girth	0.60	3.76	0.37	m2	**4.13**
Eaves gutters					
over 300 mm girth	0.65	4.07	0.50	m2	**4.57**
not exceeding 300 mm girth	0.26	1.63	0.17	m	**1.80**
Pipes or conduits					
over 300 mm girth	0.53	3.32	0.35	m2	**3.67**
not exceeding 300 mm girth	0.21	1.32	0.10	m	**1.42**
Touch up primer; two undercoats and one finishing coat of gloss oil paint; on iron or steel surfaces					
General surfaces					
over 300 mm girth	0.50	3.13	0.49	m2	**3.62**
isolated surfaces not exceeding 300 mm girth	0.20	1.25	0.15	m	**1.40**
isolated surfaces not exceeding 0.50 m2	0.37	2.32	0.25	nr	**2.57**
Windows and the like					
panes over 1.00 m2	0.50	3.13	0.18	m2	**3.31**
panes 0.50 - 1.00 m2	0.58	3.64	0.25	m2	**3.89**
panes 0.10 - 0.50 m2	0.67	4.20	0.31	m2	**4.51**
panes not exceeding 0.10 m2	0.83	5.20	0.37	m2	**5.57**
Structural steelwork					
over 300 mm girth	0.57	3.57	0.49	m2	**4.06**
Members of roof trusses					
over 300 mm girth	0.75	4.70	0.52	m2	**5.22**
Ornamental railings and the like; each side measured overall					
over 300 mm girth	0.85	5.33	0.55	m2	**5.88**
Eaves gutters					
over 300 mm girth	0.90	5.64	0.74	m2	**6.38**
not exceeding 300 mm girth	0.36	2.26	0.25	m	**2.51**
Pipes or conduits					
over 300 mm girth	0.75	4.70	0.52	m2	**5.22**
not exceeding 300 mm girth	0.30	1.88	0.15	m	**2.03**
One coat primer; one undercoat and one finishing coat of gloss oil paint; on iron or steel surfaces					
General surfaces					
over 300 mm girth	0.46	2.88	0.33	m2	**3.21**
isolated surfaces not exceeding 300 mm girth	0.19	1.19	0.10	m	**1.29**
isolated surfaces not exceeding 0.50 m2	0.35	2.19	0.17	nr	**2.36**
Windows and the like					
panes over 1.00 m2	0.46	2.88	0.12	m2	**3.00**
panes 0.50 - 1.00 m2	0.53	3.32	0.17	m2	**3.49**
panes 0.10 - 0.50 m2	0.62	3.89	0.21	m2	**4.10**
panes not exceeding 0.10 m2	0.77	4.83	0.25	m2	**5.08**

M SURFACE FINISHES Including overheads and profit at 5.00%	Labour hours	Labour £	Material £	Unit	Total rate £
Structural steelwork					
over 300 mm girth	0.52	3.26	0.33	m2	**3.59**
Members of roof trusses					
over 300 mm girth	0.68	4.26	0.35	m2	**4.61**
Ornamental railings and the like; each side measured overall					
over 300 mm girth	0.78	4.89	0.37	m2	**5.26**
Eaves gutters					
over 300 mm girth	0.82	5.14	0.50	m2	**5.64**
not exceeding 300 mm girth	0.33	2.07	0.17	m	**2.24**
Pipes or conduits					
over 300 mm girth	0.68	4.26	0.35	m2	**4.61**
not exceeding 300 mm girth	0.27	1.69	0.10	m	**1.79**
One coat primer; two undercoats and one finishing coat of gloss oil paint; on iron or steel surfaces					
General surfaces					
over 300 mm girth	0.60	3.76	0.49	m2	**4.25**
isolated surfaces not exceeding 300 mm girth	0.24	1.50	0.15	m	**1.65**
isolated surfaces not exceeding 0.50 m2	0.45	2.82	0.25	nr	**3.07**
Windows and the like					
panes over 1.00 m2	0.60	3.76	0.18	m2	**3.94**
panes 0.50 - 1.00 m2	0.70	4.39	0.25	m2	**4.64**
panes 0.10 - 0.50 m2	0.80	5.01	0.31	m2	**5.32**
panes not exceeding 0.10 m2	1.00	6.27	0.37	m2	**6.64**
Structural steelwork					
over 300 mm girth	0.68	4.26	0.49	m2	**4.75**
Members of roof trusses					
over 300 mm girth	0.90	5.64	0.52	m2	**6.16**
Ornamental railings and the like; each side measured overall					
over 300 mm girth	1.02	6.39	0.55	m2	**6.94**
Eaves gutters					
over 300 mm girth	1.08	6.77	0.74	m2	**7.51**
not exceeding 300 mm girth	0.43	2.70	0.25	m	**2.95**
Pipes or conduits					
over 300 mm girth	0.90	5.64	0.52	m2	**6.16**
not exceeding 300 mm girth	0.36	2.26	0.15	m	**2.41**
Two coats of creosote; PC £2.80/5L; on wood surfaces					
General surfaces					
over 300 mm girth	0.25	1.57	0.13	m2	**1.70**
isolated surfaces not exceeding 300 mm girth	0.10	0.63	0.04	m	**0.67**
Two coats of 'Solignum' wood preservative; PC £10.43/5L; on wood surfaces					
General surfaces					
over 300 mm girth	0.25	1.57	0.48	m2	**2.05**
isolated surfaces not exceeding 300 mm girth	0.10	0.63	0.15	m	**0.78**
Three coats of polyurethane; PC £14.34/5L; on wood surfaces					
General surfaces					
over 300 mm girth	0.50	3.13	0.72	m2	**3.85**
isolated surfaces not exceeding 300 mm girth	0.20	1.25	0.23	m	**1.48**
isolated surfaces not exceeding 0.50 m2	0.38	2.38	0.36	nr	**2.74**

M SURFACE FINISHES Including overheads and profit at 5.00%	Labour hours	Labour £	Material £	Unit	Total rate £
M60 PAINTING/CLEAR FINISHING - EXTERNALLY - cont'd					
Two coats of 'Sadovac 35'primer; PC £14.18/5L; and two coats of 'Classic'; PC £22.76/5L; pigmented; on wood surfaces					
General surfaces					
over 300 mm girth	0.68	4.26	2.03	m2	**6.29**
isolated surfaces not exceeding 300 mm girth	0.27	1.69	0.65	m	**2.34**
Windows and the like					
panes over 1.00 m2	0.68	4.26	0.74	m2	**5.00**
panes 0.50 - 1.00 m2	0.79	4.95	1.02	m2	**5.97**
panes 0.10 - 0.50 m2	0.91	5.70	1.29	m2	**6.99**
panes not exceeding 0.10 m2	1.14	7.15	1.53	m2	**8.68**
Body in with French polish; one coat of lacquer or varnish; on wood surfaces					
General surfaces					
over 300 mm girth	-	-	-	m2	**12.59**
isolated surfaces not exceeding 300 mm girth	-	-	-	m	**5.63**
isolated surfaces not exceeding 0.50 m2	-	-	-	nr	**8.81**

N FURNITURE/EQUIPMENT Including overheads and profit at 5.00%		Labour hours	Labour £	Material £	Unit	Total rate £
N10/11 GENERAL FIXTURES/KITCHEN FITTINGS ETC.						
Proprietary items						
Closed stove vitreous enamelled finish; setting in position						
571 x 606 x 267 mm	PC £253.00	2.00	20.27	265.65	nr	285.92
Closed stove; vitreous enamelled finish fitted with mild steel barffed boiler; setting in position						
571 x 606 x 267 mm	PC £330.00	2.00	20.27	346.50	nr	366.77
Tile surround preslabbed; 406 mm wide firebrick back; cast iron stool bottom; black vitreous enamelled fret; assembling, setting and pointing firebrick back in fireclay mortar and backing with fine concrete finished to splay at top, laying hearth tiles in cement mortar and pointing in white cement						
1372 x 864 x 152 mm	PC £242.06	5.00	50.66	273.22	nr	323.88
Fitting components; blockboard						
Backs, fronts, sides or divisions; over 300 mm wide						
12 mm thick	PC £23.08	1.30	10.28	24.23	m2	34.51
19 mm thick	PC £29.79	1.30	10.28	31.28	m2	41.56
25 mm thick	PC £39.18	1.30	10.28	41.14	m2	51.42
Shelves or worktops; over 300 mm wide						
19 mm thick	PC £30.83	1.30	10.28	32.37	m2	42.65
25 mm thick	PC £39.17	1.30	10.28	41.13	m2	51.41
Flush doors; lipped on four edges						
450 x 750 x 19 mm	PC £16.59	0.35	2.77	17.41	nr	20.18
450 x 750 x 25 mm	PC £20.29	0.35	2.77	21.31	nr	24.08
600 x 900 x 19 mm	PC £24.68	0.50	3.95	25.91	nr	29.86
600 x 900 x 25 mm	PC £30.03	0.50	3.95	31.54	nr	35.49
Fitting components; chipboard						
Backs, fronts, sides or divisions; over 300 mm wide						
6 mm thick	PC £8.00	1.30	10.28	8.40	m2	18.68
9 mm thick	PC £10.98	1.30	10.28	11.53	m2	21.81
12 mm thick	PC £13.44	1.30	10.28	14.11	m2	24.39
19 mm thick	PC £19.10	1.30	10.28	20.06	m2	30.34
25 mm thick	PC £25.78	1.30	10.28	27.07	m2	37.35
Shelves or worktops; over 300 mm wide						
19 mm thick	PC £20.14	1.30	10.28	21.14	m2	31.42
25 mm thick	PC £25.77	1.30	10.28	27.06	m2	37.34
Flush doors; lipped on four edges						
450 x 750 x 19 mm	PC £13.44	0.35	2.77	14.11	nr	16.88
450 x 750 x 25 mm	PC £16.39	0.35	2.77	17.21	nr	19.98
600 x 900 x 19 mm	PC £19.75	0.50	3.95	20.74	nr	24.69
600 x 900 x 25 mm	PC £23.90	0.50	3.95	25.10	nr	29.05
Fitting components; Melamine faced chipboard						
Backs, fronts, sides or divisions; over 300 mm wide						
12 mm thick	PC £18.46	1.30	10.28	19.39	m2	29.67
19 mm thick	PC £25.31	1.30	10.28	26.57	m2	36.85
Shelves or worktops; over 300 mm wide						
19 mm thick	PC £26.56	1.30	10.28	27.88	m2	38.16
Flush doors; lipped on four edges						
450 x 750 x 19 mm	PC £16.86/m2	0.35	2.77	17.71	nr	20.48
600 x 900 x 19 mm	PC £24.85/m2	0.50	3.95	26.10	nr	30.05

N FURNITURE/EQUIPMENT Including overheads and profit at 5.00%		Labour hours	Labour £	Material £	Unit	Total rate £
N10/11 GENERAL FIXTURES/KITCHEN FITTINGS ETC. - cont'd						
Fitting components; 'Warerite Xcel' standard colour laminated chipboard type LD2; PC £187.35/10m2						
Backs, fronts, sides or divisions; over 300 mm wide						
13.2 mm thick	PC £50.48	1.30	10.28	53.00	m2	63.28
Shelves or worktops; over 300 mm wide						
13.2 mm thick	PC £50.48	1.30	10.28	53.00	m2	63.28
Flush doors; lipped on four edges						
450 x 750 x 13.2 mm	PC £24.12/m2	0.35	2.77	25.33	nr	28.10
600 x 900 x 13.2 mm	PC £36.33/m2	0.50	3.95	38.15	nr	42.10
Fitting components; plywood						
Backs, fronts, sides or divisions; over 300 mm wide						
6 mm thick	PC £13.01	1.30	10.28	13.67	m2	23.95
9 mm thick	PC £17.32	1.30	10.28	18.18	m2	28.46
12 mm thick	PC £21.94	1.30	10.28	23.04	m2	33.32
19 mm thick	PC £31.82	1.30	10.28	33.41	m2	43.69
25 mm thick	PC £42.28	1.30	10.28	44.40	m2	54.68
Shelves or worktops; over 300 mm wide						
19 mm thick	PC £32.86	1.30	10.28	34.50	m2	44.78
25 mm thick	PC £42.27	1.30	10.28	44.39	m2	54.67
Flush doors; lipped on four edges						
450 x 750 x 19 mm	PC £17.02	0.35	2.77	17.87	nr	20.64
450 x 750 x 25 mm	PC £21.02	0.35	2.77	22.07	nr	24.84
600 x 900 x 19 mm	PC £25.41	0.50	3.95	26.68	nr	30.63
600 x 900 x 25 mm	PC £31.22	0.50	3.95	32.78	nr	36.73
Fitting components; wrought softwood						
Backs, fronts, sides or divisions; cross-tongued joints; over 300 mm wide						
25 mm thick		1.30	10.28	34.61	m2	44.89
Shelves or worktops; cross-tongued joints; over 300 mm wide						
25 mm thick		1.30	10.28	34.61	m2	44.89
Bearers						
19 x 38 mm		0.10	0.79	1.98	m	2.77
25 x 50 mm		0.10	0.79	2.52	m	3.31
50 x 50 mm		0.10	0.79	3.76	m	4.55
50 x 75 mm		0.10	0.79	5.05	m	5.84
Bearers; framed						
19 x 38 mm		0.13	1.03	3.72	m	4.75
25 x 50 mm		0.13	1.03	4.27	m	5.30
50 x 50 mm		0.13	1.03	5.50	m	6.53
50 x 75 mm		0.13	1.03	6.79	m	7.82
Framing to backs, fronts or sides						
19 x 38 mm		0.15	1.19	3.72	m	4.91
25 x 50 mm		0.15	1.19	4.27	m	5.46
50 x 50 mm		0.15	1.19	5.50	m	6.69
50 x 75 mm		0.15	1.19	6.79	m	7.98
Flush doors; softwood skeleton or cellular core; plywood facing both sides; lipped on four edges						
450 x 750 x 35 mm		0.35	2.77	18.75	nr	21.52
600 x 900 x 35 mm		0.50	3.95	29.67	nr	33.62
Add +5% to the above 'Material £ prices' for 'selected' softwood for staining						

N FURNITURE/EQUIPMENT Including overheads and profit at 5.00%		Labour hours	Labour £	Material £	Unit	Total rate £
Fitting components; selected West African Mahogany; PC £470.00/m3						
Bearers						
19 x 38 mm	PC £2.66	0.15	1.19	2.92	m	**4.11**
25 x 50 mm	PC £3.44	0.15	1.19	3.76	m	**4.95**
50 x 50 mm	PC £5.22	0.15	1.19	5.67	m	**6.86**
50 x 75 mm	PC £7.07	0.15	1.19	7.66	m	**8.85**
Bearers; framed						
19 x 38 mm	PC £5.12	0.20	1.58	5.51	m	**7.09**
25 x 50 mm	PC £5.90	0.20	1.58	6.35	m	**7.93**
50 x 50 mm	PC £7.67	0.20	1.58	8.25	m	**9.83**
50 x 75 mm	PC £9.52	0.20	1.58	10.25	m	**11.83**
Framing to backs, fronts or sides						
19 x 38 mm	PC £5.12	0.25	1.98	5.51	m	**7.49**
25 x 50 mm	PC £5.90	0.25	1.98	6.35	m	**8.33**
50 x 50 mm	PC £7.67	0.25	1.98	8.25	m	**10.23**
50 x 75 mm	PC £9.52	0.25	1.98	10.25	m	**12.23**
Fitting components; Iroko; PC £500.00/m3						
Backs, fronts, sides or divisions; cross-tongued joints; over 300 mm wide						
25 mm thick	PC £58.22	1.75	13.84	61.13	m2	**74.97**
Shelves or worktops; cross-tongued joints; over 300 mm wide						
25 mm thick	PC £58.22	1.75	13.84	61.13	m2	**74.97**
Draining boards; cross-tongued joints; over 300 mm wide						
25 mm thick	PC £62.19	1.75	13.84	65.30	m2	**79.14**
Stopped flutes		-	-	-	m	**2.24**
Grooves; cross-grain		-	-	-	m	**0.53**
Bearers						
19 x 38 mm	PC £3.20	0.15	1.19	3.50	m	**4.69**
25 x 50 mm	PC £4.02	0.15	1.19	4.38	m	**5.57**
50 x 50 mm	PC £5.98	0.15	1.19	6.50	m	**7.69**
50 x 75 mm		0.15	1.19	8.70	m	**9.89**
Bearers; framed						
19 x 38 mm	PC £6.47	0.20	1.58	6.96	m	**8.54**
25 x 50 mm	PC £7.31	0.20	1.58	7.86	m	**9.44**
50 x 50 mm	PC £9.26	0.20	1.58	9.96	m	**11.54**
50 x 75 mm	PC £11.32	0.20	1.58	12.18	m	**13.76**
Framing to backs, fronts or sides						
19 x 38 mm	PC £6.47	0.25	1.98	6.96	m	**8.94**
25 x 50 mm	PC £7.31	0.25	1.98	7.86	m	**9.84**
50 x 50 mm	PC £9.26	0.25	1.98	9.96	m	**11.94**
50 x 75 mm	PC £11.32	0.25	1.98	12.18	m	**14.16**
Fitting components; Teak; PC £1500.00/m3						
Backs, fronts, sides or divisions; cross-tongued joints; over 300 mm wide						
25 mm thick	PC £128.34	2.00	15.81	134.76	m2	**150.57**
Shelves or worktops; cross-tongued joints; over 300 mm wide						
25 mm thick	PC £128.34	2.00	15.81	134.76	m2	**150.57**
Draining boards; cross-tongued joints; over 300 mm wide						
25 mm thick	PC £131.32	2.00	15.81	137.89	m2	**153.70**
Stopped flutes		-	-	-	m2	**2.24**
Grooves; cross-grain		-	-	-	m	**0.53**

N FURNITURE/EQUIPMENT Including overheads and profit at 5.00%		Labour hours	Labour £	Material £	Unit	Total rate £
N10/11 GENERAL FIXTURES/KITCHEN FITTINGS ETC. - cont'd						
Fitting components; 'Formica' plastic laminated plastics coverings fixed with adhesive						
1.3 mm thick horizontal grade marbled finish over 300 mm wide	PC £15.00	1.60	12.65	23.39	m2	**36.04**
1.0 mm thick universal backing over 300 mm wide	PC £3.60	1.20	9.49	7.47	m2	**16.96**
1.3 mm thick edging strip						
18 mm wide		0.10	0.79	1.36	m	**2.15**
25 mm wide		0.10	0.79	1.36	m	**2.15**
Fixing kitchen fittings						
(Kitchen fittings are largely a matter of selection and prices vary considerably. PC supply prices for reasonable quantities for 'Standard' (moderately-priced) kitchen fittings have been shown but not extended).						
Fixing only to backgrounds requiring plugging; including any pre-assembly						
Wall units						
600 x 600 x 300 mm	PC £37.17	1.40	11.07	0.15	nr	**11.22**
600 x 900 x 300 mm	PC £46.23	1.60	12.65	0.15	nr	**12.80**
1200 x 600 x 300 mm	PC £67.50	1.85	14.63	0.20	nr	**14.83**
1200 x 900 x 300 mm	PC £80.88	2.10	16.60	0.20	nr	**16.80**
Floor units with drawers						
600 x 900 x 500 mm	PC £62.91	1.25	9.88	0.10	nr	**9.98**
600 x 900 x 600 mm	PC £65.67	1.40	11.07	0.10	nr	**11.17**
1200 x 900 x 600 mm	PC £116.31	1.70	13.44	0.10	nr	**13.54**
Laminated plastics worktops to suit last						
500 x 600 mm	PC £12.24	0.40	3.16	-	nr	**3.16**
600 x 600 mm	PC £13.72	0.45	3.56	-	nr	**3.56**
1200 x 600 mm	PC £27.36	0.70	5.53	-	nr	**5.53**
Larder units						
600 x 1950 x 600 mm	PC £105.81	2.45	19.37	0.15	nr	**19.52**
Sink units (excluding sink top)						
1200 x 900 x 600 mm	PC £112.11	1.80	14.23	0.10	nr	**14.33**
1500 x 900 x 600	PC £163.05	2.15	17.00	0.15	nr	**17.15**
6 mm thick rectangular glass mirrors; silver backed; fixed with chromium plated domed headed screws; to background requiring plugging						
Mirror with polished edges						
356 x 254 mm	PC £2.97	0.80	5.01	5.55	nr	**10.56**
400 x 300 mm	PC £3.90	0.80	5.01	6.52	nr	**11.53**
560 x 380 mm	PC £7.10	0.90	5.64	9.88	nr	**15.52**
640 x 460 mm	PC £8.66	1.00	6.27	11.52	nr	**17.79**
Mirror with bevelled edges						
356 x 254 mm	PC £5.09	0.80	5.01	7.77	nr	**12.78**
400 x 300 mm	PC £5.66	0.80	5.01	8.37	nr	**13.38**
560 x 380 mm	PC £10.06	0.90	5.64	12.99	nr	**18.63**
640 x 460 mm	PC £12.53	1.00	6.27	15.58	nr	**21.85**

N FURNITURE/EQUIPMENT Including overheads and profit at 5.00%		Labour hours	Labour £	Material £	Unit	Total rate £
N10 GENERAL FIXTURES/FURNISHINGS/EQUIPMENT						
Matwells						
Mild steel matwell; galvanized after manufacture; comprising 30 x 30 x 3 mm angle rim; with welded angles and lugs welded on; to suit mat size						
914 x 560 mm	PC £20.62	1.00	8.96	21.65	nr	**30.61**
1067 x 610 mm	PC £21.51	1.25	11.20	22.59	nr	**33.79**
1219 x 762 mm	PC £23.85	1.50	13.43	25.04	nr	**38.47**
Polished aluminium matwell; comprising 32 x 32 x 5 mm angle rim; with brazed angles and lugs brazed on; to suit mat size						
914 x 560 mm	PC £27.36	1.00	8.96	28.73	nr	**37.69**
1067 x 610 mm	PC £29.16	1.25	11.20	30.62	nr	**41.82**
1219 x 762 mm	PC £33.61	1.50	13.43	35.30	nr	**48.73**
Polished brass matwell; comprising 38 x 38 x 6 mm angle rim; with brazed angles and lugs welded on; to suit mat size						
914 x 560 mm	PC £80.87	1.00	8.96	84.92	nr	**93.88**
1067 x 610 mm	PC £86.24	1.25	11.20	90.55	nr	**101.75**
1219 x 762 mm	PC £99.38	1.50	13.43	104.35	nr	**117.78**
N13 SANITARY APPLIANCES/FITTINGS						
NOTE: Sanitary fittings are largely a matter of selection and material prices vary considerably; the PC values given below are for average quality						
Sink; white glazed fireclay; BS 1206; cast iron cantilever brackets						
610 x 455 x 205 mm	PC £70.68	3.00	26.40	79.11	nr	**105.51**
610 x 455 x 255 mm	PC £78.77	3.00	26.40	87.61	nr	**114.01**
760 x 455 x 255 mm	PC £119.20	3.00	26.40	130.06	nr	**156.46**
Sink; stainless steel combined bowl and draining board; chain and self colour plug; to BS 3380						
1050 x 500 mm with bowl						
420 x 350 x 175 mm	PC £84.43	1.75	15.40	88.65	nr	**104.05**
Sink; stainless steel combined bowl and double draining board; chain and self colour plug to BS 3380						
1550 x 500 mm with bowl						
420 x 350 x 200 mm	PC £104.67	2.00	17.60	109.91	nr	**127.51**
Lavatory basin; vitreous china; BS 1188; 32 mm chromium plated waste; chain, stay and plug; pair 13 mm chromium plated easy clean pillar taps to BS 1010; painted cantilever brackets; plugged and screwed						
560 x 405 mm; white	PC £51.68	2.30	20.24	54.26	nr	**74.50**
560 x 405 mm; coloured	PC £59.05	2.30	20.24	62.00	nr	**82.24**
635 x 455 mm; white	PC £75.14	2.30	20.24	78.90	nr	**99.14**
635 x 455 mm; coloured	PC £86.06	2.30	20.24	90.36	nr	**110.60**
Lavatory basin and pedestal; vitreous china; BS 1188; 32 mm chromium plated waste, chain, stay and plug; pair 13 mm chromium plated easy clean pillar taps to BS 1010; pedestal; wall brackets; plugged and screwed						
560 x 405 mm; white	PC £67.46	2.50	22.00	70.83	nr	**92.83**
560 x 405 mm; coloured	PC £78.89	2.50	22.00	82.83	nr	**104.83**
635 x 455 mm; white	PC £90.92	2.50	22.00	95.47	nr	**117.47**
635 x 455 mm; coloured	PC £110.13	2.50	22.00	115.64	nr	**137.64**

N FURNITURE/EQUIPMENT Including overheads and profit at 5.00%		Labour hours	Labour £	Material £	Unit	Total rate £
N13 SANITARY APPLIANCES/FITTINGS - cont'd						
Lavatory basin range; overlap joints; white glazed fireclay; 32 mm chromium plated waste, chain, stay and plug; pair 13 mm chromium plated easy clean pillar taps to BS 1010; painted cast iron cantilever brackets; plugged and screwed						
range of four 560 x 405 mm	PC £257.75	8.80	77.43	270.64	nr	348.07
Add for each additional basin in the range		2.00	17.60	72.54	nr	90.14
Drinking fountain; white glazed fireclay; 19 mm chromium plated waste; self-closing non-concussive tap; regulating valve; plugged and screwed with chromium plated screws	PC £110.00	2.50	22.00	115.50	nr	137.50
Bath; reinforced acrylic rectangular pattern; 40 mm chromium plated overflow chain and plug; 40 mm chromium plated waste; cast brass 'P' trap with plain outlet and overflow connection to BS 1184; pair 20 mm chromium plated easy clean pillar taps to BS 1010						
1700 mm long; white	PC £115.00	3.50	30.80	120.75	nr	151.55
1700 mm long; coloured	PC £115.00	3.50	30.80	120.75	nr	151.55
Bath; enamelled steel; medium gauge rectangular pattern; 40 mm chromium plated overflow chain and plug; 40 mm chromium plated waste; cast brass 'P' trap with plain outlet and overflow connection to BS 1184; pair 20 mm chromium plated easy clean pillar taps to BS 1010						
1700 mm long; white	PC £130.00	3.50	30.80	136.50	nr	167.30
1700 mm long; coloured	PC £138.50	3.50	30.80	145.43	nr	176.23
Shower tray; glazed fireclay with outlet and grated waste; chain and plug; bedding and pointing in waterproof cement mortar						
760 x 760 x 180 mm; white	PC £107.35	3.00	26.40	112.72	nr	139.12
760 x 760 x 180 mm; coloured	PC £151.08	3.00	26.40	158.63	nr	185.03
Shower fitting; riser pipe with mixing valve and shower rose; chromium plated; plugging and screwing mixing valve and pipe bracket						
15 mm dia riser pipe; 127 mm dia shower rose	PC £163.88	5.00	44.00	172.07	nr	216.07
WC suite; high level; vitreous china pan; black plastic seat; 9 litre white vitreous china cistern and brackets; low pressure ball valve; galvanized steel flush pipe and clip; plugged and screwed; mastic joint to drain						
WC suite; white	PC £127.39	3.30	29.04	133.76	nr	162.80
WC suite; coloured	PC £175.94	3.30	29.04	184.74	nr	213.78
WC suite; low level; vitreous china pan; black plastic seat; 9 litre white vitreous china cistern and brackets; low pressure ball valve and plastic flush pipe; plugged and screwed; mastic joint to drain						
WC suite; white	PC £87.78	3.00	26.40	92.17	nr	118.57
WC suite; coloured	PC £111.96	3.00	26.40	117.56	nr	143.96

N FURNITURE/EQUIPMENT Including overheads and profit at 5.00%		Labour hours	Labour £	Material £	Unit	Total rate £
Slop sink; white glazed fireclay with hardwood pad; aluminium bucket grating; vitreous china cistern and porcelain-enamelled brackets; galvanized steel flush pipe and clip; plugged and screwed; mastic joint to drain						
slop sink	PC £385.00	3.50	30.80	404.25	nr	435.05
Bowl type wall urinal; white glazed vitreous china; white vitreous china automatic flushing cistern and brackets; 38 mm chromium plated waste; chromium plated flush pipes and spreaders; cistern brackets and flush pipes plugged and screwed with chromium plated screws						
single; 455 x 380 x 330 mm	PC £121.52	4.00	35.20	127.60	nr	**162.80**
range of two	PC £201.34	7.50	65.99	211.41	nr	**277.40**
range of three	PC £268.96	11.00	96.79	282.41	nr	**379.20**
Add for each additional urinal	PC £79.82	3.20	28.16	83.81	nr	**111.97**
Add for divisions between	PC £32.25	0.75	6.60	33.86	nr	**40.46**

N15 SIGNS/NOTICES

	Labour hours	Labour £	Material £	Unit	Total rate £
Plain script; in gloss oil paint; on painted or varnished surfaces					
Capital letters; lower case letters or numerals					
per coat; per 25 mm high	0.10	0.63	-	nr	**0.63**
Stops					
per coat	0.03	0.19	-	nr	**0.19**

P BUILDING FABRIC SUNDRIES
Including overheads and profit at 5.00%

ALTERNATIVE INSULATION PRICES

Insulation (£/m2)
Expanded polystyrene
self-extinguishing grade

12 mm	0.96	25 mm	1.98	38 mm	3.02	50 mm	3.98
18 mm	1.43						

'Fibreglass'
'Crown Building Roll'

60 mm	1.73	80 mm	2.17	100 mm	2.56

'Crown Wool'

150 mm	3.33	200 mm	4.49

'Frametherm' - unfaced

60 mm	1.36	80 mm	1.82	90 mm	2.03	100 mm	2.21

Sound-deadening quilt type PF; 13 mm - £2.01/m2

		Labour hours	Labour £	Material £	Unit	Total rate £
P10 SUNDRY INSULATION/PROOFING WORK/ FIRE STOPS						
'Sisalkraft' building papers/vapour barriers						
Building paper; 150 mm laps; fixed to softwood						
'Moistop' grade 728 (class A1F)		0.10	0.79	0.62	m2	**1.41**
Vapour barrier/reflective insulation 150 mm laps; fixed to softwood						
'Insulex' grade 714; single sided		0.10	0.79	0.94	m2	**1.73**
'Insulex' grade 714; double sided		0.10	0.79	1.37	m2	**2.16**
Mat or quilt insulation						
Glass fibre quilt; Pilkingtons 'Crown Wool'; laid over ceiling joists						
60 mm thick	PC £1.38	0.10	0.79	1.52	m2	**2.31**
80 mm thick	PC £1.82	0.11	0.87	2.01	m2	**2.88**
100 mm thick	PC £2.17	0.12	0.95	2.40	m2	**3.35**
150 mm thick	PC £3.33	0.13	1.03	3.68	m2	**4.71**
200 mm thick	PC £4.49	0.14	1.11	4.95	m2	**6.06**
Raking cutting		0.05	0.40	-	m	**0.40**
Glass fibre quilt; 'Gypglass 1000'; laid loose between members at 600 mm centres						
60 mm thick	PC £1.84	0.21	1.66	2.03	m2	**3.69**
80 mm thick	PC £2.43	0.23	1.82	2.68	m2	**4.50**
100 mm thick	PC £2.90	0.25	1.98	3.19	m2	**5.17**
150 mm thick	PC £4.44	0.30	2.37	4.89	m2	**7.26**
Mineral fibre quilt; 'Gypglass 1200'; pinned vertically to softwood						
25 mm thick	PC £0.96	0.14	1.11	1.06	m2	**2.17**
50 mm thick	PC £1.92	0.15	1.19	2.12	m2	**3.31**
Glass fibre building roll; pinned vertically to softwood						
60 mm thick	PC £1.73	0.15	1.19	1.92	m2	**3.11**
80 mm thick	PC £2.17	0.16	1.27	2.40	m2	**3.67**
100 mm thick	PC £2.56	0.17	1.34	2.82	m2	**4.16**

P BUILDING FABRIC SUNDRIES Including overheads and profit at 5.00%		Labour hours	Labour £	Material £	Unit	Total rate £
Glass fibre flanged building roll; paper faces; pinned vertically or to slope between timber framing						
60 mm thick	PC £1.86	0.18	1.42	2.05	m2	3.47
80 mm thick	PC £2.30	0.19	1.50	2.55	m2	4.05
100 mm thick	PC £0.00	0.20	1.58	2.95	m2	4.53
Board or slab insulation						
Expanded polystyrene board standard grade PC £60.63/m3; fixed with adhesive						
12 mm thick		0.40	3.16	1.05	m2	4.21
25 mm thick		0.42	3.32	1.61	m2	4.93
50 mm thick		0.45	3.56	2.72	m2	6.28
75 mm thick		0.50	3.95	3.84	m2	7.79
Fire stops						
'Monolux' TRADA firecheck channel; intumescent coatings on cut mitres; fixing with brass cups and screws						
19 x 44 mm or 19 x 50 mm	PC £198.54/36m	0.60	4.74	6.64	m	11.38
'Sealmaster' intumescent fire and smoke seals; pinned into groove in timber						
type N30; for single leaf half hour door;	PC £4.89	0.30	2.37	5.13	m	7.50
type N60; for single leaf one hour door	PC £7.44	0.33	2.61	7.81	m	10.42
type IMN or IMP; for meeting or pivot styles of pair of one hour doors; per style	PC £7.44	0.33	2.61	7.81	m	10.42
Intumescent plugs in timber; including boring		0.10	0.79	0.24	nr	1.03
Rockwool fire stops; between top of brick/ block wall and concrete soffit						
25 mm deep x 112 mm wide		0.08	0.63	0.83	m	1.46
25 mm deep x 150 mm wide		0.10	0.79	1.11	m	1.90
50 mm deep x 225 mm wide		0.15	1.19	2.80	m	3.99
Fire barriers						
Rockwool fire barrier between top of suspended ceiling and concrete soffit						
one 50 mm layer x 900 mm wide; half-hour		0.60	4.74	15.75	m	20.49
two 50 mm layers x 900 mm wide; one hour		0.90	7.12	30.41	m	37.53
Dow Chemicals 'Styrofoam 1B'; cold bridging insulation fixed with adhesive to brick, block or concrete base						
Insulation to walls						
25 mm thick		0.34	3.38	4.24	m2	7.62
50 mm thick		0.36	3.58	7.32	m2	10.90
75 mm thick		0.38	3.78	10.40	m2	14.18
Insulation to isolated columns						
25 mm thick		0.41	4.08	4.24	m2	8.32
50 mm thick		0.44	4.38	7.32	m2	11.70
75 mm thick		0.47	4.67	10.40	m2	15.07
Insulation to ceilings						
25 mm thick		0.36	3.58	4.24	m2	7.82
50 mm thick		0.39	3.88	7.32	m2	11.20
75 mm thick		0.42	4.18	10.40	m2	14.58
Insulation to isolated beams						
25 mm thick		0.44	4.38	4.24	m2	8.62
50 mm thick		0.47	4.67	7.32	m2	11.99
75 mm thick		0.50	4.97	10.40	m2	15.37

P BUILDING FABRIC SUNDRIES Including overheads and profit at 5.00%	Labour hours	Labour £	Material £	Unit	Total rate £
P11 FOAMED/FIBRE/BEAD CAVITY WALL INSULATION					
Cavity wall insulation; injecting 65 mm cavity with					
UF foam	-	-	-	m2	2.63
blown EPS granules	-	-	-	m2	3.15
blown mineral wool	-	-	-	m2	3.36
P20 UNFRAMED ISOLATED TRIMS/SKIRTINGS/ SUNDRY ITEMS					
Blockboard (Birch faced); 18 mm thick PC £86.36/10m2					
Window boards and the like; rebated; hardwood lipped on one edge					
18 x 200 mm	0.25	1.98	2.87	m	4.85
18 x 250 mm	0.28	2.21	3.37	m	5.58
18 x 300 mm	0.31	2.45	3.85	m	6.30
18 x 350 mm	0.34	2.69	4.34	m	7.03
Returned and fitted ends	0.22	1.74	0.30	nr	2.04
Blockboard (Sapele veneered one side); 18 mm thick PC £75.15/10m2					
Window boards and the like; rebated; hardwood lipped on one edge					
18 x 200 mm	0.27	2.13	2.63	m	4.76
18 x 250 mm	0.30	2.37	3.04	m	5.41
18 x 300 mm	0.33	2.61	3.47	m	6.08
18 x 350 mm	0.36	2.85	3.89	m	6.74
Returned and fitted ends	0.22	1.74	0.30	nr	2.04
Blockboard (Afrormosia veneered one side); 18 mm thick PC £92.05/10m2					
Window boards and the like; rebated; hardwood lipped on one edge					
18 x 200 mm	0.27	2.13	3.05	m	5.18
18 x 250 mm	0.30	2.37	3.58	m	5.95
18 x 300 mm	0.33	2.61	4.09	m	6.70
18 x 350 mm	0.36	2.85	4.62	m	7.47
Returned and fitted ends	0.22	1.74	0.32	nr	2.06
Wrought softwood					
Skirtings, picture rails, dado rails and the like; splayed or moulded					
19 x 50 mm; splayed	0.10	0.79	1.22	m	2.01
19 x 50 mm; moulded	0.10	0.79	1.29	m	2.08
19 x 75 mm; splayed	0.10	0.79	1.47	m	2.26
19 x 75 mm; moulded	0.10	0.79	1.56	m	2.35
19 x 100 mm; splayed	0.10	0.79	1.73	m	2.52
19 x 100 mm; moulded	0.10	0.79	1.82	m	2.61
19 x 150 mm; moulded	0.12	0.95	2.40	m	3.35
19 x 175 mm; moulded	0.12	0.95	2.65	m	3.60
22 x 100 mm; splayed	0.10	0.79	1.92	m	2.71
25 x 50 mm; moulded	0.10	0.79	1.49	m	2.28
25 x 75 mm; splayed	0.10	0.79	1.73	m	2.52
25 x 100 mm; splayed	0.10	0.79	2.09	m	2.88
25 x 150 mm; splayed	0.12	0.95	2.80	m	3.75
25 x 150 mm; moulded	0.12	0.95	2.90	m	3.85
25 x 175 mm; moulded	0.12	0.95	3.24	m	4.19
25 x 225 mm; moulded	0.14	1.11	3.96	m	5.07
Returned end	0.15	1.19	-	nr	1.19
Mitres	0.10	0.79	-	nr	0.79

P BUILDING FABRIC SUNDRIES Including overheads and profit at 5.00%		Labour hours	Labour £	Material £	Unit	Total rate £
Architraves, cover fillets and the like; half round; splayed or moulded						
13 x 25 mm; half round		0.12	0.95	0.85	m	**1.80**
13 x 50 mm; moulded		0.12	0.95	1.14	m	**2.09**
16 x 32 mm; half round		0.12	0.95	0.96	m	**1.91**
16 x 38 mm; moulded		0.12	0.95	1.11	m	**2.06**
16 x 50 mm; moulded		0.12	0.95	1.23	m	**2.18**
19 x 50 mm; splayed		0.12	0.95	1.22	m	**2.17**
19 x 63 mm; splayed		0.12	0.95	1.34	m	**2.29**
19 x 75 mm; splayed		0.12	0.95	1.47	m	**2.42**
25 x 44 mm; splayed		0.12	0.95	1.38	m	**2.33**
25 x 50 mm; moulded		0.12	0.95	1.49	m	**2.44**
25 x 63 mm; splayed		0.12	0.95	1.55	m	**2.50**
25 x 75 mm; splayed		0.12	0.95	1.73	m	**2.68**
32 x 88 mm; moulded		0.12	0.95	2.48	m	**3.43**
38 x 38 mm; moulded		0.12	0.95	1.51	m	**2.46**
50 x 50 mm; moulded		0.12	0.95	2.09	m	**3.04**
Returned end		0.15	1.19	-	nr	**1.19**
Mitres		0.10	0.79	-	nr	**0.79**
Stops; screwed on						
16 x 38 mm		0.10	0.79	0.90	m	**1.69**
16 x 50 mm		0.10	0.79	1.02	m	**1.81**
19 x 38 mm		0.10	0.79	0.95	m	**1.74**
25 x 38 mm		0.10	0.79	1.09	m	**1.88**
25 x 50 mm		0.10	0.79	1.28	m	**2.07**
Glazing beads and the like						
13 x 16 mm	PC £0.58	-	-	0.65	m	**0.65**
13 x 19 mm	PC £0.58	-	-	0.65	m	**0.65**
13 x 25 mm	PC £0.62	-	-	0.70	m	**0.70**
13 x 25 mm; screwed		0.05	0.40	0.79	m	**1.19**
13 x 25 mm; fixing with brass cups and screws		0.10	0.79	1.07	m	**1.86**
16 x 25 mm	PC £0.68	-	-	0.76	m	**0.76**
16 mm; quadrant		0.05	0.40	0.74	m	**1.14**
19 mm; quadrant or scotia		0.05	0.40	0.84	m	**1.24**
19 x 36 mm		0.05	0.40	0.92	m	**1.32**
25 x 38 mm		0.05	0.40	1.05	m	**1.45**
25 mm; quadrant or scotia		0.05	0.40	0.96	m	**1.36**
38 mm; scotia		0.05	0.40	1.46	m	**1.86**
50 mm; scotia		0.05	0.40	2.04	m	**2.44**
Isolated shelves, worktops, seats and the like						
19 x 150 mm		0.16	1.27	2.29	m	**3.56**
19 x 200 mm		0.22	1.74	2.79	m	**4.53**
25 x 150 mm		0.16	1.27	2.79	m	**4.06**
25 x 200 mm		0.22	1.74	3.47	m	**5.21**
32 x 150 mm		0.16	1.27	3.47	m	**4.74**
32 x 200 mm		0.22	1.74	4.27	m	**6.01**
Isolated shelves, worktops, seats and the like; cross-tongued joints						
19 x 300 mm		0.28	2.21	9.58	m	**11.79**
19 x 450 mm		0.34	2.69	11.23	m	**13.92**
19 x 600 mm		0.40	3.16	16.20	m	**19.36**
25 x 300 mm		0.28	2.21	10.83	m	**13.04**
25 x 450 mm		0.34	2.69	12.94	m	**15.63**
25 x 600 mm		0.40	3.16	18.50	m	**21.66**
32 x 300 mm		0.28	2.21	12.09	m	**14.30**
32 x 450 mm		0.34	2.69	14.80	m	**17.49**
32 x 600 mm		0.40	3.16	21.02	m	**24.18**

P BUILDING FABRIC SUNDRIES Including overheads and profit at 5.00%		Labour hours	Labour £	Material £	Unit	Total rate £
P20 UNFRAMED ISOLATED TRIMS/SKIRTINGS/ SUNDRY ITEMS - cont'd						
Wrought softwood - cont'd						
Isolated shelves, worktops, seats and the like; slatted with 50 mm wide slats at 75 mm centres						
19 mm thick		1.33	10.52	7.76	m2	18.28
25 mm thick		1.33	10.52	9.10	m2	19.62
32 mm thick		1.33	10.52	10.26	m2	20.78
Window boards, nosings, bed moulds and the like; rebated and rounded						
19 x 75 mm		0.18	1.42	1.64	m	3.06
19 x 150 mm		0.20	1.58	2.48	m	4.06
19 x 225 mm; in one width		0.26	2.06	3.26	m	5.32
19 x 300 mm; cross-tongued joints		0.30	2.37	9.63	m	12.00
25 x 75 mm		0.18	1.42	1.89	m	3.31
25 x 150 mm		0.20	1.58	2.98	m	4.56
25 x 225 mm; in one width		0.26	2.06	4.03	m	6.09
25 x 300 mm; cross-tongued joints		0.30	2.37	10.83	m	13.20
32 x 75 mm		0.18	1.42	2.20	m	3.62
32 x 150 mm		0.20	1.58	3.56	m	5.14
32 x 225 mm; in one width		0.26	2.06	4.92	m	6.98
32 x 300 mm; cross-tongued joints		0.30	2.37	12.10	m	14.47
38 x 75 mm		0.18	1.42	2.46	m	3.88
38 x 150 mm		0.20	1.58	4.08	m	5.66
38 x 225 mm; in one width		0.26	2.06	5.68	m	7.74
38 x 300 mm; cross-tongued joints		0.30	2.37	13.20	m	15.57
Returned and fitted ends		0.15	1.19	-	nr	1.19
Handrails; mopstick						
50 mm dia		0.25	1.98	2.85	m	4.83
Handrails; rounded						
44 x 50 mm		0.25	1.98	2.85	m	4.83
50 x 75 mm		0.27	2.13	3.59	m	5.72
63 x 87 mm		0.30	2.37	5.01	m	7.38
75 x 100 mm		0.35	2.77	5.69	m	8.46
Handrails; moulded						
44 x 50 mm		0.20	1.58	3.14	m	4.72
50 x 75 mm		0.22	1.74	3.89	m	5.63
63 x 87 mm		0.24	1.90	5.29	m	7.19
75 x 100 mm		0.26	2.06	5.97	m	8.03
Add +5% to the above 'Material £ prices' for 'selected' softwood for staining						
Selected West African Mahogany; PC £470.00/m3						
Skirtings, picture rails, dado rails and the like; splayed or moulded						
19 x 50 mm; splayed	PC £2.13	0.14	1.11	2.40	m	3.51
19 x 50 mm; moulded	PC £2.26	0.14	1.11	2.54	m	3.65
19 x 75 mm; splayed	PC £2.50	0.14	1.11	2.80	m	3.91
19 x 75 mm; moulded	PC £2.63	0.14	1.11	2.95	m	4.06
19 x 100 mm; splayed	PC £2.90	0.14	1.11	3.26	m	4.37
19 x 100 mm; moulded	PC £3.03	0.14	1.11	3.40	m	4.51
19 x 150 mm; moulded	PC £3.81	0.16	1.27	4.25	m	5.52
19 x 175 mm; moulded	PC £4.38	0.16	1.27	4.88	m	6.15
22 x 100 mm; splayed	PC £3.14	0.14	1.11	3.51	m	4.62
25 x 50 mm; moulded	PC £2.50	0.14	1.11	2.80	m	3.91
25 x 75 mm; splayed	PC £2.89	0.14	1.11	3.23	m	4.34
25 x 100 mm; splayed	PC £3.39	0.14	1.11	3.80	m	4.91
25 x 150 mm; splayed	PC £4.61	0.16	1.27	5.13	m	6.40

P BUILDING FABRIC SUNDRIES Including overheads and profit at 5.00%		Labour hours	Labour £	Material £	Unit	Total rate £
25 x 150 mm; moulded	PC £4.74	0.16	1.27	5.28	m	6.55
25 x 175 mm; moulded	PC £5.24	0.16	1.27	5.83	m	7.10
25 x 225 mm; moulded	PC £6.27	0.18	1.42	6.96	m	8.38
Returned end		0.22	1.74	-	nr	1.74
Mitres		0.15	1.19	-	nr	1.19
Architraves, cover fillets and the like; half round; splayed or moulded						
13 x 25 mm; half round	PC £1.62	0.16	1.27	1.84	m	3.11
13 x 50 mm; moulded	PC £2.00	0.16	1.27	2.26	m	3.53
16 x 32 mm; half round	PC £1.76	0.16	1.27	1.99	m	3.26
16 x 38 mm; moulded	PC £1.95	0.16	1.27	2.21	m	3.48
16 x 50 mm; moulded	PC £2.13	0.16	1.27	2.40	m	3.67
19 x 50 mm; splayed	PC £2.13	0.16	1.27	2.40	m	3.67
19 x 63 mm; splayed	PC £2.32	0.16	1.27	2.62	m	3.89
19 x 75 mm; splayed	PC £2.50	0.16	1.27	2.80	m	4.07
25 x 44 mm; splayed	PC £2.37	0.16	1.27	2.67	m	3.94
25 x 50 mm; moulded	PC £2.50	0.16	1.27	2.80	m	4.07
25 x 63 mm; splayed	PC £2.63	0.16	1.27	2.95	m	4.22
25 x 75 mm; splayed	PC £2.89	0.16	1.27	3.23	m	4.50
32 x 88 mm; moulded	PC £3.98	0.16	1.27	4.44	m	5.71
38 x 38 mm; moulded	PC £2.53	0.16	1.27	2.84	m	4.11
50 x 50 mm; moulded	PC £3.39	0.16	1.27	3.80	m	5.07
Returned end		0.22	1.74	-	nr	1.74
Mitres		0.15	1.19	-	nr	1.19
Stops; screwed on						
16 x 38 mm	PC £1.71	0.15	1.19	1.93	m	3.12
16 x 50 mm	PC £1.87	0.15	1.19	2.09	m	3.28
19 x 38 mm	PC £1.81	0.15	1.19	2.03	m	3.22
25 x 38 mm	PC £2.00	0.15	1.19	2.24	m	3.43
25 x 50 mm	PC £2.26	0.15	1.19	2.52	m	3.71
Glazing beads and the like						
13 x 16 mm	PC £1.42	-	-	1.56	m	1.56
13 x 19 mm	PC £1.42	-	-	1.56	m	1.56
13 x 25 mm	PC £1.49	-	-	1.64	m	1.64
13 x 25 mm; screwed	PC £1.49	0.08	0.63	1.73	m	2.36
13 x 25 mm; fixing with brass cups and screws	PC £1.49	0.15	1.19	2.01	m	3.20
16 x 25 mm	PC £1.54	-	-	1.71	m	1.71
16 mm; quadrant	PC £1.55	0.07	0.55	1.72	m	2.27
19 mm; quadrant or scotia	PC £1.63	0.07	0.55	1.79	m	2.34
19 x 36 mm	PC £1.81	0.07	0.55	2.00	m	2.55
25 x 38 mm	PC £2.00	0.07	0.55	2.20	m	2.75
25 mm; quadrant or scotia	PC £1.85	0.07	0.55	2.04	m	2.59
38 mm; scotia	PC £2.53	0.07	0.55	2.79	m	3.34
50 mm; scotia	PC £3.39	0.07	0.55	3.74	m	4.29
Isolated shelves, worktops, seats and the like						
19 x 150 mm	PC £3.95	0.22	1.74	4.36	m	6.10
19 x 200 mm	PC £4.69	0.30	2.37	5.17	m	7.54
25 x 150 mm	PC £4.69	0.22	1.74	5.17	m	6.91
25 x 200 mm	PC £5.67	0.30	2.37	6.25	m	8.62
32 x 150 mm	PC £5.50	0.22	1.74	6.07	m	7.81
32 x 200 mm	PC £6.81	0.30	2.37	7.51	m	9.88
Isolated shelves, worktops, seats and the like; cross-tongued joints						
19 x 300 mm	PC £13.66	0.38	3.00	15.06	m	18.06
19 x 450 mm	PC £16.05	0.45	3.56	17.69	m	21.25
19 x 600 mm	PC £23.38	0.55	4.35	25.78	m	30.13
25 x 300 mm	PC £15.40	0.38	3.00	16.98	m	19.98
25 x 450 mm	PC £18.52	0.45	3.56	20.42	m	23.98
25 x 600 mm	PC £26.68	0.55	4.35	29.41	m	33.76
32 x 300 mm	PC £17.12	0.38	3.00	18.88	m	21.88
32 x 450 mm	PC £21.05	0.45	3.56	23.21	m	26.77
32 x 600 mm	PC £30.20	0.55	4.35	33.30	m	37.65

P BUILDING FABRIC SUNDRIES Including overheads and profit at 5.00%		Labour hours	Labour £	Material £	Unit	Total rate £
P20 UNFRAMED ISOLATED TRIMS/SKIRTINGS/ SUNDRY ITEMS - cont'd						
Selected West African Mahogany; PC £470.00/m3 - cont'd						
Isolated shelves, worktops, seats and the like; slatted with 50 mm wide slats at 75 mm centres						
19 mm thick	PC £15.36	1.75	13.84	17.53	m2	31.37
25 mm thick	PC £17.15	1.75	13.84	19.56	m2	33.40
32 mm thick	PC £18.79	1.75	13.84	21.41	m2	35.25
Window boards, nosings, bed moulds and the like; rebated and rounded						
19 x 75 mm	PC £2.81	0.24	1.90	3.22	m	5.12
19 x 150 mm	PC £3.99	0.27	2.13	4.52	m	6.65
19 x 225 mm; in one width	PC £5.15	0.36	2.85	5.80	m	8.65
19 x 300 mm; cross-tongued joints	PC £13.57	0.40	3.16	15.08	m	18.24
25 x 75 mm	PC £3.20	0.24	1.90	3.65	m	5.55
25 x 150 mm	PC £4.78	0.27	2.13	5.39	m	7.52
25 x 225 mm; in one width	PC £6.27	0.36	2.85	7.03	m	9.88
25 x 300 mm; cross-tongued joints	PC £15.23	0.40	3.16	16.91	m	20.07
32 x 75 mm	PC £3.63	0.24	1.90	4.12	m	6.02
32 x 150 mm	PC £5.60	0.27	2.13	6.29	m	8.42
32 x 225 mm; in one width	PC £7.56	0.36	2.85	8.46	m	11.31
32 x 300 mm; cross-tongued joints	PC £16.96	0.40	3.16	18.82	m	21.98
38 x 75 mm	PC £4.63	0.24	1.90	5.23	m	7.13
38 x 150 mm	PC £6.35	0.27	2.13	7.12	m	9.25
38 x 225 mm; in one width	PC £8.68	0.36	2.85	9.68	m	12.53
38 x 300 mm; cross-tongued joints	PC £18.52	0.40	3.16	20.54	m	23.70
Returned and fitted ends		0.23	1.82	-	nr	1.82
Handrails; rounded						
44 x 50 mm	PC £4.69	0.33	2.61	5.17	m	7.78
50 x 75 mm	PC £5.76	0.36	2.85	6.36	m	9.21
63 x 87 mm	PC £7.32	0.40	3.16	8.07	m	11.23
75 x 100 mm	PC £8.80	0.45	3.56	9.70	m	13.26
Handrails; moulded						
44 x 50 mm	PC £5.06	0.33	2.61	5.58	m	8.19
50 x 75 mm	PC £6.14	0.36	2.85	6.77	m	9.62
63 x 87 mm	PC £7.70	0.40	3.16	8.48	m	11.64
75 x 100 mm	PC £9.17	0.45	3.56	10.11	m	13.67
Afrormosia; PC £710.00/m3						
Skirtings, picture rails, dado rails and the like; splayed or moulded						
19 x 50 mm; splayed	PC £2.52	0.14	1.11	2.83	m	3.94
19 x 50 mm; moulded	PC £2.65	0.14	1.11	2.97	m	4.08
19 x 75 mm; splayed	PC £3.12	0.14	1.11	3.50	m	4.61
19 x 75 mm; moulded	PC £3.25	0.14	1.11	3.63	m	4.74
19 x 100 mm; splayed	PC £3.70	0.14	1.11	4.14	m	5.25
19 x 100 mm; moulded	PC £3.83	0.14	1.11	4.27	m	5.38
19 x 150 mm; moulded	PC £5.00	0.16	1.27	5.56	m	6.83
19 x 175 mm; moulded	PC £5.77	0.16	1.27	6.42	m	7.69
22 x 100 mm; splayed	PC £4.07	0.14	1.11	4.54	m	5.65
25 x 50 mm; moulded	PC £3.03	0.14	1.11	3.40	m	4.51
25 x 75 mm; splayed	PC £3.66	0.14	1.11	4.09	m	5.20
25 x 100 mm; splayed	PC £4.44	0.14	1.11	4.95	m	6.06
25 x 150 mm; splayed	PC £6.19	0.16	1.27	6.88	m	8.15
25 x 150 mm; moulded	PC £6.31	0.16	1.27	7.01	m	8.28
25 x 175 mm; moulded	PC £7.07	0.16	1.27	7.84	m	9.11
25 x 225 mm; moulded	PC £8.64	0.18	1.42	9.58	m	11.00
Returned end		0.22	1.74	-	nr	1.74
Mitres		0.15	1.19	-	nr	1.19

P BUILDING FABRIC SUNDRIES Including overheads and profit at 5.00%		Labour hours	Labour £	Material £	Unit	Total rate £
Architraves, cover fillets and the like; half round; splayed or moulded						
13 x 25 mm; half round	PC £1.75	0.16	1.27	1.98	m	3.25
13 x 50 mm; moulded	PC £2.27	0.16	1.27	2.55	m	3.82
16 x 32 mm; half round	PC £1.97	0.16	1.27	2.23	m	3.50
16 x 38 mm; moulded	PC £2.23	0.16	1.27	2.51	m	3.78
16 x 50 mm; moulded	PC £2.46	0.16	1.27	2.76	m	4.03
19 x 50 mm; splayed	PC £2.52	0.16	1.27	2.83	m	4.10
19 x 63 mm; splayed	PC £2.84	0.16	1.27	3.18	m	4.45
19 x 75 mm; splayed	PC £3.12	0.16	1.27	3.50	m	4.77
25 x 44 mm; splayed	PC £2.90	0.16	1.27	3.26	m	4.53
25 x 50 mm; moulded	PC £3.03	0.16	1.27	3.40	m	4.67
25 x 63 mm; splayed	PC £3.29	0.16	1.27	3.68	m	4.95
25 x 75 mm; splayed	PC £3.66	0.16	1.27	4.09	m	5.36
32 x 88 mm; moulded	PC £5.31	0.16	1.27	5.91	m	7.18
38 x 38 mm; moulded	PC £3.14	0.16	1.27	3.51	m	4.78
50 x 50 mm; moulded	PC £4.44	0.16	1.27	4.95	m	6.22
Returned end		0.22	1.74	-	nr	1.74
Mitres		0.15	1.19	-	nr	1.19
Stops; screwed on						
16 x 38 mm	PC £1.97	0.15	1.19	2.21	m	3.40
16 x 50 mm	PC £2.21	0.15	1.19	2.47	m	3.66
19 x 38 mm	PC £2.10	0.15	1.19	2.35	m	3.54
25 x 38 mm	PC £2.39	0.15	1.19	2.67	m	3.86
25 x 50 mm	PC £2.78	0.15	1.19	3.10	m	4.29
Glazing beads and the like						
13 x 16 mm	PC £1.52	-	-	1.68	m	1.68
13 x 19 mm	PC £1.52	-	-	1.68	m	1.68
13 x 25 mm	PC £1.63	-	-	1.79	m	1.79
13 x 25 mm; screwed	PC £1.63	0.08	0.63	1.88	m	2.51
13 x 25 mm; fixing with brass cups and screws	PC £1.63	0.15	1.19	2.16	m	3.35
16 x 25 mm	PC £1.71	-	-	1.89	m	1.89
16 mm; quadrant	PC £1.67	0.07	0.55	1.84	m	2.39
19 mm; quadrant or scotia	PC £1.80	0.07	0.55	1.99	m	2.54
19 x 36 mm	PC £2.10	0.07	0.55	2.31	m	2.86
25 x 38 mm	PC £2.65	0.07	0.55	2.92	m	3.47
25 mm; quadrant or scotia	PC £2.13	0.07	0.55	2.35	m	2.90
38 mm; scotia	PC £3.14	0.07	0.55	3.46	m	4.01
50 mm; scotia	PC £4.44	0.07	0.55	4.89	m	5.44
Isolated shelves, worktops, seats and the like						
19 x 150 mm	PC £5.17	0.22	1.74	5.70	m	7.44
19 x 200 mm	PC £6.29	0.30	2.37	6.93	m	9.30
25 x 150 mm	PC £6.29	0.22	1.74	6.93	m	8.67
25 x 200 mm	PC £7.77	0.30	2.37	8.57	m	10.94
32 x 150 mm	PC £7.52	0.22	1.74	8.30	m	10.04
32 x 200 mm	PC £9.51	0.30	2.37	10.49	m	12.86
Isolated shelves, worktops, seats and the like; cross-tongued joints						
19 x 300 mm	PC £16.32	0.38	3.00	17.99	m	20.99
19 x 450 mm	PC £19.92	0.45	3.56	21.96	m	25.52
19 x 600 mm	PC £28.68	0.55	4.35	31.61	m	35.96
25 x 300 mm	PC £18.92	0.38	3.00	20.86	m	23.86
25 x 450 mm	PC £23.62	0.45	3.56	26.04	m	29.60
25 x 600 mm	PC £33.63	0.55	4.35	37.08	m	41.43
32 x 300 mm	PC £21.52	0.38	3.00	23.73	m	26.73
32 x 450 mm	PC £27.48	0.45	3.56	30.29	m	33.85
32 x 600 mm	PC £38.97	0.55	4.35	42.97	m	47.32

P BUILDING FABRIC SUNDRIES Including overheads and profit at 5.00%		Labour hours	Labour £	Material £	Unit	Total rate £
P20 UNFRAMED ISOLATED TRIMS/SKIRTINGS/ SUNDRY ITEMS - cont'd						
Afrormosia; PC £710.00/m3 - cont'd						
Isolated shelves, worktops, seats and the like; slatted with 50 mm wide slats at 75 mm centres						
19 mm thick	PC £18.29	1.75	13.84	20.84	m2	**34.68**
25 mm thick	PC £20.88	1.75	13.84	23.77	m2	**37.61**
32 mm thick	PC £23.61	1.75	13.84	26.97	m2	**40.81**
Window boards, nosings, bed moulds and the like; rebated and rounded						
19 x 75 mm	PC £3.40	0.24	1.90	3.87	m	**5.77**
19 x 150 mm	PC £5.18	0.27	2.13	5.83	m	**7.96**
19 x 225 mm; in one width	PC £6.94	0.36	2.85	7.77	m	**10.62**
19 x 300 mm; cross-tongued joints	PC £16.25	0.40	3.16	18.04	m	**21.20**
25 x 75 mm	PC £3.99	0.24	1.90	4.52	m	**6.42**
25 x 150 mm	PC £6.37	0.27	2.13	7.15	m	**9.28**
25 x 225 mm; in one width	PC £8.61	0.36	2.85	9.62	m	**12.47**
25 x 300 mm; cross-tongued joints	PC £18.75	0.40	3.16	20.80	m	**23.96**
32 x 75 mm	PC £4.63	0.24	1.90	5.23	m	**7.13**
32 x 150 mm	PC £7.61	0.27	2.13	8.51	m	**10.64**
32 x 225 mm; in one width	PC £10.61	0.36	2.85	11.82	m	**14.67**
32 x 300 mm; cross-tongued joints	PC £21.36	0.40	3.16	23.67	m	**26.83**
38 x 75 mm	PC £5.18	0.24	1.90	5.83	m	**7.73**
38 x 150 mm	PC £8.74	0.27	2.13	9.76	m	**11.89**
38 x 225 mm; in one width	PC £12.27	0.36	2.85	13.65	m	**16.50**
38 x 300 mm; cross-tongued joints	PC £23.73	0.40	3.16	26.28	m	**29.44**
Returned and fitted ends		0.23	1.82	-	nr	**1.82**
Handrails; rounded						
44 x 50 mm	PC £5.74	0.33	2.61	6.33	m	**8.94**
50 x 75 mm	PC £7.29	0.36	2.85	8.03	m	**10.88**
63 x 87 mm	PC £10.46	0.40	3.16	11.53	m	**14.69**
75 x 100 mm	PC £11.94	0.45	3.56	13.17	m	**16.73**
Handrails; moulded						
44 x 50 mm	PC £6.11	0.33	2.61	6.75	m	**9.36**
50 x 75 mm	PC £7.68	0.36	2.85	8.46	m	**11.31**
63 x 87 mm	PC £10.84	0.40	3.16	11.95	m	**15.11**
75 x 100 mm	PC £12.31	0.45	3.56	13.58	m	**17.14**
Pin-boards; medium board						
Sundeala 'A' pin-board; fixed with adhesive to backing (measured elsewhere); over 300 mm wide						
6.4 mm thick		0.60	4.74	7.26	m2	**12.00**
Sundries on softwood/hardwood						
Extra over fixing with nails for						
gluing and pinning		0.02	0.16	0.03	m	**0.19**
masonry nails		0.02	0.16	0.04	m	**0.20**
steel screws		0.02	0.16	0.05	m	**0.21**
self-tapping screws		0.02	0.16	0.07	m	**0.23**
steel screws; gluing		0.04	0.32	0.05	m	**0.37**
steel screws; sinking; filling heads		0.05	0.40	0.05	m	**0.45**
steel screws; sinking; pellating over		0.10	0.79	0.39	m	**1.18**
brass cups and screws		0.15	1.19	0.05	m	**1.24**
Extra over for						
countersinking		0.02	0.16	-	m	**0.16**
pellating		0.10	0.79	0.03	m	**0.82**
Head or nut in softwood						
let in flush		0.05	0.40	-	nr	**0.40**
Head or nut; in hardwood						
let in flush		0.08	0.63	-	nr	**0.63**
let in over; pellated		0.20	1.58	-	nr	**1.58**

P BUILDING FABRIC SUNDRIES Including overheads and profit at 5.00%	Labour hours	Labour £	Material £	Unit	Total rate £
P21 IRONMONGERY					
Metalwork; mild steel; galvanized					
Water bars; groove in timber					
6 x 30 mm	0.50	3.95	3.28	m	**7.23**
6 x 40 mm	0.50	3.95	5.15	m	**9.10**
6 x 50 mm	0.50	3.95	4.49	m	**8.44**
Dowels; mortice in timber					
8 mm dia x 100 mm long	0.05	0.40	0.04	nr	**0.44**
10 mm dia x 50 mm long	0.05	0.40	0.13	nr	**0.53**
Cramps					
25 x 3 x 230 mm girth; one end bent, holed and screwed to softwood; other end fishtailed for building in	0.07	0.55	0.28	nr	**0.83**

IRONMONGERY - TYPICAL 'SUPPLY ONLY' PRICES

NOTE: Ironmongery is largely a matter of selection and prices vary considerably; indicative prices for reasonable quantities of standard quality ironmongery are given below. The prices for doors include for fixing with and including ordinary butt hinges, and only the extra cost of fixing hinges are given below.

	£		£		£		£
Bolts (£/each)							
barrel							
152 mm	1.13	203 mm	1.35	254 mm	1.67	305 mm	1.89
straight tower							
102 mm	0.54	152 mm	0.72	203 mm	1.04	254 mm	1.40

	£		£
Other bolts			
flush - 203x19 mm brass	5.67	panic - single	38.25
- 203x19 mm SAA	4.46	- double	45.90
garage bolt foot action	6.84	security hinge	2.61
monkey tail - 380 mm	7.16	security mortice	2.56
necked tower - 152 mm	1.17	WC indicator - SAA	9.63
- 203 mm	1.53	- zinc alloy	4.14

	£		£		£		£
Butts and hinges (£/pair)							
back flap; steel; medium							
25x73 mm	0.41	32x82 mm	0.54	38x89 mm	0.58	51x108 mm	0.85
bands and hooks; wrought iron							
305 mm	3.38	406 mm	4.28	457 mm	5.36	609 mm	8.33
butt hinges; brass; washered							
76x51 mm	4.28	102x66 mm	8.68	127x76 mm	11.07		
butt hinges; steel							
extra strong							
76 mm	1.71	102 mm	2.43				
medium							
50 mm	0.23	60 mm	0.23	75 mm	0.27	100 mm	0.54
parliament							
76 mm	2.61	102 mm	4.23				
rising							
76 mm	1.21	102 mm	2.16				
spring; single action							
76 mm	10.13	102 mm	11.97	127 mm	14.13	152 mm	17.28
spring; double action							
76 mm	15.84	102 mm	18.90	127 mm	21.60	152 mm	26.28

P BUILDING FABRIC SUNDRIES
Including overheads and profit at 5.00%

Butts and hinges (£/pair) - cont'd

	£		£		£		£
strong single flap							
76 mm	0.99	102 mm	1.35				
tee; medium							
230 mm	0.99	300 mm	1.04	375 mm	1.35	450 mm	1.94
washered							
76x49 mm	3.06	101x70 mm	5.58				

Other butts and hinges

	£		£
collinge - 457 mm	31.41	double strap field gate	17.73

Catches (£/each)

	£		£		£
magnetic					
cupboard	1.53	door	1.71		
roller and ball					
bales	0.90	door	1.40		

Door closers (£/each)

	£		£		£
floor springs					
single action	150.30	double action	168.75		
overhead door closer; concealed fixing					
50.8 kg	78.12	61.6 kg	82.35		
overhead door closer; surface fixing					
45 kg	43.65	67 kg	51.98	91 kg	67.05
other items					
'Perko' closer	4.77	selector	35.64		

Door furniture

	£		£		£
finger plates; 305x76 mm; quality (£/each)					
'modest'	2.34	'average'	7.65	'expensive'	13.95
kicking plates; quality (£/each)					
760x150 mm					
'modest'	2.70	'average'	8.55	'expensive'	14.85
lever latch furniture (£/set)					
'modest'	5.71	'average'	14.76	'expensive'	57.15
lever lock furniture					
'modest'	5.94	'average'	12.24	'expensive'	59.31

Other door furniture (£/each)

	£		£
cabin hook and eye	3.42	postal knocker	7.56
centre knob	16.11	pull handles	15.39 - 26.91
door chain	1.76	push pad handle; 150x150 mm	15.75
door viewer	3.55	padlock; close shackle	29.25

Latches (£/each)

cylinder rim night - 'Yale'	10.71	rim	3.20
deadlocking rim night- 'Yale'	22.68	Suffolk	2.84
mortice	3.47		

Locks (£/each)

budget	3.96	rim	9.09
cupboard lock	3.33	'Yale' deadlock	12.78
mortice deadlock	5.94	'Waterloo' mortice	7.88

Shelving; adjustable 'Spur' (£/each)

bracket - 22 cm	1.40	shelf - 1 m x 36 cm	9.09
- 36 cm	2.93	upright - 122 cm	4.23
- 61 cm	4.73	- 240 cm	7.74

P BUILDING FABRIC SUNDRIES
Including overheads and profit at 5.00%

Sliding door gear; 'Hillaldam; (£/each - unless otherwise described)

'Commercial for top hung doors; 365 kg max. weight; 3.6 m max. height

Item	£	Item	£
bow handle	4.74	door stop (rubber buffers)	13.69
det. locking bar/padlock	10.84	drop bolt	21.67
door guide - metal	3.77	flush handle	2.66
- timber	4.55	galv. steel top tack	13.78
door hanger - metal	16.42	open side wall bracket	2.49
door hanger - timber	22.27	steel bottom guide	8.79
'House One' for internal domestic doors			
pelmet	8.76	set size 2	11.84
'Twin fold' for 4 door wardrobes (1.50 m opening)			
set size TF60/4	17.84		

Window furniture (£/each)

Item	£	Item	£
casement fastener; SAA	3.24	metal window locks	17.64
casement stay; SAA	3.11	security mortice bolt	2.56
locking window catch	3.74		

Sundries

Item	£	Item	£
hat and coat hooks	2.16	pictograms; stainless steel	3.38
door stops; rubber;to concrete	2.29	numerals; stainless steel	0.54
; to wood	0.36	shelf brackets; japanned	0.18

	Labour hours	Labour £	Material £	Unit	Total rate £
P21 IRONMONGERY - cont'd					
Fixing only ironmongery to softwood					
Bolts					
barrel; not exceeding 150 mm long	0.33	2.61	-	nr	**2.61**
barrel; 150-300 mm long	0.42	3.32	-	nr	**3.32**
cylindrical mortice; not exceeding					
150 mm long	0.50	3.95	-	nr	**3.95**
150 - 300 mm long	0.60	4.74	-	nr	**4.74**
flush; not exceeding 150 mm long	0.50	3.95	-	nr	**3.95**
flush; 150 - 300 mm long	0.60	4.74	-	nr	**4.74**
monkey tail; 380 mm long	0.67	5.30	-	nr	**5.30**
necked; 150 mm long	0.33	2.61	-	nr	**2.61**
panic; single; locking	2.50	19.77	-	nr	**19.77**
panic; double; locking	3.50	27.67	-	nr	**27.67**
WC indicator	0.67	5.30	-	nr	**5.30**
Butts; extra over for					
rising	0.17	1.34	-	pr	**1.34**
skew	0.17	1.34	-	pr	**1.34**
spring; single action	1.30	10.28	-	pr	**10.28**
spring; double action	1.50	11.86	-	pr	**11.86**
Catches					
surface mounted	0.17	1.34	-	nr	**1.34**
mortice	0.33	2.61	-	nr	**2.61**
Door closers and furniture					
cabin hooks and eyes	0.17	1.34	-	nr	**1.34**
door selector	0.50	3.95	-	nr	**3.95**
finger plate	0.17	1.34	-	nr	**1.34**
floor spring	2.50	19.77	-	nr	**19.77**
lever furniture	0.25	1.98	-	nr	**1.98**
handle; not exceeding 150 mm long	0.17	1.34	-	nr	**1.34**
handle; 150 - 300 mm long	0.25	1.98	-	nr	**1.98**
handle; flush	0.33	2.61	-	nr	**2.61**
holder	0.33	2.61	-	nr	**2.61**

P BUILDING FABRIC SUNDRIES Including overheads and profit at 5.00%	Labour hours	Labour £	Material £	Unit	Total rate £
P21 IRONMONGERY - cont'd					
Fixing only ironmongery to softwood - cont'd					
Door closers and furniture					
kicking plate	0.33	2.61	-	nr	**2.61**
letter plate; including perforation	1.33	10.52	-	nr	**10.52**
overhead door closer; surface fixing	1.25	9.88	-	nr	**9.88**
overhead door closer; concealed fixing	1.75	13.84	-	nr	**13.84**
top centre	0.50	3.95	-	nr	**3.95**
'Perko' door closer	0.67	5.30	-	nr	**5.30**
rod door closers; 457 mm long	1.00	7.91	-	nr	**7.91**
Latches					
cylinder rim night latch	0.75	5.93	-	nr	**5.93**
mortice	0.67	5.30	-	nr	**5.30**
Norfolk	0.67	5.30	-	nr	**5.30**
rim	0.50	3.95	-	nr	**3.95**
Locks					
cupboard	0.42	3.32	-	nr	**3.32**
mortice	0.83	6.56	-	nr	**6.56**
mortice budget	0.75	5.93	-	nr	**5.93**
mortice dead	0.75	5.93	-	nr	**5.93**
rebated mortice	1.25	9.88	-	nr	**9.88**
rim	0.50	3.95	-	nr	**3.95**
rim budget	0.42	3.32	-	nr	**3.32**
rim dead	0.42	3.32	-	nr	**3.32**
Sliding door gear for top hung softwood timber doors; weight not exceeding 365 kg					
bottom guide; fixed to concrete in groove	0.50	3.95	-	m	**3.95**
top track	0.25	1.98	-	m	**1.98**
detachable locking bar and padlock	0.33	2.61	-	nr	**2.61**
hangers; fixed flush to timber	0.75	5.93	-	nr	**5.93**
head brackets; bolted to concrete	0.42	3.32	-	nr	**3.32**
Window furniture					
casement stay and pin	0.17	1.34	-	nr	**1.34**
catch; fanlight	0.25	1.98	-	nr	**1.98**
fastener; cockspur	0.33	2.61	-	nr	**2.61**
fastener; sash	0.25	1.98	-	nr	**1.98**
quadrant stay	0.25	1.98	-	nr	**1.98**
ring catch	0.33	2.61	-	nr	**2.61**
Sundries					
drawer pull	0.08	0.63	-	nr	**0.63**
hat and coat hook	0.08	0.63	-	nr	**0.63**
numerals	0.08	0.63	-	nr	**0.63**
rubber door stop	0.08	0.63	-	nr	**0.63**
shelf bracket	0.17	1.34	-	nr	**1.34**
skirting type door stop	0.17	1.34	-	nr	**1.34**
Fixing only ironmongery to hardwood					
Bolts					
barrel; not exceeding 150 mm long	0.44	3.48	-	nr	**3.48**
barrel; 150-300 mm long	0.56	4.43	-	nr	**4.43**
cylindrical mortice; not exceeding					
150 mm long	0.67	5.30	-	nr	**5.30**
150 - 300 mm long	0.80	6.33	-	nr	**6.33**
flush; not exceeding 150 mm long	0.67	5.30	-	nr	**5.30**
flush; 150 - 300 mm long	0.80	6.33	-	nr	**6.33**
monkey tail; 380 mm long	0.89	7.04	-	nr	**7.04**
necked; 150 mm long	0.44	3.48	-	nr	**3.48**
panic; single; locking	3.33	26.33	-	nr	**26.33**
panic; double; locking	4.67	36.92	-	nr	**36.92**
WC indicator	0.89	7.04	-	nr	**7.04**

P BUILDING FABRIC SUNDRIES Including overheads and profit at 5.00%	Labour hours	Labour £	Material £	Unit	Total rate £
Butts; extra over for					
rising	0.23	1.82	-	pr	1.82
skew	0.23	1.82	-	pr	1.82
spring; single action	1.73	13.68	-	pr	13.68
spring; double action	2.00	15.81	-	pr	15.81
Catches					
surface mounted	0.23	1.82	-	nr	1.82
mortice	0.44	3.48	-	nr	3.48
Door closers and furniture					
cabin hooks and eyes	0.23	1.82	-	nr	1.82
door selector	0.67	5.30	-	nr	5.30
finger plate	0.23	1.82	-	nr	1.82
floor spring	3.33	26.33	-	nr	26.33
lever furniture	0.33	2.61	-	nr	2.61
handle; not exceeding 150 mm long	0.23	1.82	-	nr	1.82
handle; 150 - 300 mm long	0.33	2.61	-	nr	2.61
handle; flush	0.44	3.48	-	nr	3.48
holder	0.44	3.48	-	nr	3.48
kicking plate	0.44	3.48	-	nr	3.48
letter plate; including perforation	1.77	13.99	-	nr	13.99
overhead door closer; surface fixing	1.67	13.20	-	nr	13.20
overhead door closer; concealed fixing	2.33	18.42	-	nr	18.42
top centre	0.67	5.30	-	nr	5.30
'Perko' door closer	0.89	7.04	-	nr	7.04
rod door closers; 457 mm long	1.33	10.52	-	nr	10.52
Latches					
cylinder rim night latch	1.00	7.91	-	nr	7.91
mortice	0.89	7.04	-	nr	7.04
Norfolk	0.89	7.04	-	nr	7.04
rim	0.67	5.30	-	nr	5.30
Locks					
cupboard	0.56	4.43	-	nr	4.43
mortice	1.11	8.78	-	nr	8.78
mortice budget	1.00	7.91	-	nr	7.91
mortice dead	1.00	7.91	-	nr	7.91
rebated mortice	1.67	13.20	-	nr	13.20
rim	0.67	5.30	-	nr	5.30
rim budget	0.56	4.43	-	nr	4.43
rim dead	0.56	4.43	-	nr	4.43
Sliding door gear for top hung softwood timber doors; weight not exceeding 365 kg					
bottom guide; fixed to concrete in groove	0.67	5.30	-	m	5.30
top track	0.33	2.61	-	m	2.61
detachable locking bar and padlock	0.44	3.48	-	nr	3.48
hangers; fixed flush to timber	1.00	7.91	-	nr	7.91
head brackets; bolted to concrete	0.56	4.43	-	nr	4.43
Window furniture					
casement stay and pin	0.23	1.82	-	nr	1.82
catch; fanlight	0.33	2.61	-	nr	2.61
fastener; cockspur	0.44	3.48	-	nr	3.48
fastener; sash	0.33	2.61	-	nr	2.61
quadrant stay	0.33	2.61	-	nr	2.61
ring catch	0.44	3.48	-	nr	3.48

P BUILDING FABRIC SUNDRIES Including overheads and profit at 5.00%	Labour hours	Labour £	Material £	Unit	Total rate £
P21 IRONMONGERY - cont'd					
Fixing only ironmongery to hardwood - cont'd					
Sundries					
drawer pull	0.11	0.87	-	nr	**0.87**
hat and coat hook	0.11	0.87	-	nr	**0.87**
numerals	0.11	0.87	-	nr	**0.87**
rubber door stop	0.11	0.87	-	nr	**0.87**
shelf bracket	0.23	1.82	-	nr	**1.82**
skirting type door stop	0.11	0.87	-	nr	**0.87**
Sundries					
Rubber door stop plugged and screwed to concrete	0.10	0.79	2.43	nr	**3.22**
P30 TRENCHES/PIPEWAYS/PITS FOR BURIED ENGINEERING SERVICES					
Mechanical excavation of trenches to receive pipes; grading bottoms; earthwork support; filling with excavated material and compacting; disposal of surplus soil; spreading on site average 50 m					
Pipes not exceeding 200 mm; average depth					
0.50 m deep	0.20	1.17	1.07	m	**2.24**
0.75 m deep	0.30	1.76	1.79	m	**3.55**
1.00 m deep	0.60	3.52	3.29	m	**6.81**
1.25 m deep	0.90	5.28	4.48	m	**9.76**
1.50 m deep	1.15	6.75	5.84	m	**12.59**
1.75 m deep	1.40	8.22	7.48	m	**15.70**
2.00 m deep	1.65	9.68	8.56	m	**18.24**
Hand excavation of trenches to receive pipes; grading bottoms; earthwork support; filling with excavated material and compacting; disposal of surplus soil; spreading on site average 50 m					
Pipes not exceeding 200 mm; average depth					
0.50 m deep	1.00	5.87	-	m	**5.87**
0.75 m deep	1.50	8.80	-	m	**8.80**
1.00 m deep	2.20	12.91	0.96	m	**13.87**
1.25 m deep	3.10	18.20	1.45	m	**19.65**
1.50 m deep	4.25	24.95	1.74	m	**26.69**
1.75 m deep	5.60	32.87	2.12	m	**34.99**
2.00 m deep	6.40	37.56	2.32	m	**39.88**
Pits for underground stop valves and the like; half brick thick walls in common bricks in cement mortar (1:3); on in situ concrete mix 21.00 N/mm2-20 mm aggregate (1:2:4) bed; 100 mm thick; 100 x 100 x 750 mm deep; internal holes for one small pipe; cast iron hinged box cover; bedding in cement mortar (1:3)	3.00	30.40	14.84	nr	**45.24**

P BUILDING FABRIC SUNDRIES Including overheads and profit at 5.00%	Labour hours	Labour £	Material £	Unit	Total rate £
P31 HOLES/CHASES/COVERS/SUPPORTS FOR SERVICES					
Builders' work for electrical installations					
Cutting away for and making good after electrician; including cutting or leaving all holes, notches, mortices, sinkings and chases, in both the structure and its coverings, for the following electrical points					
Exposed installation					
lighting points	0.30	2.69	-	nr	**2.69**
socket outlet points	0.50	4.48	-	nr	**4.48**
fitting outlet points	0.50	4.48	-	nr	**4.48**
equipment points or control gear points	0.70	6.27	-	nr	**6.27**
Concealed installation					
lighting points	0.40	3.58	-	nr	**3.58**
socket outlet points	0.70	6.27	-	nr	**6.27**
fitting outlet points	0.70	6.27	-	nr	**6.27**
equipment points or control gear points	1.00	8.96	-	nr	**8.96**
Builders' work for other services installations					
Cutting chases in brickwork					
for one pipe; not exceeding 55 mm nominal size; vertical	0.40	4.05	-	m	**4.05**
for one pipe; 55 - 110 nominal size; vertical	0.70	7.09	-	m	**7.09**
Cutting and pinning to brickwork or blockwork; ends of supports					
for pipes not exceeding 55 mm	0.20	2.03	-	m	**2.03**
for cast iron pipes 55 - 110 mm	0.33	3.34	-	nr	**3.34**
radiator stays or brackets	0.25	2.53	-	nr	**2.53**
Cutting holes for pipes or the like; not exceeding 55 mm nominal size					
102 mm brickwork	0.33	2.01	-	nr	**2.01**
215 mm brickwork	0.55	3.35	-	nr	**3.35**
327 mm brickwork	0.90	5.48	-	nr	**5.48**
100 mm blockwork	0.30	1.83	-	nr	**1.83**
150 mm blockwork	0.40	2.44	-	nr	**2.44**
215 mm blockwork	0.50	3.05	-	nr	**3.05**
Cutting holes for pipes or the like; 55 - 110 mm nominal size					
102 mm brickwork	0.40	2.44	-	nr	**2.44**
215 mm brickwork	0.70	4.26	-	nr	**4.26**
327 mm brickwork	1.10	6.70	-	nr	**6.70**
100 mm blockwork	0.35	2.13	-	nr	**2.13**
150 mm blockwork	0.50	3.05	-	nr	**3.05**
215 mm blockwork	0.60	3.65	-	nr	**3.65**
Cutting holes for pipes or the like; over 110 mm nominal size					
102 mm brickwork	0.50	3.05	-	nr	**3.05**
215 mm brickwork	0.85	5.18	-	nr	**5.18**
327 mm brickwork	1.35	8.22	-	nr	**8.22**
100 mm blockwork	0.45	4.56	-	nr	**4.56**
150 mm blockwork	0.60	3.65	-	nr	**3.65**
215 mm blockwork	0.75	4.57	-	nr	**4.57**
Add for making good fair face or facings one side					
pipe; not exceeding 55 mm nominal size	0.08	0.81	-	nr	**0.81**
pipe; 55 - 110 mm nominal size	0.10	1.01	-	nr	**1.01**
pipe; over 110 mm nominal size	0.12	1.22	-	nr	**1.22**

P BUILDING FABRIC SUNDRIES Including overheads and profit at 5.00%	Labour hours	Labour £	Material £	Unit	Total rate £
P31 HOLES/CHASES/COVERS/SUPPORTS FOR SERVICES - cont'd					
Builders' work for other services installations - cont'd					
Add for fixing sleeve (supply included elsewhere)					
for pipe; small	0.15	1.52	-	nr	1.52
for pipe; large	0.20	2.03	-	nr	2.03
for pipe; extra large	0.30	3.04	-	nr	3.04
Cutting or forming holes for ducts; girth not exceeding 1.00 m					
102 mm brickwork	0.60	3.65	-	nr	3.65
215 mm brickwork	1.00	6.09	-	nr	6.09
327 mm brickwork	1.60	9.74	-	nr	9.74
100 mm blockwork	0.50	3.05	-	nr	3.05
150 mm blockwork	0.70	4.26	-	nr	4.26
215 mm blockwork	0.90	5.48	-	nr	5.48
Cutting or forming holes for ducts; girth 1.00 - 2.00 m					
102 mm brickwork	0.70	4.26	-	nr	4.26
215 mm brickwork	1.20	7.31	-	nr	7.31
327 mm brickwork	1.90	11.57	-	nr	11.57
100 mm blockwork	0.60	3.65	-	nr	3.65
150 mm blockwork	0.80	4.87	-	nr	4.87
215 mm blockwork	1.00	6.09	-	nr	6.09
Cutting or forming holes for ducts; girth 2.00 - 3.00 m					
102 mm brickwork	1.10	6.70	-	nr	6.70
215 mm brickwork	1.90	11.57	-	nr	11.57
327 mm brickwork	3.00	18.27	-	nr	18.27
100 mm blockwork	0.95	5.79	-	nr	5.79
150 mm blockwork	1.30	7.92	-	nr	7.92
215 mm blockwork	1.65	10.05	-	nr	10.05
Cutting or forming holes for ducts; girth 3.00 - 4.00 m					
102 mm brickwork	1.50	9.13	-	nr	9.13
215 mm brickwork	2.50	15.23	-	nr	15.23
327 mm brickwork	4.00	24.36	-	nr	24.36
100 mm blockwork	1.10	6.70	-	nr	6.70
150 mm blockwork	1.50	9.13	-	nr	9.13
215 mm blockwork	1.90	11.57	-	nr	11.57
Mortices in brickwork					
for expansion bolt	0.20	1.22	-	nr	1.22
for 20 mm dia bolt 75 mm deep	0.15	0.91	-	nr	0.91
for 20 mm dia bolt 150 mm deep	0.25	1.52	-	nr	1.52
Mortices in brickwork; grouting with cement mortar (1:1)					
75 x 75 x 200 mm deep	0.30	1.83	0.09	nr	1.92
75 x 75 x 300 mm deep	0.40	2.44	0.13	nr	2.57
Holes in softwood for pipes, bars cables and the like					
12 mm thick	0.04	0.32	-	nr	0.32
25 mm thick	0.06	0.47	-	nr	0.47
50 mm thick	0.10	0.79	-	nr	0.79
100 mm thick	0.15	1.19	-	nr	1.19
Holes in hardwood for pipes, bars, cables and the like					
12 mm thick	0.06	0.47	-	nr	0.47
25 mm thick	0.09	0.71	-	nr	0.71
50 mm thick	0.15	1.19	-	nr	1.19
100 mm thick	0.22	1.74	-	nr	1.74

P BUILDING FABRIC SUNDRIES Including overheads and profit at 5.00%		Labour hours	Labour £	Material £	Unit	Total rate £
Duct covers with frames; cast iron; Brickhouse Dudley 'Trucast' or similar; bedding and pointing frame in cement mortar (1:3); fixing with 10 mm dia anchor bolts to concrete at 500 mm centres; including cutting and pinning anchor bolts						
Medium duty; ref 702						
300 mm wide	PC £117.82	2.50	22.39	123.71	m	**146.10**
Extra for						
ends	PC £20.00	-	-	21.00	nr	**21.00**
right angle frame corner	PC £9.60	-	-	10.08	nr	**10.08**
450 mm wide	PC £119.76	2.80	25.08	125.75	m	**150.83**
Extra for						
ends	PC £20.00	-	-	21.00	nr	**21.00**
right angle frame corner	PC £9.60	-	-	10.08	nr	**10.08**
600 mm wide	PC £189.05	3.10	27.77	198.50	m	**226.27**
Extra for						
ends	PC £20.00	-	-	21.00	nr	**21.00**
right angle frame corner	PC £9.60	-	-	10.08	nr	**10.08**
750 mm wide	PC £229.25	3.40	30.45	240.71	m	**271.16**
Extra for						
ends	PC £20.00	-	-	21.00	nr	**21.00**
right angle frame corner	PC £9.60	-	-	10.08	nr	**10.08**
900 mm wide	PC £260.82	3.70	33.14	273.86	m	**307.00**
Extra for						
ends	PC £20.00	-	-	21.00	nr	**21.00**
right angle frame corner	PC £9.60	-	-	10.08	nr	**10.08**
Heavy duty; ref 704						
300 mm wide	PC £195.22	-	-	204.99	m	**204.99**
Extra for						
ends	PC £27.20	-	-	28.56	nr	**28.56**
right angle frame corner	PC £9.60	-	-	10.08	nr	**10.08**
450 mm wide	PC £219.59	3.40	30.45	230.57	m	**261.02**
Extra for						
ends	PC £27.20	-	-	28.56	nr	**28.56**
right angle frame corner	PC £9.60	-	-	10.08	nr	**10.08**
600 mm wide	PC £242.40	3.80	34.03	254.52	m	**288.55**
Extra for						
ends	PC £27.20	-	-	28.56	nr	**28.56**
right angle frame corner	PC £9.60	-	-	10.08	nr	**10.08**
750 mm wide	PC £261.28	4.20	37.62	274.34	m	**311.96**
Extra for						
ends	PC £27.20	-	-	28.56	nr	**28.56**
right angle frame corner	PC £9.60	-	-	10.08	nr	**10.08**
900 mm wide	PC £316.29	4.60	41.20	332.10	m	**373.30**
Extra for						
ends	PC £27.20	-	-	28.56	nr	**28.56**
right angle frame corner	PC £9.60	-	-	10.08	nr	**10.08**

Q PAVING/PLANTING/FENCING/SITE FURNITURE Including overheads and profit at 5.00%	Labour hours	Labour £	Material £	Unit	Total rate £
Q10 STONE/CONCRETE/BRICK KERBS/EDGINGS/ CHANNELS					
Mechanical excavation using a wheeled hydraulic excavator with a 0.24 m3 bucket					
Excavating trenches to receive kerb foundation; average size					
300 x 100 mm	0.02	0.12	0.27	m	0.39
450 x 150 mm	0.03	0.18	0.45	m	0.63
600 x 200 mm	0.04	0.23	0.62	m	0.85
Excavating curved trenches to receive kerb foundation; average size					
300 x 100 mm	0.02	0.12	0.36	m	0.48
450 x 150 mm	0.03	0.18	0.54	m	0.72
600 x 200 mm	0.04	0.23	0.71	m	0.94
Hand excavation					
Excavating trenches to receive kerb foundation; average size					
150 x 50 mm	0.03	0.18	-	m	0.18
200 x 75 mm	0.07	0.41	-	m	0.41
250 x 100 mm	0.11	0.65	-	m	0.65
300 x 100 mm	0.14	0.82	-	m	0.82
Excavating curved trenches to receive kerb foundation; average size					
150 x 50 mm	0.04	0.23	-	m	0.23
200 x 75 mm	0.08	0.47	-	m	0.47
250 x 100 mm	0.12	0.70	-	m	0.70
300 x 100 mm	0.15	0.88	-	m	0.88
Plain in situ ready mixed concrete; 7.50 N/mm2 - 40 mm aggregate (1:8); PC £39.71/m3; poured on or against earth or unblinded hardcore					
Foundations	1.25	7.61	43.78	m3	51.39
Blinding bed					
not exceeding 150 mm thick	1.85	11.27	43.78	m3	55.05
Plain in situ ready mixed concrete; 11.50 N/mm2 - 40 mm aggregate (1:3:6); PC £40.48/m3; poured on or against earth or unblinded hardcore					
Foundations	1.25	7.61	44.63	m3	52.24
Blinding bed					
not exceeding 150 mm thick	1.85	11.27	44.63	m3	55.90
Plain in situ ready mixed concrete; 11.50 N/mm2 - 40 mm aggregate (1:3:6); PC £40.48/m3; poured on or against earth or unblinded hardcore					
Foundations	1.25	7.61	47.52	m3	55.13
Blinding bed					
not exceeding 150 mm thick	1.85	11.27	47.52	m3	58.79
Precast concrete kerbs, channels, edgings, etc.; BS 340; bedded, jointed and pointed in cement mortar (1:3); including haunching up one side with in situ concrete mix 11.50 N/mm2 - 40 mm aggregate (1:3:6); to concrete base					
Edging; straight; fig 12					
51 x 152 mm	0.25	2.24	2.20	m	4.44
51 x 203 mm	0.25	2.24	2.53	m	4.77
51 x 254 mm	0.25	2.24	2.53	m	4.77

Q PAVING/PLANTING/FENCING/SITE FURNITURE Including overheads and profit at 5.00%	Labour hours	Labour £	Material £	Unit	Total rate £
Kerb; straight					
127 x 254 mm; fig 7	0.33	2.96	4.37	m	7.33
152 x 305 mm; fig 6	0.33	2.96	5.91	m	8.87
Kerb; curved					
127 x 254 mm; fig 7	0.50	4.48	5.52	m	10.00
152 x 305 mm; fig 6	0.50	4.48	7.06	m	11.54
Channel; 255 x 125 mm; fig 8					
straight	0.33	2.96	4.54	m	7.50
curved	0.50	4.48	5.72	m	10.20
Quadrant; fig 14					
305 x 305 x 152 mm	0.35	3.13	3.84	nr	6.97
305 x 305 x 254 mm	0.35	3.13	3.84	nr	6.97
457 x 457 x 152 mm	0.40	3.58	3.84	nr	7.42
457 x 457 x 254 mm	0.40	3.58	3.84	nr	7.42
Precast concrete drainage channels; Charcon 'Safeticurb' or similar; channels jointed with plastic rings and bedded; jointed and pointed in cement mortar (1:3); including haunching up one side with in situ concrete mix 11.50 N/mm2 - 40 mm aggregate (1:3:6); to concrete base					
Channel; straight; type DBA/3					
248 x 248 mm	0.60	5.37	16.24	m	21.61
End	0.20	1.79	-	nr	1.79
Inspection unit; with cast iron lid					
248 x 248 x 914 mm	0.65	5.82	46.36	nr	52.18
Silt box top; with concrete frame and cast iron lid; set over gully					
500 x 448 x 269 mm; type A	2.00	17.91	80.97	nr	98.88
Q20 HARDCORE/GRANULAR/CEMENT BOUND BASES/ SUB-BASES TO ROADS/PAVINGS					
Mechanical filling with hardcore; PC £5.00/m3					
Filling to make up levels over 250 mm thick; depositing; compacting in layers with a 5 tonne roller	0.26	1.53	7.62	m3	9.15
Filling to make up levels not exceeding 250 mm thick; compacting	0.30	1.76	10.25	m3	12.01
Mechanical filling with granular fill; type 1; PC £9.15/t (PC £15.55/m3)					
Filling to make up levels over 250 mm thick; depositing; compacting in layers with a 5 tonne roller	0.26	1.53	21.84	m3	23.37
Filling to make up levels not exceeding 250 mm thick; compacting	0.30	1.76	27.97	m3	29.73
Mechanical filling with granular fill; type 2; PC £8.75/t (PC £15.00/m3)					
Filling to make up levels over 250 mm thick; depositing; compacting in layers with a 5 tonne roller	0.26	1.53	21.12	m3	22.65
Filling to make up levels not exceeding 250 mm thick; compacting	0.30	1.76	27.05	m3	28.81
Hand filling with hardcore; PC £5.00/m3					
Filling to make up levels over 250 mm thick; depositing; compacting	0.55	3.23	11.53	m3	14.76
Filling to make up levels not exceeding 250 mm thick; compacting	0.66	3.87	14.36	m3	18.23

Q PAVING/PLANTING/FENCING/SITE FURNITURE Including overheads and profit at 5.00%		Labour hours	Labour £	Material £	Unit	Total rate £
Filling to make up levels over 250 mm thick; depositing; compacting in layers with a 2 tonne roller		0.50	2.93	12.89	m3	**15.82**
Filling to make up levels not exceeding 250 mm thick; compacting		0.60	3.52	16.13	m3	**19.65**
Hand filling with sand; PC £8.40/t (PC £13.45/m3)						
Filling to make up levels over 250 mm thick; depositing; compacting in layers with a 2 tonne roller		0.65	3.82	23.52	m3	**27.34**
Filling to make up levels not exceeding 250 mm thick; compacting		0.77	4.52	29.55	m3	**34.07**
Surface treatments						
Compacting						
surfaces of ashes		0.04	0.23	0.11	m2	**0.34**
filling; blinding with ashes		0.12	0.57	0.38	m2	**0.95**
Q21 IN SITU CONCRETE ROADS/PAVINGS/BASES						
Reinforced in situ ready mixed concrete; normal Portland cement; mix 11.5 N/mm2 - 20 mm aggregate (1:2:4); PC £43.10/m3						
Roads; to hardcore base						
150 - 450 mm thick		1.50	9.13	42.50	m3	**51.63**
not exceeding 150 mm thick		2.20	13.40	42.50	m3	**55.90**
Reinforced in situ ready mixed concrete; normal Portland cement; mix 21.00 N/mm2						
Roads; to hardcore base						
150 - 450 mm thick		1.50	9.13	45.26	m3	**54.39**
not exceeding 150 mm thick		2.20	13.40	45.26	m3	**58.66**
Reinforced in situ ready mixed concrete; normal Portland cement; mix 26.00 N/mm2 - 20 mm aggregate (1:1:5:3); PC £43.10/m3						
Roads; to hardcore base						
150 - 450 mm thick		1.50	9.13	47.70	m3	**56.83**
not exceeding 150 mm thick		2.20	13.40	47.70	m3	**61.10**
Formwork for in situ concrete						
Sides of foundations						
not exceeding 250 mm wide		0.40	2.85	1.14	m	**3.99**
250 - 500 mm wide		0.60	4.27	1.94	m	**6.21**
500 mm - 1 m wide		0.90	6.41	3.61	m	**10.02**
Add to above for curved radius 6 m		0.04	0.28	0.14	m	**0.42**
Steel road forms to in situ concrete						
Sides of foundations						
150 mm wide		0.20	1.42	0.49	m	**1.91**
Reinforcement; fabric; BS 4483; lapped; in roads, footpaths or pavings						
Ref A142 (2.22 kg/m2)	PC £0.69	0.12	0.85	0.87	m2	**1.72**
Ref A193 (3.02 kg/m2)	PC £0.94	0.12	0.85	1.18	m2	**2.03**
Designed joints in in situ concrete						
Formed joint; 12.5 mm thick Expandite 'Flexcell' or similar						
not exceeding 150 mm wide		0.15	1.07	0.96	m	**2.03**
150 - 300 mm wide		0.20	1.42	1.64	m	**3.06**
300 - 450 mm wide		0.25	1.78	2.20	m	**3.98**

Q PAVING/PLANTING/FENCING/SITE FURNITURE Including overheads and profit at 5.00%	Labour hours	Labour £	Material £	Unit	Total rate £
Formed joint; 25 mm thick Expandite 'Flexcell' or similar					
not exceeding 150 mm wide	0.20	1.42	1.52	m	2.94
150 - 300 mm wide	0.25	1.78	2.68	m	4.46
300 - 450 mm wide	0.30	2.14	3.52	m	5.66
Sealing top 25 mm of joint with rubberized bituminous compound	0.21	1.49	0.87	m	2.36
Concrete sundries					
Treating surfaces of unset concrete; grading to cambers; tamping with a 75 mm thick steel shod tamper	0.25	1.52	-	m2	1.52
Q22 COATED MACADAM/ASPHALT ROADS/PAVINGS					
In situ finishings					
NOTE: The prices for all in situ finishings to roads and footpaths include for work to falls, crossfalls or slopes not exceeding 15 degrees from horizontal; for laying on prepared bases (priced elsewhere) and for rolling with an appropriate roller					
The following rates are based on black bitumen macadam. Red bitumen macadam rates are approximately 50% dearer.					
Fine graded wearing course; BS 4987:88; clause 2.7.7, tables 34 - 36; 14 mm pre-coated igneous rock chippings; tack coat of bitumen emulsion					
19 mm work to roads; one coat					
limestone aggregate	-	-	-	m2	3.05
igneous aggregate	-	-	-	m2	3.36
Close graded bitumen macadam; BS 4987:88; 10 mm graded aggregate to clause 2.7.4 tables 34 - 36; tack coat of bitumen emulsion					
30 mm work to roads; one coat					
limestone aggregate	-	-	-	m2	2.95
igneous aggregate	-	-	-	m2	3.07
Bitumen macadam; BS 4987:88; 45 mm thick base course of 20 mm open graded aggregate to clause 2.6.1 tables 5 - 7; 20 mm thick wearing course of 6 mm medium graded aggregate to clause 2.7.6 tables 32 - 33					
65 mm work to pavements/footpaths; two coats					
limestone aggregate	-	-	-	m2	5.40
igneous aggregate	-	-	-	m2	5.58
Add to last for 14 mm chippings; sprinkled into wearing course	-	-	-	m2	0.21
Bitumen macadam; BS 4987:88; 50 mm graded aggregate to clause 2.6.2 tables 8 - 10					
75 mm work to roads; one coat					
limestone aggregate	-	-	-	m2	4.98
igneous aggregate	-	-	-	m2	6.21

Q PAVING/PLANTING/FENCING/SITE FURNITURE Including overheads and profit at 5.00%		Labour hours	Labour £	Material £	Unit	Total rate £
Q22 COATED MACADAM/ASPHALT ROADS/PAVINGS - cont'd						
Dense bitumen macadam; BS 4987:88; 50 mm thick base course of 20 mm graded aggregate to clause 2.6.5 tables 15 - 16; 200 pen. binder; 30 mm wearing course of 10 mm graded aggregate to clause 2.7.4 tables 26 - 28						
75 mm work to roads; two coats						
limestone aggregate		-	-	-	m2	7.00
igneous aggregate		-	-	-	m2	7.34
Bitumen macadam; BS 4987:88; 50 mm thick base course of 20 mm graded aggregate to clause 2.6.1 tables 5 - 7; 25 mm thick wearing course of 10 mm graded aggregate to clause 2.7.2 tables 20 - 22						
75 mm work to roads; two coats						
limestone aggregate		-	-	-	m2	5.93
igneous aggregate		-	-	-	m2	6.21
Asphalt; BS 594 table 5; 100 mm thick roadbase; 60 mm thick dense base course; 40 mm thick hot rolled wearing course to table 1 binder 3; 14 mm chippings spread and rolled in						
200 mm work to roads; three coats						
limestone base/hardstone wearing course		-	-	-	m2	16.31
Q23 GRAVEL/HOGGIN ROADS/PAVINGS						
Two coat gravel paving; level and to falls; first layer course clinker aggregate and wearing layer fine gravel aggregate						
50 mm work to paths		0.08	0.72	1.19	m2	1.91
63 mm work to paths		0.10	0.90	1.58	m2	2.48
Q25 SLAB/BRICK/BLOCK/SETT/COBBLE PAVINGS						
NOTE: Unless otherwise described, prices for pavings do not include for ash or sand beds or bases under.						
Artificial stone paving; Redland Aggregates' 'Texitone' or similar; to falls or crossfalls; bedding 25 mm thick in lime mortar (1:4) staggered joints; jointing in coloured cement mortar (1:3); brushed in; to sand base						
Work to paved areas; over 300 mm wide						
450 x 600 x 50 mm; grey or coloured	PC £2.30/each	0.45	4.03	11.23	m2	15.26
600 x 600 x 50 mm; grey or coloured	PC £2.71/each	0.42	3.76	10.12	m2	13.88
750 x 600 x 50 mm; grey or coloured	PC £3.22/each	0.39	3.49	9.77	m2	13.26
900 x 750 x 50 mm; grey or coloured	PC £3.76/each	0.36	3.22	8.02	m2	11.24

Q PAVING/PLANTING/FENCING/SITE FURNITURE Including overheads and profit at 5.00%		Labour hours	Labour £	Material £	Unit	Total rate £
Brick paviors; 215 x 103 x 65 mm rough stock bricks; PC £430.00/1000; to falls or crossfalls; bedding 10 mm thick in cement mortar (1:3); jointing in cement mortar (1:3); as work proceeds; to concrete base						
Work to paved areas; over 300 mm wide; straight joints both ways						
bricks laid flat		0.80	8.11	18.53	m2	**26.64**
bricks laid on edge		1.12	11.35	29.17	m2	**40.52**
Work to paved areas; over 300 mm wide; laid to herringbone pattern						
bricks laid flat		1.00	10.13	18.53	m2	**28.66**
bricks laid on edge		1.40	14.19	29.17	m2	**43.36**
Add or deduct for variation of 1.00/1000 in PC of brick paviors						
bricks laid flat		-	-	-	m2	**0.42**
bricks laid on edge		-	-	-	m2	**0.67**
Cobble paving; 50 - 75 mm PC £41.00/t; to falls or crossfalls; bedding 13 mm thick in cement mortar (1:3); jointing to a height of two thirds of cobbles in dry mortar (1:3); tightly butted, washed and brushed; to concrete base						
Work to paved areas; over 300 mm wide						
regular		4.00	35.83	11.37	m2	**47.20**
laid to pattern		5.00	44.78	11.37	m2	**56.15**
Concrete paving flags; BS 368; to falls or crossfalls; bedding 25 mm thick in lime and sand mortar (1:4); butt joints straight both ways; jointing in cement mortar (1:3); brushed in; to sand base						
Work to paved areas; over 300 mm wide						
300 x 300 x 60 mm; grey	PC £0.86/each	0.55	4.93	11.32	m2	**16.25**
300 x 300 x 60 mm; coloured	PC £1.02/each	0.55	4.93	13.26	m2	**18.19**
450 x 600 x 50 mm; grey	PC £1.43/each	0.45	4.03	6.70	m2	**10.73**
450 x 600 x 50 mm; coloured	PC £2.08/each	0.45	4.03	9.35	m2	**13.38**
600 x 600 x 50 mm; grey	PC £1.59/each	0.42	3.76	5.73	m2	**9.49**
600 x 600 x 50 mm; coloured	PC £2.50/each	0.42	3.76	8.50	m2	**12.26**
750 x 600 x 50 mm; grey	PC £1.90/each	0.39	3.49	5.54	m2	**9.03**
750 x 600 x 50 mm; coloured	PC £3.01/each	0.39	3.49	8.25	m2	**11.74**
900 x 600 x 50 mm; grey	PC £2.22/each	0.36	3.22	4.51	m2	**7.73**
900 x 600 x 50 mm; coloured	PC £3.49/each	0.36	3.22	6.59	m2	**9.81**
Concrete rectangular paving blocks; to falls or crossfalls; bedding 50 mm thick in dry sharp sand; filling joints with sharp sand brushed in; on earth base						
'Keyblok' paving; over 300 mm wide; straight joints both ways						
200 x 100 x 65 mm; grey	PC £6.29	0.75	6.72	7.99	m2	**14.71**
200 x 100 x 65 mm; coloured	PC £7.29	0.75	6.72	9.09	m2	**15.81**
200 x 100 x 80 mm; grey	PC £7.29	0.80	7.17	9.18	m2	**16.35**
200 x 100 x 80 mm; coloured	PC £8.53	0.80	7.17	10.55	m2	**17.72**
'Keyblok' paving; over 300 mm wide; laid to herringbone pattern						
200 x 100 x 65 mm; grey		0.95	8.51	8.32	m2	**16.83**
200 x 100 x 65 mm; coloured		0.95	8.51	9.48	m2	**17.99**
200 x 100 x 80 mm; grey		1.00	8.96	9.57	m2	**18.53**
200 x 100 x 80 mm; coloured		1.00	8.96	11.00	m2	**19.96**

Q PAVING/PLANTING/FENCING/SITE FURNITURE Including overheads and profit at 5.00%		Labour hours	Labour £	Material £	Unit	Total rate £
Q25 SLAB/BRICK/BLOCK/SETT/COBBLE PAVINGS - cont'd						
Concrete rectangular paving blocks; to falls or crossfalls; bedding 50 mm thick in dry sharp sand; filling joints with sharp sand brushed in; on earth base - cont'd						
'Keyblok' paving; over 300 mm wide; laid to herringbone pattern						
Extra for two row boundary edging to herringbone paved areas; 200 mm wide; including a 150 mm high in situ concrete mix 11.5 N/mm2 - 40 mm aggregate (1:3:6) haunching to one side; blocks laid breaking joint						
200 x 100 x 65 mm; coloured		0.30	2.69	1.46	m	4.15
200 x 100 x 80 mm; coloured		0.30	2.69	1.55	m	4.24
'Mount Sorrel' paving; over 300 mm wide; straight joints both ways						
200 x 100 x 65 mm; grey	PC £6.42	0.75	6.72	8.14	m2	14.86
200 x 100 x 65 mm; coloured	PC £7.17	0.75	6.72	8.96	m2	15.68
200 x 100 x 80 mm; grey	PC £7.28	0.80	7.17	9.17	m2	16.34
200 x 100 x 80 mm; coloured	PC £8.23	0.80	7.17	10.22	m2	17.39
'Pedesta' paving; over 300 mm wide; straight joints both ways						
200 x 100 x 60 mm; grey	PC £7.28	0.75	6.72	9.08	m2	15.80
200 x 100 x 60 mm; coloured	PC £8.08	0.75	6.72	9.96	m2	16.68
200 x 100 x 80 mm; grey	PC £8.25	0.80	7.17	10.24	m2	17.41
200 x 100 x 80 mm; coloured	PC £9.22	0.80	7.17	11.31	m2	18.48
'Luttersett' paving; over 300 mm wide; straight joints both ways						
200 x 100 x 65 mm; grey	PC £6.72	0.75	6.72	8.47	m2	15.19
200 x 100 x 65 mm; coloured	PC £7.76	0.75	6.72	9.62	m2	16.34
200 x 100 x 80 mm; grey	PC £7.93	0.80	7.17	9.89	m2	17.06
200 x 100 x 80 mm; coloured	PC £9.21	0.80	7.17	11.30	m2	18.47
Granite setts; BS 435; 200 x 100 x 100 mm; PC £81.00/t; standard 'C' dressing; tightly butted to falls or crossfalls; bedding 25 mm thick in cement mortar (1:3); filling joints with dry mortar (1:6); washed and brushed; on concrete base						
Work to paved areas; over 300 mm wide						
straight joints		1.60	14.33	24.65	m2	38.98
laid to pattern		2.00	17.91	24.65	m2	42.56
Two rows of granite setts as boundary edging; 200 mm wide; including a 150 mm high in situ concrete mix 11.5 N/mm2 - 40 mm aggregate (1:3:6) haunching to one side; blocks laid breaking joint		0.70	6.27	6.12	m	12.39
Q26 SPECIAL SURFACINGS/PAVINGS FOR SPORT						
Sundries						
Painted line on road; one coat						
75 mm wide		0.05	0.31	0.08	m	0.39

Q PAVING/PLANTING/FENCING/SITE FURNITURE Including overheads and profit at 5.00%	Labour hours	Labour £	Material £	Unit	Total rate £
Q30 SEEDING/TURFING					
Vegetable soil					
Selected from spoil heaps; grading; preparing for turfing or seeding; to general surfaces					
average 75 mm thick	0.33	1.94	-	m2	**1.94**
average 100 mm thick	0.35	2.05	-	m2	**2.05**
average 125 mm thick	0.38	2.23	-	m2	**2.23**
average 150 mm thick	0.40	2.35	-	m2	**2.35**
average 175 mm thick	0.42	2.47	-	m2	**2.47**
average 200 mm thick	0.44	2.58	-	m2	**2.58**
Selected from spoil heaps; grading; preparing for turfing or seeding; to cutting or embankments					
average 75 mm thick	0.37	2.17	-	m2	**2.17**
average 100 mm thick	0.40	2.35	-	m2	**2.35**
average 125 mm thick	0.43	2.52	-	m2	**2.52**
average 150 mm thick	0.45	2.64	-	m2	**2.64**
average 175 mm thick	0.47	2.76	-	m2	**2.76**
average 200 mm thick	0.50	2.93	-	m2	**2.93**
Imported vegetable soil; PC £10.70/m3					
Grading; preparing for turfing or seeding; to general surfaces					
average 75 mm thick	0.30	1.76	1.12	m2	**2.88**
average 100 mm thick	0.32	1.88	1.46	m2	**3.34**
average 125 mm thick	0.34	2.00	2.13	m2	**4.13**
average 150 mm thick	0.36	2.11	2.81	m2	**4.92**
average 175 mm thick	0.38	2.23	3.15	m2	**5.38**
average 200 mm thick	0.40	2.35	3.48	m2	**5.83**
Grading; preparing for turfing or seeding; to cuttings or embankments					
average 75 mm thick	0.33	1.94	1.12	m2	**3.06**
average 100 mm thick	0.36	2.11	1.46	m2	**3.57**
average 125 mm thick	0.38	2.23	2.13	m2	**4.36**
average 150 mm thick	0.40	2.35	2.81	m2	**5.16**
average 175 mm thick	0.42	2.47	3.15	m2	**5.62**
average 200 mm thick	0.44	2.58	3.48	m2	**6.06**
Fertilizer; PC £0.34/kg					
Fertilizer 0.07 kg/m2; raking in					
general surfaces	0.04	0.23	0.02	m2	**0.25**
Selected grass seed; PC £3.60/kg					
Grass seed; sowing at a rate of 0.042 kg/m2 two applications; raking in					
general surfaces	0.07	0.41	0.32	m2	**0.73**
cuttings or embankments	0.08	0.47	0.32	m2	**0.79**
Preserved turf from stack on site					
Selected turf					
general surfaces	0.20	1.17	-	m2	**1.17**
cuttings or embankments; shallow	0.22	1.29	0.14	m2	**1.43**
cuttings or embankments; steep; pegged	0.30	1.76	0.21	m2	**1.97**
Imported turf; PC £1.20/m2					
Selected meadow turf					
general surfaces	0.20	1.17	1.26	m2	**2.43**
cuttings or embankments; shallow	0.22	1.29	1.40	m2	**2.69**
cuttings or embankments; steep; pegged	0.30	1.76	1.47	m2	**3.23**

Q PAVING/PLANTING/FENCING/SITE FURNITURE Including overheads and profit at 5.00%	Labour hours	Labour £	Material £	Unit	Total rate £
Q31 PLANTING					
Planting only					
Hedge or shrub plants					
not exceeding 750 mm high	0.25	1.47	-	nr	1.47
750 mm - 1.5 m high	0.60	3.52	-	nr	3.52
Saplings					
not exceeding 3 m high	1.70	9.98	-	nr	9.98
Q40 FENCING					
NOTE: The prices for all fencing are to include for setting posts in position, to a depth of 0.6 m for fences not exceeding 1.4 m high and of 0.76 m for fences over 1.4 m high. The prices allow for excavating post holes; filling to within 150 mm of ground level with concrete and all necessary back filling					
Strained wire fences; BS 1722 Part 3; 4 mm galvanized mild steel plain wire threaded through posts and strained with eye bolts					
900 mm fencing; three line; concrete posts at 2750 mm centres	-	-	-	m	5.35
Extra for					
end concrete straining post; one strut	-	-	-	nr	26.63
angle concrete straining post; two struts	-	-	-	nr	33.00
1.07 m fencing; five line; concrete posts at 2750 mm centres	-	-	-	m	7.02
Extra for					
end concrete straining post; one strut	-	-	-	nr	46.53
angle concrete straining post; two struts	-	-	-	nr	61.85
1.20 m fencing; six line; concrete posts at 2750 mm centres	-	-	-	m	7.26
Extra for					
end concrete straining post; one strut	-	-	-	nr	46.78
angle concrete straining post; two struts	-	-	-	nr	63.67
1.4 m fencing; seven line; concrete posts at 2750 mm centres	-	-	-	m	7.54
Extra for					
end concrete straining post; one strut	-	-	-	nr	48.34
angle concrete straining post; two struts	-	-	-	nr	64.97
Chainlink fences; BS 1722 Part 1; 3 mm; 50 mm galvanized mild steel mesh; galvanized mild steel tying and line wire; three line wires threaded through posts and strained with eye bolts and winding brackets					
900 mm fencing; galvanized mild steel angle posts at 3 m centres	-	-	-	m	7.95
Extra for					
end steel straining post; one strut	-	-	-	nr	33.50
angle steel straining post; two struts	-	-	-	nr	44.18
900 mm fencing; concrete posts at 3 m centres	-	-	-	m	8.27
Extra for					
end concrete straining post; one strut	-	-	-	nr	33.26
angle concrete straining post; two struts	-	-	-	nr	43.80

Q PAVING/PLANTING/FENCING/SITE FURNITURE Including overheads and profit at 5.00%	Labour hours	Labour £	Material £	Unit	Total rate £
1.2 m fencing; galvanized mild steel angle posts at 3 m centres	-	-	-	m	9.19
Extra for					
end steel straining post; one strut	-	-	-	nr	35.80
angle steel straining post; two struts	-	-	-	nr	47.32
1.2 m fencing; concrete posts at 3 m centres	-	-	-	m	9.68
Extra for					
end concrete straining post; one strut	-	-	-	nr	39.15
angle concrete straining post; two struts	-	-	-	nr	47.56
1.8 m fencing; galvanized mild steel angle posts at 3 m centres	-	-	-	m	12.87
Extra for					
end steel straining post; one strut	-	-	-	nr	53.64
angle steel straining post; two struts	-	-	-	nr	71.92
1.8 m fencing; concrete posts at 3 m centres	-	-	-	m	13.52
Extra for					
end concrete straining post; one strut	-	-	-	nr	55.70
angle concrete straining post; two struts	-	-	-	nr	72.46
Pair of gates and gate posts; gates to match galvanized chain link fencing, with angle framing, braces, etc., complete with hinges, locking bar, lock and bolts; two 100 x 100 mm angle section gate posts; each with one strut					
2.44 x 0.9 m high	-	-	-	nr	351.49
2.44 x 1.2 m high	-	-	-	nr	378.53
2.44 x 1.8 m high	-	-	-	nr	459.64
Chainlink fences; BS 1722 Part 1; 3 mm; 50 mm plastic coated mild steel mesh; plastic coated mild steel tying and line wire; three line wires threaded through posts and strained with eye bolts and winding brackets					
900 mm fencing; galvanized mild steel angle posts at 3 m centres	-	-	-	m	8.76
Extra for					
end steel straining post; one strut	-	-	-	nr	33.50
angle steel straining post; two struts	-	-	-	nr	44.18
900 mm fencing; concrete posts at 3 m centres	-	-	-	m	9.05
Extra for					
end concrete straining post; one strut	-	-	-	nr	33.26
angle concrete straining post; two struts	-	-	-	nr	43.80
1.2 m fencing; galvanized mild steel angle posts at 3 m centres	-	-	-	m	10.43
Extra for					
end steel straining post; one strut	-	-	-	nr	35.88
angle steel straining post; two struts	-	-	-	nr	47.32
1.2 m fencing; concrete posts at 3 m centres	-	-	-	m	10.91
Extra for					
end concrete straining post; one strut	-	-	-	nr	38.07
angle concrete straining post; two struts	-	-	-	nr	47.56
1.8 m fencing; galvanized mild steel angle posts at 3 m centres	-	-	-	m	14.18
Extra for					
end steel straining post; one strut	-	-	-	nr	53.64
angle steel straining post; two struts	-	-	-	nr	71.92
1.8 m fencing; concrete posts at 3 m centres	-	-	-	m	14.83
Extra for					
end concrete straining post; one strut	-	-	-	nr	55.70
angle concrete straining post; two struts	-	-	-	nr	72.46

Q PAVING/PLANTING/FENCING/SITE FURNITURE Including overheads and profit at 5.00%	Labour hours	Labour £	Material £	Unit	Total rate £
Q40 FENCING - cont'd					
Chainlink fences; BS 1722 Part 1; 3 mm; 50 mm plastic coated mild steel mesh; plastic coated mild steel tying and line wire; three line wires threaded through posts and strained with eye bolts and winding brackets - cont'd					
Pair of gates and gate posts; gates to match plastic chain link fencing; with angle framing, braces, etc., complete with hinges, locking bar, lock and bolts; two 100 x 100 mm angle section gate posts; each with one strut					
2.44 x 0.9 m high	-	-	-	nr	380.36
2.44 x 1.2 m high	-	-	-	nr	405.72
2.44 x 1.8 m high	-	-	-	nr	492.82
Chain link fences for tennis courts; BS 1722 Part 13; 2.5 mm; 45 mm mesh galvanized mild steel mesh; line and tying wires threaded through 45 x 45 x 5 mm galvanized mild steel angle standards, posts and struts; 60 x 60 x 6 mm angle straining posts and gate posts; straining posts and struts strained with eye bolts and winding brackets					
Fencing to tennis court 36 x 18 m; including gate 1070 x 1980 mm complete with hinges, locking bar, lock and bolts					
2745 mm fencing; standards at 3 m centres	-	-	-	nr	1727.25
3660 mm fencing; standards at 2.5 m centres	-	-	-	nr	2344.13
Cleft chestnut pale fences; BS 1722 Part 4; pales spaced 51 mm apart; on two lines of galvanized wire; 64 mm dia posts; 76 x 51 mm struts					
900 mm fences; posts at 2.50 m centres	-	-	-	m	6.02
Extra for					
straining post; one strut	-	-	-	nr	10.99
corner straining post; two struts	-	-	-	nr	14.65
1.05 m fences; posts at 2.50 m centres	-	-	-	m	6.62
Extra for					
straining post; one strut	-	-	-	nr	12.11
corner straining post; two struts	-	-	-	nr	16.12
1.20 m fences; posts at 2.25 m centres	-	-	-	m	7.09
Extra for					
straining post; one strut	-	-	-	nr	13.29
corner straining post; two struts	-	-	-	nr	17.72
1.35 m fences; posts at 2.25 m centres	-	-	-	m	7.50
Extra for					
straining post; one strut	-	-	-	nr	14.88
corner straining post; two struts	-	-	-	nr	19.73
Close boarded fencing; BS 1722 Part 5; 76 x 38 mm softwood rails; 89 x 19 mm softwood pales lapped 13 mm; 152 x 25 mm softwood gravel boards; all softwood 'treated'; posts at 3 m centres					
Fences; two rail; concrete posts					
1 m	-	-	-	m	25.16
1.2 m	-	-	-	m	25.99

Q PAVING/PLANTING/FENCING/SITE FURNITURE Including overheads and profit at 5.00%	Labour hours	Labour £	Material £	Unit	Total rate £
Fences; three rail; concrete posts					
1.4 m	-	-	-	m	33.78
1.6 m	-	-	-	m	35.08
1.8 m	-	-	-	m	36.44
Fences; two rail; oak posts					
1 m	-	-	-	m	18.19
1.2 m	-	-	-	m	20.79
Fences; three rail; oak posts					
1.4 m	-	-	-	m	23.39
1.6 m	-	-	-	m	26.58
1.8 m	-	-	-	m	29.65
Precast concrete slab fencing; 305 x 38 x 1753 mm slabs; fitted into twice grooved concrete posts at 1830 mm centres					
Fences					
1.2 m	-	-	-	m	38.63
1.5 m	-	-	-	m	45.77
1.8 m	-	-	-	m	57.17
Mild steel unclimbable fencing; in rivetted panels 2440 mm long; 44 x 13 mm flat section top and bottom rails; two 44 x 19 mm flat section standards, one with foot plate, and 38 x 13 mm raking stay with foot plate; 20 mm dia pointed verticals at 120 mm centres; two 44 x 19 mm supports 760 mm long with ragged ends to bottom rail; the whole bolted together; coated with red oxide primer; setting standards and stays in ground at 2440 mm centres and supports at 815 mm centres					
Fences					
1.67 m	-	-	-	m	61.19
2.13 m	-	-	-	m	69.13
Pair of gates and gate posts, to match mild steel unclimbable fencing; with flat section framing, braces, etc., complete with locking bar, lock, handles, drop bolt, gate stop and holding back catches; two 102 x 102 mm hollow section gate posts with cap and foot plates					
2.44 x 1.67 m	-	-	-	nr	512.66
2.44 x 2.13 m	-	-	-	nr	579.91
4.88 x 1.67 m	-	-	-	nr	1019.81
4.88 x 2.13 m	-	-	-	nr	1157.63

R DISPOSAL SYSTEMS Including overheads and profit at 5.00%		Labour hours	Labour £	Material £	Unit	Total rate £
R10 RAINWATER PIPEWORK/GUTTERS						
Aluminium pipes and fittings; BS 2997; ears cast on; powder coated finish						
63 mm pipes; plugged and nailed		0.37	3.26	7.35	m	**10.61**
Extra for						
fittings with one end		0.22	1.94	3.65	nr	**5.59**
fittings with two ends		0.42	3.70	3.93	nr	**7.63**
fittings with three ends		0.60	5.28	5.54	nr	**10.82**
shoe	PC £4.05	0.22	1.94	3.65	nr	**5.59**
bend	PC £4.32	0.42	3.70	3.93	nr	**7.63**
single branch	PC £5.63	0.60	5.28	5.54	nr	**10.82**
offset 229 mm projection	PC £9.96	0.42	3.70	8.74	nr	**12.44**
offset 305 mm projection	PC £11.12	0.42	3.70	9.95	nr	**13.65**
access pipe	PC £12.31	-	-	10.38	nr	**10.38**
connection to clay pipes; cement and sand (1:2) joint		0.15	1.32	0.09	nr	**1.41**
75 mm pipes; plugged and nailed	PC £13.79/1.8m	0.40	3.52	8.58	m	**12.10**
Extra for						
shoe	PC £5.56	0.25	2.20	5.19	nr	**7.39**
bend	PC £5.45	0.45	3.96	5.08	nr	**9.04**
single branch	PC £6.78	0.65	5.72	6.86	nr	**12.58**
offset 229 mm projection	PC £11.01	0.45	3.96	9.61	nr	**13.57**
offset 305 mm projection	PC £12.19	0.45	3.96	10.85	nr	**14.81**
access pipe	PC £13.45	-	-	11.24	nr	**11.24**
connection to clay pipes; cement and sand (1:2) joint		0.17	1.50	0.09	nr	**1.59**
100 mm pipes; plugged and nailed	PC £23.53/1.8m	0.45	3.96	14.52	m	**18.48**
Extra for						
shoe	PC £6.70	0.28	2.46	5.80	nr	**8.26**
bend	PC £7.61	0.50	4.40	6.75	nr	**11.15**
single branch	PC £9.09	0.75	6.60	8.73	nr	**15.33**
offset 229 mm projection	PC £12.74	0.50	4.40	9.92	nr	**14.32**
offset 305 mm projection	PC £14.16	0.50	4.40	11.40	nr	**15.80**
access pipe	PC £15.94	-	-	11.64	nr	**11.64**
connection to clay pipes; cement and sand (1:2) joint		0.20	1.76	0.09	nr	**1.85**
Roof outlets; circular aluminium; with flat or domed grate; joint to pipe						
50 mm dia	PC £30.59	0.60	5.28	32.76	nr	**38.04**
75 mm dia	PC £40.47	0.65	5.72	43.29	nr	**49.01**
100 mm dia	PC £53.06	0.70	6.16	56.64	nr	**62.80**
150 mm dia	PC £68.33	0.75	6.60	73.07	nr	**79.67**
Roof outlets; d-shaped; balcony; with flat or domed grate; joint to pipe						
50 mm dia	PC £37.66	0.60	5.28	40.18	nr	**45.46**
75 mm dia	PC £43.29	0.65	5.72	46.25	nr	**51.97**
100 mm dia	PC £53.17	0.70	6.16	56.76	nr	**62.92**
Galvanized wire balloon grating; BS 416 for pipes or outlets						
50 mm dia	PC £0.81	0.07	0.62	0.85	nr	**1.47**
63 mm dia	PC £0.81	0.07	0.62	0.85	nr	**1.47**
75 mm dia	PC £0.89	0.07	0.62	0.94	nr	**1.56**
100 mm dia	PC £1.08	0.08	0.70	1.13	nr	**1.83**
Aluminium gutters and fittings; BS 2997; powder coated finish						
100 mm half round gutters; on brackets screwed to timber	PC £10.78/1.8m	0.35	3.08	7.54	m	**10.62**
Extra for						
stop end	PC £1.59	0.16	1.41	2.29	nr	**3.70**
running outlet	PC £3.54	0.33	2.90	3.06	nr	**5.96**
stop end outlet	PC £3.14	0.16	1.41	3.01	nr	**4.42**
angle	PC £3.26	0.33	2.90	2.56	nr	**5.46**

R DISPOSAL SYSTEMS Including overheads and profit at 5.00%		Labour hours	Labour £	Material £	Unit	Total rate £
112 mm half round gutters; on brackets						
screwed to timber	PC £11.30/1.8m	0.35	3.08	7.86	m	**10.94**
Extra for						
stop end	PC £1.67	0.16	1.41	2.38	nr	**3.79**
running outlet	PC £3.85	0.33	2.90	3.36	nr	**6.26**
stop end outlet	PC £3.61	0.16	1.41	3.48	nr	**4.89**
angle	PC £3.68	0.33	2.90	2.94	nr	**5.84**
125 mm half round gutters; on brackets						
screwed to timber	PC £12.69/1.8m	0.40	3.52	9.51	m	**13.03**
Extra for						
stop end	PC £2.04	0.18	1.58	3.18	nr	**4.76**
running outlet	PC £4.17	0.35	3.08	3.60	nr	**6.68**
stop end outlet	PC £3.83	0.18	1.58	3.99	nr	**5.57**
angle	PC £4.09	0.35	3.08	3.89	nr	**6.97**
100 mm ogee gutters; on brackets						
screwed to timber	PC £13.45/1.8m	0.37	3.26	9.08	m	**12.34**
Extra for						
stop end	PC £1.68	0.17	1.50	2.38	nr	**3.88**
running outlet	PC £4.14	0.35	3.08	3.48	nr	**6.56**
stop end outlet	PC £3.22	0.17	1.50	2.87	nr	**4.37**
angle	PC £3.49	0.35	3.08	2.33	nr	**5.41**
112 mm ogee gutters; on						
brackets screwed to timber	PC £14.95/1.8m	0.42	3.70	10.02	m	**13.72**
Extra for						
stop end	PC £1.86	0.17	1.50	2.60	nr	**4.10**
running outlet	PC £4.21	0.35	3.08	3.43	nr	**6.51**
stop end outlet	PC £3.60	0.17	1.50	3.16	nr	**4.66**
angle	PC £4.17	0.35	3.08	2.80	nr	**5.88**
125 mm ogee gutters; on						
brackets screwed to timber	PC £17.02/1.8m	0.42	3.70	12.07	m	**15.77**
Extra for						
stop end	PC £2.02	0.19	1.67	3.16	nr	**4.83**
running outlet	PC £4.74	0.37	3.26	3.83	nr	**7.09**
stop end outlet	PC £4.21	0.19	1.67	4.03	nr	**5.70**
angle	PC £5.01	0.37	3.26	4.08	nr	**7.34**
Cast iron pipes and fittings; BS 460;						
ears cast on; joints						
50 mm pipes; primed; plugged						
and nailed	PC £15.55/1.8m	0.50	4.40	9.69	m	**14.09**
Extra for						
fittings with one end		0.30	2.64	6.91	nr	**9.55**
fittings with two ends		0.55	4.84	3.86	nr	**8.70**
shoe	PC £7.51	0.30	2.64	6.91	nr	**9.55**
bend	PC £4.61	0.55	4.84	3.86	nr	**8.70**
connection to clay pipes;						
cement and sand (1:2) joint		0.13	1.14	0.09	nr	**1.23**
63 mm pipes; primed; plugged						
and nailed	PC £15.55/1.8m	0.52	4.58	9.76	m	**14.34**
Extra for						
fittings with one end		0.32	2.82	6.97	nr	**9.79**
fittings with two ends		0.57	5.02	3.92	nr	**8.94**
fittings with three ends		0.72	6.34	6.22	nr	**12.56**
shoe	PC £7.51	0.32	2.82	6.97	nr	**9.79**
bend	PC £4.61	0.57	5.02	3.92	nr	**8.94**
single branch	PC £7.11	0.72	6.34	6.22	nr	**12.56**
offset 229 mm projection	PC £8.18	0.57	5.02	6.69	nr	**11.71**
offset 305 mm projection	PC £9.58	0.57	5.02	7.84	nr	**12.86**
connection to clay pipes;						
cement and sand (1:2) joint		0.15	1.32	0.09	nr	**1.41**

R DISPOSAL SYSTEMS Including overheads and profit at 5.00%		Labour hours	Labour £	Material £	Unit	Total rate £
R10 RAINWATER PIPEWORK/GUTTERS - cont'd						
Cast iron pipes and fittings; BS 460; ears cast on; joints - cont'd						
75 mm pipes; primed, plugged and nailed	PC £15.55/1.8m	0.55	4.84	9.85	m	**14.69**
Extra for						
shoe	PC £7.51	0.35	3.08	7.09	nr	**10.17**
bend	PC £4.61	0.60	5.28	4.04	nr	**9.32**
single branch	PC £7.11	0.75	6.60	6.46	nr	**13.06**
offset 229 mm projection	PC £8.18	0.60	5.28	6.81	nr	**12.09**
offset 305 mm projection	PC £9.58	0.60	5.28	7.96	nr	**13.24**
connection to clay pipes; cement and sand (1:2) joint		0.17	1.50	0.09	nr	**1.59**
100 mm pipes; primed, plugged and nailed	PC £20.87/1.8m	0.60	5.28	13.25	m	**18.53**
Extra for						
shoe	PC £9.76	0.40	3.52	9.23	nr	**12.75**
bend	PC £7.16	0.65	5.72	6.50	nr	**12.22**
single branch	PC £9.29	0.80	7.04	8.53	nr	**15.57**
offset 229 mm projection	PC £12.95	0.65	5.72	11.27	nr	**16.99**
offset 305 mm projection	PC £15.20	0.65	5.72	13.19	nr	**18.91**
connection to clay pipes; cement and sand (1:2) joint		0.20	1.76	0.09	nr	**1.85**
100 x 75 mm rectangular pipes; primed, plugged and nailed	PC £74.88/1.8m	0.60	5.28	46.00	m	**51.28**
Extra for					nr	
shoe	PC £29.87	0.40	3.52	25.81	nr	**29.33**
bend	PC £24.25	0.65	5.72	19.91	nr	**25.63**
offset 229 mm projection	PC £33.73	0.65	5.72	25.15	nr	**30.87**
offset 305 mm projection	PC £39.90	0.65	5.72	30.06	nr	**35.78**
connection to clay pipes; cement and sand (1:2) joint		0.20	1.76	0.09	nr	**1.85**
Rainwater head; flat; for pipes						
50 mm dia	PC £5.85	0.55	4.84	6.47	nr	**11.31**
63 mm dia	PC £5.85	0.57	5.02	6.53	nr	**11.55**
75 mm dia	PC £5.85	0.60	5.28	6.65	nr	**11.93**
100 mm dia	PC £13.49	0.65	5.72	14.91	nr	**20.63**
Rainwater head; rectangular, for pipes						
50 mm dia	PC £12.88	0.55	4.84	13.86	nr	**18.70**
63 mm dia	PC £12.88	0.57	5.02	13.92	nr	**18.94**
75 mm dia	PC £12.88	0.60	5.28	14.04	nr	**19.32**
100 mm dia	PC £26.12	0.65	5.72	28.16	nr	**33.88**
Roof outlets; cast iron; circular; with flat grate; joint to pipe						
50 mm dia	PC £37.48	0.75	6.60	39.99	nr	**46.59**
75 mm dia	PC £39.29	0.85	7.48	42.04	nr	**49.52**
100 mm dia	PC £47.15	0.90	7.92	50.43	nr	**58.35**
Copper wire balloon grating; BS 416 for pipes or outlets						
50 mm dia	PC £1.22	0.07	0.62	1.28	nr	**1.90**
63 mm dia	PC £1.22	0.07	0.62	1.28	nr	**1.90**
75 mm dia	PC £1.42	0.07	0.62	1.49	nr	**2.11**
100 mm dia	PC £1.61	0.08	0.70	1.69	nr	**2.39**
Cast iron gutters and fittings; BS 460						
100 mm half round gutters; primed; on brackets; screwed to timber	PC £7.91/1.8m	0.40	3.52	5.83	m	**9.35**
Extra for						
stop end	PC £01.0	0.17	1.50	1.61	nr	**3.11**
running outlet	PC £3.07	0.35	3.08	2.72	nr	**5.80**
angle	PC £3.07	0.35	3.08	2.78	nr	**5.86**

R DISPOSAL SYSTEMS Including overheads and profit at 5.00%		Labour hours	Labour £	Material £	Unit	Total rate £
115 mm half round gutters; primed;						
on brackets; screwed to timber	PC £8.24/1.8m	0.40	3.52	6.09	m	**9.61**
Extra for						
stop end	PC £1.45	0.17	1.50	2.12	nr	**3.62**
running outlet	PC £3.39	0.35	3.08	3.03	nr	**6.11**
angle	PC £3.39	0.35	3.08	3.10	nr	**6.18**
125 mm half round gutters; primed;						
on brackets; screwed to timber	PC £9.63/1.8m	0.45	3.96	6.93	m	**10.89**
Extra for						
stop end	PC £1.45	0.20	1.76	2.13	nr	**3.89**
running outlet	PC £4.00	0.40	3.52	3.57	nr	**7.09**
angle	PC £4.00	0.40	3.52	3.51	nr	**7.03**
150 mm half round gutters; primed;						
on brackets; screwed to timber	PC £16.48/1.8m	0.50	4.40	11.22	m	**15.62**
Extra for						
stop end	PC £1.93	0.22	1.94	2.69	nr	**4.63**
running outlet	PC £5.50	0.45	3.96	4.57	nr	**8.53**
angle	PC £5.50	0.45	3.96	3.97	nr	**7.93**
100 mm ogee gutters; primed;						
on brackets; screwed to timber	PC £8.63/1.8m	0.42	3.70	6.15	m	**9.85**
Extra for						
stop end	PC £0.93	0.18	1.58	1.90	nr	**3.48**
running outlet	PC £2.73	0.37	3.26	2.32	nr	**5.58**
angle	PC £2.73	0.37	3.26	2.26	nr	**5.52**
115 mm ogee gutters; primed;						
on brackets; screwed to timber	PC £9.70/1.8m	0.42	3.70	6.89	m	**10.59**
Extra for						
stop end	PC £1.33	0.18	1.58	2.35	nr	**3.93**
running outlet	PC £3.12	0.37	3.26	2.64	nr	**5.90**
angle	PC £3.12	0.37	3.26	2.49	nr	**5.75**
125 mm ogee gutters; primed;						
on brackets; screwed to timber	PC £10.17/1.8m	0.47	4.14	7.26	m	**11.40**
Extra for						
stop end	PC £1.33	0.21	1.85	2.45	nr	**4.30**
running outlet	PC £3.72	0.42	3.70	3.25	nr	**6.95**
angle	PC £4.95	0.42	3.70	4.42	nr	**8.12**
3 mm galvanised heavy pressed steel gutters and fittings; joggle joints; BS 1091						
200 x 100 mm (400 mm girth) box gutter;						
screwed to timber		0.65	5.72	13.08	m	**18.80**
Extra for						
stop end		0.35	3.08	10.30	nr	**13.38**
running outlet		0.70	6.16	26.48	nr	**32.64**
stop end outlet		0.35	3.08	34.13	nr	**37.21**
angle		0.70	6.16	19.82	nr	**25.98**
381 mm boundary wall gutters;						
screwed to timber		0.65	5.72	11.89	m	**17.61**
Extra for						
stop end		0.40	3.52	9.75	nr	**13.27**
running outlet		0.70	6.16	28.74	nr	**34.90**
stop end outlet		0.35	3.08	36.62	nr	**39.70**
angle		0.70	6.16	21.76	nr	**27.92**
457 mm boundary wall gutters;						
screwed to timber		0.75	6.60	13.42	m	**20.02**
Extra for						
stop end		0.35	3.08	11.78	nr	**14.86**
running outlet		0.80	7.04	29.57	nr	**36.61**
stop end outlet		0.40	3.52	38.21	nr	**41.73**
angle		0.80	7.04	23.89	nr	**30.93**

R DISPOSAL SYSTEMS Including overheads and profit at 5.00%		Labour hours	Labour £	Material £	Unit	Total rate £
R10 RAINWATER PIPEWORK/GUTTERS - cont'd						
uPVC external rainwater pipes and fittings; BS 4576; slip-in joints						
50 mm pipes; fixing with pipe or socket brackets; plugged and screwed	PC £2.77/2m	0.30	2.64	2.02	m	**4.66**
Extra for						
fittings with one end		0.20	1.76	1.05	nr	**2.81**
fittings with two ends		0.30	2.64	1.33	nr	**3.97**
fittings with three ends		0.40	3.52	1.79	nr	**5.31**
shoe	PC £0.82	0.20	1.76	1.05	nr	**2.81**
bend	PC £1.09	0.30	2.64	1.33	nr	**3.97**
single branch	PC £1.52	0.40	3.52	1.79	nr	**5.31**
two bends to form offset 229 mm projection	PC £1.90	0.30	2.64	1.69	nr	**4.33**
connection to clay pipes; cement and sand (1:2) joint		0.13	1.14	0.09	nr	**1.23**
68 mm pipes; fixing with pipe or socket brackets; plugged and screwed	PC £2.23/2m	0.33	2.90	1.84	m	**4.74**
Extra for						
shoe	PC £0.82	0.22	1.94	1.16	nr	**3.10**
bend	PC £1.43	0.33	2.90	1.80	nr	**4.70**
single branch	PC £2.50	0.44	3.87	2.93	nr	**6.80**
two bends to form offset 229 mm projection	PC £1.78	0.33	2.90	1.78	nr	**4.68**
loose drain connector; cement and sand (1:2) joint		0.15	1.32	0.09	nr	**1.41**
110 mm pipes; fixing with pipe or socket brackets; plugged and screwed	PC £7.22/3m	0.36	3.17	3.69	m	**6.86**
Extra for						
shoe	PC £3.05	0.24	2.11	3.52	nr	**5.63**
bend	PC £4.04	0.36	3.17	4.55	nr	**7.72**
single branch	PC £5.26	0.48	4.22	5.83	nr	**10.05**
two bends to form offset 229 mm projection	PC £8.08	0.36	3.17	7.73	nr	**10.90**
loose drain connector; cement and sand (1:2) joint		0.35	3.08	3.46	nr	**6.54**
68.5 mm square pipes; fixing with pipe brackets; plugged and screwed	PC £3.96/2.5m	0.33	2.90	2.37	m	**5.27**
Extra for						
shoe	PC £01.0	0.22	1.94	1.27	nr	**3.21**
bend	PC £1.01	0.33	2.90	1.29	nr	**4.19**
single branch	PC £2.59	0.44	3.87	2.95	nr	**6.82**
two bends to form offset 229 mm projection	PC £3.26	0.33	2.90	1.44	nr	**4.34**
drain connector; square to round; cement and sand (1:2) joint		0.20	1.76	2.53	nr	**4.29**
Rainwater head; rectangular; for pipes						
50 mm dia	PC £4.44	0.45	3.96	5.30	nr	**9.26**
68 mm dia	PC £3.59	0.47	4.14	4.59	nr	**8.73**
110 mm dia	PC £9.12	0.55	4.84	10.71	nr	**15.55**
68.5 mm square	PC £3.69	0.47	4.14	4.32	nr	**8.46**
uPVC gutters and fittings; BS 4576						
76 mm half round gutters; on brackets; screwed to timber	PC £2.23/2m	0.30	2.64	1.66	m	**4.30**
Extra for						
stop end	PC £0.37	0.13	1.14	0.49	nr	**1.63**
running outlet	PC £1.04	0.25	2.20	0.91	nr	**3.11**
stop end outlet	PC £1.04	0.13	1.14	1.01	nr	**2.15**
angle	PC £0.96	0.25	2.20	1.03	nr	**3.23**

R DISPOSAL SYSTEMS Including overheads and profit at 5.00%		Labour hours	Labour £	Material £	Unit	Total rate £
112 half round gutters; on brackets screwed to timber	PC £2.17/2m	0.33	2.90	2.02	m	**4.92**
Extra for						
stop end	PC £0.59	0.13	1.14	0.80	nr	**1.94**
running outlet	PC £1.15	0.28	2.46	1.03	nr	**3.49**
stop end outlet	PC £1.15	0.13	1.14	1.21	nr	**2.35**
angle	PC £1.29	0.28	2.46	1.53	nr	**3.99**
170 mm half round gutters; on brackets screwed to timber	PC £7.12/2m	0.36	3.17	6.18	m	**9.35**
Extra for						
stop end	PC £1.60	0.16	1.41	2.26	nr	**3.67**
running outlet	PC £3.56	0.31	2.73	3.14	nr	**5.87**
stop end outlet	PC £3.39	0.16	1.41	3.54	nr	**4.95**
angle	PC £4.64	0.31	2.73	5.44	nr	**8.17**
114 mm rectangular gutters; on brackets; screwed to timber	PC £2.78/2m	0.33	2.90	2.71	m	**5.61**
Extra for						
stop end	PC £0.60	0.13	1.14	0.85	nr	**1.99**
running outlet	PC £1.43	0.28	2.46	1.27	nr	**3.73**
stop end outlet	PC £1.52	0.13	1.14	1.58	nr	**2.72**
angle	PC £1.38	0.28	2.46	1.66	nr	**4.12**

ALTERNATIVE WASTE PIPE AND FITTING PRICES

ABS waste system (£/each)	£		£		£
4 m pipe -32 mm	2.24	access plug-32 mm	0.38	cn.to copper-32 mm	0.74
-40 mm	2.77	-40 mm	0.41	-40 mm	0.88
-50 mm	3.45	-50 mm	0.70	sweep bend -32 mm	0.49
pipe brkt-32 mm	0.10	reducer -32 mm	0.34	sweep bend -40 mm	0.59
-40 mm	0.11	-40 mm	0.38	-50 mm	0.89
-50 mm	0.20	-50 mm	0.50	sweep tee -32 mm	0.70
dl.socket-32 mm	0.31	str.tank cn-32 mm	0.84	-40 mm	0.85
-40 mm	0.36	-40 mm	0.95	-50 mm	1.22
-50 mm	0.59				

R11 FOUL DRAINAGE ABOVE GROUND

Cast iron pipes and fittings; BS 416

		Labour hours	Labour £	Material £	Unit	Total rate £
50 mm pipes						
primed, eared, plugged and nailed	PC £20.23/1.8m	0.75	6.60	13.57	m	20.17
primed, uneared; fixing with holderbats plugged and screwed; (PC £6.08/nr);	PC £19.01/1.8m	0.85	7.48	15.99	m	23.47
Extra for						
fittings with two ends		0.75	6.60	9.85	nr	16.45
fittings with three ends		1.00	8.80	14.76	nr	23.56
fittings with four ends		1.25	11.00	28.48	nr	39.48
bend; short radius	PC £9.25	0.75	6.60	9.85	nr	16.45
access bend; short radius	PC £22.92	0.75	6.60	24.21	nr	30.81
boss; 38 mm BSP	PC £22.32	0.75	6.60	19.19	nr	25.79
single branch	PC £13.83	1.00	8.80	14.76	nr	23.56
double branch	PC £27.05	1.25	11.00	28.48	nr	39.48
offset 229 mm projection	PC £14.66	0.75	6.60	15.53	nr	22.13
offset 305 mm projection	PC £16.76	0.75	6.60	17.74	nr	24.34

R DISPOSAL SYSTEMS Including overheads and profit at 5.00%		Labour hours	Labour £	Material £	Unit	Total rate £
R11 FOUL DRAINAGE ABOVE GROUND - cont'd						
Cast iron pipes and fittings; BS 416 - cont'd						
50 mm pipes						
access pipe	PC £19.46	0.75	6.60	16.18	nr	**22.78**
roof connector; for asphalt	PC £20.45	0.75	6.60	19.21	nr	**25.81**
roof connector; for roofing felt	PC £39.93	0.80	7.04	41.67	nr	**48.71**
isolated caulked lead joints		0.65	5.72	1.53	nr	**7.25**
connection to clay pipes; cement and sand (1:2) joint		0.13	1.14	0.09	nr	**1.23**
75 mm pipes						
primed, eared, plugged and nailed	PC £20.23/1.8m	0.80	7.04	13.76	m	**20.80**
primed, uneared; fixing with holderbats plugged and screwed; (PC £6.08/nr);	PC £19.01/1.8m	0.90	7.92	16.14	m	**24.06**
Extra for						
bend; short radius	PC £9.25	0.80	7.04	10.09	nr	**17.13**
access bend; short radius	PC £22.92	0.80	7.04	24.45	nr	**31.49**
boss; 38 mm BSP	PC £22.32	0.80	7.04	19.43	nr	**26.47**
single branch	PC £13.83	1.05	9.24	15.10	nr	**24.34**
double branch	PC £27.05	1.33	11.70	28.97	nr	**40.67**
offset 229 mm projection	PC £14.66	0.80	7.04	15.77	nr	**22.81**
offset 305 mm projection	PC £16.76	0.80	7.04	17.98	nr	**25.02**
access pipe	PC £19.46	0.80	7.04	16.42	nr	**23.46**
roof connector; for asphalt	PC £20.45	0.80	7.04	19.46	nr	**26.50**
roof connector; for roofing felt	PC £39.93	0.85	7.48	41.91	nr	**49.39**
isolated caulked lead joints		0.70	6.16	1.75	nr	**7.91**
connection to clay pipes; cement and sand (1:2) joint		0.17	1.50	0.09	nr	**1.59**
100 mm pipes						
primed, eared, plugged and nailed	PC £27.21/1.8m	0.90	7.92	18.53	m	**26.45**
primed; uneared; fixing with holderbats plugged and screwed; (PC £6.91/nr);	PC £26.07/1.8m	1.00	8.80	21.28	m	**30.08**
Extra for						
W.C. bent connector; 450 mm long tail	PC £22.50	0.70	6.16	20.26	nr	**26.42**
'Multikwik' W.C.connector	PC £1.88	0.10	0.88	1.98	nr	**2.86**
W.C. straight connector; 300 mm long tail	PC £9.96	0.70	6.16	10.92	nr	**17.08**
W.C. straight connector; 450 mm long tail	PC £12.41	0.70	6.16	13.50	nr	**19.66**
bend; short radius	PC £13.76	0.90	7.92	14.91	nr	**22.83**
access bend; short radius	PC £28.25	0.90	7.92	30.12	nr	**38.04**
boss; 38 mm BSP	PC £26.68	0.90	7.92	22.46	nr	**30.38**
single branch	PC £21.34	1.20	10.56	23.10	nr	**33.66**
double branch	PC £30.21	1.50	13.20	32.38	nr	**45.58**
offset 229 mm projection	PC £19.38	0.90	7.92	20.82	nr	**28.74**
offset 305 mm projection	PC £22.17	0.90	7.92	23.74	nr	**31.66**
access pipe	PC £21.34	0.90	7.92	16.85	nr	**24.77**
roof connector; for asphalt	PC £24.50	0.90	7.92	22.90	nr	**30.82**
roof connector; for roofing felt	PC £45.67	0.95	8.36	47.87	nr	**56.23**
isolated caulked lead joints		0.80	7.04	2.38	nr	**9.42**
connection to clay pipes; cement and sand (1:2) joint		0.20	1.76	0.09	nr	**1.85**

R DISPOSAL SYSTEMS Including overheads and profit at 5.00%		Labour hours	Labour £	Material £	Unit	Total rate £
150 mm pipes						
primed, eared, plugged and nailed	PC £55.16/1.8m	1.15	10.12	36.76	m	**46.88**
primed; uneared; fixing with holderbats plugged and screwed; (PC £11.88/nr);	PC £53.74/1.8m	1.25	11.00	41.92	m	**52.92**
Extra for						
bend; short radius	PC £24.66	1.15	10.12	25.92	nr	**36.04**
access bend; short radius	PC £35.45	1.15	10.12	37.25	nr	**47.37**
boss; 38 mm BSP	PC £45.32	1.15	10.12	35.20	nr	**45.32**
single branch	PC £38.18	1.55	13.64	40.13	nr	**53.77**
double branch	PC £54.95	1.90	16.72	57.19	nr	**73.91**
offset 229 mm projection	PC £39.89	1.15	10.12	41.91	nr	**52.03**
offset 305 mm projection	PC £45.00	1.15	10.12	47.28	nr	**57.40**
access pipe	PC £34.50	1.15	10.12	23.84	nr	**33.96**
roof connector; for asphalt	PC £52.22	1.15	10.12	48.09	nr	**58.21**
roof connector; for roofing felt	PC £76.61	1.20	10.56	79.34	nr	**89.90**
isolated caulked lead joints	PC £1.21/Kg	1.05	9.24	4.00	nr	**13.24**
connection to clay pipes; cement and sand (1:2) joint	PC £54.72/M3	0.26	2.29	0.12	nr	**2.41**
Cast iron 'Timesaver' pipes and fittings BS 416						
50 mm pipes						
primed; 2 m lengths; fixing with holderbats plugged and screwed;(PC £4.26/nr)	PC £16.60/2m	0.55	4.84	13.28	m	**18.12**
Extra for						
fittings with two ends		0.55	4.84	9.10	nr	**13.94**
fittings with three ends		0.75	6.60	15.59	nr	**22.19**
fittings with four ends		0.95	8.36	25.41	nr	**33.77**
bend; short radius	PC £6.38	0.55	4.84	9.10	nr	**13.94**
access bend; short radius	PC £15.73	0.55	4.84	18.92	nr	**23.76**
boss; 38 mm BSP	PC £16.24	0.55	4.84	15.63	nr	**20.47**
single branch	PC £9.61	0.75	6.60	15.59	nr	**22.19**
double branch	PC £16.17	0.95	8.36	25.41	nr	**33.77**
offset 229 mm projection	PC £10.19	0.55	4.84	13.10	nr	**17.94**
offset 305 mm projection	PC £11.61	0.55	4.84	14.59	nr	**19.43**
access pipe	PC £15.36	0.55	4.84	14.69	nr	**19.53**
roof connector; for asphalt	PC £15.32	0.55	4.84	16.40	nr	**21.24**
roof connector; for roofing felt	PC £36.13	0.60	5.28	39.99	nr	**45.27**
isolated 'Timesaver' coupling joint	PC £3.62	0.30	2.64	3.80	nr	**6.44**
connection to clay pipes; cement and sand (1:2) joint		0.13	1.14	0.09	nr	**1.23**
75 mm pipes						
primed; 3 m lengths; fixing with holderbats plugged and screwed;(PC £4.26/nr)	PC £23.99/3m	0.55	4.84	11.94	m	**16.78**
primed; 2 m lengths; fixing with holderbats plugged and screwed;(PC £4.26/nr)	PC £16.60/2m	0.60	5.28	13.48	m	**18.76**
Extra for						
bend; short radius	PC £6.38	0.60	5.28	9.51	nr	**14.79**
access bend; short radius	PC £15.73	0.60	5.28	19.33	nr	**24.61**
boss; 38 mm BSP	PC £16.24	0.60	5.28	16.03	nr	**21.31**
single branch	PC £9.61	0.85	7.48	16.41	nr	**23.89**
double branch	PC £16.17	1.10	9.68	26.63	nr	**36.31**
offset 229 mm projection	PC £10.19	0.60	5.28	13.51	nr	**18.79**
offset 305 mm projection	PC £11.61	0.60	5.28	15.00	nr	**20.28**
access pipe	PC £15.36	0.60	5.28	15.10	nr	**20.38**
roof connector; for asphalt	PC £15.32	0.60	5.28	16.81	nr	**22.09**
roof connector; for roofing felt	PC £36.13	0.65	5.72	40.40	nr	**46.12**

R DISPOSAL SYSTEMS Including overheads and profit at 5.00%		Labour hours	Labour £	Material £	Unit	Total rate £
R11 FOUL DRAINAGE ABOVE GROUND - cont'd						
Cast iron 'Timesaver' pipes and fittings BS 416 - cont'd						
75 mm pipes						
isolated 'Timesaver' coupling joint	PC £4.00	0.35	3.08	4.20	nr	**7.28**
connection to clay pipes; cement and sand (1:2) joint		0.15	1.32	0.09	nr	**1.41**
100 mm pipes						
primed; 3 m lengths; fixing with holderbats plugged and screwed;(PC £4.65/nr)	PC £28.96/3m	0.60	5.28	14.38	m	**19.66**
primed; 2 m lengths; fixing with holderbats plugged and screwed; (PC £4.65/nr)	PC £20.00/2m	0.67	5.90	16.21	m	**22.11**
Extra for						
W.C. bent connector; 450 mm long tail	PC £20.86	0.60	5.28	25.71	nr	**30.99**
'Multikwik' W.C.connector	PC £1.88	0.10	0.88	1.98	nr	**2.86**
W.C. straight connector; 300 mm long tail	PC £9.24	0.60	5.28	13.51	nr	**18.79**
bend; short radius		0.67	5.90	13.09	nr	**18.99**
access bend; short radius	PC £18.70	0.67	5.90	23.44	nr	**29.34**
boss; 38 mm BSP	PC £20.37	0.67	5.90	20.58	nr	**26.48**
single branch	PC £13.68	1.00	8.80	22.82	nr	**31.62**
double branch	PC £16.91	1.30	11.44	30.65	nr	**42.09**
offset 229 mm projection	PC £12.71	0.67	5.90	17.15	nr	**23.05**
offset 305 mm projection	PC £14.33	0.67	5.90	18.85	nr	**24.75**
access pipe	PC £16.17	0.67	5.90	16.17	nr	**22.07**
roof connector; for asphalt	PC £18.13	0.67	5.90	20.32	nr	**26.22**
roof connector; for roofing felt	PC £43.97	0.72	6.34	49.56	nr	**55.90**
isolated 'Timesaver' coupling joint	PC £5.23	0.42	3.70	5.49	nr	**9.19**
transitional clayware socket; cement and sand (1:2) joint		0.40	3.52	12.81	nr	**16.33**
150 mm pipes						
primed; 3 m lengths; fixing with holderbats plugged and screwed;(PC £7.99/nr)	PC £60.68/3m	0.75	6.60	29.33	m	**35.93**
primed; 2 m lengths; fixing with holderbats plugged and screwed;(PC £7.99/nr)	PC £41.06/2m	0.83	7.30	32.32	m	**39.62**
Extra for						
bend; short radius	PC £15.80	0.83	7.30	24.10	nr	**31.40**
access bend; short radius	PC £26.55	0.83	7.30	35.40	nr	**42.70**
boss; 38 mm BSP	PC £32.89	0.83	7.30	32.57	nr	**39.87**
single branch	PC £28.14	1.20	10.56	46.31	nr	**56.87**
double branch	PC £43.19	1.60	14.08	70.92	nr	**85.00**
offset 229 mm projection	PC £26.11	0.83	7.30	34.93	nr	**42.23**
offset 305 mm projection	PC £33.53	0.83	7.30	42.72	nr	**50.02**
roof connector; for asphalt	PC £33.59	0.83	7.30	37.61	nr	**44.91**
roof connector; for roofing felt	PC £67.32	0.88	7.74	77.34	nr	**85.08**
isolated 'Timesaver' coupling joint	PC £10.45	0.50	4.40	10.97	nr	**15.37**
transitional clayware socket; cement and sand (1:2) joint		0.52	4.58	14.57	nr	**19.15**

R DISPOSAL SYSTEMS Including overheads and profit at 5.00%		Labour hours	Labour £	Material £	Unit	Total rate £
Polypropylene (PP) waste pipes and fittings; BS 5254; push fit 'O' - ring joints						
32 mm pipes; fixing with pipe clips; plugged and screwed	PC £1.37/4m	0.22	1.94	0.68	m	**2.62**
Extra for						
fittings with one end		0.16	1.41	0.40	nr	**1.81**
fittings with two ends		0.22	1.94	0.59	nr	**2.53**
fittings with three ends		0.30	2.64	0.86	nr	**3.50**
access plug	PC £0.38	0.16	1.41	0.40	nr	**1.81**
double socket	PC £0.42	0.15	1.32	0.44	nr	**1.76**
male iron to PP coupling	PC £0.71	0.28	2.46	0.75	nr	**3.21**
sweep bend	PC £0.56	0.22	1.94	0.59	nr	**2.53**
sweep tee	PC £0.82	0.30	2.64	0.86	nr	**3.50**
40 mm pipes; fixing with pipe clips; plugged and screwed	PC £1.72/4m	0.27	2.38	0.79	m	**3.17**
Extra for						
fittings with one end		0.18	1.58	0.43	nr	**2.01**
fittings with two ends		0.27	2.38	0.65	nr	**3.03**
fittings with three ends		0.36	3.17	1.00	nr	**4.17**
access plug	PC £0.41	0.18	1.58	0.43	nr	**2.01**
double socket	PC £0.46	0.18	1.58	0.48	nr	**2.06**
reducing set	PC £0.36	0.09	0.79	0.38	nr	**1.17**
male iron to PP coupling	PC £0.77	0.34	2.99	0.82	nr	**3.81**
sweep bend	PC £0.62	0.27	2.38	0.65	nr	**3.03**
sweep tee	PC £0.95	0.36	3.17	1.00	nr	**4.17**
50 mm pipes; fixing with pipe clips; plugged and screwed	PC £2.47/4m	0.32	2.82	1.20	m	**4.02**
Extra for						
fittings with one end		0.20	1.76	0.73	nr	**2.49**
fittings with two ends		0.32	2.82	1.18	nr	**4.00**
fittings with three ends		0.42	3.70	1.44	nr	**5.14**
access plug	PC £0.70	0.20	1.76	0.73	nr	**2.49**
double socket	PC £0.85	0.21	1.85	0.89	nr	**2.74**
reducing set	PC £0.56	0.10	0.88	0.59	nr	**1.47**
male iron to PP coupling	PC £0.97	0.40	3.52	1.02	nr	**4.54**
sweep bend	PC £1.13	0.32	2.82	1.18	nr	**4.00**
sweep tee	PC £1.37	0.42	3.70	1.44	nr	**5.14**
muPVC waste pipes and fittings; BS 5255; solvent welded joints						
32 mm pipes; fixing with pipe clips; plugged and screwed	PC £4.31/4m	0.25	2.20	1.59	m	**3.79**
Extra for						
fittings with one end		0.17	1.50	0.98	nr	**2.48**
fittings with two ends		0.25	2.20	0.94	nr	**3.14**
fittings with three ends		0.33	2.90	1.35	nr	**4.25**
access plug	PC £0.85	0.17	1.50	0.98	nr	**2.48**
straight coupling	PC £0.56	0.17	1.50	0.67	nr	**2.17**
expansion coupling	PC £0.72	0.25	2.20	0.84	nr	**3.04**
male iron to PVC coupling	PC £0.70	0.31	2.73	0.78	nr	**3.51**
union coupling	PC £1.71	0.25	2.20	1.88	nr	**4.08**
sweep bend	PC £0.82	0.25	2.20	0.94	nr	**3.14**
spigot socket bend	PC £0.82	0.25	2.20	0.94	nr	**3.14**
sweep tee	PC £1.17	0.33	2.90	1.35	nr	**4.25**
caulking bush	PC £0.97	0.50	4.40	1.57	nr	**5.97**
40 mm pipes; fixing with pipe clips; plugged and screwed	PC £5.28/4m	0.30	2.64	1.92	m	**4.56**
Extra for						
fittings with one end		0.19	1.67	1.14	nr	**2.81**
fittings with two ends		0.30	2.64	1.05	nr	**3.69**
fittings with three ends		0.40	3.52	1.65	nr	**5.17**
fittings with four ends		0.53	4.66	3.98	nr	**8.64**
access plug	PC £1.00	0.19	1.67	1.14	nr	**2.81**

R DISPOSAL SYSTEMS Including overheads and profit at 5.00%		Labour hours	Labour £	Material £	Unit	Total rate £
R11 FOUL DRAINAGE ABOVE GROUND - cont'd						
muPVC waste pipes and fittings; BS 5255; solvent welded joints - cont'd						
40 mm pipes; fixing with pipe clips; plugged and screwed						
straight coupling	PC £0.68	0.20	1.76	0.80	nr	**2.56**
expansion coupling	PC £0.87	0.30	2.64	1.00	nr	**3.64**
male iron to PVC coupling	PC £0.82	0.38	3.34	0.91	nr	**4.25**
union coupling	PC £2.25	0.30	2.64	2.45	nr	**5.09**
level invert taper	PC £0.70	0.30	2.64	0.82	nr	**3.46**
sweep bend	PC £0.92	0.30	2.64	1.05	nr	**3.69**
spigot socket bend	PC £0.91	0.30	2.64	1.04	nr	**3.68**
sweep tee	PC £1.45	0.40	3.52	1.65	nr	**5.17**
sweep cross	PC £3.63	0.53	4.66	3.98	nr	**8.64**
50 mm pipes; fixing with pipe clips; plugged and screwed	PC £7.77/4m	0.35	3.08	2.90	m	**5.98**
Extra for						
fittings with one end		0.21	1.85	1.80	nr	**3.65**
fittings with two ends		0.35	3.08	1.48	nr	**4.56**
fittings with three ends		0.47	4.14	2.79	nr	**6.93**
fittings with four ends		0.62	5.46	4.56	nr	**10.02**
access plug	PC £1.63	0.21	1.85	1.80	nr	**3.65**
straight coupling	PC £0.84	0.23	2.02	0.97	nr	**2.99**
expansion coupling	PC £1.18	0.35	3.08	1.33	nr	**4.41**
male iron to PVC coupling	PC £1.18	0.45	3.96	1.29	nr	**5.25**
union coupling	PC £3.33	0.35	3.08	3.58	nr	**6.66**
level invert taper	PC £0.93	0.35	3.08	1.06	nr	**4.14**
sweep bend	PC £1.33	0.35	3.08	1.48	nr	**4.56**
spigot socket bend	PC £2.15	0.35	3.08	2.35	nr	**5.43**
sweep tee	PC £2.54	0.47	4.14	2.79	nr	**6.93**
sweep cross	PC £4.18	0.62	5.46	4.56	nr	**10.02**
uPVC overflow pipes and fittings; solvent						
19 mm pipes; fixing with pipe clips; plugged and screwed	PC £1.82/4m	0.22	1.94	0.71	m	**2.65**
Extra for						
splay cut end		0.02	0.18	-	nr	**0.18**
fittings with one end		0.17	1.50	0.34	nr	**1.84**
fittings with two ends		0.17	1.50	0.47	nr	**1.97**
fittings with three ends		0.22	1.94	0.56	nr	**2.50**
straight connector	PC £0.28	0.17	1.50	0.34	nr	**1.84**
female iron to PVC coupling	PC £0.53	0.20	1.76	0.59	nr	**2.35**
bend	PC £0.41	0.17	1.50	0.47	nr	**1.97**
tee	PC £0.47	0.22	1.94	0.56	nr	**2.50**
bent tank connector	PC £0.65	0.20	1.76	0.71	nr	**2.47**
uPVC pipes and fittings; BS 4514; with solvent welded joints (unless otherwise described)						
82 mm pipes; fixing with holderbats; plugged and screwed (PC £1.04/nr)	PC £11.05/4m	0.40	3.52	4.16	m	**7.68**
Extra for						
socket plug	PC £1.91	0.20	1.76	2.25	nr	**4.01**
slip coupling (push-fit)	PC £3.70	0.37	3.26	4.13	nr	**7.39**
expansion coupling	PC £2.41	0.40	3.52	2.77	nr	**6.29**
sweep bend	PC £3.38	0.40	3.52	3.79	nr	**7.31**
boss connector	PC £1.76	0.27	2.38	2.09	nr	**4.47**
single branch	PC £4.70	0.53	4.66	5.31	nr	**9.97**
access door	PC £4.08	0.60	5.28	4.40	nr	**9.68**
connection to clay pipes; caulking ring and cement and sand (1:2) joint		0.37	3.26	0.91	nr	**4.17**

R DISPOSAL SYSTEMS Including overheads and profit at 5.00%		Labour hours	Labour £	Material £	Unit	Total rate £
110 mm pipes; fixing with holderbats; plugged and screwed (PC £1.08/nr)	PC £13.26/4m	0.44	3.87	4.90	m	**8.77**
Extra for						
socket plug	PC £2.32	C.22	1.94	2.74	nr	**4.68**
slip coupling (push-fit)	PC £4.63	0.40	3.52	5.17	nr	**8.69**
expansion coupling	PC £2.42	0.44	3.87	2.86	nr	**6.73**
WC connector	PC £3.31	0.29	2.55	3.63	nr	**6.18**
sweep bend	PC £4.63	0.44	3.87	5.17	nr	**9.04**
WC connecting bend	PC £4.93	0.29	2.55	5.33	nr	**7.88**
access bend	PC £10.37	0.46	4.05	11.20	nr	**15.25**
boss connector	PC £1.76	0.29	2.55	2.16	nr	**4.71**
single branch	PC £6.12	0.58	5.10	6.89	nr	**11.99**
single branch with access door	PC £11.86	0.60	5.28	12.92	nr	**18.20**
double branch	PC £14.91	0.73	6.42	16.28	nr	**22.70**
WC manifold	PC £20.72	0.29	2.55	22.23	nr	**24.78**
access door	PC £4.08	0.60	5.28	4.40	nr	**9.68**
access pipe connector	PC £7.30	0.50	4.40	7.98	nr	**12.38**
connection to clay pipes; caulking ring and cement and sand (1:2) joint		0.42	3.70	1.25	nr	**4.95**
160 mm pipes; fixing with holderbats; plugged and screwed (PC £2.61/nr)	PC £28.24/4m	0.50	4.40	10.97	m	**15.37**
Extra for						
socket plug	PC £4.27	0.25	2.20	5.15	nr	**7.35**
slip coupling (push-fit)	PC £11.84	0.45	3.96	13.10	nr	**17.06**
expansion coupling	PC £7.33	0.50	4.40	8.36	nr	**12.76**
sweep bend	PC £9.97	0.50	4.40	11.13	nr	**15.53**
boss connector	PC £2.41	0.33	2.90	3.19	nr	**6.09**
single branch	PC £14.04	0.66	5.81	15.74	nr	**21.55**
double branch	PC £24.86	0.83	7.30	27.43	nr	**34.73**
access door		0.60	5.28	7.76	nr	**13.04**
connection to clay pipes; caulking ring and cement and sand (1:2) joint		0.50	4.40	2.09	nr	**6.49**
Weathering apron; for pipe						
82 mm dia	PC £0.95	0.34	2.99	1.12	nr	**4.11**
110 mm dia	PC £1.12	0.38	3.34	1.33	nr	**4.67**
160 mm dia	PC £3.35	0.42	3.70	3.86	nr	**7.56**
Weathering slate; for pipe						
110 mm dia	PC £15.60	0.90	7.92	16.54	nr	**24.46**
Vent cowl; for pipe						
82 mm dia	PC £0.95	0.33	2.90	1.12	nr	**4.02**
110 mm dia	PC £0.96	0.33	2.90	1.16	nr	**4.06**
160 mm dia	PC £2.50	0.33	2.90	2.95	nr	**5.85**
Roof outlets; circular PVC; with flat or domed grate; jointed to pipe						
82 mm dia	PC £11.84	0.55	4.84	12.56	nr	**17.40**
110 mm dia	PC £11.84	0.60	5.28	12.59	nr	**17.87**
Copper, brass and gunmetal ancillaries; screwed joints to fittings						
Brass trap; 'P'; 45 degree outlet; 38 mm seal						
35 mm	PC £10.46	0.48	4.22	10.98	nr	**15.20**
Brass bath trap; 88.5 degree outlet; shallow seal						
42 mm	PC £7.84	0.54	4.75	8.23	nr	**12.98**
Copper trap 'P'; two piece						
35 mm with 38 mm seal	PC £5.34	0.48	4.22	5.61	nr	**9.83**
35 mm with 76 mm seal	PC £5.75	0.48	4.22	6.04	nr	**10.26**
42 mm with 38 mm seal	PC £7.93	0.54	4.75	8.32	nr	**13.07**
42 mm with 76 mm seal	PC £8.21	0.54	4.75	8.62	nr	**13.37**

R DISPOSAL SYSTEMS Including overheads and profit at 5.00%		Labour hours	Labour £	Material £	Unit	Total rate £
R11 FOUL DRAINAGE ABOVE GROUND - cont'd						
Copper, brass and gunmetal ancillaries; screwed joints to fittings - cont'd						
Copper trap; 'S'; two piece						
35 mm with 38 mm seal	PC £5.75	0.48	4.22	6.04	nr	**10.26**
35 mm with 76 mm seal	PC £5.99	0.48	4.22	6.29	nr	**10.51**
42 mm with 38 mm seal	PC £8.25	0.54	4.75	8.66	nr	**13.41**
42 mm with 76 mm seal	PC £8.53	0.54	4.75	8.95	nr	**13.70**
Polypropylene ancillaries; screwed joint to waste fitting						
Tubular 'S' trap; bath; shallow seal						
40 mm	PC £6.29	0.55	4.84	6.60	nr	**11.44**
Trap 'P' two piece; 76 mm seal						
32 mm	PC £1.72	0.38	3.34	1.80	nr	**5.14**
40 mm	PC £1.98	0.45	3.96	2.08	nr	**6.04**
Trap 'S' two piece; 76 mm seal						
32 mm	PC £2.18	0.38	3.34	2.29	nr	**5.63**
40 mm	PC £2.56	0.45	3.96	2.68	nr	**6.64**
Bottle trap 'P'; 76 mm seal						
32 mm	PC £1.91	0.38	3.34	2.00	nr	**5.34**
40 mm	PC £2.29	0.45	3.96	2.40	nr	**6.36**
Bottle trap; 'S'; 76 mm seal						
32 mm	PC £2.30	0.38	3.34	2.42	nr	**5.76**
40 mm	PC £2.80	0.45	3.96	2.94	nr	**6.90**
R12 DRAINAGE BELOW GROUND						
NOTE: Prices for drain trenches are for excavation in 'firm' soil and it has been assumed that earthwork support will only be required for trenches 1 m or more in depth						
Mechanical excavation of trenches to receive pipes; grading bottoms; earthwork support; filling with excavated material and compacting; disposal of surplus soil; spreading on site average 50 m						
Pipes not exceeding 200 mm; average depth						
0.50 m deep		0.20	1.17	1.07	m	**2.24**
0.75 m deep		0.30	1.76	1.79	m	**3.55**
1.00 m deep		0.60	3.52	3.29	m	**6.81**
1.25 m deep		0.90	5.28	4.48	m	**9.76**
1.50 m deep		1.15	6.75	5.84	m	**12.59**
1.75 m deep		1.40	8.22	7.48	m	**15.70**
2.00 m deep		1.65	9.68	8.56	m	**18.24**
2.25 m deep		2.03	11.92	10.58	m	**22.50**
2.50 m deep		2.40	14.09	12.23	m	**26.32**
2.75 m deep		2.65	15.55	13.69	m	**29.24**
3.00 m deep		2.90	17.02	15.15	m	**32.17**
3.25 m deep		3.13	18.37	16.07	m	**34.44**
3.50 m deep		3.35	19.66	16.99	m	**36.65**
Pipes; 225 mm; average depth						
0.50 m deep		0.20	1.17	1.07	m	**2.24**
0.75 m deep		0.30	1.76	1.79	m	**3.55**
1.00 m deep		0.60	3.52	3.29	m	**6.81**
1.25 m deep		0.90	5.28	4.48	m	**9.76**
1.50 m deep		1.15	6.75	5.84	m	**12.59**
1.75 m deep		1.40	8.22	7.48	m	**15.70**
2.00 m deep		1.65	9.68	8.56	m	**18.24**

R DISPOSAL SYSTEMS Including overheads and profit at 5.00%	Labour hours	Labour £	Material £	Unit	Total rate £
2.25 m deep	2.03	11.92	10.58	m	**22.50**
2.50 m deep	2.40	14.09	12.23	m	**26.32**
2.75 m deep	2.65	15.55	13.69	m	**29.24**
3.00 m deep	2.90	17.02	15.15	m	**32.17**
3.25 m deep	3.13	18.37	16.07	m	**34.44**
3.50 m deep	3.35	19.66	16.99	m	**36.65**
Pipes; 300 mm; average depth					
0.75 m deep	0.34	2.00	1.96	m	**3.96**
1.00 m deep	0.70	4.11	3.46	m	**7.57**
1.25 m deep	0.95	5.58	4.66	m	**10.24**
1.50 m deep	1.25	7.34	6.20	m	**13.54**
1.75 m deep	1.45	8.51	7.83	m	**16.34**
2.00 m deep	1.65	9.68	9.10	m	**18.78**
2.25 m deep	2.03	11.92	10.94	m	**22.86**
2.50 m deep	2.40	14.09	12.59	m	**26.68**
2.75 m deep	2.65	15.55	14.05	m	**29.60**
3.00 m deep	2.90	17.02	15.50	m	**32.52**
3.25 m deep	3.13	18.37	16.78	m	**35.15**
3.50 m deep	3.35	19.66	17.35	m	**37.01**
Pipes; 375 mm; average depth					
0.75 m deep	0.37	2.17	2.32	m	**4.49**
1.00 m deep	0.75	4.40	4.00	m	**8.40**
1.25 m deep	1.05	6.16	5.55	m	**11.71**
1.50 m deep	1.33	7.81	7.09	m	**14.90**
1.75 m deep	1.55	9.10	8.37	m	**17.47**
2.00 m deep	1.75	10.27	9.46	m	**19.73**
2.25 m deep	2.18	12.80	11.66	m	**24.46**
2.50 m deep	2.60	15.26	13.48	m	**28.74**
2.75 m deep	2.85	16.73	14.76	m	**31.49**
3.00 m deep	3.10	18.20	16.04	m	**34.24**
3.25 m deep	3.35	19.66	17.32	m	**36.98**
3.50 m deep	3.60	21.13	18.60	m	**39.73**
Pipes; 450 mm; average depth					
0.75 m deep	0.40	2.35	2.50	m	**4.85**
1.00 m deep	0.80	4.70	4.18	m	**8.88**
1.25 m deep	1.15	6.75	5.91	m	**12.66**
1.50 m deep	1.45	8.51	7.63	m	**16.14**
1.75 m deep	1.65	9.68	8.91	m	**18.59**
2.00 m deep	1.90	11.15	10.17	m	**21.32**
2.25 m deep	2.35	13.79	12.37	m	**26.16**
2.50 m deep	2.80	16.43	14.38	m	**30.81**
2.75 m deep	3.05	17.90	15.83	m	**33.73**
3.00 m deep	3.30	19.37	17.29	m	**36.66**
3.25 m deep	3.58	21.01	18.92	m	**39.93**
3.50 m deep	3.85	22.60	20.38	m	**42.98**
Pipes; 600 mm; average depth					
1.00 m deep	0.85	4.99	4.53	m	**9.52**
1.25 m deep	1.20	7.04	6.27	m	**13.31**
1.50 m deep	1.55	9.10	8.52	m	**17.62**
1.75 m deep	1.80	10.57	9.98	m	**20.55**
2.00 m deep	2.10	12.33	11.24	m	**23.57**
2.25 m deep	2.55	14.97	13.80	m	**28.77**
2.50 m deep	3.00	17.61	16.16	m	**33.77**
2.75 m deep	3.33	19.55	17.97	m	**37.52**
3.00 m deep	3.65	21.42	19.61	m	**41.03**
3.25 m deep	3.93	23.07	21.07	m	**44.14**
3.50 m deep	4.20	24.65	22.35	m	**47.00**
Pipes; 900 mm; average depth					
1.25 m deep	1.45	8.51	7.87	m	**16.38**
1.50 m deep	1.85	10.86	10.30	m	**21.16**
1.75 m deep	2.15	12.62	12.12	m	**24.74**
2.00 m deep	2.40	14.09	13.74	m	**27.83**

R DISPOSAL SYSTEMS Including overheads and profit at 5.00%	Labour hours	Labour £	Material £	Unit	Total rate £
R12 DRAINAGE BELOW GROUND - cont'd					
Mechanical excavation of trenches to receive pipes; grading bottoms; earthwork support; filling with excavated material and compacting; disposal of surplus soil; spreading on site average 50 m - cont'd					
Pipes; 900 mm; average depth					
2.25 m deep	2.95	17.32	16.65	m	**33.97**
2.50 m deep	3.50	20.54	19.20	m	**39.74**
2.75 m deep	3.85	22.60	21.19	m	**43.79**
3.00 m deep	4.20	24.65	23.18	m	**47.83**
3.25 m deep	4.55	26.71	24.99	m	**51.70**
3.50 m deep	4.90	28.76	26.63	m	**55.39**
Pipes; 1200 mm; average depth					
1.50 m deep	2.10	12.33	12.09	m	**24.42**
1.75 m deep	2.45	14.38	14.08	m	**28.46**
2.00 m deep	2.75	16.14	16.06	m	**32.20**
2.25 m deep	3.34	19.60	19.33	m	**38.93**
2.50 m deep	4.00	23.48	22.41	m	**45.89**
2.75 m deep	4.40	25.83	24.58	m	**50.41**
3.00 m deep	4.80	28.17	26.75	m	**54.92**
3.25 m deep	5.20	30.52	28.92	m	**59.44**
3.50 m deep	5.60	32.87	31.09	m	**63.96**
Extra for breaking up					
brick	1.40	8.22	6.33	m3	**14.55**
concrete	1.95	11.45	8.68	m3	**20.13**
reinforced concrete	2.80	16.43	12.65	m3	**29.08**
concrete 150 mm thick	0.30	1.76	1.44	m2	**3.20**
tarmacadam 75 mm thick	0.15	0.88	0.76	m2	**1.64**
tarmacadam and hardcore 150 mm thick	0.20	1.17	1.04	m2	**2.21**
Hand excavation of trenches to receive pipes; grading bottoms; earthwork support; filling with excavated material and compacting; disposal of surplus soil; spreading on site average 50 m					
Pipes not exceeding 200 mm; average depth					
0.50 m deep	1.00	5.87	-	m	**5.87**
0.75 m deep	1.50	8.80	-	m	**8.80**
1.00 m deep	2.20	12.91	0.96	m	**13.87**
1.25 m deep	3.10	18.20	1.45	m	**19.65**
1.50 m deep	4.25	24.95	1.74	m	**26.69**
1.75 m deep	5.60	32.87	2.12	m	**34.99**
2.00 m deep	6.40	37.56	2.32	m	**39.88**
2.25 m deep	8.00	46.96	3.09	m	**50.05**
2.50 m deep	9.60	56.35	3.67	m	**60.02**
2.75 m deep	10.55	61.92	4.05	m	**65.97**
3.00 m deep	11.50	67.50	4.44	m	**71.94**
3.25 m deep	12.45	73.08	4.82	m	**77.90**
3.50 m deep	13.40	78.65	5.21	m	**83.86**
Pipes; 225 mm; average depth					
0.50 m deep	1.00	5.87	-	m	**5.87**
0.75 m deep	1.50	8.80	-	m	**8.80**
1.00 m deep	2.20	12.91	0.96	m	**13.87**
1.25 m deep	3.10	18.20	1.45	m	**19.65**
1.50 m deep	4.25	24.95	1.74	m	**26.69**
1.75 m deep	5.60	32.87	2.12	m	**34.99**
2.00 m deep	6.40	37.56	2.32	m	**39.88**

R DISPOSAL SYSTEMS Including overheads and profit at 5.00%	Labour hours	Labour £	Material £	Unit	Total rate £
2.25 m deep	8.00	46.96	3.09	m	**50.05**
2.50 m deep	9.60	56.35	3.67	m	**60.02**
2.75 m deep	10.55	61.92	4.05	m	**65.97**
3.00 m deep	11.50	67.50	4.44	m	**71.94**
3.25 m deep	12.45	73.08	4.82	m	**77.90**
3.50 m deep	13.40	78.65	5.21	m	**83.86**
Pipes; 300 mm; average depth					
0.75 m deep	1.75	10.27	-	m	**10.27**
1.00 m deep	2.55	14.97	0.96	m	**15.93**
1.25 m deep	3.60	21.13	1.45	m	**22.58**
1.50 m deep	4.80	28.17	1.74	m	**29.91**
1.75 m deep	5.60	32.87	2.12	m	**34.99**
2.00 m deep	6.40	37.56	2.32	m	**39.88**
2.25 m deep	8.00	46.96	3.09	m	**50.05**
2.50 m deep	9.60	56.35	3.67	m	**60.02**
2.75 m deep	10.55	61.92	4.05	m	**65.97**
3.00 m deep	11.50	67.50	4.44	m	**71.94**
3.25 m deep	12.45	73.08	4.82	m	**77.90**
3.50 m deep	13.40	78.65	5.21	m	**83.86**
Pipes; 375 mm; average depth					
0.75 m deep	1.95	11.45	-	m	**11.45**
1.00 m deep	2.85	16.73	0.96	m	**17.69**
1.25 m deep	4.00	23.48	1.45	m	**24.93**
1.50 m deep	5.33	31.28	1.74	m	**33.02**
1.75 m deep	6.20	36.39	2.12	m	**38.51**
2.00 m deep	7.10	41.67	2.32	m	**43.99**
2.25 m deep	8.90	52.24	3.09	m	**55.33**
2.50 m deep	10.70	62.80	3.67	m	**66.47**
2.75 m deep	11.75	68.97	4.05	m	**73.02**
3.00 m deep	12.80	75.13	4.44	m	**79.57**
3.25 m deep	13.90	81.59	4.82	m	**86.41**
3.50 m deep	15.00	88.04	5.21	m	**93.25**
Pipes; 450 mm; average depth					
0.75 m deep	2.20	12.91	-	m	**12.91**
1.00 m deep	3.18	18.67	0.96	m	**19.63**
1.25 m deep	4.47	26.24	1.45	m	**27.69**
1.50 m deep	5.85	34.34	1.74	m	**36.08**
1.75 m deep	6.82	40.03	2.12	m	**42.15**
2.00 m deep	7.80	45.78	2.32	m	**48.10**
2.25 m deep	9.78	57.40	3.09	m	**60.49**
2.50 m deep	11.75	68.97	3.67	m	**72.64**
2.75 m deep	12.93	75.89	4.05	m	**79.94**
3.00 m deep	14.10	82.76	4.44	m	**87.20**
3.25 m deep	15.25	89.51	4.82	m	**94.33**
3.50 m deep	16.40	96.26	5.21	m	**101.47**
Pipes; 600 mm; average depth					
1.00 m deep	3.50	20.54	0.96	m	**21.50**
1.25 m deep	5.00	29.35	1.45	m	**30.80**
1.50 m deep	6.70	39.33	1.74	m	**41.07**
1.75 m deep	7.75	45.49	2.12	m	**47.61**
2.00 m deep	8.85	51.95	2.32	m	**54.27**
2.25 m deep	9.95	58.40	3.09	m	**61.49**
2.50 m deep	11.10	65.15	3.67	m	**68.82**
2.75 m deep	13.35	78.36	4.05	m	**82.41**
3.00 m deep	16.00	93.91	4.44	m	**98.35**
3.25 m deep	17.33	101.72	4.82	m	**106.54**
3.50 m deep	18.65	109.47	5.21	m	**114.68**
Pipes; 900 mm; average depth					
1.25 m deep	6.25	36.68	1.45	m	**38.13**
1.50 m deep	8.25	48.42	1.74	m	**50.16**
1.75 m deep	9.60	56.35	2.12	m	**58.47**
2.00 m deep	10.95	64.27	2.32	m	**66.59**

R DISPOSAL SYSTEMS Including overheads and profit at 5.00%	Labour hours	Labour £	Material £	Unit	Total rate £
R12 DRAINAGE BELOW GROUND - cont'd					
Hand excavation of trenches to receive pipes; grading bottoms; earthwork support; filling with excavated material and compacting; disposal of surplus soil; spreading on site average 50 m - cont'd					
Pipes; 900 mm; average depth					
2.25 m deep	13.75	80.71	3.09	m	**83.80**
2.50 m deep	16.55	97.14	3.67	m	**100.81**
2.75 m deep	18.20	106.82	4.05	m	**110.87**
3.00 m deep	19.80	116.22	4.44	m	**120.66**
3.25 m deep	21.45	125.90	4.82	m	**130.72**
3.50 m deep	23.10	135.59	5.21	m	**140.80**
Pipes; 1200 mm; average depth					
1.50 m deep	9.85	57.81	1.74	m	**59.55**
1.75 m deep	11.45	67.21	2.12	m	**69.33**
2.00 m deep	13.10	76.89	2.32	m	**79.21**
2.25 m deep	16.43	96.44	3.09	m	**99.53**
2.50 m deep	19.75	115.92	3.67	m	**119.59**
2.75 m deep	21.70	127.37	4.05	m	**131.42**
3.00 m deep	23.65	138.81	4.44	m	**143.25**
3.25 m deep	25.58	150.14	4.82	m	**154.96**
3.50 m deep	27.50	161.41	5.21	m	**166.62**
Extra for breaking up					
brick	3.00	17.61	3.31	m3	**20.92**
concrete	4.50	26.41	5.51	m3	**31.92**
reinforced concrete	6.00	35.22	7.72	m3	**42.94**
concrete 150 mm thick	0.70	4.11	0.77	m2	**4.88**
tarmacadam 75 mm thick	0.40	2.35	0.44	m2	**2.79**
tarmacadam and hardcore 150 mm thick	0.50	2.93	0.55	m2	**3.48**
Extra for taking up precast concrete paving slabs	0.30	1.76	-	m2	**1.76**
Sand filling; PC £8.40/t; (PC £13.45/m3)					
Beds; to receive pitch fibre pipes					
600 x 50 mm	0.08	0.47	0.53	m	**1.00**
700 x 50 mm	0.10	0.59	0.62	m	**1.21**
800 x 50 mm	0.12	0.70	0.71	m	**1.41**
Granular (shingle) filling; PC £9.40/t; (PC £17.00/m3)					
Beds; 100 mm thick; to pipes size					
100 mm	0.10	0.59	1.07	m	**1.66**
150 mm	0.10	0.59	1.25	m	**1.84**
225 mm	0.12	0.70	1.43	m	**2.13**
300 mm	0.14	0.82	1.61	m	**2.43**
375 mm	0.16	0.94	1.79	m	**2.73**
450 mm	0.18	1.06	1.96	m	**3.02**
600 mm	0.20	1.17	2.14	m	**3.31**
Beds; 150 mm thick; to pipes size					
100 mm	0.14	0.82	1.61	m	**2.43**
150 mm	0.16	0.94	1.79	m	**2.73**
225 mm	0.18	1.06	1.96	m	**3.02**
300 mm	0.20	1.17	2.14	m	**3.31**
375 mm	0.24	1.41	2.68	m	**4.09**
450 mm	0.26	1.53	2.86	m	**4.39**
600 mm	0.30	1.76	3.39	m	**5.15**
Beds and benchings; beds 100 mm thick; to pipes size					
100 mm	0.23	1.35	1.96	m	**3.31**
150 mm	0.25	1.47	1.96	m	**3.43**
225 mm	0.30	1.76	2.68	m	**4.44**

R DISPOSAL SYSTEMS Including overheads and profit at 5.00%	Labour hours	Labour £	Material £	Unit	Total rate £
300 mm	0.35	2.05	3.03	m	**5.08**
375 mm	0.45	2.64	4.11	m	**6.75**
450 mm	0.52	3.05	4.64	m	**7.69**
600 mm	0.67	3.93	6.07	m	**10.00**
Beds and benchings; beds 150 mm thick; to pipes size					
100 mm	0.25	1.47	2.14	m	**3.61**
150 mm	0.28	1.64	2.32	m	**3.96**
225 mm	0.35	2.05	3.21	m	**5.26**
300 mm	0.45	2.64	3.93	m	**6.57**
375 mm	0.52	3.05	4.64	m	**7.69**
450 mm	0.62	3.64	5.53	m	**9.17**
600 mm	0.74	4.34	7.14	m	**11.48**
Beds and coverings; 100 mm thick; to pipes size					
100 mm	0.36	2.11	2.68	m	**4.79**
150 mm	0.45	2.64	3.21	m	**5.85**
225 mm	0.60	3.52	4.46	m	**7.98**
300 mm	0.72	4.23	5.36	m	**9.59**
375 mm	0.87	5.11	6.43	m	**11.54**
450 mm	1.02	5.99	7.68	m	**13.67**
600 mm	1.32	7.75	9.82	m	**17.57**
Beds and coverings; 150 mm thick; to pipes size					
100 mm	0.54	3.17	3.93	m	**7.10**
150 mm	0.60	3.52	4.46	m	**7.98**
225 mm	0.78	4.58	5.71	m	**10.29**
300 mm	0.93	5.46	6.78	m	**12.24**
375 mm	1.08	6.34	8.03	m	**14.37**
450 mm	1.29	7.57	9.64	m	**17.21**
600 mm	1.56	9.16	11.60	m	**20.76**
In situ ready mixed concrete; normal Portland cement; mix 11.50 N/mm2 - 40 mm aggregate (1:3:6); PC £40.48/m3;					
Beds; 100 mm thick; to pipes size					
100 mm	0.20	1.22	2.13	m	**3.35**
150 mm	0.20	1.22	2.13	m	**3.35**
225 mm	0.24	1.46	2.55	m	**4.01**
300 mm	0.28	1.71	2.98	m	**4.69**
375 mm	0.32	1.95	3.40	m	**5.35**
450 mm	0.36	2.19	3.83	m	**6.02**
600 mm	0.40	2.44	4.25	m	**6.69**
900 mm	0.48	2.92	5.10	m	**8.02**
1200 mm	0.64	3.90	6.80	m	**10.70**
Beds; 150 mm thick; to pipes size					
100 mm	0.28	1.71	2.98	m	**4.69**
150 mm	0.32	1.95	3.40	m	**5.35**
225 mm	0.36	2.19	3.83	m	**6.02**
300 mm	0.40	2.44	4.25	m	**6.69**
375 mm	0.48	2.92	5.10	m	**8.02**
450 mm	0.52	3.17	5.53	m	**8.70**
600 mm	0.60	3.65	6.38	m	**10.03**
900 mm	0.76	4.63	8.08	m	**12.71**
1200 mm	0.92	5.60	9.78	m	**15.38**
Beds and benchings; beds 100 mm thick; to pipes size					
100 mm	0.40	2.44	3.83	m	**6.27**
150 mm	0.45	2.74	4.25	m	**6.99**
225 mm	0.54	3.29	5.10	m	**8.39**
300 mm	0.63	3.84	5.95	m	**9.79**
375 mm	0.81	4.93	7.65	m	**12.58**

R DISPOSAL SYSTEMS Including overheads and profit at 5.00%	Labour hours	Labour £	Material £	Unit	Total rate £
R12 DRAINAGE BELOW GROUND - cont'd					
In situ ready mixed concrete; normal Portland cement; mix 11.50 N/mm2 - 40 mm aggregate (1:3:6); PC £40.48/m3; - cont'd					
Beds and benchings; beds 100 mm thick; to pipes size					
450 mm	0.95	5.79	8.93	m	**14.72**
600 mm	1.22	7.43	11.48	m	**18.91**
900 mm	1.98	12.06	18.70	m	**30.76**
1200 mm	2.93	17.84	27.63	m	**45.47**
Beds and benchings; beds 150 mm thick; to pipes size					
100 mm	0.45	2.74	4.25	m	**6.99**
150 mm	0.50	3.05	4.68	m	**7.73**
225 mm	0.63	3.84	5.95	m	**9.79**
300 mm	0.81	4.93	7.65	m	**12.58**
375 mm	0.95	5.79	8.93	m	**14.72**
450 mm	1.13	6.88	10.63	m	**17.51**
600 mm	1.44	8.77	13.60	m	**22.37**
900 mm	2.30	14.01	21.68	m	**35.69**
1200 mm	3.24	19.73	30.60	m	**50.33**
Beds and coverings; 100 mm thick; to pipes size					
100 mm	0.60	3.65	5.10	m	**8.75**
150 mm	0.70	4.26	5.95	m	**10.21**
225 mm	1.00	6.09	8.50	m	**14.59**
300 mm	1.20	7.31	10.20	m	**17.51**
375 mm	1.45	8.83	12.33	m	**21.16**
450 mm	1.70	10.35	14.45	m	**24.80**
600 mm	2.20	13.40	18.70	m	**32.10**
900 mm	3.35	20.40	28.48	m	**48.88**
1200 mm	4.60	28.01	39.10	m	**67.11**
Beds and coverings; 150 mm thick; to pipes size					
100 mm	0.90	5.48	7.65	m	**13.13**
150 mm	1.00	6.09	8.50	m	**14.59**
225 mm	1.30	7.92	11.05	m	**18.97**
300 mm	1.55	9.44	13.18	m	**22.62**
375 mm	1.80	10.96	15.30	m	**26.26**
450 mm	2.15	13.09	18.28	m	**31.37**
600 mm	2.60	15.83	22.10	m	**37.93**
900 mm	4.25	25.88	36.13	m	**62.01**
1200 mm	6.00	36.54	51.00	m	**87.54**
In situ ready mixed concrete; normal Portland cement; mix 21.00 N/mm2 - 40 mm aggregate (1:2:4); PC £43.10/m3					
Beds; 100 mm thick; to pipes size					
100 mm	0.20	1.22	2.26	m	**3.48**
150 mm	0.20	1.22	2.26	m	**3.48**
225 mm	0.24	1.46	2.72	m	**4.18**
300 mm	0.28	1.71	3.17	m	**4.88**
375 mm	0.32	1.95	3.62	m	**5.57**
450 mm	0.36	2.19	4.07	m	**6.26**
600 mm	0.40	2.44	4.53	m	**6.97**
900 mm	0.48	2.92	5.43	m	**8.35**
1200 mm	0.64	3.90	7.24	m	**11.14**
Beds; 150 mm thick; to pipes size					
100 mm	0.28	1.71	3.17	m	**4.88**
150 mm	0.32	1.95	3.62	m	**5.57**
225 mm	0.36	2.19	4.07	m	**6.26**
300 mm	0.40	2.44	4.53	m	**6.97**

R DISPOSAL SYSTEMS Including overheads and profit at 5.00%	Labour hours	Labour £	Material £	Unit	Total rate £
375 mm	0.48	2.92	5.43	m	**8.35**
450 mm	0.52	3.17	5.88	m	**9.05**
600 mm	0.60	3.65	6.79	m	**10.44**
900 mm	0.76	4.63	8.60	m	**13.23**
1200 mm	0.92	5.60	10.41	m	**16.01**
Beds and benchings; beds 100 mm thick; to pipes size					
100 mm	0.40	2.44	4.07	m	**6.51**
150 mm	0.45	2.74	4.53	m	**7.27**
225 mm	0.54	3.29	5.43	m	**8.72**
300 mm	0.63	3.84	6.34	m	**10.18**
375 mm	0.81	4.93	8.15	m	**13.08**
450 mm	0.95	5.79	9.50	m	**15.29**
600 mm	1.22	7.43	12.22	m	**19.65**
900 mm	1.98	12.06	19.91	m	**31.97**
1200 mm	2.93	17.84	29.42	m	**47.26**
Beds and benchings; beds 150 mm thick; to pipes size					
100 mm	0.45	2.74	4.53	m	**7.27**
150 mm	0.50	3.05	4.98	m	**8.03**
225 mm	0.63	3.84	6.34	m	**10.18**
300 mm	0.81	4.93	8.15	m	**13.08**
375 mm	0.95	5.79	9.50	m	**15.29**
450 mm	1.13	6.88	11.31	m	**18.19**
600 mm	1.44	8.77	14.48	m	**23.25**
900 mm	2.30	14.01	23.08	m	**37.09**
1200 mm	3.24	19.73	32.58	m	**52.31**
Beds and coverings; 100 mm thick; to pipes size					
100 mm	0.60	3.65	5.43	m	**9.08**
150 mm	0.70	4.26	6.34	m	**10.60**
225 mm	1.00	6.09	9.05	m	**15.14**
300 mm	1.20	7.31	10.86	m	**18.17**
375 mm	1.45	8.83	13.12	m	**21.95**
450 mm	1.70	10.35	15.39	m	**25.74**
600 mm	2.20	13.40	19.91	m	**33.31**
900 mm	3.35	20.40	30.32	m	**50.72**
1200 mm	4.60	28.01	41.63	m	**69.64**
Beds and coverings; 150 mm thick; to pipes size					
100 mm	0.90	5.48	8.15	m	**13.63**
150 mm	1.00	6.09	9.05	m	**15.14**
225 mm	1.30	7.92	11.77	m	**19.69**
300 mm	1.55	9.44	14.03	m	**23.47**
375 mm	1.80	10.96	16.29	m	**27.25**
450 mm	2.15	13.09	19.46	m	**32.55**
600 mm	2.60	15.83	23.53	m	**39.36**
900 mm	4.25	25.88	38.47	m	**64.35**
1200 mm	6.00	36.54	54.31	m	**90.85**

R DISPOSAL SYSTEMS Including overheads and profit at 5.00%		Labour hours	Labour £	Material £	Unit	Total rate £
NOTE: The following items unless otherwise described include for all appropriate joints/couplings in the running length. The prices for gullies and rainwater shoes, etc., include for appropriate joints to pipes and for setting on and surrounding accessory with site mixed in situ concrete 11.50 N/mm2-40 aggregate (1:3:6)						
Cast iron drain pipes and fittings; BS 437; coated; with caulked lead joints						
75 mm pipes						
laid straight; grey iron	PC £52.68/3m	0.60	3.96	19.74	m	23.70
laid straight; grey iron	PC £34.27/1.8m	0.70	4.62	21.71	m	26.33
in runs not exceeding 3 m long	PC £25.89/.91m	0.95	6.26	33.08	m	39.34
Extra for						
bend; short radius	PC £13.13	0.70	4.62	14.04	nr	18.66
bend; short radius; with round access door	PC £21.70	0.70	4.62	23.49	nr	28.11
bend; long radius	PC £18.65	0.70	4.62	19.09	nr	23.71
level invert taper	PC £14.81	0.70	4.62	14.86	nr	19.48
access pipe	PC £45.02	0.70	4.62	48.17	nr	52.79
single branch	PC £24.94	0.95	6.26	25.75	nr	32.01
100 mm pipes						
laid straight; grey iron	PC £55.52/3m	0.70	4.62	20.95	m	25.57
laid straight; grey iron	PC £35.06/1.8m	0.80	5.28	22.51	m	27.79
in runs not exceeding 3 m long	PC £28.30/.91m	1.10	7.25	36.63	m	43.88
in ducts; supported on piers (measured elsewhere)		1.40	9.23	21.22	m	30.45
supported on wall brackets (measured elsewhere)		1.15	7.58	21.22	m	28.80
supported on ceiling hangers (measured elsewhere)		1.40	9.23	21.22	m	30.45
Extra for						
bend; short radius	PC £17.20	0.80	5.28	19.02	nr	24.30
bend; short radius; with round access door	PC £25.80	0.80	5.28	28.50	nr	33.78
bend; long radius	PC £26.81	0.80	5.28	28.51	nr	33.79
bend; long radius; with rectangular access door	PC £49.99	0.80	5.28	54.07	nr	59.35
rest bend	PC £21.81	0.80	5.28	24.10	nr	29.38
level invert taper	PC £17.41	0.80	5.28	18.15	nr	23.43
access pipe	PC £49.29	0.80	5.28	53.29	nr	58.57
single branch	PC £27.20	1.10	7.25	28.91	nr	36.16
single branch; with rectangular access door	PC £62.78	1.10	7.25	68.15	nr	75.40
'Y' branch	PC £61.83	1.10	7.25	67.09	nr	74.34
double branch	PC £74.70	1.35	8.90	81.36	nr	90.26
double branch; with rectangular access door	PC £113.78	1.35	8.90	124.45	nr	133.35
WC connector; 450 mm long	PC £13.16	0.60	3.96	11.88	nr	15.84
WC connector; 600 mm long	PC £26.12	0.60	3.96	24.32	nr	28.28
SW connector	PC £14.48	0.80	5.28	14.40	nr	19.68
150 mm pipes						
laid straight; grey iron	PC £95.36/3m	0.90	5.93	35.94	m	41.87
laid straight; grey iron	PC £53.66/1.8m	1.00	6.59	34.62	m	41.21
in runs not exceeding 3 m long	PC £42.43/.91m	1.35	8.90	55.38	m	64.28

R DISPOSAL SYSTEMS Including overheads and profit at 5.00%		Labour hours	Labour £	Material £	Unit	Total rate £
Extra for						
bend; short radius	PC £37.08	1.00	6.59	40.89	nr	**47.48**
bend; short radius; with round access door	PC £49.19	1.00	6.59	54.25	nr	**60.84**
bend; long radius	PC £46.03	1.00	6.59	48.87	nr	**55.46**
bend; long radius; with rectangular access door	PC £89.93	1.00	6.59	97.27	nr	**103.86**
rest bend	PC £48.81	1.00	6.59	53.83	nr	**60.42**
level invert taper	PC £26.81	1.00	6.59	27.68	nr	**34.27**
access pipe	PC £91.30	1.00	6.59	98.79	nr	**105.38**
single branch	PC £59.10	1.35	8.90	63.18	nr	**72.08**
single branch; with rectangular access door	PC £123.71	1.35	8.90	134.42	nr	**143.32**
double branch	PC £116.10	1.70	11.21	126.14	nr	**137.35**
SW connector	PC £23.48	1.00	6.59	23.17	nr	**29.76**
225 mm pipes						
laid straight; grey iron	PC £269.21/1.8m	1.60	10.55	246.64	m	**257.19**
in runs not exceeding 3 m long	PC £155.72/.91m	2.15	14.18	194.52	m	**208.70**
Extra for						
bend; short radius	PC £141.95	1.60	10.55	162.45	nr	**173.00**
bend; long radius; with rectangular access door	PC £260.25	1.60	10.55	292.86	nr	**303.41**
rest bend	PC £170.81	1.60	10.55	194.26	nr	**204.81**
level invert taper	PC £49.74	1.60	10.55	60.79	nr	**71.34**
access pipe	PC £191.65	1.60	10.55	217.23	nr	**227.78**
single branch	PC £184.73	2.15	14.18	212.58	nr	**226.76**
single branch; with rectangular access door	PC £243.56	2.15	14.18	277.44	nr	**291.62**
SW connector		1.60	10.55	93.34	nr	**103.89**
Accessories in cast iron; with caulked lead joints to pipes (unless otherwise described)						
Rainwater shoes; horizontal inlet						
100 mm	PC £42.40	0.75	4.95	47.78	nr	**52.73**
150 mm	PC £67.74	0.85	5.60	76.42	nr	**82.02**
Gully fittings; comprising low invert gully trap and round hopper						
75 mm outlet	PC £15.97	1.10	7.25	20.93	nr	**28.18**
100 mm outlet	PC £26.81	1.20	7.91	32.91	nr	**40.82**
150 mm outlet	PC £67.99	1.60	10.55	78.20	nr	**88.75**
Add to above for						
bellmouth 300 mm high; circular plain grating						
75 mm; 175 mm grating	PC £15.25	0.60	3.96	17.52	nr	**21.48**
100 mm; 200 mm grating	PC £22.31	0.70	4.62	25.49	nr	**30.11**
150 mm; 250 mm grating	PC £32.38	1.00	6.59	37.80	nr	**44.39**
bellmouth 300 mm high; circular plain grating as above; one horizontal inlet						
75 x 50 mm	PC £29.46	0.60	3.96	32.60	nr	**36.56**
100 x 75 mm	PC £28.70	0.70	4.62	32.40	nr	**37.02**
150 x 100 mm	PC £60.79	1.00	6.59	67.63	nr	**74.22**
bellmouth 300 mm high; circular plain grating as above; one vertical inlet						
75 x 50 mm	PC £27.68	0.60	3.96	30.74	nr	**34.70**
100 x 75 mm	PC £32.68	0.70	4.62	36.58	nr	**41.20**
150 x 100 mm	PC £69.22	1.00	6.59	76.48	nr	**83.07**
raising piece; 200 mm dia						
75 mm high	PC £12.75	1.20	7.91	18.62	nr	**26.53**
150 mm high	PC £15.27	1.20	7.91	21.26	nr	**29.17**
225 mm high	PC £19.40	1.20	7.91	25.60	nr	**33.51**
300 mm high	PC £22.74	1.20	7.91	29.11	nr	**37.02**

R DISPOSAL SYSTEMS Including overheads and profit at 5.00%		Labour hours	Labour £	Material £	Unit	Total rate £
R12 DRAINAGE BELOW GROUND - cont'd						
Accessories in cast iron; with caulked lead joints to pipes (unless otherwise described) - cont'd						
Yard gully (Deans); trapped; galvanized sediment pan; 267 mm round heavy grating						
100 mm outlet	PC £142.33	3.20	21.10	154.71	nr	**175.81**
Yard gully (garage); trapless; galvanized sediment pan; 267 mm round heavy grating						
100 mm outlet	PC £124.61	3.00	19.78	136.11	nr	**155.89**
Yard gully (garage); trapped; with rodding eye; galvanized perforated sediment pan; stopper; round heavy grating						
150 mm outlet; 347 mm grating	PC £461.11	4.00	26.38	491.97	nr	**518.35**
Grease trap; with internal access; insert galvanized perforated bucket; lid and frame						
450 x 300 x 525 mm deep; 100 mm outlet	PC £293.43	3.60	23.74	315.37	nr	**339.11**
Disconnecting trap; trapped with inspection arm; bridle plate and screw; building in and cutting and fitting brickwork around						
100 mm outlet; 100 mm inlet	PC £119.73	2.50	16.49	131.98	nr	**148.47**
150 mm outlet; 150 mm inlet	PC £178.63	3.20	21.10	197.37	nr	**218.47**
Cast iron 'Timesaver' drain pipes and fittings; BS 437; coated; with mechanical coupling joints						
75 mm pipes						
laid straight	PC £39.81/3m	0.45	2.97	17.40	m	**20.37**
in runs not exceeding 3 m long	PC £24.84/1m	0.60	3.96	34.67	m	**38.63**
Extra for						
bend; medium radius	PC £12.45	0.50	3.30	18.81	nr	**22.11**
single branch	PC £17.30	0.70	4.62	30.14	nr	**34.76**
isolated 'Timesaver' joint	PC £6.94	0.30	1.98	7.65	nr	**9.63**
100 mm pipes						
laid straight	PC £38.47/3m	0.50	3.30	17.34	m	**20.64**
in runs not exceeding 3 m long	PC £24.33/1m	0.68	4.48	35.44	m	**39.92**
Extra for						
bend; medium radius	PC £15.33	0.60	3.96	23.39	nr	**27.35**
bend; medium radius with access	PC £40.29	0.60	3.96	50.91	nr	**54.87**
bend; long radius	PC £18.79	0.60	3.96	26.35	nr	**30.31**
rest bend	PC £17.60	0.60	3.96	25.89	nr	**29.85**
diminishing pipe	PC £13.04	0.60	3.96	22.13	nr	**26.09**
single branch	PC £20.36	0.75	4.95	36.27	nr	**41.22**
single branch; with access	PC £46.94	0.85	5.60	56.97	nr	**62.57**
double branch	PC £30.66	0.95	6.26	55.39	nr	**61.65**
double branch; with access	PC £56.62	0.95	6.26	89.49	nr	**95.75**
isolated 'Timesaver' joint	PC £8.20	0.35	2.31	9.04	nr	**11.35**
transitional pipe; for WC	PC £11.68	0.50	3.30	19.37	nr	**22.67**
150 mm pipes						
laid straight	PC £72.46/3m	0.60	3.96	30.71	m	**34.67**
in runs not exceeding 3 m long	PC £45.24/1m	0.82	5.41	60.31	m	**65.72**
Extra for						
bend; medium radius	PC £30.94	0.70	4.62	40.56	nr	**45.18**
bend; medium radius with access	PC £74.82	0.70	4.62	88.92	nr	**93.54**
bend; long radius	PC £40.29	0.70	4.62	49.26	nr	**53.88**
diminishing pipe	PC £19.99	0.70	4.62	30.88	nr	**35.50**
single branch	PC £44.05	0.85	5.60	63.04	nr	**68.64**
isolated 'Timesaver' joint	PC £9.94	0.42	2.77	10.96	nr	**13.73**

R DISPOSAL SYSTEMS Including overheads and profit at 5.00%		Labour hours	Labour £	Material £	Unit	Total rate £
Accessories in 'Timesaver' cast iron; with mechanical coupling joints						
Rainwater shoes; horizontal inlet						
100 mm	PC £42.40	0.50	3.30	54.13	nr	**57.43**
150 mm	PC £67.74	0.60	3.96	83.05	nr	**87.01**
Gully fittings; comprising low invert gully trap and round hopper						
75 mm outlet	PC £13.04	0.90	5.93	23.48	nr	**29.41**
100 mm outlet	PC £20.36	0.95	6.26	32.49	nr	**38.75**
150 mm outlet	PC £50.64	1.30	8.57	66.61	nr	**75.18**
Add to above for						
bellmouth 300 mm high; circular plain grating						
100 mm; 200 mm grating	PC £22.31	0.45	2.97	32.04	nr	**35.01**
bellmouth 300 mm high; circular plain grating as above; one horizontal inlet						
100 x 100 mm	PC £28.70	0.45	2.97	38.75	nr	**41.72**
bellmouth 300 mm high; circular plain grating as above; one vertical inlet						
100 x 100 mm	PC £30.42	0.45	2.97	42.93	nr	**45.90**
Yard gully (Deans); trapped; galvanized sediment pan; 267 mm round heavy grating						
100 mm outlet	PC £155.38	2.90	19.12	166.16	nr	**185.28**
Yard gully (garage); trapless; galvanized sediment pan; 267 mm round heavy grating						
100 mm outlet	PC £132.81	2.70	17.80	142.46	nr	**160.26**
Yard gully (garage); trapped; with rodding eye, galvanised perforated sediment pan; stopper; 267 mm round heavy grating						
100 mm outlet; 267 mm grating	PC £295.45	3.00	19.78	313.73	nr	**333.51**
Grease trap; internal access; galvanized perforated bucket; lid and frame						
20 gal. capacity	PC £567.35	4.00	26.38	466.16	nr	**492.54**
Extra strength vitrified clay pipes and fittings; 'Hepworths' 'SuperSleve'/'HepSleve' or similar; plain ends with push-fit polypropylene flexible couplings						
100 mm pipes						
laid straight	PC £2.46	0.25	1.37	2.71	m	**4.08**
in runs not exceeding 3 m long		0.33	1.81	3.16	m	**4.97**
Extra for						
bend	PC £2.09	0.20	1.10	3.89	nr	**4.99**
access bend	PC £13.20	0.20	1.10	16.14	nr	**17.24**
rest bend	PC £3.46	0.20	1.10	5.40	nr	**6.50**
access pipe	PC £11.44	0.20	1.10	14.00	nr	**15.10**
socket adaptor	PC £2.10	0.17	0.93	3.21	nr	**4.14**
adaptor to flexible pipe	PC £2.10	0.17	0.93	3.21	nr	**4.14**
saddle	PC £4.32	0.75	4.12	5.86	nr	**9.98**
single junction	PC £4.40	0.25	1.37	7.32	nr	**8.69**
single access junction	PC £15.30	0.25	1.37	19.35	nr	**20.72**
150 mm pipes						
laid straight	PC £5.61	0.30	1.65	6.18	m	**7.83**
in runs not exceeding 3 m long		0.40	2.20	7.23	m	**9.43**
Extra for						
bend	PC £5.38	0.24	1.32	9.64	nr	**10.96**
access bend	PC £22.64	0.24	1.32	28.94	nr	**30.26**
rest bend	PC £6.90	0.24	1.32	11.59	nr	**12.91**
taper pipe	PC £5.08	0.24	1.32	8.00	nr	**9.32**
access pipe	PC £19.38	0.24	1.32	24.90	nr	**26.22**

R DISPOSAL SYSTEMS Including overheads and profit at 5.00%		Labour hours	Labour £	Material £	Unit	Total rate £
R12 DRAINAGE BELOW GROUND - cont'd						
Extra strength vitrified clay pipes and fittings; 'Hepworths' 'SuperSleve'/'HepSleve' or similar; plain ends with push-fit polypropylene flexible couplings - cont'd						
150 mm pipes						
socket adaptor	PC £5.19	0.20	1.10	7.94	nr	**9.04**
adaptor to flexible pipe	PC £3.70	0.20	1.10	6.29	nr	**7.39**
saddle	PC £7.87	0.90	4.94	11.35	nr	**16.29**
single junction	PC £7.89	0.30	1.65	14.25	nr	**15.90**
single access junction	PC £28.21	0.30	1.65	37.31	nr	**38.96**
Extra strength vitrified clay pipes and fittings; 'Hepworths' 'HepSeal'; or similar; socketted; with push-fit flexible joints						
100 mm pipes						
laid straight	PC £4.96	0.30	1.65	5.47	m	**7.12**
in runs not exceeding 3 m long		0.40	2.20	5.99	m	**8.19**
Extra for						
bend	PC £6.39	0.24	1.32	5.41	nr	**6.73**
access bend	PC £16.25	0.24	1.32	16.28	nr	**17.60**
rest bend	PC £8.12	0.24	1.32	7.31	nr	**8.63**
stopper	PC £2.58	0.15	0.82	2.85	nr	**3.67**
access pipe	PC £14.71	0.24	1.32	14.03	nr	**15.35**
socket adaptor	PC £2.10	0.26	1.43	4.32	nr	**5.75**
saddle	PC £6.08	0.75	4.12	6.70	nr	**10.82**
single junction	PC £9.50	0.30	1.65	8.29	nr	**9.94**
single access junction	PC £17.76	0.30	1.65	17.39	nr	**19.04**
double junction	PC £18.10	0.45	2.47	17.22	nr	**19.69**
double collar	PC £6.08	0.20	1.10	6.70	nr	**7.80**
150 mm pipes						
laid straight	PC £6.42	0.35	1.92	7.08	m	**9.00**
in runs not exceeding 3 m long		0.47	2.58	7.75	m	**10.33**
Extra for						
bend	PC £11.04	0.28	1.54	10.05	nr	**11.59**
access bend	PC £25.37	0.28	1.54	25.84	nr	**27.38**
rest bend	PC £13.45	0.28	1.54	12.71	nr	**14.25**
stopper	PC £3.84	0.18	0.99	4.23	nr	**5.22**
taper reducer	PC £16.90	0.28	1.54	16.51	nr	**18.05**
access pipe	PC £23.06	0.28	1.54	22.59	nr	**24.13**
socket adaptor	PC £5.19	0.30	1.65	9.15	nr	**10.80**
saddle	PC £10.73	0.90	4.94	11.83	nr	**16.77**
single access junction	PC £30.55	0.35	1.92	30.85	nr	**32.77**
double junction	PC £30.12	0.53	2.91	29.67	nr	**32.58**
double collar	PC £10.04	0.23	1.26	11.07	nr	**12.33**
225 mm pipes						
laid straight	PC £12.57	0.45	2.47	13.86	m	**16.33**
in runs not exceeding 3 m long		0.60	3.29	15.18	m	**18.47**
Extra for						
bend	PC £23.12	0.36	1.98	21.33	nr	**23.31**
access bend	PC £58.40	0.36	1.98	60.23	nr	**62.21**
rest bend	PC £32.20	0.36	1.98	31.34	nr	**33.32**
stopper	PC £8.33	0.23	1.26	9.19	nr	**10.45**
taper reducer	PC £26.73	0.36	1.98	25.31	nr	**27.29**
access pipe	PC £58.40	0.36	1.98	58.84	nr	**60.82**
saddle	PC £26.73	1.20	6.59	29.47	nr	**36.06**
single junction	PC £34.75	0.45	2.47	32.77	nr	**35.24**
single access junction	PC £69.83	0.45	2.47	71.44	nr	**73.91**
double junction	PC £67.54	0.68	3.73	67.53	nr	**71.26**
double collar	PC £22.04	0.30	1.65	24.30	nr	**25.95**

R DISPOSAL SYSTEMS Including overheads and profit at 5.00%		Labour hours	Labour £	Material £	Unit	Total rate £
Extra strength vitrified clay pipes and fittings 'HepSeal' or similar						
300 mm pipes						
laid straight	PC £19.68	0.60	3.29	21.69	m	**24.98**
in runs not exceeding 3 m long		0.80	4.39	23.76	m	**28.15**
Extra for						
bend	PC £45.60	0.48	2.64	43.76	nr	**46.40**
access bend	PC £97.14	0.48	2.64	100.59	nr	**103.23**
rest bend	PC £68.30	0.48	2.64	68.79	nr	**71.43**
stopper	PC £18.35	0.30	1.65	20.23	nr	**21.88**
taper reducer	PC £54.67	0.48	2.64	53.77	nr	**56.41**
access pipe	PC £97.14	0.48	2.64	98.41	nr	**101.05**
saddle	PC £54.67	1.60	8.79	60.27	nr	**69.06**
single junction	PC £71.64	0.60	3.29	70.31	nr	**73.60**
single access junction	PC £116.74	0.60	3.29	120.02	nr	**123.31**
double junction	PC £143.37	0.90	4.94	147.22	nr	**152.16**
double collar	PC £35.82	0.40	2.20	39.49	nr	**41.69**
400 mm pipes						
laid straight	PC £38.46	0.80	4.39	42.40	m	**46.79**
in runs not exceeding 3 m long		1.06	5.82	46.44	m	**52.26**
Extra for						
bend	PC £143.89	0.64	3.51	145.92	nr	**149.43**
single junction	PC £134.82	0.80	4.39	131.68	nr	**136.07**
450 mm pipes						
laid straight	PC £49.84	1.00	5.49	54.95	m	**60.44**
in runs not exceeding 3 m long		1.33	7.30	60.19	m	**67.49**
Extra for						
bend	PC £189.46	0.80	4.39	192.40	nr	**196.79**
single junction	PC £161.28	1.00	5.49	155.83	nr	**161.32**
British Standard quality vitrified clay pipes and fittings; socketted; cement and sand (1:2) joints						
100 mm pipes						
laid straight	PC £3.17	0.40	2.20	3.58	m	**5.78**
in runs not exceeding 3 m long		0.53	2.91	3.92	m	**6.83**
Extra for						
bend (short/medium/knuckle)	PC £2.99	0.32	1.76	2.33	nr	**4.09**
bend (long/rest/elbow)	PC £5.49	0.32	1.76	5.09	nr	**6.85**
access bend	PC £16.39	0.32	1.76	17.11	nr	**18.87**
taper	PC £7.60	0.32	1.76	7.31	nr	**9.07**
access pipe	PC £14.88	0.32	1.76	15.09	nr	**16.85**
single junction	PC £6.13	0.40	2.20	5.53	nr	**7.73**
single access junction	PC £18.19	0.40	2.20	18.83	nr	**21.03**
double junction	PC £10.96	0.60	3.29	10.59	nr	**13.88**
double collar	PC £4.02	0.27	1.48	4.52	nr	**6.00**
double access junction	PC £23.22	0.60	3.29	24.11	nr	**27.40**
150 mm pipes						
laid straight	PC £5.45	0.45	2.47	6.09	m	**8.56**
in runs not exceeding 3 m long		0.60	3.29	6.67	m	**9.96**
Extra for						
bend (short/medium/knuckle)	PC £5.16	0.36	1.98	3.98	nr	**5.96**
bend (long/rest/elbow)	PC £9.29	0.36	1.98	8.52	nr	**10.50**
access bend	PC £27.58	0.36	1.98	28.69	nr	**30.67**
taper	PC £12.13	0.36	1.98	11.48	nr	**13.46**
access pipe	PC £22.97	0.36	1.98	23.01	nr	**24.99**
single junction	PC £10.82	0.45	2.47	9.70	nr	**12.17**
single access junction	PC £30.55	0.45	2.47	31.45	nr	**33.92**
double junction	PC £25.27	0.68	3.73	25.12	nr	**28.85**
double collar	PC £6.70	0.30	1.65	7.44	nr	**9.09**
double access junction	PC £39.17	0.68	3.73	40.44	nr	**44.17**

R DISPOSAL SYSTEMS Including overheads and profit at 5.00%		Labour hours	Labour £	Material £	Unit	Total rate £
R12 DRAINAGE BELOW GROUND - cont'd						
British Standard quality vitrified clay pipes and fittings; socketted; cement and sand (1:2) joints - cont'd						
225 mm pipes						
laid straight	PC £10.75	0.55	3.02	12.03	m	15.05
in runs not exceeding 3 m long		0.73	4.01	13.15	m	17.16
Extra for						
bend (short/medium/knuckle)	PC £16.16	0.44	2.42	14.44	nr	16.86
bend (long/rest/elbow)	PC £26.54	0.44	2.42	25.88	nr	28.30
access bend	PC £58.40	0.44	2.42	61.00	nr	63.42
taper	PC £26.33	0.44	2.42	25.29	nr	27.71
access pipe	PC £58.40	0.44	2.42	59.82	nr	62.24
single junction	PC £26.54	0.55	3.02	24.86	nr	27.88
single access junction	PC £69.83	0.55	3.02	72.59	nr	75.61
double junction	PC £52.90	0.83	4.56	52.92	nr	57.48
double collar	PC £15.77	0.36	1.98	17.56	nr	19.54
double access junction	PC £78.62	0.83	4.56	81.27	nr	85.83
300 mm pipes						
laid straight	PC £17.53	0.75	4.12	19.50	m	23.62
in runs not exceeding 3 m long		1.00	5.49	21.34	m	26.83
Extra for						
bend (short/medium/knuckle)	PC £28.61	0.60	3.29	25.92	nr	29.21
bend (long/rest/elbow)	PC £52.51	0.60	3.29	52.26	nr	55.55
access bend	PC £93.07	0.60	3.29	96.98	nr	100.27
taper	PC £47.76	0.60	3.29	46.45	nr	49.74
access pipe	PC £93.07	0.60	3.29	95.05	nr	98.34
single junction	PC £52.51	0.75	4.12	50.50	nr	54.62
single access junction	PC £103.38	0.75	4.12	106.59	nr	110.71
double junction	PC £104.93	1.13	6.21	106.54	nr	112.75
double access junction	PC £155.84	1.13	6.21	162.66	nr	168.87
400 mm pipes						
laid straight	PC £33.50	1.00	5.49	37.20	m	42.69
in runs not exceeding 3 m long	PC £33.50	1.33	7.30	40.70	m	48.00
Extra for						
bend (short/medium/knuckle)	PC £94.57	0.80	4.39	93.44	nr	97.83
taper	PC £86.00	0.80	4.39	82.89	nr	87.28
single junction		1.00	5.49	90.01	nr	95.50
450 mm pipes						
laid straight	PC £42.62	1.25	6.86	47.24	m	54.10
in runs not exceeding 3 m long		1.66	9.12	51.72	m	60.84
Extra for						
bend (short/medium/knuckle)	PC £116.49	1.00	5.49	114.60	nr	120.09
taper	PC £105.90	1.00	5.49	101.50	nr	106.99
single junction	PC £116.49	1.25	6.86	110.15	nr	117.01
500 mm pipes						
laid straight	PC £53.34	1.45	7.96	59.15	m	67.11
in runs not exceeding 3 m long		1.95	10.71	64.75	m	75.46
Extra for						
bend (short/medium/knuckle)	PC £138.49	1.15	6.32	135.38	nr	141.70
taper	PC £104.38	1.15	6.32	97.78	nr	104.10
single junction	PC £138.49	1.45	7.96	129.85	nr	137.81

R DISPOSAL SYSTEMS Including overheads and profit at 5.00%		Labour hours	Labour £	Material £	Unit	Total rate £
Accessories in vitrified clay; set in concrete; with polypropylene coupling joints to pipes						
Rodding point; with oval aluminium plate						
100 mm	PC £11.50	0.50	2.75	14.09	nr	**16.84**
Gully fittings; comprising low back trap and square hopper; 150 x 150 mm square gully grid						
100 mm outlet	PC £10.33	0.85	4.67	15.42	nr	**20.09**
Add to above for						
100 mm back inlet		-	-	-	nr	**4.75**
100 mm raising pieces	PC £4.43	0.25	1.37	5.74	nr	**7.11**
Access gully; trapped with rodding eye and integral vertical back inlet; stopper; 150 x 150 mm square gully grid						
100 mm outlet	PC £14.73	0.65	3.57	17.48	nr	**21.05**
Inspection chamber; comprising base; 300 or 450 mm raising piece; integral alloy cover and frame; 100 mm inlets						
straight through; 2 nr inlets	PC £40.05	2.00	10.98	47.65	nr	**58.63**
single junction; 3 nr inlets	PC £43.37	2.20	12.08	52.23	nr	**64.31**
double junction; 4 nr inlets	PC £47.04	2.40	13.18	57.18	nr	**70.36**
Accessories in propylene; cover set in concrete; with coupling joints to pipes						
Inspection chamber; 5 nr 100 mm inlets; cast iron cover and frame						
475 x 585 mm deep	PC £65.41	2.30	12.63	72.83	nr	**85.46**
475 x 930 mm deep	PC £76.81	2.50	13.73	84.80	nr	**98.53**
Accessories in vitrified clay; set in concrete; with cement and sand (1:2) joints to pipes						
Rainwater shoes; with 250 x 100 mm oval access						
100 mm	PC £12.05	0.50	2.75	13.66	nr	**16.41**
150 mm	PC £20.30	0.60	3.29	22.78	nr	**26.07**
Gully fittings; comprising low back trap and square hopper; square gully grid						
100 mm outlet; 150 x 150 mm grid	PC £13.25	1.00	5.49	22.64	nr	**28.13**
150 mm outlet; 225 x 225 mm grid	PC £27.80	1.30	7.14	40.00	nr	**47.14**
Add to above for						
100 mm back inlet	PC £8.92	-	-	9.37	nr	**9.37**
100 mm vertical back inlet	PC £8.92	-	-	9.37	nr	**9.37**
100 mm raising pieces	PC £2.59	0.30	1.65	2.81	nr	**4.46**
Yard gully (mud); trapped with rodding eye; galvanized square bucket; stopper; square hinged grate and frame						
100 mm outlet; 225 x 225 mm grate	PC £52.63	3.00	16.47	59.03	nr	**75.50**
150 mm outlet; 300 x 300 mm grate	PC £95.36	4.00	21.97	105.28	nr	**127.25**
Yard gully (garage); trapped with rodding eye; galvanized perforated round bucket; stopper; round hinged grate and frame						
100 mm outlet; 273 mm grate	PC £64.38	3.00	16.47	72.29	nr	**88.76**
150 mm outlet; 368 mm grate	PC £111.15	4.00	21.97	122.32	nr	**144.29**
Road gully; trapped with rodding eye and stopper (grate measured elsewhere)						
300 x 600 x 100 mm outlet	PC £33.42	3.30	18.12	45.78	nr	**63.90**
300 x 600 x 150 mm outlet	PC £33.42	3.30	18.12	45.78	nr	**63.90**
400 x 750 x 150 mm outlet	PC £40.50	4.00	21.97	58.74	nr	**80.71**
450 x 900 x 150 mm outlet	PC £55.26	5.00	27.46	77.47	nr	**104.93**

R DISPOSAL SYSTEMS Including overheads and profit at 5.00%		Labour hours	Labour £	Material £	Unit	Total rate £
R12 DRAINAGE BELOW GROUND - cont'd						
Accessories in vitrified clay; set in concrete; with cement and sand (1:2) joints to pipes - cont'd						
Grease trap; with internal access; galvanized perforated bucket; lid and frame						
450 x 300 x 525 mm deep; 100 mm outlet	PC £247.98	3.50	19.22	272.07	nr	**291.29**
600 x 450 x 600 mm deep; 100 mm outlet	PC £314.74	4.20	23.06	346.32	nr	**369.38**
Interceptor; trapped with inspection arm; lever locking stopper; chain and staple; cement and sand (1:2) joints to pipes; building in, and cutting and fitting brickwork around						
100 mm outlet; 100 mm inlet	PC £42.03	4.00	21.97	53.52	nr	**75.49**
150 mm outlet; 150 mm inlet	PC £59.60	4.50	24.71	73.81	nr	**98.52**
225 mm outlet; 225 mm inlet	PC £142.17	5.00	27.46	163.45	nr	**190.91**
Accessories: grates and covers						
Aluminium alloy gully grids; set in position						
125 x 125 mm	PC £1.52	0.10	0.55	1.60	nr	**2.15**
150 x 150 mm	PC £1.52	0.10	0.55	1.60	nr	**2.15**
225 x 225 mm	PC £4.53	0.10	0.55	4.76	nr	**5.31**
140 mm dia (for 100 mm)	PC £1.52	0.10	0.55	1.60	nr	**2.15**
197 mm dia (for 150 mm)	PC £2.36	0.10	0.55	2.48	nr	**3.03**
284 mm dia (for 225 mm)		0.10	0.55	5.32	nr	**5.87**
Aluminium alloy sealing plates and frames; set in cement and sand (1:3)						
150 x 150 mm	PC £5.88	0.25	1.37	6.26	nr	**7.63**
225 x 225 mm	PC £10.70	0.25	1.37	11.32	nr	**12.69**
254 x 150 mm; for access fittings	PC £7.47	0.25	1.37	7.93	nr	**9.30**
140 mm dia (for 100 mm)	PC £4.76	0.25	1.37	5.08	nr	**6.45**
190 mm dia (for 150 mm)	PC £6.85	0.25	1.37	7.28	nr	**8.65**
273 mm dia (for 225 mm)	PC £10.96	0.25	1.37	11.68	nr	**13.05**
Coated cast iron heavy duty road gratings and frames; BS 497 Tables 6 and 7; bedding and pointing in cement and sand (1:3); one course half brick thick wall in semi-engineering bricks in cement mortar (1:3)						
475 x 475 mm; grade A, ref GA1-450 (131 kg)	PC £59.75	2.50	13.73	64.84	nr	**78.57**
400 x 350 mm; grade A, ref GA2-325 (99 kg)	PC £50.27	2.50	13.73	54.52	nr	**68.25**
500 x 350 mm; grade A, ref GA2-325 (124 kg)	PC £71.30	2.50	13.73	76.78	nr	**90.51**
White vitreous clay floor channels; bedded, floor channel						
100 mm half round section floor channel	PC £32.32	0.60	3.29	36.09	m	**39.38**
Extra for						
stop end	PC £19.64	0.40	2.20	20.63	nr	**22.83**
angle	PC £29.47	0.60	3.29	30.94	nr	**34.23**
tee piece	PC £29.47	0.80	4.39	30.94	nr	**35.33**
stop end outlet	PC £25.39	0.50	2.75	26.66	nr	**29.41**
150 mm half round section floor channel	PC £35.87	0.75	4.12	40.22	m	**44.34**
Extra for						
stop end	PC £21.80	0.50	2.75	22.89	nr	**25.64**
angle	PC £32.74	0.75	4.12	34.38	nr	**38.50**
tee piece	PC £21.64	1.00	5.49	34.38	nr	**39.87**
stop end outlet	PC £40.36	0.60	3.29	42.38	nr	**45.67**

R DISPOSAL SYSTEMS Including overheads and profit at 5.00%		Labour hours	Labour £	Material £	Unit	Total rate £
Channel sump outlet for 150 mm channel						
one inlet	PC £46.03	1.20	6.59	48.33	nr	**54.92**
two inlets	PC £49.04	1.50	8.24	51.49	nr	**59.73**
230 mm half round section floor channel	PC £57.16	1.00	5.49	64.59	m	**70.08**
Extra for						
stop end	PC £34.90	0.65	3.57	36.64	nr	**40.21**
angle	PC £52.30	1.00	5.49	54.92	nr	**60.41**
tee piece	PC £58.39	1.35	7.41	61.31	nr	**68.72**
stop end outlet	PC £40.92	0.80	4.39	42.97	nr	**47.36**
100 mm block floor channel	PC £37.46	0.60	3.29	41.52	m	**44.81**
Extra for						
stop end	PC £21.64	0.40	2.20	22.72	nr	**24.92**
angle	PC £34.27	0.60	3.29	35.99	nr	**39.28**
tee piece	PC £34.27	0.80	4.39	35.99	nr	**40.38**
stop end outlet	PC £28.33	0.50	2.75	29.74	nr	**32.49**
150 mm block floor channel	PC £41.68	0.75	4.12	46.39	m	**50.51**
Extra for						
stop end	PC £25.37	0.50	2.75	26.64	nr	**29.39**
angle	PC £38.08	0.75	4.12	39.98	nr	**44.10**
tee piece	PC £38.08	1.00	5.49	39.98	nr	**45.47**
stop end outlet	PC £31.49	0.60	3.29	33.06	nr	**36.35**
150 mm rebated block floor channel		0.75	4.12	50.50	m	**54.62**
Extra for						
stop end	PC £27.86	0.50	2.75	29.25	nr	**32.00**
angle	PC £41.84	0.75	4.12	43.94	nr	**48.06**
tee piece	PC £41.84	1.00	5.49	43.94	nr	**49.43**
stop end outlet	PC £33.87	0.60	3.29	35.56	nr	**38.85**
Accessories; channel gratings and connectors						
Galvanized cast iron medium duty square mesh gratings; bedding and pointing in cement and sand (1:3)						
138 x 13 mm; to suit 100 mm wide channel	PC £41.01	0.70	3.84	43.06	m	**46.90**
180 x 13 mm; to suit 150 mm wide channel	PC £41.01	0.90	4.94	43.06	m	**48.00**
Galvanized cast iron medium duty square mesh gratings and frame; galvanized cast iron angle bearers; bedding and pointing in cement and sand (1:3); cutting and pinning lugs to concrete						
148 x 13 mm; to suit 100 mm wide channel	PC £66.49	1.70	9.34	69.82	m	**79.16**
190 x 13 mm; to suit 150 mm wide channel	PC £66.49	1.90	10.43	69.82	m	**80.25**
Chromium plated brass domed outlet grating; threaded joint to connector (measured elsewhere)						
50 mm	PC £15.14	0.30	1.65	15.89	nr	**17.54**
63 mm	PC £20.07	0.30	1.65	21.08	nr	**22.73**
Cast iron connector; cement and sand (1:2) joint to drain pipe						
75 x 300 mm long; screwed 50 mm	PC £11.82	0.40	2.20	12.49	nr	**14.69**
75 x 300 mm long; screwed 63 mm	PC £13.81	0.40	2.20	14.58	nr	**16.78**
Class M tested concrete centrifugally spun pipes and fittings; with flexible joints; BS 5911 Part 1						
300 mm pipes; laid straight	PC £11.75	0.70	3.84	12.96	m	**16.80**
Extra for						
bend	PC £58.77	0.70	3.84	48.59	nr	**52.43**
300 x 100 mm single junction	PC £15.30	0.50	2.75	16.86	nr	**19.61**

R DISPOSAL SYSTEMS Including overheads and profit at 5.00%		Labour hours	Labour £	Material £	Unit	Total rate £
R12 DRAINAGE BELOW GROUND - cont'd						
Class M tested concrete centrifugally spun pipes and fittings; with flexible joints; BS 5911 Part 1 - cont'd						
450 mm pipes; laid straight	PC £17.58	1.10	6.04	19.38	m	**25.42**
Extra for						
bend	PC £87.88	1.10	6.04	72.67	nr	**78.71**
450 x 150 mm single junction	PC £18.90	0.70	3.84	20.83	nr	**24.67**
600 mm pipes; laid straight	PC £23.78	1.60	8.79	26.22	m	**35.01**
Extra for						
bend	PC £118.89	1.60	8.79	98.31	nr	**107.10**
600 x 150 mm single junction	PC £19.80	0.90	4.94	21.83	nr	**26.77**
900 mm pipes; laid straight	PC £47.87	2.80	15.38	52.77	m	**68.15**
Extra for						
bend	PC £239.36	2.80	15.38	197.92	nr	**213.30**
900 x 225 mm single junction	PC £36.00	1.10	6.04	39.69	nr	**45.73**
1200 mm pipes; laid straight	PC £82.36	4.00	21.97	90.80	m	**112.77**
Extra for						
bend	PC £411.80	4.00	21.97	340.50	nr	**362.47**
1200 x 300 mm single junction	PC £42.30	1.60	8.79	46.63	nr	**55.42**
Accessories in precast concrete; top set in with rodding eye and stopper; cement and sand (1:2) joint to pipe						
Concrete road gully; BS 556 Part 2; trapped with rodding eye and stopper; cement and sand (1:2) joint to pipe						
450 mm dia x 1050 mm deep; 100 or 150 mm outlet		4.75	26.08	36.73	nr	**62.81**
uPVC pipes and fittings; BS 4660; with lip seal coupling joints						
110 mm pipes						
laid straight	PC £12.15/6m	0.20	1.10	2.56	m	**3.66**
in runs not exceeding 3 m long	PC £6.07/3m	0.27	1.48	3.10	m	**4.58**
Extra for						
bend; short radius	PC £5.55	0.16	0.88	6.01	nr	**6.89**
bend; long radius	PC £8.77	0.16	0.88	9.00	nr	**9.88**
spigot/socket bend	PC £4.76	0.20	1.10	6.94	nr	**8.04**
access bend	PC £8.69	0.16	0.88	8.92	nr	**9.80**
socket plug	PC £2.20	0.05	0.27	2.43	nr	**2.70**
variable bend	PC £6.07	0.16	0.88	8.28	nr	**9.16**
inspection pipe	PC £10.54	0.16	0.88	13.20	nr	**14.08**
adaptor to clay	PC £4.48	0.16	0.88	4.92	nr	**5.80**
WC connector	PC £4.00	0.20	1.10	6.19	nr	**7.29**
single junction	PC £7.42	0.20	1.10	7.51	nr	**8.61**
inspection junction	PC £15.40	0.20	1.10	16.30	nr	**17.40**
slip coupling	PC £3.31	0.10	0.55	3.65	nr	**4.20**
160 mm pipes						
laid straight	PC £23.74/6m	0.24	1.32	5.02	m	**6.34**
in runs not exceeding 3 m long	PC £11.87/3m	0.32	1.76	6.07	m	**7.83**
Extra for						
bend; short radius	PC £10.64	0.20	1.10	11.52	nr	**12.62**
spigot/socket bend	PC £10.13	0.25	1.37	14.52	nr	**15.89**
socket plug	PC £3.82	0.06	0.33	4.21	nr	**4.54**
inspection pipe	PC £13.69	0.20	1.10	18.22	nr	**19.32**
adaptor to clay	PC £7.61	0.20	1.10	8.26	nr	**9.36**
level invert taper	PC £6.49	0.24	1.32	10.28	nr	**11.60**
single junction	PC £19.54	0.24	1.32	20.24	nr	**21.56**
inspection junction	PC £25.08	0.24	1.32	26.34	nr	**27.66**
slip coupling	PC £7.06	0.12	0.66	7.78	nr	**8.44**

R DISPOSAL SYSTEMS Including overheads and profit at 5.00%		Labour hours	Labour £	Material £	Unit	Total rate £
Accessories in uPVC; with lip seal coupling joints to pipes (unless otherwise described)						
Access cap assembly						
110 mm	PC £5.40	0.10	0.55	7.88	nr	**8.43**
Rodding eye 200 mm; sealed cover	PC £13.87	0.50	2.75	23.71	nr	**26.46**
Gully fitting; comprising 'P' trap, square hopper 154 x 154 mm grate						
110 mm outlet	PC £9.40	0.90	4.94	24.07	nr	**29.01**
Shallow access pipe assembly; 2 nr 110 mm inlets; light duty screw down cover and frame						
110 x 600 mm deep	PC £34.48	1.00	5.49	41.54	nr	**47.03**
Shallow branch access assembly; 3 nr 110 mm inlets; light duty screw down cover and frame						
110 x 600 mm deep	PC £38.11	1.20	6.59	43.41	nr	**50.00**
Inspection chamber; 450 mm dia; 940 mm deep; heavy duty screw down cover and frame						
4 nr 110 mm outlet/inlets	PC £73.27	1.80	9.88	124.58	nr	**134.46**
Kerb to gullies; class B engineering bricks on edge to three sides in cement mortar (1:3) rendering in cement mortar (1:3) to top and two sides and skirting to brickwork 230 mm high; dishing in cement mortar (1:3) to gully; steel trowelled						
230 x 230 mm internally		1.40	7.69	1.37	nr	**9.06**
Mechanical excavation						
Excavating manholes; not exceeding						
1 m deep		0.21	1.23	3.75	m3	**4.98**
2 m deep		0.23	1.35	4.11	m3	**5.46**
4 m deep		0.27	1.58	4.82	m3	**6.40**
Hand excavation						
Excavating manholes; not exceeding						
1 m deep		3.30	19.37	-	m3	**19.37**
2 m deep		3.90	22.89	-	m3	**22.89**
4 m deep		5.00	29.35	-	m3	**29.35**
Earthwork support (average 'risk' prices)						
Not exceeding 2 m between opposing faces; not exceeding						
1 m deep		0.15	0.88	0.48	m2	**1.36**
2 m deep		0.19	1.12	0.58	m2	**1.70**
4 m deep		0.24	1.41	0.73	m2	**2.14**
Disposal (mechanical)						
Excavated material; depositing on site in spoil heaps						
average 50 m distant		0.07	0.41	1.25	m3	**1.66**
Removing from site to tip not exceeding 13 km (using lorries)		-	-	-	m3	**7.87**
Disposal (hand)						
Excavated material; depositing on site in spoil heaps						
average 50 m distant		1.30	7.63	-	m3	**7.63**
Removing from site to tip not exceeding 13 km (using lorries)		1.00	5.87	9.84	m3	**15.71**

R DISPOSAL SYSTEMS Including overheads and profit at 5.00%		Labour hours	Labour £	Material £	Unit	Total rate £
R12 DRAINAGE BELOW GROUND - cont'd						
Mechanical filling						
Excavated material filling to excavations		0.15	0.88	1.79	m3	**2.67**
Hand filling						
Excavated material filling to excavations		1.00	5.87	-	m3	**5.87**
In situ ready mixed concrete; normal Portland cement; mix 11.50 N/mm2 - 40 mm aggregate (1:3:6); PC £40.48/m3;						
Beds						
not exceeding 150 mm thick		3.10	18.88	44.63	m3	**63.51**
150 - 450 mm thick		2.30	14.01	44.63	m3	**58.64**
over 450 mm thick		1.90	11.57	44.63	m3	**56.20**
Ready mixed in situ concrete; normal Portland cement; mix 21.00 N/mm2-20 mm aggregate (1:2:4); PC £43.10/m3						
Beds						
not exceeding 150 mm thick		3.10	18.88	47.52	m3	**66.40**
150 - 450 mm thick		2.30	14.01	47.52	m3	**61.53**
over 450 mm thick		1.90	11.57	47.52	m3	**59.09**
Site mixed in situ concrete; normal Portland cement; mix 26.00 N/mm2-20 mm aggregate (1:1.5:3); (small quantities); PC £50.57/m3						
Benching in bottoms						
150 - 450 mm average thick		7.00	50.72	53.10	m3	**103.82**
Reinforced site mixed in situ concrete; normal Portland cement; mix 21.00 N/mm2-20mm aggregate (1:2:4); (small quantities); PC £47.71/m3						
Isolated cover slabs						
not exceeding 150 mm thick		6.00	36.54	50.10	m3	**86.64**
Reinforcement; fabric to BS 4483; lapped; in beds or suspended slabs						
Ref A98 (1.54 kg/m2)	PC £0.55	0.12	0.85	0.69	m2	**1.54**
Ref A142 (2.22 kg/m2)	PC £0.69	0.12	0.85	0.87	m2	**1.72**
Ref A193 (3.02 kg/m2)	PC £0.94	0.12	0.85	1.18	m2	**2.03**
Formwork to in situ concrete						
Soffits of isolated cover slabs						
horizontal		2.85	20.29	5.46	m2	**25.75**
Edges of isolated cover slabs						
not exceeding 250 mm high		0.80	5.70	2.01	m	**7.71**
Precast concrete rectangular access and inspection chambers; 'Brooklyns' chambers or similar; comprising cover frame to receive manhole cover (priced elsewhere) intermediate wall sections and base section with cut outs;bedding;jointing and pointing in cement mortar (1:3) on prepared bed						
Drainage chamber; size 600 x 450 mm internally; depth to invert						
600 mm deep		4.50	24.71	21.81	nr	**46.52**
900 mm deep		6.00	32.95	27.75	nr	**60.70**

R DISPOSAL SYSTEMS Including overheads and profit at 5.00%		Labour hours	Labour £	Material £	Unit	Total rate £
Drainage chamber; 1200 x 750 mm reducing to 600 x 600 mm; no base unit; depth to invert						
1050 mm deep		7.50	41.19	79.30	nr	**120.49**
1650 mm deep		9.00	49.42	122.39	nr	**171.81**
2250 mm deep		11.00	64.56	165.47	nr	**230.03**
Precast concrete circular manhole rings; BS 5911 Part 1; bedding, jointing and pointing in cement mortar (1:3) on prepared bed						
Chamber or shaft rings; plain						
675 mm	PC £19.98	5.00	27.46	21.38	m	**48.84**
900 mm	PC £29.16	5.50	30.20	31.25	m	**61.45**
1050 mm	PC £37.91	6.50	35.69	40.66	m	**76.35**
1200 mm	PC £50.13	7.50	41.19	53.81	m	**95.00**
Chamber or shaft rings; reinforced						
1350 mm	PC £77.85	8.50	46.68	83.63	m	**130.31**
1500 mm	PC £86.90	9.50	52.17	93.85	m	**146.02**
1800 mm	PC £110.25	12.00	65.90	119.76	m	**185.66**
2100 mm	PC £197.10	15.00	82.37	212.35	m	**294.72**
Extra for step irons built in	PC £2.07	0.15	0.82	2.17	nr	**2.99**
Taper pieces; 675 mm high; from 675 mm to						
900 mm	PC £29.61	4.50	24.71	31.45	nr	**56.16**
1050 mm	PC £39.38	5.00	27.46	41.88	nr	**69.34**
1200 mm	PC £52.83	6.00	32.95	56.19	nr	**89.14**
Taper pieces; 900 mm high; from 675 mm to						
1350 mm	PC £93.29	9.50	52.17	99.66	nr	**151.83**
1500 mm	PC £109.13	11.00	60.41	116.92	nr	**177.33**
1800 mm	PC £141.48	14.00	76.88	152.15	nr	**229.03**
Heavy duty cover slabs; to suit rings						
675 mm	PC £15.88	2.75	15.10	16.90	nr	**32.00**
900 mm	PC £26.64	3.00	16.47	28.33	nr	**44.80**
1050 mm	PC £34.88	3.50	19.22	37.16	nr	**56.38**
1200 mm	PC £48.38	4.00	21.97	51.51	nr	**73.48**
1350 mm	PC £59.67	4.50	24.71	63.78	nr	**88.49**
1500 mm	PC £82.94	5.00	27.46	88.66	nr	**116.12**
1800 mm	PC £125.55	6.00	32.95	134.26	nr	**167.21**
2100 mm	PC £260.28	7.00	38.44	276.53	nr	**314.97**
Common bricks; PC £120.00/1000; in cement mortar (1:3)						
Walls to manholes						
one brick thick		2.25	22.80	18.62	m2	**41.42**
one and a half brick thick		3.20	32.42	28.00	m2	**60.42**
Projections of footings						
two brick thick		4.35	44.08	37.24	m2	**81.32**
Class A engineering bricks; PC £340.00/1000 in cement mortar (1:3)						
Walls to manholes						
one brick thick		2.50	25.33	47.72	m2	**73.05**
one and a half brick thick		3.50	35.46	71.65	m2	**107.11**
Projections of footings						
two brick thick		4.80	48.64	95.45	m2	**144.09**
Class B engineering bricks; PC £175.00/1000 in cement mortar (1:3)						
Walls to manholes						
one brick thick		2.50	25.33	25.90	m2	**51.23**
one and a half brick thick		3.50	35.46	38.92	m2	**74.38**
Projections of footings						
two brick thick		4.80	48.64	51.79	m2	**100.43**

R DISPOSAL SYSTEMS Including overheads and profit at 5.00%		Labour hours	Labour £	Material £	Unit	Total rate £
R12 DRAINAGE BELOW GROUND - cont'd						
Brickwork sundries						
Extra over for fair face; flush smooth pointing						
manhole walls		0.20	2.03	-	m2	2.03
Building ends of pipes into brickwork; making good fair face or rendering						
not exceeding 55 mm nominal size		0.10	1.01	-	nr	1.01
55 - 110 mm nominal size		0.15	1.52	-	nr	1.52
over 110 mm nominal size		0.20	2.03	-	nr	2.03
Step irons; BS 1247; malleable; galvanized; building into joints						
general purpose pattern		0.15	1.52	4.01	nr	5.53
Cement and sand (1:3) in situ finishings; steel trowelled						
13 mm work to manhole walls; one coat; to brickwork base over 300 mm wide		0.70	6.27	0.76	m2	7.03
Manhole accessories in cast iron						
Petrol trapping bend; coated; 375 x 750 mm; building into brickwork						
100 mm	PC £52.78	1.35	8.90	57.68	nr	66.58
150 mm	PC £85.26	1.75	11.54	93.32	nr	104.86
225 mm	PC £157.98	2.85	18.79	171.82	nr	190.61
Cast iron inspection chambers; with bolted flat covers; BS 437; bedded in cement mortar (1:3); caulked lead joints to pipes						
100 x 100 mm; ref 010; no branches	PC £75.05	1.00	6.59	81.51	nr	88.10
100 x 100 mm; ref 110; one branch	PC £95.55	1.20	7.91	104.39	nr	112.30
100 x 100 mm; ref 111; one branch either side	PC £123.65	1.40	9.23	135.03	nr	144.26
100 x 100 mm; ref 212; two branches either side	PC £200.41	2.05	13.52	217.88	nr	231.40
100 x 100 mm; ref 313; three branches either side	PC £272.79	2.70	17.80	296.15	nr	313.95
150 x 100 mm; ref 110; one branch	PC £129.55	1.60	10.55	142.16	nr	152.71
150 x 100 mm; ref 111; one branch either side	PC £166.20	1.80	11.87	181.61	nr	193.48
150 x 100 mm; ref 212; two branches either side	PC £238.53	2.55	16.81	259.81	nr	276.62
150 x 100 mm; ref 313; three branches either side	PC £323.85	3.30	21.76	351.66	nr	373.42
150 x 150 mm; ref 212; two branches either side	PC £332.42	2.75	18.13	361.48	nr	379.61
225 x 100 mm; ref 212; two branches either side	PC £477.46	3.30	21.76	514.04	nr	535.80
225 x 100 mm; ref 313; three branches either side	PC £600.00	4.20	27.69	644.97	nr	672.66
Cast iron inspection chambers; with bolted flat covers; BS 437; bedded in cement mortar (1:3); with mechanical coupling joints						
100 x 100 mm; ref 110; one branch	PC £55.67	1.05	6.92	76.35	nr	83.27
100 x 100 mm; ref 111; one branch either side	PC £71.71	1.55	10.22	101.80	nr	112.02
150 x 100 mm; ref 110; one branch	PC £83.50	1.25	8.24	107.75	nr	115.99
150 x 100 mm; ref 111; one branch either side		1.80	11.87	126.23	nr	138.10
150 x 150 mm; ref 110; one branch	PC £106.90	1.35	8.90	134.14	nr	143.04
150 x 150 mm; ref 111; one branch either side	PC £117.91	1.90	12.53	156.14	nr	168.67

R DISPOSAL SYSTEMS Including overheads and profit at 5.00%		Labour hours	Labour £	Material £	Unit	Total rate £
Access covers and frames; coated; BS 497 tables 1-5; bedding frame in cement and sand (1:3); cover in grease and sand						
Grade C; light duty; rectangular single seal solid top						
450 x 450 mm; ref MC1-45/45 (31 kg)	PC £21.87	1.50	8.24	24.14	nr	**32.38**
600 x 450 mm; ref MC1-60/45 (32 kg)	PC £22.01	1.50	8.24	24.28	nr	**32.52**
600 x 600 mm; ref MC1-60/60 (61 kg)	PC £48.87	1.50	8.24	52.48	nr	**60.72**
Grade C; light duty; rectangular single seal recessed						
450 x 450 mm; ref MC1R-45/45 (43 kg)	PC £35.79	1.50	8.24	38.75	nr	**46.99**
600 x 450 mm; ref MC1R-60/45 (43 kg)	PC £45.50	1.50	8.24	48.95	nr	**57.19**
600 x 600 mm; ref MC1R-60/60 (53 kg)	PC £61.67	1.50	8.24	65.92	nr	**74.16**
Grade C; light duty; rectangular double seal solid top						
450 x 450 mm; ref MC2-45/45 (51 kg)	PC £33.50	1.50	8.24	36.35	nr	**44.59**
600 x 450 mm; ref MC2-60/45 (51 kg)	PC £32.67	1.50	8.24	35.47	nr	**43.71**
600 x 600 mm; ref MC2-60/60 (83 kg)	PC £61.88	1.50	8.24	66.15	nr	**74.39**
Grade C; light duty; rectangular double seal recessed						
450 x 450 mm; ref MC2R-45/45 56 kg)	PC £55.66	1.50	8.24	59.61	nr	**67.85**
600 x 450 mm; ref MC2R-60/45 (71 kg)	PC £77.44	1.50	8.24	82.49	nr	**90.73**
600 x 600 mm; ref MC2R-60/60 (75 kg)	PC £74.56	1.50	8.24	79.46	nr	**87.70**
Grade B; medium duty; circular single seal solid top						
500 mm; ref MB2-50 (106 kg)	PC £61.84	2.00	10.98	66.10	nr	**77.08**
550 mm; ref MB2-55 (112 kg)	PC £66.11	2.00	10.98	70.59	nr	**81.57**
600 mm; ref MB2-60 (134 kg)	PC £69.59	2.00	10.98	74.24	nr	**85.22**
Grade B; medium duty; rectangular single seal solid top						
600 x 450 mm; ref MB2-60/45 (135 kg)	PC £60.03	2.00	10.98	64.20	nr	**75.18**
600 x 600 mm; ref MB2-60/60 (170 kg)	PC £78.05	2.00	10.98	83.12	nr	**94.10**
Grade B; medium duty; rectangular single seal recessed						
600 x 450 mm; ref MB2R-60/45 (145 kg)	PC £82.45	2.00	10.98	87.75	nr	**98.73**
600 x 600 mm; ref MB2R-60/60 (171 kg)	PC £105.27	2.00	10.98	111.71	nr	**122.69**
Grade B; 'Chevron'; medium duty; double triangular solid top						
550 mm; ref MB1-55 (125 kg)	PC £60.15	2.00	10.98	64.33	nr	**75.31**
600 mm; ref MB1-60 (140 kg)	PC £71.94	2.00	10.98	76.70	nr	**87.68**
Grade A; heavy duty; single triangular solid top						
550 x 455 mm; ref MA-T (196 kg)	PC £86.08	2.50	13.73	91.56	nr	**105.29**
Grade A; 'Chevron'; heavy duty double triangular solid top						
500 mm; ref MA-50 (164 kg)	PC £100.57	3.00	16.47	106.77	nr	**123.24**
550 mm; ref MA-55 (176 kg)	PC £98.06	3.00	16.47	104.14	nr	**120.61**
600 mm; ref MA-60 (230 kg)		3.00	16.47	109.84	nr	**126.31**

R DISPOSAL SYSTEMS Including overheads and profit at 5.00%		Labour hours	Labour £	Material £	Unit	Total rate £
R12 DRAINAGE BELOW GROUND - cont'd						
British Standard best quality vitrified clay channels; bedding and jointing in cement and sand (1:2)						
Half section straight						
100 mm x 1.00 m long	PC £2.06	0.80	4.39	2.27	nr	**6.66**
150 mm x 1.00 m long	PC £3.42	1.00	5.49	3.77	nr	**9.26**
225 mm x 1.00 m long	PC £7.70	1.30	7.14	8.48	nr	**15.62**
300 mm x 1.00 m long	PC £15.27	1.60	8.79	16.83	nr	**25.62**
Half section bend						
100 mm	PC £2.11	0.60	3.29	2.33	nr	**5.62**
150 mm	PC £3.50	0.75	4.12	3.86	nr	**7.98**
225 mm	PC £11.78	1.00	5.49	12.99	nr	**18.48**
300 mm	PC £23.39	1.20	6.59	25.79	nr	**32.38**
Half section taper straight						
150 mm	PC £8.86	0.70	3.84	9.76	nr	**13.60**
225 mm	PC £19.76	0.90	4.94	21.79	nr	**26.73**
300 mm	PC £39.08	1.10	6.04	43.08	nr	**49.12**
Half section taper bend						
150 mm	PC £13.47	0.90	4.94	14.85	nr	**19.79**
225 mm	PC £38.79	1.15	6.32	42.77	nr	**49.09**
300 mm	PC £76.76	1.40	7.69	84.63	nr	**92.32**
Three quarter section branch bend						
100 mm	PC £4.80	0.50	2.75	5.29	nr	**8.04**
150 mm	PC £8.09	0.75	4.12	8.91	nr	**13.03**
225 mm	PC £29.54	1.00	5.49	32.57	nr	**38.06**
300 mm	PC £58.63	1.33	7.30	64.64	nr	**71.94**
uPVC channels; with solvent weld or lip seal coupling joints; bedding in cement and sand						
Half section cut away straight; with coupling either end						
110 mm	PC £14.76	0.30	1.65	16.19	nr	**17.84**
160 mm	PC £19.81	0.40	2.20	21.71	nr	**23.91**
Half section cut away long radius bend; with coupling either end						
110 mm	PC £16.76	0.30	1.65	18.66	nr	**20.31**
160 mm	PC £24.82	0.40	2.20	27.76	nr	**29.96**
Half section straight channel adaptor; with one coupling						
110 mm	PC £5.79	0.25	1.37	7.64	nr	**9.01**
160 mm	PC £7.69	0.33	1.81	10.95	nr	**12.76**
Half section cut away bend						
110 mm	PC £10.99	0.40	2.20	12.26	nr	**14.46**
Half section bend						
110 mm	PC £2.69	0.33	1.81	3.11	nr	**4.92**
160 mm	PC £5.23	0.50	2.75	6.09	nr	**8.84**
Half section channel connector						
110 mm	PC £1.30	0.08	0.44	1.70	nr	**2.14**
160 mm	PC £3.04	0.10	0.55	4.00	nr	**4.55**
Half section channel junction						
110 mm	PC £4.38	0.50	2.75	4.97	nr	**7.72**
160 mm	PC £8.27	0.60	3.29	9.44	nr	**12.73**
polypropylene slipper bend						
110 mm	PC £5.80	0.40	2.20	6.54	nr	**8.74**

R DISPOSAL SYSTEMS Including overheads and profit at 5.00%		Labour hours	Labour £	Material £	Unit	Total rate £
R13 LAND DRAINAGE						
Hand excavation of trenches to receive land drain pipes; grading bottoms; earthwork support; filling to within 150 mm of surface filling to within 150 mm of surface with gravel rejects; remainder filled with excavated material and compacting; disposal of surplus soil; spreading on site average 50 m						
Pipes not exceeding 200 mm; average depth						
0.75 m deep		1.70	9.98	8.91	m	**18.89**
1.00 m deep		2.25	13.21	11.86	m	**25.07**
1.25 m deep		3.15	18.49	15.07	m	**33.56**
1.50 m deep		5.40	31.70	18.99	m	**50.69**
1.75 m deep		6.40	37.56	22.47	m	**60.03**
2.00 m deep		7.40	43.43	25.93	m	**69.36**
Surplus excavated material						
Removing from site to tip not exceeding 13 km (using lorries)						
machine loaded		-	-	-	m3	**7.87**
hand loaded		1.00	5.87	9.84	m3	**15.71**
Clay field drain pipes; BS 1196; unjointed						
Pipes; laid straight						
75 mm	PC £23.00/100	0.20	1.10	0.84	m	**1.94**
100 mm	PC £41.00/100	0.25	1.37	1.49	m	**2.86**
150 mm	PC £84.00/100	0.30	1.65	3.06	m	**4.71**
Vitrified clay perforated sub-soil pipes; BS 65; 'Hepworths' 'Hepline' or similar						
Pipes; laid straight						
100 mm	PC £3.15	0.22	1.21	3.48	m	**4.69**
150 mm	PC £5.77	0.27	1.48	6.36	m	**7.84**
225 mm pipes	PC £10.60	0.36	1.98	11.69	m	**13.67**

S PIPED SUPPLY SYSTEMS
Including overheads and profit at 5.00%

ALTERNATIVE SERVICE PIPE AND FITTING PRICES

Copper pipes to BS 2871 (£/100 m)

	£		£		£		£
Table X							
6 mm	58.68	10 mm	99.90	67 mm	1442.70	108 mm	2929.50
8 mm	79.74	12 mm	124.20	76 mm	2051.10	133 mm	3618.00
Table Y							
6 mm	78.12	10 mm	135.00	67 mm	2405.70	108 mm	4824.90
8 mm	104.40	12 mm	167.40	76 mm	2714.40		
Table Z							
15 mm	108.00	35 mm	569.70	54 mm	911.70	76 mm	1826.10
22 mm	208.80	42 mm	699.30	67 mm	1288.80	108 mm	2607.30
28 mm	270.90						

PVC Class E cold water pressure system (£/m/each)
(sizes shown as nearest equivalent metric sizes)

	£		£		£
pipe-per m-13 mm	0.77	end cap-13 mm	0.38	elbow-13 mm	0.56
-19 mm	1.07	-19 mm	0.42	-19 mm	0.66
-25 mm	1.39	-25 mm	0.49	-25 mm	0.83
-32 mm	2.24	-32 mm	0.76	-32 mm	1.58
-40 mm	2.95	-40 mm	1.18	-40 mm	1.98
coupling -13 mm	0.40	reducer-32 mm	1.13	tee -13 mm	0.63
-19 mm	0.45	-40 mm	1.34	-19 mm	0.78
-25 mm	0.52	MI.conn-13 mm	0.67	-25 mm	1.17
-32 mm	0.86	-19 mm	0.71	-32 mm	1.71
-40 mm	1.08	-25 mm	1.37	-40 mm	2.52
s.tank.cn.-19 mm	1.84	-32 mm	1.82		
-25 mm	2.03	-40 mm	1.98		

		Labour hours	Labour £	Material £	Unit	**Total rate £**
S10/S11 HOT AND COLD WATER						
Copper pipes; BS 2871 table X; capillary fittings; BS 864						
15 mm pipes; fixing with pipe clips; plugged and screwed	PC £118.80/100m	0.25	2.20	1.56	m	**3.76**
Extra for						
made bend		0.15	1.32	-	nr	**1.32**
fittings with one end		0.11	0.97	0.51	nr	**1.48**
fittings with two ends		0.17	1.50	0.26	nr	**1.76**
fittings with three ends		0.25	2.20	0.48	nr	**2.68**
fittings with four ends		0.35	3.08	3.37	nr	**6.45**
stop end	PC £0.49	0.11	0.97	0.51	nr	**1.48**
straight union coupling	PC £2.18	0.17	1.50	2.28	nr	**3.78**
copper to iron connector	PC £0.80	0.22	1.94	0.84	nr	**2.78**
elbow	PC £0.25	0.17	1.50	0.26	nr	**1.76**
backplate elbow	PC £1.59	0.35	3.08	1.67	nr	**4.75**
slow bend	PC £0.85	0.17	1.50	0.89	nr	**2.39**
tee; equal	PC £0.46	0.25	2.20	0.48	nr	**2.68**
tee; reducing	PC £1.80	0.25	2.20	1.89	nr	**4.09**
cross	PC £3.21	0.35	3.08	3.37	nr	**6.45**
straight tap connector	PC £0.72	0.13	1.14	0.76	nr	**1.90**
bent tap connector	PC £0.88	0.13	1.14	0.92	nr	**2.06**
straight tank connector; backnut	PC £1.55	0.25	2.20	1.62	nr	**3.82**

S PIPED SUPPLY SYSTEMS Including overheads and profit at 5.00%		Labour hours	Labour £	Material £	Unit	Total rate £
22 mm pipes; fixing with pipe clips; plugged and screwed	PC £234.00/100m	0.26	2.29	2.89	m	**5.18**
Extra for						
made bend		0.20	1.76	-	nr	**1.76**
fittings with one end		0.13	1.14	0.85	nr	**1.99**
fittings with two ends		0.22	1.94	0.53	nr	**2.47**
fittings with three ends		0.33	2.90	0.98	nr	**3.88**
fittings with four ends		0.44	3.87	4.35	nr	**8.22**
stop end	PC £0.81	0.13	1.14	0.85	nr	**1.99**
reducing coupling	PC £0.61	0.22	1.94	0.64	nr	**2.58**
straight union coupling	PC £3.36	0.22	1.94	3.53	nr	**5.47**
copper to iron connector	PC £1.33	0.31	2.73	1.40	nr	**4.13**
elbow	PC £0.50	0.22	1.94	0.53	nr	**2.47**
backplate elbow	PC £3.31	0.44	3.87	3.47	nr	**7.34**
slow bend	PC £1.43	0.22	1.94	1.50	nr	**3.44**
tee; equal	PC £0.93	0.33	2.90	0.98	nr	**3.88**
tee; reducing	PC £0.91	0.33	2.90	0.95	nr	**3.85**
cross	PC £4.15	0.44	3.87	4.35	nr	**8.22**
straight tap connector	PC £1.11	0.17	1.50	1.17	nr	**2.67**
bent tap connector	PC £1.68	0.17	1.50	1.76	nr	**3.26**
straight tank connector; backnut	PC £2.29	0.33	2.90	2.41	nr	**5.31**
28 mm pipes; fixing with pipe clips; plugged and screwed	PC £302.40/100m	0.29	2.55	3.72	m	**6.27**
Extra for						
made bend		0.25	2.20	-	nr	**2.20**
fittings with one end		0.15	1.32	1.66	nr	**2.98**
fittings with two ends		0.28	2.46	0.97	nr	**3.43**
fittings with three ends		0.41	3.61	2.05	nr	**5.66**
fittings with four ends		0.56	4.93	6.24	nr	**11.17**
stop end	PC £1.58	0.15	1.32	1.66	nr	**2.98**
reducing coupling	PC £1.36	0.28	2.46	1.43	nr	**3.89**
straight union coupling	PC £4.61	0.28	2.46	4.84	nr	**7.30**
copper to iron connector	PC £2.09	0.39	3.43	2.19	nr	**5.62**
elbow	PC £0.92	0.28	2.46	0.97	nr	**3.43**
slow bend	PC £2.39	0.28	2.46	2.51	nr	**4.97**
tee; equal	PC £1.96	0.41	3.61	2.05	nr	**5.66**
tee; reducing	PC £2.10	0.41	3.61	2.21	nr	**5.82**
cross	PC £5.94	0.56	4.93	6.24	nr	**11.17**
straight tank connector; backnut	PC £3.16	0.41	3.61	3.32	nr	**6.93**
35 mm pipes; fixing with pipe clips; plugged and screwed	PC £659.70/100m	0.33	2.90	7.94	m	**10.84**
Extra for						
made bend		0.30	2.64	-	nr	**2.64**
fittings with one end		0.17	1.50	2.55	nr	**4.05**
fittings with two ends		0.33	2.90	2.57	nr	**5.47**
fittings with three ends		0.46	4.05	4.41	nr	**8.46**
stop end	PC £2.43	0.17	1.50	2.55	nr	**4.05**
reducing coupling	PC £2.00	0.33	2.90	2.10	nr	**5.00**
straight union coupling	PC £6.61	0.33	2.90	6.94	nr	**9.84**
copper to iron connector	PC £3.26	0.44	3.87	3.42	nr	**7.29**
elbow	PC £2.45	0.33	2.90	2.57	nr	**5.47**
bend; 91.5 degrees	PC £4.30	0.33	2.90	4.51	nr	**7.41**
tee; equal	PC £4.20	0.46	4.05	4.41	nr	**8.46**
tee; reducing	PC £3.99	0.46	4.05	4.19	nr	**8.24**
pitcher tee; equal or reducing	PC £6.98	0.46	4.05	7.33	nr	**11.38**
straight tank connector; backnut	PC £4.25	0.46	4.05	4.46	nr	**8.51**

S PIPED SUPPLY SYSTEMS Including overheads and profit at 5.00%		Labour hours	Labour £	Material £	Unit	Total rate £
S10/S11 HOT AND COLD WATER - cont'd						
Copper pipes; BS 2871 table X; capillary fittings; BS 864 - cont'd						
42 mm pipes; fixing with pipe clips; plugged and screwed	PC £800.10/100m	0.38	3.34	9.82	m	**13.16**
Extra for						
made bend		0.40	3.52	-	nr	**3.52**
fittings with one end		0.19	1.67	3.32	nr	**4.99**
fittings with two ends		0.39	3.43	3.83	nr	**7.26**
fittings with three ends		0.52	4.58	6.49	nr	**11.07**
stop end	PC £3.16	0.19	1.67	3.32	nr	**4.99**
reducing coupling	PC £3.01	0.39	3.43	3.16	nr	**6.59**
straight union coupling	PC £9.37	0.39	3.43	9.84	nr	**13.27**
copper to iron connector	PC £3.94	0.50	4.40	4.14	nr	**8.54**
elbow	PC £3.65	0.39	3.43	3.83	nr	**7.26**
bend; 91.5 degrees	PC £6.86	0.39	3.43	7.21	nr	**10.64**
tee; equal	PC £6.18	0.52	4.58	6.49	nr	**11.07**
tee; reducing	PC £7.31	0.52	4.58	7.68	nr	**12.26**
pitcher tee; equal or reducing	PC £9.81	0.52	4.58	10.30	nr	**14.88**
straight tank connector; backnut	PC £5.39	0.52	4.58	5.66	nr	**10.24**
54 mm pipes; fixing with pipe clips; plugged and screwed	PC £1034.10/100m	0.45	3.96	13.07	m	**17.03**
Extra for						
made bend		0.55	4.84	-	nr	**4.84**
fittings with one end		0.21	1.85	4.91	nr	**6.76**
fittings with two ends		0.44	3.87	8.90	nr	**12.77**
fittings with three ends		0.57	5.02	11.96	nr	**16.98**
stop end	PC £4.67	0.21	1.85	4.91	nr	**6.76**
reducing coupling	PC £4.41	0.44	3.87	4.63	nr	**8.50**
straight union coupling	PC £15.24	0.44	3.87	16.00	nr	**19.87**
copper to iron connector	PC £6.54	0.55	4.84	6.87	nr	**11.71**
elbow	PC £8.48	0.44	3.87	8.90	nr	**12.77**
bend; 91.5 degrees	PC £10.25	0.44	3.87	10.76	nr	**14.63**
tee; equal	PC £11.39	0.57	5.02	11.96	nr	**16.98**
tee; reducing	PC £11.56	0.57	5.02	12.14	nr	**17.16**
pitcher tee; equal or reducing	PC £14.13	0.57	5.02	14.83	nr	**19.85**
straight tank connector; backnut	PC £8.22	0.57	5.02	8.63	nr	**13.65**
Copper pipes; BS 2871 table X; compression fittings; BS 864						
15 mm pipes; fixing with pipe clips; plugged and screwed	PC £118.80/100m	0.24	2.11	1.65	m	**3.76**
Extra for						
made bend		0.15	1.32	-	nr	**1.32**
fittings with one end		0.10	0.88	0.71	nr	**1.59**
fittings with two ends		0.15	1.32	0.57	nr	**1.89**
fittings with three ends		0.22	1.94	0.81	nr	**2.75**
stop end	PC £0.67	0.10	0.88	0.71	nr	**1.59**
straight coupling	PC £0.47	0.15	1.32	0.49	nr	**1.81**
female coupling	PC £0.47	0.20	1.76	0.49	nr	**2.25**
elbow	PC £0.55	0.15	1.32	0.57	nr	**1.89**
female wall elbow	PC £1.19	0.30	2.64	1.25	nr	**3.89**
slow bend	PC £2.22	0.15	1.32	2.33	nr	**3.65**
bent radiator union; chrome finish	PC £1.67	0.20	1.76	1.75	nr	**3.51**
tee; equal	PC £0.77	0.22	1.94	0.81	nr	**2.75**
straight swivel connector	PC £0.95	0.12	1.06	0.99	nr	**2.05**
bent swivel connector	PC £01.0	0.12	1.06	1.05	nr	**2.11**
tank coupling; locknut	PC £1.14	0.22	1.94	1.20	nr	**3.14**

S PIPED SUPPLY SYSTEMS Including overheads and profit at 5.00%		Labour hours	Labour £	Material £	Unit	Total rate £
22 mm pipes; fixing with pipe clips; plugged and screwed	PC £234.00/100m	0.25	2.20	3.02	m	**5.22**
Extra for						
made bend		0.20	1.76	-	nr	**1.76**
fittings with one end		0.12	1.06	0.92	nr	**1.98**
fittings with two ends		0.20	1.76	0.98	nr	**2.74**
fittings with three ends		0.30	2.64	1.41	nr	**4.05**
stop end	PC £0.87	0.12	1.06	0.92	nr	**1.98**
straight coupling	PC £0.78	0.20	1.76	0.82	nr	**2.58**
reducing set	PC £0.57	0.06	0.53	0.60	nr	**1.13**
female coupling	PC £0.70	0.28	2.46	0.74	nr	**3.20**
elbow	PC £0.93	0.20	1.76	0.98	nr	**2.74**
female wall elbow	PC £2.74	0.40	3.52	2.87	nr	**6.39**
slow bend	PC £3.65	0.20	1.76	3.83	nr	**5.59**
tee; equal	PC £1.34	0.30	2.64	1.41	nr	**4.05**
tee; reducing	PC £1.91	0.30	2.64	2.00	nr	**4.64**
straight swivel connector	PC £1.48	0.16	1.41	1.55	nr	**2.96**
bent swivel connector	PC £2.15	0.16	1.41	2.26	nr	**3.67**
tank coupling; locknut	PC £1.04	0.30	2.64	1.09	nr	**3.73**
28 mm pipes; fixing with pipe clips; plugged and screwed	PC £302.40/100m	0.28	2.46	4.08	m	**6.54**
Extra for						
made bend		0.25	2.20	-	nr	**2.20**
fittings with one end		0.14	1.23	1.69	nr	**2.92**
fittings with two ends		0.25	2.20	2.35	nr	**4.55**
fittings with three ends		0.37	3.26	3.41	nr	**6.67**
stop end	PC £1.61	0.14	1.23	1.69	nr	**2.92**
straight coupling	PC £1.81	0.25	2.20	1.90	nr	**4.10**
reducing set	PC £0.86	0.07	0.62	0.90	nr	**1.52**
female coupling	PC £1.15	0.35	3.08	1.21	nr	**4.29**
elbow	PC £2.23	0.25	2.20	2.35	nr	**4.55**
slow bend	PC £4.89	0.25	2.20	5.14	nr	**7.34**
tee; equal	PC £3.25	0.37	3.26	3.41	nr	**6.67**
tee; reducing	PC £3.37	0.37	3.26	3.54	nr	**6.80**
tank coupling; locknut	PC £2.09	0.37	3.26	2.19	nr	**5.45**
35 mm pipes; fixing with pipe clips; plugged and screwed	PC £659.70/100m	0.32	2.82	8.53	m	**11.35**
Extra for						
made bend		0.30	2.64	-	nr	**2.64**
fittings with one end		0.16	1.41	2.78	nr	**4.19**
fittings with two ends		0.30	2.64	4.59	nr	**7.23**
fittings with three ends		0.42	3.70	6.22	nr	**9.92**
stop end	PC £2.65	0.16	1.41	2.78	nr	**4.19**
straight coupling	PC £3.37	0.30	2.64	3.54	nr	**6.18**
reducing set	PC £1.47	0.08	0.70	1.54	nr	**2.24**
female coupling	PC £2.96	0.40	3.52	3.11	nr	**6.63**
elbow	PC £4.37	0.30	2.64	4.59	nr	**7.23**
tee; equal	PC £5.92	0.42	3.70	6.22	nr	**9.92**
tee; reducing	PC £5.92	0.42	3.70	6.22	nr	**9.92**
tank coupling; locknut	PC £4.40	0.42	3.70	4.62	nr	**8.32**
42 mm pipes; fixing with pipe clips; plugged and screwed	PC £800.10/100m	0.37	3.26	10.54	m	**13.80**
Extra for						
made bend		0.40	3.52	-	nr	**3.52**
fittings with one end		0.18	1.58	4.57	nr	**6.15**
fittings with two ends		0.35	3.08	6.47	nr	**9.55**
fittings with three ends		0.47	4.14	10.35	nr	**14.49**
stop end	PC £4.35	0.18	1.58	4.57	nr	**6.15**
straight coupling	PC £4.34	0.35	3.08	4.55	nr	**7.63**
reducing set	PC £2.31	0.09	0.79	2.42	nr	**3.21**
female coupling	PC £3.94	0.45	3.96	4.13	nr	**8.09**
elbow	PC £6.16	0.35	3.08	6.47	nr	**9.55**
tee; equal	PC £9.86	0.47	4.14	10.35	nr	**14.49**
tee; reducing	PC £9.23	0.47	4.14	9.69	nr	**13.83**

S PIPED SUPPLY SYSTEMS Including overheads and profit at 5.00%		Labour hours	Labour £	Material £	Unit	Total rate £
S10/S11 HOT AND COLD WATER - cont'd						
Copper pipes; BS 2871 table X; compression fittings; BS 864 - cont'd						
54 mm pipes; fixing with pipe clips; plugged and screwed	PC £1034.10/100m	0.44	3.87	13.85	m	**17.72**
Extra for						
made bend		0.55	4.84	-	nr	**4.84**
fittings with two ends		0.40	3.52	10.63	nr	**14.15**
fittings with three ends		0.52	4.58	16.19	nr	**20.77**
straight coupling	PC £6.62	0.40	3.52	6.95	nr	**10.47**
reducing set	PC £3.87	0.10	0.88	4.06	nr	**4.94**
female wall elbow	PC £5.70	0.50	4.40	5.98	nr	**10.38**
elbow	PC £10.12	0.40	3.52	10.63	nr	**14.15**
tee; equal	PC £15.42	0.52	4.58	16.19	nr	**20.77**
tee; reducing	PC £15.66	0.52	4.58	16.44	nr	**21.02**
Black MDPE pipes; BS 6730; plastic compression fittings						
20 mm pipes; fixing with pipe clips; plugged and screwed	PC £29.85/100m	0.25	2.20	1.08	m	**3.28**
Extra for						
fittings with two ends		0.20	1.76	1.71	nr	**3.47**
fittings with three ends		0.30	2.64	2.20	nr	**4.84**
straight coupling	PC £1.44	0.20	1.76	1.51	nr	**3.27**
male adaptor	PC £0.90	0.28	2.46	0.95	nr	**3.41**
elbow	PC £1.62	0.20	1.76	1.71	nr	**3.47**
tee; equal	PC £2.09	0.30	2.64	2.20	nr	**4.84**
end cap	PC £1.11	0.12	1.06	1.17	nr	**2.23**
straight swivel connector	PC £0.98	0.16	1.41	1.03	nr	**2.44**
bent swivel connector	PC £1.20	0.16	1.41	1.26	nr	**2.67**
25 mm pipes; fixing with pipe clips; plugged and screwed	PC £36.40/100m	0.28	2.46	1.35	m	**3.81**
Extra for						
fittings with two ends		0.25	2.20	2.07	nr	**4.27**
fittings with three ends		0.37	3.26	2.92	nr	**6.18**
straight coupling	PC £1.77	0.25	2.20	1.86	nr	**4.06**
reducer	PC £1.76	0.25	2.20	1.84	nr	**4.04**
male adaptor	PC £1.06	0.35	3.08	1.12	nr	**4.20**
elbow	PC £1.97	0.25	2.20	2.07	nr	**4.27**
tee; equal	PC £2.78	0.37	3.26	2.92	nr	**6.18**
end cap	PC £1.29	0.14	1.23	1.35	nr	**2.58**
32 mm pipes; fixing with pipe clips; plugged and screwed	PC £59.70/100m	0.32	2.82	1.68	m	**4.50**
Extra for						
fittings with two ends		0.30	2.64	2.57	nr	**5.21**
fittings with three ends		0.42	3.70	3.81	nr	**7.51**
straight coupling	PC £2.48	0.30	2.64	2.60	nr	**5.24**
reducer	PC £2.31	0.30	2.64	2.43	nr	**5.07**
male adaptor	PC £1.32	0.40	3.52	1.39	nr	**4.91**
elbow	PC £2.44	0.30	2.64	2.57	nr	**5.21**
tee; equal	PC £3.63	0.42	3.70	3.81	nr	**7.51**
end cap	PC £1.44	0.16	1.41	1.51	nr	**2.92**
50 mm pipes; fixing with pipe clips; plugged and screwed	PC £152.88/100m	0.35	3.08	2.84	m	**5.92**
Extra for						
fittings with two ends		0.35	3.08	6.28	nr	**9.36**
fittings with three ends		0.47	4.14	8.53	nr	**12.67**
straight coupling	PC £5.57	0.35	3.08	5.84	nr	**8.92**
reducer	PC £5.13	0.35	3.08	5.39	nr	**8.47**
male adaptor	PC £3.30	0.45	3.96	3.46	nr	**7.42**
elbow	PC £5.99	0.35	3.08	6.28	nr	**9.36**
tee; equal	PC £8.13	0.47	4.14	8.53	nr	**12.67**
end cap	PC £3.68	0.18	1.58	3.87	nr	**5.45**

S PIPED SUPPLY SYSTEMS Including overheads and profit at 5.00%		Labour hours	Labour £	Material £	Unit	Total rate £
63 mm pipes; fixing with pipe clips; plugged and screwed	PC £222.04/100m	0.40	3.52	3.90	m	**7.42**
Extra for						
fittings with two ends		0.40	3.52	7.64	nr	**11.16**
fittings with three ends		0.52	4.58	12.13	nr	**16.71**
straight coupling	PC £8.41	0.40	3.52	8.83	nr	**12.35**
reducer	PC £7.27	0.40	3.52	7.64	nr	**11.16**
male adaptor	PC £4.70	0.50	4.40	4.94	nr	**9.34**
elbow	PC £7.27	0.40	3.52	7.64	nr	**11.16**
tee; equal	PC £11.55	0.52	4.58	12.13	nr	**16.71**
end cap	PC £5.09	0.20	1.76	5.34	nr	**7.10**
Stainless steel pipes; BS 4127; stainless steel capillary fittings						
15 mm pipes; fixing with pipe clips; plugged and screwed	PC £112.10/100m	0.30	2.64	2.32	m	**4.96**
Extra for						
fittings with two ends		0.20	1.76	3.57	nr	**5.33**
fittings with three ends		0.30	2.64	5.29	nr	**7.93**
bend	PC £3.40	0.20	1.76	3.57	nr	**5.33**
tee; equal	PC £5.04	0.30	2.64	5.29	nr	**7.93**
tap connector	PC £13.14	0.27	2.38	13.80	nr	**16.18**
22 mm pipes; fixing with pipe clips; plugged and screwed	PC £177.65/100m	0.32	2.82	3.48	m	**6.30**
Extra for						
fittings with two ends		0.27	2.38	4.25	nr	**6.63**
fittings with three ends		0.40	3.52	7.21	nr	**10.73**
reducer	PC £14.07	0.27	2.38	14.77	nr	**17.15**
bend	PC £4.05	0.27	2.38	4.25	nr	**6.63**
tee; equal	PC £6.87	0.40	3.52	7.21	nr	**10.73**
tee; reducing	PC £15.61	0.40	3.52	16.39	nr	**19.91**
tap connector	PC £24.68	0.38	3.34	25.92	nr	**29.26**
28 mm pipes; fixing with pipe clips; plugged and screwed	PC £259.35/100m	0.35	3.08	4.74	m	**7.82**
Extra for						
fittings with two ends		0.34	2.99	6.10	nr	**9.09**
fittings with three ends		0.50	4.40	10.73	nr	**15.13**
reducer	PC £31.26	0.34	2.99	32.82	nr	**35.81**
bend	PC £5.81	0.34	2.99	6.10	nr	**9.09**
tee; equal	PC £10.22	0.50	4.40	10.73	nr	**15.13**
tee; reducing	PC £18.20	0.50	4.40	19.11	nr	**23.51**
tap connector	PC £49.66	0.48	4.22	52.14	nr	**56.36**
Copper, brass and gunmetal ancillaries; screwed joints to fittings						
Bibtaps; brass						
chromium plated; capstan head						
15 mm	PC £8.47	0.15	1.32	8.99	nr	**10.31**
22 mm	PC £11.36	0.20	1.76	12.03	nr	**13.79**
self closing; 15 mm	PC £11.31	0.15	1.32	11.97	nr	**13.29**
crutch head with hose union; 15 mm	PC £5.26	0.15	1.32	5.62	nr	**6.94**
draincock 15 mm	PC £1.75	0.10	0.88	1.90	nr	**2.78**
Stopcock; brass/gunmetal						
capillary joints to copper						
15 mm	PC £1.75	0.20	1.76	1.84	nr	**3.60**
22 mm	PC £3.18	0.27	2.38	3.33	nr	**5.71**
28 mm	PC £9.04	0.34	2.99	9.50	nr	**12.49**
compression joints to copper						
15 mm	PC £1.75	0.18	1.58	1.84	nr	**3.42**
22 mm	PC £3.08	0.24	2.11	3.23	nr	**5.34**
28 mm	PC £8.01	0.30	2.64	8.41	nr	**11.05**

S PIPED SUPPLY SYSTEMS Including overheads and profit at 5.00%		Labour hours	Labour £	Material £	Unit	Total rate £
S10/S11 HOT AND COLD WATER - cont'd						
Copper, brass and gunmetal ancillaries; screwed joints to fittings - cont'd						
compression joints to polyethylene						
13 mm	PC £4.78	0.26	2.29	5.02	nr	**7.31**
19 mm	PC £7.21	0.33	2.90	7.57	nr	**10.47**
25 mm	PC £10.41	0.40	3.52	10.93	nr	**14.45**
Gunmetal 'Fullway' gate valve; capillary joints to copper						
15 mm	PC £5.34	0.20	1.76	5.61	nr	**7.37**
22 mm	PC £6.30	0.27	2.38	6.61	nr	**8.99**
28 mm	PC £8.61	0.34	2.99	9.04	nr	**12.03**
35 mm	PC £19.18	0.41	3.61	20.14	nr	**23.75**
42 mm	PC £22.91	0.47	4.14	24.06	nr	**28.20**
54 mm	PC £33.22	0.53	4.66	34.88	nr	**39.54**
Gunmetal stopcock; screwed joints to iron						
15 mm	PC £5.50	0.30	2.64	5.78	nr	**8.42**
20 mm	PC £8.97	0.40	3.52	9.41	nr	**12.93**
25 mm	PC £12.87	0.50	4.40	13.51	nr	**17.91**
Bronze gate valve; screwed joints to iron						
15 mm	PC £11.73	0.30	2.64	12.32	nr	**14.96**
20 mm	PC £15.60	0.40	3.52	16.38	nr	**19.90**
25 mm	PC £20.48	0.50	4.40	21.50	nr	**25.90**
35 mm	PC £29.44	0.60	5.28	30.91	nr	**36.19**
42 mm	PC £38.61	0.70	6.16	40.54	nr	**46.70**
54 mm	PC £55.95	0.80	7.04	58.75	nr	**65.79**
Chromium plated; pre-setting radiator valve; compression joint; union outlet 15 mm	PC £2.97	0.22	1.94	3.12	nr	**5.06**
Chromium plated; lockshield radiator valve; compression joint; union outlet 15 mm	PC £2.97	0.22	1.94	3.12	nr	**5.06**
Brass ball valves; BS 1212 Part 1; piston type; high pressure; copper float; screwed joint to cistern						
15 mm	PC £2.80	0.25	2.20	5.14	nr	**7.34**
22 mm	PC £5.18	0.30	2.64	8.20	nr	**10.84**
25 mm	PC £11.87	0.35	3.08	16.13	nr	**19.21**
Water tanks/cisterns						
Polyethylene cold water feed and expansion cistern; BS 4213; with covers						
ref PC15; 68 litres	PC £18.46	1.25	11.00	19.39	nr	**30.39**
ref PC25; 114 litres	PC £24.49	1.45	12.76	25.72	nr	**38.48**
ref PC40; 182 litres	PC £41.92	1.75	15.40	44.02	nr	**59.42**
ref PC50; 227 litres	PC £44.88	1.95	17.16	47.13	nr	**64.29**
GRP cold water storage cistern; with covers						
ref 899.10; 27 litres	PC £21.12	1.10	9.68	22.18	nr	**31.86**
ref 899.25; 68 litres	PC £34.29	1.25	11.00	36.00	nr	**47.00**
ref 899.40; 114 litres	PC £40.47	1.45	12.76	42.49	nr	**55.25**
ref 899.70; 227 litres	PC £80.61	1.95	17.16	84.64	nr	**101.80**
Storage cylinders/calorifiers						
Copper cylinders; direct; BS 699; grade 3						
ref 1; 350 x 900 mm; 74 litres	PC £46.54	1.50	13.20	48.87	nr	**62.07**
ref 2; 450 x 750 mm; 98 litres	PC £48.02	1.75	15.40	50.42	nr	**65.82**
ref 7; 450 x 900 mm; 120 litres	PC £52.61	2.00	17.60	55.24	nr	**72.84**
ref 8; 450 x 1050 mm; 144 litres	PC £57.42	2.80	24.64	60.29	nr	**84.93**
ref 9; 450 x 1200 mm; 166 litres	PC £67.49	3.60	31.68	70.86	nr	**102.54**

S PIPED SUPPLY SYSTEMS Including overheads and profit at 5.00%		Labour hours	Labour £	Material £	Unit	Total rate £
Copper cylinders; single feed coil indirect; BS 1566 Part 2; grade 3						
ref 2; 300 x 1500 mm; 96 litres	PC £58.09	2.00	17.60	61.00	nr	**78.60**
ref 3; 400 x 1050 mm; 114 litres	PC £61.85	2.25	19.80	64.95	nr	**84.75**
ref 7; 450 x 900 mm; 117 litres	PC £61.05	2.50	22.00	64.10	nr	**86.10**
ref 8; 450 x 1050 mm; 140 litres	PC £69.11	3.00	26.40	72.57	nr	**98.97**
ref 9; 450 x 1200 mm; 162 litres	PC £88.27	3.50	30.80	92.69	nr	**123.49**
Combination copper hot water storage units; coil direct; BS 3198; (hot/cold)						
400 x 900 mm; (65/20 litres)	PC £71.85	2.80	24.64	75.44	nr	**100.08**
450 x 900 mm; (85/25 litres)	PC £74.00	3.90	34.32	77.71	nr	**112.03**
450 x 1050 mm; (115/25 litres)	PC £81.41	4.90	43.12	85.48	nr	**128.60**
450 x 1200 mm; (115/45 litres)	PC £86.65	5.50	48.39	90.98	nr	**139.37**
Combination copper hot water storage						
450 x 900 mm; (85/25 litres)	PC £92.80	4.40	38.72	97.44	nr	**136.16**
450 x 1200 mm; (115/45 litres)	PC £100.94	6.00	52.79	105.98	nr	**158.77**
Galvanized mild steel cylinders; direct; BS 417 Part 2; grade C; welded construction						
ref YM114; 100 litres	PC £93.34	2.50	22.00	98.01	nr	**120.01**
ref YM141; 123 litres	PC £98.26	3.30	29.04	103.17	nr	**132.21**
ref YM218; 195 litres	PC £123.12	4.80	42.24	129.28	nr	**171.52**
Galvanized mild steel cylinders; indirect; BS 1565 Part 2; annular heaters; for vertical fixing						
ref BSG.2m; 136 litres	PC £165.00	4.00	35.20	173.25	nr	**208.45**
ref BSG.3m; 159 litres	PC £183.50	4.75	41.80	192.68	nr	**234.48**
ref BSG.3m; 227 litres	PC £272.30	5.50	48.39	285.92	nr	**334.31**
ref BSG.5m; 364 litres	PC £435.40	6.50	57.19	457.17	nr	**514.36**
ref BSG.5m; 455 litres	PC £586.03	8.00	70.39	615.33	nr	**685.72**
Galvanized mild steel cisterns; BS 417 Part 2; grade A						
ref SCM270; 191 litres	PC £134.96	2.00	17.60	141.71	nr	**159.31**
ref SCM450/1; 327 litres	PC £175.22	3.00	26.40	183.98	nr	**210.38**
ref SCM1600; 1227 litres	PC £460.54	6.00	52.79	483.57	nr	**536.36**
ref SCM2720; 2137 litres	PC £595.50	12.00	105.59	625.28	nr	**730.87**
ref SCM4540; 3364 litres	PC £795.58	22.00	193.58	835.36	nr	**1028.94**
Galvanized mild steel tanks; BS 417 Part 2; grade A;						
ref T25/1; 95 litres	PC £114.84	1.75	15.40	120.58	nr	**135.98**
ref T30/A; 114 litres	PC £123.13	2.75	24.20	129.29	nr	**153.49**
ref T40; 155 litres	PC £131.41	4.00	35.20	137.98	nr	**173.18**
Thermal insulation						
19 mm thick rigid mineral glass fibre sectional pipe lagging; plain finish; fixed with aluminium bands to steel or copper pipework; including working over pipe fittings						
around 15/15 mm pipes	PC £1.81	0.07	0.62	1.90	m	**2.52**
around 20/22 mm pipes	PC £1.91	0.10	0.88	2.00	m	**2.88**
around 25/28 mm pipes	PC £2.11	0.11	0.97	2.21	m	**3.18**
around 32/35 mm pipes	PC £2.33	0.12	1.06	2.45	m	**3.51**
around 40/42 mm pipes	PC £2.48	0.13	1.14	2.60	m	**3.74**
around 50/54 mm pipes	PC £2.87	0.15	1.32	3.01	m	**4.33**

S PIPED SUPPLY SYSTEMS Including overheads and profit at 5.00%		Labour hours	Labour £	Material £	Unit	Total rate £
S10/S11 HOT AND COLD WATER - cont'd						
Thermal insulation - cont'd						
19 mm thick rigid mineral glass fibre sectional pipe lagging; canvas or class 0 laquered aluminium finish; fixed with aluminium bands to steel or copper pipework; including working over pipe fittings						
around 15/15 mm pipes	PC £2.27	0.07	0.62	2.38	m	**3.00**
around 20/22 mm pipes	PC £2.46	0.10	0.88	2.58	m	**3.46**
around 25/28 mm pipes	PC £2.70	0.11	0.97	2.84	m	**3.81**
around 32/35 mm pipes	PC £2.94	0.12	1.06	3.09	m	**4.15**
around 40/42 mm pipes	PC £3.17	0.13	1.14	3.33	m	**4.47**
around 50/54 mm pipes	PC £3.68	0.15	1.32	3.87	m	**5.19**
25 mm thick expanded polystyrene lagging sets; class 0 finish; for mild steel cisterns to BS 417; complete with fixing bands; for cisterns size (ref)						
762 x 584 x 610 mm; (SCM270)	PC £3.50	0.80	7.04	3.67	nr	**10.71**
1219 x 610 x 610 mm; (SCM450/1)	PC £11.78	0.90	7.92	12.37	nr	**20.29**
1524 x 1143 x 914 mm; (SCM1600)	PC £26.23	1.10	9.68	27.54	nr	**37.22**
1829 x 1219 x 1219 mm; (SCM2270)	PC £33.31	1.30	11.44	34.97	nr	**46.41**
2438 x 1524 x 1219 mm; (SCM4540)	PC £53.04	1.50	13.20	55.69	nr	**68.89**
50 mm thick glass-fibre filled polyethylene insulating jackets for GRP or polyethylene cold water cisterns; complete with fixing bands; for cisterns size						
445 x 305 x 300 mm; (18 litres)	PC £2.33	0.40	3.52	2.45	nr	**5.97**
495 x 368 x 362 mm; (27 litres)	PC £4.71	0.50	4.40	4.95	nr	**9.35**
630 x 450 x 420 mm; (68 litres)	PC £4.71	0.60	5.28	4.95	nr	**10.23**
665 x 490 x 515 mm; (91 litres)	PC £5.26	0.70	6.16	5.52	nr	**11.68**
700 x 540 x 535 mm; (114 litres)	PC £6.08	0.80	7.04	6.38	nr	**13.42**
955 x 605 x 595 mm; (182 litres)	PC £7.90	0.85	7.48	8.29	nr	**15.77**
1155 x 640 x 595 mm; (227 litres)	PC £9.03	0.90	7.92	9.48	nr	**17.40**
80 mm thick glass-fibre filled insulating jackets in flame retardant PVC to BS 1763; type 1B; segmental type for hot water cylinders; complete with fixing bands; for cylinders size (ref)						
400 x 900 mm; (2)	PC £6.82	0.33	2.90	7.16	nr	**10.06**
450 x 750 mm; (5)	PC £5.98	0.33	2.90	6.28	nr	**9.18**
450 x 900 mm; (7)	PC £6.13	0.33	2.90	6.44	nr	**9.34**
450 x 1050 mm; (8)	PC £7.67	0.40	3.52	8.05	nr	**11.57**
500 x 1200 mm; (-)	PC £10.75	0.50	4.40	11.29	nr	**15.69**
S13 PRESSURISED WATER						
Copper pipes; BS 2871 Part 1 table Y; annealed; mains pipework; no joints in the running length; laid in trenches						
Pipes						
15 mm	PC £234.00/100m	0.10	0.88	2.58	m	**3.46**
22 mm	PC £409.50/100m	0.11	0.97	4.51	m	**5.48**
28 mm	PC £587.70/100m	0.12	1.06	6.48	m	**7.54**
35 mm	PC £922.50/100m	0.13	1.14	10.17	m	**11.31**
42 mm	PC £1115.10/100m	0.15	1.32	12.30	m	**13.62**
54 mm	PC £1904.40/100m	0.18	1.58	20.99	m	**22.57**

S PIPED SUPPLY SYSTEMS Including overheads and profit at 5.00%		Labour hours	Labour £	Material £	Unit	Total rate £
Blue MDPE pipes; BS 6527; mains pipework; no joints in the running length; laid in trenches						
Pipes						
20 mm	PC £29.85/100m	0.11	0.97	0.32	m	1.29
25 mm	PC £36.40/100m	0.12	1.06	0.40	m	1.46
32 mm	PC £59.70/100m	0.13	1.14	0.66	m	1.80
50 mm	PC £152.88/100m	0.15	1.32	1.69	m	3.01
60 mm	PC £222.04/100m	0.16	1.41	2.45	m	3.86
Steel pipes; BS 1387; heavy weight; galvanized; mains pipework; screwed joints in the running length; laid in trenches						
Pipes						
15 mm	PC £133.87/100m	0.13	1.14	1.48	m	2.62
20 mm	PC £153.46/100m	0.14	1.23	1.70	m	2.93
25 mm	PC £219.64/100m	0.16	1.41	2.42	m	3.83
32 mm	PC £275.17/100m	0.18	1.58	3.04	m	4.62
40 mm	PC £321.03/100m	0.20	1.76	3.54	m	5.30
50 mm	PC £444.83/100m	0.25	2.20	4.90	m	7.10
Ductile iron bitumen coated pipes and fittings; BS 4772; class K9; Stanton's 'Tyton' water main pipes or similar; flexible joints						
100 mm pipes; laid straight	PC £43.76/5.5m	0.60	3.96	9.12	m	13.08
Extra for						
bend; 45 degrees	PC £16.31	0.60	3.96	20.86	nr	24.82
branch; 45 degrees; socketted	PC £87.13	0.90	5.93	97.08	nr	103.01
tee	PC £27.26	0.90	5.93	34.22	nr	40.15
flanged spigot	PC £12.37	0.60	3.96	14.85	nr	18.81
flanged socket	PC £15.35	0.60	3.96	17.98	nr	21.94
150 mm pipes; laid straight	PC £64.52/5.5m	0.70	4.62	13.32	m	17.94
Extra for						
bend; 45 degrees	PC £28.13	0.70	4.62	33.66	nr	38.28
branch; 45 degrees; socketted	PC £98.53	1.05	6.92	109.64	nr	116.56
tee	PC £42.98	1.05	6.92	51.32	nr	58.24
flanged spigot	PC £21.21	0.70	4.62	24.33	nr	28.95
flanged socket	PC £24.06	0.70	4.62	27.32	nr	31.94
200 mm pipes; laid straight	PC £100.98/5.5m	1.00	6.59	20.73	m	27.32
Extra for						
bend; 45 degrees	PC £55.22	1.00	6.59	63.08	nr	69.67
branch; 45 degrees; socketted	PC £137.95	1.50	9.89	152.50	nr	162.39
tee	PC £83.47	1.50	9.89	95.30	nr	105.19
flanged spigot	PC £33.63	1.00	6.59	37.86	nr	44.45
flanged socket	PC £35.80	1.00	6.59	40.14	nr	46.73
S32 NATURAL GAS						
Ductile iron bitumen coated pipes and fittings; BS 4772; class K9; Stanton's 'Stanlock' gas main pipes or similar; bolted gland joints						
100 mm pipes; laid straight	PC £48.67/5.5m	0.70	4.62	11.07	m	15.69
Extra for						
bend; 45 degrees	PC £17.16	0.70	4.62	28.77	nr	33.39
tee	PC £26.43	1.05	6.92	45.66	nr	52.58
flanged spigot	PC £12.37	0.70	4.62	20.15	nr	24.77
flanged socket	PC £14.92	0.70	4.62	22.83	nr	27.45
isolated 'Stanlock' joint	PC £6.83	0.35	2.31	7.17	nr	9.48

S PIPED SUPPLY SYSTEMS Including overheads and profit at 5.00%		Labour hours	Labour £	Material £	Unit	Total rate £
S32 NATURAL GAS - cont'd						
Ductile iron bitumen coated pipes and fittings; BS 4772; class K9; Stanton's 'Stanlock' gas main pipes or similar; bolted gland joints - cont'd						
150 mm pipes; laid straight	PC £73.07/5.5m	0.90	5.93	16.53	m	**22.46**
Extra for						
bend; 45 degrees	PC £29.61	0.90	5.93	46.56	nr	**52.49**
tee	PC £41.88	1.35	8.90	69.74	nr	**78.64**
flanged spigot	PC £21.21	0.90	5.93	32.58	nr	**38.51**
flanged socket	PC £23.39	0.90	5.93	34.87	nr	**40.80**
isolated 'Stanlock' joint	PC £9.82	0.45	2.97	10.31	nr	**13.28**
200 mm pipes; laid straight	PC £106.92/5.5m	1.30	8.57	23.95	m	**32.52**
Extra for						
bend; 45 degrees	PC £53.69	1.30	8.57	77.00	nr	**85.57**
tee	PC £80.39	1.95	12.86	118.77	nr	**131.63**
flanged spigot	PC £33.63	1.30	8.57	49.05	nr	**57.62**
flanged socket	PC £34.77	1.30	8.57	50.26	nr	**58.83**
isolated 'Stanlock' joint	PC £13.09	0.65	4.29	13.74	nr	**18.03**

T MECHANICAL HEATING SYSTEMS ETC. Including overheads and profit at 5.00%		Labour hours	Labour £	Material £	Unit	Total rate £
T10 GAS/OIL FIRED BOILERS						
Gas fired domestic boilers; cream or white enamelled casing; 32 mm BSPT female flow and return tappings; 102 mm flue socket 13 mm BSPT male draw-off outlet; electric controls						
13.19 kW output	PC £290.88	5.00	44.00	305.67	nr	349.67
23.45 kW output	PC £439.23	5.50	48.39	461.44	nr	509.83
Smoke flue pipework; light quality 'Duracem' pipes and fittings; including asbestos yarn and composition joints in the running length						
75 mm pipes; fixing with wall clips; plugged and screwed	PC £7.89/1.8m	0.35	3.08	5.03	m	8.11
Extra for						
loose sockets	PC £2.38	0.40	3.52	2.95	nr	6.47
bend; square and obtuse	PC £3.42	0.40	3.52	4.04	nr	7.56
Terminal cone caps; asbestos yarn and composition joint	PC £13.10	0.40	3.52	14.20	nr	17.72
100 mm pipes; fixing with wall clips; plugged and screwed	PC £10.16/1.8m	0.40	3.52	6.53	m	10.05
Extra for						
loose sockets	PC £3.09	0.45	3.96	3.90	nr	7.86
bend; square and obtuse	PC £4.28	0.45	3.96	5.15	nr	9.11
Terminal cone caps; asbestos yarn and composition joint	PC £13.91	0.45	3.96	15.26	nr	19.22
150 mm pipes; fixing with wall clips; plugged and screwed	PC £17.00/1.8m	0.50	4.40	10.91	m	15.31
Extra for						
loose sockets	PC £5.30	0.55	4.84	6.65	nr	11.49
bend; square and obtuse	PC £6.60	0.55	4.84	8.02	nr	12.86
Terminal cone caps; asbestos yarn and composition joint	PC £20.99	0.55	4.84	23.13	nr	27.97
Smoke flue pipework; heavy quality 'Duracem' pipes and fittings; including asbestos yarn and composition joints in the running length						
125 mm pipes; fixing with wall clips; plugged and screwed	PC £17.47/1.8m	0.45	3.96	11.07	m	15.03
Extra for						
loose sockets	PC £5.59	0.50	4.40	6.73	nr	11.13
bend; square and obtuse	PC £7.53	0.50	4.40	8.78	nr	13.18
Terminal cone caps; asbestos yarn and composition joint	PC £15.29	0.50	4.40	16.92	nr	21.32
175 mm pipes; fixing with wall clips; plugged and screwed	PC £32.18/1.8m	0.55	4.84	20.23	m	25.07
Extra for						
loose sockets	PC £8.53	0.60	5.28	10.20	nr	15.48
bend; square and obtuse	PC £13.79	0.60	5.28	15.73	nr	21.01
Terminal cone caps; asbestos yarn and composition joint	PC £38.88	0.60	5.28	42.07	nr	47.35
225 mm pipes; fixing with wall clips; plugged and screwed	PC £42.17/1.8m	0.65	5.72	26.44	m	32.16
Extra for						
loose sockets	PC £12.00	0.70	6.16	14.14	nr	20.30
bend; square and obtuse	PC £23.50	0.70	6.16	26.23	nr	32.39
Terminal cone caps; asbestos yarn and composition joint	PC £50.24	0.70	6.16	54.30	nr	60.46

T MECHANICAL HEATING SYSTEMS ETC. Including overheads and profit at 5.00%		Labour hours	Labour £	Material £	Unit	Total rate £
T30 MEDIUM TEMPERATURE HOT WATER HEATING						
Steel pipes; BS 1387; black; screwed joints; malleable iron fittings; BS 143						
15 mm pipes						
medium weight; fixing with pipe brackets; plugged and screwed	PC £73.96/100m	0.30	2.64	1.21	m	**3.85**
heavy weight; fixing with pipe brackets; plugged and screwed	PC £86.81/100m	0.30	2.64	1.34	m	**3.98**
Extra for						
fittings with one end		0.13	1.14	0.25	nr	**1.39**
fittings with two ends		0.25	2.20	0.32	nr	**2.52**
fittings with three ends		0.37	3.26	0.39	nr	**3.65**
fittings with four ends		0.50	4.40	0.89	nr	**5.29**
cap	PC £0.24	0.13	1.14	0.25	nr	**1.39**
socket; equal	PC £0.26	0.25	2.20	0.27	nr	**2.47**
socket; reducing	PC £0.31	0.25	2.20	0.32	nr	**2.52**
bend; 90 degree long radius M/F	PC £0.54	0.25	2.20	0.57	nr	**2.77**
elbow; 90 degree M/F	PC £0.31	0.25	2.20	0.32	nr	**2.52**
tee; equal	PC £0.37	0.37	3.26	0.39	nr	**3.65**
cross	PC £0.85	0.50	4.40	0.89	nr	**5.29**
union	PC £0.82	0.25	2.20	0.86	nr	**3.06**
isolated screwed joint	PC £0.26	0.30	2.64	0.37	nr	**3.01**
tank connection; longscrew and backnuts; lead washers; joint	PC £1.17	0.37	3.26	1.33	nr	**4.59**
20 mm pipes						
medium weight; fixing with pipe brackets; plugged and screwed	PC £87.92/100m	0.32	2.82	1.39	m	**4.21**
heavy weight; fixing with pipe brackets; plugged and screwed	PC £104.03/100m	0.32	2.82	1.57	m	**4.39**
Extra for						
fittings with one end		0.16	1.41	0.29	nr	**1.70**
fittings with two ends		0.33	2.90	0.43	nr	**3.33**
fittings with three ends		0.47	4.14	0.57	nr	**4.71**
fittings with four ends		0.66	5.81	1.36	nr	**7.17**
cap	PC £0.27	0.16	1.41	0.29	nr	**1.70**
socket; equal	PC £0.31	0.33	2.90	0.32	nr	**3.22**
socket; reducing	PC £0.37	0.33	2.90	0.39	nr	**3.29**
bend; 90 degree long radius M/F	PC £0.88	0.33	2.90	0.93	nr	**3.83**
elbow; 90 degree M/F	PC £0.41	0.33	2.90	0.43	nr	**3.33**
tee; equal	PC £0.54	0.47	4.14	0.57	nr	**4.71**
cross	PC £1.29	0.66	5.81	1.36	nr	**7.17**
union	PC £0.95	0.33	2.90	1.00	nr	**3.90**
isolated screwed joint	PC £0.31	0.40	3.52	0.42	nr	**3.94**
tank connection; longscrew and backnuts; lead washers; joint	PC £1.36	0.47	4.14	1.53	nr	**5.67**
25 mm pipes						
medium weight; fixing with pipe brackets; plugged and screwed	PC £126.75/100m	0.36	3.17	1.87	m	**5.04**
heavy weight; fixing with pipe brackets; plugged and screwed	PC £152.62/100m	0.36	3.17	2.16	m	**5.33**
Extra for						
fittings with one end		0.21	1.85	0.36	nr	**2.21**
fittings with two ends		0.42	3.70	0.71	nr	**4.41**
fittings with three ends		0.57	5.02	0.82	nr	**5.84**
fittings with four ends		0.84	7.39	1.71	nr	**9.10**
cap	PC £0.34	0.21	1.85	0.36	nr	**2.21**
socket; equal	PC £0.41	0.42	3.70	0.43	nr	**4.13**
socket; reducing	PC £0.49	0.42	3.70	0.52	nr	**4.22**
bend; 90 degree long radius M/F	PC £1.26	0.42	3.70	1.32	nr	**5.02**
elbow; 90 degree M/F	PC £0.68	0.42	3.70	0.71	nr	**4.41**
tee; equal	PC £0.78	0.57	5.02	0.82	nr	**5.84**

T MECHANICAL HEATING SYSTEMS ETC. Including overheads and profit at 5.00%		Labour hours	Labour £	Material £	Unit	Total rate £
cross	PC £1.63	0.84	7.39	1.71	nr	9.10
union	PC £1.12	0.42	3.70	1.18	nr	4.88
isolated screwed joint	PC £0.41	0.50	4.40	0.55	nr	4.95
tank connection; longscrew and backnuts; lead washers; joint	PC £1.90	0.57	5.02	2.11	nr	7.13
32 mm pipes						
medium weight; fixing with pipe brackets; plugged and screwed	PC £158.38/100m	0.42	3.70	2.37	m	6.07
heavy weight; fixing with pipe brackets; plugged and screwed	PC £191.17/100m	0.42	3.70	2.73	m	6.43
Extra for						
fittings with one end		0.25	2.20	0.52	nr	2.72
fittings with two ends		0.50	4.40	1.18	nr	5.58
fittings with three ends		0.67	5.90	1.36	nr	7.26
fittings with four ends		1.00	8.80	2.25	nr	11.05
cap	PC £0.49	0.25	2.20	0.52	nr	2.72
socket; equal	PC £0.68	0.50	4.40	0.71	nr	5.11
socket; reducing	PC £0.78	0.50	4.40	0.82	nr	5.22
bend; 90 degree long radius M/F	PC £2.11	0.50	4.40	2.21	nr	6.61
elbow; 90 degree M/F	PC £1.12	0.50	4.40	1.18	nr	5.58
tee; equal	PC £1.29	0.67	5.90	1.36	nr	7.26
cross	PC £2.14	1.00	8.80	2.25	nr	11.05
union	PC £1.90	0.50	4.40	2.00	nr	6.40
isolated screwed joint	PC £0.68	0.60	5.28	0.84	nr	6.12
tank connection; longscrew and backnuts; lead washers; joint	PC £2.56	0.67	5.90	2.81	nr	8.71
40 mm pipes						
medium weight; fixing with pipe brackets; plugged and screwed	PC £184.12/100m	0.50	4.40	2.83	m	7.23
heavy weight; fixing with pipe brackets; plugged and screwed	PC £222.79/100m	0.50	4.40	3.25	m	7.65
Extra for						
fittings with one end		0.29	2.55	0.66	nr	3.21
fittings with two ends		0.58	5.10	1.75	nr	6.85
fittings with three ends		0.78	6.86	1.86	nr	8.72
fittings with four ends		1.16	10.21	3.03	nr	13.24
cap	PC £0.63	0.29	2.55	0.66	nr	3.21
socket; equal	PC £0.92	0.58	5.10	0.96	nr	6.06
socket; reducing	PC £1.02	0.58	5.10	1.07	nr	6.17
bend; 90 degree long radius M/F	PC £2.79	0.58	5.10	2.93	nr	8.03
elbow; 90 degree M/F	PC £1.67	0.58	5.10	1.75	nr	6.85
tee; equal	PC £1.77	0.78	6.86	1.86	nr	8.72
cross	PC £2.89	1.16	10.21	3.03	nr	13.24
union	PC £2.40	0.58	5.10	2.52	nr	7.62
isolated screwed joint	PC £0.92	0.70	6.16	1.11	nr	7.27
tank connection; longscrew and backnuts; lead washers; joint	PC £3.09	0.78	6.86	3.38	nr	10.24
50 mm pipes						
medium weight; fixing with pipe brackets; plugged and screwed	PC £259.09/100m	0.60	5.28	3.89	m	9.17
heavy weight; fixing with pipe brackets; plugged and screwed	PC £309.54/100m	0.60	5.28	4.44	m	9.72
Extra for						
fittings with one end		0.33	2.90	1.25	nr	4.15
fittings with two ends		0.67	5.90	2.25	nr	8.15
fittings with three ends		0.88	7.74	2.68	nr	10.42
fittings with four ends		1.34	11.79	4.71	nr	16.50
cap	PC £1.19	0.33	2.90	1.25	nr	4.15
socket; equal	PC £1.43	0.67	5.90	1.50	nr	7.40
socket; reducing	PC £1.43	0.67	5.90	1.50	nr	7.40
bend; 90 degree long radius M/F	PC £4.76	0.67	5.90	5.00	nr	10.90
elbow; 90 degree M/F	PC £2.14	0.67	5.90	2.25	nr	8.15

T MECHANICAL HEATING SYSTEMS ETC. Including overheads and profit at 5.00%		Labour hours	Labour £	Material £	Unit	Total rate £
T30 MEDIUM TEMPERATURE HOT WATER HEATING - cont'd						
Steel pipes; BS 1387; black; screwed joints; malleable iron fittings; BS 143 - cont'd						
50 mm pipes						
tee; equal	PC £2.55	0.88	7.74	2.68	nr	**10.42**
cross	PC £4.49	1.34	11.79	4.71	nr	**16.50**
union	PC £3.57	0.67	5.90	3.75	nr	**9.65**
isolated screwed joint	PC £1.43	0.80	7.04	1.66	nr	**8.70**
tank connection; longscrew and backnuts; lead washers; joint	PC £4.84	0.88	7.74	5.25	nr	**12.99**
Steel pipes; BS 1387; galvanized; screwed joints; galvanized malleable iron fittings; BS 143						
15 mm pipes						
medium weight; fixing with pipe brackets; plugged and screwed	PC £114.81/100m	0.30	2.64	1.76	m	**4.40**
heavy weight; fixing with pipe brackets; plugged and screwed	PC £133.87/100m	0.30	2.64	1.97	m	**4.61**
Extra for						
fittings with one end		0.13	1.14	0.34	nr	**1.48**
fittings with two ends		0.25	2.20	0.43	nr	**2.63**
fittings with three ends		0.37	3.26	0.53	nr	**3.79**
fittings with four ends		0.50	4.40	1.21	nr	**5.61**
cap	PC £0.32	0.13	1.14	0.34	nr	**1.48**
socket; equal	PC £0.35	0.25	2.20	0.36	nr	**2.56**
socket; reducing	PC £0.41	0.25	2.20	0.43	nr	**2.63**
bend; 90 degree long radius M/F	PC £0.74	0.25	2.20	0.77	nr	**2.97**
elbow; 90 degree M/F	PC £0.41	0.25	2.20	0.43	nr	**2.63**
tee; equal	PC £0.51	0.37	3.26	0.53	nr	**3.79**
cross	PC £1.15	0.50	4.40	1.21	nr	**5.61**
union	PC £1.10	0.25	2.20	1.16	nr	**3.36**
isolated screwed joint	PC £0.35	0.30	2.64	0.46	nr	**3.10**
tank connection; longscrew and backnuts; lead washers; joint	PC £1.58	0.37	3.26	1.76	nr	**5.02**
20 mm pipes						
medium weight; fixing with pipe brackets; plugged and screwed	PC £130.80/100m	0.32	2.82	1.99	m	**4.81**
heavy weight; fixing with pipe brackets; plugged and screwed	PC £153.46/100m	0.32	2.82	2.24	m	**5.06**
Extra for						
fittings with one end		0.16	1.41	0.39	nr	**1.80**
fittings with two ends		0.33	2.90	0.58	nr	**3.48**
fittings with three ends		0.47	4.14	0.77	nr	**4.91**
fittings with four ends		0.66	5.81	1.84	nr	**7.65**
cap	PC £0.37	0.16	1.41	0.39	nr	**1.80**
socket; equal	PC £0.41	0.33	2.90	0.43	nr	**3.33**
socket; reducing	PC £0.51	0.33	2.90	0.53	nr	**3.43**
bend; 90 degree long radius M/F	PC £1.20	0.33	2.90	1.26	nr	**4.16**
elbow; 90 degree M/F	PC £0.55	0.33	2.90	0.58	nr	**3.48**
tee; equal	PC £0.74	0.47	4.14	0.77	nr	**4.91**
cross	PC £1.75	0.66	5.81	1.84	nr	**7.65**
union	PC £1.29	0.33	2.90	1.35	nr	**4.25**
isolated screwed joint	PC £0.41	0.40	3.52	0.54	nr	**4.06**
tank connection; longscrew and backnuts; lead washers; joint	PC £1.83	0.47	4.14	2.03	nr	**6.17**

T MECHANICAL HEATING SYSTEMS ETC. Including overheads and profit at 5.00%		Labour hours	Labour £	Material £	Unit	Total rate £
25 mm pipes						
medium weight; fixing with pipe brackets; plugged and screwed	PC £183.64/100m	0.36	3.17	2.64	m	**5.81**
heavy weight; fixing with pipe brackets; plugged and screwed	PC £219.64/100m	0.36	3.17	3.04	m	**6.21**
Extra for						
fittings with one end		0.21	1.85	0.48	nr	**2.33**
fittings with two ends		0.42	3.70	0.97	nr	**4.67**
fittings with three ends		0.57	5.02	1.11	nr	**6.13**
fittings with four ends		0.84	7.39	2.32	nr	**9.71**
cap	PC £0.46	0.21	1.85	0.48	nr	**2.33**
socket; equal	PC £0.55	0.42	3.70	0.58	nr	**4.28**
socket; reducing	PC £0.67	0.42	3.70	0.70	nr	**4.40**
bend; 90 degree long radius M/F	PC £1.70	0.42	3.70	1.79	nr	**5.49**
elbow; 90 degree M/F	PC £0.92	0.42	3.70	0.97	nr	**4.67**
tee; equal	PC £1.06	0.57	5.02	1.11	nr	**6.13**
cross	PC £2.21	0.84	7.39	2.32	nr	**9.71**
union	PC £1.52	0.42	3.70	1.59	nr	**5.29**
isolated screwed joint	PC £0.55	0.50	4.40	0.70	nr	**5.10**
tank connection; longscrew and backnuts; lead washers; joint	PC £2.56	0.57	5.02	2.81	nr	**7.83**
32 mm pipes						
medium weight; fixing with pipe brackets; plugged and screwed	PC £229.42/100m	0.42	3.70	3.33	m	**7.03**
heavy weight; fixing with pipe brackets; plugged and screwed	PC £275.17/100m	0.42	3.70	3.84	m	**7.54**
Extra for						
fittings with three ends		0.25	2.20	0.70	nr	**2.90**
fittings with two ends		0.50	4.40	1.59	nr	**5.99**
fittings with three ends		0.67	5.90	1.84	nr	**7.74**
fittings with four ends		1.00	8.80	3.04	nr	**11.84**
cap	PC £0.67	0.25	2.20	0.70	nr	**2.90**
socket; equal	PC £0.92	0.50	4.40	0.97	nr	**5.37**
socket; reducing	PC £1.06	0.50	4.40	1.11	nr	**5.51**
bend; 90 degree long radius M/F	PC £2.85	0.50	4.40	2.99	nr	**7.39**
elbow; 90 degree M/F	PC £1.52	0.50	4.40	1.59	nr	**5.99**
tee; equal	PC £1.75	0.67	5.90	1.84	nr	**7.74**
cross	PC £2.90	1.00	8.80	3.04	nr	**11.84**
union	PC £2.58	0.50	4.40	2.70	nr	**7.10**
isolated screwed joint	PC £0.92	0.60	5.28	1.09	nr	**6.37**
tank connection; longscrew and backnuts; lead washers; joint	PC £3.47	0.67	5.90	3.76	nr	**9.66**
40 mm pipes						
medium weight; fixing with pipe brackets; plugged and screwed	PC £266.59/100m	0.50	4.40	3.99	m	**8.39**
heavy weight; fixing with pipe brackets; plugged and screwed	PC £321.03/100m	0.50	4.40	4.59	m	**8.99**
Extra for						
fittings with one end		0.29	2.55	0.89	nr	**3.44**
fittings with two ends		0.58	5.10	2.37	nr	**7.47**
fittings with three ends		0.78	6.86	2.51	nr	**9.37**
fittings with four ends		1.16	10.21	4.11	nr	**14.32**
cap	PC £0.85	0.29	2.55	0.89	nr	**3.44**
socket; equal	PC £1.24	0.58	5.10	1.30	nr	**6.40**
socket; reducing	PC £1.38	0.58	5.10	1.45	nr	**6.55**
bend; 90 degree long radius M/F	PC £3.77	0.58	5.10	3.96	nr	**9.06**
elbow; 90 degree M/F	PC £2.25	0.58	5.10	2.37	nr	**7.47**
tee; equal	PC £2.39	0.78	6.86	2.51	nr	**9.37**
cross	PC £3.91	1.16	10.21	4.11	nr	**14.32**
union	PC £3.24	0.58	5.10	3.41	nr	**8.51**
isolated screwed joint	PC £1.24	0.70	6.16	1.45	nr	**7.61**
tank connection; longscrew and backnuts; lead washers; joint	PC £4.16	0.78	6.86	4.51	nr	**11.37**

T MECHANICAL HEATING SYSTEMS ETC. Including overheads and profit at 5.00%		Labour hours	Labour £	Material £	Unit	Total rate £
T30 MEDIUM TEMPERATURE HOT WATER HEATING - cont'd						
Steel pipes; BS 1387; galvanized; screwed joints; galvanized malleable iron fittings; BS 143 - cont'd						
50 mm pipes						
medium weight; fixing with pipe brackets; plugged and screwed	PC £374.08/100m	0.60	5.28	5.49	m	**10.77**
heavy weight; fixing with pipe brackets; plugged and screwed	PC £444.83/100m	0.60	5.28	6.27	m	**11.55**
Extra for						
fittings with one end		0.33	2.90	1.69	nr	**4.59**
fittings with two ends		0.67	5.90	3.04	nr	**8.94**
fittings with three ends		0.88	7.74	3.62	nr	**11.36**
fittings with four ends		1.34	11.79	6.38	nr	**18.17**
cap	PC £1.61	0.33	2.90	1.69	nr	**4.59**
socket; equal	PC £1.93	0.67	5.90	2.03	nr	**7.93**
socket; reducing	PC £1.93	0.67	5.90	2.03	nr	**7.93**
bend; 90 degree long radius M/F	PC £6.44	0.67	5.90	6.76	nr	**12.66**
elbow; 90 degree M/F	PC £2.90	0.67	5.90	3.04	nr	**8.94**
tee; equal	PC £3.45	0.88	7.74	3.62	nr	**11.36**
cross	PC £6.07	1.34	11.79	6.38	nr	**18.17**
union	PC £4.83	0.67	5.90	5.07	nr	**10.97**
isolated screwed joint	PC £1.93	0.80	7.04	2.19	nr	**9.23**
tank connection; longscrew and backnuts; lead washers; joint	PC £6.55	0.88	7.74	7.03	nr	**14.77**
Radiators						
Radiators; pressed steel panel type, 590 mm high; 3 mm chromium plated air valve; 15 mm chromium plated easy clean straight valve with union; 15 mm chromium plated lockshield valve with union						
1.69 m2 single surface	PC £64.57	2.00	17.60	72.85	nr	**90.45**
2.12 m2 single surface	PC £73.58	2.25	19.80	82.35	nr	**102.15**
2.75 m2 single surface	PC £97.11	2.50	22.00	107.06	nr	**129.06**
3.39 m2 single surgace	PC £114.81	2.75	24.20	125.64	nr	**149.84**
Vibration isolation mountings						
'Tico' anti-vibration rubber pads; type CV/LF/N; fixed with Tico A/WB epoxy mortar; on concrete base						
150 x 75 x 32 mm thick		0.75	6.60	10.91	nr	**17.51**
150 x 150 x 64 mm thick		1.00	8.80	31.19	nr	**39.99**

V ELECTRICAL SYSTEMS Including overheads and profit at 5.00%	Labour hours	Labour £	Material £	Unit	Total rate £
V21 - 22 GENERAL LIGHTING AND LV POWER					
NOTE: The following items indicate approximate prices for wiring of lighting and power points complete, including accessories and socket outlets but excluding lighting fittings. Consumer control units are shown separately. For a more detailed breakdown of these costs and specialist costs for a complete range of electrical items reference should be made to Spon's Mechanical and Electrical Services Price Book.					
Consumer control units					
8-way 60 amp SP&N surface mounted insulated consumer control units fitted with miniature circuit breakers including 2 m long 32 mm screwed welded conduit with three runs of 16 mm2 PVC cables ready for final connections by the supply authority	-	-	-	nr	117.50
Extra for current operated ELCB of 30 mA tripping current	-	-	-	nr	52.50
as above but 100 amp metal cased consumer unit and 25 mm2 PVC cables	-	-	-	nr	133.00
Extra for current operated ELCB of 30 mA tripping current	-	-	-	nr	110.00
Final circuits					
Lighting points					
Wired in PVC insulated and PVC sheathed cable in flats and houses; insulated in cavities and roof space; protected where buried by heavy gauge PVC conduit	-	-	-	nr	37.50
As above but in commercial property	-	-	-	nr	41.00
Wired in PVC insulated cable in screwed welded conduit in flats and houses	-	-	-	nr	77.50
As above but in commercial property	-	-	-	nr	95.00
As above but in industrial property	-	-	-	nr	108.00
Wired in MICC cable in flats and houses	-	-	-	nr	66.00
As above but in commercial property	-	-	-	nr	80.00
As above but in industrial property with PVC sheathed cable	-	-	-	nr	90.00
Single 13 amp switched socket outlet points					
Wired in PVC insulated and PVC sheathed cable in flats and houses on a ring main circuit; protected where buried by heavy gauge PVC conduit	-	-	-	nr	40.00
As above but in commercial property	-	-	-	nr	50.00
Wired in PVC insulated cable in screwed welded conduit throughout on a ring main circuit in flats and houses	-	-	-	nr	60.00
As above but in commercial property	-	-	-	nr	70.00
As above but in industrial property	-	-	-	nr	80.00
Wired in MICC cable on a ring main circuit in flats and houses	-	-	-	nr	60.00
As above but in commercial property	-	-	-	nr	70.00
As above but in industrial property with PVC sheathed cable	-	-	-	nr	88.00

V ELECTRICAL SYSTEMS Including overheads and profit at 5.00%	Labour hours	Labour £	Material £	Unit	Total rate £
V21 - 22 GENERAL LIGHTING AND LV POWER - cont'd					
Cooker control units					
45 amp circuit including unit wired in PVC insulated and PVC sheathed cable; protected where buried by heavy gauge PVC conduit	-	-	-	nr	80.00
As above but wired in PVC insulated cable in screwed welded conduit	-	-	-	nr	117.50
As above but wired in MICC cable	-	-	-	nr	132.50

W SECURITY SYSTEMS Including overheads and profit at 5.00%	Labour hours	Labour £	Material £	Unit	Total rate £
W20 LIGHTNING PROTECTION					
Flag staff terminal	-	-	-	nr	55.47
Copper strip roof or down conductors fixed with bracket or saddle clips					
20 x 3 mm	-	-	-	m	12.29
25 x 3 mm	-	-	-	m	13.70
Aluminium strip roof or down conductors fixed with bracket or saddle clips					
20 x 3 mm	-	-	-	m	9.69
25 x 3 mm	-	-	-	m	10.16
Joints in tapes	-	-	-	nr	7.56
Bonding connections to roof and structural metalwork	-	-	-	nr	42.53
Testing points	-	-	-	nr	20.79
Earth electrodes					
16 mm driven copper electrodes in 1220 mm sectional lengths (2440 mm minimum)	-	-	-	nr	113.40
First 2440 mm driven and tested					
25 x 3 mm copper strip electrode in 457 mm deep prepared trench	-	-	-	m	8.22

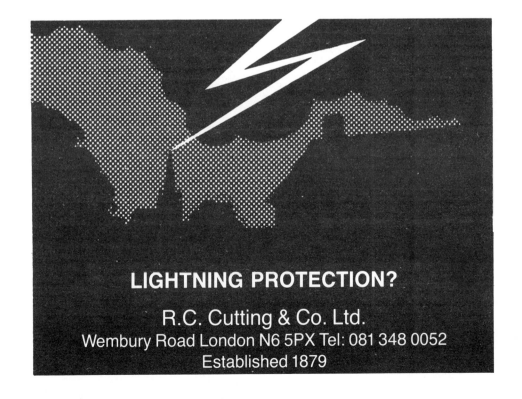

Prices for Measured Work – Minor Works

INTRODUCTION

The 'Prices for Measured Work - Minor Works' are intended to apply to a small project costing (excluding Preliminaries) about £75 000 in the outer London area.

The format of this section follows that of the 'Major Works' section with minor variations because of the different nature of the work, and reference should be made to the 'Introduction' to that section on page 161.

It has been assumed that reasonable quantities of work are involved, equivalent to quantities for two houses, although clearly this would not apply to all trades and descriptions of work in a project of this value. Where smaller quantities of work are involved it will be necessary to adjust the prices accordingly.

For C Demolition/Alteration/Renovation work, even smaller quantities have been assumed, as can be seen from the stated 'P.C.' of the materials involved.

Where work in an existing building is concerned it has been assumed that the building is vacated and that in all cases there is reasonable access and adequate storage space. Should this not be the case and if any abnormal circumstances have to be taken into account an allowance can be made either by a lump sum addition or by suitably modifying the percentage factor for overheads and profit.

Because of the different nature of the work for which prices are given in this section changes have been made to the percentage additions allowed in 'Major Works'. For those items where detailed breakdowns are given an allowance of 7.5% has been made to labour, plant and material costs for overhead charges and profit. In other cases 2.5% has been included for the General Contractor's attendance overhead charges and profit.

Labour rates are based upon typical gang costs divided by the number of primary working operatives for the trade concerned; and for general building work include an allowance for trade supervision, but exclude overheads and profit (included in the 'Labour £' column). The 'Labour hours' column gives the total hours allocated to a particular item and the 'Labour £' the consolidated cost of such labour including overhead charges and profit.

'Labour hours' have not always been given for 'spot' items because of the inclusion of Sub-Contractor's labour.

The 'Material Plant £' column includes the cost of removal of debris by skips or lorries and also overhead charges and profit.

No allowance has been made for any Value Added Tax which will probably be payable on the majority of work of this nature.

A PRELIMINARIES/CONTRACT CONDITIONS FOR MINOR WORKS

When pricing 'Preliminaries' all factors affecting the execution of the works must be considered; some of the more obvious have already been mentioned above.

As mentioned in 'Preliminaries' in the 'Prices for Measured Work - Major Works' section (page 183), the current trend is for 'Preliminaries' to be priced at approximately 12 to 15%, but for alterations and additions work in particular, care must be exercised in ensuring that all adverse factors are covered. The reader is advised to identify systematically and separately price all preliminary items with cost/time implications in order to reflect as accurately as possible preliminary costs likely to stem from any particular scheme.

Where the Standard Form of Contract applies two clauses which will affect the pricing of Preliminaries should be noted.

(a) Insurance of the works against Clause 22 Perils
Clause 22C will apply whereby the Employer and not the Contractor effects the insurance.

(b) Fluctuations
An allowance for any shortfall in recovery of increased costs under whichever clause is contained in the Contract may be covered by the inclusion of a lump sum in the Preliminaries or by increasing the prices by a suitable percentage

ADDITIONS AND NEW WORKS WITHIN EXISTING BUILDINGS

Depending upon the contract size, either the prices in 'Prices for Measured Work - Major Works' or those prices in 'Prices for Measured Work - Minor Works' will best apply.

It is likely, however, that the excavations for foundations might preclude the use of mechanical plant, and that it will be necessary to restrict prices to those applicable to hand excavation.

If, in any circumstances, less than what might be termed 'normal quantities' are likely to be involved, it is stressed that actual quotations should be invited from specialist Sub-contractors for these works.

JOBBING WORK

Jobbing work is outside the scope of this section and no attempt has been made to include prices for such work.

C DEMOLITION/ALTERATION/RENOVATION	Labour hours	Labour £	Material Plant £	Unit	Total rate £
C10 DEMOLISHING STRUCTURES					
Demolition rates vary considerably from one scheme to another; depending on access, type of construction, method of demolition, whether any redundant materials, etc. Therefore it is advisable to obtain specific quotations for each scheme under consideration, but the following rates (excluding scaffolding costs) for simple demolition may be of some assistance for comparison purposes.					
Demolish to ground level; single-storey brick out-building; timber flat roofs; volume					
50 m3	-	-	-	m3	**6.75**
200 m3	-	-	-	m3	**5.00**
500 m3	-	-	-	m3	**2.75**
Demolish down to ground level; two-storey brick out building; timber joisted suspended floor and timber flat roof; volume					
200 m3	-	-	-	m3	**3.75**
C20 ALTERATIONS - SPOT ITEMS					
Composite 'spot' items					
Few exactly similar composite items of alteration work are encountered on different schemes; for this reason it is considered more accurate for the reader to build up the value of such items from individual prices in the following section. However, for estimating purposes, the following 'spot' items have been prepared. Prices do not include for shoring, scaffolding or redecoration.					
Form openings					
Form opening through 100 mm thick softwood stud partition including framing studwork around, making good boarding and any plaster either side and extending floor finish through opening (new door and frame measured elsewhere)					
for single door and frame	-	145.00	40.00	nr	**185.00**
for pair of doors and frame	-	190.00	55.00	nr	**245.00**
Form opening through internal plastered wall for single door and frame; including cutting structure, quoining or making good jambs, cutting and pinning in suitable precast concrete plate lintel(s), making good plasterwork up to new frame both sides and extending floor finish through opening (new door and frame measured elsewhere)					
150 mm reinforced concrete wall	-	180.00	50.00	nr	**230.00**
225 mm reinforced concrete wall	-	235.00	65.00	nr	**300.00**
half brick thick wall	-	172.00	43.00	nr	**215.00**
one brick thick wall or two half brick thick skins	-	220.00	65.00	nr	**285.00**
one and a half brick thick wall	-	267.00	86.00	nr	**353.00**
two brick thick wall	-	313.00	107.00	nr	**420.00**
100 mm block wall	-	150.00	50.00	nr	**200.00**
215 mm block wall	-	185.00	82.50	nr	**267.50**

C DEMOLITION/ALTERATION/RENOVATION	Labour hours	Labour £	Material Plant £	Unit	Total rate £
C20 ALTERATIONS - SPOT ITEMS					
Form openings - cont'd					
Form opening through internal plastered wall for pair of doors and frame; including cutting structure, quoining or making good jambs, cutting and pinning in suitable precast concrete plate lintel(s), making good plasterwork up to new frame both sides and extending floor finish through opening (new door and frame measured elsewhere)					
150 mm reinforced concrete wall	-	260.00	72.50	nr	**332.50**
225 mm reinforced concrete wall	-	327.50	102.50	nr	**430.00**
half brick thick wall	-	206.50	55.50	nr	**262.00**
one brick thick wall or two half brick thick skins	-	275.00	82.00	nr	**357.00**
one and a half brick thick wall	-	342.00	108.00	nr	**450.00**
two brick thick wall	-	406.00	134.00	nr	**540.00**
100 mm block wall	-	180.00	60.00	nr	**240.00**
215 mm block wall	-	227.50	97.50	nr	**325.00**
Form opening through faced wall 1200 x 1200 mm (1.44 m2) for new window; including cutting structure, quoining up jambs, cutting and pinning in suitable precast concrete boot lintel with galvanized steel angle bolted on to support, outer brick soldier course in facing bricks to match existing (new window and frame measured elsewhere)					
one brick thick wall or two half brick thick skins	-	270.00	117.00	nr	**387.00**
one and a half brick thick wall	-	275.00	110.00	nr	**385.00**
two brick thick wall	-	325.00	125.00	nr	**450.00**
Form opening through slated, boarded and timbered roof; 700 x 1100 mm; for new rooflight; including cutting structure and finishings; trimming timbers in rafters and making good roof coverings (kerb and rooflight measured separately)	-	180.00	85.00	nr	**265.00**
Fill openings					
Take out door and frame; make good plaster and skirtings across reveals and head, and leave as blank opening					
single door	-	61.50	22.50	nr	**84.00**
pair of doors	-	72.50	23.50	nr	**96.00**
Take out door and frame in 100 mm thick softwood stud partition; block up opening with timber covered on both sides with boarding or lining to match existing and extend skirting both sides					
single doors	-	90.00	40.00	nr	**130.00**
pair of doors	-	120.00	50.00	nr	**170.00**
Take out single door and frame in internal wall; brick or block up opening; plastering walls and extend skirting both sides					
half brick thick	-	110.00	43.00	nr	**153.00**
one brick thick	-	140.00	68.00	nr	**208.00**
one and a half brick thick	-	172.00	93.00	nr	**265.00**
two brick thick	-	208.00	118.00	nr	**326.00**
100 mm blockwork	-	87.50	37.50	nr	**125.00**
215 mm blockwork	-	105.00	65.00	nr	**170.00**
Take out pair of doors and frame in internal wall; brick or block up opening; plastering walls and extend skirting both sides					

C DEMOLITION/ALTERATION/RENOVATION	Labour hours	Labour £	Material Plant £	Unit	Total rate £
half brick thick	-	172.50	73.50	nr	**246.00**
one brick thick	-	222.50	117.50	nr	**340.00**
one and a half brick thick	-	275.00	160.00	nr	**435.00**
two brick thick	-	325.00	200.00	nr	**525.00**
100 mm blockwork	-	139.00	67.00	nr	**206.00**
215 mm blockwork	-	166.50	110.00	nr	**276.50**
Take out 825 x 1406 mm (1.16 m2) sliding sash window and frame in external faced wall; brick up opening with facing bricks on outside to match existing and common bricks on inside; plastered internally					
one brick thick or two half brick thick skins	-	126.00	67.50	nr	**193.50**
one and a half brick thick	-	135.00	85.00	nr	**220.00**
two brick thick	-	157.50	102.50	nr	**260.00**
Take out 825 x 1406 mm (overall) (1.16 m2) curved headed sliding sash window in external stuccoed wall; brick up opening with common bricks; stucco on outside and plaster on inside to match existing					
one brick thick or two half brick thick skins	-	153.00	63.00	nr	**216.00**
one and a half brick thick	-	180.00	80.00	nr	**260.00**
two brick thick	-	207.00	103.00	nr	**310.00**
Take out 825 x 1406 mm (overall) (1.16 m2) curved headed sliding sash window in external masonry faced brick wall; brick up opening with masonry on outside and common bricks on inside; plastered internally					
350 mm wall	-	295.00	290.00	nr	**585.00**
500 mm wall	-	315.00	315.00	nr	**630.00**
600 mm wall	-	360.00	340.00	nr	**700.00**
Other 'spot' items					
Pull down brick chimney to 300 mm below roof level; seal off flues with slates; piece in 'treated' sawn softwood rafters and make good roof coverings over to match existing (scaffolding excluded)					
680 x 680 x 900 mm high above roof	-	117.00	33.00	nr	**150.00**
Add for each additional 300 mm high	-	17.00	5.50	nr	**22.50**
680 x 1030 x 900 mm high above roof	-	170.00	50.00	nr	**220.00**
Add for each additional 300 mm high	-	36.00	10.00	nr	**46.00**
1030 x 1030 x 900 mm high above roof	-	257.50	72.50	nr	**330.00**
Add for each additional 300 mm high	-	85.00	23.50	nr	**108.50**
Carefully take off existing chimney pots and set aside; pull down defective chimney stack to roof level and rebuild using 25% new facing bricks to match existing; provide new lead flashings; parge and core flues, and reset chimney pots including flaunching in cement mortar (scaffolding existing)					
680 x 680 x 900 mm high above roof	-	270.00	80.00	nr	**350.00**
Add for each additional 300 mm high	-	40.00	12.50	nr	**52.50**
680 x 1030 x 900 mm high above roof	-	400.00	125.00	nr	**525.00**
Add for each additional 300 mm high	-	60.00	18.00	nr	**78.00**
1030 x 1030 x 900 mm high above roof	-	590.00	185.00	nr	**775.00**
Add for each additional 300 mm high	-	87.50	27.50	nr	**115.00**
Take out fireplace; block up opening, plaster and extend skirtings; fix air brick, break up hearth and re-screed					
tiled	-	90.00	40.00	nr	**130.00**
cast iron; and set aside	-	95.00	25.00	nr	**120.00**
stone; and set aside	-	155.00	40.00	nr	**195.00**

C DEMOLITION/ALTERATION/RENOVATION Including overheads and profit at 7.50%	Labour hours	Labour £	Material £	Unit	Total rate £
C20 ALTERATIONS - SPOT ITEMS					
NOTE: All items of removal include for removing debris from site					
Removing/cutting plain/reinforced concrete work					
Break up concrete bed 100 mm thick	0.54	3.37	1.83	m2	5.20
Break up concrete bed					
150 mm thick	0.80	4.99	3.59	m2	8.58
200 mm thick	1.08	6.73	3.66	m2	10.39
300 mm thick	1.60	9.98	5.38	m2	15.36
Break up reinforced concrete bed					
100 mm thick	0.60	3.74	2.09	m2	5.83
150 mm thick	0.90	5.61	3.09	m2	8.70
200 mm thick	1.20	7.48	4.18	m2	11.66
300 mm thick	1.80	11.22	6.26	m2	17.48
Pull down reinforced concrete column or cut away casing to steel column	12.00	74.82	27.65	m3	102.47
Pull down reinforced concrete beam or cut away casing to steel beam	14.00	87.29	29.91	m3	117.20
Pull down reinforced concrete wall					
100 mm thick	1.20	7.48	2.77	m2	10.25
150 mm thick	1.80	11.22	4.10	m2	15.32
225 mm thick	2.70	16.83	6.15	m2	22.98
300 mm thick	3.60	22.45	8.30	m2	30.75
Pull down reinforced concrete suspended slabs					
100 mm thick	1.00	6.24	2.54	m2	8.78
150 mm thick	1.50	9.35	3.76	m2	13.11
225 mm thick	2.25	14.03	5.66	m2	19.69
300 mm thick	3.00	18.71	7.62	m2	26.33
Cut through reinforced concrete walls to form openings					
150 mm thick	6.00	37.90	7.50	m2	45.40
225 mm thick	8.20	51.74	11.19	m2	62.93
300 mm thick	10.40	65.56	15.09	m2	80.65
Cut through reinforced concrete suspended slabs to form openings					
150 mm thick	5.55	35.12	6.11	m2	41.23
225 mm thick	7.20	45.47	9.20	m2	54.67
300 mm thick	8.80	55.51	12.29	m2	67.80
Break up concrete plinth and make good floor beneath	4.60	28.68	16.79	m3	45.47
Remove precast concrete kerb	0.50	3.12	0.56	m	3.68
Remove precast concrete window sill and set aside for re-use	1.60	9.98	-	m	9.98
Break up concrete hearth	1.80	11.22	0.94	nr	12.16
Removing/cutting brick/blockwork					
Pull down external brick walls; in gauged mortar					
half brick thick	0.70	4.36	1.41	m2	5.77
two half brick thick skins	1.20	7.48	3.01	m2	10.49
one brick thick	1.20	7.48	3.01	m2	10.49
one and a half brick thick	1.70	10.60	4.70	m2	15.30
two brick thick	2.20	13.72	6.02	m2	19.74
Add for plaster, render or pebbledash per side	0.10	0.62	0.28	m2	0.90

C DEMOLITION/ALTERATION/RENOVATION Including overheads and profit at 7.50%	Labour hours	Labour £	Material £	Unit	Total rate £
Pull down external brick walls; in cement mortar					
half brick thick	1.05	6.55	1.41	m2	7.96
two half brick thick skins	1.75	10.91	3.01	m2	13.92
one brick thick	1.75	10.91	3.01	m2	13.92
one and a half brick thick	2.45	15.28	4.70	m2	19.98
two brick thick	3.15	19.64	6.02	m2	25.66
Add for plaster, render or pebbledash per side	0.10	0.62	0.28	m2	0.90
Pull down internal partitions; in gauged mortar					
half brick thick	1.05	6.55	1.41	m2	7.96
one brick thick	1.80	11.22	3.01	m2	14.23
one and a half brick thick	2.55	15.90	4.70	m2	20.60
75 mm blockwork	0.70	4.36	1.03	m2	5.39
90 mm blockwork	0.75	4.68	1.22	m2	5.90
100 mm blockwork	0.80	4.99	1.41	m2	6.40
115 mm blockwork	0.85	5.30	1.60	m2	6.90
125 mm blockwork	0.90	5.61	1.79	m2	7.40
140 mm blockwork	0.95	5.92	1.98	m2	7.90
150 mm blockwork	1.00	6.24	2.16	m2	8.40
190 mm blockwork	1.18	7.36	2.73	m2	10.09
215 mm blockwork	1.30	8.11	3.03	m2	11.14
255 mm blockwork	1.50	9.35	3.57	m2	12.92
Add for plaster per side	0.10	0.62	0.28	m2	0.90
Pull down internal partitions; in cement mortar					
half brick thick	1.60	9.98	1.41	m2	11.39
one brick thick	2.65	16.52	3.01	m2	19.53
one and a half brick thick	3.70	23.07	4.70	m2	27.77
Add for plaster per side	0.10	0.62	0.28	m2	0.90
Break up brick plinths	4.00	24.94	9.41	m3	34.35
Pull down bund walls or piers in cement mortar					
one brick thick	1.40	8.73	3.01	m2	11.74
Pull down walls to roof ventilator housing					
one brick thick	1.60	9.98	3.01	m2	12.99
Pull down brick chimney; to 300 mm below roof level; seal off flues with slates					
680 x 680 x 900 mm high above roof	12.50	78.07	21.34	nr	99.41
Add for each additional 300 mm high	2.50	15.61	4.22	nr	19.83
600 x 1030 x 900 mm high above roof	18.75	117.11	32.05	nr	149.16
Add for each additional 300 mm high	3.74	23.38	6.42	nr	29.80
1030 x 1030 x 900 mm high above roof	28.81	179.95	49.17	nr	229.12
Add for each additional 300 mm high	5.65	35.28	10.18	nr	45.46
Remove fireplace surround and hearth					
interior tiled	1.85	11.53	2.16	nr	13.69
cast iron; and set aside	3.10	19.33	-	nr	19.33
stone; and set aside	8.10	50.50	-	nr	50.50
Remove brick-on-edge coping; prepare walls for raising					
one brick thick	0.45	4.67	0.19	m	4.86
one and a half brick thick	0.60	6.22	0.28	m	6.50
Cut back brick projections flush with adjacent wall					
225 x 112 mm	0.30	3.11	0.09	m	3.20
225 x 225 mm	0.50	5.19	0.19	m	5.38
337 x 225 mm	0.70	7.26	0.28	m	7.54
450 x 225 mm	0.90	9.34	0.38	m	9.72
Cut back brick chimney breast flush with adjacent wall					
half brick thick	1.75	18.15	3.01	m2	21.16
one brick thick	2.35	24.38	4.70	m2	29.08

C DEMOLITION/ALTERATION/RENOVATION Including overheads and profit at 7.50%	Labour hours	Labour £	Material £	Unit	Total rate £
C20 ALTERATIONS - SPOT ITEMS - cont'd					
Removing/cutting brick/blockwork - cont'd					
Cut through brick or block walls or partitions; for lintels or beams above openings; in gauged mortar					
half brick thick	2.65	27.49	1.60	m2	29.09
one brick thick	4.40	45.64	3.20	m2	48.84
one and a half brick thick	6.15	63.80	4.80	m2	68.60
two brick thick	7.90	81.95	6.40	m2	88.35
75 mm blockwork	1.60	16.60	1.03	m2	17.63
90 mm blockwork	1.80	18.67	1.22	m2	19.89
100 mm blockwork	1.95	20.23	1.41	m2	21.64
115 mm blockwork	2.10	21.78	1.60	m2	23.38
125 mm blockwork	2.20	22.82	1.79	m2	24.61
140 mm blockwork	2.35	24.38	1.98	m2	26.36
150 mm blockwork	2.45	25.42	2.16	m2	27.58
190 mm blockwork	2.73	28.32	2.73	m2	31.05
215 mm blockwork	2.90	30.08	3.01	m2	33.09
255 mm blockwork	3.18	32.99	3.57	m2	36.56
Cut through brick walls or partitions; for lintels or beams above openings; in cement mortar					
half brick thick	3.80	39.42	1.60	m2	41.02
one brick thick	6.30	65.35	3.20	m2	68.55
one and a half brick thick	8.80	91.29	4.80	m2	96.09
two brick thick	11.30	117.22	6.40	m2	123.62
Cut through brick or block walls or partitions; for door or window openings; in gauged mortar					
half brick thick	1.35	14.00	1.60	m2	15.60
one brick thick	2.20	22.82	3.20	m2	26.02
one and a half brick thick	3.05	31.64	4.80	m2	36.44
two brick thick	3.95	40.98	6.40	m2	47.38
75 mm blockwork	0.80	8.30	1.03	m2	9.33
90 mm blockwork	0.92	9.54	1.22	m2	10.76
100 mm blockwork	1.00	10.37	1.41	m2	11.78
115 mm blockwork	1.06	11.00	1.60	m2	12.60
125 mm blockwork	1.10	11.41	1.79	m2	13.20
140 mm blockwork	1.16	12.03	1.98	m2	14.01
150 mm blockwork	1.20	12.45	2.16	m2	14.61
190 mm blockwork	1.32	13.69	2.73	m2	16.42
215 mm blockwork	1.45	15.04	3.01	m2	18.05
255 mm blockwork	1.60	16.60	3.57	m2	20.17
Cut through brick or block walls or partitions; for door or window openings; in cement mortar					
half brick thick	1.90	19.71	1.60	m2	21.31
one brick thick	3.15	32.68	3.20	m2	35.88
one and a half brick thick	4.35	45.13	4.80	m2	49.93
two brick thick	5.65	58.61	6.40	m2	65.01
Quoin up jambs in common bricks; PC £16.50/100; in gauged mortar (1:1:6); as the work proceeds					
half brick thick or skin of hollow wall	1.00	10.37	2.96	m	13.33
one brick thick	1.50	15.56	5.92	m	21.48
one and a half brick thick	1.95	20.23	8.88	m	29.11
two brick thick	2.40	24.90	11.84	m	36.74
75 mm blockwork	0.63	6.54	3.77	m	10.31
90 mm blockwork	0.67	6.95	4.06	m	11.01
100 mm blockwork	0.70	7.26	4.54	m	11.80
115 mm blockwork	0.76	7.88	5.56	m	13.44
125 mm blockwork	0.80	8.30	6.09	m	14.39

C DEMOLITION/ALTERATION/RENOVATION Including overheads and profit at 7.50%		Labour hours	Labour £	Material £	Unit	Total rate £
140 mm blockwork		0.86	8.92	6.62	m	15.54
150 mm blockwork		0.90	9.34	7.49	m	16.83
190 mm blockwork		1.01	10.48	8.70	m	19.18
215 mm blockwork		1.08	11.20	10.56	m	21.76
255 mm blockwork		1.19	12.34	11.60	m	23.94
Closing at jambs with common brickwork half brick thick						
50 mm cavity; including lead-lined hessian based vertical damp proof course		0.40	4.15	5.25	m	9.40
Quoin up jambs in sand faced facings; PC £45.38/100; in gauged mortar (1:1:6); facing and pointing one side to match existing						
half brick thick or skin of hollow wall		1.25	12.97	3.71	m	16.68
one brick thick		1.50	15.56	7.25	m	22.81
one and a half brick thick		2.30	23.86	10.83	m	34.69
two brick thick		2.80	29.05	14.37	m	43.42
Quoin up jambs in machine made facings; PC £45.38/100; in gauged mortar (1:1:6); facing and pointing one side to match existing						
half brick thick or skin of hollow wall		1.25	12.97	8.06	m	21.03
one brick thick		1.50	15.56	15.94	m	31.50
one and a half brick thick		2.30	23.86	23.87	m	47.73
two brick thick		2.80	29.05	31.75	m	60.80
Quoin up jambs in hand made facings; PC £45.38/100; in gauged mortar (1:1:6); facing and pointing one side to match existing						
half brick thick or skin of hollow wall		1.25	12.97	12.71	m	25.68
one brick thick		1.50	15.56	25.25	m	40.81
one and a half brick thick		2.30	23.86	37.84	m	61.70
two brick thick		2.80	29.05	50.38	m	79.43
Fill existing openings with common brickwork or blockwork in gauged mortar (1:1:6); any cutting and bonding measured separately						
half brick thick	PC £16.50/100	1.85	19.19	12.27	m2	31.46
one brick thick		3.05	31.64	25.26	m2	56.90
one and a half brick thick		4.20	43.57	37.97	m2	81.54
two brick thick		5.25	54.46	50.53	m2	104.99
75 mm blockwork	PC £6.66	0.92	9.54	7.90	m2	17.44
90 mm blockwork	PC £7.20	1.00	10.37	8.56	m2	18.93
100 mm blockwork	PC £8.01	1.05	10.89	9.52	m2	20.41
115 mm blockwork	PC £9.90	1.13	11.72	11.74	m2	23.46
125 mm blockwork	PC £10.80	1.18	12.24	12.81	m2	25.05
140 mm blockwork	PC £11.70	1.25	12.97	13.87	m2	26.84
150 mm blockwork	PC £13.28	1.30	13.49	15.70	m2	29.19
190 mm blockwork	PC £15.30	1.48	15.35	18.17	m2	33.52
215 mm blockwork	PC £18.67	1.60	16.60	22.13	m2	38.73
255 mm blockwork	PC £20.43	1.78	18.47	24.30	m2	42.77
Cutting and bonding ends to existing;						
half brick thick		0.40	4.15	0.81	m	4.96
one brick thick		0.58	6.02	1.62	m	7.64
one and a half brick thick		0.86	8.92	2.44	m	11.36
two brick thick		1.25	12.97	3.25	m	16.22
75 mm blockwork		0.17	1.76	0.57	m	2.33
90 mm blockwork		0.21	2.18	0.62	m	2.80
100 mm blockwork		0.23	2.39	0.69	m	3.08
115 mm blockwork		0.25	2.59	0.85	m	3.44
125 mm blockwork		0.26	2.70	0.93	m	3.63
140 mm blockwork		0.28	2.90	1.01	m	3.91
150 mm blockwork		0.29	3.01	1.14	m	4.15

C DEMOLITION/ALTERATION/RENOVATION Including overheads and profit at 7.50%	Labour hours	Labour £	Material £	Unit	Total rate £
C20 ALTERATIONS - SPOT ITEMS - cont'd					
Removing/cutting brick/blockwork - cont'd					
Cutting and bonding ends to existing;					
190 mm blockwork	0.36	3.73	1.32	m	5.05
215 mm blockwork	0.41	4.25	1.61	m	5.86
255 mm blockwork	0.48	4.98	1.76	m	6.74
half brick thick in facings; to match existing; PC £45.38/100	0.60	6.22	3.51	m2	9.73
Extra over common brickwork for fair face; flush pointing					
walls and the like	0.20	2.07	-	m2	2.07
Extra over common bricks; PC £45.38/100; for facing bricks in Flemish bond; facing and pointing one side					
sand faced facings PC £18.42/100	1.00	10.37	1.55	m2	11.92
machine made facings PC £45.38/100	1.00	10.37	23.28	m2	33.65
hand made facings PC £74.25/100	1.00	10.37	46.56	m2	56.93
ADD or DEDUCT for variation of £10.00/1000 in PC of facing bricks; in flemish bond					
half brick thick	-	-	-	m2	0.70
Fill existing openings in hollow wall with inner skin of common bricks; PC £16.50/100; 50 mm cavity and galvanized steel butter-fly ties; and outer skin of facings; all in gauged mortar (1:1:6); facing and pointing one side					
two half brick thick skins; outer skin sand faced facings	4.60	47.72	25.24	m2	72.96
two half brick thick skins; outer skin machine made facings	4.60	47.72	42.62	m2	90.34
two half brick thick skins; outer skin hand made facings	4.60	47.72	61.25	m2	108.97
Removing stone rubble walling					
Pull down external stone walls in lime mortar					
300 mm thick	1.20	7.48	2.82	m2	10.30
400 mm thick	1.60	9.98	3.76	m2	13.74
600 mm thick	2.40	14.96	5.64	m2	20.60
Pull down stone walls in lime mortar; clean off; set aside for re-use					
300 mm thick	1.80	11.22	0.94	m2	12.16
400 mm thick	2.40	14.96	1.22	m2	16.18
600 mm thick	3.60	22.45	1.88	m2	24.33
Removing metal					
Remove partitions					
corrugated metal partition	0.35	2.18	0.28	m2	2.46
lightweight steel mesh security screen	0.50	3.12	0.47	m2	3.59
solid steel demountable partition	0.75	4.68	0.66	m2	5.34
glazed sheet demountable partition; including removal of glass	1.00	6.24	0.94	m2	7.18
Remove handrails and balustrades					
tubular handrailing and brackets	0.30	1.87	0.09	m	1.96
metal balustrades	0.50	3.12	0.28	m	3.40
Remove					
metal window bars	0.30	1.87	0.09	nr	1.96
plates and bolts located in concrete bases	1.00	6.24	0.09	nr	6.33
steel cat ladder 4 m long	3.00	18.71	0.94	nr	19.65
Remove metal shutter door and track					
6.2 x 4.6 m (12.6 m long track)	12.00	74.82	14.11	nr	88.93
12.4 x 4.6 m (16.4 m long track)	15.00	93.53	28.22	nr	121.75

C DEMOLITION/ALTERATION/RENOVATION Including overheads and profit at 7.50%	Labour hours	Labour £	Material £	Unit	Total rate £
Removing timber					
Remove roof timbers complete; including rafters, purlins, ceiling joists, plates, etc, (measured flat on plan)	0.33	2.07	1.03	m2	3.10
Remove softwood floor construction					
100 mm deep joists at ground level	0.25	1.56	0.19	m2	1.75
175 mm deep joists at first floor level	0.50	3.12	0.38	m2	3.50
125 mm deep joists at roof level	0.70	4.36	0.28	m2	4.64
Remove individual floor or roof members	0.27	1.70	0.19	m	1.89
Remove infected or decayed floor plates	0.37	2.33	0.19	m	2.52
Remove boarding; withdrawing nails					
25 mm thick softwood flooring; at ground level	0.37	2.32	0.28	m2	2.60
25 mm thick softwood flooring; at first floor level	0.63	3.96	0.28	m2	4.24
25 mm thick softwood roof boarding	0.74	4.65	0.28	m2	4.93
25 mm thick softwood gutter boarding	0.80	5.02	0.28	m2	5.30
22 mm thick chipboard flooring; at first floor level	0.37	2.32	0.28	m2	2.60
Remove tilting fillet or roll	0.15	0.94	0.03	m	0.97
Remove fascia or barge boards	0.60	3.77	0.10	m	3.87
Pull down softwood stud partition; including finishings both sides etc					
solid	0.45	2.81	0.94	m2	3.75
glazed; including removal of glass	0.60	3.74	0.94	m2	4.68
Remove wall linings; including battening behind					
plain sheeting	0.30	1.87	0.38	m2	2.25
matchboarding	0.40	2.49	0.56	m2	3.05
Remove ceiling linings; including battening behind					
plain sheeting	0.45	2.81	0.38	m2	3.19
matchboarding	0.60	3.74	0.56	m2	4.30
Remove skirtings, picture rails, dado rails, architraves and the like	0.10	0.62	-	m	0.62
Carefully remove skirtings; and set aside for re-use in making good	0.25	1.56	-	m	1.56
Remove shelves, window boards and the like	0.33	2.06	0.09	m	2.15
Carefully remove oak dado panelling; and stack for subsequent re-use	0.65	7.75	-	m2	7.75
Remove; and set aside or clear away					
single door	0.40	4.77	0.28	nr	5.05
single door and frame or lining	0.80	9.54	0.47	nr	10.01
pair of doors	0.70	8.35	0.56	nr	8.91
pair of doors and frame or lining	1.20	14.31	0.94	nr	15.25
Extra for taking out floor spring box	0.75	8.94	0.19	nr	9.13
casement window and frame	1.20	14.31	0.47	nr	14.78
double hung sash window and frame	1.70	20.27	0.94	nr	21.21
pair of French windows and frame	4.00	47.69	1.41	nr	49.10
Carefully dismantle double hung sash window and frame; remove and store for re-use elsewhere	2.40	28.61	-	nr	28.61
Remove staircase; including balustrades					
single straight flight	3.50	41.73	9.41	nr	51.14
dogleg flight	5.00	59.61	14.11	m	73.72
Remove handrails and brackets	0.10	1.19	0.28	m	1.47
Remove sloping timber ramp in corridor; at change of levels	2.00	23.84	1.41	nr	25.25
Remove bath panels and bearers	0.40	4.77	0.47	nr	5.24
Remove kitchen fittings					
wall units	0.45	5.36	1.27	nr	6.63
floor units	0.30	3.58	2.12	nr	5.70
larder units	0.40	4.77	4.70	nr	9.47
built in cupboards	1.50	17.88	9.41	nr	27.29

C DEMOLITION/ALTERATION/RENOVATION Including overheads and profit at 7.50%	Labour hours	Labour £	Material £	Unit	Total rate £
C20 ALTERATIONS - SPOT ITEMS - cont'd					
Removing timber - cont'd					
Remove pipe casings	0.30	3.58	0.38	m	**3.96**
Remove ironmongery; in preparation for redecoration; and subsequently re-fix; including providing any new screws necessary	0.25	2.98	0.09	nr	**3.07**
Remove, withdrawing nails, etc; making good holes					
carpet fixing strip from floors	0.05	0.31	-	m	**0.31**
curtain track from head of window	0.25	1.56	-	m	**1.56**
nameplates or numerals from face of door	0.50	3.12	-	nr	**3.12**
fly screen and frame from window	0.90	5.61	-	nr	**5.61**
small notice board from walls	0.89	5.55	-	nr	**5.55**
fire extinguisher and bracket from walls	1.25	7.79	-	nr	**7.79**
Removing claddings/coverings					
Remove roof coverings					
slates	0.50	3.12	0.19	m2	**3.31**
slates; set aside for re-use	0.60	3.74	-	m2	**3.74**
nibbed tiles	0.40	2.49	0.19	m2	**2.68**
nibbed tiles; set aside for re-use	0.50	3.12	-	m2	**3.12**
corrugated asbestos sheeting	0.40	2.49	0.19	m2	**2.68**
corrugated metal sheeting	0.40	2.49	0.19	m2	**2.68**
underfelt; and nails	0.05	0.31	0.09	m2	**0.40**
three layer felt roofing; cleaning base off for new coverings	0.25	1.56	0.19	m2	**1.75**
sheet metal covering	0.50	3.12	0.19	m2	**3.31**
Remove defective metal flashings					
horizontal	0.20	1.25	0.19	m	**1.44**
stepped	0.25	1.56	0.09	m	**1.65**
Turn back bitumen felt and later dress up face of new brickwork as skirting; not exceeding 150 mm girth	1.00	6.33	0.39	m	**6.72**
Remove bitumen felt roofing and boarding to allow access for work to top of walls or beams beneath	0.80	5.06	-	m	**5.06**
Remove tiling battens; withdraw nails	0.08	0.50	0.09	m2	**0.59**
Removing waterproofing finishes					
Break up					
asphalt paving	0.60	3.74	-	m2	**3.74**
asphalt roofing	1.00	6.24	-	m2	**6.24**
Hack off asphalt skirtings from walls	0.15	0.94	-	m	**0.94**
Removing surface finishes					
Remove floor finishings					
carpet and underfelt	0.12	0.83	-	m2	**0.83**
linoleum sheet flooring	0.10	0.69	-	m2	**0.69**
carpet gripper	0.03	0.21	-	m2	**0.21**
Remove floor finishings; prepare screed to receive new					
carpet and underfelt	0.60	4.14	0.38	m2	**4.52**
vinyl or thermoplastic tiles	0.80	5.52	0.38	m2	**5.90**
Remove woodblock flooring; clean off and set aside for re-use	0.75	5.18	-	m2	**5.18**
Break up flooring					
floor screed	0.45	3.11	0.10	m2	**3.21**
granolithic flooring and screed	0.60	4.14	0.10	m2	**4.24**
terrazzo or ceramic floor tiles and screed	1.00	6.90	0.20	m2	**7.10**

C DEMOLITION/ALTERATION/RENOVATION Including overheads and profit at 7.50%	Labour hours	Labour £	Material £	Unit	Total rate £
Hack off in situ or tile skirtings from walls	0.25	1.73	-	m	1.73
Remove plasterboard wall finishing	0.40	2.76	-	m2	2.76
Hack off wall finishings					
plaster	0.20	1.38	0.47	m2	1.85
cement rendering or pebbledash	0.40	2.76	0.47	m2	3.23
wall tiling and screed	0.50	3.45	0.66	m2	4.11
Remove ceiling finishings					
plasterboard and skim; withdraw nails	0.30	2.07	0.28	m2	2.35
wood lath and plaster; withdraw nails	0.50	3.45	0.47	m2	3.92
suspended ceilings	0.75	5.18	0.47	m2	5.65
plaster moulded cornice; per 25 mm girth	0.15	1.04	0.09	m	1.13
Remove part of plasterboard ceiling to facilitate insertion of new steel beam	1.10	7.59	0.66	m	8.25
Removing gutterwork/pipework and equipment					
Remove gutterwork and supports					
PVC or asbestos	0.35	1.93	0.09	m	2.02
cast iron	0.40	2.20	0.19	m	2.39
Remove rainwater heads and supports					
PVC or asbestos	0.33	1.82	0.09	nr	1.91
cast iron	0.40	2.20	0.19	nr	2.39
Remove pipework and supports					
PVC or asbestos rainwater stack	0.30	1.65	0.09	m	1.74
cast iron rainwater stack	0.35	1.93	0.19	m	2.12
cast iron jointed soil stack	0.60	3.30	0.28	m	3.58
copper or steel water or gas pipework	0.15	0.83	0.09	m	0.92
cast iron rainwater shoe	0.08	0.44	-	nr	0.44
Remove sanitary fittings and supports; temporarily cap off services; to receive new (measured separately)					
sink or lavatory basin	1.00	7.26	5.64	nr	12.90
bath	2.00	14.51	8.47	nr	22.98
WC suite	1.50	10.88	5.64	nr	16.52
Remove sanitary fittings and supports, complete with associated services, overflows and waste pipes; making good all holes and other works disturbed; bring forward all surfaces ready for redecoration					
sink or lavatory basin	4.00	29.03	9.72	nr	38.75
range of three lavatory basins	8.00	58.05	18.81	nr	76.86
bath	6.00	43.54	12.45	nr	55.99
WC suite	8.00	58.05	10.40	nr	68.45
2 stall urinal	16.00	116.10	20.97	nr	137.07
3 stall urinal	24.00	174.15	38.42	nr	212.57
4 stall urinal	32.00	232.20	55.92	nr	288.12
Remove bathroom toilet fittings; making good works disturbed					
toilet roll holder or soap dispenser	0.30	1.65	-	nr	1.65
towel holder	0.60	3.30	-	nr	3.30
mirror	0.75	4.13	-	nr	4.13
Remove tap	0.10	0.73	-	nr	0.73
Remove the following equipment and ancillaries; cap off services; making good works disturbed (excluding any draining down of system)					
expansion tank 900 x 500 x 900 mm	1.80	13.06	3.76	nr	16.82
hot water cylinder 450 mm dia x 1050 mm high	1.20	8.71	1.60	nr	10.31
cold water tank 1540 x 900 x 900 mm	2.40	17.42	11.57	nr	28.99
cast iron radiator	2.00	14.51	3.01	nr	17.52
gas water heater	4.00	29.03	1.88	nr	30.91
gas fire	2.00	14.51	2.45	nr	16.96

C DEMOLITION/ALTERATION/RENOVATION Including overheads and profit at 7.50%	Labour hours	Labour £	Material £	Unit	Total rate £
C20 ALTERATIONS - SPOT ITEMS - cont'd					
Removing gutterwork/pipework and equipment - cont'd					
Remove cold water tank and housing on roof; strip out and cap off all associated piping; making good works disturbed and roof finishings					
1540 x 900 x 900 mm	12.00	87.08	14.39	nr	**101.47**
Remove lagging from pipes					
up to 42 mm dia	0.10	0.55	0.09	m	**0.64**

C30 SHORING/TEMPORARY SCREENS ETC.

Shoring and strutting for large openings

The requirements for shoring and strutting for the formation of large openings are dependent on a number of factors; for example, the weight of the superimposed structure to be supported, number (if any) of windows above; number of floors and the roof to be strutted, whether raking shores are required, and the depth to a load-bearing surface.

Prices would best be built up by assessing the use and waste of materials and the labour involved, including getting timber from and returning to yard, cutting away and making good, overheads and profit. This method is considered a more practical way of pricing than endeavouring to price the work on a cubic metre basis of timber used, and has been adopted in preparing the prices of the examples which follow.

The shoring and strutting for the formation of an opening 6100 mm wide x 2440 mm high through a one brick external wall at ground floor level, timber joisted suspended floor and timber pitched roof would cost:

Item	Labour hours	Labour £	Material £	Unit	Total rate £
Strutting to window openings over proposed new openings	0.60	4.86	1.54	nr	**6.40**
Plates, struts, braces and hardwood wedges in supports to floors and roof of opening	1.00	8.09	4.71	m	**12.80**
Other shores and associated work					
Dead shore and needle using die square timber with sole plates, braces, hardwood wedges and steel dogs	20.00	161.90	45.75	nr	**207.65**
Set of two raking shores using die square timber with 50 mm thick wall piece; hardwood wedges and steel dogs; including forming holes for needles and making good	34.00	275.22	25.13	nr	**300.35**
Cut holes through one brick wall for die square needle and make good; including facings externally and plaster internally	6.00	48.57	9.23	nr	**57.80**

C DEMOLITION/ALTERATION/RENOVATION Including overheads and profit at 7.50%	Labour hours	Labour £	Material £	Unit	Total rate £
Temporary screens and hoardings					
Provide, erect and maintain temporary dust-proof screen; with 50 x 75 mm sawn softwood framing; covered one side with 12 mm plywood					
over 300 mm wide	0.80	6.48	6.32	m2	**12.80**
Provide, erect and maintain temporary screen; with 50 x 100 mm sawn softwood framing; covered one side with 13 mm insulating board and other side with single layer of polythene sheet					
over 300 mm wide	1.00	8.09	7.17	m2	**15.26**
Provide, erect and maintain temporary hoardings; with 50 x 100 mm sawn softwood framing; covered one side with 19 mm exterior quality plywood; capped with softwood capping; including three coats of oil paint and afterwards clear away on completion					
over 300 mm wide	1.50	12.14	9.76	m2	**21.90**
C40 REPAIRING/RENOVATING CONCRETE/BRICK/ BLOCK/STONE					
Repairing/renovating plain/reinforced concrete work					
Reinstate plain concrete bed with site-mixed in situ concrete; mix 21.00 N/mm2 - 20 mm aggregate (1:2:4), where opening no longer required					
100 mm thick	0.48	3.03	6.20	m2	**9.23**
150 mm thick	0.78	4.90	9.30	m2	**14.20**
Reinstate reinforced concrete bed with site-mixed in situ concrete; mix 21.00 N/mm2 - 20 mm aggregate (1:2:4); including mesh reinforcement; where opening no longer required					
100 mm thick	0.71	4.46	7.53	m2	**11.99**
150 mm thick	0.98	6.14	10.63	m2	**16.77**
Reinstate reinforced concrete suspended floor; with site-mixed in situ concrete; mix 26.00 N/mm2 - 20 mm aggregate (1:1:5:3); including mesh reinforcement and formwork; where opening no longer required					
150 mm thick	3.20	20.34	12.25	m2	**32.59**
225 mm thick	3.75	23.80	17.16	m2	**40.96**
300 mm thick	4.15	26.30	22.07	m2	**48.37**
Reinstate 150 x 150 x 150 mm perforation through concrete suspended slab; with site-mixed in situ concrete; mix 21.00 N/mm2 - 20 mm aggregate (1:2:4); including formwork; where opening no longer required	0.92	5.77	0.24	nr	**6.01**
Clean surfaces of concrete to receive new damp proof membrane	0.15	0.94	-	m2	**0.94**
Clean out existing minor crack and fill with cement mortar mixed with bonding agent	0.33	2.06	0.45	m	**2.51**
Cut out existing crack to form 20 x 20 mm groove and fill with fine cement mixed with bonding agent	0.66	4.12	1.82	m	**5.94**

C DEMOLITION/ALTERATION/RENOVATION Including overheads and profit at 7.50%		Labour hours	Labour £	Material £	Unit	Total rate £
C40 REPAIRING/RENOVATING CONCRETE/BRICK/ BLOCK/STONE - cont'd						
Repairing/renovating plain/reinforced concrete work - cont'd						
Make good hole where pipe removed; 150 mm deep						
50 mm dia		0.42	2.62	0.33	nr	**2.95**
100 mm dia		0.55	3.43	0.50	nr	**3.93**
150 mm dia		0.70	4.36	0.66	nr	**5.02**
Add for each additional 25 mm thick up to 300 mm thick						
50 mm dia		0.09	0.56	0.03	nr	**0.59**
100 mm dia		0.12	0.75	0.03	nr	**0.78**
150 mm dia		0.15	0.94	0.05	nr	**0.99**
Repairing/renovating brick/blockwork						
Cut out small areas of defective brickwork; replace with new common bricks in gauged mortar (1:1:6)						
half brick thick	PC £16.50/100	5.25	54.46	13.87	m2	**68.33**
one brick thick		9.60	99.59	28.46	m2	**128.05**
one and a half brick thick		13.60	141.08	41.50	m2	**182.58**
two brick thick		17.50	181.54	55.51	m2	**237.05**
Cut out small areas of defective brickwork; replace with new facing brickwork in gauged mortar (1:1:6); half brick thick; facing and pointing one side						
sand faced facings	PC £18.42/100	7.30	75.73	15.17	m2	**90.90**
machine made facings	PC £45.38/100	7.30	75.73	33.43	m2	**109.16**
hand made facings	PC £74.25/100	7.30	75.73	52.99	m2	**128.72**
ADD or DEDUCT for variation of £10.00/1000 in PC of facing bricks; in flemish bond						
half brick thick		-	-	-	m2	**0.70**
Cut out individual bricks; replace with new common bricks in gauged mortar (1:1:6)						
half brick thick		0.30	3.11	0.27	nr	**3.38**
Cut out individual bricks; replace with new facing bricks in gauged mortar (1:1:6); half brick thick; facing and pointing one side						
sand faced facings	PC £18.42/100	0.45	4.67	0.29	nr	**4.96**
machine made facings	PC £45.38/100	0.45	4.67	0.58	nr	**5.25**
hand made facings	PC £74.25/100	0.45	4.67	0.89	nr	**5.56**
Cut out staggered cracks and repoint to match existing along brick joints		0.40	4.15	-	m	**4.15**
Cut out raking crack in brickwork; stitch in new common bricks and repoint to match existing						
half brick thick	PC £16.50/100	3.20	33.20	6.26	m2	**39.46**
one brick thick		5.85	60.69	12.76	m2	**73.45**
one and a half brick thick		8.75	90.77	19.26	m2	**110.03**
Cut out raking crack in brickwork; stitch in new facing bricks; half brick thick; facing and pointing one side to match existing						
sand faced facings	PC £18.42/100	4.80	49.79	6.85	m2	**56.64**
machine made facings	PC £45.38/100	4.80	49.79	15.06	m2	**64.85**
hand made facings	PC £74.25/100	4.80	49.79	23.86	m2	**73.65**

C DEMOLITION/ALTERATION/RENOVATION Including overheads and profit at 7.50%		Labour hours	Labour £	Material £	Unit	Total rate £
Cut out raking crack in cavity brickwork; stitch in new common bricks; PC £18.42/100 one side; and facing bricks the other side; both skins half brick thick; facing and pointing one side to match existing						
sand faced facings	PC £18.42/100	8.20	85.06	13.59	m2	98.65
machine made facings	PC £45.38/100	8.20	85.06	21.81	m2	106.87
hand made facings	PC £74.25/100	8.20	85.06	30.61	m2	115.67
Make good adjacent work; where intersecting wall removed						
half brick thick		0.30	3.11	0.48	m	3.59
one brick thick		0.40	4.15	0.96	m	5.11
100 mm blockwork		0.25	2.59	0.38	m	2.97
150 mm blockwork		0.29	3.01	0.57	m	3.58
215 mm blockwork		0.35	3.63	0.86	m	4.49
255 mm blockwork		0.39	4.05	0.96	m	5.01
Cut out defective brick soldier arch; renew and repoint to match existing						
sand faced facings	PC £18.42/100	1.95	20.23	3.64	m	23.87
machine made facings	PC £45.38/100	1.95	20.23	8.51	m	28.74
hand made facings	PC £74.25/100	1.95	20.23	13.72	m	33.95
Take down defective parapet wall; 600 mm high; with two courses of tiles and brick coping over; re-build in new facing bricks, tiles and coping stones						
one brick thick		6.66	69.09	59.64	m	128.73
Clean off moss and lichen from walls		0.30	3.11	-	m2	3.11
Clean off lime mortar, sort and stack old bricks for re-use		10.00	103.74	-	lt	103.74
Rake out decayed joints of brickwork and point in cement mortar (1:2); to match existing						
walls		0.75	7.78	0.62	m2	8.40
chimney stacks		1.20	12.45	0.62	m2	13.07
Rake out joint in brickwork; re-wedge flashing; and point in cement mortar (1:2) to match existing						
horizontal flashing		0.25	2.59	0.31	m	2.90
stepped flashing		0.37	3.84	0.31	m	4.15
Cut away and renew cement mortar (1:3) angle fillets; 50 mm face width		0.25	2.59	1.56	m	4.15
Cut out ends of joists and plates from walls; make good in common bricks; PC £16.50/100; in cement mortar (1:3)						
175 mm deep joists; 400 mm centres		0.65	6.74	4.32	m	11.06
225 mm deep joists; 400 mm centres		0.80	8.30	5.22	m	13.52
Cut and pin to existing brickwork						
ends of joists		0.40	4.15	-	nr	4.15
Make good hole where small pipe removed						
102 mm brickwork		0.20	1.25	0.05	nr	1.30
215 mm brickwork		0.20	1.25	0.05	nr	1.30
327 mm brickwork		0.20	1.25	0.05	nr	1.30
440 mm brickwork		0.20	1.25	0.05	nr	1.30
100 mm blockwork		0.20	1.25	0.05	nr	1.30
150 mm blockwork		0.20	1.25	0.05	nr	1.30
215 mm blockwork		0.20	1.25	0.05	nr	1.30
255 mm blockwork		0.20	1.25	0.05	nr	1.30
Make good hole and facings one side where small pipe removed						
102 mm brickwork		0.20	2.07	0.36	nr	2.43
215 mm brickwork		0.20	2.07	0.36	nr	2.43
327 mm brickwork		0.20	2.07	0.36	nr	2.43
440 mm brickwork		0.20	2.07	0.36	nr	2.43

C DEMOLITION/ALTERATION/RENOVATION Including overheads and profit at 7.50%	Labour hours	Labour £	Material £	Unit	Total rate £
C40 REPAIRING/RENOVATING CONCRETE/BRICK BLOCK/STONE - cont'd					
Repairing/renovating brick/blockwork - cont'd					
Make good hole where large pipe removed					
102 mm brickwork	0.30	1.87	0.10	nr	1.97
215 mm brickwork	0.45	2.81	0.23	nr	3.04
327 mm brickwork	0.60	3.74	0.39	nr	4.13
440 mm brickwork	0.75	4.68	0.52	nr	5.20
100 mm blockwork	0.30	1.87	0.08	nr	1.95
150 mm blockwork	0.35	2.18	0.11	nr	2.29
215 mm blockwork	0.40	2.49	0.16	nr	2.65
255 mm blockwork	0.45	2.81	0.19	nr	3.00
Make good hole and facings one side where large pipe removed					
102 mm brickwork	0.27	2.80	0.65	nr	3.45
215 mm brickwork	0.36	3.73	0.78	nr	4.51
327 mm brickwork	0.45	4.67	0.88	nr	5.55
440 mm brickwork	0.54	5.60	1.14	nr	6.74
Make good hole where extra large pipe removed					
102 mm brickwork	0.40	2.49	0.26	nr	2.75
215 mm brickwork	0.60	3.74	0.52	nr	4.26
327 mm brickwork	0.80	4.99	0.88	nr	5.87
440 mm brickwork	1.00	6.24	1.14	nr	7.38
100 mm blockwork	0.40	2.49	0.23	nr	2.72
150 mm blockwork	0.45	2.81	0.36	nr	3.17
215 mm blockwork	0.50	3.12	0.52	nr	3.64
255 mm blockwork	0.55	3.43	0.62	nr	4.05
Make good hole and facings one side where extra large pipe removed					
102 mm brickwork	0.36	3.73	0.93	nr	4.66
215 mm brickwork	0.48	4.98	1.14	nr	6.12
327 mm brickwork	0.60	6.22	1.45	nr	7.67
440 mm brickwork	0.72	7.47	1.97	nr	9.44
Cut out small areas of defective uncoursed stonework; rebuild in cement mortar to match existing					
300 mm thick wall	5.60	58.09	11.30	m2	69.39
400 mm thick wall	7.00	72.62	15.29	m2	87.91
600 mm thick wall	10.00	103.74	23.27	m2	127.01
Remove defective capping stones and haunching; replace stones and re-haunch in cement mortar to match existing					
300 mm thick wall	1.35	14.00	2.99	m2	16.99
400 mm thick wall	1.50	15.56	3.99	m2	19.55
600 mm thick wall	1.75	18.15	5.98	m2	24.13
Rake out decayed joints of uncoursed stonework and repoint in cement mortar to match existing walls	1.20	12.45	0.78	m2	13.23
C50 REPAIRING/RENOVATING METAL					
Overhaul and repair metal casement window; adjust and oil ironmongery; bring forward affected parts for redecoration	1.50	9.35	2.54	nr	11.89

C DEMOLITION/ALTERATION/RENOVATION Including overheads and profit at 7.50%	Labour hours	Labour £	Material £	Unit	Total rate £
C51 REPAIRING/RENOVATING TIMBER					
Cut out infected or decayed structural members; shore up adjacent work; provide and fix new 'treated' sawn softwood members pieced in					
Floors or flat roofs					
50 x 125 mm	0.40	3.24	2.12	m	**5.36**
50 x 150 mm	0.44	3.56	2.53	m	**6.09**
50 x 175 mm	0.48	3.89	2.95	m	**6.84**
50 x 200 mm	0.52	4.21	3.51	m	**7.72**
50 x 225 mm	0.56	4.53	4.07	m	**8.60**
Pitched roofs					
38 x 100 mm	0.36	2.91	1.46	m	**4.37**
50 x 100 mm	0.45	3.64	1.69	m	**5.33**
50 x 125 mm	0.50	4.05	2.12	m	**6.17**
50 x 150 mm	0.55	4.45	2.53	m	**6.98**
Kerbs, bearers and the like					
50 x 75 mm	0.45	3.64	1.28	m	**4.92**
50 x 100 mm	0.56	4.53	1.69	m	**6.22**
75 x 100 mm	0.68	5.50	2.86	m	**8.36**
Scarfed joint; new to existing; over 450 mm2	1.00	8.09	-	nr	**8.09**
Scarfed and bolted joint; new existing; including bolt; let in flush; over 450 mm2	1.66	13.44	1.61	nr	**15.05**
Other repairs					
Remove or punch in projecting nails; re-fix softwood or hardwood flooring					
loose floor boards	0.15	1.21	-	m2	**1.21**
floorboards previously set a side	0.80	6.48	0.28	m2	**6.76**
Remove damaged softwood flooring; provide and fix new 25 mm thick plain edge softwood boarding					
small areas	1.15	9.31	12.96	m2	**22.27**
individual boards 150 mm wide	0.30	2.43	1.40	m	**3.83**
Sand down and re-surface existing flooring; prepare body in with shellac and wax polish					
softwood	-	-	-	m2	**9.14**
hardwood	-	-	-	m2	**8.60**
Fit existing softwood skirting to new frame or architrave					
75 mm high	0.10	0.81	-	m	**0.81**
150 mm high	0.13	1.05	-	m	**1.05**
225 mm high	0.16	1.30	-	m	**1.30**
Piece in a new 25 x 150 mm moulded softwood skirting to match existing where old removed; bring forward for redecoration	0.38	2.74	5.00	m	**7.74**
Fit short length of salvaged softwood skirting to new reveal; including mitre with existing; fit to new frame; sawn softwood grounds plugged to brickwork; bring forward affected parts for redecoration					
75 mm high	0.34	2.60	0.59	nr	**3.19**
150 mm high	0.52	4.01	0.67	nr	**4.68**
225 mm high	0.65	5.01	0.73	nr	**5.74**
Piece in new 25 x 150 mm moulded softwood skirting to match existing where socket outlet removed; bring forward for redecoration	0.22	1.58	2.96	nr	**4.54**
Ease and adjust softwood door, oil ironmongery bring forward affected parts for redecoration	0.77	6.03	0.52	nr	**6.55**

C DEMOLITION/ALTERATION/RENOVATION Including overheads and profit at 7.50%	Labour hours	Labour £	Material £	Unit	Total rate £
C51 REPAIRING/RENOVATING TIMBER - cont'd					
Other repairs - cont'd					
Remove softwood door; ease and adjust; oil ironmongery; re-hang; bring forward affected parts for redecoration	1.20	9.71	0.49	nr	10.20
Remove mortice lock, piece in softwood door, bring forward affected parts for redecoration	1.10	8.74	1.26	nr	10.00
Fix only for salvaged softwood door	1.60	12.95	-	nr	12.95
Remove softwood door; plane 12 mm off bottom edge and re-hang	1.20	9.71	-	nr	9.71
Remove softwood door; alter ironmongery; piece in and rebate door and frame; re-hang on opposite style; bring forward affected parts for redecoration	2.65	21.03	2.08	nr	23.11
Remove softwood panelled door to prepare for fire upgrading; remove ironmongery; replace existing beads with new 25 x 38 mm hardwood beads screwed on; repair where damaged; piece out and re-hang on wider butts; adjust all ironmongery; seal around frame in cement mortar; bring forward affected parts for redecoration (any new glass measured separately)	5.35	42.89	21.90	nr	64.79
Upgrade and face up one side of flush door with 9 mm thick 'Supalux', screwed on	1.25	10.12	21.50	nr	31.62
Upgrade and face up one side of softwood panelled door with 9 mm thick 'Supalux', screwed on; plasterboard infilling in recesses	2.70	21.86	24.92	nr	46.78
Take off existing softwood doorstop; provide and screw on new 25 x 38 mm doorstop; bring forward for redecoration	0.22	1.75	1.74	m	3.49
Cut away defective 75 x 100 mm softwood external door frame; provide and splice in new piece 300 mm long; bed in cement mortar (1:3); point one side; bring forward for redecoration	1.40	11.00	6.56	nr	17.56
Seal roof trap flush with ceiling	0.60	4.86	3.54	nr	8.40
Form opening 762 x 762 mm in existing ceiling for new standard roof trap comprising softwood linings, architraves and 9 mm plywood cover; including all necessary trimming of ceiling joists (making good ceiling plaster measured separately)	2.70	21.86	33.60	nr	55.46
Ease and adjust softwood casement window; oil ironmongery; bring forward affected parts for redecoration	0.52	4.01	0.30	nr	4.31
Remove softwood casement window; ease and adjust; oil ironmongery; re-hang; bring forward affected parts for redecoration	0.77	6.03	0.52	nr	6.55
Remove softwood casement window; repair corner with angle plate; let in flush; oil ironmongery; re-hang; bring forward affected parts for redecoration	2.10	16.83	-	nr	16.83
Splice in or replace top or bottom rail or style of softwood casement window to match existing; bring forward for redecoration (taking off and re-hanging measured separately)	0.92	7.25	8.56	nr	15.81

C DEMOLITION/ALTERATION/RENOVATION Including overheads and profit at 7.50%	Labour hours	Labour £	Material £	Unit	Total rate £
Renew solid mullion jamb or transom of softwood casement window to match existing; bring forward for redecoration (taking off and re-hanging adjoining casements measured separately)	2.80	22.67	12.14	nr	**34.81**
Temporary lining 6 mm plywood infill to window whilst casement under repair	0.80	6.48	7.95	nr	**14.43**
Overhaul softwood double hung sash window; ease, adjust and oil pulley wheels; replace	2.65	21.03	5.90	nr	**26.93**
Cut away defective part of softwood window sill; provide and splice in new 83 x 65 mm weathered and throated piece 300 mm long; bring forward for redecoration	2.05	16.09	8.04	nr	**24.13**
Cut off broken stair nosing to tread or landing	0.40	3.24	5.32	nr	**8.56**
Renew isolated broken softwood tread; including additional framing beneath	1.20	9.71	21.50	nr	**31.21**
Remove damaged formica; provide and fix new to match existing					
worktop; over 300 mm wide	2.00	16.19	5.39	m2	**21.58**
edging; 25 mm wide	0.20	1.62	1.51	m	**3.13**
Renew 6 x 50 mm galvanized water bar	0.50	4.05	1.23	m	**5.28**
C52 FUNGUS/BEETLE ERADICATION					
Treat timbers with two coats of 'Rentokil' or similar proprietary insecticide and fungicide; PC £13.69/5ltr					
Remove cobwebs, dust and roof insulation; de-frass; treat exposed joists/rafters at 400 mm centres by spray application	-	-	-	m2	**4.52**
Treat boarding by spray application	-	-	-	m2	**1.29**
Treat individual timbers by brush application					
boarding	-	-	-	m2	**4.03**
structural members	-	-	-	m2	**9.14**
skirtings	-	-	-	m	**1.29**
Lift necessary floorboards; treat floors by spray application; later re-fix boards	-	-	-	m2	**3.00**
Treat surfaces of adjoining concrete or brickwork with two coats of 'Rentokil' dry rot fluid by spray application	0.20	1.62	0.88	m2	**2.50**
C53 REPAIRING/RENOVATING GLASS					
NOTE: All items of removal include for removing debris from site. The following prices for re-glazing are for re-glazing total windows. Prices for renewing single squares in repairs will be much higher and dependent on the quantity involved.					
Cut back putty; clean exposed rebate and re-putty					
to wood	-	-	-	m	**3.27**
to metal	-	-	-	m	**3.27**

C DEMOLITION/ALTERATION/RENOVATION Including overheads and profit at 7.50%		Labour hours	Labour £	Material £	Unit	Total rate £
C53 REPAIRING/RENOVATING GLASS - cont'd						
Hack out old glass; clean out putty and prepare rebates for new (measured elsewhere)						
3 mm thick float glass		0.55	3.53	-	m	**3.53**
4 mm thick float glass		0.63	4.04	-	m	**4.04**
6 mm thick float glass		0.80	5.13	-	m	**5.13**
Hack out old glass; clean out putty in rebates; prime for and re-glazing to wood with putty; 0.1 - 0.5 m2 panes						
3 mm thick clear float glass OQ		3.33	21.37	17.56	m2	**38.93**
4 mm thick clear float glass		3.55	22.78	19.91	m2	**42.69**
6 mm thick clear float glass		3.80	24.39	32.29	m2	**56.68**
4 mm thick patterned glass		3.55	22.78	28.60	m2	**51.38**
7 mm thick Georgian wired cast glass		3.80	24.39	25.39	m2	**49.78**
6 mm thick Georgian wired polished plate glass		4.00	25.67	55.36	m2	**81.03**
C54 REPAIRING/RENOVATING CLADDINGS/COVERINGS						
Carefully take off tiles or slates; select and re-fix; including providing 25% new; including nails, sundries, etc.						
Coverings, sloping						
asbestos cement 'blue/black' slates; 600 x 300 mm	PC £108.75/100	1.00	11.92	5.14	m2	**17.06**
asbestos-free artificial 'blue/black' slates 500 x 250 mm	PC £86.70/100	1.10	13.11	5.96	m2	**19.07**
asbestos-free artificial 'blue/black' slates 600 x 300 mm	PC £120.75/100	1.00	11.92	5.61	m2	**17.53**
natural slates; Welsh blue 510 x 255 mm	PC £204.00/100	1.20	14.31	11.98	m2	**26.29**
natural slates; Welsh blue 600 x 300 mm	PC £348.00/100	1.05	12.52	13.40	m2	**25.92**
second hand natural slates; Welsh blue 510 x 255 mm	PC £101.25/100	1.20	14.31	6.39	m2	**20.70**
second hand natural slates; Welsh blue 600 x 300 mm	PC £135.00/100	1.05	12.52	5.84	m2	**18.36**
clay plain tiles; 'Dreadnought' machine made; 265 x 165 mm	PC £41.40/100	1.10	13.11	7.91	m2	**21.02**
concrete interlocking tiles; Marley 'Ludlow Major' 413 x 330 mm	PC £84.00/100	0.70	8.35	2.91	m2	**11.26**
concrete interlocking tiles; Redland 'Renown' 417 x 330 mm	PC £84.15/100	0.70	8.35	2.92	m2	**11.27**
concrete plain tiles; 267 x 165 mm	PC £22.10/100	1.10	13.11	4.49	m2	**17.60**
Carefully take off damaged tiles or slates; provide and fix new; including nails, sundries, etc.						
Coverings; sloping; areas less than 10 m2						
asbestos cement 'blue/black' slates; 600 x 300 mm	PC £108.75/100	1.25	14.90	17.71	m2	**32.61**
asbestos-free artificial 'blue/black' slates 500 x 250 mm	PC £86.70/100	1.35	16.09	20.21	m2	**36.30**
asbestos-free artificial 'blue/black' slates 600 x 300 mm	PC £120.75/100	1.25	14.90	19.56	m2	**34.46**
natural slates; Welsh blue 510 x 255 mm	PC £204.00/100	1.45	17.29	45.72	m2	**63.01**

C DEMOLITION/ALTERATION/RENOVATION Including overheads and profit at 7.50%		Labour hours	Labour £	Material £	Unit	Total rate £
natural slates; Welsh blue 600 x 300 mm	PC £348.00/100	1.30	15.50	54.20	m2	**69.70**
second hand natural slates; Welsh blue 510 x 255 mm	PC £101.25/100	1.45	17.29	23.36	m2	**40.65**
second hand natural slates; Welsh blue 600 x 300 mm	PC £135.00/100	1.30	15.50	20.77	m2	**36.27**
clay plain tiles; 'Dreadnought' machine made; 265 x 165 mm	PC £41.40/100	1.35	16.09	30.12	m2	**46.21**
concrete interlocking tiles; Marley 'Ludlow Major' 413 x 330 mm	PC £84.00/100	0.90	10.73	10.65	m2	**21.38**
concrete interlocking tiles; Redland 'Renown' 417 x 330 mm	PC £84.15/100	0.90	10.73	10.67	m2	**21.40**
concrete plain tiles; 267 x 165 mm	PC £22.10/100	1.35	16.09	16.44	m2	**32.53**
Individual tiles or slates; isolated; rate for each						
asbestos cement 'blue/black' slates; 600 x 300 mm	PC £1.17	0.40	2.57	1.54	nr	**4.11**
asbestos-free artificial 'blue/black' slates 500 x 250 mm	PC £0.93	0.45	2.89	1.27	nr	**4.16**
asbestos-free artificial 'blue/black' slates 600 x 300 mm	PC £1.30	0.40	2.57	1.69	nr	**4.26**
natural slates; Welsh blue 510 x 255 mm	PC £2.18	0.50	3.21	2.75	nr	**5.96**
natural slates; Welsh blue 600 x 300 mm	PC £3.71	0.40	2.57	4.57	nr	**7.14**
second hand natural slates; Welsh blue 510 x 255 mm	PC £1.09	0.50	3.21	1.47	nr	**4.68**
second hand natural slates; Welsh blue 600 x 300 mm	PC £1.44	0.40	2.57	1.86	nr	**4.43**
clay plain tiles; 'Dreadnought' machine made; 265 x 165 mm	PC £0.45	0.20	1.28	0.68	nr	**1.96**
concrete interlocking tiles; Marley 'Ludlow Major' 413 x 330 mm	PC £0.90	0.25	1.60	1.22	nr	**2.82**
concrete interlocking tiles; Redland 'Renown' 417 x 330 mm	PC £0.90	0.25	1.60	1.22	nr	**2.82**
concrete plain tiles; 267 x 165 mm	PC £0.22	0.20	1.28	0.42	nr	**1.70**
Remove half round ridge or hip tile 300 mm long; provide and fix new		0.75	4.81	2.98	nr	**7.79**
Make good crack in roofing (any type), with mastic filling		0.40	2.57	1.36	m	**3.93**
Examine roof battens; renail where loose; and provide and fix 25% new						
19 x 50 mm slating battens at 262 mm centres		0.08	0.95	0.33	m2	**1.28**
19 x 38 mm tiling battens at 100 mm centres		0.20	2.38	0.79	m2	**3.17**
Remove roof battens and nails; provide and fix new 'treated' softwood battens throughout						
19 x 50 mm slating battens at 262 mm centres		0.12	1.43	1.38	m2	**2.81**
19 x 38 mm tiling battens at 100 mm centres		0.25	2.98	3.05	m2	**6.03**
Remove underfelt and nails; provide and fix new						
unreinforced felt	PC £0.87	0.10	1.19	1.24	m2	**2.43**
reinforced felt	PC £1.13	0.10	1.19	1.59	m2	**2.78**

C DEMOLITION/ALTERATION/RENOVATION Including overheads and profit at 7.50%	Labour hours	Labour £	Material £	Unit	Total rate £
C55 REPAIRING/RENOVATING WATERPROOFING FINISHES					
Cut out crack in asphalt paving; make good to match existing					
13 mm one coat	1.15	7.38	-	m	7.38
20 mm two coats	1.55	9.95	-	m	9.95
30 mm three coats	1.95	12.51	-	m	12.51
Cut out crack in asphalt roof coverings; make good to match existing					
20 mm two coats	1.90	12.19	-	m	12.19
C56 REPAIRING/RENOVATING SURFACE FINISHES					
Remove defective or damaged wall plaster; re-plaster walls with two coats of gypsum plaster; including dubbing out, joint of new to existing					
small areas	1.60	16.29	3.24	m2	19.53
isolated areas not exceeding 0.50 m2	1.15	11.71	1.79	nr	13.50
Re-plaster walls with two coats of gypsum plaster; including dubbing out, where wall or partition removed; including trimming back existing and joint of new to existing					
150 mm wide	0.63	6.41	0.57	m	6.98
225 mm wide	0.78	7.94	0.79	m	8.73
300 mm wide	0.93	9.47	1.00	m	10.47
Cut back finishings to allow new partition to be built; later re-plaster walls with two coats of gypsum plaster both sides; joints of new to existing	1.20	12.22	1.70	m	13.92
Re-plaster walls with two coats of gypsum plaster; including dubbing out; to face and returns of jamb to new opening; joints of new to existing plaster					
75 mm girth	0.65	6.62	2.16	m	8.78
150 mm girth	0.75	7.64	2.39	m	10.03
225 mm girth	0.85	8.65	2.60	m	11.25
300 mm girth	0.95	9.67	2.80	m	12.47
Re-plaster walls with two coats of gypsum plaster; including bonding agent; to faces and soffits of concrete lintel to new opening; joints of new to existing plaster					
75 mm girth	0.70	7.13	1.93	m	9.06
150 mm girth	0.80	8.14	2.07	m	10.21
225 mm girth	0.90	9.16	2.21	m	11.37
300 mm girth	1.00	10.18	2.33	m	12.51
Remove defective or damaged damp wall plaster; investigate and treat wall; re-plaster with two coats of 'Limelite' renovating plaster; including dubbing out and joint of new to existing					
small areas	1.70	17.31	4.83	m2	22.14
Dubbing out in cement and sand (1:3); average 13 mm thick					
over 300 mm wide	-	-	-	m	8.51
Remove defective or damaged damp wall					
isolated areas not exceeding 0.50 m2	1.22	12.42	2.68	nr	15.10
Remove defective or damaged ceiling plaster; remove laths or cut back boarding; prepare and fix new plasterboard; re-skim with one coat of gypsum; joint of new to existing					
small areas	1.70	17.31	2.92	m2	20.23
isolated areas not exceeding 0.50 m2	1.22	12.42	1.61	nr	14.03

C DEMOLITION/ALTERATION/RENOVATION Including overheads and profit at 7.50%	Labour hours	Labour £	Material £	Unit	Total rate £
Prepare and fix new plasterboard; re-skim with one coat of gypsum where wall or partition removed; including trimming back existing and joint of new to existing					
150 mm wide	0.73	7.43	0.78	m	**8.21**
225 mm wide	0.88	8.96	0.94	m	**9.90**
300 mm wide	1.03	10.49	1.09	m	**11.58**
Cut out; repair existing plaster cracks					
in walls	0.25	2.55	0.26	m	**2.81**
in ceilings	0.33	3.36	0.17	m	**3.53**
Make good plaster where items removed or holes left					
small pipe or conduit	0.07	0.71	0.03	nr	**0.74**
large pipe	0.10	1.02	0.06	nr	**1.08**
extra large pipe	0.13	1.32	0.09	nr	**1.41**
light switch	0.08	0.81	0.07	nr	**0.88**
Make good plasterboard and skim where items removed or holes left					
small pipe or conduit	0.15	1.53	0.28	nr	**1.81**
large pipe	0.23	2.34	0.43	nr	**2.77**
extra large pipe	0.30	3.05	0.57	nr	**3.62**
ceiling rose	0.18	1.83	0.43	nr	**2.26**
Level and repair floor screed with 'Ardit K10'					
small areas	0.50	5.09	5.18	m2	**10.27**
isolated areas not exceeding 0.50 m2	0.35	3.56	2.83	nr	**6.39**
Refix individual loose tiles or blocks					
floor tiles or blocks	0.15	1.53	0.14	nr	**1.67**
suspended ceiling tiles	0.30	3.05	-	nr	**3.05**
C57 REPAIRING/RENOVATING GUTTERWORK AND PIPEWORK					
Clean out existing					
rainwater gutters	0.08	0.44	-	m	**0.44**
rainwater gully	0.20	1.10	-	nr	**1.10**
rainwater stack; including head, swannecks and shoe (not exceeding 10 m long)	0.75	4.13	-	nr	**4.13**
Overhaul and re-make leaking joint; including new gutter bolt					
100 mm PVC	0.20	1.45	0.09	nr	**1.54**
100 mm cast iron	0.80	5.81	0.18	nr	**5.99**
Remove defective balloon grating; provide and fix in new outlet					
75 mm dia	0.10	0.90	1.41	nr	**2.31**
100 mm dia	0.10	0.90	1.69	nr	**2.59**
Overhaul section of rainwater guttering; cut out existing joints; adjust brackets to correct falls; re-make joints; including new gutter bolts					
100 mm PVC	0.25	2.25	0.03	m	**2.28**
100 mm cast iron	0.90	6.53	0.14	m	**6.67**
Clean and repair existing crack in lead gutter with wiped soldered joint	0.66	4.79	3.49	m	**8.28**
Clear blocked wastes without dismantling					
sink	0.50	2.75	-	nr	**2.75**
WC trap	0.60	3.30	-	nr	**3.30**
Turn off supplies; dismantle the following fittings; replace washers; re-assemble and test					
15 mm tap	0.25	2.25	0.03	nr	**2.28**
15 mm ball valve	0.35	3.15	0.05	nr	**3.20**
Turn off supplies; remove the following fitting; test and replace					
15 mm ball valve	0.50	4.50	4.98	nr	**9.48**

D GROUNDWORK
Including overheads and profit at 7.50%

Prices are applicable to excavation in firm soil. Multiplying factors for other soils are as follows:-

	Mechanical	Hand
Clay	x 2.00	x 1.20
Compact gravel	x 3.00	x 1.50
Soft chalk	x 4.00	x 2.00
Hard rock	x 5.00	x 6.00
Running sand or silt	x 6.00	x 2.00

	Labour hours	Labour £	Material Plant £	Unit	Total rate £
D20 EXCAVATION AND FILLING					
Site preparation					
Removing trees					
600 mm - 1.50 m girth	22.00	132.20	-	nr	**132.20**
1.50 - 3.00 m girth	38.50	231.36	-	nr	**231.36**
over 3.00 m girth	55.00	330.51	-	nr	**330.51**
Removing tree stumps					
600 mm - 1.50 m girth	1.10	6.61	29.56	nr	**36.17**
1.50 - 3.00 m girth	1.65	9.92	44.34	nr	**54.26**
over 3.00 m girth	2.20	13.22	59.13	nr	**72.35**
Clearing site vegetation	0.04	0.24	-	m2	**0.24**
Lifting turf for preservation; stacking	0.39	2.34	-	m2	**2.34**
Excavation using a wheeled hydraulic excavator with a 0.24 m3 bucket					
Excavating topsoil to be preserved					
average 150 mm deep	0.03	0.18	0.65	m2	**0.83**
Add or deduct for each 25 mm variation in depth	0.01	0.03	0.11	m2	**0.14**
Excavating to reduce levels; not exceeding					
0.25 m deep	0.10	0.60	2.15	m3	**2.75**
1 m deep	0.08	0.48	1.72	m3	**2.20**
2 m deep	0.10	0.60	2.15	m3	**2.75**
4 m deep	0.13	0.78	2.80	m3	**3.58**
Excavating basements and the like; starting from reduced level; not exceeding					
1 m deep	0.10	0.60	2.37	m3	**2.97**
2 m deep	0.08	0.48	1.89	m3	**2.37**
4 m deep	0.10	0.60	2.37	m3	**2.97**
6 m deep	0.12	0.72	2.84	m3	**3.56**
8 m deep	0.14	0.84	3.31	m3	**4.15**
Excavating pits to receive bases; not exceeding					
0.25 m deep	0.30	1.80	7.10	m3	**8.90**
1 m deep	0.26	1.56	6.15	m3	**7.71**
2 m deep	0.30	1.80	7.10	m3	**8.90**
4 m deep	0.32	1.92	7.57	m3	**9.49**
6 m deep	0.34	2.04	8.04	m3	**10.08**
Extra over pit excavation for commencing below ground or reduced level					
1 m below	0.03	0.18	0.71	m3	**0.89**
2 m below	0.05	0.30	0.95	m3	**1.25**
3 m below	0.06	0.36	1.18	m3	**1.54**
4 m below	0.08	0.48	1.66	m3	**2.14**

D GROUNDWORK Including overheads and profit at 7.50%	Labour hours	Labour £	Material Plant £	Unit	Total rate £
Excavating trenches; not exceeding 0.30 m wide; not exceeding					
0.25 m deep	0.25	1.50	5.91	m3	**7.41**
1 m deep	0.22	1.32	5.20	m3	**6.52**
2 m deep	0.25	1.50	5.91	m3	**7.41**
4 m deep	0.30	1.80	7.10	m3	**8.90**
6 m deep	0.35	2.10	8.28	m3	**10.38**
Excavating trenches; over 0.30 m wide; not exceeding					
0.25 m deep	0.23	1.38	5.44	m3	**6.82**
1 m deep	0.20	1.20	4.73	m3	**5.93**
2 m deep	0.23	1.38	5.44	m3	**6.82**
4 m deep	0.26	1.56	6.15	m3	**7.71**
6 m deep	0.32	1.92	7.57	m3	**9.49**
Extra over trench excavation for commencing below ground or reduced level					
1 m below	0.03	0.18	0.71	m3	**0.89**
2 m below	0.05	0.30	0.95	m3	**1.25**
3 m below	0.06	0.36	1.18	m3	**1.54**
4 m below	0.08	0.48	1.66	m3	**2.14**
Extra over trench excavation for curved trench	0.01	0.06	0.24	m3	**0.30**
Excavating for pile caps and ground beams between piles; not exceeding					
0.25 m deep	0.38	2.28	7.80	m3	**10.08**
1 m deep	0.35	2.10	7.10	m3	**9.20**
2 m deep	0.38	2.28	7.80	m3	**10.08**
Excavating to bench sloping ground to receive filling; not exceeding					
0.25 m deep	0.10	0.60	2.37	m3	**2.97**
1 m deep	0.08	0.48	1.89	m3	**2.37**
2 m deep	0.10	0.60	2.37	m3	**2.97**
Extra over any type of excavation at any depth for excavating					
alongside services	2.25	12.34		m3	**12.34**
around services crossing excavation	5.20	27.46		m3	**27.46**
below ground water level	0.11	0.66	2.37	m3	**3.03**
Extra over excavation for breaking out existing materials					
rock	2.65	15.92	10.97	m3	**26.89**
concrete	2.25	13.52	8.25	m3	**21.77**
reinforced concrete	3.20	19.23	12.09	m3	**31.32**
brickwork; blockwork or stonework	1.60	9.61	6.05	m3	**15.66**
Extra over excavation for breaking out existing hard pavings					
tarmacadam 75 mm thick	0.17	1.02	0.51	m2	**1.53**
tarmacadam and hardcore 150 mm thick	0.23	1.38	0.55	m2	**1.93**
concrete 150 mm thick	0.35	2.10	1.21	m2	**3.31**
reinforced concrete 150 mm thick	0.52	3.12	1.76	m2	**4.88**
Working space allowance to excavations					
reduced levels; basements and the like	0.09	0.54	1.89	m2	**2.43**
pits	0.20	1.20	4.02	m2	**5.22**
trenches	0.18	1.08	3.78	m2	**4.86**
pile caps and ground beams between piles	0.21	1.26	4.26	m2	**5.52**
Extra over excavation for working space for filling in with					
hardcore	0.11	0.66	6.13	m2	**6.79**
coarse ashes	0.11	0.66	7.31	m2	**7.97**
sand	0.11	0.66	13.38	m2	**14.04**
40 - 20 mm gravel	0.11	0.66	16.61	m2	**17.27**
in situ concrete - 7.50 N/mm2 - 40 mm aggregate (1:8)	0.83	5.18	26.90	m2	**32.08**

D GROUNDWORK Including overheads and profit at 7.50%	Labour hours	Labour £	Material Plant £	Unit	Total rate £
D20 EXCAVATION AND FILLING - cont'd					
Hand excavation					
Excavating topsoil to be preserved					
average 150 mm deep	0.28	1.68	-	m2	**1.68**
Add or deduct for each 25 mm variation in depth	0.04	0.24	-	m2	**0.24**
Excavating to reduce levels; not exceeding					
0.25 m deep	2.00	12.02	-	m3	**12.02**
1 m deep	2.20	13.22	-	m3	**13.22**
2 m deep	2.40	14.42	-	m3	**14.42**
4 m deep	2.65	15.92	-	m3	**15.92**
Excavating basements and the like; starting from reduced level; not exceeding					
1 m deep	2.65	15.92	-	m3	**15.92**
2 m deep	2.95	17.73	-	m3	**17.73**
4 m deep	3.85	23.14	-	m3	**23.14**
6 m deep	4.75	28.54	-	m3	**28.54**
8 m deep	5.70	34.25	-	m3	**34.25**
Excavating pits to receive bases; not exceeding					
0.25 m deep	3.10	18.63	-	m3	**18.63**
1 m deep	3.30	19.83	-	m3	**19.83**
2 m deep	3.85	23.14	-	m3	**23.14**
4 m deep	4.95	29.75	-	m3	**29.75**
6 m deep	6.25	37.56	-	m3	**37.56**
Extra over pit excavation for commencing below ground or reduced level					
1 m below	0.55	3.31	-	m3	**3.31**
2 m below	1.10	6.61	-	m3	**6.61**
3 m below	1.65	9.92	-	m3	**9.92**
4 m below	2.20	13.22	-	m3	**13.22**
Excavating trenches; not exceeding 0.30 m wide; not exceeding					
0.25 m deep	2.75	16.53	-	m3	**16.53**
1 m deep	3.05	18.33	-	m3	**18.33**
2 m deep	3.50	21.03	-	m3	**21.03**
4 m deep	4.45	26.74	-	m3	**26.74**
6 m deep	5.65	33.95	-	m3	**33.95**
Excavating trenches; over 0.30 m wide; not exceeding					
0.25 m deep	2.55	15.32	-	m3	**15.32**
1 m deep	2.75	16.53	-	m3	**16.53**
2 m deep	3.20	19.23	-	m3	**19.23**
4 m deep	4.05	24.34	-	m3	**24.34**
6 m deep	5.15	30.95	-	m3	**30.95**
Extra over trench excavation for commencing below ground or reduced level					
1 m below	0.55	3.31	-	m3	**3.31**
2 m below	1.10	6.61	-	m3	**6.61**
3 m below	1.65	9.92	-	m3	**9.92**
4 m below	2.20	13.22	-	m3	**13.22**
Extra over trench excavation for curved trench	0.55	3.31	-	m3	**3.31**
Excavating for pile caps and ground beams between piles; not exceeding					
0.25 m deep	3.65	21.93	-	m3	**21.93**
1 m deep	3.95	23.74	-	m3	**23.74**
2 m deep	4.60	27.64	-	m3	**27.64**
Excavating to bench sloping ground to receive filling; not exceeding					
0.25 m deep	1.65	9.92	-	m3	**9.92**
1 m deep	2.00	12.02	-	m3	**12.02**
2 m deep	2.20	13.22	-	m3	**13.22**

D GROUNDWORK Including overheads and profit at 7.50%	Labour hours	Labour £	Material Plant £	Unit	Total rate £
Extra over any type of excavation at any depth for excavating					
alongside services	1.10	6.61	-	m3	**6.61**
around services crossing excavation	2.20	13.22	-	m3	**13.22**
below ground water level	0.39	2.34	-	m3	**2.34**
Extra over excavation for breaking out existing materials					
rock	5.50	33.05	6.77	m3	**39.82**
concrete	4.95	29.75	5.64	m3	**35.39**
reinforced concrete	6.60	39.66	7.90	m3	**47.56**
brickwork; blockwork or stonework	3.30	19.83	3.39	m3	**23.22**
Extra over excavation for breaking out existing hard pavings					
tarmacadam 75 mm thick	0.44	2.64	0.45	m2	**3.09**
tarmacadam and hardcore 150 mm thick	0.55	3.31	0.56	m2	**3.87**
concrete 150 mm thick	0.77	4.63	0.79	m2	**5.42**
reinforced concrete 150 mm thick	0.99	5.95	1.13	m2	**7.08**
Extra for taking up precast concrete paving slabs	0.33	1.98	-	m2	**1.98**
Working space allowance to excavations					
reduced levels; basements and the like	2.85	17.13	-	m2	**17.13**
pits	2.95	17.73	-	m2	**17.73**
trenches	2.60	15.62	-	m2	**15.62**
pile caps and ground beams between piles	3.05	18.33	-	m2	**18.33**
Extra over excavation for working space for filling in with					
hardcore	0.88	5.29	4.84	m2	**10.13**
coarse ashes	0.79	4.75	6.02	m2	**10.77**
sand	0.88	5.29	12.09	m2	**17.38**
40 - 20 mm gravel	0.88	5.29	15.32	m2	**20.61**
in situ concrete - 7.50 N/mm2 - 40 mm aggregate (1:8)	1.20	7.48	25.61	m2	**33.09**
Hand excavation inside existing buildings					
Excavating basements and the like; starting from reduced level; not exceeding					
1 m deep	3.30	19.83	-	m3	**19.83**
2 m deep	3.70	22.23	-	m3	**22.23**
4 m deep	4.80	28.84	-	m3	**28.84**
6 m deep	5.90	35.45	-	m3	**35.45**
8 m deep	7.15	42.97	-	m3	**42.97**
Excavating pits to receive bases; not exceeding					
0.25 m deep	3.85	23.14	-	m3	**23.14**
1 m deep	4.15	24.94	-	m3	**24.94**
2 m deep	4.80	28.84	-	m3	**28.84**
4 m deep	6.20	37.26	-	m3	**37.26**
Extra over pit excavation for commencing below ground or reduced level					
1 m below	0.65	3.91	-	m3	**3.91**
2 m below	1.35	8.11	-	m3	**8.11**
3 m below	2.05	12.32	-	m3	**12.32**
4 m below	2.75	16.53	-	m3	**16.53**
Excavating trenches; not exceeding 0.30 m wide; not exceeding					
0.25 m deep	3.15	18.93	-	m3	**18.93**
1 m deep	3.45	20.73	-	m3	**20.73**
2 m deep	4.00	24.04	-	m3	**24.04**
4 m deep	5.10	30.65	-	m3	**30.65**
6 m deep	6.45	38.76	-	m3	**38.76**

D GROUNDWORK Including overheads and profit at 7.50%	Labour hours	Labour £	Material Plant £	Unit	Total rate £
D20 EXCAVATION AND FILLING - cont'd					
Hand excavation inside existing buildings - cont'd					
Excavating trenches; over 0.30 m wide; not exceeding					
0.25 m deep	3.15	18.93	-	m3	**18.93**
1 m deep	3.45	20.73	-	m3	**20.73**
2 m deep	4.00	24.04	-	m3	**24.04**
4 m deep	5.10	30.65	-	m3	**30.65**
6 m deep	6.45	38.76	-	m3	**38.76**
Extra over trench excavation for commencing below ground or reduced level					
1 m below	0.65	3.91	-	m3	**3.91**
2 m below	1.35	8.11	-	m3	**8.11**
3 m below	2.05	12.32	-	m3	**12.32**
4 m below	2.75	16.53	-	m3	**16.53**
Extra over trench excavation for curved trench	0.70	4.21	-	m3	**4.21**
Extra over any type of excavation at any depth for excavating					
below ground water level	0.07	0.42	1.10	m3	**1.52**
Extra over excavation for breaking out existing materials					
concrete	6.20	37.26	5.64	m3	**42.90**
reinforced concrete	8.25	49.58	7.90	m3	**57.48**
brickwork; blockwork or stonework	4.15	24.94	3.39	m3	**28.33**
Extra over excavation for breaking out existing hard pavings					
concrete 150 mm thick	0.96	5.77	0.79	m3	**6.56**
reinforced concrete 150 mm thick	0.96	5.77	0.79	m3	**6.56**
Working space allowance to excavations					
pits	0.31	1.86	5.67	m2	**7.53**
trenches	0.06	0.36	0.91	m2	**1.27**
Earthwork support (average 'risk' prices)					
Not exceeding 2 m between opposing faces; not exceeding					
1 m deep	0.11	0.66	0.34	m2	**1.00**
2 m deep	0.14	0.84	0.38	m2	**1.22**
4 m deep	0.18	1.08	0.45	m2	**1.53**
6 m deep	0.21	1.26	0.53	m2	**1.79**
2 - 4 m between opposing faces; not exceeding					
1 m deep	0.12	0.72	0.38	m2	**1.10**
2 m deep	0.15	0.90	0.45	m2	**1.35**
4 m deep	0.19	1.14	0.53	m2	**1.67**
6 m deep	0.23	1.38	0.64	m2	**2.02**
Over 4 m between opposing faces; not exceeding					
1 m deep	0.13	0.78	0.45	m2	**1.23**
2 m deep	0.17	1.02	0.53	m2	**1.55**
4 m deep	0.21	1.26	0.64	m2	**1.90**
6 m deep	0.26	1.56	0.77	m2	**2.33**
Earthwork support (open boarded)					
Not exceeding 2 m between opposing faces; not exceeding					
1 m deep	0.33	1.98	1.06	m2	**3.04**
2 m deep	0.42	2.52	1.28	m2	**3.80**
4 m deep	0.53	3.18	1.59	m2	**4.77**
6 m deep	0.66	3.97	1.83	m2	**5.80**

D GROUNDWORK Including overheads and profit at 7.50%	Labour hours	Labour £	Material Plant £	Unit	Total rate £
2 - 4 m between opposing faces; not exceeding					
1 m deep	0.37	2.22	1.28	m2	**3.50**
2 m deep	0.46	2.76	1.62	m2	**4.38**
4 m deep	0.59	3.55	1.91	m2	**5.46**
6 m deep	0.73	4.39	2.25	m2	**6.64**
Over 4 m between opposing faces; not exceeding					
1 m deep	0.42	2.52	1.49	m2	**4.01**
2 m deep	0.53	3.18	1.87	m2	**5.05**
4 m deep	0.66	3.97	2.25	m2	**6.22**
6 m deep	0.84	5.05	2.59	m2	**7.64**
Earthwork support (close boarded)					
Not exceeding 2 m between opposing faces; not exceeding					
1 m deep	0.88	5.29	1.91	m2	**7.20**
2 m deep	1.10	6.61	2.34	m2	**8.95**
4 m deep	1.40	8.41	2.87	m2	**11.28**
6 m deep	1.70	10.22	3.40	m2	**13.62**
2 - 4 m between opposing faces; not exceeding					
1 m deep	0.97	5.83	2.13	m2	**7.96**
2 m deep	1.20	7.21	2.66	m2	**9.87**
4 m deep	1.40	8.41	3.19	m2	**11.60**
6 m deep	1.70	10.22	3.72	m2	**13.94**
Over 4 m between opposing faces; not exceeding					
1 m deep	1.05	6.31	2.34	m2	**8.65**
2 m deep	1.30	7.81	2.98	m2	**10.79**
4 m deep	1.55	9.31	3.51	m2	**12.82**
6 m deep	1.85	11.12	4.04	m2	**15.16**
Extra over earthwork support for					
extending into running silt or the like	0.55	3.31	0.64	m2	**3.95**
extending below ground water level	0.33	1.98	0.32	m2	**2.30**
support next to roadways	0.44	2.64	0.53	m3	**3.17**
curved support	0.03	0.18	0.36	m2	**0.54**
support left in	0.72	4.33	12.51	m2	**16.84**
Earthwork support (average 'risk' prices - inside existing buildings)					
Not exceeding 2 m between opposing faces; not exceeding					
1 m deep	0.21	1.26	0.53	m2	**1.79**
2 m deep	0.26	1.56	0.64	m2	**2.20**
4 m deep	0.33	1.98	0.81	m2	**2.79**
6 m deep	0.41	2.46	0.91	m2	**3.37**
2 - 4 m between opposing faces; not exceeding					
1 m deep	0.23	1.38	0.64	m2	**2.02**
2 m deep	0.29	1.74	0.81	m2	**2.55**
4 m deep	0.37	2.22	0.96	m2	**3.18**
6 m deep	0.45	2.70	1.13	m2	**3.83**
Over 4 m between opposing faces; not exceeding					
1 m deep	0.26	1.56	0.74	m2	**2.30**
2 m deep	0.33	1.98	0.94	m2	**2.92**
4 m deep	0.41	2.46	1.13	m2	**3.59**
6 m deep	0.52	3.12	1.34	m2	**4.46**

D GROUNDWORK Including overheads and profit at 7.50%	Labour hours	Labour £	Material Plant £	Unit	Total rate £
D20 EXCAVATION AND FILLING - cont'd					
Mechanical disposal of excavated materials					
Removing from site to tip not exceeding 13 km (using lorries)	-	-	-	m3	**8.86**
Depositing on site in spoil heaps					
average 25 m distant	-	-	-	m3	**0.65**
average 50 m distant	-	-	-	m3	**1.08**
average 100 m distant	-	-	-	m3	**1.72**
average 200 m distant	-	-	-	m3	**2.58**
Spreading on site					
average 25 m distant	0.23	1.38	0.65	m3	**2.03**
average 50 m distant	0.23	1.38	1.08	m3	**2.46**
average 100 m distant	0.23	1.38	1.72	m3	**3.10**
average 200 m distant	0.23	1.38	2.58	m3	**3.96**
Hand disposal of excavated materials					
Removing from site to tip not exceeding 13 km (using lorries)	1.10	6.61	11.09	m3	**17.70**
Depositing on site in spoil heaps					
average 25 m distant	1.10	6.61	-	m3	**6.61**
average 50 m distant	1.45	8.71	-	m3	**8.71**
average 100 m distant	2.10	12.62	-	m3	**12.62**
average 200 m distant	3.10	18.63	-	m3	**18.63**
Spreading on site					
average 25 m distant	1.45	8.71	-	m3	**8.71**
average 50 m distant	1.75	10.52	-	m3	**10.52**
average 100 m distant	2.40	14.42	-	m3	**14.42**
average 200 m distant	3.40	20.43	-	m3	**20.43**
Hand disposal of excavated materials inside existing buildings					
Removing from site to tip not exceeding 13 km (using lorries)	1.40	8.41	11.09	m3	**19.50**
Depositing on site in spoil heaps					
average 25 m distant	1.40	8.41	-	m3	**8.41**
Spreading on site					
average 25 m distant	1.80	10.82	-	m3	**10.82**
Mechanical filling with excavated material					
Filling to excavations	0.17	1.02	1.72	m3	**2.74**
Filling to make up levels over 250 mm thick; depositing; compacting	0.25	1.50	1.29	m3	**2.79**
Filling to make up levels not exceeding 250 mm thick; compacting	0.30	1.80	1.72	m3	**3.52**
Mechanical filling with imported soil					
Filling to make up levels over 250 mm thick; depositing; compacting in layers	0.25	1.50	13.38	m3	**14.88**
Filling to make up levels not exceeding 250 mm thick; compacting	0.30	1.80	17.20	m3	**19.00**
Mechanical filling with hardcore; PC £6.00/m3					
Filling to excavations	0.20	1.20	10.21	m3	**11.41**
Filling to make up levels over 250 mm thick; depositing; compacting	0.30	1.80	9.78	m3	**11.58**
Filling to make up levels not exceeding 250 mm thick; compacting	0.35	2.10	12.47	m3	**14.57**

D GROUNDWORK Including overheads and profit at 7.50%	Labour hours	Labour £	Material Plant £	Unit	Total rate £
Mechanical filling with granular fill type 1; PC £10.15/t (PC £17.25/m3)					
Filling to excavations	0.20	1.20	25.33	m3	**26.53**
Filling to make up levels over 250 mm thick; depositing; compacting	0.30	1.80	24.90	m3	**26.70**
Filling to make up levels not exceeding 250 mm thick; compacting	0.35	2.10	31.82	m3	**33.92**
Mechanical filling with granular fill type 2; PC £9.75/t (PC £16.60/m3)					
Filling to excavations	0.20	1.20	24.46	m3	**25.66**
Filling to make up levels over 250 mm thick; depositing; compacting	0.30	1.80	24.03	m3	**25.83**
Filling to make up levels not exceeding 250 mm thick; compacting	0.35	2.10	30.70	m3	**32.80**
Mechanical filling with coarse ashes; PC £7.00/m3					
Filling to make up levels over 250 mm thick; depositing; compacting	0.28	1.68	11.51	m3	**13.19**
Filling to make up levels not exceeding 250 mm thick; compacting	0.32	1.92	14.73	m3	**16.65**
Mechanical filling with sand; PC £9.40/t (PC £15.00/m3)					
Filling to make up levels over 250 mm thick; depositing; compacting	0.30	1.80	21.88	m3	**23.68**
Filling to make up levels not exceeding 250 mm thick; compacting	0.35	2.10	27.95	m3	**30.05**
Hand filling with excavated material					
Filling to excavations	1.10	6.61	-	m3	**6.61**
Filling to make up levels over 250 mm thick; depositing; compacting	1.20	7.21	-	m3	**7.21**
Filling to make up levels exceeding 250 mm thick; multiple handling via spoil heap average - 25 m distant	2.65	15.92	-	m3	**15.92**
Filling to make up levels not exceeding 250 mm thick; compacting	1.50	9.01	-	m3	**9.01**
Hand filling with imported soil					
Filling to make up levels over 250 mm thick; depositing; compacting in layers	1.20	7.21	12.09	m3	**19.30**
depositing; compacting in layers	1.50	9.01	15.48	m3	**24.49**
Hand filling with hardcore; PC £6.00/m3					
Filling to excavations	1.45	8.71	8.06	m3	**16.77**
Filling to make up levels over 250 mm thick; depositing; compacting	0.61	3.67	13.15	m3	**16.82**
Filling to make up levels not exceeding 250 mm thick; compacting	0.73	4.39	16.42	m3	**20.81**
Hand filling with granular fill type 1; PC £10.15/t (PC £17.25/m3)					
Filling to excavations	1.45	8.71	23.18	m3	**31.89**
Filling to make up levels over 250 mm thick; depositing; compacting	0.61	3.67	28.26	m3	**31.93**
Filling to make up levels not exceeding 250 mm thick; compacting	0.73	4.39	35.77	m3	**40.16**

D GROUNDWORK Including overheads and profit at 7.50%	Labour hours	Labour £	Material Plant £	Unit	Total rate £
D20 EXCAVATION AND FILLING - cont'd					
Hand filling with granular fill type 2; PC £9.75/t (PC £16.60/m3)					
Filling to excavations	1.45	8.71	22.31	m3	**31.02**
Filling to make up levels over 250 mm thick; depositing; compacting	0.61	3.67	27.39	m3	**31.06**
Filling to make up levels not exceeding 250 mm thick; compacting	0.73	4.39	34.65	m3	**39.04**
Hand filling with coarse ashes; PC £7.00/m3					
Filling to make up levels over 250 mm thick; depositing; compacting	0.55	3.31	14.63	m3	**17.94**
Filling to make up levels not exceeding 250 mm thick; compacting	0.66	3.97	18.34	m3	**22.31**
Hand filling with sand; PC £9.40/t (PC £15.00/m3)					
Filling to excavations	0.55	3.31	24.78	m3	**28.09**
Filling to make up levels over 250 mm thick; depositing; compacting	0.72	4.33	26.17	m3	**30.50**
Filling to make up levels not exceeding 250 mm thick; compacting	0.85	5.11	32.92	m3	**38.03**
Hand filling with excavated material inside existing buildings					
Filling to excavations	1.40	8.41	-	m3	**8.41**
Hand filling with hardcore; PC £6.00/m3; inside existing buildings					
Filling to excavations	1.80	10.82	8.06	m3	**18.88**
Filling to make up levels over 250 mm thick; depositing; compacting	0.96	5.77	14.53	m3	**20.30**
Filling to make up levels not exceeding 250 mm thick; compacting	0.96	5.77	14.53	m3	**20.30**
Surface packing to filling					
To vertical or battered faces	0.20	1.20	0.48	m2	**1.68**
Surface treatments					
Compacting					
surfaces of ashes	0.04	0.24	-	m2	**0.24**
filling; blinding with ashes	0.09	0.54	0.53	m2	**1.07**
filling; blinding with sand	0.09	0.54	1.21	m2	**1.75**
bottoms of excavations	0.06	0.36	-	m2	**0.36**
Trimming					
sloping surfaces	0.20	1.20	-	m2	**1.20**
sloping surfaces; in rock	1.10	6.05	2.26	m2	**8.31**
Filter membrane; one layer; laid on earth to receive granular material					
'Terram 500'	0.06	0.36	0.44	m2	**0.80**
'Terram 700'	0.06	0.36	0.44	m2	**0.80**
'Terram 1000'	0.07	0.42	0.59	m2	**1.01**
D50 UNDERPINNING					
Mechanical excavation using a wheeled hydraulic off-centre excavator with a 0.24 m3 bucket					
Excavating preliminary trenches; not exceeding					
1 m deep	0.29	1.74	5.11	m3	**6.85**
2 m deep	0.35	2.10	6.13	m3	**8.23**
4 m deep	0.40	2.40	7.15	m3	**9.55**

D GROUNDWORK Including overheads and profit at 7.50%	Labour hours	Labour £	Material £	Unit	Total rate £
Extra for breaking up					
concrete 150 mm thick	0.81	4.87	0.79	m2	**5.66**
Hand excavation					
Excavating preliminary trenches; not exceeding					
1 m deep	3.20	19.23	-	m3	**19.23**
2 m deep	3.65	21.93	-	m3	**21.93**
4 m deep	4.70	28.24	-	m3	**28.24**
Extra for breaking up					
concrete 150 mm thick	0.33	1.98	1.59	m2	**3.57**
Excavating underpinning pits starting from 1 m below ground level; not exceeding					
0.25 m deep	4.85	29.14	-	m3	**29.14**
1 m deep	5.30	31.85	-	m3	**31.85**
2 m deep	6.35	38.16	-	m3	**38.16**
Excavating underpinning pits starting from 2 m below ground level; not exceeding					
0.25 m deep	5.95	35.76	-	m3	**35.76**
1 m deep	6.40	38.46	-	m3	**38.46**
2 m deep	7.45	44.77	-	m3	**44.77**
Excavating underpinning pits starting from 4 m below ground level; not exceeding					
0.25 m deep	7.05	42.37	-	m3	**42.37**
1 m deep	7.50	45.07	-	m3	**45.07**
2 m deep	8.55	51.38	-	m3	**51.38**
Extra over any type of excavation at any depth for excavating below ground water level	0.39	2.34	-	m3	**2.34**
Earthwork support (open boarded) in 3 m lengths					
To preliminary trenches; not exceeding 2 m between opposing faces; not exceeding					
1 m deep	0.44	2.64	1.38	m2	**4.02**
2 m deep	0.55	3.31	1.70	m2	**5.01**
4 m deep	0.70	4.21	2.13	m2	**6.34**
To underpinning pits; not exceeding 2 m between opposing faces; not exceeding					
1 m deep	0.48	2.88	1.49	m2	**4.37**
2 m deep	0.61	3.67	1.91	m2	**5.58**
4 m deep	0.77	4.63	2.34	m2	**6.97**
Earthwork support (closed boarded) in 3 m lengths					
To preliminary trenches; not exceeding 2 m between opposing faces; not exceeding					
1 m deep	1.10	6.61	2.34	m2	**8.95**
2 m deep	1.40	8.41	2.98	m2	**11.39**
4 m deep	1.70	10.22	3.61	m2	**13.83**
To underpinning pits; not exceeding 2 m between opposing faces; not exceeding					
1 m deep	1.20	7.21	2.55	m2	**9.76**
2 m deep	1.50	9.01	3.19	m2	**12.20**
4 m deep	1.85	11.12	4.04	m2	**15.16**
Extra over all types of earthwork support for earthwork support left in	0.83	4.99	12.51	m2	**17.50**
Preparation/cutting away					
Cutting away projecting plain concrete foundations					
150 x 150 mm	0.18	1.08	0.36	m	**1.44**
150 x 225 mm	0.26	1.56	0.54	m	**2.10**
150 x 300 mm	0.35	2.10	0.72	m	**2.82**
300 x 300 mm	0.69	4.15	1.42	m	**5.57**

D GROUNDWORK Including overheads and profit at 7.50%		Labour hours	Labour £	Material £	Unit	Total rate £
D50 UNDERPINNING - cont'd						
Preparation/cutting away - cont'd						
Cutting away masonry						
one course high		0.06	0.36	0.11	m	**0.47**
two courses high		0.15	0.90	0.32	m	**1.22**
three courses high		0.30	1.80	0.61	m	**2.41**
four courses high		0.50	3.00	1.02	m	**4.02**
Preparing underside of existing work to receive new underpinning						
380 mm wide		0.66	3.97	-	m	**3.97**
600 mm wide		0.88	5.29	-	m	**5.29**
900 mm wide		1.10	6.61	-	m	**6.61**
1200 mm wide		1.30	7.81	-	m	**7.81**
Hand disposal of excavated materials						
Removing from site to tip not exceeding 13 km (using lorries)		1.10	6.61	11.09	m3	**17.70**
Hand filling with excavated material						
Filling to excavations		1.10	6.61	-	m3	**6.61**
Surface treatments						
Compacting bottoms of excavations		0.06	0.36	-	m2	**0.36**
Plain insitu ready mixed concrete; 11.50 N/mm2 - 40 mm aggregate (1:3:6); poured against faces of excavation						
Underpinning						
over 450 mm thick		3.10	19.33	49.06	m3	**68.39**
150 - 450 mm thick		3.55	22.13	49.06	m3	**71.19**
not exceeding 150 mm thick		4.25	26.50	49.06	m3	**75.56**
Extra for working around reinforcement		0.35	2.18	-	m3	**2.18**
Plain insitu ready mixed concrete; 21.00 N/mm2 - 20 mm aggregate (1:2:4); poured against faces of excavation						
Underpinning						
over 450 mm thick		3.10	19.33	52.25	m3	**71.58**
150 - 450 mm thick		3.55	22.13	52.25	m3	**74.38**
not exceeding 150 mm thick		4.25	26.50	52.25	m3	**78.75**
Extra for working around reinforcement		0.35	2.18	-	m3	**2.18**
Sawn formwork						
Sides of foundations in underpinning						
over 1 m high		1.85	13.48	5.23	m2	**18.71**
not exceeding 250 mm high		0.63	4.59	1.47	m	**6.06**
250 - 500 mm high		0.98	7.14	2.78	m	**9.92**
500 mm - 1 m high		1.50	10.93	5.23	m	**16.16**
Reinforcement						
Reinforcement bars; BS 4449; hot rolled plain round mild steel bars; bent						
20 mm	PC £364.05	18.70	135.29	445.84	t	**581.13**
16 mm	PC £369.45	22.00	159.16	455.20	t	**614.36**
12 mm	PC £396.45	25.30	183.04	488.96	t	**672.00**
10 mm	PC £418.05	29.70	214.87	516.61	t	**731.48**
8 mm	PC £445.95	34.10	246.71	551.37	t	**798.08**
6 mm	PC £485.55	39.60	286.50	599.35	t	**885.85**
8 mm; links or the like	PC £445.95	46.20	334.25	561.20	t	**895.45**
6 mm; links or the like	PC £485.55	53.90	389.95	612.44	t	**1002.39**

D GROUNDWORK Including overheads and profit at 7.50%		Labour hours	Labour £	Material £	Unit	Total rate £
Reinforcement bars; BS 4461; cold worked deformed square high yield steel bars; bent						
20 mm	PC £367.65	18.70	135.29	449.90	t	**585.19**
16 mm	PC £373.05	22.00	159.16	459.27	t	**618.43**
12 mm	PC £400.05	25.30	183.04	493.02	t	**676.06**
10 mm	PC £421.65	29.70	214.87	520.67	t	**735.54**
8 mm	PC £449.55	34.10	246.71	555.44	t	**802.15**
6 mm	PC £489.15	39.60	286.50	603.41	t	**889.91**
Common bricks; PC £124.00/1000; in cement mortar (1:3)						
Walls in underpinning						
one brick thick		2.75	28.53	19.96	m2	**48.49**
one and a half brick thick		3.75	38.90	30.03	m2	**68.93**
two brick thick		4.75	49.28	39.93	m2	**89.21**
Add or deduct for variation of £10.00/1000 in PC of common bricks						
one brick thick		-	-	-	m2	**1.35**
one and a half brick thick		-	-	-	m2	**2.03**
two brick thick		-	-	-	m2	**2.71**
Class A engineering bricks; PC £350.00/1000; in cement mortar (1:3)						
Walls in underpinning						
one brick thick		3.00	31.12	58.25	m2	**89.37**
one and a half brick thick		4.00	41.50	87.45	m2	**128.95**
two brick thick		5.00	51.87	116.50	m2	**168.37**
Add or deduct for variation of £10.00/1000 in PC of common bricks						
one brick thick		-	-	-	m2	**1.35**
one and a half brick thick		-	-	-	m2	**2.03**
two brick thick		-	-	-	m2	**2.71**
Class B engineering bricks; PC £180.00/1000; in cement mortar (1:3)						
Walls in underpinning						
one brick thick		3.00	31.12	27.55	m2	**58.67**
one and a half brick thick		4.00	41.50	41.40	m2	**82.90**
two brick thick		5.00	51.87	55.10	m2	**106.97**
Add or deduct for variation of £10.00/1000 in PC of common bricks						
one brick thick		-	-	-	m2	**1.35**
one and a half brick thick		-	-	-	m2	**2.03**
two brick thick		-	-	-	m2	**2.71**
'Pluvex' (hessian based) damp proof course or similar; PC £3.95/m2; 200 mm laps; in cement mortar (1:3)						
Horizontal						
over 225 mm wide		0.29	3.01	4.57	m2	**7.58**
not exceeding 225 mm wide		0.58	6.02	4.66	m2	**10.68**
'Hyload' (pitch polymer) damp proof course or similar; PC £4.43/m2; 150 mm laps; in cement mortar (1:3)						
Horizontal						
over 225 mm wide		0.29	3.01	5.11	m2	**8.12**
not exceeding 225 mm wide		0.58	6.02	5.23	m2	**11.25**

D GROUNDWORK Including overheads and profit at 7.50%	Labour hours	Labour £	Material £	Unit	Total rate £
D50 UNDERPINNING - cont'd					
'Ledkore' grade A (bitumen based lead cored) damp proof course or similar; PC £12.21/m2; 200 mm laps; in cement mortar (1:3)					
Horizontal					
over 225 mm wide	0.38	3.94	14.11	m2	**18.05**
not exceeding 225 mm wide	0.76	7.88	14.44	m2	**22.32**
Two courses of slates in cement mortar (1:3)					
Horizontal					
over 225 mm wide	1.70	17.64	25.80	m2	**43.44**
not exceeding 225 mm wide	2.90	30.08	26.25	m2	**56.33**
Wedging and pinning					
To underside of existing construction with slates in cement mortar (1:3)					
102 mm wall	1.25	12.97	4.87	m	**17.84**
215 mm wall	1.50	15.56	7.37	m	**22.93**
317 mm wall	1.70	17.64	9.79	m	**27.43**

E IN SITU CONCRETE/LARGE PRECAST CONCRETE
Including overheads and profit at 7.50%

BASIC CONCRETE PRICES

	£		£		£
Concrete aggregates (£/tonne)					
40 mm all-in	10.70	40 mm shingle	10.70	10 mm shingle	10.80
20 mm all-in	11.80	20 mm shingle	10.80	sharp sand	9.40

	£		£
Formwork items			
plywood (£/m2)	7.43	timber (£/m3)	197.75
Lightweight aggregates			
'Lytag' (£/m3)			
'fines'	23.00	6-12 mm granular	21.79
Portland cement (£/tonne)			
normal - in bags	65.43	normal - in bulk to silos	58.75
sulphate - resisting - plus	9.92	rapid - hardening - plus	3.85

Tying wire for reinforcement - £0.56/kg

MIXED CONCRETE PRICES (£/m3)

The following prices are for ready or site mixed concrete ready for placing including 2.5% for waste and 7.5% or overheads and profit	Mix 7.50 N/mm2 - 40mm aggregate (1:8) £	Mix 11.50 N/mm2 - 40mm aggregate (1:3:6) £	Mix 15.00 N/mm2 - 40mm aggregate £	Mix 21.00 N/mm2 - 20mm aggregate (1:2:4) £	Mix 26.00 N/mm2 - 20mm aggregate (1:1.5:3) £	Mix 31.00 N/mm2 - 20mm aggregate (1:1:2) £	Mix 40.00 N/mm2 - 20mm aggregate £
Ready mixed concrete							
Normal Portland cement	43.75	44.60	46.00	47.50	50.00	52.50	55.50
Sulphate - resistant cement	45.50	46.50	48.00	50.00	53.00	56.00	60.00
Normal Portland cement with water-repellent additive	43.80	46.75	48.15	49.65	52.15	54.65	57.65
Normal Portland cement; air-entrained	43.80	46.75	48.15	49.65	52.15	54.65	57.65
Lightweight concrete using Lytag medium and natural sand	-	-	64.00	62.60	64.70	66.60	-
Site mixed concrete							
Normal Portland cement	-	50.00	52.00	54.00	56.00	61.00	66.00
Sulphate - resistant cement	-	52.00	54.50	57.00	60.00	65.00	71.00

E IN SITU CONCRETE/LARGE PRECAST CONCRETE Including overheads and profit at 7.50%	Labour hours	Labour £	Material £	Unit	Total rate £
E10 IN SITU CONCRETE					
Plain in situ ready mixed concrete; 11.50 N/mm2 - 40 mm aggregate (1:3:6)					
Foundations	1.85	11.53	44.60	m3	56.13
Isolated foundations	2.20	13.72	44.60	m3	58.32
Beds					
over 450 mm thick	1.45	9.04	44.60	m3	53.64
150 - 450 mm thick	1.90	11.85	44.60	m3	56.45
not exceeding 150 mm thick	2.75	17.15	44.60	m3	61.75
Filling to hollow walls					
not exceeding 150 mm thick	4.90	30.55	44.60	m3	75.15
Plain in situ ready mixed concrete; 11.50 N/mm2 - 40 mm aggregate (1:3:6); poured on or against earth or unblinded hardcore					
Foundations	1.95	12.16	46.83	m3	58.99
Isolated foundations	2.30	14.34	46.83	m3	61.17
Beds					
over 450 mm thick	1.50	9.35	46.83	m3	56.18
150 - 450 mm thick	2.00	12.47	46.83	m3	59.30
not exceeding 150 mm thick	2.90	18.08	46.83	m3	64.91
Plain in situ ready mixed concrete; 21.00 N/mm2 - 20 mm aggregate (1:2:4)					
Foundations	1.85	11.53	47.50	m3	59.03
Isolated foundations	2.20	13.72	47.50	m3	61.22
Beds					
over 450 mm thick	1.45	9.04	47.50	m3	56.54
150 - 450 mm thick	1.90	11.85	47.50	m3	59.35
not exceeding 150 mm thick	2.75	17.15	47.50	m3	64.65
Filling to hollow walls					
not exceeding 150 mm thick	4.90	30.55	47.50	m3	78.05
Plain in situ ready mixed concrete; 21.00 N/mm2 - 20 mm aggregate (1:2:4); poured on or against earth or unblinded hardcore					
Foundations	1.95	12.16	49.88	m3	62.04
Isolated foundations	2.30	14.34	49.88	m3	64.22
Beds					
over 450 mm thick	1.50	9.35	49.88	m3	59.23
150 - 450 mm thick	2.00	12.47	49.88	m3	62.35
not exceeding 150 mm thick	2.90	18.08	49.88	m3	67.96
Reinforced in situ ready mixed concrete; 21.00 N/mm2 - 20 mm aggregate (1:2:4)					
Foundations	2.30	14.34	47.50	m3	61.84
Ground beams	4.05	25.25	47.50	m3	72.75
Isolated foundations	2.65	16.52	47.50	m3	64.02
Beds					
over 450 mm thick	1.85	11.53	47.50	m3	59.03
150 - 450 mm thick	2.30	14.34	47.50	m3	61.84
not exceeding 150 mm thick	3.15	19.64	47.50	m3	67.14
Slabs					
over 450 mm thick	3.55	22.13	47.50	m3	69.63
150 - 450 mm thick	4.05	25.25	47.50	m3	72.75
not exceeding 150 mm thick	5.05	31.49	47.50	m3	78.99
Coffered or troughed slabs					
over 450 mm thick	4.05	25.25	47.50	m3	72.75
150 - 450 mm thick	4.60	28.68	47.50	m3	76.18

E IN SITU CONCRETE/LARGE PRECAST CONCRETE Including overheads and profit at 7.50%	Labour hours	Labour £	Material £	Unit	Total rate £
Extra over for laying to slopes					
not exceeding 15 degrees	0.35	2.18	-	m3	2.18
over 15 degrees	0.69	4.30	-	m3	4.30
Walls					
over 450 mm thick	3.75	23.38	47.50	m3	70.88
150 - 450 mm thick	4.25	26.50	47.50	m3	74.00
not exceeding 150 mm thick	5.30	33.05	47.50	m3	80.55
Isolated beams	5.75	35.85	47.50	m3	83.35
Isolated deep beams	6.35	39.59	47.50	m3	87.09
Attached deep beams	5.75	35.85	47.50	m3	83.35
Isolated beam casings	6.35	39.59	47.50	m3	87.09
Isolated deep beam casings	6.90	43.02	47.50	m3	90.52
Attached deep beam casings	6.35	39.59	47.50	m3	87.09
Columns	6.90	43.02	47.50	m3	90.52
Column casings	7.60	47.39	47.50	m3	94.89
Staircases	8.65	53.93	47.50	m3	101.43
Upstands	5.50	34.29	47.50	m3	81.79
Reinforced in situ ready mixed concrete; 26.00 N/mm2 - 20 mm aggregate (1:1.5:3)					
Foundations	2.30	14.34	50.00	m3	64.34
Ground beams	4.05	25.25	50.00	m3	75.25
Isolated foundations	2.65	16.52	50.00	m3	66.52
Beds					
over 450 mm thick	1.85	11.53	50.00	m3	61.53
150 - 450 mm thick	2.30	14.34	50.00	m3	64.34
not exceeding 150 mm thick	3.15	19.64	50.00	m3	69.64
Slabs					
over 450 mm thick	3.55	22.13	50.00	m3	72.13
150 - 450 mm thick	4.05	25.25	50.00	m3	75.25
not exceeding 150 mm thick	5.05	31.49	50.00	m3	81.49
Coffered or troughed slabs					
over 450 mm thick	4.05	25.25	50.00	m3	75.25
150 - 450 mm thick	4.60	28.68	50.00	m3	78.68
Extra over for laying to slopes					
not exceeding 15 degrees	0.35	2.18	-	m3	2.18
over 15 degrees	0.69	4.30	-	m3	4.30
Walls					
over 450 mm thick	3.75	23.38	50.00	m3	73.38
150 - 450 mm thick	4.25	26.50	50.00	m3	76.50
not exceeding 150 mm thick	5.30	33.05	50.00	m3	83.05
Isolated beams	5.75	35.85	50.00	m3	85.85
Isolated deep beams	6.35	39.59	50.00	m3	89.59
Attached deep beams	5.75	35.85	50.00	m3	85.85
Isolated beam casings	6.35	39.59	50.00	m3	89.59
Isolated deep beam casings	6.90	43.02	50.00	m3	93.02
Attached deep beam casings	6.35	39.59	50.00	m3	89.59
Columns	6.90	43.02	50.00	m3	93.02
Column casings	7.60	47.39	50.00	m3	97.39
Staircases	8.65	53.93	50.00	m3	103.93
Upstands	5.50	34.29	50.00	m3	84.29
Reinforced in situ ready mixed concrete; 31.00 N/mm2 - 20 mm aggregate (1:1:2)					
Beds					
over 450 mm thick	1.85	11.53	52.50	m3	64.03
150 - 450 mm thick	2.30	14.34	52.50	m3	66.84
not exceeding 150 mm thick	3.15	19.64	52.50	m3	72.14
Slabs					
over 450 mm thick	3.55	22.13	52.50	m3	74.63
150 - 450 mm thick	4.05	25.25	52.50	m3	77.75
not exceeding 150 mm thick	5.05	31.49	52.50	m3	83.99

E IN SITU CONCRETE/LARGE PRECAST CONCRETE Including overheads and profit at 7.50%	Labour hours	Labour £	Material £	Unit	Total rate £
E10 IN SITU CONCRETE - cont'd					
Reinforced in situ ready mixed concrete; 31.00 N/mm2 - 20 mm aggregate (1:1:2) - cont'd					
Coffered or troughed slabs					
over 450 mm thick	4.05	25.25	52.50	m3	**77.75**
150 - 450 mm thick	4.60	28.68	52.50	m3	**81.18**
Extra over for laying to slopes					
not exceeding 15 degrees	0.35	2.18	-	m3	**2.18**
over 15 degrees	0.69	4.30	-	m3	**4.30**
Walls					
over 450 mm thick	3.75	23.38	52.50	m3	**75.88**
150 - 450 mm thick	4.25	26.50	52.50	m3	**79.00**
not exceeding 150 mm thick	5.30	33.05	52.50	m3	**85.55**
Isolated beams	5.75	35.85	52.50	m3	**88.35**
Isolated deep beams	6.35	39.59	52.50	m3	**92.09**
Attached deep beams	5.75	35.85	52.50	m3	**88.35**
Isolated beam casings	6.35	39.59	52.50	m3	**92.09**
Isolated deep beam casings	6.90	43.02	52.50	m3	**95.52**
Attached deep beam casings	6.35	39.59	52.50	m3	**92.09**
Columns	6.90	43.02	52.50	m3	**95.52**
Column casings	7.60	47.39	52.50	m3	**99.89**
Staircases	8.65	53.93	52.50	m3	**106.43**
Upstands	5.50	34.29	52.50	m3	**86.79**
Extra over vibrated concrete for reinforcement content over 5%	0.63	3.93	-	m3	**3.93**
Grouting with cement mortar (1:1)					
Stanchion bases					
10 mm thick	1.10	6.86	0.44	nr	**7.30**
25 mm thick	1.40	8.73	1.06	nr	**9.79**
Grouting with epoxy resin					
Stanchion bases					
10 mm thick	1.40	8.73	0.31	nr	**9.04**
25 mm thick	1.65	10.29	0.74	nr	**11.03**
Filling; plain in situ concrete; 21.00 N/mm2 - 20 mm aggregate (1:2:4)					
Mortices	0.11	0.69	0.28	nr	**0.97**
Holes	0.28	1.75	1.93	m3	**3.68**
Chases					
over 0.01 m2	0.22	1.37	1.13	m3	**2.50**
not exceeding 0.01 m2	0.17	1.06	0.56	m	**1.62**
Sheeting to prevent moisture loss					
Building paper; lapped joints					
subsoil grade; horizontal on foundations	0.03	0.19	0.53	m2	**0.72**
standard grade; horizontal on slabs	0.06	0.37	0.85	m2	**1.22**
Polyethylene sheeting; lapped joints; horizontal on slabs					
250 microns; 0.25 mm thick	0.06	0.37	0.33	m2	**0.70**
'Visqueen' sheeting; lapped joints; horizontal on slabs					
1000 Grade; 0.25 mm thick	0.06	0.37	0.28	m2	**0.65**
1200 Super; 0.30 mm thick	0.07	0.44	0.38	m2	**0.82**

E IN SITU CONCRETE/LARGE PRECAST CONCRETE Including overheads and profit at 7.50%	Labour hours	Labour £	Material £	Unit	Total rate £
E20 FORMWORK FOR IN SITU CONCRETE					
Note: Generally all formwork based on four uses unless otherwise stated					
Sides of foundations					
over 1 m high	1.60	11.66	4.11	m2	15.77
not exceeding 250 mm high	0.52	3.79	1.30	m	5.09
250 - 500 mm high	0.86	6.27	2.21	m	8.48
500 mm - 1 m high	1.30	9.48	4.11	m	13.59
Sides of foundations; left in					
over 1 m high	1.60	11.66	13.83	m2	25.49
not exceeding 250 mm high	0.52	3.79	3.59	m	7.38
250 - 500 mm high	0.86	6.27	7.00	m	13.27
500 mm - 1 m high	1.30	9.48	13.83	m	23.31
Sides of ground beams and edges of beds					
over 1 m high	1.90	13.85	5.65	m2	19.50
not exceeding 250 mm high	0.58	4.23	1.58	m	5.81
250 - 500 mm high	1.05	7.65	2.91	m	10.56
500 mm - 1 m high	1.45	10.57	5.65	m	16.22
Edges of suspended slabs					
not exceeding 250 mm high	0.86	6.27	1.87	m	8.14
250 - 500 mm high	1.25	9.11	3.73	m	12.84
500 mm - 1 m high	2.00	14.58	7.11	m	21.69
Sides of upstands					
over 1 m high	2.30	16.76	7.47	m2	24.23
not exceeding 250 mm high	0.72	5.25	1.94	m	7.19
250 - 500 mm high	1.15	8.38	3.90	m	12.28
500 mm - 1 m high	2.00	14.58	7.47	m	22.05
Steps in top surfaces					
not exceeding 250 mm high	0.58	4.23	2.03	m	6.26
250 - 500 mm high	0.92	6.71	4.21	m	10.92
Steps in soffits					
not exceeding 250 mm high	0.63	4.59	2.03	m	6.62
250 - 500 mm high	1.00	7.29	4.21	m	11.50
Machine bases and plinths					
over 1 m high	1.85	13.48	5.65	m2	19.13
not exceeding 250 mm high	0.58	4.23	1.58	m	5.81
250 - 500 mm high	0.98	7.14	2.91	m	10.05
500 mm - 1 m high	1.45	10.57	5.65	m	16.22
Soffits of slabs; 1.5 - 3 m height to soffit					
not exceeding 200 mm thick	1.95	14.21	5.52	m2	19.73
not exceeding 200 mm thick (5 uses)	1.90	13.85	5.05	m2	18.90
not exceeding 200 mm thick (6 uses)	1.85	13.48	4.71	m2	18.19
200 - 300 mm thick	2.05	14.94	7.45	m2	22.39
300 - 400 mm thick	2.10	15.31	7.90	m2	23.21
400 - 500 mm thick	2.25	16.40	8.36	m2	24.76
500 - 600 mm thick	2.40	17.49	8.82	m2	26.31
Soffits of slabs; not exceeding 200 mm thick					
not exceeding 1.5 m height to soffit	2.05	14.94	6.89	m2	21.83
3 - 4.5 m height to soffit	1.95	14.21	6.72	m2	20.93
4.5 - 6 m height to soffit	2.05	14.94	7.87	m2	22.81
Soffits of landings; 1.5 - 3m height to soffit					
not exceeding 200 mm thick	2.05	14.94	5.73	m2	20.67
200 - 300 mm thick	2.20	16.03	7.73	m2	23.76
300 - 400 mm thick	2.25	16.40	8.19	m2	24.59
400 - 500 mm thick	2.35	17.13	8.65	m2	25.78
500 - 600 mm thick	2.55	18.59	9.10	m2	27.69
Extra over for sloping					
not exceeding 15 degrees	0.23	1.68	-	m2	1.68
over 15 degrees	0.46	3.35	-	m2	3.35

E IN SITU CONCRETE/LARGE PRECAST CONCRETE Including overheads and profit at 7.50%	Labour hours	Labour £	Material £	Unit	Total rate £
E20 FORMWORK FOR IN SITU CONCRETE - cont'd					
Soffits of coffered or troughed slabs; including 'Cordek' troughed forms; 300 mm deep; ribs at 600 mm centres and cross ribs at centres of bay; 300 - 400 mm thick					
1.5 - 3 m height to soffit	2.85	20.77	8.32	m2	**29.09**
3 - 4.5 m height to soffit	3.00	21.87	9.11	m2	**30.98**
4.5 - 6 m height to soffit	3.10	22.59	10.27	m2	**32.86**
Soffits of bands and margins to troughed slabs					
horizontal; 300 - 400 mm thick	2.30	16.76	6.27	m2	**23.03**
Top formwork	1.70	12.39	4.27	m2	**16.66**
Walls					
vertical	2.30	16.76	6.55	m2	**23.31**
vertical; interrupted	2.40	17.49	6.85	m2	**24.34**
vertical; exceeding 3 m high; inside stairwells	2.55	18.59	6.93	m2	**25.52**
vertical; exceeding 3 m high; inside lift shaft	2.75	20.04	7.47	m2	**27.51**
battered	3.20	23.32	7.72	m2	**31.04**
Beams attached to insitu slabs					
square or rectangular; 1.5 - 3 m height to soffit	2.55	18.59	8.28	m2	**26.87**
square or rectangular; 3 - 4.5 m height to soffit	2.65	19.31	9.45	m2	**28.76**
square or rectangular; 4.5 - 6 m height to soffit	2.75	20.04	10.62	m2	**30.66**
Beams attached to walls					
square or rectangular; 1.5 - 3 m height to soffit	2.65	19.31	8.28	m2	**27.59**
Isolated beams					
square or rectangular; 1.5 - 3 m height to soffit	2.75	20.04	8.28	m2	**28.32**
square or rectangular; 3 - 4.5 m height to soffit	2.90	21.14	9.45	m2	**30.59**
square or rectangular; 4.5 - 6 m height to soffit	3.00	21.87	10.62	m2	**32.49**
Beam casings attached to insitu slabs					
square or rectangular; 1.5 - 3 m height to soffit	2.65	19.31	8.28	m2	**27.59**
square or rectangular; 3 - 4.5 m height to soffit	2.75	20.04	9.45	m2	**29.49**
Beam casings attached to walls					
square or rectangular; 1.5 - 3 m height to soffit	2.75	20.04	8.28	m2	**28.32**
Isolated beam casings					
square or rectangular; 1.5 - 3 m height to soffit	2.90	21.14	8.28	m2	**29.42**
square or rectangular; 3 - 4.5 m height to soffit	3.00	21.87	9.45	m2	**31.32**
Extra over for sloping					
not exceeding 15 degrees	0.33	2.41	0.80	m2	**3.21**
over 15 degrees	0.66	4.81	1.60	m2	**6.41**
Columns attached to walls					
square or rectangular	2.55	18.59	6.55	m2	**25.14**
Isolated columns					
square or rectangular	2.65	19.31	6.55	m2	**25.86**
Column casings attached to walls					
square or rectangular	2.65	19.31	6.55	m2	**25.86**
Isolated column casings					
square or rectangular	2.75	20.04	6.55	m2	**26.59**

E IN SITU CONCRETE/LARGE PRECAST CONCRETE Including overheads and profit at 7.50%	Labour hours	Labour £	Material £	Unit	Total rate £
Extra over for					
Throat	0.06	0.44	0.17	m	0.61
Chamfer					
30 mm wide	0.07	0.51	0.24	m	0.75
60 mm wide	0.08	0.58	0.40	m	0.98
90 mm wide	0.09	0.66	0.83	m	1.49
Rebate or horizontal recess					
12 x 12 mm	0.08	0.58	0.10	m	0.68
25 x 25 mm	0.08	0.58	0.17	m	0.75
25 x 50 mm	0.08	0.58	0.17	m	0.75
50 x 50 mm	0.08	0.58	0.58	m	1.16
Nibs					
50 x 50 mm	0.63	4.59	1.28	m	5.87
100 x 100 mm	0.90	6.56	2.49	m	9.05
100 x 200 mm	1.20	8.75	3.31	m	12.06
Extra over basic formwork for rubbing down, filling and leaving face of concrete smooth					
general surfaces	0.38	2.77	0.10	m2	2.87
edges	0.58	4.23	0.12	m2	4.35
Add to prices for basic formwork for					
add +27.5% for curved radius 6 m					
add +50% for curved radius 2 m					
coating with retardant agent	0.02	0.15	0.49	m2	0.64
Wall kickers to both sides					
150 mm high	0.58	4.23	1.29	m	5.52
225 mm high	0.75	5.47	1.68	m	7.15
150 mm high; one side suspended	0.72	5.25	2.06	m	7.31
Wall ends, soffits and steps in walls					
over 1 m wide	2.20	16.03	6.89	m	22.92
not exceeding 250 mm wide	0.69	5.03	1.96	m	6.99
250 - 500 mm wide	1.10	8.02	3.77	m	11.79
500 mm - 1 m wide	1.70	12.39	6.89	m	19.28
Openings in walls					
over 1 m wide	2.40	17.49	8.29	m	25.78
not exceeding 250 mm wide	0.75	5.47	2.10	m	7.57
250 - 500 mm wide	1.25	9.11	4.20	m	13.31
500 mm - 1 m wide	1.95	14.21	8.29	m	22.50
Stair flights					
1 m wide; 150 mm waist; 150 mm undercut risers	5.75	41.91	21.04	m	62.95
2 m wide; 200 mm waist; 150 mm vertical risers	10.35	75.44	33.94	m	109.38
Mortices; not exceeding 250 mm deep					
not exceeding 500 mm girth	0.17	1.24	1.00	nr	2.24
Holes; not exceeding 250 mm deep					
not exceeding 500 mm girth	0.23	1.68	1.04	nr	2.72
500 mm - 1 m girth	0.29	2.11	2.00	nr	4.11
1 - 2 m girth	0.52	3.79	3.99	nr	7.78
2 - 3 m girth	0.69	5.03	5.99	nr	11.02
Holes; 250 - 500 mm deep					
not exceeding 500 mm girth	0.35	2.55	2.00	nr	4.55
500 mm - 1 m girth	0.44	3.21	3.99	nr	7.20
1 - 2 m girth	0.77	5.61	7.98	nr	13.59
2 - 3 m girth	1.05	7.65	11.97	nr	19.62
Permanent shuttering; left in					
Dufaylite 'Clayboard' shuttering; type KN30; horizontal; under concrete beds; left in					
50 mm thick	0.17	1.24	7.69	m2	8.93
75 mm thick	0.18	1.31	8.22	m2	9.53
100 mm thick	0.20	1.46	8.92	m2	10.38
150 mm thick	0.23	1.68	9.55	m2	11.23

E IN SITU CONCRETE/LARGE PRECAST CONCRETE Including overheads and profit at 7.50%		Labour hours	Labour £	Material £	Unit	Total rate £
E20 FORMWORK FOR IN SITU CONCRETE - cont'd						
Permanent shuttering; left in - cont'd						
Dufaylite 'Clayboard' shuttering; type KN30; horizontal or vertical (including temporary supports); beneath or to sides of foundations; left in						
50 mm thick; vertical		0.29	2.11	8.33	m2	**10.44**
400 x 50 mm thick; horizontal		0.10	0.73	3.14	m	**3.87**
600 x 50 mm thick; horizontal		0.14	1.02	4.73	m	**5.75**
800 x 50 mm thick; horizontal		0.17	1.24	6.30	m	**7.54**
75 mm thick; vertical		0.31	2.26	8.89	m2	**11.15**
400 x 75 mm thick; horizontal		0.11	0.80	3.37	m	**4.17**
600 x 75 mm thick; horizontal		0.15	1.09	5.06	m	**6.15**
800 x 75 mm thick; horizontal		0.18	1.31	6.74	m	**8.05**
100 mm thick; vertical		0.35	2.55	9.58	m2	**12.13**
400 x 100 mm thick; horizontal		0.13	0.95	3.66	m	**4.61**
600 x 100 mm thick; horizontal		0.16	1.17	5.48	m	**6.65**
800 x 100 mm thick; horizontal		0.20	1.46	7.30	m	**8.76**
Hyrib permanent shuttering and reinforcement ref 2411 to soffits of slabs; left in						
horizontal		1.75	12.75	14.49	m2	**27.24**
E30 REINFORCEMENT FOR IN SITU CONCRETE						
Reinforcement bars; BS 4449; hot rolled plain round mild steel bars; straight						
40 mm	PC £355.95	12.10	87.54	426.88	t	**514.42**
32 mm	PC £344.25	12.70	91.88	416.95	t	**508.83**
25 mm	PC £337.95	14.90	107.80	413.11	t	**520.91**
20 mm	PC £336.15	17.10	123.71	414.34	t	**538.05**
16 mm	PC £337.05	19.80	143.25	418.63	t	**561.88**
12 mm	PC £355.05	23.10	167.12	442.23	t	**609.35**
10 mm	PC £363.15	26.40	191.00	454.65	t	**645.65**
8 mm	PC £373.05	29.70	214.87	469.09	t	**683.96**
6 mm	PC £408.15	35.20	254.66	511.99	t	**766.65**
Reinforcement bars; BS 4449; hot rolled plain round mild steel bars; bent						
40 mm	PC £378.45	12.10	87.54	452.27	t	**539.81**
32 mm	PC £367.65	14.30	103.46	443.36	t	**546.82**
25 mm	PC £365.85	16.50	119.37	444.59	t	**563.96**
20 mm	PC £364.05	18.70	135.29	445.84	t	**581.13**
16 mm	PC £369.45	22.00	159.16	455.20	t	**614.36**
12 mm	PC £396.45	25.30	183.04	488.96	t	**672.00**
10 mm	PC £418.05	29.70	214.87	516.61	t	**731.48**
8 mm	PC £445.95	34.10	246.71	551.37	t	**798.08**
6 mm	PC £485.55	39.60	286.50	599.35	t	**885.85**
8 mm; links or the like	PC £445.95	46.20	334.25	561.20	t	**895.45**
6 mm; links or the like	PC £485.55	53.90	389.95	612.44	t	**1002.39**
Reinforcement bars; BS 4461; cold worked deformed square high steel bars; straight						
40 mm	PC £359.55	11.00	79.58	430.94	t	**510.52**
32 mm	PC £348.75	12.70	91.88	422.02	t	**513.90**
25 mm	PC £342.45	14.90	107.80	418.18	t	**525.98**
20 mm	PC £340.65	17.10	123.71	419.43	t	**543.14**
16 mm	PC £341.55	19.80	143.25	423.72	t	**566.97**
12 mm	PC £359.55	23.10	167.12	447.30	t	**614.42**
10 mm	PC £367.65	26.40	191.00	459.72	t	**650.72**
8 mm	PC £377.55	29.70	214.87	474.17	t	**689.04**
6 mm	PC £412.65	35.20	254.66	517.06	t	**771.72**

E IN SITU CONCRETE/LARGE PRECAST CONCRETE Including overheads and profit at 7.50%		Labour hours	Labour £	Material £	Unit	Total rate £
Reinforcement bars; BS 4461; cold worked deformed square high steel bars; bent						
40 mm	PC £382.05	12.10	87.54	456.33	t	**543.87**
32 mm	PC £344.25	14.30	103.46	416.95	t	**520.41**
25 mm	PC £369.45	16.50	119.37	448.66	t	**568.03**
20 mm	PC £367.65	18.70	135.29	449.90	t	**585.19**
16 mm	PC £373.05	22.00	159.16	459.27	t	**618.43**
12 mm	PC £400.05	25.30	183.04	493.02	t	**676.06**
10 mm	PC £421.65	29.70	214.87	520.67	t	**735.54**
8 mm	PC £449.55	34.10	246.71	555.44	t	**802.15**
6 mm	PC £489.15	39.60	286.50	603.41	t	**889.91**
Reinforcement fabric; BS 4483; lapped; in beds or suspended slabs						
Ref A98 (1.54 kg/m2)	PC £0.57	0.13	0.94	0.73	m2	**1.67**
Ref A142 (2.22 kg/m2)	PC £0.71	0.13	0.94	0.91	m2	**1.85**
Ref A193 (3.02 kg/m2)	PC £0.97	0.13	0.94	1.24	m2	**2.18**
Ref A252 (3.95 kg/m2)	PC £1.25	0.14	1.01	1.61	m2	**2.62**
Ref A393 (6.16 kg/m2)	PC £2.00	0.17	1.23	2.58	m2	**3.81**
Ref B196 (3.05 kg/m2)	PC £1.08	0.13	0.94	1.40	m2	**2.34**
Ref B283 (3.73 kg/m2)	PC £1.27	0.13	0.94	1.63	m2	**2.57**
Ref B385 (4.53 kg/m2)	PC £1.52	0.14	1.01	1.95	m2	**2.96**
Ref B503 (5.93 kg/m2)	PC £1.87	0.17	1.23	2.41	m2	**3.64**
Ref B785 (8.14 kg/m2)	PC £2.65	0.19	1.37	3.42	m2	**4.79**
Ref B1131 (10.90 kg/m2)	PC £3.59	0.21	1.52	4.63	m2	**6.15**
Ref C283 (2.61 kg/m2)	PC £0.93	0.13	0.94	1.20	m2	**2.14**
Ref C385 (3.41 kg/m2)	PC £1.18	0.13	0.94	1.52	m2	**2.46**
Ref C503 (4.34 kg/m2)	PC £1.42	0.14	1.01	1.83	m2	**2.84**
Ref C636 (5.55 kg/m2)	PC £1.84	0.15	1.09	2.38	m2	**3.47**
Ref C785 (6.72 kg/m2)	PC £2.23	0.17	1.23	2.88	m2	**4.11**
Reinforcement fabric; BS 4483; lapped; in casings to steel columns or beams						
Ref D49 (0.77 kg/m2)	PC £0.56	0.28	2.03	0.72	m2	**2.75**
Ref D98 (1.54 kg/m2)	PC £0.56	0.28	2.03	0.72	m2	**2.75**
E40 DESIGNED JOINTS IN IN SITU CONCRETE						
Expandite 'Flexcell' impregnated fibreboard joint filler; or similar						
Formed joint; 10 mm thick						
not exceeding 150 mm wide		0.17	1.24	1.04	m	**2.28**
150 - 300 mm wide		0.22	1.60	1.80	m	**3.40**
300 - 450 mm wide		0.27	1.97	2.36	m	**4.33**
Formed joint; 12.5 mm thick						
not exceeding 150 mm wide		0.17	1.24	1.10	m	**2.34**
150 - 300 mm wide		0.22	1.60	1.26	m	**2.86**
300 - 450 mm wide		0.27	1.97	2.52	m	**4.49**
Formed joint; 20 mm thick						
not exceeding 150 mm wide		0.22	1.60	1.54	m	**3.14**
150 - 300 mm wide		0.27	1.97	2.01	m	**3.98**
300 - 450 mm wide		0.33	2.41	3.56	m	**5.97**
Formed joint; 25 mm thick						
not exceeding 150 mm wide		0.22	1.60	1.76	m	**3.36**
150 - 300 mm wide		0.27	1.97	3.10	m	**5.07**
300 - 450 mm wide		0.33	2.41	4.07	m	**6.48**
Sealing top of joint with Expandite 'Pliastic' hot poured rubberized bituminous compound						
10 x 25 mm		0.20	1.46	0.40	m	**1.86**
12.5 x 25 mm		0.21	1.53	0.48	m	**2.01**
20 x 25 mm		0.22	1.60	0.67	m	**2.27**
25 x 25 mm		0.23	1.68	0.94	m	**2.62**

E IN SITU CONCRETE/LARGE PRECAST CONCRETE Including overheads and profit at 7.50%		Labour hours	Labour £	Material £	Unit	Total rate £
E40 DESIGNED JOINTS IN IN SITU CONCRETE - cont'd						
Expandite 'Flexcell' impregnated fibreboard joint filler; or similar - cont'd						
Sealing top of joint with Expandite 'Thioflex 600' cold poured polysulphide rubberized compound						
10 x 25 mm		0.07	0.51	2.66	m	**3.17**
12.5 x 25 mm		0.08	0.58	3.21	m	**3.79**
20 x 25 mm		0.09	0.66	4.45	m	**5.11**
25 x 25 mm		0.10	0.73	6.40	m	**7.13**
Servicised 'Kork-pak' waterproof bonded cork joint filler board; or similar						
Formed joint; 10 mm thick						
not exceeding 150 mm wide		0.17	1.24	2.76	m	**4.00**
150 - 300 mm wide		0.22	1.60	4.29	m	**5.89**
300 - 450 mm wide		0.28	2.04	6.84	m	**8.88**
Formed joint; 13 mm thick						
not exceeding 150 mm wide		0.17	1.24	3.19	m	**4.43**
150 - 300 mm wide		0.22	1.60	4.42	m	**6.02**
300 - 450 mm wide		0.28	2.04	6.51	m	**8.55**
Formed joint; 19 mm thick						
not exceeding 150 mm wide		0.22	1.60	3.36	m	**4.96**
150 - 300 mm wide		0.28	2.04	5.72	m	**7.76**
300 - 450 mm wide		0.33	2.41	9.09	m	**11.50**
Formed joint; 25 mm thick						
not exceeding 150 mm wide		0.22	1.60	3.96	m	**5.56**
150 - 300 mm wide		0.28	2.04	6.76	m	**8.80**
300 - 450 mm wide		0.33	2.41	11.01	m	**13.42**
Sealing top of joint with 'Paraseal' polysulphide sealing compound						
10 x 25 mm		0.07	0.51	2.63	m	**3.14**
13 x 25 mm		0.08	0.58	3.04	m	**3.62**
19 x 25 mm		0.09	0.66	4.58	m	**5.24**
25 x 25 mm		0.10	0.73	5.57	m	**6.30**
Servicised water stops or similar						
Formed joint; PVC water stop; flat dumbell type; heat welded joints						
100 mm wide	PC £44.40/15m	0.24	1.75	3.36	m	**5.11**
Flat angle	PC £2.82	0.30	2.19	3.18	nr	**5.37**
Vertical angle	PC £4.91	0.30	2.19	5.55	nr	**7.74**
Flat three way intersection	PC £5.56	0.41	2.99	6.28	nr	**9.27**
Vertical three way intersection	PC £5.98	0.41	2.99	6.75	nr	**9.74**
Four way intersection	PC £6.98	0.52	3.79	7.88	nr	**11.67**
170 mm wide	PC £62.55/15m	0.28	2.04	4.73	m	**6.77**
Flat angle	PC £2.86	0.33	2.41	3.23	nr	**5.64**
Vertical angle	PC £4.91	0.33	2.41	5.55	nr	**7.96**
Flat three way intersection	PC £5.61	0.44	3.21	6.33	nr	**9.54**
Vertical three way intersection	PC £6.96	0.44	3.21	7.86	nr	**11.07**
Four way intersection	PC £7.61	0.55	4.01	8.59	nr	**12.60**
210 mm wide	PC £73.95/15m	0.31	2.26	5.59	m	**7.85**
Flat angle	PC £4.77	0.35	2.55	5.39	nr	**7.94**
Vertical angle	PC £5.45	0.35	2.55	6.15	nr	**8.70**
Flat three way intersection	PC £6.95	0.46	3.35	7.85	nr	**11.20**
Vertical three way intersection	PC £8.60	0.46	3.35	9.71	nr	**13.06**
Four way intersection	PC £8.72	0.57	4.15	9.85	nr	**14.00**
250 mm wide	PC £108.15/15m	0.33	2.41	8.18	m	**10.59**
Flat angle	PC £5.80	0.37	2.70	6.55	nr	**9.25**
Vertical angle	PC £6.13	0.37	2.70	6.92	nr	**9.62**
Flat three way intersection	PC £7.79	0.48	3.50	8.79	nr	**12.29**
Vertical three way intersection	PC £9.69	0.48	3.50	10.93	nr	**14.43**
Four way intersection	PC £10.03	0.59	4.30	11.32	nr	**15.62**

E IN SITU CONCRETE/LARGE PRECAST CONCRETE Including overheads and profit at 7.50%		Labour hours	Labour £	Material £	Unit	Total rate £
Formed joint; PVC water stop; centre bulb type; heat welded joints						
160 mm wide	PC £60.00/15m	0.28	2.04	4.54	m	**6.58**
Flat angle	PC £3.74	0.33	2.41	4.22	nr	**6.63**
Vertical angle	PC £6.16	0.33	2.41	6.96	nr	**9.37**
Flat three way intersection	PC £7.89	0.44	3.21	8.90	nr	**12.11**
Vertical three way intersection	PC £9.76	0.44	3.21	11.02	nr	**14.23**
Four way intersection	PC £9.40	0.55	4.01	10.61	nr	**14.62**
210 mm wide	PC £86.55/15m	0.31	2.26	6.55	m	**8.81**
Flat angle	PC £5.56	0.35	2.55	6.28	nr	**8.83**
Vertical angle	PC £7.32	0.35	2.55	8.27	nr	**10.82**
Flat three way intersection	PC £8.94	0.46	3.35	10.09	nr	**13.44**
Vertical three way intersection	PC £11.08	0.46	3.35	12.50	nr	**15.85**
Four way intersection	PC £10.73	0.57	4.15	12.12	nr	**16.27**
260 mm wide	PC £123.00/15m	0.33	2.41	9.30	m	**11.71**
Flat angle	PC £7.26	0.37	2.70	8.19	nr	**10.89**
Vertical angle	PC £7.97	0.37	2.70	9.00	nr	**11.70**
Flat three way intersection	PC £10.90	0.48	3.50	12.30	nr	**15.80**
Vertical three way intersection	PC £13.51	0.48	3.50	15.25	nr	**18.75**
Four way intersection	PC £13.06	0.59	4.30	14.74	nr	**19.04**
325 mm wide	PC £206.70/15m	0.36	2.62	15.63	m	**18.25**
Flat angle	PC £10.52	0.40	2.92	11.88	nr	**14.80**
Vertical angle	PC £10.86	0.40	2.92	12.26	nr	**15.18**
Flat three way intersection	PC £15.45	0.51	3.72	17.44	nr	**21.16**
Vertical three way intersection	PC £15.79	0.51	3.72	17.82	nr	**21.54**
Four way intersection	PC £17.26	0.62	4.52	19.48	nr	**24.00**
Formed joint; rubber water stop; flat dumbell type; sleeved joints						
150 mm wide	PC £107.87/9m	0.22	1.60	14.88	m	**16.48**
Flat angle	PC £30.70	0.22	1.60	34.66	nr	**36.26**
Vertical angle	PC £30.70	0.22	1.60	34.66	nr	**36.26**
Flat three way intersection	PC £33.87	0.28	2.04	38.23	nr	**40.27**
Vertical three way intersection	PC £33.87	0.28	2.04	38.23	nr	**40.27**
Four way intersection	PC £37.42	0.33	2.41	42.24	nr	**44.65**
230 mm wide	PC £163.18/9m	0.28	2.04	22.18	m	**24.22**
Flat angle	PC £37.15	0.24	1.75	41.94	nr	**43.69**
Vertical angle	PC £37.15	0.24	1.75	41.94	nr	**43.69**
Flat three way intersection	PC £40.47	0.30	2.19	45.68	nr	**47.87**
Vertical three way intersection	PC £40.47	0.30	2.19	45.68	nr	**47.87**
Four way intersection	PC £43.60	0.36	2.62	49.21	nr	**51.83**
Formed joint; rubber water stop; centre bulb type; sleeved joints						
150 mm wide	PC £124.12/9m	0.22	1.60	16.90	m	**18.50**
Flat angle	PC £33.63	0.22	1.60	37.96	nr	**39.56**
Vertical angle	PC £33.63	0.22	1.60	37.96	nr	**39.56**
Flat three way intersection	PC £37.11	0.28	2.04	41.89	nr	**43.93**
Vertical three way intersection	PC £37.11	0.28	2.04	41.89	nr	**43.93**
Four way intersection	PC £40.60	0.33	2.41	45.83	nr	**48.24**
230 mm wide	PC £185.45/9m	0.28	2.04	25.08	m	**27.12**
Flat angle	PC £39.50	0.24	1.75	44.59	nr	**46.34**
Vertical angle	PC £39.50	0.24	1.75	44.59	nr	**46.34**
Flat three way intersection	PC £41.32	0.30	2.19	46.64	nr	**48.83**
Vertical three way intersection	PC £41.32	0.30	2.19	46.64	nr	**48.83**
Four way intersection	PC £47.89	0.36	2.62	54.05	nr	**56.67**
305 mm wide	PC £305.83/9m	0.33	2.41	41.04	m	**43.45**
Flat angle	PC £62.07	0.26	1.90	70.06	nr	**71.96**
Vertical angle	PC £62.07	0.26	1.90	70.06	nr	**71.96**
Flat three way intersection	PC £75.04	0.33	2.41	84.70	nr	**87.11**
Vertical three way intersection	PC £75.04	0.33	2.41	84.70	nr	**87.11**
Four way intersection	PC £93.97	0.40	2.92	106.07	nr	**108.99**

E IN SITU CONCRETE/LARGE PRECAST CONCRETE Including overheads and profit at 7.50%	Labour hours	Labour £	Material £	Unit	Total rate £
E41 WORKED FINISHES/CUTTING ON IN SITU CONCRETE					
Tamping by mechanical means	0.03	0.19	0.13	m2	0.32
Power floating	0.19	1.18	0.43	m2	1.61
Trowelling	0.36	2.24	-	m2	2.24
Lightly shot blast surface of concrete to receive finishes	0.45	2.81	-	m2	2.81
Hacking					
by mechanical means	0.36	2.24	0.37	m2	2.61
by hand	0.77	4.80	-	m2	4.80
Wood float finish	0.14	0.87	-	m2	0.87
Tamped finish	0.06	0.37	-	m2	0.37
to falls	0.08	0.50	-	m2	0.50
to crossfalls	0.11	0.69	-	m2	0.69
Spade finish	0.17	1.06	-	m2	1.06
Cutting chases					
not exceeding 50 mm deep; 10 mm wide	0.36	2.24	0.16	m	2.40
not exceeding 50 mm deep; 50 mm wide	0.55	3.43	0.24	m	3.67
not exceeding 50 mm deep; 75 mm wide	0.73	4.55	0.32	m	4.87
50 - 100 mm deep; 75 mm wide	0.99	6.17	0.44	m	6.61
50 - 100 mm deep; 100 mm wide	1.10	6.86	0.48	m	7.34
100 - 150 mm deep; 100 mm wide	1.45	9.04	0.63	m	9.67
100 - 150 mm deep; 150 mm wide	1.75	10.91	0.77	m	11.68
Cutting chases in reinforced concrete					
50 - 100 mm deep; 100 mm wide	1.65	10.29	0.73	m	11.02
100 - 150 mm deep; 100 mm wide	2.20	13.72	0.97	m	14.69
100 - 150 mm deep; 150 mm wide	2.65	16.52	1.16	m	17.68
Cutting rebates					
not exceeding 50 mm deep; 50 mm wide	0.55	3.43	0.24	m	3.67
50 - 100 mm deep; 100 mm wide	1.10	6.86	0.48	m	7.34
Cutting mortices; not exceeding 100 mm deep; making good					
20 mm dia	0.17	1.06	0.07	nr	1.13
50 mm dia	0.19	1.18	0.08	nr	1.26
150 x 150 mm	0.39	2.43	0.19	nr	2.62
300 x 300 mm	0.77	4.80	0.40	nr	5.20
Cutting mortices in reinforced concrete; not exceeding 100 mm deep; making good					
150 x 150 mm	0.61	3.80	0.26	nr	4.06
300 x 300 mm	1.15	7.17	0.55	nr	7.72
Cutting holes; not exceeding 100 mm deep					
50 mm dia	0.39	2.43	0.40	nr	2.83
100 mm dia	0.44	2.74	0.45	nr	3.19
150 x 150 mm	0.50	3.12	0.51	nr	3.63
300 x 300 mm	0.61	3.80	0.62	nr	4.42
Cutting holes; 100 - 200 mm deep					
50 mm dia	0.55	3.43	0.56	nr	3.99
100 mm dia	0.66	4.12	0.68	nr	4.80
150 x 150 mm	0.83	5.18	0.85	nr	6.03
300 x 300 mm	1.05	6.55	1.07	nr	7.62
Cutting holes; 200 - 300 mm deep					
50 mm dia	0.83	5.18	0.85	nr	6.03
100 mm dia	0.99	6.17	1.02	nr	7.19
150 x 150 mm	1.20	7.48	1.24	nr	8.72
300 x 300 mm	1.55	9.66	1.58	nr	11.24
Add for making good fair finish one side					
50 mm dia	0.06	0.37	0.02	nr	0.39
100 mm dia	0.13	0.81	0.03	nr	0.84
150 x 150 mm	0.22	1.37	0.04	nr	1.41
300 x 300 mm	0.44	2.74	0.09	nr	2.83

E IN SITU CONCRETE/LARGE PRECAST CONCRETE Including overheads and profit at 7.50%	Labour hours	Labour £	Material £	Unit	Total rate £
Add for fixing only sleeve					
50 mm dia	0.11	0.69	-	nr	0.69
100 mm dia	0.24	1.50	-	nr	1.50
150 x 150 mm	0.36	2.24	-	nr	2.24
300 x 300 mm	0.66	4.12	-	nr	4.12
Cutting holes in reinforced concrete; not exceeding 100 mm deep					
50 mm dia	0.61	3.80	0.62	nr	4.42
100 mm dia	0.66	4.12	0.68	nr	4.80
150 x 150 mm dia	0.77	4.80	0.79	nr	5.59
300 x 300 mm dia	0.94	5.86	0.96	nr	6.82
Cutting holes in reinforced concrete; 100 - 200 mm deep					
50 mm dia	0.83	5.18	0.85	nr	6.03
100 mm dia	0.99	6.17	1.02	nr	7.19
150 x 150 mm dia	1.25	7.79	1.30	nr	9.09
300 x 300 mm dia	1.60	9.98	1.64	nr	11.62
Cutting holes in reinforced concrete; 200 - 300 mm deep					
50 mm dia	1.25	7.79	1.30	nr	9.09
100 mm dia	1.50	9.35	1.52	nr	10.87
150 x 150 mm dia	1.80	11.22	1.86	nr	13.08
300 x 300 mm dia	2.30	14.34	2.37	nr	16.71

E42 ACCESSORIES CAST INTO IN SITU CONCRETE

Item	Labour hours	Labour £	Material £	Unit	Total rate £
Temporary plywood foundation bolt boxes					
75 x 75 x 150 mm	0.50	3.12	0.40	nr	3.52
75 x 75 x 250 mm	0.55	3.43	0.64	nr	4.07
'Expamet' cylindrical expanded steel foundation boxes					
76 mm dia x 152 mm high	0.33	2.06	3.40	nr	5.46
76 mm dia x 305 mm high	0.22	1.37	1.08	nr	2.45
102 mm dia x 457 mm high	0.28	1.75	1.95	nr	3.70
10 mm dia x 100 mm long	0.28	1.75	1.29	nr	3.04
12 mm dia x 120 mm long	0.28	1.75	1.47	nr	3.22
16 mm dia x 160 mm long	0.33	2.06	3.37	nr	5.43
20 mm dia x 200 mm long	0.33	2.06	3.40	nr	5.46
'Abbey' galvanized steel masonry slots; 18 G (1.22 mm)					
3.048 m lengths	0.39	2.43	1.08	m	3.51
76 mm long	0.09	0.56	0.15	nr	0.71
102 mm long	0.09	0.56	0.17	nr	0.73
152 mm long	0.10	0.62	0.22	nr	0.84
229 mm long	0.11	0.69	0.31	nr	1.00
'Unistrut' galvanized steel slotted metal inserts; 2.5 mm thick; end caps and foam filling					
41 x 41 mm; ref P3270	0.44	2.74	4.89	m	7.63
41 x 41 x 75 mm; ref P3249	0.11	0.69	1.84	nr	2.53
41 x 41 x 100 mm; ref P3250	0.11	0.69	1.95	nr	2.64
41 x 41 x 150 mm; ref P3251	0.11	0.69	2.21	nr	2.90
Butterfly type wall ties; casting one end into concrete; other end built into joint of brickwork					
galvanized steel	0.11	0.69	0.08	nr	0.77
stainless steel	0.11	0.69	0.10	nr	0.79
Mild steel fixing cramp; once bent; one end shot fired into concrete; other end fanged and built into joint of brickwork					
200 mm girth	0.17	1.06	0.59	nr	1.65

E IN SITU CONCRETE/LARGE PRECAST CONCRETE Including overheads and profit at 7.50%	Labour hours	Labour £	Material £	Unit	Total rate £
E42 ACCESSORIES CAST INTO IN SITU CONCRETE - cont'd					
Sherardized steel floor clips; pinned to surface of concrete					
50 mm wide; standard type	0.09	0.56	0.10	nr	0.66
50 mm wide; direct fix acoustic type	0.11	0.69	0.69	nr	1.38
Hardwood dovetailed fillets					
50 x 50/40 x 1000 mm	0.11	0.69	1.21	nr	1.90
50 x 50/40 x 100 mm	0.09	0.56	0.17	nr	0.73
50 x 50/40 x 200 mm	0.09	0.56	0.24	nr	0.80
'Rigifix' galvanized steel plate column guard; 1 m long					
75 mm x 75 mm x 3 mm	0.66	4.12	8.89	nr	13.01
75 mm x 75 mm x 4.5 mm	0.66	4.12	11.89	nr	16.01
'Rigifix' white nylon coated steel plate corner guard; plugged and screwed to concrete with chromium plated domed headed screws					
75 x 75 x 1.5 mm x 1 m long	0.88	5.49	13.10	nr	18.59
E60 PRECAST/COMPOSITE CONCRETE DECKING					
Prestressed precast flooring planks; Bison 'Drycast' or similar; cement and sand (1:3) grout between planks and on prepared bearings					
100 mm thick suspended slabs; horizontal					
400 mm wide planks	-	-	-	m2	34.83
1200 mm wide planks	-	-	-	m2	32.25
150 mm thick suspended slabs; horizontal					
400 mm wide planks	-	-	-	m2	35.47
1200 mm wide planks	-	-	-	m2	32.90
Prestressed precast concrete beam and block floor; Bison 'Housefloor' or similar; in situ concrete 30 N/mm2 - 10 mm aggregate in filling at wall abutments; cement and sand (1:6) grout brushed in between beams and blocks					
155 mm thick suspended slab at ground level; 440 x 215 x 100 mm blocks; horizontal					
beams at 520 mm centres; up to 3.30 m span with a superimposed load of 5 kN/m2	-	-	-	m2	20.64
beams at 295 mm centres; up to 4.35 m span with a superimposed load of 5 kN/m2	-	-	-	m2	23.87

F MASONRY
Including overheads and profit at 7.50%

BASIC MORTAR PRICES

	£		£		£		£
Coloured mortar materials (£/tonne); (excluding cement)							
light	29.69	medium	31.19	dark	33.94	extra dark	33.94
Mortar materials (£/tonne)							
cement	65.43	lime	100.72	sand	9.86	white cement	77.94

Mortar plasticizer - £1.86/Litre

	Labour hours	Labour £	Material £	Unit	Total rate £
F10 BRICK/BLOCK WALLING					
Common bricks; PC £124.00/1000; in cement mortar (1:3)					
Walls					
half brick thick	1.45	15.04	9.59	m2	**24.63**
one brick thick	2.40	24.90	19.96	m2	**44.86**
one and a half brick thick	3.25	33.71	30.03	m2	**63.74**
two brick thick	4.00	41.50	39.93	m2	**81.43**
Walls; facework one side					
half brick thick	1.60	16.60	9.59	m2	**26.19**
one brick thick	2.60	26.97	19.96	m2	**46.93**
one and a half brick thick	3.45	35.79	30.03	m2	**65.82**
two brick thick	4.20	43.57	39.93	m2	**83.50**
Walls; facework both sides					
half brick thick	1.70	17.64	9.59	m2	**27.23**
one brick thick	2.70	28.01	19.96	m2	**47.97**
one and a half brick thick	3.55	36.83	30.03	m2	**66.86**
two brick thick	4.30	44.61	39.93	m2	**84.54**
Walls; built curved mean radius 6 m					
half brick thick	1.90	19.71	10.33	m2	**30.04**
one brick thick	3.15	32.68	21.43	m2	**54.11**
Walls; built curved mean radius 1.50 m					
half brick thick	2.40	24.90	10.86	m2	**35.76**
one brick thick	4.00	41.50	22.50	m2	**64.00**
Walls; built overhand					
half brick thick	1.80	18.67	9.59	m2	**28.26**
Walls; building up against concrete including flushing up at back					
half brick thick	1.55	16.08	10.89	m2	**26.97**
Walls; backing to masonry; cutting and bonding					
one brick thick	2.90	30.08	20.37	m2	**50.45**
one and a half brick thick	3.85	39.94	30.62	m2	**70.56**
Honeycomb walls					
half brick thick	1.15	11.93	6.79	m2	**18.72**
Dwarf support wall					
half brick thick	1.80	18.67	9.59	m2	**28.26**
one brick thick	2.90	30.08	19.96	m2	**50.04**
Battering walls					
one and a half brick thick	3.80	39.42	30.62	m2	**70.04**
two brick thick	4.70	48.76	40.72	m2	**89.48**
Walls; tapering one side; average					
337 mm thick	4.20	43.57	31.22	m2	**74.79**
450 mm thick	5.40	56.02	41.53	m2	**97.55**

F MASONRY Including overheads and profit at 7.50%	Labour hours	Labour £	Material £	Unit	Total rate £
F10 BRICK/BLOCK WALLING - cont'd					
Common bricks; PC £124.00/1000; in cement mortar (1:3) - cont'd					
Walls; tapering both sides; average					
337 mm thick	4.80	49.79	31.22	m2	**81.01**
450 mm thick	6.00	62.24	41.53	m2	**103.77**
Isolated piers					
one brick thick	3.70	38.38	20.37	m2	**58.75**
two brick thick	5.75	59.65	40.72	m2	**100.37**
three brick thick	7.25	75.21	61.09	m2	**136.30**
Isolated casings to steel columns					
half brick thick	1.85	19.19	9.79	m2	**28.98**
one brick thick	3.15	32.68	20.37	m2	**53.05**
Chimney stacks					
one brick thick	3.70	38.38	20.37	m2	**58.75**
two brick thick	5.75	59.65	40.72	m2	**100.37**
three brick thick	7.25	75.21	61.09	m2	**136.30**
Projections; vertical					
225 x 112 mm	0.45	4.67	2.27	m	**6.94**
225 x 225 mm	0.85	8.82	4.39	m	**13.21**
337 x 225 mm	1.25	12.97	7.37	m	**20.34**
440 x 225 mm	1.45	15.04	8.63	m	**23.67**
Bonding ends to existing					
half brick thick	0.45	4.67	0.71	m	**5.38**
one brick thick	0.65	6.74	1.42	m	**8.16**
one and a half brick thick	1.00	10.37	2.13	m	**12.50**
two brick thick	1.40	14.52	2.84	m	**17.36**
ADD or DEDUCT to walls for variation of £10.00/1000 in PC of common bricks					
half brick thick	-	-	-	m2	**0.70**
one brick thick	-	-	-	m2	**1.39**
one and a half brick thick	-	-	-	m2	**2.07**
two brick thick	-	-	-	m2	**2.71**
Extra over walls for sulphate-resisting cement mortar (1:3) in lieu of cement mortar (1:3)					
half brick thick	-	-	-	m2	**0.11**
one brick thick	-	-	-	m2	**0.30**
one and a half brick thick	-	-	-	m2	**0.45**
two brick thick	-	-	-	m2	**0.59**
Common bricks; PC £124.00/1000; in gauged mortar (1:1:6)					
Walls					
half brick thick	1.45	15.04	9.49	m2	**24.53**
one brick thick	2.40	24.90	19.71	m2	**44.61**
one and a half brick thick	3.25	33.71	29.64	m2	**63.35**
two brick thick	4.00	41.50	39.42	m2	**80.92**
Walls; facework one side					
half brick thick	1.60	16.60	9.49	m2	**26.09**
one brick thick	2.60	26.97	19.71	m2	**46.68**
one and a half brick thick	3.45	35.79	29.64	m2	**65.43**
two brick thick	4.20	43.57	39.42	m2	**82.99**
Walls; facework both sides					
half brick thick	1.70	17.64	9.49	m2	**27.13**
one brick thick	2.70	28.01	19.71	m2	**47.72**
one and a half brick thick	3.55	36.83	29.64	m2	**66.47**
two brick thick	4.30	44.61	39.42	m2	**84.03**
Walls; built curved mean radius 6 m					
half brick thick	1.90	19.71	10.23	m2	**29.94**
one brick thick	3.15	32.68	21.16	m2	**53.84**

F MASONRY Including overheads and profit at 7.50%	Labour hours	Labour £	Material £	Unit	Total rate £
Walls; built curved mean radius 1.50 m					
half brick thick	2.40	24.90	10.74	m2	**35.64**
one brick thick	4.00	41.50	22.21	m2	**63.71**
Walls; built overhand					
half brick thick	1.80	18.67	9.49	m2	**28.16**
Walls; built up against concrete including flushing up at back					
half brick thick	1.55	16.08	10.69	m2	**26.77**
Walls; backing to masonry; cutting and bonding					
one brick thick	2.90	30.08	20.11	m2	**50.19**
one and a half brick thick	3.85	39.94	30.23	m2	**70.17**
Honeycomb walls					
half brick thick	1.15	11.93	6.70	m2	**18.63**
Dwarf support wall					
half brick thick	1.80	18.67	9.49	m2	**28.16**
one brick thick	2.90	30.08	19.71	m2	**49.79**
Battering walls					
one and a half brick thick	3.80	39.42	30.23	m2	**69.65**
two brick thick	4.70	48.76	40.22	m2	**88.98**
Walls; tapering one side; average					
337 mm thick	4.20	43.57	30.83	m2	**74.40**
450 mm thick	5.40	56.02	41.02	m2	**97.04**
Walls; tapering both sides; average					
337 mm thick	4.80	49.79	30.83	m2	**80.62**
450 mm thick	6.00	62.24	41.02	m2	**103.26**
Isolated piers					
one brick thick	3.70	38.38	20.11	m2	**58.49**
two brick thick	5.75	59.65	40.22	m2	**99.87**
three brick thick	7.25	75.21	60.33	m2	**135.54**
Isolated casings to steel columns					
half brick thick	1.85	19.19	9.70	m2	**28.89**
one brick thick	3.15	32.68	20.11	m2	**52.79**
Chimney stacks					
one brick thick	3.70	38.38	20.11	m2	**58.49**
two brick thick	5.75	59.65	40.22	m2	**99.87**
three brick thick	7.25	75.21	60.33	m2	**135.54**
Projections; vertical					
225 x 112 mm	0.45	4.67	2.24	m	**6.91**
225 x 225 mm	0.85	8.82	4.35	m	**13.17**
337 x 225 mm	1.25	12.97	7.31	m	**20.28**
440 x 225 mm	1.45	15.04	8.55	m	**23.59**
Bonding ends to existing					
half brick thick	0.45	4.67	0.69	m	**5.36**
one brick thick	0.65	6.74	1.39	m	**8.13**
one and a half brick thick	1.00	10.37	2.08	m	**12.45**
two brick thick	1.40	14.52	2.77	m	**17.29**
ADD or DEDUCT to walls for variation of £10.00/1000 in PC of common bricks					
half brick thick	-	-	-	m2	**0.68**
one brick thick	-	-	-	m2	**1.35**
one and a half brick thick	-	-	-	m2	**2.03**
two brick thick	-	-	-	m2	**2.71**
Segmental arches; one ring, 102 mm high on face					
102 mm wide on exposed soffit	2.55	23.21	2.74	m	**25.95**
215 mm wide on exposed soffit	3.05	26.86	5.70	m	**32.56**
Segmental arches; two ring, 215 mm high on face					
102 mm wide on exposed soffit	3.25	30.48	4.18	m	**34.66**
215 mm wide on exposed soffit	3.75	34.12	8.56	m	**42.68**

F MASONRY Including overheads and profit at 7.50%	Labour hours	Labour £	Material £	Unit	Total rate £
F10 BRICK/BLOCK WALLING - cont'd					
Common bricks; PC £124.00/1000; in gauged mortar (1:1:6) - cont'd					
Semi-circular arches; one ring, 102 mm high on face					
102 mm wide on exposed soffit	3.25	29.40	3.32	m	**32.72**
215 mm wide on exposed soffit	3.70	32.68	6.64	m	**39.32**
Semi-circular arches; two ring, 215 mm high on face					
102 mm wide on exposed soffit	4.15	38.73	5.04	m	**43.77**
215 mm wide on exposed soffit	4.60	42.01	9.13	m	**51.14**
Labours on brick fairface					
Fair returns					
half brick wide	0.03	0.31	-	m	**0.31**
one brick wide	0.06	0.62	-	m	**0.62**
Class A engineering bricks; PC £350.00/1000; in cement mortar (1:3)					
Walls					
half brick thick	1.55	16.08	24.34	m2	**40.42**
one brick thick	2.60	26.97	49.45	m2	**76.42**
one and a half brick thick	3.45	35.79	74.25	m2	**110.04**
two brick thick	4.30	44.61	98.89	m2	**143.50**
Walls; facework one side					
half brick thick	1.70	17.64	24.90	m2	**42.54**
one brick thick	2.75	28.53	50.57	m2	**79.10**
one and a half brick thick	3.60	37.35	75.94	m2	**113.29**
two brick thick	4.50	46.68	101.15	m2	**147.83**
Walls; facework both sides					
half brick thick	1.80	18.67	24.90	m2	**43.57**
one brick thick	2.85	29.57	50.57	m2	**80.14**
one and a half brick thick	3.70	38.38	75.94	m2	**114.32**
two brick thick	4.60	47.72	101.15	m2	**148.87**
Walls; built curved mean radius 6 m					
one brick thick	3.45	35.79	53.15	m2	**88.94**
Walls; backing to masonry; cutting and bonding					
one brick thick	3.10	32.16	51.70	m2	**83.86**
one and a half brick thick	4.15	43.05	77.63	m2	**120.68**
Walls; tapering one side; average					
337 mm thick	4.50	46.68	77.63	m2	**124.31**
450 mm thick	5.75	59.65	103.41	m2	**163.06**
Walls; tapering both sides; average					
337 mm thick	5.20	53.94	79.33	m2	**133.27**
450 mm thick	6.55	67.95	105.66	m2	**173.61**
Isolated piers					
one brick thick	4.00	41.50	50.57	m2	**92.07**
two brick thick	6.30	65.35	101.15	m2	**166.50**
three brick thick	7.75	80.40	151.83	m2	**232.23**
Isolated casings to steel columns					
half brick thick	2.00	20.75	25.47	m2	**46.22**
one brick thick	3.45	35.79	51.70	m2	**87.49**
Projections; vertical					
225 x 112 mm	0.50	5.19	5.92	m	**11.11**
225 x 225 mm	0.90	9.34	11.44	m	**20.78**
337 x 225 mm	1.40	14.52	19.39	m	**33.91**
440 x 225 mm	1.55	16.08	22.47	m	**38.55**
Bonding ends to existing					
half brick thick	0.50	5.19	1.59	m	**6.78**
one brick thick	0.70	7.26	3.18	m	**10.44**
one and a half brick thick	1.05	10.89	4.77	m	**15.66**
two brick thick	1.50	15.56	6.37	m	**21.93**

F MASONRY Including overheads and profit at 7.50%	Labour hours	Labour £	Material £	Unit	Total rate £
ADD or DEDUCT to walls for variation of £10.00/1000 in PC of engineering bricks					
half brick thick	-	-	-	m2	**0.68**
one brick thick	-	-	-	m2	**1.35**
one and a half brick thick	-	-	-	m2	**2.03**
two brick thick	-	-	-	m2	**2.71**
Class B engineering bricks; PC £180.00/1000; in cement mortar (1:3)					
Walls					
half brick thick	1.55	16.08	13.09	m2	**29.17**
one brick thick	2.60	26.97	26.97	m2	**53.94**
one and a half brick thick	3.45	35.79	40.53	m2	**76.32**
two brick thick	4.30	44.61	53.94	m2	**98.55**
Walls; facework one side					
half brick thick	1.70	17.64	13.38	m2	**31.02**
one brick thick	2.75	28.53	27.55	m2	**56.08**
one and a half brick thick	3.60	37.35	41.40	m2	**78.75**
two brick thick	4.50	46.68	55.10	m2	**101.78**
Walls; facework both sides					
half brick thick	1.80	18.67	13.38	m2	**32.05**
one brick thick	2.85	29.57	27.55	m2	**57.12**
one and a half brick thick	3.70	38.38	41.40	m2	**79.78**
two brick thick	4.60	47.72	55.10	m2	**102.82**
Walls; built curved mean radius 6 m					
one brick thick	3.45	35.79	28.97	m2	**64.76**
Walls; backing to masonry; cutting and bonding					
one brick thick	3.10	32.16	28.13	m2	**60.29**
one and a half brick thick	4.15	43.05	42.27	m2	**85.32**
Walls; tapering one side; average					
337 mm thick	4.50	46.68	42.27	m2	**88.95**
450 mm thick	5.75	59.65	56.26	m2	**115.91**
Walls; tapering both sides; average					
337 mm thick	5.20	53.94	43.14	m2	**97.08**
450 mm thick	6.55	67.95	57.42	m2	**125.37**
Isolated piers					
one brick thick	4.00	41.50	27.55	m2	**69.05**
two brick thick	6.30	65.35	55.10	m2	**120.45**
three brick thick	7.75	80.40	82.75	m2	**163.15**
Isolated casings to steel columns					
half brick thick	2.00	20.75	13.67	m2	**34.42**
one brick thick	3.45	35.79	28.13	m2	**63.92**
Projections					
225 x 112 mm	0.50	5.19	3.17	m	**8.36**
225 x 225 mm	0.90	9.34	6.13	m	**15.47**
337 x 225 mm	1.40	14.52	10.35	m	**24.87**
440 x 225 mm	1.55	16.08	12.07	m	**28.15**
Bonding ends to existing					
half brick thick	0.50	5.19	0.92	m	**6.11**
one brick thick	0.70	7.26	1.83	m	**9.09**
one and a half brick thick	1.05	10.89	2.75	m	**13.64**
two brick thick	1.50	15.56	3.68	m	**19.24**
ADD or DEDUCT to walls for variation of £10.00/1000 in PC of engineering bricks					
half brick thick	-	-	-	m2	**0.68**
one brick thick	-	-	-	m2	**1.35**
one and a half brick thick	-	-	-	m2	**2.03**
two brick thick	-	-	-	m2	**2.71**

F MASONRY Including overheads and profit at 7.50%	Labour hours	Labour £	Material £	Unit	Total rate £
F10 BRICK/BLOCK WALLING - cont'd					
Refractory bricks; PC £587.00/1000; stretcher bond lining to flue; in fireclay cement mortar (1:4); built 50 mm clear of flues; one header per m2					
Walls; vertical; facework one side					
half brick thick	1.95	20.23	39.20	m2	**59.43**

ALTERNATIVE FACING BRICK PRICES (£/1000)

	£		£
Ibstock facing bricks; 215 x 102.5 x 65 mm			
Aldridge brown blend	311.06	Leicester Anglican Red Rustic	302.82
Cattybrook Gloucester Golden	309.00	Leicester Red Stock	335.78
Himley Dark Brown Rustic	355.35	Roughdales Red Multi Rustic	339.90
Himley Mixed Russet	305.91	Roughdales Trafford Buff Multi	350.20
London Brick Company facing bricks; 215 x 102.5 x 65 mm			
Brecken Grey	124.90	Orton Multi Buff	136.85
Chiltern	137.59	Regency	126.38
Claydon Red Multi	127.66	Sandfaced	137.54
Delph Autumn	125.25	Saxon Gold	132.76
Edwardian	131.08	Tudor	140.12
Georgian Red Multi	128.29	Victorian	138.54
Heather	135.67	Wansford Multi	125.39
Ironstone	127.37	Windsor	127.85
Milton Buff	131.60		
Redland facing bricks; 215 x 102.5 x 65 mm			
Arun	421.27	Southwater class B	290.00
Beare Green restoration red	589.16	Sheppy 'matured' yellow	432.60
Chailey yellow multicoloured	425.39	Stourbridge Sherbourne range	326.51
Cottage mixed multicoloured	312.09	Stourbridge Henley range	305.91
Crowborough multicoloured	389.34	Stourbridge Pennine range	294.58
Dorking	323.42	Stourbridge Stratford range	248.23
Funton yellow London	309.00	Surrey bronze multicoloured	335.78
Hamsey multicoloured	408.91	Tonbridge handmade	533.54
Holbrook Sherbourne textured	320.33	Tonbridge handmade (50 mm deep)	533.54
Holbrook smooth red	365.65	Tudor (53 mm deep)	676.71
Nutbourne sandfaced	296.64	Wealden	421.27
Pevensey red multicoloured	392.43	Wealdmade	463.50
Pluckley multicoloured	407.88		

	Labour hours	Labour £	Material £	Unit	Total rate £
Facing bricks; sand faced; PC £134.00/1000 (unless otherwise stated); in gauged mortar (1:1:6)					
Extra over common bricks; PC £124.00/1000; for facing bricks in					
stretcher bond	0.45	4.67	0.69	m2	**5.36**
flemish bond with snapped headers	0.55	5.71	3.79	m2	**9.50**
english bond with snapped headers	0.55	5.71	3.70	m2	**9.41**
ADD or DEDUCT for variation of £10.00/1000 in PC of facing bricks	-	-	-	m2	**0.92**

F MASONRY Including overheads and profit at 7.50%	Labour hours	Labour £	Material £	Unit	Total rate £
Half brick thick; stretcher bond; facework one side					
walls	1.90	19.71	10.39	m2	**30.10**
walls; building curved mean radius 6 m	2.75	28.53	11.17	m2	**39.70**
walls; building curved mean radius 1.50 m	3.45	35.79	11.74	m2	**47.53**
walls; building overhand	2.30	23.86	10.39	m2	**34.25**
walls; building up against concrete including flushing up at back	2.00	20.75	11.79	m2	**32.54**
walls; as formwork; temporary strutting	2.75	28.53	12.73	m2	**41.26**
walls; panels and aprons; not exceeding 1 m2	2.40	24.90	10.43	m2	**35.33**
isolated casings to steel columns	2.90	30.08	10.39	m2	**40.47**
bonding ends to existing	0.75	7.78	1.56	m	**9.34**
projections; vertical					
225 x 112 mm	0.46	4.77	2.41	m	**7.18**
337 x 112 mm	0.86	8.92	3.73	m	**12.65**
440 x 112 mm	1.25	12.97	5.05	m	**18.02**
Half brick thick; flemish bond with snapped headers; facework one side					
walls	2.20	22.82	11.40	m2	**34.22**
walls; building curved mean radius 6 m	3.10	32.16	11.97	m2	**44.13**
walls; building curved mean radius 1.50 m	4.00	41.50	12.54	m2	**54.04**
walls; building overhand	2.60	26.97	11.40	m2	**38.37**
walls; building up against concrete including flushing up at back	2.30	23.86	12.80	m2	**36.66**
walls; as formwork; temporary strutting	3.00	31.12	13.74	m2	**44.86**
walls; panels and aprons; not exceeding 1 m2	2.70	28.01	11.40	m2	**39.41**
isolated casings to steel columns	3.15	32.68	10.39	m2	**43.07**
bonding ends to existing	0.75	7.78	1.56	m	**9.34**
projections; vertical					
225 x 112 mm	0.58	6.02	2.72	m	**8.74**
337 x 112 mm	0.98	10.17	4.04	m	**14.21**
440 x 112 mm	1.40	14.52	5.52	m	**20.04**
One brick thick; two stretcher skins tied together; facework both sides					
walls	3.20	33.20	21.71	m2	**54.91**
walls; building curved mean radius 6 m	4.50	46.68	23.27	m2	**69.95**
walls; building curved mean radius 1.50 m	5.50	57.06	26.69	m2	**83.75**
isolated piers	3.80	39.42	23.55	m2	**62.97**
bonding ends to existing	0.98	10.17	3.11	m	**13.28**
One brick thick; flemish bond; facework both sides					
walls	3.35	34.75	21.50	m2	**56.25**
walls; building curved mean radius 6 m	4.60	47.72	23.07	m2	**70.79**
walls; building curved mean radius 1.50 m	5.75	59.65	26.50	m2	**86.15**
isolated piers	3.90	40.46	23.36	m2	**63.82**
bonding ends to existing	0.98	10.17	3.11	m	**13.28**
projections; vertical					
225 x 225 mm	0.92	9.54	3.19	m	**12.73**
337 x 225 mm	1.70	17.64	4.66	m	**22.30**
440 x 225 mm	2.55	26.45	6.23	m	**32.68**
ADD or DEDUCT for variation of £10.00/1000 in PC of facing bricks; in stretcher bond					
half brick thick	-	-	-	m2	**0.70**
one brick thick	-	-	-	m2	**1.40**
ADD or DEDUCT for variation of £10.00/1000 in PC of facing bricks; in flemish bond					
half brick thick	-	-	-	m2	**0.87**
one brick thick	-	-	-	m2	**1.74**

F MASONRY Including overheads and profit at 7.50%	Labour hours	Labour £	Material £	Unit	Total rate £
F10 BRICK/BLOCK WALLING - cont'd					
Facing bricks; sand faced; PC £134.00/1000 (unless otherwise stated); in gauged mortar (1:1:6) - cont'd					
Extra over facing bricks for					
recessed joints	0.03	0.31	-	m2	**0.31**
raking out joints and pointing in black mortar	0.58	6.02	0.18	m2	**6.20**
bedding and pointing half brick wall in black mortar	-	-	-	m2	**0.87**
bedding and pointing one brick wall in black mortar	-	-	-	m2	**2.31**
flush plain bands; 225 mm wide stretcher bond; horizontal; bricks; PC £154.00/1000	0.29	3.01	0.32	m	**3.33**
flush quoins; average 320 mm girth; block bond vertical; facing bricks; PC £154.00/1000	0.46	4.77	0.30	m	**5.07**
Flat arches; 215 mm high on face					
102 mm wide exposed soffit	1.32	12.27	1.61	m	**13.88**
215 mm wide exposed soffit	1.99	18.51	3.22	m	**21.73**
Flat arches; 215 mm high on face; bullnosed specials; PC £51.20/100					
102 mm wide exposed soffit	1.38	12.90	9.27	m	**22.17**
215 mm wide exposed soffit	2.09	19.55	18.55	m	**38.10**
Segmental arches; one ring; 215 mm high on face					
102 mm wide exposed soffit	2.40	21.35	2.24	m	**23.59**
215 mm wide exposed soffit	3.60	32.10	4.70	m	**36.80**
Segmental arches; two ring; 215 mm high on face					
102 mm wide exposed soffit	3.10	28.61	2.24	m	**30.85**
215 mm wide exposed soffit	4.65	42.99	4.70	m	**47.69**
Segmental arches; 215 mm high on face; cut voussoirs; PC £71.10/100					
102 mm wide exposed soffit	2.55	22.91	12.82	m	**35.73**
215 mm wide exposed soffit	3.75	33.66	25.84	m	**59.50**
Segmental arches; one and a half ring; 320 mm high on face; cut voussoirs; PC £71.10/100					
102 mm wide exposed soffit	3.45	32.24	24.31	m2	**56.55**
215 mm wide exposed soffit	5.15	48.18	48.85	m2	**97.03**
Semi circular arches; one ring; 215 mm high on face					
102 mm wide exposed soffit	4.15	38.73	5.34	m	**44.07**
215 mm wide exposed soffit	6.00	56.53	10.67	m	**67.20**
Semi circular arches; two ring; 215 mm high on face					
102 mm wide exposed soffit	5.55	53.25	5.34	m	**58.59**
215 mm wide exposed soffit	6.00	56.53	10.67	m	**67.20**
Semi circular arches; one ring; 215 mm high on face; cut voussoirs PC £71.10/100					
102 mm wide exposed soffit	3.10	27.84	13.11	m	**40.95**
215 mm wide exposed soffit	4.45	40.46	41.51	m	**81.97**
Bullseye window 600 mm dia; two rings; 215 mm high on face					
102 mm wide exposed soffit	6.90	68.03	3.37	nr	**71.40**
215 mm wide exposed soffit	10.35	102.12	6.54	nr	**108.66**
Bullseye window 1200 mm dia; two rings; 215 mm high on face					
102 mm wide exposed soffit	12.05	117.91	5.15	nr	**123.06**
215 mm wide exposed soffit	18.15	177.64	10.30	m	**187.94**

F MASONRY Including overheads and profit at 7.50%	Labour hours	Labour £	Material £	Unit	Total rate £
Bullseye window 600 mm dia; one ring; 215 mm high on face; cut voussoirs PC £71.10/100					
102 mm wide exposed soffit	5.75	56.10	30.28	nr	86.38
215 mm wide exposed soffit	8.60	83.97	60.33	nr	144.30
Bullseye window 1200 mm dia; one ring; 215 mm high on face; cut voussoirs					
102 mm wide exposed soffit	10.35	100.27	51.96	nr	152.23
215 mm wide exposed soffit	15.55	150.67	103.92	nr	254.59
ADD or DEDUCT for variation of £10.00/1000 in PC of facing bricks	-	-	-	m	0.29
Sills; horizontal; headers on edge; pointing top and one side; set weathering					
150 x 102 mm	0.81	8.40	2.46	m	10.86
150 x 102 mm; cant headers; PC £66.10/100	0.86	8.92	10.98	m	19.90
Sills; horizontal; headers on flat; pointing top and one side					
150 x 102 mm; bullnosed specials; PC £51.20/100	0.75	7.78	6.10	m	13.88
Coping; horizontal; headers on edge; pointing top and both sides					
215 x 102 mm	0.64	6.64	2.41	m	9.05
260 x 102 mm	1.05	10.89	3.73	m	14.62
215 x 102 mm; double bullnosed specials; PC £52.30/100	0.69	7.16	8.70	m	15.86
260 x 120 mm; single bullnosed specials; PC £51.20/100	1.05	10.89	13.94	m	24.83
ADD or DEDUCT for variation of £10.00/1000 in PC of facing bricks	-	-	-	m	0.29
Facing bricks; white sandlime; PC £138.00/1000 in gauged mortar (1:1:6)					
Extra over common bricks; PC £124.00/1000; for facing bricks in					
stretcher bond	0.46	4.77	1.22	m2	5.99
flemish bond with snapped headers	0.58	6.02	4.49	m2	10.51
ADD or DEDUCT for variation of £10.00/1000 in PC of facing bricks	-	-	-	m2	0.92
Half brick thick; stretcher bond; facework one side					
walls	1.90	19.71	10.92	m2	30.63
One brick thick; flemish bond; facework both sides					
walls	3.35	34.75	22.55	m2	57.30
ADD or DEDUCT for variation of £10.00/1000 in PC of facing bricks; in flemish bond					
half brick thick	-	-	-	m2	0.87
one brick thick	-	-	-	m2	1.74
Facing bricks; machine made facings; PC £330.00/1000 (unless otherwise stated; in gauged mortar (1:1:6)					
Extra over common bricks; PC £124.00/1000; for facing bricks in					
stretcher bond	0.45	4.67	14.29	m2	18.96
flemish bond with snapped headers	0.55	5.71	17.86	m2	23.57
english bond with snapped headers	0.55	5.71	20.23	m2	25.94
ADD or DEDUCT for variation of £10.00/1000 in PC of facing bricks	-	-	-	m2	0.87

F MASONRY Including overheads and profit at 7.50%	Labour hours	Labour £	Material £	Unit	Total rate £
F10 BRICK/BLOCK WALLING - cont'd					
Facing bricks; machine made facings; PC £330.00/1000 (unless otherwise stated; in gauged mortar (1:1:6) - cont'd					
Half brick thick; stretcher bond; facework one side					
walls	1.90	19.71	23.97	m2	**43.68**
walls; building curved mean radius 6 m	2.75	28.53	25.78	m2	**54.31**
walls; building curved mean radius 1.50 m	3.45	35.79	27.05	m2	**62.84**
walls; building overhand	2.30	23.86	23.97	m2	**47.83**
walls; building up against concrete including flushing up at back	2.00	20.75	25.38	m2	**46.13**
walls; as formwork; temporary strutting	2.75	28.53	26.31	m2	**54.84**
walls; panels and aprons; not exceeding 1 m2	2.40	24.90	24.51	m2	**49.41**
isolated casings to steel columns	2.90	30.08	23.97	m2	**54.05**
bonding ends to existing	0.75	7.78	1.56	m	**9.34**
projections; vertical					
225 x 112 mm	0.46	4.77	5.58	m	**10.35**
337 x 112 mm	0.86	8.92	8.49	m	**17.41**
440 x 112 mm	1.25	12.97	11.39	m	**24.36**
Half brick thick; flemish bond with snapped headers; facework one side					
walls	2.20	22.82	26.47	m2	**49.29**
walls; building curved mean radius 6 m	3.10	32.16	27.73	m2	**59.89**
walls; building curved mean radius 1.50 m	4.00	41.50	28.99	m2	**70.49**
walls; building overhand	2.60	26.97	26.47	m2	**53.44**
walls; building up against concrete including flushing up at back	2.30	23.86	28.81	m2	**52.67**
walls; as formwork; temporary strutting	3.00	31.12	27.88	m2	**59.00**
walls; panels and aprons; not exceeding 1 m2	2.70	28.01	26.47	m2	**54.48**
isolated casings to steel columns	3.15	32.68	23.97	m2	**56.65**
bonding ends to existing	0.75	7.78	1.56	m	**9.34**
projections; vertical					
225 x 112 mm	0.58	6.02	6.34	m	**12.36**
337 x 112 mm	0.98	10.17	9.25	m	**19.42**
440 x 112 mm	1.40	14.52	12.54	m	**27.06**
One brick thick; two stretcher skins tied together; facework both sides					
walls	3.20	33.20	48.88	m2	**82.08**
walls; building curved mean radius 6 m	4.50	46.68	52.28	m2	**98.96**
walls; building curved mean radius 1.50 m	5.50	57.06	55.00	m2	**112.06**
isolated piers	3.80	39.42	48.87	m2	**88.29**
bonding ends to existing	0.98	10.17	3.11	m	**13.28**
One brick thick; flemish bond; facework both sides					
walls	3.35	34.75	48.68	m2	**83.43**
walls; building curved mean radius 6 m	4.60	47.72	52.28	m2	**100.00**
walls; building curved mean radius 1.50 m	5.75	59.65	54.81	m2	**114.46**
isolated piers	3.90	40.46	48.88	m2	**89.34**
bonding ends to existing	0.98	10.17	3.11	m	**13.28**
projections; vertical					
225 x 225 mm	0.92	9.54	10.78	m	**20.32**
337 x 225 mm	1.70	17.64	18.26	m	**35.90**
440 x 225 mm	2.55	26.45	21.17	m	**47.62**
ADD or DEDUCT for variation of £10.00/1000 in PC of facing bricks; in stretcher bond					
half brick thick	-	-	-	m2	**0.70**
one brick thick	-	-	-	m2	**1.40**
ADD or DEDUCT for variation of £10.00/1000 in PC of facing bricks; in flemish bond					
half brick thick	-	-	-	m2	**0.87**
one brick thick	-	-	-	m2	**1.74**

F MASONRY Including overheads and profit at 7.50%	Labour hours	Labour £	Material £	Unit	Total rate £
Extra over facing bricks for					
recessed joints	0.03	0.31	-	m2	**0.31**
raking out joints and pointing in black mortar	0.58	6.02	0.18	m2	**6.20**
bedding and pointing half brick wall in black mortar	-	-	-	m2	**0.87**
bedding and pointing one brick wall in black mortar	-	-	-	m2	**2.31**
flush plain bands; 225 mm wide stretcher bond; horizontal; bricks PC £360.00/1000	0.29	3.01	0.48	m	**3.49**
flush quoins; average 320 mm girth; black bond vertical; bricks PC £360.00/1000	0.46	4.77	0.45	m	**5.22**
Flat arches; 215 mm high on face					
102 mm wide exposed soffit	1.32	12.27	2.52	m	**14.79**
215 mm wide exposed soffit	1.99	18.51	5.04	m	**23.55**
Flat arches; 215 mm high on face; bullnosed specials; PC £124.30/100					
102 mm wide exposed soffit	1.38	12.90	21.10	m	**34.00**
215 mm wide exposed soffit	2.09	19.55	42.19	m	**61.74**
Segmental arches; one ring; 215 mm high on face					
102 mm wide exposed soffit	2.40	21.35	3.61	m	**24.96**
215 mm wide exposed soffit	3.60	32.10	7.43	m	**39.53**
Segmental arches; two ring; 215 mm high on face					
102 mm wide exposed soffit	3.10	28.61	3.61	m	**32.22**
215 mm wide exposed soffit	4.65	42.99	7.43	m	**50.42**
Segmental arches; 215 mm high on face; cut voussoirs; PC £269.50/100					
102 mm wide exposed soffit	2.55	22.91	44.91	m	**67.82**
215 mm wide exposed soffit	3.75	33.66	90.04	m	**123.70**
Segmental arches; one and a half ring; 320 mm high on face; cut voussoirs PC £269.50/100					
102 mm wide exposed soffit	3.45	32.24	88.51	m	**120.75**
215 mm wide exposed soffit	5.15	48.18	177.24	m	**225.42**
Semi circular arches; one ring; 215 mm high on face					
102 mm wide exposed soffit	4.15	38.73	10.52	m	**49.25**
215 mm wide exposed soffit	6.00	56.53	21.04	m	**77.57**
Semi circular arches; two ring; 215 mm high on face					
102 mm wide exposed soffit	5.55	53.25	10.52	m	**63.77**
215 mm wide exposed soffit	6.00	56.53	21.04	m	**77.57**
Semi circular arches; one ring; 215 mm high on face; cut voussoirs PC £269.50/100					
102 mm wide exposed soffit	3.10	27.84	45.20	m	**73.04**
215 mm wide exposed soffit	4.45	40.46	105.71	m	**146.17**
Bullseye window 600 mm dia; two rings; 215 mm high on face					
102 mm wide exposed soffit	6.90	68.03	47.49	nr	**115.52**
215 mm wide exposed soffit	10.35	102.12	94.78	nr	**196.90**
Bullseye window 1200 mm dia; two rings; 215 mm high on face					
102 mm wide exposed soffit	12.05	117.91	8.79	nr	**126.70**
215 mm wide exposed soffit	18.15	177.64	17.58	nr	**195.22**
Bullseye window 600 mm dia; one ring; 215 mm high on face; cut voussoirs PC £269.50/100					
102 mm wide exposed soffit	5.75	56.10	110.52	nr	**166.62**
215 mm wide exposed soffit	8.60	83.97	220.83	nr	**304.80**

F MASONRY Including overheads and profit at 7.50%	Labour hours	Labour £	Material £	Unit	Total rate £
F10 BRICK/BLOCK WALLING - cont'd					
Facing bricks; machine made facings; PC £330.00/1000 (unless otherwise stated; in gauged mortar (1:1:6) - cont'd					
Bullseye window 1200 mm dia; one ring; 215 mm high on face; cut voussoirs PC £269.50/100					
102 mm wide exposed soffit	10.35	100.27	190.40	nr	**290.67**
215 mm wide exposed soffit	15.55	150.67	380.79	nr	**531.46**
ADD or DEDUCT for variation of £10.00/1000 in PC of facing bricks	-	-	-	m2	**0.29**
Sills; horizontal; headers on edge; pointing top and one side; set weathering					
150 x 102 mm;	0.81	8.40	5.63	m	**14.03**
150 x 102 mm; cant headers; PC £124.30/100	0.86	8.92	20.40	m	**29.32**
Sills; horizontal; headers on flat; pointing top and one side					
150 x 102 mm; bullnosed specials; PC £124.30/100	0.75	7.78	14.62	m	**22.40**
Coping; horizontal; headers on edge; pointing top and both sides					
215 x 102 mm	0.64	6.64	12.99	m	**19.63**
260 x 102 mm	1.05	10.89	8.49	m	**19.38**
215 x 102 mm; double bullnosed specials; PC £124.30/100	0.69	7.16	20.35	m	**27.51**
260 x 120 mm; single bullnosed specials; PC £124.30/100	1.05	10.89	33.37	m	**44.26**
ADD or DEDUCT for variation of £10.00/1000 in PC of facing bricks	-	-	-	m	**0.29**
Facing bricks; hand made; PC £540.00/1000 (unless otherwise stated); in gauged mortar (1:1:6)					
Extra over common bricks; PC £124.00/1000; for facing bricks in					
stretcher bond	0.45	4.67	28.84	m2	**33.51**
flemish bond with snapped headers	0.55	5.71	36.06	m2	**41.77**
english bond with snapped headers	0.55	5.71	40.86	m2	**46.57**
ADD or DEDUCT for variation of £10.00/1000 in PC of facing bricks	-	-	-	m2	**0.87**
Half brick thick; stretcher bond; facework one side					
walls	1.90	19.71	38.54	m2	**58.25**
walls; building curved mean radius 6 m	2.75	28.53	41.42	m2	**69.95**
walls; building curved mean radius 1.50 m	3.45	35.79	43.43	m2	**79.22**
walls; building overhand	2.30	23.86	38.54	m2	**62.40**
walls; building up against concrete including flushing up at back	2.00	20.75	39.95	m2	**60.70**
walls; as formwork; temporary strutting	2.75	28.53	40.88	m2	**69.41**
walls; panels and aprons; not exceeding 1 m2	2.40	24.90	38.54	m2	**63.44**
isolated casings to steel columns	2.90	30.08	38.54	m2	**68.62**
bonding ends to existing	0.75	7.78	1.56	m	**9.34**
projections; vertical					
225 x 112 mm	0.46	4.77	8.98	m	**13.75**
337 x 112 mm	0.86	8.92	13.58	m	**22.50**
440 x 112 mm	1.25	12.97	18.19	m	**31.16**
Half brick thick; flemish bond with snapped headers; facework one side					
walls	2.20	22.82	42.60	m2	**65.42**
walls; building curved mean radius 6 m	3.10	32.16	44.61	m2	**76.77**
walls; building curved mean radius 1.50 m	4.00	41.50	46.62	m2	**88.12**
walls; building overhand	2.60	26.97	42.60	m2	**69.57**

F MASONRY Including overheads and profit at 7.50%	Labour hours	Labour £	Material £	Unit	Total rate £
walls; building up against concrete including flushing up at back	2.30	23.86	44.01	m2	**67.87**
walls; as formwork; temporary strutting	3.00	31.12	44.94	m2	**76.06**
walls; panels and aprons; not exceeding 1 m2	2.70	28.01	42.60	m2	**70.61**
isolated casings to steel columns	3.15	32.68	38.54	m2	**71.22**
bonding ends to existing	0.75	7.78	1.56	m	**9.34**
projections; vertical					
225 x 112 mm	0.58	6.02	10.23	m	**16.25**
337 x 112 mm	0.98	10.17	14.83	m	**25.00**
440 x 112 mm	1.40	14.52	20.07	m	**34.59**
One brick thick; two stretcher skins tied together; facework both sides					
walls	3.20	33.20	78.36	m2	**111.56**
walls; building curved mean radius 6 m	4.50	46.68	82.31	m2	**128.99**
walls; building curved mean radius 1.50 m	5.50	57.06	87.59	m2	**144.65**
isolated piers	3.80	39.42	78.34	m2	**117.76**
bonding ends to existing	0.98	10.17	3.11	m	**13.28**
One brick thick; flemish bond; facework both sides					
walls	3.35	34.75	77.80	m2	**112.55**
walls; building curved mean radius 6 m	4.60	47.72	83.56	m2	**131.28**
walls; building curved mean radius 1.50 m	5.75	59.65	87.59	m2	**147.24**
isolated piers	3.90	40.46	77.80	m2	**118.26**
bonding ends to existing	0.98	10.17	3.11	m	**13.28**
projections; vertical					
225 x 225 mm	0.92	9.54	17.32	m	**26.86**
337 x 225 mm	1.70	17.64	29.42	m	**47.06**
440 x 225 mm	2.55	26.45	34.03	m	**60.48**
ADD or DEDUCT for variation of £10.00/1000 in PC of facing bricks; in stretcher bond					
half brick thick	-	-	-	m2	**0.70**
one brick thick	-	-	-	m2	**1.40**
ADD or DEDUCT for variation of £10.00/1000 in PC of facing bricks; in flemish bond					
half brick thick	-	-	-	m2	**0.87**
one brick thick	-	-	-	m2	**1.74**
Extra over facing bricks for					
recessed joints	0.03	0.31	-	m2	**0.31**
raking out joints and pointing in black mortar	0.58	6.02	0.18	m2	**6.20**
bedding and pointing half brick wall in black mortar	-	-	-	m2	**0.87**
bedding and pointing one brick wall in black mortar	-	-	-	m2	**2.31**
flush plain bands; 225 mm wide stretcher bond; horizontal; bricks PC £580.00/1000	0.29	3.01	0.64	m	**3.65**
flush quoins; average 320 mm girth; block bond vertical; bricks PC £580.00/1000	0.46	4.77	0.60	m	**5.37**
Flat arches; 215 mm high on face					
102 mm wide exposed soffit	1.32	12.27	3.49	m	**15.76**
215 mm wide exposed soffit	1.99	18.51	6.99	m	**25.50**
Flat arches; 215 mm high on face; bullnosed specials; PC £124.30/100					
102 mm wide exposed soffit	1.38	12.90	21.10	m	**34.00**
215 mm wide exposed soffit	2.09	19.55	42.19	m	**61.74**
Segmental arches; one ring; 215 mm high on face					
102 mm wide exposed soffit	2.40	21.35	5.07	m	**26.42**
215 mm wide exposed soffit	3.60	32.10	10.36	m	**42.46**
Segmental arches; two ring; 215 mm high on face					
102 mm wide exposed soffit	3.10	28.61	5.07	m	**33.68**
215 mm wide exposed soffit	4.65	42.99	10.36	m	**53.35**

F MASONRY Including overheads and profit at 7.50%	Labour hours	Labour £	Material £	Unit	Total rate £
F10 BRICK/BLOCK WALLING - cont'd					
Facing bricks; hand made; PC £540.00/1000 (unless otherwise stated); in gauged mortar (1:1:6) - cont'd					
Segmental arches; 215 mm high on face; cut voussoirs; PC £269.50/100					
102 mm wide exposed soffit	2.55	22.91	44.91	m	**67.82**
215 mm wide exposed soffit	3.75	33.66	90.04	m	**123.70**
Segmental arches; one and a half ring; 320 mm high on face; cut voussoirs PC £269.50/100					
102 mm wide exposed soffit	3.45	32.24	88.51	m	**120.75**
215 mm wide exposed soffit	5.15	48.18	177.24	m	**225.42**
Semi circular arches; one ring; 215 mm high on face					
102 mm wide exposed soffit	4.15	38.73	16.65	m	**55.38**
215 mm wide exposed soffit	6.00	56.53	33.30	m	**89.83**
Semi circular arches; two ring; 215 mm high on face					
102 mm wide exposed soffit	5.55	53.26	16.65	m	**69.91**
215 mm wide exposed soffit	6.00	56.53	33.30	m	**89.83**
Semi circular arches; one ring; 215 mm high on face; cut voussoirs PC £269.50/100					
102 mm wide exposed soffit	3.10	27.84	45.20	m	**73.04**
215 mm wide exposed soffit	4.45	40.46	105.71	m	**146.17**
Bullseye window 600 mm dia; two rings; 215 mm high on face					
102 mm wide exposed soffit	6.90	68.03	76.75	nr	**144.78**
215 mm wide exposed soffit	10.35	102.12	153.29	nr	**255.41**
Bullseye window 1200 mm dia; two rings; 215 mm high on face					
102 mm wide exposed soffit	12.05	117.91	12.69	nr	**130.60**
215 mm wide exposed soffit	18.15	177.64	25.38	nr	**203.02**
Bullseye window 600 mm dia; one ring; 215 mm high on face; cut voussoirs; PC £269.50/100					
102 mm wide exposed soffit	5.75	56.10	110.44	nr	**166.54**
215 mm wide exposed soffit	8.60	83.97	220.83	nr	**304.80**
Bullseye window 1200 mm dia; one ring; 215 mm high on face; cut voussoirs; PC £269.50/100					
102 mm wide exposed soffit	10.35	100.27	190.40	nr	**290.67**
215 mm wide exposed soffit	15.55	150.67	380.79	nr	**531.46**
ADD or DEDUCT for variation of £10.00/1000 in PC of facing bricks	-	-	-	m	**0.29**
Sills; horizontal; headers on edge; pointing top and one side; set weathering					
150 x 102 mm;	0.81	8.40	9.03	m	**17.43**
150 x 102 mm; cant headers; PC £126.50/100	0.86	8.92	20.75	m	**29.67**
Sills; horizontal; headers on flat; pointing top and one side					
150 x 102 mm; bullnosed specials; PC £124.30/100	0.75	7.78	14.62	m	**22.40**
Coping; horizontal; headers on edge; pointing top and both sides					
215 x 102 mm	0.64	6.64	16.38	m	**23.02**
260 x 102 mm	1.05	10.89	13.58	m	**24.47**
215 x 102 mm; double bullnosed specials; PC £124.30/100	0.69	7.16	20.35	m	**27.51**
260 x 120 mm; single bullnosed specials; PC £124.30/100	1.05	10.89	33.37	m	**44.26**
ADD or DEDUCT for variation of £10.00/1000 in PC of facing bricks	-	-	-	m2	**0.29**

F MASONRY Including overheads and profit at 7.50%	Labour hours	Labour £	Material £	Unit	Total rate £
50 mm facing bricks slips; PC £110.00/100; in gauged mortar (1:1:6) built up against concrete including flushing up at back (ties measured elsewhere)					
Walls	2.90	30.08	78.22	m2	**108.30**
Edges of suspended slabs 200 mm wide	0.86	8.92	15.75	m	**24.67**
Columns 400 mm wide	1.70	17.64	31.39	m	**49.03**
Engineering bricks; PC £350.00/1000; and specials at PC £124.30/100; in cement mortar (1:3)					
Steps; all headers-on-edge; edges set with					
215 x 102 mm; horizontal; set weathering	0.81	8.40	19.90	m	**28.30**
Returned ends pointed	0.23	2.39	3.37	nr	**5.76**
430 x 102 mm; horizontal; set weathering	1.15	11.93	25.85	m	**37.78**
Returned ends pointed	0.29	3.01	6.17	nr	**9.18**
Labours on brick facework					
Fair cutting					
to curve	0.23	2.39	1.77	m	**4.16**
Fair returns					
half brick wide	0.06	0.62	-	m	**0.62**
one brick wide	0.08	0.83	-	m	**0.83**
one and a half brick wide	0.10	1.04	-	m	**1.04**
Fair angles formed by cutting					
squint	0.92	9.54	-	m	**9.54**
birdsmouth	0.81	8.40	-	m	**8.40**
external; chamfered 25 mm wide	1.15	11.93	-	m	**11.93**
external; rounded 100 mm radius	1.40	14.52	-	m	**14.52**
Fair chases					
100 x 50 mm; horizontal	1.70	17.64	-	m	**17.64**
100 x 50 mm; vertical	1.70	17.64	-	m	**17.64**
300 x 50 mm; vertical	3.10	32.16	-	m	**32.16**
Bonding ends to existing					
half brick thick	0.75	7.78	1.56	m	**9.34**
one brick thick	0.98	10.17	3.11	m	**13.28**
Centering to brickwork soffits; (prices included within arches rates)					
Flat soffits; not exceeding 2 m span					
over 0.3 m wide	2.30	16.76	6.05	m2	**22.81**
102 mm wide	0.46	3.35	1.08	m	**4.43**
215 mm wide	0.69	5.03	2.17	m	**7.20**
Segmental soffits; 1500 mm span 200 mm rise					
102 mm wide	1.55	11.30	2.23	nr	**13.53**
215 mm wide	2.30	16.76	4.46	nr	**21.22**
Semicircular soffits; 1500 mm span					
102 mm wide	2.05	14.94	2.63	nr	**17.57**
215 mm wide	2.75	20.04	5.28	nr	**25.32**
Bullseye window; 265 mm wide					
600 mm dia	1.70	12.39	3.10	nr	**15.49**
1200 mm dia	3.45	25.15	5.86	nr	**31.01**

ALTERNATIVE BLOCK PRICES (£/m2)

Aerated Concrete Durox 'Supablocs'; 630 x 225 mm

	£		£		£		£
75 mm	4.32	125 mm	8.11	175 mm	11.35	225 mm	14.60
90 mm	5.55	130 mm	8.44	190 mm	11.72	250 mm	16.22
100 mm	6.17	140 mm	9.08	200 mm	12.34	280 mm	18.17
115 mm	7.46	150 mm	9.73	215 mm	13.26		

F MASONRY
Including overheads and profit at 7.50%

F10 BRICK/BLOCK WALLING - cont'd

ALTERNATIVE BLOCK PRICES (£/m2) - cont'd

	£		£		£
ARC Conbloc blocks; 450 x 225 mm					
Cream fair faced					
75 mm solid	5.82	190 mm solid	14.53	190 mm hollow	11.83
100 mm solid	6.96	100 mm hollow	6.74	215 mm hollow	13.28
140 mm solid	9.99	140 mm hollow	9.16		
Fenlite					
90 mm solid	4.02	100 mm solid	4.48	140 mm solid	6.44
Leca					
75 mm solid	7.11	100 mm solid	13.70	140 mm solid	9.37
Standard facing					
100 mm solid	6.13	190 mm solid	12.57	190 mm hollow	10.35
140 mm solid	9.06	140 mm hollow	8.45	215 mm hollow	11.79

	£		£		£		£
Celcon 'Standard' blocks; 450 x 225 mm							
75 mm	4.70	125 mm	7.83	190 mm	11.90	230 mm	14.42
90 mm	5.63	140 mm	8.78	200 mm	12.52	250 mm	15.66
100 mm	6.26	150 mm	9.39	215 mm	13.46	300 mm	18.79

'Solar' blocks are also available at the same price, in a limited range

	£		£		£
Forticrete painting quality blocks; 450 x 225 mm					
100 mm solid	6.75	215 mm solid	14.53	190 mm hollow	10.97
140 mm solid	10.03	100 mm hollow	5.81	215 mm hollow	11.29
190 mm solid	13.41	140 mm hollow	8.08		
Lytag blocks; 450 x 225 mm					
3.5 N/mm2 Insulating blocks					
75 mm solid	3.42	140 mm solid	6.43	100 mm cellular	7.06
90 mm solid	6.68	190 mm solid	8.71	215 mm cellular	9.09
100 mm solid	4.56				
7.0 N/mm2 Insulating extra strength blocks					
100 mm solid	4.72	140 mm solid	6.67	190 mm solid	9.04
10.5 N/mm2 High strength blocks					
100 mm solid	5.21	140 mm solid	7.41	190 mm solid	10.02
3.5 N/mm2 and 7.0 N/mm2 Close textured blocks					
100 mm solid (3.5)	5.88	140 mm solid (3.5)	8.26	190 mm solid (3.5)	11.16
100 mm solid (7.0)	6.09	140 mm solid (7.0)	8.59	190 mm solid (7.0)	11.60
Tarmac 'Topblocks'; 450 x 225 mm					
3.5 N/mm2 'Hemelite' blocks					
70/75 mm solid	4.27	100 mm solid	4.12	215 mm solid	10.04
90 mm solid	7.00	190 mm solid	8.65		
7.0 N/mm2 'Hemelite' blocks					
90 mm solid	6.59	140 mm solid	7.49	215 mm solid	10.88
100 mm solid	4.85	190 mm solid	9.35		

	£		£		£		£
'Toplite' standard blocks							
75 mm	4.21	100 mm	5.61	150 mm	8.43	215 mm	12.07
90 mm	5.06	140 mm	7.86	200 mm	11.23		
'Toplite' GTI (thermal) blocks							
115 mm	6.63	130 mm	7.50	150 mm	8.65	215 mm	12.40
125 mm	7.21	140 mm	8.08	200 mm	11.54		

Discounts of 0 - 7.5% available depending on quantity/status

F MASONRY Including overheads and profit at 7.50%		Labour hours	Labour £	Material £	Unit	Total rate £
F10 BRICK/BLOCK WALLING - cont'd						
Lightweight aerated concrete blocks; Thermalite 'Shield'/'Turbo' blocks or similar; in gauged mortar (1:2:9)						
Walls or partitions or skins of hollow walls						
75 mm thick	PC £4.79	0.69	7.16	6.15	m2	13.31
90 mm thick	PC £5.75	0.76	7.88	7.34	m2	15.22
100 mm thick	PC £6.39	0.81	8.40	8.19	m2	16.59
115 mm thick	PC £7.73	0.85	8.82	9.87	m2	18.69
125 mm thick	PC £8.39	0.89	9.23	10.77	m2	20.00
130 mm thick	PC £8.73	0.92	9.54	11.17	m2	20.71
140 mm thick	PC £8.94	0.94	9.75	11.46	m2	21.21
150 mm thick	PC £9.59	0.98	10.17	12.23	m2	22.40
190 mm thick	PC £12.13	1.15	11.93	15.49	m2	27.42
200 mm thick	PC £12.77	1.20	12.45	16.24	m2	28.69
215 mm thick	PC £13.73	1.25	12.97	17.48	m2	30.45
255 mm thick	PC £16.28	1.40	14.52	20.66	m2	35.18
Isolated piers or chimney stacks						
190 mm thick		1.60	16.60	15.49	m2	32.09
215 mm thick		1.70	17.64	17.37	m2	35.01
255 mm thick		1.95	20.23	20.66	m2	40.89
Isolated casings						
75 mm thick		0.81	8.40	6.15	m2	14.55
90 mm thick		0.87	9.03	7.40	m2	16.43
100 mm thick		0.92	9.54	8.19	m2	17.73
115 mm thick		0.97	10.06	9.87	m2	19.93
125 mm thick		1.00	10.37	10.77	m2	21.14
140 mm thick		1.05	10.89	11.46	m2	22.35
Extra over for fair face; flush pointing						
walls; one side		0.11	1.14	-	m2	1.14
walls; both sides		0.20	2.07	-	m2	2.07
Bonding ends to common brickwork						
75 mm blockwork		0.17	1.76	0.50	m	2.26
90 mm blockwork		0.17	1.76	0.59	m	2.35
100 mm blockwork		0.23	2.39	0.65	m	3.04
115 mm blockwork		0.23	2.39	0.77	m	3.16
125 mm blockwork		0.26	2.70	0.89	m	3.59
130 mm blockwork		0.26	2.70	0.92	m	3.62
140 mm blockwork		0.29	3.01	0.93	m	3.94
150 mm blockwork		0.29	3.01	1.00	m	4.01
190 mm blockwork		0.35	3.63	1.29	m	4.92
200 mm blockwork		0.38	3.94	1.34	m	5.28
215 mm blockwork		0.41	4.25	1.44	m	5.69
255 mm blockwork		0.46	4.77	1.72	m	6.49
Lightweight smooth face aerated concrete blocks; Thermalite 'Smooth Face' blocks or similar; in gauged mortar (1:2:9); flush pointing one side						
Walls or partitions or skins of hollow walls						
100 mm thick	PC £9.03	1.01	10.10	11.04	m2	21.14
140 mm thick	PC £12.65	1.16	11.58	15.50	m2	27.08
150 mm thick	PC £13.01	1.21	12.10	15.93	m2	28.03
190 mm thick	PC £17.17	1.39	13.84	21.03	m2	34.87
200 mm thick	PC £18.07	1.44	14.36	22.09	m2	36.45
215 mm thick	PC £19.43	1.56	15.52	23.74	m2	39.26
Isolated piers or chimney stacks						
190 mm thick		1.84	18.51	21.03	m2	39.54
200 mm thick		1.99	20.06	22.09	m2	42.15
215 mm thick		2.21	22.26	23.74	m2	46.00

F MASONRY
Including overheads and profit at 7.50%

		Labour hours	Labour £	Material £	Unit	Total rate £
F10 BRICK/BLOCK WALLING - cont'd						
Lightweight smooth face aerated concrete blocks; Thermalite 'Smooth Face' blocks or similar; in gauged mortar (1:2:9); flush pointing one side - cont'd						
Isolated casings						
100 mm thick		1.14	11.45	11.04	m2	**22.49**
140 mm thick		1.26	12.62	15.50	m2	**28.12**
Extra over for flush pointing						
walls; both sides		0.08	0.83	-	m2	**0.83**
Bonding ends to common brickwork						
100 mm blockwork		0.29	3.01	0.90	m	**3.91**
140 mm blockwork		0.32	3.32	1.29	m	**4.61**
150 mm blockwork		0.35	3.63	1.32	m	**4.95**
190 mm blockwork		0.40	4.15	1.76	m	**5.91**
200 mm blockwork		0.44	4.56	1.84	m	**6.40**
215 mm blockwork		0.46	4.77	1.98	m	**6.75**
Lightweight aerated high strength concrete blocks (7 N/mm2); Thermalite 'High Strength' blocks or similar; in cement mortar (1:3)						
Walls or partitions or skins of hollow walls						
100 mm thick	PC £8.24	0.86	8.71	10.10	m2	**18.81**
140 mm thick	PC £11.54	0.98	9.92	14.18	m2	**24.10**
150 mm thick	PC £11.82	1.04	10.54	14.52	m2	**25.06**
190 mm thick	PC £15.00	1.22	12.37	18.46	m2	**30.83**
200 mm thick	PC £15.78	1.27	12.88	18.66	m2	**31.54**
215 mm thick	PC £16.96	1.33	13.47	20.83	m2	**34.30**
Isolated piers or chimney stacks						
190 mm thick		1.62	16.52	16.85	m2	**33.37**
200 mm thick		1.72	17.55	19.39	m2	**36.94**
215 mm thick		1.98	20.21	20.79	m2	**41.00**
Isolated casings						
100 mm thick		0.97	9.86	10.10	m2	**19.96**
140 mm thick		1.11	11.27	14.18	m2	**25.45**
150 mm thick		1.16	11.79	14.52	m2	**26.31**
190 mm thick		1.32	13.40	18.46	m2	**31.86**
200 mm thick		1.47	14.96	19.39	m2	**34.35**
215 mm thick		1.53	15.54	20.83	m2	**36.37**
Extra over for fair face; flush pointing						
walls; one side		0.11	1.14	-	m2	**1.14**
walls; both sides		0.20	2.07	-	m2	**2.07**
Bonding ends to common brickwork						
100 mm blockwork		0.29	3.01	0.83	m	**3.84**
140 mm blockwork		0.32	3.32	1.18	m	**4.50**
150 mm blockwork		0.35	3.63	1.20	m	**4.83**
190 mm blockwork		0.40	4.15	1.55	m	**5.70**
200 mm blockwork		0.44	4.56	1.63	m	**6.19**
215 mm blockwork		0.46	4.77	1.75	m	**6.52**
Dense aggregate concrete blocks; 'ARC Conbloc' or similar; in gauged mortar (1:2:9)						
Walls or partitions or skins of hollow walls						
75 mm thick; solid	PC £4.15	0.81	9.77	5.17	m2	**14.94**
100 mm thick; solid	PC £4.15	0.92	11.10	5.26	m2	**16.36**
140 mm thick; hollow	PC £6.64	1.05	12.66	8.35	m2	**21.01**
140 mm thick; solid	PC £7.10	1.15	13.87	8.89	m2	**22.76**
190 mm thick; hollow	PC £8.97	1.30	15.68	11.33	m2	**27.01**
215 mm thick; hollow	PC £9.19	1.45	17.49	11.63	m2	**29.12**

F MASONRY Including overheads and profit at 7.50%		Labour hours	Labour £	Material £	Unit	Total rate £
Isolated piers or chimney stacks						
140 mm thick; hollow		1.45	17.49	8.35	m2	**25.84**
190 mm thick; hollow		1.85	22.31	11.33	m2	**33.64**
215 mm thick; hollow		2.05	24.73	11.63	m2	**36.36**
Isolated casings						
75 mm thick; solid		0.92	11.10	5.17	m2	**16.27**
100 mm thick; solid		1.05	12.66	5.26	m2	**17.92**
140 mm thick; solid		1.25	15.08	8.89	m2	**23.97**
Extra over for fair face; flush pointing						
walls; one side		0.11	1.33	-	m2	**1.33**
walls; both sides		0.20	2.41	-	m2	**2.41**
Bonding ends to common brickwork						
75 mm blockwork		0.23	2.77	0.43	m	**3.20**
100 mm blockwork		0.30	3.62	0.43	m	**4.05**
140 mm blockwork		0.37	4.46	0.77	m	**5.23**
190 mm blockwork		0.44	5.31	0.98	m	**6.29**
215 mm blockwork		0.51	6.15	1.00	m	**7.15**
Dense aggregate concrete blocks; (7 N/mm2) Forticrete 'Leicester Common' blocks or similar; in cement mortar (1:3)						
Walls or partitions or skins of hollow walls						
75 mm thick; solid	PC £4.77	0.81	9.77	5.95	m2	**15.72**
100 mm thick; hollow	PC £5.27	0.92	11.10	6.65	m2	**17.75**
100 mm thick; solid	PC £6.12	0.92	11.10	7.65	m2	**18.75**
140 mm thick; hollow	PC £7.30	1.05	12.66	9.21	m2	**21.87**
140 mm thick; solid	PC £9.13	1.15	13.87	11.36	m2	**25.23**
190 mm thick; hollow	PC £9.90	1.30	15.68	12.54	m2	**28.22**
190 mm thick; solid	PC £12.22	1.45	17.49	15.27	m2	**32.76**
215 mm thick; hollow	PC £10.25	1.45	17.49	13.00	m2	**30.49**
215 mm thick; solid	PC £13.16	1.55	18.70	16.45	m2	**35.15**
Dwarf support wall						
140 mm thick; solid		1.60	19.30	11.36	m2	**30.66**
190 mm thick; solid		1.85	22.31	15.27	m2	**37.58**
215 mm thick; solid		2.05	24.73	16.45	m2	**41.18**
Isolated piers or chimney stacks						
140 mm thick; hollow		1.45	17.49	9.21	m2	**26.70**
190 mm thick; hollow		1.85	22.31	12.54	m2	**34.85**
215 mm thick; hollow		2.05	24.73	13.00	m2	**37.73**
Isolated casings						
75 mm thick; solid		0.92	11.10	5.95	m2	**17.05**
100 mm thick; solid		1.05	12.66	7.65	m2	**20.31**
140 mm thick; solid		1.25	15.08	11.36	m2	**26.44**
Extra over for fair face; flush pointing						
walls; one side		0.11	1.33	-	m2	**1.33**
walls; both sides		0.20	2.41	-	m2	**2.41**
Bonding ends to common brickwork						
75 mm blockwork		0.23	2.77	0.51	m	**3.28**
100 mm blockwork		0.30	3.62	0.63	m	**4.25**
140 mm blockwork		0.37	4.46	0.96	m	**5.42**
190 mm blockwork		0.44	5.31	1.31	m	**6.62**
215 mm blockwork		0.51	6.15	1.41	m	**7.56**
Dense aggregate coloured concrete blocks; Forticrete 'Leicester Bathstone'; in coloured gauged mortar (1:1:6); flush pointing one side						
Walls or partitions or skins of hollow walls						
100 mm thick; hollow	PC £11.80	1.05	12.66	14.60	m2	**27.26**
100 mm thick; solid	PC £14.17	1.05	12.66	17.41	m2	**30.07**
140 mm thick; hollow	PC £15.64	1.15	13.87	19.37	m2	**33.24**
140 mm thick; solid	PC £20.96	1.25	15.08	25.68	m2	**40.76**

F MASONRY Including overheads and profit at 7.50%		Labour hours	Labour £	Material £	Unit	Total rate £
F10 BRICK/BLOCK WALLING - cont'd						
Dense aggregate coloured concrete blocks; Forticrete 'Leicester Bathstone'; in coloured gauged mortar (1:1:6); flush pointing one side - cont'd						
Walls or partitions or skins of hollow walls						
190 mm thick; hollow	PC £19.52	1.45	17.49	24.37	m2	**41.86**
190 mm thick; solid	PC £28.55	1.55	18.70	35.06	m2	**53.76**
215 mm thick; hollow	PC £21.07	1.55	18.70	26.30	m2	**45.00**
215 mm thick; solid	PC £30.47	1.65	19.90	37.40	m2	**57.30**
Isolated piers or chimney stacks						
140 mm thick; solid		1.70	20.50	25.68	m2	**46.18**
190 mm thick; solid		1.95	23.52	35.06	m2	**58.58**
215 mm thick; solid		2.20	26.54	37.57	m2	**64.11**
Extra over for flush pointing						
walls; both sides		0.08	0.96	-	m2	**0.96**
Extra over blocks for						
100 mm thick lintol blocks; ref D14		0.29	3.50	8.33	m	**11.83**
140 mm thick lintol blocks; ref H14		0.35	4.22	9.19	m	**13.41**
140 mm thick quoin blocks; ref H16		0.46	5.55	19.40	m	**24.95**
140 mm thick cavity closer blocks; ref H17		0.46	5.55	22.34	m	**27.89**
140 mm thick cill blocks; ref H21		0.35	4.22	13.51	m	**17.73**
190 mm thick lintol blocks; ref A14		0.46	5.55	10.65	m	**16.20**
190 mm thick cill blocks; ref A21		0.40	4.82	14.48	m	**19.30**
F20 NATURAL STONE RUBBLE WALLING						
Cotswold Guiting limestone; laid dry						
Uncoursed random rubble walling						
275 mm thick		2.05	24.44	31.94	m2	**56.38**
350 mm thick		2.40	28.61	40.41	m2	**69.02**
425 mm thick		2.70	32.19	49.42	m2	**81.61**
500 mm thick		3.00	35.77	57.88	m2	**93.65**
Cotswold Guiting limestone; bedded; jointed and pointed in cement - lime mortar (1:2:9)						
Uncoursed random rubble walling; faced and pointed; both sides						
275 mm thick		1.95	23.25	33.75	m2	**57.00**
350 mm thick		2.05	24.44	42.67	m2	**67.11**
425 mm thick		2.20	26.23	52.14	m2	**78.37**
500 mm thick		2.30	27.42	61.51	m2	**88.93**
Coursed random rubble walling; rough dressed; faced and pointed one side						
114 mm thick		1.70	17.64	48.31	m2	**65.95**
150 mm thick		2.00	20.75	55.01	m2	**75.76**
Fair returns on walling						
114 mm wide		0.03	0.31	-	m	**0.31**
150 mm wide		0.05	0.52	-	m	**0.52**
275 mm wide		0.08	0.83	-	m	**0.83**
350 mm wide		0.10	1.04	-	m	**1.04**
425 mm wide		0.13	1.35	-	m	**1.35**
500 mm wide		0.15	1.56	-	m	**1.56**
Fair raking cutting on walling						
114 mm thick		0.23	2.39	6.82	m	**9.21**
150 mm thick		0.28	2.90	7.77	m	**10.67**
Level uncoursed rubble walling for damp proof courses and the like						
275 mm wide		0.24	2.86	2.07	m	**4.93**
350 mm wide		0.25	2.98	2.33	m	**5.31**
425 mm wide		0.26	3.10	2.60	m	**5.70**
500 mm wide		0.28	3.34	3.14	m	**6.48**

F MASONRY Including overheads and profit at 7.50%	Labour hours	Labour £	Material £	Unit	Total rate £
Copings formed of rough stones; faced and pointed all round					
275 x 200 mm (average) high	0.59	7.03	7.23	m	14.26
350 x 250 mm (average) high	0.77	9.18	10.89	m	20.07
425 x 300 mm (average) high	0.98	11.68	15.20	m	26.88
500 x 350 mm (average) high	1.20	14.31	20.75	m	35.06
F22 CAST STONE WALLING/DRESSINGS					
Reconstructed limestone walling; 'Bradstone' 100 mm bed weathered Cotswold or North Cerney masonry blocks or similar; laid to pattern or course recommended; bedded, jointed and pointed in approved coloured cement - lime mortar (1:2:9)					
Walls; facing and pointing one side					
masonry blocks; random uncoursed	1.30	13.49	21.45	m2	34.94
Extra for					
Return ends	0.46	4.77	1.25	m	6.02
Plain 'L' shaped quoins	0.15	1.56	6.67	m	8.23
traditional walling; coursed squared	1.60	16.60	20.63	m2	37.23
squared random rubble	1.60	16.60	21.34	m2	37.94
squared coursed rubble (large module)	1.50	15.56	21.17	m2	36.73
squared coursed rubble (small module)	1.55	16.08	21.50	m2	37.58
squared and pitched rock faced walling; coursed	1.65	17.12	20.63	m2	37.75
rough hewn rockfaced walling; random	1.70	17.64	20.46	m2	38.10
Extra for return ends	0.18	1.87	-	m	1.87
Isolated piers or chimney stacks; facing and pointing one side					
masonry blocks; random uncoursed	1.80	18.67	22.47	m2	41.14
traditional walling; coursed squared	2.25	23.34	21.62	m2	44.96
squared random rubble	2.25	23.34	22.35	m2	45.69
squared coursed rubble (large module)	2.05	21.27	22.18	m2	43.45
squared coursed rubble (small module)	2.20	22.82	22.53	m2	45.35
squared and pitched rock faced walling; coursed	2.35	24.38	21.62	m2	46.00
rough hewn rockfaced walling; random	2.40	24.90	21.43	m2	46.33
Isolated casings; facing and pointing one side					
masonry blocks; random uncoursed	1.55	16.08	22.47	m2	38.55
traditional walling; coursed squared	1.95	20.23	21.62	m2	41.85
squared random rubble	1.95	20.23	22.35	m2	42.58
squared coursed rubble (large module)	1.80	18.67	22.18	m2	40.85
squared coursed rubble (small module)	1.90	19.71	22.53	m2	42.24
squared and pitched rock faced walling; coursed	2.00	20.75	21.62	m2	42.37
rough hewn rockfaced walling; random	2.05	21.27	21.43	m2	42.70
Fair returns 100 mm wide					
masonry blocks; random uncoursed	0.14	1.45	-	m	1.45
traditional walling; coursed squared	0.17	1.76	-	m	1.76
squared random rubble	0.17	1.76	-	m	1.76
squared coursed rubble (large module)	0.16	1.66	-	m	1.66
squared coursed rubble (small module)	0.16	1.66	-	m	1.66
squared and pitched rock faced walling; coursed	0.17	1.76	-	m	1.76
rough hewn rockfaced walling; random	0.18	1.87	-	m	1.87
Fair raking cutting on masonry blocks					
100 mm thick	0.21	2.18	-	m	2.18

F MASONRY Including overheads and profit at 7.50%		Labour hours	Labour £	Material £	Unit	Total rate £
F22 CAST STONE WALLING/DRESSINGS - cont'd						
Reconstructed limestone dressings; 'Bradstone Architectural' dressings in weathered Cotswold or North Cerney shades or similar; bedded, jointed and pointed in approved coloured cement - lime mortar (1:2:9)						
Copings; twice weathered and throated						
152 x 76 mm; type A		0.38	3.94	8.87	m	**12.81**
178 x 64 mm; type B		0.38	3.94	9.54	m	**13.48**
305 x 76 mm; type A		0.46	4.77	16.95	m	**21.72**
Extra for						
Fair end		-	-	-	nr	**3.35**
Returned mitred fair end		-	-	-	nr	**3.35**
Copings; once weathered and throated						
191 x 76 mm		0.38	3.94	10.33	m	**14.27**
305 x 76 mm		0.46	4.77	16.65	m	**21.42**
365 x 76 mm		0.46	4.77	17.91	m	**22.68**
Extra for						
Fair end		-	-	-	nr	**3.35**
Returned mitred fair end		-	-	-	nr	**3.35**
Chimney caps; four times weathered and throated; once holed						
553 x 533 x 76 mm		0.46	4.77	22.93	nr	**27.70**
686 x 686 x 76 mm		0.46	4.77	37.68	nr	**42.45**
Pier caps; four times weathered and throated						
305 x 305 mm		0.29	3.01	7.61	nr	**10.62**
381 x 381 mm		0.29	3.01	10.70	nr	**13.71**
457 x 457 mm		0.35	3.63	14.87	nr	**18.50**
533 x 533 mm		0.35	3.63	20.64	nr	**24.27**
Splayed corbels						
457 x 102 x 229 mm		0.17	1.76	9.58	nr	**11.34**
686 x 102 x 229 mm		0.23	2.39	15.27	nr	**17.66**
Air bricks						
229 x 142 x 76 mm		0.09	0.93	3.74	nr	**4.67**
102 x 152 mm lintels; rectangular; reinforced with mild steel bars						
not exceeding 1.22 m long		0.28	2.90	15.57	m	**18.47**
1.37 - 1.67 m long		0.30	3.11	15.86	m	**18.97**
1.83 - 1.98 m long		0.32	3.32	17.35	m	**20.67**
102 x 229 mm lintels; rectangular; reinforced with mild steel bars						
not exceeding 1.67 m long		0.30	3.11	19.43	m	**22.54**
1.83 - 1.98 m long		0.32	3.32	19.89	m	**23.21**
2.13 - 2.44 m long		0.35	3.63	20.03	m	**23.66**
2.59 x 2.90 m long		0.37	3.84	21.83	m	**25.67**
197 x 67 mm sills to suit standard softwood windows; stooled at ends						
0.56 - 2.50 m long		0.35	3.63	24.47	m	**28.10**
Window surround; traditional with label moulding; for single light; sill 146 x 133 mm; jambs 146 x 146 mm; head 146 x 105 mm; including all dowels and anchors						
window size 508 x 1479 mm	PC £110.74	1.05	10.89	119.53	nr	**130.42**
Window surround; traditional with label moulding; three light; for windows 508 x 1219 mm; sill 146 x 133 mm; jambs 146 x 146 mm; head 146 x 103 mm; mullions 146 x 108 mm; including all dowels and anchors						
overall size 1975 x 1479 mm	PC £241.19	2.70	28.01	260.09	nr	**288.10**

F MASONRY Including overheads and profit at 7.50%	Labour hours	Labour £	Material £	Unit	Total rate £
Door surround; moulded continuous jambs and head with label moulding; including all dowels and anchors					
door 839 x 1981 mm in 102 x 64 mm frame	1.90	19.71	315.77	nr	**335.48**
F30 ACCESSORIES/SUNDRY ITEMS FOR BRICK/BLOCK/ STONE WALLING					
Sundries - brick/block walling					
Forming cavities in hollow walls; three wall ties per m2					
50 mm cavity; polypropylene ties	0.07	0.73	0.17	m2	**0.90**
50 mm cavity; galvanized steel butterfly wall ties	0.07	0.73	0.19	m2	**0.92**
50 mm cavity; galvanized steel twisted wall ties	0.07	0.73	0.26	m2	**0.99**
50 mm cavity; stainless steel butterfly wall ties	0.07	0.73	0.25	m2	**0.98**
50 mm cavity; stainless steel twisted wall ties	0.07	0.73	0.51	m2	**1.24**
75 mm cavity; polypropylene ties	0.07	0.73	0.17	m2	**0.90**
75 mm cavity; galvanised steel butterfly wall ties	0.07	0.73	0.19	m2	**0.92**
75 mm cavity; galvanised steel twisted wall ties	0.07	0.73	0.26	m2	**0.99**
75 mm cavity; stainless steel butterfly wall ties	0.07	0.73	0.25	m2	**0.98**
75 mm cavity; stainless steel twisted wall ties	0.07	0.73	0.51	m2	**1.24**
Closing at jambs with common brickwork half brick thick					
50 mm cavity	0.35	3.63	0.96	m	**4.59**
50 mm cavity; including damp proof course	0.46	4.77	1.52	m	**6.29**
75 mm cavity	0.35	3.63	1.24	m	**4.87**
75 mm cavity; including damp proof course	0.46	4.77	1.80	m	**6.57**
Closing at jambs with blockwork 100 mm thick					
50 mm cavity	0.29	3.01	0.47	m	**3.48**
50 mm cavity; including damp proof course	0.35	3.63	1.04	m	**4.67**
75 mm cavity	0.29	3.01	0.66	m	**3.67**
75 mm cavity; including damp proof course	0.35	3.63	1.22	m	**4.85**
Closing at sill with one course common brickwork					
50 mm cavity	0.35	3.63	2.10	m	**5.73**
50 mm cavity; including damp proof course	0.40	4.15	2.66	m	**6.81**
75 mm cavity	0.35	3.63	2.10	m	**5.73**
75 mm cavity; including damp proof course	0.40	4.15	2.66	m	**6.81**
Closing at top with					
single course of slates	0.17	1.76	2.55	m	**4.31**
'Westbrick' cavity closer	0.17	1.76	3.44	m	**5.20**
Cavity wall insulation; 'Dritherm'					
filling 50 mm cavity	0.17	1.76	2.45	m2	**4.21**
filling 75 mm cavity	0.20	2.07	3.23	m2	**5.30**
Cavity wall insulation; fixing with insulation retaining ties					
30 mm Celotex RR cavity insulation	0.23	2.39	4.07	m2	**6.46**
30 mm cavity wall batts	0.23	2.39	3.21	m2	**5.60**
25 mm Plaschem 'Aerobuild' foil faced polyurethene cavity slabs	0.23	2.39	5.60	m2	**7.99**
25 mm Styrofoam cavity wall insulation	0.23	2.39	3.23	m2	**5.62**

F MASONRY
Including overheads and profit at 7.50%

ALTERNATIVE DAMP PROOF COURSE PRICES (£/m2)

	£		£		£
Asbestos based					
'Astos'	4.33	'Barchester'	4.33		
Fibre based					
'Challenge'	2.52	'Stormax'	2.64		
Hessian based					
'Callendrite'	3.53	'Nubit'	3.91	'Permaseal'	3.91
Leadcored					
'Astos'	10.54	'Ledumite' grade 2	14.01	'Permalead'	11.59
'Ledkore' grade B	14.03	'Ledumite' grade 3	17.63	'Pluvex'	9.40
'Ledumite' grade 1	11.33	'Nuled'	9.30	'Trindos'	10.46
Pitch polymer					
'Aquagard'	4.43	'Permaflex'	4.57		

Further discounts may be available depending on quantity/status

	Labour hours	Labour £	Material £	Unit	Total rate £
'Pluvex' (hessian based) damp proof course or similar; PC £3.95/m2; 200 mm laps; in gauged mortar (1:1:6)					
Horizontal					
over 225 mm wide	0.29	3.01	4.59	m2	**7.60**
over 225 mm wide; forming cavity gutters in hollow walls	0.46	4.77	4.59	m2	**9.36**
not exceeding 225 mm wide	0.58	6.02	4.59	m2	**10.61**
Vertical					
not exceeding 225 mm wide	0.86	8.92	4.59	m2	**13.51**
'Pluvex' (fibre based) damp proof course or similar; PC £2.64/m2; 200 mm laps; in gauged mortar (1:1:6)					
Horizontal					
over 225 mm wide	0.29	3.01	3.06	m2	**6.07**
over 225 mm wide; forming cavity gutters in hollow walls	0.46	4.77	3.05	m2	**7.82**
not exceeding 225 mm wide	0.58	6.02	3.05	m2	**9.07**
Vertical					
not exceeding 225 mm wide	0.86	8.92	3.05	m2	**11.97**
'Asbex' (asbestos based) damp proof course or similar; PC £4.33/m2; 200 mm laps; in gauged mortar (1:1:6)					
Horizontal					
over 225 mm wide	0.29	3.01	5.00	m2	**8.01**
over 225 mm wide; forming cavity gutters in hollow walls	0.46	4.77	5.00	m2	**9.77**
not exceeding 225 mm wide	0.58	6.02	5.00	m2	**11.02**
Vertical					
not exceeding 225 mm wide	0.86	8.92	5.00	m2	**13.92**

F MASONRY Including overheads and profit at 7.50%	Labour hours	Labour £	Material £	Unit	Total rate £
Polyethelene damp proof course; PC £0.89/m2; 200 mm laps; in gauged mortar (1:1:6)					
Horizontal					
over 225 mm wide	0.29	3.01	1.03	m2	4.04
over 225 mm wide; forming cavity gutters in hollow walls	0.46	4.77	1.03	m2	5.80
not exceeding 225 mm wide	0.58	6.02	1.03	m2	7.05
Vertical					
not exceeding 225 mm wide	0.86	8.92	1.03	m2	9.95
'Permabit' bitumen polymer damp proof course or similar; PC £2.67/m2; 150 mm laps; in gauged mortar (1:1:6)					
Horizontal					
over 225 mm wide	0.29	3.01	3.09	m2	6.10
over 225 mm wide; forming cavity gutters in hollow walls	0.46	4.77	3.09	m2	7.86
not exceeding 225 mm wide	0.58	6.02	3.09	m2	9.11
Vertical					
not exceeding 225 mm wide	0.86	8.92	3.08	m2	12.00
'Hyload' (pitch polymer) damp proof course or similar; PC £4.43/m2; 150 mm laps; in gauged mortar (1:1:6)					
Horizontal					
over 225 mm wide	0.29	3.01	5.14	m2	8.15
over 225 mm wide; forming cavity gutters in hollow walls	0.46	4.77	5.14	m2	9.91
not exceeding 225 mm wide	0.58	6.02	5.14	m2	11.16
Vertical					
not exceeding 225 mm wide	0.86	8.92	5.14	m2	14.06
'Ledkore' grade A (bitumen based lead cored); damp proof course or similar; PC £12.21/m2; 200 mm laps; in gauged mortar (1:1:6)					
Horizontal					
over 225 mm wide	0.38	3.94	14.18	m2	18.12
over 225 mm wide; forming cavity gutters in hollow walls	0.61	6.33	14.18	m2	20.51
not exceeding 225 mm wide	0.58	6.02	14.18	m2	20.20
Vertical					
not exceeding 225 mm wide	0.86	8.92	14.11	m2	23.03
Milled lead damp proof course; PC £26.52/m2; BS 1178; 1.80 mm (code 4), 175 mm laps; in cement-lime mortar (1:2:9)					
Horizontal					
over 225 mm wide	2.30	23.86	30.79	m2	54.65
not exceeding 225 mm wide	3.45	35.79	30.79	m2	66.58
Two courses slates in cement mortar (1:3)					
Horizontal					
over 225 mm wide	1.70	17.64	19.04	m2	36.68
Vertical					
over 225 mm wide	2.60	26.97	19.50	m2	46.47
Silicone injection damp-proofing					
Horizontal; 450 mm centres; make good brickwork					
half brick thick	-	-	-	m	10.43
one brick thick	-	-	-	m	11.18
one and a half brick thick	-	-	-	m	13.06

F MASONRY Including overheads and profit at 7.50%	Labour hours	Labour £	Material £	Unit	Total rate £
F30 ACCESSORIES/SUNDRY ITEMS FOR BRICK/BLOCK/ - cont'd					
STONE WALLING - cont'd					
'Synthaprufe' damp proof membrane; PC £30.04/25L; three coats brushed on					
Vertical					
not exceeding 150 mm wide	0.09	0.93	0.51	m2	**1.44**
150 - 225 mm wide	0.11	1.14	0.78	m2	**1.92**
225 - 300 mm wide	0.15	1.56	1.02	m2	**2.58**
over 300 mm wide	0.32	3.32	3.39	m2	**6.71**
'Type X' polypropylene abutment cavity tray by Cavity Trays Ltd; built into facing brickwork as the work proceeds; complete with Code 4 lead flashing					
Intermediate tray with short leads (requiring soakers); to suit roof of					
17 - 20 degrees pitch	0.07	0.73	4.53	nr	**5.26**
21 - 25 degrees pitch	0.07	0.73	4.20	nr	**4.93**
26 - 45 degrees pitch	0.07	0.73	3.71	nr	**4.44**
Intermediate tray with long leads (suitable only for corrugated roof tiles); to suit roof of					
17 - 20 degrees pitch	0.07	0.73	5.64	nr	**6.37**
21 - 25 degrees pitch	0.07	0.73	5.19	nr	**5.92**
26 - 45 degrees pitch	0.07	0.73	4.80	nr	**5.53**
Extra for					
ridge tray	0.07	0.73	7.33	nr	**8.06**
catchment end tray	0.07	0.73	9.95	nr	**10.68**
Servicised 'Bitu-thene' self-adhesive cavity flashing; type 'CA'; well lapped at joints; in gauged mortar (1:1:6)					
Horizontal					
over 225 mm wide	0.98	10.17	0.59	m2	**10.76**
'Brickforce' galvanised steel joint reinforcement					
In walls					
40 mm wide; ref. GBF 40	0.03	0.31	0.19	m	**0.50**
60 mm wide; ref. GBF 60	0.04	0.41	0.20	m	**0.61**
100 mm wide; ref. GBF 100	0.05	0.52	0.27	m	**0.79**
160 mm wide; ref. GBF 160	0.06	0.62	0.37	m	**0.99**
'Brickforce' stainless steel joint reinforcement					
In walls					
40 mm wide; ref. SBF 40	0.03	0.31	1.94	m	**2.25**
60 mm wide; ref. SBF 60	0.04	0.41	2.12	m	**2.53**
100 mm wide; ref. SBF 100	0.05	0.52	2.17	m	**2.69**
160 mm wide; ref. SBF 160	0.06	0.62	2.26	m	**2.88**
Fillets/pointing/wedging and pinning etc.					
Weather or angle fillets in cement mortar (1:3)					
50 mm face width	0.14	1.45	0.08	m	**1.53**
100 mm face width	0.23	2.39	0.31	m	**2.70**
Pointing wood frames or sills with mastic					
one side	0.08	0.83	0.44	m	**1.27**
each side	0.16	1.66	0.89	m	**2.55**
Pointing wood frames or sills with polysulphide sealant					
one side	0.08	0.83	0.93	m	**1.76**
each side	0.16	1.66	1.86	m	**3.52**

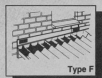

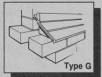

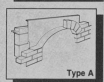

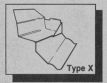

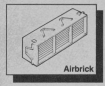

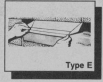

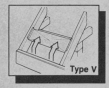

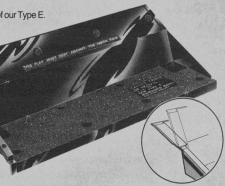

F MASONRY Including overheads and profit at 7.50%	Labour hours	Labour £	Material £	Unit	Total rate £
Bedding wood plates in cement mortar (1:3)					
100 mm wide	0.07	0.73	0.05	m	**0.78**
Bedding wood frame in cement mortar (1:3) and point					
one side	0.09	0.93	0.05	m	**0.98**
each side	0.11	1.14	0.07	m	**1.21**
one side in mortar; other side in mastic	0.17	1.76	0.49	m	**2.25**
Wedging and pinning up to underside of existing construction with slates in cement mortar (1:3)					
102 mm wall	0.92	9.54	2.66	m	**12.20**
215 mm wall	1.15	11.93	4.88	m	**16.81**
327 mm wall	1.40	14.52	7.11	m	**21.63**
Raking out joint in brickwork or blockwork for turned-in edge of flashing					
horizontal	0.17	1.76	0.05	m	**1.81**
stepped	0.23	2.39	0.05	m	**2.44**
Raking out and enlarging joint in brickwork or blockwork for nib of asphalt					
horizontal	0.23	2.39	-	m	**2.39**
Cutting grooves in brickwork or blockwork					
for water bars and the like	0.35	3.63	-	m	**3.63**
for nib of asphalt; horizontal	0.35	3.63	-	m	**3.63**
Preparing to receive new walls					
top existing 215 mm brick wall	0.23	2.39	-	m	**2.39**
Expansion joints					
Cleaning and priming both faces; filling with pre-formed closed cell joint filler and pointing one side with polysulphide sealant; 12 mm deep					
12 mm joint	0.29	3.01	2.57	m	**5.58**
19 mm joint	0.31	3.22	3.30	m	**6.52**
25 mm joint	0.35	3.63	4.01	m	**7.64**
Fire resisting horizontal expansion joint; filling with 'Nullifire System J' joint filler; fixed with high temparature slip adhesive; between top of wall and soffit					
10 mm joint with 30 mm deep ref. J60/10 filler (one hour fire seal)					
Wall 100 mm wide	0.29	3.01	5.39	m	**8.40**
Wall 150 mm wide	0.29	3.01	5.39	m	**8.40**
Wall 200 mm wide	0.29	3.01	5.39	m	**8.40**
10 mm joint with 30 mm deep ref. J120/10 filler (two hour fire seal)					
Wall 100 mm wide	0.29	3.01	5.39	m	**8.40**
Wall 150 mm wide	0.29	3.01	5.39	m	**8.40**
Wall 200 mm wide	0.29	3.01	5.39	m	**8.40**
20 mm joint with 45 mm deep ref. J120/20 filler (two hour fire seal)					
Wall 100 mm wide	0.35	3.63	8.82	m	**12.45**
Wall 150 mm wide	0.35	3.63	8.82	m	**12.45**
Wall 200 mm wide	0.35	3.63	8.82	m	**12.45**
30 mm joint with 75 mm deep ref. J180/30 filler (three hour fire seal)					
Wall 100 mm wide	0.40	4.15	27.62	m	**31.77**
Wall 150 mm wide	0.40	4.15	27.62	m	**31.77**
Wall 200 mm wide	0.40	4.15	27.62	m	**31.77**

F MASONRY Including overheads and profit at 7.50%	Labour hours	Labour £	Material £	Unit	Total rate £
F30 ACCESSORIES/SUNDRY ITEMS FOR BRICK/BLOCK/ - cont'd STONE WALLING - cont'd					
Fire resisting vertical expansion joint; filling with 'Nullifire System J' joint filler; fixed with high temparature slip adhesive; with polysulphide sealant one side; between end of wall and concrete					
20 mm joint with 45 mm deep ref. J120/20 filler (two hour fire seal)					
Wall 100 mm wide	0.40	4.15	12.37	m	**16.52**
Wall 150 mm wide	0.40	4.15	12.37	m	**16.52**
Wall 200 mm wide	0.40	4.15	12.37	m	**16.52**
Sills and tile creasings					
Sills; two courses of machine made plain roofing tiles; set weathering; bedded and pointed	0.69	7.16	3.39	m	**10.55**
Extra over brickwork for two courses of machine made tile creasing; bedded and pointed; projecting 25 mm each side; horizontal					
215 mm wide copings	0.58	6.02	5.12	m	**11.14**
260 mm wide copings	0.81	8.40	6.57	m	**14.97**
Galvanised steel coping cramp; built in	0.11	1.14	0.47	nr	**1.61**
Flue linings etc					
Flue linings; True Flue 200 mm refractory concrete square flue linings; rebated joints in refractory mortar (1:2.5)					
linings	0.21	2.18	11.96	m	**14.14**
bottom svivel unit; ref 2u	0.11	1.14	9.23	nr	**10.37**
45 deg. bend; ref 4u	0.11	1.14	9.23	nr	**10.37**
offset unit; ref 5u	0.11	1.14	9.41	nr	**10.55**
offset unit; ref 6u	0.11	1.14	8.45	nr	**9.59**
off set unit; ref 7u	0.11	1.14	7.90	nr	**9.04**
pot; ref 8u	0.11	1.14	5.24	nr	**6.38**
single cap unit; ref 10u	0.29	3.01	21.12	nr	**24.13**
double cap unit; ref 11u	0.35	3.63	18.82	nr	**22.45**
U-type lintol with U1 attachment	0.35	3.63	22.32	nr	**25.95**
Gas flue system; True Flue 'Typex HP'; concrete blocks built in; in refractory mortar (1:2.5); cutting brickwork or blockwork around					
recess; ref HP.1	0.11	1.14	3.66	nr	**4.80**
cover ref HP.2	0.11	1.14	5.19	nr	**6.33**
222 mm standard block; ref HP.3	0.11	1.14	3.41	nr	**4.55**
112 mm standard block; ref HP.3	0.11	1.14	3.39	nr	**4.53**
72 mm standard block; ref HP.3	0.11	1.14	3.41	nr	**4.55**
vent block; ref HP.3/H	0.11	1.14	6.94	nr	**8.08**
222 mm standard block; ref HP.4	0.11	1.14	3.28	nr	**4.42**
112 mm standard block; ref HP.4	0.11	1.14	3.28	nr	**4.42**
72 mm standard block; ref HP.4	0.11	1.14	3.28	nr	**4.42**
120 mm side offset; ref HP.5	0.11	1.14	4.07	nr	**5.21**
70 mm back offset; ref HP.6	0.11	1.14	10.08	nr	**11.22**
vertical exit; ref HP.7	0.11	1.14	6.77	nr	**7.91**
angled entry/exit; ref HP.8	0.11	1.14	6.77	nr	**7.91**
reverse rebate; ref HP.9	0.11	1.14	5.33	nr	**6.47**
corbel block; ref HP.10	0.11	1.14	6.41	nr	**7.55**
lintol block; ref HP.11	0.11	1.14	6.08	nr	**7.22**

F MASONRY Including overheads and profit at 7.50%		Labour hours	Labour £	Material £	Unit	Total rate £
Parging and coring flues with refractory mortar (1:2.5); sectional area not exceeding 0.25 m2						
900 x 900 mm		0.69	7.16	5.70	m	**12.86**
Ancillaries						
Forming openings; one ring arch over 225 x 225 mm; one brick facing brickwork making good facings both sides		1.70	17.64	-	nr	**17.64**
Forming opening through hollow wall; slate lintel over; sealing 50 cavity with slates in cement mortar (1:3); making good fair face or facings one side						
225 x 75 mm		0.23	2.39	1.84	nr	**4.23**
225 x 150 mm		0.29	3.01	2.71	nr	**5.72**
225 x 225 mm		0.35	3.63	3.57	nr	**7.20**
Air bricks; red terracotta; building into prepared openings						
215 x 65 mm		0.09	0.93	1.05	nr	**1.98**
215 x 140 mm		0.09	0.93	1.72	nr	**2.65**
215 x 215 mm		0.09	0.93	3.64	nr	**4.57**
Air bricks; cast iron; building into prepared openings						
215 x 65 mm		0.09	0.93	3.08	nr	**4.01**
215 x 140 mm		0.09	0.93	5.62	nr	**6.55**
215 x 215 mm		0.09	0.93	7.40	nr	**8.33**
Bearers; mild steel						
51 x 10 mm flat bar		0.29	2.66	3.90	m	**6.56**
64 x 13 mm flat bar		0.29	2.66	6.24	m	**8.90**
50 x 50 x 6 mm angle section		0.40	3.67	3.18	m	**6.85**
Ends fanged for building in		-	-	-	nr	**2.37**
Proprietary items						
Ties in walls; 150 mm long butterfly type; building into joints of brickwork or blockwork						
galvanized steel or polypropylene		0.03	0.31	0.08	nr	**0.39**
stainless steel		0.03	0.31	0.10	nr	**0.41**
copper		0.03	0.31	0.15	nr	**0.46**
Ties in walls; 20 x 3 x 150 mm long twisted wall type; building into joints of brickwork or blockwork						
galvanized steel		0.05	0.52	0.11	nr	**0.63**
stainless steel		0.05	0.52	0.21	nr	**0.73**
copper		0.05	0.52	0.38	nr	**0.90**
Anchors in walls; 25 x 3 x 100 mm long; one end dovetailed; other end building into joints of brickwork or blockwork						
galvanized steel		0.07	0.73	0.24	nr	**0.97**
stainless steel		0.07	0.73	0.38	nr	**1.11**
copper		0.07	0.73	0.59	nr	**1.32**
Fixing cramp 25 x 3 x 250 mm long; once bent; fixed to back of frame; other end building into joints of brickwork or blockwork						
galvanized steel		0.07	0.73	0.12	nr	**0.85**
Chimney pots; red terracotta; plain or cannon-head; setting and flaunching in cement mortar (1:3)						
185 mm dia x 300 mm long	PC £12.75	2.05	21.27	14.57	nr	**35.84**
185 mm dia x 600 mm long	PC £22.00	2.30	23.86	24.76	nr	**48.62**
185 mm dia x 900 mm long	PC £29.25	2.55	26.45	32.75	nr	**59.20**

F MASONRY Including overheads and profit at 7.50%		Labour hours	Labour £	Material £	Unit	Total rate £
Galvanised steel lintels; Catnic or similar; built into brickwork or blockwork						
'CN7' combined lintel; 143 mm high; for standard cavity walls						
750 mm long	PC £12.09	0.28	2.57	13.65	nr	**16.22**
900 mm long	PC £14.55	0.33	3.03	16.43	nr	**19.46**
1200 mm long	PC £19.34	0.39	3.58	21.84	nr	**25.42**
1500 mm long	PC £25.13	0.44	4.03	28.37	nr	**32.40**
1800 mm long	PC £30.77	0.50	4.58	34.73	nr	**39.31**
2100 mm long	PC £35.88	0.55	5.04	40.50	nr	**45.54**
'CN8' combined lintel; 219 mm high; for standard cavity walls						
2400 mm long	PC £48.72	0.66	6.05	54.99	nr	**61.04**
2700 mm long	PC £54.97	0.77	7.06	62.05	nr	**69.11**
3000 mm long	PC £67.03	0.88	8.07	75.66	nr	**83.73**
3300 mm long	PC £74.61	0.99	9.08	84.22	nr	**93.30**
3600 mm long	PC £81.71	1.10	10.09	92.23	nr	**102.32**
3900 mm long	PC £108.81	1.20	11.00	122.82	nr	**133.82**
4200 mm long	PC £117.14	1.30	11.92	132.23	nr	**144.15**
'CN92' single lintel; for 75 mm internal walls						
900 mm long	PC £2.17	0.33	3.03	2.45	nr	**5.48**
1050 mm long	PC £2.55	0.33	3.03	2.88	nr	**5.91**
1200 mm long	PC £2.86	0.39	3.58	3.22	nr	**6.80**
'CN102' single lintel; for 100 mm internal walls						
900 mm long	PC £2.67	0.33	3.03	3.01	nr	**6.04**
1050 mm long	PC £3.11	0.33	3.03	3.51	nr	**6.54**
1200 mm long	PC £3.39	0.39	3.58	3.83	nr	**7.41**
F31 PRECAST CONCRETE SILLS/LINTELS/COPINGS/ FEATURES						
Mix 21.00 N/mm2 - 20 mm aggregate (1:2:4)						
Lintels; plate; prestressed; bedded						
100 x 65 x 750 mm	PC £2.50	0.45	4.67	3.21	nr	**7.88**
100 x 65 x 900 mm	PC £2.75	0.45	4.67	3.48	nr	**8.15**
100 x 65 x 1050 mm	PC £3.25	0.45	4.67	4.01	nr	**8.68**
100 x 65 x 1200 mm	PC £3.70	0.45	4.67	4.50	nr	**9.17**
150 x 65 x 900 mm	PC £3.80	0.60	6.22	4.60	nr	**10.82**
150 x 65 x 1050 mm	PC £4.70	0.60	6.22	5.57	nr	**11.79**
150 x 65 x 1200 mm	PC £5.10	0.60	6.22	6.00	nr	**12.22**
220 x 65 x 900 mm	PC £5.90	0.70	7.26	6.86	nr	**14.12**
220 x 65 x 1200 mm	PC £7.80	0.70	7.26	8.90	nr	**16.16**
220 x 65 x 1500 mm	PC £9.70	0.80	8.30	10.95	nr	**19.25**
265 x 65 x 900 mm	PC £7.25	0.70	7.26	8.31	nr	**15.57**
265 x 65 x 1200 mm	PC £9.90	0.70	7.26	11.16	nr	**18.42**
265 x 65 x 1500 mm	PC £12.35	0.80	8.30	13.80	nr	**22.10**
265 x 65 x 1800 mm	PC £13.75	0.90	9.34	15.30	nr	**24.64**
Lintels; rectangular; reinforced with mild steel bars; bedded						
100 x 150 x 900 mm	PC £8.00	0.70	7.26	9.12	nr	**16.38**
100 x 150 x 1050 mm	PC £9.00	0.70	7.26	10.19	nr	**17.45**
100 x 150 x 1200 mm	PC £14.98	0.70	7.26	16.62	nr	**23.88**
225 x 150 x 1200 mm	PC £18.90	0.90	9.34	20.84	nr	**30.18**
225 x 225 x 1800 mm	PC £35.00	1.70	17.64	38.14	nr	**55.78**
Lintels; boot; reinforced with mild steel bars; bedded						
250 x 225 x 1200 mm	PC £28.00	1.40	14.52	30.62	nr	**45.14**
275 x 225 x 1800 mm	PC £37.80	2.00	20.75	41.67	nr	**62.42**
Padstones						
300 x 100 x 75 mm	PC £2.80	0.35	3.63	3.53	nr	**7.16**
225 x 225 x 150 mm	PC £4.20	0.45	4.67	5.03	nr	**9.70**
450 x 450 x 150 mm	PC £11.20	0.70	7.26	12.56	nr	**19.82**

F MASONRY Including overheads and profit at 7.50%		Labour hours	Labour £	Material £	Unit	Total rate £
Mix 31.00 N/mm2 - 20 mm aggregate (1:1:2)						
Copings; once weathered; once throated bedded and pointed						
152 x 76 mm	PC £9.79	0.80	8.30	11.04	m	**19.34**
178 x 64 mm	PC £9.79	0.80	8.30	11.04	m	**19.34**
305 x 76 mm	PC £12.69	0.90	9.34	14.16	m	**23.50**
Extra for						
fair end	PC £3.57	-	-	3.84	nr	**3.84**
angles	PC £7.80	-	-	8.39	nr	**8.39**
Copings; twice weathered; twice throated; bedded and pointed						
152 x 76 mm	PC £9.05	0.80	8.30	10.25	m	**18.55**
178 x 64 mm	PC £9.05	0.80	8.30	10.25	m	**18.55**
305 x 76 mm	PC £11.95	0.90	9.34	13.36	m	**22.70**
Extra for						
fair end	PC £3.57	-	-	3.84	nr	**3.84**
angles	PC £7.80	-	-	8.39	nr	**8.39**

G STRUCTURAL/CARCASSING METAL/TIMBER Including overheads and profit at 7.50%		Labour hours	Labour £	Material £	Unit	Total rate £
G10 STRUCTURAL STEEL FRAMING						
Fabricated steelwork; BS 4360; grade 50						
Single beams; universal beams, joists or channels	PC £440.40	-	-	-	t	1135.00
Single beams; castellated universal beams		-	-	-	t	2760.00
Single beams; rectangular hollow sections		-	-	-	t	1380.00
Single cranked beams; rectangular hollow sections		-	-	-	t	1575.00
Built-up beams; universal beams, joists or channels and plates	PC £440.40	-	-	-	t	1645.00
Latticed beams; angle sections	PC £331.20	-	-	-	t	1310.00
Latticed beams; circular hollow sections		-	-	-	t	1475.00
Single columns; universal beams, joists or channels	PC £440.40	-	-	-	t	1025.00
Single columns; circular hollow sections	PC £1163.80	-	-	-	t	1990.00
Single columns; rectangular and square hollow sections	PC £1035.00	-	-	-	t	1845.00
Built-up columns; universal beams, joists or channels and plates	PC £440.40	-	-	-	t	1700.00
Roof trusses; angle sections	PC £331.20	-	-	-	t	1600.00
Roof trusses; circular hollow sections	PC £568.10	-	-	-	t	1775.00
Bolted fittings; other than in connections; consisting of cleats, brackets etc.		-	-	-	t	1810.00
Welded fittings; other than in connections; consisting of cleats, brackets, etc.		-	-	-	t	1635.00
Erection of fabricated steelwork						
Erection of steelwork on site		-	-	-	t	175.00
Wedging						
stanchion bases		-	-	-	nr	8.17
Holes for other trades; made on site						
16 mm dia; 6 mm thick metal		-	-	-	nr	3.57
16 mm dia; 10 mm thick metal		-	-	-	nr	5.00
16 mm dia; 20 mm thick metal		-	-	-	nr	7.01
Anchorage for stanchion base; including plate; holding down bolts; nuts and washers; for stanchion size						
152 x 152 mm		-	-	-	nr	10.17
254 x 254 mm		-	-	-	nr	15.19
356 x 406 mm		-	-	-	nr	22.73
Surface treatments off site						
On steelwork; general surfaces						
galvanizing		-	-	-	t	516.00
shot blasting		-	-	-	m2	0.90
grit blast and one coat zinc chromate primer		-	-	-	m2	2.43
touch up primer and one coat of two pack epoxy zinc phosphate primer		-	-	-	m2	2.30
G12 ISOLATED STRUCTURAL METAL MEMBERS						
Unfabricated steelwork; BS 4360; grade 43A						
Single beams; joists or channels	PC £440.40	-	-	-	t	980.00
Erection of steelwork on site		-	-	-	t	205.00

G STRUCTURAL/CARCASSING METAL/TIMBER Including overheads and profit at 7.50%	Labour hours	Labour £	Material £	Unit	Total rate £
G12 ISOLATED STRUCTURAL METAL MEMBERS - cont'd					
Unfabricated steelwork; BS 4360; grade 50 inside existing buildings					
Single beams; joists or channels	-	-	-	t	1200.00
Erection of steelwork within existing buildings	-	-	-	t	900.00
Metsec open web steel lattice beams; in single members; raised 3.50 m above ground; ends built in					
Beams; one coat zinc phosphate primer at works					
20 cm dp; to span 5m (7.69kg/m); ref NA20	0.22	2.02	22.19	m	24.21
25 cm dp; to span 6m (7.64kg/m); ref NA25	0.22	2.02	22.73	m	24.75
30 cm dp; to span 7m (10.26kg/m); ref NBs30	0.28	2.57	24.71	m	27.28
35 cm dp; to span 8.5m (10.6kg/m); ref NBs35	0.28	2.57	25.24	m	27.81
35 cm dp; to span 10m (12.76kg/m); ref NCs35	0.33	3.03	26.77	m	29.80
40 cm dp; to span 11.5m (17.08kg/m); ref E40	0.39	3.58	38.09	m	41.67
45 cm dp; to span 13m (25.44kg/m); ref G45	0.55	5.04	46.90	m	51.94
Beams; galvanised					
20 cm dp; to span 5m (7.69kg/m); ref NA20	0.22	2.02	23.88	m	25.90
25 cm dp; to span 6m (7.64kg/m); ref NA25	0.22	2.02	24.43	m	26.45
30 cm dp; to span 7m (10.26kg/m); ref NBs30	0.28	2.57	27.04	m	29.61
35 cm dp; to span 8.5m (10.6kg/m); ref NBs35	0.28	2.57	27.53	m	30.10
35 cm dp; to span 10m (12.76kg/m); ref NCs35	0.33	3.03	29.59	m	32.62
40 cm dp; to span 11.5m (17.08kg/m); ref E40	0.39	3.58	42.08	m	45.66
45 cm dp; to span 13m (25.44kg/m); ref G45	0.55	5.04	52.51	m	57.55

BASIC TIMBER PRICES

	£		£		£
Hardwood; fair average (£/m3 ex-wharf)					
African Walnut	618.13	European Oak	1290.00	Sapele	559.00
Afrormosia	763.25	Iroko	537.50	Teak	1612.50
Agba	618.13	Maple	752.50	Utile	736.38
Beech	537.50	Obeche	365.50	W.A.Mahogany	505.25
Brazil. Mahogany	682.63				
Softwood					
Carcassing quality (£/m3)					
2-4.8 m lengths	220.00	G.S.grade	13.98		
4.8-6 m lengths	209.00	S.S.grade	26.88		
6-9 m lenths	231.00				
Joinery quality - £333.25/m3					
'Treatment'(£/m3)					
Pre-treatment of timber by vacuum/pressure impregnation, excluding transport costs and any subsequent seasoning:-					
interior work; min. salt ret. 4 kg/m3			32.25		
exterior work; min. salt ret. 5.30 kg/m3			37.63		
Pre-treatment of timber including flame proofing					
all purposes; min. salt ret. 36 kg/m3			114.00		

G STRUCTURAL/CARCASSING METAL/TIMBER Including overheads and profit at 7.50%	Labour hours	Labour £	Material £	Unit	Total rate £
G20 CARPENTRY/TIMBER FRAMING/FIRST FIXING					
Sawn softwood; untreated					
Floor members					
38 x 75 mm	0.13	1.05	0.78	m	1.83
38 x 100 mm	0.13	1.05	1.00	m	2.05
38 x 125 mm	0.14	1.13	1.23	m	2.36
38 x 150 mm	0.15	1.21	1.46	m	2.67
50 x 75 mm	0.13	1.05	0.88	m	1.93
50 x 100 mm	0.15	1.21	1.16	m	2.37
50 x 125 mm	0.15	1.21	1.43	m	2.64
50 x 150 mm	0.17	1.38	1.71	m	3.09
50 x 175 mm	0.17	1.38	2.00	m	3.38
50 x 200 mm	0.18	1.46	2.36	m	3.82
50 x 225 mm	0.18	1.46	2.74	m	4.20
50 x 250 mm	0.19	1.54	3.22	m	4.76
75 x 125 mm	0.18	1.46	2.27	m	3.73
75 x 150 mm	0.18	1.46	2.67	m	4.13
75 x 175 mm	0.18	1.46	3.20	m	4.66
75 x 200 mm	0.19	1.54	3.73	m	5.27
75 x 225 mm	0.19	1.54	4.26	m	5.80
75 x 250 mm	0.20	1.62	4.82	m	6.44
100 x 150 mm	0.24	1.94	4.09	m	6.03
100 x 200 mm	0.25	2.02	5.45	m	7.47
100 x 250 mm	0.28	2.27	6.80	m	9.07
100 x 300 mm	0.30	2.43	8.17	m	10.60
Wall or partition members					
25 x 25 mm	0.08	0.65	0.21	m	0.86
25 x 38 mm	0.08	0.65	0.30	m	0.95
25 x 75 mm	0.10	0.81	0.54	m	1.35
38 x 38 mm	0.10	0.81	0.45	m	1.26
38 x 50 mm	0.10	0.81	0.53	m	1.34
38 x 75 mm	0.13	1.05	0.78	m	1.83
38 x 100 mm	0.17	1.38	1.00	m	2.38
50 x 50 mm	0.13	1.05	0.66	m	1.71
50 x 75 mm	0.17	1.38	0.89	m	2.27
50 x 100 mm	0.20	1.62	1.17	m	2.79
50 x 125 mm	0.21	1.70	1.45	m	3.15
75 x 75 mm	0.20	1.62	1.44	m	3.06
75 x 100 mm	0.23	1.86	1.97	m	3.83
100 x 100 mm	0.23	1.86	2.72	m	4.58
Flat roof members					
38 x 75 mm	0.15	1.21	0.80	m	2.01
38 x 100 mm	0.15	1.21	1.00	m	2.21
38 x 125 mm	0.15	1.21	1.23	m	2.44
38 x 150 mm	0.15	1.21	1.46	m	2.67
50 x 100 mm	0.15	1.21	1.16	m	2.37
50 x 125 mm	0.15	1.21	1.43	m	2.64
50 x 150 mm	0.17	1.38	1.71	m	3.09
50 x 175 mm	0.17	1.38	2.00	m	3.38
50 x 200 mm	0.18	1.46	2.36	m	3.82
50 x 225 mm	0.18	1.46	2.74	m	4.20
50 x 250 mm	0.19	1.54	3.22	m	4.76
75 x 150 mm	0.18	1.46	2.65	m	4.11
75 x 175 mm	0.18	1.46	3.18	m	4.64
75 x 200 mm	0.19	1.54	3.73	m	5.27
75 x 225 mm	0.19	1.54	4.26	m	5.80
75 x 250 mm	0.20	1.62	4.82	m	6.44
Pitched roof members					
25 x 100 mm	0.13	1.05	0.66	m	1.71
25 x 125 mm	0.13	1.05	0.81	m	1.86
25 x 150 mm	0.17	1.38	0.96	m	2.34
25 x 150 mm; notching over trussed rafters	0.33	2.67	0.96	m	3.63

G STRUCTURAL/CARCASSING METAL/TIMBER Including overheads and profit at 7.50%	Labour hours	Labour £	Material £	Unit	Total rate £
25 x 175 mm	0.17	1.38	1.16	m	2.54
25 x 175 mm; notching over trussed rafters	0.33	2.67	1.16	m	3.83
25 x 200 mm	0.20	1.62	1.35	m	2.97
32 x 150 mm; notching over trussed rafters	0.36	2.91	1.36	m	4.27
32 x 175 mm; notching over trussed rafters	0.36	2.91	1.63	m	4.54
32 x 200 mm; notching over trussed rafters	0.36	2.91	1.89	m	4.80
38 x 100 mm	0.17	1.38	1.00	m	2.38
38 x 125 mm	0.17	1.38	1.23	m	2.61
38 x 150 mm	0.17	1.38	1.46	m	2.84
50 x 50 mm	0.13	1.05	0.65	m	1.70
50 x 75 mm	0.17	1.38	0.88	m	2.26
50 x 100 mm	0.20	1.62	1.16	m	2.78
50 x 125 mm	0.20	1.62	1.43	m	3.05
50 x 150 mm	0.23	1.86	1.71	m	3.57
50 x 175 mm	0.23	1.86	2.00	m	3.86
50 x 200 mm	0.23	1.86	2.36	m	4.22
50 x 225 mm	0.23	1.86	2.74	m	4.60
75 x 100 mm	0.28	2.27	1.96	m	4.23
75 x 125 mm	0.28	2.27	2.27	m	4.54
75 x 150 mm	0.28	2.27	2.67	m	4.94
100 x 150 mm	0.33	2.67	4.11	m	6.78
100 x 175 mm	0.33	2.67	4.79	m	7.46
100 x 200 mm	0.33	2.67	5.46	m	8.13
100 x 225 mm	0.36	2.91	6.13	m	9.04
100 x 250 mm	0.36	2.91	6.80	m	9.71
Kerbs, bearers and the like					
19 x 100 mm	0.04	0.32	0.53	m	0.85
19 x 125 mm	0.04	0.32	0.66	m	0.98
19 x 150 mm	0.04	0.32	0.78	m	1.10
25 x 75 mm	0.06	0.49	0.54	m	1.03
25 x 100 mm	0.06	0.49	0.66	m	1.15
38 x 75 mm	0.07	0.57	0.78	m	1.35
38 x 100 mm	0.07	0.57	1.00	m	1.57
50 x 75 mm	0.07	0.57	0.88	m	1.45
50 x 100 mm	0.08	0.65	1.16	m	1.81
75 x 100 mm	0.08	0.65	1.94	m	2.59
75 x 125 mm	0.09	0.73	2.25	m	2.98
75 x 150 mm	0.09	0.73	2.65	m	3.38
75 x 150 mm; splayed and rounded	0.11	0.89	4.64	m	5.53
Kerbs, bearers and the like; fixing by bolting					
19 x 100 mm	0.09	0.73	0.53	m	1.26
19 x 125 mm	0.09	0.73	0.66	m	1.39
19 x 150 mm	0.09	0.73	0.78	m	1.51
25 x 75 mm	0.11	0.89	0.54	m	1.43
25 x 100 mm	0.11	0.89	0.66	m	1.55
38 x 75 mm	0.13	1.05	0.78	m	1.83
38 x 100 mm	0.13	1.05	1.00	m	2.05
50 x 75 mm	0.13	1.05	0.88	m	1.93
50 x 100 mm	0.15	1.21	1.16	m	2.37
75 x 100 mm	0.15	1.21	1.96	m	3.17
75 x 125 mm	0.18	1.46	2.27	m	3.73
75 x 150 mm	0.18	1.46	2.65	m	4.11
Herringbone strutting 50 x 50 mm					
to 150 mm deep joists	0.55	4.45	1.54	m	5.99
to 175 mm deep joists	0.55	4.45	1.56	m	6.01
to 200 mm deep joists	0.55	4.45	1.59	m	6.04
to 225 mm deep joists	0.55	4.45	1.63	m	6.08
to 250 mm deep joists	0.55	4.45	1.66	m	6.11

G STRUCTURAL/CARCASSING METAL/TIMBER Including overheads and profit at 7.50%	Labour hours	Labour £	Material £	Unit	Total rate £
G20 CARPENTRY/TIMBER FRAMING/FIRST FIXING - cont'd					
Sawn softwood; untreated - cont'd					
Solid strutting to joists					
50 x 150 mm	0.33	2.67	1.98	m	4.65
50 x 175 mm	0.33	2.67	2.26	m	4.93
50 x 200 mm	0.33	2.67	2.65	m	5.32
50 x 225 mm	0.33	2.67	3.02	m	5.69
50 x 250 mm	0.33	2.67	3.53	m	6.20
Cleats					
225 x 100 x 75 mm	0.22	1.78	0.69	nr	2.47
Sprockets					
50 x 50 x 200 mm	0.17	1.38	0.69	nr	2.07
Extra for stress grading to above timbers					
general structural (GS) grade	-	-	-	m3	15.02
special structural (SS) grade	-	-	-	m3	28.89
Extra for protecting and flameproofing timber with 'Celcure F' protection					
small sections	-	-	-	m3	129.00
large sections	-	-	-	m3	122.55
Wrought faces					
generally	0.33	2.67	-	m2	2.67
50 mm wide	0.04	0.32	-	m	0.32
75 mm wide	0.06	0.49	-	m	0.49
100 mm wide	0.07	0.57	-	m	0.57
Raking cutting					
50 mm thick	0.22	1.78	0.53	m	2.31
75 mm thick	0.28	2.27	0.81	m	3.08
100 mm thick	0.33	2.67	1.06	m	3.73
Curved cutting					
50 mm thick	0.28	2.27	0.74	m	3.01
Scribing					
50 mm thick	0.33	2.67	-	m	2.67
Notching and fitting ends to metal	0.13	1.05	-	nr	1.05
Trimming to openings					
760 x 760 mm; joists 38 x 150 mm	2.40	19.43	0.85	nr	20.28
760 x 760 mm; joists 50 x 175 mm	2.55	20.64	1.49	nr	22.13
1500 x 500 mm; joists 38 x 200 mm	2.75	22.26	1.70	nr	23.96
1500 x 500 mm; joists 50 x 200 mm	2.75	22.26	2.13	nr	24.39
3000 x 1000 mm; joists 50 x 225 mm	4.95	40.07	4.68	nr	44.75
3000 x 1000 mm; joists 75 x 225 mm	5.50	44.52	7.02	nr	51.54
Sawn softwood; 'Tanalised'					
Floor members					
38 x 75 mm	0.13	1.05	0.97	m	2.02
38 x 100 mm	0.13	1.05	1.24	m	2.29
38 x 125 mm	0.14	1.13	1.46	m	2.59
38 x 150 mm	0.15	1.21	1.74	m	2.95
50 x 75 mm	0.13	1.05	1.03	m	2.08
50 x 100 mm	0.15	1.21	1.43	m	2.64
50 x 125 mm	0.15	1.21	1.79	m	3.00
50 x 150 mm	0.17	1.38	2.13	m	3.51
50 x 175 mm	0.17	1.38	2.50	m	3.88
50 x 200 mm	0.18	1.46	2.94	m	4.40
50 x 225 mm	0.18	1.46	3.41	m	4.87
50 x 250 mm	0.19	1.54	4.03	m	5.57
75 x 125 mm	0.18	1.46	2.83	m	4.29
75 x 150 mm	0.18	1.46	3.31	m	4.77
75 x 175 mm	0.18	1.46	3.97	m	5.43
75 x 200 mm	0.19	1.54	4.65	m	6.19
75 x 225 mm	0.19	1.54	5.30	m	6.84
75 x 250 mm	0.20	1.62	6.01	m	7.63

G STRUCTURAL/CARCASSING METAL/TIMBER Including overheads and profit at 7.50%	Labour hours	Labour £	Material £	Unit	Total rate £
100 x 150 mm	0.24	1.94	5.10	m	7.04
100 x 200 mm	0.25	2.02	6.79	m	8.81
100 x 250 mm	0.28	2.27	8.46	m	10.73
100 x 300 mm	0.30	2.43	10.18	m	12.61
Wall or partition members					
25 x 25 mm	0.08	0.65	0.26	m	0.91
25 x 38 mm	0.08	0.65	0.36	m	1.01
25 x 75 mm	0.10	0.81	0.67	m	1.48
38 x 38 mm	0.10	0.81	0.55	m	1.36
38 x 50 mm	0.10	0.81	0.66	m	1.47
38 x 75 mm	0.13	1.05	0.97	m	2.02
38 x 100 mm	0.17	1.38	1.24	m	2.62
50 x 50 mm	0.13	1.05	0.77	m	1.82
50 x 75 mm	0.17	1.38	1.05	m	2.43
50 x 100 mm	0.20	1.62	1.45	m	3.07
50 x 125 mm	0.21	1.70	1.81	m	3.51
75 x 75 mm	0.20	1.62	1.84	m	3.46
75 x 100 mm	0.23	1.86	2.44	m	4.30
100 x 100 mm	0.23	1.86	3.45	m	5.31
Flat roof members					
38 x 75 mm	0.15	1.21	0.97	m	2.18
38 x 100 mm	0.15	1.21	1.24	m	2.45
38 x 125 mm	0.15	1.21	1.46	m	2.67
38 x 150 mm	0.15	1.21	1.74	m	2.95
50 x 100 mm	0.15	1.21	1.43	m	2.64
50 x 125 mm	0.15	1.21	1.79	m	3.00
50 x 150 mm	0.17	1.38	2.13	m	3.51
50 x 175 mm	0.17	1.38	2.50	m	3.88
50 x 200 mm	0.18	1.46	2.94	m	4.40
50 x 225 mm	0.18	1.46	3.41	m	4.87
50 x 250 mm	0.19	1.54	4.03	m	5.57
75 x 150 mm	0.18	1.46	3.31	m	4.77
75 x 175 mm	0.18	1.46	3.96	m	5.42
75 x 200 mm	0.19	1.54	4.65	m	6.19
75 x 225 mm	0.19	1.54	5.30	m	6.84
75 x 250 mm	0.20	1.62	6.01	m	7.63
Pitched roof members					
25 x 100 mm	0.13	1.05	0.81	m	1.86
25 x 125 mm	0.13	1.05	1.00	m	2.05
25 x 150 mm	0.17	1.38	1.19	m	2.57
25 x 150 mm; notching over trussed rafters	0.33	2.67	1.19	m	3.86
25 x 175 mm	0.17	1.38	1.43	m	2.81
25 x 175 mm; notching over trussed rafters	0.33	2.67	1.43	m	4.10
25 x 200 mm	0.20	1.62	1.69	m	3.31
32 x 150 mm; notching over trussed rafters	0.36	2.91	1.69	m	4.60
32 x 175 mm; notching over trussed rafters	0.36	2.91	2.02	m	4.93
32 x 200 mm; notching over trussed rafters	0.36	2.91	2.36	m	5.27
38 x 100 mm	0.17	1.38	1.24	m	2.62
38 x 125 mm	0.17	1.38	1.46	m	2.84
38 x 150 mm	0.17	1.38	1.74	m	3.12
50 x 50 mm	0.13	1.05	0.76	m	1.81
50 x 75 mm	0.17	1.38	1.03	m	2.41
50 x 100 mm	0.20	1.62	1.43	m	3.05
50 x 125 mm	0.20	1.62	1.79	m	3.41
50 x 150 mm	0.23	1.86	2.13	m	3.99
50 x 175 mm	0.23	1.86	2.50	m	4.36
50 x 200 mm	0.23	1.86	2.94	m	4.80
50 x 225 mm	0.23	1.86	3.41	m	5.27
75 x 100 mm	0.28	2.27	2.43	m	4.70
75 x 125 mm	0.28	2.27	2.83	m	5.10
75 x 150 mm	0.28	2.27	3.31	m	5.58

G STRUCTURAL/CARCASSING METAL/TIMBER Including overheads and profit at 7.50%	Labour hours	Labour £	Material £	Unit	Total rate £
G20 CARPENTRY/TIMBER FRAMING/FIRST FIXING - cont'd					
Sawn softwood; 'Tanalised' - cont'd					
Pitched roof members - cont'd					
100 x 150 mm	0.33	2.67	5.11	m	7.78
100 x 175 mm	0.33	2.67	5.96	m	8.63
100 x 200 mm	0.33	2.67	6.80	m	9.47
100 x 225 mm	0.36	2.91	7.65	m	10.56
100 x 250 mm	0.36	2.91	8.48	m	11.39
Kerbs, bearers and the like					
19 x 100 mm	0.04	0.32	0.66	m	0.98
19 x 125 mm	0.04	0.32	0.81	m	1.13
19 x 150 mm	0.04	0.32	0.97	m	1.29
25 x 75 mm	0.06	0.49	0.67	m	1.16
25 x 100 mm	0.06	0.49	0.81	m	1.30
38 x 75 mm	0.07	0.57	0.97	m	1.54
38 x 100 mm	0.07	0.57	1.24	m	1.81
50 x 75 mm	0.07	0.57	1.03	m	1.60
50 x 100 mm	0.08	0.65	1.43	m	2.08
75 x 100 mm	0.08	0.65	2.41	m	3.06
75 x 125 mm	0.09	0.73	2.81	m	3.54
75 x 150 mm	0.09	0.73	3.29	m	4.02
75 x 150 mm; splayed and rounded	0.11	0.89	5.78	m	6.67
Kerbs, bearers and the like; fixing by bolting					
19 x 100 mm	0.09	0.73	0.66	m	1.39
19 x 125 mm	0.09	0.73	0.81	m	1.54
19 x 150 mm	0.09	0.73	0.97	m	1.70
25 x 75 mm	0.11	0.89	0.68	m	1.57
25 x 100 mm	0.11	0.89	0.81	m	1.70
38 x 75 mm	0.13	1.05	0.97	m	2.02
38 x 100 mm	0.13	1.05	1.24	m	2.29
50 x 75 mm	0.13	1.05	1.03	m	2.08
50 x 100 mm	0.15	1.21	1.43	m	2.64
75 x 100 mm	0.15	1.21	2.43	m	3.64
75 x 125 mm	0.18	1.46	2.83	m	4.29
75 x 150 mm	0.18	1.46	3.29	m	4.75
Herringbone strutting 50 x 50 mm					
to 150 mm deep joists	0.55	4.45	1.77	m	6.22
to 175 mm deep joists	0.55	4.45	1.81	m	6.26
to 200 mm deep joists	0.55	4.45	1.84	m	6.29
to 225 mm deep joists	0.55	4.45	1.89	m	6.34
to 250 mm deep joists	0.55	4.45	1.92	m	6.37
Solid strutting to joists					
50 x 150 mm	0.33	2.67	2.40	m	5.07
50 x 175 mm	0.33	2.67	2.76	m	5.43
50 x 200 mm	0.33	2.67	3.22	m	5.89
50 x 225 mm	0.33	2.67	3.70	m	6.37
50 x 250 mm	0.33	2.67	4.34	m	7.01
Cleats					
225 x 100 x 75 mm	0.22	1.78	0.82	nr	2.60
Sprockets					
50 x 50 x 200 mm	0.17	1.38	0.82	nr	2.20
Extra for stress grading to above timbers					
general structural (GS) grade	-	-	-	m3	15.02
special structural (SS) grade	-	-	-	m3	28.89
Extra for protecting and flameproofing timber with 'Celcure F' protection					
small sections	-	-	-	m3	129.00
large sections	-	-	-	m3	122.55

G STRUCTURAL/CARCASSING METAL/TIMBER Including overheads and profit at 7.50%	Labour hours	Labour £	Material £	Unit	Total rate £
Wrought faces					
generally	0.33	2.67	-	m2	2.67
50 mm wide	0.04	0.32	-	m	0.32
75 mm wide	0.06	0.49	-	m	0.49
100 mm wide	0.07	0.57	-	m	0.57
Raking cutting					
50 mm thick	0.22	1.78	0.64	m	2.42
75 mm thick	0.28	2.27	0.97	m	3.24
100 mm thick	0.33	2.67	1.27	m	3.94
Curved cutting					
50 mm thick	0.28	2.27	0.89	m	3.16
Scribing					
50 mm thick	0.33	2.67	-	m	2.67
Notching and fitting ends to metal	0.13	1.05	-	nr	1.05
Trimming to openings					
760 x 760 mm; joists 38 x 150 mm	2.40	19.43	1.02	nr	20.45
760 x 760 mm; joists 50 x 175 mm	2.55	20.64	1.78	nr	22.42
1500 x 500 mm; joists 38 x 200 mm	2.75	22.26	2.03	nr	24.29
1500 x 500 mm; joists 50 x 200 mm	2.75	22.26	2.54	nr	24.80
3000 x 1000 mm; joists 50 x 225 mm	4.95	40.07	5.59	nr	45.66
3000 x 1000 mm; joists 75 x 225 mm	5.50	44.52	8.39	nr	52.91
Trussed rafters, stress graded sawn softwood pressure impregnated; raised through two storeys and fixed in position					
'W' type truss (Fink); 22.5 degree pitch; 450 mm eaves overhang					
5.00 m span	1.75	14.17	20.08	nr	34.25
7.60 m span	1.95	15.78	29.40	nr	45.18
10.00 m span	2.20	17.81	46.45	nr	64.26
'W' type truss (Fink); 30 degree pitch; 450 mm eaves overhang					
5.00 m span	1.75	14.17	21.05	nr	35.22
7.60 m span	1.95	15.78	30.65	nr	46.43
10.00 m span	2.20	17.81	48.84	nr	66.65
'W' type truss (Fink); 45 degree pitch; 450 mm eaves overhang					
4.60 m span	1.75	14.17	23.61	nr	37.78
7.00 m span	1.95	15.78	35.94	nr	51.72
'Mono' type truss; 17.5 degree pitch; 450 mm eaves overhang					
3.30 m span	1.55	12.55	17.89	nr	30.44
5.60 m span	1.75	14.17	28.20	nr	42.37
7.00 m span	2.05	16.59	38.61	nr	55.20
'Mono' type truss; 30 degree pitch; 450 mm eaves overhang					
3.30 m span	1.55	12.55	19.77	nr	32.32
5.60 m span	1.75	14.17	31.93	nr	46.10
7.00 m span	2.05	16.59	45.53	nr	62.12
Attic type truss; 45 degree pitch; 450 mm eaves overhang					
5.00 m span	3.45	27.93	41.11	nr	69.04
7.60 m span	3.65	29.55	57.64	nr	87.19
9.00 m span	3.85	31.16	70.58	nr	101.74

G STRUCTURAL/CARCASSING METAL/TIMBER Including overheads and profit at 7.50%	Labour hours	Labour £	Material £	Unit	Total rate £
G20 CARPENTRY/TIMBER FRAMING/FIRST FIXING - cont'd					
Standard 'Toreboda' glulam timber beams; Moelven (UK) Ltd.; LB grade whitewood; pressure impregnated; phenol resorcinal adhesive; clean planed finish; fixed					
Laminated roof beams					
56 x 255 mm	0.61	4.94	16.03	m	**20.97**
66 x 315 mm	0.77	6.23	21.53	m	**27.76**
90 x 315 mm	0.99	8.01	27.00	m	**35.01**
90 x 405 mm	1.25	10.12	33.79	m	**43.91**
115 x 405 mm	1.60	12.95	40.72	m	**53.67**
115 x 495 mm	2.00	16.19	54.03	m	**70.22**
115 x 630 mm	2.40	19.43	67.58	m	**87.01**

ALTERNATIVE FIRST FIXING MATERIAL PRICES

	£		£		£
Chipboard roofing (£/10 m2)					
12 mm	29.02	18 mm	40.23	25 mm	55.88
Non-asbestos boards (£/10 m2)					
'Masterboard'					
6 mm	45.30	9 mm	82.30	12 mm	108.50
'Masterclad'; sanded finish					
4.5 mm	36.90	6 mm	47.60	9 mm	79.60
Plywood (£/10 m2)					
External quality					
12 mm	75.58	15 mm	92.76	25 mm	135.62
Marine quality					
12 mm	62.71	15 mm	79.54	25 mm	127.20

Discounts of 0 - 10% available depending on quantity/status

	Labour hours	Labour £	Material £	Unit	Total rate £
'Masterboard'; 6 mm thick; PC £45.30/10m2					
Boarding to eaves, verges, fascias and the like					
over 300 mm wide	0.77	6.23	5.56	m2	**11.79**
75 mm wide	0.23	1.86	0.50	m	**2.36**
150 mm wide	0.26	2.10	0.94	m	**3.04**
200 mm wide	0.30	2.43	1.21	m	**3.64**
225 mm wide	0.31	2.51	1.38	m	**3.89**
250 mm wide	0.32	2.59	1.55	m	**4.14**
Raking cutting	0.06	0.49	0.26	m	**0.75**
Plywood; external quality; 15 mm thick; PC £93.83/10m2					
Boarding to eaves, verges, fascias and the like					
over 300 mm wide	0.90	7.29	11.16	m2	**18.45**
75 mm wide	0.28	2.27	1.02	m	**3.29**
150 mm wide	0.32	2.59	1.90	m	**4.49**
225 mm wide	0.36	2.91	2.80	m	**5.71**

G STRUCTURAL/CARCASSING METAL/TIMBER Including overheads and profit at 7.50%	Labour hours	Labour £	Material £	Unit	Total rate £
Plywood; external quality; 18 mm thick; PC £110.21/10m2					
Boarding to eaves, verges, fascias and the like					
over 300 mm wide	0.90	7.29	13.06	m2	20.35
75 mm wide	0.28	2.27	1.18	m	3.45
150 mm wide	0.32	2.59	2.22	m	4.81
225 mm wide	0.36	2.91	3.27	m	6.18
Plywood; marine quality; 18 mm thick; PC £91.59/10m2					
Boarding to gutter bottoms or sides; butt joints					
over 300 mm wide	1.00	8.09	10.91	m2	19.00
150 mm wide	0.36	2.91	1.85	m	4.76
225 mm wide	0.41	3.32	2.84	m	6.16
300 mm wide	0.46	3.72	3.60	m	7.32
Boarding to eaves, verges, fascias and the like					
over 300 mm wide	0.90	7.29	10.91	m2	18.20
75 mm wide	0.28	2.27	0.98	m	3.25
150 mm wide	0.32	2.59	1.85	m	4.44
225 mm wide	0.34	2.75	2.39	m	5.14
Plywood; marine quality; 25 mm thick; PC £136.97/10m2					
Boarding to gutter bottoms or sides; butt joints					
150 mm wide	1.10	8.90	16.16	m	25.06
150 mm wide	0.39	3.16	2.76	m	5.92
225 mm wide	0.44	3.56	4.20	m	7.76
300 mm wide	0.50	4.05	5.34	m	9.39
Boarding to eaves, verges, fascias and the like					
over 300 mm wide	0.97	7.85	16.16	m2	24.01
75 mm wide	0.29	2.35	1.45	m	3.80
150 mm wide	0.34	2.75	2.76	m	5.51
225 mm wide	0.39	3.16	3.55	m	6.71
Sawn softwood; untreated					
Boarding to gutter bottoms or sides; butt joints					
19 mm thick; sloping	1.40	11.33	5.63	m2	16.96
19 mm thick x 75 mm wide	0.39	3.16	0.45	m	3.61
19 mm thick x 150 mm wide	0.44	3.56	0.84	m	4.40
19 mm thick x 225 mm wide	0.50	4.05	1.31	m	5.36
25 mm thick; sloping	1.40	11.33	6.83	m2	18.16
25 mm thick x 75 mm wide	0.39	3.16	0.57	m	3.73
25 mm thick x 150 mm wide	0.44	3.56	1.02	m	4.58
25 mm thick x 225 mm wide	0.50	4.05	1.62	m	5.67
Cesspools with 25 mm thick sides and bottom					
225 x 225 x 150 mm	1.30	10.52	3.23	nr	13.75
300 x 300 x 150 mm	1.55	12.55	4.57	nr	17.12
Firrings					
50 mm wide x 36 mm average depth	0.17	1.38	0.70	m	2.08
50 mm wide x 50 mm average depth	0.17	1.38	0.83	m	2.21
50 mm wide x 75 mm average depth	0.17	1.38	1.23	m	2.61
Bearers					
25 x 50 mm	0.11	0.89	0.41	m	1.30
38 x 50 mm	0.11	0.89	0.57	m	1.46
50 x 50 mm	0.11	0.89	0.69	m	1.58
50 x 75 mm	0.11	0.89	0.93	m	1.82
Angle fillets					
38 x 38 mm	0.11	0.89	0.33	m	1.22
50 x 50 mm	0.11	0.89	0.50	m	1.39
75 x 75 mm	0.13	1.05	0.96	m	2.01

G STRUCTURAL/CARCASSING METAL/TIMBER Including overheads and profit at 7.50%	Labour hours	Labour £	Material £	Unit	Total rate £
G20 CARPENTRY/TIMBER FRAMING/FIRST FIXING - cont'd					
Sawn softwood; untreated - cont'd					
Tilting fillets					
19 x 38 mm	0.11	0.89	0.20	m	1.09
25 x 50 mm	0.11	0.89	0.28	m	1.17
38 x 75 mm	0.11	0.89	0.53	m	1.42
50 x 75 mm	0.11	0.89	0.69	m	1.58
75 x 100 mm	0.17	1.38	1.11	m	2.49
Grounds or battens					
13 x 19 mm	0.06	0.49	0.14	m	0.63
13 x 32 mm	0.06	0.49	0.18	m	0.67
25 x 50 mm	0.06	0.49	0.38	m	0.87
Grounds or battens; plugged and screwed					
13 x 19 mm	0.17	1.38	0.20	m	1.58
13 x 32 mm	0.17	1.38	0.23	m	1.61
25 x 50 mm	0.17	1.38	0.44	m	1.82
Open-spaced grounds or battens; at 300 mm centres one way					
25 x 50 mm	0.17	1.38	1.28	m2	2.66
25 x 50 mm; plugged and screwed	0.50	4.05	1.48	m2	5.53
Framework to walls; at 300 mm centres one way and 600 mm centres the other					
25 x 50 mm	0.83	6.72	2.12	m2	8.84
38 x 50 mm	0.83	6.72	2.97	m2	9.69
50 x 50 mm	0.83	6.72	3.63	m2	10.35
50 x 75 mm	0.83	6.72	4.93	m2	11.65
75 x 75 mm	0.83	6.72	8.04	m2	14.76
Framework to walls; at 300 mm centres one way and 600 mm centres the other way; plugged and screwed					
25 x 50 mm	1.40	11.33	2.41	m2	13.74
38 x 50 mm	1.40	11.33	3.28	m2	14.61
50 x 50 mm	1.40	11.33	3.94	m2	15.27
50 x 75 mm	1.40	11.33	5.24	m2	16.57
75 x 75 mm	1.40	11.33	8.35	m2	19.68
Framework to bath panel; at 500 mm centres both ways					
25 x 50 mm	0.99	8.01	2.31	m2	10.32
Framework as bracketing and cradling around steelwork					
25 x 50 mm	1.55	12.55	2.75	m2	15.30
50 x 50 mm	1.65	13.36	4.72	m2	18.08
50 x 75 mm	1.75	14.17	6.40	m2	20.57
Blockings wedged between flanges of steelwork					
50 x 50 x 150 mm	0.13	1.05	0.31	nr	1.36
50 x 75 x 225 mm	0.14	1.13	0.47	nr	1.60
50 x 100 x 300 mm	0.15	1.21	0.63	nr	1.84
Sawn softwood; 'Tanalised'					
Boarding to gutter bottoms or sides; butt joints					
19 mm thick; sloping	1.40	11.33	6.89	m2	18.22
19 mm thick x 75 mm wide	0.39	3.16	0.52	m	3.68
19 mm thick x 150 mm wide	0.44	3.56	1.03	m	4.59
19 mm thick x 225 mm wide	0.50	4.05	1.55	m	5.60
25 mm thick; sloping	1.40	11.33	8.48	m2	19.81
25 mm thick x 75 mm wide	0.39	3.16	0.70	m	3.86
25 mm thick x 150 mm wide	0.44	3.56	1.26	m	4.82
25 mm thick x 225 mm wide	0.50	4.05	1.92	m	5.97
Firrings					
50 mm wide x 36 mm average depth	0.17	1.38	0.83	m	2.21
50 mm wide x 50 mm average depth	0.17	1.38	1.00	m	2.38
50 mm wide x 75 mm average depth	0.17	1.38	1.50	m	2.88

G STRUCTURAL/CARCASSING METAL/TIMBER Including overheads and profit at 7.50%	Labour hours	Labour £	Material £	Unit	Total rate £
Bearers					
25 x 50 mm	0.11	0.89	0.51	m	1.40
38 x 50 mm	0.11	0.89	0.70	m	1.59
50 x 50 mm	0.11	0.89	0.81	m	1.70
50 x 75 mm	0.11	0.89	1.10	m	1.99
Cesspools with 25 mm thick sides and bottom					
225 x 225 x 150 mm	1.30	10.52	4.02	nr	14.54
300 x 300 x 150 mm	1.55	12.55	5.71	nr	18.26
Angle fillets					
38 x 38 mm	0.11	0.89	0.39	m	1.28
50 x 50 mm	0.11	0.89	0.60	m	1.49
75 x 75 mm	0.13	1.05	1.17	m	2.22
Tilting fillets					
19 x 38 mm	0.11	0.89	0.23	m	1.12
25 x 50 mm	0.11	0.89	0.34	m	1.23
38 x 75 mm	0.11	0.89	0.65	m	1.54
50 x 75 mm	0.11	0.89	0.85	m	1.74
75 x 100 mm	0.17	1.38	1.37	m	2.75
Grounds or battens					
13 x 19 mm	0.06	0.49	0.16	m	0.65
13 x 32 mm	0.06	0.49	0.21	m	0.70
25 x 50 mm	0.06	0.49	0.48	m	0.97
Grounds or battens; plugged and screwed					
13 x 19 mm	0.17	1.38	0.20	m	1.58
13 x 32 mm	0.17	1.38	0.27	m	1.65
25 x 50 mm	0.17	1.38	0.54	m	1.92
Open-spaced grounds or battens; at 300 mm centres one way					
25 x 50 mm	0.17	1.38	1.58	m2	2.96
25 x 50 mm; plugged and screwed	0.50	4.05	1.78	m2	5.83
Framework to walls; at 300 mm centres one way and 600 mm centres the other					
25 x 50 mm	0.83	6.72	2.61	m2	9.33
38 x 50 mm	0.83	6.72	3.70	m2	10.42
50 x 50 mm	0.83	6.72	4.28	m2	11.00
50 x 75 mm	0.83	6.72	5.87	m2	12.59
75 x 75 mm	0.83	6.72	10.29	m2	17.01
Framework to walls; at 300 mm centres one way and 600 mm centres the other way; plugged and screwed					
25 x 50 mm	1.40	11.33	2.92	m2	14.25
38 x 50 mm	1.40	11.33	4.00	m2	15.33
50 x 50 mm	1.40	11.33	4.59	m2	15.92
50 x 75 mm	1.40	11.33	6.17	m2	17.50
75 x 75 mm	1.40	11.33	10.60	m2	21.93
Framework to bath panel; at 500 mm centres both ways					
25 x 50 mm	0.99	8.01	2.86	m2	10.87
Framework as bracketing and cradling around steelwork					
25 x 50 mm	1.55	12.55	3.41	m2	15.96
50 x 50 mm	1.65	13.36	5.56	m2	18.92
50 x 75 mm	1.75	14.17	7.61	m2	21.78
Blockings wedged between flanges of steelwork					
50 x 50 x 150 mm	0.13	1.05	0.40	nr	1.45
50 x 75 x 225 mm	0.14	1.13	0.59	nr	1.72
50 x 100 x 300 mm	0.15	1.21	0.78	nr	1.99
Floor fillets set in or on concrete					
38 x 50 mm	0.13	1.05	0.64	m	1.69
50 x 50 mm	0.13	1.05	0.78	m	1.83
Floor fillets fixed to floor clips (priced elsewhere)					
38 x 50 mm	0.11	0.89	0.67	m	1.56
50 x 50 mm	0.11	0.89	0.82	m	1.71

G STRUCTURAL/CARCASSING METAL/TIMBER Including overheads and profit at 7.50%	Labour hours	Labour £	Material £	Unit	Total rate £
G20 CARPENTRY/TIMBER FRAMING/FIRST FIXING - cont'd					
Wrought softwood					
Boarding to gutter bottoms or sides; tongued and grooved joints					
19 mm thick; sloping	1.65	13.36	7.61	m2	20.97
19 mm thick x 75 mm wide	0.44	3.56	0.74	m	4.30
19 mm thick x 150 mm wide	0.50	4.05	1.13	m	5.18
19 mm thick x 225 mm wide	0.55	4.45	1.80	m	6.25
25 mm thick; sloping	1.65	13.36	9.86	m2	23.22
25 mm thick x 75 mm wide	0.44	3.56	0.76	m	4.32
25 mm thick x 150 mm wide	0.50	4.05	1.44	m	5.49
25 mm thick x 225 mm wide	0.55	4.45	2.32	m	6.77
Boarding to eaves, verges, fascias and the like					
19 mm thick x over 300 mm wide	1.35	10.93	7.43	m2	18.36
19 mm thick x 150 mm wide; once grooved	0.22	1.78	2.83	m	4.61
25 mm thick x 150 mm wide; once grooved	0.22	1.78	3.45	m	5.23
25 mm thick x 175 mm wide; once grooved	0.24	1.94	3.88	m	5.82
32 mm thick x 225 mm wide; moulded	0.28	2.27	5.87	m	8.14
Mitred angles	0.11	0.89	-	nr	0.89
Rolls					
32 x 44 mm	0.13	1.05	0.63	m	1.68
50 x 50 mm	0.13	1.05	1.00	m	2.05
50 x 75 mm	0.14	1.13	1.56	m	2.69
75 x 75 mm	0.15	1.21	2.31	m	3.52
Wrought softwood; 'Tanalised'					
Boarding to gutter bottoms or sides; tongued and grooved joints					
19 mm thick; sloping	1.65	13.36	9.37	m2	22.73
19 mm thick x 75 mm wide	0.44	3.56	0.91	m	4.47
19 mm thick x 150 mm wide	0.50	4.05	1.39	m	5.44
19 mm thick x 225 mm wide	0.55	4.45	2.25	m	6.70
25 mm thick; sloping	1.65	13.36	12.14	m2	25.50
25 mm thick x 75 mm wide	0.44	3.56	0.93	m	4.49
25 mm thick x 150 mm wide	0.50	4.05	1.79	m	5.84
25 mm thick x 225 mm wide	0.55	4.45	2.89	m	7.34
Boarding to eaves, verges, fascias and the like					
19 mm thick x over 300 mm wide	1.35	10.93	9.19	m2	20.12
19 mm thick x 150 mm wide; once grooved	0.22	1.78	3.51	m	5.29
25 mm thick x 150 mm wide; once grooved	0.22	1.78	4.28	m	6.06
25 mm thick x 175 mm wide; once grooved	0.24	1.94	4.86	m	6.80
32 mm thick x 225 mm wide; moulded	0.28	2.27	7.33	m	9.60
Mitred angles	0.11	0.89	-	nr	0.89
Rolls					
32 x 44 mm	0.13	1.05	0.80	m	1.85
50 x 50 mm	0.13	1.05	1.29	m	2.34
50 x 75 mm	0.14	1.13	2.04	m	3.17
75 x 75 mm	0.15	1.21	3.00	m	4.21
Labours on softwood boarding					
Raking cutting					
12 mm thick	0.07	0.57	0.34	m	0.91
19 mm thick	0.09	0.73	0.30	m	1.03
25 mm thick	0.11	0.89	0.42	m	1.31
Boundary cutting					
19 mm thick	0.10	0.81	0.30	m	1.11
Curved cutting					
19 mm thick	0.13	1.05	0.42	m	1.47
Tongued edges and mitred angle					
19 mm thick	0.15	1.21	0.42	m	1.63

G STRUCTURAL/CARCASSING METAL/TIMBER Including overheads and profit at 7.50%	Labour hours	Labour £	Material £	Unit	Total rate £
Plugging					
Plugging blockwork					
300 mm centres; both ways	0.13	1.05	0.06	m2	**1.11**
300 mm centres; one way	0.07	0.57	0.03	m	**0.60**
isolated	0.03	0.24	0.01	nr	**0.25**
Plugging brickwork					
300 mm centres; both ways	0.22	1.78	0.06	m2	**1.84**
300 mm centres; one way	0.11	0.89	0.03	m	**0.92**
isolated	0.06	0.49	0.01	nr	**0.50**
Plugging concrete					
300 mm centres; both ways	0.40	3.24	0.06	m2	**3.30**
300 mm centres; one way	0.20	1.62	0.03	m	**1.65**
isolated	0.10	0.81	0.01	nr	**0.82**
Shot-firing					
Concrete					
1 m centres; one way	0.09	0.73	0.25	m	**0.98**
isolated	0.11	0.89	0.25	nr	**1.14**
Steel					
1 m centres; one way	0.09	0.73	0.25	m	**0.98**
isolated	0.11	0.89	0.25	nr	**1.14**

BASIC BOLT PRICES

	£		£		£		£
Black bolts, nuts and washers (£/100)							
Mild steel hex. hdd. bolts/nuts							
M6x50 mm	6.46	M10x50 mm	15.48	M12x100 mm	31.95	M16x140 mm	85.50
M6x80 mm	10.82	M10x80 mm	22.22	M12x140 mm	48.43	M16x180 mm	124.33
M6x100 mm	11.76	M10x100 mm	27.38	M12x180 mm	92.07	M20x80 mm	87.55
M8x50 mm	10.17	M10x140 mm	37.65	M16x50 mm	39.04	M20x100 mm	102.15
M8x80 mm	14.19	M12x50 mm	22.00	M16x80 mm	50.00	M20x140 mm	138.09
M8x100 mm	17.71	M12x80 mm	28.04	M16x100 mm	55.37	M20x180 mm	186.04
Mild steel cup. hdd. bolts/nuts							
M6x50 mm	5.95	M8x75 mm	10.76	M10x100 mm	24.63	M12x100 mm	34.79
M6x75 mm	7.28	M8x100 mm	17.70	M10x150 mm	35.34	M12x150 mm	48.56
M6x100 mm	11.33	M8x150 mm	24.20	M10x200 mm	73.93	M12x200 mm	119.74
M6x150 mm	17.25	M10x50 mm	14.39	M12x50 mm	22.76		
M8x50 mm	9.02	M10x75 mm	16.50	M12x75 mm	28.56		
Mild steel washers; round							
M6	1.65	M10	2.04	M16	4.41	M20	5.09
M8	1.43	M12	3.16				
Mild steel washers; square							
38x38 mm	9.43	50x50 mm	6.53				

	Labour hours	Labour £	Material £	Unit	Total rate £
Metalwork; mild steel; galvanized					
Fix only bolts; 50-200 mm long					
6 mm dia	0.04	0.32	-	nr	**0.32**
8 mm dia	0.04	0.32	-	nr	**0.32**
10 mm dia	0.06	0.49	-	nr	**0.49**
12 mm dia	0.06	0.49	-	nr	**0.49**
16 mm dia	0.07	0.57	-	nr	**0.57**
20 mm dia	0.07	0.57	-	nr	**0.57**

G STRUCTURAL/CARCASSING METAL/TIMBER Including overheads and profit at 7.50%		Labour hours	Labour £	Material £	Unit	Total rate £
G20 CARPENTRY/TIMBER FRAMING/FIRST FIXING - cont'd						
Metalwork; mild steel; galvanized - cont'd						
Straps; standard twisted vertical restraint; fixing to softwood and brick or blockwork						
30 x 2.5 x 400 mm girth		0.28	2.27	0.73	nr	3.00
30 x 2.5 x 600 mm girth		0.29	2.35	0.94	nr	3.29
30 x 2.5 x 800 mm girth		0.30	2.43	1.25	nr	3.68
30 x 2.5 x 1000 mm girth		0.33	2.67	1.53	nr	4.20
30 x 2.5 x 1200 mm girth		0.34	2.75	1.78	nr	4.53
Timber connectors; round toothed plate; for 10 mm or 12 mm dia bolts						
38 mm dia; single sided		0.02	0.16	0.18	nr	0.34
38 mm dia; double sided		0.02	0.16	0.21	nr	0.37
50 mm dia; single sided		0.02	0.16	0.21	nr	0.37
50 mm dia; double sided		0.02	0.16	0.23	nr	0.39
63 mm dia; single sided		0.02	0.16	0.29	nr	0.45
63 mm dia; double sided		0.02	0.16	0.32	nr	0.48
75 mm dia; single sided		0.02	0.16	0.41	nr	0.57
75 mm dia; double sided		0.02	0.16	0.46	nr	0.62
Framing anchor		0.17	1.38	0.53	nr	1.91
Joist hangers 1.0 mm thick; for fixing to softwood; joint sizes						
50 x 100 mm	PC £0.68	0.13	1.05	0.89	nr	1.94
50 x 125 mm	PC £0.68	0.13	1.05	0.89	nr	1.94
50 x 150 mm	PC £0.68	0.14	1.13	0.89	nr	2.02
50 x 175 mm	PC £0.68	0.14	1.13	0.89	nr	2.02
50 x 200 mm	PC £0.68	0.15	1.21	0.89	nr	2.10
50 x 225 mm	PC £0.68	0.15	1.21	0.89	nr	2.10
50 x 250 mm	PC £0.68	0.17	1.38	0.89	nr	2.27
75 x 150 mm	PC £0.72	0.14	1.13	0.94	nr	2.07
75 x 175 mm	PC £0.72	0.14	1.13	0.94	nr	2.07
75 x 200 mm	PC £0.72	0.15	1.21	0.94	nr	2.15
75 x 225 mm	PC £0.72	0.15	1.21	0.94	nr	2.15
75 x 250 mm	PC £0.72	0.17	1.38	0.94	nr	2.32
100 x 200 mm	PC £0.78	0.17	1.38	1.01	nr	2.39
Joist hangers 2.7 mm thick; for building in; joist sizes						
50 x 100 mm	PC £1.23	0.09	0.73	1.45	nr	2.18
50 x 125 mm	PC £1.24	0.09	0.73	1.46	nr	2.19
50 x 150 mm	PC £1.27	0.10	0.81	1.49	nr	2.30
50 x 175 mm	PC £1.30	0.10	0.81	1.52	nr	2.33
50 x 200 mm	PC £1.43	0.11	0.89	1.68	nr	2.57
50 x 225 mm	PC £1.53	0.11	0.89	1.79	nr	2.68
50 x 250 mm	PC £2.00	0.12	0.97	2.32	nr	3.29
75 x 150 mm	PC £1.91	0.10	0.81	2.22	nr	3.03
75 x 175 mm	PC £1.76	0.10	0.81	2.05	nr	2.86
75 x 200 mm	PC £1.93	0.11	0.89	2.24	nr	3.13
75 x 225 mm	PC £2.01	0.11	0.89	2.33	nr	3.22
75 x 250 mm	PC £2.23	0.12	0.97	2.58	nr	3.55
100 x 200 mm	PC £2.24	0.11	0.89	2.59	nr	3.48
Herringbone joist struts; to suit joists at						
400 mm centres	PC £24.21/100	0.33	2.67	1.30	m	3.97
450 mm centres	PC £27.35/100	0.30	2.43	1.29	m	3.72
600 mm centres	PC £30.50/100	0.26	2.10	1.31	m	3.41
Expanding bolts; 'Rawlbolt' projecting type; plated; one nut; one washer						
6 mm dia; ref M6 10P		0.09	0.73	0.55	nr	1.28
6 mm dia; ref M6 25P		0.09	0.73	0.62	nr	1.35
6 mm dia; ref M6 60P		0.09	0.73	0.65	nr	1.38
8 mm dia; ref M8 25P		0.09	0.73	0.74	nr	1.47
8 mm dia; ref M8 60P		0.09	0.73	0.79	nr	1.52

G STRUCTURAL/CARCASSING METAL/TIMBER Including overheads and profit at 7.50%		Labour hours	Labour £	Material £	Unit	Total rate £
10 mm dia; ref M10 15P		0.11	0.89	0.96	nr	1.85
10 mm dia; ref M10 30P		0.11	0.89	1.01	nr	1.90
10 mm dia; ref M10 60P		0.11	0.89	1.05	nr	1.94
12 mm dia; ref M12 15P		0.11	0.89	1.52	nr	2.41
12 mm dia; ref M12 30P		0.11	0.89	1.63	nr	2.52
12 mm dia; ref M12 75P		0.11	0.89	2.04	nr	2.93
16 mm dia; ref M16 35P		0.13	1.05	3.76	nr	4.81
16 mm dia; ref M16 75P		0.13	1.05	3.95	nr	5.00
Expanding bolts; 'Rawlbolt' loose bolt type; plated; one bolt; one washer						
6 mm dia; ref M6 10L		0.11	0.89	0.55	nr	1.44
6 mm dia; ref M6 25L		0.11	0.89	0.59	nr	1.48
6 mm dia; ref M6 40L		0.11	0.89	0.59	nr	1.48
8 mm dia; ref M8 25L		0.11	0.89	0.72	nr	1.61
8 mm dia; ref M8 40L		0.11	0.89	0.77	nr	1.66
10 mm dia; ref M10 10L		0.11	0.89	0.93	nr	1.82
10 mm dia; ref M10 25L		0.11	0.89	0.96	nr	1.85
10 mm dia; ref M10 50L		0.11	0.89	1.01	nr	1.90
10 mm dia; ref M10 75L		0.11	0.89	1.05	nr	1.94
12 mm dia; ref M12 10L		0.11	0.89	1.38	nr	2.27
12 mm dia; ref M12 25L		0.11	0.89	1.52	nr	2.41
12 mm dia; ref M12 40L		0.11	0.89	1.59	nr	2.48
12 mm dia; ref M12 60L		0.11	0.89	1.67	nr	2.56
16 mm dia; ref M16 30L		0.13	1.05	3.58	nr	4.63
16 mm dia; ref M16 60L		0.13	1.05	3.87	nr	4.92
Metalwork; mild steel; galvanised						
Ragbolts; mild steel; one nut; one washer						
M10 x 120 mm long		0.11	1.01	1.13	nr	2.14
M12 x 160 mm long		0.14	1.28	1.40	nr	2.68
M20 x 200 mm long		0.17	1.56	3.07	nr	4.63
G32 EDGE SUPPORTED/REINFORCED WOODWOOL SLAB DECKING						
Woodwool interlocking reinforced slabs; Torvale 'Woodcelip' or similar; natural finish; fixing to timber or steel with galvanized nails or clips; flat or sloping						
50 mm slabs; type 503; max. span 2100 mm						
1800 - 2100 mm lengths	PC £11.89	0.55	4.45	13.66	m2	18.11
2400 mm lengths	PC £12.56	0.55	4.45	14.41	m2	18.86
2700 - 3000 mm lengths	PC £12.88	0.55	4.45	14.95	m2	19.40
75 mm slabs; type 751; max. span 2100 mm						
1800 - 2400 mm lengths	PC £17.28	0.61	4.94	19.78	m2	24.72
2700 - 3000 mm lengths	PC £17.37	0.61	4.94	19.88	m2	24.82
75 mm slabs; type 752; max. span 2100 mm						
1800 - 2400 mm lengths	PC £17.28	0.61	4.94	19.78	m2	24.72
2700 - 3000 mm lengths	PC £16.99	0.61	4.94	19.66	m2	24.60
75 mm slabs; type 753; max. span 3600 mm						
2400 mm lengths	PC £17.27	0.61	4.94	19.77	m2	24.71
2700 - 3000 mm lengths	PC £18.04	0.61	4.94	20.64	m2	25.58
3300 - 3900 mm lengths		0.61	4.94	24.24	m2	29.18
Raking cutting; including additional trim		0.24	1.94	4.36	m	6.30
Holes for pipes and the like		0.13	1.05	-	nr	1.05
100 mm slabs; type 1001; max. span 3600 mm						
3000 mm lengths	PC £25.60	0.66	5.34	29.22	m2	34.56
3300 - 3600 mm lengths	PC £26.29	0.66	5.34	29.99	m2	35.33

G STRUCTURAL/CARCASSING METAL/TIMBER Including overheads and profit at 7.50%		Labour hours	Labour £	Material £	Unit	Total rate £
G32 EDGE SUPPORTED/REINFORCED WOODWOOL SLAB DECKING - cont'd						
Woodwool interlocking reinforced slabs; Torvale 'Woodcelip' or similar; natural finish; fixing to timber or steel with galvanized nails or clips; flat or sloping - cont'd						
100 mm slabs; type 1002; max. span 3600 mm						
3000 mm lengths	PC £26.42	0.66	5.34	30.14	m2	35.48
3300 - 3600 mm lengths	PC £24.23	0.66	5.34	27.91	m2	33.25
100 mm slabs; type 1003; max. span 4000 mm						
3000 - 3600 mm lengths	PC £23.80	0.66	5.34	27.19	m2	32.53
3900 - 4000 mm lengths	PC £23.80	0.66	5.34	27.19	m2	32.53
125 mm slabs; type 1252; max. span 3000 mm						
2400 - 3000 mm lengths	PC £26.71	0.66	5.34	30.48	m2	35.82
Extra over slabs for						
pre-screeded deck		-	-	-	m2	1.05
pre-screeded soffit		-	-	-	m2	2.39
pre-screeded deck and soffit		-	-	-	m2	3.03
pre-screeded and proofed deck		-	-	-	m2	1.93
pre-screeded and proofed deck plus pre-screeded soffit		-	-	-	m2	4.51
pre-felted deck (glass fibre)		-	-	-	m2	2.53
pre-felted deck plus pre-screeded soffit		-	-	-	m2	4.84
'Weatherdeck'		-	-	-	m2	1.64

H CLADDING/COVERING Including overheads and profit at 7.50%		Labour hours	Labour £	Material £	Unit	Total rate £
H10 PATENT GLAZING						
Patent glazing; aluminium alloy bars 2.44 mm long at 622 mm centres; fixed to supports						
Roof cladding; glazing with 7 mm thick Georgian wired cast glass		-	-	-	m2	88.02
Extra for associated code 4 lead flashings						
top flashing; 210 mm girth		-	-	-	m	26.91
bottom flashing; 240 mm girth		-	-	-	m	34.99
end flashing; 300 mm girth		-	-	-	m	49.77
Wall cladding; glazing with 7 mm thick Georgian wired cast glass		-	-	-	m2	85.53
Wall cladding; glazing with 6 mm thick plate glass		-	-	-	m2	98.12
Extra for aluminium alloy members						
38 x 38 x 3 mm angle jamb		-	-	-	m	21.63
extruded cill member		-	-	-	m	24.21
extruded channel head and PVC came		-	-	-	m	13.50
H20 RIGID SHEET CLADDING						
Eternit 2000 'Glasal' sheet; Eternit TAC Ltd; flexible neoprene gasket joints; fixing with stainless steel screws and coloured caps						
7.5 mm thick cladding to walls						
over 300 mm wide		2.40	28.61	48.07	m2	76.68
not exceeding 300 mm wide		0.80	9.54	22.40	m2	31.94
External angle		0.11	1.31	7.17	m	8.48
7.5 mm thick cladding to eaves; verges fascias or the like						
100 mm wide		0.58	6.91	12.18	m	19.09
150 mm wide		0.63	7.51	14.40	m	21.91
200 mm wide		0.69	8.23	16.61	m	24.84
250 mm wide		0.75	8.94	18.82	m	27.76
300 mm wide		0.81	9.66	21.78	m	31.44
H30 FIBRE CEMENT PROFILED SHEET CLADDING/ COVERING/SIDING						
Asbestos-free corrugated sheets; Eternit '2000' or similar						
Roof cladding; sloping not exceeding 50 degrees; fixing to timber purlins with drive screws						
'Profile 3'; natural	PC £7.49	0.23	2.74	12.04	m2	14.78
'Profile 3'; coloured	PC £8.24	0.23	2.74	12.85	m2	15.59
'Profile 6'; natural	PC £7.67	0.29	3.46	11.98	m2	15.44
'Profile 6'; coloured	PC £8.24	0.29	3.46	12.76	m2	16.22
'Profile 6'; natural; insulated; 60 mm glass fibre infill; lining panel	PC £7.67	0.52	6.20	27.64	m2	33.84
Roof cladding; sloping not exceeding 50 degrees; fixing to steel purlins with hook bolts						
'Profile 3'; natural		0.29	3.46	12.62	m2	16.08
'Profile 3'; coloured		0.29	3.46	13.41	m2	16.87
'Profile 6'; natural		0.35	4.17	12.48	m2	16.65
'Profile 6'; coloured		0.35	4.17	13.26	m2	17.43
'Profile 6'; natural; insulated; 60 mm glass fibre infill; lining panel		0.58	6.91	25.63	m2	32.54

H CLADDING/COVERING Including overheads and profit at 7.50%	Labour hours	Labour £	Material £	Unit	Total rate £
H30 FIBRE CEMENT PROFILED SHEET CLADDING/ COVERING/SIDING - cont'd					
Asbestos-free corrugated sheets; Eternit '2000' or similar - cont'd					
Wall cladding; vertical; fixing to steel rails with hook bolts					
'Profile 3'; natural	0.35	4.17	12.62	m2	16.79
'Profile 3'; coloured	0.35	4.17	13.64	m2	17.81
'Profile 6'; natural	0.40	4.77	12.48	m2	17.25
'Profile 6'; coloured	0.40	4.77	13.26	m2	18.03
'Profile 6'; natural; insulated; 60 mm glass fibre infill; lining panel	0.63	7.51	25.63	m2	33.14
Raking cutting	0.17	2.03	1.86	m	3.89
Holes for pipes and the like	0.17	2.03	-	nr	2.03
Accessories; to 'Profile 3' cladding; natural					
eaves filler	0.11	1.31	8.23	m	9.54
vertical corrugation closure	0.14	1.67	8.23	m	9.90
apron flashing	0.14	1.67	9.28	m	10.95
underglazing flashing	0.14	1.67	8.91	m	10.58
plain wing or close fitting two-piece adjustable capping to ridge	0.20	2.38	20.67	m	23.05
ventilating two-piece adjustable capping to ridge	0.20	2.38	20.67	m	23.05
Accessories; to 'Profile 3' cladding; coloured					
eaves filler	0.11	1.31	9.90	m	11.21
vertical corrugation closure	0.14	1.67	9.90	m	11.57
apron flashing	0.14	1.67	13.96	m	15.63
underglazing flashing	0.14	1.67	10.72	m	12.39
plain wing or close fitting two-piece adjustable capping to ridge	0.20	2.38	24.70	m	27.08
ventilating two-piece adjustable capping to ridge	0.20	2.38	24.70	m	27.08
Accessories; to 'Profile 6' cladding; natural					
eaves filler	0.11	1.31	7.00	m	8.31
vertical corrugation closure	0.14	1.67	7.00	m	8.67
apron flashing	0.14	1.67	6.02	m	7.69
plain cranked crown to ridge	0.20	2.38	13.86	m	16.24
plain wing, close fitting or north light two-piece adjustable capping to ridge	0.20	2.38	13.33	m	15.71
ventilating two-piece adjustable capping to ridge	0.20	2.38	18.11	m	20.49
Accessories; to 'Profile 6' cladding; coloured					
eaves filler	0.11	1.31	8.40	m	9.71
vertical corrugation closure	0.14	1.67	8.40	m	10.07
apron flashing	0.14	1.67	7.23	m	8.90
plain cranked crown to ridge	0.20	2.38	16.55	m	18.93
plain wing, close fitting or north light two-piece adjustable capping to ridge	0.20	2.38	15.89	m	18.27
ventilating two-piece adjustable capping to ridge	0.20	2.38	21.63	m	24.01

H CLADDING/COVERING Including overheads and profit at 7.50%		Labour hours	Labour £	Material £	Unit	Total rate £
H41 GLASS REINFORCED PLASTICS CLADDING/ FEATURES						
Glass fibre translucent sheeting grade AB class 3						
Roof cladding; sloping not exceeding 50 degrees; fixing to timber purlins with drive screws; to suit						
'Profile 3'	PC £6.93	0.23	2.74	10.97	m2	**13.71**
'Profile 6'	PC £9.64	0.29	3.46	14.64	m2	**18.10**
Roof cladding; sloping not exceeding 50 degrees; fixing to steel purlins with hook bolts; to suit						
'Profile 3'		0.29	3.46	11.55	m2	**15.01**
'Profile 6'		0.35	4.17	15.22	m2	**19.39**
'Longrib 1000'		0.35	4.17	15.74	m2	**19.91**

ALTERNATIVE TILE PRICES (£/1000)

	£		£		£
Clay tiles; plain, interlocking and pantile					
'Langleys' 'Sterreberg' pantiles					
anthracite	1524.00	natural red	1248.00	rustic	1248.00
deep brown	1650.00				
Sandtoft pantiles					
Bold roll 'Roman'	861.00	'Gaelic'	565.95	'County' i'locking	519.75
William Blyth pantiles					
'Barco' bold roll	459.90	'Celtic' (French)	512.40		
Concrete tiles, plain and interlocking					
Marley roof tiles					
'Anglia'	403.00	plain	227.00	'Roman'	566.00
'Ludlow +'	325.00				
Redland roof tiles					
'49'-granule	315.00	'Grovebury'	624.00	'Stonewold Mk 1	846.00
'50 Roman'	561.00				

Discounts of 2.5 - 15% available depending on quantity/status

H CLADDING/COVERING Including overheads and profit at 7.50%	Labour hours	Labour £	Material £	Unit	Total rate £
NOTE: The following items of tile roofing unless otherwise described, include for conventional fixing assuming 'normal exposure' with appropriate nails and/or rivets or clips to pressure impregnated softwood battens fixed with galvanized nails; Prices also include for all bedding and pointing at verges; beneath ridge tiles, etc.					
H60 CLAY/CONCRETE ROOF TILING					
Clay interlocking pantiles; Sandtoft Goxhill 'Tudor' red sand faced; PC £892.50/1000; 470 x 285 mm; to 100 mm lap; on 25 x 38 mm battens and type 1F reinforced underlay					
Roof coverings	0.46	5.48	14.30	m2	**19.78**
Extra over coverings for					
fixing every tile	0.02	0.24	0.33	m2	**0.57**
eaves course with plastic filler	0.35	4.17	5.21	m	**9.38**
verges; extra single undercloak course of plain tiles	0.35	4.17	2.30	m	**6.47**
open valleys; cutting both sides	0.21	2.50	5.18	m	**7.68**
ridge tiles	0.69	8.23	10.53	m	**18.76**
hip tiles; cutting both sides	0.86	10.25	15.71	m	**25.96**
Holes for pipes and the like	0.23	2.74	-	nr	**2.74**
Clay interlocking pantiles; Langley's 'Sterreberg' black glazed; or similar; PC £1650.00/1000; 355 x 240 mm: to 75 mm lap; on 25 x 38 mm battens and type 1F reinforced underlay					
Roof coverings	0.52	6.20	35.13	m2	**41.33**
Extra over coverings for					
double course at eaves	0.38	4.53	5.49	m	**10.02**
verges; extra single undercloak course of plain tiles	0.35	4.17	7.13	m	**11.30**
open valleys; cutting both sides	0.21	2.50	13.48	m	**15.98**
saddleback ridge tiles	0.69	8.23	18.77	m	**27.00**
saddleback hip tiles; cutting both sides	0.86	10.25	32.25	m	**42.50**
Holes for pipes and the like	0.23	2.74	-	nr	**2.74**
Clay pantiles; Sandtoft Goxhill 'Old English' red sand faced; PC £514.50/1000 342 x 241 mm; to 75 mm lap; on 25 x 38 mm battens and type 1F reinforced underlay					
Roof coverings	0.52	6.20	14.26	m2	**20.46**
Extra over coverings for					
fixing every tile	0.03	0.36	0.54	m2	**0.90**
other colours	-	-	-	m2	**0.84**
double course at eaves	0.38	4.53	3.61	m	**8.14**
verges; extra single undercloak course of plain tiles	0.35	4.17	2.30	m	**6.47**
open valleys; cutting both sides	0.21	2.50	4.20	m	**6.70**
ridge tiles; tile slips	0.69	8.23	11.30	m	**19.53**
hip tiles; tile slips; cutting both sides	0.86	10.25	15.50	m	**25.75**
Holes for pipes and the like	0.23	2.74	-	nr	**2.74**

H CLADDING/COVERING Including overheads and profit at 7.50%	Labour hours	Labour £	Material £	Unit	Total rate £
Clay pantiles; William Blyth's 'Lincoln' natural; 343 x 280 mm; to 75 mm lap; PC £622.88/1000; on 19 x 38 mm battens and type 1F reinforced underlay					
Roof coverings	0.52	6.20	16.39	m2	**22.59**
Extra over coverings for					
fixing every tile	0.03	0.36	0.54	m2	**0.90**
other colours	-	-	-	m2	**0.68**
double course at eaves	0.38	4.53	3.45	m	**7.98**
verges; extra single undercloak course of plain tiles	0.35	4.17	6.09	m	**10.26**
open valleys; cutting both sides	0.21	2.50	5.36	m	**7.86**
ridge tiles; tile slips	0.69	8.23	11.64	m	**19.87**
hip tiles; tile slips; cutting both sides	0.86	10.25	17.00	m	**27.25**
Holes for pipes and the like	0.23	2.74	-	nr	**2.74**
Clay plain tiles; Hinton, Perry and Davenhill 'Dreadnought' smooth red machine-made; PC £276.00/1000; 265 x 165 mm; on 19 x 38 mm battens and type 1F reinforced underlay					
Roof coverings; to 64 mm lap	1.20	14.31	22.90	m2	**37.21**
Wall coverings; to 38 mm lap	1.45	17.29	20.47	m2	**37.76**
Extra over coverings for					
25 x 38 mm battens in lieu	-	-	-	m2	**0.36**
other colours	-	-	-	m2	**3.87**
ornamental tiles in lieu	-	-	-	m2	**8.19**
double course at eaves	0.29	3.46	2.02	m	**5.48**
verges; extra single undercloak course	0.38	4.53	2.91	m	**7.44**
valley tiles; cutting both sides	0.75	8.94	31.40	m	**40.34**
bonnet hip tiles; cutting both sides	0.92	10.97	32.57	m	**43.54**
external vertical angle tiles; supplementary nail fixings	0.46	5.48	23.41	m	**28.89**
half round ridge tiles	0.58	6.91	10.87	m	**17.78**
Holes for pipes and the like	0.23	2.74	-	nr	**2.74**
Clay plain tiles; Keymer best hand-made sand-faced tiles; PC £534.00/1000; 265 x 165 mm; on 19 x 38 mm battens and type 1F reinforced underlay					
Roof coverings; to 64 mm lap	1.20	14.31	40.38	m2	**54.69**
Wall coverings; to 38 mm lap	1.45	17.29	35.90	m2	**53.19**
Extra over coverings for					
25 x 38 mm battens in lieu	-	-	-	m2	**0.36**
ornamental tiles in lieu	-	-	-	m2	**3.87**
double course at eaves	0.29	3.46	3.68	m	**7.14**
verges; extra single undercloak course	0.38	4.53	5.41	m	**9.94**
valley tiles; cutting both sides	0.75	8.94	21.75	m	**30.69**
bonnet hip tiles; cutting both sides	0.92	10.97	32.11	m	**43.08**
external vertical angle tiles; supplementary nail fixings	0.46	5.48	18.94	m	**24.42**
half round ridge tiles	0.58	6.91	9.60	m	**16.51**
Holes for pipes and the like	0.23	2.74	-	nr	**2.74**
Concrete interlocking tiles; Marley 'Bold Roll' granule finish tiles or similar; PC £627.00/1000; 419 x 330 mm; to 75 mm lap; on 22 x 38 mm battens and type 1F reinforced underlay					
Roof coverings	0.40	4.77	9.31	m2	**14.08**
Extra over coverings for					
fixing every tile	0.03	0.36	0.65	m2	**1.01**
25 x 38 mm battens in lieu	-	-	-	m2	**0.04**
eaves; eave filler	0.06	0.72	0.78	m	**1.50**

H CLADDING/COVERING Including overheads and profit at 7.50%	Labour hours	Labour £	Material £	Unit	Total rate £
H60 CLAY/CONCRETE ROOF TILING - cont'd					
Concrete interlocking tiles; Marley 'Bold Roll' granule finish tiles or similar; PC £627.00/1000; 419 x 330 mm; to 75 mm lap; on 22 x 38 mm battens and type 1F reinforced underlay - cont'd					
Extra over coverings for					
verges; 150 mm asbestos cement strip undercloak	0.26	3.10	1.72	m	4.82
valley trough tiles; cutting both sides	0.63	7.51	13.60	m	21.11
segmental ridge tiles; tile slips	0.63	7.51	10.40	m	17.91
segmental ridge tiles; tile slips; cutting both sides	0.81	9.66	13.10	m	22.76
dry ridge tiles; segmental including batten sections; unions and filler pieces	0.35	4.17	10.25	m	14.42
segmental monoridge tiles	0.63	7.51	11.94	m	19.45
gas ridge terminal	0.58	6.91	41.87	nr	48.78
Holes for pipes and the like	0.23	2.74	-	nr	2.74
Concrete interlocking tiles; Marley 'Ludlow Major' granule finish tiles or similar; PC £560.00/1000; 413 x 330 mm; to 75 mm lap; on 22 x 38 mm battens					
Roof coverings	0.40	4.77	8.78	m2	13.55
Extra over coverings for					
fixing every tile	0.03	0.36	0.39	m2	0.75
25 x 38 mm battens in lieu	-	-	-	m2	0.04
verges; 150 mm asbestos cement strip undercloak	0.26	3.10	1.72	m	4.82
dry verge system; extruded white pvc	0.17	2.03	5.88	m	7.91
Segmental ridge cap	0.03	0.36	1.81	nr	2.17
valley trough tiles; cutting both sides	0.63	7.51	15.33	m	22.84
segmental ridge tiles	0.58	6.91	5.28	m	12.19
segmental hip tiles; cutting both sides	0.75	8.94	7.81	m	16.75
dry ridge tiles; segmental including batten sections; unions and filler pieces	0.35	4.17	10.25	m	14.42
segmental monoridge tiles	0.58	6.91	10.56	m	17.47
gas ridge terminal	0.58	6.91	41.87	nr	48.78
Holes for pipes and the like	0.23	2.74	-	nr	2.74
Concrete interlocking tiles; Marley 'Mendip' granule finish double pantiles or similar; PC £627.00/1000; 413 x 330 mm; to 75 mm lap; on 22 x 38 mm battens and type 1F reinforced underlay					
Roof coverings	0.40	4.77	9.56	m2	14.33
Extra over coverings for					
fixing every tile	0.03	0.36	0.39	m2	0.75
25 x 38 mm battens in lieu	-	-	-	m2	0.04
eaves; eave filler	0.03	0.36	0.07	m	0.43
verges; 150 mm asbestos cement strip undercloak	0.26	3.10	1.72	m	4.82
dry verge system; extruded white pvc	0.17	2.03	5.88	m	7.91
valley trough tiles; cutting both sides	0.63	7.51	15.63	m	23.14
segmental ridge tiles	0.63	7.51	8.45	m	15.96
segmental hip tiles; cutting both sides	0.81	9.66	11.29	m	20.95
dry ridge tiles; segmental including batten sections; unions and filler pieces	0.35	4.17	10.25	m	14.42
segmental monoridge tiles	0.58	6.91	10.56	m	17.47
gas ridge terminal	0.58	6.91	41.87	nr	48.78
Holes for pipes and the like	0.23	2.74	-	nr	2.74

H CLADDING/COVERING Including overheads and profit at 7.50%	Labour hours	Labour £	Material £	Unit	Total rate £
Concrete interlocking tiles; Marley 'Modern' smooth finish tiles or similar; PC £643.00/1000; 413 x 330 mm; to 75 mm lap; on 22 x 38 mm battens and type 1F reinforced underlay					
Roof coverings	0.40	4.77	9.83	m2	**14.60**
Extra over coverings for					
fixing every tile	0.05	0.60	0.33	m2	**0.93**
25 x 38 mm battens in lieu	-	-	-	m2	**0.04**
verges; 150 mm asbestos cement strip undercloak	0.32	3.81	2.76	m	**6.57**
dry verge system, extruded white pvc	0.23	2.74	7.10	m	**9.84**
'Modern' ridge cap	0.03	0.36	1.81	nr	**2.17**
valley trough tiles; cutting both sides	0.63	7.51	15.70	m	**23.21**
'Modern' ridge tiles	0.58	6.91	5.54	m	**12.45**
'Modern' hip tiles; cutting both sides	0.75	8.94	8.44	m	**17.38**
dry ridge tiles; 'Modern'; including batten sections, unions and filler pieces	0.35	4.17	9.58	m	**13.75**
monoridge tiles	0.58	6.91	10.56	m	**17.47**
gas ridge terminal	0.58	6.91	41.87	nr	**48.78**
Holes for pipes and the like	0.23	2.74	-	nr	**2.74**
Concrete interlocking tiles; Marley 'Wessex' smooth finish tiles or similar; PC £725.00/1000; 413 x 330 mm; to 75 mm lap; on 22 x 38 mm battens and type 1F reinforced underlay					
Roof coverings	0.40	4.77	10.78	m2	**15.55**
Extra over coverings for					
fixing every tile	0.05	0.60	0.33	m2	**0.93**
25 x 38 mm battens in lieu	-	-	-	m2	**0.04**
verges; 150 mm asbestos cement strip undercloak	0.26	3.10	1.72	m	**4.82**
dry verge system, extruded white pvc	0.17	2.03	5.88	m	**7.91**
'Modern' ridge cap	0.03	0.36	1.81	nr	**2.17**
valley trough tiles; cutting both sides	0.63	7.51	16.07	m	**23.58**
'Modern' ridge tiles	0.63	7.51	7.11	m	**14.62**
'Modern' hip tiles; cutting both sides	0.81	9.66	10.38	m	**20.04**
dry ridge tiles; 'Modern'; including batten sections, unions and filler pieces	0.35	4.17	10.52	m	**14.69**
monoridge tiles	0.58	6.91	10.56	m	**17.47**
gas ridge terminal	0.58	6.91	41.87	nr	**48.78**
Holes for pipes and the like	0.23	2.74	-	nr	**2.74**
Concrete interlocking tiles; Redland 'Delta' smooth finish tiles or similar; PC £931.00/1000; 430 x 380 mm; to 75 mm lap; on 22 x 38 mm battens and type 1F reinforced underlay					
Roof coverings	0.40	4.77	10.90	m2	**15.67**
Extra over coverings for					
fixing every tile	0.03	0.36	0.35	m2	**0.71**
25 x 38 mm battens in lieu	-	-	-	m2	**0.03**
eaves; eave filler	0.03	0.36	0.11	m	**0.47**
verges; extra single undercloak course of plain tiles	0.29	3.46	3.69	m	**7.15**
dry verge system; extruded white pvc	0.23	2.74	9.07	m	**11.81**
Ridge end unit	0.03	0.36	2.15	nr	**2.51**
valley trough tiles; cutting both sides	0.63	7.51	16.88	m	**24.39**
universal 'Delta' ridge tiles	0.58	6.91	5.37	m	**12.28**
universal 'Delta' hip tiles; cutting both sides	0.75	8.94	8.98	m	**17.92**
universal 'Delta' mono-pitch ridge tiles	0.58	6.91	9.56	m	**16.47**
gas flue terminal; 'Delta' type	0.58	6.91	41.87	nr	**48.78**
Holes for pipes and the like	0.23	2.74	-	nr	**2.74**

H CLADDING/COVERING Including overheads and profit at 7.50%	Labour hours	Labour £	Material £	Unit	Total rate £
H60 CLAY/CONCRETE ROOF TILING - cont'd					
Concrete interlocking tiles; Redland 'Norfolk' smooth finish pantiles or similar; PC £408.00/1000; 381 x 229 mm; to 75 mm lap; on 22 x 38 mm battens and type 1F reinforced underlay					
Roof coverings	0.52	6.20	9.87	m2	16.07
Extra over coverings for					
fixing every tile	0.06	0.72	0.32	m2	1.04
25 x 38 mm battens in lieu	-	-	-	m2	0.04
eaves; eave filler	0.06	0.72	0.63	m	1.35
verges; extra single undercloak course of plain tiles	0.35	4.17	1.14	m	5.31
valley trough tiles; cutting both sides	0.69	8.23	16.61	m	24.84
segmental ridge tiles	0.69	8.23	8.03	m	16.26
segmental hip tiles; cutting both sides	0.86	10.25	11.36	m	21.61
Holes for pipes and the like	0.23	2.74	-	nr	2.74
Concrete interlocking tiles; Redland 'Regent' granule finish bold roll tiles or similar; PC £624.00/1000; 418 x 332 mm; to 75 mm lap; on 22 x 38 mm battens and type 1F reinforced underlay					
Roof coverings	0.40	4.77	9.32	m2	14.09
Extra over coverings for					
fixing every tile	0.05	0.60	0.54	m2	1.14
25 x 38 mm battens in lieu	-	-	-	m2	0.03
eaves; eave filler	0.06	0.72	0.77	m	1.49
verges; extra single undercloak course of plain tiles	0.29	3.46	2.25	m	5.71
dry verge system; extruded white pvc	0.17	2.03	7.29	m	9.32
Ridge end unit	0.03	0.36	2.15	nr	2.51
cloaked verge system	0.17	2.03	3.64	m	5.67
Blocked end ridge unit	0.03	0.36	3.87	nr	4.23
valley trough tiles; cutting both sides	0.63	7.51	15.96	m	23.47
segmental ridge tiles; tile slips	0.63	7.51	8.03	m	15.54
segmental hip tiles; tile slips; cutting both sides	0.81	9.66	10.71	m	20.37
dry ridge system; segmental ridge tiles; including fixing straps; 'Nuralite' fillets and seals	0.29	3.46	13.36	m	16.82
half round mono-pitch ridge tiles	0.63	7.51	13.90	m	21.41
gas flue terminal; half round type	0.58	6.91	39.67	nr	46.58
Holes for pipes and the like	0.23	2.74	-	nr	2.74
Concrete interlocking tiles; Redland 'Renown' granule finish tiles or similar; PC £561.00/1000; 418 x 330 mm; to 75 mm lap; on 22 x 38 mm battens and type 1F reinforced underlay					
Roof coverings	0.40	4.77	8.46	m2	13.23
Extra over coverings for					
fixing every tile	0.03	0.36	0.45	m2	0.81
25 x 38 mm battens in lieu	-	-	-	m2	0.04
verges; extra single undercloak course of plain tiles	0.29	3.46	1.14	m	4.60
dry verge system; extruded white pvc	0.17	2.03	7.29	m	9.32
Ridge end unit	0.03	0.36	2.15	nr	2.51
cloaked verge system	0.17	2.03	3.86	m	5.89
Blocked end ridge unit	0.03	0.36	3.87	nr	4.23

H CLADDING/COVERING Including overheads and profit at 7.50%	Labour hours	Labour £	Material £	Unit	Total rate £
valley trough tiles; cutting both sides	0.63	7.51	15.69	m	23.20
segmental ridge tiles	0.58	6.91	5.07	m	11.98
segmental hip tiles; cutting both sides	0.75	8.94	7.48	m	16.42
dry ridge system; segmental ridge tiles; including fixing straps; 'Nuralite' fillets and seals	0.29	3.46	13.36	m	16.82
half round mono-pitch ridge tiles	0.58	6.91	9.16	m	16.07
gas flue terminal; half round type	0.58	6.91	39.67	nr	46.58
Holes for pipes and the like	0.23	2.74	-	nr	2.74
Concrete interlocking tiles; Redland 'Stonewold' smooth finish tiles or similar; PC £776.00/1000; 430 x 380 mm; to 75 mm lap; on 22 x 38 mm battens and type 1F reinforced underlay					
Roof coverings	0.40	4.77	9.40	m2	14.17
Extra over coverings for					
fixing every tile	0.03	0.36	0.35	m2	0.71
25 x 38 mm battens in lieu	-	-	-	m2	0.03
verges; extra single undercloak course of plain tiles	0.35	4.17	3.44	m	7.61
dry verge system; extruded white pvc	0.23	2.74	8.80	m	11.54
Ridge end unit	0.03	0.36	2.15	nr	2.51
valley trough tiles; cutting both sides	0.63	7.51	16.28	m	23.79
universal 'Stonewold' ridge tiles	0.58	6.91	5.37	m	12.28
universal 'Stonewold' hip tiles; cutting both sides	0.75	8.94	8.38	m	17.32
dry ridge system; universal 'Stonewold' ridge tiles; including fixing straps; 'Nuralite' fillets and seals	0.29	3.46	13.89	m	17.35
universal 'Stonewold' mono-pitch ridge tiles	0.58	6.91	9.56	m	16.47
gas flue terminal; 'Stonewold' type	0.58	6.91	41.87	nr	48.78
Holes for pipes and the like	0.23	2.74	-	nr	2.74
Concrete plain tiles; BS 473 and 550 group A; PC £221.00/1000; 267 x 165 mm; on 19 x 38 mm battens and type 1F reinforced underlay					
Roof coverings; to 64 mm lap	1.20	14.31	19.18	m2	33.49
Wall coverings; to 38 mm lap	1.45	17.29	17.17	m2	34.46
Extra over coverings for					
25 x 38 mm battens in lieu	-	-	-	m2	0.36
ornamental tiles in lieu	-	-	-	m2	7.48
double course at eaves	0.29	3.46	1.66	m	5.12
verges; extra single undercloak course	0.38	4.53	2.38	m	6.91
valley tiles; cutting both sides	0.75	8.94	17.49	m	26.43
bonnet hip tiles; cutting both sides	0.92	10.97	18.66	m	29.63
external vertical angle tiles; supplementary nail fixings	0.46	5.48	11.15	m	16.63
segmental ridge tiles	0.58	6.91	7.37	m	14.28
segmental hip tiles; cutting both sides	0.86	10.25	8.04	m	18.29
Holes for pipes and the like	0.23	2.74	-	nr	2.74
Sundries					
Hip irons					
galvanized mild steel; fixing with screws	0.11	1.31	1.98	nr	3.29
Fixing					
lead soakers (supply included elsewhere)	0.09	1.07	-	nr	1.07
Pressure impregnated softwood counter battens; 25 x 50 mm					
450 mm centres	0.08	0.95	0.81	m2	1.76
600 mm centres	0.06	0.72	0.61	m2	1.33

H CLADDING/COVERING Including overheads and profit at 7.50%	Labour hours	Labour £	Material £	Unit	Total rate £
H60 CLAY/CONCRETE ROOF TILING - cont'd					
Underlay; BS 747 type 1B; bitumen felt weighing 14 kg/10 m2; PC £16.01/20m2; 75 mm laps					
To sloping or vertical surfaces	0.03	0.36	0.92	m2	1.28
Underlay; BS 747 type 1F; reinforced bitumen felt; weighing 22.5 kg/15 m2; PC £17.00/15m2; 75 mm laps					
To sloping or vertical surfaces	0.03	0.36	1.31	m2	1.67
H61 FIBRE CEMENT SLATING					
Asbestos-cement slates; Eternit or similar; to 75 mm lap; on 19 x 50 mm battens and type 1F reinforced underlay					
Coverings; 600 x 300 mm 'blue/black' slates					
roof coverings	0.58	6.91	17.28	m2	24.19
wall coverings	0.75	8.94	17.28	m2	26.22
Extra over slate coverings for					
double course at eaves	0.29	3.46	2.92	m	6.38
verges; extra single undercloak course	0.38	4.53	2.29	m	6.82
open valleys; cutting both sides	0.23	2.74	6.08	m	8.82
valley gutters; cutting both sides	0.63	7.51	21.15	m	28.66
half round ridge tiles	0.58	6.91	15.92	m	22.83
Stop end	0.11	1.31	5.49	nr	6.80
roll top ridge tiles	0.58	6.91	19.89	m	26.80
Stop end	0.11	1.31	8.23	nr	9.54
mono-pitch ridges	0.58	6.91	22.92	m	29.83
Stop end	0.11	1.31	24.96	nr	26.27
duo-pitch ridges	0.58	6.91	19.59	m	26.50
Stop end	0.11	1.31	18.31	nr	19.62
mitred hips; cutting both sides	0.23	2.74	6.08	m	8.82
half round hip tiles; cutting both sides	0.75	8.94	21.99	m	30.93
Holes for pipes and the like	0.23	2.74	-	nr	2.74
Asbestos-free artificial slates; Eternit '2000' or similar; to 75 mm lap; on 19 x 50 mm battens and type 1F reinforced underlay					
Coverings; 400 x 200 mm 'blue/black' slates					
roof coverings	0.92	10.97	21.93	m2	32.90
wall coverings	1.20	14.31	21.93	m2	36.24
Coverings; 500 x 250 mm 'blue/black' slates					
roof coverings	0.75	8.94	19.67	m2	28.61
wall coverings	0.98	11.68	19.67	m2	31.35
Coverings; 600 x 300 mm 'blue/black' slates					
roof coverings	0.58	6.91	18.81	m2	25.72
wall coverings	0.75	8.94	18.81	m2	27.75
Coverings; 600 x 300 mm 'brown' or 'rose nuit' slates					
roof coverings	0.58	6.91	18.81	m2	25.72
wall coverings	0.75	8.94	18.81	m2	27.75
Extra over slate coverings for					
double course at eaves	0.29	3.46	3.21	m	6.67
verges; extra single undercloak course	0.38	4.53	2.51	m	7.04
open valleys; cutting both sides	0.23	2.74	6.75	m	9.49
valley gutters; cutting both sides	0.63	7.51	21.82	m	29.33

H CLADDING/COVERING Including overheads and profit at 7.50%	Labour hours	Labour £	Material £	Unit	Total rate £
half round ridge tiles	0.58	6.91	15.92	m	**22.83**
Stop end	0.11	1.31	5.49	nr	**6.80**
roll top ridge tiles	0.69	8.23	19.89	m	**28.12**
Stop end	0.11	1.31	8.23	nr	**9.54**
mono-pitch ridges	0.58	6.91	18.75	m	**25.66**
Stop end	0.11	1.31	24.96	nr	**26.27**
duo-pitch ridges	0.58	6.91	18.75	m	**25.66**
Stop end	0.11	1.31	18.31	nr	**19.62**
mitred hips; cutting both sides	0.23	2.74	6.75	m	**9.49**
half round hip tiles; cutting both sides	0.75	8.94	22.67	m	**31.61**
Holes for pipes and the like	0.23	2.74	-	nr	**2.74**

ALTERNATIVE SLATE PRICES (£/1000)

	£		£		£		£
Natural slates							
Greaves Portmadoc Welsh blue-grey							
Mediums (Class 1)							
305x255 mm	583.00	405x255 mm	929.50	510x255 mm	1496.00	610x305 mm	2552.00
355x255 mm	670.00	460x255 mm	1080.00	560x305 mm	1790.00	610x355 mm	2475.00
Strongs (Class 2)							
305x255 mm	515.00	405x255 mm	790.00	510x255 mm	1300.00	610x305 mm	2170.00
355x255 mm	645.00	460x255 mm	1015.00	560x305 mm	1735.00	610x355 mm	2360.00

Discounts of 2.5 - 15% available depending on quantity/status

NOTE: The following items of slate roofing unless otherwise described, include for conventional fixing assuming 'normal exposure' with appropriate nails and/or rivets or clips to pressure impregnated softwood battens fixed with galvanized nails; Prices also include for all bedding and pointing at verges; beneath ridge tiles, etc.	Labour hours	Labour £	Material £	Unit	Total rate £
H62 NATURAL SLATING					
Natural slates; BS 680 Part 2; Welsh blue; uniform size; to 75 mm lap; on 25 x 50 mm battens and type 1F reinforced underlay					
Coverings; 405 x 255 mm slates					
roof coverings	1.05	12.52	29.36	m2	**41.88**
wall coverings	1.30	15.50	29.36	m2	**44.86**
Coverings; 510 x 255 mm slates					
roof coverings	0.86	10.25	34.41	m2	**44.66**
wall coverings	1.05	12.52	34.41	m2	**46.93**
Coverings; 610 x 305 mm slates					
roof coverings	0.69	8.23	38.70	m2	**46.93**
wall coverings	0.86	10.25	38.70	m2	**48.95**
Extra over coverings for					
double course at eaves	0.35	4.17	7.01	m	**11.18**
verges; extra single undercloak course	0.48	5.72	5.75	m	**11.47**
open valleys; cutting both sides	0.25	2.98	15.36	m	**18.34**
blue/black glazed ware 152 mm half round ridge tiles	0.58	6.91	9.96	m	**16.87**
blue/black glazed ware 125 x 125 mm plain angle ridge tiles	0.58	6.91	13.30	m	**20.21**
mitred hips; cutting both sides	0.25	2.98	15.36	m	**18.34**

H CLADDING/COVERING Including overheads and profit at 7.50%	Labour hours	Labour £	Material £	Unit	Total rate £
H62 NATURAL SLATING - cont'd					
Natural slates; BS 680 Part 2; Welsh blue; uniform size; to 75 mm lap; on 25 x 50 mm battens and type 1F reinforced underlay - cont'd					
Extra over coverings for					
blue/black glazed ware 152 mm half round hip tiles; cutting both sides	0.81	9.66	25.33	m	34.99
blue/black glazed ware 125 x 125 mm plain angle hip tiles; cutting both sides	0.81	9.66	28.67	m	38.33
Holes for pipes and the like	0.23	2.74	-	nr	2.74
Natural slates; Westmorland green; PC £1430.00/t; random lengths; 457 - 229 mm proportionate widths to 75 mm lap; in diminishing courses; on 25 x 50 mm battens and type 1F underlay					
Roof coverings	1.30	15.50	101.98	m2	117.48
Wall coverings	1.65	19.67	101.98	m2	121.65
Extra over coverings for					
double course at eaves	0.76	9.06	17.54	m	26.60
verges; extra single undercloak course slates 152 mm wide	0.86	10.25	13.39	m	23.64
Holes for pipes and the like	0.35	4.17	-	nr	4.17
H63 RECONSTRUCTED STONE SLATING/TILING					
Reconstructed stone slates; 'Hardrow Slates' standard colours; or similar; PC £579.70/1000 75 mm lap; on 25 x 50 mm battens and type 1F reinforced underlay					
Coverings; 457 x 305 mm slates					
roof coverings	0.92	10.97	14.40	m2	25.37
wall coverings	1.15	13.71	14.40	m2	28.11
Coverings; 457 x 457 mm slates					
roof coverings	0.75	8.94	14.39	m2	23.33
wall coverings	0.98	11.68	14.39	m2	26.07
Extra over coverings for					
double course at eaves	0.35	4.17	2.62	m	6.79
verges; extra single undercloak course	0.48	5.72	2.69	m	8.41
open valleys; cutting both sides	0.25	2.98	5.24	m	8.22
ridge tiles	0.58	6.91	10.12	m	17.03
hip tiles	0.81	9.66	13.78	m	23.44
Holes for pipes and the like	0.23	2.74	-	nr	2.74
Reconstructed stone slates; 'Hardrow Slates' green/'oldstone' colours or similar; PC £626.87/1000 75 mm lap; on 25 x 50 mm battens and type 1F reinforced underlay					
Coverings; 457 x 305 mm slates					
roof coverings	0.92	10.97	15.25	m2	26.22
wall coverings	1.15	13.71	15.25	m2	28.96
Coverings; 457 x 457 mm slates					
roof coverings	0.75	8.94	15.25	m2	24.19
wall coverings	0.98	11.68	15.25	m2	26.93
Extra over coverings for					
double course at eaves	0.35	4.17	2.82	m	6.99
verges; extra single undercloak course	0.48	5.72	2.89	m	8.61
open valleys; cutting both sides	0.25	2.98	5.66	m	8.64
ridge tiles	0.58	6.91	10.89	m	17.80
hip tiles	0.81	9.66	14.40	m	24.06
Holes for pipes and the like	0.23	2.74	-	nr	2.74

H CLADDING/COVERING Including overheads and profit at 7.50%	Labour hours	Labour £	Material £	Unit	Total rate £
Reconstructed stone slates; Bradstone 'Cotswold' style or similar; PC £19.36/m2; random lengths 550 - 300 mm; proportional widths; to 80 mm lap; in diminishing courses; on 25 x 50 mm battens and type 1F reinforced underlay					
Roof coverings	1.20	14.31	27.29	m2	**41.60**
Wall coverings	1.55	18.48	28.89	m2	**47.37**
Extra over coverings for					
double course at eaves	0.58	6.91	4.49	m	**11.40**
verges; extra single undercloak course	0.76	9.06	3.64	m	**12.70**
open valleys; cutting both sides	0.52	6.20	9.37	m	**15.57**
ridge tile	0.76	9.06	10.94	m	**20.00**
mitred hips; cutting both sides	0.52	6.20	9.37	m	**15.57**
hip tile; cutting both sides	1.20	14.31	20.06	m	**34.37**
Holes for pipes and the like	0.35	4.17	-	nr	**4.17**
Reconstructed stone slates; Bradstone 'Moordale' style or similar; PC £19.42/m2; random lengths 550 - 450 mm; proportional widths; to 80 mm lap; in diminishing courses; on 25 x 50 mm battens and type 1F reinforced underlay					
Roof coverings	1.10	13.11	26.88	m2	**39.99**
Wall coverings	1.45	17.29	28.73	m2	**46.02**
Extra over coverings for					
double course at eaves	0.58	6.91	4.50	m	**11.41**
verges; extra single undercloak course	0.76	9.06	3.65	m	**12.71**
ridge tile	0.76	9.06	10.94	m	**20.00**
mitred hips; cutting both sides	0.52	6.20	9.39	m	**15.59**
Holes for pipes and the like	0.35	4.17	-	nr	**4.17**
H64 TIMBER SHINGLING					
Red cedar sawn shingles preservative treated; PC £34.29 per bundle (2.11 m2 cover); uniform length 450 mm; varying widths; to 125 mm lap; on 25 x 100 mm battens and type 1F reinforced underlay					
Roof coverings	1.20	14.31	25.80	m2	**40.11**
Wall coverings	1.55	18.48	25.80	m2	**44.28**
Extra over coverings for					
double course at eaves; three rows of battens	0.35	4.17	5.08	m	**9.25**
verges; extra single undercloak course	0.58	6.91	5.30	m	**12.21**
open valleys; cutting both sides	0.23	2.74	4.05	m	**6.79**
selected shingles to form hip capping	1.40	16.69	10.25	m	**26.94**
Double starter course on last	0.23	2.74	2.22	nr	**4.96**
Holes for pipes and the like	0.17	2.03	-	nr	**2.03**
H71 LEAD SHEET COVERINGS/FLASHINGS					
Milled lead; BS 1178; PC £1352.70/t					
1.25 mm (code 3) roof coverings					
flat	3.10	27.93	21.61	m2	**49.54**
sloping 10 - 50 degrees	3.45	31.08	21.61	m2	**52.69**
vertical or sloping over 50 degrees	3.80	34.23	21.61	m2	**55.84**
1.80 mm (code 4) roof coverings					
flat	3.35	30.18	31.10	m2	**61.28**
sloping 10 - 50 degrees	3.70	33.33	31.10	m2	**64.43**
vertical or sloping over 50 degrees	4.05	36.48	31.10	m2	**67.58**
1.80 mm (code 4) dormer coverings					
flat	3.90	35.13	31.85	m2	**66.98**
sloping 10 - 50 degrees	4.50	40.54	31.85	m2	**72.39**
vertical or sloping over 50 degrees	4.85	43.69	31.85	m2	**75.54**

H CLADDING/COVERING Including overheads and profit at 7.50%	Labour hours	Labour £	Material £	Unit	Total rate £
H71 LEAD SHEET COVERINGS/FLASHINGS - cont'd					
Milled lead; BS 1178; PC £1352.70/t - cont'd					
2.24 mm (code 5) roof coverings					
flat	3.55	31.98	38.70	m2	**70.68**
sloping 10 - 50 degrees	3.90	35.13	38.70	m2	**73.83**
vertical or sloping over 50 degrees	4.25	38.29	38.70	m2	**76.99**
2.24 mm (code 5) dormer coverings					
flat	4.25	38.29	39.62	m2	**77.91**
sloping 10 - 50 degrees	4.70	42.34	39.62	m2	**81.96**
vertical or sloping over 50 degrees	5.20	46.84	39.62	m2	**86.46**
2.50 mm (code 6) roof coverings					
flat	3.80	34.23	43.21	m2	**77.44**
sloping 10 - 50 degrees	4.15	37.39	43.21	m2	**80.60**
vertical or sloping over 50 degrees	4.50	40.54	43.21	m2	**83.75**
2.50 mm (code 6) dormer coverings					
flat	4.60	41.44	44.24	m2	**85.68**
sloping 10 - 50 degrees	4.95	44.59	44.24	m2	**88.83**
vertical or sloping over 50 degrees	5.40	48.65	44.24	m2	**92.89**
Dressing over glazing bars and glass	0.38	3.42	-	m	**3.42**
Soldered dot	1.45	13.06	2.41	nr	**15.47**
Copper nailing 75 mm spacing	0.23	2.07	0.33	m	**2.40**
1.80 mm (code 4) lead flashings, etc.					
Flashings; wedging into grooves					
150 mm girth	0.92	8.29	4.95	m	**13.24**
240 mm girth	1.05	9.46	7.96	m	**17.42**
Stepped flashings; wedging into grooves					
180 mm girth	1.05	9.46	5.93	m	**15.39**
270 mm girth	1.15	10.36	8.89	m	**19.25**
Linings to sloping gutters					
390 mm girth	1.40	12.61	12.90	m	**25.51**
450 mm girth	1.50	13.51	14.82	m	**28.33**
750 mm girth	1.85	16.67	23.43	m	**40.10**
Cappings to hips or ridges					
450 mm girth	1.70	15.31	14.82	m	**30.13**
600 mm girth	1.85	16.67	19.81	m	**36.48**
Soakers					
200 x 200 mm	0.17	1.53	1.42	nr	**2.95**
300 x 300 mm	0.23	2.07	3.18	nr	**5.25**
Saddle flashings; at intersections of hips and ridges; dressing and bossing					
450 x 600 mm	2.05	18.47	9.51	nr	**27.98**
Slates; with 150 mm high collar					
450 x 450 mm; to suit 50 mm pipe	1.95	17.57	13.03	nr	**30.60**
450 x 450 mm; to suit 100 mm pipe	2.30	20.72	18.37	nr	**39.09**
2.24 mm (code 5) lead flashings, etc.					
Flashings; wedging into grooves					
150 mm girth	0.92	8.29	6.16	m	**14.45**
240 mm girth	1.05	9.46	9.83	m	**19.29**
Stepped flashings; wedging into grooves					
180 mm girth	1.05	9.46	7.41	m	**16.87**
270 mm girth	1.15	10.36	11.08	m	**21.44**
Linings to sloping gutters					
390 mm girth	1.40	12.61	16.04	m	**28.65**
450 mm girth	1.50	13.51	18.46	m	**31.97**
750 mm girth	1.85	16.67	29.30	m	**45.97**
Cappings to hips or ridges					
450 mm girth	1.70	15.31	18.46	m	**33.77**
600 mm girth	1.85	16.67	24.66	m	**41.33**

H CLADDING/COVERING Including overheads and profit at 7.50%	Labour hours	Labour £	Material £	Unit	Total rate £
Soakers					
200 x 200 mm	0.17	1.53	1.76	nr	**3.29**
300 x 300 mm	0.23	2.07	3.89	nr	**5.96**
Saddle flashings; at intersections of hips and ridges; dressing and bossing					
450 x 600 mm	2.05	18.47	13.21	nr	**31.68**
Slates; with 150 mm high collar					
450 x 450 mm; to suit 50 mm pipe	1.95	17.57	14.78	nr	**32.35**
450 x 450 mm; to suit 100 mm pipe	2.30	20.72	20.11	nr	**40.83**
H72 ALUMINIUM SHEET COVERINGS/FLASHINGS					
Aluminium roofing; commercial grade; PC £2312.50/t					
0.90 mm roof coverings					
flat	3.45	31.08	6.12	m2	**37.20**
sloping 10 - 50 degrees	3.80	34.23	6.12	m2	**40.35**
vertical or sloping over 50 degrees	4.15	37.39	6.12	m2	**43.51**
0.90 mm dormer coverings					
flat	4.15	37.39	6.27	m2	**43.66**
sloping 10 - 50 degrees	4.60	41.44	6.27	m2	**47.71**
vertical or sloping over 50 degrees	5.05	45.49	6.27	m2	**51.76**
Aluminium nailing; 75 mm spacing	0.23	2.07	0.24	m	**2.31**
0.90 mm aluminium; commercial grade flashings, etc.					
Flashings; wedging into grooves					
150 mm girth	0.92	8.29	0.96	m	**9.25**
240 mm girth	1.05	9.46	1.54	m	**11.00**
300 mm girth	1.20	10.81	1.91	m	**12.72**
Stepped flashings; wedging into grooves					
180 mm girth	1.05	9.46	1.15	m	**10.61**
270 mm girth	1.15	10.36	1.76	m	**12.12**
H73 COPPER SHEET COVERINGS/FLASHINGS					
Copper roofing; BS 2870					
0.56 mm (24 swg) roof coverings					
flat PC £4000.00/t	3.70	33.33	21.98	m2	**55.31**
sloping 10 - 50 degrees	4.05	36.48	21.98	m2	**58.46**
vertical or sloping over 50 degrees	4.35	39.19	21.98	m2	**61.17**
0.56 mm (24 swg) dormer coverings					
flat	4.35	39.19	22.51	m2	**61.70**
sloping 10 - 50 degrees	4.85	43.69	22.51	m2	**66.20**
vertical or sloping over 50 degrees	5.30	47.75	22.51	m2	**70.26**
0.61 mm (23 swg) roof coverings					
flat PC £3875.00/t	3.70	33.33	25.36	m2	**58.69**
sloping 10 - 50 degrees	4.05	36.48	25.36	m2	**61.84**
vertical or sloping over 50 degrees	4.35	39.19	25.36	m2	**64.55**
0.61 mm (23 swg) dormer coverings					
flat	4.35	39.19	25.98	m2	**65.17**
sloping 10 - 50 degrees	4.85	43.69	25.98	m2	**69.67**
vertical or sloping over 50 degrees	5.30	47.75	25.98	m2	**73.73**
Copper nailing; 75 mm spacing	0.23	2.07	0.33	m	**2.40**
0.56 mm copper flashings, etc.					
Flashings; wedging into grooves					
150 mm girth	0.92	8.29	3.45	m	**11.74**
240 mm girth	1.05	9.46	5.54	m	**15.00**
300 mm girth	1.20	10.81	6.90	m	**17.71**
Stepped flashings; wedging into grooves					
180 mm girth	1.05	9.46	4.16	m	**13.62**
270 mm girth	1.15	10.36	6.25	m	**16.61**

H CLADDING/COVERING Including overheads and profit at 7.50%	Labour hours	Labour £	Material £	Unit	Total rate £
H73 COPPER SHEET COVERINGS/FLASHINGS - cont'd					
0.61 mm copper flashings, etc.					
Flashings; wedging into grooves					
150 mm girth	0.92	8.29	3.99	m	**12.28**
240 mm girth	1.05	9.46	6.41	m	**15.87**
300 mm girth	1.20	10.81	7.97	m	**18.78**
Stepped flashings; wedging into grooves					
180 mm girth	1.05	9.46	4.81	m	**14.27**
270 mm girth	1.15	10.36	7.20	m	**17.56**
H74 ZINC SHEET COVERINGS/FLASHINGS					
Zinc BS 849; PC £2500.00/t					
0.81 mm roof coverings					
flat	3.70	33.33	16.06	m2	**49.39**
sloping 10 - 50 degrees	4.05	36.48	16.06	m2	**52.54**
vertical or sloping over 50 degrees	4.35	39.19	16.06	m2	**55.25**
0.81 mm dormer coverings					
flat	4.35	39.19	16.44	m2	**55.63**
sloping 10 - 50 degrees	4.85	43.69	16.44	m2	**60.13**
vertical or sloping over 50 degrees	5.30	47.75	16.44	m2	**64.19**
0.81 mm zinc flashings, etc.					
Flashings; wedging into grooves					
150 mm girth	0.92	8.29	2.51	m	**10.80**
240 mm girth	1.05	9.46	4.05	m	**13.51**
300 mm girth	1.20	10.81	5.06	m	**15.87**
Stepped flashings; wedging into grooves					
180 mm girth	1.05	9.46	3.05	m	**12.51**
270 mm girth	1.15	10.36	4.55	m	**14.91**
H75 STAINLESS STEEL SHEET COVERINGS/ FLASHINGS					
Terne coated stainless steel roofing					
0.38 mm roof coverings					
flat	3.70	33.33	33.86	m2	**67.19**
sloping 10 - 50 degrees	4.05	36.48	33.86	m2	**70.34**
vertical or sloping over 50 degrees	4.35	39.19	33.86	m2	**73.05**
Flashings; wedging into grooves					
150 mm girth	1.15	10.36	7.10	m	**17.46**
240 mm girth	1.30	11.71	10.29	m	**22.00**
300 mm girth	1.50	13.51	12.42	m	**25.93**
Stepped flashings; wedging into grooves					
180 mm girth	1.30	11.71	8.16	m	**19.87**
270 mm girth	1.45	13.06	11.35	m	**24.41**

H CLADDING/COVERING Including overheads and profit at 7.50%	Labour hours	Labour £	Material £	Unit	Total rate £
H76 FIBRE BITUMEN THERMOPLASTIC SHEET COVERINGS/FLASHINGS					
Glass fibre reinforced bitumen strip slates; 'Langhome 1000' or similar; PC £19.11/2 m2; strip pack; 900 x 300 mm mineral finish; fixed to external plywood boarding (measured separately)					
Roof coverings	0.29	3.46	9.10	m2	12.56
Wall coverings	0.46	5.48	9.10	m2	14.58
Extra over coverings for					
double course at eaves; felt soaker	0.23	2.74	1.83	m	4.57
verges; felt soaker	0.29	3.46	1.73	m	5.19
valley slate; cut to shape; felt soaker both sides; cutting both sides	0.52	6.20	4.38	m	10.58
ridge slate; cut to shape	0.35	4.17	2.26	m	6.43
hip slate; cut to shape; cutting both sides	0.52	6.20	4.34	m	10.54
Holes for pipes and the like	0.06	0.72	-	nr	0.72
'Evode Flashband' sealing strips and flashings; special grey finish					
Flashings; wedging at top if required; pressure bonded; flashband primer before application; to walls					
100 mm girth	0.29	2.61	1.14	m	3.75
150 mm girth	0.38	3.42	1.70	m	5.12
225 mm girth	0.46	4.14	2.57	m	6.71
300 mm girth	0.52	4.68	3.35	m	8.03
450 mm girth	0.69	6.22	6.49	m	12.71
'Nuralite' semi-rigid bitumen membrane roofing DC12 jointing strip system PC £27.70/3m2					
Roof coverings					
flat	0.58	5.22	14.47	m2	19.69
sloping 10 - 50 degrees	0.63	5.68	14.47	m2	20.15
vertical or sloping over 50 degrees	0.81	7.30	14.47	m2	21.77
'Nuralite' flashings, etc.					
Flashings; wedging into grooves					
150 mm girth	0.58	5.22	1.83	m	7.05
200 mm girth	0.69	6.22	2.53	m	8.75
250 mm girth	0.69	6.22	2.88	m	9.10
Linings to sloping gutters					
450 mm girth	0.98	8.83	6.25	m	15.08
490 mm girth	1.10	9.91	6.89	m	16.80
Cappings to hips or ridges					
450 mm girth	1.05	9.46	9.24	m	18.70
600 mm girth	1.15	10.36	13.88	m	24.24
Undersoakers					
150 mm girth	0.14	1.26	0.94	nr	2.20
250 mm girth	0.21	1.89	0.99	nr	2.88
Oversoakers					
200 mm girth	0.17	1.53	9.55	nr	11.08
Cavity tray without apron					
450 mm long	0.52	4.68	1.77	nr	6.45
250 mm long	0.29	2.61	1.37	nr	3.98
350 mm long	0.40	3.60	1.66	nr	5.26

J WATERPROOFING Including overheads and profit at 7.50%	Labour hours	Labour £	Material £	Unit	Total rate £
J10 SPECIALIST WATERPROOF RENDERING					
'Sika' waterproof rendering; steel trowelled					
20 mm work to walls; three coat; to concrete base					
over 300 mm wide	-	-	-	m2	31.28
not exceeding 300 mm wide	-	-	-	m2	49.01
25 mm work to walls; three coat; to concrete base					
over 300 mm wide	-	-	-	m2	35.45
not exceeding 300 mm wide	-	-	-	m2	56.31
40 mm work to walls; four coat; to concrete base					
over 300 mm wide	-	-	-	m2	54.22
not exceeding 300 mm wide	-	-	-	m2	83.42
J20 MASTIC ASPHALT TANKING/DAMP PROOF MEMBRANES					
Mastic asphalt to BS 1097					
13 mm one coat coverings to concrete base; flat; subsequently covered					
over 300 mm wide	-	-	-	m2	8.41
225 - 300 mm wide	-	-	-	m2	15.96
150 - 225 mm wide	-	-	-	m2	18.21
not exceeding 150 mm wide	-	-	-	m2	19.08
20 mm two coat coverings to concrete base; flat; subsequently covered					
over 300 mm wide	-	-	-	m2	11.40
225 - 300 mm wide	-	-	-	m2	24.55
150 - 300 mm wide	-	-	-	m2	28.02
not exceeding 150 mm wide	-	-	-	m2	29.35
30 mm three coat coverings to concrete base; flat; subsequently covered					
over 300 mm wide	-	-	-	m2	16.29
225 - 300 mm wide	-	-	-	m2	36.83
150 - 225 mm wide	-	-	-	m2	42.06
not exceeding 150 mm wide	-	-	-	m2	44.02
13 mm two coat coverings to brickwork base; vertical; subsequently covered					
over 300 mm wide	-	-	-	m2	35.84
225 - 300 mm wide	-	-	-	m2	42.87
150 - 225 mm wide	-	-	-	m2	50.89
not exceeding 150 mm wide	-	-	-	m2	66.92
20 mm three coat coverings to brickwork base; vertical; subsequently covered					
over 300 mm wide	-	-	-	m2	58.60
225 - 300 mm wide	-	-	-	m2	59.97
150 - 225 mm wide	-	-	-	m2	70.53
not exceeding 150 mm wide	-	-	-	m2	91.74
Turning 20 mm into groove	-	-	-	m	1.01
Internal angle fillets; subsequently covered	-	-	-	m	4.53
Mastic asphalt to BS 6577					
13 mm one coat coverings to concrete base; flat; subsequently covered					
over 300 mm wide	-	-	-	m2	11.46
225 - 300 mm wide	-	-	-	m2	18.50
150 - 225 mm wide	-	-	-	m2	21.18
not exceeding 150 mm wide	-	-	-	m2	22.32

J WATERPROOFING Including overheads and profit at 7.50%	Labour hours	Labour £	Material £	Unit	Total rate £
20 mm two coat coverings to concrete base; flat; subsequently covered					
over 300 mm wide	-	-	-	m2	16.58
225 - 300 mm wide	-	-	-	m2	28.46
150 - 225 mm wide	-	-	-	m2	32.59
not exceeding 150 mm wide	-	-	-	m2	34.33
30 mm three coat coverings to concrete base; flat; subsequently covered					
over 300 mm wide	-	-	-	m2	24.08
225 - 300 mm wide	-	-	-	m2	42.70
150 - 225 mm wide	-	-	-	m2	48.91
not exceeding 150 mm wide	-	-	-	m2	51.55
13 mm two coat coverings to brickwork base; vertical; subsequently covered					
over 300 mm wide	-	-	-	m2	39.21
225 - 300 mm wide	-	-	-	m2	47.70
150 - 225 mm wide	-	-	-	m2	56.44
not exceeding 150 mm wide	-	-	-	m2	73.91
20 mm three coat coverings to brickwork base; vertical; subsequently covered					
over 300 mm wide	-	-	-	m2	54.13
225 - 300 mm wide	-	-	-	m2	67.01
150 - 225 mm wide	-	-	-	m2	78.60
not exceeding 150 mm wide	-	-	-	m2	101.76
Turning 20 mm into groove	-	-	-	m	1.01
Internal angle fillets; subsequently covered	-	-	-	m	5.08
J21 MASTIC ASPHALT ROOFING/INSULATION/ FINISHES					
Mastic asphalt to BS 988					
20 mm two coat coverings; felt isolating membrane; to concrete (or timber) base; flat or to falls or slopes not exceeding 10 degrees from horizontal					
over 300 mm wide	-	-	-	m2	11.80
225 - 300 mm wide	-	-	-	m2	23.38
150 - 225 mm wide	-	-	-	m2	24.90
not exceeding 150 mm wide	-	-	-	m2	27.95
Add to the above for covering with:					
10 mm limestone chippings in hot bitumen	-	-	-	m2	2.39
coverings with solar reflective paint	-	-	-	m2	2.47
300 x 300 x 8 mm g.r.p. tiles in hot bitumen	-	-	-	m2	34.00
Cutting to line; jointing to old asphalt	-	-	-	m	8.20
13 mm two coat skirtings to brickwork base					
not exceeding 150 mm girth	-	-	-	m	10.04
150 - 225 mm girth	-	-	-	m	11.45
225 - 300 mm girth	-	-	-	m	12.86
13 mm three coat skirtings; expanded metal lathing reinforcement nailed to timber base					
not exceeding 150 mm girth	-	-	-	m	15.74
150 - 225 mm girth	-	-	-	m	18.68
225 - 300 mm girth	-	-	-	m	21.62
13 mm two coat fascias to concrete base					
not exceeding 150 mm girth	-	-	-	m	10.04
150 - 225 mm girth	-	-	-	m	11.45
20 mm two coat linings to channels to concrete base					
not exceeding 150 mm girth	-	-	-	m	24.27
150 - 225 mm girth	-	-	-	m	24.27
225 - 300 mm girth	-	-	-	m	24.27

J WATERPROOFING Including overheads and profit at 7.50%		Labour hours	Labour £	Material £	Unit	Total rate £
J21 MASTIC ASPHALT ROOFING/INSULATION/ FINISHES - cont'd						
Mastic asphalt to BS 988 - cont'd						
20 mm two coat lining to cesspools						
250 x 150 x 150 mm deep		-	-	-	nr	28.67
Collars around pipes, standards and like members		-	-	-	nr	13.87
Mastic asphalt to BS 6577						
20 mm two coat coverings; felt isolating membrane; to concrete (or timber) base; flat or to falls or slopes not exceeding 10 degrees from horizontal						
over 300 mm wide		-	-	-	m2	16.42
225 - 300 mm wide		-	-	-	m2	27.10
150 - 225 mm wide		-	-	-	m2	28.97
not exceeding 150 mm wide		-	-	-	m2	32.69
Add to the above for covering with:						
10 mm limestone chippings in hot bitumen		-	-	-	m2	2.39
solar reflective paint		-	-	-	m2	2.47
300 x 300 x 8 mm g.r.p. tiles in hot bitumen		-	-	-	m2	34.00
Cutting to line; jointing to old asphalt		-	-	-	m	8.20
13 mm two coat skirtings to brickwork base						
not exceeding 150 mm girth		-	-	-	m	11.09
150 - 225 mm girth		-	-	-	m	12.70
225 - 300 mm girth		-	-	-	m	14.31
13 mm three coat skirtings; expanded metal lathing reinforcement nailed to timber base						
not exceeding 150 mm girth		-	-	-	m	17.03
150 - 225 mm girth		-	-	-	m	20.33
225 - 300 mm girth		-	-	-	m	23.64
13 mm two coat fascias to concrete base						
not exceeding 150 mm girth		-	-	-	m	11.09
150 - 225 mm girth		-	-	-	m	12.70
20 mm two coat linings to channels to concrete base						
not exceeding 150 mm girth		-	-	-	m	27.08
150 - 225 mm girth		-	-	-	m	27.08
225 - 300 mm girth		-	-	-	m	27.08
20 mm two coat lining to cesspools						
250 x 150 x 150 mm deep		-	-	-	nr	29.57
Collars around pipes, standards and like members		-	-	-	nr	13.87
Accessories						
Eaves trim; extruded aluminium alloy; working asphalt into trim						
'Alutrim'; type A roof edging	PC £9.09/2.5m	0.44	5.25	4.17	m	9.42
Angle	PC £3.14	0.13	1.55	3.54	nr	5.09
Roof screed ventilator - aluminium alloy						
Extr-aqua-vent; set on screed over and including dished sinking; working collar around ventilator	PC £3.71	1.20	14.31	4.19	nr	18.50
J30 LIQUID APPLIED TANKING/DAMP PROOF MEMBRANES						
'Synthaprufe'; blinding with sand; horizontal on slabs						
two coats		0.22	1.37	2.10	m2	3.47
three coats		0.31	1.93	3.07	m2	5.00

J WATERPROOFING Including overheads and profit at 7.50%	Labour hours	Labour £	Material £	Unit	Total rate £
'Tretolastex 202T'; on vertical surfaces of concrete					
two coats	0.22	1.37	2.06	m2	3.43
three coats	0.31	1.93	3.09	m2	5.02
One coat Vandex 'Super' 0.75/m2 slurry; one consolidating coat of Vandex 'Premix' 1kg/m2 slurry; horizontal on beds					
over 225 mm wide	-	-	-	m2	5.73
'Ventrot' hot applied damp proof membrane; one coat; horizontal on slabs					
over 225 mm wide	-	-	-	m2	4.05
J40 FLEXIBLE SHEET TANKING/DAMP PROOF MEMBRANES					
'Bituthene' sheeting; lapped joints; horizontal on slabs					
standard 500 grade	0.11	0.69	5.15	m2	5.84
1000 grade	0.12	0.75	5.69	m2	6.44
1200 grade	0.13	0.81	8.60	m2	9.41
heavy duty grade	0.14	0.87	6.34	m2	7.21
'Bituthene' sheeting; lapped joints; dressed up vertical face of concrete					
1000 grade	0.20	1.25	6.20	m2	7.45
'Servi-pak' protection board; butt jointed; to horizontal surfaces; including 'Bituthene' standard sheeting; fixed with adhesive dabs					
3 mm thick	0.44	2.74	8.87	m2	11.61
6 mm thick	0.50	3.12	10.60	m2	13.72
12 mm thick	0.55	3.43	14.57	m2	18.00
'Servi-pak' protection board; butt jointed; to vertical surfaces; including 'Bituthene' standard sheeting; fixed with adhesive dabs					
3 mm thick	0.55	3.43	8.87	m2	12.30
6 mm thick	0.61	3.80	10.60	m2	14.40
12 mm thick	0.66	4.12	14.57	m2	18.69
'Bituthene' fillet					
40 x 40 mm	0.11	0.69	4.25	m	4.94
'Bituthene' reinforcing strip; 300 mm wide					
1000 grade	0.11	0.69	1.92	m	2.61
Expandite 'Famflex' waterproof tanking; 150 mm laps					
horizontal; over 300 mm wide	0.44	2.74	6.71	m2	9.45
vertical; over 300 mm wide	0.72	4.49	6.71	m2	11.20
J41 BUILT UP FELT ROOF COVERINGS					
NOTE: The following items of felt roofing, unless otherwise described, include for conventional lapping, laying and bonding between layers and to base; and laying flat or to falls to cross-falls or to slopes not exceeding 10 degrees - but exclude any insulation etc. (measured separately)					
Felt roofing; BS 747; suitable for flat roofs					
Three layer coverings type 1B (two 18 kg/10 m2 and one 25 kg/10 m2) bitumen fibre based felts	-	-	-	m2	12.83
Extra over top layer type 1B for mineral surfaced layer type 1E	-	-	-	m2	1.69

J WATERPROOFING Including overheads and profit at 7.50%	Labour hours	Labour £	Material £	Unit	Total rate £
J41 BUILT UP FELT ROOF COVERINGS - cont'd					
Felt roofing; BS 747; suitable for flat roofs - cont'd					
Three layer coverings type 2B bitumen asbestos based felts	-	-	-	m2	16.92
Extra over top layer type 2B for mineral surfaced layer type 2E	-	-	-	m2	2.10
Three layer coverings first layer type 3G; subsequent layers type 3B bitumen glass fibre based felt	-	-	-	m2	13.53
Extra over top layer type 3B for mineral surfaced layer type 3E	-	-	-	m2	1.90
Extra over felt for covering with and bedding in hot bitumen					
13 mm granite chippings	-	-	-	m2	5.34
300 x 300 x 8 mm grp tiles	-	-	-	m2	34.00
Working into outlet pipes and the like	-	-	-	nr	5.34
Skirtings; three layer; top layer mineral surfaced; dressed over tilting fillet; turned into groove					
not exceeding 200 mm girth	-	-	-	m	8.49
200 - 400 mm girth	-	-	-	m	10.84
Coverings to kerbs; three layer					
400 - 600 mm girth	-	-	-	m	18.09
Linings to gutters; three layer					
400 - 600 mm (average) girth	-	-	-	m	23.53
Collars around pipes and the like; three layer mineral surfaced; 150 mm high					
not exceeding 55 mm nominal size	-	-	-	nr	6.29
55 - 110 mm nominal size	-	-	-	nr	7.69
Felt roofing; BS 747; suitable for pitched timber roofs; sloping not exceeding 50 degrees					
Two layer coverings; first layer type 2B; second layer 2E; mineral surfaced asbestos based felts	-	-	-	m2	17.95
Three layer coverings; first two layers type 2B; top layer type 2E; mineral surfaced asbestos based felts	-	-	-	m2	25.28
'Andersons' high performance polyester-based roofing system					
Two layer coverings; first layer HT 125 underlay; second layer HT 350; fully bonded to wood; fibre or cork base	0.35	4.17	10.73	m2	14.90
Extra over for					
Top layer mineral surfaced	-	-	-	m2	1.72
13 mm granite chippings	-	-	-	m2	5.34
Third layer of type 3B as underlay for concrete or screeded base	0.17	2.03	1.81	m2	3.84
Working into outlet pipes and the like	0.58	6.91	-	nr	6.91
Skirtings; two layer; top layer mineral surfaced; dressed over tilting fillet; turned into groove					
not exceeding 200 mm girth	0.17	2.03	2.86	m	4.89
200 - 400 mm girth	0.23	2.74	5.43	m	8.17
Coverings to kerbs; two layer					
400 - 600 mm girth	0.29	3.46	7.43	m	10.89
Linings to gutters; three layer					
400 - 600 mm (average) girth	0.69	8.23	9.22	m	17.45

J WATERPROOFING Including overheads and profit at 7.50%	Labour hours	Labour £	Material £	Unit	Total rate £
Collars around pipes and the like; two layer; 150 mm high					
not exceeding 55 mm nominal size	0.38	4.53	0.45	nr	4.98
55 - 110 mm nominal size	0.46	5.48	0.74	nr	6.22
'Ruberglas 120 GP' high performance roofing					
Two layer coverings; first and second layers 'Ruberglas 120 GP'; fully bonded to wood; fibre or cork base	-	-	-	m2	15.35
Extra over for					
Top layer mineral surfaced	-	-	-	m2	1.61
13 mm granite chippings	-	-	-	m2	5.34
Third layer of 'Rubervent 3G' as underlay for concrete or screeded base	-	-	-	m2	6.51
Working into outlet pipes and the like	-	-	-	nr	10.20
Skirtings; two layer; top layer mineral surfaced; dressed over tilting fillet; turned into groove					
not exceeding 200 mm girth	-	-	-	m	8.32
200 - 400 mm girth	-	-	-	m	10.86
Coverings to kerbs; two layer					
400 - 600 mm girth	-	-	-	m	18.10
Linings to gutters; three layer					
400 - 600 mm (average) girth	-	-	-	m	26.69
Collars around pipes and the like; two layer; 150 mm high					
not exceeding 55 mm nominal size	-	-	-	nr	9.06
55 - 110 mm nominal size	-	-	-	nr	10.99
'Ruberfort HP 350' high performance roofing					
Two layer coverings; first layer 'Ruberfort HP 180'; second layer 'Ruberfort HP 350'; fully bonded; to wood; fibre or cork base	-	-	-	m2	21.62
Extra over for					
Top layer mineral surfaced	-	-	-	m2	2.77
13 mm granite chippings	-	-	-	m2	5.34
Third layer of 'Rubervent 3G' as underlay for concrete or screeded base	-	-	-	m2	6.51
Working into outlet pipes and the like	-	-	-	nr	10.20
Skirtings; two layer; top layer mineral surfaced; dressed over tilting fillet; turned into groove					
not exceeding 200 mm girth	-	-	-	m	9.88
200 - 400 mm girth	-	-	-	m	14.17
Coverings to kerbs; two layer					
400 - 600 mm girth	-	-	-	m	22.84
Linings to gutters; three layer					
400 - 600 mm (average) girth	-	-	-	m	34.77
Collars around pipes and the like; two layer; 150 mm high					
not exceeding 55 mm nominal size	-	-	-	nr	9.06
55 - 110 mm nominal size	-	-	-	nr	10.99
'Polybit 350' elastomeric roofing					
Two layer coverings; first layer 'Polybit 180'; second layer 'Polybit 350'; fully bonded to wood; fibre or cork base	-	-	-	m2	23.54
Extra over for					
Top layer mineral surfaced	-	-	-	m2	2.86
13 mm granite chippings	-	-	-	m2	5.34
Third layer of 'Rubervent 3G' as underlay for concrete or screeded base	-	-	-	m2	6.51
Working into outlet pipes and the like	-	-	-	nr	10.15

J WATERPROOFING Including overheads and profit at 7.50%	Labour hours	Labour £	Material £	Unit	Total rate £
J41 BUILT UP FELT ROOF COVERINGS - cont'd					
'Polybit 350' elastomeric roofing - cont'd					
Skirtings; two layer; top layer mineral surfaced; dressed over tilting fillet; turned into groove					
not exceeding 200 mm girth	-	-	-	m	10.35
200 - 400 mm girth	-	-	-	m	15.03
Coverings to kerbs; two layer					
400 - 600 mm girth	-	-	-	m	24.34
Linings to gutters; three layer					
400 - 600 mm (average) girth	-	-	-	m	37.27
Collars around pipes and the like; two layer; 150 mm high					
not exceeding 55 mm nominal size	-	-	-	nr	10.99
55 - 110 mm nominal size	-	-	-	nr	14.88
'Hyload 150 E' elastomeric roofing					
Two layer coverings; first layer 'Ruberglas 120 GP'; second layer 'Hyload 150 E' fully bonded to wood; fibre or cork base	-	-	-	m2	22.72
Extra over for					
13 mm granite chippings	-	-	-	m2	5.34
Third layer of 'Rubervent 3G' as underlay for concrete or screeded base	-	-	-	m2	6.51
Three layer coverings; finished with 13 mm granite chippings					
Working into outlet pipes and the like	-	-	-	nr	10.15
Skirtings; two layer; dressed over tilting fillet; turned into groove					
not exceeding 200 mm girth	-	-	-	m	9.80
200 - 400 mm girth	-	-	-	m	14.01
Coverings to kerbs; two layer					
400 - 600 mm girth	-	-	-	m	23.10
Linings to gutters; three layer					
400 - 600 mm (average) girth	-	-	-	m	36.44
Collars around pipes and the like; two layer; 150 mm high					
not exceeding 55 mm nominal size	-	-	-	nr	9.06
55 - 110 mm nominal size	-	-	-	nr	10.99
Felt; 'Paradiene' elastomeric bitumen roofing; perforated crepe paper isolating membrane; first layer 'Paradiene 20'; second layer 'Paradiene 30' pre-finished surface					
Two layer coverings	-	-	-	m2	22.11
Working into outlet pipes and the like	-	-	-	nr	8.39
Skirtings; three layer; dressed over tilting fillet; turned into groove					
not exceeding 200 mm girth	-	-	-	m	6.81
200 - 400 mm girth	-	-	-	m	11.95
Coverings to kerbs; two layer					
400 - 600 mm girth	-	-	-	m	14.61
Linings to gutters; three layer					
400 - 600 mm (average) girth	-	-	-	m	22.83
Collars around pipes and the like; three layer; 150 mm high					
not exceeding 55 mm nominal size	-	-	-	nr	6.05
55 - 110 mm nominal size	-	-	-	nr	6.72

J WATERPROOFING Including overheads and profit at 7.50%	Labour hours	Labour £	Material £	Unit	Total rate £
Metal faced 'Veral' glass cloth reinforced bitumen roofing; first layer 'Veralvent' perforated underlay; second layer 'Veralglas'; third layer 'Veral' natural slate aluminium surfaced					
Three layer coverings	-	-	-	m2	**30.65**
Working into outlet pipes and the like	-	-	-	nr	**8.73**
Skirtings; three layer; dressed over tilting fillet; turned into groove					
not exceeding 200 mm girth	-	-	-	m	**6.14**
200 - 400 mm girth	-	-	-	m	**10.56**
Coverings to kerbs; three layer					
400 - 600 mm girth	-	-	-	m	**15.32**
Linings to gutters; three layer					
400 - 600 mm (average) girth	-	-	-	m	**19.91**
Collars around pipes and the like; three layer; 150 mm high					
not exceeding 55 mm nominal size	-	-	-	nr	**6.05**
55 - 110 mm nominal size	-	-	-	nr	**6.72**
Accessories					
Eaves trim; extruded aluminium alloy; working felt into trim					
'Alutrim' type F roof edging PC £9.08/2.5m	0.29	3.46	4.17	m	**7.63**
Angle PC £3.14	0.14	1.67	3.54	nr	**5.21**
Roof screed ventilator - aluminium alloy 'Extr-aqua-vent'; set on screed over and including dished sinking; working collar around ventilator PC £3.71	0.69	8.23	4.19	nr	**12.42**
Insulation board underlays					
Vapour barrier					
reinforced; metal lined	0.03	0.36	4.61	m2	**4.97**
Cork boards; density 112 - 125 kg/m3					
60 mm thick	0.35	4.17	5.61	m2	**9.78**
Foamed glass boards; density 125 - 135 kg/m2					
60 mm thick	0.35	4.17	14.41	m2	**18.58**
Glass fibre boards; density 120 - 130 kg/m2					
60 mm thick	0.35	4.17	10.12	m2	**14.29**
Perlite boards; density 170 - 180 kg/m3					
60 mm thick	0.35	4.17	10.18	m2	**14.35**
Polyurethene boards; density 32 kg/m3					
30 mm thick	0.23	2.74	4.70	m2	**7.44**
35 mm thick	0.23	2.74	4.93	m2	**7.67**
50 mm thick	0.35	4.17	6.30	m2	**10.47**
Wood fibre boards; impregnated; density 220 - 350 kg/m3					
12.7 mm thick	0.23	2.74	2.07	m2	**4.81**
Insulation board overlays					
Dow 'Roofmate SL' extruded polystyrene foam boards					
50 mm thick	0.35	4.17	10.36	m2	**14.53**
75 mm thick	0.35	4.17	14.85	m2	**19.02**
Dow 'Roofmate LG' extruded polystyrene foam boards					
50 mm thick	0.35	4.17	18.52	m2	**22.69**
75 mm thick	0.35	4.17	20.35	m2	**24.52**
100 mm thick	0.35	4.17	24.09	m2	**28.26**

J WATERPROOFING Including overheads and profit at 7.50%	Labour hours	Labour £	Material £	Unit	Total rate £
J42 SINGLE LAYER PLASTICS ROOF COVERINGS					
Felt; 'Derbigum' special polyester 4 mm roofing; first layer 'Ventilag' (partial bond) underlay; second layer 'Derbigum SF'; glass reinforced weathering surface					
Two layer coverings	-	-	-	m2	27.46
Skirtings; two layer; dressed over tilting fillet; turned into groove					
not exceeding 200 mm girth	-	-	-	m	13.94
200 - 400 mm girth	-	-	-	m	17.84
Coverings to kerbs; two layer					
400 - 600 mm girth	-	-	-	m	29.54
Collars around pipes and the like; two layer; 150 mm high					
not exceeding 55 mm nominal size	-	-	-	nr	10.99
55 - 110 mm nominal size	-	-	-	nr	14.88

K LININGS/SHEATHING/DRY PARTITIONING
Including overheads and profit at 7.50%

ALTERNATIVE SHEET LINING MATERIAL PRICES

	£		£		£		£
Asbestos cement flat sheets (£/10 m2)							
Fully compressed							
4.5 mm	59.48	6 mm	73.80	9 mm	111.66	12 mm	139.00
Semi compressed							
4.5 mm	30.24	6 mm	41.14	9 mm	61.23	12 mm	102.01
Blockboard							
Gaboon faced (£/10 m2)							
16 mm	97.38	18 mm	102.31	22 mm	115.81	25 mm	130.02

	£		£		£
18 mm Decorative faced (£/10 m2)					
Ash	126.00	Mahogany	94.48	Teak	119.68
Beech	107.08	Oak	112.00		
Edgings; self adhesive (£/25 m roll)					
19 mm Mahogany	2.47	19 mm Oak	3.46	19 mm Ash	4.00
25 mm Mahogany	3.63	25 mm Oak	4.12	19 mm Teak	4.00

	£		£		£		£
Chipboard (£/10 m2)							
Standard grade							
3.2 mm	8.34	9 mm	17.74	16 mm	25.40	22 mm	33.27
4 mm	10.37	12 mm	19.42	18 mm	26.46	25 mm	37.77
6 mm	15.31						
Melamine faced							
12 mm	34.52	18 mm	41.87				
Laminboard; Birch faced (£/10 m2)							
18 mm	132.88	22 mm	155.70	25 mm	173.12		
Medium density fibreboard (£/10 m2)							
6.5 mm	22.31	12 mm	35.64	17.5 mm	46.09	25 mm	65.25
9 mm	27.07	16 mm	42.79	19 mm	50.40		

	£		£		£		£
Plasterboard (£/100m2)							
Wallboard plank							
9.5 mm	121.41	12.5 mm	145.62	15 mm	204.07	19 mm	254.46
Lath (Thistle baseboard)							
9.5 mm	126.91	12.5	152.56				

	£		£		£
Industrial board				**Fireline Industrial board**	
9.5 mm	238.44	12.5 mm	269.56	12.5 mm	346.66
Fireline board					
12.5 mm	214.97	15 mm	250.00		

	£		£		£
Plywood (£/10 m2)					
Decorative					
6 mm Ash	49.90	9 mm Ash	63.64	12 mm Ash	78.52
6 mm Oak	52.65	9 mm Oak	66.70	12 mm Oak	80.70
6 mm Sapele	44.25	9 mm Sapele	59.64	12 mm Sapele	72.37
6 mm Teak	48.65	9 mm Teak	63.70	12 mm Teak	78.25

Discounts of 0 - 10% available depending on quantity/status

K LININGS/SHEATHING/DRY PARTITIONING Including overheads and profit at 7.50%	Labour hours	Labour £	Material £	Unit	Total rate £
K10 PLASTERBOARD DRY LINING					
Gypsum plasterboard; BS 1230; fixing with nails; joints left open to receive 'Artex'; to softwood base					
Plain grade tapered edge wallboard					
9.5 mm board to ceilings					
over 300 mm wide	0.31	3.70	1.32	m2	**5.02**
9.5 mm board to beams					
over 300 mm wide	0.39	4.65	0.83	m	**5.48**
12.5 mm board to ceilings					
total girth not exceeding 600 mm	0.50	5.96	1.60	m	**7.56**
total girth 600 - 1200 mm	0.33	3.93	1.84	m2	**5.77**
12.5 mm board to beams					
total girth not exceeding 600 mm	0.40	4.77	1.15	m	**5.92**
total girth 600 - 1200 mm	0.53	6.32	2.20	m	**8.52**
Gypsum plasterboard to BS 1230; fixing with nails; joints filled with joint filler and joint tape to receive direct decoration; to softwood base					
Plain grade tapered edge wallboard					
9.5 mm board to walls					
wall height 2.40 - 2.70	1.10	13.11	4.66	m	**17.77**
wall height 2.70 - 3.00	1.25	14.90	5.20	m	**20.10**
wall height 3.00 - 3.30	1.45	17.29	5.74	m	**23.03**
wall height 3.30 - 3.60	1.65	19.67	6.27	m	**25.94**
9.5 mm board to reveals and soffits of openings and recesses					
not exceeding 300 mm wide	0.22	2.62	0.57	m	**3.19**
width 300 - 600 mm	0.44	5.25	1.11	m	**6.36**
9.5 mm board to faces of columns					
total girth not exceeding 600 mm	0.55	6.56	1.08	m	**7.64**
total girth 600 - 1200 mm	1.10	13.11	2.24	m	**15.35**
total girth 1200 - 1800 mm	1.45	17.29	3.29	m	**20.58**
9.5 mm board to ceilings					
over 300 mm wide	0.46	5.48	1.73	m2	**7.21**
9.5 mm board to faces of beams					
total girth not exceeding 600 mm	0.58	6.91	1.11	m	**8.02**
total girth 600 - 1200 mm	1.15	13.71	2.24	m	**15.95**
total girth 1200 - 1800 mm	1.50	17.88	3.29	m	**21.17**
Add for 'Duplex' insulating grade	-	-	-	m2	**0.48**
12.5 mm board to walls					
wall height 2.40 - 2.70	1.15	13.71	5.49	m	**19.20**
wall height 2.70 - 3.00	1.30	15.50	6.08	m	**21.58**
wall height 3.00 - 3.30	1.50	17.88	6.72	m	**24.60**
wall height 3.30 - 3.60	1.75	20.86	7.35	m	**28.21**
12.5 mm board to reveals and soffits of openings and recesses					
not exceeding 300 mm wide	0.23	2.74	0.65	m	**3.39**
width 300 - 600 mm	0.46	5.48	1.30	m	**6.78**
12.5 mm board to faces of columns					
total girth not exceeding 600 mm	0.57	6.80	1.30	m	**8.10**
total girth 600 - 1200 mm	1.15	13.71	2.61	m	**16.32**
total girth 1200 - 1800 mm	1.50	17.88	3.84	m	**21.72**
12.5 mm board to ceilings					
over 300 mm wide	0.48	5.72	2.05	m2	**7.77**
12.5 mm board to faces of beams					
total girth not exceeding 600 mm	0.62	7.39	1.30	m	**8.69**
total girth 600 - 1200 mm	1.25	14.90	2.61	m	**17.51**
total girth 1200 - 1800 mm	1.60	19.07	3.84	m	**22.91**

K LININGS/SHEATHING/DRY PARTITIONING Including overheads and profit at 7.50%	Labour hours	Labour £	Material £	Unit	Total rate £
External angle; with joint tape bedded					
in joint filler; covered with joint finish	0.13	1.55	0.50	m	2.05
Add for 'Duplex' insulating grade	-	-	-	m2	0.48
Tapered edge plank					
19 mm plank to walls					
wall height 2.40 - 2.70	1.20	14.31	8.92	m	23.23
wall height 2.70 - 3.00	1.40	16.69	9.90	m	26.59
wall height 3.00 - 3.30	1.55	18.48	10.89	m	29.37
wall height 3.30 - 3.60	1.85	22.06	11.28	m	33.34
19 mm plank to reveals and soffits of openings and recesses					
not exceeding 300 mm wide	0.24	2.86	1.07	m	3.93
width 300 - 600 mm	0.48	5.72	2.11	m	7.83
19 mm plank to faces of columns					
total girth not exceeding 600 mm	0.59	7.03	2.11	m	9.14
total girth 600 - 1200 mm	1.20	14.31	4.23	m	18.54
total girth 1200 - 1800 mm	1.55	18.48	6.14	m	24.62
19 mm plank to ceilings					
over 300 mm wide	0.51	6.08	3.30	m2	9.38
19 mm plank to faces of beams					
total girth not exceeding 600 mm	0.64	7.63	2.06	m	9.69
total girth 600 - 1200 mm	1.30	15.50	4.23	m	19.73
total girth 1200 - 1800 mm	1.65	19.67	6.14	m	25.81
Thermal board					
25 mm board to walls					
wall height 2.40 - 2.70	1.25	14.90	11.90	m	26.80
wall height 2.70 - 3.00	1.45	17.29	13.18	m	30.47
wall height 3.00 - 3.30	1.60	19.07	14.53	m	33.60
wall height 3.30 - 3.60	1.95	23.25	15.89	m	39.14
25 mm board to reveals and soffits of openings and recesses					
not exceeding 300 mm wide	0.25	2.98	1.40	m	4.38
width 300 - 600 mm	0.51	6.08	2.80	m	8.88
25 mm board to faces of columns					
total girth not exceeding 600 mm	0.62	7.39	2.80	m	10.19
total girth 600 - 1200 mm	1.25	14.90	5.60	m	20.50
total girth 1200 - 1800 mm	1.60	19.07	8.21	m	27.28
25 mm board to ceilings					
over 300 mm wide	0.55	6.56	4.39	m2	10.95
25 mm board to faces of beams					
total girth not exceeding 600 mm	0.66	7.87	2.80	m	10.67
total girth 600 - 1200 mm	1.30	15.50	5.60	m	21.10
total girth 1200 - 1800 mm	1.75	20.86	8.21	m	29.07
50 mm board to walls					
wall height 2.40 - 2.70	1.40	16.69	20.09	m	36.78
wall height 2.70 - 3.00	1.55	18.48	22.35	m	40.83
wall height 3.00 - 3.30	1.70	20.27	24.57	m	44.84
wall height 3.30 - 3.60	2.05	24.44	26.81	m	51.25
50 mm board to reveals and soffits of openings and recesses					
not exceeding 300 mm wide	0.28	3.34	2.35	m	5.69
width 300 - 600 mm	0.55	6.56	4.70	m	11.26
50 mm board to faces of columns					
total girth not exceeding 600 mm	0.66	7.87	4.70	m	12.57
total girth 600 - 1200 mm	1.30	15.50	9.39	m	24.89
total girth 1200 - 1800 mm	1.70	20.27	13.80	m	34.07
50 mm board to ceilings					
over 300 mm wide	0.58	6.91	7.45	m2	14.36
50 mm board to faces of beams					
total girth not exceeding 600 mm	0.69	8.23	4.70	m	12.93
total girth 600 - 1200 mm	1.40	16.69	9.39	m	26.08
total girth 1200 - 1800 mm	1.85	22.06	13.80	m	35.86

K LININGS/SHEATHING/DRY PARTITIONING Including overheads and profit at 7.50%	Labour hours	Labour £	Material £	Unit	Total rate £
K10 PLASTERBOARD DRY LINING - cont'd					
White plastic faced gypsum plasterboard to BS 1230; fixing with screws; butt joints; to softwood base					
Insulating grade square edge wallboard					
9.5 mm board to walls					
wall height 2.40 - 2.70	1.10	13.11	8.64	m	21.75
wall height 2.70 - 3.00	1.25	14.90	9.60	m	24.50
wall height 3.00 - 3.30	1.45	17.29	10.56	m	27.85
wall height 3.30 - 3.60	1.65	19.67	11.52	m	31.19
9.5 mm board to reveals and soffits of openings and recesses					
not exceeding 300 mm wide	0.22	2.62	1.00	m	3.62
width 300 - 600 mm	0.44	5.25	1.99	m	7.24
9.5 mm board to faces of columns					
total girth not exceeding 600 mm	0.55	6.56	1.99	m	8.55
total girth 600 - 1200 mm	1.10	13.11	3.96	m	17.07
total girth 1200 - 1800 mm	1.45	17.29	5.84	m	23.13
9.5 mm board to ceilings					
over 300 mm wide	0.46	5.48	3.20	m2	8.68
9.5 mm board to faces of beams					
total girth not exceeding 600 mm	0.58	6.91	1.99	m	8.90
total girth 600 - 1200 mm	1.15	13.71	3.96	m	17.67
total girth 1200 - 1800 mm	1.50	17.88	5.84	m	23.72
12.5 mm board to walls					
wall height 2.40 - 2.70	1.15	13.71	8.64	m	22.35
wall height 2.70 - 3.00	1.30	15.50	9.60	m	25.10
wall height 3.00 - 3.30	1.50	17.88	10.56	m	28.44
wall height 3.30 - 3.60	1.75	20.86	11.52	m	32.38
12.5 mm board to reveals and soffits of openings and recesses					
not exceeding 300 mm wide	0.23	2.74	1.00	m	3.74
width 300 - 600 mm	0.46	5.48	1.99	m	7.47
12.5 mm board to faces of columns					
total girth not exceeding 600 mm	0.57	6.80	1.99	m	8.79
total girth 600 - 1200 mm	1.15	13.71	3.96	m	17.67
total girth 1200 - 1800 mm	1.50	17.88	5.84	m	23.72
12.5 mm board to ceilings					
over 300 mm wide	0.48	5.72	3.54	m2	9.26
12.5 mm board to faces of beams					
total girth not exceeding 600 mm	0.62	7.39	2.22	m	9.61
total girth 600 - 1200 mm	1.25	14.90	4.40	m	19.30
total girth 1200 - 1800 mm	1.60	19.07	6.48	m	25.55
Two layers of gypsum plasterboard to BS 1230; fixing with nails; joints filled					
19 mm two layer board to walls					
wall height 2.40 - 2.70	1.75	20.86	8.81	m	29.67
wall height 2.70 - 3.00	2.00	23.84	9.78	m	33.62
wall height 3.00 - 3.30	2.20	26.23	10.76	m	36.99
wall height 3.30 - 3.60	2.65	31.59	11.14	m	42.73
19 mm two layer board to reveals and soffits of openings and recesses					
not exceeding 300 mm wide	0.33	3.93	1.05	m	4.98
width 300 - 600 mm	0.66	7.87	2.08	m	9.95
19 mm two layer board to faces of columns					
total girth not exceeding 600 mm	0.88	10.49	2.08	m	12.57
total girth 600 - 1200 mm	1.65	19.67	4.17	m	23.84
total girth 1200 - 1800 mm	2.20	26.23	6.06	m	32.29
19 mm two layer board to ceilings					
over 300 mm wide	0.73	8.70	3.27	m2	11.97

K LININGS/SHEATHING/DRY PARTITIONING Including overheads and profit at 7.50%	Labour hours	Labour £	Material £	Unit	Total rate £
19 mm two layer board to faces of beams					
total girth not exceeding 600 mm	0.94	11.21	2.03	m	13.24
total girth 600 - 1200 mm	1.75	20.86	4.17	m	25.03
total girth 1200 - 1800 mm	2.35	28.02	6.06	m	34.08
25 mm two layer board to walls					
wall height 2.40 - 2.70	1.85	22.06	10.56	m	32.62
wall height 2.70 - 3.00	2.10	25.04	11.73	m	36.77
wall height 3.00 - 3.30	2.40	28.61	12.96	m	41.57
wall height 3.30 - 3.60	2.85	33.98	14.20	m	48.18
25 mm two layer board to reveals and soffits of openings and recesses					
not exceeding 300 mm wide	0.36	4.29	1.25	m	5.54
width 300 - 600 mm	0.72	8.58	2.48	m	11.06
25 mm two layer board to faces of columns					
total girth not exceeding 600 mm	0.95	11.33	2.48	m	13.81
total girth 600 - 1200 mm	1.75	20.86	4.97	m	25.83
total girth 1200 - 1800 mm	2.35	28.02	7.30	m	35.32
25 mm two layer board to ceilings					
over 300 mm wide	0.79	9.42	3.91	m2	13.33
25 mm two layer board to faces of beams					
total girth not exceeding 600 mm	1.00	11.92	2.48	m	14.40
total girth 600 - 1200 mm	1.95	23.25	4.97	m	28.22
total girth 1200 - 1800 mm	2.55	30.40	7.30	m	37.70
Gypsum plasterboard to BS 1230; 3 mm joints; fixed by the 'Thistleboard' system of dry linings; joints; filled with joint filler and joint tape; to receive direct decoration					
Plain grade tapered edge wallboard					
9.5 mm board to walls					
wall height 2.40 - 2.70	1.15	13.71	4.58	m	18.29
wall height 2.70 - 3.00	1.30	15.50	5.10	m	20.60
wall height 3.00 - 3.30	1.50	17.88	5.63	m	23.51
wall height 3.30 - 3.60	1.75	20.86	6.15	m	27.01
9.5 mm board to reveals and soffits of openings and recesses					
not exceeding 300 mm wide	0.23	2.74	0.53	m	3.27
width 300 - 600 mm	0.46	5.48	1.06	m2	6.54
9.5 mm board to faces of columns					
total girth not exceeding 600 mm	0.57	6.80	1.06	m	7.86
total girth 600 - 1200 mm	1.15	13.71	2.13	m	15.84
total girth 1200 - 1800 mm	1.50	17.88	3.13	m	21.01
Angle; with joint tape bedded in joint filler; covered with joint finish					
internal	0.07	0.83	0.22	m	1.05
external	0.13	1.55	0.49	m	2.04
Vermiculite gypsum cladding; 'Vicuclad' board on and including shaped noggins; fixed with nails and adhesive; joints pointed in adhesive					
25 mm column casings; 2 hour fire protection rating					
over 1 m girth	1.10	13.11	14.38	m2	27.49
150 mm girth	0.33	3.93	4.36	m	8.29
300 mm girth	0.50	5.96	6.53	m	12.49
600 mm girth	0.83	9.90	10.63	m	20.53
900 mm girth	1.05	12.52	13.67	m	26.19

K LININGS/SHEATHING/DRY PARTITIONING Including overheads and profit at 7.50%		Labour hours	Labour £	Material £	Unit	Total rate £
K10 PLASTERBOARD DRY LINING - cont'd						
Vermiculite gypsum cladding; 'Vicuclad' board on and including shaped noggins; fixed with nails and adhesive; joints pointed in adhesive - cont'd						
30 mm beam casings; 2 hour fire protection rating						
over 1 m girth		1.20	14.31	17.65	m2	31.96
150 mm girth		0.36	4.29	5.34	m	9.63
300 mm girth		0.55	6.56	8.01	m	14.57
600 mm girth		0.91	10.85	13.09	m	23.94
900 mm girth		1.15	13.71	16.77	m	30.48
55 mm column casings; 4 hour fire protection rating						
over 1 m girth		1.30	15.50	36.22	m2	51.72
150 mm girth		0.40	4.77	10.91	m	15.68
300 mm girth		0.61	7.27	16.36	m	23.63
600 mm girth		0.99	11.80	27.01	m	38.81
900 mm girth		1.25	14.90	34.42	m	49.32
60 mm beam casings; 4 hour fire protection rating						
over 1 m girth		1.45	17.29	38.19	m2	55.48
150 mm girth		0.43	5.13	11.50	m	16.63
300 mm girth		0.66	7.87	17.25	m	25.12
600 mm girth		1.05	12.52	28.49	m	41.01
900 mm girth		1.40	16.69	36.28	m	52.97
Add to the above for						
Plus 3% for work 3.5 - 5 m high						
Plus 6% for work 5 - 6.5 m high						
Plus 12% for work 6.5 - 8 m high						
Plus 18% for work over 8 m high						
Cutting and fitting around steel joists, angles, trunking, ducting, ventilators, pipes, tubes, etc						
over 2 m girth		0.33	3.93	-	m	3.93
not exceeding 0.30 m girth		0.22	2.62	-	nr	2.62
0.30 - 1 m girth		0.28	3.34	-	nr	3.34
1 - 2 m girth		0.39	4.65	-	nr	4.65
K11 RIGID SHEET FLOORING/SHEATHING/LININGS/ CASINGS						
Blockboard (Birch faced)						
12 mm lining to walls						
over 300 mm wide	PC £79.93/10m2	0.52	4.21	9.58	m2	13.79
not exceeding 300 mm wide		0.34	2.75	3.14	m	5.89
Raking cutting		0.09	0.73	0.34	m	1.07
Holes for pipes and the like		0.04	0.32	3.44	nr	3.76
18 mm lining to walls						
over 300 mm wide	PC £97.16/10m2	0.55	4.45	11.62	m2	16.07
not exceeding 300 mm wide		0.35	2.83	3.81	m	6.64
Raking cutting		0.11	0.89	0.52	m	1.41
Holes for pipes and the like		0.06	0.49	-	nr	0.49
Two-sided 18 mm thick pipe casing; 50 x 50 mm softwood framing; two members plugged to wall						
300 mm girth		1.40	11.33	5.97	m	17.30
450 mm girth		1.50	12.14	7.90	m	20.04
600 mm girth		1.60	12.95	9.82	m	22.77
750 mm girth		1.70	13.76	11.74	m	25.50

K LININGS/SHEATHING/DRY PARTITIONING Including overheads and profit at 7.50%		Labour hours	Labour £	Material £	Unit	Total rate £
Three-sided 18 mm thick pipe casing; 50 x 50 mm softwood framing; two members plugged to wall						
450 mm girth		1.85	14.98	8.61	m	23.59
600 mm girth		2.00	16.19	10.54	m	26.73
750 mm girth		2.10	17.00	12.47	m	29.47
900 mm girth		2.20	17.81	14.40	m	32.21
1050 mm girth		2.30	18.62	16.32	m	34.94
Extra for 400 x 400 mm removable access panel; brass cups and screws; additional framing		1.10	8.90	5.10	nr	14.00
25 mm lining to walls						
over 300 mm wide	PC £126.65/10m2	0.59	4.78	15.13	m2	19.91
not exceeding 300 mm wide		0.39	3.16	4.97	m	8.13
Raking cutting		0.13	1.05	0.68	m	1.73
Holes for pipes and the like		0.07	0.57	-	nr	0.57
Chipboard (plain)						
12 mm lining to walls						
over 300 mm wide	PC £19.42/10m2	0.42	3.40	2.42	m2	5.82
not exceeding 300 mm wide		0.24	1.94	0.80	m	2.74
Raking cutting		0.07	0.57	0.10	m	0.67
Holes for pipes and the like		0.03	0.24	-	nr	0.24
15 mm lining to walls						
over 300 mm wide	PC £22.63/10m2	0.44	3.56	2.81	m2	6.37
not exceeding 300 mm wide		0.26	2.10	0.93	m	3.03
Raking cutting		0.09	0.73	0.12	m	0.85
Holes for pipes and the like		0.04	0.32	-	nr	0.32
Two-sided 15 mm thick pipe casing; 50 x 50 mm softwood framing; two members plugged to wall						
300 mm girth		1.10	8.90	3.32	m	12.22
450 mm girth		1.20	9.71	3.92	m	13.63
600 mm girth		1.30	10.52	4.53	m	15.05
750 mm girth		1.35	10.93	5.13	m	16.06
Three-sided 15 mm thick pipe casing; 50 x 50 mm softwood framing; two members plugged to wall						
450 mm girth		1.65	13.36	4.65	m	18.01
600 mm girth		1.75	14.17	5.25	m	19.42
750 mm girth		1.85	14.98	5.86	m	20.84
900 mm girth		1.95	15.78	6.46	m	22.24
1050 mm girth		2.20	17.81	7.06	m	24.87
Extra for 400 x 400 mm removable access panel; brass cups and screws; additional framing		1.10	8.90	5.04	nr	13.94
18 mm lining to walls						
over 300 mm wide	PC £26.46/10m2	0.46	3.72	3.25	m2	6.97
not exceeding 300 mm wide		0.30	2.43	1.07	m	3.50
Raking cutting		0.11	0.89	0.14	m	1.03
Holes for pipes and the like		0.06	0.49	-	nr	0.49
Chipboard (Melamine faced white matt finish)						
15 mm thick; PC £37.03/10m2; laminated masking strips						
Lining to walls						
over 300 mm wide		1.15	9.31	6.76	m2	16.07
not exceeding 300 mm wide		0.75	6.07	3.11	m	9.18
Raking cutting		0.14	1.13	0.20	m	1.33
Holes for pipes and the like		0.08	0.65	-	nr	0.65
Chipboard boarding and flooring						
Boarding to floors; butt joints						
18 mm thick	PC £28.31/10m2	0.33	2.67	3.36	m2	6.03
22 mm thick	PC £33.02/10m2	0.36	2.91	3.89	m2	6.80

K LININGS/SHEATHING/DRY PARTITIONING Including overheads and profit at 7.50%		Labour hours	Labour £	Material £	Unit	Total rate £
K11 RIGID SHEET FLOORING/SHEATHING/LININGS/ CASINGS - cont'd						
Chipboard boarding and flooring - cont'd						
Boarding to floors; tongued and grooved joints						
18 mm thick	PC £30.70/10m2	0.35	2.83	3.62	m2	6.45
22 mm thick	PC £40.36/10m2	0.39	3.16	4.72	m2	7.88
Boarding to roofs; butt joints						
18 mm thick; pre-felted	PC £26.46/10m2	0.36	2.91	3.15	m2	6.06
Raking cutting on 18 mm thick chipboard		0.09	0.73	1.52	m	2.25
Durabella 'Westbourne' flooring system or similar; comprising 19 mm thick tongued and grooved chipboard panels secret nailed to softwood MK 10X-profiled foam backed battens at 600 mm centres; on concrete floor						
Flooring tongued and grooved joints						
63 mm thick 36 x 50 mm battens	PC £9.52	0.99	8.01	11.51	m2	19.52
75 mm thick 43 x 50 mm battens	PC £10.23	0.99	8.01	12.35	m2	20.36
Plywood flooring						
Boarding to floors; tongued and grooved joints						
15 mm thick	PC £124.99/10m2	0.44	3.56	14.26	m2	17.82
18 mm thick	PC £145.44/10m2	0.48	3.89	16.58	m2	20.47
Plywood; external quality; 18 mm thick; PC £110.21/10m2						
Boarding to roofs; butt joints						
flat to falls		0.44	3.56	12.60	m2	16.16
sloping		0.47	3.80	12.60	m2	16.40
vertical		0.63	5.10	12.60	m2	17.70
Hardboard to BS 1142						
3.2 mm lining to walls						
over 300 mm wide	PC £10.56/10m2	0.33	2.67	1.30	m2	3.97
not exceeding 300 mm wide		0.20	1.62	0.42	m	2.04
Raking cutting		0.02	0.16	0.06	m	0.22
Holes for pipes and the like		0.01	0.08	-	nr	0.08
6.4 mm lining to walls						
over 300 mm wide	PC £20.98/10m2	0.36	2.91	2.53	m2	5.44
not exceeding 300 mm wide		0.23	1.86	0.83	m	2.69
Raking cutting		0.03	0.24	0.11	m	0.35
Holes for pipes and the like		0.02	0.16	-	nr	0.16
Glazed hardboard to BS 1142; PC £15.86/10m2; on and including 38 x 38 mm wrought softwood framing						
3.2 mm thick panel						
to side of bath		2.00	16.19	4.90	nr	21.09
to end of bath		0.77	6.23	1.83	nr	8.06
Insulation board to BS 1142						
12.7 mm lining to walls						
over 300 mm wide	PC £18.04/10m2	0.26	2.10	2.26	m2	4.36
not exceeding 300 mm wide		0.15	1.21	0.75	m	1.96
Raking cutting		0.03	0.24	0.10	m	0.34
Holes for pipes and the like		0.01	0.08	-	nr	0.08
19 mm lining to walls						
over 300 mm wide	PC £29.43/10m2	0.29	2.35	3.60	m2	5.95
not exceeding 300 mm wide		0.18	1.46	1.19	m	2.65
Raking cutting		0.06	0.49	0.16	m	0.65
Holes for pipes and the like		0.02	0.16	-	nr	0.16

K LININGS/SHEATHING/DRY PARTITIONING Including overheads and profit at 7.50%		Labour hours	Labour £	Material £	Unit	Total rate £
25 mm lining to walls						
over 300 mm wide	PC £38.12/10m2	0.35	2.83	4.65	m2	7.48
not exceeding 300 mm wide		0.21	1.70	1.54	m	3.24
Raking cutting		0.07	0.57	0.20	m	0.77
Holes for pipes and the like		0.03	0.24	-	nr	0.24
Laminboard (Birch Faced); 18 mm thick PC £132.88/10m2						
Lining to walls						
over 300 mm wide		0.58	4.69	15.84	m2	20.53
not exceeding 300 mm wide		0.37	3.00	5.19	m	8.19
Raking cutting		0.12	0.97	0.71	m	1.68
Holes for pipes and the like		0.07	0.57	-	nr	0.57
Non-asbestos board; 'Masterboard'; sanded finish						
6 mm lining to walls						
over 300 mm wide	PC £45.30/10m2	0.36	2.91	5.45	m2	8.36
not exceeding 300 mm wide		0.22	1.78	1.78	m	3.56
6 mm lining to ceilings						
over 300 mm wide		0.48	3.89	5.45	m2	9.34
not exceeding 300 mm wide		0.30	2.43	1.78	m	4.21
Raking cutting		0.06	0.49	0.24	m	0.73
Holes for pipes and the like		0.03	0.24	-	nr	0.24
9 mm lining to walls						
over 300 mm wide	PC £82.30/10m2	0.40	3.24	9.84	m2	13.08
not exceeding 300 mm wide		0.24	1.94	3.23	m	5.17
9 mm lining to ceilings						
over 300 mm wide		0.50	4.05	9.84	m2	13.89
not exceeding 300 mm wide		0.32	2.59	3.23	m	5.82
Raking cutting		0.07	0.57	0.44	m	1.01
Holes for pipes and the like		0.04	0.32	-	nr	0.32
12 mm lining to walls						
over 300 mm wide	PC £115.90/10m2	0.44	3.56	13.84	m2	17.40
not exceeding 300 mm wide		0.26	2.10	4.54	m	6.64
12 mm lining to ceilings						
over 300 mm wide		0.58	4.69	13.84	m2	18.53
not exceeding 300 mm wide		0.35	2.83	4.54	m	7.37
Raking cutting		0.08	0.65	0.62	m	1.27
Holes for pipes and the like		0.06	0.49	-	nr	0.49
Non-asbestos board; 'Supalux'; sanded finish						
6 mm lining to walls						
over 300 mm wide	PC £57.40/10m2	0.36	2.91	6.88	m2	9.79
not exceeding 300 mm wide		0.22	1.78	2.25	m	4.03
6 mm lining to ceilings						
over 300 mm wide		0.48	3.89	6.88	m2	10.77
not exceeding 300 mm wide		0.30	2.43	2.25	m	4.68
Raking cutting		0.06	0.49	0.31	m	0.80
Holes for pipes and the like		0.03	0.24	-	nr	0.24
9 mm lining to walls						
over 300 mm wide	PC £87.60/10m2	0.40	3.24	10.48	m2	13.72
not exceeding 300 mm wide		0.24	1.94	3.44	m	5.38
9 mm lining to ceilings						
over 300 mm wide		0.50	4.05	10.48	m2	14.53
not exceeding 300 mm wide		0.32	2.59	3.44	m	6.03
Raking cutting		0.07	0.57	0.47	m	1.04
Holes for pipes and the like		0.04	0.32	-	nr	0.32
12 mm lining to walls						
over 300 mm wide	PC £115.90/10m2	0.44	3.56	13.84	m2	17.40
not exceeding 300 mm wide		0.26	2.10	4.54	m	6.64
12 mm lining to ceilings						
over 300 mm wide		0.58	4.69	13.84	m2	18.53
not exceeding 300 mm wide		0.35	2.83	4.54	m	7.37
Raking cutting		0.08	0.65	0.62	m	1.27
Holes for pipes and the like		0.06	0.49	-	nr	0.49

K LININGS/SHEATHING/DRY PARTITIONING Including overheads and profit at 7.50%		Labour hours	Labour £	Material £	Unit	Total rate £
K11 RIGID SHEET FLOORING/SHEATHING/LININGS/ CASINGS - cont'd						
Plywood (Russian Birch); internal quality						
4 mm lining to walls						
over 300 mm wide	PC £25.44/10m2	0.41	3.32	3.13	m2	6.45
not exceeding 300 mm wide		0.26	2.10	1.03	m	3.13
4 mm lining to ceilings						
over 300 mm wide		0.55	4.45	3.13	m2	7.58
not exceeding 300 mm wide		0.35	2.83	1.03	m	3.86
Raking cutting		0.06	0.49	0.14	m	0.63
Holes for pipes and the like		0.03	0.24	-	nr	0.24
6 mm lining to walls						
over 300 mm wide	PC £35.93/10m2	0.44	3.56	4.38	m2	7.94
not exceeding 300 mm wide		0.29	2.35	1.44	m	3.79
6 mm lining to ceilings						
over 300 mm wide		0.58	4.69	4.38	m2	9.07
not exceeding 300 mm wide		0.39	3.16	1.44	m	4.60
Raking cutting		0.06	0.49	0.19	m	0.68
Holes for pipes and the like		0.03	0.24	-	nr	0.24
Two-sided 6 mm thick pipe casings; 50 x 50 mm softwood framing; two members plugged to wall						
300 mm girth		1.30	10.52	3.79	m	14.31
450 mm girth		1.45	11.74	4.63	m	16.37
600 mm girth		1.55	12.55	5.47	m	18.02
750 mm girth		1.65	13.36	6.31	m	19.67
Three-sided 6 mm thick pipe casing; 50 x 50 mm softwood framing; to members plugged to wall						
450 mm girth		1.80	14.57	5.36	m	19.93
600 mm girth		1.95	15.78	6.20	m	21.98
750 mm girth		2.05	16.59	7.04	m	23.63
900 mm girth		2.15	17.40	7.88	m	25.28
1050 mm girth		2.20	17.81	8.72	m	26.53
9 mm lining to walls						
over 300 mm wide	PC £50.41/10m2	0.47	3.80	6.09	m2	9.89
not exceeding 300 mm wide		0.31	2.51	2.00	m	4.51
9 mm lining to ceilings						
over 300 mm wide		0.63	5.10	6.09	m2	11.19
not exceeding 300 mm wide		0.41	3.32	2.00	m	5.32
Raking cutting		0.07	0.57	0.27	m	0.84
Holes for pipes and the like		0.04	0.32	-	nr	0.32
12 mm lining to walls						
over 300 mm wide	PC £66.11/10m2	0.51	4.13	7.95	m2	12.08
not exceeding 300 mm wide		0.33	2.67	2.61	m	5.28
12 mm lining to ceilings						
over 300 mm wide		0.67	5.42	7.95	m2	13.37
not exceeding 300 mm wide		0.44	3.56	2.61	m	6.17
Raking cutting		0.07	0.57	0.36	m	0.93
Holes for pipes and the like		0.04	0.32	-	nr	0.32
Plywood (Finnish Birch); external quality						
4 mm lining to walls						
over 300 mm wide	PC £30.54/10m2	0.41	3.32	3.75	m2	7.07
not exceeding 300 mm wide		0.26	2.10	1.23	m	3.33
4 mm lining to ceilings						
over 300 mm wide		0.55	4.45	3.75	m2	8.20
not exceeding 300 mm wide		0.35	2.83	1.23	m	4.06
Raking cutting		0.06	0.49	0.16	m	0.65
Holes for pipes and the like		0.03	0.24	-	nr	0.24
6 mm lining to walls						
over 300 mm wide	PC £45.90/10m2	0.44	3.56	5.56	m2	9.12
not exceeding 300 mm wide		0.29	2.35	1.83	m	4.18

K LININGS/SHEATHING/DRY PARTITIONING Including overheads and profit at 7.50%		Labour hours	Labour £	Material £	Unit	Total rate £
6 mm lining to ceilings						
over 300 mm wide		0.58	4.69	5.56	m2	10.25
not exceeding 300 mm wide		0.39	3.16	1.83	m	4.99
Raking cutting		0.06	0.49	0.25	m	0.74
Holes for pipes and the like		0.03	0.24	-	nr	0.24
Two-sided 6 mm thick pipe casings; 50 x 50 mm softwood framing; two members plugged to wall						
300 mm girth		1.30	10.52	4.15	m	14.67
450 mm girth		1.45	11.74	5.17	m	16.91
600 mm girth		1.55	12.55	6.18	m	18.73
750 mm girth		1.65	13.36	7.20	m	20.56
Three-sided 6 mm thick pipe casing; 50 x 50 mm softwood framing; to members plugged to wall						
450 mm girth		1.80	14.57	5.89	m	20.46
600 mm girth		1.95	15.78	6.91	m	22.69
750 mm girth		2.05	16.59	7.92	m	24.51
900 mm girth		2.15	17.40	8.94	m	26.34
1050 mm girth		2.20	17.81	9.95	m	27.76
9 mm lining to walls						
over 300 mm wide	PC £59.25/10m2	0.47	3.80	7.13	m2	10.93
not exceeding 300 mm wide		0.31	2.51	2.35	m	4.86
9 mm lining to ceilings						
over 300 mm wide		0.63	5.10	7.13	m2	12.23
not exceeding 300 mm wide		0.41	3.32	2.35	m	5.67
Raking cutting		0.08	0.65	0.32	m	0.97
Holes for pipes and the like		0.06	0.49	-	nr	0.49
12 mm lining to walls						
over 300 mm wide	PC £75.58/10m2	0.51	4.13	9.07	m2	13.20
not exceeding 300 mm wide		0.33	2.67	2.99	m	5.66
12 mm lining to ceilings						
over 300 mm wide		0.67	5.42	9.09	m2	14.51
not exceeding 300 mm wide		0.44	3.56	2.99	m	6.55
Raking cutting		0.11	0.89	0.41	m	1.30
Holes for pipes and the like		0.09	0.73	-	nr	0.73
Extra over wall linings fixed with nails; for screwing		0.17	1.38	0.19	m2	1.57
Woodwool unreinforced slabs; Torvale 'Woodcemair' or similar; BS 1105 type SB; natural finish; fixing to timber or steel with galvanized nails or clips; flat or sloping						
50 mm slabs; type 500; max. span 600 mm						
1800 - 2400 mm lengths	PC £5.20	0.44	3.56	6.10	m2	9.66
2700 - 3000 mm lengths	PC £5.26	0.44	3.56	6.17	m2	9.73
75 mm slabs; type 750; max. span 900 mm						
2100 mm lengths	PC £6.72	0.50	4.05	7.87	m2	11.92
2400 - 2700 mm lengths	PC £6.80	0.50	4.05	8.00	m2	12.05
3000 mm lengths	PC £6.91	0.50	4.05	8.04	m2	12.09
Raking cutting		0.15	1.21	1.67	m	2.88
Holes for pipes and the like		0.13	1.05	-	nr	1.05
100 mm slabs; type 1000; max. span 1200 mm						
3000 - 3600 mm lengths	PC £9.83	0.55	4.45	11.33	m2	15.78
Internal quality West African Mahogany veneered plywood; 6 mm thick; PC £40.05/10m2; WAM cover strips						
Lining to walls						
over 300 mm wide		0.77	6.23	5.85	m2	12.08
not exceeding 300 mm wide		0.50	4.05	2.31	m	6.36

K LININGS/SHEATHING/DRY PARTITIONING Including overheads and profit at 7.50%	Labour hours	Labour £	Material £	Unit	Total rate £
K13 RIGID SHEET FINE LININGS/PANELLING					
Formica faced chipboard; 17 mm thick; white matt finish; balancer; PC £215.45/10m2; aluminium cover strips and countersunk screws					
Lining to walls					
over 300 mm wide	2.10	17.00	27.32	m2	44.32
not exceeding 300 mm wide	1.35	10.93	9.62	m	20.55
Lining to isolated columns or the like					
over 300 mm wide	3.15	25.50	29.99	m2	55.49
not exceeding 300 mm wide	1.80	14.57	10.27	m	24.84
K20 TIMBER BOARD FLOORING/SHEATHING/LININGS/ CASINGS					
Sawn softwood; untreated					
Boarding to roofs; 150 mm wide boards; butt joints					
19 mm thick; flat; over 300 mm wide PC £65.29/100m	0.50	4.05	5.38	m2	9.43
19 mm thick; flat; not exceeding 300 mm wide	0.33	2.67	1.66	m	4.33
19 mm thick; sloping; over 300 mm wide	0.55	4.45	5.38	m2	9.83
19 mm thick; sloping; not exceeding 300 mm wide	0.36	2.91	1.66	m	4.57
19 mm thick; sloping; laid diagonally; over 300 mm wide	0.69	5.59	5.50	m2	11.09
19 mm thick; sloping; laid diagonally; not exceeding 300 mm wide	0.44	3.56	1.69	m	5.25
25 mm thick; flat; over 300 mm wide; PC £82.99/100m	0.50	4.05	6.75	m2	10.80
25 mm thick; flat; not exceeding 300 mm	0.33	2.67	2.08	m	4.75
25 mm thick; sloping; over 300 mm wide	0.55	4.45	6.75	m2	11.20
25 mm thick; sloping; not exceeding 300 mm wide	0.36	2.91	2.08	m	4.99
Boarding to tops or cheeks of dormers; 150 mm wide boards; butt joints					
19 mm thick; diagonally; over 300 mm wide	0.88	7.12	5.50	m2	12.62
19 mm thick; diagonally; not exceeding 300 mm wide	0.55	4.45	1.66	m	6.11
19 mm thick; diagonally; area not exceeding 1.00 m2; irrespective of width	1.10	8.90	5.62	nr	14.52
Sawn softwood; 'Tanalised'					
Boarding to roofs; 150 mm wide boards; butt joints					
19 mm thick; flat; over 300 mm wide PC £81.61/100m	0.50	4.05	6.64	m2	10.69
19 mm thick; flat; not exceeding 300 mm wide	0.33	2.67	2.04	m	4.71
19 mm thick; sloping; over 300 mm wide	0.55	4.45	6.64	m2	11.09
19 mm thick; sloping; not exceeding 300 mm wide	0.36	2.91	2.04	m	4.95
19 mm thick; sloping; laid diagonally; over 300 mm wide	0.69	5.59	6.79	m2	12.38
19 mm thick; sloping; laid diagonally; not exceeding 300 mm wide	0.44	3.56	2.08	m	5.64
25 mm thick; flat; over 300 mm wide; PC £103.73/100m	0.50	4.05	8.36	m2	12.41
25 mm thick; flat; not exceeding 300 mm	0.33	2.67	2.57	m	5.24
25 mm thick; sloping; over 300 mm wide	0.55	4.45	8.36	m2	12.81
25 mm thick; sloping; not exceeding 300 mm wide	0.36	2.91	2.57	m	5.48

K LININGS/SHEATHING/DRY PARTITIONING Including overheads and profit at 7.50%	Labour hours	Labour £	Material £	Unit	Total rate £
Boarding to tops or cheeks of dormers; 150 mm wide boards; butt joints					
19 mm thick; diagonally; over 300 mm wide	0.88	7.12	6.79	m2	13.91
19 mm thick; diagonally; not exceeding 300 mm wide	0.55	4.45	2.08	m	6.53
19 mm thick; diagonally; area not exceeding 1.00 m2; irrespective of width	1.10	8.90	6.79	nr	15.69
Wrought softwood					
Boarding to floors; butt joints					
19 mm thick x 75 mm wide boards PC £58.73/100m	0.66	5.34	9.61	m2	14.95
19 mm thick x 125 mm wide boards PC £69.87/100m	0.55	4.45	6.85	m2	11.30
22 mm thick x 150 mm wide boards PC £82.52/100m	0.50	4.05	6.71	m2	10.76
25 mm thick x 100 mm wide boards PC £80.85/100m	0.61	4.94	9.80	m2	14.74
25 mm thick x 150 mm wide boards PC £117.92/100m	0.50	4.05	9.45	m2	13.50
25 mm boarding and bearers to floors; butt joints; in making good where partitions removed or openings formed (boards running in direction of partition)					
150 mm wide	0.30	2.43	1.67	m	4.10
225 mm wide	0.45	3.64	2.51	m	6.15
300 mm wide	0.60	4.86	3.35	m	8.21
25 mm boarding and bearers to floors; butt joints; in making good where partitions removed or openings formed (boards running at right angles to partition)					
150 mm wide	0.45	3.64	3.51	m	7.15
225 mm wide	0.70	5.67	4.46	m	10.13
300 mm wide	0.90	7.29	5.42	m	12.71
450 mm wide	1.35	10.93	7.17	m	18.10
Boarding to floors; tongued and grooved joints					
19 mm thick x 75 mm wide boards PC £58.73/100m	0.77	6.23	10.15	m2	16.38
19 mm thick x 125 mm wide boards PC £69.87/100m	0.66	5.34	7.24	m2	12.58
22 mm thick x 150 mm wide boards PC £82.52/100m	0.61	4.94	7.10	m2	12.04
25 mm thick x 100 mm wide boards PC £80.85/100m	0.72	5.83	10.36	m2	16.19
25 mm thick x 150 mm wide boards PC £117.92/100m	0.61	4.94	10.01	m2	14.95
Nosings; tongued to edge of flooring					
19 x 75 mm; once rounded	0.28	2.27	1.65	m	3.92
25 x 75 mm; once rounded	0.28	2.27	1.93	m	4.20
Boarding to internal walls; tongued and grooved and V-jointed					
12 mm thick x 100 mm wide boards PC £62.69/100m	0.88	7.12	8.14	m2	15.26
16 mm thick x 100 mm wide boards PC £73.96/100m	0.88	7.12	9.52	m2	16.64
19 mm thick x 100 mm wide boards PC £87.81/100m	0.88	7.12	11.22	m2	18.34
19 mm thick x 125 mm wide boards PC £108.55/100m	0.83	6.72	11.04	m2	17.76
19 mm thick x 125 mm wide boards; chevron pattern	1.30	10.52	11.60	m2	22.12
25 mm thick x 125 mm wide boards PC £141.10/100m	0.83	6.72	14.22	m2	20.94

K LININGS/SHEATHING/DRY PARTITIONING Including overheads and profit at 7.50%		Labour hours	Labour £	Material £	Unit	Total rate £
K20 TIMBER BOARD FLOORING/SHEATHING/LININGS/ CASINGS - cont'd						
Wrought softwood - cont'd						
Boarding to internal walls;						
12 mm thick x 100 mm wide						
Knotty Pine boards	PC £37.95/100m	0.88	7.12	5.10	m2	12.22
Boarding to internal ceilings;						
tongued and grooved and V-jointed						
12 mm thick x 100 mm wide boards		1.10	8.90	8.14	m2	17.04
16 mm thick x 100 mm wide boards		1.10	8.90	9.52	m2	18.42
19 mm thick x 100 mm wide boards		1.10	8.90	11.22	m2	20.12
19 mm thick x 125 mm wide boards		1.05	8.50	11.04	m2	19.54
19 mm thick x 125 mm wide boards;						
chevron pattern		1.55	12.55	11.60	m2	24.15
25 mm thick x 125 mm wide boards		1.05	8.50	14.22	m2	22.72
12 mm thick x 100 mm wide						
Knotty Pine boards		1.10	8.90	5.10	m2	14.00
Boarding to roofs; tongued and grooved joints						
19 mm thick; flat to falls	PC £90.92/100m	0.61	4.94	7.79	m2	12.73
19 mm thick; sloping		0.66	5.34	7.79	m2	13.13
19 mm thick; sloping; laid diagonally		0.83	6.72	7.92	m2	14.64
25 mm thick; flat to falls	PC £117.92/100m	0.61	4.94	10.01	m2	14.95
25 mm thick; sloping		0.66	5.34	10.01	m2	15.35
Boarding to tops or cheeks of dormers;						
tongued and grooved joints						
19 mm thick; laid diagonally		1.10	8.90	7.92	m2	16.82
Wrought softwood; 'Tanalised'						
Boarding to roofs; tongued and grooved joints						
19 mm thick; flat to falls	PC £113.66/100m	0.61	4.94	9.66	m2	14.60
19 mm thick; sloping		0.66	5.34	9.66	m2	15.00
19 mm thick; sloping; laid diagonally		0.83	6.72	9.82	m2	16.54
25 mm thick; flat to falls	PC £147.40/100m	0.61	4.94	12.42	m2	17.36
25 mm thick; sloping		0.66	5.34	12.42	m2	17.76
Boarding to tops or cheeks of dormers;						
tongued and grooved joints						
19 mm thick; laid diagonally		1.10	8.90	9.82	m2	18.72
Wood strip; 22 mm thick; 'Junckers' pre-treated or similar; tongued and grooved joints; pre-finished boards; level fixing to resilient battens; to cement and sand base						
Strip flooring; over 300 mm wide						
Beech; prime		-	-	-	m2	74.18
Beech; standard		-	-	-	m2	70.95
Beech; sylvia squash		-	-	-	m2	74.18
Oak; quality A		-	-	-	m2	75.25
Wrought hardwood						
Strip flooring to floors; 25 mm thick x 75 mm wide; tongued and grooved joints; secret fixing; surface sanded after laying						
American Oak		-	-	-	m2	51.06
Canadian Maple		-	-	-	m2	50.55
Gurjun		-	-	-	m2	42.89
Iroko		-	-	-	m2	50.04
Raking cutting		-	-	-	m	5.62
Curved cutting		-	-	-	m	6.64

K LININGS/SHEATHING/DRY PARTITIONING Including overheads and profit at 7.50%		Labour hours	Labour £	Material £	Unit	Total rate £
K29 TIMBER FRAMED AND PANELLED PARTITIONS						
Purpose made screen components; wrought softwood						
Panelled partitions						
32 mm thick frame; 9 mm thick plywood panels; over 300 mm wide	PC £35.07	0.99	8.01	38.65	m2	**46.66**
38 mm thick frame; 9 mm thick plywood panels over 300 mm wide	PC £37.39	1.05	8.50	41.20	m2	**49.70**
50 mm thick frame; 12 mm thick plywood panels; over 300 mm wide	PC £45.50	1.10	8.90	50.14	m2	**59.04**
Panelled partitions; mouldings worked on solid both sides						
32 mm thick frame; 9 mm thick plywood panels; over 300 mm wide	PC £40.52	0.99	8.01	44.64	m2	**52.65**
38 mm thick frame; 9 mm thick plywood panels over 300 mm wide	PC £42.39	1.05	8.50	46.71	m2	**55.21**
50 mm thick frame; 12 mm thick plywood panels; over 300 mm wide	PC £51.45	1.10	8.90	56.69	m2	**65.59**
Panelled partitions; mouldings planted on both sides						
32 mm thick frame; 9 mm thick plywood panels; over 300 mm wide	PC £41.01	0.99	8.01	45.19	m2	**53.20**
38 mm thick frame; 9 mm thick plywood panels over 300 mm wide	PC £42.76	1.05	8.50	47.12	m2	**55.62**
50 mm thick frame; 12 mm thick plywood panels; over 300 mm wide	PC £51.45	1.10	8.90	56.69	m2	**65.59**
Panelled partitions; diminishing stiles over 300 mm wide; upper portion open panels for glass; in medium panes						
32 mm thick frame; 9 mm thick plywood panels; over 300 mm wide	PC £58.25	0.99	8.01	64.19	m2	**72.20**
38 mm thick frame; 9 mm thick plywood panels over 300 mm wide	PC £65.14	1.05	8.50	71.77	m2	**80.27**
50 mm thick frame; 12 mm thick plywood panels; over 300 mm wide	PC £81.08	1.10	8.90	89.35	m2	**98.25**
Purpose made screen components; selected West African Mahogany; PC £505.25/m3						
Panelled partitions						
32 mm thick frame; 9 mm thick plywood panels; over 300 mm wide	PC £62.51	1.30	10.52	68.87	m2	**79.39**
38 mm thick frame; 9 mm thick plywood panels over 300 mm wide	PC £67.70	1.40	11.33	74.59	m2	**85.92**
50 mm thick frame; 12 mm thick plywood panels; over 300 mm wide	PC £85.23	1.50	12.14	93.91	m2	**106.05**
Panelled partitions; mouldings worked on solid both sides						
32 mm thick frame; 9 mm thick plywood panels; over 300 mm wide	PC £71.40	1.30	10.52	78.68	m2	**89.20**
38 mm thick frame; 9 mm thick plywood panels over 300 mm wide	PC £79.77	1.40	11.33	87.90	m2	**99.23**
50 mm thick frame; 12 mm thick plywood panels; over 300 mm wide	PC £88.52	1.50	12.14	97.53	m2	**109.67**
Panelled partitions; mouldings planted on both sides						
32 mm thick frame; 9 mm thick plywood panels; over 300 mm wide	PC £90.28	1.30	10.52	99.49	m2	**110.01**
38 mm thick frame; 9 mm thick plywood panels over 300 mm wide	PC £91.07	1.40	11.33	100.35	m2	**111.68**
50 mm thick frame; 12 mm thick plywood panels; over 300 mm wide	PC £99.17	1.50	12.14	109.27	m2	**121.41**

K LININGS/SHEATHING/DRY PARTITIONING Including overheads and profit at 7.50%		Labour hours	Labour £	Material £	Unit	Total rate £
K29 TIMBER FRAMED AND PANELLED PARTITIONS - cont'd						
Purpose made screen components; selected West African Mahogany; PC £505.25/m3 - cont'd						
Panelled partitions; diminishing stiles over 300 mm wide; upper portion open panels for glass; in medium panes						
32 mm thick frame; 9 mm thick plywood panels; over 300 mm wide	PC £117.23	1.30	10.52	129.17	m2	**139.69**
38 mm thick frame; 9 mm thick plywood panels over 300 mm wide	PC £125.95	1.40	11.33	138.79	m2	**150.12**
50 mm thick frame; 12 mm thick plywood panels; over 300 mm wide	PC £133.69	1.50	12.14	147.30	m2	**159.44**
Purpose made screen components; Afrormosia PC £763.25/m3						
Panelled partitions						
32 mm thick frame; 9 mm thick plywood panels; over 300 mm wide	PC £78.73	1.30	10.52	86.75	m2	**97.27**
38 mm thick frame; 9 mm thick plywood panels over 300 mm wide	PC £85.38	1.40	11.33	94.07	m2	**105.40**
50 mm thick frame; 12 mm thick plywood panels; over 300 mm wide	PC £107.61	1.50	12.14	118.58	m2	**130.72**
Panelled partitions; mouldings worked on solid both sides						
32 mm thick frame; 9 mm thick plywood panels; over 300 mm wide	PC £90.02	1.30	10.52	99.19	m2	**109.71**
38 mm thick frame; 9 mm thick plywood panels over 300 mm wide	PC £100.74	1.40	11.33	111.01	m2	**122.34**
50 mm thick frame; 12 mm thick plywood panels; over 300 mm wide	PC £111.53	1.50	12.14	122.89	m2	**135.03**
Panelled partitions; mouldings planted on both sides						
32 mm thick frame; 9 mm thick plywood panels; over 300 mm wide	PC £120.91	1.30	10.52	133.23	m2	**143.75**
38 mm thick frame; 9 mm thick plywood panels over 300 mm wide	PC £121.88	1.40	11.33	134.30	m2	**145.63**
50 mm thick frame; 12 mm thick plywood panels; over 300 mm wide	PC £125.11	1.50	12.14	137.86	m2	**150.00**
Panelled partitions; diminishing stiles over 300 mm wide; upper portion open panels for glass; in medium panes						
32 mm thick frame; 9 mm thick plywood panels; over 300 mm wide	PC £147.79	1.30	10.52	162.85	m2	**173.37**
38 mm thick frame; 9 mm thick plywood panels over 300 mm wide	PC £158.54	1.40	11.33	174.69	m2	**186.02**
50 mm thick frame; 12 mm thick plywood panels; over 300 mm wide	PC £168.76	1.50	12.14	185.95	m2	**198.09**
K31 PLASTERBOARD FIXED PARTITIONS/ INNER WALLS/ LININGS						
'Gyproc' laminated partition; comprising two skins of gypsum plasterboard bonded to a centre core of plasterboard square edge plank 19 mm thick; fixing with nails to softwood studwork (measured separately); joints filled with joint filler and joint tape; to receive direct decoration; (perimeter studwork measured separately)						
50 mm partition; two outer skins of 12.7 mm tapered edge wallboard						
height 2.10 - 2.40 m		2.35	28.02	19.08	m	**47.10**
height 2.40 - 2.70 m		2.65	31.59	21.53	m	**53.12**

K LININGS/SHEATHING/DRY PARTITIONING Including overheads and profit at 7.50%	Labour hours	Labour £	Material £	Unit	Total rate £
height 2.70 - 3.00 m	2.95	35.17	24.04	m	**59.21**
height 3.00 - 3.30 m	3.30	39.34	26.53	m	**65.87**
height 3.30 - 3.60 m	3.75	44.71	29.04	m	**73.75**
65 mm partition; two outer skins of 19 mm tapered edge plank					
height 2.10 - 2.40 m	2.60	31.00	24.94	m	**55.94**
height 2.40 - 2.70 m	2.85	33.98	28.11	m	**62.09**
height 2.70 - 3.00 m	3.30	39.34	31.32	m	**70.66**
height 3.00 - 3.30 m	3.70	44.11	34.53	m	**78.64**
height 3.30 - 3.60 m	4.15	49.48	37.74	m	**87.22**
Labours and associated additional wrought softwood studwork					
Floor, wall or ceiling battens					
25 x 38 mm	0.13	1.55	0.49	m	**2.04**
Forming openings					
25 x 38 mm framing	0.33	3.93	0.75	m	**4.68**
Fair ends	0.22	2.62	0.49	m	**3.11**
Angle	0.33	3.93	0.49	m	**4.42**
Intersection	0.22	2.62	-	m	**2.62**
Cutting and fitting around steel joists, angles, trunking, ducting, ventilators, pipes, tubes, etc					
over 2 m girth	0.10	1.19	-	m	**1.19**
not exceeding 0.30 m girth	0.06	0.72	-	nr	**0.72**
0.30 - 1 m girth	0.08	0.95	-	nr	**0.95**
1 - 2 m girth	0.12	1.43	-	nr	**1.43**
'Paramount' dry partition comprising Paramount panels with and including wrought softwood battens at vertical joints; fixing with nails to softwood studwork; (perimeter studwork measured separately)					
Square edge panels; joints filled with plaster and jute scrim cloth; to receive plaster (measured elsewhere);					
57 mm partition					
height 2.10 - 2.40 m	1.60	19.07	16.23	m	**35.30**
height 2.40 - 2.70 m	1.75	20.86	18.28	m	**39.14**
height 2.70 - 3.00 m	2.00	23.84	20.30	m	**44.14**
height 3.00 - 3.30 m	2.25	26.82	20.47	m	**47.29**
height 3.30 - 3.60 m	2.60	31.00	24.47	m	**55.47**
63 mm partition					
height 2.10 - 2.40 m	1.75	20.86	18.49	m	**39.35**
height 2.40 - 2.70 m	1.95	23.25	20.82	m	**44.07**
height 2.70 - 3.00 m	2.15	25.63	23.12	m	**48.75**
height 3.00 - 3.30 m	2.40	28.61	25.48	m	**54.09**
height 3.30 - 3.60 m	2.75	32.78	27.85	m	**60.63**
Tapered edge panels; joints filled with joint filler and joint tape to receive direct decoration;					
57 mm partition					
height 2.10 - 2.40 m	2.25	26.82	17.60	m	**44.42**
height 2.40 - 2.70 m	2.40	28.61	19.84	m	**48.45**
height 2.70 - 3.00 m	2.65	31.59	22.06	m	**53.65**
height 3.00 - 3.30 m	2.90	34.57	24.34	m	**58.91**
height 3.30 - 3.60 m	3.30	39.34	26.63	m	**65.97**
63 mm partition					
height 2.10 - 2.40 m	2.40	28.61	19.90	m	**48.51**
height 2.40 - 2.70 m	2.60	31.00	22.41	m	**53.41**
height 2.70 - 3.00 m	2.80	33.38	24.92	m	**58.30**
height 3.00 - 3.30 m	3.10	36.96	27.49	m	**64.45**
height 3.30 - 3.60 m	3.45	41.13	30.05	m	**71.18**

K LININGS/SHEATHING/DRY PARTITIONING Including overheads and profit at 7.50%	Labour hours	Labour £	Material £	Unit	Total rate £
K31 PLASTERBOARD FIXED PARTITIONS/ INNER WALLS/ LININGS - cont'd					
Labours and associated additional wrought softwood studwork on 57 mm partition					
Wall or ceiling battens					
19 x 37 mm	0.13	1.55	0.44	m	1.99
Sole plates; with 19 x 37 x 150 mm long battens spiked on at 600 mm centres					
19 x 57 mm	0.28	3.34	0.81	m	4.15
50 x 57 mm	0.33	3.93	1.29	m	5.22
Forming openings					
37 x 37 mm framing	0.33	3.93	0.44	m	4.37
Fair ends	0.22	2.62	0.44	m	3.06
Angle	0.33	3.93	0.93	m	4.86
Intersection	0.28	3.34	0.63	m	3.97
Cutting and fitting around steel joists, angles, trunking, ducting, ventilators, pipes, tubes, etc					
over 2 m girth	0.11	1.31	-	m	1.31
not exceeding 0.30 m girth	0.07	0.83	-	nr	0.83
0.30 - 1 m girth	0.09	1.07	-	nr	1.07
1 - 2 m girth	0.13	1.55	-	nr	1.55
Plugging; 300 mm centres; one way					
brickwork	0.11	1.31	0.03	m	1.34
concrete	0.20	2.38	0.03	m	2.41
'Gyroc' metal stud partition; comprising 146 mm metal stud frame; with floor channel plugged and screwed to concrete through 38 x 148 mm tanalised softwood sole plate					
Tapered edge panels; joints filled with joint filler and joint tape to receive direct decoration					
171 mm partition; one hour; one layer of 12.5 mm Fireline board each side					
height 2.10 - 2.40 m	4.60	54.84	22.81	m	77.65
height 2.40 - 2.70 m	5.35	63.78	25.20	m	88.98
height 2.70 - 3.00 m	5.95	70.93	27.73	m	98.66
height 3.00 - 3.30 m	6.90	82.26	30.40	m	112.66
height 3.30 - 3.60 m	7.55	90.01	32.51	m	122.52
height 3.60 - 3.90 m	9.00	107.30	35.55	m	142.85
height 3.90 - 4.20 m	9.70	115.64	38.09	m	153.73
Angles	0.22	2.62	1.28	m	3.90
T-junctions	0.22	2.62	1.19	m	3.81
Fair end	0.33	3.93	1.99	m	5.92
Tapered edge panels; joints filled with joint filler and joint tape to receive direct decoration					
196 mm partition; two hour; two layers of 12.5 mm Fireline board both sides					
height 2.10 - 2.40 m	6.60	78.68	36.19	m	114.87
height 2.40 - 2.70 m	7.45	88.82	40.25	m	129.07
height 2.70 - 3.00 m	8.25	98.35	44.44	m	142.79
height 3.00 - 3.30 m	9.65	115.04	48.50	m	163.54
height 3.30 - 3.60 m	10.56	125.89	52.57	m	178.46
height 3.60 - 3.90 m	12.90	153.79	57.27	m	211.06
height 3.90 - 4.20 m	13.90	165.71	61.48	m	227.19
Angles	0.22	2.62	1.28	m	3.90
T-junctions	0.22	2.62	1.19	m	3.81
Fair end	0.39	4.65	2.54	m	7.19

L WINDOWS/DOORS/STAIRS
Including overheads and profit at 7.50%

ALTERNATIVE TIMBER WINDOW PRICES (£/each)

	£		£
Softwood shallow circular bay windows			
3 lightx1200 mm; ref CSB312CV	175.77	4 lightx1350 mm; ref CSB413CV	247.59
3 lightx1350 mm; ref CSB313CV	184.17	4 lightx1500 mm; ref CSB415T	280.63
3 light glass fibre roof unit	77.56	4 light glass fibre roof unit	83.58

		Labour hours	Labour £	Material £	Unit	Total rate £
L10 TIMBER WINDOWS/ROOFLIGHTS/SCREENS/LOUVRES						
Standard windows; 'treated' wrought softwood (refs. refer to Boulton & Paul cat. nos.)						
Side hung casement windows without glazing bars; with 140 mm wide softwood sills; opening casements and ventilators hung on rustproof hinges; fitted with aluminized laquered finish casement stays and fasteners; knotting and priming by manufacturer before delivery						
500 x 750 mm; ref N07V	PC £25.94	0.77	6.23	27.88	nr	**34.11**
500 x 900 mm; ref N09V	PC £30.35	0.88	7.12	32.62	nr	**39.74**
600 x 750 mm; ref 107V	PC £32.93	0.88	7.12	35.41	nr	**42.53**
600 x 750 mm; ref 107C	PC £30.10	0.88	7.12	32.36	nr	**39.48**
600 x 900 mm; ref 109V	PC £29.96	0.99	8.01	32.21	nr	**40.22**
600 x 900 mm; ref 109C	PC £35.84	0.88	7.12	38.53	nr	**45.65**
600 x 1050 mm; ref 110V	PC £35.07	0.88	7.12	37.70	nr	**44.82**
600 x 1050 mm; ref 110C	PC £37.45	1.10	8.90	40.26	nr	**49.16**
900 x 900 mm; ref 2N09W	PC £38.64	1.20	9.71	41.54	nr	**51.25**
900 x 1050 mm; ref 2N10W	PC £39.34	1.25	10.12	42.29	nr	**52.41**
900 x 1200 mm; ref 2N12W	PC £40.50	1.30	10.52	43.53	nr	**54.05**
900 x 1350 mm; ref 2N13W	PC £41.48	1.50	12.14	44.59	nr	**56.73**
900 x 1500 mm; ref 2N15W	PC £42.21	1.55	12.55	45.38	nr	**57.93**
1200 x 750 mm; ref 207C	PC £40.22	1.25	10.12	43.23	nr	**53.35**
1200 x 750 mm; ref 207CV	PC £52.61	1.25	10.12	56.55	nr	**66.67**
1200 x 900 mm; ref 209C	PC £42.49	1.30	10.52	45.68	nr	**56.20**
1200 x 900 mm; ref 209W	PC £44.84	1.30	10.52	48.20	nr	**58.72**
1200 x 900 mm; ref 209CV	PC £54.88	1.30	10.52	59.00	nr	**69.52**
1200 x 1050 mm; ref 210C	PC £44.17	1.50	12.14	47.48	nr	**59.62**
1200 x 1050 mm; ref 210W	PC £45.78	1.50	12.14	49.21	nr	**61.35**
1200 x 1050 mm; ref 210T	PC £54.88	1.50	12.14	59.00	nr	**71.14**
1200 x 1050 mm; ref 210CV	PC £56.74	1.50	12.14	60.99	nr	**73.13**
1200 x 1200 mm; ref 212C	PC £46.27	1.60	12.95	49.74	nr	**62.69**
1200 x 1200 mm; ref 212W	PC £46.83	1.60	12.95	50.34	nr	**63.29**
1200 x 1200 mm; ref 212T	PC £56.77	1.60	12.95	61.03	nr	**73.98**
1200 x 1200 mm; ref 212CV	PC £58.91	1.60	12.95	63.32	nr	**76.27**
1200 x 1350 mm; ref 213W	PC £47.95	1.70	13.76	51.55	nr	**65.31**
1200 x 1350 mm; ref 213CV	PC £62.23	1.70	13.76	66.90	nr	**80.66**
1200 x 1500 mm; ref 215W	PC £49.28	1.85	14.98	52.98	nr	**67.96**
1770 x 750 mm; ref 307CC	PC £63.63	1.55	12.55	68.40	nr	**80.95**
1770 x 900 mm; ref 309CC	PC £66.92	1.85	14.98	71.94	nr	**86.92**
1770 x 1050 mm; ref 310C	PC £53.52	1.95	15.78	57.53	nr	**73.31**
1770 x 1050 mm; ref 310T	PC £65.52	1.85	14.98	70.43	nr	**85.41**
1770 x 1050 mm; ref 310CC	PC £69.82	1.55	12.55	75.06	nr	**87.61**
1770 x 1050 mm; ref 310WW	PC £77.17	1.55	12.55	82.96	nr	**95.51**
1770 x 1200 mm; ref 312C	PC £55.34	2.00	16.19	59.49	nr	**75.68**
1770 x 1200 mm; ref 312T	PC £67.59	2.00	16.19	72.65	nr	**88.84**

L WINDOWS/DOORS/STAIRS Including overheads and profit at 7.50%		Labour hours	Labour £	Material £	Unit	Total rate £
L10 TIMBER WINDOWS/ROOFLIGHTS/SCREENS/LOUVRES - cont'd						
Standard windows; 'treated' wrought softwood (refs. refer to Boulton & Paul cat. nos.) - cont'd						
Side hung casement windows without glazing bars; with 140 mm wide softwood sills; opening casements and ventilators hung on rustproof hinges; fitted with aluminized laquered finish casement stays and fasteners; knotting and priming by manufacturer before delivery						
1770 x 1200 mm; ref 312CC	PC £72.77	2.00	16.19	78.22	nr	94.41
1770 x 1200 mm; ref 312WW	PC £79.35	2.00	16.19	85.30	nr	101.49
1770 x 1200 mm; ref 312CVC	PC £85.40	2.00	16.19	91.81	nr	108.00
1770 x 1350 mm; ref 313CC	PC £78.30	2.10	17.00	84.17	nr	101.17
1770 x 1350 mm; ref 312CC	PC £81.38	2.10	17.00	87.48	nr	104.48
1770 x 1350 mm; ref 313CVC	PC £90.97	2.10	17.00	97.79	nr	114.79
1770 x 1500 mm; ref 315T	PC £71.26	2.20	17.81	76.60	nr	94.41
2340 x 1050 mm; ref 410CWC	PC £97.86	2.15	17.40	105.20	nr	122.60
2340 x 1200 mm; ref 412CWC	PC £101.29	2.25	18.21	108.89	nr	127.10
2340 x 1350 mm; ref 413CWC	PC £107.35	2.40	19.42	115.40	nr	134.82
Top hung casement windows; with 140 mm wide softwood sills; opening casements and ventilators hung on rustproof hinges; fitted with aluminized laquered finish casement stays; knotting and priming by manufacturer before delivery						
600 x 750 mm; ref 107A	PC £32.90	0.88	7.12	35.37	nr	42.49
600 x 900 mm; ref 109A	PC £34.23	0.99	8.01	36.80	nr	44.81
600 x 1050 mm; ref 110A	PC £35.95	1.10	8.90	38.64	nr	47.54
900 x 750 mm; ref 2N07A	PC £41.06	1.15	9.31	44.13	nr	53.44
900 x 900 mm; ref 2N09A	PC £44.63	1.20	9.71	47.97	nr	57.68
900 x 1050 mm; ref 2N10A	PC £46.55	1.25	10.12	50.04	nr	60.16
900 x 1350 mm; ref 2N13AS	PC £53.87	1.50	12.14	57.90	nr	70.04
900 x 1500 mm; ref 2N15AS	PC £55.62	1.55	12.55	59.79	nr	72.34
1200 x 750 mm; ref 207A	PC £47.22	1.25	10.12	50.76	nr	60.88
1200 x 900 mm; ref 209A	PC £50.47	1.30	10.52	54.26	nr	64.78
1200 x 1050 mm; ref 210A	PC £52.40	1.50	12.14	56.32	nr	68.46
1200 x 1050 mm; ref 210AT	PC £70.24	1.50	12.14	75.51	nr	87.65
1200 x 1200 mm; ref 212A	PC £54.22	1.60	12.95	58.28	nr	71.23
1200 x 1200 mm; ref 212AT	PC £74.13	1.60	12.95	79.69	nr	92.64
1200 x 1350 mm; ref 213AS	PC £60.06	1.70	13.76	64.56	nr	78.32
1200 x 1500 mm; ref 215AS	PC £61.91	1.85	14.98	66.56	nr	81.54
1770 x 1050 mm; ref 310A	PC £63.32	1.85	14.98	68.06	nr	83.04
1770 x 1050 mm; ref 310AV	PC £77.67	1.85	14.98	83.49	nr	98.47
1770 x 1200 mm; ref 312A	PC £65.49	2.00	16.19	70.40	nr	86.59
1770 x 1220 mm; ref 312AV	PC £79.87	2.00	16.19	85.86	nr	102.05
2340 x 1200 mm; ref 412A	PC £73.57	2.25	18.21	79.09	nr	97.30
2340 x 1200 mm; ref 412AW	PC £93.14	2.25	18.21	100.12	nr	118.33
2340 x 1500 mm; ref 415AWS	PC £99.16	2.60	21.05	106.59	nr	127.64
High performance top hung reversible windows; with 140 mm wide softwood sills; adjustable ventilators weather stripping; opening sashes and fanlights hung on rustproof hinges; fitted with aluminized laquered espagnolette bolts; knotting and priming by manufacturer before delivery						
600 x 900 mm; ref R0609	PC £82.67	0.99	8.01	88.87	nr	96.88
900 x 900 mm; ref R0909	PC £90.41	1.20	9.71	97.19	nr	106.90
900 x 1050 mm; ref R0910	PC £92.19	1.25	10.12	99.10	nr	109.22
900 x 1200 mm; ref R0912	PC £94.75	1.40	11.33	101.85	nr	113.18
900 x 1500 mm; ref R0915	PC £107.35	1.55	12.55	115.40	nr	127.95
1200 x 1050 mm; ref R1210	PC £100.56	1.50	12.14	108.10	nr	120.24

L WINDOWS/DOORS/STAIRS Including overheads and profit at 7.50%		Labour hours	Labour £	Material £	Unit	Total rate £
1200 x 1200 mm; ref R1212	PC £103.25	1.60	12.95	110.99	nr	**123.94**
1200 x 1500 mm; ref R1215	PC £117.01	1.85	14.98	125.78	nr	**140.76**
1500 x 1200 mm; ref R1512	PC £115.88	1.85	14.98	124.58	nr	**139.56**
1800 x 1050 mm; ref R1810	PC £121.03	1.85	14.98	130.11	nr	**145.09**
1800 x 1200 mm; ref R1812	PC £123.20	2.00	16.19	132.44	nr	**148.63**
1800 x 1500 mm; ref R1815	PC £132.27	2.10	17.00	142.18	nr	**159.18**
2400 x 1500 mm; ref R2415	PC £146.76	2.60	21.05	157.76	nr	**178.81**
High performance double hung sash windows with glazing bars; solid frames; 63 x 175 mm softwood sills; standard flush external linings; spiral spring balances and sash catch; knotting and priming by manufacturer before delivery						
635 x 1050 mm; ref DH0610B	PC £118.37	2.20	17.81	127.25	nr	**145.06**
635 x 1350 mm; ref DH0613B	PC £128.77	2.40	19.43	138.42	nr	**157.85**
635 x 1650 mm; ref DH0616B	PC £138.01	2.70	21.86	148.36	nr	**170.22**
860 x 1050 mm; ref DH0810B	PC £129.75	2.55	20.64	139.48	nr	**160.12**
860 x 1350 mm; ref DH0813B	PC £140.94	2.85	23.07	151.52	nr	**174.59**
860 x 1650 mm; ref DH0816B	PC £152.15	3.30	26.71	163.56	nr	**190.27**
1085 x 1050 mm; ref DH1010B	PC £141.68	2.85	23.07	152.31	nr	**175.38**
1085 x 1350 mm; ref DH1013B	PC £156.03	3.30	26.71	167.73	nr	**194.44**
1085 x 1650 mm; ref DH1016B	PC £169.12	4.05	32.78	181.80	nr	**214.58**
1699 x 1050 mm; ref DH1710B	PC £258.26	4.05	32.78	277.63	nr	**310.41**
1699 x 1350 mm; ref DH1713B	PC £282.63	5.05	40.88	303.82	nr	**344.70**
1699 x 1650 mm; ref DH1716B	PC £304.22	5.15	41.69	327.04	nr	**368.73**
Purpose made window casements; 'treated' wrought softwood						
Casements; rebated; moulded						
38 mm thick	PC £20.15	-	-	21.66	m2	**21.66**
50 mm thick	PC £23.94	-	-	25.73	m2	**25.73**
Casements; rebated; moulded; in medium panes						
38 mm thick	PC £29.42	-	-	31.63	m2	**31.63**
50 mm thick	PC £36.40	-	-	39.13	m2	**39.13**
Casements; rebated; moulded; with semi-circular head						
38 mm thick	PC £36.76	-	-	39.52	m2	**39.52**
50 mm thick	PC £45.53	-	-	48.95	m2	**48.95**
Casements; rebated; moulded; to bullseye window						
38 mm thick x 600 mm dia	PC £66.10/nr	-	-	71.06	nr	**71.06**
38 mm thick x 900 mm dia	PC £98.06/nr	-	-	105.42	nr	**105.42**
50 mm thick x 600 mm dia	PC £76.04/nr	-	-	81.74	nr	**81.74**
50 mm thick x 900 mm dia	PC £112.74/nr	-	-	121.20	nr	**121.20**
Fitting and hanging casements						
square or rectangular		0.55	4.45	-	nr	**4.45**
semicircular		1.40	11.33	-	nr	**11.33**
bullseye		2.20	17.81	-	nr	**17.81**
Purpose made window frames; 'treated' wrought softwood						
Frames; rounded; rebated check grooved						
25 x 120 mm		0.15	1.21	6.90	m	**8.11**
50 x 75 mm		0.15	1.21	7.82	m	**9.03**
50 x 100 mm		0.18	1.46	9.63	m	**11.09**
50 x 125 mm		0.18	1.46	11.72	m	**13.18**
63 x 100 mm		0.18	1.46	11.98	m	**13.44**
75 x 150 mm		0.20	1.62	20.17	m	**21.79**
90 x 140 mm		0.20	1.62	26.13	m	**27.75**

L WINDOWS/DOORS/STAIRS Including overheads and profit at 7.50%		Labour hours	Labour £	Material £	Unit	Total rate £
L10 TIMBER WINDOWS/ROOFLIGHTS/SCREENS/LOUVRES - cont'd						
Purpose made window frames; 'treated' wrought softwood - cont'd						
Mullions and transoms; twice rounded, rebated and check grooved						
50 x 75 mm		0.11	0.89	9.49	m	10.38
50 x 100 mm		0.13	1.05	11.28	m	12.33
63 x 100 mm		0.13	1.05	13.65	m	14.70
75 x 150 mm		0.15	1.21	21.84	m	23.05
Sill; sunk weathered, rebated and grooved						
75 x 100 mm		0.22	1.78	19.74	m	21.52
75 x 150 mm		0.22	1.78	26.02	m	27.80
Add +5% to the above 'Material £ prices' for 'selected' softwood for staining						
Purpose made double hung sash windows; 'treated' wrought softwood						
Cased frames of 100 x 25 mm grooved inner linings; 114 x 25 mm grooved outer linings; 125 x 38 mm twice rebated head linings; 125 x 32 mm twice rebated grooved pulley stiles; 150 x 13 mm linings; 50 x 19 mm parting slips; 25 x 13 mm parting beads; 25 x 19 mm inside beads; 150 x 75 mm Oak twice sunk weathered throated sill; 50 mm thick rebated and moulded sashes; moulded horns; over 1.25 m2 each; both sashes in medium panes; including spiral spring balances	PC £148.40	2.50	20.24	225.51	m2	245.75
As above but with cased mullions	PC £156.59	2.75	22.26	234.32	m2	256.58
'Velux' pre-glazed roof windows; 'treated' Nordic Red Pine and aluminium trimmed 'Velux' windows, or equivalent; including type U flashings and soakers (for tiles and pantiles), and sealed double glazing unit; trimming opening measured separately						
550 x 700 mm; ref GGL-9	PC £100.49	10.00	80.95	115.44	nr	196.39
550 x 980 mm; ref GGL-6	PC £112.95	10.00	80.95	128.84	nr	209.79
700 x 1180 mm; ref GGL-5	PC £134.11	11.00	89.04	151.59	nr	240.63
780 x 980 mm; ref GGL-1	PC £129.10	10.50	84.99	146.20	nr	231.19
780 x 1400 mm; ref GGL-2	PC £152.01	11.50	93.09	170.83	nr	263.92
940 x 1600 mm; ref GGL-3	PC £178.98	11.50	93.09	199.82	nr	292.91
1140 x 1180 mm; ref GGL-4	PC £170.97	12.00	97.14	191.21	nr	288.35
1340 x 980 mm; ref GGL-7	PC £174.42	12.00	97.14	194.92	nr	292.06
1340 x 1400 mm; ref GGL-8	PC £201.79	12.50	101.18	224.34	nr	325.52
Standard windows; selected Philippine Mahogany; preservative stain finish						
Side hung casement windows; with 45 x 140 mm hardwood sills; weather stripping; opening sashes on canopy hinges; fitted with fasteners; aluminized lacquered finish ironmongery						
600 x 600 mm; ref SS0606/L	PC £63.35	1.05	8.50	68.10	nr	76.60
600 x 900 mm; ref SS0609/L	PC £70.21	1.30	10.52	75.48	nr	86.00
600 x 900 mm; ref SV0609/O	PC £93.60	1.05	8.50	100.62	nr	109.12
600 x 1050 mm; ref SS0610/L	PC £84.96	1.45	11.74	91.33	nr	103.07
600 x 1050 mm; ref SV0610/O	PC £96.40	1.45	11.74	103.63	nr	115.37
900 x 900 mm; ref SV0909/O	PC £124.72	1.65	13.36	134.07	nr	147.43
900 x 1050 mm; ref SV0910/O	PC £125.60	1.75	14.17	135.02	nr	149.19
900 x 1200 mm; ref SV0912/O	PC £128.48	1.85	14.98	138.12	nr	153.10
900 x 1350 mm; ref SV0913/O	PC £131.04	2.00	16.19	140.87	nr	157.06
900 x 1500 mm; ref SV0915/O	PC £133.76	2.10	17.00	143.79	nr	160.79

L WINDOWS/DOORS/STAIRS Including overheads and profit at 7.50%		Labour hours	Labour £	Material £	Unit	Total rate £
1200 x 900 mm; ref SS1209/L	PC £131.36	1.85	14.98	141.21	nr	**156.19**
1200 x 900 mm; ref SV1209/O	PC £143.20	1.85	14.98	153.94	nr	**168.92**
1200 x 1050 mm; ref SS1210/L	PC £121.66	2.00	16.19	130.78	nr	**146.97**
1200 x 1050 mm; ref SV1210/O	PC £146.00	2.00	16.19	156.95	nr	**173.14**
1200 x 1200 mm; ref SS1212/L	PC £128.17	2.15	17.40	137.78	nr	**155.18**
1200 x 1200 mm; ref SV1212/O	PC £148.64	2.15	17.40	159.79	nr	**177.19**
1200 x 1350 mm; ref SV1213/O	PC £151.04	2.30	18.62	162.37	nr	**180.99**
1200 x 1500 mm; ref SV1215/O	PC £154.00	2.40	19.43	165.55	nr	**184.98**
1800 x 900 mm; ref SS189D/O	PC £144.48	2.50	20.24	155.32	nr	**175.56**
1800 x 1050 mm; ref SS1810/L	PC £144.48	2.50	20.24	155.32	nr	**175.56**
1800 x 1050 mm; ref SS180D/O	PC £215.60	2.50	20.24	231.77	nr	**252.01**
1800 x 1200 mm; ref SS1812/L	PC £172.72	2.65	21.45	185.67	nr	**207.12**
1800 x 1200mm; ref SS182D/O	PC £210.49	2.65	21.45	226.28	nr	**247.73**
2400 x 1200 mm; ref SS242D/O	PC £233.38	3.10	25.09	250.88	nr	**275.97**
Top hung casement windows; with 45 x 140 mm hardwood sills; weather stripping; opening sashes on canopy hinges; fitted with fasteners; aluminized lacquered finish ironmongery						
600 x 900 mm; ref ST0609/O	PC £81.92	1.05	8.50	88.06	nr	**96.56**
600 x 1050 mm; ref ST0610/O	PC £88.16	1.45	11.74	94.77	nr	**106.51**
900 x 900 mm; ref ST0909/O	PC £104.48	1.65	13.36	112.32	nr	**125.68**
900 x 1050 mm; ref ST0910/O	PC £108.00	1.75	14.17	116.10	nr	**130.27**
900 x 1200 mm; ref ST0912/O	PC £119.36	1.85	14.98	128.31	nr	**143.29**
900 x 1350 mm; ref ST0913/O	PC £125.92	2.00	16.19	135.36	nr	**151.55**
900 x 1500 mm; ref ST0915/O	PC £136.24	2.10	17.00	146.46	nr	**163.46**
1200 x 900 mm; ref ST1209/O	PC £121.60	2.10	17.00	130.72	nr	**147.72**
1200 x 1050 mm; ref ST1210/O	PC £115.64	2.00	16.19	124.31	nr	**140.50**
1200 x 1200 mm; ref ST1212/O	PC £125.65	2.15	17.40	135.07	nr	**152.47**
1200 x 1350 mm; ref ST1213/O	PC £152.24	2.30	18.62	163.66	nr	**182.28**
1200 x 1500 mm; ref ST1215/O	PC £134.75	2.35	19.02	144.86	nr	**163.88**
1500 x 1050 mm; ref ST1510/L	PC £161.84	2.30	18.62	173.98	nr	**192.60**
1500 x 1200 mm; ref ST1512/L	PC £176.24	2.40	19.43	189.46	nr	**208.89**
1500 x 1350 mm; ref ST1513/L	PC £186.80	2.55	20.64	200.81	nr	**221.45**
1500 x 1500 mm; ref ST1515/L	PC £200.00	2.75	22.26	215.00	nr	**237.26**
1800 x 1050 mm; ref ST1810/L	PC £153.23	2.50	20.24	164.72	nr	**184.96**
1800 x 1200 mm; ref ST1812/L	PC £165.76	2.65	21.45	178.19	nr	**199.64**
1800 x 1350 mm; ref ST1813/L	PC £174.86	2.75	22.26	187.97	nr	**210.23**
1800 x 1500 mm; ref ST1815/L	PC £213.20	2.95	23.88	229.19	nr	**253.07**
Purpose made window casements; selected West African Mahogany; PC £505.25/m3						
Casements; rebated; moulded						
38 mm thick	PC £29.47	-	-	31.68	m2	**31.68**
50 mm thick	PC £38.40	-	-	41.28	m2	**41.28**
Casements; rebated; moulded; in medium panes						
38 mm thick	PC £35.75	-	-	38.44	m2	**38.44**
50 mm thick	PC £47.85	-	-	51.44	m2	**51.44**
Casements; rebated; moulded; with semi-circular head						
38 mm thick	PC £44.70	-	-	48.05	m2	**48.05**
50 mm thick	PC £59.83	-	-	64.32	m2	**64.32**
Casements; rebated; moulded; to bullseye window						
38 mm thick x 600 mm dia	PC £96.74/nr	-	-	104.00	nr	**104.00**
38 mm thick x 900 mm dia	PC £144.78/nr	-	-	155.64	nr	**155.64**
50 mm thick x 600 mm dia	PC £111.42/nr	-	-	119.78	nr	**119.78**
50 mm thick x 900 mm dia	PC £165.89/nr	-	-	178.33	nr	**178.33**
Fitting and hanging casements						
square or rectangular		0.77	6.23	-	nr	**6.23**
semicircular		1.85	14.98	-	nr	**14.98**
bullseye		2.95	23.88	-	nr	**23.88**

L WINDOWS/DOORS/STAIRS Including overheads and profit at 7.50%		Labour hours	Labour £	Material £	Unit	Total rate £
L10 TIMBER WINDOWS/ROOFLIGHTS/SCREENS/LOUVRES - cont'd						
Purpose made window frames; selected West African Mahogany; PC £505.25/m3						
Frames; rounded; rebated check grooved						
25 x 120 mm	PC £9.10	0.20	1.62	10.02	m	**11.64**
50 x 75 mm	PC £10.31	0.20	1.62	11.36	m	**12.98**
50 x 100 mm	PC £12.67	0.23	1.86	13.96	m	**15.82**
50 x 125 mm	PC £15.40	0.23	1.86	16.98	m	**18.84**
63 x 100 mm	PC £15.78	0.23	1.86	17.38	m	**19.24**
75 x 150 mm	PC £26.54	0.26	2.10	29.24	m	**31.34**
90 x 140 mm	PC £34.32	0.26	2.10	37.82	m	**39.92**
Mullions and transoms; twice rounded, rebated and check grooved						
50 x 75 mm	PC £12.50	0.15	1.21	13.77	m	**14.98**
50 x 100 mm	PC £14.88	0.18	1.46	16.40	m	**17.86**
63 x 100 mm	PC £17.98	0.18	1.46	19.81	m	**21.27**
75 x 150 mm	PC £28.73	0.20	1.62	31.66	m	**33.28**
Sill; sunk weathered, rebated and grooved						
75 x 100 mm	PC £26.05	0.30	2.43	28.71	m	**31.14**
75 x 150 mm	PC £34.33	0.30	2.43	37.83	m	**40.26**
Purpose made double hung sash windows; selected West African Mahogany; PC £505.25/m3						
Cased frames of 100 x 25 mm grooved inner linings; 114 x 25 mm grooved outer linings; 125 x 38 mm twice rebated head linings; 125 x 32 mm twice rebated grooved pulley stiles; 150 x 13 mm linings; 50 x 19 mm parting slips; 25 x 13 mm parting beads; 25 x 19 mm inside beads; 150 x 75 mm Oak twice sunk weathered throated sill; 50 mm thick rebated and moulded sashes; moulded horns; over 1.25 m2 each; both sashes in medium panes; including spiral spring balances	PC £202.95	3.30	26.71	284.16	m2	**310.87**
As above but with cased mullions	PC £214.10	3.65	29.55	296.14	m2	**325.69**
Purpose made window casements; Afrormosia; PC £763.25/m3						
Casements; rebated; moulded						
38 mm thick	PC £36.23	-	-	38.95	m2	**38.95**
50 mm thick	PC £44.52	-	-	47.86	m2	**47.86**
Casements; rebated; moulded; in medium panes						
38 mm thick	PC £46.97	-	-	50.49	m2	**50.49**
50 mm thick	PC £56.43	-	-	60.67	m2	**60.67**
Casements; rebated; moulded; with semi-circular head						
38 mm thick	PC £58.72	-	-	63.13	m2	**63.13**
50 mm thick	PC £70.56	-	-	75.85	m2	**75.85**
Casements; rebated; moulded; to bullseye window						
38 mm thick x 600 mm dia	PC £119.03/nr	-	-	127.96	nr	**127.96**
38 mm thick x 900 mm dia	PC £177.85/nr	-	-	191.18	nr	**191.18**
50 mm thick x 600 mm dia	PC £136.89/nr	-	-	147.16	nr	**147.16**
50 mm thick x 900 mm dia	PC £204.69/nr	-	-	220.04	nr	**220.04**
Fitting and hanging casements						
square or rectangular		0.77	6.23	-	nr	**6.23**
semicircular		1.85	14.98	-	nr	**14.98**
bullseye		2.95	23.88	-	nr	**23.88**

L WINDOWS/DOORS/STAIRS Including overheads and profit at 7.50%		Labour hours	Labour £	Material £	Unit	Total rate £
Purpose made window frames; Afrormosia; PC £763.25/m3						
Frames; rounded; rebated check grooved						
25 x 120 mm	PC £10.62	0.20	1.62	11.70	m	**13.32**
50 x 75 mm	PC £12.13	0.20	1.62	13.36	m	**14.98**
50 x 100 mm	PC £15.08	0.23	1.86	16.62	m	**18.48**
50 x 125 mm	PC £18.47	0.23	1.86	20.35	m	**22.21**
63 x 100 mm	PC £18.83	0.23	1.86	20.75	m	**22.61**
75 x 150 mm	PC £31.96	0.26	2.10	35.21	m	**37.31**
90 x 140 mm	PC £41.56	0.26	2.10	45.80	m	**47.90**
Mullions and transoms; twice rounded, rebated and check grooved						
50 x 75 mm	PC £14.31	0.15	1.21	15.77	m	**16.98**
50 x 100 mm	PC £17.28	0.18	1.46	19.03	m	**20.49**
63 x 100 mm	PC £21.01	0.18	1.46	23.16	m	**24.62**
75 x 150 mm	PC £34.15	0.20	1.62	37.62	m	**39.24**
Sill; sunk weathered, rebated and grooved						
75 x 100 mm	PC £29.67	0.30	2.43	32.69	m	**35.12**
75 x 150 mm	PC £39.80	0.30	2.43	43.85	m	**46.28**
Purpose made double hung sash windows; Afrormosia PC £763.25/m3						
Cased frames of 100 x 25 mm grooved inner linings; 114 x 25 mm grooved outer linings; 125 x 38 mm twice rebated head linings; 125 x 32 mm twice rebated grooved pulley stiles; 150 x 13 mm linings; 50 x 19 mm parting slips; 25 x 13 mm parting beads; 25 x 19 mm inside beads; 150 x 75 mm Oak twice sunk weathered throated sill; 50 mm thick rebated and moulded sashes; moulded horns; over 1.25 m2 each; both sashes in medium panes; including spiral spring balances	PC £229.99	3.30	26.71	284.16	m2	**310.87**
As above but with cased mullions	PC £242.64	3.65	29.55	296.14	m2	**325.69**
L11 METAL WINDOWS/ROOFLIGHTS/SCREENS/LOUVRES						
Aluminium fixed and fanlight windows; Crittall 'Luminaire' stock sliders or similar; white acrylic finish; fixed in position; including lugs plugged and screwed to brickwork or blockwork; or screwed to wooden sub-frame (measured elsewhere)						
Fixed lights; factory glazed with 3, 4 or 5 mm OQ clear float glass						
600 x 900 mm; ref 6FL9A	PC £47.70	2.40	22.01	52.02	nr	**74.03**
900 x 1500 mm; ref 9FL15A	PC £67.50	3.45	31.64	73.30	nr	**104.94**
1200 x 900 mm; ref 12FL9A	PC £72.90	3.45	31.64	79.11	nr	**110.75**
1200 x 1200 mm; ref 12FL12A	PC £74.70	3.45	31.64	81.04	nr	**112.68**
1500 x 1200 mm; ref 15FL12A	PC £82.80	3.45	31.64	89.75	nr	**121.39**
1500 x 1500 mm; ref 15FL15A	PC £168.30	4.40	40.35	181.66	nr	**222.01**
Fixed lights; site double glazed with 11 mm clear float glass						
600 x 900 mm; ref 6FL9A	PC £63.90	3.30	30.26	69.43	nr	**99.69**
900 x 1500 mm; ref 9FL15A	PC £88.20	5.25	48.14	95.56	nr	**143.70**
1200 x 900 mm; ref 12FL9A	PC £102.60	5.80	53.18	111.04	nr	**164.22**
1200 x 1200 mm; ref 12FL12A	PC £101.70	6.00	55.02	110.07	nr	**165.09**
Fixed lights with fanlight over; factory glazed with 3, 4 or 5 mm OQ clear float glass						
600 x 900 mm; ref 6FV9A	PC £126.90	2.40	22.01	137.16	nr	**159.17**
600 x 1500 mm; ref 6FV15A	PC £140.40	2.40	22.01	151.67	nr	**173.68**
900 x 1200 mm; ref 9FV12A	PC £154.80	3.45	31.64	167.15	nr	**198.79**
900 x 1500 mm; ref 9FV15A	PC £165.60	3.45	31.64	178.76	nr	**210.40**

L WINDOWS/DOORS/STAIRS Including overheads and profit at 7.50%		Labour hours	Labour £	Material £	Unit	Total rate £
L11 METAL WINDOWS/ROOFLIGHTS/SCREENS/LOUVRES - cont'd						
Aluminium fixed and fanlight windows; Crittall 'Luminaire' stock sliders or similar; white acrylic finish; fixed in position; including lugs plugged and screwed to brickwork or blockwork; or screwed to wooden sub-frame (measured elsewhere) - cont'd						
Fixed lights with fanlight over; site double glazed with 11 mm clear float glass						
600 x 900 mm; ref 6FV9A	PC £152.10	3.30	30.26	164.25	nr	194.51
600 x 1500 mm; ref 6FV15A	PC £181.80	3.90	35.76	196.18	nr	231.94
900 x 1200 mm; ref 9FV12A	PC £200.70	4.75	43.56	216.49	nr	260.05
900 x 1500 mm; ref 9FV15A	PC £216.90	5.80	53.18	233.91	nr	287.09
Vertical slider; factory glazed with 3; 4 or 5 mm OQ clear float glass						
600 x 900 mm; ref 6VV9A	PC £108.90	2.95	27.05	117.81	nr	144.86
900 x 1100 mm; ref 9VV11A	PC £128.70	2.95	27.05	139.09	nr	166.14
1200 x 1500 mm; ref 12VV15A	PC £156.60	4.15	38.05	169.09	nr	207.14
Vertical slider; factory double glazed with 11 mm clear float glass						
600 x 900 mm; ref 6VV9A	PC £177.30	3.75	34.39	191.34	nr	225.73
900 x 1100 mm; ref 9VV11A	PC £208.80	3.75	34.39	225.20	nr	259.59
1200 x 1500 mm; ref 12VV15A	PC £292.50	5.25	48.14	315.18	nr	363.32
Horizontal slider; factory glazed with 3, 4 or 5 mm OQ clear float glass						
1200 x 900 mm; ref 12HH9A	PC £139.50	4.15	38.05	150.70	nr	188.75
1200 x 1200 mm; ref 12HH12A	PC £152.10	4.15	38.05	164.25	nr	202.30
1500 x 900 mm; ref 15HH9A	PC £152.10	4.15	38.05	164.25	nr	202.30
1500 x 1300 mm; ref 15HH13A	PC £174.60	4.15	38.05	188.44	nr	226.49
1800 x 1200 mm; ref 18HH12A	PC £184.50	5.80	53.18	199.08	nr	252.26
Horizontal slider; factory double glazed with 11 mm clear float glass						
1200 x 900 mm; ref 12HH9A	PC £189.90	5.25	48.14	204.88	nr	253.02
1200 x 1200 mm; ref 12HH12A	PC £208.80	5.25	48.14	225.20	nr	273.34
1500 x 900 mm; ref 15HH9A	PC £207.90	5.25	48.14	224.23	nr	272.37
1500 x 1300 mm; ref 15HH13A	PC £245.70	5.25	48.14	264.87	nr	313.01
1800 x 1200 mm; ref 18HH12A	PC £260.10	6.35	58.23	280.35	nr	338.58
Galvanized steel fixed light; casement and fanlight windows; Crittall; 'Homelight' range or similar; site glazing measured elsewhere; fixed in position; including lugs cut and pinned to brickwork or blockwork						
Basic fixed lights; including easy-glaze beads						
628 x 292 mm; ref ZNG5	PC £12.38	1.30	11.92	13.61	nr	25.53
628 x 923 mm; ref ZNC5	PC £17.64	1.30	11.92	19.27	nr	31.19
628 x 1513 mm; ref ZNDV5	PC £25.09	1.30	11.92	27.28	nr	39.20
1237 x 292 mm; ref ZNG13	PC £18.90	1.30	11.92	20.63	nr	32.55
1237 x 923 mm; ref ZNC13	PC £24.45	1.95	17.88	26.60	nr	44.48
1237 x 1218 mm; ref ZND13	PC £28.79	1.95	17.88	31.26	nr	49.14
1237 x 1513 mm; ref ZNDV13	PC £31.54	1.95	17.88	34.21	nr	52.09
1846 x 292 mm; ref ZNG14	PC £24.50	1.30	11.92	26.64	nr	38.56
1846 x 923 mm; ref ZNC14	PC £30.20	1.95	17.88	32.77	nr	50.65
1846 x 1513 mm; ref ZNDV14	PC £36.58	2.40	22.01	39.63	nr	61.64
Basic opening lights; including easy-glaze beads and weatherstripping						
628 x 292 mm; ref ZNG1	PC £28.11	1.30	11.92	30.52	nr	42.44
1237 x 292 mm; ref ZNG13G	PC £41.72	1.30	11.92	45.15	nr	57.07
1846 x 292 mm; ref ZNG4	PC £71.34	1.30	11.92	77.00	nr	88.92

L WINDOWS/DOORS/STAIRS Including overheads and profit at 7.50%		Labour hours	Labour £	Material £	Unit	Total rate £
One-piece composites; including easy-glaze beads and weatherstripping						
628 x 923 mm; ref ZNC5F	PC £39.30	1.30	11.92	42.56	nr	**54.48**
628 x 1513 mm; ref ZNDV5F	PC £47.03	1.30	11.92	50.87	nr	**62.79**
1237 x 923 mm; ref ZNC2F	PC £77.56	1.95	17.88	83.69	nr	**101.57**
1237 x 1218 mm; ref ZND2F	PC £91.26	1.95	17.88	98.41	nr	**116.29**
1237 x 1513 mm; ref ZNDV2V	PC £109.55	1.95	17.88	118.07	nr	**135.95**
1846 x 923 mm; ref NC4F	PC £119.51	1.95	17.88	128.78	nr	**146.66**
1846 x 1218 mm; ref ZND10F	PC £115.79	2.40	22.01	124.79	nr	**146.80**
Reversible windows; including easy-glaze beads and weatherstripping						
997 x 923 mm; ref NC13R	PC £104.51	1.70	15.59	112.66	nr	**128.25**
997 x 1067 mm; ref NC013R	PC £110.81	1.70	15.59	119.43	nr	**135.02**
1237 x 923 mm; ref ZNC13R	PC £115.44	2.55	23.38	124.41	nr	**147.79**
1237 x 1218 mm; ref ZND13R	PC £126.18	2.55	23.38	135.95	nr	**159.33**
1237 x 1513 mm; ref ZNDV13RS	PC £141.88	2.55	23.38	152.83	nr	**176.21**
Pressed steel sills; to suit above window widths						
628 mm	PC £5.97	0.39	3.58	7.16	nr	**10.74**
997 mm	PC £8.38	0.50	4.58	9.75	nr	**14.33**
1237 mm	PC £9.45	0.61	5.59	10.90	nr	**16.49**
1486 mm	PC £10.97	0.72	6.60	12.54	nr	**19.14**
1846 mm	PC £12.46	0.83	7.61	14.13	nr	**21.74**
Factory finished steel fixed light; casement and fanlight windows; Crittall polyester powder coated 'Homelight' range or similar; site glazing measured elsewhere; fixed in position; including lugs cut and pinned to brickwork or blockwork						
Basic fixed lights; including easy-glaze beads						
628 x 292 mm; ref ZNG5	PC £15.46	1.30	11.92	16.93	nr	**28.85**
628 x 923 mm; ref ZNC5	PC £22.04	1.30	11.92	24.00	nr	**35.92**
628 x 1513 mm; ref ZNDV5	PC £31.36	1.30	11.92	34.02	nr	**45.94**
1237 x 292 mm; ref ZNG13	PC £23.61	1.30	11.92	25.69	nr	**37.61**
1237 x 923 mm; ref ZNC13	PC £30.57	1.95	17.88	33.18	nr	**51.06**
1237 x 1218 mm; ref ZND13	PC £35.99	1.95	17.88	39.00	nr	**56.88**
1237 x 1513 mm; ref ZNDV13	PC £39.41	1.95	17.88	42.68	nr	**60.56**
1846 x 292 mm; ref ZNG14	PC £30.62	1.30	11.92	33.22	nr	**45.14**
1846 x 923 mm; ref ZNC14	PC £37.74	1.95	17.88	40.88	nr	**58.76**
1846 x 1513 mm; ref ZNDV14	PC £45.72	2.40	22.01	49.46	nr	**71.47**
Basic opening lights; including easy-glaze beads and weatherstripping						
628 x 292 mm; ref ZNG1	PC £34.18	1.30	11.92	37.05	nr	**48.97**
1237 x 292 mm; ref ZNG13G	PC £50.83	1.30	11.92	54.95	nr	**66.87**
1846 x 292 mm; ref ZNG4	PC £86.92	1.30	11.92	93.75	nr	**105.67**
One-piece composites; including easy-glaze beads and weatherstripping						
628 x 923 mm; ref ZNC5F	PC £48.12	1.30	11.92	52.04	nr	**63.96**
628 x 1513 mm; ref ZNDV5F	PC £57.79	1.30	11.92	62.43	nr	**74.35**
1237 x 923 mm; ref ZNC2F	PC £94.19	1.95	17.88	101.56	nr	**119.44**
1237 x 1218 mm; ref ZND2F	PC £110.93	1.95	17.88	119.56	nr	**137.44**
1237 x 1513 mm; ref ZNDV2V	PC £132.95	1.95	17.88	143.23	nr	**161.11**
1846 x 923 mm; ref NC4F	PC £144.80	1.95	17.88	155.97	nr	**173.85**
1846 x 1218 mm; ref ZND10F	PC £141.02	2.40	22.01	151.91	nr	**173.92**
Reversible windows; including easy-glaze beads and weatherstripping						
997 x 923 mm; ref NC13R	PC £135.86	1.70	15.59	146.35	nr	**161.94**
997 x 1067 mm; ref NC013R	PC £144.05	1.70	15.59	155.17	nr	**170.76**
1237 x 923 mm; ref ZNC13R	PC £150.08	2.55	23.38	161.64	nr	**185.02**
1237 x 1218 mm; ref ZND13R	PC £164.03	2.55	23.38	176.64	nr	**200.02**
1237 x 1513 mm; ref ZNDV13RS	PC £184.44	2.55	23.38	198.58	nr	**221.96**

L WINDOWS/DOORS/STAIRS Including overheads and profit at 7.50%		Labour hours	Labour £	Material £	Unit	Total rate £
L11 METAL WINDOWS/ROOFLIGHTS/SCREENS/LOUVRES - cont'd						
Factory finished steel fixed light; casement and fanlight windows; Crittall polyester powder coated 'Homelight' range or similar; site glazing measured elsewhere; fixed in position; including lugs cut and pinned to brickwork or blockwork - cont'd						
Pressed steel sills; to suit above window widths						
628 mm	PC £7.45	0.39	3.58	8.75	nr	**12.33**
997 mm	PC £10.47	0.50	4.58	11.99	nr	**16.57**
1237 mm	PC £11.81	0.61	5.59	13.44	nr	**19.03**
1486 mm	PC £13.72	0.72	6.60	15.49	nr	**22.09**
1846 mm	PC £15.56	0.83	7.61	17.47	nr	**25.08**
L12 PLASTICS WINDOWS/ROOFLIGHTS/SCREENS/ LOUVRES						
uPVC windows to BS 2782; 'Anglian' or similar; reinforced where appropriate with aluminium alloy; including standard ironmongery; cills and glazing; fixed in position; including lugs plugged and screwed to brickwork or blockwork						
Fixed light; including e.p.d.m. glazing gaskets and weather seals						
600 x 900 mm; single glazed	PC £104.40	3.85	35.30	112.54	nr	**147.84**
600 x 900 mm; double glazed	PC £117.60	3.85	35.30	126.73	nr	**162.03**
Casement/fixed light; including e.p.d.m. glazing gaskets and weather seals						
600 x 1200 mm; single glazed	PC £150.00	4.15	38.05	161.56	nr	**199.61**
600 x 1200 mm; double glazed	PC £163.20	4.15	38.05	175.75	nr	**213.80**
1200 x 1200 mm; single glazed	PC £268.80	4.95	45.39	289.27	nr	**334.66**
1200 x 1200 mm; double glazed	PC £295.20	4.95	45.39	317.65	nr	**363.04**
1800 x 1200 mm; single glazed	PC £384.00	5.50	50.43	413.11	nr	**463.54**
1800 x 1200 mm; double glazed	PC £422.40	5.50	50.43	454.39	nr	**504.82**
'Tilt & Turn' light; including e.p.d.m. glazing gaskets and weather seals						
1200 x 1200 mm; single glazed	PC £216.00	4.95	45.39	232.51	nr	**277.90**
1200 x 1200 mm; double glazed	PC £241.20	4.95	45.39	259.60	nr	**304.99**
uPVC Rooflights; plugged and screwed to concrete; or screwed to timber						
Rooflight; 'Coxdome Mark 2'; UPVC double skin; dome or pyramid						
619 x 619 mm	PC £109.85	1.65	10.59	123.99	nr	**134.58**
924 x 619 mm	PC £110.60	1.80	11.55	124.84	nr	**136.39**
924 x 924 mm	PC £153.35	2.00	12.84	173.10	nr	**185.94**
1229 x 924 mm	PC £241.50	2.15	13.80	272.60	nr	**286.40**
1229 x 1229 mm	PC £371.50	2.30	14.76	419.34	nr	**434.10**
1686 x 1076 mm		2.40	15.40	289.36	nr	**304.76**
Rooflight; 'Coxdome Mark 4'; UPVC double skin; GRP splayed upstand; hit and miss vent						
600 x 600 mm	PC £162.95	3.30	21.18	183.93	nr	**205.11**
900 x 600 mm	PC £256.25	3.65	23.42	289.24	nr	**312.66**
900 x 900 mm	PC £271.40	3.95	25.35	306.34	nr	**331.69**
1200 x 900 mm	PC £327.40	4.30	27.60	369.55	nr	**397.15**
1200 x 1200 mm	PC £343.05	4.60	29.52	387.22	nr	**416.74**
1800 x 1200 mm	PC £514.95	5.50	35.30	581.25	nr	**616.55**

L WINDOWS/DOORS/STAIRS
Including overheads and profit at 7.50%

ALTERNATIVE TIMBER DOOR PRICES (£/each)

	£		£		£
Hardwood doors					
Brazilian Mahogany period doors (£/each); 838 x 1981 x 44 mm					
4 panel	169.40	'Carolina'	195.80	'Gothic'	163.35
6 panel	174.24	'Elizabethan	203.28	'Kentucky'	186.12
Red Meranti period doors					
4 panel					
762x1981x44 mm	137.50	838x1981x44 mm	137.50	807x2000x44 mm	137.50
6 panel					
762x1981x44 mm	137.45	838x1981x44 mm	137.45	807x2000x44 mm	137.45
Half bow					
762x1981x44 mm	143.00	838x1981x44 mm	143.00	807x2000x44 mm	143.00
'Kentucky'					
762x1981x44 mm	138.77	838x1981x44 mm	138.77	807x2000x44 mm	138.77
Softwood doors					
Casement doors					
2 XG; two panel; beaded					
762x1981x44 mm	34.98	838x1981x44 mm	37.24	807x2000x44 mm	37.24
813x2032x44 mm	37.24	726x2040x44 mm	37.24	826x2040x44 mm	37.24
2 XGG; two panel; beaded					
762x1981x44 mm	33.83	838x1981x44 mm	36.08	807x2000x44 mm	36.08
813x2032x44 mm	36.08	726x2040x44 mm	36.08	826x2040x44 mm	36.08
10; one panel; beaded					
762x1981x44 mm	28.22	838x1981x44 mm	30.47	807x2000x44 mm	30.47
SA; 15 panel; beaded					
762x1981x44 mm	45.10	838x1981x44 mm	47.36	807x2000x44 mm	47.36
22; pair of two panel; beaded					
914x1981x44 mm	76.67	1168x1981x44 mm	76.67		
2 SA; pair of 10 panel; beaded					
914x1981x44 mm	101.48	1168x1981x44 mm	101.48		
Louvre doors					
533x1524x28 mm	13.86	610x1524x28 mm	15.18	686x1981x28 mm	20.59
533x1676x28 mm	15.01	610x1676x28 mm	16.62	762x1981x28 mm	22.83
533x1829x28 mm	15.99	610x1829x28 mm	17.65	737x1067x28 mm	
533x1981x28 mm	17.02	610x1981x28 mm	18.92	- pair	16.79
Period doors					
4 panel					
762x1981x44 mm	109.73	838x1981x44 mm	109.73	807x2000x44 mm	109.73
6 panel					
762x1981x44 mm	115.50	838x1981x44 mm	115.50	807x2000x44 mm	115.50
Half bow					
762x1981x44 mm	109.73	838x1981x44 mm	109.73	807x2000x44 mm	109.73
'Kentucky'					
762x1981x44 mm	106.43	838x1981x44 mm	106.43	807x2000x44 mm	106.43

L WINDOWS/DOORS/STAIRS Including overheads and profit at 7.50%		Labour hours	Labour £	Material £	Unit	Total rate £
L20 TIMBER DOORS/SHUTTERS/HATCHES						
Standard matchboarded doors; wrought softwood						
Matchboarded, ledged and braced doors; 25 mm ledges and braces; 19 mm tongued, grooved and V-jointed; one side vertical boarding						
762 x 1981 mm	PC £43.70	1.65	13.36	46.97	nr	**60.33**
838 x 1981 mm	PC £46.30	1.65	13.36	49.77	nr	**63.13**
Matchboarded, framed, ledged and braced doors; 44 mm framing; 25 mm intermediate and bottom rails; 19 mm tongued, grooved and V-jointed; one side vertical boarding						
762 x 1981 x 44 mm	PC £56.85	2.00	16.19	61.11	nr	**77.30**
838 x 1981 x 44 mm	PC £58.94	2.00	16.19	63.36	nr	**79.55**
Standard flush doors; softwood composition						
Flush door; internal quality; skeleton or cellular core; hardboard faced both sides;						
457 x 1981 x 35 mm	PC £11.61	1.40	11.33	12.48	nr	**23.81**
533 x 1981 x 35 mm	PC £11.61	1.40	11.33	12.48	nr	**23.81**
610 x 1981 x 35 mm	PC £13.20	1.40	11.33	14.54	nr	**25.87**
686 x 1981 x 35 mm	PC £11.61	1.40	11.33	12.48	nr	**23.81**
762 x 1981 x 35 mm	PC £11.61	1.40	11.33	12.48	nr	**23.81**
838 x 1981 x 35 mm	PC £12.71	1.40	11.33	13.66	nr	**24.99**
526 x 2040 x 40 mm	PC £13.20	1.40	11.33	14.19	nr	**25.52**
626 x 2040 x 40 mm	PC £13.20	1.40	11.33	14.19	nr	**25.52**
726 x 2040 x 40 mm	PC £13.53	1.40	11.33	14.54	nr	**25.87**
826 x 2040 x 40 mm	PC £13.53	1.40	11.33	14.54	nr	**25.87**
Flush door; internal quality; skeleton or cellular core; plywood faced both sides; lipped on two long edges						
457 x 1981 x 35 mm	PC £14.96	1.40	11.33	16.08	nr	**27.41**
533 x 1981 x 35 mm	PC £14.96	1.40	11.33	16.08	nr	**27.41**
610 x 1981 x 35 mm	PC £14.96	1.40	11.33	17.62	nr	**28.95**
686 x 1981 x 35 mm	PC £14.96	1.40	11.33	16.08	nr	**27.41**
762 x 1981 x 35 mm	PC £14.96	1.40	11.33	16.08	nr	**27.41**
838 x 1981 x 35 mm	PC £16.12	1.40	11.33	17.32	nr	**28.65**
526 x 2040 x 40 mm	PC £16.39	1.40	11.33	17.62	nr	**28.95**
626 x 2040 x 40 mm	PC £16.39	1.40	11.33	17.62	nr	**28.95**
726 x 2040 x 40 mm	PC £16.39	1.40	11.33	17.62	nr	**28.95**
826 x 2040 x 40 mm	PC £16.39	1.40	11.33	17.62	nr	**28.95**
Flush door; internal quality; skeleton or cellular core; Sapele faced both sides; lipped on all four edges						
457 x 1981 x 35 mm	PC £18.70	1.50	12.14	20.10	nr	**32.24**
533 x 1981 x 35 mm	PC £18.70	1.50	12.14	20.10	nr	**32.24**
610 x 1981 x 35 mm	PC £18.70	1.50	12.14	20.10	nr	**32.24**
686 x 1981 x 35 mm	PC £18.76	1.50	12.14	20.16	nr	**32.30**
762 x 1981 x 35 mm	PC £18.81	1.50	12.14	20.22	nr	**32.36**
838 x 1981 x 35 mm	PC £20.19	1.50	12.14	21.70	nr	**33.84**
Flush door; internal quality; skeleton or cellular core; Teak faced both sides; lipped on all four edges						
457 x 1981 x 35 mm	PC £71.50	1.50	12.14	76.86	nr	**89.00**
533 x 1981 x 35 mm	PC £62.87	1.50	12.14	67.58	nr	**79.72**
610 x 1981 x 35 mm	PC £62.87	1.50	12.14	67.58	nr	**79.72**
686 x 1981 x 35 mm	PC £62.87	1.50	12.14	67.58	nr	**79.72**
762 x 1981 x 35 mm	PC £74.80	1.50	12.14	80.41	nr	**92.55**
838 x 1981 x 35 mm	PC £64.90	1.50	12.14	69.77	nr	**81.91**
526 x 2040 x 40 mm	PC £64.13	1.50	12.14	68.94	nr	**81.08**
626 x 2040 x 40 mm	PC £64.13	1.50	12.14	68.94	nr	**81.08**
726 x 2040 x 40 mm	PC £64.15	1.50	12.14	68.96	nr	**81.10**
826 x 2040 x 40 mm	PC £64.90	1.50	12.14	69.77	nr	**81.91**

L WINDOWS/DOORS/STAIRS Including overheads and profit at 7.50%		Labour hours	Labour £	Material £	Unit	Total rate £
Flush door; half-hour fire check (30/20); hardboard faced both sides;						
457 x 1981 x 44 mm	PC £32.11	1.95	15.78	34.52	nr	**50.30**
533 x 1981 x 44 mm	PC £32.11	1.95	15.78	34.52	nr	**50.30**
610 x 1981 x 44 mm	PC £32.11	1.95	15.78	34.52	nr	**50.30**
686 x 1981 x 44 mm	PC £32.11	1.95	15.78	34.52	nr	**50.30**
762 x 1981 x 44 mm	PC £32.11	1.95	15.78	34.52	nr	**50.30**
838 x 1981 x 44 mm	PC £33.78	1.95	15.78	36.31	nr	**52.09**
526 x 2040 x 44 mm	PC £33.35	1.95	15.78	35.85	nr	**51.63**
626 x 2040 x 44 mm	PC £33.35	1.95	15.78	35.85	nr	**51.63**
726 x 2040 x 44 mm	PC £33.35	1.95	15.78	35.85	nr	**51.63**
826 x 2040 x 44 mm	PC £35.04	1.95	15.78	37.67	nr	**53.45**
Flush door; half-hour fire check (30/20); plywood faced both sides; lipped on all four edges						
457 x 1981 x 44 mm	PC £30.82	1.95	15.78	33.13	nr	**48.91**
533 x 1981 x 44 mm	PC £30.82	1.95	15.78	33.13	nr	**48.91**
610 x 1981 x 44 mm	PC £30.82	1.95	15.78	33.13	nr	**48.91**
686 x 1981 x 44 mm	PC £30.82	1.95	15.78	33.13	nr	**48.91**
762 x 1981 x 44 mm	PC £30.82	1.95	15.78	33.13	nr	**48.91**
838 x 1981 x 44 mm	PC £32.43	1.95	15.78	34.86	nr	**50.64**
526 x 2040 x 44 mm	PC £31.66	1.95	15.78	34.04	nr	**49.82**
626 x 2040 x 44 mm	PC £31.66	1.95	15.78	34.04	nr	**49.82**
726 x 2040 x 44 mm	PC £31.66	1.95	15.78	34.04	nr	**49.82**
826 x 2040 x 44 mm	PC £33.27	1.95	15.78	35.77	nr	**51.55**
Flush door; half hour fire resisting (type B30); Red Meranti veneer for painting; hardwood lipping two long edges						
526 x 2040 x 44 mm	PC £66.20	2.05	16.59	71.17	nr	**87.76**
626 x 2040 x 44 mm	PC £68.20	2.05	16.59	73.32	nr	**89.91**
726 x 2040 x 44 mm	PC £70.20	2.05	16.59	75.46	nr	**92.05**
826 x 2040 x 44 mm	PC £72.20	2.05	16.59	77.62	nr	**94.21**
Flush door; half-hour fire check (30/20); Sapele faced both sides; lipped on all four edges						
457 x 1981 x 44 mm	PC £45.62	2.05	16.59	49.04	nr	**65.63**
533 x 1981 x 44 mm	PC £45.62	2.05	16.59	49.04	nr	**65.63**
610 x 1981 x 44 mm	PC £45.62	2.05	16.59	49.04	nr	**65.63**
686 x 1981 x 44 mm	PC £45.62	2.05	16.59	49.04	nr	**65.63**
762 x 1981 x 44 mm	PC £45.62	2.05	16.59	49.04	nr	**65.63**
526 x 2040 x 44 mm	PC £46.80	2.05	16.59	50.31	nr	**66.90**
626 x 2040 x 44 mm	PC £46.80	2.05	16.59	50.31	nr	**66.90**
726 x 2040 x 44 mm	PC £46.80	2.05	16.59	50.31	nr	**66.90**
826 x 2040 x 44 mm	PC £48.56	2.05	16.59	52.20	nr	**68.79**
Flush door; half-hour fire resisting (30/30) Sapele faced both sides; lipped on all four edges						
457 x 1981 x 44 mm	PC £37.03	2.05	16.59	39.81	nr	**56.40**
533 x 1981 x 44 mm	PC £37.59	2.05	16.59	40.41	nr	**57.00**
610 x 1981 x 44 mm	PC £38.01	2.05	16.59	40.86	nr	**57.45**
686 x 1981 x 44 mm	PC £42.40	2.05	16.59	45.58	nr	**62.17**
762 x 1981 x 44 mm	PC £43.48	2.05	16.59	46.74	nr	**63.33**
838 x 1981 x 44 mm	PC £45.36	2.05	16.59	48.76	nr	**65.35**
526 x 2040 x 44 mm	PC £43.52	2.05	16.59	46.78	nr	**63.37**
626 x 2040 x 44 mm	PC £44.20	2.05	16.59	47.52	nr	**64.11**
726 x 2040 x 44 mm	PC £43.96	2.05	16.59	47.26	nr	**63.85**
826 x 2040 x 44 mm	PC £46.16	2.05	16.59	49.62	nr	**66.21**
Flush door; half hour fire resisting (type B30); American light oak veneer; hardwood lipping all edges						
526 x 2040 x 44 mm	PC £105.40	2.05	16.59	113.31	nr	**129.90**
626 x 2040 x 44 mm	PC £107.40	2.05	16.59	115.46	nr	**132.05**
726 x 2040 x 44 mm	PC £109.40	2.05	16.59	117.61	nr	**134.20**
826 x 2040 x 44 mm	PC £111.40	2.05	16.59	119.76	nr	**136.35**

L WINDOWS/DOORS/STAIRS Including overheads and profit at 7.50%		Labour hours	Labour £	Material £	Unit	Total rate £
L20 TIMBER DOORS/SHUTTERS/HATCHES - cont'd						
Standard flush doors; softwood composition - cont'd						
Flush door; half hour fire-resisting (30/30); 'Melador'; laminate faced both sides; hardwood lipped on all edges						
610 x 1981 x 47 mm	PC £210.92	2.50	20.24	226.74	nr	246.98
686 x 1981 x 47 mm	PC £210.92	2.50	20.24	226.74	nr	246.98
762 x 1981 x 47 mm	PC £210.92	2.50	20.24	226.74	nr	246.98
838 x 1981 x 47 mm	PC £210.92	2.50	20.24	226.74	nr	246.98
526 x 2040 x 47 mm	PC £210.92	2.50	20.24	226.74	nr	246.98
626 x 2040 x 47 mm	PC £210.92	2.50	20.24	226.74	nr	246.98
726 x 2040 x 47 mm	PC £210.92	2.50	20.24	226.74	nr	246.98
826 x 2040 x 47 mm	PC £210.92	2.50	20.24	226.74	nr	246.98
Flush door; one hour fire check (60/45); plywood faced both sides; lipped on all four edges						
610 x 1981 x 54 mm	PC £164.30	2.20	17.81	176.63	nr	194.44
686 x 1981 x 54 mm	PC £164.30	2.20	17.81	176.63	nr	194.44
762 x 1981 x 54 mm	PC £106.52	2.20	17.81	114.51	nr	132.32
838 x 1981 x 54 mm	PC £114.42	2.20	17.81	123.00	nr	140.81
526 x 2040 x 54 mm	PC £164.30	2.20	17.81	176.63	nr	194.44
626 x 2040 x 54 mm	PC £164.30	2.20	17.81	176.63	nr	194.44
726 x 2040 x 54 mm	PC £106.52	2.20	17.81	114.51	nr	132.32
826 x 2040 x 54 mm	PC £114.42	2.20	17.81	123.00	nr	140.81
Flush door; one hour fire check (60/45); Sapele faced both sides; lipped on all four edges						
610 x 1981 x 54 mm	PC £171.84	2.30	18.62	184.73	nr	203.35
686 x 1981 x 54 mm	PC £171.84	2.30	18.62	184.73	nr	203.35
762 x 2040 x 54 mm	PC £171.84	2.30	18.62	184.73	nr	203.35
838 x 1981 x 54 mm	PC £171.84	2.30	18.62	184.73	nr	203.35
526 x 2040 x 54 mm	PC £171.84	2.30	18.62	184.73	nr	203.35
626 x 2040 x 54 mm	PC £171.84	2.30	18.62	184.73	nr	203.35
762 x 2040 x 54 mm	PC £171.84	2.30	18.62	184.73	nr	203.35
826 x 2040 x 54 mm	PC £171.84	2.30	18.62	184.73	nr	203.35
Flush door; one hour fire resisting (60/60); Sapele faced both sides; lipped on all four edges						
610 x 1981 x 54 mm	PC £193.22	2.30	18.62	207.71	nr	226.33
686 x 1981 x 54 mm	PC £193.22	2.30	18.62	207.71	nr	226.33
762 x 1981 x 54 mm	PC £193.22	2.30	18.62	207.71	nr	226.33
838 x 1981 x 54 mm	PC £193.22	2.30	18.62	207.71	nr	226.33
526 x 2040 x 54 mm	PC £193.22	2.30	18.62	207.71	nr	226.33
626 x 2040 x 54 mm	PC £193.22	2.30	18.62	207.71	nr	226.33
726 x 2040 x 54 mm	PC £193.22	2.30	18.62	207.71	nr	226.33
826 x 2040 x 54 mm	PC £193.22	2.30	18.62	207.71	nr	226.33
Flush door; one hour fire resisting (type B60); Afrormosia veneer; hardwood lipping all edges including groove and 'Leaderseal' intumescent strip						
457 x 1981 x 54 mm	PC £150.20	2.30	18.62	161.47	nr	180.09
533 x 1981 x 54 mm	PC £152.20	2.30	18.62	163.62	nr	182.24
610 x 1981 x 54 mm	PC £152.20	2.30	18.62	163.62	nr	182.24
686 x 1981 x 54 mm	PC £154.20	2.30	18.62	165.77	nr	184.39
762 x 1981 x 54 mm	PC £156.20	2.30	18.62	167.92	nr	186.54
838 x 1981 x 54 mm	PC £158.20	2.30	18.62	170.07	nr	188.69
526 x 2040 x 54 mm	PC £150.20	2.30	18.62	161.47	nr	180.09
626 x 2040 x 54 mm	PC £152.20	2.30	18.62	163.62	nr	182.24
726 x 2040 x 54 mm	PC £154.20	2.30	18.62	165.77	nr	184.39
826 x 2040 x 54 mm	PC £156.20	2.30	18.62	167.92	nr	186.54

L WINDOWS/DOORS/STAIRS Including overheads and profit at 7.50%		Labour hours	Labour £	Material £	Unit	Total rate £
Flush door; one hour fire-resisting (60/60); 'Melador'; laminate faced both sides; hardwood lipped on all edges						
610 x 1981 x 57 mm	PC £292.43	3.05	24.69	314.36	nr	**339.05**
686 x 1981 x 57 mm	PC £292.43	3.05	24.69	314.36	nr	**339.05**
762 x 1981 x 57 mm	PC £292.43	3.05	24.69	314.36	nr	**339.05**
838 x 1981 x 57 mm	PC £292.43	3.05	24.69	314.36	nr	**339.05**
526 x 2040 x 57 mm	PC £292.43	3.05	24.69	314.36	nr	**339.05**
626 x 2040 x 57 mm	PC £292.43	3.05	24.69	314.36	nr	**339.05**
726 x 2040 x 57 mm	PC £292.43	3.05	24.69	314.36	nr	**339.05**
826 x 2040 x 57 mm	PC £292.43	3.05	24.69	314.36	nr	**339.05**
Flush door; external quality; skeleton or cellular core; plywood faced both sides; lipped on all four edges						
762 x 1981 x 44 mm	PC £28.11	1.65	13.36	30.21	nr	**43.57**
838 x 1981 x 44 mm	PC £29.37	1.65	13.36	31.57	nr	**44.93**
Flush door; external quality with standard glass opening; skeleton or cellular core; plywood faced both sides; lipped on all four edges; including glazing beads						
762 x 1981 x 44 mm	PC £34.97	1.95	15.78	37.59	nr	**53.37**
838 x 1981 x 44 mm	PC £36.26	1.95	15.78	38.98	nr	**54.76**
Purpose made panelled doors; wrought softwood						
Panelled doors; one open panel for glass; including glazing beads						
686 x 1981 x 44 mm	PC £49.36	1.95	15.78	53.06	nr	**68.84**
762 x 1981 x 44 mm	PC £50.90	1.95	15.78	54.72	nr	**70.50**
838 x 1981 x 44 mm	PC £52.43	1.95	15.78	56.37	nr	**72.15**
Panelled doors; two open panel for glass; including glazing beads						
686 x 1981 x 44 mm	PC £67.98	1.95	15.78	73.08	nr	**88.86**
762 x 1981 x 44 mm	PC £71.64	1.95	15.78	77.01	nr	**92.79**
838 x 1981 x 44 mm	PC £75.22	1.95	15.78	80.86	nr	**96.64**
Panelled doors; four 19 mm thick plywood panel; mouldings worked on solid both sides						
686 x 1981 x 44 mm	PC £95.10	1.95	15.78	102.24	nr	**118.02**
762 x 1981 x 44 mm	PC £99.08	1.95	15.78	106.51	nr	**122.29**
838 x 1981 x 44 mm	PC £104.06	1.95	15.78	111.87	nr	**127.65**
Panelled doors; six 19 mm thick panels raised and fielded; mouldings worked on solid both sides						
686 x 1981 x 50 mm	PC £208.14	2.30	18.62	223.75	nr	**242.37**
762 x 1981 x 50 mm	PC £216.80	2.30	18.62	233.06	nr	**251.68**
838 x 1981 x 50 mm	PC £227.64	2.30	18.62	244.72	nr	**263.34**
Rebated edges beaded		-	-	-	m	**1.13**
Rounded edges or heels		-	-	-	m	**0.57**
Weatherboard; fixed to bottom rail		0.28	2.27	2.40	m	**4.67**
Stopped groove for weatherboard		-	-	-	m	**0.38**
Purpose made panelled doors; selected West African Mahogany; PC £505.25/m3						
Panelled doors; one open panel for glass; mouldings worked on the solid one side; 19 x 13 mm beads one side; fixing with brass screws and cups						
686 x 1981 x 50 mm	PC £85.40	2.75	22.26	91.81	nr	**114.07**
762 x 1981 x 50 mm	PC £88.17	2.75	22.26	94.78	nr	**117.04**
838 x 1981 x 50 mm	PC £90.93	2.75	22.26	97.75	nr	**120.01**
686 x 1981 x 63 mm	PC £103.59	3.05	24.69	111.36	nr	**136.05**
762 x 1981 x 63 mm	PC £107.07	3.05	24.69	115.10	nr	**139.79**
838 x 1981 x 63 mm	PC £110.51	3.05	24.69	118.80	nr	**143.49**

L WINDOWS/DOORS/STAIRS Including overheads and profit at 7.50%		Labour hours	Labour £	Material £	Unit	Total rate £
L20 TIMBER DOORS/SHUTTERS/HATCHES - cont'd						
Purpose made panelled doors; selected West African Mahogany; PC £505.25/m3 - cont'd						
Panelled doors; 250 mm wide cross-tongued intermediate rail; two open panels for glass mouldings worked on the solid one side; 19 x 13 mm beads one side; fixing with brass screws and cups						
686 x 1981 x 50 mm	PC £116.87	2.75	22.26	125.63	nr	**147.89**
762 x 1981 x 50 mm	PC £122.56	2.75	22.26	131.75	nr	**154.01**
838 x 1981 x 50 mm	PC £128.20	2.75	22.26	137.81	nr	**160.07**
686 x 1981 x 63 mm	PC £139.96	3.05	24.69	150.46	nr	**175.15**
762 x 1981 x 63 mm	PC £146.84	3.05	24.69	157.85	nr	**182.54**
838 x 1981 x 63 mm	PC £153.69	3.05	24.69	165.21	nr	**189.90**
Panelled doors; four panels; (19 mm for 50 mm thick doors and 25 mm for 63 mm thick doors); mouldings worked on solid both sides						
686 x 1981 x 50 mm	PC £148.20	2.75	22.26	159.31	nr	**181.57**
762 x 1981 x 50 mm	PC £154.37	2.75	22.26	165.95	nr	**188.21**
838 x 1981 x 50 mm	PC £162.08	2.75	22.26	174.24	nr	**196.50**
686 x 1981 x 63 mm	PC £174.37	3.05	24.69	187.45	nr	**212.14**
762 x 1981 x 63 mm	PC £181.67	3.05	24.69	195.29	nr	**219.98**
838 x 1981 x 63 mm	PC £190.74	3.05	24.69	205.04	nr	**229.73**
Panelled doors; 150 mm wide stiles in one width; 430 mm wide cross-tongued bottom rail; six panels raised and fielded one side (19 mm thick for 50 mm thick doors and 25 mm thick for 63 mm thick doors); mouldings worked on the solid both sides						
686 x 1981 x 50 mm	PC £318.78	2.75	22.26	342.69	nr	**364.95**
762 x 1981 x 50 mm	PC £332.05	2.75	22.26	356.96	nr	**379.22**
838 x 1981 x 50 mm	PC £348.65	2.75	22.26	374.80	nr	**397.06**
686 x 1981 x 63 mm	PC £391.44	3.05	24.69	420.79	nr	**445.48**
762 x 1981 x 63 mm	PC £407.82	3.05	24.69	438.40	nr	**463.09**
838 x 1981 x 63 mm	PC £428.15	3.05	24.69	460.26	nr	**484.95**
Rebated edges beaded		-	-	-	m	**1.72**
Rounded edges or heels		-	-	-	m	**0.85**
Weatherboard; fixed to bottom rail		0.36	2.91	4.57	m	**7.48**
Stopped groove for weatherboard		-	-	-	m	**0.74**
Purpose made panelled doors; Afrormosia; PC £763.25/m3						
Panelled doors; one open panel for glass; mouldings worked on the solid one side; 19 x 13 mm beads one side; fixing with brass screws and cups						
686 x 1981 x 50 mm	PC £100.56	2.75	22.26	108.10	nr	**130.36**
762 x 1981 x 50 mm	PC £103.91	2.75	22.26	111.71	nr	**133.97**
838 x 1981 x 50 mm	PC £107.26	2.75	22.26	115.31	nr	**137.57**
686 x 1981 x 63 mm	PC £122.51	3.05	24.69	131.70	nr	**156.39**
762 x 1981 x 63 mm	PC £126.74	3.05	24.69	136.25	nr	**160.94**
838 x 1981 x 63 mm	PC £130.90	3.05	24.69	140.72	nr	**165.41**
Panelled doors; 250 mm wide cross-tongued intermediate rail; two open panels for glass mouldings worked on the solid one side; 19 x 13 mm beads one side; fixing with brass screws and cups						
686 x 1981 x 50 mm	PC £136.88	2.75	22.26	147.15	nr	**169.41**
762 x 1981 x 50 mm	PC £143.71	2.75	22.26	154.49	nr	**176.75**
838 x 1981 x 50 mm	PC £150.45	2.75	22.26	161.74	nr	**184.00**
686 x 1981 x 63 mm	PC £164.95	3.05	24.69	177.32	nr	**202.01**
762 x 1981 x 63 mm	PC £173.21	3.05	24.69	186.20	nr	**210.89**
838 x 1981 x 63 mm	PC £181.43	3.05	24.69	195.04	nr	**219.73**

L WINDOWS/DOORS/STAIRS Including overheads and profit at 7.50%		Labour hours	Labour £	Material £	Unit	Total rate £
Panelled doors; four panels; (19 mm for 50 mm thick doors and 25 mm for 63 mm thick doors); mouldings worked on solid both sides						
686 x 1981 x 50 mm	PC £173.80	2.75	22.26	186.84	nr	**209.10**
762 x 1981 x 50 mm	PC £181.03	2.75	22.26	194.60	nr	**216.86**
838 x 1981 x 50 mm	PC £190.08	2.75	22.26	204.34	nr	**226.60**
686 x 1981 x 63 mm	PC £206.23	3.05	24.69	221.70	nr	**246.39**
762 x 1981 x 63 mm	PC £214.83	3.05	24.69	230.95	nr	**255.64**
838 x 1981 x 63 mm	PC £225.59	3.05	24.69	242.51	nr	**267.20**
Panelled doors; 150 mm wide stiles in one width; 430 mm wide cross-tongued bottom rail; six panels raised and fielded one side (19 mm thick for 50 mm thick doors and 25 mm thick for 63 mm thick doors); mouldings worked on the solid both sides						
686 x 1981 x 50 mm	PC £358.75	2.75	22.26	385.65	nr	**407.91**
762 x 1981 x 50 mm	PC £373.70	2.75	22.26	401.73	nr	**423.99**
838 x 1981 x 50 mm	PC £392.37	2.75	22.26	421.80	nr	**444.06**
686 x 1981 x 63 mm	PC £442.04	3.05	24.69	475.19	nr	**499.88**
762 x 1981 x 63 mm	PC £460.47	3.05	24.69	495.00	nr	**519.69**
838 x 1981 x 63 mm	PC £483.50	3.05	24.69	519.76	nr	**544.45**
Rebated edges beaded		-	-	-	m	**1.72**
Rounded edges or heels		-	-	-	m	**0.85**
Weatherboard; fixed to bottom rail		0.36	2.91	5.24	m	**8.15**
Stopped groove for weatherboard		-	-	-	m	**0.74**
Standard joinery sets; wrought softwood						
Internal door frame or lining set for 686 x 1981 mm door; all with loose stops unless rebated; 'finished sizes'						
27 x 94 mm lining	PC £16.28	0.88	7.12	17.50	nr	**24.62**
27 x 107 mm lining	PC £17.12	0.88	7.12	18.40	nr	**25.52**
35 x 107 mm rebated lining	PC £17.74	0.88	7.12	19.08	nr	**26.20**
27 x 121 mm lining	PC £18.48	0.88	7.12	19.87	nr	**26.99**
27 x 121 mm lining with fanlight over	PC £26.11	1.05	8.50	28.07	nr	**36.57**
27 x 133 mm lining	PC £19.91	0.88	7.12	21.41	nr	**28.53**
35 x 133 mm rebated linings	PC £20.44	0.88	7.12	21.97	nr	**29.09**
27 x 133 mm lining with fanlight over	PC £28.00	1.05	8.50	30.10	nr	**38.60**
33 x 57 mm frame	PC £13.27	0.88	7.12	14.26	nr	**21.38**
33 x 57 mm storey height frame	PC £16.28	0.94	7.61	17.50	nr	**25.11**
33 x 57 mm frame with fanlight over	PC £19.95	1.05	8.50	21.45	nr	**29.95**
33 x 64 mm frame	PC £14.21	0.88	7.12	15.28	nr	**22.40**
33 x 64 mm storey height frame	PC £17.40	0.94	7.61	18.70	nr	**26.31**
33 x 64 mm frame with fanlight over	PC £20.93	1.05	8.50	22.50	nr	**31.00**
44 x 94 mm frame	PC £21.77	1.00	8.09	23.40	nr	**31.49**
44 x 94 mm storey height frame	PC £25.94	1.10	8.90	27.88	nr	**36.78**
44 x 94 mm frame with fanlight over	PC £30.42	1.20	9.71	32.70	nr	**42.41**
44 x 107 mm frame	PC £25.03	1.00	8.09	26.90	nr	**34.99**
44 x 107 mm storey height frame	PC £29.26	1.10	8.90	31.45	nr	**40.35**
44 x 107 mm frame with fanlight over	PC £33.56	1.20	9.71	36.08	nr	**45.79**
Internal door frame or lining set for 762 x 1981 mm door; all with loose stops unless rebated; 'finished sizes'						
27 x 94 mm lining	PC £16.28	0.88	7.12	17.50	nr	**24.62**
27 x 107 mm lining	PC £17.12	0.88	7.12	18.40	nr	**25.52**
35 x 107 mm rebated lining	PC £17.74	0.88	7.12	19.08	nr	**26.20**
27 x 121 mm lining	PC £18.48	0.88	7.12	19.87	nr	**26.99**
27 x 121 mm lining with fanlight over	PC £26.11	1.05	8.50	28.07	nr	**36.57**
27 x 133 mm lining	PC £19.91	0.88	7.12	21.41	nr	**28.53**

L WINDOWS/DOORS/STAIRS Including overheads and profit at 7.50%		Labour hours	Labour £	Material £	Unit	Total rate £
L20 TIMBER DOORS/SHUTTERS/HATCHES - cont'd						
Standard joinery sets; wrought softwood - cont'd						
Internal door frame or lining set for 762 x 1981 mm door; all with loose stops unless rebated; 'finished sizes'						
35 x 133 mm rebated linings	PC £20.44	0.88	7.12	21.97	nr	**29.09**
27 x 133 mm lining with fanlight over	PC £28.00	1.05	8.50	30.10	nr	**38.60**
33 x 57 mm frame	PC £13.27	0.88	7.12	14.26	nr	**21.38**
33 x 57 mm storey height frame	PC £16.28	0.94	7.61	17.50	nr	**25.11**
33 x 57 mm frame with fanlight over	PC £19.95	1.05	8.50	21.45	nr	**29.95**
33 x 64 mm frame	PC £14.21	0.88	7.12	15.28	nr	**22.40**
33 x 64 mm storey height frame	PC £17.40	0.94	7.61	18.70	nr	**26.31**
33 x 64 mm frame with fanlight over	PC £20.93	1.05	8.50	22.50	nr	**31.00**
44 x 94 mm frame	PC £21.77	1.00	8.09	23.40	nr	**31.49**
44 x 94 mm storey height frame	PC £25.94	1.10	8.90	27.88	nr	**36.78**
44 x 94 mm frame with fanlight over	PC £30.42	1.20	9.71	32.70	nr	**42.41**
44 x 107 mm frame	PC £25.03	1.00	8.09	26.90	nr	**34.99**
44 x 107 mm storey height frame	PC £29.26	1.10	8.90	31.45	nr	**40.35**
44 x 107 mm frame with fanlight over	PC £33.56	1.20	9.71	36.08	nr	**45.79**
Internal door frame or lining set for 726 x 2040 mm door; with loose stops						
30 x 94 mm lining	PC £18.90	0.88	7.12	20.32	nr	**27.44**
30 x 94 mm lining with fanlight over	PC £27.09	1.05	8.50	29.12	nr	**37.62**
30 x 107 mm lining	PC £22.26	0.88	7.12	23.93	nr	**31.05**
30 x 107 mm lining with fanlight over	PC £30.56	1.05	8.50	32.85	nr	**41.35**
30 x 133 mm lining	PC £24.92	0.88	7.12	26.79	nr	**33.91**
30 x 133 mm lining with fanlight over	PC £32.93	1.05	8.50	35.41	nr	**43.91**
Internal door frame or lining set for 826 x 2040 mm door; with loose stops						
30 x 94 mm lining	PC £18.90	0.88	7.12	20.32	nr	**27.44**
30 x 94 mm lining with fanlight over	PC £27.09	1.05	8.50	29.12	nr	**37.62**
30 x 107 mm lining	PC £22.26	0.88	7.12	23.93	nr	**31.05**
30 x 107 mm lining with fanlight over	PC £30.56	1.05	8.50	32.85	nr	**41.35**
30 x 133 mm lining	PC £24.92	0.88	7.12	26.79	nr	**33.91**
30 x 133 mm lining with fanlight over	PC £32.93	1.05	8.50	35.41	nr	**43.91**
Trap door set; 9 mm plywood in 35 x 107 mm (fin) rebated lining						
762 x 762 mm	PC £21.67	0.88	7.12	23.61	nr	**30.73**
Serving hatch set; plywood faced doors; rebated meeting stiles; hung on nylon hinges; in 35 x 140 mm (fin) rebated lining						
648 x 533 mm	PC £41.27	1.45	11.74	44.68	nr	**56.42**
Purpose made door frames and lining sets; wrought softwood						
Jambs and heads; as linings						
32 x 63 mm		0.19	1.54	3.87	m	**5.41**
32 x 100 mm		0.19	1.54	5.05	m	**6.59**
32 x 140 mm		0.19	1.54	6.56	m	**8.10**

L WINDOWS/DOORS/STAIRS Including overheads and profit at 7.50%		Labour hours	Labour £	Material £	Unit	Total rate £
Jambs and heads; as frames; rebated, rounded and grooved						
38 x 75 mm		0.19	1.54	4.99	m	**6.53**
38 x 100 mm		0.19	1.54	5.96	m	**7.50**
38 x 115 mm		0.19	1.54	6.97	m	**8.51**
38 x 140 mm		0.22	1.78	8.12	m	**9.90**
50 x 100 mm		0.22	1.78	7.49	m	**9.27**
50 x 125 mm		0.22	1.78	8.96	m	**10.74**
63 x 88 mm		0.22	1.78	8.76	m	**10.54**
63 x 100 mm		0.22	1.78	8.74	m	**10.52**
63 x 125 mm		0.22	1.78	11.06	m	**12.84**
75 x 100 mm		0.22	1.78	10.60	m	**12.38**
75 x 125 mm		0.24	1.94	13.16	m	**15.10**
75 x 150 mm		0.24	1.94	15.39	m	**17.33**
100 x 100 mm		0.24	1.94	13.85	m	**15.79**
100 x 150 mm		0.24	1.94	20.00	m	**21.94**
Mullions and transoms; in linings						
32 x 63 mm		0.13	1.05	6.26	m	**7.31**
32 x 100 mm		0.13	1.05	7.47	m	**8.52**
32 x 140 mm		0.13	1.05	8.95	m	**10.00**
Mullions and transoms; in frames; twice rebated, rounded and grooved						
38 x 75 mm		0.13	1.05	7.51	m	**8.56**
38 x 100 mm		0.13	1.05	8.51	m	**9.56**
38 x 115 mm		0.13	1.05	9.34	m	**10.39**
38 x 140 mm		0.15	1.21	10.51	m	**11.72**
50 x 100 mm		0.15	1.21	9.88	m	**11.09**
50 x 125 mm		0.15	1.21	11.50	m	**12.71**
63 x 88 mm		0.15	1.21	11.14	m	**12.35**
63 x 100 mm		0.15	1.21	11.65	m	**12.86**
75 x 100 mm		0.15	1.21	13.12	m	**14.33**
Extra for additional labours						
one		0.02	0.16	-	m	**0.16**
two		0.03	0.24	-	m	**0.24**
three		0.04	0.32	-	m	**0.32**
Add +5% to the above 'Material £ prices' for 'selected' softwood for staining						
Purpose made door frames and lining sets; selected West African Mahogany; PC £505.25/m3						
Jambs and heads; as linings						
32 x 63 mm	PC £5.45	0.25	2.02	6.01	m	**8.03**
32 x 100 mm	PC £7.01	0.25	2.02	7.73	m	**9.75**
32 x 140 mm	PC £8.94	0.25	2.02	9.84	m	**11.86**
Jambs and heads; as frames; rebated, rounded and grooved						
38 x 75 mm	PC £6.86	0.25	2.02	7.56	m	**9.58**
38 x 100 mm	PC £8.26	0.25	2.02	9.10	m	**11.12**
38 x 115 mm	PC £9.45	0.25	2.02	10.41	m	**12.43**
38 x 140 mm	PC £10.88	0.30	2.43	11.99	m	**14.42**
50 x 100 mm	PC £10.35	0.30	2.43	11.41	m	**13.84**
50 x 125 mm	PC £12.44	0.30	2.43	13.70	m	**16.13**
63 x 88 mm	PC £11.57	0.30	2.43	12.75	m	**15.18**
63 x 100 mm	PC £12.43	0.30	2.43	13.69	m	**16.12**
63 x 125 mm	PC £15.34	0.30	2.43	16.90	m	**19.33**
75 x 100 mm	PC £14.71	0.30	2.43	16.21	m	**18.64**
75 x 125 mm	PC £18.28	0.33	2.67	20.14	m	**22.81**
75 x 150 mm	PC £21.40	0.33	2.67	23.59	m	**26.26**
100 x 100 mm	PC £19.21	0.33	2.67	21.17	m	**23.84**
100 x 150 mm	PC £27.77	0.33	2.67	30.59	m	**33.26**
Mullions and transoms; in linings						
32 x 63 mm	PC £8.64	0.18	1.46	9.52	m	**10.98**
32 x 100 mm	PC £10.20	0.18	1.46	11.25	m	**12.71**
32 x 140 mm	PC £12.00	0.18	1.46	13.22	m	**14.68**

L WINDOWS/DOORS/STAIRS Including overheads and profit at 7.50%		Labour hours	Labour £	Material £	Unit	Total rate £
L20 TIMBER DOORS/SHUTTERS/HATCHES - cont'd						
Purpose made door frames and lining sets; selected West African Mahogany; PC £505.25/m3 - cont'd						
Mullions and transoms; in frames; twice rebated, rounded and grooved						
38 x 75 mm	PC £10.21	0.18	1.46	11.26	m	**12.72**
38 x 100 mm	PC £11.61	0.18	1.46	12.79	m	**14.25**
38 x 115 mm	PC £12.62	0.18	1.46	13.91	m	**15.37**
38 x 140 mm	PC £14.07	0.20	1.62	15.51	m	**17.13**
50 x 100 mm	PC £13.54	0.20	1.62	14.92	m	**16.54**
50 x 125 mm	PC £15.79	0.20	1.62	17.39	m	**19.01**
63 x 88 mm	PC £14.74	0.20	1.62	16.25	m	**17.87**
63 x 100 mm	PC £16.01	0.20	1.62	17.64	m	**19.26**
75 x 100 mm		0.20	1.62	19.90	m	**21.52**
Sills; once sunk weathered; once rebated, three times grooved						
63 x 175 mm	PC £29.55	0.36	2.91	32.56	m	**35.47**
75 x 125 mm	PC £26.89	0.36	2.91	29.63	m	**32.54**
75 x 150 mm	PC £30.07	0.36	2.91	33.14	m	**36.05**
Extra for additional labours						
one		0.03	0.24	-	m	**0.24**
two		0.06	0.49	-	m	**0.49**
three		0.08	0.65	-	m	**0.65**
Purpose made door frames and lining sets; Afrormosia; PC £763.25/m3						
Jambs and heads; as linings						
32 x 63 mm	PC £6.54	0.25	2.02	7.21	m	**9.23**
32 x 100 mm	PC £8.55	0.25	2.02	9.42	m	**11.44**
32 x 140 mm	PC £10.99	0.25	2.02	12.10	m	**14.12**
Jambs and heads; as frames; rebated, rounded and grooved						
38 x 75 mm	PC £8.23	0.25	2.02	9.08	m	**11.10**
38 x 100 mm	PC £10.10	0.25	2.02	11.12	m	**13.14**
38 x 115 mm	PC £11.65	0.25	2.02	12.83	m	**14.85**
38 x 140 mm	PC £13.44	0.30	2.43	14.81	m	**17.24**
50 x 100 mm	PC £10.35	0.30	2.43	14.07	m	**16.50**
50 x 125 mm	PC £15.45	0.30	2.43	17.02	m	**19.45**
63 x 88 mm	PC £14.31	0.30	2.43	15.77	m	**18.20**
63 x 100 mm	PC £15.12	0.30	2.43	16.66	m	**19.09**
63 x 125 mm	PC £19.15	0.30	2.43	21.10	m	**23.53**
75 x 100 mm	PC £18.33	0.30	2.43	20.20	m	**22.63**
75 x 125 mm	PC £22.80	0.33	2.67	25.12	m	**27.79**
75 x 150 mm	PC £26.82	0.33	2.67	29.55	m	**32.22**
100 x 100 mm	PC £24.04	0.33	2.67	26.49	m	**29.16**
100 x 150 mm	PC £35.01	0.33	2.67	38.58	m	**41.25**
Mullions and transoms; in linings						
32 x 63 mm	PC £9.71	0.18	1.46	10.70	m	**12.16**
32 x 100 mm	PC £11.73	0.18	1.46	12.93	m	**14.39**
32 x 140 mm	PC £14.18	0.18	1.46	15.62	m	**17.08**
Mullions and transoms; in frames; twice rebated, rounded and grooved						
38 x 75 mm	PC £11.58	0.18	1.46	12.77	m	**14.23**
38 x 100 mm	PC £13.44	0.18	1.46	14.81	m	**16.27**
38 x 115 mm	PC £14.83	0.18	1.46	16.34	m	**17.80**
38 x 140 mm	PC £14.83	0.20	1.62	16.34	m	**17.96**
50 x 100 mm	PC £15.95	0.20	1.62	17.57	m	**19.19**
50 x 125 mm	PC £18.80	0.20	1.62	20.71	m	**22.33**
63 x 88 mm	PC £17.51	0.20	1.62	19.30	m	**20.92**
63 x 100 mm	PC £19.05	0.20	1.62	21.00	m	**22.62**
75 x 100 mm	PC £21.68	0.20	1.62	23.89	m	**25.51**

L WINDOWS/DOORS/STAIRS Including overheads and profit at 7.50%		Labour hours	Labour £	Material £	Unit	Total rate £
Sills; once sunk weathered; once rebated, three times grooved						
63 x 175 mm	PC £34.85	0.36	2.91	38.40	m	41.31
75 x 125 mm	PC £31.42	0.36	2.91	34.63	m	37.54
75 x 150 mm	PC £35.73	0.36	2.91	39.37	m	42.28
Extra for additional labours						
one		0.03	0.24	-	m	0.24
two		0.06	0.49	-	m	0.49
three		0.08	0.65	-	m	0.65
Door sills; European Oak PC £1290.00/m3						
Sills; once sunk weathered; once rebated, three times grooved						
63 x 175 mm	PC £58.31	0.36	2.91	64.25	m	67.16
75 x 125 mm	PC £51.97	0.36	2.91	57.26	m	60.17
75 x 150 mm	PC £59.34	0.36	2.91	65.38	m	68.29
Extra for additional labours						
one		0.03	0.24	-	m	0.24
two		0.07	0.57	-	m	0.57
three		0.10	0.81	-	m	0.81
Bedding and pointing frames						
Pointing wood frames or sills with mastic						
one side		0.08	0.83	0.44	m	1.27
each side		0.16	1.66	0.89	m	2.55
Pointing wood frames or sills with polysulphide sealant						
one side		0.08	0.83	0.93	m	1.76
each side		0.16	1.66	1.86	m	3.52
Bedding wood plates in cement mortar (1:3)						
100 mm wide		0.07	0.73	0.05	m	0.78
Bedding wood frame in cement mortar (1:3) and point						
one side		0.09	0.93	0.05	m	0.98
each side		0.11	1.14	0.07	m	1.21
one side in mortar; other side in mastic		0.17	1.76	0.49	m	2.25
L21 METAL DOORS/SHUTTERS/HATCHES						
Aluminium double glazed sliding patio doors; Crittall 'Luminaire' or similar; white acrylic finish; with and including 18 mm annealed double glazing; fixed in position; including lugs plugged and screwed to brickwork or blockwork; or screwed to wooden sub-frame (measured elsewhere)						
Patio doors						
1800 x 2100 mm; ref D18HDL21	PC £486.00	8.25	75.65	525.17	nr	600.82
2400 x 2100 mm; ref D24HDL21	PC £575.55	9.90	90.78	621.44	nr	712.22
3000 x 2100 mm; ref D30HDF21	PC £804.60	11.60	106.37	867.66	nr	974.03
Galvanized steel 'up and over' type garage doors; Catnic 'Garador' or similar; spring counterbalanced; fixed to timber frame (measured elsewhere)						
Garage door						
2135 x 1980 mm; ref MK 3C	PC £137.70	5.50	50.43	149.51	nr	199.94
2135 x 2135 mm; ref MK 3C	PC £151.20	5.80	53.18	164.02	nr	217.20
2400 x 2125 mm; ref MK 3C	PC £178.20	6.60	60.52	193.05	nr	253.57
4270 x 2135 mm; ref 'Carlton'	PC £586.35	9.90	90.78	631.81	nr	722.59

L WINDOWS/DOORS/STAIRS Including overheads and profit at 7.50%	Labour hours	Labour £	Material £	Unit	Total rate £
L30 TIMBER STAIRS/WALKWAYS/BALUSTRADES					
Standard staircases; wrought softwood					
Stairs; 25 mm treads with rounded nosings; 12 mm plywood risers; 32 mm once rounded strings; bullnose bottom tread; 50 x 75 mm hardwood handrail; two 32 x 140 mm balustrade knee rails; 32 x 50 mm stiffeners and 100 x 100 mm newel posts with hardwood newel caps on top					
straight flight; 838 mm wide; 2688 mm going; 2600 mm rise; with two newel posts	14.90	120.61	325.31	nr	445.92
straight flight; 838 mm wide; 2600 mm rise; with two newel posts and three top treads winding	19.80	160.28	402.97	nr	563.25
dogleg staircase; 838 mm wide; 2600 mm rise; quarter space landing third riser from top; with three newel posts	20.90	169.18	415.36	nr	584.54
as last but with half space landing; one 100 x 200 mm newel post; and two 100 x 100 mm newel posts	22.00	178.08	497.17	nr	675.25
Stairs; 25 mm treads with rounded nosings; 12 mm plywood risers; 32 mm once rounded strings; with string cappings; bullnose bottom tread; 50 x 75 mm hardwood handrail; two 32 x 32 mm balusters per tread and 100 x 100 mm newel post with hardwood newel caps on top					
straight flight; 838 mm wide 2688 going 2600 mm rise with two newel posts	16.50	133.56	319.48	nr	453.04
Standard balustrades; wrought softwood					
Landing balustrade; 50 x 75 mm hardwood handrail; three 32 x 140 mm balustrades knee rails; two 32 x 50 mm stiffeners; one end jointed to newel post; other end built into wall (newel post and mortices both measured separately)					
3 m long	3.30	26.71	119.82	nr	146.53
Landing balustrade; 50 x 75 mm hardwood handrail; 32 x 32 mm balusters; one end of handrail jointed to newel post; other end built into wall; balusters housed in at bottom (newel post and mortices both measured separately)					
3 m long	4.95	40.07	97.66	nr	137.73
Purpose made staircase components; wrought softwood					
Board landings; cross-tongued joints; 100 x 50 mm sawn softwood bearers					
25 mm thick	1.10	8.90	46.50	m2	55.40
32 mm thick	1.10	8.90	55.43	m2	64.33
Treads cross-tongued joints and risers; rounded nosings; tongued, grooved, glued and blocked together; one 175 x 50 mm sawn softwood carriage					
25 mm treads; 19 mm risers	1.65	13.36	60.17	m2	73.53
Ends; quadrant	-	-	-	nr	29.22
32 mm treads; 25 mm risers	1.65	13.36	70.60	m2	83.96
Ends; quadrant	-	-	-	nr	34.43
Ends; housed to hardwood	-	-	-	nr	0.51

L WINDOWS/DOORS/STAIRS Including overheads and profit at 7.50%	Labour hours	Labour £	Material £	Unit	Total rate £
Winders; cross-tongued joints and risers in one width; rounded nosings; tongued, grooved glued and blocked together; one 175 x 50 mm sawn softwood carriage					
25 mm treads; 19 mm risers	2.75	22.26	65.29	m2	87.55
32 mm treads; 25 mm risers	2.75	22.26	76.77	m2	99.03
Wide ends; housed to hardwood	-	-	-	nr	1.04
Narrow ends; housed to hardwood	-	-	-	nr	0.79
Closed strings; in one width; 230 mm wide; rounded twice					
32 mm thick	0.55	4.45	11.09	m	15.54
38 mm thick	0.55	4.45	13.44	m	17.89
50 mm thick	0.55	4.45	17.65	m	22.10
Closed strings; cross-tongued joints; 280 mm wide; once rounded; fixing with screws; plugging 450 mm centres					
32 mm thick	0.66	5.34	20.30	m	25.64
Extra for short ramp	0.13	1.05	10.16	nr	11.21
38 mm thick	0.66	5.34	23.01	m	28.35
Extra for short ramp	0.13	1.05	11.49	nr	12.54
50 mm thick	0.66	5.34	28.33	m	33.67
Ends; fitted	0.11	0.89	0.33	nr	1.22
Ends; framed	0.13	1.05	3.32	nr	4.37
Extra for tongued heading joint	0.13	1.05	1.72	nr	2.77
Extra for short ramp	0.13	1.05	14.16	nr	15.21
Closed strings; ramped; crossed tongued joints 280 mm wide; once rounded; fixing with screws; plugging 450 mm centres					
32 mm thick	0.66	5.34	22.36	m	27.70
38 mm thick	0.66	5.34	25.30	m	30.64
50 mm thick	0.66	5.34	31.15	m	36.49
Apron linings; in one width 230 mm wide					
19 mm thick	0.36	2.91	3.69	m	6.60
25 mm thick	0.36	2.91	3.37	m	6.28
Handrails; rounded					
44 x 50 mm	0.28	2.27	3.28	m	5.55
50 x 75 mm	0.30	2.43	4.13	m	6.56
63 x 87 mm	0.33	2.67	5.77	m	8.44
75 x 100 mm	0.39	3.16	6.54	m	9.70
Handrails; moulded					
44 x 50 mm	0.28	2.27	3.61	m	5.88
50 x 75 mm	0.30	2.43	4.47	m	6.90
63 x 87 mm	0.33	2.67	6.09	m	8.76
75 x 100 mm	0.39	3.16	6.85	m	10.01
Handrails; rounded; ramped					
44 x 50 mm	0.36	2.91	6.57	m	9.48
50 x 75 mm	0.40	3.24	8.29	m	11.53
63 x 87 mm	0.44	3.56	11.53	m	15.09
75 x 100 mm	0.50	4.05	13.07	m	17.12
Handrails; moulded; ramped					
44 x 50 mm	0.36	2.91	7.22	m	10.13
50 x 75 mm	0.40	3.24	8.95	m	12.19
63 x 87 mm	0.44	3.56	12.19	m	15.75
75 x 100 mm	0.50	4.05	13.72	m	17.77
Add to above for					
grooved once	-	-	-	m	0.31
ends; framed	0.11	0.89	2.64	nr	3.53
ends; framed on rake	0.17	1.38	3.32	nr	4.70
Heading joints on rake; handrail screws					
44 x 50 mm	0.17	1.38	16.82	nr	18.20
50 x 75 mm	0.17	1.38	16.82	nr	18.20
63 x 87 mm	0.17	1.38	21.36	nr	22.74
75 x 100 mm	0.17	1.38	21.36	nr	22.74

L WINDOWS/DOORS/STAIRS Including overheads and profit at 7.50%		Labour hours	Labour £	Material £	Unit	Total rate £
L30 TIMBER STAIRS/WALKWAYS/BALUSTRADES - cont'd						
Purpose made staircase components; wrought softwood - cont'd						
Mitres; handrail screws						
44 x 50 mm		0.22	1.78	16.82	nr	**18.60**
50 x 75 mm		0.22	1.78	16.82	nr	**18.60**
63 x 87 mm		0.22	1.78	21.36	nr	**23.14**
75 x 100 mm		0.22	1.78	21.36	nr	**23.14**
Balusters; stiffeners						
25 x 25 mm		0.09	0.73	1.35	m	**2.08**
32 x 32 mm		0.09	0.73	1.57	m	**2.30**
32 x 50 mm		0.09	0.73	1.96	m	**2.69**
Ends; housed		0.03	0.24	0.64	nr	**0.88**
Sub rails						
32 x 63 mm		0.36	2.91	2.68	m	**5.59**
Ends; housed to newel		0.11	0.89	2.64	nr	**3.53**
Knee rails						
32 x 140 mm		0.44	3.56	4.40	m	**7.96**
Ends; housed to newel		0.11	0.89	2.64	nr	**3.53**
Newel posts						
50 x 100 mm; half		0.44	3.56	4.52	m	**8.08**
75 x 75 mm		0.44	3.56	4.89	m	**8.45**
100 x 100 mm		0.55	4.45	7.64	m	**12.09**
Newel caps; splayed on four sides						
62.5 x 125 x 50 mm; half		0.17	1.38	3.12	nr	**4.50**
100 x 100 x 50 mm		0.17	1.38	3.12	nr	**4.50**
125 x 125 x 50 mm		0.17	1.38	3.35	nr	**4.73**
Purpose made staircase components; selected West African Mahogany; PC £505.25/m3						
Board landings; cross-tongued joints; 100 x 50 mm sawn softwood bearers						
25 mm thick	PC £61.05	1.65	13.36	66.79	m2	**80.15**
32 mm thick	PC £74.68	1.65	13.36	81.44	m2	**94.80**
Treads cross-tongued joints and risers; rounded nosings; tongued, grooved, glued and blocked together; one 175 x 50 mm sawn softwood carriage						
25 mm treads; 19 mm risers	PC £77.16	2.20	17.81	84.71	m2	**102.52**
Ends; quadrant	PC £45.21	-	-	48.60	nr	**48.60**
32 mm treads; 25 mm risers	PC £92.30	2.20	17.81	100.98	m2	**118.79**
Ends; quadrant	PC £64.60	-	-	69.45	nr	**69.45**
Ends; housed to hardwood		-	-	-	nr	**0.79**
Winders; cross-tongued joints and risers in one width; rounded nosings; tongued, grooved glued and blocked together; one 175 x 50 mm sawn softwood carriage						
25 mm treads; 19 mm risers	PC £84.86	3.70	29.95	92.27	m2	**122.22**
32 mm treads; 25 mm risers	PC £101.53	3.70	29.95	110.19	m2	**140.14**
Wide ends; housed to hardwood		-	-	-	nr	**1.56**
Narrow ends; housed to hardwood		-	-	-	nr	**1.17**
Closed strings; in one width; 230 mm wide; rounded twice						
32 mm thick	PC £16.15/m	0.74	5.99	17.44	m2	**23.43**
38 mm thick	PC £19.40/m	0.74	5.99	20.94	m2	**26.93**
50 mm thick	PC £25.62/m	0.74	5.99	27.62	m2	**33.61**
Closed strings; cross-tongued joints; 280 mm wide; once rounded; fixing with screws; plugging 450 mm centres						
32 mm thick	PC £28.78	0.88	7.12	31.02	m	**38.14**
Extra for short ramp	PC £14.38	0.20	1.62	15.46	nr	**17.08**
38 mm thick	PC £32.57	0.88	7.12	35.10	m	**42.22**
Extra for short ramp	PC £16.31	0.20	1.62	17.53	nr	**19.15**

L WINDOWS/DOORS/STAIRS Including overheads and profit at 7.50%		Labour hours	Labour £	Material £	Unit	Total rate £
50 mm thick	PC £40.18	0.88	7.12	43.27	m	50.39
Ends; fitted	PC £0.90	0.17	1.38	0.97	nr	2.35
Ends; framed	PC £5.44	0.20	1.62	5.84	nr	7.46
Extra for tongued heading joint	PC £2.68	0.20	1.62	2.88	nr	4.50
Extra for short ramp	PC £20.12	0.20	1.62	21.63	nr	23.25
Closed strings; ramped; crossed tongued joints 280 mm wide; once rounded; fixing with screws; plugging 450 mm centres						
32 mm thick	PC £31.68	0.88	7.12	34.14	m	41.26
38 mm thick	PC £35.86	0.88	7.12	38.63	m	45.75
50 mm thick	PC £44.20	0.88	7.12	47.60	m	54.72
Apron linings; in one width 230 mm wide						
19 mm thick	PC £6.21	0.48	3.89	6.76	m	10.65
25 mm thick	PC £7.45	0.48	3.89	8.09	m	11.98
Handrails; rounded						
44 x 50 mm	PC £5.39	0.36	2.91	6.09	m	9.00
50 x 75 mm	PC £6.63	0.40	3.24	7.48	m	10.72
63 x 87 mm	PC £8.41	0.44	3.56	9.50	m	13.06
75 x 100 mm	PC £10.12	0.50	4.05	11.42	m	15.47
Handrails; moulded						
44 x 50 mm	PC £5.82	0.36	2.91	6.57	m	9.48
50 x 75 mm	PC £7.06	0.40	3.24	7.97	m	11.21
63 x 87 mm	PC £8.85	0.44	3.56	9.99	m	13.55
75 x 100 mm	PC £10.54	0.50	4.05	11.90	m	15.95
Handrails; rounded; ramped						
44 x 50 mm	PC £10.79	0.48	3.89	12.18	m	16.07
50 x 75 mm	PC £13.24	0.53	4.29	14.95	m	19.24
63 x 87 mm	PC £16.84	0.58	4.69	19.01	m	23.70
75 x 100 mm	PC £20.22	0.66	5.34	22.83	m	28.17
Handrails; moulded; ramped						
44 x 50 mm	PC £11.65	0.48	3.89	13.15	m	17.04
50 x 75 mm	PC £14.12	0.53	4.29	15.94	m	20.23
63 x 87 mm	PC £17.71	0.58	4.69	20.00	m	24.69
75 x 100 mm	PC £21.08	0.66	5.34	23.79	m	29.13
Add to above for						
grooved once		-	-	-	m	0.51
ends; framed	PC £4.11	0.17	1.38	4.41	nr	5.79
ends; framed on rake	PC £5.05	0.24	1.94	5.43	nr	7.37
Heading joints on rake; handrail screws						
44 x 50 mm	PC £19.24	0.24	1.94	20.69	nr	22.63
50 x 75 mm	PC £19.24	0.24	1.94	20.69	nr	22.63
63 x 87 mm	PC £24.66	0.24	1.94	26.51	nr	28.45
75 x 100 mm	PC £24.66	0.24	1.94	26.51	nr	28.45
Mitres; handrail screws						
44 x 50 mm	PC £19.24	0.30	2.43	20.69	nr	23.12
50 x 75 mm	PC £19.24	0.30	2.43	20.69	nr	23.12
63 x 87 mm	PC £24.66	0.30	2.43	26.51	nr	28.94
75 x 100 mm	PC £24.66	0.30	2.43	26.51	nr	28.94
Balusters; stiffeners						
25 x 25 mm	PC £2.72	0.11	0.89	2.93	m	3.82
32 x 32 mm	PC £3.10	0.11	0.89	3.33	m	4.22
32 x 50 mm	PC £3.66	0.11	0.89	3.93	m	4.82
Ends; housed	PC £0.90	0.06	0.49	0.97	nr	1.46
Sub rails						
32 x 63 mm	PC £4.69	0.48	3.89	5.04	m	8.93
Ends; housed to newel	PC £3.69	0.17	1.38	3.97	nr	5.35
Knee rails						
32 x 140 mm	PC £7.07	0.58	4.69	7.60	m	12.29
Ends; housed to newel	PC £3.69	0.17	1.38	3.97	nr	5.35
Newel posts						
50 x 100 mm; half	PC £7.53	0.58	4.69	8.10	m	12.79
75 x 75 mm	PC £8.11	0.58	4.69	8.71	m	13.40
100 x 100 mm	PC £12.25	0.74	5.99	13.17	m	19.16

L WINDOWS/DOORS/STAIRS Including overheads and profit at 7.50%		Labour hours	Labour £	Material £	Unit	Total rate £
L30 TIMBER STAIRS/WALKWAYS/BALUSTRADES - cont'd						
Purpose made staircase components; selected West African Mahogany; PC £505.25/m3 - cont'd						
Newel caps; splayed on four sides						
62.5 x 125 x 50 mm; half	PC £5.21	0.22	1.78	5.60	nr	7.38
100 x 100 x 50 mm	PC £5.21	0.22	1.78	5.60	nr	7.38
125 x 125 x 50 mm	PC £5.58	0.22	1.78	6.00	nr	7.78
Purpose made staircase components; Oak; PC £1290.00/m3						
Board landings; cross-tongued joints; 100 x 50 mm sawn softwood bearers						
25 mm thick	PC £129.73	1.65	13.36	140.62	m2	153.98
32 mm thick	PC £159.67	1.65	13.36	172.80	m2	186.16
Treads cross-tongued joints and risers; rounded nosings; tongued, grooved, glued and blocked together; one 175 x 50 mm sawn softwood carriage						
25 mm treads; 19 mm risers	PC £143.04	2.20	17.81	155.53	m2	173.34
Ends; quadrant	PC £71.53	-	-	76.89	nr	76.89
32 mm treads; 25 mm risers	PC £175.62	2.20	17.81	190.56	m2	208.37
Ends; quadrant	PC £87.83	-	-	94.41	nr	94.41
Ends; housed to hardwood		-	-	-	nr	1.04
Winders; cross-tongued joints and risers in one width; rounded nosings; tongued, grooved glued and blocked together; one 175 x 50 mm sawn softwood carriage						
25 mm treads; 19 mm risers	PC £157.36	3.70	29.95	170.20	m2	200.15
32 mm treads; 25 mm risers	PC £193.20	3.70	29.95	208.73	m2	238.68
Wide ends; housed to hardwood		-	-	-	nr	2.07
Narrow ends; housed to hardwood		-	-	-	nr	1.56
Closed strings; in one width; 230 mm wide; rounded twice						
32 mm thick	PC £34.22	0.74	5.99	36.87	m	42.86
38 mm thick	PC £40.80	0.74	5.99	43.94	m	49.93
50 mm thick	PC £53.88	0.74	5.99	58.00	m	63.99
Closed strings; cross-tongued joints; 280 mm wide; once rounded; fixing with screws; plugging 450 mm centres						
32 mm thick	PC £55.33	0.88	7.12	59.56	m	66.68
Extra for short ramp	PC £27.66	0.20	1.62	29.73	nr	31.35
38 mm thick	PC £63.34	0.88	7.12	68.17	m	75.29
Extra for short ramp	PC £31.66	0.20	1.62	34.03	nr	35.65
50 mm thick	PC £79.49	0.88	7.12	85.53	m	92.65
Ends; fitted	PC £1.04	0.17	1.38	1.12	nr	2.50
Ends; framed	PC £6.99	0.20	1.62	7.51	nr	9.13
Extra for tongued heading joint	PC £3.49	0.20	1.62	3.75	nr	5.37
Extra for short ramp	PC £39.75	0.20	1.62	42.74	nr	44.36
Closed strings; ramped; crossed tongued joints 280 mm wide; once rounded; fixing with screws; plugging 450 mm centres						
32 mm thick	PC £60.85	0.88	7.12	65.49	m	72.61
38 mm thick	PC £69.69	0.88	7.12	75.00	m	82.12
50 mm thick	PC £87.43	0.88	7.12	94.07	m	101.19
Apron linings; in one width 230 mm wide						
19 mm thick	PC £15.40	0.48	3.89	16.64	m	20.53
25 mm thick	PC £19.28	0.48	3.89	20.80	m	24.69
Handrails; rounded						
44 x 50 mm	PC £11.18	0.36	2.91	12.62	m	15.53
50 x 75 mm	PC £15.08	0.40	3.24	17.02	m	20.26
63 x 87 mm	PC £20.74	0.44	3.56	23.42	m	26.98
75 x 100 mm	PC £26.10	0.50	4.05	29.45	m	33.50

L WINDOWS/DOORS/STAIRS Including overheads and profit at 7.50%		Labour hours	Labour £	Material £	Unit	Total rate £
Handrails; moulded						
44 x 50 mm	PC £11.77	0.36	2.91	13.28	m	**16.19**
50 x 75 mm	PC £15.67	0.40	3.24	17.68	m	**20.92**
63 x 87 mm	PC £21.33	0.44	3.56	24.08	m	**27.64**
75 x 100 mm	PC £26.67	0.50	4.05	30.10	m	**34.15**
Handrails; rounded; ramped						
44 x 50 mm	PC £22.38	0.48	3.89	25.27	m	**29.16**
50 x 75 mm	PC £30.16	0.53	4.29	34.04	m	**38.33**
63 x 87 mm	PC £41.49	0.58	4.69	46.83	m	**51.52**
75 x 100 mm	PC £52.19	0.66	5.34	58.91	m	**64.25**
Handrails; moulded; ramped						
44 x 50 mm	PC £23.53	0.48	3.89	26.56	m	**30.45**
50 x 75 mm	PC £31.32	0.53	4.29	35.35	m	**39.64**
63 x 87 mm	PC £42.66	0.58	4.69	48.15	m	**52.84**
75 x 100 mm	PC £53.33	0.66	5.34	60.20	m	**65.54**
Add to above for						
grooved once		-	-	-	m	**0.61**
ends; framed	PC £5.33	0.17	1.38	5.73	nr	**7.11**
ends; framed on rake	PC £6.58	0.24	1.94	7.08	nr	**9.02**
Heading joints on rake; handrail screws						
44 x 50 mm	PC £29.68	0.24	1.94	31.91	nr	**33.85**
50 x 75 mm	PC £29.68	0.24	1.94	31.91	nr	**33.85**
63 x 87 mm	PC £36.31	0.24	1.94	39.03	nr	**40.97**
75 x 100 mm	PC £36.31	0.24	1.94	39.03	nr	**40.97**
Mitres; handrail screws						
44 x 50 mm	PC £29.68	0.30	2.43	31.91	nr	**34.34**
50 x 75 mm	PC £29.68	0.30	2.43	31.91	nr	**34.34**
63 x 87 mm	PC £36.31	0.30	2.43	39.03	nr	**41.46**
75 x 100 mm	PC £36.31	0.30	2.43	39.03	nr	**41.46**
Balusters; stiffeners						
25 x 25 mm	PC £4.39	0.11	0.89	4.72	m	**5.61**
32 x 32 mm	PC £5.57	0.11	0.89	5.99	m	**6.88**
32 x 50 mm	PC £7.34	0.11	0.89	7.89	m	**8.78**
Ends; housed	PC £1.20	0.06	0.49	1.29	nr	**1.78**
Sub rails						
32 x 63 mm	PC £9.43	0.48	3.89	10.13	m	**14.02**
Ends; housed to newel	PC £4.91	0.17	1.38	5.28	nr	**6.66**
Knee rails						
32 x 140 mm	PC £16.86	0.58	4.69	18.13	m	**22.82**
Ends; housed to newel	PC £4.91	0.17	1.38	5.28	nr	**6.66**
Newel posts						
50 x 100 mm; half	PC £18.36	0.58	4.69	19.74	m	**24.43**
75 x 75 mm	PC £20.14	0.58	4.69	21.65	m	**26.34**
100 x 100 mm	PC £33.24	0.74	5.99	35.74	m	**41.73**
Newel caps; splayed on four sides						
62.5 x 125 x 50 mm; half	PC £7.51	0.22	1.78	8.07	nr	**9.85**
100 x 100 x 50 mm	PC £7.51	0.22	1.78	8.07	nr	**9.85**
125 x 125 x 50 mm	PC £8.68	0.22	1.78	9.33	nr	**11.11**

L31 METAL STAIRS/WALKWAYS/BALUSTRADES

		Labour hours	Labour £	Material £	Unit	Total rate £
Cat ladders, balustrades and handrails, etc.; mild steel; BS 4360						
Cat ladders; welded construction; 64 x 13 mm bar strings; 19 mm rungs at 250 mm centres; fixing by bolting; 0.46 m wide; 3.05 m high		3.30	30.26	141.90	nr	**172.16**
Extra for						
ends of strings; bent once; holed once for 10 mm dia bolt		-	-	-	nr	**2.96**
ends of strings; fanged		-	-	-	nr	**1.77**
ends of strings; bent in plane		-	-	-	nr	**8.28**

L WINDOWS/DOORS/STAIRS Including overheads and profit at 7.50%	Labour hours	Labour £	Material £	Unit	Total rate £
L31 METAL STAIRS/WALKWAYS/BALUSTRADES - cont'd					
Cat ladders, balustrades and handrails, etc.; mild steel; BS 4360 - cont'd					
Chequer plate flooring; over 300 mm wide; bolted to steel supports					
6 mm thick	6.60	60.52	22.57	m2	83.09
8 mm thick	7.70	70.61	29.33	m2	99.94
Balustrades; welded construction; galvanized after manufacture; 1070 mm high; 50 x 50 x 3.2 mm r.h.s. top rail; 38 x 13 mm bottom rail, 50 x 50 x 3.2 mm r.h.s. standards at 1830 mm centres with base plate drilled and bolted to concrete; 13 x 13 mm balusters at 102 mm centres	1.65	15.13	59.13	m	74.26
Balusters; isolated; one end ragged and cemented in; one 76 x 25 x 6 mm flange plate welded on; ground to a smooth finish; countersunk drilled and tap screwed to underside of handrail					
19 x 19 x 914 mm square bar	-	-	-	nr	10.64
Core-rails; joints prepared, welded and ground to a smooth finish; fixing on brackets (measured elsewhere)					
38 x 10 mm flat bar	-	-	-	m	14.19
50 x 8 mm flat bar	-	-	-	m	13.60
Extra for					
ends fanged	-	-	-	nr	2.37
ends scrolled	-	-	-	nr	3.55
ramps in thickness	-	-	-	nr	4.73
wreaths	-	-	-	nr	5.91
Handrails; joints prepared, welded and ground to a smooth finish; fixing on brackets (measured elsewhere)					
38 x 12 mm half oval bar	-	-	-	m	18.92
44 x 13 mm half oval bar	-	-	-	m	20.40
Extra for					
ends fanged	-	-	-	nr	2.37
ends scrolled	-	-	-	nr	3.55
ramps in thickness	-	-	-	nr	4.73
wreaths	-	-	-	nr	5.91
Handrail bracket; comprising 40 x 5 mm plate with mitred and welded angle; one end welded to 100 mm dia x 5 mm backplate; three times holed and plugged and screwed to brickwork; other end scribed and welded to underside of handrail					
140 mm girth	-	-	-	nr	11.23
Holes					
Holes; countersunk; for screws or bolts					
6 mm dia wood screw; 3 mm thick	0.06	0.55	-	nr	0.55
6 mm dia wood screw; 6 mm thick	0.08	0.73	-	nr	0.73
8 mm dia bolt; 6 mm thick	0.08	0.73	-	nr	0.73
10 mm dia bolt; 6 mm thick	0.09	0.83	-	nr	0.83
12 mm dia bolt; 8 mm thick	0.11	1.01	-	nr	1.01

L WINDOWS/DOORS/STAIRS Including overheads and profit at 7.50%	Labour hours	Labour £	Material £	Unit	Total rate £
L40 GENERAL GLAZING					
Standard plain glass; BS 952; clear float; panes area 0.15 - 4.00 m2					
3 mm thick; glazed with					
putty or bradded beads	-	-	-	m2	14.57
bradded beads and butyl compound	-	-	-	m2	18.24
screwed beads	-	-	-	m2	19.86
screwed beads and butyl compound	-	-	-	m2	23.07
4 mm thick; glazed with					
putty or bradded beads	-	-	-	m2	17.44
bradded beads and butyl compound	-	-	-	m2	21.59
screwed beads	-	-	-	m2	23.26
screwed beads and butyl compound	-	-	-	m2	26.94
5 mm thick; glazed with					
putty or bradded beads	-	-	-	m2	23.03
bradded beads and butyl compound	-	-	-	m2	25.74
screwed beads	-	-	-	m2	27.41
screwed beads and butyl compound	-	-	-	m2	31.09
6 mm thick; glazed with					
putty or bradded beads	-	-	-	m2	24.25
bradded beads and butyl compound	-	-	-	m2	26.96
screwed beads	-	-	-	m2	28.64
screwed beads and butyl compound	-	-	-	m2	32.31
Standard plain glass; BS 952; white patterned; panes area 0.15 - 4.00 m2					
4 mm thick; glazed with					
putty or bradded beads	-	-	-	m2	16.52
bradded beads and butyl compound	-	-	-	m2	20.20
screwed beads	-	-	-	m2	21.81
screwed beads and butyl compound	-	-	-	m2	25.02
6 mm thick; glazed with					
putty or bradded beads	-	-	-	m2	23.88
bradded beads and butyl compound	-	-	-	m2	27.62
screwed beads	-	-	-	m2	29.12
screwed beads and butyl compound	-	-	-	m2	32.43
Standard plain glass; BS 952; rough cast; panes area 0.15 - 4.00 m2					
6 mm thick; glazed with					
putty or bradded beads	-	-	-	m2	19.83
bradded beads and butyl compound	-	-	-	m2	23.56
screwed beads	-	-	-	m2	25.08
screwed beads and butyl compound	-	-	-	m2	28.39
Standard plain glass; BS 952; Georgian wired cast; panes area 0.15 - 4.00 m2					
7 mm thick; glazed with					
putty or bradded beads	-	-	-	m2	21.29
bradded beads and butyl compound	-	-	-	m2	25.02
screwed beads	-	-	-	m2	26.53
screwed beads and butyl compound	-	-	-	m2	29.84
Extra for lining up wired glass	-	-	-	m2	2.23
Standard plain glass; BS 952; Georgian wired polished; panes area 0.15 - 4.00 m2					
6 mm thick; glazed with					
putty or bradded beads	-	-	-	m2	45.91
bradded beads and butyl compound	-	-	-	m2	48.35
screwed beads	-	-	-	m2	49.87
screwed beads and butyl compound	-	-	-	m2	53.17
Extra for lining up wired glass	-	-	-	m2	2.50

L WINDOWS/DOORS/STAIRS Including overheads and profit at 7.50%	Labour hours	Labour £	Material £	Unit	Total rate £
L40 GENERAL GLAZING - cont'd					
Special glass; BS 952; toughened clear float; panes area 0.15 - 4.00 m2					
4 mm thick; glazed with					
putty or bradded beads	-	-	-	m2	44.06
bradded beads and butyl compound	-	-	-	m2	46.49
screwed beads	-	-	-	m2	48.01
screwed beads and butyl compound	-	-	-	m2	51.32
5 mm thick; glazed with					
putty or bradded beads	-	-	-	m2	48.81
bradded beads and butyl compound	-	-	-	m2	51.23
screwed beads	-	-	-	m2	52.75
screwed beads and butyl compound	-	-	-	m2	56.06
6 mm thick; glazed with					
putty or bradded beads	-	-	-	m2	52.33
bradded beads and butyl compound	-	-	-	m2	54.76
screwed beads	-	-	-	m2	56.28
screwed beads and butyl compound	-	-	-	m2	59.59
10 mm thick; glazed with					
putty or bradded beads	-	-	-	m2	95.63
bradded beads and butyl compound	-	-	-	m2	98.06
screwed beads	-	-	-	m2	99.58
screwed beads and butyl compound	-	-	-	m2	102.89
Special glass; BS 952; clear laminated safety glass; panes area 0.15 - 4.00 m2					
4.4 mm thick; glazed with					
putty or bradded beads	-	-	-	m2	58.94
bradded beads and butyl compound	-	-	-	m2	61.37
screwed beads	-	-	-	m2	62.89
screwed beads and butyl compound	-	-	-	m2	66.20
5.4 mm thick; glazed with					
putty or bradded beads	-	-	-	m2	60.11
bradded beads and butyl compound	-	-	-	m2	62.54
screwed beads	-	-	-	m2	64.06
screwed beads and butyl compound	-	-	-	m2	67.37
6.4 mm thick; glazed with					
putty or bradded beads	-	-	-	m2	63.50
bradded beads and butyl compound	-	-	-	m2	65.93
screwed beads	-	-	-	m2	66.73
screwed beads and butyl compound	-	-	-	m2	70.76
Special glass; BS 952; 'Antisun' solar control float glass infill panels; panes area 0.15 - 4.00 m2					
4 mm thick; glazed with					
non-hardening compound to metal	-	-	-	m2	62.64
6 mm thick; glazed with					
non-hardening compound to metal	-	-	-	m2	83.50
10 mm thick; glazed with					
non-hardening compound to metal	-	-	-	m2	165.12
12 mm thick; glazed with					
non-hardening compound to metal	-	-	-	m2	204.17
Special glass; BS 952; 'Pyran' half-hour fire resisting glass					
6.5 mm thick rectangular panes; glazed with screwed hardwood beads and Interdens intumescent strip					
300 x 400 mm pane	1.55	18.48	34.59	nr	53.07
400 x 800 mm pane	2.65	31.59	83.55	nr	115.14
500 x 1400 mm pane	4.20	50.07	173.32	nr	223.39
600 x 1800 mm pane	5.30	63.19	260.47	nr	323.66

L WINDOWS/DOORS/STAIRS Including overheads and profit at 7.50%	Labour hours	Labour £	Material £	Unit	Total rate £
6.5 mm thick irregular panes; glazed with screwed hardwood beads and Intergens intumescent strip					
300 x 400 mm pane	1.95	23.25	52.34	nr	75.59
400 x 800 mm pane	3.30	39.34	130.87	nr	170.21
500 x 1400 mm pane	5.25	62.59	276.82	nr	339.41
600 x 1800 mm pane	6.60	78.68	420.16	nr	498.84
Special glass; BS 952; 'Pyrostop' one-hour fire resisting glass					
11 mm thick regular panes; glazed with screwed hardwood beads and Intergens intumescent strip					
300 x 400 mm pane	1.75	20.86	60.05	nr	80.91
400 x 800 mm pane	2.95	35.17	151.46	nr	186.63
500 x 1400 mm pane	4.60	54.84	321.86	nr	376.70
600 x 1800 mm pane	5.85	69.74	489.65	nr	559.39
Special glass; BS 952; clear laminated security glass					
7.5 mm thick regular panes; glazed with screwed hardwood beads and Intergens intumescent strip					
300 x 400 mm pane	1.55	18.48	32.72	nr	51.20
400 x 800 mm pane	2.65	31.59	51.61	nr	83.20
500 x 1400 mm pane	4.20	50.07	103.42	nr	153.49
600 x 1800 mm pane	5.30	63.19	152.63	nr	215.82
Glass louvres; BS 952; with long edges ground or smooth					
5 mm or 6 mm thick float					
100 mm wide	-	-	-	m	5.72
150 mm wide	-	-	-	m	6.36
7 mm thick Georgian wired cast					
100 mm wide	-	-	-	m	12.05
150 mm wide	-	-	-	m	12.86
6 mm thick Georgian polished wired					
100 mm wide	-	-	-	m	16.13
150 mm wide	-	-	-	m	18.53
Labours on glass/sundries					
Curved cutting on panes					
4 mm thick	-	-	-	m	1.99
6 mm thick	-	-	-	m	1.99
6 mm thick; wired	-	-	-	m	3.01
Imitation washleather or black velvet strip; as bedding to edge of glass	-	-	-	m	0.82
Intumescent paste to fire doors per side of glass	-	-	-	m	4.62
Drill hole exceeding 6 mm and not exceeding 15 mm dia. through panes					
not exceeding 6 mm thick	-	-	-	nr	2.20
not exceeding 10 mm thick	-	-	-	nr	2.85
not exceeding 12 mm thick	-	-	-	nr	3.55
not exceeding 19 mm thick	-	-	-	nr	4.41
not exceeding 25 mm thick	-	-	-	nr	5.54
Drill hole exceeding 16 mm and not exceeding 38 mm dia. through panes					
not exceeding 6 mm thick	-	-	-	nr	3.12
not exceeding 10 mm thick	-	-	-	nr	4.19
not exceeding 12 mm thick	-	-	-	nr	5.00
not exceeding 19 mm thick	-	-	-	nr	6.29
not exceeding 25 mm thick	-	-	-	nr	7.85

L WINDOWS/DOORS/STAIRS Including overheads and profit at 7.50%	Labour hours	Labour £	Material £	Unit	Total rate £
L40 GENERAL GLAZING - cont'd					
Labours on glass/sundries - cont'd					
Drill hole over 38 mm dia. through panes					
not exceeding 6 mm thick	-	-	-	nr	6.29
not exceeding 10 mm thick	-	-	-	nr	7.63
not exceeding 12 mm thick	-	-	-	nr	9.03
not exceeding 19 mm thick	-	-	-	nr	11.13
not exceeding 25 mm thick	-	-	-	nr	13.87
Add to the above					
Plus 33% for countersunk holes					
Plus 50% for wired or laminated glass					
Factory made double hermetically sealed units; Pilkington's 'Insulight' or similar; to wood or metal with butyl non-setting compound and screwed or clipped beads					
Two panes; BS 952; clear float glass; GG; 3 mm or 4 mm thick; 13 mm air space					
2 - 4 m2	-	-	-	m2	57.90
1 - 2 m2	-	-	-	m2	55.91
0.75 - 1 m2	-	-	-	m2	62.08
0.5 - 0.75 m2	-	-	-	m2	70.68
0.35 - 0.5 m2	-	-	-	m2	78.39
0.25 - 0.35 m2	-	-	-	m2	96.69
not exceeding 0.25 m2	-	-	-	m2	96.69
Two panes; BS 952; clear float glass; GG; 5 mm or 6 mm thick; 13 mm air space					
2 - 4 m2	-	-	-	m2	74.58
1 - 2 m2	-	-	-	m2	71.06
0.75 - 1 m2	-	-	-	m2	78.11
0.5 - 0.75 m2	-	-	-	m2	87.37
0.35 - 0.5 m2	-	-	-	m2	96.62
0.25 - 0.35 m2	-	-	-	m2	112.38
not exceeding 0.25 m2	-	-	-	m2	112.38

M SURFACE FINISHES Including overheads and profit at 7.50%	Labour hours	Labour £	Material £	Unit	Total rate £
M10 SAND CEMENT/CONCRETE/GRANOLITHIC SCREEDS/ FLOORING					
Cement and sand (1:3); steel trowelled					
Work to floors; one coat; level and to falls not exceeding 15 degrees from horizontal; to concrete base; over 300 mm wide					
32 mm	0.44	5.25	1.89	m2	**7.14**
40 mm	0.46	5.48	2.26	m2	**7.74**
48 mm	0.50	5.96	2.68	m2	**8.64**
50 mm	0.51	6.08	2.75	m2	**8.83**
60 mm	0.54	6.44	3.15	m2	**9.59**
65 mm	0.56	6.68	3.42	m2	**10.10**
70 mm	0.58	6.91	3.68	m2	**10.59**
75 mm	0.61	7.27	3.94	m2	**11.21**
Finishing around pipes; not exceeding 0.30 m girth	0.01	0.09	-	nr	**0.09**
Add to the above for work					
to falls and crossfalls and to slopes not exceeding 15 degrees from horizontal	0.03	0.28	-	m2	**0.28**
to slopes over 15 degrees from horizontal	0.11	1.01	-	m2	**1.01**
three coats of surface hardener brushed on	0.09	0.83	1.52	m2	**2.35**
water-repellent additive incorporated in the mix	-	-	-	m2	**0.79**
oil-repellent additive incorporated in the mix	-	-	-	m2	**3.39**
Cement and sand (1:3) beds and backings					
Work to floors; one coat; level; to concrete base; screeded; over 300 mm wide					
25 mm thick	-	-	-	m2	**6.65**
50 mm thick	-	-	-	m2	**8.34**
75 mm thick	-	-	-	m2	**11.81**
100 mm thick	-	-	-	m2	**15.68**
Work to floors; one coat; level; to concrete base; steel trowelled; over 300 mm wide					
25 mm thick	-	-	-	m2	**6.65**
50 mm thick	-	-	-	m2	**8.34**
75 mm thick	-	-	-	m2	**11.81**
100 mm thick	-	-	-	m2	**15.68**
Granolithic paving; cement and granite chippings 5 mm down (1:2.5); steel trowelled					
Work to floors; one coat; level; laid on concrete while green; over 300 mm wide					
20 mm	0.33	4.79	3.76	m2	**8.55**
25 mm	0.39	5.66	4.54	m2	**10.20**
Work to floors; two coat; laid on hacked concrete with slurry; over 300 mm wide					
38 mm	0.66	9.58	6.05	m2	**15.63**
50 mm	0.77	11.17	7.75	m2	**18.92**
75 mm	0.99	14.37	11.42	m2	**25.79**
Work to landings; one coat; level; laid on concrete while green; over 300 mm wide					
20 mm	0.50	7.26	3.76	m2	**11.02**
25 mm	0.50	7.26	4.54	m2	**11.80**
Work to landings; two coat; laid on hacked concrete with slurry; over 300 mm wide					
38 mm	0.83	12.05	6.05	m2	**18.10**
50 mm	0.94	13.64	7.75	m2	**21.39**
75 mm	1.10	15.96	11.42	m2	**27.38**
Finishing around pipes; not exceeding 0.30 m girth	0.06	0.87	-	nr	**0.87**

M SURFACE FINISHES Including overheads and profit at 7.50%	Labour hours	Labour £	Material £	Unit	Total rate £
M10 SAND CEMENT/CONCRETE/GRANOLITHIC SCREEDS/ FLOORING - cont'd					
Granolithic paving; cement and granite chippings 5 mm down (1:2.5); steel trowelled - cont'd					
Add to the above over 300 mm wide for					
1.35 kg/m2 carborundum grains trowelled in	0.07	1.02	1.16	m2	2.18
three coats of surface hardener brushed on	0.06	0.87	1.52	m2	2.39
liquid hardening additive incorporated in the mix	-	-	-	m2	1.58
oil-repellent additive incorporated in the mix	-	-	-	m2	3.39
25 mm work to treads; one coat; to concrete base					
225 mm wide	0.99	14.37	4.96	m	19.33
275 mm wide	0.99	14.37	5.03	m	19.40
Return end	0.20	2.90	-	nr	2.90
Extra for 38 mm 'Ferodo' nosing	0.22	3.19	7.15	m	10.34
13 mm skirtings; rounded top edge and coved bottom junction; to brickwork or blockwork base					
75 mm wide on face	0.61	8.85	0.71	m	9.56
150 mm wide on face	0.83	12.05	1.07	m	13.12
Ends; fair	0.06	0.87	-	nr	0.87
Angles	0.08	1.16	-	nr	1.16
13 mm outer margin to stairs; to follow profile of and with rounded nosing to treads and risers; fair edge and arris at bottom; to concrete base					
75 mm wide	0.99	14.37	1.42	m	15.79
Angles	0.08	1.16	-	nr	1.16
13 mm wall string to stairs; fair edge and arris on top; coved bottom junction with treads and risers; to brickwork or blockwork base					
275 mm (extreme) wide	0.88	12.77	1.42	m	14.19
Ends	0.06	0.87	-	nr	0.87
Angles	0.08	1.16	-	nr	1.16
Ramps	0.09	1.31	-	nr	1.31
Ramped and wreathed corners	0.11	1.60	-	nr	1.60
13 mm outer string to stairs; rounded nosing on top at junction with treads and risers; fair edge and arris at bottom; to concrete base					
300 mm (extreme) wide	0.88	12.77	2.37	m	15.14
Ends	0.06	0.87	-	nr	0.87
Angles	-	-	-	nr	1.02
Ramps	0.09	1.31	-	nr	1.31
Ramped and wreathed corners	0.11	1.60	-	nr	1.60
19 mm skirtings; rounded top edge and coved bottom junction; to brickwork or blockwork base					
75 mm wide on face	0.61	8.85	1.19	m	10.04
150 mm wide on face	0.83	12.05	1.78	m	13.83
Ends; fair	0.06	0.87	-	nr	0.87
Angles	0.08	1.16	-	nr	1.16
19 mm riser; one rounded nosing; to concrete base					
150 mm high; plain	0.99	14.37	4.02	m	18.39
150 mm high; undercut	0.99	14.37	4.02	m	18.39
180 mm high; plain	0.99	14.37	4.09	m	18.46
180 mm high; undercut	0.99	14.37	4.09	m	18.46

M SURFACE FINISHES Including overheads and profit at 7.50%	Labour hours	Labour £	Material £	Unit	Total rate £
M11 MASTIC ASPHALT FLOORING					
Mastic asphalt paving to BS 1076; black					
20 mm one coat coverings; felt isolating membrane; to concrete base; flat					
over 300 mm wide	-	-	-	m2	13.60
225 - 300 mm wide	-	-	-	m2	16.86
150 - 225 mm wide	-	-	-	m2	26.18
not exceeding 150 mm wide	-	-	-	m2	25.42
25 mm one coat coverings; felt isolating membrane; to concrete base; flat					
over 300 mm wide	-	-	-	m2	16.19
225 - 300 mm wide	-	-	-	m2	20.09
150 - 225 mm wide	-	-	-	m2	31.15
not exceeding 150 mm wide	-	-	-	m2	30.23
20 mm three coat skirtings to brickwork base					
not exceeding 150 mm girth	-	-	-	m	13.76
150 - 225 mm girth	-	-	-	m	15.87
225 - 300 mm girth	-	-	-	m	17.99
Mastic asphalt paving; acid-resisting; black					
20 mm one coat coverings; felt isolating membrane; to concrete base; flat					
over 300 mm wide	-	-	-	m2	16.04
225 - 300 mm wide	-	-	-	m2	19.89
150 - 225 mm wide	-	-	-	m2	28.62
not exceeding 150 mm wide	-	-	-	m2	30.01
25 mm one coat coverings; felt isolating membrane; to concrete base; flat					
over 300 mm wide	-	-	-	m2	18.61
225 - 300 mm wide	-	-	-	m2	23.07
150 - 225 mm wide	-	-	-	m2	35.85
not exceeding 150 mm wide	-	-	-	m2	34.80
20 mm three coat skirtings to brickwork base					
not exceeding 150 mm girth	-	-	-	m	18.01
150 - 225 mm girth	-	-	-	m	20.99
225 - 300 mm girth	-	-	-	m	23.97
Mastic asphalt paving to BS 6577; black					
20 mm one coat coverings; felt isolating membrane; to concrete base; flat					
over 300 mm wide	-	-	-	m2	21.70
225 - 300 mm wide	-	-	-	m2	28.62
150 - 225 mm wide	-	-	-	m2	37.36
not exceeding 150 mm wide	-	-	-	m2	40.58
25 mm one coat coverings; felt isolating membrane; to concrete base; flat					
over 300 mm wide	-	-	-	m2	26.27
225 - 300 mm wide	-	-	-	m2	32.57
150 - 225 mm wide	-	-	-	m2	45.26
not exceeding 150 mm wide	-	-	-	m2	49.11
20 mm three coat skirtings to brickwork base					
not exceeding 150 mm girth	-	-	-	m	15.27
150 - 225 mm girth	-	-	-	m	20.16
225 - 300 mm girth	-	-	-	m	23.27
Mastic asphalt paving to BS 1451; red					
20 mm one coat coverings; felt isolating membrane; to concrete base; flat					
over 300 mm wide	-	-	-	m2	22.44
225 - 300 mm wide	-	-	-	m2	48.25
150 - 225 mm wide	-	-	-	m2	55.20
not exceeding 150 mm wide	-	-	-	m2	57.90

M SURFACE FINISHES Including overheads and profit at 7.50%	Labour hours	Labour £	Material £	Unit	Total rate £
M11 MASTIC ASPHALT FLOORING - cont'd					
Mastic asphalt paving to BS 1451; red - cont'd					
20 mm three coat skirtings to brickwork base					
not exceeding 150 mm girth	-	-	-	m	19.03
150 - 225 mm girth	-	-	-	m	21.20
M12 TROWELLED BITUMEN/RESIN/RUBBER LATEX FLOORING					
Latex cement floor screeds; steel trowelled					
Work to floors; level; to concrete base; over 300 mm wide					
3 mm thick; one coat	0.19	1.74	1.72	m2	3.46
5 mm thick; two coats	0.24	2.20	2.69	m2	4.89
Isocrete K screeds; steel trowelled					
Work to floors; level; to concrete base; over 300 mm wide					
35 mm thick; plus polymer bonder coat	-	-	-	m2	13.48
40 mm thick	-	-	-	m2	13.99
45 mm thick	-	-	-	m2	14.60
50 mm thick	-	-	-	m2	15.32
Work to floors; to falls or cross-falls; to concrete base; over 300 mm wide					
55 mm (average) thick	-	-	-	m2	11.03
60 mm (average) thick	-	-	-	m2	11.74
65 mm (average) thick	-	-	-	m2	12.24
75 mm (average) thick	-	-	-	m2	13.64
90 mm (average) thick	-	-	-	m2	15.75
'Synthanite' floor screeds; steel trowelled					
Work to floors; one coat; level; paper felt underlay; to concrete base; over 300 mm wide					
25 mm thick	0.55	5.04	12.25	m2	17.29
50 mm thick	0.66	6.05	19.45	m2	25.50
75 mm thick	0.77	7.06	23.39	m2	30.45
Bituminous lightweight insulating roof screeds					
'Bit-Ag' or similar roof screed; to falls or cross-falls; bitumen felt vapour barrier; over 300 mm wide					
75 mm (average) thick	-	-	-	m2	14.70
100 mm (average) thick	-	-	-	m2	17.54

BASIC PLASTER PRICES

	£		£
Plaster prices (£/tonne)			
BS 1191 Part 1; class A			
CB stucco	59.20		
BS 1191 Part 1; class B			
'Thistle' board	66.42	'Thistle' finish	67.89
BS 1191 Part 1; class C			
'Limelite' - backing	151.00	'Limelite' - renovating	200.04
- finishing	119.60		

M SURFACE FINISHES
Including overheads and profit at 7.50%

Pre-mixed lightweight; BS 1191 Part 2

'Carlite'		'Carlite'	
- bonding	111.36	- finishing	88.88
- browning	109.01	- metal lathing	112.58
- browning HSB	119.18		
Projection	102.94		
Tyrolean 'Cullamix'	179.52		

Plastering sand £9.86/tonne

Sundries

ceramic tile - adhesive (£/5L)	6.02	plasterboard nails (£/25kg)	41.17
- grout (£/kg)	0.49	scrim (£/100 m roll)	3.68
impact adhesive (£/5L)	19.00		

	Labour hours	Labour £	Material £	Unit	Total rate £
M20 PLASTERED/RENDERED/ROUGHCAST COATINGS					
Prepare and brush down; two coats of 'Unibond' or similar bonding agent; PC £15.73/5 Litre					
Brick or block walls					
over 300 mm wide	0.15	1.38	0.52	m2	**1.90**
Concrete walls or ceilings					
over 300 mm wide	0.12	1.10	0.40	m2	**1.50**
Cement and sand (1:3) beds and backings					
10 mm work to walls; one coat; to brickwork or blockwork base					
over 300 mm wide	-	-	-	m2	**6.15**
not exceeding 300 mm wide	-	-	-	m	**3.70**
12 mm work to walls; one coat; to brickwork or blockwork base					
over 300 mm wide	-	-	-	m2	**6.73**
not exceeding 300 mm wide	-	-	-	m	**4.07**
13 mm work to walls; one coat; to brickwork or blockwork base; screeded					
over 300 mm wide	-	-	-	m2	**7.42**
not exceeding 300 mm wide	-	-	-	m	**3.71**
15 mm work to walls; one coat; to brickwork or blockwork base					
over 300 mm wide	-	-	-	m2	**7.92**
not exceeding 300 mm wide	-	-	-	m	**4.74**
Cement and sand (1:3); steel trowelled					
13 mm work to walls; two coats; to brickwork or blockwork base					
over 300 mm wide	-	-	-	m2	**8.81**
not exceeding 300 mm wide	-	-	-	m	**5.21**
16 mm work to walls; two coats; to brickwork or blockwork base					
over 300 mm wide	-	-	-	m2	**9.59**
not exceeding 300 mm wide	-	-	-	m	**5.74**
19 mm work to walls; two coats; to brickwork or blockwork base					
over 300 mm wide	-	-	-	m2	**10.50**
not exceeding 300 mm wide	-	-	-	m	**5.25**
Add to the above over 300 mm wide for first coat in water-repellent cement	-	-	-	m2	**1.87**
finishing coat in coloured cement	-	-	-	m2	**4.54**

M SURFACE FINISHES Including overheads and profit at 7.50%	Labour hours	Labour £	Material £	Unit	Total rate £
M20 PLASTERED/RENDERED/ROUGHCAST COATINGS - cont'd					
Cement-lime-sand (1:2:9); steel trowelled					
19 mm work to walls; two coats; to brickwork or blockwork base					
over 300 mm wide	-	-	-	m2	**9.91**
not exceeding 300 mm wide	-	-	-	m	**5.94**
Cement-lime-sand (1:1:6); steel trowelled					
13 mm work to walls; two coats; to brickwork or blockwork base					
over 300 mm wide	-	-	-	m2	**9.32**
not exceeding 300 mm wide	-	-	-	m	**5.59**
Add to the above over 300 mm wide for waterproof additive	-	-	-	m2	**1.18**
19 mm work to ceilings; three coats; to metal lathing base					
over 300 mm wide	-	-	-	m2	**13.71**
not exceeding 300 mm wide	-	-	-	m	**6.86**
Plaster; first and finishing coats of 'Carlite' pre-mixed lightweight plaster; steel trowelled					
13 mm work to walls; two coats; to brickwork or blockwork base (or 10 mm work to concrete base)					
over 300 mm wide	-	-	-	m2	**6.98**
over 300 mm wide; in staircase areas or plant rooms	-	-	-	m2	**9.23**
not exceeding 300 mm wide	-	-	-	m	**3.49**
13 mm work to isolated piers or columns; two coats over 300 mm wide	-	-	-	m2	**12.17**
not exceeding 300 mm wide	-	-	-	m	**6.08**
10 mm work to ceilings; two coats; to concrete base					
over 300 mm wide	-	-	-	m2	**7.29**
over 300 mm wide; 3.5 - 5 m high	-	-	-	m2	**7.86**
over 300 mm wide; in staircase areas or plant rooms	-	-	-	m2	**9.55**
not exceeding 300 mm wide	-	-	-	m	**4.77**
10 mm work to isolated beams; two coats; to concrete base					
over 300 mm wide	-	-	-	m2	**10.51**
over 300 mm wide; 3.5 - 5 m high	-	-	-	m2	**11.07**
not exceeding 300 mm wide	-	-	-	m	**5.26**
Plaster; one coat 'Snowplast' plaster; steel trowelled					
13 mm work to walls; one coat; to brickwork or blockwork base					
over 300 mm wide	-	-	-	m2	**7.62**
over 300 mm wide; in staircase areas or plant rooms	-	-	-	m2	**10.07**
not exceeding 300 mm wide	-	-	-	m	**3.82**
13 mm work to isolated columns; one coat					
over 300 mm wide	-	-	-	m2	**10.84**
not exceeding 300 mm wide	-	-	-	m	**5.42**

M SURFACE FINISHES Including overheads and profit at 7.50%	Labour hours	Labour £	Material £	Unit	Total rate £
Plaster; first coat of cement and sand (1:3); finishing coat of 'Thistle' class B plaster; steel trowelled					
13 mm work to walls; two coats; to brickwork or blockwork base					
over 300 mm wide	-	-	-	m2	9.22
over 300 mm wide; in staircase areas or plant rooms	-	-	-	m2	11.48
not exceeding 300 mm wide	-	-	-	m	4.62
13 mm work to isolated columns; two coats					
over 300 mm wide	-	-	-	m2	12.44
not exceeding 300 mm wide	-	-	-	m	6.22
Plaster; first coat of cement-lime-sand (1:1:6); finishing coat of 'Sirapite' class B plaster; steel trowelled					
13 mm work to walls; two coats; to brickwork or blockwork base					
over 300 mm wide	-	-	-	m2	9.22
over 300 mm wide; in staircase areas or plant rooms	-	-	-	m2	11.48
not exceeding 300 mm wide	-	-	-	m	4.62
13 mm work to isolated columns; two coats					
over 300 mm wide	-	-	-	m2	13.55
not exceeding 300 mm wide	-	-	-	m	8.13
Plaster; first coat of 'Limelite' renovating plaster; finishing coat of 'Limelite' finishing plaster; steel trowelled					
13 mm work to walls; two coats; to brickwork or blockwork base					
over 300 mm wide	-	-	-	m2	10.16
over 300 mm wide; in staricase areas or compartments under 4 m2	-	-	-	m2	12.19
not exceeding 300 mm wide	-	-	-	m	6.10
Dubbing out existing walls with undercoat plaster; average 6 mm thick					
over 300 mm wide	-	-	-	m2	4.07
not exceeding 300 mm wide	-	-	-	m	2.43
Dubbing out existing walls with undercoat plaster; average 12 mm thick					
over 300 mm wide	-	-	-	m2	5.50
not exceeding 300 mm wide	-	-	-	m	3.29
Plaster; one coat 'Thistle' projection plaster; steel trowelled					
13 mm work to walls; one coat; to brickwork or blockwork base					
over 300 mm wide	-	-	-	m2	9.87
over 300 mm wide; in staircase areas or plant rooms	-	-	-	m2	12.12
not exceeding 300 mm wide	-	-	-	m	4.93
6 mm work to isolated columns; one coat					
over 300 mm wide	-	-	-	m2	12.69
not exceeding 300 mm wide	-	-	-	m	6.34

M SURFACE FINISHES Including overheads and profit at 7.50%	Labour hours	Labour £	Material £	Unit	Total rate £
M20 PLASTERED/RENDERED/ROUGHCAST COATINGS - cont'd					
Plaster; first, second and finishing coats of 'Carlite' pre-mixed lightweight plaster; steel trowelled					
13 mm work to ceilings; three coats to metal lathing base					
over 300 mm wide	-	-	-	m2	10.84
over 300 mm wide; in staircase areas or plant rooms	-	-	-	m2	13.09
not exceeding 300 mm wide	-	-	-	m	5.42
13 mm work to swept soffit of metal lathing arch former					
not exceeding 300 mm wide	-	-	-	m	16.08
300 - 400 mm wide	-	-	-	m	17.21
13 mm work to vertical face of metal lathing arch former					
not exceeding 0.5 m2 per side	-	-	-	nr	31.04
0.5 m2 - 1 m2 per side	-	-	-	nr	36.68
Tyrolean decorative rendering; 13 mm first coat of cement-lime-sand (1:1:6); finishing three coats of 'Cullamix' applied with approved hand operated machine external					
To walls; four coats; to brickwork or blockwork base					
over 300 mm wide	-	-	-	m2	22.70
not exceeding 300 mm wide	-	-	-	m	11.36
'Mineralite' decorative rendering; first coat of cement-lime-sand (1:0.5:4.5); finishing coat of 'Mineralite'; applied by Specialist Subcontractor					
17 mm work to walls; to brickwork or blockwork base					
over 300 mm wide	-	-	-	m2	48.38
not exceeding 300 mm wide	-	-	-	m	24.19
form 9 x 6 mm expansion joint and point in one part polysulphide mastic	-	-	-	m	8.28
Plaster; one coat 'Thistle' board finish; steel trowelled (prices included within plasterboard rates)					
3 mm work to walls or ceilings; one coat; to plasterboard base					
over 300 mm wide	-	-	-	m2	5.20
over 300 mm wide; in staircase areas or plant rooms	-	-	-	m2	7.46
not exceeding 300 mm wide	-	-	-	m	2.61
Plaster; one coat 'Thistle' board finish; steel trowelled 3 mm work to walls or ceilings; one coat on and including gypsum plasterboard; BS 1230; fixing with nails; 3 mm joints filled with plaster and jute scrim cloth; to softwood base; plain grade baseboard or lath with rounded edges					
9.5 mm board to walls					
over 300 mm wide	0.55	6.56	2.67	m2	9.23
not exceeding 300 mm wide	0.28	3.34	0.85	m	4.19
9.5 mm board to walls; in staircase areas or plant rooms					
over 300 mm wide	0.66	7.87	2.67	m2	10.54
not exceeding 300 mm wide	0.33	3.93	0.85	m	4.78

M SURFACE FINISHES Including overheads and profit at 7.50%	Labour hours	Labour £	Material £	Unit	Total rate £
9.5 mm board to isolated columns					
over 300 mm wide	0.73	8.70	2.69	m2	11.39
not exceeding 300 mm wide	0.36	4.29	0.85	m	5.14
9.5 mm board to ceilings					
over 300 mm wide	0.62	7.39	2.67	m2	10.06
over 300 mm wide; 3.5 - 5 m high	0.73	8.70	2.67	m2	11.37
not exceeding 300 mm wide	0.31	3.70	0.85	m	4.55
9.5 mm board to ceilings; in staircase areas or plant rooms					
over 300 mm wide	0.73	8.70	2.67	m2	11.37
not exceeding 300 mm wide	0.36	4.29	0.85	m	5.14
9.5 mm board to isolated beams					
over 300 mm wide	0.77	9.18	2.69	m2	11.87
not exceeding 300 mm wide	0.39	4.65	0.82	m	5.47
Add for 'Duplex' insulating grade	-	-	-	m2	0.48
12.5 mm board to walls					
over 300 mm wide	0.62	7.39	3.18	m2	10.57
not exceeding 300 mm wide	0.31	3.70	1.01	m	4.71
12.5 mm board to walls; in staircase areas or plant rooms					
over 300 mm wide	0.73	8.70	3.18	m2	11.88
not exceeding 300 mm wide	0.36	4.29	1.01	m	5.30
12.5 mm board to isolated columns					
over 300 mm wide	0.77	9.18	3.20	m2	12.38
not exceeding 300 mm wide	0.39	4.65	1.01	m	5.66
12.5 mm board to ceilings					
over 300 mm wide	0.66	7.87	3.18	m2	11.05
over 300 mm wide; 3.5 - 5 m high	0.77	9.18	3.18	m2	12.36
not exceeding 300 mm wide	0.33	3.93	1.01	m	4.94
12.5 mm board to ceilings; in staircase areas or plant rooms					
over 300 mm wide	0.77	9.18	3.18	m2	12.36
not exceeding 300 mm wide	0.39	4.65	1.01	m	5.66
12.5 mm board to isolated beams					
over 300 mm wide	0.84	10.01	3.20	m2	13.21
not exceeding 300 mm wide	0.42	5.01	1.01	m	6.02
Add for 'Duplex' insulating grade	-	-	-	m2	0.48
Accessories					
'Expamet' render beads; white PVC nosings; to brickwork or blockwork base					
external stop bead; ref 1222	0.07	0.83	1.76	m	2.59
'Expamet' plaster beads; to brickwork or blockwork base					
angle bead; ref 550	0.08	0.95	0.49	m	1.44
architrave bead; ref 579	0.10	1.19	0.75	m	1.94
stop bead; ref 563	0.07	0.83	0.58	m	1.41
M22 SPRAYED MINERAL FIBRE COATINGS					
Prepare and apply by spray Mandolite P20 fire protection on structural steel/metalwork					
16 mm (one hour) fire protection					
to walls and columns	-	-	-	m2	8.87
to ceilings and beams	-	-	-	m2	9.76
to isolated metalwork	-	-	-	m2	22.17
22 mm (one and a half hour) fire protection					
to walls and columns	-	-	-	m2	11.23
to ceilings and beams	-	-	-	m2	12.36
to isolated metalwork	-	-	-	m2	25.84

M SURFACE FINISHES Including overheads and profit at 7.50%		Labour hours	Labour £	Material £	Unit	Total rate £
M22 SPRAYED MINERAL FIBRE COATINGS - cont'd						
Prepare and apply by spray Mandolite P20 fire protection on structural steel/metalwork						
28 mm (two hour) fire protection						
to walls and columns		-	-	-	m2	**12.42**
to ceilings and beams		-	-	-	m2	**13.66**
to isolated metalwork		-	-	-	m2	**27.32**
52 mm (four hour) fire protection						
to walls and columns		-	-	-	m2	**25.42**
to ceilings and beams		-	-	-	m2	**27.97**
to isolated metalwork		-	-	-	m2	**50.85**
M30 METAL MESH LATHING/ANCHORED REINFORCEMENT FOR PLASTERED COATING						
Accessories						
Pre-formed galvanised expanded steel arch-frames; 'Truline' or similar; semi-circular; to suit walls up to 230 mm thick						
375 mm radius; for 800 mm opening; ref SC 750	PC £16.63	0.28	3.34	17.88	nr	**21.22**
425 mm radius; for 850 mm opening; ref SC 850	PC £17.64	0.28	3.34	18.96	nr	**22.30**
450 mm radius; for 900 mm opening; ref SC 900	PC £20.48	0.28	3.34	22.01	nr	**25.35**
600 mm radius; for 1200 mm opening; ref SC 1200	PC £25.48	0.28	3.34	27.39	nr	**30.73**
'Newlath' damp-free lathing or similar; PC £62.96/15 m2; shot-fired to brick or block or concrete base						
Lining to walls						
over 300 mm wide		0.25	2.98	6.64	m2	**9.62**
not exceeding 300 mm wide		0.15	1.79	2.18	m	**3.97**
Lathing; Expamet 'BB' expanded metal lathing or similar; BS 1369; 50 mm laps						
6 mm mesh linings to ceilings; fixing with staples; to softwood base; over 300 mm wide						
ref BB263; 0.500 mm thick	PC £2.49	0.33	3.93	2.81	m2	**6.74**
ref BB264; 0.675 mm thick	PC £2.90	0.33	3.93	3.29	m2	**7.22**
6 mm mesh linings to ceilings; fixing with wire; to steelwork; over 300 mm wide						
ref BB263; 0.500 mm thick		0.35	4.17	2.89	m2	**7.06**
ref BB264; 0.675 mm thick		0.35	4.17	3.36	m2	**7.53**
6 mm mesh linings to ceilings; fixings with wire; to steelwork; not exceeding 300 mm wide						
ref BB263; 0.500 mm thick		0.22	2.62	0.91	m	**3.53**
ref BB264; 0.675 mm thick		0.22	2.62	1.06	m	**3.68**
Raking cutting		0.22	2.62	0.91	m	**3.53**
Cutting and fitting around pipes; not exceeding 0.30 m girth		0.22	2.62	1.07	nr	**3.69**
Lathing; Expamet 'Riblath' or 'Spraylath' stiffened expanded metal lathing or similar; 50 mm laps						
10 mm mesh lining to walls; fixing with nails; to softwood base; over 300 mm wide						
Riblath ref 269; 0.30 mm thick	PC £3.13	0.28	3.34	3.55	m2	**6.89**
Riblath ref 271; 0.50 mm thick	PC £3.61	0.28	3.34	4.08	m2	**7.42**
Spraylath ref 273; 0.50 mm thick		0.28	3.34	4.95	m2	**8.29**

M SURFACE FINISHES Including overheads and profit at 7.50%		Labour hours	Labour £	Material £	Unit	Total rate £
10 mm mesh lining to walls; fixing with nails; to softwood base; not exceeding 300 mm wide						
Riblath ref 269; 0.30 mm thick		0.17	2.03	1.11	m	**3.14**
Riblath ref 271; 0.50 mm thick		0.17	2.03	1.29	m	**3.32**
Spraylath ref 273; 0.50 mm thick		0.17	2.03	1.56	m	**3.59**
10 mm mesh lining to walls; fixing to brick or blockwork; over 300 mm wide						
Red-rib ref 274; 0.50 mm thick	PC £3.97	0.22	2.62	6.85	m2	**9.47**
Stainless steel Riblath ref 267; 0.30 mm thick	PC £8.19	0.22	2.62	11.61	m2	**14.23**
10 mm mesh lining to ceilings; fixing with wire; to steelwork; over 300 mm wide						
Riblath ref 269; 0.30 mm thick		0.35	4.17	3.65	m2	**7.82**
Riblath ref 271; 0.50 mm thick		0.35	4.17	4.19	m2	**8.36**
Spraylath ref 273; 0.50 mm thick		0.35	4.17	5.06	m2	**9.23**
Raking cutting		0.17	2.03	0.71	m	**2.74**
Cutting and fitting around pipes; not exceeding 0.30 m girth		0.06	0.72	-	nr	**0.72**
M31 FIBROUS PLASTER						
Fibrous plaster; fixing with screws; plugging; countersinking; stopping; filling and pointing joints with plaster						
16 mm plain slab coverings to ceilings						
over 300 mm wide		-	-	-	m2	**97.13**
not exceeding 300 mm wide		-	-	-	m	**31.30**
Coves; not exceeding 150 mm girth						
per 25 mm girth		-	-	-	m	**4.54**
Coves; 150 - 300 mm girth						
per 25 mm girth		-	-	-	m	**5.59**
Cornices						
per 25 mm girth		-	-	-	m	**5.59**
Cornice enrichments						
per 25 mm girth; depending on degree of enrichments		-	-	-	m	**6.85**
Fibrous plaster; fixing with plaster wadding filling and pointing joints with plaster; to steel base						
16 mm plain slab coverings to ceilings						
over 300 mm wide		-	-	-	m2	**97.13**
not exceeding 300 mm wide		-	-	-	m2	**31.30**
16 mm plain casings to stanchions						
per 25 mm girth		-	-	-	m	**2.73**
16 mm plain casings to beams						
per 25 mm girth		-	-	-	m	**2.73**
Gyproc cove; fixing with adhesive; filling and pointing joints with plaster						
Cove						
125 mm girth		0.22	2.02	1.00	m	**3.02**
Angles		0.04	0.37	0.30	nr	**0.67**

ALTERNATIVE TILE MATERIALS

	£		£
Dennis Ruabon clay floor quarries (£/1000)			
Heather brown			
150x72x12.5 mm	183.31	194x94x12.5 mm	259.90
150x150x12.5 mm	259.10	194x194x12.5 mm	502.66

M SURFACE FINISHES
Including overheads and profit at 7.50%

	£		£
Red			
150x150x12.5 mm; square	172.16	150x150x12.5 mm; hexagonal	252.20
Rustic			
150x150x12.5 mm	253.35		

Daniel Platt heavy duty floor tiles (£/m2)

'Ferrolite' flat; 150x150x9 mm			
black	11.82	steel	11.03
cream	11.03	red	9.44
'Ferrolite' flat; 150x150x12 mm			
black	15.67	chocolate	15.67
cream	13.23	red	12.94
'Ferrundum' anti-slip; 150x150x12 mm			
black	24.82	red	22.54
chocolate	24.82		

Langley wall and floor tiles (£/m2)

'Buchtal' fine grain ceramic wall and floor tiles			
240x115x11 mm - series 1	29.14	194x144x11 mm - series 1	28.31
- series 2	31.61	- series 2	30.69
'Buchtal' rustic glazed wall and floor tiles			
240x115x11 mm - series 1	29.14	194x94x11 mm - series 1	28.42
- series 2	30.92	- series 2	30.14
- series 3	27.72	- series 3	27.10
'Sinzag' glazed ceramic wall facings			
240x115x10 mm - series 1	27.13	240x52x10 mm - series 1	30.69
- series 2	35.94		
- series 3	49.07		
'Sinzag' vitrified ceramic floor tiles; 150x150x11/12 mm			
plain - grey/white speckled	29.71	textured - series 2	31.52

Marley floor tiles (£/m2); 300x300 mm

'Econoflex' - series 1/2	2.02	anti-static	20.89
- series 4	2.39	'Travertine' - 2.5 mm	3.88
'Marleyflex' - 2.0 mm	2.60	'Vylon'	2.72
- 2.5 mm	3.04		

	Labour hours	Labour £	Material £	Unit	**Total rate £**
M40 STONE/CONCRETE/QUARRY/CERAMIC TILING/ MOSAIC					
Clay floor quarries; BS 6431; class 1; Daniel Platt 'Crown' tiles or similar; level bedding 10 mm thick and jointing in cement and sand (1:3); butt joints straight both ways; flush pointing with grout; to cement and sand base					
Work to floors; over 300 mm wide					
150 x 150 x 12.5 mm; red PC £242.67/1000	0.88	8.07	13.14	m2	**21.21**
150 x 150 x 12.5 mm; brown PC £381.20/1000	0.88	8.07	20.03	m2	**28.10**
200 x 200 x 19 mm; brown PC £929.75/1000	0.72	6.60	27.28	m2	**33.88**
Works to floors; in staircase areas or plant rooms					
150 x 150 x 12.5 mm; red	0.99	9.08	13.14	m2	**22.22**
150 x 150 x 12.5 mm; brown	0.99	9.08	20.03	m2	**29.11**
200 x 200 x 19 mm; brown	0.83	7.61	27.28	m2	**34.89**

M SURFACE FINISHES Including overheads and profit at 7.50%		Labour hours	Labour £	Material £	Unit	Total rate £
Work to floors; not exceeding 300 mm wide						
150 x 150 x 12.5 mm; red		0.44	4.03	4.72	m	8.75
150 x 150 x 12.5 mm; brown		0.44	4.03	7.18	m	11.21
200 x 200 x 19 mm; brown		0.36	3.30	10.30	m	13.60
Fair square cutting against flush edges of existing finishings		0.11	1.01	0.61	m	1.62
Raking cutting		0.22	2.02	0.90	m	2.92
Cutting around pipes; not exceeding 0.30 m girth		0.17	1.56	-	nr	1.56
Extra for cutting and fitting into recessed manhole covers 600 x 600 mm; lining up with adjoining work		1.10	10.09	-	nr	10.09
Work to sills; 150 mm wide; rounded edge tiles						
150 x 150 x 12.5 mm; red	PC £453.72/1000	0.36	3.30	3.79	m	7.09
150 x 150 x 12.5 mm; brown	PC £489.08/1000	0.36	3.30	4.07	m	7.37
Fitted end		0.17	1.56	-	m	1.56
Coved skirtings; 150 mm high; rounded top edge						
150 x 150 x 12.5 mm; red	PC £453.72/1000	0.28	2.57	3.79	m	6.36
150 x 150 x 12.5 mm; brown	PC £489.08/1000	0.28	2.57	4.07	m	6.64
Ends		0.06	0.55	-	nr	0.55
Angles		0.17	1.56	1.71	nr	3.27
Glazed ceramic wall tiles; BS 6431; fixing with adhesive; butt joints straight both ways; flush pointing with white grout; to plaster base						
Work to walls; over 300 mm wide						
152 x 152 x 5.5 mm; white	PC £10.64	0.66	7.87	13.76	m2	21.63
152 x 152 x 5.5 mm; light colours	PC £14.89	0.66	7.87	18.56	m2	26.43
152 x 152 x 5.5 mm; dark colours	PC £18.41	0.66	7.87	22.54	m2	30.41
Extra for RE or REX tile		-	-	-	nr	0.08
200 x 100 x 6.5 mm; white and light colours	PC £9.68	0.66	7.87	12.67	m2	20.54
250 x 200 x 7 mm; light colours	PC £15.43	0.55	6.56	19.17	m2	25.73
Work to walls; in staircase areas or plant rooms						
152 x 152 x 5.5 mm; white		0.74	8.82	13.76	m2	22.58
Work to walls; not exceeding 300 mm wide						
152 x 152 x 5.5 mm; white		0.33	3.93	4.30	m	8.23
152 x 152 x 5.5 mm; light colours		0.33	3.93	5.81	m	9.74
152 x 152 x 5.5 mm; dark colours		0.33	3.93	7.06	m	10.99
200 x 100 x 6.5 mm; white and light colours		0.33	3.93	3.96	m	7.89
250 x 200 x 7 mm; light colours		0.28	3.34	6.00	m	9.34
Cutting around pipes; not exceeding 0.30 m girth		0.11	1.31	-	nr	1.31
Work to sills; 150 mm wide; rounded edge tiles						
152 x 152 x 5.5 mm; white		0.28	3.34	2.19	m	5.53
Fitted end		0.11	1.31	-	nr	1.31
198 x 64.5 x 6 mm Langley 'Coloursound' wall tiles; fixing with adhesive; butt joints straight both ways; flush pointing with white grout; to plaster base						
Work to walls						
over 300 mm wide	PC £21.95	2.00	23.84	26.53	m2	50.37
not exceeding 300 mm wide		0.77	9.18	7.96	m	17.14

M SURFACE FINISHES Including overheads and profit at 7.50%		Labour hours	Labour £	Material £	Unit	Total rate £
M40 STONE/CONCRETE/QUARRY/CERAMIC TILING/ MOSAIC - cont'd						
Glazed ceramic floor tiles; level; bedding 10 mm thick and jointing in cement and sand (1:3); butt joints straight both ways; flush pointing with white grout; to cement and sand base						
Work to floors; over 300 mm wide						
200 x 200 x 8 mm	PC £27.98	0.66	7.87	32.68	m2	**40.55**
Work to floors; not exceeding 300 mm wide						
200 x 200 x 8 mm		0.33	3.93	10.35	m	**14.28**
M42 WOOD BLOCK/COMPOSITION BLOCK/PARQUET FLOORING						
Wood block; Vigers, Stevens & Adams 'Feltwood'; 7.5 mm thick; level; fixing with adhesive; to cement and sand base						
Work to floors; over 300 mm wide						
Iroko		0.55	5.04	11.85	m2	**16.89**
Wood blocks 25 mm thick; tongued and grooved joints; herringbone pattern; level; fixing with adhesive; to cement and sand base						
Work to floors; over 300 mm wide						
Iroko		-	-	-	m2	**60.52**
Merbau		-	-	-	m2	**57.46**
French Oak		-	-	-	m2	**69.61**
American Oak		-	-	-	m2	**56.76**
Fair square cutting against flush edges of existing finishings		-	-	-	m	**4.30**
Extra for cutting and fitting into recessed duct covers 450 mm wide; lining up with adjoining work		-	-	-	m	**13.84**
Cutting around pipes; not exceeding 0.30 m girth		-	-	-	nr	**2.26**
Extra for cutting and fitting into recessed manhole covers 600 x 600 mm; lining up with adjoining work		-	-	-	nr	**9.46**
Add to wood block flooring over 300 mm wide for						
sanding; one coat sealer; one coat wax polish		-	-	-	m2	**5.38**
sanding; two coats sealer; buffing with steel wool		-	-	-	m2	**3.87**
sanding; three coats polyurethane lacquer; buffing down between coats		-	-	-	m2	**3.87**

ALTERNATIVE FLEXIBLE SHEET MATERIALS (£/m2)

	£		£
Forbo-Nairn flooring			
'Armourfloor' - 3.2 mm	7.99	'Amourflex' - 2.0 mm	6.00
'Armourflex' - 3.2 mm	7.25		
Marley sheet flooring			
'HD Acoustic'	5.24	'HD Safetread'	8.12
'Format +'	6.55	'Vynatred'	3.75

M SURFACE FINISHES Including overheads and profit at 7.50%	Labour hours	Labour £	Material £	Unit	Total rate £
M50 RUBBER/PLASTICS/CORK/LINO/CARPET TILING/ SHEETING					
Linoleum sheet; BS 810; Forbo-Nairn or similar; level; fixing with adhesive; butt joints; to cement and sand base					
Work to floors; over 300 mm wide					
3.2 mm; plain	0.44	4.03	7.80	m2	**11.83**
3.2 mm; marbled	0.44	4.03	9.88	m2	**13.91**
Vinyl sheet; Altro 'Safety' range or similar with welded seams; level; fixing with adhesive; to cement and sand base					
Work to floors; over 300 mm wide					
2 mm thick; Marine T20	0.66	6.05	11.22	m2	**17.27**
2.5 mm thick; Classic D25	0.77	7.06	14.72	m2	**21.78**
3.5 mm thick; stronghold	0.88	8.07	19.13	m2	**27.20**
Vinyl sheet; heavy duty; Marley 'HD' or similar; level; with welded seams; fixing with adhesive; level; to cement and sand base					
Work to floors; over 300 mm wide					
2 mm thick	0.50	4.58	6.17	m2	**10.75**
2.5 mm thick	0.55	5.04	6.92	m2	**11.96**
2 mm skirtings					
75 mm high	0.11	1.01	0.74	m	**1.75**
100 mm high	0.13	1.19	0.88	m	**2.07**
150 mm high	0.17	1.56	1.29	m	**2.85**
Vinyl sheet; 'Gerflex' standard sheet; 'Classic' range; level; with welded seams; fixing with adhesive; to cement and sand base					
Work to floors; over 300 mm wide					
2 mm thick	0.55	5.04	5.46	m2	**10.50**
Vinyl sheet; 'Armstrong Rhino Contract'; level; with welded seams; fixing with adhesive; to cement and sand base					
Work to floors; over 300 mm wide					
2.5 mm thick	0.55	5.04	10.64	m2	**15.68**
Vinyl sheet; Marmoleum'; level; with welded seams; fixing with adhesive; to cement and sand base					
Work to floors; over 300 mm wide					
2.5 mm thick	0.55	5.04	8.58	m2	**13.62**
Vinyl tiles; 'Accoflex'; level; fixing with adhesive; butt joints; straight both ways; to cement and sand base					
Work to floors; over 300 mm wide					
300 x 300 x 2.0 mm	0.28	2.57	4.13	m2	**6.70**
Vinyl semi-flexible tiles; 'Arlon' or similar; level; fixing with adhesive; butt joints; straight both ways; to cement and sand base					
Work to floors; over 300 mm wide					
250 x 250 x 2.0 mm	0.28	2.57	4.48	m2	**7.05**

M SURFACE FINISHES Including overheads and profit at 7.50%	Labour hours	Labour £	Material £	Unit	Total rate £
M50 RUBBER/PLASTICS/CORK/LINO/CARPET TILING/ SHEETING - cont'd					
Vinyl semi-flexible tiles; Marley 'Marleyflex' or similar; level; fixing with adhesive; butt joints straight both ways; to cement and sand base					
Work to floors; over 300 mm wide					
250 x 250 x 2.0 mm	0.28	2.57	3.80	m2	**6.37**
250 x 250 x 2.5 mm	0.28	2.57	4.32	m2	**6.89**
Vinyl semi-flexible tiles; 'Vylon' or similar; level; fixing with adhesive; butt joints; straight both ways; to cement and sand base					
Work to floors; over 300 mm wide					
250 x 250 x 2.0 mm	0.28	2.57	3.80	m2	**6.37**
Vinyl tiles; anti-static; level; fixing with adhesive; butt joints straight both ways; to cement and sand base					
Work to floors; over 300 mm wide					
457 x 457 x 2 mm	0.50	4.58	25.83	m2	**30.41**
Vinyl tiles; 'Polyflex'; level; fixing with adhesive; butt joints; straight both ways;to cement and sand base					
Work to floors; over 300 mm wide					
300 x 300 x 1.5 mm	0.28	2.57	3.39	m2	**5.96**
300 x 300 x 2.0 mm	0.28	2.57	3.67	m2	**6.24**
Vinyl tiles; 'Polyflor XL'; level; fixing with adhesive; butt joints; straight both ways; to cement and sand base					
Work to floors; over 300 mm wide					
300 x 300 x 2.0 mm	0.28	2.57	5.71	m2	**8.28**
Vinyl tiles; 'Marley 'HD''; level; fixing with adhesive; butt joints; straight both ways; to cement and sand base					
Work to floors; over 300 mm wide					
300 x 300 x 2.0 mm	0.28	2.57	6.57	m2	**9.14**
Thermoplastic tiles; Marley 'Econoflex' or similar; level; fixing with adhesive; butt joints straight both ways; to cement and sand base					
Work to floors; over 300 mm wide					
250 x 250 x 2.0 mm; (series 2)	0.25	2.29	3.11	m2	**5.40**
250 x 250 x 2.0 mm; (series 4)	0.25	2.29	3.56	m2	**5.85**
Linoleum tiles; BS 810; Forbo-Nairn Floors or similar; level; fixing with adhesive; butt joints straight both ways; to cement and sand base					
Work to floors; over 300 mm wide					
3.2 mm thick (marbled patterns)	0.33	3.03	10.09	m2	**13.12**
Cork tiles; Wicanders 'Corktile' or similar; level fixing with adhesive; butt joints straight both ways; to cement and sand base					
Work to floors; over 300 mm wide					
300 x 300 x 3.2 mm	0.44	4.03	14.92	m2	**18.95**

M SURFACE FINISHES Including overheads and profit at 7.50%		Labour hours	Labour £	Material £	Unit	Total rate £
Rubber studded tiles; Altro 'Mondopave' or similar; level; fixing with adhesive; butt joints; straight; to cement and sand base						
Work to floors; over 300 mm wide						
500 x 500 x 2.5 mm; type GS; black		0.66	6.05	17.71	m2	**23.76**
500 x 500 x 4 mm; type BT; black		0.66	6.05	22.52	m2	**28.57**
Work to landings; over 300 mm wide						
500 x 500 x 4 mm; type BT; black		0.88	8.07	22.52	m2	**30.59**
4 mm thick to tread						
275 mm wide		0.55	5.04	7.28	m	**12.32**
4 mm thick to riser						
185 mm wide		0.66	6.05	5.04	m	**11.09**
Sundry floor sheeting underlays						
For floor finishings; over 300 mm wide						
building paper to BS 1521; class A;						
75 mm lap		0.07	0.64	0.71	m2	**1.35**
3.2 mm thick hardboard		0.22	2.02	1.14	m2	**3.16**
6.0 mm thick plywood		0.33	3.03	3.86	m2	**6.89**
Stair nosings						
'Ferodo' or equivalent; light duty hard aluminium alloy stair tread nosings; plugged and screwed to concrete						
type SD1	PC £5.62	0.28	2.57	6.63	m	**9.20**
type SD2	PC £7.76	0.33	3.03	9.05	m	**12.08**
'Ferodo' or equivalent; heavy duty aluminium alloy stair tread nosings; plugged and screwed to concrete						
type HD1	PC £6.65	0.33	3.03	7.79	m	**10.82**
type HD2	PC £9.25	0.39	3.58	10.72	m	**14.30**
Nylon needlepunch tiles; 'Marleyflex' or similar level; fixing with adhesive; to cement and sand base						
Work to floors						
over 300 mm wide		0.24	2.20	7.51	m2	**9.71**
Heavy duty carpet tiles; 'Heuga 581 Olympic' or similar; PC £12.60/m2; to cement and sand base						
Work to floors						
over 300 mm wide		0.27	2.48	14.95	m2	**17.43**
Nylon needlepunch carpet; 'Marleyflex' or similar; fixing; with adhesive; level; to cement						
Work to floors						
over 300 mm wide		0.27	2.48	6.91	m2	**9.39**

M SURFACE FINISHES Including overheads and profit at 7.50%	Labour hours	Labour £	Material £	Unit	Total rate £
M51 EDGE FIXED CARPETING					
Fitted carpeting; Wilton wool/nylon; 80/20 velvet pile; heavy domestic plain; PC £30.00/m2					
Work to floors					
over 300 mm wide	0.45	4.13	35.38	m2	**39.51**
Work to treads and risers					
over 300 mm wide	0.90	8.25	36.18	m2	**44.43**
Raking cutting	0.09	0.83	2.48	m2	**3.31**
Underlay to carpeting; PC £3.00/m2					
Work to floors					
over 300 mm wide	0.09	0.83	3.39	m2	**4.22**
Sundries					
Carpet gripper fixed to floor; standard edging	0.06	0.55	0.17	m	**0.72**
M52 DECORATIVE PAPERS/FABRICS					
Lining paper; PC £0.91/roll; roll; and hanging					
Plaster walls or columns					
over 300 mm girth	0.21	1.35	0.30	m2	**1.65**
Plaster ceilings or beams					
over 300 mm girth	0.26	1.67	0.30	m2	**1.97**
Decorative vinyl wallpaper; PC £5.20/roll; roll; and hanging					
Plaster walls or columns					
over 300 mm girth	0.26	1.67	1.31	m2	**2.98**
M60 PREPARATION OF EXISTING SURFACES - INTERNALLY					
NOTE: The prices for preparation given hereunder assume that existing surfaces are in 'fair condition' and should be increased for badly dilapidated surfaces.					
Wash down walls; cut out and make good cracks					
emulsion painted surfaces; including bringing forward bare patches	0.10	0.64	-	m2	**0.64**
gloss painted surfaces	0.08	0.51	-	m2	**0.51**
Wash down ceilings; cut out and make good cracks					
distempered surfaces	0.12	0.77	-	m2	**0.77**
emulsion painted surfaces; including bringing forward bare patches	0.14	0.90	-	m2	**0.90**
gloss painted surfaces	0.12	0.77	-	m2	**0.77**
Wash down plaster moulded cornices; cut out and make good cracks					
distempered surfaces	0.18	1.16	-	m2	**1.16**
emulsion painted surfaces; including bringing forward bare patches	0.21	1.35	-	m2	**1.35**

M SURFACE FINISHES Including overheads and profit at 7.50%	Labour hours	Labour £	Material £	Unit	Total rate £
Wash and rub down iron or steel surfaces; bringing forward					
General surfaces					
over 300 mm girth	0.16	1.03	-	m2	**1.03**
isolated surfaces not exceeding 300 mm girth	0.07	0.45	-	m	**0.45**
isolated surfaces not exceeding 0.50 m2	0.12	0.77	-	nr	**0.77**
Windows and the like					
panes over 1.00 m2	0.16	1.03	-	m2	**1.03**
panes 0.50 - 1.00 m2	0.18	1.16	-	m2	**1.16**
panes 0.10 - 0.50 m2	0.21	1.35	-	m2	**1.35**
panes not exceeding 0.10 m2	0.27	1.73	-	m2	**1.73**
Wash and rub down wood surfaces; prime bare patches; bringing forward					
General surfaces					
over 300 mm girth	0.25	1.60	-	m2	**1.60**
isolated surfaces not exceeding 300 mm girth	0.10	0.64	-	m	**0.64**
isolated surfaces not exceeding 0.50 m2	0.19	1.22	-	nr	**1.22**
Windows and the like					
panes over 1.00 m2	0.25	1.60	-	m2	**1.60**
panes 0.50 - 1.00 m2	0.29	1.86	-	m2	**1.86**
panes 0.10 - 0.50 m2	0.33	2.12	-	m2	**2.12**
panes not exceeding 0.10 m2	0.42	2.70	-	m2	**2.70**
Wash down and remove paint with chemical stripper from iron, steel or wood surfaces					
General surfaces					
over 300 mm girth	0.75	4.81	-	m2	**4.81**
isolated surfaces not exceeding 300 mm girth	0.33	2.12	-	m	**2.12**
isolated surfaces not exceeding 0.50 m2	0.56	3.59	-	nr	**3.59**
Windows and the like					
panes over 1.00 m2	0.96	6.16	-	m2	**6.16**
panes 0.50 - 1.00 m2	1.10	7.06	-	m2	**7.06**
panes 0.10 - 0.50 m2	1.28	8.21	-	m2	**8.21**
panes not exceeding 0.10 m2	1.60	10.27	-	m2	**10.27**
Burn off and rub down to remove paint from iron, steel or wood surfaces					
General surfaces					
over 300 mm girth	0.93	5.97	0.17	m2	**6.14**
isolated surfaces not exceeding 300 mm girth	0.42	2.70	0.07	m	**2.77**
isolated surfaces not exceeding 0.50 m2	0.70	4.49	0.07	nr	**4.56**
Windows and the like					
panes over 1.00 m2	1.20	7.70	-	m2	**7.70**
panes 0.50 - 1.00 m2	1.38	8.86	-	m2	**8.86**
panes 0.10 - 0.50 m2	1.60	10.27	-	m2	**10.27**
panes not exceeding 0.10 m2	2.00	12.84	-	m2	**12.84**

BASIC PAINT PRICES (£/5 Litre tin)

	£		£		£
Bituminous	9.89				
Emulsion					
matt	9.26	silk	9.76		
Knotting solution	25.88				
Oil					
gloss	11.32	undercoat	10.82		
Primer/undercoats					
acrylic	9.76	plaster	14.52	wood	10.77
aluminium	19.75	red lead	31.99	zinc phosphate	12.18
calcium plumbate	25.10	red oxide	10.79		
Road marking	20.89				

M SURFACE FINISHES
Including overheads and profit at 7.50%

	£		£
Special paints			
aluminium	19.53	fire retardant	
anti-condensation	18.71	- undercoat	28.00
heat resisting	15.97	- top coat	37.60
Stains and preservatives			
creosote	3.15	Solignum - 'Architectural'	23.06
Cuprinol 'clear'	11.21	- brown	7.92
linseed oil	15.82	- cedar	11.67
Sadolin - 'Holdex'	28.48	- clear	11.21
- 'Classic'	24.10	- green	10.68
- 'Sadovac 35'	15.01	'Multiplus'	12.85
Varnishes			
polyurethene	16.14	yacht	17.21

NOTE: The following prices include for preparing surfaces. Painting woodwork also includes for knotting prior to applying the priming coat and for all stopping of nail holes etc.	Labour hours	Labour £	Material £	Unit	Total rate £
M60 PAINTING/CLEAR FINISHING					
One coat primer; PC £10.77/5L; on wood surfaces before fixing					
General surfaces					
over 300 mm girth	0.16	1.03	0.23	m2	**1.26**
isolated surfaces not exceeding 300 mm girth	0.06	0.39	0.07	m	**0.46**
isolated surfaces not exceeding 0.50 m2	0.12	0.77	0.11	nr	**0.88**
One coat polyurethene sealer; PC £16.14/5L; on wood surfaces before fixing					
General surfaces					
over 300 mm girth	0.19	1.22	0.28	m2	**1.50**
isolated surfaces not exceeding 300 mm girth	0.07	0.45	0.09	m	**0.54**
isolated surfaces not exceeding 0.50 m2	0.14	0.90	0.14	nr	**1.04**
One coat clear wood preservative; PC £11.21/5L; on wood surfaces before fixing					
General surfaces					
over 300 mm girth	0.14	0.90	0.27	m2	**1.17**
isolated surfaces not exceeding 300 mm girth	0.05	0.32	0.08	m	**0.40**
isolated surfaces not exceeding 0.50 m2	0.11	0.71	0.13	nr	**0.84**
Two coats emulsion paint; PC £9.26/5L;					
Brick or block walls					
over 300 mm girth	0.21	1.35	0.40	m2	**1.75**
Cement render or concrete					
over 300 mm girth	0.19	1.22	0.30	m2	**1.52**
isolated surfaces not exceeding 300 mm girth	0.08	0.51	0.09	m	**0.60**
Plaster walls or plaster/plasterboard ceilings					
over 300 mm girth	0.18	1.16	0.30	m2	**1.46**
over 300 mm girth; in multi-colours	0.28	1.80	0.35	m2	**2.15**
over 300 mm girth; in staircase areas	0.21	1.35	0.30	m2	**1.65**
cutting in edges on flush surfaces	0.08	0.51	-	m	**0.51**
Plaster/plasterboard ceilings					
over 300 mm girth; 3.5 - 5.0 m high	0.20	1.28	0.30	m2	**1.58**

M SURFACE FINISHES Including overheads and profit at 7.50%	Labour hours	Labour £	Material £	Unit	Total rate £
One mist and two coats emulsion paint					
Brick or block walls					
over 300 mm girth	0.24	1.54	0.55	m2	**2.09**
Cement render or concrete					
over 300 mm girth	0.21	1.35	0.44	m2	**1.79**
isolated surfaces not exceeding 300 mm girth	0.09	0.58	0.12	m	**0.70**
Plaster walls or plaster/plasterboard ceilings					
over 300 mm girth	0.21	1.35	0.40	m2	**1.75**
over 300 mm girth; in multi-colours	0.32	2.05	0.45	m2	**2.50**
over 300 mm girth; in staircase areas	0.24	1.54	0.40	m2	**1.94**
cutting in edges on flush surfaces	0.12	0.77	-	m	**0.77**
Plaster/plasterboard ceilings					
over 300 mm girth; 3.5 - 5.0 m high	0.23	1.48	0.40	m2	**1.88**
Textured plastic; 'Artex' finish;					
Plasterboard ceilings					
over 300 mm girth	0.26	1.67	1.24	m2	**2.91**
Concrete walls or ceilings					
over 300 mm girth	0.28	1.80	1.39	m2	**3.19**
Touch up primer; one undercoat and one finishing coat of gloss oil paint; PC £11.32/5L; on wood surfaces					
General surfaces					
over 300 mm girth	0.34	2.18	0.40	m2	**2.58**
isolated surfaces not exceeding 300 mm girth	0.14	0.90	0.12	m	**1.02**
isolated surfaces not exceeding 0.50 m2	0.25	1.60	0.20	nr	**1.80**
Windows and the like					
panes over 1.00 m2	0.34	2.18	0.15	m2	**2.33**
panes 0.50 - 1.00 m2	0.39	2.50	0.20	m2	**2.70**
panes 0.10 - 0.50 m2	0.45	2.89	0.25	m2	**3.14**
panes not exceeding 0.10 m2	0.56	3.59	0.30	m2	**3.89**
Touch up primer; two undercoats and one finishing coat of gloss oil paint; on wood surfaces					
General surfaces					
over 300 mm girth	0.47	3.02	0.59	m2	**3.61**
isolated surfaces not exceeding 300 mm girth	0.19	1.22	0.18	m	**1.40**
isolated surfaces not exceeding 0.50 m2	0.35	2.25	0.29	nr	**2.54**
Windows and the like					
panes over 1.00 m2	0.47	3.02	0.22	m2	**3.24**
panes 0.50 - 1.00 m2	0.55	3.53	0.29	m2	**3.82**
panes 0.10 - 0.50 m2	0.63	4.04	0.37	m2	**4.41**
panes not exceeding 0.10 m2	0.79	5.07	0.44	m2	**5.51**
Knot; one coat primer; stop; one undercoat and one finishing coat of gloss oil paint; on wood surfaces					
General surfaces					
over 300 mm girth	0.49	3.14	0.68	m2	**3.82**
isolated surfaces not exceeding 300 mm girth	0.20	1.28	0.21	m	**1.49**
isolated surfaces not exceeding 0.50 m2	0.37	2.37	0.34	nr	**2.71**
Windows and the like					
panes over 1.00 m2	0.49	3.14	0.25	m2	**3.39**
panes 0.50 - 1.00 m2	0.58	3.72	0.34	m2	**4.06**
panes 0.10 - 0.50 m2	0.68	4.36	0.42	m2	**4.78**
panes not exceeding 0.10 m2	0.84	5.39	0.51	m2	**5.90**

M SURFACE FINISHES Including overheads and profit at 7.50%	Labour hours	Labour £	Material £	Unit	Total rate £
M60 PAINTING/CLEAR FINISHING - cont'd					
Knot; one coat primer; stop; two undercoats and one finishing coat of gloss oil paint; on wood surfaces					
General surfaces					
over 300 mm girth	0.63	4.04	0.87	m2	**4.91**
isolated surfaces not exceeding 300 mm girth	0.25	1.60	0.27	m	**1.87**
isolated surfaces not exceeding 0.50 m2	0.47	3.02	0.43	nr	**3.45**
Windows and the like					
panes over 1.00 m2	0.63	4.04	0.32	m2	**4.36**
panes 0.50 - 1.00 m2	0.73	4.68	0.43	m2	**5.11**
panes 0.10 - 0.50 m2	0.84	5.39	0.54	m2	**5.93**
panes not exceeding 0.10 m2	1.05	6.74	0.65	m2	**7.39**
One coat primer; one undercoat and one finishing coat of gloss oil paint					
Plaster surfaces					
over 300 mm girth	0.44	2.82	0.83	m2	**3.65**
One coat primer; two undercoats and one finishing coat of gloss oil paint					
Plaster surfaces					
over 300 mm girth	0.58	3.72	1.01	m2	**4.73**
Touch up primer; one undercoat and one finishing coat of gloss paint; PC £11.32/5L; on iron or steel surfaces					
General surfaces					
over 300 mm girth	0.34	2.18	0.38	m2	**2.56**
isolated surfaces not exceeding 300 mm girth	0.14	0.90	0.12	m	**1.02**
isolated surfaces not exceeding 0.50 m2	0.25	1.60	0.19	nr	**1.79**
Windows and the like					
panes over 1.00 m2	0.34	2.18	0.14	m2	**2.32**
panes 0.50 - 1.00 m2	0.39	2.50	0.19	m2	**2.69**
panes 0.10 - 0.50 m2	0.45	2.89	0.24	m2	**3.13**
panes not exceeding 0.10 m2	0.56	3.59	0.29	m2	**3.88**
Structural steelwork					
over 300 mm girth	0.38	2.44	0.38	m2	**2.82**
Members of roof trusses					
over 300 mm girth	0.50	3.21	0.40	m2	**3.61**
Ornamental railings and the like; each side measured overall					
over 300 mm girth	0.58	3.72	0.43	m2	**4.15**
Iron or steel radiators					
over 300 mm girth	0.34	2.18	0.40	m2	**2.58**
Pipes or conduits					
over 300 mm girth	0.50	3.21	0.40	m2	**3.61**
not exceeding 300 mm girth	0.20	1.28	0.12	m	**1.40**
Touch up primer; two undercoats and one finishing coat of gloss oil paint; on iron or steel surfaces					
General surfaces					
over 300 mm girth	0.47	3.02	0.57	m2	**3.59**
isolated surfaces not exceeding 300 mm girth	0.19	1.22	0.18	m	**1.40**
isolated surfaces not exceeding 0.50 m2	0.35	2.25	0.28	nr	**2.53**
Windows and the like					
panes over 1.00 m2	0.47	3.02	0.21	m2	**3.23**
panes 0.50 - 1.00 m2	0.55	3.53	0.28	m2	**3.81**
panes 0.10 - 0.50 m2	0.63	4.04	0.35	m2	**4.39**
panes not exceeding 0.10 m2	0.79	5.07	0.41	m2	**5.48**
Structural steelwork					
over 300 mm girth	0.54	3.47	0.57	m2	**4.04**

M SURFACE FINISHES Including overheads and profit at 7.50%	Labour hours	Labour £	Material £	Unit	Total rate £
Members of roof trusses					
over 300 mm girth	0.71	4.56	0.60	m2	**5.16**
Ornamental railings and the like; each side measured overall					
over 300 mm girth	0.81	5.20	0.64	m2	**5.84**
Iron or steel radiators					
over 300 mm girth	0.47	3.02	0.60	m2	**3.62**
Pipes or conduits					
over 300 mm girth	0.71	4.56	0.60	m2	**5.16**
not exceeding 300 mm girth	0.28	1.80	0.18	m	**1.98**
One coat primer; one undercoat and one finishing coat of gloss oil paint; on iron or steel surfaces					
General surfaces					
over 300 mm girth	0.44	2.82	0.38	m2	**3.20**
isolated surfaces not exceeding 300 mm girth	0.18	1.16	0.12	m	**1.28**
isolated surfaces not exceeding 0.50 m2	0.34	2.18	0.19	nr	**2.37**
Windows and the like					
panes over 1.00 m2	0.44	2.82	0.14	m2	**2.96**
panes 0.50 - 1.00 m2	0.50	3.21	0.19	m2	**3.40**
panes 0.10 - 0.50 m2	0.59	3.79	0.24	m2	**4.03**
panes not exceeding 0.10 m2	0.73	4.68	0.29	m2	**4.97**
Structural steelwork					
over 300 mm girth	0.49	3.14	0.38	m2	**3.52**
Members of roof trusses					
over 300 mm girth	0.66	4.24	0.40	m2	**4.64**
Ornamental railings and the like; each side measured overall					
over 300 mm girth	0.76	4.88	0.43	m2	**5.31**
Iron or steel radiators					
over 300 mm girth	0.44	2.82	0.40	m2	**3.22**
Pipes or conduits					
over 300 mm girth	0.66	4.24	0.40	m2	**4.64**
not exceeding 300 mm girth	0.26	1.67	0.12	m	**1.79**
One coat primer; two undercoats and one finishing coat of gloss oil paint; on iron or steel surfaces					
General surfaces					
over 300 mm girth	0.58	3.72	0.57	m2	**4.29**
isolated surfaces not exceeding 300 mm girth	0.23	1.48	0.18	m	**1.66**
isolated surfaces not exceeding 0.50 m2	0.42	2.70	0.28	nr	**2.98**
Windows and the like					
panes over 1.00 m2	0.58	3.72	0.21	m2	**3.93**
panes 0.50 - 1.00 m2	0.68	4.36	0.28	m2	**4.64**
panes 0.10 - 0.50 m2	0.79	5.07	0.35	m2	**5.42**
panes not exceeding 0.10 m2	0.94	6.03	0.43	m2	**6.46**
Structural steelwork					
over 300 mm girth	0.65	4.17	0.57	m2	**4.74**
Members of roof trusses					
over 300 mm girth	0.86	5.52	0.60	m2	**6.12**
Ornamental railings and the like; each side measured overall					
over 300 mm girth	0.98	6.29	0.64	m2	**6.93**
Iron or steel radiators					
over 300 mm girth	0.58	3.72	0.60	m2	**4.32**
Pipes or conduits					
over 300 mm girth	0.87	5.58	0.60	m2	**6.18**
not exceeding 300 mm girth	0.35	2.25	0.18	m	**2.43**

M SURFACE FINISHES Including overheads and profit at 7.50%	Labour hours	Labour £	Material £	Unit	Total rate £
M60 PAINTING/CLEAR FINISHING - cont'd					
Two coats of bituminous paint; PC £9.89/5L; on iron or steel surfaces					
General surfaces					
over 300 mm girth	0.42	2.70	0.34	m2	**3.04**
Inside of galvanized steel cistern					
over 300 mm girth	0.63	4.04	0.34	m2	**4.38**
Two coats of boiled linseed oil; PC £15.82/5; on hardwood surfaces					
General surfaces					
over 300 mm girth	0.32	2.05	1.19	m2	**3.24**
isolated surfaces not exceeding 300 mm girth	0.13	0.83	0.36	m	**1.19**
isolated surfaces not exceeding 0.50 m2	0.24	1.54	0.60	nr	**2.14**
Two coats polyurethane; PC £16.14/5L; on wood surfaces					
General surfaces					
over 300 mm girth	0.32	2.05	0.56	m2	**2.61**
isolated surfaces not exceeding 300 mm girth	0.13	0.83	0.17	m	**1.00**
isolated surfaces not exceeding 0.50 m2	0.23	1.48	0.28	nr	**1.76**
Three coats polyurethane; on wood surfaces					
General surfaces					
over 300 mm girth	0.47	3.02	0.83	m2	**3.85**
isolated surfaces not exceeding 300 mm girth	0.19	1.22	0.26	m	**1.48**
isolated surfaces not exceeding 0.50 m2	0.36	2.31	0.42	nr	**2.73**
One undercoat; PC £28.00/5L; and one finishing coat; PC £37.60/5L; of 'Albi' clear flame retardent surface coating; on wood surfaces					
General surfaces					
over 300 mm girth	0.57	3.66	1.40	m2	**5.06**
isolated surfaces not exceeding 300 mm girth	0.23	1.48	0.44	m	**1.92**
isolated surfaces not exceeding 0.50 m2	0.44	2.82	0.68	nr	**3.50**
Two undercoats; PC £28.00/5L; and one finishing coat; PC £37.60/5L; of 'Albi' clear flame retardent surface coating; on wood surfaces					
General surfaces					
over 300 mm girth	0.57	3.66	2.15	m2	**5.81**
isolated surfaces not exceeding 300 mm girth	0.23	1.48	0.68	m	**2.16**
isolated surfaces not exceeding 0.50 m2	0.44	2.82	1.08	nr	**3.90**
Seal and wax polish; dull gloss finish on wood surfaces					
General surfaces					
over 300 mm girth	-	-	-	m2	**5.75**
isolated surfaces not exceeding 300 mm girth	-	-	-	m	**2.58**
isolated surfaces not exceeding 0.50 m2	-	-	-	nr	**4.03**

M SURFACE FINISHES Including overheads and profit at 7.50%	Labour hours	Labour £	Material £	Unit	Total rate £
One coat of 'Sadolins Classic'; PC £24.10/5L; clear or pigmented; one further coat of 'Holdex' clear interior silk matt laquer; PC £28.48/5L					
General surfaces					
over 300 mm girth	0.42	2.70	1.06	m2	**3.76**
isolated surfaces not exceeding 300 mm girth	0.17	1.09	0.33	m	**1.42**
isolated surfaces not exceeding 0.50 m2	0.32	2.05	0.53	nr	**2.58**
Windows and the like					
panes over 1.00 m2	0.42	2.70	0.39	m2	**3.09**
panes 0.50 - 1.00 m2	0.48	3.08	0.53	m2	**3.61**
panes 0.10 - 0.50 m2	0.56	3.59	0.67	m2	**4.26**
panes not exceeding 0.10 m2	0.69	4.43	0.78	m2	**5.21**
Two coats of 'Sadolins Classic'; PC £24.10/5L; clear or pigmented; two further coats of 'Holdex' clear interior silk matt laquer; PC £28.48/5L					
General surfaces					
over 300 mm girth	0.65	4.17	2.12	m2	**6.29**
isolated surfaces not exceeding 300 mm girth	0.26	1.67	0.67	m	**2.34**
isolated surfaces not exceeding 0.50 m2	0.49	3.14	1.06	nr	**4.20**
Windows and the like					
panes over 1.00 m2	0.65	4.17	0.78	m2	**4.95**
panes 0.50 - 1.00 m2	0.76	4.88	1.06	m2	**5.94**
panes 0.10 - 0.50 m2	0.87	5.58	1.34	m2	**6.92**
panes not exceeding 0.10 m2	1.10	7.06	1.59	m2	**8.65**
Body in and wax polish; dull gloss finish; on hardwood surfaces					
General surfaces					
over 300 mm girth	-	-	-	m2	**8.17**
isolated surfaces not exceeding 300 mm girth	-	-	-	m	**3.66**
isolated surfaces not exceeding 0.50 m2	-	-	-	nr	**5.72**
Stain; body in and wax polish; dull gloss finish; on hardwood surface					
General surfaces					
over 300 mm girth	-	-	-	m2	**10.21**
isolated surfaces not exceeding 300 mm girth	-	-	-	m	**4.57**
isolated surfaces not exceeding 0.50 m2	-	-	-	nr	**7.15**
Seal; two coats of synthetic resin lacquer; decorative flatted finish; wire down, wax and burnish; on wood surfaces					
General surfaces					
over 300 mm girth	-	-	-	m2	**12.90**
isolated surfaces not exceeding 300 mm girth	-	-	-	m	**5.81**
isolated surfaces not exceeding 0.50 m2	-	-	-	nr	**9.03**
Stain; body in and fully French polish; full gloss finish; on hardwood surfaces					
General surfaces					
over 300 mm girth	-	-	-	m2	**15.48**
isolated surfaces not exceeding 300 mm girth	-	-	-	m	**6.99**
isolated surfaces not exceeding 0.50 m2	-	-	-	nr	**10.84**
Stain; fill grain and fully French polish; full gloss finish; on hardwood surfaces					
General surfaces					
over 300 mm girth	-	-	-	m2	**23.65**
isolated surfaces not exceeding 300 mm girth	-	-	-	m	**10.64**
isolated surfaces not exceeding 0.50 m2	-	-	-	nr	**16.56**

M SURFACE FINISHES Including overheads and profit at 7.50%	Labour hours	Labour £	Material £	Unit	Total rate £
M60 PAINTING/CLEAR FINISHING - cont'd					
Stain black; body in and fully French polish; ebonized finish; on hardwood surfaces					
General surfaces					
over 300 mm girth	-	-	-	m2	**29.03**
isolated surfaces not exceeding 300 mm girth	-	-	-	m	**13.06**
isolated surfaces not exceeding 0.50 m2	-	-	-	nr	**20.32**
M60 PREPARATION OF EXISTING SURFACES - EXTERNALLY					
Wash and rub down iron or steel surfaces; bringing forward					
General surfaces					
over 300 mm girth	0.20	1.28	-	m2	**1.28**
isolated surfaces not exceeding 300 mm girth	0.08	0.51	-	m	**0.51**
isolated surfaces not exceeding 0.50 m2	0.15	0.96	-	nr	**0.96**
Windows and the like					
panes over 1.00 m2	0.20	1.28	-	m2	**1.28**
panes 0.50 - 1.00 m2	0.23	1.48	-	m2	**1.48**
panes 0.10 - 0.50 m2	0.27	1.73	-	m2	**1.73**
panes not exceeding 0.10 m2	0.33	2.12	-	m2	**2.12**
Wash and rub down wood surfaces; prime bare patches; bringing forward					
General surfaces					
over 300 mm girth	0.33	2.12	-	m2	**2.12**
isolated surfaces not exceeding 300 mm girth	0.13	0.83	-	m	**0.83**
isolated surfaces not exceeding 0.50 m2	0.25	1.60	-	nr	**1.60**
Windows and the like					
panes over 1.00 m2	0.33	2.12	-	m2	**2.12**
panes 0.50 - 1.00 m2	0.38	2.44	-	m2	**2.44**
panes 0.10 - 0.50 m2	0.44	2.82	-	m2	**2.82**
panes not exceeding 0.10 m2	0.55	3.53	-	m2	**3.53**
Wash down and remove paint with chemical stripper from iron, steel or wood surfaces					
General surfaces					
over 300 mm girth	1.00	6.42	-	m2	**6.42**
isolated surfaces not exceeding 300 mm girth	0.45	2.89	-	m	**2.89**
isolated surfaces not exceeding 0.50 m2	0.75	4.81	-	nr	**4.81**
Windows and the like					
panes over 1.00 m2	1.28	8.21	-	m2	**8.21**
panes 0.50 - 1.00 m2	1.47	9.43	-	m2	**9.43**
panes 0.10 - 0.50 m2	1.70	10.91	-	m2	**10.91**
panes not exceeding 0.10 m2	2.14	13.73	-	m2	**13.73**
Burn off and rub down to remove paint from iron, steel or wood surfaces					
General surfaces					
over 300 mm girth	1.25	8.02	0.22	m2	**8.24**
isolated surfaces not exceeding 300 mm girth	0.56	3.59	0.03	m	**3.62**
isolated surfaces not exceeding 0.50 m2	0.94	6.03	0.08	nr	**6.11**
Windows and the like					
panes over 1.00 m2	1.60	10.27	-	m2	**10.27**
panes 0.50 - 1.00 m2	1.84	11.81	-	m2	**11.81**
panes 0.10 - 0.50 m2	2.13	13.67	-	m2	**13.67**
panes not exceeding 0.10 m2	2.67	17.14	-	m2	**17.14**

M SURFACE FINISHES Including overheads and profit at 7.50%	Labour hours	Labour £	Material £	Unit	Total rate £
M60 PAINTING/CLEAR FINISHING - EXTERNALLY					
Two coats of cement paint, 'Sandtex Matt' or similar; PC £12.60/5L;					
Brick or block walls					
over 300 mm girth	0.47	3.02	1.63	m2	**4.65**
Cement render or concrete walls					
over 300 mm girth	0.42	2.70	1.08	m2	**3.78**
Roughcast walls					
over 300 mm girth	0.52	3.34	2.17	m2	**5.51**
One coat sealer and two coats of external grade emulsion paint, Dulux 'Weathershield' or similar					
Brick or block walls					
over 300 mm girth	0.63	4.04	2.31	m2	**6.35**
Cement render or concrete walls					
over 300 mm girth	0.52	3.34	1.54	m2	**4.88**
Concrete soffits					
over 300 mm girth	0.58	3.72	1.54	m2	**5.26**
One coat sealer (applied by brush) and two coats of external grade emulsion paint, Dulux 'Weathershield' or similar (spray applied)					
Roughcast					
over 300 mm girth	0.63	2.99	3.08	m2	**6.07**
Touch up primer; one undercoat and one finishing coat of gloss oil paint; PC £11.32/5L; on wood surfaces					
General surfaces					
over 300 mm girth	0.38	2.44	0.40	m2	**2.84**
isolated surfaces not exceeding 300 mm girth	0.16	1.03	0.12	m	**1.15**
isolated surfaces not exceeding 0.50 m2	0.28	1.80	0.20	nr	**2.00**
Windows and the like					
panes over 1.00 m2	0.38	2.44	0.15	m2	**2.59**
panes 0.50 - 1.00 m2	0.44	2.82	0.20	m2	**3.02**
panes 0.10 - 0.50 m2	0.50	3.21	0.25	m2	**3.46**
panes not exceeding 0.10 m2	0.63	4.04	0.30	m2	**4.34**
Windows and the like; casements in colours differing from frames					
panes over 1.00 m2	0.42	2.70	0.17	m2	**2.87**
panes 0.50 - 1.00 m2	0.47	3.02	0.22	m2	**3.24**
panes 0.10 - 0.50 m2	0.56	3.59	0.27	m2	**3.86**
panes not exceeding 0.10 m2	0.69	4.43	0.32	m2	**4.75**
Touch up primer; two undercoats and one finishing coat of gloss oil paint; on wood surfaces					
General surfaces					
over 300 mm girth	0.52	3.34	0.59	m2	**3.93**
isolated surfaces not exceeding 300 mm girth	0.21	1.35	0.18	m2	**1.53**
isolated surfaces not exceeding 0.50 m2	0.40	2.57	0.29	m2	**2.86**
Windows and the like					
panes over 1.00 m2	0.52	3.34	0.22	m2	**3.56**
panes 0.50 - 1.00 m2	0.61	3.91	0.29	m2	**4.20**
panes 0.10 - 0.50 m2	0.70	4.49	0.37	m2	**4.86**
panes not exceeding 0.10 m2	0.87	5.58	0.44	m2	**6.02**
Windows and the like; casements in colours differing from frames					
panes over 1.00 m2	0.61	3.91	0.25	m2	**4.16**
panes 0.50 - 1.00 m2	0.70	4.49	0.33	m2	**4.82**
panes 0.10 - 0.50 m2	0.81	5.20	0.40	m2	**5.60**
panes not exceeding 0.10 m2	1.00	6.42	0.47	m2	**6.89**

M SURFACE FINISHES Including overheads and profit at 7.50%	Labour hours	Labour £	Material £	Unit	Total rate £
M60 PAINTING/CLEAR FINISHING - EXTERNALLY - cont'd					
Knot; one coat primer; one undercoat and one finishing coat of gloss oil paint; on wood surfaces					
General surfaces					
over 300 mm girth	0.56	3.59	0.68	m2	**4.27**
isolated surfaces not exceeding 300 mm girth	0.23	1.48	0.21	m	**1.69**
isolated surfaces not exceeding 0.50 m2	0.42	2.70	0.34	nr	**3.04**
Windows and the like					
panes over 1.00 m2	0.56	3.59	0.25	m2	**3.84**
panes 0.50 - 1.00 m2	0.65	4.17	0.34	m2	**4.51**
panes 0.10 - 0.50 m2	0.73	4.68	0.42	m2	**5.10**
panes not exceeding 0.10 m2	0.92	5.90	0.51	m2	**6.41**
Windows and the like; casements in colours differing from frames					
panes over 1.00 m2	0.61	3.91	0.29	m2	**4.20**
panes 0.50 - 1.00 m2	0.71	4.56	0.38	m2	**4.94**
panes 0.10 - 0.50 m2	0.81	5.20	0.66	m2	**5.86**
panes not exceeding 0.10 m2	1.00	6.42	0.54	m2	**6.96**
Knot; one coat primer; two undercoats and one finishing coat of gloss oil paint on wood surfaces					
General surfaces					
over 300 mm girth	0.69	4.43	0.87	m2	**5.30**
isolated surfaces not exceeding 300 mm girth	0.28	1.80	0.27	m	**2.07**
isolated surfaces not exceeding 0.50 m2	0.52	3.34	0.43	nr	**3.77**
Windows and the like					
panes over 1.00 m2	0.69	4.43	0.32	m2	**4.75**
panes 0.50 - 1.00 m2	0.81	5.20	0.43	m2	**5.63**
panes 0.10 - 0.50 m2	0.92	5.90	0.54	m2	**6.44**
panes not exceeding 0.10 m2	1.15	7.38	0.65	m2	**8.03**
Windows and the like; casements in colours differing from frames					
panes over 1.00 m2	0.80	5.13	0.37	m2	**5.50**
panes 0.50 - 1.00 m2	0.94	6.03	0.48	m2	**6.51**
panes 0.10 - 0.50 m2	1.05	6.74	0.79	m2	**7.53**
panes not exceeding 0.10 m2	1.30	8.34	0.70	m2	**9.04**
Touch up primer; one undercoat and one finishing coat of gloss oil paint; PC £11.32/5L; on iron or steel surfaces					
General surfaces					
over 300 mm girth	0.38	2.44	0.38	m2	**2.82**
isolated surfaces not exceeding 300 mm girth	0.16	1.03	0.12	m	**1.15**
isolated surfaces not exceeding 0.50 m2	0.28	1.80	0.19	nr	**1.99**
Windows and the like					
panes over 1.00 m2	0.38	2.44	0.14	m2	**2.58**
panes 0.50 - 1.00 m2	0.44	2.82	0.19	m2	**3.01**
panes 0.10 - 0.50 m2	0.50	3.21	0.24	m2	**3.45**
panes not exceeding 0.10 m2	0.63	4.04	0.29	m2	**4.33**
Structural steelwork					
over 300 mm girth	0.42	2.70	0.38	m2	**3.08**
Members of roof trusses					
over 300 mm girth	0.56	3.59	0.40	m2	**3.99**
Ornamental railings and the like; each side measured overall					
over 300 mm girth	0.63	4.04	0.43	m2	**4.47**
Eaves gutters					
over 300 mm girth	0.68	4.36	0.57	m2	**4.93**
not exceeding 300 mm girth	0.27	1.73	0.19	m	**1.92**
Pipes or conduits					
over 300 mm girth	0.56	3.59	0.40	m2	**3.99**
not exceeding 300 mm girth	0.22	1.41	0.12	m	**1.53**

M SURFACE FINISHES Including overheads and profit at 7.50%	Labour hours	Labour £	Material £	Unit	Total rate £
Touch up primer; two undercoats and one finishing coat of gloss oil paint; on iron or steel surfaces					
General surfaces					
over 300 mm girth	0.52	3.34	0.57	m2	**3.91**
isolated surfaces not exceeding 300 mm girth	0.21	1.35	0.18	m	**1.53**
isolated surfaces not exceeding 0.50 m2	0.39	2.50	0.28	nr	**2.78**
Windows and the like					
panes over 1.00 m2	0.52	3.34	0.21	m2	**3.55**
panes 0.50 - 1.00 m2	0.61	3.91	0.28	m2	**4.19**
panes 0.10 - 0.50 m2	0.70	4.49	0.35	m2	**4.84**
panes not exceeding 0.10 m2	0.87	5.58	0.43	m2	**6.01**
Structural steelwork					
over 300 mm girth	0.60	3.85	0.57	m2	**4.42**
Members of roof trusses					
over 300 mm girth	0.79	5.07	0.60	m2	**5.67**
Ornamental railings and the like; each side measured overall					
over 300 mm girth	0.89	5.71	0.64	m2	**6.35**
Eaves gutters					
over 300 mm girth	0.94	6.03	0.85	m2	**6.88**
not exceeding 300 mm girth	0.38	2.44	0.28	m	**2.72**
Pipes or conduits					
over 300 mm girth	0.79	5.07	0.60	m2	**5.67**
not exceeding 300 mm girth	0.32	2.05	0.18	m	**2.23**
One coat primer; one undercoat and one finishing coat of gloss oil paint; on iron or steel surfaces					
General surfaces					
over 300 mm girth	0.48	3.08	0.38	m2	**3.46**
isolated surfaces not exceeding 300 mm girth	0.20	1.28	0.12	m	**1.40**
isolated surfaces not exceeding 0.50 m2	0.37	2.37	0.19	nr	**2.56**
Windows and the like					
panes over 1.00 m2	0.48	3.08	0.14	m2	**3.22**
panes 0.50 - 1.00 m2	0.56	3.59	0.19	m2	**3.78**
panes 0.10 - 0.50 m2	0.65	4.17	0.24	m2	**4.41**
panes not exceeding 0.10 m2	0.81	5.20	0.29	m2	**5.49**
Structural steelwork					
over 300 mm girth	0.55	3.53	0.38	m2	**3.91**
Members of roof trusses					
over 300 mm girth	0.71	4.56	0.40	m2	**4.96**
Ornamental railings and the like; each side measured overall					
over 300 mm girth	0.82	5.26	0.43	m2	**5.69**
Eaves gutters					
over 300 mm girth	0.86	5.52	0.57	m2	**6.09**
not exceeding 300 mm girth	0.35	2.25	0.19	m	**2.44**
Pipes or conduits					
over 300 mm girth	0.71	4.56	0.40	m2	**4.96**
not exceeding 300 mm girth	0.28	1.80	0.12	m	**1.92**
One coat primer; two undercoats and one finishing coat of gloss oil paint; on iron or steel surfaces					
General surfaces					
over 300 mm girth	0.63	4.04	0.57	m2	**4.61**
isolated surfaces not exceeding 300 mm girth	0.25	1.60	0.18	m	**1.78**
isolated surfaces not exceeding 0.50 m2	0.47	3.02	0.28	nr	**3.30**
Windows and the like					
panes over 1.00 m2	0.63	4.04	0.21	m2	**4.25**
panes 0.50 - 1.00 m2	0.73	4.68	0.28	m2	**4.96**
panes 0.10 - 0.50 m2	0.84	5.39	0.35	m2	**5.74**
panes not exceeding 0.10 m2	1.05	6.74	0.43	m2	**7.17**

M SURFACE FINISHES Including overheads and profit at 7.50%	Labour hours	Labour £	Material £	Unit	Total rate £
M60 PAINTING/CLEAR FINISHING - EXTERNALLY - cont'd					
One coat primer; two undercoats and one finishing coat of gloss oil paint; on iron or steel surfaces - cont'd					
Structural steelwork					
over 300 mm girth	0.71	4.56	0.57	m2	**5.13**
Members of roof trusses					
over 300 mm girth	0.94	6.03	0.60	m2	**6.63**
Ornamental railings and the like; each side measured overall					
over 300 mm girth	1.05	6.74	0.64	m2	**7.38**
Eaves gutters					
over 300 mm girth	1.15	7.38	0.85	m2	**8.23**
not exceeding 300 mm girth	0.45	2.89	0.28	m	**3.17**
Pipes or conduits					
over 300 mm girth	0.94	6.03	0.60	m2	**6.63**
not exceeding 300 mm girth	0.38	2.44	0.18	m	**2.62**
Two coats of creosote; PC £3.15/5L; on wood surfaces					
General surfaces					
over 300 mm girth	0.26	1.67	0.15	m2	**1.82**
isolated surfaces not exceeding 300 mm girth	0.11	0.71	0.05	m	**0.76**
Two coats of 'Solignum' wood preservative; PC £11.67/5L; on wood surfaces					
General surfaces					
over 300 mm girth	0.26	1.67	0.55	m2	**2.22**
isolated surfaces not exceeding 300 mm girth	0.11	0.71	0.18	m	**0.89**
Three coats of polyurethane; PC £16.14/5L; on wood surfaces					
General surfaces					
over 300 mm girth	0.52	3.34	0.83	m2	**4.17**
isolated surfaces not exceeding 300 mm girth	0.21	1.35	0.26	m	**1.61**
isolated surfaces not exceeding 0.50 m2	0.40	2.57	0.42	nr	**2.99**
Two coats of 'Sadovac 35'primer; PC £15.01/5L; and two coats of 'Classic'; PC £24.10/5L; pigmented; on wood surfaces					
General surfaces					
over 300 mm girth	0.71	4.56	2.21	m2	**6.77**
isolated surfaces not exceeding 300 mm girth	0.28	1.80	0.70	m	**2.50**
Windows and the like					
panes over 1.00 m2	0.71	4.56	0.80	m2	**5.36**
panes 0.50 - 1.00 m2	0.83	5.33	1.10	m2	**6.43**
panes 0.10 - 0.50 m2	0.96	6.16	1.40	m2	**7.56**
panes not exceeding 0.10 m2	1.20	7.70	1.66	m2	**9.36**
Body in with French polish; one coat of lacquer or varnish; on wood surfaces					
General surfaces					
over 300 mm girth	-	-	-	m2	**13.22**
isolated surfaces not exceeding 300 mm girth	-	-	-	m	**5.91**
isolated surfaces not exceeding 0.50 m2	-	-	-	nr	**9.26**

N FURNITURE/EQUIPMENT Including overheads and profit at 7.50%		Labour hours	Labour £	Material £	Unit	Total rate £
N10/11 GENERAL FIXTURES/KITCHEN FITTINGS ETC.						
Proprietary items						
Closed stove vitreous enamelled finish; setting in position						
571 x 606 x 267 mm	PC £276.00	2.30	23.86	296.70	nr	**320.56**
Closed stove; vitreous enamelled finish fitted with mild steel barffed boiler; setting in position						
571 x 606 x 267 mm	PC £360.00	2.30	23.86	387.00	nr	**410.86**
Tile surround preslabbed; 406 mm wide firebrick back; cast iron stool bottom; black vitreous enamelled fret; assembling, setting and pointing firebrick back in fireclay mortar and backing with fine concrete finished to splay at top, laying hearth tiles in cement mortar and pointing in white cement						
1372 x 864 x 152 mm	PC £264.06	5.75	59.65	305.15	nr	**364.80**
Fitting components; blockboard						
Backs, fronts, sides or divisions; over 300 mm wide						
12 mm thick	PC £26.54	1.45	11.74	28.53	m2	**40.27**
19 mm thick	PC £34.26	1.45	11.74	36.83	m2	**48.57**
25 mm thick	PC £45.06	1.45	11.74	48.44	m2	**60.18**
Shelves or worktops; over 300 mm wide						
19 mm thick	PC £35.46	1.45	11.74	38.12	m2	**49.86**
25 mm thick	PC £45.05	1.45	11.74	48.43	m2	**60.17**
Flush doors; lipped on four edges						
450 x 750 x 19 mm	PC £19.07	0.39	3.16	20.50	nr	**23.66**
450 x 750 x 25 mm	PC £23.34	0.39	3.16	25.09	nr	**28.25**
600 x 900 x 19 mm	PC £28.38	0.55	4.45	30.51	nr	**34.96**
600 x 900 x 25 mm	PC £34.54	0.55	4.45	37.13	nr	**41.58**
Fitting components; chipboard						
Backs, fronts, sides or divisions; over 300 mm wide						
6 mm thick	PC £9.20	1.45	11.74	9.89	m2	**21.63**
9 mm thick	PC £12.63	1.45	11.74	13.57	m2	**25.31**
12 mm thick	PC £15.46	1.45	11.74	16.62	m2	**28.36**
19 mm thick	PC £21.97	1.45	11.74	23.62	m2	**35.36**
25 mm thick	PC £29.65	1.45	11.74	31.87	m2	**43.61**
Shelves or worktops; over 300 mm wide						
19 mm thick	PC £23.16	1.45	11.74	24.90	m2	**36.64**
25 mm thick	PC £29.64	1.45	11.74	31.86	m2	**43.60**
Flush doors; lipped on four edges						
450 x 750 x 19 mm	PC £15.46	0.39	3.16	16.62	nr	**19.78**
450 x 750 x 25 mm	PC £18.85	0.39	3.16	20.26	nr	**23.42**
600 x 900 x 19 mm	PC £22.71	0.55	4.45	24.42	nr	**28.87**
600 x 900 x 25 mm	PC £27.49	0.55	4.45	29.55	nr	**34.00**
Fitting components; Melamine faced chipboard						
Backs, fronts, sides or divisions; over 300 mm wide						
12 mm thick	PC £21.23	1.45	11.74	22.83	m2	**34.57**
19 mm thick	PC £29.11	1.45	11.74	31.29	m2	**43.03**
Shelves or worktops; over 300 mm wide						
19 mm thick	PC £30.54	1.45	11.74	32.83	m2	**44.57**
Flush doors; lipped on four edges						
450 x 750 x 19 mm	PC £19.39/m2	0.39	3.16	20.85	nr	**24.01**
600 x 900 x 19 mm	PC £28.58/m2	0.55	4.45	30.73	nr	**35.18**

N FURNITURE/EQUIPMENT Including overheads and profit at 7.50%		Labour hours	Labour £	Material £	Unit	Total rate £
N10/11 GENERAL FIXTURES/KITCHEN FITTINGS ETC. - cont'd						
Fitting components; 'Warerite Xcel' standard colour laminated chipboard type LD2; PC £215.45/10m2						
Backs, fronts, sides or divisions; over 300 mm wide						
13.2 mm thick	PC £58.05	1.45	11.74	62.41	m2	74.15
Shelves or worktops; over 300 mm wide						
13.2 mm thick	PC £58.05	1.45	11.74	62.41	m2	74.15
Flush doors; lipped on four edges						
450 x 750 x 13.2 mm	PC £27.74/m2	0.39	3.16	29.82	nr	32.98
600 x 900 x 13.2 mm	PC £41.79/m2	0.55	4.45	44.92	nr	49.37
Fitting components; plywood						
Backs, fronts, sides or divisions; over 300 mm wide						
6 mm thick	PC £14.97	1.45	11.74	16.09	m2	27.83
9 mm thick	PC £19.91	1.45	11.74	21.41	m2	33.15
12 mm thick	PC £25.23	1.45	11.74	27.13	m2	38.87
19 mm thick	PC £36.59	1.45	11.74	39.34	m2	51.08
25 mm thick	PC £48.63	1.45	11.74	52.27	m2	64.01
Shelves or worktops; over 300 mm wide						
19 mm thick	PC £37.79	1.45	11.74	40.62	m2	52.36
25 mm thick	PC £48.62	1.45	11.74	52.26	m2	64.00
Flush doors; lipped on four edges						
450 x 750 x 19 mm	PC £19.57	0.39	3.16	21.04	nr	24.20
450 x 750 x 25 mm	PC £24.17	0.39	3.16	25.98	nr	29.14
600 x 900 x 19 mm	PC £29.22	0.55	4.45	31.41	nr	35.86
600 x 900 x 25 mm	PC £35.90	0.55	4.45	38.60	nr	43.05
Fitting components; wrought softwood						
Backs, fronts, sides or divisions; cross-tongued joints; over 300 mm wide						
25 mm thick		1.45	11.74	40.75	m2	52.49
Shelves or worktops; cross-tongued joints; over 300 mm wide						
25 mm thick		1.45	11.74	40.75	m2	52.49
Bearers						
19 x 38 mm		0.11	0.89	2.33	m	3.22
25 x 50 mm		0.11	0.89	2.97	m	3.86
50 x 50 mm		0.11	0.89	4.42	m	5.31
50 x 75 mm		0.11	0.89	5.95	m	6.84
Bearers; framed						
19 x 38 mm		0.14	1.13	4.38	m	5.51
25 x 50 mm		0.14	1.13	5.03	m	6.16
50 x 50 mm		0.14	1.13	6.48	m	7.61
50 x 75 mm		0.14	1.13	7.99	m	9.12
Framing to backs, fronts or sides						
19 x 38 mm		0.17	1.38	4.38	m	5.76
25 x 50 mm		0.17	1.38	5.03	m	6.41
50 x 50 mm		0.17	1.38	6.48	m	7.86
50 x 75 mm		0.17	1.38	7.99	m	9.37
Flush doors; softwood skeleton or cellular core; plywood facing both sides; lipped on four edges						
450 x 750 x 35 mm		0.39	3.16	22.08	nr	25.24
600 x 900 x 35 mm		0.55	4.45	34.94	nr	39.39
Add +5% to the above 'Material £ prices' for 'selected' softwood for staining						

N FURNITURE/EQUIPMENT Including overheads and profit at 7.50%		Labour hours	Labour £	Material £	Unit	Total rate £
Fitting components; selected West African Mahogany; PC £505.25/m3						
Bearers						
19 x 38 mm	PC £3.06	0.17	1.38	3.44	m	**4.82**
25 x 50 mm	PC £3.96	0.17	1.38	4.43	m	**5.81**
50 x 50 mm	PC £6.00	0.17	1.38	6.68	m	**8.06**
50 x 75 mm	PC £8.13	0.17	1.38	9.02	m	**10.40**
Bearers; framed						
19 x 38 mm	PC £5.88	0.22	1.78	6.49	m	**8.27**
25 x 50 mm	PC £6.79	0.22	1.78	7.48	m	**9.26**
50 x 50 mm	PC £8.82	0.22	1.78	9.72	m	**11.50**
50 x 75 mm	PC £10.95	0.22	1.78	12.06	m	**13.84**
Framing to backs, fronts or sides						
19 x 38 mm	PC £5.88	0.28	2.27	6.49	m	**8.76**
25 x 50 mm	PC £6.79	0.28	2.27	7.48	m	**9.75**
50 x 50 mm	PC £8.82	0.28	2.27	9.72	m	**11.99**
50 x 75 mm	PC £10.95	0.28	2.27	12.06	m	**14.33**
Fitting components; Iroko; PC £537.50/m3						
Backs, fronts, sides or divisions; cross-tongued joints; over 300 mm wide						
25 mm thick	PC £66.96	1.95	15.78	71.98	m2	**87.76**
Shelves or worktops; cross-tongued joints; over 300 mm wide						
25 mm thick	PC £66.96	1.95	15.78	71.98	m2	**87.76**
Draining boards; cross-tongued joints; over 300 mm wide						
25 mm thick	PC £71.52	1.95	15.78	76.88	m2	**92.66**
Stopped flutes		-	-	-	m	**2.64**
Grooves; cross-grain		-	-	-	m	**0.63**
Bearers						
19 x 38 mm	PC £3.68	0.17	1.38	4.12	m	**5.50**
25 x 50 mm	PC £4.63	0.17	1.38	5.17	m	**6.55**
50 x 50 mm	PC £6.88	0.17	1.38	7.65	m	**9.03**
50 x 75 mm		0.17	1.38	10.25	m	**11.63**
Bearers; framed						
19 x 38 mm	PC £7.44	0.22	1.78	8.20	m	**9.98**
25 x 50 mm	PC £8.40	0.22	1.78	9.26	m	**11.04**
50 x 50 mm	PC £10.65	0.22	1.78	11.74	m	**13.52**
50 x 75 mm	PC £13.02	0.22	1.78	14.35	m	**16.13**
Framing to backs, fronts or sides						
19 x 38 mm	PC £7.44	0.28	2.27	8.20	m	**10.47**
25 x 50 mm	PC £8.40	0.28	2.27	9.26	m	**11.53**
50 x 50 mm	PC £10.65	0.28	2.27	11.74	m	**14.01**
50 x 75 mm	PC £13.02	0.28	2.27	14.35	m	**16.62**
Fitting components; Teak; PC £1612.50/m3						
Backs, fronts, sides or divisions; cross-tongued joints; over 300 mm wide						
25 mm thick	PC £147.60	2.20	17.81	158.67	m2	**176.48**
Shelves or worktops; cross-tongued joints; over 300 mm wide						
25 mm thick	PC £147.60	2.20	17.81	158.67	m2	**176.48**
Draining boards; cross-tongued joints; over 300 mm wide						
25 mm thick	PC £151.03	2.20	17.81	162.35	m2	**180.16**
Stopped flutes		-	-	-	m2	**2.64**
Grooves; cross-grain		-	-	-	m	**0.63**

N FURNITURE/EQUIPMENT Including overheads and profit at 7.50%		Labour hours	Labour £	Material £	Unit	Total rate £
N10/11 GENERAL FIXTURES/KITCHEN FITTINGS ETC. - cont'd						
Fitting components; 'Formica' plastic laminated plastics coverings fixed with adhesive						
1.3 mm thick horizontal grade marbled finish over 300 mm wide	PC £18.00	1.75	14.17	28.51	m2	**42.68**
1.0 mm thick universal backing over 300 mm wide	PC £4.00	1.30	10.52	8.50	m2	**19.02**
1.3 mm thick edging strip						
18 mm wide		0.11	0.89	1.54	m	**2.43**
25 mm wide		0.11	0.89	1.54	m	**2.43**
Fixing kitchen fittings						
(Kitchen fittings are largely a matter of selection and prices vary considerably. PC supply prices for reasonable quantities for 'Standard' (moderately-priced) kitchen fittings have been shown but not extended).						
Fixing only to backgrounds requiring plugging; including any pre-assembly						
Wall units						
600 x 600 x 300 mm	PC £43.37	1.55	12.55	0.18	nr	**12.73**
600 x 900 x 300 mm	PC £53.94	1.75	14.17	0.18	nr	**14.35**
1200 x 600 x 300 mm	PC £78.75	2.05	16.59	0.24	nr	**16.83**
1200 x 900 x 300 mm	PC £94.36	2.30	18.62	0.24	nr	**18.86**
Floor units with drawers						
600 x 900 x 500 mm	PC £73.39	1.40	11.33	0.12	nr	**11.45**
600 x 900 x 600 mm	PC £76.62	1.55	12.55	0.12	nr	**12.67**
1200 x 900 x 600 mm	PC £135.69	1.85	14.98	0.12	nr	**15.10**
Laminated plastics worktops to suit last						
500 x 600 mm	PC £14.28	0.44	3.56	-	nr	**3.56**
600 x 600 mm	PC £16.00	0.50	4.05	-	nr	**4.05**
1200 x 600 mm	PC £31.92	0.77	6.23	-	nr	**6.23**
Larder units						
600 x 1950 x 600 mm	PC £123.45	2.70	21.86	0.18	nr	**22.04**
Sink units (excluding sink top)						
1200 x 900 x 600 mm	PC £130.80	2.00	16.19	0.12	nr	**16.31**
1500 x 900 x 600	PC £190.23	2.35	19.02	0.18	nr	**19.20**
6 mm thick rectangular glass mirrors; silver backed; fixed with chromium plated domed headed screws; to background requiring plugging						
Mirror with polished edges						
356 x 254 mm	PC £3.13	0.88	5.65	5.98	nr	**11.63**
400 x 300 mm	PC £4.11	0.88	5.65	7.03	nr	**12.68**
560 x 380 mm	PC £7.47	0.99	6.35	10.64	nr	**16.99**
640 x 460 mm	PC £9.12	1.10	7.06	12.42	nr	**19.48**
Mirror with bevelled edges						
356 x 254 mm	PC £5.36	0.88	5.65	8.37	nr	**14.02**
400 x 300 mm	PC £5.96	0.88	5.65	9.02	nr	**14.67**
560 x 380 mm	PC £10.59	0.99	6.35	14.00	nr	**20.35**
640 x 460 mm	PC £13.19	1.10	7.06	16.79	nr	**23.85**

N FURNITURE/EQUIPMENT Including overheads and profit at 7.50%		Labour hours	Labour £	Material £	Unit	Total rate £
N10 GENERAL FIXTURES/FURNISHINGS/EQUIPMENT						
Matwells						
Mild steel matwell; galvanized after manufacture; comprising 30 x 30 x 3 mm angle rim; with welded angles and lugs welded on; to suit mat size						
914 x 560 mm	PC £22.68	1.10	10.09	24.38	nr	**34.47**
1067 x 610 mm	PC £23.66	1.40	12.84	25.44	nr	**38.28**
1219 x 762 mm	PC £26.24	1.65	15.13	28.20	nr	**43.33**
Polished aluminium matwell; comprising 32 x 32 x 5 mm angle rim; with brazed angles and lugs brazed on; to suit mat size						
914 x 560 mm	PC £30.40	1.10	10.09	32.68	nr	**42.77**
1067 x 610 mm	PC £32.40	1.40	12.84	34.83	nr	**47.67**
1219 x 762 mm	PC £37.35	1.65	15.13	40.15	nr	**55.28**
Polished brass matwell; comprising 38 x 38 x 6 mm angle rim; with brazed angles and lugs welded on; to suit mat size						
914 x 560 mm	PC £89.86	1.10	10.09	96.60	nr	**106.69**
1067 x 610 mm	PC £95.82	1.40	12.84	103.01	nr	**115.85**
1219 x 762 mm	PC £110.42	1.65	15.13	118.70	nr	**133.83**
N13 SANITARY APPLIANCES/FITTINGS						
NOTE: Sanitary fittings are largely a matter of selection and material prices vary considerably; the PC values given below are for average quality						
Sink; white glazed fireclay; BS 1206; cast iron cantilever brackets						
610 x 455 x 205 mm	PC £101.00	3.60	32.43	114.12	nr	**146.55**
610 x 455 x 255 mm	PC £121.75	3.60	32.43	136.42	nr	**168.85**
760 x 455 x 255 mm	PC £170.00	3.60	32.43	188.29	nr	**220.72**
Sink; stainless steel combined bowl and draining board; chain and self colour plug; to BS 3380						
1050 x 500 mm with bowl 420 x 350 x 175 mm	PC £88.87	2.10	18.92	95.54	nr	**114.46**
Sink; stainless steel combined bowl and double draining board; chain and self colour plug to BS 3380						
1550 x 500 mm with bowl 420 x 350 x 200 mm	PC £110.18	2.40	21.62	118.45	nr	**140.07**
Lavatory basin; vitreous china; BS 1188; 32 mm chromium plated waste; chain, stay and plug; pair 13 mm chromium plated easy clean pillar taps to BS 1010; painted cantilever brackets; plugged and screwed						
560 x 405 mm; white	PC £67.85	2.75	24.77	72.94	nr	**97.71**
560 x 405 mm; coloured	PC £77.30	2.75	24.77	83.10	nr	**107.87**
635 x 455 mm; white	PC £98.64	2.75	24.77	106.04	nr	**130.81**
635 x 455 mm; coloured	PC £112.98	2.75	24.77	121.45	nr	**146.22**
Lavatory basin and pedestal; vitreous china; BS 1188; 32 mm chromium plated waste, chain, stay and plug; pair 13 mm chromium plated easy clean pillar taps to BS 1010; pedestal; wall brackets; plugged and screwed						
560 x 405 mm; white	PC £88.57	3.00	27.03	95.21	nr	**122.24**
560 x 405 mm; coloured	PC £103.57	3.00	27.03	111.34	nr	**138.37**
635 x 455 mm; white	PC £119.37	3.00	27.03	128.32	nr	**155.35**
635 x 455 mm; coloured	PC £139.02	3.00	27.03	149.45	nr	**176.48**

N FURNITURE/EQUIPMENT Including overheads and profit at 7.50%		Labour hours	Labour £	Material £	Unit	Total rate £
N13 SANITARY APPLIANCES/FITTINGS - cont'd						
Lavatory basin range; overlap joints; white glazed fireclay; 32 mm chromium plated waste, chain, stay and plug; pair 13 mm chromium plated easy clean pillar taps to BS 1010; painted cast iron cantilever brackets; plugged and screwed						
range of four 560 x 405 mm	PC £354.00	10.56	95.13	380.55	nr	**475.68**
Add for each additional basin in the range		2.40	21.62	97.50	nr	**119.12**
Drinking fountain; white glazed fireclay; 19 mm chromium plated waste; self-closing non-concussive tap; regulating valve; plugged and screwed with chromium plated screws	PC £156.80	3.00	27.03	168.56	nr	**195.59**
Bath; reinforced acrylic rectangular pattern; 40 mm chromium plated overflow chain and plug; 40 mm chromium plated waste; cast brass 'P' trap with plain outlet and overflow connection to BS 1184; pair 20 mm chromium plated easy clean pillar taps to BS 1010						
1700 mm long; white	PC £165.00	4.20	37.84	177.38	nr	**215.22**
1700 mm long; coloured	PC £165.00	4.20	37.84	177.38	nr	**215.22**
Bath; enamelled steel; medium gauge rectangular pattern; 40 mm chromium plated overflow chain and plug; 40 mm chromium plated waste; cast brass 'P' trap with plain outlet and overflow connection to BS 1184; pair 20 mm chromium plated easy clean pillar taps to BS 1010						
1700 mm long; white	PC £181.75	4.20	37.84	195.38	nr	**233.22**
1700 mm long; coloured	PC £193.50	4.20	37.84	208.01	nr	**245.85**
Shower tray; glazed fireclay with outlet and grated waste; chain and plug; bedding and pointing in waterproof cement mortar						
760 x 760 x 180 mm; white	PC £140.93	3.60	32.43	151.50	nr	**183.93**
760 x 760 x 180 mm; coloured	PC £198.35	3.60	32.43	213.23	nr	**245.66**
Shower fitting; riser pipe with mixing valve and shower rose; chromium plated; plugging and screwing mixing valve and pipe bracket						
15 mm dia riser pipe; 127 mm dia shower rose	PC £230.00	6.00	54.05	247.25	nr	**301.30**
WC suite; high level; vitreous china pan; black plastic seat; 9 litre white vitreous china cistern and brackets; low pressure ball valve; galvanized steel flush pipe and clip; plugged and screwed; mastic joint to drain						
WC suite; white	PC £166.08	3.95	35.58	178.54	nr	**214.12**
WC suite; coloured	PC £205.23	3.95	35.58	220.62	nr	**256.20**
WC suite; low level; vitreous china pan; black plastic seat; 9 litre white vitreous china cistern and brackets; low pressure ball valve and plastic flush pipe; plugged and screwed; mastic joint to drain						
WC suite; white	PC £115.24	3.60	32.43	123.88	nr	**156.31**
WC suite; coloured	PC £143.86	3.60	32.43	154.65	nr	**187.08**

N FURNITURE/EQUIPMENT Including overheads and profit at 7.50%		Labour hours	Labour £	Material £	Unit	Total rate £
Slop sink; white glazed fireclay with hardwood pad; aluminium bucket grating; vitreous china cistern and porcelain-enamelled brackets; galvanized steel flush pipe and clip; plugged and screwed; mastic joint to drain						
slop sink	PC £550.00	4.20	37.84	591.25	nr	**629.09**
Bowl type wall urinal; white glazed vitreous china; white vitreous china automatic flushing cistern and brackets; 38 mm chromium plated waste; chromium plated flush pipes and spreaders; cistern brackets and flush pipes plugged and screwed with chromium plated screws						
single; 455 x 380 x 330 mm	PC £159.53	4.80	43.24	171.49	nr	**214.73**
range of two	PC £264.32	9.00	81.08	284.14	nr	**365.22**
range of three	PC £353.10	13.20	118.91	379.58	nr	**498.49**
Add for each additional urinal	PC £104.79	3.85	34.68	112.65	nr	**147.33**
Add for divisions between	PC £42.33	0.90	8.11	45.50	nr	**53.61**

N15 SIGNS/NOTICES

	Labour hours	Labour £	Material £	Unit	Total rate £
Plain script; in gloss oil paint; on painted or varnished surfaces					
Capital letters; lower case letters or numerals					
per coat; per 25 mm high	0.11	0.71	-	nr	**0.71**
Stops					
per coat	0.03	0.19	-	nr	**0.19**

P BUILDING FABRIC SUNDRIES
Including overheads and profit at 7.50%

ALTERNATIVE INSULATION PRICES

Insulation (£/m2)

Expanded polystyrene

self-extinguishing grade

12 mm	1.02	25 mm	2.11	38 mm	3.22	50 mm	4.24
18 mm	1.52						

'Fibreglass'

'Crown Building Roll'

60 mm	2.38	80 mm	2.99	100 mm	3.51

'Crown Wool'

150 mm	5.00	200 mm	6.74

'Frametherm' - unfaced

60 mm	1.87	80 mm	2.50	90 mm	2.80	100 mm	3.04

Sound-deadening quilt type PF;
13 mm £3.02/m2

		Labour hours	Labour £	Material £	Unit	Total rate £
P10 SUNDRY INSULATION/PROOFING WORK/ FIRE STOPS						
'Sisalkraft' building papers/vapour barriers						
Building paper; 150 mm laps; fixed to softwood						
'Moistop' grade 728 (class A1F)		0.11	0.89	0.74	m2	**1.63**
Vapour barrier/reflective insulation 150 mm laps; fixed to softwood						
'Insulex' grade 714; single sided		0.11	0.89	1.14	m2	**2.03**
'Insulex' grade 714; double sided		0.11	0.89	1.66	m2	**2.55**
Mat or quilt insulation						
Glass fibre quilt; Pilkingtons 'Crown Wool'; laid over ceiling joists						
60 mm thick	PC £2.07	0.11	0.89	2.33	m2	**3.22**
80 mm thick	PC £2.74	0.12	0.97	3.09	m2	**4.06**
100 mm thick	PC £3.26	0.13	1.05	3.67	m2	**4.72**
150 mm thick	PC £5.00	0.14	1.13	5.64	m2	**6.77**
200 mm thick	PC £6.74	0.15	1.21	7.61	m2	**8.82**
Raking cutting		0.06	0.49	-	m	**0.49**
Glass fibre quilt; 'Gypglass 1000'; laid loose between members at 600 mm centres						
60 mm thick	PC £2.30	0.23	1.86	2.59	m2	**4.45**
80 mm thick	PC £3.04	0.25	2.02	3.43	m2	**5.45**
100 mm thick	PC £3.62	0.28	2.27	4.09	m2	**6.36**
150 mm thick	PC £5.55	0.33	2.67	6.27	m2	**8.94**
Mineral fibre quilt; 'Gypglass 1200'; pinned vertically to softwood						
25 mm thick	PC £1.20	0.15	1.21	1.35	m2	**2.56**
50 mm thick	PC £2.40	0.17	1.38	2.71	m2	**4.09**
Glass fibre building roll; pinned vertically to softwood						
60 mm thick	PC £2.38	0.17	1.38	2.69	m2	**4.07**
80 mm thick	PC £2.99	0.18	1.46	3.37	m2	**4.83**
100 mm thick	PC £3.51	0.19	1.54	3.97	m2	**5.51**

P BUILDING FABRIC SUNDRIES Including overheads and profit at 7.50%		Labour hours	Labour £	Material £	Unit	Total rate £
Glass fibre flanged building roll; paper faces; pinned vertically or to slope between timber framing						
60 mm thick	PC £2.56	0.20	1.62	2.89	m2	**4.51**
80 mm thick	PC £3.17	0.21	1.70	3.58	m2	**5.28**
100 mm thick	PC £3.69	0.22	1.78	4.16	m2	**5.94**
Board or slab insulation						
Expanded polystyrene board standard grade PC £64.67/m3; fixed with adhesive						
12 mm thick		0.44	3.56	1.47	m2	**5.03**
25 mm thick		0.46	3.72	2.39	m2	**6.11**
50 mm thick		0.50	4.05	4.21	m2	**8.26**
75 mm thick		0.55	4.45	6.03	m2	**10.48**
Fire stops						
'Monolux' TRADA firecheck channel; intumescent coatings on cut mitres; fixing with brass cups and screws						
19 x 44 mm or 19 x 50 mm	PC £248.18/36m	0.66	5.34	8.46	m	**13.80**
'Sealmaster' intumescent fire and smoke seals; pinned into groove in timber						
type N30; for single leaf half hour door;	PC £5.46	0.33	2.67	5.88	m	**8.55**
type N60; for single leaf one hour door	PC £8.31	0.36	2.91	8.95	m	**11.86**
type IMN or IMP; for meeting or pivot styles of pair of one hour doors; per style	PC £8.31	0.36	2.91	8.95	m	**11.86**
Intumescent plugs in timber; including boring		0.11	0.89	0.28	nr	**1.17**
Rockwool fire stops; between top of brick/ block wall and concrete soffit						
25 mm deep x 112 mm wide		0.09	0.73	0.93	m	**1.66**
25 mm deep x 150 mm wide		0.11	0.89	1.25	m	**2.14**
50 mm deep x 225 mm wide		0.17	1.38	3.16	m	**4.54**
Fire barriers						
Rockwool fire barrier between top of suspended ceiling and concrete soffit						
one 50 mm layer x 900 mm wide; half-hour		0.66	5.34	17.75	m	**23.09**
two 50 mm layers x 900 mm wide; one hour		1.00	8.09	34.25	m	**42.34**
Dow Chemicals 'Styrofoam 1B'; cold bridging insulation fixed with adhesive to brick, block or concrete base						
Insulation to walls						
25 mm thick		0.37	3.77	5.36	m2	**9.13**
50 mm thick		0.40	4.07	9.56	m2	**13.63**
75 mm thick		0.42	4.28	13.62	m2	**17.90**
Insulation to isolated columns						
25 mm thick		0.45	4.58	5.36	m2	**9.94**
50 mm thick		0.48	4.89	9.56	m2	**14.45**
75 mm thick		0.52	5.29	13.62	m2	**18.91**
Insulation to ceilings						
25 mm thick		0.40	4.07	5.36	m2	**9.43**
50 mm thick		0.43	4.38	9.56	m2	**13.94**
75 mm thick		0.46	4.68	13.62	m2	**18.30**
Insulation to isolated beams						
25 mm thick		0.48	4.89	5.36	m2	**10.25**
50 mm thick		0.52	5.29	9.56	m2	**14.85**
75 mm thick		0.55	5.60	13.62	m2	**19.22**

P BUILDING FABRIC SUNDRIES Including overheads and profit at 7.50%	Labour hours	Labour £	Material £	Unit	Total rate £
P11 FOAMED/FIBRE/BEAD CAVITY WALL INSULATION					
Cavity wall insulation; injecting 65 mm cavity with					
UF foam	-	-	-	m2	2.96
blown EPS granules	-	-	-	m2	3.23
blown mineral wool	-	-	-	m2	3.44
P20 UNFRAMED ISOLATED TRIMS/SKIRTINGS/ SUNDRY ITEMS					
Blockboard (Birch faced); 18 mm thick PC £97.16/10m2					
Window boards and the like; rebated; hardwood lipped on one edge					
18 x 200 mm	0.28	2.27	3.34	m	5.61
18 x 250 mm	0.31	2.51	3.90	m	6.41
18 x 300 mm	0.34	2.75	4.46	m	7.21
18 x 350 mm	0.37	3.00	5.03	m	8.03
Returned and fitted ends	0.24	1.94	0.35	nr	2.29
Blockboard (Sapele veneered one side); 18 mm thick PC £84.54/10m2					
Window boards and the like; rebated; hardwood lipped on one edge					
18 x 200 mm	0.30	2.43	3.05	m	5.48
18 x 250 mm	0.33	2.67	3.54	m	6.21
18 x 300 mm	0.36	2.91	4.02	m	6.93
18 x 350 mm	0.40	3.24	4.51	m	7.75
Returned and fitted ends	0.24	1.94	0.35	nr	2.29
Blockboard (Afrormosia veneered one side); 18 mm thick PC £103.56/10m2					
Window boards and the like; rebated; hardwood lipped on one edge					
18 x 200 mm	0.30	2.43	3.55	m	5.98
18 x 250 mm	0.33	2.67	4.15	m	6.82
18 x 300 mm	0.36	2.91	4.75	m	7.66
18 x 350 mm	0.40	3.24	5.35	m	8.59
Returned and fitted ends	0.24	1.94	0.38	nr	2.32
Wrought softwood					
Skirtings, picture rails, dado rails and the like; splayed or moulded					
19 x 50 mm; splayed	0.11	0.89	1.43	m	2.32
19 x 50 mm; moulded	0.11	0.89	1.52	m	2.41
19 x 75 mm; splayed	0.11	0.89	1.74	m	2.63
19 x 75 mm; moulded	0.11	0.89	1.83	m	2.72
19 x 100 mm; splayed	0.11	0.89	2.05	m	2.94
19 x 100 mm; moulded	0.11	0.89	2.14	m	3.03
19 x 150 mm; moulded	0.13	1.05	2.82	m	3.87
19 x 175 mm; moulded	0.13	1.05	3.13	m	4.18
22 x 100 mm; splayed	0.11	0.89	2.26	m	3.15
25 x 50 mm; moulded	0.11	0.89	1.75	m	2.64
25 x 75 mm; splayed	0.11	0.89	2.05	m	2.94
25 x 100 mm; splayed	0.11	0.89	2.47	m	3.36
25 x 150 mm; splayed	0.13	1.05	3.30	m	4.35
25 x 150 mm; moulded	0.13	1.05	3.41	m	4.46
25 x 175 mm; moulded	0.13	1.05	3.81	m	4.86
25 x 225 mm; moulded	0.15	1.21	4.66	m	5.87
Returned end	0.17	1.38	-	nr	1.38
Mitres	0.11	0.89	-	nr	0.89

P BUILDING FABRIC SUNDRIES Including overheads and profit at 7.50%		Labour hours	Labour £	Material £	Unit	Total rate £
Architraves, cover fillets and the like; half round; splayed or moulded						
13 x 25 mm; half round		0.13	1.05	1.01	m	2.06
13 x 50 mm; moulded		0.13	1.05	1.34	m	2.39
16 x 32 mm; half round		0.13	1.05	1.14	m	2.19
16 x 38 mm; moulded		0.13	1.05	1.31	m	2.36
16 x 50 mm; moulded		0.13	1.05	1.44	m	2.49
19 x 50 mm; splayed		0.13	1.05	1.43	m	2.48
19 x 63 mm; splayed		0.13	1.05	1.58	m	2.63
19 x 75 mm; splayed		0.13	1.05	1.74	m	2.79
25 x 44 mm; splayed		0.13	1.05	1.62	m	2.67
25 x 50 mm; moulded		0.13	1.05	1.75	m	2.80
25 x 63 mm; splayed		0.13	1.05	1.82	m	2.87
25 x 75 mm; splayed		0.13	1.05	2.05	m	3.10
32 x 88 mm; moulded		0.13	1.05	2.92	m	3.97
38 x 38 mm; moulded		0.13	1.05	1.77	m	2.82
50 x 50 mm; moulded		0.13	1.05	2.47	m	3.52
Returned end		0.17	1.38	-	nr	1.38
Mitres		0.11	0.89	-	nr	0.89
Stops; screwed on						
16 x 38 mm		0.11	0.89	1.07	m	1.96
16 x 50 mm		0.11	0.89	1.19	m	2.08
19 x 38 mm		0.11	0.89	1.12	m	2.01
25 x 38 mm		0.11	0.89	1.28	m	2.17
25 x 50 mm		0.11	0.89	1.50	m	2.39
Glazing beads and the like						
13 x 16 mm	PC £0.67	-	-	0.77	m	0.77
13 x 19 mm	PC £0.67	-	-	0.77	m	0.77
13 x 25 mm	PC £0.71	-	-	0.82	m	0.82
13 x 25 mm; screwed		0.06	0.49	0.92	m	1.41
13 x 25 mm; fixing with brass cups and screws		0.11	0.89	1.27	m	2.16
16 x 25 mm	PC £0.78	-	-	0.90	m	0.90
16 mm; quadrant		0.06	0.49	0.88	m	1.37
19 mm; quadrant or scotia		0.06	0.49	0.98	m	1.47
19 x 36 mm		0.06	0.49	1.08	m	1.57
25 x 38 mm		0.06	0.49	1.24	m	1.73
25 mm; quadrant or scotia		0.06	0.49	1.13	m	1.62
38 mm; scotia		0.06	0.49	1.71	m	2.20
50 mm; scotia		0.06	0.49	2.40	m	2.89
Isolated shelves, worktops, seats and the like						
19 x 150 mm		0.18	1.46	2.71	m	4.17
19 x 200 mm		0.24	1.94	3.28	m	5.22
25 x 150 mm		0.18	1.46	3.28	m	4.74
25 x 200 mm		0.24	1.94	4.08	m	6.02
32 x 150 mm		0.18	1.46	4.08	m	5.54
32 x 200 mm		0.24	1.94	5.03	m	6.97
Isolated shelves, worktops, seats and the like; cross-tongued joints						
19 x 300 mm		0.31	2.51	11.27	m	13.78
19 x 450 mm		0.37	3.00	13.22	m	16.22
19 x 600 mm		0.44	3.56	19.07	m	22.63
25 x 300 mm		0.31	2.51	12.75	m	15.26
25 x 450 mm		0.37	3.00	15.23	m	18.23
25 x 600 mm		0.44	3.56	21.78	m	25.34
32 x 300 mm		0.31	2.51	14.23	m	16.74
32 x 450 mm		0.37	3.00	17.43	m	20.43
32 x 600 mm		0.44	3.56	24.75	m	28.31

P BUILDING FABRIC SUNDRIES Including overheads and profit at 7.50%		Labour hours	Labour £	Material £	Unit	Total rate £
P20 UNFRAMED ISOLATED TRIMS/SKIRTINGS/ SUNDRY ITEMS - cont'd						
Wrought softwood - cont'd						
Isolated shelves, worktops, seats and the like; slatted with 50 mm wide slats at 75 mm centres						
19 mm thick		1.45	11.74	9.15	m2	**20.89**
25 mm thick		1.45	11.74	10.73	m2	**22.47**
32 mm thick		1.45	11.74	12.08	m2	**23.82**
Window boards, nosings, bed moulds and the like; rebated and rounded						
19 x 75 mm		0.20	1.62	1.92	m	**3.54**
19 x 150 mm		0.22	1.78	2.91	m	**4.69**
19 x 225 mm; in one width		0.29	2.35	3.83	m	**6.18**
19 x 300 mm; cross-tongued joints		0.33	2.67	11.35	m	**14.02**
25 x 75 mm		0.20	1.62	2.22	m	**3.84**
25 x 150 mm		0.22	1.78	3.52	m	**5.30**
25 x 225 mm; in one width		0.29	2.35	4.75	m	**7.10**
25 x 300 mm; cross-tongued joints		0.33	2.67	12.76	m	**15.43**
32 x 75 mm		0.20	1.62	2.59	m	**4.21**
32 x 150 mm		0.22	1.78	4.19	m	**5.97**
32 x 225 mm; in one width		0.29	2.35	5.79	m	**8.14**
32 x 300 mm; cross-tongued joints		0.33	2.67	14.25	m	**16.92**
38 x 75 mm		0.20	1.62	2.90	m	**4.52**
38 x 150 mm		0.22	1.78	4.80	m	**6.58**
38 x 225 mm; in one width		0.29	2.35	6.68	m	**9.03**
38 x 300 mm; cross-tongued joints		0.33	2.67	15.55	m	**18.22**
Returned and fitted ends		0.17	1.38	-	nr	**1.38**
Handrails; mopstick						
50 mm dia		0.28	2.27	3.36	m	**5.63**
Handrails; rounded						
44 x 50 mm		0.28	2.27	3.36	m	**5.63**
50 x 75 mm		0.30	2.43	4.22	m	**6.65**
63 x 87 mm		0.33	2.67	5.90	m	**8.57**
75 x 100 mm		0.39	3.16	6.69	m	**9.85**
Handrails; moulded						
44 x 50 mm		0.22	1.78	3.70	m	**5.48**
50 x 75 mm		0.24	1.94	4.58	m	**6.52**
63 x 87 mm		0.26	2.10	6.23	m	**8.33**
75 x 100 mm		0.29	2.35	7.02	m	**9.37**
Add +5% to the above 'Material £ prices' for 'selected' softwood for staining						
Selected West African Mahogany; PC £505.25/m3						
Skirtings, picture rails, dado rails and the like; splayed or moulded						
19 x 50 mm; splayed	PC £2.45	0.15	1.21	2.82	m	**4.03**
19 x 50 mm; moulded	PC £2.60	0.15	1.21	3.00	m	**4.21**
19 x 75 mm; splayed	PC £2.87	0.15	1.21	3.30	m	**4.51**
19 x 75 mm; moulded	PC £3.02	0.15	1.21	3.47	m	**4.68**
19 x 100 mm; splayed	PC £3.34	0.15	1.21	3.84	m	**5.05**
19 x 100 mm; moulded	PC £3.49	0.15	1.21	4.00	m	**5.21**
19 x 150 mm; moulded	PC £4.38	0.18	1.46	5.01	m	**6.47**
19 x 175 mm; moulded	PC £5.03	0.18	1.46	5.74	m	**7.20**
22 x 100 mm; splayed	PC £3.61	0.15	1.21	4.14	m	**5.35**
25 x 50 mm; moulded	PC £2.87	0.15	1.21	3.30	m	**4.51**
25 x 75 mm; splayed	PC £3.32	0.15	1.21	3.82	m	**5.03**

P BUILDING FABRIC SUNDRIES Including overheads and profit at 7.50%		Labour hours	Labour £	Material £	Unit	Total rate £
25 x 100 mm; splayed	PC £3.90	0.15	1.21	4.48	m	**5.69**
25 x 150 mm; splayed	PC £5.30	0.18	1.46	6.04	m	**7.50**
25 x 150 mm; moulded	PC £5.45	0.18	1.46	6.21	m	**7.67**
25 x 175 mm; moulded	PC £6.03	0.18	1.46	6.87	m	**8.33**
25 x 225 mm; moulded	PC £7.21	0.20	1.62	8.21	m	**9.83**
Returned end		0.24	1.94	-	nr	**1.94**
Mitres		0.17	1.38	-	nr	**1.38**
Architraves, cover fillets and the like; half round; splayed or moulded						
13 x 25 mm; half round	PC £1.86	0.18	1.46	2.16	m	**3.62**
13 x 50 mm; moulded	PC £2.30	0.18	1.46	2.65	m	**4.11**
16 x 32 mm; half round	PC £2.02	0.18	1.46	2.35	m	**3.81**
16 x 38 mm; moulded	PC £2.24	0.18	1.46	2.60	m	**4.06**
16 x 50 mm; moulded	PC £2.45	0.18	1.46	2.82	m	**4.28**
19 x 50 mm; splayed	PC £2.45	0.18	1.46	2.82	m	**4.28**
19 x 63 mm; splayed	PC £2.67	0.18	1.46	3.08	m	**4.54**
19 x 75 mm; splayed	PC £2.87	0.18	1.46	3.30	m	**4.76**
25 x 44 mm; splayed	PC £2.72	0.18	1.46	3.14	m	**4.60**
25 x 50 mm; moulded	PC £2.87	0.18	1.46	3.30	m	**4.76**
25 x 63 mm; splayed	PC £3.02	0.18	1.46	3.47	m	**4.93**
25 x 75 mm; splayed	PC £3.32	0.18	1.46	3.82	m	**5.28**
32 x 88 mm; moulded	PC £4.57	0.18	1.46	5.23	m	**6.69**
38 x 38 mm; moulded	PC £2.90	0.18	1.46	3.35	m	**4.81**
50 x 50 mm; moulded	PC £3.90	0.18	1.46	4.48	m	**5.94**
Returned end		0.24	1.94	-	nr	**1.94**
Mitres		0.17	1.38	-	nr	**1.38**
Stops; screwed on						
16 x 38 mm	PC £1.97	0.17	1.38	2.26	m	**3.64**
16 x 50 mm	PC £2.15	0.17	1.38	2.47	m	**3.85**
19 x 38 mm	PC £2.09	0.17	1.38	2.39	m	**3.77**
25 x 38 mm	PC £2.30	0.17	1.38	2.63	m	**4.01**
25 x 50 mm	PC £2.60	0.17	1.38	2.97	m	**4.35**
Glazing beads and the like						
13 x 16 mm	PC £1.63	-	-	1.84	m	**1.84**
13 x 19 mm	PC £1.63	-	-	1.84	m	**1.84**
13 x 25 mm	PC £1.71	-	-	1.94	m	**1.94**
13 x 25 mm; screwed	PC £1.71	0.09	0.73	2.04	m	**2.77**
13 x 25 mm; fixing with brass cups and screws	PC £1.71	0.17	1.38	2.38	m	**3.76**
16 x 25 mm	PC £1.78	-	-	2.01	m	**2.01**
16 mm; quadrant	PC £1.79	0.08	0.65	2.02	m	**2.67**
19 mm; quadrant or scotia	PC £1.87	0.08	0.65	2.11	m	**2.76**
19 x 36 mm	PC £2.09	0.08	0.65	2.35	m	**3.00**
25 x 38 mm	PC £2.30	0.08	0.65	2.59	m	**3.24**
25 mm; quadrant or scotia	PC £2.13	0.08	0.65	2.41	m	**3.06**
38 mm; scotia	PC £2.90	0.08	0.65	3.28	m	**3.93**
50 mm; scotia	PC £3.90	0.08	0.65	4.41	m	**5.06**
Isolated shelves, worktops, seats and the like						
19 x 150 mm	PC £4.54	0.24	1.94	5.13	m	**7.07**
19 x 200 mm	PC £5.39	0.33	2.67	6.09	m	**8.76**
25 x 150 mm	PC £5.39	0.24	1.94	6.09	m	**8.03**
25 x 200 mm	PC £6.52	0.33	2.67	7.36	m	**10.03**
32 x 150 mm	PC £6.33	0.24	1.94	7.15	m	**9.09**
32 x 200 mm	PC £7.83	0.33	2.67	8.84	m	**11.51**
Isolated shelves, worktops, seats and the like; cross-tongued joints						
19 x 300 mm	PC £15.71	0.42	3.40	17.74	m	**21.14**
19 x 450 mm	PC £18.46	0.50	4.05	20.83	m	**24.88**
19 x 600 mm	PC £26.89	0.61	4.94	30.35	m	**35.29**
25 x 300 mm	PC £17.71	0.42	3.40	20.00	m	**23.40**
25 x 450 mm	PC £21.30	0.50	4.05	24.03	m	**28.08**

P BUILDING FABRIC SUNDRIES Including overheads and profit at 7.50%		Labour hours	Labour £	Material £	Unit	Total rate £
P20 UNFRAMED ISOLATED TRIMS/SKIRTINGS/ SUNDRY ITEMS - cont'd						
Selected West African Mahogany; PC £505.25/m3 - cont'd						
Isolated shelves, worktops, seats and the like; cross-tongued joints						
25 x 600 mm	PC £30.68	0.61	4.94	34.63	m	39.57
32 x 300 mm	PC £19.69	0.42	3.40	22.22	m	25.62
32 x 450 mm	PC £24.21	0.50	4.05	27.33	m	31.38
32 x 600 mm	PC £34.73	0.61	4.94	39.21	m	44.15
Isolated shelves, worktops, seats and the like; slatted with 50 mm wide slats at 75 mm centres						
19 mm thick	PC £17.66	1.95	15.78	20.65	m2	36.43
25 mm thick	PC £19.72	1.95	15.78	23.04	m2	38.82
32 mm thick	PC £21.61	1.95	15.78	25.21	m2	40.99
Window boards, nosings, bed moulds and the like; rebated and rounded						
19 x 75 mm	PC £3.23	0.26	2.10	3.79	m	5.89
19 x 150 mm	PC £4.58	0.30	2.43	5.32	m	7.75
19 x 225 mm; in one width	PC £5.93	0.40	3.24	6.84	m	10.08
19 x 300 mm; cross-tongued joints	PC £15.61	0.44	3.56	17.76	m	21.32
25 x 75 mm	PC £3.68	0.26	2.10	4.30	m	6.40
25 x 150 mm	PC £5.50	0.30	2.43	6.35	m	8.78
25 x 225 mm; in one width	PC £7.21	0.40	3.24	8.29	m	11.53
25 x 300 mm; cross-tongued joints	PC £17.51	0.44	3.56	19.92	m	23.48
32 x 75 mm	PC £4.17	0.26	2.10	4.85	m	6.95
32 x 150 mm	PC £6.44	0.30	2.43	7.41	m	9.84
32 x 225 mm; in one width	PC £8.69	0.40	3.24	9.95	m	13.19
32 x 300 mm; cross-tongued joints	PC £19.50	0.44	3.56	22.15	m	25.71
38 x 75 mm	PC £4.58	0.26	2.10	5.32	m	7.42
38 x 150 mm	PC £7.30	0.30	2.43	8.38	m	10.81
38 x 225 mm; in one width	PC £9.98	0.40	3.24	11.41	m	14.65
38 x 300 mm; cross-tongued joints	PC £21.30	0.44	3.56	24.18	m	27.74
Returned and fitted ends		0.25	2.02	-	nr	2.02
Handrails; rounded						
44 x 50 mm	PC £5.39	0.36	2.91	6.09	m	9.00
50 x 75 mm	PC £6.63	0.40	3.24	7.48	m	10.72
63 x 87 mm	PC £8.41	0.44	3.56	9.50	m	13.06
75 x 100 mm	PC £10.12	0.50	4.05	11.42	m	15.47
Handrails; moulded						
44 x 50 mm	PC £5.82	0.36	2.91	6.57	m	9.48
50 x 75 mm	PC £7.06	0.40	3.24	7.97	m	11.21
63 x 87 mm	PC £8.85	0.44	3.56	9.99	m	13.55
75 x 100 mm	PC £10.54	0.50	4.05	11.90	m	15.95
Afrormosia; PC £763.25/m3						
Skirtings, picture rails, dado rails and the like; splayed or moulded						
19 x 50 mm; splayed	PC £2.89	0.15	1.21	3.33	m	4.54
19 x 50 mm; moulded	PC £3.04	0.15	1.21	3.50	m	4.71
19 x 75 mm; splayed	PC £3.59	0.15	1.21	4.11	m	5.32
19 x 75 mm; moulded	PC £3.73	0.15	1.21	4.28	m	5.49
19 x 100 mm; splayed	PC £4.26	0.15	1.21	4.87	m	6.08
19 x 100 mm; moulded	PC £4.40	0.15	1.21	5.04	m	6.25
19 x 150 mm; moulded	PC £5.74	0.18	1.46	6.55	m	8.01
19 x 175 mm; moulded	PC £6.64	0.18	1.46	7.56	m	9.02
22 x 100 mm; splayed	PC £4.68	0.15	1.21	5.34	m	6.55
25 x 50 mm; moulded	PC £3.49	0.15	1.21	4.00	m	5.21
25 x 75 mm; splayed	PC £4.21	0.15	1.21	4.82	m	6.03

P BUILDING FABRIC SUNDRIES Including overheads and profit at 7.50%		Labour hours	Labour £	Material £	Unit	Total rate £
25 x 100 mm; splayed	PC £5.11	0.15	1.21	5.83	m	**7.04**
25 x 150 mm; splayed	PC £7.12	0.18	1.46	8.10	m	**9.56**
25 x 150 mm; moulded	PC £7.26	0.18	1.46	8.25	m	**9.71**
25 x 175 mm; moulded	PC £8.13	0.18	1.46	9.24	m	**10.70**
25 x 225 mm; moulded	PC £9.94	0.20	1.62	11.28	m	**12.90**
Returned end		0.24	1.94	-	nr	**1.94**
Mitres		0.17	1.38	-	nr	**1.38**
Architraves, cover fillets and the like; half round; splayed or moulded						
13 x 25 mm; half round	PC £2.01	0.18	1.46	2.33	m	**3.79**
13 x 50 mm; moulded	PC £2.61	0.18	1.46	3.01	m	**4.47**
16 x 32 mm; half round	PC £2.27	0.18	1.46	2.62	m	**4.08**
16 x 38 mm; moulded	PC £2.56	0.18	1.46	2.96	m	**4.42**
16 x 50 mm; moulded	PC £2.83	0.18	1.46	3.26	m	**4.72**
19 x 50 mm; splayed	PC £2.89	0.18	1.46	3.33	m	**4.79**
19 x 63 mm; splayed	PC £3.27	0.18	1.46	3.75	m	**5.21**
19 x 75 mm; splayed	PC £3.59	0.18	1.46	4.11	m	**5.57**
25 x 44 mm; splayed	PC £3.34	0.18	1.46	3.84	m	**5.30**
25 x 50 mm; moulded	PC £3.49	0.18	1.46	4.00	m	**5.46**
25 x 63 mm; splayed	PC £3.79	0.18	1.46	4.34	m	**5.80**
25 x 75 mm; splayed	PC £4.21	0.18	1.46	4.82	m	**6.28**
32 x 88 mm; moulded	PC £6.11	0.18	1.46	6.96	m	**8.42**
38 x 38 mm; moulded	PC £3.61	0.18	1.46	4.14	m	**5.60**
50 x 50 mm; moulded	PC £5.11	0.18	1.46	5.83	m	**7.29**
Returned end		0.24	1.94	-	nr	**1.94**
Mitres		0.17	1.38	-	nr	**1.38**
Stops; screwed on						
16 x 38 mm	PC £2.27	0.17	1.38	2.60	m	**3.98**
16 x 50 mm	PC £2.54	0.17	1.38	2.91	m	**4.29**
19 x 38 mm	PC £2.41	0.17	1.38	2.77	m	**4.15**
25 x 38 mm	PC £2.74	0.17	1.38	3.14	m	**4.52**
25 x 50 mm	PC £3.19	0.17	1.38	3.64	m	**5.02**
Glazing beads and the like						
13 x 16 mm	PC £1.74	-	-	1.97	m	**1.97**
13 x 19 mm	PC £1.74	-	-	1.97	m	**1.97**
13 x 25 mm	PC £1.87	-	-	2.11	m	**2.11**
13 x 25 mm; screwed	PC £1.87	0.09	0.73	2.21	m	**2.94**
13 x 25 mm; fixing with brass cups and screws	PC £1.87	0.17	1.38	2.56	m	**3.94**
16 x 25 mm	PC £1.97	-	-	2.22	m	**2.22**
16 mm; quadrant	PC £1.93	0.08	0.65	2.18	m	**2.83**
19 mm; quadrant or scotia	PC £2.07	0.08	0.65	2.34	m	**2.99**
19 x 36 mm	PC £2.41	0.08	0.65	2.72	m	**3.37**
25 x 38 mm	PC £3.05	0.08	0.65	3.44	m	**4.09**
25 mm; quadrant or scotia	PC £2.45	0.08	0.65	2.76	m	**3.41**
38 mm; scotia	PC £3.61	0.08	0.65	4.07	m	**4.72**
50 mm; scotia	PC £5.11	0.08	0.65	5.77	m	**6.42**
Isolated shelves, worktops, seats and the like						
19 x 150 mm	PC £5.95	0.24	1.94	6.72	m	**8.66**
19 x 200 mm	PC £7.23	0.33	2.67	8.16	m	**10.83**
25 x 150 mm	PC £7.23	0.24	1.94	8.16	m	**10.10**
25 x 200 mm	PC £8.94	0.33	2.67	10.09	m	**12.76**
32 x 150 mm	PC £8.65	0.24	1.94	9.76	m	**11.70**
32 x 200 mm	PC £10.94	0.33	2.67	12.35	m	**15.02**
Isolated shelves, worktops, seats and the like; cross-tongued joints						
19 x 300 mm	PC £18.77	0.42	3.40	21.18	m	**24.58**
19 x 450 mm	PC £22.90	0.50	4.05	25.86	m	**29.91**
19 x 600 mm	PC £32.98	0.61	4.94	37.22	m	**42.16**
25 x 300 mm	PC £21.75	0.42	3.40	24.56	m	**27.96**
25 x 450 mm	PC £27.17	0.50	4.05	30.67	m	**34.72**
25 x 600 mm	PC £38.68	0.61	4.94	43.66	m	**48.60**

P BUILDING FABRIC SUNDRIES Including overheads and profit at 7.50%		Labour hours	Labour £	Material £	Unit	Total rate £
P20 UNFRAMED ISOLATED TRIMS/SKIRTINGS/ SUNDRY ITEMS - cont'd						
Afrormosia; PC £763.25/m3 - cont'd						
Isolated shelves, worktops, seats and the like; cross-tongued joints						
32 x 300 mm	PC £24.75	0.42	3.40	27.94	m	**31.34**
32 x 450 mm	PC £31.61	0.50	4.05	35.67	m	**39.72**
32 x 600 mm	PC £44.82	0.61	4.94	50.59	m	**55.53**
Isolated shelves, worktops, seats and the like; slatted with 50 mm wide slats at 75 mm centres						
19 mm thick	PC £21.03	1.95	15.78	24.55	m2	**40.33**
25 mm thick	PC £24.01	1.95	15.78	27.99	m2	**43.77**
32 mm thick	PC £27.15	1.95	15.78	31.76	m2	**47.54**
Window boards, nosings, bed moulds and the like; rebated and rounded						
19 x 75 mm	PC £3.91	0.26	2.10	4.57	m	**6.67**
19 x 150 mm	PC £5.96	0.30	2.43	6.87	m	**9.30**
19 x 225 mm; in one width	PC £7.98	0.40	3.24	9.15	m	**12.39**
19 x 300 mm; cross-tongued joints	PC £18.69	0.44	3.56	21.24	m	**24.80**
25 x 75 mm	PC £4.58	0.26	2.10	5.32	m	**7.42**
25 x 150 mm	PC £7.33	0.30	2.43	8.42	m	**10.85**
25 x 225 mm; in one width	PC £9.90	0.40	3.24	11.33	m	**14.57**
25 x 300 mm; cross-tongued joints	PC £21.56	0.44	3.56	24.49	m	**28.05**
32 x 75 mm	PC £5.33	0.26	2.10	6.17	m	**8.27**
32 x 150 mm	PC £8.76	0.30	2.43	10.03	m	**12.46**
32 x 225 mm; in one width	PC £12.20	0.40	3.24	13.92	m	**17.16**
32 x 300 mm; cross-tongued joints	PC £24.56	0.44	3.56	27.87	m	**31.43**
38 x 75 mm	PC £5.96	0.26	2.10	6.87	m	**8.97**
38 x 150 mm	PC £10.05	0.30	2.43	11.49	m	**13.92**
38 x 225 mm; in one width	PC £14.12	0.40	3.24	16.08	m	**19.32**
38 x 300 mm; cross-tongued joints	PC £27.29	0.44	3.56	30.94	m	**34.50**
Returned and fitted ends		0.25	2.02	-	nr	**2.02**
Handrails; rounded						
44 x 50 mm	PC £6.60	0.36	2.91	7.44	m	**10.35**
50 x 75 mm	PC £8.38	0.40	3.24	9.46	m	**12.70**
63 x 87 mm	PC £12.03	0.44	3.56	13.58	m	**17.14**
75 x 100 mm	PC £13.73	0.50	4.05	15.51	m	**19.56**
Handrails; moulded						
44 x 50 mm	PC £7.03	0.36	2.91	7.94	m	**10.85**
50 x 75 mm	PC £8.83	0.40	3.24	9.96	m	**13.20**
63 x 87 mm	PC £12.47	0.44	3.56	14.07	m	**17.63**
75 x 100 mm	PC £14.16	0.50	4.05	15.98	m	**20.03**
Pin-boards; medium board						
Sundeala 'A' pin-board; fixed with adhesive to backing (measured elsewhere); over 300 mm wide						
6.4 mm thick		0.66	5.34	8.34	m2	**13.68**
Sundries on softwood/hardwood						
Extra over fixing with nails for						
gluing and pinning		0.02	0.16	0.03	m	**0.19**
masonry nails		0.02	0.16	0.04	m	**0.20**
steel screws		0.02	0.16	0.06	m	**0.22**
self-tapping screws		0.02	0.16	0.08	m	**0.24**
steel screws; gluing		0.04	0.32	0.06	m	**0.38**
steel screws; sinking; filling heads		0.06	0.49	0.06	m	**0.55**
steel screws; sinking; pellating over		0.11	0.89	0.47	m	**1.36**
brass cups and screws		0.17	1.38	0.06	m	**1.44**
Extra over for						
countersinking		0.02	0.16	-	m	**0.16**
pellating		0.11	0.89	0.03	m	**0.92**

P BUILDING FABRIC SUNDRIES Including overheads and profit at 7.50%	Labour hours	Labour £	Material £	Unit	Total rate £
Head or nut in softwood					
let in flush	0.06	0.49	-	nr	**0.49**
Head or nut; in hardwood					
let in flush	0.09	0.73	-	nr	**0.73**
let in over; pellated	0.22	1.78	-	nr	**1.78**
P21 IRONMONGERY					
Metalwork; mild steel; galvanized					
Water bars; groove in timber					
6 x 30 mm	0.55	4.45	3.76	m	**8.21**
6 x 40 mm	0.55	4.45	5.90	m	**10.35**
6 x 50 mm	0.55	4.45	5.15	m	**9.60**
Dowels; mortice in timber					
8 mm dia x 100 mm long	0.06	0.49	0.05	nr	**0.54**
10 mm dia x 50 mm long	0.06	0.49	0.14	nr	**0.63**
Cramps					
25 x 3 x 230 mm girth; one end bent, holed and screwed to softwood; other end fishtailed for building in	0.08	0.65	0.34	nr	**0.99**
Fixing only ironmongery to softwood					
Bolts					
barrel; not exceeding 150 mm long	0.36	2.91	-	nr	**2.91**
barrel; 150-300 mm long	0.46	3.72	-	nr	**3.72**
cylindrical mortice; not exceeding					
150 mm long	0.55	4.45	-	nr	**4.45**
150 - 300 mm long	0.66	5.34	-	nr	**5.34**
flush; not exceeding 150 mm long	0.55	4.45	-	nr	**4.45**
flush; 150 - 300 mm long	0.66	5.34	-	nr	**5.34**
monkey tail; 380 mm long	0.74	5.99	-	nr	**5.99**
necked; 150 mm long	0.36	2.91	-	nr	**2.91**
panic; single; locking	2.75	22.26	-	nr	**22.26**
panic; double; locking	3.85	31.16	-	nr	**31.16**
WC indicator	0.74	5.99	-	nr	**5.99**
Butts; extra over for					
rising	0.19	1.54	-	pr	**1.54**
skew	0.19	1.54	-	pr	**1.54**
spring; single action	1.45	11.74	-	pr	**11.74**
spring; double action	1.65	13.36	-	pr	**13.36**
Catches					
surface mounted	0.19	1.54	-	nr	**1.54**
mortice	0.36	2.91	-	nr	**2.91**
Door closers and furniture					
cabin hooks and eyes	0.19	1.54	-	nr	**1.54**
door selector	0.55	4.45	-	nr	**4.45**
finger plate	0.19	1.54	-	nr	**1.54**
floor spring	2.75	22.26	-	nr	**22.26**
lever furniture	0.28	2.27	-	nr	**2.27**
handle; not exceeding 150 mm long	0.19	1.54	-	nr	**1.54**
handle; 150 - 300 mm long	0.28	2.27	-	nr	**2.27**
handle; flush	0.36	2.91	-	nr	**2.91**
holder	0.36	2.91	-	nr	**2.91**
kicking plate	0.36	2.91	-	nr	**2.91**
letter plate; including perforation	1.45	11.74	-	nr	**11.74**
overhead door closer; surface fixing	1.40	11.33	-	nr	**11.33**
overhead door closer; concealed fixing	1.95	15.78	-	nr	**15.78**
top centre	0.55	4.45	-	nr	**4.45**
'Perko' door closer	0.74	5.99	-	nr	**5.99**
rod door closers; 457 mm long	1.10	8.90	-	nr	**8.90**

P BUILDING FABRIC SUNDRIES Including overheads and profit at 7.50%	Labour hours	Labour £	Material £	Unit	Total rate £
P21 IRONMONGERY - cont'd					
Fixing only ironmongery to softwood - cont'd					
Latches					
cylinder rim night latch	0.83	6.72	-	nr	6.72
mortice	0.74	5.99	-	nr	5.99
Norfolk	0.74	5.99	-	nr	5.99
rim	0.55	4.45	-	nr	4.45
Locks					
cupboard	0.46	3.72	-	nr	3.72
mortice	0.91	7.37	-	nr	7.37
mortice budget	0.83	6.72	-	nr	6.72
mortice dead	0.83	6.72	-	nr	6.72
rebated mortice	1.40	11.33	-	nr	11.33
rim	0.55	4.45	-	nr	4.45
rim budget	0.46	3.72	-	nr	3.72
rim dead	0.46	3.72	-	nr	3.72
Sliding door gear for top hung softwood timber doors; weight not exceeding 365 kg					
bottom guide; fixed to concrete in groove	0.55	4.45	-	m	4.45
top track	0.28	2.27	-	m	2.27
detachable locking bar and padlock	0.36	2.91	-	nr	2.91
hangers; fixed flush to timber	0.83	6.72	-	nr	6.72
head brackets; bolted to concrete	0.46	3.72	-	nr	3.72
Window furniture					
casement stay and pin	0.19	1.54	-	nr	1.54
catch; fanlight	0.28	2.27	-	nr	2.27
fastener; cockspur	0.36	2.91	-	nr	2.91
fastener; sash	0.28	2.27	-	nr	2.27
quadrant stay	0.28	2.27	-	nr	2.27
ring catch	0.36	2.91	-	nr	2.91
Sundries					
drawer pull	0.09	0.73	-	nr	0.73
hat and coat hook	0.09	0.73	-	nr	0.73
numerals	0.09	0.73	-	nr	0.73
rubber door stop	0.09	0.73	-	nr	0.73
shelf bracket	0.19	1.54	-	nr	1.54
skirting type door stop	0.19	1.54	-	nr	1.54
Fixing only ironmongery to hardwood					
Bolts					
barrel; not exceeding 150 mm long	0.48	3.89	-	nr	3.89
barrel; 150-300 mm long	0.62	5.02	-	nr	5.02
cylindrical mortice; not exceeding					
150 mm long	0.74	5.99	-	nr	5.99
150 - 300 mm long	0.88	7.12	-	nr	7.12
flush; not exceeding 150 mm long	0.74	5.99	-	nr	5.99
flush; 150 - 300 mm long	0.88	7.12	-	nr	7.12
monkey tail; 380 mm long	0.98	7.93	-	nr	7.93
necked; 150 mm long	0.48	3.89	-	nr	3.89
panic; single; locking	3.65	29.55	-	nr	29.55
panic; double; locking	5.15	41.69	-	nr	41.69
WC indicator	0.98	7.93	-	nr	7.93
Butts; extra over for					
rising	0.25	2.02	-	pr	2.02
skew	0.25	2.02	-	pr	2.02
spring; single action	1.90	15.38	-	pr	15.38
spring; double action	2.20	17.81	-	pr	17.81
Catches					
surface mounted	0.25	2.02	-	nr	2.02
mortice	0.48	3.89	-	nr	3.89

P BUILDING FABRIC SUNDRIES Including overheads and profit at 7.50%	Labour hours	Labour £	Material £	Unit	Total rate £
Door closers and furniture					
cabin hooks and eyes	0.25	2.02	-	nr	2.02
door selector	0.74	5.99	-	nr	5.99
finger plate	0.25	2.02	-	nr	2.02
floor spring	3.65	29.55	-	nr	29.55
lever furniture	0.36	2.91	-	nr	2.91
handle; not exceeding 150 mm long	0.25	2.02	-	nr	2.02
handle; 150 - 300 mm long	0.36	2.91	-	nr	2.91
handle; flush	0.48	3.89	-	nr	3.89
holder	0.48	3.89	-	nr	3.89
kicking plate	0.48	3.89	-	nr	3.89
letter plate; including perforation	1.95	15.78	-	nr	15.78
overhead door closer; surface fixing	1.85	14.98	-	nr	14.98
overhead door closer; concealed fixing	2.55	20.64	-	nr	20.64
top centre	0.74	5.99	-	nr	5.99
'Perko' door closer	0.98	7.93	-	nr	7.93
rod door closers; 457 mm long	1.45	11.74	-	nr	11.74
Latches					
cylinder rim night latch	1.10	8.90	-	nr	8.90
mortice	0.98	7.93	-	nr	7.93
Norfolk	0.98	7.93	-	nr	7.93
rim	0.74	5.99	-	nr	5.99
Locks					
cupboard	0.62	5.02	-	nr	5.02
mortice	1.20	9.71	-	nr	9.71
mortice budget	1.10	8.90	-	nr	8.90
mortice dead	1.10	8.90	-	nr	8.90
rebated mortice	1.85	14.98	-	nr	14.98
rim	0.74	5.99	-	nr	5.99
rim budget	0.62	5.02	-	nr	5.02
rim dead	0.62	5.02	-	nr	5.02
Sliding door gear for top hung softwood timber doors; weight not exceeding 365 kg					
bottom guide; fixed to concrete in groove	0.74	5.99	-	m	5.99
top track	0.36	2.91	-	m	2.91
detachable locking bar and padlock	0.48	3.89	-	nr	3.89
hangers; fixed flush to timber	1.10	8.90	-	nr	8.90
head brackets; bolted to concrete	0.62	5.02	-	nr	5.02
Window furniture					
casement stay and pin	0.25	2.02	-	nr	2.02
catch; fanlight	0.36	2.91	-	nr	2.91
fastener; cockspur	0.48	3.89	-	nr	3.89
fastener; sash	0.36	2.91	-	nr	2.91
quadrant stay	0.36	2.91	-	nr	2.91
ring catch	0.48	3.89	-	nr	3.89
Sundries					
drawer pull	0.12	0.97	-	nr	0.97
hat and coat hook	0.12	0.97	-	nr	0.97
numerals	0.12	0.97	-	nr	0.97
rubber door stop	0.12	0.97	-	nr	0.97
shelf bracket	0.25	2.02	-	nr	2.02
skirting type door stop	0.12	0.97	-	nr	0.97
Sundries					
Rubber door stop plugged and screwed to concrete	0.11	0.89	2.77	nr	3.66

P BUILDING FABRIC SUNDRIES Including overheads and profit at 7.50%	Labour hours	Labour £	Material £	Unit	Total rate £
P30 TRENCHES/PIPEWAYS/PITS FOR BURIED ENGINEERING SERVICES					
Mechanical excavation of trenches to receive pipes; grading bottoms; earthwork support; filling with excavated material and compacting; disposal of surplus soil; spreading on site average 50 m					
Pipes not exceeding 200 mm; average depth					
0.50 m deep	0.23	1.38	1.10	m	**2.48**
0.75 m deep	0.35	2.10	1.83	m	**3.93**
1.00 m deep	0.69	4.15	3.44	m	**7.59**
1.25 m deep	1.05	6.31	4.70	m	**11.01**
1.50 m deep	1.30	7.81	6.12	m	**13.93**
1.75 m deep	1.60	9.61	7.82	m	**17.43**
2.00 m deep	1.90	11.42	8.95	m	**20.37**
Hand excavation of trenches to receive pipes; grading bottoms; earthwork support; filling with excavated material and compacting; disposal of surplus soil; spreading on site average 50 m					
Pipes not exceeding 200 mm; average depth					
0.50 m deep	1.10	6.61	-	m	**6.61**
0.75 m deep	1.65	9.92	-	m	**9.92**
1.00 m deep	2.40	14.42	1.06	m	**15.48**
1.25 m deep	3.40	20.43	1.59	m	**22.02**
1.50 m deep	4.70	28.24	1.91	m	**30.15**
1.75 m deep	6.15	36.96	2.34	m	**39.30**
2.00 m deep	7.05	42.37	2.55	m	**44.92**
Pits for underground stop valves and the like; half brick thick walls in common bricks in cement mortar (1:3); on in situ concrete mix 21.00 N/mm2-20 mm aggregate (1:2:4) bed; 100 mm thick; 100 x 100 x 750 mm deep; internal holes for one small pipe; cast iron hinged box cover; bedding in cement mortar (1:3)	3.45	35.79	16.38	nr	**52.17**
P31 HOLES/CHASES/COVERS/SUPPORTS FOR SERVICES					
Builders' work for electrical installations Cutting away for and making good after electrician; including cutting or leaving all holes, notches, mortices, sinkings and chases, in both the structure and its coverings, for the following electrical points					
Exposed installation					
lighting points	0.35	3.21	-	nr	**3.21**
socket outlet points	0.58	5.32	-	nr	**5.32**
fitting outlet points	0.58	5.32	-	nr	**5.32**
equipment points or control gear points	0.81	7.43	-	nr	**7.43**
Concealed installation					
lighting points	0.46	4.22	-	nr	**4.22**
socket outlet points	0.81	7.43	-	nr	**7.43**
fitting outlet points	0.81	7.43	-	nr	**7.43**
equipment points or control gear points	1.15	10.55	-	nr	**10.55**

P BUILDING FABRIC SUNDRIES Including overheads and profit at 7.50%	Labour hours	Labour £	Material £	Unit	Total rate £
Builders' work for other services installations					
Cutting chases in brickwork					
for one pipe; not exceeding 55 mm nominal size; vertical	0.46	4.77	-	m	**4.77**
for one pipe; 55 - 110 nominal size; vertical	0.81	8.40	-	m	**8.40**
Cutting and pinning to brickwork or blockwork; ends of supports					
for pipes not exceeding 55 mm	0.23	2.39	-	m	**2.39**
for cast iron pipes 55 - 110 mm	0.38	3.94	-	nr	**3.94**
radiator stays or brackets	0.29	3.01	-	nr	**3.01**
Cutting holes for pipes or the like; not exceeding 55 mm nominal size					
102 mm brickwork	0.38	2.37	-	nr	**2.37**
215 mm brickwork	0.63	3.93	-	nr	**3.93**
327 mm brickwork	1.05	6.55	-	nr	**6.55**
100 mm blockwork	0.35	2.18	-	nr	**2.18**
150 mm blockwork	0.46	2.87	-	nr	**2.87**
215 mm blockwork	0.58	3.62	-	nr	**3.62**
Cutting holes for pipes or the like; 55 - 110 mm nominal size					
102 mm brickwork	0.46	2.87	-	nr	**2.87**
215 mm brickwork	0.81	5.05	-	nr	**5.05**
327 mm brickwork	1.25	7.79	-	nr	**7.79**
100 mm blockwork	0.40	2.49	-	nr	**2.49**
150 mm blockwork	0.58	3.62	-	nr	**3.62**
215 mm blockwork	0.69	4.30	-	nr	**4.30**
Cutting holes for pipes or the like; over 110 mm nominal size					
102 mm brickwork	0.58	3.62	-	nr	**3.62**
215 mm brickwork	0.98	6.11	-	nr	**6.11**
327 mm brickwork	1.55	9.66	-	nr	**9.66**
100 mm blockwork	0.52	5.39	-	nr	**5.39**
150 mm blockwork	0.69	4.30	-	nr	**4.30**
215 mm blockwork	0.86	5.36	-	nr	**5.36**
Add for making good fair face or facings one side					
pipe; not exceeding 55 mm nominal size	0.09	0.93	-	nr	**0.93**
pipe; 55 - 110 mm nominal size	0.11	1.14	-	nr	**1.14**
pipe; over 110 mm nominal size	0.14	1.45	-	nr	**1.45**
Add for fixing sleeve (supply included elsewhere)					
for pipe; small	0.17	1.76	-	nr	**1.76**
for pipe; large	0.23	2.39	-	nr	**2.39**
for pipe; extra large	0.35	3.63	-	nr	**3.63**
Cutting or forming holes for ducts; girth not exceeding 1.00 m					
102 mm brickwork	0.69	4.30	-	nr	**4.30**
215 mm brickwork	1.15	7.17	-	nr	**7.17**
327 mm brickwork	1.85	11.53	-	nr	**11.53**
100 mm blockwork	0.58	3.62	-	nr	**3.62**
150 mm blockwork	0.81	5.05	-	nr	**5.05**
215 mm blockwork	1.05	6.55	-	nr	**6.55**
Cutting or forming holes for ducts; girth 1.00 - 2.00 m					
102 mm brickwork	0.81	5.05	-	nr	**5.05**
215 mm brickwork	1.40	8.73	-	nr	**8.73**
327 mm brickwork	2.20	13.72	-	nr	**13.72**
100 mm blockwork	0.69	4.30	-	nr	**4.30**
150 mm blockwork	0.92	5.74	-	nr	**5.74**
215 mm blockwork	1.15	7.17	-	nr	**7.17**

P BUILDING FABRIC SUNDRIES Including overheads and profit at 7.50%		Labour hours	Labour £	Material £	Unit	Total rate £
P31 HOLES/CHASES/COVERS/SUPPORTS FOR SERVICES - cont'd						
Builders' work for other services installations - cont'd						
Cutting or forming holes for ducts; girth 2.00 - 3.00 m						
102 mm brickwork		1.25	7.79	-	nr	**7.79**
215 mm brickwork		2.20	13.72	-	nr	**13.72**
327 mm brickwork		3.45	21.51	-	nr	**21.51**
100 mm blockwork		1.10	6.86	-	nr	**6.86**
150 mm blockwork		1.50	9.35	-	nr	**9.35**
215 mm blockwork		1.90	11.85	-	nr	**11.85**
Cutting or forming holes for ducts; girth 3.00 - 4.00 m						
102 mm brickwork		1.70	10.60	-	nr	**10.60**
215 mm brickwork		2.90	18.08	-	nr	**18.08**
327 mm brickwork		4.60	28.68	-	nr	**28.68**
100 mm blockwork		1.25	7.79	-	nr	**7.79**
150 mm blockwork		1.70	10.60	-	nr	**10.60**
215 mm blockwork		2.20	13.72	-	nr	**13.72**
Mortices in brickwork						
for expansion bolt		0.23	1.43	-	nr	**1.43**
for 20 mm dia bolt 75 mm deep		0.17	1.06	-	nr	**1.06**
for 20 mm dia bolt 150 mm deep		0.29	1.81	-	nr	**1.81**
Mortices in brickwork; grouting with cement mortar (1:1)						
75 x 75 x 200 mm deep		0.35	2.18	0.11	nr	**2.29**
75 x 75 x 300 mm deep		0.46	2.87	0.15	nr	**3.02**
Holes in softwood for pipes, bars cables and the like						
12 mm thick		0.04	0.32	-	nr	**0.32**
25 mm thick		0.07	0.57	-	nr	**0.57**
50 mm thick		0.11	0.89	-	nr	**0.89**
100 mm thick		0.17	1.38	-	nr	**1.38**
Holes in hardwood for pipes, bars, cables and the like						
12 mm thick		0.07	0.57	-	nr	**0.57**
25 mm thick		0.10	0.81	-	nr	**0.81**
50 mm thick		0.17	1.38	-	nr	**1.38**
100 mm thick		0.24	1.94	-	nr	**1.94**
Duct covers with frames; cast iron; Brickhouse Dudley 'Trucast' or similar; bedding and pointing frame in cement mortar (1:3); fixing with 10 mm dia anchor bolts to concrete at 500 mm centres; including cutting and pinning anchor bolts						
Medium duty; ref 702						
300 mm wide	PC £147.27	2.75	25.22	158.32	m	**183.54**
Extra for						
ends	PC £25.00	-	-	26.88	nr	**26.88**
right angle frame corner	PC £12.00	-	-	12.90	nr	**12.90**
450 mm wide	PC £149.70	3.10	28.43	160.93	m	**189.36**
Extra for						
ends	PC £25.00	-	-	26.88	nr	**26.88**
right angle frame corner	PC £12.00	-	-	12.90	nr	**12.90**
600 mm wide	PC £236.31	3.40	31.18	254.03	m	**285.21**
Extra for						
ends	PC £25.00	-	-	26.88	nr	**26.88**
right angle frame corner	PC £12.00	-	-	12.90	nr	**12.90**
750 mm wide	PC £286.56	3.75	34.39	308.05	m	**342.44**
Extra for						
ends	PC £25.00	-	-	26.88	nr	**26.88**
right angle frame corner	PC £12.00	-	-	12.90	nr	**12.90**

P BUILDING FABRIC SUNDRIES Including overheads and profit at 7.50%		Labour hours	Labour £	Material £	Unit	Total rate £
900 mm wide	PC £326.02	4.05	37.14	350.47	m	**387.61**
Extra for						
ends	PC £25.00	-	-	26.88	nr	**26.88**
right angle frame corner	PC £12.00	-	-	12.90	nr	**12.90**
Heavy duty; ref 704						
300 mm wide	PC £244.03	-	-	262.33	m	**262.33**
Extra for						
ends	PC £34.00	-	-	36.55	nr	**36.55**
right angle frame corner	PC £12.00	-	-	12.90	nr	**12.90**
450 mm wide	PC £274.49	3.75	34.39	295.08	m	**329.47**
Extra for						
ends	PC £34.00	-	-	36.55	nr	**36.55**
right angle frame corner	PC £12.00	-	-	12.90	nr	**12.90**
600 mm wide	PC £303.00	4.20	38.51	325.73	m	**364.24**
Extra for						
ends	PC £34.00	-	-	36.55	nr	**36.55**
right angle frame corner	PC £12.00	-	-	12.90	nr	**12.90**
750 mm wide	PC £326.60	4.60	42.18	351.10	m	**393.28**
Extra for						
ends	PC £34.00	-	-	36.55	nr	**36.55**
right angle frame corner	PC £12.00	-	-	12.90	nr	**12.90**
900 mm wide	PC £395.36	5.05	46.31	425.01	m	**471.32**
Extra for						
ends	PC £34.00	-	-	36.55	nr	**36.55**
right angle frame corner	PC £12.00	-	-	12.90	nr	**12.90**

Q PAVING/PLANTING/FENCING/SITE FURNITURE Including overheads and profit at 7.50%	Labour hours	Labour £	Material £	Unit	Total rate £
Q10 STONE/CONCRETE/BRICK KERBS/EDGINGS/ CHANNELS					
Mechanical excavation using a wheeled hydraulic excavator with a 0.24 m3 bucket					
Excavating trenches to receive kerb foundation; average size					
300 x 100 mm	0.02	0.12	0.27	m	0.39
450 x 150 mm	0.03	0.18	0.46	m	0.64
600 x 200 mm	0.05	0.30	0.64	m	0.94
Excavating curved trenches to receive kerb foundation; average size					
300 x 100 mm	0.02	0.12	0.37	m	0.49
450 x 150 mm	0.03	0.18	0.55	m	0.73
600 x 200 mm	0.05	0.30	0.73	m	1.03
Hand excavation					
Excavating trenches to receive kerb foundation; average size					
150 x 50 mm	0.03	0.18	-	m	0.18
200 x 75 mm	0.08	0.48	-	m	0.48
250 x 100 mm	0.12	0.72	-	m	0.72
300 x 100 mm	0.15	0.90	-	m	0.90
Excavating curved trenches to receive kerb foundation; average size					
150 x 50 mm	0.04	0.24	-	m	0.24
200 x 75 mm	0.09	0.54	-	m	0.54
250 x 100 mm	0.13	0.78	-	m	0.78
300 x 100 mm	0.17	1.02	-	m	1.02
Plain in situ ready mixed concrete; 7.50 N/mm2 - 40 mm aggregate (1:8); PC £40.70/m3; poured on or against earth or unblinded hardcore					
Foundations	1.45	9.04	45.94	m3	54.98
Blinding bed					
not exceeding 150 mm thick	2.15	13.41	45.94	m3	59.35
Plain in situ ready mixed concrete; 11.50 N/mm2 - 40 mm aggregate (1:3:6); PC £41.49/m3; poured on or against earth or unblinded hardcore					
Foundations	1.45	9.04	46.83	m3	55.87
Blinding bed					
not exceeding 150 mm thick	2.15	13.41	46.83	m3	60.24
Plain in situ ready mixed concrete; 11.50 N/mm2 - 40 mm aggregate (1:3:6); PC £41.49/m3; poured on or against earth or unblinded hardcore					
Foundations	1.45	9.04	49.88	m3	58.92
Blinding bed					
not exceeding 150 mm thick	2.15	13.41	49.88	m3	63.29
Precast concrete kerbs, channels, edgings, etc.; BS 340; bedded, jointed and pointed in cement mortar (1:3); including haunching up one side with in situ concrete mix 11.50 N/mm2 - 40 mm aggregate (1:3:6); to concrete base					
Edging; straight; fig 12					
51 x 152 mm	0.29	2.66	2.63	m	5.29
51 x 203 mm	0.29	2.66	3.03	m	5.69
51 x 254 mm	0.29	2.66	3.03	m	5.69

Q PAVING/PLANTING/FENCING/SITE FURNITURE Including overheads and profit at 7.50%	Labour hours	Labour £	Material £	Unit	Total rate £
Kerb; straight					
127 x 254 mm; fig 7	0.38	3.48	5.26	m	**8.74**
152 x 305 mm; fig 6	0.38	3.48	7.15	m	**10.63**
Kerb; curved					
127 x 254 mm; fig 7	0.58	5.32	6.68	m	**12.00**
152 x 305 mm; fig 6	0.58	5.32	8.56	m	**13.88**
Channel; 255 x 125 mm; fig 8					
straight	0.38	3.48	5.46	m	**8.94**
curved	0.58	5.32	6.92	m	**12.24**
Quadrant; fig 14					
305 x 305 x 152 mm	0.40	3.67	4.67	nr	**8.34**
305 x 305 x 254 mm	0.40	3.67	4.67	nr	**8.34**
457 x 457 x 152 mm	0.46	4.22	4.67	nr	**8.89**
457 x 457 x 254 mm	0.46	4.22	4.67	nr	**8.89**
Precast concrete drainage channels; Charcon 'Safeticurb' or similar; channels jointed with plastic rings and bedded; jointed and pointed in cement mortar (1:3); including haunching up one side with in situ concrete mix 11.50 N/mm2 - 40 mm aggregate (1:3:6); to concrete base					
Channel; straight; type DBA/3					
248 x 248 mm	0.69	6.33	19.52	m	**25.85**
End	0.23	2.11	-	nr	**2.11**
Inspection unit; with cast iron lid					
248 x 248 x 914 mm	0.75	6.88	56.52	nr	**63.40**
Silt box top; with concrete frame and cast iron lid; set over gully					
500 x 448 x 269 mm; type A	2.30	21.09	99.06	nr	**120.15**
Q20 HARDCORE/GRANULAR/CEMENT BOUND BASES/ SUB-BASES TO ROADS/PAVINGS					
Mechanical filling with hardcore; PC £6.00/m3					
Filling to make up levels over 250 mm thick; depositing; compacting in layers with a 5 tonne roller	0.30	1.80	9.15	m3	**10.95**
Filling to make up levels not exceeding 250 mm thick; compacting	0.35	2.10	12.21	m3	**14.31**
Mechanical filling with granular fill; type 1; PC £10.15/t (PC £17.25/m3)					
Filling to make up levels over 250 mm thick; depositing; compacting in layers with a 5 tonne roller	0.30	1.80	24.64	m3	**26.44**
Filling to make up levels not exceeding 250 mm thick; compacting	0.35	2.10	31.56	m3	**33.66**
Mechanical filling with granular fill; type 2; PC £9.75/t (PC £16.60/m3)					
Filling to make up levels over 250 mm thick; depositing; compacting in layers with a 5 tonne roller	0.30	1.80	23.77	m3	**25.57**
Filling to make up levels not exceeding 250 mm thick; compacting	0.35	2.10	30.44	m3	**32.54**
Hand filling with hardcore; PC £6.00/m3					
Filling to make up levels over 250 mm thick; depositing; compacting	0.61	3.67	13.15	m3	**16.82**
Filling to make up levels not exceeding 250 mm thick; compacting	0.73	4.39	16.42	m3	**20.81**

Q PAVING/PLANTING/FENCING/SITE FURNITURE Including overheads and profit at 7.50%		Labour hours	Labour £	Material £	Unit	Total rate £
Filling to make up levels over 250 mm thick; depositing; compacting in layers with a 2 tonne roller		0.55	3.31	14.63	m3	17.94
Filling to make up levels not exceeding 250 mm thick; compacting		0.66	3.97	18.34	m3	22.31
Hand filling with sand; PC £9.40/t (PC £15.00/m3)						
Filling to make up levels over 250 mm thick; depositing; compacting in layers with a 2 tonne roller		0.72	4.33	26.17	m3	30.50
Filling to make up levels not exceeding 250 mm thick; compacting		0.85	5.11	32.92	m3	38.03
Surface treatments						
Compacting						
surfaces of ashes		0.05	0.30	0.11	m2	0.41
filling; blinding with ashes		0.09	0.54	0.56	m2	1.10
Q21 IN SITU CONCRETE ROADS/PAVINGS/BASES						
Reinforced in situ ready mixed concrete; normal Portland cement; mix 11.5 N/mm2 - 20 mm aggregate (1:2:4); PC £44.19/m3						
Roads; to hardcore base						
150 - 450 mm thick		1.70	10.60	44.60	m3	55.20
not exceeding 150 mm thick		2.55	15.90	44.60	m3	60.50
Reinforced in situ ready mixed concrete; normal Portland cement; mix 21.00 N/mm2						
Roads; to hardcore base						
150 - 450 mm thick		1.70	10.60	47.50	m3	58.10
not exceeding 150 mm thick		2.55	15.90	47.50	m3	63.40
Reinforced in situ ready mixed concrete; normal Portland cement; mix 26.00 N/mm2 - 20 mm aggregate (1:1:5:3); PC £44.19/m3						
Roads; to hardcore base						
150 - 450 mm thick		1.70	10.60	50.00	m3	60.60
not exceeding 150 mm thick		2.55	15.90	50.00	m3	65.90
Formwork for in situ concrete						
Sides of foundations						
not exceeding 250 mm wide		0.46	3.35	1.30	m	4.65
250 - 500 mm wide		0.69	5.03	2.21	m	7.24
500 mm - 1 m wide		1.05	7.65	4.11	m	11.76
Add to above for curved radius 6 m		0.05	0.36	0.16	m	0.52
Steel road forms to in situ concrete						
Sides of foundations						
150 mm wide		0.22	1.60	0.50	m	2.10
Reinforcement; fabric; BS 4483; lapped; in roads, footpaths or pavings						
Ref A142 (2.22 kg/m2)	PC £0.71	0.13	0.94	0.91	m2	1.85
Ref A193 (3.02 kg/m2)	PC £0.97	0.13	0.94	1.24	m2	2.18
Designed joints in in situ concrete						
Formed joint; 12.5 mm thick Expandite 'Flexcell' or similar						
not exceeding 150 mm wide		0.17	1.24	1.10	m	2.34
150 - 300 mm wide		0.22	1.60	1.89	m	3.49
300 - 450 mm wide		0.28	2.04	2.52	m	4.56

Q PAVING/PLANTING/FENCING/SITE FURNITURE Including overheads and profit at 7.50%	Labour hours	Labour £	Material £	Unit	Total rate £
Formed joint; 25 mm thick Expandite 'Flexcell' or similar					
not exceeding 150 mm wide	0.22	1.60	1.76	m	3.36
150 - 300 mm wide	0.28	2.04	3.10	m	5.14
300 - 450 mm wide	0.33	2.41	4.07	m	6.48
Sealing top 25 mm of joint with rubberized bituminous compound	0.23	1.68	0.92	m	2.60
Concrete sundries					
Treating surfaces of unset concrete; grading to cambers; tamping with a 75 mm thick steel shod tamper	0.28	1.75	-	m2	1.75
Q22 COATED MACADAM/ASPHALT ROADS/PAVINGS					
In situ finishings					
NOTE: The prices for all in situ finishings to roads and footpaths include for work to falls, crossfalls or slopes not exceeding 15 degrees from horizontal; for laying on prepared bases (priced elsewhere) and for rolling with an appropriate roller					
The following rates are based on black bitumen macadam. Red bitumen macadam rates are approximately 50% dearer.					
Fine graded wearing course; BS 4987:88; clause 2.7.7, tables 34 - 36; 14 mm pre-coated igneous rock chippings; tack coat of bitumen emulsion					
19 mm work to roads; one coat					
limestone aggregate	-	-	-	m2	9.00
igneous aggregate	-	-	-	m2	9.14
Close graded bitumen macadam; BS 4987:88; 10 mm graded aggregate to clause 2.7.4 tables 34 - 36; tack coat of bitumen emulsion					
30 mm work to roads; one coat					
limestone aggregate	-	-	-	m2	8.87
igneous aggregate	-	-	-	m2	8.93
Bitumen macadam; BS 4987:88; 45 mm thick base course of 20 mm open graded aggregate to clause 2.6.1 tables 5 - 7; 20 mm thick wearing course of 6 mm medium graded aggregate to clause 2.7.6 tables 32 - 33					
65 mm work to pavements/footpaths; two coats					
limestone aggregate	-	-	-	m2	9.38
igneous aggregate	-	-	-	m2	9.75
Add to last for 14 mm chippings; sprinkled into wearing course	-	-	-	m2	0.43
Bitumen macadam; BS 4987:88; 50 mm graded aggregate to clause 2.6.2 tables 8 - 10					
75 mm work to roads; one coat					
limestone aggregate	-	-	-	m2	10.33
igneous aggregate	-	-	-	m2	10.98

Q PAVING/PLANTING/FENCING/SITE FURNITURE Including overheads and profit at 7.50%		Labour hours	Labour £	Material £	Unit	Total rate £
Q22 COATED MACADAM/ASPHALT ROADS/PAVINGS - cont'd						
Dense bitumen macadam; BS 4987:88; 50 mm thick base course of 20 mm graded aggregate to clause 2.6.5 tables 15 - 16; 200 pen. binder; 30 mm wearing course of 10 mm graded aggregate to clause 2.7.4 tables 26 - 28						
75 mm work to roads; two coats						
limestone aggregate		-	-	-	m2	**11.95**
igneous aggregate		-	-	-	m2	**12.33**
Bitumen macadam; BS 4987:88; 50 mm thick base course of 20 mm graded aggregate to clause 2.6.1 tables 5 - 7; 25 mm thick wearing course of 10 mm graded aggregate to clause 2.7.2 tables 20 - 22						
75 mm work to roads; two coats						
limestone aggregate		-	-	-	m2	**10.70**
igneous aggregate		-	-	-	m2	**10.98**
Asphalt; BS 594 table 5; 100 mm thick roadbase; 60 mm thick dense base course; 40 mm thick hot rolled wearing course to table 1 binder 3; 14 mm chippings spread and rolled in						
200 mm work to roads; three coats						
limestone base/hardstone wearing course		-	-	-	m2	**29.40**
Q25 SLAB/BRICK/BLOCK/SETT/COBBLE PAVINGS						
NOTE: Unless otherwise described, prices for pavings do not include for ash or sand beds or bases under.						
Artificial stone paving; Redland Aggregates' 'Texitone' or similar; to falls or crossfalls; bedding 25 mm thick in lime mortar (1:4) staggered joints; jointing in coloured cement mortar (1:3); brushed in; to sand base						
Work to paved areas; over 300 mm wide						
450 x 600 x 50 mm;						
grey or coloured	PC £2.76/each	0.52	4.77	13.58	m2	**18.35**
600 x 600 x 50 mm;						
grey or coloured	PC £3.25/each	0.48	4.40	12.22	m2	**16.62**
750 x 600 x 50 mm;						
grey or coloured	PC £3.86/each	0.45	4.13	11.77	m2	**15.90**
900 x 750 x 50 mm;						
grey or coloured	PC £4.51/each	0.41	3.76	9.62	m2	**13.38**
Brick paviors; 215 x 103 x 65 mm rough stock bricks; PC £443.00/1000; to falls or crossfalls; bedding 10 mm thick in cement mortar (1:3); jointing in cement mortar (1:3); as work proceeds; to concrete base						
Work to paved areas; over 300 mm wide; straight joints both ways						
bricks laid flat		0.92	9.54	19.73	m2	**29.27**
bricks laid on edge		1.30	13.49	30.99	m2	**44.48**

Q PAVING/PLANTING/FENCING/SITE FURNITURE Including overheads and profit at 7.50%		Labour hours	Labour £	Material £	Unit	Total rate £
Work to paved areas; over 300 mm wide; laid to herringbone pattern						
bricks laid flat		1.15	11.93	19.73	m2	**31.66**
bricks laid on edge		1.60	16.60	30.99	m2	**47.59**
Add or deduct for variation of 1.00/1000 in PC of brick paviors						
bricks laid flat		-	-	-	m2	**0.43**
bricks laid on edge		-	-	-	m2	**0.69**
Cobble paving; 50 - 75 mm PC £58.00/t; to falls or crossfalls; bedding 13 mm thick in cement mortar (1:3); jointing to a height of two thirds of cobbles in dry mortar (1:3); tightly butted, washed and brushed; to concrete base						
Work to paved areas; over 300 mm wide						
regular		4.60	42.18	15.55	m2	**57.73**
laid to pattern		5.75	52.73	15.55	m2	**68.28**
Concrete paving flags; BS 368; to falls or crossfalls; bedding 25 mm thick in lime and sand mortar (1:4); butt joints straight both ways; jointing in cement mortar (1:3); brushed in; to sand base						
Work to paved areas; over 300 mm wide						
300 x 300 x 60 mm; grey	PC £1.03/each	0.63	5.78	13.81	m2	**19.59**
300 x 300 x 60 mm; coloured	PC £1.22/each	0.63	5.78	16.19	m2	**21.97**
450 x 600 x 50 mm; grey	PC £1.72/each	0.52	4.77	8.15	m2	**12.92**
450 x 600 x 50 mm; coloured	PC £2.50/each	0.52	4.77	11.39	m2	**16.16**
600 x 600 x 50 mm; grey	PC £1.91/each	0.48	4.40	6.94	m2	**11.34**
600 x 600 x 50 mm; coloured	PC £3.00/each	0.48	4.40	10.34	m2	**14.74**
750 x 600 x 50 mm; grey	PC £2.28/each	0.45	4.13	6.71	m2	**10.84**
750 x 600 x 50 mm; coloured	PC £3.61/each	0.45	4.13	10.05	m2	**14.18**
900 x 600 x 50 mm; grey	PC £2.66/each	0.41	3.76	5.45	m2	**9.21**
900 x 600 x 50 mm; coloured	PC £4.19/each	0.41	3.76	7.99	m2	**11.75**
Concrete rectangular paving blocks; to falls or crossfalls; bedding 50 mm thick in dry sharp sand; filling joints with sharp sand brushed in; on earth base						
'Keyblok' paving; over 300 mm wide; straight joints both ways						
200 x 100 x 65 mm; grey	PC £7.55	0.86	7.89	9.74	m2	**17.63**
200 x 100 x 65 mm; coloured	PC £8.75	0.86	7.89	11.09	m2	**18.98**
200 x 100 x 80 mm; grey	PC £8.75	0.92	8.44	11.19	m2	**19.63**
200 x 100 x 80 mm; coloured	PC £10.24	0.92	8.44	12.87	m2	**21.31**
'Keyblok' paving; over 300 mm wide; laid to herringbone pattern						
200 x 100 x 65 mm; grey		1.10	10.09	10.13	m2	**20.22**
200 x 100 x 65 mm; coloured		1.10	10.09	11.55	m2	**21.64**
200 x 100 x 80 mm; grey		1.15	10.55	11.65	m2	**22.20**
200 x 100 x 80 mm; coloured		1.15	10.55	13.41	m2	**23.96**
Extra for two row boundary edging to herringbone paved areas; 200 mm wide; including a 150 mm high in situ concrete mix 11.5 N/mm2 - 40 mm aggregate (1:3:6) haunching to one side; blocks laid breaking joint						
200 x 100 x 65 mm; coloured		0.35	3.21	1.73	m	**4.94**
200 x 100 x 80 mm; coloured		0.35	3.21	1.84	m	**5.05**

Q PAVING/PLANTING/FENCING/SITE FURNITURE Including overheads and profit at 7.50%		Labour hours	Labour £	Material £	Unit	Total rate £
Q25 SLAB/BRICK/BLOCK/SETT/COBBLE PAVINGS - cont'd						
Concrete rectangular paving blocks; to falls or crossfalls; bedding 50 mm thick in dry sharp sand; filling joints with sharp sand brushed in; on earth base - cont'd						
'Mount Sorrel' paving; over 300 mm wide; straight joints both ways						
200 x 100 x 65 mm; grey	PC £7.70	0.86	7.89	9.91	m2	17.80
200 x 100 x 65 mm; coloured	PC £8.60	0.86	7.89	10.92	m2	18.81
200 x 100 x 80 mm; grey	PC £8.74	0.92	8.44	11.18	m2	19.62
200 x 100 x 80 mm; coloured	PC £9.88	0.92	8.44	12.46	m2	20.90
'Pedesta' paving; over 300 mm wide; straight joints both ways						
200 x 100 x 60 mm; grey	PC £8.74	0.86	7.89	11.08	m2	18.97
200 x 100 x 60 mm; coloured	PC £9.70	0.86	7.89	12.15	m2	20.04
200 x 100 x 80 mm; grey	PC £9.90	0.92	8.44	12.49	m2	20.93
200 x 100 x 80 mm; coloured	PC £11.06	0.92	8.44	13.80	m2	22.24
'Luttersett' paving; over 300 mm wide; straight joints both ways						
200 x 100 x 65 mm; grey	PC £8.06	0.86	7.89	10.31	m2	18.20
200 x 100 x 65 mm; coloured	PC £9.31	0.86	7.89	11.73	m2	19.62
200 x 100 x 80 mm; grey	PC £9.52	0.92	8.44	12.06	m2	20.50
200 x 100 x 80 mm; coloured	PC £11.05	0.92	8.44	13.79	m2	22.23
Granite setts; BS 435; 200 x 100 x 100 mm; PC £100.00/t; standard 'C' dressing; tightly butted to falls or crossfalls; bedding 25 mm thick in cement mortar (1:3); filling joints with dry mortar (1:6); washed and brushed; on concrete base						
Work to paved areas; over 300 mm wide						
straight joints		1.85	16.96	30.86	m2	47.82
laid to pattern		2.30	21.09	30.86	m2	51.95
Two rows of granite setts as boundary edging; 200 mm wide; including a 150 mm high in situ concrete mix 11.5 N/mm2 - 40 mm aggregate (1:3:6) haunching to one side; blocks laid breaking joint		0.81	7.43	7.40	m	14.83
Q26 SPECIAL SURFACINGS/PAVINGS FOR SPORT						
Sundries						
Painted line on road; one coat						
75 mm wide		0.06	0.39	0.09	m	0.48
Q30 SEEDING/TURFING						
Vegetable soil						
Selected from spoil heaps; grading; preparing for turfing or seeding; to general surfaces						
average 75 mm thick		0.36	2.16	-	m2	2.16
average 100 mm thick		0.39	2.34	-	m2	2.34
average 125 mm thick		0.42	2.52	-	m2	2.52
average 150 mm thick		0.44	2.64	-	m2	2.64
average 175 mm thick		0.46	2.76	-	m2	2.76
average 200 mm thick		0.48	2.88	-	m2	2.88

Q PAVING/PLANTING/FENCING/SITE FURNITURE Including overheads and profit at 7.50%	Labour hours	Labour £	Material £	Unit	Total rate £
Selected from spoil heaps; grading; preparing for turfing or seeding; to cutting or embankments					
average 75 mm thick	0.41	2.46	-	m2	2.46
average 100 mm thick	0.44	2.64	-	m2	2.64
average 125 mm thick	0.47	2.82	-	m2	2.82
average 150 mm thick	0.50	3.00	-	m2	3.00
average 175 mm thick	0.52	3.12	-	m2	3.12
average 200 mm thick	0.55	3.31	-	m2	3.31
Imported vegetable soil; PC £11.70/m3					
Grading; preparing for turfing or seeding; to general surfaces					
average 75 mm thick	0.33	1.98	1.26	m2	3.24
average 100 mm thick	0.35	2.10	1.64	m2	3.74
average 125 mm thick	0.37	2.22	2.39	m2	4.61
average 150 mm thick	0.40	2.40	3.14	m2	5.54
average 175 mm thick	0.42	2.52	3.52	m2	6.04
average 200 mm thick	0.44	2.64	3.90	m2	6.54
Grading; preparing for turfing or seeding; to cuttings or embankments					
average 75 mm thick	0.36	2.16	1.26	m2	3.42
average 100 mm thick	0.40	2.40	1.64	m2	4.04
average 125 mm thick	0.42	2.52	2.39	m2	4.91
average 150 mm thick	0.44	2.64	3.14	m2	5.78
average 175 mm thick	0.46	2.76	3.52	m2	6.28
average 200 mm thick	0.48	2.88	3.90	m2	6.78
Fertilizer; PC £0.47/kg					
Fertilizer 0.07 kg/m2; raking in					
general surfaces	0.04	0.24	0.04	m2	0.28
Selected grass seed; PC £4.70/kg					
Grass seed; sowing at a rate of 0.042 kg/m2 two applications; raking in					
general surfaces	0.08	0.48	0.42	m2	0.90
cuttings or embankments	0.09	0.54	0.42	m2	0.96
Preserved turf from stack on site					
Selected turf					
general surfaces	0.22	1.32	-	m2	1.32
cuttings or embankments; shallow	0.24	1.44	0.15	m2	1.59
cuttings or embankments; steep; pegged	0.33	1.98	0.23	m2	2.21
Imported turf; PC £1.45/m2					
Selected meadow turf					
general surfaces	0.22	1.32	1.56	m2	2.88
cuttings or embankments; shallow	0.24	1.44	1.71	m2	3.15
cuttings or embankments; steep; pegged	0.33	1.98	1.79	m2	3.77
Q31 PLANTING					
Planting only					
Hedge or shrub plants					
not exceeding 750 mm high	0.28	1.68	-	nr	1.68
750 mm - 1.5 m high	0.66	3.97	-	nr	3.97
Saplings					
not exceeding 3 m high	1.85	11.12	-	nr	11.12

Q PAVING/PLANTING/FENCING/SITE FURNITURE Including overheads and profit at 7.50%	Labour hours	Labour £	Material £	Unit	Total rate £
Q40 FENCING					
NOTE: The prices for all fencing are to include for setting posts in position, to a depth of 0.6 m for fences not exceeding 1.4 m high and of 0.76 m for fences over 1.4 m high. The prices allow for excavating post holes; filling to within 150 mm of ground level with concrete and all necessary back filling					
Strained wire fences; BS 1722 Part 3; 4 mm galvanized mild steel plain wire threaded through posts and strained with eye bolts					
900 mm fencing; three line; concrete posts at 2750 mm centres	-	-	-	m	6.17
Extra for					
end concrete straining post; one strut	-	-	-	nr	27.49
angle concrete straining post; two struts	-	-	-	nr	34.13
1.07 m fencing; five line; concrete posts at 2750 mm centres	-	-	-	m	7.47
Extra for					
end concrete straining post; one strut	-	-	-	nr	47.46
angle concrete straining post; two struts	-	-	-	nr	63.18
1.20 m fencing; six line; concrete posts at 2750 mm centres	-	-	-	m	7.52
Extra for					
end concrete straining post; one strut	-	-	-	nr	48.24
angle concrete straining post; two struts	-	-	-	nr	64.32
1.4 m fencing; seven line; concrete posts at 2750 mm centres	-	-	-	m	7.74
Extra for					
end concrete straining post; one strut	-	-	-	nr	48.76
angle concrete straining post; two struts	-	-	-	nr	65.35
Chainlink fences; BS 1722 Part 1; 3 mm; 50 mm galvanized mild steel mesh; galvanized mild steel tying and line wire; three line wires threaded through posts and strained with eye bolts and winding brackets					
900 mm fencing; galvanized mild steel angle posts at 3 m centres	-	-	-	m	8.31
Extra for					
end steel straining post; one strut	-	-	-	nr	34.31
angle steel straining post; two struts	-	-	-	nr	45.60
900 mm fencing; concrete posts at 3 m centres	-	-	-	m	8.58
Extra for					
end concrete straining post; one strut	-	-	-	nr	34.20
angle concrete straining post; two struts	-	-	-	nr	45.33
1.2 m fencing; galvanized mild steel angle posts at 3 m centres	-	-	-	m	9.72
Extra for					
end steel straining post; one strut	-	-	-	nr	36.48
angle steel straining post; two struts	-	-	-	nr	49.08
1.2 m fencing; concrete posts at 3 m centres	-	-	-	m	9.99
Extra for					
end concrete straining post; one strut	-	-	-	nr	40.06
angle concrete straining post; two struts	-	-	-	nr	48.75
1.8 m fencing; galvanized mild steel angle posts at 3 m centres	-	-	-	m	13.57
Extra for					
end steel straining post; one strut	-	-	-	nr	55.92
angle steel straining post; two struts	-	-	-	nr	73.72

Q PAVING/PLANTING/FENCING/SITE FURNITURE Including overheads and profit at 7.50%	Labour hours	Labour £	Material £	Unit	Total rate £
1.8 m fencing; concrete posts at 3 m centres	-	-	-	m	14.06
Extra for					
end concrete straining post; one strut	-	-	-	nr	57.27
angle concrete straining post; two struts	-	-	-	nr	74.37
Pair of gates and gate posts; gates to match galvanized chain link fencing, with angle framing, braces, etc., complete with hinges, locking bar, lock and bolts; two 100 x 100 mm angle section gate posts; each with one strut					
2.44 x 0.9 m high	-	-	-	nr	388.70
2.44 x 1.2 m high	-	-	-	nr	418.01
2.44 x 1.8 m high	-	-	-	nr	508.13
Chainlink fences; BS 1722 Part 1; 3 mm; 50 mm plastic coated mild steel mesh; plastic coated mild steel tying and line wire; three line wires threaded through posts and strained with eye bolts and winding brackets					
900 mm fencing; galvanized mild steel angle posts at 3 m centres	-	-	-	m	9.12
Extra for					
end steel straining post; one strut	-	-	-	nr	34.31
angle steel straining post; two struts	-	-	-	nr	45.60
900 mm fencing; concrete posts at 3 m centres	-	-	-	m	9.39
Extra for					
end concrete straining post; one strut	-	-	-	nr	34.20
angle concrete straining post; two struts	-	-	-	nr	45.33
1.2 m fencing; galvanized mild steel angle posts at 3 m centres	-	-	-	m	10.94
Extra for					
end steel straining post; one strut	-	-	-	nr	36.48
angle steel straining post; two struts	-	-	-	nr	49.08
1.2 m fencing; concrete posts at 3 m centres	-	-	-	m	11.23
Extra for					
end concrete straining post; one strut	-	-	-	nr	40.06
angle concrete straining post; two struts	-	-	-	nr	48.75
1.8 m fencing; galvanized mild steel angle posts at 3 m centres	-	-	-	m	14.89
Extra for					
end steel straining post; one strut	-	-	-	nr	55.92
angle steel straining post; two struts	-	-	-	nr	73.72
1.8 m fencing; concrete posts at 3 m centres	-	-	-	m	15.37
Extra for					
end concrete straining post; one strut	-	-	-	nr	57.27
angle concrete straining post; two struts	-	-	-	nr	73.83
Pair of gates and gate posts; gates to match plastic chain link fencing; with angle framing, braces, etc., complete with hinges, locking bar, lock and bolts; two 100 x 100 mm angle section gate posts; each with one strut					
2.44 x 0.9 m high	-	-	-	nr	418.71
2.44 x 1.2 m high	-	-	-	nr	446.26
2.44 x 1.8 m high	-	-	-	nr	542.12

Q PAVING/PLANTING/FENCING/SITE FURNITURE Including overheads and profit at 7.50%	Labour hours	Labour £	Material £	Unit	Total rate £
Q40 FENCING - cont'd					
Chain link fences for tennis courts; BS 1722 Part 13; 2.5 mm; 45 mm mesh galvanized mild steel mesh; line and tying wires threaded through 45 x 45 x 5 mm galvanized mild steel angle standards, posts and struts; 60 x 60 x 6 mm angle straining posts and gate posts; straining posts and struts strained with eye bolts and winding brackets					
Fencing to tennis court 36 x 18 m; including gate 1070 x 1980 mm complete with hinges, locking bar, lock and bolts					
2745 mm fencing; standards at 3 m centres	-	-	-	nr	1881.25
3660 mm fencing; standards at 2.5 m centres	-	-	-	nr	2553.13
Cleft chestnut pale fences; BS 1722 Part 4; pales spaced 51 mm apart; on two lines of galvanized wire; 64 mm dia posts; 76 x 51 mm struts					
900 mm fences; posts at 2.50 m centres	-	-	-	m	6.58
Extra for					
straining post; one strut	-	-	-	nr	12.00
corner straining post; two struts	-	-	-	nr	16.00
1.05 m fences; posts at 2.50 m centres	-	-	-	m	7.22
Extra for					
straining post; one strut	-	-	-	nr	13.22
corner straining post; two struts	-	-	-	nr	17.61
1.20 m fences; posts at 2.25 m centres	-	-	-	m	7.74
Extra for					
straining post; one strut	-	-	-	nr	14.51
corner straining post; two struts	-	-	-	nr	19.35
1.35 m fences; posts at 2.25 m centres	-	-	-	m	8.19
Extra for					
straining post; one strut	-	-	-	nr	16.25
corner straining post; two struts	-	-	-	nr	21.54
Close boarded fencing; BS 1722 Part 5; 76 x 38 mm softwood rails; 89 x 19 mm softwood pales lapped 13 mm; 152 x 25 mm softwood gravel boards; all softwood 'treated'; posts at 3 m centres					
Fences; two rail; concrete posts					
1 m	-	-	-	m	27.48
1.2 m	-	-	-	m	28.38
Fences; three rail; concrete posts					
1.4 m	-	-	-	m	36.89
1.6 m	-	-	-	m	38.31
1.8 m	-	-	-	m	39.80
Fences; two rail; oak posts					
1 m	-	-	-	m	19.87
1.2 m	-	-	-	m	22.70
Fences; three rail; oak posts					
1.4 m	-	-	-	m	25.54
1.6 m	-	-	-	m	29.03
1.8 m	-	-	-	m	32.38
Precast concrete slab fencing; 305 x 38 x 1753 mm slabs; fitted into twice grooved concrete posts at 1830 mm centres					
Fences					
1.2 m	-	-	-	m	42.18
1.5 m	-	-	-	m	49.99
1.8 m	-	-	-	m	62.44

Q PAVING/PLANTING/FENCING/SITE FURNITURE Including overheads and profit at 7.50%	Labour hours	Labour £	Material £	Unit	Total rate £
Mild steel unclimbable fencing; in rivetted panels 2440 mm long; 44 x 13 mm flat section top and bottom rails; two 44 x 19 mm flat section standards, one with foot plate, and 38 x 13 mm raking stay with foot plate; 20 mm dia pointed verticals at 120 mm centres; two 44 x 19 mm supports 760 mm long with ragged ends to bottom rail; the whole bolted together; coated with red oxide primer; setting standards and stays in ground at 2440 mm centres and supports at 815 mm centres					
Fences					
1.67 m	-	-	-	m	62.81
2.13 m	-	-	-	m	72.17
Pair of gates and gate posts, to match mild steel unclimbable fencing; with flat section framing, braces, etc., complete with locking bar, lock, handles, drop bolt, gate stop and holding back catches; two 102 x 102 mm hollow section gate posts with cap and foot plates					
2.44 x 1.67 m	-	-	-	nr	564.16
2.44 x 2.13 m	-	-	-	nr	637.99
4.88 x 1.67 m	-	-	-	nr	1121.71
4.88 x 2.13 m	-	-	-	nr	1272.67

R DISPOSAL SYSTEMS Including overheads and profit at 7.50%		Labour hours	Labour £	Material £	Unit	Total rate £
R10 RAINWATER PIPEWORK/GUTTERS						
Aluminium pipes and fittings; BS 2997; ears cast on; powder coated finish						
63 mm pipes; plugged and nailed		0.44	3.96	8.96	m	**12.92**
Extra for						
fittings with one end		0.26	2.34	4.42	nr	**6.76**
fittings with two ends		0.50	4.50	4.77	nr	**9.27**
fittings with three ends		0.72	6.49	6.71	nr	**13.20**
shoe	PC £4.83	0.26	2.34	4.42	nr	**6.76**
bend	PC £5.16	0.50	4.50	4.77	nr	**9.27**
single branch	PC £6.72	0.72	6.49	6.71	nr	**13.20**
offset 229 mm projection	PC £11.90	0.50	4.50	10.65	nr	**15.15**
offset 305 mm projection	PC £13.27	0.50	4.50	12.13	nr	**16.63**
access pipe	PC £14.70	-	-	12.65	nr	**12.65**
connection to clay pipes; cement and sand (1:2) joint		0.18	1.62	0.10	nr	**1.72**
75 mm pipes; plugged and nailed	PC £16.46/1.8m	0.48	4.32	10.47	m	**14.79**
Extra for						
shoe	PC £6.64	0.30	2.70	6.30	nr	**9.00**
bend	PC £6.51	0.54	4.86	6.16	nr	**11.02**
single branch	PC £8.10	0.78	7.03	8.30	nr	**15.33**
offset 229 mm projection	PC £13.15	0.54	4.86	11.71	nr	**16.57**
offset 305 mm projection	PC £14.55	0.54	4.86	13.21	nr	**18.07**
access pipe	PC £16.06	-	-	13.70	nr	**13.70**
connection to clay pipes; cement and sand (1:2) joint		0.20	1.80	0.10	nr	**1.90**
100 mm pipes; plugged and nailed	PC £28.10/1.8m	0.54	4.86	17.72	m	**22.58**
Extra for						
shoe	PC £8.00	0.60	5.41	7.02	nr	**12.43**
bend	PC £9.09	0.34	3.06	8.19	nr	**11.25**
single branch	PC £10.85	-	-	10.55	nr	**10.55**
offset 229 mm projection	PC £15.22	0.90	8.11	12.06	nr	**20.17**
offset 305 mm projection	PC £16.90	0.60	5.41	13.87	nr	**19.28**
access pipe	PC £19.03	-	-	14.17	nr	**14.17**
connection to clay pipes; cement and sand (1:2) joint		0.24	2.16	0.10	nr	**2.26**
Roof outlets; circular aluminium; with flat or domed grate; joint to pipe						
50 mm dia	PC £33.99	0.72	6.49	37.26	nr	**43.75**
75 mm dia	PC £44.97	0.78	7.03	49.24	nr	**56.27**
100 mm dia	PC £58.95	0.84	7.57	64.43	nr	**72.00**
150 mm dia	PC £75.92	0.90	8.11	83.12	nr	**91.23**
Roof outlets; d-shaped; balcony; with flat or domed grate; joint to pipe						
50 mm dia	PC £41.84	0.72	6.49	45.70	nr	**52.19**
75 mm dia	PC £48.10	0.78	7.03	52.61	nr	**59.64**
100 mm dia	PC £59.08	0.84	7.57	64.57	nr	**72.14**
Galvanized wire balloon grating; BS 416 for pipes or outlets						
50 mm dia	PC £0.95	0.08	0.72	1.02	nr	**1.74**
63 mm dia	PC £0.95	0.08	0.72	1.02	nr	**1.74**
75 mm dia	PC £1.05	0.08	0.72	1.13	nr	**1.85**
100 mm dia	PC £1.27	0.10	0.90	1.37	nr	**2.27**
Aluminium gutters and fittings; BS 2997; powder coated finish						
100 mm half round gutters; on brackets screwed to timber	PC £13.27/1.8m	0.42	3.78	9.49	m	**13.27**
Extra for						
stop end	PC £1.96	0.19	1.71	2.88	nr	**4.59**
running outlet	PC £4.35	0.40	3.60	3.85	nr	**7.45**
stop end outlet	PC £3.86	0.19	1.71	3.78	nr	**5.49**
angle	PC £4.01	0.40	3.60	3.21	nr	**6.81**

R DISPOSAL SYSTEMS Including overheads and profit at 7.50%		Labour hours	Labour £	Material £	Unit	Total rate £
112 mm half round gutters; on brackets						
screwed to timber	PC £13.91/1.8m	0.42	3.78	9.89	m	13.67
Extra for						
stop end	PC £2.06	0.19	1.71	2.99	nr	4.70
running outlet	PC £4.74	0.40	3.60	4.22	nr	7.82
stop end outlet	PC £4.44	0.19	1.71	4.37	nr	6.08
angle	PC £4.53	0.40	3.60	3.69	nr	7.29
125 mm half round gutters; on brackets						
screwed to timber	PC £15.62/1.8m	0.48	4.32	11.98	m	16.30
Extra for						
stop end	PC £2.51	0.22	1.98	3.99	nr	5.97
running outlet	PC £5.13	0.42	3.78	4.52	nr	8.30
stop end outlet	PC £4.72	0.22	1.98	5.02	nr	7.00
angle	PC £5.03	0.42	3.78	4.88	nr	8.66
100 mm ogee gutters; on brackets						
screwed to timber	PC £16.55/1.8m	0.44	3.96	11.44	m	15.40
Extra for						
stop end	PC £2.06	0.20	1.80	2.99	nr	4.79
running outlet	PC £5.10	0.42	3.78	4.37	nr	8.15
stop end outlet	PC £3.96	0.20	1.80	3.60	nr	5.40
angle	PC £4.30	0.42	3.78	2.92	nr	6.70
112 mm ogee gutters; on						
brackets screwed to timber	PC £18.40/1.8m	0.50	4.50	12.61	m	17.11
Extra for						
stop end	PC £2.22	0.20	1.80	3.18	nr	4.98
running outlet	PC £5.18	0.42	3.78	4.31	nr	8.09
stop end outlet	PC £4.43	0.20	1.80	3.98	nr	5.78
angle	PC £5.13	0.42	3.78	3.52	nr	7.30
125 mm ogee gutters; on						
brackets screwed to timber	PC £20.32/1.8m	0.50	4.50	14.83	m	19.33
Extra for						
stop end	PC £2.42	0.23	2.07	3.88	nr	5.95
running outlet	PC £5.66	0.44	3.96	4.68	nr	8.64
stop end outlet	PC £5.03	0.23	2.07	4.95	nr	7.02
angle	PC £5.98	0.44	3.96	5.05	nr	9.01
Cast iron pipes and fittings; BS 460; ears cast on; joints						
50 mm pipes; primed; plugged						
and nailed	PC £18.44/1.8m	0.60	5.41	11.75	m	17.16
Extra for						
fittings with one end		0.36	3.24	8.36	nr	11.60
fittings with two ends		0.66	5.95	4.67	nr	10.62
shoe	PC £8.90	0.36	3.24	8.36	nr	11.60
bend	PC £5.47	0.66	5.95	4.67	nr	10.62
connection to clay pipes;						
cement and sand (1:2) joint		0.16	1.44	0.10	nr	1.54
63 mm pipes; primed; plugged						
and nailed	PC £18.44/1.8m	0.62	5.59	11.83	m	17.42
Extra for						
fittings with one end		0.38	3.42	8.43	nr	11.85
fittings with two ends		0.68	6.13	4.73	nr	10.86
fittings with three ends		0.86	7.75	7.49	nr	15.24
shoe	PC £8.90	0.38	3.42	8.43	nr	11.85
bend	PC £5.47	0.68	6.13	4.73	nr	10.86
single branch	PC £8.44	0.86	7.75	7.49	nr	15.24
offset 229 mm projection	PC £9.70	0.68	6.13	8.09	nr	14.22
offset 305 mm projection	PC £11.36	0.68	6.13	9.48	nr	15.61
connection to clay pipes;						
cement and sand (1:2) joint		0.18	1.62	0.10	nr	1.72

R DISPOSAL SYSTEMS Including overheads and profit at 7.50%		Labour hours	Labour £	Material £	Unit	Total rate £
R10 RAINWATER PIPEWORK/GUTTERS - cont'd						
Cast iron pipes and fittings; BS 460; ears cast on; joints - cont'd						
75 mm pipes; primed, plugged and nailed	PC £18.44/1.8m	0.66	5.95	11.94	m	17.89
Extra for						
shoe	PC £8.90	0.42	3.78	8.56	nr	12.34
bend	PC £5.47	0.72	6.49	4.87	nr	11.36
single branch	PC £8.44	0.90	8.11	7.76	nr	15.87
offset 229 mm projection	PC £9.70	0.72	6.49	8.23	nr	14.72
offset 305 mm projection	PC £11.36	0.72	6.49	9.62	nr	16.11
connection to clay pipes; cement and sand (1:2) joint		0.20	1.80	0.10	nr	1.90
100 mm pipes; primed, plugged and nailed	PC £24.76/1.8m	0.72	6.49	16.06	m	22.55
Extra for						
shoe	PC £11.58	0.48	4.32	11.15	nr	15.47
bend	PC £8.49	0.78	7.03	7.83	nr	14.86
single branch	PC £11.02	0.96	8.65	10.24	nr	18.89
offset 229 mm projection	PC £15.36	0.78	7.03	13.62	nr	20.65
offset 305 mm projection	PC £18.02	0.78	7.03	15.95	nr	22.98
connection to clay pipes; cement and sand (1:2) joint		0.24	2.16	0.10	nr	2.26
100 x 75 mm rectangular pipes; primed, plugged and nailed	PC £88.81/1.8m	0.72	6.49	55.83	m	62.32
Extra for					nr	
shoe	PC £35.42	0.48	4.32	31.28	nr	35.60
bend	PC £28.76	0.78	7.03	24.12	nr	31.15
offset 229 mm projection	PC £40.00	0.78	7.03	30.47	nr	37.50
offset 305 mm projection	PC £47.33	0.78	7.03	36.44	nr	43.47
connection to clay pipes; cement and sand (1:2) joint		0.24	2.16	0.10	nr	2.26
Rainwater head; flat; for pipes						
50 mm dia	PC £6.94	0.66	5.95	7.83	nr	13.78
63 mm dia	PC £6.94	0.68	6.13	7.90	nr	14.03
75 mm dia	PC £6.94	0.72	6.49	8.03	nr	14.52
100 mm dia	PC £16.00	0.78	7.03	18.04	nr	25.07
Rainwater head; rectangular, for pipes						
50 mm dia	PC £15.28	0.66	5.95	16.80	nr	22.75
63 mm dia	PC £15.28	0.68	6.13	16.87	nr	23.00
75 mm dia	PC £15.28	0.72	6.49	17.00	nr	23.49
100 mm dia	PC £30.98	0.78	7.03	34.14	nr	41.17
Roof outlets; cast iron; circular; with flat grate; joint to pipe						
50 mm dia		0.90	8.11	48.51	nr	56.62
75 mm dia	PC £46.60	1.00	9.01	50.98	nr	59.99
100 mm dia	PC £55.92	1.10	9.91	61.17	nr	71.08
Copper wire balloon grating; BS 416 for pipes or outlets						
50 mm dia	PC £1.45	0.08	0.72	1.56	nr	2.28
63 mm dia	PC £1.45	0.08	0.72	1.56	nr	2.28
75 mm dia	PC £1.68	0.08	0.72	1.81	nr	2.53
100 mm dia		0.10	0.90	2.05	nr	2.95
Cast iron gutters and fittings; BS 460						
100 mm half round gutters; primed; on brackets; screwed to timber	PC £9.38/1.8m	0.48	4.32	7.07	m	11.39
Extra for						
stop end	PC £1.18	0.20	1.80	1.94	nr	3.74
running outlet	PC £3.64	0.42	3.78	3.28	nr	7.06
angle	PC £3.64	0.42	3.78	3.36	nr	7.14

R DISPOSAL SYSTEMS Including overheads and profit at 7.50%		Labour hours	Labour £	Material £	Unit	Total rate £
115 mm half round gutters; primed;						
on brackets; screwed to timber	PC £9.77/1.8m	0.48	4.32	7.38	m	11.70
Extra for						
stop end	PC £1.72	0.20	1.80	2.55	nr	4.35
running outlet	PC £4.02	0.42	3.78	3.66	nr	7.44
angle	PC £4.02	0.42	3.78	3.75	nr	7.53
125 mm half round gutters; primed;						
on brackets; screwed to timber	PC £11.42/1.8m	0.54	4.86	8.40	m	13.26
Extra for						
stop end	PC £1.72	0.24	2.16	2.56	nr	4.72
running outlet	PC £4.74	0.48	4.32	4.31	nr	8.63
angle	PC £4.74	0.48	4.32	4.24	nr	8.56
150 mm half round gutters; primed;						
on brackets; screwed to timber	PC £19.54/1.8m	0.60	5.41	13.62	m	19.03
Extra for						
stop end	PC £2.28	0.26	2.34	3.24	nr	5.58
running outlet	PC £6.52	0.54	4.86	5.53	nr	10.39
angle	PC £6.52	0.54	4.86	4.80	nr	9.66
100 mm ogee gutters; primed;						
on brackets; screwed to timber	PC £10.24/1.8m	0.50	4.50	7.45	m	11.95
Extra for						
stop end	PC £1.10	0.22	1.98	2.29	nr	4.27
running outlet	PC £3.24	0.44	3.96	2.80	nr	6.76
angle	PC £3.24	0.44	3.96	2.72	nr	6.68
115 mm ogee gutters; primed;						
on brackets; screwed to timber	PC £11.51/1.8m	0.50	4.50	8.35	m	12.85
Extra for						
stop end	PC £1.58	0.22	1.98	2.84	nr	4.82
running outlet	PC £3.70	0.44	3.96	3.19	nr	7.15
angle	PC £3.70	0.44	3.96	3.00	nr	6.96
125 mm ogee gutters; primed;						
on brackets; screwed to timber	PC £12.06/1.8m	0.56	5.04	8.80	m	13.84
Extra for						
stop end	PC £1.58	0.25	2.25	2.95	nr	5.20
running outlet	PC £4.41	0.50	4.50	3.92	nr	8.42
angle	PC £5.87	0.50	4.50	5.35	nr	9.85
3 mm galvanised heavy pressed steel gutters and fittings; joggle joints; BS 1091						
200 x 100 mm (400 mm girth) box gutter;						
screwed to timber		0.78	7.03	14.60	m	21.63
Extra for						
stop end		0.42	3.78	11.51	nr	15.29
running outlet		0.84	7.57	29.57	nr	37.14
stop end outlet		0.42	3.78	38.08	nr	41.86
angle		0.84	7.57	22.22	nr	29.79
381 mm boundary wall gutters;						
screwed to timber		0.78	7.03	13.28	m	20.31
Extra for						
stop end		0.48	4.32	10.90	nr	15.22
running outlet		0.84	7.57	32.11	nr	39.68
stop end outlet		0.42	3.78	40.86	nr	44.64
angle		0.84	7.57	24.39	nr	31.96
457 mm boundary wall gutters;						
screwed to timber		0.90	8.11	15.00	m	23.11
Extra for						
stop end		0.42	3.78	13.16	nr	16.94
running outlet		0.96	8.65	33.03	nr	41.68
stop end outlet		0.48	4.32	42.64	nr	46.96
angle		0.96	8.65	26.78	nr	35.43

R DISPOSAL SYSTEMS Including overheads and profit at 7.50%		Labour hours	Labour £	Material £	Unit	Total rate £
R10 RAINWATER PIPEWORK/GUTTERS - cont'd						
uPVC external rainwater pipes and fittings; BS 4576; slip-in joints						
50 mm pipes; fixing with pipe or socket brackets; plugged and screwed	PC £3.23/2m	0.36	3.24	2.41	m	**5.65**
Extra for						
fittings with one end		0.24	2.16	1.25	nr	**3.41**
fittings with two ends		0.36	3.24	1.59	nr	**4.83**
fittings with three ends		0.48	4.32	2.14	nr	**6.46**
shoe	PC £0.95	0.24	2.16	1.25	nr	**3.41**
bend	PC £1.27	0.36	3.24	1.59	nr	**4.83**
single branch	PC £1.78	0.48	4.32	2.14	nr	**6.46**
two bends to form offset 229 mm projection	PC £2.22	0.36	3.24	2.02	nr	**5.26**
connection to clay pipes; cement and sand (1:2) joint		0.16	1.44	0.10	nr	**1.54**
68 mm pipes; fixing with pipe or socket brackets; plugged and screwed	PC £2.60/2m	0.40	3.60	2.19	m	**5.79**
Extra for						
shoe	PC £0.95	0.26	2.34	1.38	nr	**3.72**
bend	PC £1.67	0.40	3.60	2.16	nr	**5.76**
single branch	PC £2.92	0.53	4.77	3.50	nr	**8.27**
two bends to form offset 229 mm projection	PC £2.08	0.40	3.60	2.13	nr	**5.73**
loose drain connector; cement and sand (1:2) joint		0.18	1.62	0.10	nr	**1.72**
110 mm pipes; fixing with pipe or socket brackets; plugged and screwed	PC £8.42/3m	0.43	3.87	4.41	m	**8.28**
Extra for						
shoe	PC £3.56	0.29	2.61	4.20	nr	**6.81**
bend	PC £4.71	0.43	3.87	5.44	nr	**9.31**
single branch	PC £6.13	0.58	5.22	6.96	nr	**12.18**
two bends to form offset 229 mm projection	PC £9.42	0.43	3.87	9.23	nr	**13.10**
loose drain connector; cement and sand (1:2) joint		0.42	3.78	4.12	nr	**7.90**
68.5 mm square pipes; fixing with pipe brackets; plugged and screwed	PC £4.62/2.5m	0.40	3.60	2.83	m	**6.43**
Extra for						
shoe	PC £1.16	0.26	2.34	1.52	nr	**3.86**
bend	PC £1.18	0.40	3.60	1.54	nr	**5.14**
single branch	PC £3.02	0.53	4.77	3.52	nr	**8.29**
two bends to form offset 229 mm projection	PC £3.80	0.40	3.60	1.72	nr	**5.32**
drain connector; square to round; cement and sand (1:2) joint		0.24	2.16	3.02	nr	**5.18**
Rainwater head; rectangular; for pipes						
50 mm dia	PC £5.18	0.54	4.86	6.33	nr	**11.19**
68 mm dia	PC £4.19	0.56	5.04	5.48	nr	**10.52**
110 mm dia	PC £10.64	0.66	5.95	12.80	nr	**18.75**
68.5 mm square	PC £4.30	0.56	5.04	5.16	nr	**10.20**
uPVC gutters and fittings; BS 4576						
76 mm half round gutters; on brackets; screwed to timber	PC £2.60/2m	0.36	3.24	1.98	m	**5.22**
Extra for						
stop end	PC £0.43	0.16	1.44	0.59	nr	**2.03**
running outlet	PC £1.22	0.30	2.70	1.09	nr	**3.79**
stop end outlet	PC £1.22	0.16	1.44	1.21	nr	**2.65**
angle	PC £1.12	0.30	2.70	1.23	nr	**3.93**

R DISPOSAL SYSTEMS Including overheads and profit at 7.50%		Labour hours	Labour £	Material £	Unit	Total rate £
112 half round gutters; on brackets screwed to timber	PC £2.53/2m	0.40	3.60	2.42	m	**6.02**
Extra for						
stop end	PC £0.69	0.16	1.44	0.95	nr	**2.39**
running outlet	PC £1.34	0.34	3.06	1.23	nr	**4.29**
stop end outlet	PC £1.34	0.16	1.44	1.44	nr	**2.88**
angle	PC £1.51	0.34	3.06	1.83	nr	**4.89**
170 mm half round gutters; on brackets screwed to timber	PC £8.31/2m	0.43	3.87	7.38	m	**11.25**
Extra for						
stop end	PC £1.87	0.19	1.71	2.70	nr	**4.41**
running outlet	PC £4.16	0.37	3.33	3.76	nr	**7.09**
stop end outlet	PC £3.96	0.19	1.71	4.23	nr	**5.94**
angle	PC £5.42	0.37	3.33	6.49	nr	**9.82**
114 mm rectangular gutters; on brackets; screwed to timber	PC £3.24/2m	0.40	3.60	3.24	m	**6.84**
Extra for						
stop end	PC £0.70	0.16	1.44	1.02	nr	**2.46**
running outlet	PC £1.67	0.34	3.06	1.52	nr	**4.58**
stop end outlet	PC £1.77	0.16	1.44	1.89	nr	**3.33**
angle	PC £1.61	0.34	3.06	1.99	nr	**5.05**

ALTERNATIVE WASTE PIPE AND FITTING PRICES

ABS waste system (£/each)	£		£		£
4 m pipe -32 mm	2.62	access plug-32 mm	0.45	cn.to copper-32 mm	0.86
-40 mm	3.23	-40 mm	0.48	-40 mm	1.02
-50 mm	4.03	-50 mm	0.81	sweep bend -32 mm	0.57
pipe brkt-32 mm	0.12	reducer -32 mm	0.40	sweep bend -40 mm	0.69
-40 mm	0.13	-40 mm	0.45	-50 mm	1.04
-50 mm	0.24	-50 mm	0.59	sweep tee -32 mm	0.81
dl.socket-32 mm	0.36	str.tank cn-32 mm	0.98	-40 mm	0.99
-40 mm	0.42	-40 mm	1.11	-50 mm	1.43
-50 mm	0.69				

		Labour hours	Labour £	Material £	Unit	Total rate £
R11 FOUL DRAINAGE ABOVE GROUND						
Cast iron pipes and fittings; BS 416						
50 mm pipes						
primed, eared, plugged and nailed	PC £24.01/1.8m	0.90	8.11	16.44	m	**24.55**
primed, uneared; fixing with holderbats plugged and screwed; (PC £7.21/nr);	PC £22.56/1.8m	1.00	9.01	19.36	m	**28.37**
Extra for						
fittings with two ends		0.90	8.11	11.83	nr	**19.94**
fittings with three ends		1.20	10.81	17.73	nr	**28.54**
fittings with four ends		1.50	13.51	34.34	nr	**47.85**
bend; short radius	PC £10.97	0.90	8.11	11.83	nr	**19.94**
access bend; short radius	PC £27.21	0.90	8.11	29.28	nr	**37.39**
boss; 38 mm BSP	PC £26.49	0.90	8.11	23.18	nr	**31.29**
single branch	PC £16.42	1.20	10.81	17.73	nr	**28.54**
double branch	PC £32.11	1.50	13.51	34.34	nr	**47.85**
offset 229 mm projection	PC £17.40	0.90	8.11	18.74	nr	**26.85**
offset 305 mm projection	PC £19.89	0.90	8.11	21.42	nr	**29.53**

R DISPOSAL SYSTEMS Including overheads and profit at 7.50%		Labour hours	Labour £	Material £	Unit	Total rate £
R11 FOUL DRAINAGE ABOVE GROUND - cont'd						
Cast iron pipes and fittings; BS 416 - cont'd						
50 mm pipes						
access pipe	PC £23.09	0.90	8.11	19.52	nr	27.63
roof connector; for asphalt	PC £24.27	0.90	8.11	23.21	nr	31.32
roof connector; for roofing felt	PC £47.39	0.96	8.65	50.49	nr	59.14
isolated caulked lead joints		0.78	7.03	1.74	nr	8.77
connection to clay pipes; cement and sand (1:2) joint		0.16	1.44	0.10	nr	1.54
75 mm pipes						
primed, eared, plugged and nailed	PC £24.01/1.8m	0.96	8.65	16.66	m	25.31
primed, uneared; fixing with holderbats plugged and screwed; (PC £7.21/nr);	PC £22.56/1.8m	1.10	9.91	19.53	m	29.44
Extra for						
bend; short radius	PC £10.97	0.96	8.65	12.10	nr	20.75
access bend; short radius	PC £27.21	0.96	8.65	29.55	nr	38.20
boss; 38 mm BSP	PC £26.49	0.96	8.65	23.45	nr	32.10
single branch	PC £16.42	1.25	11.26	18.12	nr	29.38
double branch	PC £32.11	1.60	14.41	34.89	nr	49.30
offset 229 mm projection	PC £17.40	0.96	8.65	19.01	nr	27.66
offset 305 mm projection	PC £19.89	0.96	8.65	21.69	nr	30.34
access pipe	PC £23.09	0.96	8.65	19.80	nr	28.45
roof connector; for asphalt	PC £24.27	0.96	8.65	23.49	nr	32.14
roof connector; for roofing felt	PC £47.39	1.00	9.01	50.77	nr	59.78
isolated caulked lead joints		0.84	7.57	1.99	nr	9.56
connection to clay pipes; cement and sand (1:2) joint		0.20	1.80	0.10	nr	1.90
100 mm pipes						
primed, eared, plugged and nailed	PC £32.29/1.8m	1.10	9.91	22.45	m	32.36
primed; uneared; fixing with holderbats plugged and screwed; (PC £8.20/nr);	PC £30.94/1.8m	1.20	10.81	25.74	m	36.55
Extra for						
W.C. bent connector; 450 mm long tail	PC £26.71	0.84	7.57	24.41	nr	31.98
'Multikwik' W.C.connector	PC £2.24	0.12	1.08	2.40	nr	3.48
W.C. straight connector; 300 mm long tail	PC £11.82	0.84	7.57	13.06	nr	20.63
W.C. straight connector; 450 mm long tail	PC £14.73	0.84	7.57	16.19	nr	23.76
bend; short radius	PC £16.33	1.10	9.91	17.91	nr	27.82
access bend; short radius	PC £33.52	1.10	9.91	36.40	nr	46.31
boss; 38 mm BSP	PC £31.67	1.10	9.91	27.08	nr	36.99
single branch	PC £25.33	1.45	13.06	27.76	nr	40.82
double branch	PC £35.86	1.80	16.22	38.93	nr	55.15
offset 229 mm projection	PC £23.00	1.10	9.91	25.09	nr	35.00
offset 305 mm projection	PC £26.31	1.10	9.91	28.64	nr	38.55
access pipe	PC £25.33	1.10	9.91	20.27	nr	30.18
roof connector; for asphalt	PC £29.07	1.10	9.91	27.62	nr	37.53
roof connector; for roofing felt	PC £54.21	1.15	10.36	57.96	nr	68.32
isolated caulked lead joints		0.96	8.65	2.71	nr	11.36
connection to clay pipes; cement and sand (1:2) joint		0.24	2.16	0.10	nr	2.26

R DISPOSAL SYSTEMS Including overheads and profit at 7.50%		Labour hours	Labour £	Material £	Unit	Total rate £
150 mm pipes						
primed, eared, plugged and nailed	PC £65.46/1.8m	1.40	12.61	44.54	m	57.15
primed; uneared; fixing with holderbats plugged and screwed; (PC £14.09/nr);	PC £63.77/1.8m	1.50	13.51	50.74	m	64.25
Extra for						
bend; short radius	PC £29.27	1.40	12.61	31.14	nr	43.75
access bend; short radius	PC £42.08	1.40	12.61	44.91	nr	57.52
boss; 38 mm BSP	PC £53.78	1.40	12.61	42.42	nr	55.03
single branch	PC £45.32	1.85	16.67	48.24	nr	64.91
double branch	PC £65.22	2.30	20.72	68.79	nr	89.51
offset 229 mm projection	PC £47.34	1.40	12.61	50.57	nr	63.18
offset 305 mm projection	PC £53.41	1.40	12.61	57.09	nr	69.70
access pipe	PC £40.94	1.40	12.61	28.61	nr	41.22
roof connector; for asphalt	PC £61.98	1.40	12.61	58.08	nr	70.69
roof connector; for roofing felt	PC £90.93	1.45	13.06	96.06	nr	109.12
isolated caulked lead joints		1.25	11.26	4.55	nr	15.81
connection to clay pipes; cement and sand (1:2) joint		0.31	2.79	0.14	nr	2.93
Cut into existing soil stack; provide and insert new single branch						
50 mm	PC £19.76	3.00	27.03	47.71	nr	74.74
50 mm; with access door	PC £39.29	3.00	27.03	68.70	nr	95.73
75 mm	PC £19.76	3.00	27.03	47.38	nr	74.41
75 mm; with access door	PC £39.29	3.00	27.03	68.37	nr	95.40
100 mm	PC £30.48	3.85	34.68	69.32	nr	104.00
100 mm; with access door	PC £51.22	3.85	34.68	91.61	nr	126.29
150 mm	PC £54.55	4.70	42.34	127.20	nr	169.54
150 mm; with access door	PC £72.07	4.70	42.34	146.04	nr	188.38
Cast iron 'Timesaver' pipes and fittings BS 416						
50 mm pipes						
primed; 2 m lengths; fixing with holderbats plugged and screwed;(PC £5.03/nr)	PC £19.61/2m	0.66	5.95	16.07	m	22.02
Extra for						
fittings with two ends		0.66	5.95	11.01	nr	16.96
fittings with three ends		0.90	8.11	18.87	nr	26.98
fittings with four ends		1.15	10.36	30.75	nr	41.11
bend; short radius	PC £7.54	0.66	5.95	11.01	nr	16.96
access bend; short radius	PC £18.60	0.66	5.95	22.90	nr	28.85
boss; 38 mm BSP	PC £19.20	0.66	5.95	18.91	nr	24.86
single branch	PC £11.36	0.90	8.11	18.87	nr	26.98
double branch	PC £19.12	1.15	10.36	30.75	nr	41.11
offset 229 mm projection	PC £12.04	0.66	5.95	15.85	nr	21.80
offset 305 mm projection	PC £13.72	0.66	5.95	17.65	nr	23.60
access pipe	PC £18.15	0.66	5.95	17.78	nr	23.73
roof connector; for asphalt	PC £18.11	0.66	5.95	19.84	nr	25.79
roof connector; for roofing felt	PC £42.70	0.72	6.49	48.39	nr	54.88
isolated 'Timesaver' coupling joint	PC £4.27	0.36	3.24	4.59	nr	7.83
connection to clay pipes; cement and sand (1:2) joint	PC £61.85/M3	0.16	1.44	0.10	nr	1.54
75 mm pipes						
primed; 3 m lengths; fixing with holderbats plugged and screwed;(PC £5.03/nr)	PC £28.35/3m	0.66	5.95	14.45	m	20.40

R DISPOSAL SYSTEMS Including overheads and profit at 7.50%		Labour hours	Labour £	Material £	Unit	Total rate £
R11 FOUL DRAINAGE ABOVE GROUND - cont'd						
Cast iron 'Timesaver' pipes and fittings BS 416 - cont'd						
75 mm pipes						
primed; 2 m lengths; fixing with holderbats plugged and screwed;(PC £5.03/nr)	PC £19.61/2m	0.72	6.49	16.32	m	22.81
Extra for						
bend; short radius	PC £7.54	0.72	6.49	11.51	nr	18.00
access bend; short radius	PC £18.60	0.72	6.49	23.39	nr	29.88
boss; 38 mm BSP	PC £19.20	0.72	6.49	19.40	nr	25.89
single branch	PC £11.36	1.00	9.01	19.85	nr	28.86
double branch	PC £19.12	1.30	11.71	32.23	nr	43.94
offset 229 mm projection	PC £12.04	0.72	6.49	16.35	nr	22.84
offset 305 mm projection	PC £13.72	0.72	6.49	18.15	nr	24.64
access pipe	PC £18.15	0.72	6.49	18.27	nr	24.76
roof connector; for asphalt	PC £18.11	0.72	6.49	20.33	nr	26.82
roof connector; for roofing felt	PC £42.70	0.78	7.03	48.88	nr	55.91
isolated 'Timesaver' coupling joint	PC £4.73	0.42	3.78	5.09	nr	8.87
connection to clay pipes; cement and sand (1:2) joint		0.18	1.62	0.10	nr	1.72
100 mm pipes						
primed; 3 m lengths; fixing with holderbats plugged and screwed;(PC £5.49/nr)	PC £34.23/3m	0.72	6.49	17.40	m	23.89
primed; 2 m lengths; fixing with holderbats plugged and screwed; (PC £5.49/nr)	PC £23.64/2m	0.80	7.21	19.61	m	26.82
Extra for						
W.C. bent connector; 450 mm long tail	PC £24.65	0.72	6.49	31.10	nr	37.59
'Multikwik' W.C.connector	PC £2.24	0.12	1.08	2.40	nr	3.48
W.C. straight connector; 300 mm long tail	PC £10.92	0.72	6.49	16.35	nr	22.84
bend; short radius	PC £10.44	0.80	7.21	15.83	nr	23.04
access bend; short radius	PC £22.10	0.80	7.21	28.37	nr	35.58
boss; 38 mm BSP	PC £24.08	0.80	7.21	24.90	nr	32.11
single branch	PC £16.16	1.20	10.81	27.61	nr	38.42
double branch	PC £19.99	1.55	13.96	37.09	nr	51.05
offset 229 mm projection	PC £15.02	0.80	7.21	20.75	nr	27.96
offset 305 mm projection	PC £16.93	0.80	7.21	22.81	nr	30.02
access pipe	PC £19.12	0.80	7.21	19.57	nr	26.78
roof connector; for asphalt	PC £21.42	0.80	7.21	24.59	nr	31.80
roof connector; for roofing felt	PC £51.97	0.86	7.75	59.97	nr	67.72
isolated 'Timesaver' coupling joint	PC £6.18	0.50	4.50	6.64	nr	11.14
transitional clayware socket; cement and sand (1:2) joint		0.48	4.32	15.49	nr	19.81
150 mm pipes						
primed; 3 m lengths; fixing with holderbats plugged and screwed;(PC £9.44/nr)	PC £71.72/3m	0.90	8.11	35.49	m	43.60
primed; 2 m lengths; fixing with holderbats plugged and screwed;(PC £9.44/nr)	PC £48.53/2m	1.00	9.01	39.09	m	48.10
Extra for						
bend; short radius	PC £18.67	1.00	9.01	29.17	nr	38.18
access bend; short radius	PC £31.38	1.00	9.01	42.83	nr	51.84
boss; 38 mm BSP	PC £38.88	1.00	9.01	39.41	nr	48.42
single branch	PC £33.26	1.45	13.06	56.04	nr	69.10
double branch		1.90	17.12	85.82	nr	102.94

R DISPOSAL SYSTEMS Including overheads and profit at 7.50%		Labour hours	Labour £	Material £	Unit	Total rate £
offset 229 mm projection	PC £30.86	1.00	9.01	42.27	nr	**51.28**
offset 305 mm projection	PC £39.62	1.00	9.01	51.69	nr	**60.70**
roof connector; for asphalt	PC £39.70	1.00	9.01	45.51	nr	**54.52**
roof connector; for roofing felt	PC £79.56	1.05	9.46	93.58	nr	**103.04**
isolated 'Timesaver' coupling joint	PC £12.34	0.60	5.41	13.27	nr	**18.68**
transitional clayware socket; cement and sand (1:2) joint		0.62	5.59	17.63	nr	**23.22**
Polypropylene (PP) waste pipes and fittings; BS 5254; push fit 'O' - ring joints						
32 mm pipes; fixing with pipe clips; plugged and screwed	PC £1.60/4m	0.26	2.34	0.82	m	**3.16**
Extra for						
fittings with one end		0.19	1.71	0.48	nr	**2.19**
fittings with two ends		0.26	2.34	0.71	nr	**3.05**
fittings with three ends		0.36	3.24	1.03	nr	**4.27**
access plug	PC £0.45	0.19	1.71	0.48	nr	**2.19**
double socket	PC £0.49	0.18	1.62	0.53	nr	**2.15**
male iron to PP coupling	PC £0.83	0.34	3.06	0.89	nr	**3.95**
sweep bend	PC £0.66	0.26	2.34	0.71	nr	**3.05**
sweep tee	PC £0.96	0.36	3.24	1.03	nr	**4.27**
40 mm pipes; fixing with pipe clips; plugged and screwed	PC £2.01/4m	0.32	2.88	0.96	m	**3.84**
Extra for						
fittings with one end		0.22	1.98	0.52	nr	**2.50**
fittings with two ends		0.32	2.88	0.78	nr	**3.66**
fittings with three ends		0.43	3.87	1.20	nr	**5.07**
access plug	PC £0.48	0.22	1.98	0.52	nr	**2.50**
double socket	PC £0.53	0.22	1.98	0.57	nr	**2.55**
reducing set	PC £0.42	0.11	0.99	0.45	nr	**1.44**
male iron to PP coupling		0.41	3.69	0.98	nr	**4.67**
sweep bend	PC £0.72	0.32	2.88	0.78	nr	**3.66**
sweep tee	PC £1.11	0.43	3.87	1.20	nr	**5.07**
50 mm pipes; fixing with pipe clips; plugged and screwed	PC £2.88/4m	0.38	3.42	1.45	m	**4.87**
Extra for						
fittings with one end		0.24	2.16	0.87	nr	**3.03**
fittings with two ends		0.38	3.42	1.41	nr	**4.83**
fittings with three ends		0.50	4.50	1.72	nr	**6.22**
access plug	PC £0.81	0.24	2.16	0.87	nr	**3.03**
double socket	PC £0.99	0.25	2.25	1.07	nr	**3.32**
reducing set	PC £0.66	0.12	1.08	0.71	nr	**1.79**
male iron to PP coupling	PC £1.13	0.48	4.32	1.22	nr	**5.54**
sweep bend	PC £1.32	0.38	3.42	1.41	nr	**4.83**
sweep tee	PC £1.60	0.50	4.50	1.72	nr	**6.22**
muPVC waste pipes and fittings; BS 5255; solvent welded joints						
32 mm pipes; fixing with pipe clips; plugged and screwed	PC £5.03/4m	0.30	2.70	1.90	m	**4.60**
Extra for						
fittings with one end		0.20	1.80	1.17	nr	**2.97**
fittings with two ends		0.30	2.70	1.12	nr	**3.82**
fittings with three ends		0.40	3.60	1.62	nr	**5.22**
access plug	PC £0.99	0.20	1.80	1.17	nr	**2.97**
straight coupling	PC £0.65	0.20	1.80	0.80	nr	**2.60**
expansion coupling	PC £0.84	0.30	2.70	1.00	nr	**3.70**
male iron to PVC coupling	PC £0.82	0.37	3.33	0.93	nr	**4.26**
union coupling	PC £2.00	0.30	2.70	2.24	nr	**4.94**
sweep bend	PC £0.95	0.30	2.70	1.12	nr	**3.82**
spigot socket bend	PC £0.95	0.30	2.70	1.12	nr	**3.82**
sweep tee	PC £1.37	0.40	3.60	1.62	nr	**5.22**
caulking bush	PC £1.13	0.60	5.41	1.85	nr	**7.26**

R DISPOSAL SYSTEMS Including overheads and profit at 7.50%		Labour hours	Labour £	Material £	Unit	Total rate £
R11 FOUL DRAINAGE ABOVE GROUND - cont'd						
muPVC waste pipes and fittings; BS 5255; solvent welded joints - cont'd						
40 mm pipes; fixing with pipe clips; plugged and screwed	PC £6.16/4m	0.36	3.24	2.30	m	5.54
Extra for						
fittings with one end		0.23	2.07	1.36	nr	3.43
fittings with two ends		0.36	3.24	1.26	nr	4.50
fittings with three ends		0.48	4.32	1.97	nr	6.29
fittings with four ends		0.64	5.77	4.75	nr	10.52
access plug	PC £1.17	0.23	2.07	1.36	nr	3.43
straight coupling	PC £0.80	0.24	2.16	0.96	nr	3.12
expansion coupling	PC £1.02	0.36	3.24	1.19	nr	4.43
male iron to PVC coupling	PC £0.96	0.46	4.14	1.09	nr	5.23
union coupling	PC £2.63	0.36	3.24	2.92	nr	6.16
level invert taper	PC £0.82	0.36	3.24	0.98	nr	4.22
sweep bend	PC £1.08	0.36	3.24	1.26	nr	4.50
spigot socket bend	PC £1.06	0.36	3.24	1.24	nr	4.48
sweep tee	PC £1.69	0.48	4.32	1.97	nr	6.29
sweep cross	PC £4.24	0.64	5.77	4.75	nr	10.52
50 mm pipes; fixing with pipe clips; plugged and screwed	PC £9.06/4m	0.42	3.78	3.47	m	7.25
Extra for						
fittings with one end		0.25	2.25	2.15	nr	4.40
fittings with two ends		0.42	3.78	1.77	nr	5.55
fittings with three ends		0.56	5.04	3.33	nr	8.37
fittings with four ends		0.74	6.67	5.45	nr	12.12
access plug	PC £1.90	0.25	2.25	2.15	nr	4.40
straight coupling	PC £0.98	0.28	2.52	1.15	nr	3.67
expansion coupling	PC £1.38	0.42	3.78	1.58	nr	5.36
male iron to PVC coupling	PC £1.38	0.54	4.86	1.54	nr	6.40
union coupling	PC £3.89	0.42	3.78	4.28	nr	8.06
level invert taper	PC £1.09	0.42	3.78	1.27	nr	5.05
sweep bend	PC £1.55	0.42	3.78	1.77	nr	5.55
spigot socket bend	PC £2.51	0.42	3.78	2.80	nr	6.58
sweep tee	PC £2.96	0.56	5.04	3.33	nr	8.37
sweep cross	PC £4.88	0.74	6.67	5.45	nr	12.12
uPVC overflow pipes and fittings; solvent						
19 mm pipes; fixing with pipe clips; plugged and screwed	PC £2.13/4m	0.26	2.34	0.86	m	3.20
Extra for						
splay cut end		0.02	0.18	-	nr	0.18
fittings with one end		0.20	1.80	0.40	nr	2.20
fittings with two ends		0.20	1.80	0.56	nr	2.36
fittings with three ends		0.26	2.34	0.67	nr	3.01
straight connector	PC £0.33	0.20	1.80	0.40	nr	2.20
female iron to PVC coupling	PC £0.62	0.24	2.16	0.70	nr	2.86
bend	PC £0.48	0.20	1.80	0.56	nr	2.36
tee	PC £0.55	0.26	2.34	0.67	nr	3.01
bent tank connector	PC £0.76	0.24	2.16	0.84	nr	3.00
uPVC pipes and fittings; BS 4514; with solvent welded joints (unless otherwise described)						
82 mm pipes; fixing with holderbats; plugged and screwed (PC £1.22/nr)	PC £12.89/4m	0.48	4.32	4.97	m	9.29
Extra for						
socket plug	PC £2.23	0.24	2.16	2.69	nr	4.85
slip coupling (push-fit)	PC £4.32	0.44	3.96	4.94	nr	8.90
expansion coupling	PC £2.81	0.48	4.32	3.31	nr	7.63
sweep bend	PC £3.94	0.48	4.32	4.53	nr	8.85

R DISPOSAL SYSTEMS Including overheads and profit at 7.50%		Labour hours	Labour £	Material £	Unit	Total rate £
boss connector	PC £2.05	0.32	2.88	2.50	nr	5.38
single branch	PC £5.48	0.64	5.77	6.34	nr	12.11
access door	PC £4.76	0.72	6.49	5.26	nr	11.75
connection to clay pipes; caulking ring and cement and sand (1:2) joint		0.44	3.96	1.08	nr	5.04
110 mm pipes; fixing with holderbats; plugged and screwed (PC £1.26/nr)	PC £15.47/4m	0.53	4.77	5.84	m	10.61
Extra for						
socket plug	PC £2.70	0.26	2.34	3.27	nr	5.61
slip coupling (push-fit)	PC £5.40	0.48	4.32	6.17	nr	10.49
expansion coupling	PC £2.83	0.53	4.77	3.41	nr	8.18
WC connector	PC £3.86	0.35	3.15	4.33	nr	7.48
sweep bend	PC £5.40	0.53	4.77	6.17	nr	10.94
WC connecting bend	PC £5.75	0.35	3.15	6.36	nr	9.51
access bend	PC £12.10	0.55	4.95	13.38	nr	18.33
boss connector	PC £2.05	0.35	3.15	2.58	nr	5.73
single branch	PC £7.14	0.70	6.31	8.23	nr	14.54
single branch with access door	PC £13.83	0.72	6.49	15.43	nr	21.92
double branch	PC £17.40	0.88	7.93	19.44	nr	27.37
WC manifold	PC £24.18	0.35	3.15	26.55	nr	29.70
access door	PC £4.76	0.72	6.49	5.26	nr	11.75
access pipe connector	PC £8.52	0.60	5.41	9.53	nr	14.94
connection to clay pipes; caulking ring and cement and sand (1:2) joint		0.50	4.50	1.49	nr	5.99
160 mm pipes; fixing with holderbats; plugged and screwed (PC £3.04/nr)	PC £32.95/4m	0.60	5.41	13.11	m	18.52
Extra for						
socket plug	PC £4.98	0.30	2.70	6.15	nr	8.85
slip coupling (push-fit)	PC £13.82	0.54	4.86	15.65	nr	20.51
expansion coupling	PC £8.55	0.60	5.41	9.98	nr	15.39
sweep bend	PC £11.63	0.60	5.41	13.30	nr	18.71
boss connector	PC £2.81	0.40	3.60	3.81	nr	7.41
single branch	PC £16.38	0.79	7.12	18.81	nr	25.93
double branch	PC £29.00	1.00	9.01	32.76	nr	41.77
access door	PC £8.50	0.72	6.49	9.27	nr	15.76
connection to clay pipes; caulking ring and cement and sand (1:2) joint		0.60	5.41	2.49	nr	7.90
Weathering apron; for pipe						
82 mm dia	PC £1.11	0.41	3.69	1.34	nr	5.03
110 mm dia	PC £1.30	0.46	4.14	1.59	nr	5.73
160 mm dia	PC £3.91	0.50	4.50	4.60	nr	9.10
Weathering slate; for pipe						
110 mm dia	PC £18.20	1.10	9.91	19.75	nr	29.66
Vent cowl; for pipe						
82 mm dia	PC £1.11	0.40	3.60	1.34	nr	4.94
110 mm dia	PC £1.12	0.40	3.60	1.39	nr	4.99
160 mm dia	PC £2.91	0.40	3.60	3.53	nr	7.13
Roof outlets; circular PVC; with flat or domed grate; jointed to pipe						
82 mm dia	PC £13.82	0.66	5.95	15.00	nr	20.95
110 mm dia	PC £13.82	0.72	6.49	15.04	nr	21.53
Cut into existing soil stack; provide and insert new single branch						
82 mm	PC £6.66	1.00	9.01	7.60	nr	16.61
110 mm	PC £8.57	1.10	9.91	9.77	nr	19.68
110 mm; with access door	PC £16.80	1.10	9.91	18.61	nr	28.52
160 mm	PC £19.89	1.20	10.81	22.58	nr	33.39
Copper, brass and gunmetal ancillaries; screwed joints to fittings						
Brass trap; 'P'; 45 degree outlet; 38 mm seal						
35 mm	PC £12.47	0.58	5.22	13.40	nr	18.62

R DISPOSAL SYSTEMS Including overheads and profit at 7.50%		Labour hours	Labour £	Material £	Unit	Total rate £
R11 FOUL DRAINAGE ABOVE GROUND - cont'd						
Copper, brass and gunmetal ancillaries; screwed joints to fittings - cont'd						
Brass bath trap; 88.5 degree outlet; shallow seal						
42 mm	PC £9.34	0.65	5.86	10.04	nr	15.90
Copper trap 'P'; two piece						
35 mm with 38 mm seal	PC £6.37	0.58	5.22	6.85	nr	12.07
35 mm with 76 mm seal	PC £6.86	0.58	5.22	7.37	nr	12.59
42 mm with 38 mm seal	PC £9.45	0.65	5.86	10.16	nr	16.02
42 mm with 76 mm seal	PC £9.79	0.65	5.86	10.52	nr	16.38
Copper trap; 'S'; two piece						
35 mm with 38 mm seal	PC £6.86	0.58	5.22	7.37	nr	12.59
35 mm with 76 mm seal	PC £7.14	0.58	5.22	7.68	nr	12.90
42 mm with 38 mm seal	PC £9.83	0.65	5.86	10.57	nr	16.43
42 mm with 76 mm seal	PC £10.16	0.65	5.86	10.92	nr	16.78
Polypropylene ancillaries; screwed joint to waste fitting						
Tubular 'S' trap; bath; shallow seal						
40 mm	PC £7.34	0.66	5.95	7.89	nr	13.84
Trap 'P' two piece; 76 mm seal						
32 mm	PC £2.00	0.46	4.14	2.15	nr	6.29
40 mm	PC £2.31	0.54	4.86	2.48	nr	7.34
Trap 'S' two piece; 76 mm seal						
32 mm	PC £2.54	0.46	4.14	2.73	nr	6.87
40 mm	PC £2.98	0.54	4.86	3.21	nr	8.07
Bottle trap 'P'; 76 mm seal						
32 mm	PC £2.23	0.46	4.14	2.39	nr	6.53
40 mm	PC £2.67	0.54	4.86	2.87	nr	7.73
Bottle trap; 'S'; 76 mm seal						
32 mm	PC £2.69	0.46	4.14	2.89	nr	7.03
40 mm	PC £3.27	0.54	4.86	3.51	nr	8.37
R12 DRAINAGE BELOW GROUND						
Mechanical excavation of trenches to receive pipes; grading bottoms; earthwork support; filling with excavated material and compacting; disposal of surplus soil; spreading on site average 50 m						
Pipes not exceeding 200 mm; average depth						
0.50 m deep		0.23	1.38	1.10	m	2.48
0.75 m deep		0.35	2.10	1.83	m	3.93
1.00 m deep		0.69	4.15	3.44	m	7.59
1.25 m deep		1.05	6.31	4.70	m	11.01
1.50 m deep		1.30	7.81	6.12	m	13.93
1.75 m deep		1.60	9.61	7.82	m	17.43
2.00 m deep		1.90	11.42	8.95	m	20.37
2.25 m deep		2.35	14.12	11.08	m	25.20
2.50 m deep		2.75	16.53	12.81	m	29.34
2.75 m deep		3.05	18.33	14.33	m	32.66
3.00 m deep		3.35	20.13	15.85	m	35.98
3.25 m deep		3.60	21.63	16.83	m	38.46
3.50 m deep		3.85	23.14	17.80	m	40.94
Pipes; 225 mm; average depth						
0.50 m deep		0.23	1.38	1.10	m	2.48
0.75 m deep		0.35	2.10	1.83	m	3.93
1.00 m deep		0.69	4.15	3.44	m	7.59
1.25 m deep		1.05	6.31	4.70	m	11.01
1.50 m deep		1.30	7.81	6.12	m	13.93
1.75 m deep		1.60	9.61	7.82	m	17.43

R DISPOSAL SYSTEMS Including overheads and profit at 7.50%	Labour hours	Labour £	Material £	Unit	Total rate £
2.00 m deep	1.90	11.42	8.95	m	**20.37**
2.25 m deep	2.35	14.12	11.08	m	**25.20**
2.50 m deep	2.75	16.53	12.81	m	**29.34**
2.75 m deep	3.05	18.33	14.33	m	**32.66**
3.00 m deep	3.35	20.13	15.85	m	**35.98**
3.25 m deep	3.60	21.63	16.83	m	**38.46**
3.50 m deep	3.85	23.14	17.80	m	**40.94**
Pipes; 300 mm; average depth					
0.75 m deep	0.39	2.34	2.01	m	**4.35**
1.00 m deep	0.81	4.87	3.62	m	**8.49**
1.25 m deep	1.10	6.61	4.88	m	**11.49**
1.50 m deep	1.45	8.71	6.48	m	**15.19**
1.75 m deep	1.65	9.92	8.19	m	**18.11**
2.00 m deep	1.90	11.42	9.50	m	**20.92**
2.25 m deep	2.35	14.12	11.44	m	**25.56**
2.50 m deep	2.75	16.53	13.18	m	**29.71**
2.75 m deep	3.05	18.33	14.70	m	**33.03**
3.00 m deep	3.35	20.13	16.22	m	**36.35**
3.25 m deep	3.60	21.63	17.56	m	**39.19**
3.50 m deep	3.85	23.14	18.17	m	**41.31**
Extra for breaking up					
brick	1.60	9.61	6.48	m3	**16.09**
concrete	2.25	13.52	8.88	m3	**22.40**
reinforced concrete	3.20	19.23	12.95	m3	**32.18**
concrete 150 mm thick	0.35	2.10	1.48	m2	**3.58**
tarmacadam 75 mm thick	0.17	1.02	0.77	m2	**1.79**
tarmacadam and hardcore 150 mm thick	0.23	1.38	1.07	m2	**2.45**
Hand excavation of trenches to receive pipes; grading bottoms; earthwork support; filling with excavated material and compacting; disposal of surplus soil; spreading on site average 50 m					
Pipes not exceeding 200 mm; average depth					
0.50 m deep	1.10	6.61	-	m	**6.61**
0.75 m deep	1.65	9.92	-	m	**9.92**
1.00 m deep	2.40	14.42	1.06	m	**15.48**
1.25 m deep	3.40	20.43	1.59	m	**22.02**
1.50 m deep	4.70	28.24	1.91	m	**30.15**
1.75 m deep	6.15	36.96	2.34	m	**39.30**
2.00 m deep	7.05	42.37	2.55	m	**44.92**
2.25 m deep	8.80	52.88	3.40	m	**56.28**
2.50 m deep	10.56	63.46	4.04	m	**67.50**
2.75 m deep	11.60	69.71	4.46	m	**74.17**
3.00 m deep	12.70	76.32	4.89	m	**81.21**
3.25 m deep	13.70	82.33	5.31	m	**87.64**
3.50 m deep	14.70	88.34	5.74	m	**94.08**
Pipes; 225 mm; average depth					
0.50 m deep	1.10	6.61	-	m	**6.61**
0.75 m deep	1.65	9.92	-	m	**9.92**
1.00 m deep	2.40	14.42	1.06	m	**15.48**
1.25 m deep	3.40	20.43	1.59	m	**22.02**
1.50 m deep	4.70	28.24	1.91	m	**30.15**
1.75 m deep	6.15	36.96	2.34	m	**39.30**
2.00 m deep	7.05	42.37	2.55	m	**44.92**
2.25 m deep	8.80	52.88	3.40	m	**56.28**
2.50 m deep	10.56	63.46	4.04	m	**67.50**
2.75 m deep	11.60	69.71	4.46	m	**74.17**
3.00 m deep	12.70	76.32	4.89	m	**81.21**
3.25 m deep	13.70	82.33	5.31	m	**87.64**
3.50 m deep	14.70	88.34	5.74	m	**94.08**

R DISPOSAL SYSTEMS Including overheads and profit at 7.50%	Labour hours	Labour £	Material £	Unit	Total rate £
R12 DRAINAGE BELOW GROUND - cont'd					
Hand excavation of trenches to receive pipes; grading bottoms; earthwork support; filling with excavated material and compacting; disposal of surplus soil; spreading on site average 50 m - cont'd					
Pipes; 300 mm; average depth					
0.75 m deep	1.95	11.72	-	m	11.72
1.00 m deep	2.80	16.83	1.06	m	17.89
1.25 m deep	3.95	23.74	1.59	m	25.33
1.50 m deep	5.30	31.85	1.91	m	33.76
1.75 m deep	6.15	36.96	2.34	m	39.30
2.00 m deep	7.05	42.37	2.55	m	44.92
2.25 m deep	8.80	52.88	3.40	m	56.28
2.50 m deep	10.56	63.46	4.04	m	67.50
2.75 m deep	11.60	69.71	4.46	m	74.17
3.00 m deep	12.70	76.32	4.89	m	81.21
3.25 m deep	13.70	82.33	5.31	m	87.64
3.50 m deep	14.70	88.34	5.74	m	94.08
Extra for breaking up					
brick	3.30	19.83	3.39	m3	23.22
concrete	4.95	29.75	5.64	m3	35.39
reinforced concrete	6.60	39.66	7.90	m3	47.56
concrete 150 mm thick	0.77	4.63	0.79	m2	5.42
tarmacadam 75 mm thick	0.44	2.64	0.45	m2	3.09
tarmacadam and hardcore 150 mm thick	0.55	3.31	0.56	m2	3.87
Extra for taking up precast concrete paving slabs	0.33	1.98	-	m2	1.98
Sand filling; PC £9.40/t; (PC £15.00/m3)					
Beds; to receive pitch fibre pipes					
600 x 50 mm	0.09	0.54	0.61	m	1.15
700 x 50 mm	0.11	0.66	0.71	m	1.37
800 x 50 mm	0.13	0.78	0.81	m	1.59
Granular (shingle) filling; PC £10.40/t; (PC £18.70/m3)					
Beds; 100 mm thick; to pipes size					
100 mm	0.11	0.66	1.21	m	1.87
150 mm	0.11	0.66	1.41	m	2.07
225 mm	0.13	0.78	1.61	m	2.39
300 mm	0.15	0.90	1.81	m	2.71
Beds; 150 mm thick; to pipes size					
100 mm	0.15	0.90	1.81	m	2.71
150 mm	0.18	1.08	2.01	m	3.09
225 mm	0.20	1.20	2.21	m	3.41
300 mm	0.22	1.32	2.41	m	3.73
Beds and benchings; beds 100 mm thick; to pipes size					
100 mm	0.25	1.50	2.21	m	3.71
150 mm	0.28	1.68	2.21	m	3.89
225 mm	0.33	1.98	3.02	m	5.00
300 mm	0.39	2.34	3.42	m	5.76
Beds and benchings; beds 150 mm thick; to pipes size					
100 mm	0.28	1.68	2.41	m	4.09
150 mm	0.31	1.86	2.61	m	4.47
225 mm	0.39	2.34	3.62	m	5.96
300 mm	0.50	3.00	4.42	m	7.42

R DISPOSAL SYSTEMS Including overheads and profit at 7.50%	Labour hours	Labour £	Material £	Unit	Total rate £
Beds and coverings; 100 mm thick; to pipes size					
100 mm	0.40	2.40	3.02	m	5.42
150 mm	0.50	3.00	3.62	m	6.62
225 mm	0.66	3.97	5.03	m	9.00
300 mm	0.79	4.75	6.03	m	10.78
Beds and coverings; 150 mm thick; to pipes size					
100 mm	0.59	3.55	4.42	m	7.97
150 mm	0.66	3.97	5.03	m	9.00
225 mm	0.86	5.17	6.43	m	11.60
300 mm	1.00	6.01	7.64	m	13.65
In situ ready mixed concrete; normal Portland cement; mix 11.50 N/mm2 - 40 mm aggregate (1:3:6); PC £41.49/m3;					
Beds; 100 mm thick; to pipes size					
100 mm	0.23	1.43	2.23	m	3.66
150 mm	0.23	1.43	2.23	m	3.66
225 mm	0.28	1.75	2.68	m	4.43
300 mm	0.32	2.00	3.12	m	5.12
Beds; 150 mm thick; to pipes size					
100 mm	0.32	2.00	3.12	m	5.12
150 mm	0.37	2.31	3.57	m	5.88
225 mm	0.41	2.56	4.01	m	6.57
300 mm	0.46	2.87	4.46	m	7.33
Beds and benchings; beds 100 mm thick; to pipes size					
100 mm	0.46	2.87	4.01	m	6.88
150 mm	0.52	3.24	4.46	m	7.70
225 mm	0.62	3.87	5.35	m	9.22
300 mm	0.72	4.49	6.24	m	10.73
Beds and benchings; beds 150 mm thick; to pipes size					
100 mm	0.52	3.24	4.46	m	7.70
150 mm	0.58	3.62	4.91	m	8.53
225 mm	0.72	4.49	6.24	m	10.73
300 mm	0.93	5.80	8.03	m	13.83
Beds and coverings; 100 mm thick; to pipes size					
100 mm	0.69	4.30	5.35	m	9.65
150 mm	0.81	5.05	6.24	m	11.29
225 mm	1.15	7.17	8.92	m	16.09
300 mm	1.40	8.73	10.70	m	19.43
Beds and coverings; 150 mm thick; to pipes size					
100 mm	1.05	6.55	8.03	m	14.58
150 mm	1.15	7.17	8.92	m	16.09
225 mm	1.50	9.35	11.60	m	20.95
300 mm	1.80	11.22	13.83	m	25.05
In situ ready mixed concrete; normal Portland cement; mix 21.00 N/mm2 - 40 mm aggregate (1:2:4); PC £44.19/m3					
Beds; 100 mm thick; to pipes size					
100 mm	0.23	1.43	2.38	m	3.81
150 mm	0.23	1.43	2.38	m	3.81
225 mm	0.28	1.75	2.85	m	4.60
300 mm	0.32	2.00	3.33	m	5.33
Beds; 150 mm thick; to pipes size					
100 mm	0.32	2.00	3.33	m	5.33
150 mm	0.37	2.31	3.80	m	6.11
225 mm	0.41	2.56	4.28	m	6.84
300 mm	0.46	2.87	4.75	m	7.62

R DISPOSAL SYSTEMS Including overheads and profit at 7.50%	Labour hours	Labour £	Material £	Unit	Total rate £
R12 DRAINAGE BELOW GROUND - cont'd					
In situ ready mixed concrete; normal Portland cement; mix 21.00 N/mm2 - 40 mm aggregate (1:2:4); PC £44.19/m3 - cont'd					
Beds and benchings; beds 100 mm thick; to pipes size					
100 mm	0.46	2.87	4.28	m	7.15
150 mm	0.52	3.24	4.75	m	7.99
225 mm	0.62	3.87	5.70	m	9.57
300 mm	0.72	4.49	6.65	m	11.14
Beds and benchings; beds 150 mm thick; to pipes size					
100 mm	0.52	3.24	4.75	m	7.99
150 mm	0.58	3.62	5.23	m	8.85
225 mm	0.72	4.49	6.65	m	11.14
300 mm	0.93	5.80	8.55	m	14.35
Beds and coverings; 100 mm thick; to pipes size					
100 mm	0.69	4.30	5.70	m	10.00
150 mm	0.81	5.05	6.65	m	11.70
225 mm	1.15	7.17	9.50	m	16.67
300 mm	1.40	8.73	11.40	m	20.13
Beds and coverings; 150 mm thick; to pipes size					
100 mm	1.05	6.55	8.55	m	15.10
150 mm	1.15	7.17	9.50	m	16.67
225 mm	1.50	9.35	12.35	m	21.70
300 mm	1.80	11.22	14.73	m	25.95

NOTE: The following items unless otherwise described include for all appropriate joints/couplings in the running length
The prices for gullies and rainwater shoes, etc., include for appropriate joints to pipes and for setting on and surrounding accessory with site mixed in situ concrete 11.50 N/mm2-40 aggregate (1:3:6)

	Labour hours	Labour £	Material £	Unit	Total rate £
Cast iron drain pipes and fittings; BS 437; coated; with caulked lead joints					
75 mm pipes					
laid straight; grey iron PC £62.56/3m	0.69	4.66	23.95	m	28.61
laid straight; grey iron PC £40.70/1.8m	0.81	5.47	26.32	m	31.79
in runs not exceeding 3 m long	1.10	7.43	40.07	m	47.50
Extra for					
bend; short radius	0.81	5.47	16.95	nr	22.42
bend; short radius; with round access door	0.81	5.47	28.44	nr	33.91
bend; long radius	0.81	5.47	23.08	nr	28.55
level invert taper	0.81	5.47	17.93	nr	23.40
access pipe	0.81	5.47	58.42	nr	63.89
single branch	1.10	7.43	31.11	nr	38.54
100 mm pipes					
laid straight; grey iron PC £65.93/3m	0.81	5.47	25.41	m	30.88
laid straight; grey iron PC £41.63/1.8m	0.92	6.21	27.27	m	33.48
in runs not exceeding 3 m long PC £33.60/.91m	1.25	8.44	44.34	m	52.78
in ducts; supported on piers (measured elsewhere)	1.60	10.80	25.78	m	36.58
supported on wall brackets (measured elsewhere)	1.30	8.78	25.78	m	34.56

R DISPOSAL SYSTEMS Including overheads and profit at 7.50%		Labour hours	Labour £	Material £	Unit	Total rate £
supported on ceiling hangers (measured elsewhere)		1.60	10.80	25.78	m	36.58
Extra for						
bend; short radius	PC £20.42	0.92	6.21	22.94	nr	29.15
bend; short radius; with round access door	PC £30.64	0.92	6.21	34.47	nr	40.68
bend; long radius	PC £31.83	0.92	6.21	34.48	nr	40.69
bend; long radius; with rectangular access door	PC £59.37	0.92	6.21	65.56	nr	71.77
rest bend	PC £25.90	0.92	6.21	29.13	nr	35.34
level invert taper	PC £20.67	0.92	6.21	21.89	nr	28.10
access pipe	PC £58.53	0.92	6.21	64.62	nr	70.83
single branch	PC £32.30	1.25	8.44	34.88	nr	43.32
single branch; with rectangular access door	PC £74.56	1.25	8.44	82.58	nr	91.02
'Y' branch	PC £73.43	1.25	8.44	81.30	nr	89.74
double branch	PC £88.70	1.55	10.46	98.57	nr	109.03
double branch; with rectangular access door	PC £135.11	1.55	10.46	150.95	nr	161.41
WC connector; 450 mm long	PC £15.63	0.69	4.66	14.26	nr	18.92
WC connector; 600 mm long	PC £31.02	0.69	4.66	29.40	nr	34.06
SW connector	PC £17.20	0.92	6.21	17.41	nr	23.62
150 mm pipes						
laid straight; grey iron	PC £113.24/3m	1.05	7.09	43.60	m	50.69
laid straight; grey iron	PC £63.73/1.8m	1.15	7.76	41.92	m	49.68
in runs not exceeding 3 m long	PC £50.39/.91m	1.55	10.46	67.01	m	77.47
Extra for						
bend; short radius	PC £44.03	1.15	7.76	49.43	nr	57.19
bend; short radius; with round access door	PC £58.42	1.15	7.76	65.66	nr	73.42
bend; long radius	PC £54.66	1.15	7.76	59.13	nr	66.89
bend; long radius; with rectangular access door	PC £106.79	1.15	7.76	117.96	nr	125.72
rest bend	PC £57.96	1.15	7.76	65.14	nr	72.90
level invert taper	PC £31.83	1.15	7.76	33.36	nr	41.12
access pipe	PC £108.42	1.15	7.76	119.81	nr	127.57
single branch	PC £70.18	1.55	10.46	76.37	nr	86.83
single branch; with rectangular access door	PC £146.91	1.55	10.46	162.97	nr	173.43
double branch	PC £137.86	1.95	13.16	152.75	nr	165.91
SW connector	PC £27.88	1.15	7.76	28.02	nr	35.78
225 mm pipes						
laid straight; grey iron	PC £319.68/1.8m	1.85	12.49	299.60	m	312.09
in runs not exceeding 3 m long	PC £184.92/.91m	2.45	16.54	235.97	m	252.51
Extra for						
bend; short radius	PC £168.57	1.85	12.49	197.03	nr	209.52
bend; long radius; with rectangular access door	PC £309.04	1.85	12.49	355.59	nr	368.08
rest bend	PC £202.83	1.85	12.49	235.71	nr	248.20
level invert taper	PC £59.07	1.85	12.49	73.43	nr	85.92
access pipe	PC £227.58	1.85	12.49	263.64	nr	276.13
single branch	PC £219.36	2.45	16.54	257.75	nr	274.29
single branch; with rectangular access door	PC £289.23	2.45	16.54	336.60	nr	353.14
SW connector		1.85	12.49	113.25	nr	125.74
Accessories in cast iron; with caulked lead joints to pipes (unless otherwise described)						
Rainwater shoes; horizontal inlet						
100 mm	PC £50.22	0.86	5.81	57.68	nr	63.49
150 mm	PC £80.30	0.98	6.62	92.34	nr	98.96

R DISPOSAL SYSTEMS Including overheads and profit at 7.50%		Labour hours	Labour £	Material £	Unit	Total rate £
R12 DRAINAGE BELOW GROUND - cont'd						
Accessories in cast iron; with caulked lead joints to pipes (unless otherwise described) - cont'd						
Gully fittings; comprising low invert gully trap and round hopper						
75 mm outlet	PC £18.96	1.25	8.44	25.09	nr	33.53
100 mm outlet	PC £31.83	1.40	9.45	39.61	nr	49.06
150 mm outlet	PC £80.74	1.85	12.49	94.50	nr	106.99
Add to above for						
bellmouth 300 mm high; circular plain grating						
75 mm; 175 mm grating	PC £18.12	0.69	4.66	21.18	nr	25.84
100 mm; 200 mm grating	PC £26.50	0.81	5.47	30.83	nr	36.30
150 mm; 250 mm grating	PC £38.45	1.15	7.76	45.65	nr	53.41
bellmouth 300 mm high; circular plain grating as above; one horizontal inlet						
75 x 50 mm	PC £34.99	0.69	4.66	39.51	nr	44.17
100 x 75 mm	PC £34.09	0.81	5.47	39.21	nr	44.68
150 x 100 mm	PC £72.19	1.15	7.76	81.93	nr	89.69
bellmouth 300 mm high; circular plain grating as above; one vertical inlet						
75 x 50 mm	PC £32.88	0.69	4.66	37.24	nr	41.90
100 x 75 mm	PC £38.82	0.81	5.47	44.30	nr	49.77
150 x 100 mm	PC £82.19	1.15	7.76	92.68	nr	100.44
raising piece; 200 mm dia						
75 mm high	PC £15.14	1.40	9.45	22.23	nr	31.68
150 mm high	PC £18.14	1.40	9.45	25.44	nr	34.89
225 mm high	PC £23.04	1.40	9.45	30.71	nr	40.16
300 mm high		1.40	9.45	34.98	nr	44.43
Yard gully (Deans); trapped; galvanized sediment pan; 267 mm round heavy grating						
100 mm outlet	PC £166.68	3.70	24.98	185.13	nr	210.11
Yard gully (garage); trapless; galvanized sediment pan; 267 mm round heavy grating						
100 mm outlet	PC £147.87	3.45	23.29	164.92	nr	188.21
Yard gully (garage); trapped; with rodding eye; galvanized perforated sediment pan; stopper; round heavy grating						
150 mm outlet; 347 mm grating	PC £542.48	4.60	31.05	592.00	nr	623.05
Grease trap; with internal access; insert galvanized perforated bucket; lid and frame						
450 x 300 x 525 mm deep; 100 mm outlet	PC £348.45	4.15	28.02	382.79	nr	410.81
Disconnecting trap; trapped with inspection arm; bridle plate and screw; building in and cutting and fitting brickwork around						
100 mm outlet; 100 mm inlet	PC £142.17	2.90	19.58	159.92	nr	179.50
150 mm outlet; 150 mm inlet	PC £212.12	3.70	24.98	239.11	nr	264.09
Cast iron 'Timesaver' drain pipes and fittings; BS 437; coated; with mechanical coupling joints						
75 mm pipes						
laid straight	PC £47.27/3m	0.52	3.51	21.15	m	24.66
in runs not exceeding 3 m long	PC £29.50/1m	0.69	4.66	42.14	m	46.80
Extra for						
bend; medium radius	PC £14.78	0.58	3.92	22.87	nr	26.79
single branch	PC £20.55	0.81	5.47	36.64	nr	42.11
isolated 'Timesaver' joint	PC £8.24	0.35	2.36	9.29	nr	11.65

R DISPOSAL SYSTEMS Including overheads and profit at 7.50%		Labour hours	Labour £	Material £	Unit	Total rate £
100 mm pipes						
laid straight	PC £45.69/3m	0.58	3.92	21.10	m	**25.02**
in runs not exceeding 3 m long	PC £28.89/1m	0.78	5.27	43.07	m	**48.34**
Extra for						
bend; medium radius	PC £18.20	0.69	4.66	28.44	nr	**33.10**
bend; medium radius with access	PC £47.84	0.69	4.66	61.89	nr	**66.55**
bend; long radius	PC £22.32	0.69	4.66	32.05	nr	**36.71**
rest bend	PC £20.90	0.69	4.66	31.48	nr	**36.14**
diminishing pipe	PC £15.49	0.69	4.66	26.92	nr	**31.58**
single branch	PC £24.18	0.86	5.81	44.10	nr	**49.91**
single branch; with access	PC £55.74	0.98	6.62	69.25	nr	**75.87**
double branch	PC £36.40	1.10	7.43	67.34	nr	**74.77**
double branch; with access	PC £67.24	1.10	7.43	108.79	nr	**116.22**
isolated 'Timesaver' joint	PC £9.74	0.40	2.70	10.99	nr	**13.69**
transitional pipe; for WC	PC £13.87	0.58	3.92	23.55	nr	**27.47**
150 mm pipes						
laid straight	PC £86.04/3m	0.69	4.66	37.33	m	**41.99**
in runs not exceeding 3 m long	PC £53.72/1m	0.94	6.35	73.33	m	**79.68**
Extra for						
bend; medium radius	PC £36.75	0.81	5.47	49.30	nr	**54.77**
bend; medium radius with access	PC £88.84	0.81	5.47	108.11	nr	**113.58**
bend; long radius	PC £47.84	0.81	5.47	59.89	nr	**65.36**
diminishing pipe	PC £23.74	0.81	5.47	37.54	nr	**43.01**
single branch	PC £52.31	0.98	6.62	76.64	nr	**83.26**
isolated 'Timesaver' joint	PC £11.80	0.48	3.24	13.32	nr	**16.56**
Accessories in 'Timesaver' cast iron; with mechanical coupling joints						
Rainwater shoes; horizontal inlet						
100 mm	PC £50.22	0.58	3.92	65.58	nr	**69.50**
150 mm	PC £80.30	0.69	4.66	100.70	nr	**105.36**
Gully fittings; comprising low invert gully trap and round hopper						
75 mm outlet	PC £15.49	1.05	7.09	28.32	nr	**35.41**
100 mm outlet	PC £24.18	1.10	7.43	39.28	nr	**46.71**
150 mm outlet	PC £60.14	1.50	10.13	80.71	nr	**90.84**
Add to above for						
bellmouth 300 mm high; circular plain grating						
100 mm; 200 mm grating	PC £26.50	0.52	3.51	38.95	nr	**42.46**
bellmouth 300 mm high; circular plain grating as above; one horizontal inlet						
100 x 100 mm	PC £34.09	0.52	3.51	47.11	nr	**50.62**
bellmouth 300 mm high; circular plain grating as above; one vertical inlet						
100 x 100 mm	PC £36.13	0.52	3.51	52.20	nr	**55.71**
Yard gully (Deans); trapped; galvanized sediment pan; 267 mm round heavy grating						
100 mm outlet	PC £184.41	3.35	22.62	201.62	nr	**224.24**
Yard gully (garage); trapless; galvanized sediment pan; 267 mm round heavy grating						
100 mm outlet	PC £157.61	3.10	20.93	172.81	nr	**193.74**
Yard gully (garage); trapped; with rodding eye, galvanised perforated sediment pan; stopper; 267 mm round heavy grating						
100 mm outlet; 267 mm grating	PC £350.40	3.45	23.29	380.63	nr	**403.92**
Grease trap; internal access; galvanized perforated bucket; lid and frame						
20 gal. capacity	PC £664.15	4.60	31.05	566.07	nr	**597.12**

R DISPOSAL SYSTEMS Including overheads and profit at 7.50%		Labour hours	Labour £	Material £	Unit	Total rate £
R12 DRAINAGE BELOW GROUND - cont'd						
Extra strength vitrified clay pipes and fittings; 'Hepworths' 'SuperSleeve'/'HepSleve' or similar; plain ends with push-fit polypropylene flexible couplings						
100 mm pipes						
laid straight	PC £2.71	0.29	1.63	3.06	m	4.69
in runs not exceeding 3 m long		0.38	2.14	3.56	m	5.70
Extra for						
bend	PC £2.30	0.23	1.29	4.39	nr	5.68
access bend	PC £14.52	0.23	1.29	18.18	nr	19.47
rest bend	PC £3.81	0.23	1.29	6.08	nr	7.37
access pipe	PC £12.58	0.23	1.29	15.77	nr	17.06
socket adaptor	PC £2.31	0.20	1.12	3.62	nr	4.74
adaptor to flexible pipe	PC £2.31	0.20	1.12	3.62	nr	4.74
saddle	PC £4.75	0.86	4.84	6.60	nr	11.44
single junction	PC £4.84	0.29	1.63	8.25	nr	9.88
single access junction	PC £16.83	0.29	1.63	21.79	nr	23.42
150 mm pipes						
laid straight	PC £6.17	0.35	1.97	6.97	m	8.94
in runs not exceeding 3 m long		0.46	2.59	8.15	m	10.74
Extra for						
bend	PC £5.92	0.28	1.57	10.86	nr	12.43
access bend	PC £24.90	0.28	1.57	32.60	nr	34.17
rest bend	PC £7.59	0.28	1.57	13.05	nr	14.62
taper pipe	PC £5.59	0.28	1.57	9.01	nr	10.58
access pipe	PC £21.32	0.28	1.57	28.03	nr	29.60
socket adaptor	PC £5.71	0.23	1.29	8.94	nr	10.23
adaptor to flexible pipe	PC £4.07	0.23	1.29	7.09	nr	8.38
saddle	PC £8.66	1.05	5.90	12.78	nr	18.68
single junction	PC £8.68	0.35	1.97	16.06	nr	18.03
single access junction	PC £31.03	0.35	1.97	42.01	nr	43.98
Extra strength vitrified clay pipes and fittings; 'Hepworths' 'HepSeal'; or similar; socketted; with push-fit flexible joints						
100 mm pipes						
laid straight	PC £5.46	0.35	1.97	6.16	m	8.13
in runs not exceeding 3 m long		0.46	2.59	6.75	m	9.34
Extra for						
bend	PC £7.03	0.28	1.57	6.09	nr	7.66
access bend	PC £17.88	0.28	1.57	18.33	nr	19.90
rest bend	PC £8.93	0.28	1.57	8.23	nr	9.80
stopper	PC £2.84	0.17	0.96	3.20	nr	4.16
access pipe	PC £16.18	0.28	1.57	15.80	nr	17.37
socket adaptor	PC £2.31	0.30	1.69	4.87	nr	6.56
saddle	PC £6.69	0.86	4.84	7.54	nr	12.38
single junction	PC £10.45	0.35	1.97	9.33	nr	11.30
single access junction	PC £19.54	0.35	1.97	19.59	nr	21.56
double junction	PC £19.91	0.52	2.92	19.40	nr	22.32
double collar	PC £6.69	0.23	1.29	7.54	nr	8.83
150 mm pipes						
laid straight	PC £7.06	0.40	2.25	7.97	m	10.22
in runs not exceeding 3 m long		0.54	3.04	8.73	m	11.77
Extra for						
bend	PC £12.14	0.32	1.80	11.31	nr	13.11
access bend	PC £27.91	0.32	1.80	29.11	nr	30.91
rest bend	PC £14.80	0.32	1.80	14.30	nr	16.10
stopper	PC £4.22	0.21	1.18	4.77	nr	5.95
taper reducer	PC £18.59	0.32	1.80	18.59	nr	20.39
access pipe	PC £25.37	0.32	1.80	25.45	nr	27.25

R DISPOSAL SYSTEMS Including overheads and profit at 7.50%		Labour hours	Labour £	Material £	Unit	Total rate £
socket adaptor	PC £5.71	0.35	1.97	10.30	nr	12.27
saddle	PC £11.80	1.05	5.90	13.32	nr	19.22
single access junction	PC £33.61	0.40	2.25	34.74	nr	36.99
double junction	PC £33.13	0.61	3.43	33.41	nr	36.84
double collar	PC £11.04	0.26	1.46	12.46	nr	13.92
225 mm pipes						
laid straight	PC £13.83	0.52	2.92	15.61	m	18.53
in runs not exceeding 3 m long		0.69	3.88	17.09	m	20.97
Extra for						
bend	PC £25.43	0.41	2.31	24.02	nr	26.33
access bend	PC £64.24	0.41	2.31	67.82	nr	70.13
rest bend	PC £35.42	0.41	2.31	35.29	nr	37.60
stopper	PC £9.16	0.26	1.46	10.34	nr	11.80
taper reducer	PC £29.40	0.41	2.31	28.50	nr	30.81
access pipe	PC £64.24	0.41	2.31	66.27	nr	68.58
saddle	PC £29.40	1.40	7.87	33.19	nr	41.06
single junction	PC £38.23	0.52	2.92	36.90	nr	39.82
single access junction	PC £76.81	0.52	2.92	80.46	nr	83.38
double junction	PC £74.29	0.78	4.39	76.06	nr	80.45
double collar	PC £24.24	0.35	1.97	27.36	nr	29.33
Extra strength vitrified clay pipes and fittings 'HepSeal' or similar						
300 mm pipes						
laid straight	PC £21.65	0.69	3.88	24.43	m	28.31
in runs not exceeding 3 m long		0.92	5.17	26.77	m	31.94
Extra for						
bend	PC £50.16	0.55	3.09	49.28	nr	52.37
access bend	PC £106.85	0.55	3.09	113.28	nr	116.37
rest bend	PC £75.13	0.55	3.09	77.47	nr	80.56
stopper	PC £20.19	0.35	1.97	22.78	nr	24.75
taper reducer	PC £60.14	0.55	3.09	60.55	nr	63.64
access pipe	PC £106.85	0.55	3.09	110.84	nr	113.93
saddle	PC £60.14	1.85	10.40	67.88	nr	78.28
single junction	PC £78.80	0.69	3.88	79.18	nr	83.06
single access junction	PC £128.41	0.69	3.88	135.18	nr	139.06
double junction	PC £157.71	1.05	5.90	165.79	nr	171.69
double collar	PC £39.40	0.46	2.59	44.47	nr	47.06
British Standard quality vitrified clay pipes and fittings; socketted; cement and sand (1:2) joints						
100 mm pipes						
laid straight	PC £3.49	0.46	2.59	4.03	m	6.62
in runs not exceeding 3 m long		0.61	3.43	4.41	m	7.84
Extra for						
bend (short/medium/knuckle)	PC £3.29	0.37	2.08	2.63	nr	4.71
bend (long/rest/elbow)	PC £6.04	0.37	2.08	5.73	nr	7.81
access bend	PC £18.03	0.37	2.08	19.27	nr	21.35
taper	PC £8.36	0.37	2.08	8.24	nr	10.32
access pipe	PC £16.37	0.37	2.08	17.00	nr	19.08
single junction	PC £6.74	0.46	2.59	6.24	nr	8.83
single access junction	PC £20.01	0.46	2.59	21.21	nr	23.80
double junction	PC £12.06	0.69	3.88	11.94	nr	15.82
double collar	PC £4.42	0.31	1.74	5.09	nr	6.83
double access junction	PC £25.54	0.69	3.88	27.16	nr	31.04

R DISPOSAL SYSTEMS Including overheads and profit at 7.50%		Labour hours	Labour £	Material £	Unit	Total rate £
R12 DRAINAGE BELOW GROUND - cont'd						
British Standard quality vitrified clay pipes and fittings; socketted; cement and sand (1:2) joints - cont'd						
150 mm pipes						
laid straight	PC £6.00	0.52	2.92	6.87	m	9.79
in runs not exceeding 3 m long		0.69	3.88	7.51	m	11.39
Extra for						
bend (short/medium/knuckle)	PC £5.68	0.41	2.31	4.47	nr	6.78
bend (long/rest/elbow)	PC £10.22	0.41	2.31	9.60	nr	11.91
access bend	PC £30.34	0.41	2.31	32.32	nr	34.63
taper	PC £13.34	0.41	2.31	12.93	nr	15.24
access pipe	PC £25.27	0.41	2.31	25.91	nr	28.22
single junction	PC £11.90	0.52	2.92	10.93	nr	13.85
single access junction	PC £33.61	0.52	2.92	35.42	nr	38.34
double junction	PC £27.80	0.78	4.39	28.29	nr	32.68
double collar	PC £7.37	0.35	1.97	8.39	nr	10.36
double access junction	PC £43.09	0.78	4.39	45.54	nr	49.93
225 mm pipes						
laid straight	PC £11.83	0.63	3.54	13.54	m	17.08
in runs not exceeding 3 m long		0.84	4.72	14.81	m	19.53
Extra for						
bend (short/medium/knuckle)	PC £17.78	0.51	2.87	16.26	nr	19.13
bend (long/rest/elbow)	PC £29.19	0.51	2.87	29.14	nr	32.01
access bend	PC £64.24	0.51	2.87	68.70	nr	71.57
taper	PC £28.96	0.51	2.87	28.48	nr	31.35
access pipe	PC £64.24	0.51	2.87	67.37	nr	70.24
single junction	PC £29.19	0.63	3.54	28.01	nr	31.55
single access junction	PC £76.81	0.63	3.54	81.75	nr	85.29
double junction	PC £58.19	0.95	5.34	59.60	nr	64.94
double collar	PC £17.35	0.41	2.31	19.78	nr	22.09
double access junction	PC £86.48	0.95	5.34	91.54	nr	96.88
300 mm pipes						
laid straight	PC £19.28	0.86	4.84	21.96	m	26.80
in runs not exceeding 3 m long		1.15	6.47	24.03	m	30.50
Extra for						
bend (short/medium/knuckle)	PC £31.47	0.69	3.88	29.19	nr	33.07
bend (long/rest/elbow)	PC £57.76	0.69	3.88	58.87	nr	62.75
access bend	PC £102.38	0.69	3.88	109.23	nr	113.11
taper	PC £52.54	0.69	3.88	52.32	nr	56.20
access pipe	PC £102.38	0.69	3.88	107.05	nr	110.93
single junction	PC £57.76	0.86	4.84	56.88	nr	61.72
single access junction	PC £113.72	0.86	4.84	120.05	nr	124.89
double junction	PC £115.42	1.30	7.31	119.99	nr	127.30
double access junction	PC £171.42	1.30	7.31	183.21	nr	190.52
Accessories in vitrified clay; set in concrete; with polypropylene coupling joints to pipes						
Rodding point; with oval aluminium plate						
100 mm	PC £12.65	0.58	3.26	15.90	nr	19.16
Gully fittings; comprising low back trap and square hopper; 150 x 150 mm square gully grid						
100 mm outlet	PC £11.36	0.98	5.51	17.44	nr	22.95
Add to above for						
100 mm back inlet		-	-	-	nr	5.34
100 mm raising pieces		0.29	1.63	6.47	nr	8.10

R DISPOSAL SYSTEMS Including overheads and profit at 7.50%		Labour hours	Labour £	Material £	Unit	Total rate £
Access gully; trapped with rodding eye and integral vertical back inlet; stopper; 150 x 150 mm square gully grid						
100 mm outlet	PC £16.20	0.75	4.22	19.71	nr	**23.93**
Inspection chamber; comprising base; 300 or 450 mm raising piece; integral alloy cover and frame; 100 mm inlets						
straight through; 2 nr inlets	PC £44.05	2.30	12.93	53.66	nr	**66.59**
single junction; 3 nr inlets	PC £47.70	2.55	14.34	58.82	nr	**73.16**
double junction; 4 nr inlets	PC £51.73	2.75	15.46	64.39	nr	**79.85**
Accessories in propylene; cover set in concrete; with coupling joints to pipes						
Inspection chamber; 5 nr 100 mm inlets; cast iron cover and frame						
475 x 585 mm deep	PC £71.95	2.65	14.90	82.15	nr	**97.05**
475 x 930 mm deep	PC £84.49	2.90	16.30	95.63	nr	**111.93**
Accessories in vitrified clay; set in concrete; with cement and sand (1:2) joints to pipes						
Rainwater shoes; with 250 x 100 mm oval access						
100 mm	PC £13.26	0.58	3.26	15.41	nr	**18.67**
150 mm	PC £22.33	0.69	3.88	25.70	nr	**29.58**
Gully fittings; comprising low back trap and square hopper; square gully grid						
100 mm outlet; 150 x 150 mm grid	PC £15.60	1.15	6.47	25.57	nr	**32.04**
150 mm outlet; 225 x 225 mm grid	PC £32.70	1.50	8.43	45.14	nr	**53.57**
Add to above for						
100 mm back inlet	PC £9.81	-	-	10.55	nr	**10.55**
100 mm vertical back inlet	PC £9.81	-	-	10.55	nr	**10.55**
100 mm raising pieces	PC £2.85	0.35	1.97	3.16	nr	**5.13**
Yard gully (mud); trapped with rodding eye; galvanized square bucket; stopper; square hinged grate and frame						
100 mm outlet; 225 x 225 mm grate	PC £57.89	3.45	19.40	66.60	nr	**86.00**
150 mm outlet; 300 x 300 mm grate	PC £104.90	4.60	25.86	118.73	nr	**144.59**
Yard gully (garage); trapped with rodding eye; galvanized perforated round bucket; stopper; round hinged grate and frame						
100 mm outlet; 273 mm grate	PC £70.82	3.45	19.40	81.56	nr	**100.96**
150 mm outlet; 368 mm grate	PC £122.27	4.60	25.86	137.93	nr	**163.79**
Road gully; trapped with rodding eye and stopper (grate measured elsewhere)						
300 x 600 x 100 mm outlet	PC £36.76	3.80	21.36	51.89	nr	**73.25**
300 x 600 x 150 mm outlet	PC £36.76	3.80	21.36	51.89	nr	**73.25**
400 x 750 x 150 mm outlet	PC £44.55	4.60	25.86	66.66	nr	**92.52**
450 x 900 x 150 mm outlet	PC £60.79	5.75	32.33	87.85	nr	**120.18**
Grease trap; with internal access; galvanized perforated bucket; lid and frame						
450 x 300 x 525 mm deep; 100 mm outlet	PC £272.78	4.05	22.77	306.77	nr	**329.54**
600 x 450 x 600 mm deep; 100 mm outlet	PC £346.21	4.85	27.27	390.51	nr	**417.78**
Interceptor; trapped with inspection arm; lever locking stopper; chain and staple; cement and sand (1:2) joints to pipes; building in, and cutting and fitting brickwork around						
100 mm outlet; 100 mm inlet	PC £46.24	4.60	25.86	60.57	nr	**86.43**
150 mm outlet; 150 mm inlet	PC £65.57	5.20	29.24	83.48	nr	**112.72**
225 mm outlet; 225 mm inlet	PC £156.38	5.75	32.33	184.51	nr	**216.84**

R DISPOSAL SYSTEMS Including overheads and profit at 7.50%		Labour hours	Labour £	Material £	Unit	Total rate £
R12 DRAINAGE BELOW GROUND - cont'd						
Accessories: grates and covers						
Aluminium alloy gully grids; set in position						
125 x 125 mm	PC £1.67	0.11	0.62	1.80	nr	2.42
150 x 150 mm	PC £1.67	0.11	0.62	1.80	nr	2.42
225 x 225 mm	PC £4.98	0.11	0.62	5.36	nr	5.98
140 mm dia (for 100 mm)	PC £1.67	0.11	0.62	1.80	nr	2.42
197 mm dia (for 150 mm)	PC £2.60	0.11	0.62	2.79	nr	3.41
284 mm dia (for 225 mm)	PC £5.58	0.11	0.62	6.00	nr	6.62
Aluminium alloy sealing plates and frames; set in cement and sand (1:3)						
150 x 150 mm	PC £6.47	0.29	1.63	7.05	nr	8.68
225 x 225 mm	PC £11.77	0.29	1.63	12.75	nr	14.38
254 x 150 mm; for access fittings	PC £8.22	0.29	1.63	8.93	nr	10.56
140 mm dia (for 100 mm)	PC £5.24	0.29	1.63	5.73	nr	7.36
190 mm dia (for 150 mm)	PC £7.54	0.29	1.63	8.20	nr	9.83
273 mm dia (for 225 mm)	PC £12.06	0.29	1.63	13.16	nr	14.79
Coated cast iron heavy duty road gratings and frames; BS 497 Tables 6 and 7; bedding and pointing in cement and sand (1:3); one course half brick thick wall in semi-engineering bricks in cement mortar (1:3)						
475 x 475 mm; grade A, ref GA1-450 (131 kg)	PC £70.29	2.90	16.30	77.82	nr	94.12
400 x 350 mm; grade A, ref GA2-325 (99 kg)	PC £59.14	2.90	16.30	65.45	nr	81.75
500 x 350 mm; grade A, ref GA2-325 (124 kg)	PC £83.88	2.90	16.30	92.24	nr	108.54
White vitreous clay floor channels; bedded, floor channel						
100 mm half round section floor channel	PC £45.79	0.69	3.88	52.20	m	56.08
Extra for						
stop end	PC £27.83	0.46	2.59	29.92	nr	32.51
angle	PC £41.74	0.69	3.88	44.87	nr	48.75
tee piece	PC £41.74	0.92	5.17	44.87	nr	50.04
stop end outlet	PC £35.96	0.58	3.26	38.66	nr	41.92
150 mm half round section floor channel	PC £50.81	0.86	4.84	58.13	m	62.97
Extra for						
stop end	PC £30.88	0.58	3.26	33.20	nr	36.46
angle	PC £46.38	0.86	4.84	49.86	nr	54.70
tee piece	PC £30.66	1.15	6.47	49.86	nr	56.33
stop end outlet	PC £57.18	0.69	3.88	61.47	nr	65.35
Channel sump outlet for 150 mm channel						
one inlet	PC £65.20	1.40	7.87	70.09	nr	77.96
two inlets	PC £69.47	1.70	9.56	74.68	nr	84.24
230 mm half round section floor channel	PC £80.97	1.15	6.47	93.21	m	99.68
Extra for						
stop end	PC £49.44	0.75	4.22	53.14	nr	57.36
angle	PC £74.09	1.15	6.47	79.65	nr	86.12
tee piece	PC £82.72	1.55	8.71	88.93	nr	97.64
stop end outlet	PC £57.97	0.92	5.17	62.32	nr	67.49
100 mm block floor channel	PC £53.07	0.69	3.88	60.15	m	64.03
Extra for						
stop end	PC £30.66	0.46	2.59	32.96	nr	35.55
angle	PC £48.55	0.69	3.88	52.19	nr	56.07
tee piece	PC £48.55	0.92	5.17	52.19	nr	57.36
stop end outlet	PC £40.13	0.58	3.26	43.14	nr	46.40

R DISPOSAL SYSTEMS Including overheads and profit at 7.50%		Labour hours	Labour £	Material £	Unit	Total rate £
150 mm block floor channel	PC £59.04	0.86	4.84	67.16	m	72.00
Extra for						
stop end	PC £35.94	0.58	3.26	38.63	nr	41.89
angle	PC £53.94	0.86	4.84	57.99	nr	62.83
tee piece	PC £53.94	1.15	6.47	57.99	nr	64.46
stop end outlet	PC £44.61	0.69	3.88	47.95	nr	51.83
150 mm rebated block floor channel		0.86	4.84	73.12	m	77.96
Extra for						
stop end	PC £39.47	0.58	3.26	42.43	nr	45.69
angle	PC £59.28	0.86	4.84	63.72	nr	68.56
tee piece	PC £59.28	1.15	6.47	63.72	nr	70.19
stop end outlet	PC £47.98	0.69	3.88	51.58	nr	55.46
Accessories; channel gratings and connectors						
Galvanized cast iron medium duty square mesh gratings; bedding and pointing in cement and sand (1:3)						
138 x 13 mm; to suit 100 mm wide channel	PC £50.13	0.81	4.55	53.89	m	58.44
180 x 13 mm; to suit 150 mm wide channel	PC £50.13	1.05	5.90	53.89	m	59.79
Galvanized cast iron medium duty square mesh gratings and frame; galvanized cast iron angle bearers; bedding and pointing in cement and sand (1:3); cutting and pinning lugs to concrete						
148 x 13 mm; to suit 100 mm wide channel	PC £81.27	1.95	10.96	87.36	m	98.32
190 x 13 mm; to suit 150 mm wide channel	PC £81.27	2.20	12.37	87.36	m	99.73
Chromium plated brass domed outlet grating; threaded joint to connector (measured elsewhere)						
50 mm	PC £17.97	0.35	1.97	19.32	nr	21.29
63 mm	PC £23.84	0.35	1.97	25.62	nr	27.59
Cast iron connector; cement and sand (1:2) joint to drain pipe						
75 x 300 mm long; screwed 50 mm	PC £14.03	0.46	2.59	15.18	nr	17.77
75 x 300 mm long; screwed 63 mm	PC £16.40	0.46	2.59	17.73	nr	20.32
Accessories in precast concrete; top set in with rodding eye and stopper; cement and sand (1:2) joint to pipe						
Concrete road gully; BS 556 Part 2; trapped with rodding eye and stopper; cement and sand (1:2) joint to pipe						
450 mm dia x 1050 mm deep; 100 or 150 mm outlet		5.45	30.64	42.37	nr	73.01
uPVC pipes and fittings; BS 4660; with lip seal coupling joints						
110 mm pipes						
laid straight	PC £14.18/6m	0.23	1.29	3.06	m	4.35
in runs not exceeding 3 m long	PC £7.08/3m	0.31	1.74	3.70	m	5.44
Extra for						
bend; short radius	PC £6.48	0.18	1.01	7.18	nr	8.19
bend; long radius	PC £10.23	0.18	1.01	10.75	nr	11.76
spigot/socket bend	PC £5.55	0.23	1.29	8.29	nr	9.58
access bend	PC £10.14	0.18	1.01	10.66	nr	11.67
socket plug	PC £2.57	0.06	0.34	2.90	nr	3.24

R DISPOSAL SYSTEMS Including overheads and profit at 7.50%		Labour hours	Labour £	Material £	Unit	Total rate £
R12 DRAINAGE BELOW GROUND - cont'd						
uPVC pipes and fittings; BS 4660; with lip seal coupling joints - cont'd						
110 mm pipes						
variable bend	PC £7.08	0.18	1.01	9.89	nr	**10.90**
inspection pipe	PC £12.29	0.18	1.01	15.77	nr	**16.78**
adaptor to clay	PC £5.23	0.18	1.01	5.88	nr	**6.89**
WC connector	PC £4.66	0.23	1.29	7.40	nr	**8.69**
single junction	PC £8.66	0.23	1.29	8.98	nr	**10.27**
inspection junction	PC £17.96	0.23	1.29	19.47	nr	**20.76**
slip coupling	PC £3.86	0.11	0.62	4.35	nr	**4.97**
160 mm pipes						
laid straight	PC £27.70/6m	0.28	1.57	5.99	m	**7.56**
in runs not exceeding 3 m long	PC £13.85/3m	0.37	2.08	7.25	m	**9.33**
Extra for						
bend; short radius	PC £12.42	0.23	1.29	13.77	nr	**15.06**
spigot/socket bend	PC £11.82	0.29	1.63	17.35	nr	**18.98**
socket plug	PC £4.46	0.07	0.39	5.03	nr	**5.42**
inspection pipe	PC £15.97	0.23	1.29	21.77	nr	**23.06**
adaptor to clay	PC £8.88	0.23	1.29	9.86	nr	**11.15**
level invert taper	PC £7.57	0.28	1.57	12.29	nr	**13.86**
single junction	PC £22.80	0.28	1.57	24.17	nr	**25.74**
inspection junction	PC £29.26	0.28	1.57	31.46	nr	**33.03**
slip coupling	PC £8.24	0.14	0.79	9.30	nr	**10.09**
Accessories in uPVC; with lip seal coupling joints to pipes (unless otherwise described)						
Access cap assembly						
110 mm	PC £6.30	0.11	0.62	9.40	nr	**10.02**
Rodding eye 200 mm; sealed cover	PC £16.18	0.58	3.26	28.23	nr	**31.49**
Gully fitting; comprising 'P' trap, square hopper 154 x 154 mm grate						
110 mm outlet	PC £10.97	1.05	5.90	28.57	nr	**34.47**
Shallow access pipe assembly; 2 nr 110 mm inlets; light duty screw down cover and frame						
110 x 600 mm deep	PC £40.23	1.15	6.47	49.51	nr	**55.98**
Shallow branch access assembly; 3 nr 110 mm inlets; light duty screw down cover and frame						
110 x 600 mm deep	PC £44.46	1.40	7.87	51.79	nr	**59.66**
Inspection chamber; 450 mm dia; 940 mm deep; heavy duty screw down cover and frame						
4 nr 110 mm outlet/inlets	PC £85.48	2.05	11.53	148.68	nr	**160.21**
Kerb to gullies; class B engineering bricks on edge to three sides in cement mortar (1:3) rendering in cement mortar (1:3) to top and two sides and skirting to brickwork 230 mm high; dishing in cement mortar (1:3) to gully; steel trowelled						
230 x 230 mm internally		1.60	9.00	1.49	nr	**10.49**
Mechanical excavation						
Excavating manholes; not exceeding						
1 m deep		0.24	1.44	3.84	m3	**5.28**
2 m deep		0.26	1.56	4.20	m3	**5.76**
4 m deep		0.31	1.86	4.93	m3	**6.79**

R DISPOSAL SYSTEMS Including overheads and profit at 7.50%	Labour hours	Labour £	Material £	Unit	Total rate £
Hand excavation					
Excavating manholes; not exceeding					
1 m deep	3.65	21.93	-	m3	**21.93**
2 m deep	4.30	25.84	-	m3	**25.84**
4 m deep	5.50	33.05	-	m3	**33.05**
Earthwork support (average 'risk' prices)					
Not exceeding 2 m between opposing faces; not exceeding					
1 m deep	0.17	1.02	0.53	m2	**1.55**
2 m deep	0.21	1.26	0.64	m2	**1.90**
4 m deep	0.26	1.56	0.81	m2	**2.37**
Disposal (mechanical)					
Excavated material; depositing on site in spoil heaps					
average 50 m distant	0.08	0.48	1.28	m3	**1.76**
Removing from site to tip not exceeding 13 km (using lorries)	-	-	-	m3	**8.86**
Disposal (hand)					
Excavated material; depositing on site in spoil heaps					
average 50 m distant	1.45	8.71	-	m3	**8.71**
Removing from site to tip not exceeding 13 km (using lorries)	1.10	6.61	11.09	m3	**17.70**
Mechanical filling					
Excavated material filling to excavations	0.17	1.02	1.83	m3	**2.85**
Hand filling					
Excavated material filling to excavations	1.10	6.61	-	m3	**6.61**
In situ ready mixed concrete; normal Portland cement; mix 11.50 N/mm2 - 40 mm aggregate (1:3:6); PC £41.49/m3;					
Beds					
not exceeding 150 mm thick	3.55	22.13	46.83	m3	**68.96**
150 - 450 mm thick	2.65	16.52	46.83	m3	**63.35**
over 450 mm thick	2.20	13.72	46.83	m3	**60.55**
Ready mixed in situ concrete; normal Portland cement; mix 21.00 N/mm2-20 mm aggregate (1:2:4); PC £44.19/m3					
Beds					
not exceeding 150 mm thick	3.55	22.13	49.88	m3	**72.01**
150 - 450 mm thick	2.65	16.52	49.88	m3	**66.40**
over 450 mm thick	2.20	13.72	49.88	m3	**63.60**
Site mixed in situ concrete; normal Portland cement; mix 26.00 N/mm2-20 mm aggregate (1:1.5:3); (small quantities); PC £55.33/m3					
Benching in bottoms					
150 - 450 mm average thick	8.05	59.71	59.48	m3	**119.19**
Reinforced site mixed in situ concrete; normal Portland cement; mix 21.00 N/mm2-20mm aggregate (1:2:4); (small quantities); PC £52.43/m3					
Isolated cover slabs					
not exceeding 150 mm thick	6.90	43.02	56.36	m3	**99.38**

R DISPOSAL SYSTEMS Including overheads and profit at 7.50%		Labour hours	Labour £	Material £	Unit	Total rate £
R12 DRAINAGE BELOW GROUND - cont'd						
Reinforcement; fabric to BS 4483; lapped; in beds or suspended slabs						
Ref A98 (1.54 kg/m2)	PC £0.57	0.13	0.94	0.73	m2	**1.67**
Ref A142 (2.22 kg/m2)	PC £0.71	0.13	0.94	0.91	m2	**1.85**
Ref A193 (3.02 kg/m2)	PC £0.97	0.13	0.94	1.24	m2	**2.18**
Formwork to in situ concrete						
Soffits of isolated cover slabs						
horizontal		3.30	24.05	6.16	m2	**30.21**
Edges of isolated cover slabs						
not exceeding 250 mm high		0.92	6.71	2.26	m	**8.97**
Precast concrete rectangular access and inspection chambers; 'Brooklyns' chambers or similar; comprising cover frame to receive manhole cover (priced elsewhere) intermediate wall sections and base section with cut outs;bedding;jointing and pointing in cement mortar (1:3) on prepared bed						
Drainage chamber; size 600 x 450 mm internally; depth to invert						
600 mm deep		5.20	29.24	24.84	nr	**54.08**
900 mm deep		6.90	38.79	31.59	nr	**70.38**
Drainage chamber; 1200 x 750 mm reducing to 600 x 600 mm; no base unit; depth to invert						
1050 mm deep		8.65	48.63	90.25	nr	**138.88**
1650 mm deep		10.35	58.19	139.28	nr	**197.47**
2250 mm deep		12.70	76.32	188.31	nr	**264.63**
Common bricks; PC £124.00/1000; in cement mortar (1:3)						
Walls to manholes						
one brick thick		2.60	26.97	19.96	m2	**46.93**
one and a half brick thick		3.70	38.38	30.03	m2	**68.41**
Projections of footings						
two brick thick		5.00	51.87	39.93	m2	**91.80**
Class A engineering bricks; PC £350.00/1000 in cement mortar (1:3)						
Walls to manholes						
one brick thick		2.90	30.08	50.57	m2	**80.65**
one and a half brick thick		4.05	42.01	75.94	m2	**117.95**
Projections of footings						
two brick thick		5.50	57.06	101.15	m2	**158.21**
Class B engineering bricks; PC £180.00/1000 in cement mortar (1:3)						
Walls to manholes						
one brick thick		2.90	30.08	27.55	m2	**57.63**
one and a half brick thick		4.05	42.01	41.40	m2	**83.41**
Projections of footings						
two brick thick		5.50	57.06	55.10	m2	**112.16**
Brickwork sundries						
Extra over for fair face; flush smooth pointing						
manhole walls		0.23	2.39	-	m2	**2.39**

R DISPOSAL SYSTEMS Including overheads and profit at 7.50%		Labour hours	Labour £	Material £	Unit	Total rate £
Building ends of pipes into brickwork; making good fair face or rendering						
not exceeding 55 mm nominal size		0.11	1.14	-	nr	1.14
55 - 110 mm nominal size		0.17	1.76	-	nr	1.76
over 110 mm nominal size		0.23	2.39	-	nr	2.39
Step irons; BS 1247; malleable; galvanized; building into joints						
general purpose pattern		0.17	1.76	4.76	nr	6.52
Cement and sand (1:3) in situ finishings; steel trowelled						
13 mm work to manhole walls; one coat; to brickwork base over 300 mm wide		0.77	7.06	0.88	m2	7.94
Manhole accessories in cast iron						
Petrol trapping bend; coated; 375 x 750 mm; building into brickwork						
100 mm	PC £62.68	1.55	10.46	69.95	nr	80.41
150 mm	PC £101.24	2.00	13.50	113.16	nr	126.66
225 mm	PC £187.61	3.30	22.28	208.44	nr	230.72
Cast iron inspection chambers; with bolted flat covers; BS 437; bedded in cement mortar (1:3); caulked lead joints to pipes						
100 x 100 mm; ref 010; no branches	PC £89.12	1.15	7.76	98.89	nr	106.65
100 x 100 mm; ref 110; one branch	PC £113.47	1.40	9.45	126.61	nr	136.06
100 x 100 mm; ref 111; one branch either side	PC £146.83	1.60	10.80	163.77	nr	174.57
100 x 100 mm; ref 212; two branches either side	PC £237.98	2.35	15.86	264.33	nr	280.19
100 x 100 mm; ref 313; three branches either side	PC £323.94	3.10	20.93	359.30	nr	380.23
150 x 100 mm; ref 110; one branch	PC £153.84	1.85	12.49	172.38	nr	184.87
150 x 100 mm; ref 111; one branch either side	PC £197.36	2.05	13.84	220.25	nr	234.09
150 x 100 mm; ref 212; two branches either side	PC £283.25	2.95	19.92	315.16	nr	335.08
150 x 100 mm; ref 313; three branches either side	PC £384.57	3.80	25.65	426.64	nr	452.29
150 x 150 mm; ref 212; two branches either side	PC £394.75	3.15	21.27	438.53	nr	459.80
225 x 100 mm; ref 212; two branches either side	PC £566.98	3.80	25.65	624.00	nr	649.65
225 x 100 mm; ref 313; three branches either side	PC £712.50	4.85	32.74	783.01	nr	815.75
Cast iron inspection chambers; with bolted flat covers; BS 437; bedded in cement mortar (1:3); with mechanical coupling joints						
100 x 100 mm; ref 110; one branch	PC £66.11	1.20	8.10	92.78	nr	100.88
100 x 100 mm; ref 111; one branch either side	PC £85.16	1.80	12.15	123.73	nr	135.88
150 x 100 mm; ref 110; one branch	PC £99.15	1.45	9.79	130.93	nr	140.72
150 x 100 mm; ref 111; one branch either side		2.05	13.84	153.40	nr	167.24
150 x 150 mm; ref 110; one branch	PC £126.94	1.55	10.46	163.02	nr	173.48
150 x 150 mm; ref 111; one branch either side	PC £140.02	2.20	14.85	189.77	nr	204.62

R DISPOSAL SYSTEMS Including overheads and profit at 7.50%		Labour hours	Labour £	Material £	Unit	Total rate £
R12 DRAINAGE BELOW GROUND - cont'd						
Access covers and frames; coated; BS 497 tables 1-5; bedding frame in cement and sand (1:3); cover in grease and sand						
Grade C; light duty; rectangular single seal solid top						
450 x 450 mm; ref MC1-45/45 (31 kg)	PC £25.73	1.70	9.56	29.03	nr	**38.59**
600 x 450 mm; ref MC1-60/45 (32 kg)	PC £25.89	1.70	9.56	29.20	nr	**38.76**
600 x 600 mm; ref MC1-60/60 (61 kg)	PC £57.49	1.70	9.56	63.17	nr	**72.73**
Grade C; light duty; rectangular single seal recessed						
450 x 450 mm; ref MC1R-45/45 (43 kg)	PC £42.10	1.70	9.56	46.63	nr	**56.19**
600 x 450 mm; ref MC1R-60/45 (43 kg)	PC £53.53	1.70	9.56	58.92	nr	**68.48**
600 x 600 mm; ref MC1R-60/60 (53 kg)	PC £72.55	1.70	9.56	79.36	nr	**88.92**
Grade C; light duty; rectangular double seal solid top						
450 x 450 mm; ref MC2-45/45 (51 kg)	PC £39.41	1.70	9.56	43.74	nr	**53.30**
600 x 450 mm; ref MC2-60/45 (51 kg)	PC £38.43	1.70	9.56	42.68	nr	**52.24**
600 x 600 mm; ref MC2-60/60 (83 kg)	PC £72.80	1.70	9.56	79.63	nr	**89.19**
Grade C; light duty; rectangular double seal recessed						
450 x 450 mm; ref MC2R-45/45 56 kg)	PC £65.48	1.70	9.56	71.76	nr	**81.32**
600 x 450 mm; ref MC2R-60/45 (71 kg)	PC £91.11	1.70	9.56	99.31	nr	**108.87**
600 x 600 mm; ref MC2R-60/60 (75 kg)	PC £66.67	1.70	9.56	73.04	nr	**82.60**
Grade B; medium duty; circular single seal solid top						
500 mm; ref MB2-50 (106 kg)	PC £72.75	2.30	12.93	79.58	nr	**92.51**
550 mm; ref MB2-55 (112 kg)	PC £77.78	2.30	12.93	84.98	nr	**97.91**
600 mm; ref MB2-60 (134 kg)	PC £81.87	2.30	12.93	89.38	nr	**102.31**
Grade B; medium duty; rectangular single seal solid top						
600 x 450 mm; ref MB2-60/45 (135 kg)	PC £70.62	2.30	12.93	77.29	nr	**90.22**
600 x 600 mm; ref MB2-60/60 (170 kg)	PC £91.82	2.30	12.93	100.08	nr	**113.01**
Grade B; medium duty; rectangular single seal recessed						
600 x 450 mm; ref MB2R-60/45 (145 kg)	PC £97.00	2.30	12.93	105.65	nr	**118.58**
600 x 600 mm; ref MB2R-60/60 (171 kg)	PC £123.85	2.30	12.93	134.51	nr	**147.44**
Grade B; 'Chevron'; medium duty; double triangular solid top						
550 mm; ref MB1-55 (125 kg)	PC £70.76	2.30	12.93	77.44	nr	**90.37**
600 mm; ref MB1-60 (140 kg)	PC £84.63	2.30	12.93	92.35	nr	**105.28**
Grade A; heavy duty; single triangular solid top						
550 x 455 mm; ref MA-T (196 kg)	PC £101.27	2.90	16.30	110.24	nr	**126.54**

R DISPOSAL SYSTEMS Including overheads and profit at 7.50%		Labour hours	Labour £	Material £	Unit	Total rate £
Grade A; 'Chevron'; heavy duty double triangular solid top						
500 mm; ref MA-50 (164 kg)	PC £118.32	3.45	19.40	128.56	nr	**147.96**
550 mm; ref MA-55 (176 kg)	PC £115.37	3.45	19.40	125.39	nr	**144.79**
600 mm; ref MA-60 (230 kg)	PC £121.76	3.45	19.40	132.26	nr	**151.66**
British Standard best quality vitrified clay channels; bedding and jointing in cement and sand (1:2)						
Half section straight						
100 mm x 1.00 m long	PC £2.27	0.92	5.17	2.55	nr	**7.72**
150 mm x 1.00 m long	PC £3.76	1.15	6.47	4.25	nr	**10.72**
225 mm x 1.00 m long	PC £8.47	1.50	8.43	9.56	nr	**17.99**
300 mm x 1.00 m long	PC £16.80	1.85	10.40	18.96	nr	**29.36**
Half section bend						
100 mm	PC £2.32	0.69	3.88	2.62	nr	**6.50**
150 mm	PC £3.85	0.86	4.84	4.34	nr	**9.18**
225 mm	PC £12.96	1.15	6.47	14.63	nr	**21.10**
300 mm	PC £25.73	1.40	7.87	29.05	nr	**36.92**
Half section taper straight						
150 mm	PC £9.75	0.81	4.55	11.00	nr	**15.55**
225 mm	PC £21.74	1.05	5.90	24.54	nr	**30.44**
300 mm	PC £42.99	1.25	7.03	48.52	nr	**55.55**
Half section taper bend						
150 mm	PC £14.82	1.05	5.90	16.72	nr	**22.62**
225 mm	PC £42.67	1.30	7.31	48.16	nr	**55.47**
300 mm	PC £84.44	1.60	9.00	95.31	nr	**104.31**
Three quarter section branch bend						
100 mm	PC £5.28	0.58	3.26	5.96	nr	**9.22**
150 mm	PC £8.90	0.86	4.84	10.04	nr	**14.88**
225 mm	PC £32.49	1.15	6.47	36.67	nr	**43.14**
300 mm	PC £64.49	1.55	8.71	72.79	nr	**81.50**
uPVC channels; with solvent weld or lip seal coupling joints; bedding in cement and sand						
Half section cut away straight; with coupling either end						
110 mm	PC £17.22	0.35	1.97	19.33	nr	**21.30**
160 mm	PC £23.11	0.46	2.59	25.93	nr	**28.52**
Half section cut away long radius bend; with coupling either end						
110 mm	PC £19.55	0.35	1.97	22.29	nr	**24.26**
160 mm	PC £28.96	0.46	2.59	33.16	nr	**35.75**
Half section straight channel adaptor; with one coupling						
110 mm	PC £6.76	0.29	1.63	9.12	nr	**10.75**
160 mm	PC £8.97	0.38	2.14	13.09	nr	**15.23**
Half section cut away bend						
110 mm	PC £12.82	0.46	2.59	14.64	nr	**17.23**
Half section bend						
110 mm	PC £3.14	0.38	2.14	3.72	nr	**5.86**
160 mm	PC £6.10	0.58	3.26	7.27	nr	**10.53**
Half section channel connector						
110 mm	PC £1.51	0.09	0.51	2.05	nr	**2.56**
160 mm	PC £3.55	0.11	0.62	4.79	nr	**5.41**
Half section channel junction						
110 mm	PC £5.11	0.58	3.26	5.94	nr	**9.20**
160 mm	PC £9.65	0.69	3.88	11.28	nr	**15.16**
polypropylene slipper bend						
110 mm	PC £6.77	0.46	2.59	7.81	nr	**10.40**

R DISPOSAL SYSTEMS Including overheads and profit at 7.50%		Labour hours	Labour £	Material £	Unit	Total rate £
R13 LAND DRAINAGE						
Hand excavation of trenches to receive land drain pipes; grading bottoms; earthwork support; filling to within 150 mm of surface filling to within 150 mm of surface with gravel rejects; remainder filled with excavated material and compacting; disposal of surplus soil; spreading on site average 50 m						
Pipes not exceeding 200 mm; average depth						
0.75 m deep		1.85	11.12	10.0	m	21.12
1.00 m deep		2.50	15.02	13.32	m	28.34
1.25 m deep		3.45	20.73	16.91	m	37.64
1.50 m deep		5.95	35.76	21.32	m	57.08
1.75 m deep		7.05	42.37	25.21	m	67.58
2.00 m deep		8.15	48.98	29.10	m	78.08
Surplus excavated material						
Removing from site to tip not exceeding 13 km (using lorries)						
machine loaded		-	-	-	m3	8.86
hand loaded		1.10	6.61	11.09	m3	17.70
Clay field drain pipes; BS 1196; unjointed						
Pipes; laid straight						
75 mm	PC £25.30/100	0.23	1.29	0.94	m	2.23
100 mm	PC £45.10/100	0.29	1.63	1.68	m	3.31
150 mm	PC £92.40/100	0.35	1.97	3.44	m	5.41
Vitrified clay perforated sub-soil pipes; BS 65; 'Hepworths' 'Hepline' or similar						
Pipes; laid straight						
100 mm	PC £3.47	0.25	1.41	3.91	m	5.32
150 mm	PC £6.35	0.31	1.74	7.17	m	8.91
225 mm pipes	PC £11.66	0.41	2.31	13.16	m	15.47

S PIPED SUPPLY SYSTEMS
Including overheads and profit at 7.50%

ALTERNATIVE SERVICE PIPE AND FITTING PRICES

Copper pipes to BS 2871 (£/100 m)

	£		£		£		£
Table X							
6 mm	65.20	10 mm	111.00	67 mm	1603.00	108 mm	3255.00
8 mm	88.60	12 mm	138.00	76 mm	2279.00	133 mm	4020.00
Table Y							
6 mm	86.80	10 mm	150.00	67 mm	2673.00	108 mm	5361.00
8 mm	116.00	12 mm	186.00	76 mm	3016.00		
Table Z							
15 mm	120.00	35 mm	633.00	54 mm	1013.00	76 mm	2029.00
22 mm	232.00	42 mm	777.00	67 mm	1432.00	108 mm	2897.00
28 mm	301.00						

PVC Class E cold water pressure system (£/m/each)
(sizes shown as nearest equivalent metric sizes)

	£		£		£
pipe-per m-13 mm	0.86	end cap-13 mm	0.42	elbow-13 mm	0.62
-19 mm	1.19	-19 mm	0.47	-19 mm	0.73
-25 mm	1.54	-25 mm	0.55	-25 mm	0.92
-32 mm	2.49	-32 mm	0.84	-32 mm	1.75
-40 mm	3.28	-40 mm	1.31	-40 mm	2.20
coupling -13 mm	0.44	reducer-32 mm	1.25	tee -13 mm	0.70
-19 mm	0.50	-40 mm	1.49	-19 mm	0.87
-25 mm	0.58	MI.conn-13 mm	0.74	-25 mm	1.30
-32 mm	0.96	-19 mm	0.79	-32 mm	1.90
-40 mm	1.20	-25 mm	1.52	-40 mm	2.80
s.tank.cn.-19 mm	2.05	-32 mm	2.02		
-25 mm	2.26	-40 mm	2.20		

		Labour hours	Labour £	Material £	Unit	Total rate £
S10/S11 HOT AND COLD WATER						
Copper pipes; BS 2871 table X; capillary fittings; BS 864						
15 mm pipes; fixing with pipe clips; plugged and screwed	PC £132.00/100m	0.30	2.70	1.80	m	4.50
Extra for						
made bend		0.18	1.62	-	nr	1.62
fittings with one end		0.13	1.17	0.63	nr	1.80
fittings with two ends		0.20	1.80	0.32	nr	2.12
fittings with three ends		0.30	2.70	0.59	nr	3.29
fittings with four ends		0.42	3.78	4.12	nr	7.90
stop end	PC £0.58	0.13	1.17	0.63	nr	1.80
straight union coupling	PC £2.59	0.20	1.80	2.79	nr	4.59
copper to iron connector	PC £0.95	0.26	2.34	1.02	nr	3.36
elbow	PC £0.30	0.20	1.80	0.32	nr	2.12
backplate elbow	PC £1.90	0.42	3.78	2.04	nr	5.82
slow bend	PC £1.01	0.20	1.80	1.08	nr	2.88
tee; equal	PC £0.55	0.30	2.70	0.59	nr	3.29
tee; reducing	PC £2.14	0.30	2.70	2.30	nr	5.00
cross	PC £3.83	0.42	3.78	4.12	nr	7.90
straight tap connector	PC £0.86	0.16	1.44	0.93	nr	2.37
bent tap connector	PC £1.04	0.16	1.44	1.12	nr	2.56
straight tank connector; backnut	PC £1.84	0.30	2.70	1.98	nr	4.68

S PIPED SUPPLY SYSTEMS Including overheads and profit at 7.50%		Labour hours	Labour £	Material £	Unit	Total rate £
S10/S11 HOT AND COLD WATER - cont'd						
Copper pipes; BS 2871 table X; capillary fittings; BS 864 - cont'd						
22 mm pipes; fixing with pipe clips; plugged and screwed	PC £260.00/100m	0.31	2.79	3.31	m	6.10
Extra for						
made bend		0.24	2.16	-	nr	2.16
fittings with one end		0.16	1.44	1.04	nr	2.48
fittings with two ends		0.26	2.34	0.65	nr	2.99
fittings with three ends		0.40	3.60	1.20	nr	4.80
fittings with four ends		0.53	4.77	5.31	nr	10.08
stop end	PC £0.97	0.16	1.44	1.04	nr	2.48
reducing coupling	PC £0.73	0.26	2.34	0.79	nr	3.13
straight union coupling	PC £4.00	0.26	2.34	4.30	nr	6.64
copper to iron connector	PC £1.58	0.37	3.33	1.70	nr	5.03
elbow	PC £0.60	0.26	2.34	0.65	nr	2.99
backplate elbow	PC £3.94	0.53	4.77	4.24	nr	9.01
slow bend	PC £1.71	0.26	2.34	1.83	nr	4.17
tee; equal	PC £1.11	0.40	3.60	1.20	nr	4.80
tee; reducing	PC £1.08	0.40	3.60	1.16	nr	4.76
cross	PC £4.94	0.53	4.77	5.31	nr	10.08
straight tap connector	PC £1.32	0.20	1.80	1.42	nr	3.22
bent tap connector	PC £2.00	0.20	1.80	2.15	nr	3.95
straight tank connector; backnut	PC £2.73	0.40	3.60	2.94	nr	6.54
28 mm pipes; fixing with pipe clips; plugged and screwed	PC £336.00/100m	0.35	3.15	4.28	m	7.43
Extra for						
made bend		0.30	2.70	-	nr	2.70
fittings with one end		0.18	1.62	2.02	nr	3.64
fittings with two ends		0.34	3.06	1.18	nr	4.24
fittings with three ends		0.49	4.41	2.51	nr	6.92
fittings with four ends		0.67	6.04	7.61	nr	13.65
stop end	PC £1.88	0.18	1.62	2.02	nr	3.64
reducing coupling	PC £1.62	0.34	3.06	1.74	nr	4.80
straight union coupling	PC £5.50	0.34	3.06	5.91	nr	8.97
copper to iron connector	PC £2.49	0.47	4.23	2.67	nr	6.90
elbow	PC £1.10	0.34	3.06	1.18	nr	4.24
slow bend	PC £2.85	0.34	3.06	3.07	nr	6.13
tee; equal	PC £2.33	0.49	4.41	2.51	nr	6.92
tee; reducing	PC £2.51	0.49	4.41	2.69	nr	7.10
cross	PC £7.08	0.67	6.04	7.61	nr	13.65
straight tank connector; backnut	PC £3.77	0.49	4.41	4.05	nr	8.46
35 mm pipes; fixing with pipe clips; plugged and screwed	PC £733.00/100m	0.40	3.60	9.09	m	12.69
Extra for						
made bend		0.36	3.24	-	nr	3.24
fittings with one end		0.20	1.80	3.11	nr	4.91
fittings with two ends		0.40	3.60	3.13	nr	6.73
fittings with three ends		0.55	4.95	5.38	nr	10.33
stop end	PC £2.90	0.20	1.80	3.11	nr	4.91
reducing coupling	PC £2.38	0.40	3.60	2.56	nr	6.16
straight union coupling	PC £7.87	0.40	3.60	8.46	nr	12.06
copper to iron connector	PC £3.88	0.53	4.77	4.17	nr	8.94
elbow	PC £2.91	0.40	3.60	3.13	nr	6.73
bend; 91.5 degrees	PC £5.12	0.40	3.60	5.51	nr	9.11
tee; equal	PC £5.00	0.55	4.95	5.38	nr	10.33
tee; reducing	PC £4.75	0.55	4.95	5.11	nr	10.06
pitcher tee; equal or reducing	PC £8.32	0.55	4.95	8.94	nr	13.89
straight tank connector; backnut	PC £5.06	0.55	4.95	5.44	nr	10.39

S PIPED SUPPLY SYSTEMS Including overheads and profit at 7.50%		Labour hours	Labour £	Material £	Unit	Total rate £
42 mm pipes; fixing with pipe clips; plugged and screwed	PC £889.00/100m	0.46	4.14	11.26	m	15.40
Extra for						
made bend		0.48	4.32	-	nr	4.32
fittings with one end		0.23	2.07	4.05	nr	6.12
fittings with two ends		0.47	4.23	4.68	nr	8.91
fittings with three ends		0.62	5.59	7.92	nr	13.51
stop end	PC £3.77	0.23	2.07	4.05	nr	6.12
reducing coupling	PC £3.58	0.47	4.23	3.85	nr	8.08
straight union coupling	PC £11.17	0.47	4.23	12.01	nr	16.24
copper to iron connector	PC £4.70	0.60	5.41	5.05	nr	10.46
elbow	PC £4.35	0.47	4.23	4.68	nr	8.91
bend; 91.5 degrees	PC £8.18	0.47	4.23	8.79	nr	13.02
tee; equal	PC £7.37	0.62	5.59	7.92	nr	13.51
tee; reducing	PC £8.72	0.62	5.59	9.37	nr	14.96
pitcher tee; equal or reducing	PC £11.69	0.62	5.59	12.57	nr	18.16
straight tank connector; backnut	PC £6.43	0.62	5.59	6.91	nr	12.50
54 mm pipes; fixing with pipe clips; plugged and screwed	PC £1149.00/100m	0.54	4.86	15.00	m	19.86
Extra for						
made bend		0.66	5.95	-	nr	5.95
fittings with one end		0.25	2.25	5.99	nr	8.24
fittings with two ends		0.53	4.77	10.86	nr	15.63
fittings with three ends		0.68	6.13	14.59	nr	20.72
stop end	PC £5.57	0.25	2.25	5.99	nr	8.24
reducing coupling	PC £5.25	0.53	4.77	5.65	nr	10.42
straight union coupling	PC £18.17	0.53	4.77	19.53	nr	24.30
copper to iron connector	PC £7.80	0.66	5.95	8.38	nr	14.33
elbow	PC £10.10	0.53	4.77	10.86	nr	15.63
bend; 91.5 degrees	PC £12.21	0.53	4.77	13.13	nr	17.90
tee; equal	PC £13.57	0.68	6.13	14.59	nr	20.72
tee; reducing	PC £13.78	0.68	6.13	14.81	nr	20.94
pitcher tee; equal or reducing	PC £16.83	0.68	6.13	18.10	nr	24.23
straight tank connector; backnut	PC £9.80	0.68	6.13	10.53	nr	16.66
Cut into existing copper services; provide and insert new capillary jointed tee; equal or reducing						
15 mm	PC £0.55	0.50	4.50	0.59	nr	5.09
22 mm	PC £1.08	0.58	5.22	1.16	nr	6.38
28 mm	PC £2.51	0.66	5.95	2.69	nr	8.64
35 mm	PC £4.75	0.74	6.67	5.11	nr	11.78
42 mm	PC £8.72	0.80	7.21	9.37	nr	16.58
54 mm	PC £13.78	0.86	7.75	14.81	nr	22.56
Copper pipes; BS 2871 table X; compression fittings; BS 864						
15 mm pipes; fixing with pipe clips; plugged and screwed	PC £132.00/100m	0.29	2.61	1.91	m	4.52
Extra for						
made bend		0.18	1.62	-	nr	1.62
fittings with one end		0.12	1.08	0.87	nr	1.95
fittings with two ends		0.18	1.62	0.71	nr	2.33
fittings with three ends		0.26	2.34	1.00	nr	3.34
stop end	PC £0.81	0.12	1.08	0.87	nr	1.95
straight coupling	PC £0.56	0.18	1.62	0.60	nr	2.22
female coupling	PC £0.56	0.24	2.16	0.60	nr	2.76
elbow	PC £0.66	0.18	1.62	0.71	nr	2.33
female wall elbow	PC £1.43	0.36	3.24	1.54	nr	4.78
slow bend	PC £2.67	0.18	1.62	2.87	nr	4.49
bent radiator union; chrome finish	PC £2.00	0.24	2.16	2.15	nr	4.31
tee; equal	PC £0.93	0.26	2.34	1.00	nr	3.34
straight swivel connector	PC £1.14	0.14	1.26	1.22	nr	2.48
bent swivel connector	PC £1.20	0.14	1.26	1.29	nr	2.55
tank coupling; locknut	PC £1.37	0.26	2.34	1.47	nr	3.81

S PIPED SUPPLY SYSTEMS Including overheads and profit at 7.50%		Labour hours	Labour £	Material £	Unit	Total rate £
S10/S11 HOT AND COLD WATER - cont'd						
Copper pipes; BS 2871 table X; compression fittings; BS 864 - cont'd						
22 mm pipes; fixing with pipe clips; plugged and screwed	PC £260.00/100m	0.30	2.70	3.48	m	6.18
Extra for						
made bend		0.24	2.16	-	nr	2.16
fittings with one end		0.14	1.26	1.13	nr	2.39
fittings with two ends		0.24	2.16	1.21	nr	3.37
fittings with three ends		0.36	3.24	1.73	nr	4.97
stop end	PC £1.05	0.14	1.26	1.13	nr	2.39
straight coupling	PC £0.93	0.24	2.16	1.00	nr	3.16
reducing set	PC £0.69	0.07	0.63	0.74	nr	1.37
female coupling	PC £0.85	0.34	3.06	0.91	nr	3.97
elbow	PC £1.12	0.24	2.16	1.21	nr	3.37
female wall elbow	PC £3.29	0.48	4.32	3.54	nr	7.86
slow bend	PC £4.39	0.24	2.16	4.72	nr	6.88
tee; equal	PC £1.61	0.36	3.24	1.73	nr	4.97
tee; reducing	PC £2.30	0.36	3.24	2.47	nr	5.71
straight swivel connector	PC £1.78	0.19	1.71	1.91	nr	3.62
bent swivel connector	PC £2.59	0.19	1.71	2.78	nr	4.49
tank coupling; locknut	PC £1.25	0.36	3.24	1.34	nr	4.58
28 mm pipes; fixing with pipe clips; plugged and screwed	PC £336.00/100m	0.34	3.06	4.72	m	7.78
Extra for						
made bend		0.30	2.70	-	nr	2.70
fittings with one end		0.17	1.53	2.09	nr	3.62
fittings with two ends		0.30	2.70	2.89	nr	5.59
fittings with three ends		0.44	3.96	4.20	nr	8.16
stop end	PC £1.94	0.17	1.53	2.09	nr	3.62
straight coupling	PC £2.17	0.30	2.70	2.33	nr	5.03
reducing set	PC £1.03	0.08	0.72	1.11	nr	1.83
female coupling	PC £1.39	0.42	3.78	1.49	nr	5.27
elbow	PC £2.69	0.30	2.70	2.89	nr	5.59
slow bend	PC £5.88	0.30	2.70	6.32	nr	9.02
tee; equal	PC £3.91	0.44	3.96	4.20	nr	8.16
tee; reducing	PC £4.06	0.44	3.96	4.36	nr	8.32
tank coupling; locknut	PC £2.51	0.44	3.96	2.70	nr	6.66
35 mm pipes; fixing with pipe clips; plugged and screwed	PC £733.00/100m	0.38	3.42	9.82	m	13.24
Extra for						
made bend		0.36	3.24	-	nr	3.24
fittings with one end		0.19	1.71	3.43	nr	5.14
fittings with two ends		0.36	3.24	5.65	nr	8.89
fittings with three ends		0.50	4.50	7.65	nr	12.15
stop end	PC £3.19	0.19	1.71	3.43	nr	5.14
straight coupling	PC £4.06	0.36	3.24	4.36	nr	7.60
reducing set	PC £1.76	0.10	0.90	1.89	nr	2.79
female coupling	PC £3.56	0.48	4.32	3.83	nr	8.15
elbow	PC £5.26	0.36	3.24	5.65	nr	8.89
tee; equal	PC £7.12	0.50	4.50	7.65	nr	12.15
tee; reducing	PC £7.12	0.50	4.50	7.65	nr	12.15
tank coupling; locknut	PC £5.30	0.50	4.50	5.69	nr	10.19
42 mm pipes; fixing with pipe clips; plugged and screwed	PC £889.00/100m	0.44	3.96	12.15	m	16.11
Extra for						
made bend		0.48	4.32	-	nr	4.32
fittings with one end		0.22	1.98	5.63	nr	7.61
fittings with two ends		0.42	3.78	7.97	nr	11.75
fittings with three ends		0.56	5.04	12.74	nr	17.78
stop end	PC £5.23	0.22	1.98	5.63	nr	7.61
straight coupling	PC £5.22	0.42	3.78	5.61	nr	9.39

S PIPED SUPPLY SYSTEMS Including overheads and profit at 7.50%		Labour hours	Labour £	Material £	Unit	Total rate £
reducing set	PC £2.78	0.11	0.99	2.99	nr	3.98
female coupling	PC £4.73	0.54	4.86	5.09	nr	9.95
elbow	PC £7.41	0.42	3.78	7.97	nr	11.75
tee; equal	PC £11.85	0.56	5.04	12.74	nr	17.78
tee; reducing	PC £11.10	0.56	5.04	11.93	nr	16.97
54 mm pipes; fixing with pipe clips; plugged and screwed	PC £1149.00/100m	0.53	4.77	15.97	m	20.74
Extra for						
made bend		0.66	5.95	-	nr	5.95
fittings with two ends		0.48	4.32	13.09	nr	17.41
fittings with three ends		0.62	5.59	19.94	nr	25.53
straight coupling	PC £7.97	0.48	4.32	8.56	nr	12.88
reducing set	PC £4.65	0.12	1.08	5.00	nr	6.08
female wall elbow	PC £6.85	0.60	5.41	7.37	nr	12.78
elbow	PC £12.18	0.48	4.32	13.09	nr	17.41
tee; equal	PC £18.55	0.62	5.59	19.94	nr	25.53
tee; reducing	PC £18.83	0.62	5.59	20.24	nr	25.83
Black MDPE pipes; BS 6730; plastic compression fittings						
20 mm pipes; fixing with pipe clips; plugged and screwed	PC £34.11/100m	0.30	2.70	1.29	m	3.99
Extra for						
fittings with two ends		0.24	2.16	2.00	nr	4.16
fittings with three ends		0.36	3.24	2.57	nr	5.81
straight coupling	PC £1.64	0.24	2.16	1.76	nr	3.92
male adaptor	PC £1.03	0.34	3.06	1.11	nr	4.17
elbow	PC £1.86	0.24	2.16	2.00	nr	4.16
tee; equal	PC £2.39	0.36	3.24	2.57	nr	5.81
end cap	PC £1.27	0.14	1.26	1.37	nr	2.63
straight swivel connector	PC £1.12	0.19	1.71	1.20	nr	2.91
bent swivel connector	PC £1.37	0.19	1.71	1.47	nr	3.18
25 mm pipes; fixing with pipe clips; plugged and screwed	PC £41.60/100m	0.34	3.06	1.60	m	4.66
Extra for						
fittings with two ends		0.30	2.70	2.42	nr	5.12
fittings with three ends		0.44	3.96	3.41	nr	7.37
straight coupling	PC £2.02	0.30	2.70	2.18	nr	4.88
reducer	PC £2.01	0.30	2.70	2.16	nr	4.86
male adaptor	PC £1.22	0.42	3.78	1.31	nr	5.09
elbow	PC £2.25	0.30	2.70	2.42	nr	5.12
tee; equal	PC £3.18	0.44	3.96	3.41	nr	7.37
end cap	PC £1.47	0.17	1.53	1.58	nr	3.11
32 mm pipes; fixing with pipe clips; plugged and screwed	PC £68.22/100m	0.38	3.42	1.99	m	5.41
Extra for						
fittings with two ends		0.36	3.24	3.00	nr	6.24
fittings with three ends		0.50	4.50	4.46	nr	8.96
straight coupling	PC £2.83	0.36	3.24	3.04	nr	6.28
reducer	PC £2.64	0.36	3.24	2.84	nr	6.08
male adaptor	PC £1.51	0.48	4.32	1.63	nr	5.95
elbow	PC £2.79	0.36	3.24	3.00	nr	6.24
tee; equal	PC £4.15	0.50	4.50	4.46	nr	8.96
end cap	PC £1.64	0.19	1.71	1.76	nr	3.47
50 mm pipes; fixing with pipe clips; plugged and screwed	PC £174.72/100m	0.42	3.78	3.34	m	7.12
Extra for						
fittings with two ends		0.42	3.78	7.35	nr	11.13
fittings with three ends		0.56	5.04	9.98	nr	15.02
straight coupling	PC £6.36	0.42	3.78	6.84	nr	10.62
reducer	PC £5.86	0.42	3.78	6.30	nr	10.08
male adaptor	PC £3.77	0.54	4.86	4.05	nr	8.91

S PIPED SUPPLY SYSTEMS Including overheads and profit at 7.50%		Labour hours	Labour £	Material £	Unit	Total rate £
S10/S11 HOT AND COLD WATER - cont'd						
Black MDPE pipes; BS 6730; plastic compression fittings - cont'd						
50 mm pipes; fixing with pipe clips; plugged and screwed						
elbow	PC £6.84	0.42	3.78	7.35	nr	11.13
tee; equal	PC £9.29	0.56	5.04	9.98	nr	15.02
end cap	PC £4.21	0.22	1.98	4.52	nr	6.50
63 mm pipes; fixing with pipe clips; plugged and screwed	PC £253.76/100m	0.48	4.32	4.60	m	8.92
Extra for						
fittings with two ends		0.48	4.32	8.94	nr	13.26
fittings with three ends		0.62	5.59	14.19	nr	19.78
straight coupling	PC £9.61	0.48	4.32	10.33	nr	14.65
reducer	PC £8.31	0.48	4.32	8.94	nr	13.26
male adaptor	PC £5.38	0.60	5.41	5.78	nr	11.19
elbow	PC £8.31	0.48	4.32	8.94	nr	13.26
tee; equal	PC £13.20	0.62	5.59	14.19	nr	19.78
end cap	PC £5.82	0.24	2.16	6.25	nr	8.41
Stainless steel pipes; BS 4127; stainless steel capillary fittings						
15 mm pipes; fixing with pipe clips; plugged and screwed	PC £129.80/100m	0.36	3.24	2.75	m	5.99
Extra for						
fittings with two ends		0.24	2.16	4.23	nr	6.39
fittings with three ends		0.36	3.24	6.27	nr	9.51
bend	PC £3.94	0.24	2.16	4.23	nr	6.39
tee; equal	PC £5.83	0.36	3.24	6.27	nr	9.51
tap connector	PC £15.21	0.32	2.88	16.35	nr	19.23
22 mm pipes; fixing with pipe clips; plugged and screwed	PC £205.70/100m	0.38	3.42	4.13	m	7.55
Extra for						
fittings with two ends		0.32	2.88	5.04	nr	7.92
fittings with three ends		0.48	4.32	8.55	nr	12.87
reducer	PC £16.29	0.32	2.88	17.51	nr	20.39
bend	PC £4.69	0.32	2.88	5.04	nr	7.92
tee; equal	PC £7.95	0.48	4.32	8.55	nr	12.87
tee; reducing	PC £18.07	0.48	4.32	19.43	nr	23.75
tap connector	PC £28.58	0.46	4.14	30.72	nr	34.86
28 mm pipes; fixing with pipe clips; plugged and screwed	PC £300.30/100m	0.42	3.78	5.62	m	9.40
Extra for						
fittings with two ends		0.41	3.69	7.24	nr	10.93
fittings with three ends		0.60	5.41	12.72	nr	18.13
reducer	PC £36.19	0.41	3.69	38.90	nr	42.59
bend	PC £6.73	0.41	3.69	7.24	nr	10.93
tee; equal	PC £11.84	0.60	5.41	12.72	nr	18.13
tee; reducing	PC £21.08	0.60	5.41	22.66	nr	28.07
tap connector	PC £57.50	0.58	5.22	61.81	nr	67.03
Copper, brass and gunmetal ancillaries; screwed joints to fittings						
Bibtaps; brass						
chromium plated; capstan head						
15 mm	PC £9.87	0.18	1.62	10.72	nr	12.34
22 mm	PC £13.23	0.24	2.16	14.34	nr	16.50
self closing; 15 mm	PC £13.17	0.18	1.62	14.28	nr	15.90
crutch head with hose union; 15 mm	PC £6.13	0.18	1.62	6.70	nr	8.32
draincock 15 mm	PC £2.09	0.12	1.08	2.32	nr	3.40

S PIPED SUPPLY SYSTEMS Including overheads and profit at 7.50%		Labour hours	Labour £	Material £	Unit	Total rate £
Stopcock; brass/gunmetal						
capillary joints to copper						
15 mm	PC £2.09	0.24	2.16	2.24	nr	4.40
22 mm	PC £3.78	0.32	2.88	4.07	nr	6.95
28 mm	PC £10.78	0.41	3.69	11.59	nr	15.28
compression joints to copper						
15 mm	PC £2.11	0.22	1.98	2.27	nr	4.25
22 mm	PC £3.70	0.29	2.61	3.98	nr	6.59
28 mm	PC £9.63	0.36	3.24	10.35	nr	13.59
compression joints to polyethylene						
13 mm	PC £5.75	0.30	2.70	6.18	nr	8.88
19 mm	PC £8.67	0.40	3.60	9.32	nr	12.92
25 mm	PC £12.52	0.48	4.32	13.46	nr	17.78
Gunmetal 'Fullway' gate valve; capillary joints to copper						
15 mm	PC £6.37	0.24	2.16	6.85	nr	9.01
22 mm	PC £7.51	0.32	2.88	8.07	nr	10.95
28 mm	PC £10.26	0.41	3.69	11.03	nr	14.72
35 mm	PC £22.86	0.49	4.41	24.58	nr	28.99
42 mm	PC £27.31	0.56	5.04	29.36	nr	34.40
54 mm	PC £39.59	0.64	5.77	42.56	nr	48.33
Gunmetal stopcock; screwed joints to iron						
15 mm	PC £6.00	0.36	3.24	6.45	nr	9.69
20 mm	PC £9.78	0.48	4.32	10.51	nr	14.83
25 mm	PC £14.04	0.60	5.41	15.09	nr	20.50
Bronze gate valve; screwed joints to iron						
15 mm	PC £12.90	0.36	3.24	13.87	nr	17.11
20 mm	PC £17.16	0.48	4.32	18.45	nr	22.77
25 mm	PC £22.53	0.60	5.41	24.22	nr	29.63
35 mm	PC £32.38	0.72	6.49	34.81	nr	41.30
42 mm	PC £42.47	0.84	7.57	45.66	nr	53.23
54 mm	PC £61.55	0.96	8.65	66.16	nr	74.81
Chromium plated; pre-setting radiator valve; compression joint; union outlet 15 mm	PC £3.54	0.26	2.34	3.81	nr	6.15
Chromium plated; lockshield radiator valve; compression joint; union outlet 15 mm	PC £3.54	0.26	2.34	3.81	nr	6.15
Brass ball valves; BS 1212 Part 1; piston type; high pressure; copper float; screwed joint to cistern						
15 mm	PC £3.08	0.30	2.70	5.79	nr	8.49
22 mm	PC £5.88	0.36	3.24	9.52	nr	12.76
25 mm	PC £13.06	0.42	3.78	18.17	nr	21.95
Water tanks/cisterns						
Polyethylene cold water feed and expansion cistern; BS 4213; with covers						
ref PC15; 68 litres	PC £20.77	1.50	13.51	22.33	nr	35.84
ref PC25; 114 litres	PC £27.56	1.75	15.76	29.62	nr	45.38
ref PC40; 182 litres	PC £47.16	2.10	18.92	50.70	nr	69.62
ref PC50; 227 litres	PC £50.50	2.35	21.17	54.29	nr	75.46
GRP cold water storage cistern; with covers						
ref 899.10; 27 litres	PC £24.64	1.30	11.71	26.49	nr	38.20
ref 899.25; 68 litres	PC £40.01	1.50	13.51	43.01	nr	56.52
ref 899.40; 114 litres	PC £47.22	1.75	15.76	50.76	nr	66.52
ref 899.70; 227 litres	PC £94.05	2.35	21.17	101.10	nr	122.27
Storage cylinders/calorifiers						
Copper cylinders; direct; BS 699; grade 3						
ref 1; 350 x 900 mm; 74 litres	PC £48.99	1.80	16.22	52.66	nr	68.88
ref 2; 450 x 750 mm; 98 litres	PC £50.55	2.10	18.92	54.34	nr	73.26
ref 7; 450 x 900 mm; 120 litres	PC £55.38	2.40	21.62	59.53	nr	81.15
ref 8; 450 x 1050 mm; 144 litres	PC £60.44	3.35	30.18	64.97	nr	95.15
ref 9; 450 x 1200 mm; 166 litres	PC £71.04	4.30	38.74	76.37	nr	115.11

S PIPED SUPPLY SYSTEMS Including overheads and profit at 7.50%		Labour hours	Labour £	Material £	Unit	Total rate £
S10/S11 HOT AND COLD WATER - cont'd						
Storage cylinders/calorifiers - cont'd						
Copper cylinders; single feed coil indirect; BS 1566 Part 2; grade 3						
ref 2; 300 x 1500 mm; 96 litres	PC £61.15	2.40	21.62	65.74	nr	87.36
ref 3; 400 x 1050 mm; 114 litres	PC £65.11	2.70	24.32	69.99	nr	94.31
ref 7; 450 x 900 mm; 117 litres	PC £64.26	3.00	27.03	69.08	nr	96.11
ref 8; 450 x 1050 mm; 140 litres	PC £72.75	3.60	32.43	78.21	nr	110.64
ref 9; 450 x 1200 mm; 162 litres	PC £92.92	4.20	37.84	99.89	nr	137.73
Combination copper hot water storage units; coil direct; BS 3198; (hot/cold)						
400 x 900 mm; (65/20 litres)	PC £75.63	3.35	30.18	81.30	nr	111.48
450 x 900 mm; (85/25 litres)	PC £77.90	4.70	42.34	83.74	nr	126.08
450 x 1050 mm; (115/25 litres)	PC £85.69	5.90	53.15	92.12	nr	145.27
450 x 1200 mm; (115/45 litres)	PC £91.21	6.60	59.46	98.05	nr	157.51
Combination copper hot water storage						
450 x 900 mm; (85/25 litres)	PC £97.68	5.30	47.75	105.01	nr	152.76
450 x 1200 mm; (115/45 litres)	PC £106.25	7.20	64.86	114.22	nr	179.08
Thermal insulation						
19 mm thick rigid mineral glass fibre sectional pipe lagging; plain finish; fixed with aluminium bands to steel or copper pipework; including working over pipe fittings						
around 15/15 mm pipes	PC £2.11	0.08	0.72	2.27	m	2.99
around 20/22 mm pipes	PC £2.23	0.12	1.08	2.39	m	3.47
around 25/28 mm pipes	PC £2.46	0.13	1.17	2.64	m	3.81
around 32/35 mm pipes	PC £2.72	0.14	1.26	2.93	m	4.19
around 40/42 mm pipes	PC £2.89	0.16	1.44	3.11	m	4.55
around 50/54 mm pipes	PC £3.35	0.18	1.62	3.60	m	5.22
19 mm thick rigid mineral glass fibre sectional pipe lagging; canvas or class O laquered aluminium finish; fixed with aluminium bands to steel or copper pipework; including working over pipe fittings						
around 15/15 mm pipes	PC £2.65	0.08	0.72	2.84	m	3.56
around 20/22 mm pipes	PC £2.87	0.12	1.08	3.09	m	4.17
around 25/28 mm pipes	PC £3.15	0.13	1.17	3.39	m	4.56
around 32/35 mm pipes	PC £3.43	0.14	1.26	3.69	m	4.95
around 40/42 mm pipes	PC £3.70	0.16	1.44	3.98	m	5.42
around 50/54 mm pipes	PC £4.30	0.18	1.62	4.62	m	6.24
25 mm thick expanded polystyrene lagging sets; class O finish; for mild steel cisterns to BS 417; complete with fixing bands; for cisterns size (ref)						
762 x 584 x 610 mm; (SCM270)	PC £4.08	0.96	8.65	4.39	nr	13.04
1219 x 610 x 610 mm; (SCM450/1)	PC £13.74	1.10	9.91	14.77	nr	24.68
1524 x 1143 x 914 mm; (SCM1600)	PC £30.60	1.30	11.71	32.90	nr	44.61
1829 x 1219 x 1219 mm; (SCM2270)	PC £38.86	1.55	13.96	41.77	nr	55.73
2438 x 1524 x 1219 mm; (SCM4540)	PC £61.88	1.80	16.22	66.52	nr	82.74
50 mm thick glass-fibre filled polyethylene insulating jackets for GRP or polyethylene cold water cisterns; complete with fixing bands; for cisterns size						
445 x 305 x 300 mm; (18 litres)	PC £2.64	0.48	4.32	2.84	nr	7.16
495 x 368 x 362 mm; (27 litres)	PC £5.34	0.60	5.41	5.74	nr	11.15
630 x 450 x 420 mm; (68 litres)	PC £5.34	0.72	6.49	5.74	nr	12.23
665 x 490 x 515 mm; (91 litres)	PC £5.96	0.84	7.57	6.41	nr	13.98
700 x 540 x 535 mm; (114 litres)	PC £6.89	0.96	8.65	7.40	nr	16.05
955 x 605 x 595 mm; (182 litres)	PC £8.95	1.00	9.01	9.62	nr	18.63
1155 x 640 x 595 mm; (227 litres)	PC £10.23	1.10	9.91	11.00	nr	20.91

S PIPED SUPPLY SYSTEMS Including overheads and profit at 7.50%		Labour hours	Labour £	Material £	Unit	Total rate £
80 mm thick glass-fibre filled insulating jackets in flame retardant PVC to BS 1763; type 1B; segmental type for hot water cylinders; complete with fixing bands; for cylinders size (ref)						
400 x 900 mm; (2)	PC £7.50	0.40	3.60	8.06	nr	11.66
450 x 750 mm; (5)	PC £6.58	0.40	3.60	7.07	nr	10.67
450 x 900 mm; (7)	PC £6.74	0.40	3.60	7.25	nr	10.85
450 x 1050 mm;(8)	PC £8.44	0.48	4.32	9.07	nr	13.39
500 x 1200 mm;(-)	PC £11.83	0.60	5.41	12.71	nr	18.12
S13 PRESSURISED WATER						
Copper pipes; BS 2871 Part 1 table Y; annealed; mains pipework; no joints in the running length; laid in trenches						
Pipes						
15 mm	PC £260.00/100m	0.12	1.08	2.93	m	4.01
22 mm	PC £455.00/100m	0.13	1.17	5.14	m	6.31
28 mm	PC £653.00/100m	0.14	1.26	7.37	m	8.63
35 mm	PC £1025.00/100m	0.16	1.44	11.57	m	13.01
42 mm	PC £1239.00/100m	0.18	1.62	13.99	m	15.61
54 mm	PC £2116.00/100m	0.22	1.98	23.89	m	25.87
Blue MDPE pipes; BS 6527; mains pipework; no joints in the running length; laid in trenches						
Pipes						
20 mm	PC £34.11/100m	0.13	1.17	0.39	m	1.56
25 mm	PC £41.60/100m	0.14	1.26	0.47	m	1.73
32 mm	PC £68.22/100m	0.16	1.44	0.77	m	2.21
50 mm	PC £174.72/100m	0.18	1.62	1.97	m	3.59
60 mm	PC £253.76/100m	0.19	1.71	2.87	m	4.58
Steel pipes; BS 1387; heavy weight; galvanized; mains pipework; screwed joints in the running length; laid in trenches						
Pipes						
15 mm	PC £147.11/100m	0.16	1.44	1.66	m	3.10
20 mm	PC £168.64/100m	0.17	1.53	1.90	m	3.43
25 mm	PC £241.36/100m	0.19	1.71	2.72	m	4.43
32 mm	PC £302.38/100m	0.22	1.98	3.41	m	5.39
40 mm	PC £352.78/100m	0.24	2.16	3.99	m	6.15
50 mm	PC £488.82/100m	0.30	2.70	5.51	m	8.21
Ductile iron bitumen coated pipes and fittings; BS 4772; class K9; Stanton's 'Tyton' water main pipes or similar; flexible joints						
100 mm pipes; laid straight	PC £51.48/5.5m	0.69	4.66	10.99	m	15.65
Extra for						
bend; 45 degrees	PC £19.19	0.69	4.66	25.12	nr	29.78
branch; 45 degrees; socketted	PC £102.50	1.05	7.09	116.93	nr	124.02
tee	PC £32.07	1.05	7.09	41.22	nr	48.31
flanged spigot	PC £14.55	0.69	4.66	17.89	nr	22.55
flanged socket	PC £18.06	0.69	4.66	21.66	nr	26.32
150 mm pipes; laid straight	PC £75.90/5.5m	0.81	5.47	16.04	m	21.51
Extra for						
bend; 45 degrees	PC £33.09	0.81	5.47	40.54	nr	46.01
branch; 45 degrees; socketted	PC £115.92	1.20	8.10	132.06	nr	140.16
tee	PC £50.57	1.20	8.10	61.81	nr	69.91
flanged spigot	PC £24.95	0.81	5.47	29.30	nr	34.77
flanged socket	PC £28.30	0.81	5.47	32.91	nr	38.38

S PIPED SUPPLY SYSTEMS Including overheads and profit at 7.50%		Labour hours	Labour £	Material £	Unit	Total rate £
S13 PRESSURISED WATER - cont'd						
Ductile iron bitumen coated pipes and fittings; BS 4772; class K9; Stanton's 'Tyton' water main pipes or similar; flexible joints - cont'd						
200 mm pipes; laid straight	PC £118.80/5.5m	1.15	7.76	24.96	m	32.72
Extra for						
bend; 45 degrees	PC £64.96	1.15	7.76	75.98	nr	83.74
branch; 45 degrees; socketted	PC £162.29	1.70	11.48	183.69	nr	195.17
tee	PC £98.20	1.70	11.48	114.79	nr	126.27
flanged spigot	PC £39.56	1.15	7.76	45.60	nr	53.36
flanged socket	PC £42.12	1.15	7.76	48.35	nr	56.11
S32 NATURAL GAS						
Ductile iron bitumen coated pipes and fittings; BS 4772; class K9; Stanton's 'Stanlock' gas main pipes or similar; bolted gland joints						
100 mm pipes; laid straight	PC £57.26/5.5m	0.81	5.47	13.33	m	18.80
Extra for						
bend; 45 degrees	PC £20.19	0.81	5.47	34.65	nr	40.12
tee	PC £31.09	1.20	8.10	55.00	nr	63.10
flanged spigot	PC £14.55	0.81	5.47	24.27	nr	29.74
flanged socket	PC £17.55	0.81	5.47	27.50	nr	32.97
isolated 'Stanlock' joint	PC £8.03	0.40	2.70	8.63	nr	11.33
150 mm pipes; laid straight	PC £85.97/5.5m	1.05	7.09	19.92	m	27.01
Extra for						
bend; 45 degrees	PC £34.84	1.05	7.09	56.08	nr	63.17
tee	PC £49.27	1.55	10.46	84.01	nr	94.47
flanged spigot	PC £24.95	1.05	7.09	39.24	nr	46.33
flanged socket	PC £27.52	1.05	7.09	42.00	nr	49.09
isolated 'Stanlock' joint	PC £11.55	0.52	3.51	12.42	nr	15.93
200 mm pipes; laid straight	PC £125.79/5.5m	1.50	10.13	28.85	m	38.98
Extra for						
bend; 45 degrees	PC £63.17	1.50	10.13	92.74	nr	102.87
tee	PC £94.58	2.25	15.19	143.06	nr	158.25
flanged spigot	PC £39.56	1.50	10.13	59.08	nr	69.21
flanged socket	PC £40.91	1.50	10.13	60.53	nr	70.66
isolated 'Stanlock' joint	PC £15.40	0.75	5.06	16.56	nr	21.62

T MECHANICAL HEATING SYSTEMS ETC. Including overheads and profit at 7.50%		Labour hours	Labour £	Material £	Unit	Total rate £
T10 GAS/OIL FIRED BOILERS						
Gas fired domestic boilers; cream or white enamelled casing; 32 mm BSPT female flow and return tappings; 102 mm flue socket 13 mm BSPT male draw-off outlet; electric controls						
13.19 kW output	PC £319.97	6.00	54.05	344.26	nr	398.31
23.45 kW output	PC £483.15	6.60	59.46	519.68	nr	579.14
Smoke flue pipework; light quality 'Duracem' pipes and fittings; including asbestos yarn and composition joints in the running length						
75 mm pipes; fixing with wall clips; plugged and screwed	PC £9.56/1.8m	0.42	3.78	6.21	m	9.99
Extra for						
loose sockets	PC £2.89	0.48	4.32	3.62	nr	7.94
bend; square and obtuse	PC £4.15	0.48	4.32	4.97	nr	9.29
Terminal cone caps; asbestos yarn and composition joint	PC £15.88	0.48	4.32	17.58	nr	21.90
100 mm pipes; fixing with wall clips; plugged and screwed	PC £12.32/1.8m	0.48	4.32	8.06	m	12.38
Extra for						
loose sockets	PC £3.75	0.54	4.86	4.77	nr	9.63
bend; square and obtuse	PC £5.19	0.54	4.86	6.32	nr	11.18
Terminal cone caps; asbestos yarn and composition joint	PC £16.86	0.54	4.86	18.87	nr	23.73
150 mm pipes; fixing with wall clips; plugged and screwed	PC £20.61/1.8m	0.60	5.41	13.48	m	18.89
Extra for						
loose sockets	PC £6.42	0.66	5.95	8.14	nr	14.09
bend; square and obtuse	PC £8.00	0.66	5.95	9.84	nr	15.79
Terminal cone caps; asbestos yarn and composition joint	PC £25.44	0.66	5.95	28.59	nr	34.54
Smoke flue pipework; heavy quality 'Duracem' pipes and fittings; including asbestos yarn and composition joints in the running length						
125 mm pipes; fixing with wall clips; plugged and screwed	PC £21.18/1.8m	0.54	4.86	13.69	m	18.55
Extra for						
loose sockets	PC £6.77	0.60	5.41	8.27	nr	13.68
bend; square and obtuse	PC £9.13	0.60	5.41	10.81	nr	16.22
Terminal cone caps; asbestos yarn and composition joint	PC £18.53	0.60	5.41	20.91	nr	26.32
175 mm pipes; fixing with wall clips; plugged and screwed	PC £39.01/1.8m	0.66	5.95	25.02	m	30.97
Extra for						
loose sockets	PC £10.34	0.72	6.49	12.54	nr	19.03
bend; square and obtuse	PC £16.72	0.72	6.49	19.40	nr	25.89
Terminal cone caps; asbestos yarn and composition joint	PC £47.13	0.72	6.49	52.09	nr	58.58
225 mm pipes; fixing with wall clips; plugged and screwed	PC £51.12/1.8m	0.78	7.03	32.73	m	39.76
Extra for						
loose sockets	PC £14.54	0.84	7.57	17.40	nr	24.97
bend; square and obtuse	PC £28.49	0.84	7.57	32.39	nr	39.96
Terminal cone caps; asbestos yarn and composition joint	PC £60.90	0.84	7.57	67.23	nr	74.80

T MECHANICAL HEATING SYSTEMS ETC. Including overheads and profit at 7.50%		Labour hours	Labour £	Material £	Unit	Total rate £
T30 MEDIUM TEMPERATURE HOT WATER HEATING						
Steel pipes; BS 1387; black; screwed joints; malleable iron fittings; BS 143						
15 mm pipes						
medium weight; fixing with pipe brackets; plugged and screwed	PC £81.27/100m	0.36	3.24	1.39	m	4.63
heavy weight; fixing with pipe brackets; plugged and screwed	PC £95.40/100m	0.36	3.24	1.55	m	4.79
Extra for						
fittings with one end		0.16	1.44	0.30	nr	1.74
fittings with two ends		0.30	2.70	0.39	nr	3.09
fittings with three ends		0.44	3.96	0.47	nr	4.43
fittings with four ends		0.60	5.41	1.08	nr	6.49
cap	PC £0.28	0.16	1.44	0.30	nr	1.74
socket; equal	PC £0.30	0.30	2.70	0.32	nr	3.02
socket; reducing	PC £0.36	0.30	2.70	0.39	nr	3.09
bend; 90 degree long radius M/F	PC £0.64	0.30	2.70	0.69	nr	3.39
elbow; 90 degree M/F	PC £0.36	0.30	2.70	0.39	nr	3.09
tee; equal	PC £0.44	0.44	3.96	0.47	nr	4.43
cross	PC £1.00	0.60	5.41	1.08	nr	6.49
union	PC £0.96	0.30	2.70	1.03	nr	3.73
isolated screwed joint	PC £0.30	0.36	3.24	0.44	nr	3.68
tank connection; longscrew and backnuts; lead washers; joint	PC £1.53	0.44	3.96	1.77	nr	5.73
20 mm pipes						
medium weight; fixing with pipe brackets; plugged and screwed	PC £96.62/100m	0.38	3.42	1.61	m	5.03
heavy weight; fixing with pipe brackets; plugged and screwed	PC £114.32/100m	0.38	3.42	1.81	m	5.23
Extra for						
fittings with one end		0.19	1.71	0.34	nr	2.05
fittings with two ends		0.40	3.60	0.52	nr	4.12
fittings with three ends		0.56	5.04	0.69	nr	5.73
fittings with four ends		0.79	7.12	1.63	nr	8.75
cap	PC £0.32	0.19	1.71	0.34	nr	2.05
socket; equal	PC £0.36	0.40	3.60	0.39	nr	3.99
socket; reducing	PC £0.44	0.40	3.60	0.47	nr	4.07
bend; 90 degree long radius M/F	PC £1.04	0.40	3.60	1.12	nr	4.72
elbow; 90 degree M/F	PC £0.48	0.40	3.60	0.52	nr	4.12
tee; equal	PC £0.64	0.56	5.04	0.69	nr	5.73
cross	PC £1.52	0.79	7.12	1.63	nr	8.75
union	PC £1.12	0.40	3.60	1.20	nr	4.80
isolated screwed joint	PC £0.36	0.48	4.32	0.51	nr	4.83
tank connection; longscrew and backnuts; lead washers; joint	PC £1.78	0.56	5.04	2.03	nr	7.07
25 mm pipes						
medium weight; fixing with pipe brackets; plugged and screwed	PC £139.29/100m	0.43	3.87	2.15	m	6.02
heavy weight; fixing with pipe brackets; plugged and screwed	PC £167.71/100m	0.43	3.87	2.47	m	6.34
Extra for						
fittings with one end		0.25	2.25	0.43	nr	2.68
fittings with two ends		0.50	4.50	0.86	nr	5.36
fittings with three ends		0.68	6.13	0.99	nr	7.12
fittings with four ends		1.00	9.01	2.06	nr	11.07
cap	PC £0.40	0.25	2.25	0.43	nr	2.68
socket; equal	PC £0.48	0.50	4.50	0.52	nr	5.02
socket; reducing	PC £0.58	0.50	4.50	0.62	nr	5.12
bend; 90 degree long radius M/F	PC £1.48	0.50	4.50	1.59	nr	6.09
elbow; 90 degree M/F	PC £0.80	0.50	4.50	0.86	nr	5.36
tee; equal	PC £0.92	0.68	6.13	0.99	nr	7.12
cross	PC £1.92	1.00	9.01	2.06	nr	11.07

T MECHANICAL HEATING SYSTEMS ETC. Including overheads and profit at 7.50%		Labour hours	Labour £	Material £	Unit	Total rate £
union	PC £1.32	0.50	4.50	1.42	nr	5.92
isolated screwed joint	PC £0.48	0.60	5.41	0.66	nr	6.07
tank connection; longscrew and backnuts; lead washers; joint	PC £2.49	0.68	6.13	2.82	nr	8.95
32 mm pipes						
medium weight; fixing with pipe brackets; plugged and screwed	PC £174.04/100m	0.50	4.50	2.72	m	7.22
heavy weight; fixing with pipe brackets; plugged and screwed	PC £210.08/100m	0.50	4.50	3.13	m	7.63
Extra for						
fittings with one end		0.30	2.70	0.62	nr	3.32
fittings with two ends		0.60	5.41	1.42	nr	6.83
fittings with three ends		0.80	7.21	1.63	nr	8.84
fittings with four ends		1.20	10.81	2.71	nr	13.52
cap	PC £0.58	0.30	2.70	0.62	nr	3.32
socket; equal	PC £0.80	0.60	5.41	0.86	nr	6.27
socket; reducing	PC £0.92	0.60	5.41	0.99	nr	6.40
bend; 90 degree long radius M/F	PC £2.48	0.60	5.41	2.67	nr	8.08
elbow; 90 degree M/F	PC £1.32	0.60	5.41	1.42	nr	6.83
tee; equal	PC £1.52	0.80	7.21	1.63	nr	8.84
cross	PC £2.52	1.20	10.81	2.71	nr	13.52
union	PC £2.24	0.60	5.41	2.41	nr	7.82
isolated screwed joint	PC £0.80	0.72	6.49	1.01	nr	7.50
tank connection; longscrew and backnuts; lead washers; joint	PC £3.36	0.80	7.21	3.76	nr	10.97
40 mm pipes						
medium weight; fixing with pipe brackets; plugged and screwed	PC £202.33/100m	0.60	5.41	3.25	m	8.66
heavy weight; fixing with pipe brackets; plugged and screwed	PC £244.82/100m	0.60	5.41	3.73	m	9.14
Extra for						
fittings with one end		0.35	3.15	0.80	nr	3.95
fittings with two ends		0.70	6.31	2.11	nr	8.42
fittings with three ends		0.94	8.47	2.24	nr	10.71
fittings with four ends		1.40	12.61	3.66	nr	16.27
cap	PC £0.74	0.35	3.15	0.80	nr	3.95
socket; equal	PC £1.08	0.70	6.31	1.16	nr	7.47
socket; reducing	PC £1.20	0.70	6.31	1.29	nr	7.60
bend; 90 degree long radius M/F	PC £3.28	0.70	6.31	3.53	nr	9.84
elbow; 90 degree M/F	PC £1.96	0.70	6.31	2.11	nr	8.42
tee; equal	PC £2.08	0.94	8.47	2.24	nr	10.71
cross	PC £3.40	1.40	12.61	3.66	nr	16.27
union	PC £2.82	0.70	6.31	3.03	nr	9.34
isolated screwed joint	PC £1.08	0.84	7.57	1.33	nr	8.90
tank connection; longscrew and backnuts; lead washers; joint	PC £4.04	0.94	8.47	4.51	nr	12.98
50 mm pipes						
medium weight; fixing with pipe brackets; plugged and screwed	PC £284.71/100m	0.72	6.49	4.46	m	10.95
heavy weight; fixing with pipe brackets; plugged and screwed	PC £340.15/100m	0.72	6.49	5.09	m	11.58
Extra for						
fittings with one end		0.40	3.60	1.51	nr	5.11
fittings with two ends		0.80	7.21	2.71	nr	9.92
fittings with three ends		1.05	9.46	3.23	nr	12.69
fittings with four ends		1.60	14.41	5.68	nr	20.09
cap	PC £1.40	0.40	3.60	1.51	nr	5.11
socket; equal	PC £1.68	0.80	7.21	1.81	nr	9.02
socket; reducing	PC £1.68	0.80	7.21	1.81	nr	9.02
bend; 90 degree long radius M/F	PC £5.60	0.80	7.21	6.02	nr	13.23
elbow; 90 degree M/F	PC £2.52	0.80	7.21	2.71	nr	9.92
tee; equal	PC £3.00	1.05	9.46	3.23	nr	12.69
cross	PC £5.28	1.60	14.41	5.68	nr	20.09

T MECHANICAL HEATING SYSTEMS ETC. Including overheads and profit at 7.50%		Labour hours	Labour £	Material £	Unit	Total rate £
T30 MEDIUM TEMPERATURE HOT WATER HEATING - cont'd						
Steel pipes; BS 1387; black; screwed joints; malleable iron fittings; BS 143 - cont'd						
50 mm pipes						
union	PC £4.20	0.80	7.21	4.52	nr	11.73
isolated screwed joint	PC £1.68	0.96	8.65	2.00	nr	10.65
tank connection; longscrew and backnuts; lead washers; joint	PC £6.32	1.05	9.46	6.99	nr	16.45
Steel pipes; BS 1387; galvanized; screwed joints; galvanized malleable iron fittings; BS 143						
15 mm pipes						
medium weight; fixing with pipe brackets; plugged and screwed	PC £126.17/100m	0.36	3.24	2.03	m	5.27
heavy weight; fixing with pipe brackets; plugged and screwed	PC £147.11/100m	0.36	3.24	2.27	m	5.51
Extra for						
fittings with one end		0.16	1.44	0.41	nr	1.85
fittings with two ends		0.30	2.70	0.53	nr	3.23
fittings with three ends		0.44	3.96	0.65	nr	4.61
fittings with four ends		0.60	5.41	1.48	nr	6.89
cap	PC £0.39	0.16	1.44	0.41	nr	1.85
socket; equal	PC £0.41	0.30	2.70	0.44	nr	3.14
socket; reducing	PC £0.50	0.30	2.70	0.53	nr	3.23
bend; 90 degree long radius M/F	PC £0.88	0.30	2.70	0.95	nr	3.65
elbow; 90 degree M/F	PC £0.50	0.30	2.70	0.53	nr	3.23
tee; equal	PC £0.61	0.44	3.96	0.65	nr	4.61
cross	PC £1.38	0.60	5.41	1.48	nr	6.89
union	PC £1.32	0.30	2.70	1.42	nr	4.12
isolated screwed joint	PC £0.41	0.36	3.24	0.56	nr	3.80
tank connection; longscrew and backnuts; lead washers; joint	PC £1.90	0.44	3.96	2.15	nr	6.11
20 mm pipes						
medium weight; fixing with pipe brackets; plugged and screwed	PC £143.74/100m	0.38	3.42	2.29	m	5.71
heavy weight; fixing with pipe brackets; plugged and screwed	PC £168.64/100m	0.38	3.42	2.57	m	5.99
Extra for						
fittings with one end		0.19	1.71	0.47	nr	2.18
fittings with two ends		0.40	3.60	0.71	nr	4.31
fittings with three ends		0.56	5.04	0.95	nr	5.99
fittings with four ends		0.79	7.12	2.25	nr	9.37
cap	PC £0.44	0.19	1.71	0.47	nr	2.18
socket; equal	PC £0.50	0.40	3.60	0.53	nr	4.13
socket; reducing	PC £0.61	0.40	3.60	0.65	nr	4.25
bend; 90 degree long radius M/F	PC £1.43	0.40	3.60	1.54	nr	5.14
elbow; 90 degree M/F	PC £0.66	0.40	3.60	0.71	nr	4.31
tee; equal	PC £0.88	0.56	5.04	0.95	nr	5.99
cross	PC £2.09	0.79	7.12	2.25	nr	9.37
union	PC £1.54	0.40	3.60	1.66	nr	5.26
isolated screwed joint	PC £0.50	0.48	4.32	0.65	nr	4.97
tank connection; longscrew and backnuts; lead washers; joint	PC £2.20	0.56	5.04	2.48	nr	7.52

T MECHANICAL HEATING SYSTEMS ETC. Including overheads and profit at 7.50%		Labour hours	Labour £	Material £	Unit	Total rate £
25 mm pipes						
medium weight; fixing with pipe brackets; plugged and screwed	PC £201.80/100m	0.43	3.87	3.03	m	6.90
heavy weight; fixing with pipe brackets; plugged and screwed	PC £241.36/100m	0.43	3.87	3.48	m	7.35
Extra for						
fittings with one end		0.25	2.25	0.59	nr	2.84
fittings with two ends		0.50	4.50	1.18	nr	5.68
fittings with three ends		0.68	6.13	1.36	nr	7.49
fittings with four ends		1.00	9.01	2.84	nr	11.85
cap	PC £0.55	0.25	2.25	0.59	nr	2.84
socket; equal	PC £0.66	0.50	4.50	0.71	nr	5.21
socket; reducing	PC £0.80	0.50	4.50	0.86	nr	5.36
bend; 90 degree long radius M/F	PC £2.04	0.50	4.50	2.19	nr	6.69
elbow; 90 degree M/F	PC £1.10	0.50	4.50	1.18	nr	5.68
tee; equal	PC £1.27	0.68	6.13	1.36	nr	7.49
cross	PC £2.64	1.00	9.01	2.84	nr	11.85
union	PC £1.82	0.50	4.50	1.95	nr	6.45
isolated screwed joint	PC £0.66	0.60	5.41	0.85	nr	6.26
tank connection; longscrew and backnuts; lead washers; joint	PC £3.08	0.68	6.13	3.45	nr	9.58
32 mm pipes						
medium weight; fixing with pipe brackets; plugged and screwed	PC £252.11/100m	0.50	4.50	3.84	m	8.34
heavy weight; fixing with pipe brackets; plugged and screwed	PC £302.38/100m	0.50	4.50	4.41	m	8.91
Extra for						
fittings with three ends		0.30	2.70	0.86	nr	3.56
fittings with two ends		0.60	5.41	1.95	nr	7.36
fittings with three ends		0.80	7.21	2.25	nr	9.46
fittings with four ends		1.20	10.81	3.72	nr	14.53
cap	PC £0.80	0.30	2.70	0.86	nr	3.56
socket; equal	PC £1.10	0.60	5.41	1.18	nr	6.59
socket; reducing	PC £1.27	0.60	5.41	1.36	nr	6.77
bend; 90 degree long radius M/F	PC £3.41	0.60	5.41	3.67	nr	9.08
elbow; 90 degree M/F	PC £1.82	0.60	5.41	1.95	nr	7.36
tee; equal	PC £2.09	0.80	7.21	2.25	nr	9.46
cross	PC £3.47	1.20	10.81	3.72	nr	14.53
union	PC £3.08	0.60	5.41	3.31	nr	8.72
isolated screwed joint	PC £1.10	0.72	6.49	1.33	nr	7.82
tank connection; longscrew and backnuts; lead washers; joint	PC £4.16	0.80	7.21	4.61	nr	11.82
40 mm pipes						
medium weight; fixing with pipe brackets; plugged and screwed	PC £292.96/100m	0.60	5.41	4.60	m	10.01
heavy weight; fixing with pipe brackets; plugged and screwed	PC £352.78/100m	0.60	5.41	5.28	m	10.69
Extra for						
fittings with one end		0.35	3.15	1.09	nr	4.24
fittings with two ends		0.70	6.31	2.90	nr	9.21
fittings with three ends		0.94	8.47	3.07	nr	11.54
fittings with four ends		1.40	12.61	5.03	nr	17.64
cap	PC £1.02	0.35	3.15	1.09	nr	4.24
socket; equal	PC £1.49	0.70	6.31	1.60	nr	7.91
socket; reducing	PC £1.65	0.70	6.31	1.77	nr	8.08
bend; 90 degree long radius M/F	PC £4.51	0.70	6.31	4.85	nr	11.16
elbow; 90 degree M/F	PC £2.70	0.70	6.31	2.90	nr	9.21
tee; equal	PC £2.86	0.94	8.47	3.07	nr	11.54
cross	PC £4.68	1.40	12.61	5.03	nr	17.64
union	PC £3.88	0.70	6.31	4.17	nr	10.48
isolated screwed joint	PC £1.49	0.84	7.57	1.77	nr	9.34
tank connection; longscrew and backnuts; lead washers; joint	PC £4.99	0.94	8.47	5.53	nr	14.00

T MECHANICAL HEATING SYSTEMS ETC. Including overheads and profit at 7.50%		Labour hours	Labour £	Material £	Unit	Total rate £
T30 MEDIUM TEMPERATURE HOT WATER HEATING - cont'd						
Steel pipes; BS 1387; galvanized; screwed joints; galvanized malleable iron fittings; BS 143 - cont'd						
50 mm pipes						
medium weight; fixing with pipe brackets; plugged and screwed	PC £411.08/100m	0.72	6.49	6.32	m	12.81
heavy weight; fixing with pipe brackets; plugged and screwed	PC £488.82/100m	0.72	6.49	7.19	m	13.68
Extra for						
fittings with one end		0.40	3.60	1.73	nr	5.33
fittings with two ends		0.80	7.21	3.72	nr	10.93
fittings with three ends		1.05	9.46	4.43	nr	13.89
fittings with four ends		1.60	14.41	7.80	nr	22.21
cap	PC £1.61	0.40	3.60	1.73	nr	5.33
socket; equal	PC £2.31	0.80	7.21	2.48	nr	9.69
socket; reducing	PC £2.31	0.80	7.21	2.48	nr	9.69
bend; 90 degree long radius M/F	PC £7.70	0.80	7.21	8.28	nr	15.49
elbow; 90 degree M/F	PC £3.47	0.80	7.21	3.72	nr	10.93
tee; equal	PC £4.13	1.05	9.46	4.43	nr	13.89
cross	PC £7.26	1.60	14.41	7.80	nr	22.21
union	PC £5.78	0.80	7.21	6.21	nr	13.42
isolated screwed joint	PC £2.31	0.96	8.65	2.68	nr	11.33
tank connection; longscrew and backnuts; lead washers; joint	PC £7.85	1.05	9.46	8.63	nr	18.09
Radiators						
Radiators; pressed steel panel type, 590 mm high; 3 mm chromium plated air valve; 15 mm chromium plated easy clean straight valve with union; 15 mm chromium plated lockshield valve with union						
1.69 m2 single surface	PC £71.68	2.40	21.62	83.24	nr	104.86
2.12 m2 single surface	PC £81.59	2.70	24.32	93.94	nr	118.26
2.75 m2 single surface	PC £107.47	3.00	27.03	121.77	nr	148.80
3.39 m2 single surgace	PC £126.94	3.30	29.73	142.70	nr	172.43
Vibration isolation mountings						
'Tico' anti-vibration rubber pads; type CV/LF/N; fixed with Tico A/WB epoxy mortar; on concrete base						
150 x 75 x 32 mm thick		0.90	8.11	12.42	nr	20.53
150 x 150 x 64 mm thick		1.20	10.81	35.47	nr	46.28

V ELECTRICAL SYSTEMS Including overheads and profit at 7.50%	Labour hours	Labour £	Material £	Unit	Total rate £
V21 - 22 GENERAL LIGHTING AND LV POWER					
NOTE: The following items indicate approximate prices for wiring of lighting and power points complete, including accessories and socket outlets but excluding lighting fittings. Consumer control units are shown separately. For a more detailed breakdown of these costs and specialist costs for a complete range of electrical items reference should be made to Spon's Mechanical and Electrical Services Price Book.					
Consumer control units					
8-way 60 amp SP&N surface mounted insulated consumer control units fitted with miniature circuit breakers including 2 m long 32 mm screwed welded conduit with three runs of 16 mm2 PVC cables ready for final connections by the supply authority	-	-	-	nr	130.00
Extra for current operated ELCB of 30 mA tripping current	-	-	-	nr	57.50
as above but 100 amp metal cased consumer unit and 25 mm2 PVC cables	-	-	-	nr	146.00
Extra for current operated ELCB of 30 mA tripping current	-	-	-	nr	120.00
Final circuits					
Lighting points					
Wired in PVC insulated and PVC sheathed cable in flats and houses; insulated in cavities and roof space; protected where buried by heavy gauge PVC conduit	-	-	-	nr	41.50
As above but in commercial property	-	-	-	nr	50.00
Wired in PVC insulated cable in screwed welded conduit in flats and houses	-	-	-	nr	85.00
As above but in commercial property	-	-	-	nr	105.00
As above but in industrial property	-	-	-	nr	117.50
Wired in MICC cable in flats and houses	-	-	-	nr	72.50
As above but in commercial property	-	-	-	nr	87.50
As above but in industrial property with PVC sheathed cable	-	-	-	nr	97.50
Single 13 amp switched socket outlet points					
Wired in PVC insulated and PVC sheathed cable in flats and houses on a ring main circuit; protected where buried by heavy gauge PVC conduit	-	-	-	nr	45.00
As above but in commercial property	-	-	-	nr	55.00
Wired in PVC insulated cable in screwed welded conduit throughout on a ring main circuit in flats and houses	-	-	-	nr	65.00
As above but in commercial property	-	-	-	nr	77.50
As above but in industrial property	-	-	-	nr	88.00
Wired in MICC cable on a ring main circuit in flats and houses	-	-	-	nr	66.00
As above but in commercial property	-	-	-	nr	77.00
As above but in industrial property with PVC sheathed cable	-	-	-	nr	95.00

V ELECTRICAL SYSTEMS Including overheads and profit at 7.50%	Labour hours	Labour £	Material £	Unit	Total rate £
V21 - 22 GENERAL LIGHTING AND LV POWER - cont'd					
Cooker control units					
45 amp circuit including unit wired in PVC insulated and PVC sheathed cable; protected where buried by heavy gauge PVC conduit	-	-	-	nr	87.50
As above but wired in PVC insulated cable in screwed welded conduit	-	-	-	nr	130.00
As above but wired in MICC cable	-	-	-	nr	145.00

W SECURITY SYSTEMS Including overheads and profit at 7.50%	Labour hours	Labour £	Material £	Unit	Total rate £
W20 LIGHTNING PROTECTION					
Flag staff terminal	-	-	-	nr	59.95
Copper strip roof or down conductors fixed with bracket or saddle clips					
20 x 3 mm	-	-	-	m	13.28
25 x 3 mm	-	-	-	m	14.81
Aluminium strip roof or down conductors fixed with bracket or saddle clips					
20 x 3 mm	-	-	-	m	10.47
25 x 3 mm	-	-	-	m	10.98
Joints in tapes	-	-	-	nr	8.17
Bonding connections to roof and structural metalwork	-	-	-	nr	45.96
Testing points	-	-	-	nr	22.47
Earth electrodes					
16 mm driven copper electrodes in 1220 mm sectional lengths (2440 mm minimum)	-	-	-	nr	122.55
First 2440 mm driven and tested					
25 x 3 mm copper strip electrode in 457 mm deep prepared trench	-	-	-	m	8.88

PART IV

Approximate Estimating

This part of the book contains the following sections:

Building Costs and Tender Prices Index

The tables which follow show the changes in building costs and tender prices since 1976.

To avoid confusion it is essential that the term 'building costs' and 'tender price' are clearly defined and understood.

'Building costs' are the costs actually incurred by the builder in the course of his business the major ones being those for labour and materials.

'Tender price' is the price for which a builder offers to erect a building. This includes 'building costs' but also takes into account market considerations such as the availability of labour and materials and the prevailing economic situation. This means that in 'boom' periods when there is a surfeit of building work to be done 'tender prices' may increase at a greater rate than 'building costs' whilst in a period when work is scarce 'tender prices' may actually fall when building costs are rising.

Building costs

This table reflects the fluctuations since 1976 in wages and materials costs to the builder. In compiling the table the proportion of labour to material has been assumed to be 40:60. The wages element has been assessed from a contract wages sheet revalued for each variation in labour costs whilst the changes in the cost of materials have been based on the indices prepared by the Department of Trade and Industry. No allowance has been made for changes in productivity, plus rates or hours worked which may occur in particular conditions and localities.

1976 = 100

Year	First quarter	Second quarter	Third quarter	Fourth quarter	Annual average
1976	93	97	104	107	100
1977	109	112	116	117	114
1978	118	120	127	129	124
1979	131	135	149	153	142
1980	157	161	180	181	170
1981	182	185	195	199	190
1982	203	206	214	216	210
1983	217	219	227	229	223
1984	230	232	239	241	236
1985	243	245	252	254	249
1986	256	258	266	267	262
1987	270	272	281	282	276
1988	284	286	299	302	293
1989	305	307	322	323	314
1990	325 (F)	328 (F)	346 (F)	349 (F)	

Note: P = Provisional F = Forecast

Tender prices

This table reflects the changes in tender prices since 1976. It indicates the level of pricing contained in the lowest competitive fluctuating tenders for new work in the outer London area (over £300 000 in value) compared with a common base.

1976 = 100

Year	First quarter	Second quarter	Third quarter	Fourth quarter	Annual average
1976	97	98	102	103	100
1977	105	105	109	110	107
1978	113	116	126	139	124
1979	142	146	160	167	154
1980	179	200	192	188	190
1981	199	193	190	195	194
1982	191	188	195	195	192
1983	198	200	198	200	199
1984	205	206	214	215	210
1985	215	219	219	220	218
1986	221	226	234	234	229
1987	242	249	265	279	258
1988	289	299	321	328	309
1989	341	335	340	345	340
1990	330 (P)	325 (F)	330 (F)	330 (F)	

Regional variations

As well as being aware of inflationary trends when preparing an estimate, it is also important to establish the appropriate price level. Regional variations in price levels can be significant. Variations between London and other regions have narrowed over the past year because following an abrupt halt to the London property markets, Greater London prices have been more effected by the current slump than prices in other regions. Location adjustment factors to assist with the preparation of initial estimates are shown in the following table. For ease of reference the table shows both the forecast fourth quarter 1990 tender price indices for each region of the country and the percentage adjustment to the Prices for Measured Work. In addition, a band of percentage adjustments is shown to reflect the range of price levels within each region.

Region	Forecast fourth quarter 1990 tender price index	Percentage adjustment to Major Works section
East Anglia	300	-9%
East Midlands	274	-17%
Inner London	350	+ 6%
Northern	277	-16%
Northern Ireland	231	-30%
North West	281	-15%
Outer London	330	0%
Scotland	281	-15%
South East	312	- 6%
South West	284	-14%
Wales	281	-15%
West Midlands	281	-15%
Yorkshire and Humberside	274	-17%

Special further adjustment to the above percentages may be necessary when considering city centre or very isolated locations.

Note: P = Provisional F = Forecast

The following example illustrates the adjustment of an estimate prepared using Spon's A&B 1991, to a price level that reflects the forecast Outer London market conditions for competitive tenders in the fourth quarter 1990:

		£
A Value of items priced using Spon's A&B 1991 i.e. Tender price index 330		1 050 000
B Adjustment to increase value of A to forecast price level for fourth quarter 1990 i.e. Forecast tender price index 330 $\frac{(330 - 330)}{330} \times 100 = 0\%$		-
		1 050 000
C Value of items priced using competitive quotations that reflect the market conditions in the fourth quarter 1990		600 000
		1 650 000
D Allowance for preliminaries	+12.5% say	206 000
E Total value of estimate at fourth quarter 1990 price levels		£ 1 856 000

Alternatively, for a similar estimate in East Anglia:

		£
A Value of items priced using Spon's A&B 1991 i.e. Tender price index 330		1 050 000
B Adjustment to reduce value of A to forecast price level for fourth quarter 1991 for East Anglia (from Regional variation table) ie Tender price index 330 $\frac{(330 - 300)}{300} \times 100 = +10\%$	deduct 10%	105 000
		945 000
C Value of items priced using competitive quotations that reflect the market conditions in the fourth quarter 1990		600 000
		1 545 000
D Allowance for preliminaries	+12.5% say	193 000
E Total value of estimate at fourth quarter 1990 price levels		£ 1 738 000

Building Prices per Square Metre

Prices given under this heading are average prices, on a 'fluctuating basis', for typical buildings during the fourth quarter 1990 with a tender price level index of 330 (1976 = 100). Unless otherwise stated, prices do not allow for external works, other than those adjacent to the building, furniture, loose or special equipment and are, of course, exclusive of fees for professional services.

Prices are based upon the total floor area of all storeys, measured between external walls and without deduction for internal walls, columns, stairwells, liftwells and the like.

As in previous editions it is emphasized that the prices must be treated with reserve in that they represent the average of prices from our records and cannot provide more than a rough guide to the probable cost of a building.

In many instances normal commercial pressures together with a limited range of available specifications ensure that a single rate is sufficient to indicate the prevailing average price. However, where such restrictions do not apply a range has been given; this is not to suggest that figures outside this range will not be encountered, but simply that the calibre of such a type of building can itself vary significantly.

For assistance with the compilation of a closer estimate, or of a 'Cost Plan' the reader is directed to the 'Approximate Estimates' sections.

As elsewhere in this edition, prices do not include Value Added Tax, which should be applied at the current rate to all non-domestic building.

Utilities, civil engineering facilities (CI/SFB 1)	Square metre excluding VAT £
Surface car parking	35 to 50
Multi-storey car parks	
split level	190 to 210
flat slab	200 to 250
warped	220 to 265
Underground car parks	
partially underground under buildings	325 to 390
completely underground under buildings	390 to 475
completely underground with landscaped roof	430 to 515
Railway stations	1200 to 2000
Bus and coach stations	600 to 1000
Bus garages	575 to 625
Petrol stations	750 to 1100
Garage showrooms	450 to 700
Garages, domestic	250 to 400
Airport passenger terminal buildings (excluding aprons)	
national standard	1000 to 1200
international standard	1400 to 1800
Airport facility buildings	
large hangars	1000 to 1200
workshops and small hangers	630 to 720
TV, radio and video studios	725 to 1175
Telephone exchanges	600 to 950
Telephone engineering centres	500 to 625
Branch Post Offices	625 to 850

Utilities, civil engineering facilities (CI/SfB 1)	Square metre excluding VAT £
Postal Delivery Offices/Sorting Offices	500 to 750
Mortuaries	1150 to 1600
Sub-stations	850 to 1000
Industrial facilities (CI/SfB 2)	
Agricultural storage buildings	300 to 400
Factories	
for letting (incoming services only)	215 to 295
for letting (including lighting, power and heating)	285 to 390
nursery units (including lighting, power and heating)	355 to 515
workshops	400 to 515
maintenance/motor transport workshops	400 to 675
owner occupation-for light industrial use	360 to 515
owner occupation-for heavy industrial use	675 to 765
Factory/office buildings - high technology production	
for letting (shell and core only)	390 to 515
for letting (ground floor shell, first floor offices)	630 to 810
for owner occupation (controlled environment, fully finished)	810 to 1080
Laboratory workshops and offices	745 to 925
High technology laboratory workshop centres, air conditioned	1700 to 2150
Warehouses	
low bay (6 - 8 m high) for letting (no heating)	215 to 260
low bay for owner occupation (including heating)	300 to 380
high bay (9 - 18 m high) for owner occupation (including heating)	400 to 515
Cold stores, refrigerated stores	450 to 575
Administrative, commercial protective service facilities (CI/SfB 3)	
Embassies	1175 to 1700
County Courts	1050 to 1300
Magistrates Courts	800 to 1000
Civic offices	
non air conditioned	800 to 1000
fully air conditioned	1000 to 1175
Probation/Registrar Offices	565 to 675
Offices for letting	
low rise, non air conditioned	550 to 800
low rise, air conditioned	700 to 950
medium rise, non air conditioned	675 to 875
medium rise, air conditioned	825 to 1100
high rise, non air conditioned	850 to 1100
high rise, air conditioned	1050 to 1400
Offices for owner occupation	
low rise, non air conditioned	675 to 850
low rise, air conditioned	800 to 1100
medium rise, non air conditioned	800 to 1000
medium rise, air conditioned	1050 to 1300
high rise, air conditioned	1350 to 1650
Offices, prestige	
medium rise	1350 to 1650
high rise	1650 to 2250

	Square metre excluding VAT £
Administrative, commercial protective service facilities (CI/SfB 3) - cont'd	
Large trading floors in medium rise offices	1800 to 2100
Two storey ancillary office accommodation to warehouses/factories	565 to 630
Fitting out offices	
basic fitting out including carpets, decorations, partitions and services	180 to 225
good quality fitting out including carpets, decorations, partitions, ceilings, furniture and services	360 to 435
high quality fitting out including raised floors and carpets, decorations, partitions, ceilings, furniture, air conditioning and electrical services	630 to 785
Office refurbishment	
basic refurbishment	300 to 450
good quality, including air conditioning	600 to 800
high quality, including air conditioning	1100 to 1450
Banks	
local	925 to 1100
city centre/head office	1400 to 2000
Building Society Branch Offices	800 to 1050
refurbishments	450 to 750
Shop shells	
small	390 to 495
large including department stores and supermarkets	360 to 405
Fitting out shell for small shop (including shop fittings)	
simple store	400 to 465
fashion store	750 to 925
Fitting out shell for department store or supermarket	
excluding shop fittings	450 to 585
including shop fittings	675 to 925
Retail Warehouses	
shell	250 to 370
fitting out	180 to 210
Shopping centres	
Malls including fitting out	
comfort cooled	1350 to 1500
air-conditioned	1500 to 1700
Retail area shells, capped off services	400 to 500
Landlord's back-up areas, management offices, plant rooms non air conditioned	560 to 630
*Ambulance stations	520 to 750
Ambulance controls centre	750 to 1100
Fire stations	750 to 1050
Police stations	775 to 1150
Prisons	1050 to 1250
Health and welfare facilities (CI/SfB 4)	
*District general hospitals	825 to 1100
Refurbishment	400 to 800
Hospice	850 to 1050
Private hospitals	900 to 1250
Hospital laboratories	1050 to 1500

	Square metre excluding VAT £
Health and welfare facilities (CI/SfB 4) - cont'd	
Ward blocks	800 to 1000
Refurbishment	375 to 600
Geriatric units	825 to 1100
Psychiatric units	825 to 950
Psycho-geriatric units	775 to 1125
Maternity units	825 to 1100
Operating theatres	875 to 1350
Outpatients/casualty units	900 to 1175
Hospital teaching centres	675 to 950
*Health centres	625 to 850
Welfare centres	825 to 975
*Day centres	700 to 975
Group practice surgeries	625 to 775
*Homes for the physically handicapped	
houses	800 to 950
*Homes for the mentally handicapped	650 to 925
Geriatric day hospital	700 to 1000
Nursing homes, convalescent homes	650 to 950
*Children's homes	525 to 800
*Homes for the aged	600 to 800
Refurbishments	225 to 550
*Observation and assessment units	600 to 1000
Recreational facilities (CI/SfB 5)	
Public houses	725 to 975
Kitchen blocks (including fitting out)	1175 to 1350
Dining blocks and canteens in shop and factory	700 to 1000
Restaurants	900 to 1100
Community centres	600 to 825
General purpose halls	600 to 825
Visitors' centres	800 to 1200
Youth clubs	550 to 825
Arts and drama centres	800 to 925
Theatres, including seating and stage equipment	
large-over 500 seats	1250 to 1500
workshop - less than 500 seats	900 to 1125
Concert halls, including seating and stage equipment	1450 to 2400
Cinema	
shell	385 to 435
fitting out including all equipment, air-conditioned	800 to 925
Exhibition centres	925 to 1250
Swimming pools	
international standard	1050 to 1250
local authority standard	925 to 1175
school standard	750 to 880
leisure pools, including wave making equipment	1175 to 1550
Ice rinks	850 to 975
Rifle ranges	650 to 825
Leisure centre	
dry	525 to 775
wet	775 to 1125
Sports halls including changing	550 to 775

	Square metre excluding VAT £
Recreational facilities (CI/SfB 5) - cont'd	
School gymnasiums	550 to 650
Squash courts	550 to 775
Indoor bowls halls	450 to 600
Bowls pavilions	550 to 675
Sports pavilions	
changing only	700 to 950
social and changing	600 to 1025
Grandstands	
simple stands	450 to 525
first class stands with ancillary accommodation	800 to 1000
Clubhouses	600 to 800
Golf clubhouses	750 to 1050
Religious facilities (CI/SfB 6)	
Temples, mosques, synagogues	850 to 1200
Churches	725 to 1025
Mission halls, meeting houses	850 to 1100
Convents	775 to 875
Crematoria	975 to 1125
Educational, scientific, information facilities (CI/SfB 7)	
Nursery Schools	725 to 1050
*Primary/junior school	600 to 775
*Secondary/middle schools	550 to 700
*Extensions to schools	
classrooms	650 to 725
residential	650 to 725
laboratories	900 to 975
School refurbishment	175 to 550
Sixth form colleges	675 to 850
*Special schools	600 to 850
*Polytechnics	
Students Union buildings	650 to 750
arts buildings	575 to 675
scientific laboratories	775 to 975
*Training colleges	550 to 850
Management training centres	800 to 1075
*Universities	
arts buildings	725 to 825
science buildings	800 to 1025
College/University Libraries	675 to 950
Laboratories and offices, low level servicing	725 to 900
Laboratories (specialist, controlled environment)	1100 to 2100
Computer buildings	950 to 1375
Museums	
national, including full air conditioning and standby generator	2175 to 2575
regional, including full air conditioning	1000 to 1675
local, air conditioned	800 to 1125
conversion of existing warehouse to regional standard	775 to 1175
conversion of existing warehouse to local standard	675 to 1000
Libraries	
city centre	900 to 1050
branch	675 to 900

	Square metre excluding VAT £
Residential facilities (CI/SfB 8)	
*Local authority and housing association schemes	
Bungalows	
semi-detached	475 to 575
terraced	400 to 500
Two storey housing	
detached	450 to 550
semi-detached	400 to 500
terraced	350 to 450
Three storey housing	
semi-detached	375 to 475
terraced	325 to 425
Flats	
low rise excluding lifts	475 to 625
medium rise excluding lifts	500 to 675
Sheltered housing with wardens' accommodation	500 to 650
Terraced blocks of garages	325 to 375
Private developments	
Single detached houses	575 to 800
Houses - two or three storey	375 to 525
Flats	
standard	475 to 600
luxury	775 to 975
Warehouse conversions to apartments	575 to 850
Rehabilitation	
housing	225 to 375
flats	350 to 550
Hotels	
5 star city centre hotel	1350 to 1800
4 star city/provincial centre hotel	1075 to 1625
3 star city/provincial hotel	950 to 1500
3/2 star, provincial hotel	675 to 1125
3/2 star, provincial hotel bedroom extension	625 to 900
*Students' residences	550 to 725
Hostels	575 to 800
Common facilities, other facilities (CI/SfB 9)	
Conference centres	1125 to 1525
Public conveniences	1225 to 1750

* Refer also to 'Cost Limits and Allowances' in following section

Approximate Estimates
(incorporating Comparative Prices)

Estimating by means of priced approximate quantities is always more accurate than by using overall prices per square metre. Prices given in this section, which is arranged in elemental order, are derived from 'Prices for Measured Work - Major Works' section, but also include for all incidental items and labours which are normally measured separately in Bills of Quantities. As in other sections, they have been established with a tender price index level of 330 (1976 = 100). They do not include for preliminaries, details of which are given in Part II and which may amount to approximately 12.50% of the value of the measured work or fees for professional services.

In 1989, the **Approximate Estimates** section was re-drafted to include a considerable expansion of items and a consolidation of items from the previous **Approximate Estimates and Comparative Prices** sections.

The Comparative prices section previously consisted of schedules of commonly used forms of construction and types and qualities of materials, but provided no guidance as to the types of buildings in which certain materials were used.

The authors have therefore taken this opportunity to include not only these items but have also extracted an extensive range of structural and specification alternatives from their Cost Information library to indicate typical specifications and ranges of cost for a number of building types.

Price ranges have been completed for those building types for commonly-used specifications and where they vary indicate for different building types, the effect of different structural or specifications standards/ranges for each building type.

Whilst every effort has been made to ensure the accuracy of these figures, they have been prepared for approximate estimate purposes and on no account should be used for the preparation of tenders.

Unless otherwise described, units denoted as m2 refer to appropriate area unit (rather than gross floor) areas.

As elsewhere in this edition, prices do not include Value Added Tax, which should be applied at the current rate to all non-domestic building.

Item nr.	SPECIFICATIONS	Unit	RESIDENTIAL £ range £
1.0	**SUBSTRUCTURE** **ground floor plan area** (unless otherwise described)		
	comprising:-		
	Trench fill foundations		
	Strip foundations		
	Strip or base foundations		
	Raft foundations		
	Piled foundations		
	Other foundations/Extras		
	Basements		
	Trench fill foundations		
	Foundations; concrete bed and thickening under partitions; for two storey residential		
1.0.1	1 m deep	m2	65.00 - 75.00
	Extra for		
1.0.2	each additional 0.25 m deep	m2	11.00 - 12.50
1.0.3	each additional storey	m2	9.00 - 11.00
	Foundations; hollow ground floor; timber and boarding for two storey residential		
1.0.4	1 m deep	m2	80.00 -100.00
	Strip foundations		
	Foundations; brickwork; concrete bed and thickening under partitions; for two storey residential		
1.0.5	1 m deep; 265 mm cavity brickwork	m2	75.00 - 85.00
	Extra for		
1.0.6	additional 0.25 m deep	m2	11.00 - 12.50
1.0.7	each additional storey	m2	9.00 - 11.00
1.0.8	1 m deep; 322.5 mm solid brickwork	m2	80.00 - 90.00
	Extra for		
1.0.9	additional 0.25 m deep	m2	11.50 - 13.50
	Foundations; brickwork; hollow ground floor; timber and boarding; for two storey residential		
1.0.10	1 m deep; 265 mm cavity brickwork	m2	90.00 -110.00
	Extra for		
1.0.11	each additional 0.25 m deep	m2	11.00 - 12.50
	Strip or base foundations		
	Foundations in good ground; reinforced concrete bed; for one storey development		
1.0.12	shallow foundations	m2	- -
1.0.13	deep foundations	m2	- -
	Foundations in good ground; reinforced concrete bed; for two storey development		
1.0.14	shallow foundations	m2	- -
1.0.15	deep foundations	m2	- -
	Extra for		
1.0.16	each additional storey	m2	- -
	Raft foundations		
	Raft on poor ground for development		
1.0.17	one storey	m2	- -
1.0.18	two storey	m2	- -
	Extra for		
1.0.19	each additional storey	m2	- -

INDUSTRIAL £ range £	RETAILING £ range £	LEISURE £ range £	OFFICES £ range £	HOTELS £ range £	Item nr.
- -	- -	- -	- -	- -	1.0.1
- -	- -	- -	- -	- -	1.0.2
- -	- -	- -	- -	- -	1.0.3
- -	- -	- -	- -	- -	1.0.4
- -	- -	- -	- -	- -	1.0.5
- -	- -	- -	- -	- -	1.0.6
- -	- -	- -	- -	- -	1.0.7
- -	- -	- -	- -	- -	1.0.8
- -	- -	- -	- -	- -	1.0.9
- -	- -	- -	- -	- -	1.0.10
- -	- -	- -	- -	- -	1.0.11
40.00 - 55.00	45.00 - 65.00	45.00 - 65.00	- -	- -	1.0.12
65.00 - 80.00	70.00 - 90.00	65.00 - 80.00	- -	- -	1.0.13
50.00 - 70.00	55.00 - 80.00	50.00 - 75.00	55.00 - 80.00	55.00 - 80.00	1.0.14
70.00 -110.00	80.00 -115.00	75.00 -110.00	75.00 -110.00	75.00 -110.00	1.0.15
15.00 - 18.00	13.00 - 16.00	13.00 - 16.00	13.00 - 16.00	13.00 - 16.00	1.0.16
55.00 - 90.00	60.00 - 90.00	65.00 -100.00	- -	- -	1.0.17
90.00 -125.00	100.00 -135.00	100.00 -135.00	100.00 -135.00	100.00 -135.00	1.0.18
15.00 - 18.00	13.00 - 16.00	13.00 - 16.00	13.00 - 16.00	13.00 - 16.00	1.0.19

Item nr.	SPECIFICATIONS	Unit	RESIDENTIAL £ range £
1.0	**SUBSTRUCTURE - cont'd** **ground floor plan area** (unless otherwise described)		
	Piled foundations		
	Foundations in poor ground; reinforced concrete slab and ground beams; for two storey residential		
1.0.20	short bore piled	m2	110.00 -145.00
1.0.21	fully piled	m2	135.00 -170.00
	Foundations in poor ground; hollow ground floor; timber and boarding; for two storey residential		
1.0.22	short bore piled	m2	130.00 -160.00
	Foundations in poor ground; reinforced concrete slab; for one storey commercial development		
1.0.23	short bore piles to columns only	m2	- -
1.0.24	short bore piles	m2	- -
1.0.25	fully piled	m2	- -
	Foundations in poor ground with reinforced concrete slab and ground beams; for two storey commercial development		
1.0.26	short bore piles	m2	- -
1.0.27	fully piled	m2	- -
	Extra for		
1.0.28	each additional storey	m2	- -
	Foundations in bad ground; inner city redevelopment; reinforced concrete slab and ground beams; for two storey commercial development		
1.0.29	fully piled	m2	- -
	Other foundations/alternative slabs/extras		
	Cantilevered foundations in good ground; reinforced		
1.0.30	concrete slab for two storey commercial development	m2	- -
	Underpinning foundations of existing buildings		
1.0.31	abutting site	m	- -
	Extra to substructure rates for		
1.0.32	watertight pool construction	m2	- -
1.0.33	ice pad	m2	- -
	Reinforced concrete bed including excavation and hardcore under		
1.0.34	150 mm thick	m2	18.00 - 23.00
1.0.35	200 mm thick	m2	24.00 - 30.00
1.0.36	300 mm thick	m2	- -
	Hollow ground floor with timber and boarding, including excavation, concrete and hardcore under		
1.0.37	300 mm deep	m2	29.00 - 36.00
	Extra for		
1.0.38	sound reducing quilt in screed	m2	2.75 - 3.50
1.0.39	50 mm insulation under slab and at edges	m2	2.75 - 4.50
1.0.40	75 mm insulation under slab and at edges	m2	- -
	suspended precast concrete slabs in lieu of insitu		
1.0.41	slab	m2	4.50 - 6.75
	Basement (excluding bulk excavation costs) **basement floor/wall area (as appropriate)**		
	Reinforced concrete basement floors		
1.0.42	non-waterproofed	m2	- -
1.0.43	waterproofed	m2	- -
	Reinforced concrete basement walls		
1.0.44	non-waterproofed	m2	- -
1.0.45	waterproofed	m2	- -
1.0.46	sheet piled	m2	- -
1.0.47	diaphragm walling	m2	- -
	Extra for		
1.0.48	each additional basement level	%	- -

INDUSTRIAL £ range £	RETAILING £ range £	LEISURE £ range £	OFFICES £ range £	HOTELS £ range £	Item nr.
- -	- -	- -	- -	- -	1.0.20
- -	- -	- -	- -	- -	1.0.21
- -	- -	- -	- -	- -	1.0.22
65.00 -100.00	70.00 -100.00	65.00 -100.00	- -	- -	1.0.23
85.00 -115.00	90.00 -120.00	90.00 -120.00	90.00 -120.00	90.00 -120.00	1.0.24
110.00 -155.00	125.00 -170.00	125.00 -170.00	125.00 -170.00	125.00 -170.00	1.0.25
- -	100.00 -125.00	100.00 -125.00	100.00 -125.00	100.00 -125.00	1.0.26
- -	160.00 -220.00	160.00 -220.00	160.00 -220.00	160.00 -220.00	1.0.27
15.00 - 18.00	9.00 - 11.00	9.00 - 11.00	9.00 - 11.00	9.00 - 11.00	1.0.28
- -	180.00 -270.00	- -	180.00 -315.00	180.00 -270.00	1.0.29
- -	225.00 -290.00	225.00 -290.00	225.00 -290.00	225.00 -290.00	1.0.30
- -	540.00 -720.00	540.00 -720.00	540.00 -720.00	540.00 -720.00	1.0.31
- -	- -	90.00 -125.00	- -	- -	1.0.32
- -	- -	100.00 -150.00	- -	- -	1.0.33
25.00 - 30.00	25.00 - 32.00	25.00 - 30.00	25.00 - 30.00	25.00 - 30.00	1.0.34
32.00 - 40.00	30.00 - 38.00	30.00 - 36.00	30.00 - 36.00	30.00 - 36.00	1.0.35
40.00 - 54.00	36.00 - 50.00	36.00 - 45.00	36.00 - 45.00	36.00 - 45.00	1.0.36
- -	- -	- -	- -	- -	1.0.37
2.75 - 4.50	2.75 - 4.50	2.75 - 4.50	2.75 - 4.50	2.75 - 4.50	1.0.38
4.50 - 6.25	4.50 - 6.25	4.50 - 6.25	4.50 - 6.25	4.50 - 6.25	1.0.39
5.50 - 8.00	5.50 - 8.00	5.50 - 8.00	5.50 - 8.00	5.50 - 8.00	1.0.40
11.00 - 14.00	9.00 - 11.00	9.00 - 11.00	9.00 - 11.00	9.00 - 11.00	1.0.41
- -	45.00 - 55.00	45.00 - 55.00	45.00 - 55.00	45.00 - 55.00	1.0.42
- -	55.00 - 70.00	55.00 - 70.00	55.00 - 70.00	55.00 - 70.00	1.0.43
- -	120.00 -145.00	120.00 -145.00	120.00 -145.00	120.00 -145.00	1.0.44
- -	135.00 -170.00	135.00 -170.00	135.00 -170.00	135.00 -170.00	1.0.45
- -	245.00 -295.00	245.00 -295.00	245.00 -295.00	245.00 -295.00	1.0.46
- -	270.00 -315.00	270.00 -315.00	270.00 -315.00	270.00 -315.00	1.0.47
- -	+ 20%	+ 20%	+ 20%	+ 20%	1.0.48

Item nr.	SPECIFICATIONS	Unit	RESIDENTIAL £ range £	
2.0	**SUPERSTRUCTURE**			
2.1	**FRAME TO ROOF** — **roof plan area** (unless otherwise described)			
	comprising:-			
	Reinforced concrete frame			
	Precast concrete frame			
	Steel frame			
	Other frames/extras			
	Reinforced concrete frame			
2.1.1	Columns	m2	-	-
2.1.2	Columns and beams; 7.2 x 7.2 m grid	m2	-	-
	Extra for			
2.1.3	spans 7.5 - 15 m	m2	-	-
	Precast concrete frame			
	Portal frame			
2.1.4	industrial	m2	-	-
2.1.5	industrial; to small units	m2	-	-
2.1.6	retailing	m2	-	-
	Steel frame			
	Columns			
2.1.7	unprotected	m2	-	-
2.1.8	fully protected	m2	-	-
	Portal frame			
2.1.9	unprotected	m2	-	-
2.1.10	unprotected; to small units	m2	-	-
2.1.11	protected columns only	m2	-	-
2.1.12	fully protected	m2	-	-
2.1.13	protected; to small units	m2	-	-
	Columns and beams; to flat roofs			
2.1.14	unprotected	m2	-	-
2.1.15	unprotected; to small units	m2	-	-
2.1.16	unprotected; large span	m2	-	-
2.1.17	unprotected; very large span	m2	-	-
2.1.18	protected columns only	m2	-	-
2.1.19	fully protected	m2	-	-
2.1.20	fully protected; insulation	m2	-	-
	Other frames/extras			
	Space deck on steel frame			
2.1.21	unprotected	m2	-	-
2.1.22	Exposed steel frame for tent/mast structures	m2	-	-
	Columns and beams to 18 m high bay warehouse			
2.1.23	unprotected	m2	-	-
	Columns and beams to mansard			
2.1.24	protected	m2	-	-
	Feature columns and beams to glazed atrium roof			
2.1.25	unprotected	m2	-	-

INDUSTRIAL £ range £	RETAILING £ range £	LEISURE £ range £	OFFICES £ range £	HOTELS £ range £	Item nr.
22.50 - 31.50	25.00 - 36.00	25.00 - 36.00	25.00 - 31.50	25.00 - 31.50	2.1.1
40.00 - 65.00	45.00 - 67.50	45.00 - 65.00	45.00 - 65.00	45.00 - 67.50	2.1.2
13.50 - 36.00	18.00 - 40.00	13.50 - 36.00	13.50 - 36.00	13.50 - 36.00	2.1.3
40.00 - 55.00	- -	- -	- -	- -	2.1.4
55.00 - 75.00	- -	- -	- -	- -	2.1.5
- -	45.00 - 65.00	- -	- -	- -	2.1.6
27.50 - 40.00	32.50 - 50.00	32.50 - 45.00	32.50 - 45.00	- -	2.1.7
32.50 - 45.00	40.00 - 55.00	40.00 - 50.00	40.00 - 50.00	- -	2.1.8
40.00 - 55.00	45.00 - 67.50	- -	- -	- -	2.1.9
50.00 - 72.50	- -	- -	- -	- -	2.1.10
45.00 - 67.50	54.00 - 80.00	50.00 - 67.50	- -	- -	2.1.11
60.00 - 80.00	67.50 - 95.00	60.00 - 80.00	- -	- -	2.1.12
85.00 -100.00	- -	- -	- -	- -	2.1.13
45.00 - 55.00	50.00 - 60.00	50.00 - 60.00	55.00 - 67.50	55.00 - 67.50	2.1.14
80.00 - 90.00	- -	- -	- -	- -	2.1.15
80.00 - 95.00	85.00 -100.00	87.50 -112.50	- -	- -	2.1.16
85.00 -100.00	90.00 -112.50	95.00 -117.50	- -	- -	2.1.17
67.50 - 80.00	60.00 - 85.00	57.50 - 77.50	- -	- -	2.1.18
67.50 - 85.00	67.50 - 90.00	72.50 -100.00	67.50 - 95.00	67.50 -100.00	2.1.19
- -	72.50 -100.00	85.00 -112.50	80.00 -105.00	80.00 -112.50	2.1.20
90.00 -170.00	95.00 -170.00	95.00 -170.00	95.00 -170.00	- -	2.1.21
150.00 -225.00	150.00 -225.00	150.00 -225.00	- -	- -	2.1.22
95.00 -130.00	- -	- -	- -	- -	2.1.23
- -	65.00 - 90.00	65.00 - 90.00	65.00 - 90.00	65.00 - 90.00	2.1.24
- -	- -	- -	70.00 -110.00	- -	2.1.25

Item nr.	SPECIFICATIONS	Unit	RESIDENTIAL £ range £
2.2	**FRAME & UPPER FLOORS (COMBINED)**	**upper floor area** (unless otherwise described)	
	comprising:-		
	Softwood floors; no frame		
	Softwood floors; steel frame		
	Reinforced concrete floors; no frame		
	Reinforced concrete floors and frame		
	Reinforced concrete floors; steel frame		
	Precast concrete floors; no frame		
	Precast concrete floors; reinforced concrete frame		
	Precast concrete floors and frame		
	Precast concrete floors; steel frame		
	Other floor and frame constructions/extras		
	Softwood floors; no frame		
	Joisted floor; supported on layers; 22 mm chipboard t & g flooring; herring bone strutting; no coverings or finishes		
2.2.1	150 x 50 mm joists	m2	18.00 - 21.00
2.2.2	175 x 50 mm joists	m2	20.00 - 23.00
2.2.3	200 x 50 mm joists	m2	21.00 - 24.00
2.2.4	225 x 50 mm joists	m2	23.00 - 25.00
2.2.5	250 x 50 mm joists	m2	24.00 - 26.00
2.2.6	275 x 50 mm joists	m2	26.00 - 29.00
2.2.7	Joisted floor; average depth; plasterboard; skim; emulsion; vinyl flooring and painted softwood skirtings	m2	45.00 - 55.00
	Softwood construction; steel frame		
2.2.8	Joisted floor; average depth; plasterboard; skim; emulsion; vinyl flooring and painted softwood skirtings	m2	- -
	Reinforced concrete floors; no frame		
	Suspended slab; no coverings or finishes		
2.2.9	3.65 m span; 3.00 KN/m2 loading	m2	34.00 - 36.00
2.2.10	4.25 m span; 3.00 KN/m2 loading	m2	40.00 - 42.00
2.2.11	2.75 m span; 8.00 KN/m2 loading	m2	- -
2.2.12	3.35 m span; 8.00 KN/m2 loading	m2	- -
2.2.13	4.25 m span; 8.00 KN/m2 loading	m2	- -
	Suspended slab; no coverings or finishes		
2.2.14	150 mm thick	m2	40.00 - 50.00
2.2.15	225 mm thick	m2	60.00 - 70.00
	Reinforced concrete floors and frame		
	Suspended slab; average depth; no coverings or finishes		
2.2.16	up to six storeys	m2	- -
2.2.17	seven to twelve storeys	m2	- -
2.2.18	thirteen to eighteen storeys	m2	- -
	Extra for		
2.2.19	section 20 fire regulations	m2	- -
	Wide span suspended slab		
2.2.20	up to six storeys	m2	- -
	Reinforced concrete floors; steel frame		
	Suspended slab; average depth; 'Holorib' permanent steel shuttering; protected steel frame; no coverings or finishes		
2.2.21	up to six storeys	m2	- -
	Extra for		
2.2.22	spans 7.5 to 15 m	m2	- -
2.2.23	seven to twelve storeys	m2	- -
	Extra for		
2.2.24	section 20 fire regulations	m2	- -

INDUSTRIAL £ range £	RETAILING £ range £	LEISURE £ range £	OFFICES £ range £	HOTELS £ range £	Item nr.
- -	- -	- -	- -	- -	2.2.1
- -	- -	- -	- -	- -	2.2.2
- -	- -	- -	- -	- -	2.2.3
- -	- -	- -	- -	- -	2.2.4
- -	- -	- -	- -	- -	2.2.5
- -	- -	- -	- -	- -	2.2.6
- -	- -	- -	- -	- -	2.2.7
60.00 - 65.00	65.00 - 90.00	65.00 - 85.00	65.00 - 85.00	65.00 - 85.00	2.2.8
- -	- -	34.00 - 40.00	34.00 - 40.00	34.00 - 40.00	2.2.9
- -	- -	40.00 - 42.00	40.00 - 47.00	40.00 - 47.00	2.2.10
34.00 - 40.00	34.00 - 40.00	34.00 - 40.00	34.00 - 40.00	34.00 - 40.00	2.2.11
38.00 - 45.00	38.00 - 45.00	38.00 - 45.00	38.00 - 45.00	38.00 - 45.00	2.2.12
49.00 - 54.00	49.00 - 54.00	49.00 - 54.00	49.00 - 56.00	49.00 - 56.00	2.2.13
45.00 - 65.00	40.00 - 65.00	40.00 - 60.00	40.00 - 55.00	40.00 - 65.00	2.2.14
70.00 - 80.00	60.00 - 80.00	60.00 - 75.00	60.00 - 72.50	60.00 - 80.00	2.2.15
85.00 -115.00	70.00 -100.00	75.00 -105.00	70.00 - 90.00	70.00 - 95.00	2.2.16
- -	- -	- -	90.00 -140.00	- -	2.2.17
- -	- -	- -	140.00 -175.00	- -	2.2.18
- -	- -	- -	7.00 - 9.00	- -	2.2.19
95.00 -120.00	75.00 -110.00	85.00 -115.00	75.00 -100.00	75.00 -105.00	2.2.20
115.00 -140.00	95.00 -130.00	100.00 -130.00	95.00 -120.00	95.00 -125.00	2.2.21
13.50 - 35.00	17.50 - 40.00	13.50 - 35.00	13.50 - 35.00	13.50 - 35.00	2.2.22
- -	- -	- -	110.00 -165.00	- -	2.2.23
- -	- -	- -	13.50 - 16.25	- -	2.2.24

Item nr.	SPECIFICATIONS	Unit	RESIDENTIAL £ range £
2.2	**FRAME & UPPER FLOORS (COMBINED) - cont'd** upper floor area (unless otherwise described)		
	Reinforced concrete floors; steel frame - cont'd		
	Suspended slab; average depth; protected steel frame; no coverings or finishes		
2.2.25	up to six storeys	m2	- -
2.2.26	seven to twelve storeys	m2	- -
	Extra for		
2.2.27	section 20 fire regulations	m2	- -
	Precast concrete floors; no frame		
	Suspended slab; 75 mm screed; no coverings or finishes		
2.2.28	3 m span; 5.00 KN/m2 loading	m2	29.00 - 36.00
2.2.29	6 m span; 5.00 KN/m2 loading	m2	30.00 - 37.00
2.2.30	7.5 m span; 5.00 KN/m2 loading	m2	31.00 - 38.00
2.2.31	3 m span; 8.50 KN/m2 loading	m2	- -
2.2.32	6 m span; 8.50 KN/m2 loading	m2	- -
2.2.33	7.5 m span; 8.50 KN/m2 loading	m2	- -
2.2.34	3 m span; 12.50 KN/m2 loading	m2	- -
2.2.35	6 m span; 12.50 KN/m2 loading	m2	- -
2.2.36	Suspended slab; average depth; no coverings or finishes	m2	27.50 - 40.00
	Precast concrete floors; reinforced concrete frame		
2.2.37	Suspended slab; average depth; no coverings or finishes	m2	- -
	Precast concrete floors and frame		
2.2.38	Suspended slab; average depth; no coverings or finishes	m2	- -
	Precast concrete floors; steel frame		
	Suspended slabs; average depth; unprotected steel frame; no coverings or finishes		
2.2.39	up to three storeys	m2	- -
	Suspended slabs; average depth; protected steel frame; no coverings or finishes		
2.2.40	up to six storeys	m2	- -
2.2.41	seven to twelve storeys	m2	- -
	Other floor and frame construction/extras		
2.2.42	Reinforced concrete cantilevered balcony	nr	1200 -1600
2.2.43	Reinforced concrete cantilevered walkways	m2	- -
2.2.44	Reinforced concrete walkways and supporting frame	m2	- -
	Reinforced concrete core with steel umbrella frame		
2.2.45	twelve to twenty four storeys	m2	- -
	Extra for		
2.2.46	wrought formwork	m2	2.40 - 3.00
2.2.47	sound reducing quilt in screed	m2	2.75 - 3.50
2.2.48	insulation to avoid cold bridging	m2	2.75 - 4.50

INDUSTRIAL £ range £	RETAILING £ range £	LEISURE £ range £	OFFICES £ range £	HOTELS £ range £	Item nr.
115.00 -150.00	100.00 -140.00	105.00 -140.00	100.00 -130.00	100.00 -130.00	2.2.25
- -	- -	- -	115.00 -175.00	- -	2.2.26
- -	- -	- -	13.50 - 16.25	- -	2.2.27
- -	34.00 - 39.00	34.00 - 39.00	34.00 - 39.00	34.00 - 39.00	2.2.28
- -	35.00 - 41.00	35.00 - 41.00	35.00 - 40.00	39.00 - 41.00	2.2.29
- -	36.00 - 45.00	36.00 - 45.00	36.00 - 41.00	36.00 - 45.00	2.2.30
36.00 - 40.00	- -	- -	- -	- -	2.2.31
38.00 - 43.00	- -	- -	- -	- -	2.2.32
40.00 - 45.00	- -	- -	- -	- -	2.2.33
38.00 - 42.50	- -	- -	- -	- -	2.2.34
42.50 - 47.50	- -	- -	- -	- -	2.2.35
35.00 - 50.00	35.00 - 60.00	35.00 - 55.00	35.00 - 50.00	35.00 - 60.00	2.2.36
55.00 - 75.00	50.00 - 80.00	50.00 - 75.00	50.00 - 70.00	50.00 - 80.00	2.2.37
55.00 -105.00	50.00 -110.00	50.00 -105.00	50.00 -105.00	50.00 -120.00	2.2.38
62.50 -112.50	60.00 -100.00	60.00 - 95.00	50.00 - 95.00	60.00 -100.00	2.2.39
100.00 -130.00	85.00 -120.00	95.00 -120.00	85.00 -115.00	85.00 -115.00	2.2.40
- -	- -	- -	110.00 -165.00	- -	2.2.41
- -	- -	- -	- -	- -	2.2.42
- -	90.00 -110.00	- -	- -	- -	2.2.43
- -	100.00 -120.00	- -	- -	- -	2.2.44
- -	- -	- -	205.00 -270.00	- -	2.2.45
2.40 - 3.00	2.40 - 5.00	2.40 - 5.00	2.40 - 5.00	2.40 - 6.50	2.2.46
2.75 - 4.50	2.75 - 4.50	2.75 - 4.50	2.75 - 4.50	2.75 - 4.50	2.2.47
4.50 - 6.25	4.50 - 6.25	4.50 - 6.25	4.50 - 6.25	4.50 - 6.25	2.2.48

Item nr.	SPECIFICATIONS	Unit	RESIDENTIAL £ range £
2.3	**ROOF** **roof plan area** (unless otherwise described)		
	comprising:-		
	Softwood flat roofs		
	Softwood trussed pitched roofs		
	Steel trussed pitched roofs		
	Concrete flat roofs		
	Flatroof decking and finishes		
	Roof claddings		
	Rooflights/patent glazing and glazed roofs		
	Comparative over/underlays		
	Comparative tiling and slating finishes/perimeter treatments		
	Comparative cladding finishes/perimeter treatments		
	Comparative waterproofing finishes/perimeter treatments		
	Softwood Flat roofs		
	Structure only comprising roof joists; 100 x 50 mm wall plates; herring-bone strutting; 50 mm woodwool slabs; no coverings or finishes		
2.3.1	150 x 50 mm joists	m2	24.00 - 27.00
2.3.2	200 x 50 mm joists	m2	27.00 - 30.00
2.3.3	250 x 50 mm joists	m2	30.00 - 33.00
	Structure only comprising roof joists; 100 x 50 mm wall plates; herring-bone strutting; 25 mm softwood boarding; no coverings or finishes		
2.3.4	150 x 50 mm joists	m2	26.00 - 29.00
2.3.5	200 x 50 mm joists	m2	29.00 - 32.00
2.3.6	250 x 50 mm joists	m2	32.00 - 35.00
	Roof joists; average depth; 25 mm softwood boarding; PVC rainwater goods; plasterboard; skim and emulsion		
2.3.7	three layer felt and chippings	m2	56.00 - 72.00
2.3.8	two coat asphalt and chippings	m2	54.00 - 80.00
	Softwood trussed pitched roofs		
	Structure only comprising 75 x 50 mm Fink roof trusses @ 600 mm centres (measured on plan)		
2.3.9	22.5° pitch	m2	11.00 - 13.50
	Structure only comprising 100 x 38 mm Fink roof trusses @ 600 mm centres (measured on plan)		
2.3.10	30° pitch	m2	12.50 - 15.50
2.3.11	35° pitch	m2	13.00 - 16.00
2.3.12	40° pitch	m2	14.50 - 18.00
	Structure only comprising 100 x 50 mm Fink roof trusses @ 375 mm centres (measured on plan)		
2.3.13	30° pitch	m2	22.00 - 24.50
2.3.14	35° pitch	m2	22.00 - 25.00
2.3.15	40° pitch	m2	23.50 - 27.00
	Extra for		
2.3.16	forming dormers	nr	320.00 -450.00
	Structure only for Mansard roof comprising 100 x 50 mm roof trusses @ 600 mm centres		
2.3.17	70° pitch	m2	16.00 - 22.00
	Fink roof trusses; narrow span; 100 mm insulation; PVC rainwater goods; plasterboard; skim and emulsion		
2.3.18	concrete interlocking tile coverings	m2	55.00 - 70.00
2.3.19	clay pantile coverings	m2	60.00 - 75.00
2.3.20	composition slate coverings	m2	63.00 - 78.00
2.3.21	plain clay tile coverings	m2	75.00 - 90.00
2.3.22	natural slate coverings	m2	80.00 -100.00
2.3.23	reconstructed stone coverings	m2	65.00 -105.00

INDUSTRIAL £ range £		RETAILING £ range £	LEISURE £ range £		OFFICES £ range £	HOTELS £ range £	Item nr.
-	-	24.00 - 27.00	-	-	24.00 - 27.00	24.00 - 27.00	2.3.1
-	-	27.00 - 30.00	-	-	27.00 - 30.00	27.00 - 30.00	2.3.2
-	-	30.00 - 33.00	-	-	30.00 - 33.00	30.00 - 33.00	2.3.3
-	-	26.00 - 29.00	-	-	26.00 - 29.00	26.00 - 29.00	2.3.4
-	-	29.00 - 32.00	-	-	29.00 - 32.00	29.00 - 32.00	2.3.5
-	-	31.00 - 35.00	-	-	31.00 - 35.00	31.00 - 35.00	2.3.6
-	-	56.00 - 72.00	-	-	56.00 - 72.00	56.00 - 72.00	2.3.7
-	-	55.00 - 80.00	-	-	55.00 - 80.00	55.00 - 80.00	2.3.8
-	-	12.00 - 14.50	-	-	12.00 - 14.50	12.00 - 15.50	2.3.9
-	-	13.50 - 16.00	-	-	13.50 - 16.00	13.50 - 17.00	2.3.10
-	-	14.50 - 16.50	-	-	14.50 - 16.50	14.50 - 17.50	2.3.11
-	-	15.50 - 19.00	-	-	15.50 - 19.00	15.50 - 20.00	2.3.12
-	-	22.00 - 25.00	-	-	22.00 - 25.00	22.00 - 26.00	2.3.13
-	-	23.00 - 26.00	-	-	23.00 - 26.00	23.00 - 27.00	2.3.14
-	-	24.00 - 28.00	-	-	24.00 - 28.00	24.00 - 29.00	2.3.15
-	-	320.00 -450.00	-	-	320.00 -450.00	320.00 -450.00	2.3.16
-	-	17.00 - 23.00	-	-	17.00 - 23.00	17.00 - 24.00	2.3.17
-	-	60.00 - 80.00	-	-	60.00 - 85.00	65.00 - 90.00	2.3.18
-	-	65.00 - 85.00	-	-	65.00 - 90.00	70.00 - 95.00	2.3.19
-	-	68.00 - 88.00	-	-	68.00 - 93.00	72.50 - 98.00	2.3.20
-	-	80.00 -100.00	-	-	80.00 -105.00	85.00 -110.00	2.3.21
-	-	85.00 -105.00	-	-	85.00 -110.00	90.00 -115.00	2.3.22
-	-	70.00 -110.00	-	-	70.00 -115.00	75.00 -120.00	2.3.23

Item nr.	SPECIFICATIONS	Unit	RESIDENTIAL £ range £
2.3	**ROOF - cont'd** roof plan area (unless otherwise described)		
	Softwood trussed pitched roofs - cont'd		
	Monopitch roof trusses; 100 mm insulation; PVC rainwater goods; plasterboard; skim and emulsion		
2.3.24	concrete interlocking tile coverings	m2	60.00 - 75.00
2.3.25	clay pantile coverings	m2	65.00 - 80.00
2.3.26	composition slate coverings	m2	68.00 - 83.00
2.3.27	plain clay tile coverings	m2	80.00 - 95.00
2.3.28	natural slate coverings	m2	85.00 -100.00
2.3.29	reconstructed stone coverings	m2	70.00 -110.00
	Dormer roof trusses; 100 mm insulation; PVC rainwater goods; plasterboard; skim and emulsion		
2.3.30	concrete interlocking tile coverings	m2	80.00 -100.00
2.3.31	clay pantile coverings	m2	85.00 -105.00
2.3.32	composition slate coverings	m2	90.00 -120.00
2.3.33	plain clay tile coverings	m2	100.00 -120.00
2.3.34	natural slate coverings	m2	105.00 -125.00
2.3.35	reconstructed stone coverings	m2	90.00 -135.00
	Extra for		
2.3.36	end of terrace/semi-detached configeration	m2	22.00 - 24.00
2.3.37	hipped roof configeration	m2	22.00 - 27.00
	Steel trussed pitched roofs		
	Fink roof trusses; wide span; 100 mm insulation		
2.3.38	concrete interlocking tile coverings	m2	- -
2.3.39	clay pantile coverings	m2	- -
2.3.40	composition slate coverings	m2	- -
2.3.41	clay plain tile coverings	m2	- -
2.3.42	natural slate coverings	m2	- -
2.3.43	reconstructed stone coverings	m2	- -
	Steel roof trusses and beams; thermal and accoustic insulation		
2.3.44	aluminium profiled composite cladding	m2	- -
2.3.45	copper roofing on boarding	m2	- -
	Steel roof and glulam beams; thermal and accoustic insulation		
2.3.46	aluminium profiled composite cladding	m2	- -
	Concrete flat roofs		
	Structure only comprising reinforced concrete suspended slab; no coverings or finishes		
2.3.47	3.65 m span; 3.00 KN/m2 loading	m2	34.00- 36.00
2.3.48	4.25 m span; 3.00 KN/m2 loading	m2	40.00 - 42.00
2.3.49	3.65 m span; 8.00 KN/m2 loading	m2	- -
2.3.50	4.25 m span; 8.00 KN/m2 loading	m2	- -
	Precast concrete suspended slab;average depth; 100 mm insulation; PVC rainwater goods; no coverings or finishes		
2.3.51	two coat asphalt coverings and chippings	m2	- -
2.3.52	polyester roofing	m2	- -
	Reinforced concrete or waffle suspended slabs; average depth; 100 mm insulation; PVC rainwater goods; no coverings or finishes		
2.3.53	two coat asphalt coverings and chippings	m2	- -
2.3.54	two coat asphalt coverings and paving slabs	m2	- -
	Reinforced concrete suspended slabs; on 'Holorib' permanent steel shuttering; average depth; 100 mm insulation; PVC rainwater goods		
2.3.55	two coat asphalt coverings and chippings	m2	- -

INDUSTRIAL £ range £	RETAILING £ range £	LEISURE £ range £	OFFICES £ range £	HOTELS £ range £	Item nr.
- -	65.00 - 85.00	- -	65.00 - 90.00	70.00 - 95.00	2.3.24
- -	70.00 - 90.00	- -	70.00 - 95.00	75.00 -100.00	2.3.25
- -	73.00 - 95.00	- -	73.00 -100.00	78.00 -105.00	2.3.26
- -	85.00 -105.00	- -	85.00 -110.00	90.00 -115.00	2.3.27
- -	90.00 -110.00	- -	90.00 -115.00	95.00 -120.00	2.3.28
- -	75.00 -115.00	- -	75.00 -120.00	80.00 -125.00	2.3.29
- -	85.00 -110.00	- -	85.00 -115.00	90.00 -120.00	2.3.30
- -	90.00 -115.00	- -	90.00 -120.00	95.00 -125.00	2.3.31
- -	93.00 -120.00	- -	93.00 -125.00	98.00 -130.00	2.3.32
- -	105.00 -130.00	- -	105.00 -135.00	110.00 -140.00	2.3.33
- -	110.00 -135.00	- -	110.00 -140.00	115.00 -145.00	2.3.34
- -	95.00 -140.00	- -	95.00 -145.00	100.00 -150.00	2.3.35
- -	- -	- -	- -	- -	2.3.36
- -	- -	- -	- -	- -	2.3.37
- -	90.00 -115.00	85.00 -110.00	90.00 -120.00	95.00 -125.00	2.3.38
- -	95.00 -120.00	90.00 -115.00	95.00 -125.00	100.00 -130.00	2.3.39
- -	98.00 -123.00	95.00 -120.00	98.00 -128.00	103.00 -133.00	2.3.40
- -	110.00 -135.00	105.00 -130.00	110.00 -140.00	115.00 -145.00	2.3.41
- -	115.00 -140.00	110.00 -135.00	115.00 -145.00	120.00 -150.00	2.3.42
- -	105.00 -150.00	100.00 -145.00	105.00 -155.00	110.00 -160.00	2.3.43
135.00 -160.00	125.00 -150.00	120.00 -145.00	125.00 -140.00	- -	2.3.44
- -	135.00 -165.00	140.00 -165.00	145.00 -160.00	- -	2.3.45
- -	135.00 -180.00	130.00 -175.00	135.00 -170.00	- -	2.3.46
- -	- -	34.00 - 40.00	34.00 - 40.00	34.00 - 40.00	2.3.47
- -	- -	40.00 - 42.00	40.00 - 47.00	40.00 - 47.00	2.3.48
38.00 - 45.00	38.00 - 45.00	38.00 - 45.00	38.00 - 45.00	38.00 - 45.00	2.3.49
48.00 - 54.00	48.00 - 54.00	48.00 - 54.00	48.00 - 54.00	48.00 - 54.00	2.3.50
65.00 - 90.00	60.00 - 97.50	60.00 - 92.50	60.00 - 90.00	60.00 - 97.50	2.3.51
68.00 - 80.00	65.00 - 90.00	65.00 - 85.00	65.00 - 80.00	65.00 - 90.00	2.3.52
70.00 - 92.50	65.00 -100.00	65.00 -97.50	65.00 - 92.50	65.00 -100.00	2.3.53
97.50 -120.00	92.50 -130.00	92.50 -125.00	92.50 -120.00	92.50 -133.00	2.3.54
65.00 - 80.00	60.00 - 90.00	60.00 - 85.00	60.00 - 80.00	60.00 - 90.00	2.3.55

Item nr.	SPECIFICATIONS	Unit	RESIDENTIAL £ range £	
2.3	**ROOF - cont'd** **roof plan area** (unless otherwise described)			
	Flat roof decking and finishes			
	Woodwool roof decking			
2.3.56	50 mm thick; two coat asphalt coverings to BS 988 and chippings	m2	-	-
	Galvanised steel roof decking; 100 mm insulation; three layer felt roofing and chippings			
2.3.57	0.7 mm thick; 2.38 m span	m2	-	-
2.3.58	0.7 mm thick; 2.96 m span	m2	-	-
2.3.59	0.7 mm thick; 3.74 m span	m2	-	-
2.3.60	0.7 mm thick; 5.13 m span	m2	-	-
	Aluminium roof decking; 100 mm insulation; three layer felt roofing and chippings			
2.3.61	0.9 mm thick; 1.79 m span	m2	-	-
2.3.62	0.9 mm thick; 2.34 m span	m2	-	-
	Metal decking; 100 mm insulation; on wood/steel open lattice beams			
2.3.63	three layer felt roofing and chippings	m2	-	-
2.3.64	two layer high performance felt roofing and chippings	m2	-	-
2.3.65	two coats asphalt coverings and chippings	m2	-	-
	Metal decking to mansard; excluding frame; 50 mm insulation;			
2.3.66	two coat asphalt coverings and chippings on decking; natural slate covering to mansard faces	m2	-	-
	Roof claddings			
	Non-asbestos profiled cladding			
2.3.67	'profile 3'; natural	m2	-	-
2.3.68	'profile 3'; coloured	m2	-	-
2.3.69	'profile 6'; natural	m2	-	-
2.3.70	'profile 6'; coloured	m2	-	-
2.3.71	'profile 6'; natural; insulated; inner lining panel	m2	-	-
	Non-asbestos profiled cladding on steel purlins			
2.3.72	insulated	m2	-	-
2.3.73	insulated; with 10% transluscent sheets	m2	-	-
2.3.74	insulated; plasterboard inner lining on metal tees	m2	-	-
	Asbestos cement profiled cladding on steel purlins			
2.3.75	insulated	m2	-	-
2.3.76	insulated; with 10% transluscent sheets	m2	-	-
2.3.77	insulated; plasterboard inner lining on metal tees	m2	-	-
2.3.78	insulated; steel sheet liner on metal tees	m2	-	-
	PVF2 coated galvanised steel profiled cladding			
2.3.79	0.72 mm thick; 'profile 20B'	m2	-	-
2.3.80	0.72 mm thick; 'profile TOP 40'	m2	-	-
2.3.81	0.72 mm thick; 'profile 45'	m2	-	-
	Extra for			
2.3.82	80 mm insulation and 0.4 mm thick coated inner lining sheet	m2	-	-
	PVF2 coated galvanised steel profiled cladding on steel purlins			
2.3.83	insulated	m2	-	-
2.3.84	insulated; plasterboard inner lining on metal tees	m2	-	-
2.3.85	insulated; plasterboard inner lining on metal tees; with 1% fire vents	m2	-	-
2.3.86	insulated; plasterboard inner lining on metal tees; with 2.5% fire vents	m2	-	-
2.3.87	insulated; coloured inner lining panel	m2	-	-
2.3.88	insulated; coloured inner lining panel; with 1% fire vents	m2	-	-
2.3.89	insulated; coloured inner lining panel; with 2.5% fire vents	m2	-	-
2.3.90	insulated; sandwich panel	m2	-	-

INDUSTRIAL £ range £	RETAILING £ range £	LEISURE £ range £	OFFICES £ range £	HOTELS £ range £	Item nr.
31.50 - 45.00	34.00 - 43.00	36.00 - 45.00	31.50 - 40.50	36.00 - 45.00	2.3.56
36.00 - 45.00	36.00 - 45.00	36.00 - 45.00	36.00 - 45.00	36.00 - 45.00	2.3.57
37.00 - 46.00	37.00 - 46.00	37.00 - 46.00	37.00 - 46.00	37.00 - 46.00	2.3.58
38.00 - 47.00	38.00 - 47.00	38.00 - 47.00	38.00 - 47.00	38.00 - 47.00	2.3.59
39.00 - 48.00	39.00 - 48.00	39.00 - 48.00	39.00 - 48.00	39.00 - 48.00	2.3.60
43.00 - 52.00	43.00 - 52.00	43.00 - 52.00	43.00 - 52.00	43.00 - 52.00	2.3.61
45.00 - 54.00	45.00 - 54.00	45.00 - 54.00	45.00 - 54.00	45.00 - 54.00	2.3.62
65.00 - 80.50	70.00 - 82.50	70.00 - 82.50	70.00 - 82.50	70.00 - 85.00	2.3.63
67.50 - 82.50	72.50 - 82.50	72.50 - 87.50	72.50 - 87.50	72.50 - 90.00	2.3.64
65.00 - 85.00	65.00 - 90.00	65.00 - 90.00	65.00 - 90.00	70.00 - 95.00	2.3.65
- -	120.00 -195.00	110.00 -180.00	110.00 -180.00	120.00 -210.00	2.3.66
11.50 - 13.50	- -	- -	- -	- -	2.3.67
12.50 - 14.50	- -	- -	- -	- -	2.3.68
12.50 - 14.50	- -	- -	- -	- -	2.3.69
13.50 - 15.50	- -	- -	- -	- -	2.3.70
25.00 - 30.00	- -	- -	- -	- -	2.3.71
23.00 - 27.00	- -	- -	- -	- -	2.3.72
26.00 - 30.00	- -	- -	- -	- -	2.3.73
36.00 - 43.00	- -	- -	- -	- -	2.3.74
22.00 - 26.00	- -	- -	- -	- -	2.3.75
25.00 - 29.00	- -	- -	- -	- -	2.3.76
35.00 - 42.00	- -	- -	- -	- -	2.3.77
34.00 - 42.00	- -	- -	- -	- -	2.3.78
15.50 - 20.00	20.50 - 27.00	20.50 - 27.00	- -	- -	2.3.79
15.50 - 19.00	20.00 - 26.00	20.00 - 26.00	- -	- -	2.3.80
17.00 - 22.00	22.00 - 29.00	22.00 - 29.00	- -	- -	2.3.81
10.00 - 11.00	10.00 - 11.00	10.00 - 11.00	- -	- -	2.3.82
27.00 - 36.00	31.00 - 40.00	31.00 - 40.00	- -	- -	2.3.83
40.00 - 52.00	45.00 - 56.00	45.00 - 56.00	- -	- -	2.3.84
50.00 - 65.00	55.00 - 70.00	55.00 - 70.00	- -	- -	2.3.85
60.00 - 75.00	65.00 - 80.00	65.00 - 80.00	- -	- -	2.3.86
40.00 - 50.00	45.00 - 60.00	45.00 - 60.00	- -	- -	2.3.87
50.00 - 60.00	55.00 - 65.00	55.00 - 65.00	- -	- -	2.3.88
60.00 - 75.00	65.00 - 75.00	65.00 - 75.00	- -	- -	2.3.89
90.00 -135.00	110.00 -155.00	110.00 -155.00	- -	- -	2.3.90

Item nr.	SPECIFICATIONS	Unit	RESIDENTIAL £ range £	
2.3	**ROOF - cont'd**	**roof plan area** (unless otherwise described)		
	Roof claddings - cont'd			
	Pre-painted 'Rigidal' aluminium profiled cladding			
2.3.91	0.7 mm thick; type WA6	m2	-	-
2.3.92	0.9 mm thick; type A7	m2	-	-
	PVF2 coated aluminium profiled cladding on steel purlins			
2.3.93	insulated; plasterboard inner lining on metal tees	m2	-	-
2.3.94	insulated; coloured inner lining panel	m2	-	-
	Rooflights/patent glazing and glazed roofs			
	Rooflights			
2.3.95	standard pvc	m2	-	-
2.3.96	feature/ventilating	m2	-	-
	Patent glazing; including flashings			
	standard aluminium georgian wired			
2.3.97	single glazed	m2	-	-
2.3.98	double glazed	m2	-	-
	purpose made polyester powder coated aluminium;			
2.3.99	double glazed low emissivity glass	m2	-	-
2.3.100	feature; to covered walkways	m2	-	-
	Glazed roofing on framing; to covered walkways			
2.3.101	feature; single glazed	m2	-	-
2.3.102	feature; double glazed barrel vault	m2	-	-
2.3.103	feature; very expensive	m2	-	-
	Comparative over/underlays			
	Roofing felt; unreinforced			
2.3.104	sloping (measured on face)	m2	1.10 -	1.35
	Roofing felt; reinforced			
2.3.105	sloping (measured on face)	m2	1.35 -	1.60
	sloping (measured on plan)			
2.3.106	20° pitch	m2	1.60 -	2.00
2.3.107	30° pitch	m2	1.80 -	2.20
2.3.108	35° pitch	m2	2.20 -	2.40
2.3.109	40° pitch	m2	2.25 -	2.50
2.3.110	Building paper	m2	1.00 -	2.00
2.3.111	Vapour barrier	m2	1.50 -	4.50
	Insulation quilt; laid over ceiling joists			
2.3.112	80 mm thick	m2	2.75 -	3.00
2.3.113	100 mm thick	m2	3.25 -	3.50
2.3.114	150 mm thick	m2	4.50 -	5.00
2.3.115	200 mm thick	m2	6.00 -	6.50
	Wood fibre insulation boards; impregnated; density 220 - 350 kg/m3			
2.3.116	12.7 mm thick	m2	-	-
	Polystyrene insulation boards; fixed vertically with adhesive			
2.3.117	12 mm thick	m2	-	-
2.3.118	25 mm thick	m2	-	-
2.3.119	50 mm thick	m2	-	-
2.3.120	Ballast	m2	-	-
	Polyurethene insulation boards; density 32 kg/m3			
2.3.121	30 mm thick	m2	-	-
2.3.122	35 mm thick	m2	-	-
2.3.123	50 mm thick	m2	-	-
	Cork insulation boards; density 112 - 125 kg/m3			
2.3.124	60 mm thick	m2	-	-
	Glass fibre insulation boards; density 120 - 130 kg/m2			
2.3.125	60 mm thick	m2	-	-

INDUSTRIAL £ range £	RETAILING £ range £	LEISURE £ range £	OFFICES £ range £	HOTELS £ range £	Item nr.
20.00 - 25.00	25.00 - 32.50	25.00 - 32.50	- -	- -	2.3.91
25.00 - 29.00	29.00 - 36.00	29.00 - 36.00	- -	- -	2.3.92
43.00 - 52.00	48.00 - 59.00	48.00 - 59.00	- -	- -	2.3.93
45.00 - 54.00	51.00 - 61.00	51.00 - 61.00	- -	- -	2.3.94
90.00 -160.00	90.00 -160.00	90.00 -160.00	90.00 -160.00	90.00 -160.00	2.3.95
- -	160.00 -290.00	160.00 -290.00	160.00 -290.00	160.00 -290.00	2.3.96
110.00 -155.00	120.00 -160.00	120.00 -160.00	120.00 -160.00	125.00 -180.00	2.3.97
135.00 -170.00	145.00 -180.00	145.00 -180.00	145.00 -180.00	155.00 -200.00	2.3.98
160.00 -190.00	170.00 -200.00	170.00 -200.00	170.00 -200.00	180.00 -225.00	2.3.99
- -	180.00 -315.00	180.00 -315.00	180.00 -315.00	180.00 -340.00	2.3.100
- -	250.00 -360.00	250.00 -360.00	250.00 -360.00	270.00 -400.00	2.3.101
- -	360.00 -540.00	360.00 -540.00	360.00 -540.00	360.00 -590.00	2.3.102
- -	540.00 -675.00	540.00 -675.00	540.00 -675.00	540.00 -720.00	2.3.103
- -	1.10 - 1.40	- -	1.10 - 1.40	1.10 - 1.40	2.3.104
- -	1.40 - 1.60	- -	1.40 - 1.60	1.40 - 1.60	2.3.105
- -	1.60 - 2.00	- -	1.60 - 2.00	1.60 - 2.00	2.3.106
- -	1.80 - 2.20	- -	1.80 - 2.20	1.80 - 2.20	2.3.107
- -	2.20 - 2.40	- -	2.20 - 2.40	2.20 - 2.40	2.3.108
- -	2.25 - 2.40	- -	2.25 - 2.40	2.25 - 2.40	2.3.109
- -	1.00 - 2.00	- -	1.00 - 2.00	1.00 - 2.00	2.3.110
- -	1.50 - 4.50	- -	1.50 - 4.50	1.50 - 4.50	2.3.111
- -	2.75 - 3.00	- -	2.75 - 3.00	2.75 - 3.00	2.3.112
- -	3.25 - 3.50	- -	3.25 - 3.50	3.25 - 3.50	2.3.113
- -	4.50 - 5.00	- -	4.50 - 5.00	4.50 - 5.00	2.3.114
- -	6.00 - 6.50	- -	6.00 - 6.50	6.00 - 6.50	2.3.115
- -	4.00 - 5.50	- -	- -	- -	2.3.116
4.50 - 5.50	4.50 - 5.50	4.50 - 5.50	4.50 - 5.50	4.50 - 5.50	2.3.117
5.50 - 6.25	5.50 - 6.25	5.50 - 6.25	5.50 - 6.25	5.50 - 6.25	2.3.118
6.25 - 7.25	6.25 - 7.25	6.25 - 7.25	6.25 - 7.25	6.25 - 7.25	2.3.119
4.50 - 7.25	4.50 - 7.25	4.50 - 7.25	4.50 - 7.25	4.50 - 7.25	2.3.120
6.25 - 7.25	6.25 - 7.25	6.25 - 7.25	6.25 - 7.25	6.25 - 7.25	2.3.121
7.25 - 8.00	7.25 - 8.00	7.25 - 8.00	7.25 - 8.00	7.25 - 8.00	2.3.122
9.00 - 10.00	9.00 - 10.00	9.00 - 10.00	9.00 - 10.00	9.00 - 10.00	2.3.123
8.00 - 11.00	8.00 - 11.00	8.00 - 11.00	8.00 - 11.00	8.00 - 11.00	2.3.124
11.50 - 13.50	11.50 - 13.50	11.50 - 13.50	11.50 - 13.50	11.50 - 13.50	2.3.125

Item nr.	SPECIFICATIONS	Unit	RESIDENTIAL £ range £
2.3	**ROOF - cont'd** roof plan area (unless otherwise described)		
	Comparative over/underlays - cont'd		
	Extruded polystyrene foam boards		
2.3.126	50 mm thick	m2	- -
2.3.127	50 mm thick; with cement topping	m2	- -
2.3.128	75 mm thick	m2	- -
	Perlite insulation board; density 170 - 180 kg/m3		
2.3.129	60 mm thick	m2	- -
	Foam glass insulation board; density 125 - 135 kg/m2		
2.3.130	60 mm thick	m2	- -
	Screeds to receive roof coverings		
2.3.131	50 mm cement and sand screed	m2	7.50 - 8.25
2.3.132	60 mm (av.) 'Isocrete K' screed; density 500 kg	m2	8.00 - 8.50
2.3.133	75 mm lightweight bituminous screed and vapour barrier	m2	12.50 - 14.00
2.3.134	100 mm lightweight bituminous screed and vapour barrier	m2	14.50 - 16.50
	50 mm Woodwool slabs; unreinforced		
2.3.135	sloping (measured on face)	m2	8.00 - 10.00
	sloping (measured on plan)		
2.3.136	20° pitch	m2	9.00 - 11.00
2.3.137	30° pitch	m2	11.00 - 13.00
2.3.138	35° pitch	m2	12.25 - 14.25
2.3.139	40° pitch	m2	12.75 - 14.50
2.3.140	50 mm Woodwool slabs; unreinforced; on and including steel purlins @ 600 mm centres	m2	13.50 - 16.25
	19 mm 'Tanalised' softwood boarding		
2.3.141	sloping (measured on face)	m2	9.00 - 11.00
	sloping (measured on plan)		
2.3.142	20° pitch	m2	10.00 - 11.50
2.3.143	30° pitch	m2	12.00 - 14.00
2.3.144	35° pitch	m2	13.50 - 15.50
2.3.145	40° pitch	m2	14.00 - 15.75
	25 mm 'Tanalised' softwood boarding		
2.3.146	sloping (measured on face)	m2	10.50 - 12.00
	sloping (measured on plan)		
2.3.147	20° pitch	m2	11.50 - 13.25
2.3.148	30° pitch	m2	14.00 - 15.75
2.3.149	35° pitch	m2	15.50 - 17.50
2.3.150	40° pitch	m2	16.25 - 18.00
	18 mm External quality plywood boarding		
2.3.151	sloping (measured on face)	m2	13.50 - 16.25
	sloping (measured on plan)		
2.3.152	20° pitch	m2	15.00 - 17.50
2.3.153	30° pitch	m2	18.25 - 21.25
2.3.154	35° pitch	m2	20.75 - 23.50
2.3.155	40° pitch	m2	21.25 - 24.50
	Comparative tiling and slating finishes/perimeter treatments (including underfelt, battening, eaves courses and ridges)		
	Concrete troughed interlocking tiles; 413 x 300 mm; 75 mm lap		
2.3.156	sloping (measured on face)	m2	13.00 - 15.75
	sloping (measured on plan)		
2.3.157	30° pitch	m2	17.00 - 19.75
2.3.158	35° pitch	m2	19.00 - 21.50
2.3.159	40° pitch	m2	20.00 - 22.50

INDUSTRIAL £ range £	RETAILING £ range £	LEISURE £ range £	OFFICES £ range £	HOTELS £ range £	Item nr.
11.00 - 12.50	11.00 - 12.50	11.00 - 12.50	11.00 - 12.50	11.00 - 12.50	2.3.126
18.00 - 20.00	18.00 - 20.00	18.00 - 20.00	18.00 - 20.00	18.00 - 20.00	2.3.127
14.50 - 16.00	14.50 - 16.00	14.50 - 16.00	14.50 - 16.00	14.50 - 16.00	2.3.128
12.00 - 13.00	12.00 - 13.00	12.00 - 13.00	12.00 - 13.00	12.00 - 13.00	2.3.129
15.00 - 17.00	15.00 - 17.00	15.00 - 17.00	15.00 - 17.00	15.00 - 17.00	2.3.130
- -	7.50 - 8.25	- -	7.50 - 8.25	7.50 - 8.25	2.3.131
- -	8.00 - 8.50	- -	8.00 - 8.50	8.00 - 8.50	2.3.132
- -	12.50 - 14.00	- -	12.50 - 14.00	12.50 - 14.00	2.3.133
- -	14.50 - 16.50	- -	14.50 - 16.50	14.00 - 16.50	2.3.134
8.00 - 10.00	- -	- -	8.00 - 10.00	- -	2.3.135
9.00 - 11.00	- -	- -	9.00 - 11.00	- -	2.3.136
11.25 - 13.00	- -	- -	11.25 - 13.00	- -	2.3.137
12.25 - 14.25	- -	- -	12.25 - 14.25	- -	2.3.138
12.75 - 14.50	- -	- -	12.75 - 14.50	- -	2.3.139
13.50 - 16.25	- -	- -	13.50 - 16.25	- -	2.3.140
- -	9.00 - 11.00	- -	9.00 - 11.00	9.00 - 11.00	2.3.141
- -	10.00 - 11.75	- -	10.00 - 11.75	10.00 - 11.75	2.3.142
- -	12.00 - 14.00	- -	12.00 - 14.00	12.00 - 14.00	2.3.143
- -	13.50 - 15.25	- -	13.50 - 15.25	13.50 - 15.25	2.3.144
- -	14.00 - 15.75	- -	14.00 - 15.75	14.00 - 15.75	2.3.145
- -	10.25 - 12.00	- -	10.25 - 12.00	10.25 - 12.00	2.3.146
- -	11.50 - 13.25	- -	11.50 - 13.25	11.50 - 13.25	2.3.147
- -	14.00 - 15.75	- -	14.00 - 15.75	14.00 - 15.75	2.3.148
- -	15.50 - 17.50	- -	15.50 - 17.50	15.50 - 17.50	2.3.149
- -	16.25 - 18.00	- -	16.25 - 18.00	16.25 - 18.00	2.3.150
- -	13.50 - 16.25	- -	13.50 - 16.25	13.50 - 16.25	2.3.151
- -	15.00 - 17.50	- -	15.00 - 17.50	15.00 - 17.50	2.3.152
- -	18.25 - 21.00	- -	18.25 - 21.00	18.25 - 21.00	2.3.153
- -	20.75 - 23.50	- -	20.75 - 23.50	20.75 - 23.50	2.3.154
- -	21.00 - 24.25	- -	21.00 - 24.25	21.00 - 24.25	2.3.155
- -	13.00 - 15.75	- -	13.00 - 15.75	13.00 - 15.75	2.3.156
- -	17.00 - 20.00	- -	17.00 - 20.00	17.00 - 20.00	2.3.157
- -	19.00 - 21.50	- -	19.00 - 21.50	19.00 - 21.50	2.3.158
- -	20.00 - 22.50	- -	20.00 - 22.50	20.00 - 22.50	2.3.159

Item nr.	SPECIFICATIONS	Unit	RESIDENTIAL £ range £
2.3	**ROOF - cont'd** roof plan area (unless otherwise described)		
	Comparative tiling and slating finishes/perimeter treatments - cont'd		
	Concrete interlocking slates; 430 x 330 mm; 75 mm lap		
2.3.160	sloping (measured on face)	m2	13.50 - 16.25
	sloping (measured on plan)		
2.3.161	30° pitch	m2	17.50 - 20.25
2.3.162	35° pitch	m2	19.50 - 22.25
2.3.163	40° pitch	m2	20.00 - 22.75
	Glass fibre reinforced bitumen slates; 900 x 300 mm; fixed to boarding (measured elsewhere)		
2.3.164	sloping (measured on face)	m2	12.50 - 15.00
	sloping (measured on plan)		
2.3.165	30° pitch	m2	16.00 - 18.00
2.3.166	35° pitch	m2	17.50 - 21.00
2.3.167	40° pitch	m2	19.00 - 22.50
	Concrete bold roll interlocking tiles; 418 x 332 mm; 75 mm lap		
2.3.168	sloping (measured on face)	m2	13.00 - 15.75
	sloping (measured on plan)		
2.3.169	30° pitch	m2	17.00 - 20.00
2.3.170	35° pitch	m2	19.00 - 21.50
2.3.171	40° pitch	m2	19.25 - 22.00
	Tudor clay pantiles; 470 x 285 mm; 100 mm lap		
2.3.172	sloping (measured on face)	m2	18.50 - 22.00
	sloping (measured on plan)		
2.3.173	30° pitch	m2	23.00 - 27.50
2.3.174	35° pitch	m2	26.00 - 31.00
2.3.175	40° pitch	m2	27.50 - 33.00
	Natural red pantiles; 337 x 241 mm; 76 mm head and 38 mm side laps		
2.3.176	sloping (measured on face)	m2	21.50 - 26.00
	sloping (measured on plan)		
2.3.177	30° pitch	m2	28.00 - 32.50
2.3.178	35° pitch	m2	30.50 - 35.00
2.3.179	40° pitch	m2	31.50 - 36.00
	Blue composition (non-asbestos) slates; 600 x 300 mm; 75 mm lap		
2.3.180	sloping (measured on face)	m2	22.50 - 27.00
2.3.181	sloping to mansard (measured on face)	m2	31.50 - 36.00
	sloping (measured on plan)		
2.3.182	30° pitch	m2	30.00 - 34.25
2.3.183	35° pitch	m2	32.50 - 37.00
2.3.184	40° pitch	m2	33.25 - 37.75
	vertical to mansard; including 18 mm blockboard		
2.3.185	(measured on face)	m2	45.00 - 54.00
	Concrete plain tiles; 267 x 165 mm; 64 mm lap		
2.3.186	sloping (measured on face)	m2	30.00 - 36.00
	sloping (measured on plan)		
2.3.187	30° pitch	m2	37.50 - 45.00
2.3.188	35° pitch	m2	42.00 - 50.00
2.3.189	40° pitch	m2	45.00 - 54.00
	Machine made clay plain tiles; 267 x 165 mm; 64 mm lap		
2.3.190	sloping (measured on face)	m2	34.00 - 40.00
	sloping (measured on plan)		
2.3.191	30° pitch	m2	42.50 - 50.00
2.3.192	35° pitch	m2	47.50 - 56.00
2.3.193	40° pitch	m2	51.00 - 60.00

INDUSTRIAL £ range £	RETAILING £ range £	LEISURE £ range £	OFFICES £ range £	HOTELS £ range £	Item nr.
- -	13.50 - 16.00	- -	13.50 - 16.00	13.50 - 16.00	2.3.160
- -	17.50 - 20.25	- -	17.50 - 20.25	17.50 - 20.25	2.3.161
- -	19.50 - 22.00	- -	19.50 - 22.00	19.50 - 22.00	2.3.162
- -	20.00 - 22.75	- -	20.00 - 22.75	20.00 - 22.75	2.3.163
- -	12.50 - 15.00	- -	12.50 - 15.00	12.50 - 15.00	2.3.164
- -	16.00 - 18.00	- -	16.00 - 18.00	16.00 - 18.00	2.3.165
- -	17.50 - 21.00	- -	17.50 - 21.00	17.50 - 21.00	2.3.166
- -	19.00 - 22.50	- -	19.00 - 22.50	19.00 - 22.50	2.3.167
- -	13.00 - 15.75	- -	13.00 - 15.75	13.00 - 15.75	2.3.168
- -	17.00 - 20.00	- -	17.00 - 20.00	17.00 - 20.00	2.3.169
- -	19.00 - 21.50	- -	19.00 - 21.50	19.00 - 21.50	2.3.170
- -	19.25 - 22.00	- -	19.25 - 22.00	19.25 - 22.00	2.3.171
- -	18.50 - 22.00	- -	18.50 - 22.00	18.50 - 22.00	2.3.172
- -	23.00 - 27.50	- -	23.00 - 27.50	23.00 - 27.50	2.3.173
- -	26.00 - 31.00	- -	26.00 - 31.00	26.00 - 31.00	2.3.174
- -	27.50 - 33.00	- -	27.50 - 33.00	27.50 - 33.00	2.3.175
- -	21.50 - 26.00	- -	21.50 - 26.00	21.50 - 26.00	2.3.176
- -	28.00 - 32.50	- -	28.00 - 32.50	28.00 - 32.50	2.3.177
- -	30.50 - 35.00	- -	30.50 - 35.00	30.50 - 35.00	2.3.178
- -	31.50 - 36.00	- -	31.50 - 36.00	31.50 - 36.00	2.3.179
- -	22.50 - 27.00	- -	22.50 - 27.00	22.50 - 27.00	2.3.180
- -	31.50 - 36.00	- -	31.50 - 36.00	31.50 - 36.00	2.3.181
- -	29.75 - 34.25	- -	29.75 - 34.25	29.75 - 34.25	2.3.182
- -	32.50 - 37.00	- -	32.50 - 37.00	32.50 - 37.00	2.3.183
- -	33.25 - 37.75	- -	33.25 - 37.75	33.25 - 37.75	2.3.184
- -	45.00 - 54.00	- -	45.00 - 54.00	45.00 - 54.00	2.3.185
- -	30.00 - 36.00	- -	30.00 - 36.00	30.00 - 36.00	2.3.186
- -	37.50 - 45.00	- -	37.50 - 45.00	37.50 - 45.00	2.3.187
- -	42.00 - 50.00	- -	42.00 - 50.00	42.00 - 50.00	2.3.188
- -	45.00 - 54.00	- -	45.00 - 54.00	45.00 - 54.00	2.3.189
- -	34.00 - 40.00	- -	34.00 - 40.00	34.00 - 40.00	2.3.190
- -	42.50 - 50.00	- -	42.50 - 50.00	42.50 - 50.00	2.3.191
- -	47.50 - 56.00	- -	47.50 - 56.00	47.50 - 56.00	2.3.192
- -	51.00 - 60.00	- -	51.00 - 60.00	51.00 - 60.00	2.3.193

Item nr.	SPECIFICATIONS	Unit	RESIDENTIAL £ range £
2.3	**ROOF - cont'd** **roof plan area (unless otherwise described)**		
	Comparative tiling and slating finishes/perimeter treatments - cont'd		
	Black 'Sterreberg' glazed interlocking pantiles; 355 x 240 mm; 76 mm head and 38 mm side laps		
2.3.194	sloping (measured on face)	m2	31.00 - 37.50
	sloping (measured on plan)		
2.3.195	30° pitch	m2	38.00 - 46.00
2.3.196	35° pitch	m2	43.00 - 52.50
2.3.197	40° pitch	m2	46.00 - 56.00
	Red cedar sawn shingles; 450 mm long; 125 mm lap		
2.3.198	sloping (measured on face)	m2	35.00 - 42.00
	sloping (measured on plan)		
2.3.199	30° pitch	m2	43.00 - 52.50
2.3.200	35° pitch	m2	49.00 - 59.00
2.3.201	40° pitch	m2	52.50 - 63.00
	Welsh natural slates; 510 x 255 mm; 76 mm lap		
2.3.202	sloping (measured on face)	m2	38.00 - 44.00
	sloping (measured on plan)		
2.3.203	30° pitch	m2	50.00 - 58.00
2.3.204	35° pitch	m2	55.00 - 61.00
2.3.205	40° pitch	m2	57.00 - 63.00
	Welsh slates; 610 x 305 mm; 76 mm lap		
2.3.206	sloping (measured on face)	m2	39.00 - 45.00
	sloping (measured on plan)		
2.3.207	30° pitch	m2	50.00 - 57.00
2.3.208	35° pitch	m2	56.00 - 62.00
2.3.209	40° pitch	m2	58.00 - 64.00
	Reconstructed stone slates; random lengths; 80 mm lap		
2.3.210	sloping (measured on face)	m2	23.00 - 45.00
	sloping (measured on plan)		
2.3.211	30° pitch	m2	29.00 - 58.00
2.3.212	35° pitch	m2	32.00 - 63.00
2.3.213	40° pitch	m2	35.00 - 65.00
	Handmade sandfaced plain tiles; 267 x 165 mm; 64 mm lap		
2.3.214	sloping (measured on face)	m2	50.00 - 60.00
	sloping (measured on plan)		
2.3.215	30° pitch	m2	62.50 - 75.00
2.3.216	35° pitch	m2	70.00 - 84.00
2.3.217	40° pitch	m2	75.00 - 90.00
	Westmorland green slates; random sizes; 76 mm lap		
2.3.218	sloping (measured on face)	m2	100.00 -115.00
	sloping (measured on plan)		
2.3.219	30° pitch	m2	125.00 -142.50
2.3.220	35° pitch	m2	140.00 -160.00
2.3.221	40° pitch	m2	150.00 -172.50
	Verges to sloping roofs; 250 x 25 mm painted softwood bargeboard **perimeter length**		
2.3.222	6 mm 'Masterboard' soffit lining 150 mm wide	m	12.50 - 14.50
2.3.223	19 x 150 mm painted softwood soffit	m	14.50 - 16.25
	Eaves to sloping roofs; 200 x 25 mm painted softwood fascia; 6 mm 'Masterboard' soffit lining 225 mm wide		
2.3.224	100 mm PVC gutter	m	17.00 - 22.50
2.3.225	150 mm PVC gutter	m	21.50 - 27.00
2.3.226	100 mm cast iron gutter; decorated	m	27.00 - 31.50
2.3.227	150 mm cast iron gutter; decorated	m	32.50 - 37.75

INDUSTRIAL £ range £		RETAILING £ range £	LEISURE £ range £		OFFICES £ range £	HOTELS £ range £	Item nr.
-	-	31.00 - 37.50	-	-	31.00 - 37.50	31.00 - 37.50	2.3.194
-	-	38.00 - 46.00	-	-	38.00 - 46.00	38.00 - 46.00	2.3.195
-	-	43.00 - 52.50	-	-	43.00 - 52.50	43.00 - 52.50	2.3.196
-	-	46.00 - 56.00	-	-	46.00 - 56.00	46.00 - 56.00	2.3.197
-	-	35.00 - 42.00	-	-	35.00 - 42.00	35.00 - 42.00	2.3.198
-	-	43.00 - 52.50	-	-	43.00 - 52.50	43.00 - 52.50	2.3.199
-	-	49.00 - 59.00	-	-	49.00 - 59.00	49.00 - 59.00	2.3.200
-	-	52.50 - 63.00	-	-	52.50 - 63.00	52.50 - 63.00	2.3.201
-	-	38.00 - 44.00	-	-	42.00 - 49.00	42.00 - 49.00	2.3.202
-	-	50.00 - 58.00	-	-	50.00 - 58.00	50.00 - 58.00	2.3.203
-	-	55.00 - 61.00	-	-	55.00 - 61.00	55.00 - 61.00	2.3.204
-	-	57.00 - 63.00	-	-	57.00 - 63.00	57.00 - 63.00	2.3.205
-	-	39.00 - 45.00	-	-	39.00 - 45.00	39.00 - 45.00	2.3.206
-	-	50.00 - 57.00	-	-	50.00 - 57.00	50.00 - 57.00	2.3.207
-	-	56.00 - 62.00	-	-	56.00 - 62.00	56.00 - 62.00	2.3.208
-	-	58.00 - 64.00	-	-	58.00 - 64.00	58.00 - 64.00	2.3.209
-	-	23.00 - 45.00	-	-	23.00 - 45.00	23.00 - 45.00	2.3.210
-	-	29.00 - 58.00	-	-	29.00 - 58.00	29.00 - 58.00	2.3.211
-	-	32.00 - 63.00	-	-	32.00 - 63.00	32.00 - 63.00	2.3.212
-	-	35.00 - 65.00	-	-	35.00 - 65.00	35.00 - 65.00	2.3.213
-	-	50.00 - 60.00	-	-	50.00 - 60.00	50.00 - 60.00	2.3.214
-	-	62.50 - 75.00	-	-	62.50 - 75.00	62.50 - 75.00	2.3.215
-	-	70.00 - 84.00	-	-	70.00 - 84.00	70.00 - 84.00	2.3.216
-	-	75.00 - 90.00	-	-	75.00 - 90.00	75.00 - 90.00	2.3.217
-	-	100.00 -115.00	-	-	100.00 -115.00	100.00 -115.00	2.3.218
-	-	125.00 -142.50	-	-	125.00 -142.50	125.00 -142.50	2.3.219
-	-	140.00 -160.00	-	-	140.00 -160.00	140.00 -160.00	2.3.220
-	-	150.00 -172.50	-	-	150.00 -172.50	150.00 -172.50	2.3.221
-	-	12.50 - 14.50	-	-	12.50 - 14.50	12.50 - 14.50	2.3.222
-	-	14.50 - 16.25	-	-	14.50 - 16.25	14.50 - 16.25	2.3.223
-	-	17.00 - 22.50	-	-	17.00 - 22.50	17.00 - 22.50	2.3.224
-	-	21.50 - 27.00	-	-	21.50 - 27.00	21.50 - 27.00	2.3.225
-	-	27.00 - 31.50	-	-	27.00 - 31.50	27.00 - 31.50	2.3.226
-	-	32.50 - 37.75	-	-	32.50 - 37.75	32.50 - 37.75	2.3.227

Item nr.	SPECIFICATIONS	Unit	RESIDENTIAL £ range £
2.3	**ROOF** - cont'd	roof plan area (unless otherwise described)	
	Comparative tiling and slating finishes/perimeter treatments - cont'd		
	Eaves to sloping roofs; 200 x 25 mm painted softwood fascia; 19 x 225 mm painted softwood soffit		
2.3.228	100 mm PVC gutter	m	20.00 - 25.00
2.3.229	150 mm PVC gutter	m	24.50 - 30.00
2.3.230	100 mm cast iron gutter; decorated	m	29.00 - 34.00
2.3.231	150 mm cast iron gutter; decorated	m	35.00 - 40.50
	Rainwater pipes; fixed to backgrounds; including offsets and shoe		
2.3.232	68 mm PVC	m	6.00 - 8.00
2.3.233	110 mm PVC	m	9.00 - 11.00
2.3.234	75 mm cast iron; decorated	m	21.00 - 24.50
2.3.235	100 mm cast iron; decorated	m	25.00 - 29.00
	Ridges		
2.3.236	concrete half round tiles	m	12.00 - 14.50
2.3.237	machine-made clay half round tiles	m	15.00 - 17.00
2.3.238	hand-made clay half round tiles	m	15.00 - 24.00
	Hips; including mitreing roof tiles		
2.3.239	concrete half round tiles	m	15.50 - 20.00
2.3.240	machine-made clay half round tiles	m	22.00 - 24.00
2.3.241	hand-made clay half round tiles	m	22.00 - 26.00
2.3.242	hand-made clay bonnet hip tiles	m	35.00 - 40.00
	Comparative cladding finishes (including underfelt, labours etc.)		
	0.91 mm Aluminium roofing; commercial grade		
2.3.243	flat	m2	- -
	0.91 mm Aluminium roofing; commercial grade; fixed to boarding (included)		
2.3.244	sloping (measured on face)	m2	- -
	sloping (measured on plan)		
2.3.245	20° pitch	m2	- -
2.3.246	30° pitch	m2	- -
2.3.247	35° pitch	m2	- -
2.3.248	40° pitch	m2	- -
	0.81 mm Zinc roofing		
2.3.249	flat	m2	- -
	0.81 mm Zinc roofing; fixed to boarding (included)		
2.3.250	sloping (measured on face)	m2	- -
	sloping (measured on plan)		
2.3.251	20° pitch	m2	- -
2.3.252	30° pitch	m2	- -
2.3.253	35° pitch	m2	- -
2.3.254	40° pitch	m2	- -
	Copper roofing		
2.3.255	0.56 mm thick; flat	m2	- -
2.3.256	0.61 mm thick; flat	m2	- -
	Copper roofing; fixed to boarding (included)		
2.3.257	0.56 mm thick; sloping (measured on face)	m2	- -
	0.56 mm thick; sloping (measured on plan)		
2.3.258	20° pitch	m2	- -
2.3.259	30° pitch	m2	- -
2.3.260	35° pitch	m2	- -
2.3.261	40° pitch	m2	- -
2.3.262	0.61 mm thick; sloping (measured on face)	m2	- -
	0.61 mm thick; sloping (measured on plan)		
2.3.263	20° pitch	m2	- -
2.3.264	30° pitch	m2	- -
2.3.265	35° pitch	m2	- -
2.3.266	40° pitch	m2	- -

INDUSTRIAL £ range £	RETAILING £ range £	LEISURE £ range £	OFFICES £ range £	HOTELS £ range £	Item nr.
- -	20.00 - 25.00	- -	20.00 - 25.00	20.00 - 25.00	2.3.228
- -	24.50 - 30.00	- -	24.50 - 30.00	24.50 - 30.00	2.3.229
- -	29.00 - 34.00	- -	29.00 - 34.00	29.00 - 34.00	2.3.230
- -	35.00 - 40.50	- -	35.00 - 40.50	35.00 - 40.50	2.3.231
- -	6.00 - 8.00	- -	6.00 - 8.00	6.00 - 8.00	2.3.232
- -	9.00 - 11.00	- -	9.00 - 11.00	9.00 - 11.00	2.3.233
- -	21.00 - 24.50	- -	21.00 - 24.50	21.00 - 24.50	2.3.234
- -	25.00 - 29.00	- -	25.00 - 29.00	25.00 - 29.00	2.3.235
- -	12.00 - 14.50	- -	12.00 - 14.50	12.00 - 14.50	2.3.236
- -	15.00 - 17.00	- -	15.00 - 17.00	15.00 - 17.00	2.3.237
- -	15.00 - 24.00	- -	15.00 - 24.00	15.00 - 24.00	2.3.238
- -	15.50 - 20.00	- -	15.50 - 20.00	15.50 - 20.00	2.3.239
- -	22.00 - 24.00	- -	22.00 - 24.00	22.00 - 24.00	2.3.240
- -	22.00 - 26.00	- -	22.00 - 26.00	22.00 - 26.00	2.3.241
- -	35.00 - 40.00	- -	35.00 - 40.00	35.00 - 40.00	2.3.242
- -	34.00 - 39.00	34.00 - 39.00	34.00 - 39.00	34.00 - 39.00	2.3.243
- -	37.00 - 42.00	37.00 - 42.00	37.00 - 42.00	37.00 - 42.00	2.3.244
- -	40.00 - 46.00	40.00 - 46.00	40.00 - 46.00	40.00 - 46.00	2.3.245
- -	50.00 - 55.00	50.00 - 55.00	50.00 - 55.00	50.00 - 55.00	2.3.246
- -	55.00 - 62.50	55.00 - 62.50	55.00 - 62.50	55.00 - 62.50	2.3.247
- -	58.00 - 62.50	58.00 - 62.50	58.00 - 62.50	58.00 - 62.50	2.3.248
- -	45.00 - 50.00	45.00 - 50.00	45.00 - 50.00	45.00 - 50.00	2.3.249
- -	48.00 - 54.00	48.00 - 54.00	48.00 - 54.00	48.00 - 54.00	2.3.250
- -	53.00 - 59.00	53.00 - 59.00	53.00 - 59.00	53.00 - 59.00	2.3.251
- -	65.00 - 70.00	65.00 - 70.00	65.00 - 70.00	65.00 - 70.00	2.3.252
- -	72.00 - 78.00	72.00 - 78.00	72.00 - 78.00	72.00 - 78.00	2.3.253
- -	76.00 - 81.00	76.00 - 81.00	76.00 - 81.00	76.00 - 81.00	2.3.254
- -	51.00 - 57.00	51.00 - 57.00	51.00 - 57.00	51.00 - 57.00	2.3.255
- -	54.00 - 60.00	54.00 - 60.00	54.00 - 60.00	54.00 - 60.00	2.3.256
- -	54.00 - 60.00	54.00 - 60.00	54.00 - 60.00	54.00 - 60.00	2.3.257
- -	60.00 - 65.00	60.00 - 65.00	60.00 - 65.00	60.00 - 65.00	2.3.258
- -	72.00 - 78.00	72.00 - 78.00	72.00 - 78.00	72.00 - 78.00	2.3.259
- -	81.00 - 87.00	81.00 - 87.00	81.00 - 87.00	81.00 - 87.00	2.3.260
- -	83.00 - 89.00	83.00 - 89.00	83.00 - 89.00	83.00 - 89.00	2.3.261
- -	57.00 - 63.00	57.00 - 63.00	57.00 - 63.00	57.00 - 63.00	2.3.262
- -	63.00 - 68.00	63.00 - 68.00	63.00 - 68.00	63.00 - 68.00	2.3.263
- -	77.00 - 83.00	77.00 - 83.00	77.00 - 83.00	77.00 - 83.00	2.3.264
- -	86.00 - 92.00	86.00 - 92.00	86.00 - 92.00	86.00 - 92.00	2.3.265
- -	89.00 - 95.00	89.00 - 95.00	89.00 - 95.00	89.00 - 95.00	2.3.266

Item nr.	SPECIFICATIONS	Unit	RESIDENTIAL £ range £	
2.3	**ROOF** - cont'd **roof plan area** (unless otherwise described)			
	Comparative cladding finishes (including underfelt, labours etc.) - cont'd			
	Lead roofing			
2.3.267	code 4 sheeting; flat	m2	-	-
2.3.268	code 5 sheeting; flat	m2	-	-
2.3.269	code 6 sheeting; flat	m2	-	-
	Lead roofing; fixed to boarding (included)			
2.3.270	code 4 sheeting; sloping (measured on face)	m2	-	-
	code 4 sheeting; sloping (measured on plan)			
2.3.271	20° pitch	m2	-	-
2.3.272	30° pitch	m2	-	-
2.3.273	35° pitch	m2	-	-
2.3.274	40° pitch	m2	-	-
2.3.275	code 6 sheeting; sloping (measured on face)	m2	-	-
	code 6 sheeting; sloping (measured on plan)			
2.3.276	20° pitch	m2	-	-
2.3.277	30° pitch	m2	-	-
2.3.278	35° pitch	m2	-	-
2.3.279	40° pitch	m2	-	-
2.3.280	code 6 sheeting; vertical to mansard; including insulation (measured on face)	m2	-	-
	Comparative waterproof finishes/perimeter treatments			
	Liquid applied coatings			
2.3.281	solar reflective paint	m2	-	-
2.3.282	spray applied bitumen	m2	-	-
2.3.283	spray applied co-polymer	m2	-	-
2.3.284	spray applied polyurethene	m2	-	-
	20 mm Two coat asphalt roofing; laid flat; on felt underlay			
2.3.285	to BS 988	m2	-	-
2.3.286	to BS 6577	m2	-	-
	Extra for			
2.3.287	solar reflective paint	m2	-	-
2.3.288	limestone chipping finish	m2	-	-
2.3.289	grp tiles in hot bitumen	m2	-	-
	20 mm Two coat reinforced asphaltic compound; laid flat; on felt underlay			
2.3.290	to BS 6577	m2	-	-
	Built-up bitumen felt roofing; laid flat			
2.3.291	three layer glass fibre roofing	m2	-	-
2.3.292	three layer asbestos based roofing	m2	-	-
	Extra for			
2.3.293	granite chipping finish	m2	-	-
	Built-up self-finished asbestos based bitumen felt roofing; laid sloping			
2.3.294	two layer roofing (measured on face)	m2	19.00	22.00
	two layer roofing (measured on plan)			
2.3.295	35° pitch	m2	28.00	31.00
2.3.296	40° pitch	m2	29.50	32.50
2.3.297	three layer roofing (measured on face)	m2	26.00	29.00
	three layer roofing (measured on plan)			
2.3.298	20° pitch	m2	38.00	41.00
2.3.299	30° pitch	m2	39.00	42.50
	Elastomeric single ply roofing; laid flat			
2.3.300	EPDM membrane; laid loose	m2	-	-
2.3.301	Bytyl rubber membrane; laid loose	m2	-	-
	Extra for			
2.3.302	ballast	m2	-	-

INDUSTRIAL £ range £	RETAILING £ range £	LEISURE £ range £	OFFICES £ range £	HOTELS £ range £	Item nr.
- -	56.00 - 61.00	56.00 - 61.00	56.00 - 61.00	56.00 - 61.00	2.3.267
- -	65.00 - 70.00	65.00 - 70.00	65.00 - 70.00	65.00 - 70.00	2.3.268
- -	70.00 - 77.00	70.00 - 77.00	70.00 - 77.00	70.00 - 77.00	2.3.269
- -	59.00 - 64.00	59.00 - 64.00	59.00 - 64.00	59.00 - 64.00	2.3.270
- -	64.00 - 69.00	64.00 - 69.00	64.00 - 69.00	64.00 - 69.00	2.3.271
- -	78.00 - 84.00	78.00 - 84.00	78.00 - 84.00	78.00 - 84.00	2.3.272
- -	88.00 - 94.00	88.00 - 94.00	88.00 - 94.00	88.00 - 94.00	2.3.273
- -	90.00 - 96.00	90.00 - 96.00	90.00 - 96.00	90.00 - 96.00	2.3.274
- -	74.00 - 81.00	74.00 - 81.00	74.00 - 81.00	74.00 - 81.00	2.3.275
- -	81.00 - 88.00	81.00 - 88.00	81.00 - 88.00	81.00 - 88.00	2.3.276
- -	100.00 -106.00	100.00 -106.00	100.00 -106.00	100.00 -106.00	2.3.277
- -	110.00 -117.00	110.00 -117.00	110.00 -117.00	110.00 -117.00	2.3.278
- -	114.00 -122.00	114.00 -122.00	114.00 -122.00	114.00 -122.00	2.3.279
- -	100.00 -145.00	100.00 -145.00	100.00 -145.00	100.00 -145.00	2.3.280
- -	1.25 - 2.25	1.25 - 2.25	1.25 - 2.25	1.25 - 2.25	2.3.281
- -	5.00 - 8.00	5.00 - 8.00	5.00 - 8.00	5.00 - 8.00	2.3.282
- -	6.00 - 9.00	6.00 - 9.00	6.00 - 9.00	6.00 - 9.00	2.3.283
- -	11.00 - 13.00	11.00 - 13.00	11.00 - 13.00	11.00 - 13.00	2.3.284
- -	10.50 - 13.50	10.50 - 13.50	10.50 - 13.50	10.50 - 13.50	2.3.285
- -	14.50 - 18.00	14.50 - 18.00	14.50 - 18.00	14.50 - 18.00	2.3.286
- -	1.75 - 2.25	1.75 - 2.25	1.75 - 2.25	1.75 - 2.25	2.3.287
- -	2.00 - 5.50	2.00 - 5.50	2.00 - 5.50	2.00 - 5.50	2.3.288
- -	23.00 - 28.00	23.00 - 28.00	23.00 - 28.00	23.00 - 28.00	2.3.289
- -	16.00 - 20.00	16.00 - 20.00	16.00 - 20.00	16.00 - 20.00	2.3.290
- -	14.00 - 18.00	14.00 - 18.00	14.00 - 18.00	14.00 - 18.00	2.3.291
- -	17.50 - 20.00	17.50 - 20.00	17.50 - 20.00	17.50 - 20.00	2.3.292
- -	2.00 - 5.50	2.00 - 5.50	2.00 - 5.50	2.00 - 5.50	2.3.293
- -	- -	- -	- -	- -	2.3.294
- -	- -	- -	- -	- -	2.3.295
- -	- -	- -	- -	- -	2.3.296
- -	- -	- -	- -	- -	2.3.297
- -	- -	- -	- -	- -	2.3.298
- -	- -	- -	- -	- -	2.3.299
16.00 - 19.00	16.00 - 19.00	16.00 - 19.00	16.00 - 19.00	16.00 - 19.00	2.3.300
18.00 - 21.00	18.00 - 21.00	18.00 - 21.00	18.00 - 21.00	18.00 - 21.00	2.3.301
5.00 - 8.00	5.00 - 8.00	5.00 - 8.00	5.00 - 8.00	5.00 - 8.00	2.3.302

Item nr.	SPECIFICATIONS	Unit	RESIDENTIAL £ range £
2.3	**ROOF - cont'd** **roof plan area (unless otherwise described)**		
	Comparative waterproof finishes/perimeter treatments - cont'd		
	Thermoplastic single ply roofing; laid flat		
	PVC membrane		
2.3.303	laid loose	m2	- -
2.3.304	mechanically fixed	m2	- -
2.3.305	fully adhered	m2	- -
2.3.306	CPE membrane; laid loose	m2	- -
2.3.307	CSPG membrane; fully adhered	m2	- -
2.3.308	PIB membrane; laid loose	m2	- -
	Extra for		
2.3.309	ballast	m2	- -
	High performance built-up felt roofing; laid flat		
2.3.310	two layer self-finished 'Paradiene' elastomeric bitumen roofing	m2	- -
2.3.311	three layer 'Ruberglas 120 GP' felt roofing; granite chipping finish	m2	- -
2.3.312	'Andersons' three layer self-finished polyester based bitumen felt roofing	m2	- -
2.3.313	three layer polyester based modified bitumen felt roofing	m2	- -
2.3.314	'Anderson' three layer polyester based bitumen felt roofing; granite chipping finish	m2	- -
2.3.315	two layer 'Derbigum' self-finished polyester roofing	m2	- -
2.3.316	three layer metal faced 'Veral' glass cloth reinforced bitumen roofing	m2	- -
2.3.317	three layer 'Ruberfort HP 350' felt roofing; granite chipping finish	m2	- -
2.3.318	three layer 'Hyload 150E' elastomeric roofing; granite chipping finish	m2	- -
2.3.319	three layer 'Polybit 350' elastomeric roofing; granite chipping finish	m2	- -
	Torch on roofing; laid flat		
2.3.320	three layer polyester-based modified bitumen roofing	m2	- -
2.3.321	two layer polymeric isotropic roofing	m2	- -
	Extra for		
2.3.322	granite chipping finish	m2	- -
	Edges to flat felt roofs; softwood splayed fillet; 280 x 25 mm painted softwood fascia; no gutter		
2.3.323	aluminium edge trim	m	23.50 - 25.00
	Edges to flat roofs; code 4 lead drip dressed into gutter; 230 x 25 mm painted softwood fascia		
2.3.324	100 mm PVC gutter	m	22.50 - 29.00
2.3.325	150 mm PVC gutter	m	27.00 - 33.50
2.3.326	100 mm cast iron gutter; decorated	m	33.50 - 40.50
2.3.327	150 mm cast iron gutter; decorated	m	40.50 - 49.50

INDUSTRIAL £ range £	RETAILING £ range £	LEISURE £ range £	OFFICES £ range £	HOTELS £ range £	Item nr.
16.00 - 19.00	16.00 - 19.00	16.00 - 19.00	16.00 - 19.00	16.00 - 19.00	2.3.303
20.00 - 23.00	20.00 - 23.00	20.00 - 23.00	20.00 - 23.00	20.00 - 23.00	2.3.304
22.00 - 25.00	22.00 - 25.00	22.00 - 25.00	22.00 - 25.00	22.00 - 25.00	2.3.305
18.50 - 22.00	18.50 - 22.00	18.50 - 22.00	18.50 - 22.00	18.50 - 22.00	2.3.306
18.50 - 22.00	18.50 - 22.00	18.50 - 22.00	18.50 - 22.00	18.50 - 22.00	2.3.307
20.50 - 25.00	20.50 - 25.00	20.50 - 25.00	20.50 - 25.00	20.50 - 25.00	2.3.308
5.00 - 8.00	5.00 - 8.00	5.00 - 8.00	5.00 - 8.00	5.00 - 8.00	2.3.309
19.50 - 21.50	19.50 - 21.50	19.50 - 21.50	19.50 - 21.50	19.50 - 21.50	2.3.310
21.50 - 24.00	21.50 - 24.00	21.50 - 24.00	21.50 - 24.00	21.50 - 24.00	2.3.311
22.00 - 25.00	22.00 - 25.00	22.00 - 25.00	22.00 - 25.00	22.00 - 25.00	2.3.312
23.00 - 25.00	23.00 - 25.00	23.00 - 25.00	23.00 - 25.00	23.00 - 25.00	2.3.313
25.00 - 27.50	25.00 - 27.50	25.00 - 27.50	25.00 - 27.50	25.00 - 27.50	2.3.314
26.00 - 29.00	26.00 - 29.00	26.00 - 29.00	26.00 - 29.00	26.00 - 29.00	2.3.315
28.00 - 31.00	28.00 - 31.00	28.00 - 31.00	28.00 - 31.00	28.00 - 31.00	2.3.316
28.00 - 31.00	28.00 - 31.00	28.00 - 31.00	28.00 - 31.00	28.00 - 31.00	2.3.317
30.00 - 32.50	30.00 - 32.50	30.00 - 32.50	30.00 - 32.50	30.00 - 32.50	2.3.318
31.50 - 35.00	31.50 - 35.00	31.50 - 35.00	31.50 - 35.00	31.50 - 35.00	2.3.319
- -	20.00 - 23.00	20.00 - 23.00	20.00 - 23.00	20.00 - 23.00	2.3.320
- -	20.00 - 23.00	20.00 - 23.00	20.00 - 23.00	20.00 - 23.00	2.3.321
- -	2.00 - 5.00	2.00 - 5.00	2.00 - 5.00	2.00 - 5.00	2.3.322
- -	23.50 - 25.00	23.50 - 25.00	23.50 - 25.00	23.50 - 25.00	2.3.323
- -	22.50 - 29.00	- -	22.50 - 29.00	22.50 - 29.00	2.3.324
- -	27.00 - 33.50	- -	27.00 - 33.50	27.00 - 33.50	2.3.325
- -	33.50 - 40.50	- -	33.50 - 40.50	33.50 - 40.50	2.3.326
- -	40.50 - 49.50	- -	40.50 - 49.50	40.50 - 49.50	2.3.327

Item nr.	SPECIFICATIONS	Unit	RESIDENTIAL £ range £	
2.4	**STAIRS**	storey (unless otherwise described)		
	comprising:-			
	Timber construction **Reinforced concrete construction** **Metal construction** **Comparative finishes/balustrading**			
	Timber construction			
	Softwood staircase; softwood balustrades and hardwood handrail; plasterboard; skim and emulsion to soffit			
2.4.1	2.6 m rise; standard; straight flight	nr	450.00	-675.00
2.4.2	2.6 m rise; standard; top three treads winding	nr	550.00	-750.00
2.4.3	2.6 m rise; standard; dogleg	nr	625.00	-800.00
	Oak staircase; balusters and handrails; plasterboard; skim and emulsion to soffit			
2.4.4	2.6 m rise; purpose-made; dogleg	nr	-	-
	Plus or minus for			
2.4.5	each 300 mm variation in storey height	nr	-	-
	Reinforced concrete construction			
	Escape staircase; granolithic finish; mild steel balustrades and handrails			
2.4.6	3 m rise; dogleg	nr	2250	-2750
	Plus or minus for			
2.4.7	each 300 mm variation in storey height	nr	225.00	-275.00
	Staircase; terrazzo finish; mild steel balustrades and handrails; plastered soffit; balustrades and staircase soffit decorated			
2.4.8	3 m rise; dogleg	nr	-	-
	Plus or minus for			
2.4.9	each 300 mm variation in storey height	nr	-	-
	Staircase; terrazzo finish; stainless steel balustrades and handrails; plastered and decorated soffit			
2.4.10	3 m rise; dogleg	nr	-	-
	Plus or minus for			
2.4.11	each 300 mm variation in storey height	nr	-	-
	Staircase; high quality finishes; stainless steel and glass balustrades; plastered and decorated soffit			
2.4.12	3 m rise; dogleg	nr	-	-
	Plus or minus for			
2.4.13	each 300 mm variation in storey height	nr	-	-
	Metal construction			
	Steel access/fire ladder			
2.4.14	3 m high	nr	-	-
2.4.15	4 m high; epoxide finished	nr	-	-
	Light duty metal staircase; galvanised finish; perforated treads; no risers; balustrades and handrails; decorated			
2.4.16	3 m rise; spiral; 1548 mm diameter	nr	-	-
	Plus or minus for			
2.4.17	each 300 mm variation in storey height	nr	-	-
2.4.18	3 m rise; spiral; 1936 mm diameter	nr	-	-
	Plus or minus for			
2.4.19	each 300 mm variation in storey height	nr	-	-
2.4.20	3 m rise; spiral; 2072 mm diameter	nr	-	-
	Plus or minus for			
2.4.21	each 300 mm variation in storey height	nr	-	-

INDUSTRIAL £ range £		RETAILING £ range £		LEISURE £ range £		OFFICES £ range £		HOTELS £ range £		Item nr.
-	-	-	-	-	-	-	-	-	-	2.4.1
-	-	-	-	-	-	-	-	-	-	2.4.2
-	-	-	-	-	-	-	-	-	-	2.4.3
-	-	-	-	-	-	-	-	4000	-5500	2.4.4
-	-	-	-	-	-	-	-	650.00	-725.00	2.4.5
3250	-4000	2250	-4000	3000	-3500	3000	-3500	2250	-4500	2.4.6
325.00	-400.00	225.00	-400.00	300.00	-350.00	300.00	-350.00	225.00	-450.00	2.4.7
4500	-5500	4500	-6000	4250	-5500	4250	-5750	3500	-6500	2.4.8
450.00	-550.00	450.00	-600.00	425.00	-550.00	425.00	-575.00	350.00	-650.00	2.4.9
-	-	4500	-7250	5000-6750		5000	-7250	4500	-7750	2.4.10
-	-	450.00	-725.00	500.00	-675.00	500.00	-725.00	450.00	-775.00	2.4.11
9000	-12250	10000	-13500	9000	-12250	8500	-11750	10750	-14500	2.4.12
900	-1225	1000	-1350	900	-1225	850	-1175	1075	-1450	2.4.13
325.00	-450.00	-	-	325.00	-450.00	-	-	-	-	2.4.14
-	-	-	-	450.00	-750.00	-	-	-	-	2.4.15
1550	-1900	-	-	-	-	-	-	-	-	2.4.16
155.00	-190.00	-	-	-	-	-	-	-	-	2.4.17
1800	-2200	-	-	-	-	-	-	-	-	2.4.18
180.00	-220.00	-	-	-	-	-	-	-	-	2.4.19
1900	-2250	-	-	-	-	-	-	-	-	2.4.20
190.00	-225.00	-	-	-	-	-	-	-	-	2.4.21

Item nr.	SPECIFICATIONS	Unit	RESIDENTIAL £ range £	
2.4	**STAIRS - cont'd**	**Storey** (unless otherwise described)		
	Metal construction - cont'd			
	Light duty metal staircase; galvanised finish; perforated treads; no risers; balustrades and handrails; decorated - cont'd			
2.4.22	3 m rise; straight; 760 mm wide	nr	-	-
	Plus or minus for			
2.4.23	each 300 mm variation in storey height	nr	-	-
2.4.24	3 m rise; straight; 900 mm wide	nr	-	-
	Plus or minus for			
2.4.25	each 300 mm variation in storey height	nr	-	-
2.4.26	3 m rise; straight; 1070 mm wide	nr	-	-
	Plus or minus for			
2.4.27	each 300 mm variation in storey height	nr	-	-
	Heavy duty cast iron staircase; perforated treads; no risers; balustrades and handrails; decorated			
2.4.28	3 m rise; spiral; 1548 mm diameter	nr	-	-
	Plus or minus for			
2.4.29	each 300 mm variation in storey height	nr	-	-
2.4.30	3 m rise; straight	nr	-	-
	Plus or minus for			
2.4.31	each 300 mm variation in storey height	nr	-	-
	Feature metal staircase; galvanised finish; perforated treads; no risers; decorated			
2.4.32	3 m rise; spiral; balustrades and handrails	nr	-	-
	Plus or minus for			
2.4.33	each 300 mm variation in storey height	nr	-	-
2.4.34	3 m rise; dogleg; hardwood balustrades and handrails	nr	-	-
2.4.35	3 m rise; dogleg; stainless steel balustrades and handrails	nr	-	-
	Feature metal staircase; galvanised finish; concrete treads; balustrades and handrails; decorated			
2.4.36	3 m rise; dogleg	nr	-	-
	Feature metal staircase to water chute; steel springers; stainless steel treads in-filled with tiling; landings every one metre rise			
2.4.37	9 m rise; spiral	nr	-	-
	Galvanised steel catwalk; nylon coated balustrading			
2.4.38	450 mm wide	m	-	-
	Comparative finishes/balustrading			
	Finishes to treads and risers			
2.4.39	PVC floor tiles including screeds	per flight	-	-
2.4.40	granolithic	per flight	-	-
2.4.41	heavy duty carpet	per flight	-	-
2.4.42	terrazzo	per flight	-	-
	Wall handrails			
2.4.43	PVC covered mild steel rail on brackets	per flight	-	-
2.4.44	hardwood handrail on brackets	per flight	-	-
2.4.45	stainless steel handrail on brackets	per flight	-	-
	Balustrading and handrails			
2.4.46	mild steel balustrades and PVC covered handrails	per flight	-	-
2.4.47	mild steel balustrades and hardwood handrails	per flight	-	-
2.4.48	stainless steel balustrades and handrails	per flight	-	-
2.4.49	stainless steel and glass balustrades	per flight	-	-

INDUSTRIAL £ range £	RETAILING £ range £	LEISURE £ range £	OFFICES £ range £	HOTELS £ range £	Item nr.
1800 -2250	- -	- -	- -	- -	2.4.22
180.00 -225.00	- -	- -	- -	- -	2.4.23
1900 -2400	- -	- -	- -	- -	2.4.24
190.00 -240.00	- -	- -	- -	- -	2.4.25
2000 -2400	- -	- -	- -	- -	2.4.26
200.00 -240.00	- -	- -	- -	- -	2.4.27
2500 -3250	- -	- -	- -	- -	2.4.28
250.00 -325.00	- -	- -	- -	- -	2.4.29
2700 -3600	- -	- -	- -	- -	2.4.30
270.00 -360.00	- -	- -	- -	- -	2.4.31
- -	- -	3500 -3800	- -	- -	2.4.32
- -	- -	350.00 -380.00	- -	- -	2.4.33
- -	- -	3600 -5000	- -	- -	2.4.34
- -	- -	4100 -6750	- -	-	2.4.35
- -	- -	6300 -8500	- -	- -	2.4.36
- -	- -	20250 -24750	- -	- -	2.4.37
180.00 -220.00	- -	190.00 -230.00	- -	- -	2.4.38
- -	550.00 -725.00	550.00 -725.00	550.00 -725.00	550.00 -725.00	2.4.39
- -	800.00 -900.00	800.00 -900.00	800.00 -900.00	800.00 -900.00	2.4.40
- -	1100 -1350	1100 -1350	1100 -1350	1100 -1350	2.4.41
- -	2250 -2750	2250 -2750	2250 -2750	2250 -2750	2.4.42
- -	180.00 -270.00	180.00 -270.00	180.00 -270.00	180.00 -270.00	2.4.43
- -	550.00 -900.00	550.00 -900.00	550.00 -900.00	550.00 -900.00	2.4.44
- -	2250 -2750	2250 -2750	2250 -2750	2250 -2750	2.4.45
- -	550 -675	550 -675	550 -675	550 -675	2.4.46
- -	990 -1400	990 -1400	990 -1400	990 -1400	2.4.47
- -	4050 -4950	4050 -4950	4050 -4950	4050 -4950	2.4.48
- -	5400 -9000	5400 -9000	5400 -9000	5400 -9000	2.4.49

Item nr.	SPECIFICATIONS	Unit	RESIDENTIAL £ range £
2.5	**EXTERNAL WALLS**	**external wall area** (unless otherwise described)	
	comprising:-		
	Timber framed walling		
	Brick/block walling		
	Reinforced concrete walling		
	Panelled walling		
	Wall claddings		
	Curtain/glazed walling		
	Comparative external finishes		
	Timber framed walling		
	Structure only comprising softwood studs at 400 x 600 mm centres; head and sole plates		
2.5.1	125 x 50 mm	m2	11.50 - 14.50
2.5.2	125 x 50 mm; one layer of double sided building paper	m2	13.50 - 17.00
	Softwood stud wall; vapour barrier and plasterboard inner lining; decorated		
2.5.3	PVC weatherboard outer lining	m2	47.50 - 62.50
2.5.4	tile hanging on battens outer lining	m2	57.50 - 67.50
	Brick/block walling		
	Autoclaved aerated light weight block walls		
2.5.5	100 mm thick	m2	16.00 - 18.00
2.5.6	140 mm thick	m2	19.00 - 22.00
2.5.7	190 mm thick	m2	26.00 - 29.00
	Dense aggregate block walls		
2.5.8	100 mm thick	m2	15.00 - 17.00
2.5.9	140 mm thick	m2	21.00 - 23.00
	Coloured dense aggregate masonry block walls		
2.5.10	100 mm thick; hollow	m2	27.00 - 30.00
2.5.11	100 mm thick; solid	m2	30.00 - 32.00
2.5.12	140 mm thick; hollow	m2	31.00 - 36.00
2.5.13	140 mm thick; solid	m2	41.00 - 45.00
	Common brick solid walls; bricks PC £120.00/1000		
2.5.14	half brick thick	m2	23.00 - 25.00
2.5.15	one brick thick	m2	41.00 - 45.00
2.5.16	one and a half brick thick	m2	58.00 - 63.00
	Add or deduct for each variation of £10.00/1000 in PC value		
2.5.17	half brick thick	m2	0.60 - 0.90
2.5.18	one brick thick	m2	1.30 - 1.50
2.5.19	one and a half brick thick	m2	1.90 - 2.20
	Extra for		
2.5.20	fair face one side	m2	1.30 - 1.80
	Engineering brick walls; class B; bricks PC £175.00/1000		
2.5.21	half brick thick	m2	27.00 - 31.00
2.5.22	one brick thick	m2	51.00 - 57.00
	Facing brick walls; sand faced facings; bricks PC £130.00/1000		
2.5.23	half brick thick; pointed one side	m2	27.00 - 31.00
2.5.24	one brick thick; pointed both sides	m2	50.00 - 56.00
	Facing brick walls; machine-made facings; bricks PC £320.00/1000		
2.5.25	half brick thick; pointed one side	m2	41.00 - 45.00
2.5.26	half brick thick; built against concrete	m2	42.50 - 47.50
2.5.27	one brick thick; pointed both sides	m2	79.00 - 88.00
	Facing bricks solid walls; hand-made facings; bricks PC £525.00/1000		
2.5.28	half brick thick; pointed one side	m2	54.00 - 59.50
2.5.29	one brick thick; pointed both sides	m2	107.50 -117.50

INDUSTRIAL £ range £	RETAILING £ range £	LEISURE £ range £	OFFICES £ range £	HOTELS £ range £	Item nr.
- -	- -	- -	- -	- -	2.5.1
- -	- -	- -	- -	- -	2.5.2
- -	- -	- -	- -	- -	2.5.3
- -	- -	- -	- -	- -	2.5.4
15.00 - 17.00	15.00 - 17.00	15.00 - 17.00	15.00 - 17.00	17.00 - 19.00	2.5.5
18.00 - 21.00	18.00 - 21.00	18.00 - 21.00	18.00 - 21.00	20.00 - 23.00	2.5.6
25.00 - 29.00	25.00 - 29.00	25.00 - 29.00	25.00 - 29.00	27.00 - 30.00	2.5.7
14.00 - 16.00	14.00 - 16.00	14.00 - 16.00	14.00 - 16.00	16.00 - 18.00	2.5.8
20.00 - 23.00	20.00 - 23.00	20.00 - 23.00	20.00 - 23.00	22.00 - 24.00	2.5.9
25.50 - 28.50	25.50 - 28.50	25.50 - 28.50	25.50 - 28.50	28.00 - 30.50	2.5.10
28.00 - 30.50	28.00 - 30.50	28.00 - 30.50	28.00 - 30.50	30.50 - 33.50	2.5.11
31.50 - 34.00	31.50 - 34.00	31.50 - 34.00	31.50 - 34.00	34.00 - 38.00	2.5.12
38.50 - 43.00	38.50 - 43.00	38.50 - 43.00	38.50 - 43.00	41.50 - 47.00	2.5.13
22.00 - 24.00	22.00 - 24.00	22.00 - 24.00	22.00 - 24.00	24.00 - 26.00	2.5.14
39.00 - 43.00	39.00 - 43.00	39.00 - 43.00	39.00 - 43.00	42.00 - 47.00	2.5.15
56.00 - 61.00	56.00 - 61.00	56.00 - 61.00	56.00 - 61.00	59.00 - 65.00	2.5.16
0.60 - 0.90	0.60 - 0.90	0.60 - 0.90	0.60 - 0.90	0.60 - 0.90	2.5.17
1.30 - 1.50	1.30 - 1.50	1.30 - 1.50	1.30 - 1.50	1.30 - 1.50	2.5.18
1.90 - 2.20	1.90 - 2.20	1.90 - 2.20	1.90 - 2.20	1.90 - 2.20	2.5.19
1.40 - 1.80	1.40 - 1.80	1.40 - 1.80	1.40 - 1.80	1.40 - 1.80	2.5.20
26.00 - 30.00	26.00 - 30.00	26.00 - 30.00	26.00 - 30.00	28.00 - 32.00	2.5.21
50.00 - 55.00	50.00 - 55.00	50.00 - 55.00	50.00 - 55.00	54.00 - 59.00	2.5.22
26.00 - 30.00	26.00 - 30.00	26.00 - 30.00	26.00 - 30.00	28.00 - 32.00	2.5.23
49.00 - 54.00	49.00 - 54.00	49.00 - 54.00	49.00 - 54.00	52.00 - 58.00	2.5.24
39.00 - 43.00	41.00 - 45.00	41.00 - 45.00	41.00 - 45.00	42.00 - 47.00	2.5.25
41.00 - 46.00	43.00 - 47.50	43.00 - 47.50	43.00 - 47.50	45.00 - 49.50	2.5.26
77.50 - 88.00	77.00 - 88.00	77.00 - 88.00	77.00 - 88.00	80.00 - 90.00	2.5.27
52.00 - 57.50	54.00 - 59.50	54.00 - 59.50	54.00 - 59.50	56.50 - 63.00	2.5.28
103.00 -112.00	105.00 -115.00	105.00 -115.00	105.00 -115.00	110.00 -120.00	2.5.29

Item nr.	SPECIFICATIONS	Unit	RESIDENTIAL £ range £
2.5	**EXTERNAL WALLS** - cont'd **external wall area** (unless otherwise described)		
	Brick/block walling - cont'd		
	Add or deduct for		
	each variation of £10.00/1000 in PC value		
2.5.30	half brick thick	m2	0.60 - 0.90
2.5.31	one brick thick	m2	1.30 - 1.50
	Composite solid walls; facing brick on outside; bricks PC £320.00/1000 and common brick on inside; bricks PC £120.00/1000		
2.5.32	one brick thick; pointed one side	m2	63.00 - 69.50
	Extra for		
2.5.33	weather pointing as a separate operation	m2	3.50 - 5.50
2.5.34	one and a half brick thick; pointed one side	m2	81.00 - 90.00
	Composite cavity wall; block outer skin; 50 mm insulation; lightweight block inner skin		
2.5.35	outer block rendered	m2	45.00 - 56.00
	Extra for		
2.5.36	heavyweight block inner skin	m2	0.90 - 1.80
2.5.37	fair face one side	m2	0.40 - 1.80
2.5.38	75 mm cavity insulation	m2	0.90 - 1.80
2.5.39	100 mm cavity insulation	m2	1.80 - 2.70
2.5.40	plaster and emulsion	m2	9.00 - 10.80
	outer block rendered; no insulation; inner skin		
2.5.41	insulating	m2	45.00 - 58.50
2.5.42	outer block roughcast	m2	49.50 - 63.00
2.5.43	coloured masonry outer block	m2	54.00 - 67.50
	Composite cavity wall; facing brick outer skin; 50 mm insulation; plasterboard on stud inner skin; emulsion		
2.5.44	sand-faced facings; PC £130.00/1000	m2	52.00 - 61.00
2.5.45	machine-made facings; PC £320.00/1000	m2	65.50 - 74.50
2.5.46	hand-made facings; PC £525.00/1000	m2	77.50 - 86.50
	Composite cavity wall; facing brick outer skin; lightweight block inner skin; plaster and emulsion		
2.5.47	sand-faced facings; PC £130.00/1000	m2	47.50 - 58.50
2.5.48	machine-made facings; PC £320.00/1000	m2	61.50 - 72.00
2.5.49	hand-made facings; PC £525.00/1000	m2	72.00 - 85.50
	Add or deduct for		
2.5.50	each variation of £10.00/1000 in PC value	m2	0.60 - 0.90
	Extra for		
2.5.51	heavyweight block inner skin	m2	0.90 - 1.80
2.5.52	insulating block inner skin	m2	1.80 - 4.50
2.5.53	30 mm cavity wall slab	m2	2.00 - 5.00
2.5.54	50 mm cavity insulation	m2	2.50 - 3.00
2.5.55	75 mm cavity insulation	m2	3.50 - 4.00
2.5.56	100 mm cavity insulation	m2	4.50 - 5.00
2.5.57	weather-pointing as a separate operation	m2	3.50 - 5.50
2.5.58	purpose made feature course to windows	m2	4.50 - 9.00
	Composite cavity wall; facing brick outer skin; 50 mm insulation; common brick inner skin; fair face on inside		
2.5.59	sand-faced facings; PC £130.00/1000	m2	50.50 - 63.00
2.5.60	machine-made facings; PC £320.00/1000	m2	66.50 - 76.50
2.5.61	hand-made facings; PC £525.00/1000	m2	79.00 - 90.00
	Composite cavity wall; facing brick outer skin; 50 mm insulation; common brick inner skin; plaster and emulsion		
2.5.62	sand-faced facings; PC £130.00/1000	m2	57.50 - 70.00
2.5.63	machine-made facings; PC £320.00/1000	m2	74.00 - 83.50
2.5.64	hand-made facings; PC £525.00/1000	m2	86.50 - 99.00
	Composite cavity wall; coloured masonry block; outer and		
2.5.65	inner skins; fair faced both sides	m2	81.00 -108.00

INDUSTRIAL £ range £	RETAILING £ range £	LEISURE £ range £	OFFICES £ range £	HOTELS £ range £	Item nr.
0.60 - 0.90	0.60 - 0.90	0.60 - 0.90	0.60 - 0.90	0.60 - 0.90	2.5.30
1.30 - 1.50	1.30 - 1.50	1.30 - 1.50	1.30 - 1.50	1.30 - 1.50	2.5.31
61.50 - 67.50	63.00 - 70.00	63.00 - 70.00	63.00 - 70.00	66.00 - 72.00	2.5.32
3.50 - 5.50	3.50 - 5.50	3.50 - 5.50	3.50 - 5.50	3.50 - 5.50	2.5.33
77.00 - 86.00	81.00 - 90.00	81.00 - 90.00	81.00 - 90.00	86.00 - 95.00	2.5.34
43.00 - 54.00	45.00 - 56.50	45.00 - 56.50	45.00 - 56.50	47.50 - 58.50	2.5.35
0.90 - 1.80	0.90 - 1.80	0.90 - 1.80	0.90 - 1.80	0.90 - 1.80	2.5.36
0.40 - 1.80	0.40 - 1.80	0.40 - 1.80	0.40 - 1.80	0.40 - 1.80	2.5.37
0.90 - 1.80	0.90 - 1.80	0.90 - 1.80	0.90 - 1.80	0.90 - 1.80	2.5.38
1.80 - 2.70	1.80 - 2.70	1.80 - 2.70	1.80 - 2.70	1.80 - 2.70	2.5.39
8.50 - 10.50	9.00 - 11.00	9.00 - 11.00	9.00 - 11.00	9.50 - 11.50	2.5.40
43.00 - 57.00	45.00 - 59.00	45.00 - 59.00	45.00 - 59.00	47.00 - 60.00	2.5.41
48.00 - 61.00	49.50 - 63.00	49.50 - 63.00	49.50 - 63.00	51.50 - 65.00	2.5.42
52.00 - 65.00	54.00 - 67.50	54.00 - 67.50	54.00 - 67.50	56.50 - 70.00	2.5.43
50.50 - 59.50	52.00 - 61.00	52.00 - 61.00	52.00 - 61.00	54.00 - 63.00	2.5.44
63.00 - 72.00	65.50 - 74.50	65.50 - 74.50	65.50 - 74.50	67.50 - 76.50	2.5.45
74.00 - 83.00	77.50 - 86.50	77.50 - 86.50	77.50 - 86.50	81.00 - 90.00	2.5.46
46.00 - 56.50	47.50 - 58.50	47.50 - 58.50	47.50 - 58.50	50.50 - 61.00	2.5.47
58.50 - 69.50	61.50 - 72.00	61.50 - 72.00	61.50 - 72.00	65.50 - 75.50	2.5.48
69.50 - 85.50	72.00 - 85.50	72.00 - 85.50	72.00 - 85.50	75.50 - 90.00	2.5.49
0.60 - 0.90	0.60 - 0.90	0.60 - 0.90	0.60 - 0.90	0.60 - 0.90	2.5.50
0.90 - 1.80	0.90 - 1.80	0.90 - 1.80	0.90 - 1.80	0.90 - 1.80	2.5.51
1.80 - 4.50	1.80 - 4.50	1.80 - 4.50	1.80 - 4.50	1.80 - 4.50	2.5.52
2.00 - 5.00	2.00 - 5.00	2.00 - 5.00	2.00 - 5.00	2.00 - 5.00	2.5.53
2.70 - 3.20	2.70 - 3.20	2.70 - 3.20	2.70 - 3.20	2.70 - 3.20	2.5.54
3.50 - 4.00	3.50 - 4.00	3.50 - 4.00	3.50 - 4.00	3.50 - 4.00	2.5.55
4.50 - 5.00	4.50 - 5.00	4.50 - 5.00	4.50 - 5.00	4.50 - 5.00	2.5.56
3.50 - 5.50	3.50 - 5.50	3.50 - 5.50	3.50 - 5.50	3.50 - 5.50	2.5.57
4.50 - 9.00	4.50 - 9.00	4.50 - 9.00	4.50 - 9.00	4.50 - 9.00	2.5.58
49.00 - 61.00	50.00 - 63.00	50.00 - 63.00	50.00 - 63.00	52.00 - 65.00	2.5.59
64.00 - 74.00	67.00 - 77.00	67.00 - 77.00	67.00 - 77.00	69.00 - 79.00	2.5.60
76.00 - 86.00	79.00 - 90.00	79.00 - 90.00	79.00 - 90.00	83.00 - 95.00	2.5.61
56.00 - 67.50	57.50 - 70.00	57.50 - 70.00	57.50 - 70.00	60.50 - 74.00	2.5.62
72.00 - 81.00	74.00 - 83.50	74.00 - 83.50	74.00 - 83.50	77.50 - 86.50	2.5.63
82.00 - 94.50	86.50 - 99.00	86.50 - 99.00	86.50 - 99.00	90.00 -103.50	2.5.64
77.50 -103.50	81.00 -108.00	81.00 -108.00	81.00 -108.00	85.50 -112.50	2.5.65

Item nr.	SPECIFICATIONS	Unit	RESIDENTIAL £ range £
2.5	**EXTERNAL WALLS** - cont'd	**external wall area** (unless otherwise described)	
	Reinforced concrete walling		
	Insitu reinforced concrete 25.5 N/mm2; 13 kg/m2 reinforcement; formwork both sides		
2.5.66	150 mm thick	m2	60.00 - 70.00
2.5.67	225 mm thick	m2	70.00 - 80.00
	Panelled walling		
	Precast concrete panels; including insulation; lining and fixings		
2.5.68	standard panels	m2	- -
2.5.69	standard panels; exposed aggregate finish	m2	- -
2.5.70	brick clad panels	m2	- -
2.5.71	reconstructed stone faced panels	m2	- -
2.5.72	natural stone faced panels	m2	- -
2.5.73	marble or granite faced panels	m2	- -
	GRP/laminate panels; including battens; back-up walls; plaster and emulsion		
2.5.74	melamined finished solid laminate panels	m2	- -
2.5.75	GRP single skin panels	m2	- -
2.5.76	GRP double skin panels	m2	- -
2.5.77	GRP insulated sandwich panels	m2	- -
	Wall claddings		
	Non-asbestos profiled cladding		
2.5.78	'profile 3'; natural	m2	- -
2.5.79	'profile 3'; coloured	m2	- -
2.5.80	'profile 6'; natural	m2	- -
2.5.81	'profile 6'; coloured	m2	- -
2.5.82	insulated; inner lining of plasterboard	m2	- -
2.5.83	'profile 6'; natural; insulated; inner lining panel	m2	- -
2.5.84	insulated; with 2.8 m high block inner skin; emulsion	m2	- -
2.5.85	insulated; with 2.8 m high block inner skin; plasterboard lining on metal tees; emulsion	m2	- -
	Asbestos cement profiled cladding on steel rails		
2.5.86	insulated; with 2.8 m high block inner skin; emulsion	m2	- -
2.5.87	insulated; with 2.8 m high block inner skin; plasterboard lining on metal tees; emulsion	m2	- -
	PVF2 coated galvanised steel profiled cladding		
2.5.88	0.60 mm thick; 'profile 20B'; corrugations vertical	m2	- -
2.5.89	0.60 mm thick; 'profile 30'; corrugations vertical	m2	- -
2.5.90	0.60 mm thick; 'profile TOP 40'; corrugations vertical	m2	- -
2.5.91	0.60 mm thick; 'profile 60B'; corrugations vertical	m2	- -
2.5.92	0.60 mm thick; 'profile 30'; corrugations horizontal	m2	- -
2.5.93	0.60 mm thick; 'profile 60B; corrugations horizontal	m2	- -
	Extra for		
2.5.94	80 mm insulation and 0.4 mm thick coated inner lining sheet	m2	- -
	PVF2 coated galvanised steel profiled cladding on steel rails		
2.5.95	insulated	m2	- -
2.5.96	2.8 m high insulating block inner skin; emulsion	m2	- -
2.5.97	2.8 m high insulated block inner skin; plasterboard lining on metal tees; emulsion	m2	- -
2.5.98	insulated; coloured inner lining panel	m2	- -
2.5.99	insulated; full-height insulating block inner skin; plaster and emulsion	m2	- -
2.5.100	insulated; metal sandwich panel system	m2	- -

INDUSTRIAL £ range £	RETAILING £ range £	LEISURE £ range £	OFFICES £ range £	HOTELS £ range £	Item nr.
60.00 - 70.00	60.00 - 70.00	60.00 - 70.00	60.00 - 70.00	60.00 - 70.00	2.5.66
70.00 - 80.00	70.00 - 80.00	70.00 - 80.00	70.00 - 80.00	70.00 - 80.00	2.5.67
75.00 -115.00	80.00 -115.00	80.00 -115.00	80.00 -115.00	80.00 -125.00	2.5.68
115.00 -160.00	125.00 -160.00	125.00 -160.00	125.00 -160.00	125.00 -170.00	2.5.69
155.00 -190.00	160.00 -190.00	160.00 -190.00	160.00 -190.00	160.00 -200.00	2.5.70
- -	250.00 -315.00	250.00 -315.00	250.00 -315.00	250.00 -340.00	2.5.71
- -	360.00 -450.00	360.00 -450.00	360.00 -450.00	360.00 -475.00	2.5.72
- -	405.00 -495.00	405.00 -495.00	405.00 -495.00	405.00 -540.00	2.5.73
100.00 -125.00	110.00 -125.00	110.00 -125.00	- -	- -	2.5.74
105.00 -135.00	115.00 -135.00	115.00 -135.00	90.00 -110.00	- -	2.5.75
145.00 -180.00	155.00 -180.00	155.00 -180.00	135.00 -170.00	- -	2.5.76
- -	180.00 -225.00	180.00 -225.00	170.00 -215.00	170.00 -215.00	2.5.77
15.00 - 17.00	- -	- -	- -	- -	2.5.78
16.00 - 18.00	- -	- -	- -	- -	2.5.79
16.00 - 18.00	- -	- -	- -	- -	2.5.80
17.00 - 19.00	- -	- -	- -	- -	2.5.81
30.00 - 36.00	- -	- -	- -	- -	2.5.82
30.00 - 36.00	- -	- -	- -	- -	2.5.83
26.00 - 28.00	- -	- -	- -	- -	2.5.84
34.00 - 40.00	- -	- -	- -	- -	2.5.85
25.00 - 27.00	- -	- -	- -	- -	2.5.86
33.00 - 39.00	- -	- -	- -	- -	2.5.87
20.00 - 25.00	25.00 - 32.00	25.00 - 32.00	- -	- -	2.5.88
20.00 - 25.00	25.00 - 32.00	25.00 - 32.00	- -	- -	2.5.89
19.00 - 24.00	24.00 - 31.00	24.00 - 31.00	- -	- -	2.5.90
23.00 - 28.00	28.00 - 34.00	28.00 - 34.00	- -	- -	2.5.91
21.00 - 26.00	26.00 - 33.00	26.00 - 33.00	- -	- -	2.5.92
24.00 - 29.00	29.00 - 36.00	29.00 - 36.00	- -	- -	2.5.93
11.00 - 12.00	11.00 - 12.00	11.00 - 12.00	- -	- -	2.5.94
33.00 - 43.00	- -	- -	- -	- -	2.5.95
40.00 - 50.00	- -	- -	- -	- -	2.5.96
47.50 - 57.50	- -	- -	- -	- -	2.5.97
47.50 - 57.50	55.00 - 70.00	55.00 - 70.00	- -	- -	2.5.98
60.00 - 80.00	65.00 - 85.00	65.00 - 85.00	65.00 - 90.00	- -	2.5.99
120.00 -180.00	140.00 -190.00	130.00 -180.00	150.00 -180.00	- -	2.5.100

Item nr.	SPECIFICATIONS	Unit	RESIDENTIAL £ range £	
2.5	**EXTERNAL WALLS - cont'd** **external wall area** (unless otherwise described)			
	Wall claddings - cont'd			
	PVF2 coated aluminium profiled cladding on steel rails			
2.5.101	insulated	m2	-	-
2.5.102	insulated; plasterboard lining on metal tees; emulsion	m2	-	-
2.5.103	insulated; coloured inner lining panel	m2	-	-
2.5.104	insulated; full-height insulating block inner skin; plaster and emulsion	m2	-	-
	Extra for			
2.5.105	heavyweight block inner skin	m2	-	-
2.5.106	insulated; aluminium sandwich panel system	m2	-	-
2.5.107	insulated; aluminium sandwich panel system; on framing; on block inner skin; fair face one side	m2	-	-
	Other cladding systems			
2.5.108	vitreous enamelled insulated steel sandwich panel system; with non-asbestos fibre insulating board on inner face	m2	-	-
2.5.109	Formalux sandwich panel system; with coloured lining tray; on steel cladding rails	m2	-	-
2.5.110	aluminium over-cladding system rain screen	m2	-	-
2.5.111	natural stone cladding on full-height insulating block inner skin; plaster and emulsion	m2	-	-
	Curtain/glazed walling			
	Single glazed anodised aluminium curtain walling			
2.5.112	economical; including part-height block back-up wall; plaster and emulsion	m2	-	-
	Extra over single 6 mm float glass for			
2.5.113	double glazing unit with two 6 mm float glass skins	m2	-	-
2.5.114	double glazing unit with one 6 mm 'Antisun' skin and one 6 mm float glass skin	m2	-	-
2.5.115	economical; including part-height block back-up wall; plaster and emulsion	m2	-	-
2.5.116	economical; including infill panels	m2	-	-
	Extra for			
2.5.117	50 mm insulation	m2	-	-
2.5.118	polyester powder coating in lieu of anodised finish	m2	-	-
2.5.119	bronze anodising in lieu of anodised finish	m2	-	-
2.5.120	good quality	m2	-	-
2.5.121	good quality; 35% opening lights	m2	-	-
2.5.122	high quality	m2	-	-
2.5.123	high quality; fire rated	m2	-	-
	Suspended panelled glazing system (e.g. Planar)			
2.5.124	10 mm toughened float glass panels	m2	-	-
	Extra for			
2.5.125	solar control glass	m2	-	-
2.5.126	supporting structure of aluminium; stainless steel or glass fins	m2	-	-
2.5.127	High quality structural glazing to entrance elevation	m2	-	-
	Comparative external finishes			
	Comparative concrete wall finishes			
2.5.128	wrought formwork one side including rubbing down	m2	-	-
2.5.129	shotblasting to expose aggregate	m2	-	-
2.5.130	bush hammering to expose aggregate	m2	-	-
	Comparative insitu finishes			
2.5.131	two coats cement paint	m2	-	-
2.5.132	cement and sand plain face rendering	m2	-	-
2.5.133	three-coat Tyrolean rendering; including backing	m2	-	-
2.5.134	'Mineralite' decorative rendering; including backing	m2	-	-

INDUSTRIAL £ range £	RETAILING £ range £	LEISURE £ range £	OFFICES £ range £	HOTELS £ range £	Item nr.
37.50 - 47.50	- -	- -	- -	- -	2.5.101
50.00 - 60.00	- -	- -	- -	- -	2.5.102
55.00 - 65.00	60.00 - 75.00	60.00 - 75.00	- -	- -	2.5.103
70.00 - 90.00	75.00 - 95.00	75.00 - 95.00	75.00 -100.00	- -	2.5.104
1.00 - 2.00	1.00 - 2.00	1.00 - 2.00	1.00 - 2.00	- -	2.5.105
100.00 -130.00	130.00 -170.00	130.00 -170.00	140.00 -180.00	- -	2.5.106
- -	160.00 -250.00	- -	- -	- -	2.5.107
110.00 -135.00	- -	- -	- -	- -	2.5.108
135.00 -160.00	145.00 -170.00	145.00 -170.00	145.00 -170.00	- -	2.5.109
170.00 -190.00	- -	- -	- -	- -	2.5.110
- -	- -	- -	300.00 -430.00	315.00 -450.00	2.5.111
180.00 -245.00	190.00 -250.00	190.00 -250.00	190.00 -250.00	200.00 -270.00	2.5.112
55.00 - 65.00	55.00 - 65.00	55.00 - 65.00	55.00 - 65.00	55.00 - 65.00	2.5.113
90.00 -100.00	90.00 -100.00	90.00 -100.00	90.00 -100.00	90.00 -100.00	2.5.114
205.00 -250.00	215.00 -270.00	220.00 -270.00	215.00 -270.00	225.00 -300.00	2.5.115
205.00 -300.00	225.00 -315.00	225.00 -315.00	225.00 -315.00	250.00 -360.00	2.5.116
10.00 - 12.00	10.00 - 12.00	10.00 - 12.00	10.00 - 12.00	10.00 - 12.00	2.5.117
5.00 - 12.00	5.00 - 12.00	5.00 - 12.00	5.00 - 12.00	5.00 - 12.00	2.5.118
11.00 - 18.00	11.00 - 18.00	11.00 - 18.00	11.00 - 18.00	11.00 - 18.00	2.5.119
305.00 -390.00	315.00 -405.00	315.00 -415.00	325.00 -415.00	325.00 -415.00	2.5.120
370.00 -440.00	380.00 -460.00	380.00 -460.00	380.00 -470.00	380.00 -475.00	2.5.121
405.00 -540.00	435.00 -600.00	415.00 -570.00	435.00 -575.00	415.00 -585.00	2.5.122
- -	- -	- -	720.00 -1080	- -	2.5.123
- -	250.00 -315.00	250.00 -315.00	250.00 -315.00	- -	2.5.124
- -	45.00 -135.00	45.00 -135.00	45.00 -135.00	- -	2.5.125
- -	25.00 -105.00	25.00 -105.00	25.00 -105.00	- -	2.5.126
- -	405.00 -630.00	405.00 -630.00	405.00 -630.00	- -	2.5.127
- -	2.00 - 4.00	2.00 - 4.00	2.00 - 4.00	2.00 - 4.00	2.5.128
- -	2.50 - 5.00	2.50 - 5.00	2.50 - 5.00	2.50 - 5.00	2.5.129
- -	8.00 - 11.00	8.00 - 11.00	8.00 - 11.00	8.00 - 11.00	2.5.130
- -	5.00 - 7.00	5.00 - 7.00	5.00 - 7.00	5.00 - 7.00	2.5.131
- -	9.00 - 13.00	9.00 - 13.00	9.00 - 13.00	9.00 - 13.00	2.5.132
- -	22.00 - 25.00	22.00 - 25.00	22.00 - 25.00	22.00 - 28.00	2.5.133
- -	43.00 - 48.00	43.00 - 50.00	43.00 - 50.00	45.00 - 55.00	2.5.134

Item nr.	SPECIFICATIONS	Unit	RESIDENTIAL £ range £
2.5	**EXTERNAL WALLS - cont'd**	**external wall area** (unless otherwise described)	
	Comparative external finishes - cont'd		
	Comparative claddings		
2.5.135	25 mm tongued and grooved 'tanalised' softwood boarding; including battens	m2	- -
2.5.136	25 mm tongued and grooved Western Red Cedar boarding including battens	m2	- -
2.5.137	machine-made tiles; including battens	m2	- -
2.5.138	best hand-made sand-faced tiles; including battens	m2	- -
2.5.139	20 x 20 mm mosaic glass or ceramic; in common colours; fixed on prepared surface	m2	- -
2.5.140	75 mm Portland stone facing slabs and fixing; including clamps	m2	- -
2.5.141	75 mm Ancaster stone facing slabs and fixing; including clamps	m2	- -
	Comparative curtain wall finishes; extra over aluminium mill finish for		
2.5.142	natural anodising	m2	- -
2.5.143	polyester powder coating	m2	- -
2.5.144	bronze anodising	m2	- -
2.6	**WINDOWS AND EXTERNAL DOORS**	**window and external door area** (unless otherwise described)	
	comprising:-		
	Softwood windows and external doors **Steel windows and external doors** **Steel roller shutters** **Hardwood windows and external doors** **UPVC windows and external doors** **Aluminium windows, entrance screens and doors** **Stainless steel entrance screens and doors** **Shop fronts, shutters and grilles**		
	Softwood windows and external doors		
	Standard windows; painted		
2.6.1	single glazed	m2	125.00 -160.00
2.6.2	double glazed	m2	160.00 -195.00
	Purpose-made windows; painted		
2.6.3	single glazed	m2	160.00 -200.00
2.6.4	double glazed	m2	200.00 -250.00
	Standard external softwood doors and hardwood frames; doors painted; including ironmongery		
2.6.5	two panelled door; plywood panels	nr	210.00 -250.00
2.6.6	solid flush door	nr	235.00 -270.00
2.6.7	two panelled door; glazed panels	nr	450.00 -500.00
	heavy-duty solid flush door		
2.6.8	single leaf	nr	- -
2.6.9	double leaf	nr	- -
	Extra for		
2.6.10	emergency fire exit door	nr	- -
	Steel windows and external doors		
	Standard windows		
2.6.11	single glazed; galvanised; painted	m2	135.00 -170.00
2.6.12	single glazed; powder-coated	m2	140.00 -175.00
2.6.13	double glazed; galvanised; painted	m2	170.00 -205.00
2.6.14	double glazed; powder coated	m2	175.00 -210.00
	Purpose-made windows		
2.6.15	double glazed; powder coated	m2	- -

INDUSTRIAL £ range £	RETAILING £ range £	LEISURE £ range £	OFFICES £ range £	HOTELS £ range £	Item nr.
- -	22.00 - 26.00	22.00 - 26.00	22.00 - 26.00	22.00 - 26.00	2.5.135
- -	25.00 - 30.00	25.00 - 30.00	25.00 - 30.00	25.00 - 30.00	2.5.136
- -	30.00 - 33.00	30.00 - 33.00	30.00 - 33.00	30.00 - 33.00	2.5.137
- -	34.00 - 38.00	34.00 - 38.00	34.00 - 38.00	34.00 - 38.00	2.5.138
- -	65.00 - 75.00	65.00 - 75.00	65.00 - 75.00	65.00 - 75.00	2.5.139
- -	- -	- -	260.00 -325.00	275.00 -350.00	2.5.140
- -	- -	- -	285.00 -350.00	- -	2.5.141
- -	13.50 - 18.00	13.50 - 18.00	13.50 - 18.00	13.50 - 18.00	2.5.142
- -	18.00 - 29.50	18.00 - 29.50	18.00 - 29.50	18.00 - 29.50	2.5.143
- -	24.50 - 36.00	24.50 - 36.00	24.50 - 36.00	24.50 - 36.00	2.5.144
120.00 -150.00	115.00 -155.00	125.00 -160.00	125.00 -160.00	125.00 -170.00	2.6.1
150.00 -190.00	155.00 -190.00	160.00 -200.00	160.00 -200.00	160.00 -205.00	2.6.2
160.00 -200.00	160.00 -190.00	170.00 -205.00	165.00 -205.00	165.00 -215.00	2.6.3
200.00 -235.00	200.00 -235.00	200.00 -230.00	200.00 -230.00	205.00 -245.00	2.6.4
- -	- -	- -	- -	- -	2.6.5
315.00 -540.00	245.00 -280.00	245.00 -280.00	245.00 -280.00	250.00 -290.00	2.6.6
- -	- -	- -	- -	- -	2.6.7
360.00 -540.00	405.00 -500.00	405.00 -495.00	- -	- -	2.6.8
630.00 -900.00	765.00 -855.00	765.00 -855.00	- -	- -	2.6.9
- -	135.00 -205.00	135.00 -205.00	- -	- -	2.6.10
125.00 -160.00	125.00 -165.00	135.00 -170.00	135.00 -170.00	135.00 -180.00	2.6.11
130.00 -165.00	130.00 -165.00	140.00 -175.00	140.00 -175.00	140.00 -190.00	2.6.12
160.00 -200.00	160.00 -200.00	165.00 -205.00	165.00 -205.00	170.00 -215.00	2.6.13
160.00 -200.00	160.00 -205.00	170.00 -210.00	170.00 -210.00	175.00 -225.00	2.6.14
- -	225.00 -280.00	225.00 -280.00	225.00 -280.00	245.00 -300.00	2.6.15

Item nr.	SPECIFICATIONS	Unit	RESIDENTIAL £ range £	
2.6	**WINDOWS AND EXTERNAL DOORS - cont'd** **Window and external door area (unless otherwise described)**			
	Steel roller shutters			
	Shutters; galvanised			
2.6.16	manual	m2	-	-
2.6.17	electric	m2	-	-
2.6.18	manual; insulated	m2	-	-
2.6.19	electric; insulated	m2	-	-
2.6.20	electric; insulated; fire-resistant	m2	-	-
	Hardwood windows and external doors			
	Standard windows; stained or UPVC coated			
2.6.21	single glazed	m2	180.00	-245.00
2.6.22	double glazed	m2	225.00	-290.00
	Purpose-made windows; stained or UPVC coated			
2.6.23	single glazed	m2	210.00	-275.00
2.6.24	double glazed	m2	255.00	-325.00
	Upvc windows and external doors			
	Purpose-made windows			
2.6.25	double glazed	m2	360.00	-450.00
	Extra for			
2.6.26	tinted glass	m2	-	-
	Aluminium windows, entrance screens and doors			
	Standard windows; anodised finish			
2.6.27	single glazed; horizontal sliding sash	m2	-	-
2.6.28	single glazed; vertical sliding sash	m2	-	-
2.6.29	single glazed; casement; in hardwood sub-frame	m2	190.00	-255.00
2.6.30	double glazed; vertical sliding sash	m2	-	-
2.6.31	double glazed; casement; in hardwood sub-frame	m2	225.00	-305.00
	Purpose-made windows			
2.6.32	single glazed	m2	-	-
2.6.33	double glazed	m2	-	-
2.6.34	double glazed; feature; with precast concrete surrounds	m2	-	-
	Purpose-made entrance screens and doors			
2.6.35	double glazed	m2	-	-
	Purpose-made revolving door			
2.6.36	2000 mm dia.; double glazed	nr	-	-
	Stainless steel entrance screens and doors			
	Purpose-made screen; double glazed			
2.6.37	with manual doors	m2	-	-
2.6.38	with automatic doors	m2	-	-
	Purpose-made revolving door			
2.6.39	2000 mm dia; double glazed	nr	-	-
	Shop fronts, shutters and grilles **shop front length**			
2.6.40	Temporary timber shop fronts	m	-	-
2.6.41	Grilles or shutters	m	-	-
2.6.42	Fire shutters; power-operated	m	-	-
	Shop front			
2.6.43	flat facade; glass in aluminium framing; manual centre doors only	m	-	-
2.6.44	flat facade; glass in aluminium framing; automatic centre doors only	m	-	-
2.6.45	hardwood and glass; including high enclosed window beds	m	-	-
2.6.46	high quality; marble or granite plasters and stair risers; window beds and backings; illuminated signs	m	-	-

INDUSTRIAL £ range £	RETAILING £ range £	LEISURE £ range £	OFFICES £ range £	HOTELS £ range £	Item nr.
135.00 -170.00	125.00 -160.00	125.00 -165.00	- -	- -	2.6.16
160.00 -225.00	155.00 -215.00	155.00 -215.00	- -	- -	2.6.17
215.00 -260.00	205.00 -250.00	205.00 -250.00	- -	- -	2.6.18
250.00 -315.00	235.00 -305.00	235.00 -305.00	- -	- -	2.6.19
600.00 -720.00	585.00 -720.00	585.00 -720.00	- -	- -	2.6.20
225.00 -270.00	225.00 -270.00	225.00 -270.00	225.00 -270.00	235.00 -290.00	2.6.21
290.00 -345.00	290.00 -345.00	290.00 -345.00	290.00 -345.00	290.00 -360.00	2.6.22
270.00 -325.00	270.00 -325.00	270.00 -325.00	270.00 -325.00	270.00 -345.00	2.6.23
325.00 -390.00	325.00 -390.00	325.00 -390.00	325.00 -390.00	325.00 -400.00	2.6.24
340.00 -405.00	340.00 -405.00	340.00 -405.00	340.00 -405.00	340.00 -405.00	2.6.25
17.00 - 22.00	17.00 - 22.00	17.00 - 22.00	17.00 - 22.00	17.00 - 22.00	2.6.26
160.00 -195.00	160.00 -195.00	160.00 -195.00	160.00 -195.00	160.00 -205.00	2.6.27
250.00 -295.00	250.00 -295.00	250.00 -295.00	250.00 -295.00	250.00 -305.00	2.6.28
- -	- -	- -	- -	- -	2.6.29
270.00 -325.00	315.00 -390.00	315.00 -390.00	315.00 -390.00	315.00 -405.00	2.6.30
- -	- -	- -	- -	- -	2.6.31
160.00 -225.00	160.00 -225.00	160.00 -225.00	160.00 -225.00	160.00 -225.00	2.6.32
360.00 -430.00	360.00 -430.00	360.00 -430.00	360.00 -430.00	360.00 -450.00	2.6.33
- -	- -	900.00 -1350	- -	- -	2.6.34
- -	450.00 -720.00	450.00 -720.00	450.00 -720.00	- -	2.6.35
- -	15750 -20250	15750 -20250	15750 -20250	15750 -20250	2.6.36
- -	720.00 -1080	- -	- -	- -	2.6.37
- -	900.00 -1260	- -	- -	- -	2.6.38
- -	21600 -27000	21600 -27000	21600 -27000	21600 -27000	2.6.39
- -	30.00 - 40.00	- -	- -	- -	2.6.40
- -	360.00 -720.00	- -	- -	- -	2.6.41
- -	630.00 -900.00	- -	- -	- -	2.6.42
- -	720.00 -1620	- -	- -	- -	2.6.43
- -	900.00 -1980	- -	- -	- -	2.6.44
- -	2610 -3060	- -	- -	- -	2.6.45
- -	2970 -3960	- -	- -	- -	2.6.46

Item nr.	SPECIFICATIONS	Unit	RESIDENTIAL £ range £
2.7	**INTERNAL WALLS, PARTITIONS AND DOORS**	**internal wall area** (unless otherwise described)	
	comprising:-		
	Timber or metal stud partitions and doors		
	Brick/block partitions and doors		
	Reinforced concrete walls		
	Solid partitioning and doors		
	Glazed partitioning and doors		
	Special partitioning and doors		
	WC/Changing cubicles		
	Comparative doors/door linings/frames		
	Perimeter treatments		
	Timber or metal stud partitions and doors		
	Structure only comprising softwood studs at 400 x 600 mm centres; head and sole plates		
2.7.1	100 x 38 mm	m2	9.00 - 11.00
	Softwood stud and plasterboard partitions		
2.7.2	57 mm Paramount dry partition	m2	16.50 - 18.00
2.7.3	65 mm Paramount dry partition	m2	18.00 - 20.00
2.7.4	50 mm laminated partition	m2	17.50 - 19.50
2.7.5	65 mm laminated partition	m2	20.50 - 22.50
2.7.6	65 mm laminated partition; emulsioned both sides	m2	24.50 - 27.00
2.7.7	100 mm partition; taped joints; emulsioned both sides	m2	27.00 - 31.50
2.7.8	100 mm partition; skim and emulsioned both sides	m2	31.50 - 36.00
2.7.9	150 mm partition as party wall; skim and emulsioned both sides	m2	38.00 - 45.00
	Metal stud and plasterboard partitions		
2.7.10	170 mm partition; one hour; taped joints; emulsioned both sides	m2	- -
2.7.11	200 mm partition; two hour; taped joints; emulsioned both sides	m2	- -
2.7.12	325 mm two layer partition; cavity insulation	m2	- -
	Extra for		
2.7.13	curved work		- -
	Metal stud and plasterboard partitions; emulsioned both sides; softwood doors and frames; painted		
2.7.14	170 mm partition	m2	- -
2.7.15	200 mm partition; insulated	m2	- -
	Stud or plasterboard partitions; softwood doors and frames; painted		
2.7.16	partition; plastered and emulsioned both sides	m2	54.00 - 67.50
	Extra for		
2.7.17	vinyl paper in lieu of emulsion	m2	3.50 - 5.50
2.7.18	hardwood doors and frames in lieu of softwood	m2	15.50 - 18.00
2.7.19	partition; plastered and vinyled both sides	m2	63.00 - 76.50
	Stud or plasterboard partitions; hardwood doors and frames		
2.7.20	partition; plastered and emulsioned both sides	m2	69.00 - 86.00
2.7.21	partition; plastered and vinyled both sides	m2	78.00 - 95.00
	Brick/block partitions and doors		
	Autoclaved aerated/lightweight block partitions		
2.7.22	75 mm thick	m2	12.00 - 13.50
2.7.23	100 mm thick	m2	15.50 - 18.00
2.7.24	130 mm thick; insulating	m2	19.00 - 20.50
2.7.25	150 mm thick	m2	20.50 - 22.00
2.7.26	190 mm thick	m2	24.50 - 28.00
	Extra for		
2.7.27	fair face both sides	m2	2.00 - 4.00
2.7.28	curved work		+10% to +20%

INDUSTRIAL £ range £	RETAILING £ range £	LEISURE £ range £	OFFICES £ range £	HOTELS £ range £	Item nr.
- -	- -	- -	- -	- -	2.7.1
- -	- -	- -	- -	- -	2.7.2
- -	- -	- -	- -	- -	2.7.3
- -	- -	- -	- -	- -	2.7.4
- -	- -	- -	- -	- -	2.7.5
- -	- -	- -	- -	- -	2.7.6
- -	- -	- -	- -	- -	2.7.7
- -	- -	- -	- -	- -	2.7.8
- -	- -	- -	- -	- -	2.7.9
32.00 - 39.00	34.00 - 41.00	34.00 - 41.00	34.00 - 41.00	35.00 - 41.00	2.7.10
45.00 - 50.00	47.00 - 51.00	47.00 - 51.00	47.00 - 51.00	48.00 - 52.00	2.7.11
75.00 -105.00	80.00 -110.00	80.00 -110.00	80.00 -110.00	80.00 -110.00	2.7.12
+50%	+50%	+50%	+50%	+50%	2.7.13
43.00 - 57.00	45.00 - 59.00	45.00 - 59.00	45.00 - 59.00	45.00 - 61.00	2.7.14
57.00 - 70.00	59.00 - 72.00	59.00 - 72.00	59.00 - 72.00	59.00 - 77.00	2.7.15
52.00 - 66.00	54.00 - 68.00	54.00 - 68.00	54.00 - 68.00	54.00 - 72.00	2.7.16
- -	4.00 - 5.50	4.00 - 5.50	4.00 - 5.50	4.00 - 9.00	2.7.17
15.00 - 18.00	15.00 - 18.00	15.00 - 18.00	15.00 - 18.00	15.00 - 23.00	2.7.18
61.00 - 74.50	63.00 - 77.00	63.00 - 77.00	63.00 - 77.00	63.00 - 81.00	2.7.19
67.00 - 81.00	69.00 - 86.00	69.00 - 86.00	69.00 - 86.00	69.00 - 90.00	2.7.20
77.00 - 90.00	78.00 - 95.00	78.00 - 95.00	78.00 - 95.00	78.00 -108.00	2.7.21
11.00 - 13.00	12.00 - 13.50	12.00 - 13.50	12.00 - 13.50	13.00 - 14.50	2.7.22
14.00 - 17.00	15.00 - 18.00	15.00 - 18.00	15.00 - 18.00	16.00 - 19.00	2.7.23
18.00 - 20.00	19.00 - 21.00	19.00 - 21.00	19.00 - 21.00	20.00 - 22.00	2.7.24
19.50 - 21.00	20.00 - 22.00	20.00 - 22.00	20.00 - 22.00	21.00 - 22.00	2.7.25
23.50 - 27.00	24.50 - 28.00	24.50 - 28.00	24.50 - 28.00	25.00 - 29.00	2.7.26
2.00 - 4.50	2.00 - 4.50	2.00 - 4.50	2.00 - 4.50	2.00 - 4.50	2.7.27
+10% to +20%	+10% to +20%	+10% to +20%	+10% to +20%	+10% to +20%	2.7.28

Item nr.	SPECIFICATIONS	Unit	RESIDENTIAL £ range £
2.7	**INTERNAL WALLS, PARTITIONS AND DOORS - cont'd** internal wall area (unless otherwise described)		
	Brick/block partitions and doors - cont'd		
	Autoclaved aerated/lightweight block partitions - cont'd		
2.7.29	average thickness; fair face both sides	m2	20.00 - 25.00
2.7.30	average thickness; fair face and emulsioned both sides	m2	24.00 - 30.00
2.7.31	average thickness; plastered and emulsioned both sides	m2	37.00 - 42.00
	Concrete block partitions		
2.7.32	to retail units	m2	- -
	Dense aggregate block partitions		
2.7.33	average thickness; fair face both sides	m2	23.00 - 28.00
2.7.34	average thickness; fair face and emulsioned both sides	m2	27.00 - 32.00
2.7.35	average thickness; plastered and emulsioned both sides	m2	40.00 - 45.00
	Coloured dense aggregate masonry block partition		
2.7.36	fair face both sides	m2	- -
	Common brick partitions; bricks PC £120.00/1000		
2.7.37	half brick thick	m2	21.00 - 24.00
2.7.38	half brick thick; fair face both sides	m2	23.00 - 28.00
2.7.39	half brick thick; fair face and emulsioned both sides	m2	27.00 - 32.00
2.7.40	half brick thick; plastered and emulsioned both sides	m2	39.00 - 50.00
2.7.41	one brick thick	m2	40.00 - 45.00
2.7.42	one brick thick; fair face both sides	m2	43.00 - 49.00
2.7.43	one brick thick; fair face and emulsioned both sides	m2	47.00 - 53.00
2.7.44	one brick thick; plastered and emulsioned both sides	m2	59.00 - 68.00
	Block partitions; softwood doors and frames; painted		
2.7.45	partition	m2	34.00 - 45.00
2.7.46	partition; fair face both sides	m2	36.00 - 47.00
2.7.47	partition; fair face and emulsioned both sides	m2	40.00 - 51.00
2.7.48	partition; plastered and emulsioned both sides	m2	52.00 - 66.00
	Block partitions; hardwood doors and frames		
2.7.49	partition	m2	50.00 - 63.00
2.7.50	partition; plastered and emulsioned both sides	m2	68.00 - 86.00
	Reinforced concrete walls		
	Walls		
2.7.51	150 mm thick	m2	60.00 - 70.00
2.7.52	150 mm thick; plastered and emulsioned both sides	m2	80.00 - 95.00
	Solid partitioning and doors		
	Patent partitioning; softwood doors		
2.7.53	frame and sheet	m2	- -
2.7.54	frame and panel	m2	- -
2.7.55	panel to panel	m2	- -
2.7.56	economical	m2	- -
	Patent partitioning; hardwood doors		
2.7.57	economical	m2	- -
	Demountable partitioning; hardwood doors		
2.7.58	medium quality; vinyl-faced	m2	- -
2.7.59	high quality; vinyl-faced	m2	- -
	Glazed partitioning and doors		
	Aluminium internal patent glazing		
2.7.60	single glazed	m2	- -
2.7.61	double glazed	m2	- -
	Demountable steel partitioning and doors		
2.7.62	medium quality	m2	- -
2.7.63	high quality	m2	- -
	Demountable aluminium/steel partitioning and doors		
2.7.64	high quality	m2	- -
2.7.65	high quality; sliding	m2	- -
	Stainless steel glazed manual doors and screens		
2.7.66	high quality; to inner lobby of malls	m2	- -

INDUSTRIAL £ range £	RETAILING £ range £	LEISURE £ range £	OFFICES £ range £	HOTELS £ range £	Item nr.
20.00 - 24.00	21.00 - 25.00	21.00 - 25.00	21.00 - 25.00	22.00 - 26.00	2.7.29
23.00 - 29.00	24.00 - 30.00	24.00 - 30.00	24.00 - 30.00	25.00 - 31.00	2.7.30
36.00 - 41.00	37.00 - 42.00	37.00 - 42.00	37.00 - 42.00	38.00 - 45.00	2.7.31
- -	32.00 - 36.00	- -	- -	- -	2.7.32
23.00 - 27.00	23.00 - 28.00	23.00 - 28.00	23.00 - 28.00	24.00 - 29.00	2.7.33
26.00 - 32.00	27.00 - 32.00	27.00 - 32.00	27.00 - 32.00	28.00 - 33.00	2.7.34
39.00 - 43.00	40.00 - 45.00	40.00 - 45.00	40.00 - 45.00	40.00 - 47.00	2.7.35
- -	- -	45.00 - 54.00	- -	- -	2.7.36
20.00 - 23.00	21.00 - 24.00	21.00 - 24.00	21.00 - 24.00	22.00 - 25.00	2.7.37
22.00 - 27.00	23.00 - 28.00	23.00 - 28.00	23.00 - 28.00	24.00 - 29.00	2.7.38
26.00 - 32.00	27.00 - 32.00	27.00 - 32.00	27.00 - 32.00	28.00 - 33.00	2.7.39
38.00 - 48.00	39.00 - 50.00	39.00 - 50.00	39.00 - 50.00	40.00 - 54.00	2.7.40
39.00 - 43.00	40.00 - 45.00	40.00 - 45.00	40.00 - 45.00	42.00 - 47.00	2.7.41
41.00 - 47.00	43.00 - 49.00	43.00 - 49.00	43.00 - 49.00	45.00 - 50.00	2.7.42
45.00 - 51.00	47.00 - 53.00	47.00 - 53.00	47.00 - 53.00	49.00 - 55.00	2.7.43
57.00 - 65.00	58.00 - 68.00	58.00 - 68.00	58.00 - 68.00	61.50 - 72.00	2.7.44
32.00 - 43.00	34.00 - 45.00	34.00 - 45.00	34.00 - 45.00	36.00 - 45.00	2.7.45
34.00 - 45.00	36.00 - 47.00	36.00 - 47.00	36.00 - 47.00	38.00 - 47.00	2.7.46
38.00 - 49.00	40.00 - 51.00	40.00 - 51.00	40.00 - 51.00	41.00 - 51.00	2.7.47
51.00 - 63.00	52.00 - 66.00	52.00 - 66.00	52.00 - 66.00	52.00 - 70.00	2.7.48
48.00 - 60.00	50.00 - 63.00	50.00 - 63.00	50.00 - 63.00	53.50 - 65.00	2.7.49
66.00 - 83.00	68.00 - 86.00	68.00 - 86.00	68.00 - 86.00	69.00 - 90.00	2.7.50
60.00 - 70.00	60.00 - 70.00	60.00 - 70.00	60.00 - 70.00	60.00 - 70.00	2.7.51
80.00 - 95.00	80.00 - 95.00	80.00 - 95.00	80.00 - 95.00	80.00 - 95.00	2.7.52
45.00 - 86.00	50.00 - 90.00	50.00 - 90.00	50.00 - 90.00	50.00 -108.00	2.7.53
38.00 - 70.00	40.00 - 72.00	40.00 - 72.00	40.00 - 72.00	40.00 - 90.00	2.7.54
56.00 -117.00	59.00 -122.00	59.00 -122.00	59.00 -122.00	59.00 -135.00	2.7.55
59.00 - 76.00	63.00 - 81.00	63.00 - 81.00	63.00 - 81.00	63.00 -100.00	2.7.56
65.00 - 80.00	65.00 - 85.00	65.00 - 85.00	65.00 - 85.00	65.00 - 90.00	2.7.57
85.00 -110.00	90.00 -110.00	90.00 -110.00	90.00 -110.00	90.00 -120.00	2.7.58
110.00 -155.00	110.00 -160.00	110.00 -160.00	110.00 -160.00	120.00 -170.00	2.7.59
65.00 - 90.00	- -	- -	- -	- -	2.7.60
110.00 -135.00	- -	- -	- -	- -	2.7.61
120.00 -145.00	- -	- -	120.00 -150.00	- -	2.7.62
145.00 -180.00	- -	- -	160.00 -190.00	- -	2.7.63
- -	- -	180.00 -315.00	- -	250.00 -315.00	2.7.64
- -	- -	425.00 -520.00	- -	- -	2.7.65
- -	250.00 -675.00	- -	- -	- -	2.7.66

Item nr.	SPECIFICATIONS	Unit	RESIDENTIAL £ range £
2.7	**INTERNAL WALLS, PARTITION AND DOORS - cont'd**	**internal wall area** (unless otherwise described)	
	Special partitioning and doors		
	Demountable fire partitions		
2.7.67	enamelled steel; half hour	m2	- -
2.7.68	stainless steel; half hour	m2	- -
	Soundproof partitions; hardwood doors		
2.7.69	luxury veneered	m2	- -
	Folding screens		
2.7.70	gym divider; electronically operated	m2	- -
2.7.71	bar divider	m2	- -
2.7.72	Squash court glass back wall and door	m2	- -
	WC/Changing cubicles	**each**	
2.7.73	WC cubicles	nr	- -
	Changing cubicles		
2.7.74	aluminium	nr	- -
2.7.75	aluminium; textured glass and bench seating	nr	- -
	Comparative doors/door linings/frames		
	Standard softwood doors; excluding ironmongery; linings and frames		
	40 mm flush; hollow core; painted		
2.7.76	726 x 2040 mm	nr	30.00 - 35.00
2.7.77	826 x 2040 mm	nr	30.00 - 35.00
	40 mm flush; hollow core; plywood faced; painted		
2.7.78	726 x 2040 mm	nr	35.00 - 45.00
2.7.79	826 x 2040 mm	nr	35.00 - 45.00
	40 mm flush; hollow core; Sapele veneered hard board faced		
2.7.80	726 x 2040 mm	nr	30.00 - 35.00
2.7.81	826 x 2040 mm	nr	30.00 - 35.00
	40 mm flush; hollow core; Teak veneered hard board faced		
2.7.82	726 x 2040 mm	nr	70.00 - 85.00
2.7.83	826 x 2040 mm	nr	70.00 - 85.00
	Standard softwood fire-doors; excluding ironmongery; linings and frames		
	44 mm flush; half hour fire check; plywood faced; painted		
2.7.84	726 x 2040 mm	nr	45.00 - 50.00
2.7.85	826 x 2040 mm	nr	47.50 - 52.50
	44 mm flush; half hour fire check; Sapele veneered hardboard faced		
2.7.86	726 x 2040 mm	nr	57.50 - 65.00
2.7.87	826 x 2040 mm	nr	60.00 - 67.50
	54 mm flush; one hour fire check; Sapele veneered hardboard faced		
2.7.88	726 x 2040 mm	nr	175.00 -200.00
2.7.89	826 x 2040 mm	nr	180.00 -205.00
	54 mm flush; one hour fire resisting; Sapele veneered hardboard faced		
2.7.90	726 x 2040 mm	nr	195.00 -215.00
2.7.91	826 x 2040 mm	nr	200.00 -225.00
	Purpose-made softwood doors; excluding ironmongery; linings and frames		
	44 mm four panel door; painted		
2.7.92	726 x 2040 mm	nr	100.00 -110.00
2.7.93	826 x 2040 mm	nr	105.00 -115.00

INDUSTRIAL £ range £	RETAILING £ range £	LEISURE £ range £	OFFICES £ range £	HOTELS £ range £	Item nr.
270.00 -405.00	315.00 -430.00	315.00 -430.00	315.00 -430.00	- -	2.7.67
- -	540.00 -675.00	540.00 -675.00	540.00 -675.00	- -	2.7.68
135.00 -180.00	145.00 -205.00	145.00 -205.00	145.00 -205.00	155.00 -225.00	2.7.69
- -	- -	100.00 -110.00	- -	- -	2.7.70
- -	- -	270.00 -300.00	- -	- -	2.7.71
- -	- -	145.00 -180.00	- -	- -	2.7.72
225.00 -340.00	230.00 -340.00	230.00 -340.00	225.00 -495.00	280.00 -495.00	2.7.73
- -	- -	270.00 -495.00	- -	- -	2.7.74
- -	- -	450.00 -585.00	- -	- -	2.7.75
30.00 - 35.00	30.00 - 35.00	30.00 - 35.00	30.00 - 35.00	30.00 - 35.00	2.7.76
30.00 - 35.00	30.00 - 35.00	30.00 - 35.00	30.00 - 35.00	30.00 - 35.00	2.7.77
35.00 - 45.00	35.00 - 45.00	35.00 - 45.00	35.00 - 45.00	35.00 - 45.00	2.7.78
35.00 - 45.00	35.00 - 45.00	35.00 - 45.00	35.00 - 45.00	35.00 - 45.00	2.7.79
30.00 - 35.00	30.00 - 35.00	30.00 - 35.00	30.00 - 35.00	30.00 - 35.00	2.7.80
30.00 - 35.00	30.00 - 35.00	30.00 - 35.00	30.00 - 35.00	30.00 - 35.00	2.7.81
70.00 - 85.00	70.00 - 85.00	70.00 - 85.00	70.00 - 85.00	70.00 - 85.00	2.7.82
70.00 - 85.00	70.00 - 85.00	70.00 - 85.00	70.00 - 85.00	70.00 - 85.00	2.7.83
45.00 - 50.00	45.00 - 50.00	45.00 - 50.00	45.00 - 50.00	45.00 - 50.00	2.7.84
47.50 - 52.50	47.50 - 52.50	47.50 - 52.50	47.50 - 52.50	47.50 - 52.50	2.7.85
57.50 - 65.00	57.50 - 65.00	57.50 - 65.00	57.50 - 65.00	57.50 - 65.00	2.7.86
60.00 - 67.50	60.00 - 67.50	60.00 - 67.50	60.00 - 67.50	60.00 - 67.50	2.7.87
175.00 -200.00	175.00 -200.00	175.00 -200.00	175.00 -200.00	175.00 -200.00	2.7.88
180.00 -200.00	180.00 -200.00	180.00 -200.00	180.00 -200.00	180.00 -200.00	2.7.89
195.00 -215.00	195.00 -215.00	195.00 -215.00	195.00 -215.00	195.00 -215.00	2.7.90
200.00 -225.00	200.00 -225.00	200.00 -225.00	200.00 -225.00	200.00 -225.00	2.7.91
100.00 -110.00	100.00 -110.00	100.00 -110.00	100.00 -110.00	100.00 -110.00	2.7.92
105.00 -115.00	105.00 -115.00	105.00 -115.00	105.00 -115.00	105.00 -115.00	2.7.93

Item nr.	SPECIFICATIONS	Unit	RESIDENTIAL £ range £
2.7	**INTERNAL WALLS, PARTITION AND DOORS - cont'd**	each (unless otherwise described)	
	Comparative doors/door linings/frames - cont'd		
	Purpose-made Mahogany doors; excluding ironmongery; linings and frames		
	50 mm four panel door; wax polished		
2.7.94	726 x 2040 mm	nr	195.00 -215.00
2.7.95	826 x 2040 mm	nr	205.00 -225.00
	Purpose-made softwood door frames/linings; painted including grounds		
	32 x 100 mm lining		
2.7.96	726 x 2040 mm opening	nr	60.00 - 67.50
2.7.97	826 x 2040 mm opening	nr	62.00 - 69.00
	32 x 140 mm lining		
2.7.98	726 x 2040 mm opening	nr	67.00 - 77.00
2.7.99	826 x 2040 mm opening	nr	68.00 - 78.00
	32 x 250 mm Cross-tongued lining		
2.7.100	726 x 2040 mm opening	nr	89.00 -100.00
2.7.101	826 x 2040 mm opening	nr	90.00 -101.00
	32 x 375 mm Cross-tongued lining		
2.7.102	726 x 2040 mm opening	nr	115.00 -125.00
2.7.103	826 x 2040 mm opening	nr	119.00 -128.00
	75 x 100 mm rebated frame		
2.7.104	726 x 2040 mm opening	nr	78.50 - 83.50
2.7.105	826 x 2040 mm opening	nr	83.00 - 87.50
	Purpose-made Mahogany door frames/linings; wax polished including grounds		
	32 x 100 mm lining		
2.7.106	726 x 2040 mm opening	nr	81.00 - 87.50
2.7.107	826 x 2040 mm opening	nr	83.00 - 90.00
	32 x 140 mm lining		
2.7.108	726 x 2040 mm opening	nr	90.00 -100.00
2.7.109	826 x 2040 mm opening	nr	92.00 -101.00
	32 x 250 mm cross-tongued lining		
2.7.110	726 x 2040 mm opening	nr	101.00 -110.00
2.7.111	826 x 2040 mm opening	nr	104.00 -113.00
	32 x 375 mm cross-tongued lining		
2.7.112	726 x 2040 mm opening	nr	155.00 -165.00
2.7.113	826 x 2040 mm opening	nr	155.00 -170.00
	75 x 100 mm rebated frame		
2.7.114	726 x 2040 mm opening	nr	110.00 -120.00
2.7.115	826 x 2040 mm opening	nr	110.00 -120.00
	Standard softwood doors and frames; including ironmongery and painting		
2.7.116	flush; hollow core	nr	145.00 -180.00
2.7.117	flush; hollow core; hardwood faced	nr	150.00 -200.00
	flush; solid core		
2.7.118	single leaf	nr	170.00 -225.00
2.7.119	double leaf	nr	250.00 -340.00
2.7.120	flush; solid core; hardwood faced	nr	180.00 -235.00
2.7.121	four panel door	nr	250.00 -315.00
	Purpose-made softwood doors and hardwood frames; including ironmongery; painting and polishing		
	flush; solid core; heavy duty		
2.7.122	single leaf	nr	- -
2.7.123	double leaf	nr	- -
	flush; solid core; heavy duty; plastic laminate faced		
2.7.124	single leaf	nr	- -
2.7.125	double leaf	nr	- -

INDUSTRIAL £ range £	RETAILING £ range £	LEISURE £ range £	OFFICES £ range £	HOTELS £ range £	Item nr.
195.00 -215.00	195.00 -215.00	195.00 -215.00	195.00 -215.00	195.00 -215.00	2.7.94
200.00 -225.00	200.00 -225.00	200.00 -225.00	200.00 -225.00	200.00 -225.00	2.7.95
60.50 - 67.50	60.50 - 67.50	60.50 - 67.50	60.50 - 67.50	60.50 - 67.50	2.7.96
62.00 - 69.00	62.00 - 69.00	62.00 - 69.00	62.00 - 69.00	62.00 - 69.00	2.7.97
66.50 - 76.50	66.50 - 76.70	66.50 - 76.70	66.50 - 76.50	66.50 - 76.50	2.7.98
68.50 - 78.50	68.50 - 78.50	68.50 - 78.50	68.50 - 78.50	68.50 - 78.50	2.7.99
88.50 -100.00	88.50 -100.00	88.50 -100.00	88.50 -100.00	88.50 -100.00	2.7.100
90.00 -101.00	90.00 -101.00	90.00 -101.00	90.00 -101.00	90.00 -101.00	2.7.101
120.00 -125.00	120.00 -125.00	120.00 -125.00	120.00 -125.00	120.00 -125.00	2.7.102
120.00 -128.00	120.00 -128.00	120.00 -128.00	120.00 -128.00	120.00 -128.00	2.7.103
78.50 - 83.50	78.50 - 83.50	78.50 - 83.50	78.50 - 83.50	78.50 - 83.50	2.7.104
83.50 - 88.00	83.50 - 88.00	83.50 - 88.00	83.50 - 88.00	83.50 - 88.00	2.7.105
81.00 - 88.00	81.00 - 88.00	81.00 - 88.00	81.00 - 88.00	81.00 - 88.00	2.7.106
83.00 - 90.00	83.00 - 90.00	83.00 - 90.00	83.00 - 90.00	83.00 - 90.00	2.7.107
90.00 -100.00	90.00 -100.00	90.00 -100.00	90.00 -100.00	90.00 -100.00	2.7.108
92.50 -101.50	92.50 -101.50	92.50 -101.50	92.50 -101.50	92.50 -101.50	2.7.109
100.00 -110.00	100.00 -110.00	100.00 -110.00	100.00 -110.00	100.00 -110.00	2.7.110
105.00 -115.00	105.00 -115.00	105.00 -115.00	105.00 -115.00	105.00 -115.00	2.7.111
150.00 -165.00	150.00 -165.00	150.00 -165.00	150.00 -165.00	150.00 -165.00	2.7.112
155.00 -170.00	155.00 -170.00	155.00 -170.00	155.00 -170.00	155.00 -170.00	2.7.113
110.00 -120.00	110.00 -120.00	110.00 -120.00	110.00 -120.00	110.00 -120.00	2.7.114
110.00 -120.00	110.00 -120.00	110.00 -120.00	110.00 -120.00	110.00 -120.00	2.7.115
145.00 -180.00	145.00 -180.00	145.00 -180.00	145.00 -180.00	145.00 -180.00	2.7.116
150.00 -200.00	150.00 -200.00	150.00 -200.00	150.00 -200.00	150.00 -200.00	2.7.117
170.00 -225.00	170.00 -225.00	170.00 -225.00	170.00 -225.00	170.00 -225.00	2.7.118
250.00 -340.00	250.00 -340.00	250.00 -340.00	250.00 -340.00	250.00 -340.00	2.7.119
180.00 -235.00	180.00 -235.00	180.00 -235.00	180.00 -235.00	180.00 -235.00	2.7.120
250.00 -315.00	250.00 -315.00	250.00 -315.00	250.00 -315.00	250.00 -315.00	2.7.121
430.00 -500.00	430.00 -500.00	430.00 -500.00	430.00 -500.00	430.00 -500.00	2.7.122
580.00 -740.00	580.00 -740.00	580.00 -740.00	580.00 -740.00	580.00 -740.00	2.7.123
520.00 -590.00	520.00 -590.00	520.00 -590.00	520.00 -590.00	520.00 -590.00	2.7.124
720.00 -810.00	720.00 -810.00	720.00 -810.00	720.00 -810.00	720.00 -810.00	2.7.125

Item nr.	SPECIFICATIONS	Unit	RESIDENTIAL £ range £
2.7	**INTERNAL WALLS, PARTITION AND DOORS - cont'd**	**each** (unless otherwise described)	
	Comparative doors/door linings/frames - cont'd		
	Purpose-made softwood fire doors and hardwood frames; including ironmongery; painting and polishing		
	flush; one hour fire resisting		
2.7.126	single leaf	nr	- -
2.7.127	double leaf	nr	- -
	flush; one hour fire resisting; plastic laminate faced		
2.7.128	single leaf	nr	- -
2.7.129	double leaf	nr	- -
	Purpose-made softwood doors and pressed steel frames;		
2.7.130	flush; half hour fire check; plastic laminate faced	nr	- -
	Purpose-made Mahogany doors and frames; including ironmongery and polishing		
2.7.131	four panel door	nr	- -
	Perimeter treatments		
	Precast concrete lintels; in block walls		
2.7.132	75 mm wide	m	7.50 - 12.50
2.7.133	100 mm wide	m	12.50 - 15.00
	Precast concrete lintels; in brick walls		
2.7.134	half brick thick	m	12.50 - 15.00
2.7.135	one brick thick	m	20.00 - 25.00
	Purpose-made Softwood architraves; painted; including grounds		
	25 x 50 mm; to both sides of openings		
2.7.136	726 x 2040 mm opening	nr	58.50 - 63.00
2.7.137	826 x 2040 mm opening	nr	60.00 - 64.50
	Purpose-made Mahogany architraves; wax polished; including grounds		
	25 x 50 mm; to both sides of openings		
2.7.138	726 x 2040 mm opening	nr	95.00 -105.00
2.7.139	826 x 2040 mm opening	nr	96.00 -105.00
3.1	**WALL FINISHES**	**wall finish area** (unless otherwise described)	
	comprising:-		
	Sheet/board finishes		
	In situ wall finishes		
	Rigid tile/panel finishes		
	Sheet/board finishes		
	Dry plasterboard lining; taped joints; for direct decoration		
3.1.1	9.5 mm Gyproc Wallboard	m2	6.50 - 9.00
	Extra for		
3.1.2	insulating grade	m2	0.40 - 0.50
3.1.3	insulating grade; plastic faced	m2	1.40 - 1.60
3.1.4	12.5 mm Gyproc Wallboard (half-hour fire-resisting)	m2	7.00 - 10.00
	Extra for		
3.1.5	insulating grade	m2	0.45 - 0.55
3.1.6	insulating grade; plastic faced	m2	1.30 - 1.45
3.1.7	two layers of 12.5 mm Gyproc Wallboard (one hour fire-resisting)	m2	12.50 - 16.00
3.1.8	9 mm Supalux (half-hour fire-resisting)	m2	12.50 - 16.00
	Dry plasterboard lining; taped joints; for direct decoration; fixed to wall on dabs		
3.1.9	9.5 mm Gyproc Wallboard	m2	7.00 - 10.00

INDUSTRIAL £ range £	RETAILING £ range £	LEISURE £ range £	OFFICES £ range £	HOTELS £ range £	Item nr.
565.00 -630.00	565.00 -630.00	565.00 -630.00	565.00 -630.00	565.00 -630.00	2.7.126
720.00 -875.00	720.00 -875.00	720.00 -875.00	720.00 -875.00	720.00 -875.00	2.7.127
700.00 -765.00	700.00 -765.00	700.00 -765.00	700.00 -765.00	700.00 -765.00	2.7.128
900.00 -975.00	900.00 -975.00	900.00 -975.00	900.00 -975.00	900.00 -975.00	2.7.129
650.00 -785.00	650.00 -785.00	650.00 -785.00	650.00 -785.00	650.00 -785.00	2.7.130
540.00 -630.00	540.00 -630.00	540.00 -630.00	540.00 -630.00	540.00 -630.00	2.7.131
6.00 - 9.00	6.00 - 9.00	6.00 - 9.00	6.00 - 9.00	6.00 - 9.00	2.7.132
7.00 - 10.00	7.00 - 10.00	7.00 - 10.00	7.00 - 10.00	7.00 - 10.00	2.7.133
7.00 - 10.00	7.00 - 10.00	7.00 - 10.00	7.00 - 10.00	7.00 - 10.00	2.7.134
11.00 - 13.50	11.00 - 13.50	11.00 - 13.50	11.00 - 13.50	11.00 - 13.50	2.7.135
58.50 - 63.00	58.50 - 63.00	58.50 - 63.00	58.50 - 63.00	58.50 - 63.00	2.7.136
60.00 - 64.50	60.00 - 64.50	60.00 - 64.50	60.00 - 64.50	60.00 - 64.50	2.7.137
95.00 -105.00	95.00 -105.00	95.00 -105.00	95.00 -105.00	95.00 -105.00	2.7.138
96.50 -105.50	96.50 -105.50	96.50 -105.50	96.50 -105.50	96.50 -105.50	2.7.139
6.00 - 8.50	6.50 - 9.00	6.50 - 9.00	6.50 - 9.00	6.50 - 9.00	3.1.1
0.40 - 0.50	0.40 - 0.50	0.40 - 0.50	0.40 - 0.50	0.40 - 0.50	3.1.2
1.45 - 1.60	1.45 - 1.60	1.45 - 1.60	1.45 - 1.60	1.45 - 1.60	3.1.3
7.00 - 9.50	7.00 - 10.00	7.00 - 10.00	7.00 - 10.00	7.50 - 11.50	3.1.4
0.45 - 0.55	0.45 - 0.55	0.45 - 0.55	0.45 - 0.55	0.45 - 0.55	3.1.5
1.25 - 1.45	1.25 - 1.45	1.25 - 1.45	1.25 - 1.45	1.25 - 1.45	3.1.6
12.00 - 16.00	12.50 - 16.00	12.50 - 16.00	12.50 - 16.00	13.50 - 18.00	3.1.7
12.00 - 16.00	12.50 - 16.00	12.50 - 16.00	12.50 - 16.00	13.50 - 18.00	3.1.8
6.50 - 9.50	7.00 - 10.00	7.00 - 10.00	7.00 - 10.00	7.50 - 10.50	3.1.9

Item nr.	SPECIFICATIONS	Unit	RESIDENTIAL £ range £
3.1	**WALL FINISHES - cont'd**	wall finish area (unless otherwise described)	
	Sheet/board finishes - cont'd		
	Dry plasterboard lining; taped joints; for direct decoration; including metal tees		
3.1.10	9.5 mm Gyproc Wallboard	m2	- -
3.1.11	12.5 mm Gyproc Wallboard	m2	- -
	Dry lining/sheet panelling; including battens; plugged to wall		
3.1.12	6.4 mm hardboard	m2	8.00 - 9.00
3.1.13	9.5 mm Gyproc Wallboard	m2	11.50 - 15.00
3.1.14	6 mm birch faced plywood	m2	12.50 - 14.50
3.1.15	6 mm WAM plywood	m2	15.50 - 18.00
3.1.16	15 mm chipboard	m2	11.50 - 12.50
3.1.17	15 mm melamine faced chipboard	m2	20.00 - 24.00
3.1.18	13.2 mm 'Formica' faced chipboard	m2	27.00 - 40.00
	Timber boarding/panelling; on and including battens; plugged to wall		
3.1.19	12 mm softwood boarding	m2	18.00 - 22.50
3.1.20	25 mm softwood boarding	m2	25.00 - 27.50
3.1.21	hardwood panelling; t, g & v-jointed	m2	40.00 - 90.00
	In situ wall finishes		
	Extra over common brickwork for		
3.1.22	fair face and pointing both sides	m2	2.50 - 3.50
	Comparative finishes		
3.1.23	one mist and two coats emulsion paint	m2	2.00 - 3.00
3.1.24	multi-coloured gloss paint	m2	3.50 - 4.50
3.1.25	two coats of lightweight plaster	m2	7.00 - 9.00
3.1.26	9.5 mm Gyproc Wallboard and skim coat	m2	9.00 - 11.00
3.1.27	12.5 mm Gyproc Wallboard and skim coat	m2	10.00 - 12.00
3.1.28	two coats of 'Thistle' plaster	m2	9.00 - 11.50
3.1.29	plaster and emulsion	m2	9.00 - 13.50
	Extra for		
3.1.30	gloss paint in lieu of emulsion	m2	1.50 - 1.80
3.1.31	two coat render and emulsion	m2	15.50 - 20.50
3.1.32	plaster and vinyl	m2	12.50 - 18.00
3.1.33	plaster and fabric	m2	12.50 - 27.00
3.1.34	squash court plaster 'including markings'	m2	- -
3.1.35	6 mm terrazzo wall lining; including backing	m2	- -
3.1.36	glass reinforced gypsum	m2	- -
	Rigid tile/panel finishes		
	Ceramic wall tiles; including backing		
3.1.37	economical	m2	17.00 - 30.00
3.1.38	medium quality	m2	30.00 - 54.00
3.1.39	high quality; to toilet blocks, kitchens and first aid rooms	m2	- -
3.1.40	high quality; to changing areas, toilets, showers and fitness areas	m2	- -
	Porcelain mosaic tiling; including backing		
3.1.41	to swimming pool lining; walls and floors	m2	- -
	'Roman Travertine' marble wall linings; polished		
3.1.42	19 mm thick	m2	- -
3.1.43	40 mm thick	m2	- -
3.1.44	Metal mirror cladding panels	m2	- -

INDUSTRIAL £ range £	RETAILING £ range £	LEISURE £ range £	OFFICES £ range £	HOTELS £ range £	Item nr.
15.00 - 18.00	- -	- -	- -	- -	3.1.10
16.00 - 19.00	- -	- -	- -	- -	3.1.11
7.50 - 8.50	8.00 - 9.00	8.00 - 9.00	8.00 - 9.00	8.50 - 10.00	3.1.12
11.00 - 14.50	11.00 - 15.00	11.00 - 15.00	11.00 - 15.00	11.50 - 16.00	3.1.13
12.00 - 14.00	12.50 - 14.50	12.50 - 14.50	12.50 - 14.50	13.00 - 16.00	3.1.14
15.00 - 17.50	15.50 - 18.00	15.50 - 18.00	15.50 - 18.00	16.00 - 19.00	3.1.15
11.00 - 12.00	11.00 - 12.50	11.00 - 12.50	11.00 - 12.50	11.50 - 13.50	3.1.16
19.00 - 23.00	20.00 - 24.00	20.00 - 24.00	20.00 - 24.00	21.00 - 25.00	3.1.17
26.00 - 38.50	27.00 - 40.50	27.00 - 40.50	27.00 - 40.50	27.00 - 45.00	3.1.18
17.00 - 21.50	18.00 - 22.50	18.00 - 22.50	18.00 - 22.50	19.00 - 23.50	3.1.19
24.00 - 26.50	25.00 - 27.50	25.00 - 27.50	25.00 - 27.50	26.00 - 29.00	3.1.20
40.00 - 90.00	40.00 - 90.00	40.00 - 90.00	40.00 - 90.00	45.00 -100.00	3.1.21
2.50 - 3.50	2.50 - 3.50	2.50 - 3.50	2.50 - 3.50	2.50 - 3.50	3.1.22
2.00 - 3.00	2.00 - 3.00	2.00 - 3.00	2.00 - 3.00	2.00 - 3.00	3.1.23
3.50 - 4.50	3.50 - 4.50	3.50 - 4.50	3.50 - 4.50	3.50 - 4.50	3.1.24
6.50 - 8.50	7.00 - 9.00	7.00 - 9.00	7.00 - 9.00	7.50 - 10.00	3.1.25
8.50 - 10.50	9.00 - 11.00	9.00 - 11.00	9.00 - 11.00	9.50 - 11.50	3.1.26
9.50 - 11.50	10.00 - 12.00	10.00 - 12.00	10.00 - 12.00	10.50 - 12.00	3.1.27
9.00 - 11.50	9.00 - 11.50	9.00 - 11.50	9.00 - 11.50	9.50 - 11.50	3.1.28
8.50 - 12.50	9.00 - 13.50	9.00 - 13.50	9.00 - 13.50	10.00 - 16.00	3.1.29
1.50 - 1.80	1.50 - 1.80	1.50 - 1.80	1.50 - 1.80	1.50 - 1.80	3.1.30
15.00 - 20.00	15.50 - 20.50	15.50 - 20.50	15.50 - 20.50	16.00 - 21.50	3.1.31
- -	12.50 - 18.00	12.50 - 18.00	12.50 - 18.00	13.50 - 19.00	3.1.32
- -	12.50 - 27.00	12.50 - 27.00	12.50 - 27.00	13.50 - 29.00	3.1.33
- -	- -	18.00 - 22.50	- -	- -	3.1.34
- -	125.00 -160.00	125.00 -160.00	125.00 -160.00	135.00 -170.00	3.1.35
- -	115.00 -170.00	115.00 -170.00	115.00 -170.00	125.00 -180.00	3.1.36
16.00 - 28.00	17.00 - 30.00	17.00 - 30.00	17.00 - 30.00	20.00 - 35.00	3.1.37
28.00 - 50.00	30.00 - 55.00	30.00 - 55.00	30.00 - 55.00	35.00 - 60.00	3.1.38
- -	49.50 - 63.00	49.50 - 63.00	49.50 - 63.00	49.50 - 63.00	3.1.39
- -	- -	63.00 - 76.50	- -	- -	3.1.40
- -	- -	40.50 - 54.00	- -	- -	3.1.41
- -	200.00 -270.00	200.00 -270.00	200.00 -270.00	200.00 -270.00	3.1.42
- -	275.00 -350.00	275.00 -350.00	275.00 -350.00	275.00 -350.00	3.1.43
- -	215.00 -350.00	200.00 -340.00	- -	225.00 -360.00	3.1.44

Item nr.	SPECIFICATIONS	Unit	RESIDENTIAL £ range £
3.2	**FLOOR FINISHES** **floor finish area** (unless otherwise described)		
	comprising:-		
	Sheet/board flooring		
	Insitu screed and floor finishes		
	Rigid tile/slab finishes		
	Parquet/Wood block finishes		
	Flexible tiling/sheet finishes		
	Carpet tiles/Carpetting		
	Access floors and finishes		
	Perimeter treatments and sundries		
	Sheet/board flooring		
	Chipboard flooring; t & g joints		
3.2.1	18 mm thick	m2	6.50 - 7.50
3.2.2	22 mm thick	m2	8.00 - 9.00
	Wrought softwood flooring		
3.2.3	25 mm thick; butt joints	m2	13.50 - 15.50
3.2.4	25 mm thick; butt joints; cleaned off and polished	m2	16.00 - 19.00
3.2.5	25 mm thick; t & g joints	m2	15.00 - 18.00
3.2.6	25 mm thick; t & g joints; cleaned off and polished	m2	18.00 - 21.50
3.2.7	Wrought softwood t & g strip flooring; 25 mm thick; polished; including fillets	m2	21.50 - 27.00
	Wrought hardwood t & g strip flooring; 25 mm thick; polished		
3.2.8	Maple	m2	- -
3.2.9	Gurjun	m2	- -
3.2.10	Iroko	m2	- -
3.2.11	American Oak	m2	- -
	Wrought hardwood t & g strip flooring; 25 mm thick; polished; including fillets		
3.2.12	Maple	m2	- -
3.2.13	Gurjun	m2	- -
3.2.14	Iroko	m2	- -
3.2.15	American Oak	m2	- -
	Wrought hardwood t & g strip flooring; 25 mm thick; polished; including rubber pads		
3.2.16	Maple	m2	- -
	In situ screed and floor finishes		
	Extra over concrete floor for		
3.2.17	power floating	m2	- -
3.2.18	power floating; surface hardener	m2	- -
	Latex cement screeds		
3.2.19	3 mm thick; one coat	m2	- -
3.2.20	5 mm thick; two coat	m2	- -
3.2.21	Rubber latex non-slip solution and epoxy sealant	m2	- -
	Cement and sand (1:3) screeds		
3.2.22	25 mm thick	m2	7.00 - 7.50
3.2.23	50 mm thick	m2	8.50 - 9.50
3.2.24	75 mm thick	m2	12.00 - 13.00
	Cement and sand (1:3) paving		
3.2.25	paving	m2	6.50 - 8.00
3.2.26	32 mm thick; surface hardener	m2	8.00 - 12.00
3.2.27	Screed only (for subsequent finish)	m2	10.00 - 14.50
3.2.28	Screed only (for subsequent finish); allowance for skirtings	m2	11.50 - 17.00
	Mastic asphalt paving		
3.2.29	20 mm thick; BS 1076; black	m2	- -
3.2.30	20 mm thick; BS 1451; red	m2	- -

INDUSTRIAL £ range £	RETAILING £ range £	LEISURE £ range £	OFFICES £ range £	HOTELS £ range £	Item nr.
- -	- -	- -	- -	- -	3.2.1
- -	- -	- -	- -	- -	3.2.2
- -	- -	- -	- -	- -	3.2.3
- -	- -	- -	- -	- -	3.2.4
- -	- -	- -	- -	- -	3.2.5
- -	- -	- -	- -	- -	3.2.6
- -	- -	- -	- -	- -	3.2.7
- -	- -	37.50 - 42.00	37.50 - 42.00	38.00 - 44.00	3.2.8
- -	- -	- -	33.50 - 37.00	34.00 - 37.50	3.2.9
- -	- -	- -	37.50 - 42.00	38.00 - 44.00	3.2.10
- -	- -	- -	40.00 - 45.00	41.00 - 46.00	3.2.11
- -	- -	42.00 - 47.00	42.00 - 47.00	43.00 - 48.00	3.2.12
- -	- -	- -	38.00 - 42.00	39.00 - 43.00	3.2.13
- -	- -	- -	42.00 - 47.00	43.00 - 48.00	3.2.14
- -	- -	- -	44.00 - 50.00	45.00 - 52.50	3.2.15
- -	- -	50.00 - 60.00	- -	- -	3.2.16
3.50 - 7.00	2.50 - 7.00	- -	- -	- -	3.2.17
7.00 - 10.00	5.50 - 8.50	- -	- -	- -	3.2.18
3.50 - 4.00	- -	- -	3.50 - 4.00	- -	3.2.19
5.00 - 5.50	- -	- -	5.00 - 5.50	- -	3.2.20
6.00 - 13.50	- -	- -	- -	- -	3.2.21
7.00 - 7.50	7.00 - 7.50	7.00 - 7.50	7.00 - 7.50	7.00 - 7.50	3.2.22
8.50 - 9.50	8.50 - 9.50	8.50 - 9.50	8.50 - 9.50	8.50 - 9.50	3.2.23
12.00 - 13.00	12.00 - 13.00	12.00 - 13.00	12.00 - 13.00	12.00 - 13.00	3.2.24
6.00 - 7.50	6.50 - 8.00	6.50 - 8.00	6.50 - 8.00	6.50 - 8.00	3.2.25
7.00 - 11.00	8.00 - 12.00	8.00 - 12.00	8.00 - 12.00	8.00 - 12.00	3.2.26
9.00 - 13.50	10.00 - 14.50	10.00 - 14.50	10.00 - 14.50	10.00 - 14.50	3.2.27
11.00 - 16.00	11.50 - 17.00	11.50 - 17.00	11.50 - 17.00	11.50 - 18.00	3.2.28
14.50 - 16.50	15.00 - 17.00	15.00 - 17.00	- -	- -	3.2.29
17.00 - 19.00	17.50 - 20.00	17.50 - 20.00	- -	- -	3.2.30

Item nr.	SPECIFICATIONS	Unit	RESIDENTIAL £ range £	
3.2	**FLOOR FINISHES - cont'd**	**floor finish area** (unless otherwise described)		
	In situ screed and floor finishes - cont'd			
	Granolithic			
3.2.31	20 mm thick	m2	-	-
3.2.32	25 mm thick	m2	-	-
3.2.33	25 mm thick; including screed	m2	-	-
3.2.34	38 mm thick; including screed	m2	-	-
	Synthanite; on and including building paper			
3.2.35	25 mm thick	m2	-	-
3.2.36	50 mm thick	m2	-	-
3.2.37	75 mm thick	m2	-	-
	Acrylic polymer floor finish			
3.2.38	10 mm thick	m2	-	
	Epoxy floor finish			
3.2.39	1.5 - 2 mm thick	m2	-	-
3.2.40	5 - 6 mm thick	m2	-	-
	Polyester resin floor finish			
3.2.41	5 - 9 mm thick	m2	-	-
	Terrazzo paving; divided into squares with ebonite strip; polished			
3.2.42	16 mm thick	m2	-	-
3.2.43	16 mm thick; including screed	m2	-	-
	Rigid Tile/slab finishes			
	Quarry tile flooring			
3.2.44	150 x 150 x 12.5 mm thick; red	m2	-	-
3.2.45	150 x 150 x 12.5 mm thick; brown	m2	-	-
3.2.46	200 x 200 x 19 mm thick; brown	m2	-	-
3.2.47	average tiling	m2	-	-
3.2.48	tiling; including screed	m2	-	-
3.2.49	tiling; including screed and allowance for skirtings	m2	-	-
	Brick paving			
3.2.50	paving	m2	-	-
3.2.51	paving; including screed	m2	-	-
	Glazed ceramic tile flooring			
3.2.52	100 x 100 x 9 mm thick; red	m2	-	-
3.2.53	150 x 150 x 12 mm thick; red	m2	-	-
3.2.54	100 x 100 x 9 mm thick; black	m2	-	-
3.2.55	150 x 150 x 12 mm thick; black	m2	-	-
3.2.56	150 x 150 x 12 mm thick; antislip	m2	-	-
3.2.57	fully vitrified	m2	-	-
3.2.58	fully vitrified; including screed	m2	-	-
3.2.59	fully vitrified; including screed and allowance for skirtings	m2	-	-
3.2.60	high quality; to service areas; kitchen and toilet blocks; including screed	m2	-	-
3.2.61	high quality; to foyer; fitness and bar areas; including screed	m2	-	-
3.2.62	high quality; to pool surround, bottoms, steps and changing room; including screed	m2	-	-
	Porcelain mosaic paving; including screed			
3.2.63	to swimming pool lining, walls and floors	m2	-	-
	Extra for			
3.2.64	non-slip finish to pool lining, beach and changing areas and showers	m2	-	-
	Terrazzo tile flooring			
3.2.65	28 mm thick white Sicilian marble aggregate tiling	m2	-	-
3.2.66	tiling; including screed	m2	-	-
	York stone			
3.2.67	50 mm thick paving	m2	-	-
3.2.68	paving; including screed	m2	-	-

INDUSTRIAL £ range £	RETAILING £ range £	LEISURE £ range £	OFFICES £ range £	HOTELS £ range £	Item nr.
8.00 - 12.00	8.50 - 12.50	8.50 - 12.50	8.50 - 12.50	8.50 - 12.50	3.2.31
10.50 - 13.50	11.00 - 14.00	11.00 - 14.00	11.00 - 14.00	11.50 - 15.00	3.2.32
17.00 - 19.00	17.50 - 20.00	17.50 - 20.00	17.50 - 20.00	17.50 - 22.50	3.2.33
25.00 - 27.50	26.00 - 29.00	26.00 - 29.00	26.00 - 29.00	27.00 - 30.00	3.2.34
16.00 - 19.00	16.00 - 20.00	16.00 - 20.00	16.00 - 20.00	- -	3.2.35
22.50 - 25.50	23.50 - 27.00	23.50 - 27.00	23.50 - 27.00	- -	3.2.36
27.00 - 31.50	29.00 - 32.50	29.00 - 32.50	29.00 - 32.50	- -	3.2.37
16.00 - 20.00	17.00 - 21.00	17.00 - 21.00	17.00 - 21.00	- -	3.2.38
16.00 - 21.00	17.00 - 21.50	17.00 - 21.50	17.00 - 21.50	- -	3.2.39
32.50 - 37.00	33.50 - 38.00	33.50 - 38.00	33.50 - 38.00	- -	3.2.40
37.00 - 41.50	38.00 - 43.00	38.00 - 43.00	38.00 - 43.00	- -	3.2.41
- -	37.00 - 42.00	37.00 - 42.00	37.00 - 42.00	40.00 - 50.00	3.2.42
- -	54.00 - 63.00	54.00 - 63.00	54.00 - 63.00	58.00 - 72.00	3.2.43
20.00 - 22.00	21.00 - 23.00	21.00 - 23.00	21.00 - 23.00	21.50 - 24.00	3.2.44
25.00 - 27.50	26.00 - 28.00	26.00 - 28.00	26.00 - 28.00	27.00 - 30.00	3.2.45
30.00 - 33.00	31.00 - 34.00	31.00 - 34.00	31.00 - 34.00	32.00 - 35.00	3.2.46
20.00 - 33.00	21.00 - 34.00	21.00 - 34.00	21.00 - 34.00	21.50 - 35.00	3.2.47
28.50 - 43.00	30.00 - 44.00	30.00 - 44.00	30.00 - 44.00	31.00 - 45.00	3.2.48
38.50 - 53.00	40.00 - 54.00	40.00 - 54.00	40.00 - 54.00	41.00 - 55.00	3.2.49
- -	30.00 - 43.00	30.00 - 43.00	30.00 - 43.00	30.00 - 45.00	3.2.50
- -	38.00 - 54.00	38.00 - 54.00	38.00 - 54.00	38.00 - 54.00	3.2.51
- -	27.00 - 30.50	27.00 - 30.50	27.00 - 30.50	28.00 - 31.50	3.2.52
- -	23.50 - 27.00	23.50 - 27.00	23.50 - 27.00	24.00 - 28.00	3.2.53
- -	30.00 - 32.50	30.00 - 32.50	30.00 - 32.50	30.50 - 33.50	3.2.54
- -	24.50 - 30.00	24.50 - 30.00	24.50 - 30.00	25.00 - 30.50	3.2.55
- -	30.00 - 31.50	30.00 - 31.50	30.00 - 31.50	30.50 - 32.50	3.2.56
- -	31.50 - 45.00	31.50 - 45.00	31.50 - 45.00	32.50 - 47.00	3.2.57
- -	38.00 - 56.50	38.00 - 56.50	38.00 - 56.50	39.50 - 58.50	3.2.58
- -	43.00 - 67.50	43.00 - 67.50	43.00 - 67.50	43.00 - 72.00	3.2.59
- -	- -	59.50 - 70.00	- -	- -	3.2.60
- -	- -	63.00- 72.00	- -	- -	3.2.61
- -	- -	65.50 - 76.50	- -	- -	3.2.62
- -	- -	40.50 - 54.00	- -	- -	3.2.63
- -	- -	1.00 - 2.00	- -	- -	3.2.64
- -	58.00 - 65.00	58.00 - 65.00	58.00 - 65.00	60.00 - 70.00	3.2.65
- -	76.00 - 85.00	76.00 - 85.00	76.00 - 85.00	78.00 - 90.00	3.2.66
- -	77.00 -100.00	77.00 -100.00	77.00 -100.00	81.00 -108.00	3.2.67
- -	84.00 -110.00	84.00 -110.00	84.00 -110.00	90.00 -122.00	3.2.68

Item nr.	SPECIFICATIONS	Unit	RESIDENTIAL £ range £
3.2	**FLOOR FINISHES - cont'd** **floor finish area** (unless otherwise described)		
	Rigid Tile/slab finishes - cont'd		
	Slate		
3.2.69	200 x 400 x 10 mm blue-grey	m2	- -
3.2.70	Otta riven	m2	- -
3.2.71	Otta honed (polished)	m2	- -
	Portland stone		
3.2.72	50 mm thick paving	m2	- -
	Fine sanded 'Roman Travertine' marble		
3.2.73	20 mm thick paving	m2	- -
3.2.74	paving; including screed	m2	- -
3.2.75	paving; including screed and allowance for skirtings	m2	- -
	Granite		
3.2.76	20 mm thick paving	m2	- -
	Parquet/wood block finishes		
	Parquet flooring; polished		
3.2.77	8 mm Gurjun 'Feltwood'	m2	20.00 - 23.00
	Wrought hardwood block floorings; 25 mm thick; polished; tongued and grooved joints; herringbone pattern		
3.2.78	Merbau	m2	- -
3.2.79	Iroko	m2	- -
3.2.80	Iroko; including screed	m2	- -
3.2.81	Oak	m2	- -
3.2.82	Oak; including screed	m2	- -
	Composition block flooring		
3.2.83	174 x 57 mm blocks	m2	- -
	Flexible tiling/sheet finishes		
	Thermoplastic tile flooring		
3.2.84	2 mm thick (series 2)	m2	6.00 - 7.00
3.2.85	2 mm thick (series 4)	m2	6.00 - 7.00
3.2.86	2 mm thick; including screed	m2	13.50 - 15.50
	Cork tile flooring		
3.2.87	3.2 mm thick	m2	11.00 - 13.00
3.2.88	3.2 mm thick; including screed	m2	19.00 - 25.00
3.2.89	6.3 mm thick	m2	- -
3.2.90	6.3 mm thick; including screed	m2	- -
	Vinyl tile flooring		
3.2.91	2 mm thick; semi-flexible tiles	m2	6.50 - 9.00
3.2.92	2 mm thick; fully flexible tiles	m2	6.00 - 8.00
3.2.93	2.5 mm thick; semi-flexible tiles	m2	7.50 - 10.00
3.2.94	tiling; including screed	m2	16.00 - 20.00
3.2.95	tiling; including screed and allowance for skirtings	m2	18.00 - 23.50
3.2.96	tiling; antistatic	m2	- -
3.2.97	tiling; antistatic; including screed	m2	- -
	Vinyl sheet flooring; heavy duty		
3.2.98	2 mm thick	m2	- -
3.2.99	2.5 mm thick	m2	- -
3.2.100	3 mm thick; needle felt backed	m2	- -
3.2.101	3 mm thick; foam backed	m2	- -
3.2.102	sheeting; including screed and allowance for skirtings	m2	- -
	'Altro' safety flooring		
3.2.103	2 mm thick; Marine T20	m2	- -
3.2.104	2.5 mm thick; Classic D25	m2	- -
3.2.105	3.5 mm thick; stronghold	m2	- -
3.2.106	flooring	m2	- -
3.2.107	flooring; including screed	m2	- -
	Linoleum tile flooring		
3.2.108	3.2 mm thick; coloured	m2	- -
3.2.109	3.2 mm thick; coloured; including screed	m2	- -

INDUSTRIAL £ range £	RETAILING £ range £	LEISURE £ range £	OFFICES £ range £	HOTELS £ range £	Item nr.
- -	95.00 -105.00	95.00 -105.00	95.00 -105.00	- -	3.2.69
- -	105.00 -115.00	105.00 -115.00	105.00 -115.00	- -	3.2.70
- -	115.00 -125.00	115.00 -125.00	115.00 -125.00	- -	3.2.71
- -	160.00 -190.00	160.00 -190.00	160.00 -190.00	- -	3.2.72
- -	200.00 -240.00	200.00 -240.00	200.00 -240.00	210.00 -250.00	3.2.73
- -	210.00 -255.00	210.00 -255.00	210.00 -255.00	220.00 -265.00	3.2.74
- -	260.00 -320.00	260.00 -320.00	260.00 -320.00	265.00 -330.00	3.2.75
- -	275.00 -325.00	275.00 -325.00	275.00 -325.00	280.00 -350.00	3.2.76
- -	- -	- -	- -	- -	3.2.77
- -	42.00 - 48.00	42.00 - 48.00	42.00 - 48.00	43.00 - 49.00	3.2.78
- -	45.00 - 50.00	45.00 - 50.00	45.00 - 50.00	47.50 - 52.50	3.2.79
- -	53.00 - 63.00	53.00 - 63.00	53.00 - 63.00	54.00 - 64.00	3.2.80
- -	44.00 - 55.00	44.00 - 55.00	44.00 - 55.00	45.00 - 57.00	3.2.81
- -	52.00 - 62.00	52.00 - 62.00	52.00 - 62.00	53.00 - 63.00	3.2.82
- -	50.00 - 55.00	50.00 - 55.00	50.00 - 55.00	52.50 - 57.50	3.2.83
- -	- -	- -	- -	- -	3.2.84
- -	- -	- -	- -	- -	3.2.85
- -	- -	- -	- -	- -	3.2.86
- -	- -	- -	- -	- -	3.2.87
- -	- -	- -	- -	- -	3.2.88
- -	19.00 - 21.00	19.00 - 21.00	19.00 - 21.00	20.00 - 22.50	3.2.89
- -	25.00 - 32.50	25.00 - 32.50	25.00 - 32.50	26.00 - 34.00	3.2.90
- -	7.00 - 9.00	7.00 - 9.00	7.00 - 9.00	7.00 - 9.00	3.2.91
- -	6.50 - 8.00	6.50 - 8.00	6.50 - 8.00	6.50 - 8.00	3.2.92
- -	7.50 - 10.00	7.50 - 10.00	7.50 - 10.00	7.50 - 10.00	3.2.93
- -	16.00 - 20.00	16.00 - 20.00	16.00 - 20.00	16.00 - 20.00	3.2.94
- -	18.00 - 23.50	18.00 - 23.50	18.00 - 23.50	18.00 - 23.50	3.2.95
- -	29.00 - 33.50	29.00 - 33.50	29.00 - 33.50	29.00 - 33.50	3.2.96
- -	36.00 - 45.00	36.00 - 45.00	36.00 - 45.00	36.00 - 45.00	3.2.97
- -	11.00 - 12.50	11.00 - 12.50	11.00 - 12.50	11.00 - 12.50	3.2.98
- -	11.50 - 13.50	11.50 - 13.50	11.50 - 13.50	11.50 - 13.50	3.2.99
- -	8.00 - 10.00	8.00 - 10.00	8.00 - 10.00	8.00 - 10.00	3.2.100
- -	11.00 - 13.50	11.00 - 13.50	11.00 - 13.50	11.00 - 13.50	3.2.101
- -	21.50 - 24.50	21.50 - 24.50	21.50 - 24.50	21.50 - 24.50	3.2.102
- -	16.00 - 20.00	16.00 - 20.00	16.00 - 20.00	- -	3.2.103
- -	20.00 - 23.50	20.00 - 23.50	20.00 - 23.50	- -	3.2.104
- -	25.00 - 29.50	25.00 - 29.50	25.00 - 29.50	- -	3.2.105
- -	16.00 - 29.50	16.00 - 29.50	16.00 - 29.50	- -	3.2.106
- -	23.50 - 40.50	23.50 - 40.50	23.50 - 40.50	- -	3.2.107
- -	12.50 - 14.50	12.50 - 14.50	12.50 - 14.50	- -	3.2.108
- -	20.00 - 27.00	20.00 - 27.00	20.00 - 27.00	- -	3.2.109

Item nr.	SPECIFICATIONS	Unit	RESIDENTIAL £ range £	
3.2	**FLOOR FINISHES - cont'd**	**floor finish area** (unless otherwise described)		
	Flexible tiling/sheet finishes - cont'd			
	Linoleum sheet flooring			
3.2.110	3.2 mm thick; coloured	m2	-	-
3.2.111	3.2 mm thick; marbled; including screed	m2	-	-
	Rubber tile flooring; smooth; ribbed or studded tiles			
3.2.112	2.5 mm thick	m2	-	-
3.2.113	5 mm thick	m2	-	-
3.2.114	5 mm thick; including screed	m2	-	-
	Carpet tiles/Carpetting			
3.2.115	Underlay	m2	3.50 -	4.50
	Carpet tiles			
3.2.116	nylon needlepunch (stick down)	m2	9.00 -	11.00
3.2.117	80% animal hair; 20% wool cord	m2	-	-
3.2.118	100% wool	m2	-	-
3.2.119	80% wool; 20% nylon antistatic	m2	-	-
3.2.120	economical; including screed and allowance for skirtings	m2	-	-
3.2.121	good quality	m2	-	-
3.2.122	good quality; including screed	m2	-	-
3.2.123	good quality; including screed and allowance for skirtings	m2	-	-
	Carpet; including underlay			
3.2.124	nylon needlepunch	m2	11.50 -	14.50
3.2.125	100% acrylic; light duty	m2	-	-
3.2.126	80% animal hair; 20% wool cord	m2	-	-
3.2.127	open-weave matting poolside carpet (no underlay)	m2	-	-
3.2.128	80% wool; 20% acrylic; light duty	m2	-	-
3.2.129	100% acrylic; heavy duty	m2	-	-
3.2.130	cord	m2	-	-
3.2.131	100% wool	m2	-	-
3.2.132	good quality; including screed	m2	-	-
3.2.133	good quality (grade 5); including screed and allowance for skirtings	m2	-	-
3.2.134	80% wool; 20% acrylic; heavy duty	m2	-	-
3.2.135	Wilton/pile carpet	m2	-	-
3.2.136	high quality; including screed	m2	-	-
3.2.137	high quality; including screed and allowance for skirtings	m2	-	-
	Access floors and finishes			
	Shallow void block and battened floors			
3.2.138	chipboard on softwood battens; partial access	m2	16.00 -	20.00
3.2.139	chipboard on softwood cradles (for uneven floors)	m2	19.50 -	22.50
3.2.140	chipboard on softwood battens and cross battens; full access	m2	-	-
3.2.141	fibre and particle board; on lightweight concrete pedestal blocks	m2	-	-
	Shallow void block and battened floors; including carpet-tile finish			
3.2.142	fibre and particle board; on lightweight concrete pedestal blocks	m2	-	-
	Access floors; excluding finish			
	600 x 600 mm chipboard panels; faced both sides with galvanised steel sheet; on adjustable steel/aluminium pedestals; cavity height 100 - 300 mm high			
3.2.143	light grade duty	m2	-	-
3.2.144	medium grade duty	m2	-	-
3.2.145	heavy grade duty	m2	-	-
3.2.146	extra heavy grade duty	m2	-	-

INDUSTRIAL £ range £		RETAILING £ range £		LEISURE £ range £		OFFICES £ range £		HOTELS £ range £		Item nr.
-	-	13.00 -	14.50	13.00 -	14.50	13.00 -	14.50	-	-	3.2.110
-	-	21.00 -	27.00	21.00 -	27.00	21.00 -	27.00	-	-	3.2.111
-	-	20.00 -	23.50	20.00 -	23.50	20.00 -	23.50	-	-	3.2.112
-	-	23.50 -	27.00	23.50 -	27.00	23.50 -	27.00	-	-	3.2.113
-	-	32.00 -	40.00	32.00 -	40.00	32.00 -	40.00	-	-	3.2.114
-	-	3.50 -	4.50	3.50 -	4.50	3.50 -	4.50	-	-	3.2.115
-	-	9.00 -	11.00	9.00 -	11.00	9.00 -	11.00	-	-	3.2.116
-	-	16.00 -	18.00	16.00 -	18.00	16.00 -	18.00	-	-	3.2.117
-	-	22.50 -	27.00	22.50 -	27.00	22.50 -	27.00	-	-	3.2.118
-	-	23.50 -	32.50	23.50 -	32.50	23.50 -	32.50	-	-	3.2.119
-	-	25.00 -	27.00	25.00 -	27.00	25.00 -	27.00	-	-	3.2.120
-	-	22.50 -	32.50	22.50 -	32.50	22.50 -	32.50	22.50 -	36.00	3.2.121
-	-	30.00 -	40.50	30.00 -	40.50	30.00 -	40.50	30.00 -	45.00	3.2.122
-	-	33.00 -	45.00	33.00 -	45.00	33.00 -	45.00	34.00 -	49.50	3.2.123
-	-	11.50 -	14.50	11.50 -	14.50	11.50 -	14.50	-	-	3.2.124
-	-	15.50 -	18.00	15.50 -	18.00	15.50 -	18.00	-	-	3.2.125
-	-	19.00 -	23.50	19.00 -	23.50	19.00 -	23.50	-	-	3.2.126
-	-	-	-	16.00 -	20.00	-	-	-	-	3.2.127
-	-	23.50 -	30.50	23.50 -	30.50	23.50 -	30.50	-	-	3.2.128
-	-	24.00 -	29.00	24.00 -	29.00	24.00 -	29.00	-	-	3.2.129
-	-	-	-	29.00 -	32.50	-	-	-	-	3.2.130
-	-	29.00 -	38.00	29.00 -	38.00	29.00 -	38.00	29.00 -	40.50	3.2.131
-	-	36.00 -	49.50	36.00 -	49.50	36.00 -	49.50	38.00 -	54.00	3.2.132
-	-	38.00 -	52.00	38.00 -	52.00	38.00 -	52.00	40.50 -	56.00	3.2.133
-	-	38.50 -	45.00	38.50 -	45.00	38.50 -	45.00	40.50 -	47.50	3.2.134
-	-	42.00 -	46.00	42.00 -	46.00	42.00 -	46.00	43.00 -	47.50	3.2.135
-	-	50.00 -	57.50	50.00 -	57.50	50.00 -	57.50	52.50 -	60.00	3.2.136
-	-	52.50 -	60.00	52.50 -	60.00	52.50 -	60.00	55.00 -	65.00	3.2.137
-	-	-	-	-	-	16.00 -	20.00	-	-	3.2.138
-	-	-	-	-	-	19.50 -	22.50	-	-	3.2.139
-	-	-	-	-	-	23.50 -	26.00	-	-	3.2.140
-	-	-	-	-	-	22.50 -	27.00	-	-	3.2.141
-	-	-	-	-	-	34.00 -	38.00	-	-	3.2.142
-	-	-	-	-	-	30.00 -	35.00	-	-	3.2.143
-	-	-	-	-	-	35.00 -	40.00	-	-	3.2.144
-	-	-	-	-	-	43.50 -	54.00	-	-	3.2.145
-	-	-	-	-	-	48.50 -	54.00	-	-	3.2.146

Item nr.	SPECIFICATIONS	Unit	RESIDENTIAL £ range £
3.2	**FLOOR FINISHES - cont'd**	**floor finish area** (unless otherwise described)	
	Access floors and finishes - cont'd		
	600 x 600 mm chipboard panels; faced both sides with galvanised steel sheet; on adjustable steel/ aluminium pedestals; cavity height 300 - 600 mm high		
3.2.147	medium grade duty	m2	- -
3.2.148	heavy grade duty	m2	- -
3.2.149	extra heavy grade duty	m2	- -
	Access floor with medium quality carpetting		
3.2.150	'Durabella' suspended floors	m2	- -
3.2.151	'Buroplan' partial access raised floor	m2	- -
3.2.152	'Pedestal' partial access raised floor	m2	- -
3.2.153	Modular floor; 100% access raised floor	m2	-
	Access floor with high quality carpetting		
3.2.154	computer loading; 100% access raised floor	m2	- -
	Common floor coverings bonded to access floor panels		
3.2.155	heavy-duty fully flexible vinyl to BS 3261, type A	m2	- -
3.2.156	fibre-bonded carpet	m2	- -
3.2.157	high-pressure laminate to BS 3794 class D	m2	- -
3.2.158	anti-static grade fibre-bonded carpet	m2	- -
3.2.159	anti-static grade sheet PVC to BS 3261	m2	- -
3.2.160	low loop tufted carpet	m2	- -
	Perimeter treatments and sundries		
	Comparative skirtings		
3.2.161	25 x 75 mm softwood skirting; painted; including grounds	m	7.00 - 8.00
3.2.162	25 x 100 mm Mahogany skirting; polished; including grounds	m	10.00 - 11.00
3.2.163	12.5 x 150 mm Quarry tile skirting; including backing	m	9.50 - 11.50
3.2.164	13 x 75 mm granolithic skirting; including backing	m	13.50 - 16.00
3.2.165	6 x 75 mm terrazzo; including backing	m	23.50 - 27.00
	Entrance matting in aluminium-framed		
3.2.166	matwell	m2	- -
3.3	**CEILING FINISHES**	**ceiling finish area** (unless otherwise described)	
	comprising:-		
	In situ/board finishes		
	Suspended and integrated ceilings		
	In situ/board finishes		
	Decoration only to soffits		
3.3.1	to exposed steelwork	m2	- -
3.3.2	to concrete soffits	m2	2.00 - 3.00
3.3.3	one mist and two coats emulsion paint; to plaster/ plasterboard	m2	2.00 - 3.00
	Plaster to soffits		
3.3.4	lightweight plaster	m2	7.50 - 9.50
3.3.5	plaster and emulsion	m2	9.50 - 14.00
	Extra for		
3.3.6	gloss paint in lieu of emulsion	m2	1.50 - 1.80
	Plasterboard to soffits		
3.3.7	9.5 mm Gyproc lath and skim coat	m2	10.00 - 12.00
3.3.8	9.5 mm Gyproc insulating lath and skim coat	m2	10.50 - 12.00
3.3.9	plasterboard, skim and emulsion	m2	11.50 - 14.50

INDUSTRIAL £ range £	RETAILING £ range £	LEISURE £ range £	OFFICES £ range £	HOTELS £ range £	Item nr.
- -	- -	- -	40.50 - 45.00	- -	3.2.147
- -	- -	- -	45.00 - 54.00	- -	3.2.148
- -	- -	- -	49.50 - 54.00	- -	3.2.149
- -	- -	- -	43.00 - 47.00	- -	3.2.150
- -	- -	- -	48.50 - 54.00	- -	3.2.151
- -	- -	- -	54.00 - 67.50	- -	3.2.152
- -	- -	- -	72.00 - 90.00	- -	3.2.153
- -	- -	- -	80.00 -115.00	- -	3.2.154
- -	- -	- -	5.50 - 15.50	- -	3.2.155
- -	- -	- -	6.00 - 11.00	- -	3.2.156
- -	- -	- -	6.00 - 17.00	- -	3.2.157
- -	- -	- -	7.50 - 12.00	- -	3.2.158
- -	- -	- -	10.00 - 15.50	- -	3.2.159
- -	- -	- -	12.00 - 17.00	- -	3.2.160
6.50 - 7.50	7.00 - 8.00	7.00 - 8.00	7.00 - 8.00	8.00 - 9.00	3.2.161
9.50 - 10.50	10.00 - 11.00	10.00 - 11.00	10.00 - 11.00	10.50 - 11.50	3.2.162
9.00 - 11.00	9.50 - 11.50	9.50 - 11.50	9.50 - 11.50	10.00 - 12.00	3.2.163
13.00 - 16.00	13.50 - 16.00	13.50 - 16.00	13.50 - 16.00	14.00 - 17.00	3.2.164
- -	23.50 - 27.00	23.50 - 27.00	23.50 - 27.00	24.50 - 29.00	3.2.165
- -	210.00 -290.00	200.00 -265.00	205.00 -270.00	205.00 -270.00	3.2.166
2.50 - 3.50	2.00 - 3.00	- -	- -	- -	3.3.1
2.00 - 3.00	2.00 - 3.00	2.00 - 3.00	2.00 - 3.00	2.00 - 3.00	3.3.2
2.00 - 3.00	2.00 - 3.00	2.00 - 3.00	2.00 - 3.00	2.00 - 3.00	3.3.3
7.00 - 9.00	7.00 - 9.00	7.00 - 9.00	7.00 - 9.00	8.00 - 10.50	3.3.4
9.00 - 13.50	9.50 - 14.00	9.50 - 14.00	9.50 - 14.00	10.50 - 16.00	3.3.5
1.50 - 1.80	1.50 - 1.80	1.50 - 1.80	1.50 - 1.80	1.50 - 1.80	3.3.6
9.50 - 11.50	10.00 - 12.00	10.00 - 12.00	10.00 - 12.00	10.50 - 12.00	3.3.7
10.00 - 11.50	10.50 - 12.00	10.50 - 12.00	10.50 - 12.00	11.00 - 12.50	3.3.8
11.00 - 14.00	11.50 - 14.50	11.50 - 14.50	11.50 - 14.50	12.00 - 15.00	3.3.9

Item nr.	SPECIFICATIONS	Unit	RESIDENTIAL £ range £
3.3	**CEILING FINISHES - cont'd** **ceiling finish area** (unless otherwise described)		
	In situ/board finishes - cont'd		
	Plasterboard to soffits - cont'd		
	Extra for		
3.3.10	gloss paint in lieu of emulsion	m2	1.50 - 1.80
3.3.11	plasterboard and Artex	m2	8.00 - 10.00
3.3.12	plasterboard, Artex and emulsion	m2	10.00 - 12.50
3.3.13	plaster and emulsion; including metal lathing	m2	16.00 - 22.50
	Other board finishes; with fire-resisting properties; excluding decoration		
3.3.14	12.5 mm Gyproc Fireline; half hour	m2	- -
3.3.15	6 mm Supalux; half hour	m2	- -
3.3.16	two layers of 12.5 mm Gyproc Wallboard; half hour	m2	- -
3.3.17	two layers of 12.5 mm Gyproc Fireline; one hour	m2	- -
3.3.18	9 mm Supalux; one hour; on fillets	m2	- -
	Specialist plasters; to soffits		
3.3.19	sprayed accoustic plaster; self-finished	m2	- -
3.3.20	rendering; 'Tyrolean' finish	m2	- -
	Other ceiling finishes		
3.3.21	50 mm wood wool slabs as permanent lining	m2	- -
3.3.22	12 mm Pine tongued and grooved boarding	m2	12.50 - 15.00
3.3.23	16 mm Softwood tongued and grooved boardings	m2	15.50 - 18.00
	Suspended and integrated ceilings		
	Suspended ceiling		
3.3.24	economical; exposed grid	m2	- -
3.3.25	jointless; plasterboard	m2	- -
3.3.26	semi-concealed grid	m2	- -
3.3.27	medium quality; 'Minatone'; concealed grid	m2	- -
3.3.28	high quality; 'Travertone'; concealed grid	m2	- -
	Other suspended ceilings		
3.3.29	metal linear strip; 'Dampa'/'Luxalon'	m2	- -
3.3.30	metal tray	m2	- -
3.3.31	egg-crate	m2	- -
3.3.32	open grid; 'Formalux'/'Dimension'	m2	- -
	Integrated ceilings		
3.3.33	coffered; with steel services	m2	- -
3.4	**DECORATIONS** **surface area** (unless otherwise described)		
	comprising:-		
	Comparative wall and ceiling finishes		
	Comparative steel/metalwork finishes		
	Comparative woodwork finishes		
	Comparative wall and ceiling finishes		
	Emulsion		
3.4.1	two coats	m2	1.50 - 2.00
3.4.2	one mist and two coats	m2	1.75 - 2.75
	Artex plastic compound		
3.4.3	one coat; textured	m2	2.50 - 3.50
3.4.4	Wall paper	m2	3.50 - 5.50
3.4.5	Hessian wall coverings	m2	- -
	Gloss		
3.4.6	primer and two coats	m2	3.00 - 4.00
3.4.7	primer and three coats	m2	4.00 - 5.00

INDUSTRIAL £ range £	RETAILING £ range £	LEISURE £ range £	OFFICES £ range £	HOTELS £ range £	Item nr.
1.50 - 1.75	1.50 - 1.75	1.50 - 1.75	1.50 - 1.75	1.50 - 1.75	3.3.10
7.50 - 9.50	8.00 - 10.00	8.00 - 10.00	8.00 - 10.00	8.50 - 10.50	3.3.11
9.50 - 12.00	10.00 - 12.50	10.00 - 12.50	10.00 - 12.50	10.50 - 13.00	3.3.12
15.50 - 21.50	16.00 - 22.50	16.00 - 22.50	16.00 - 22.50	16.50 - 23.50	3.3.13
8.00 - 10.00	8.25 - 10.25	8.25 - 10.25	8.25 - 10.25	9.00 - 11.00	3.3.14
10.00 - 11.00	10.50 - 11.50	10.50 - 11.50	10.50 - 11.50	11.00 - 11.50	3.3.15
11.00 - 13.00	11.50 - 13.50	11.50 - 13.50	11.50 - 13.50	12.00 - 14.00	3.3.16
13.00 - 15.50	13.50 - 16.00	13.50 - 16.00	13.50 - 16.00	14.00 - 16.00	3.3.17
15.00 - 17.50	15.00 - 18.00	15.00 - 18.00	15.00 - 18.00	16.00 - 19.00	3.3.18
20.00 - 27.00	- -	- -	- -	- -	3.3.19
- -	- -	21.00 - 30.00	- -	- -	3.3.20
- -	- -	- -	9.50 - 11.50	- -	3.3.21
- -	12.50 - 15.00	12.50 - 15.00	12.50 - 15.00	13.00 - 16.00	3.3.22
- -	16.00 - 18.00	16.00 - 18.00	16.00 - 18.00	16.00 - 18.00	3.3.23
16.00 - 21.00	18.00 - 22.50	18.00 - 22.50	18.00 - 22.50	18.00 - 25.00	3.3.24
20.00 - 25.00	21.50 - 27.00	21.50 - 27.00	21.50 - 27.00	21.50 - 30.00	3.3.25
22.00 - 27.00	23.50 - 29.50	23.50 - 29.50	23.50 - 29.50	23.50 - 31.50	3.3.26
23.50 - 31.50	25.00 - 34.00	25.00 - 34.00	25.00 - 34.00	25.00 - 36.00	3.3.27
27.00 - 34.50	29.50 - 36.00	29.50 - 36.00	29.50 - 36.00	29.50 - 39.50	3.3.28
30.00 - 37.50	32.00 - 40.00	32.00 - 40.00	32.00 - 40.00	- -	3.3.29
31.00 - 40.00	33.00 - 40.00	33.00 - 40.00	33.00 - 40.00	- -	3.3.30
34.00 - 70.00	36.00 - 72.00	36.00 - 72.00	36.00 - 72.00	- -	3.3.31
61.00 - 75.00	63.00 - 77.50	63.00 - 77.50	63.00 - 77.50	- -	3.3.32
70.00 -115.00	70.00 -115.00	70.00 -115.00	70.00 -115.00	- -	3.3.33
1.50 - 2.00	1.50 - 2.00	1.50 - 2.00	1.50 - 2.00	1.50 - 2.00	3.4.1
1.75 - 2.75	1.75 - 2.75	1.75 - 2.75	1.75 - 2.75	1.75 - 2.75	3.4.2
2.60 - 3.50	2.60 - 3.50	2.60 - 3.50	2.60 - 3.50	2.60 - 3.50	3.4.3
3.50 - 5.50	3.50 - 5.50	3.50 - 5.50	3.50 - 5.50	3.50 - 9.00	3.4.4
- -	8.00 - 11.00	8.00 - 11.00	8.00 - 11.00	9.00 - 13.50	3.4.5
3.00 - 4.00	3.00 - 4.00	3.00 - 4.00	3.00 - 4.00	3.00 - 4.00	3.4.6
4.00 - 5.00	4.00 - 5.00	4.00 - 5.00	4.00 - 5.00	4.00 - 5.00	3.4.7

Item nr.	SPECIFICATIONS	Unit	RESIDENTIAL £ range £
3.4	**DECORATIONS - cont'd**	**surface area** (unless otherwise described)	
	Comparative steel/metalwork finishes		
	Primer		
3.4.8	only	m2	- -
3.4.9	grit blast and one coat zinc chromate primer	m2	- -
3.4.10	touch up primer and one coat of two pack epoxy zinc phosphate primer	m2	- -
	Gloss		
3.4.11	three coats	m2	4.00 - 5.00
	Sprayed mineral fibre		
3.4.12	one hour	m2	- -
3.4.13	two hour	m2	- -
	Sprayed vermiculate cement		
3.4.14	one hour	m2	- -
3.4.15	two hour	m2	- -
	Intumescent coating with decorative top seal		
3.4.16	half hour	m2	- -
3.4.17	one hour	m2	- -
	Comparative woodwork finishes		
	Primer		
3.4.18	only	m2	1.10 - 1.25
	Gloss		
3.4.19	two coats; touch up primer	m2	2.25 - 2.50
3.4.20	three coats; touch up primer	m2	3.00 - 3.75
3.4.21	primer and two coat	m2	3.50 - 4.00
3.4.22	primer and three coat	m2	4.50 - 5.00
	Polyurethene laquer		
3.4.23	two coats	m2	2.25 - 2.75
3.4.24	three coats	m2	3.50 - 4.00
	Flame-retardent paint		
3.4.25	three coats	m2	5.20 - 6.20
	Wax polish		
3.4.26	seal	m2	5.50 - 7.50
3.4.27	stain and body-in	m2	9.50 - 10.50
	French polish		
3.4.28	stain and body-in	m2	14.50 - 16.50
4.1	**FITTINGS AND FURNISHINGS**	**gross internal area or individual units**	
	comprising:-		
	Residential fittings		
	Comparative fittings/sundries		
	Industrial/Office furniture, fittings and equipment		
	Retail furniture, fittings and equipment		
	Leisure furniture, fittings and equipment		
	Hotel furniture, fittings and equipment		
	Residential fittings		
	Kitchen fittings for residential units		
4.1.1	one person flat/bed-sit	nr	400 -800
4.1.2	two person flat/house	nr	550 -1350
4.1.3	three person flat/house	nr	650 -1900
4.1.4	four person house	nr	700 -2400
4.1.5	five person house	nr	900 -4500

INDUSTRIAL £ range £	RETAILING £ range £	LEISURE £ range £	OFFICES £ range £	HOTELS £ range £	Item nr.
0.65 - 1.10	- -	- -	- -	- -	3.4.8
1.35 - 2.00	- -	- -	- -	- -	3.4.9
1.75 - 2.25	- -	- -	- -	- -	3.4.10
- -	- -	- -	- -	- -	3.4.11
8.00 - 11.50	- -	- -	- -	- -	3.4.12
13.00 - 15.00	- -	- -	- -	- -	3.4.13
9.00 - 13.00	- -	- -	- -	- -	3.4.14
11.00 - 15.00	- -	- -	- -	- -	3.4.15
14.00 - 15.00	14.00 - 15.00	14.00 - 15.00	14.00 - 15.00	- -	3.4.16
22.50 - 27.00	22.50 - 27.00	22.50 - 27.00	22.50 - 27.00	- -	3.4.17
1.10 - 1.25	1.10 - 1.25	1.10 - 1.25	1.10 - 1.25	1.10 - 1.25	3.4.18
2.25 - 2.50	2.25 - 2.50	2.25 - 2.50	2.25 - 2.50	2.25 - 2.50	3.4.19
3.00 - 3.75	3.00 - 3.75	3.00 - 3.75	3.00 - 3.75	3.00 - 3.75	3.4.20
3.50 - 4.00	3.50 - 4.00	3.50 - 4.00	3.50 - 4.00	3.50 - 4.00	3.4.21
4.50 - 5.00	4.50 - 5.00	4.50 - 5.00	4.50 - 5.00	4.50 - 5.00	3.4.22
2.25 - 2.75	2.25 - 2.75	2.25 - 2.75	2.25 - 2.75	2.25 - 2.75	3.4.23
3.50 - 4.00	3.50 - 4.00	3.50 - 4.00	3.50 - 4.00	3.50 - 4.00	3.4.24
5.20 - 6.20	5.20 - 6.20	5.20 - 6.20	5.20 - 6.20	5.20 - 6.20	3.4.25
5.50 - 7.50	5.50 - 7.50	5.50 - 7.50	5.50 - 7.50	5.50 - 7.50	3.4.26
9.50 - 10.50	9.50 - 10.50	9.50 - 10.50	9.50 - 10.50	9.50 - 10.50	3.4.27
14.50 - 16.50	14.50 - 16.50	14.50 - 16.50	14.50 - 16.50	14.50 - 16.50	3.4.28
- -	- -	- -	- -	- -	4.1.1
- -	- -	- -	- -	- -	4.1.2
- -	- -	- -	- -	- -	4.1.3
- -	- -	- -	- -	- -	4.1.4
- -	- -	- -	- -	- -	4.1.5

Item nr.	SPECIFICATIONS	Unit	RESIDENTIAL £ range £
4.1	**FITTINGS AND FURNISHINGS - cont'd** **gross internal area or individual areas** (unless otherwise described)		
	Comparative fittings/sundries		
	Individual kitchen fittings		
4.1.6	600 x 600 x 300 mm wall unit	nr	50.00 - 60.00
4.1.7	1200 x 900 x 300 mm wall unit	nr	80.00 - 90.00
4.1.8	500 x 900 x 600 mm floor unit	nr	80.00 - 90.00
4.1.9	600 x 500 x 195 mm store cupboard	nr	120.00 -130.00
4.1.10	1200 x 900 x 600 mm sink unit (excluding top)	nr	120.00 -130.00
	Comparative wrot softwood shelving		
4.1.11	25 x 225 mm; including black japanned brackets	m	8.00 - 9.00
4.1.12	25 mm thick slatted shelving; including bearers	m2	32.00 - 35.00
4.1.13	25 mm thick cross-tongued shelving; including bearers	m2	40.00 - 45.00
	Industrial/Office furniture, fittings and equipment		
	Reception desk, shelves and cupboards for general areas		
4.1.14	economical	m2	- -
4.1.15	medium quality	m2	- -
4.1.16	high quality	m2	- -
	Extra for		
4.1.17	high quality finishes to reception areas	m2	- -
4.1.18	full kitchen equipment (one cover/20m2)	m2	- -
	Furniture and fittings to general office areas		
4.1.19	economical	m2	- -
4.1.20	medium quality	m2	- -
4.1.21	high quality	m2	- -
	Retail fitting out, furniture, fittings and equipment		
	Mall furniture etc.		
4.1.22	minimal provision	m2	- -
4.1.23	good provision	m2	- -
4.1.24	internal planting	m2	- -
4.1.25	glazed metal balustrades to voids	m	- -
4.1.26	feature pond and fountain	nr	- -
4.1.27	Fitting out a retail warehouse	m2	- -
	Fitting out shell for small shop (including shop fittings)		
4.1.28	simple store	m2	- -
4.1.29	fashion store	m2	- -
	Fitting out shell for department store or supermarket		
4.1.30	excluding shop fittings	m2	- -
4.1.31	including shop fittings	m2	- -
	Special fittings/equipment		
4.1.32	refrigerated installation for cold stores; display fittings in food stores	m2	- -
4.1.33	food court furniture; fittings and special finishes (excluding catering display units)	m2	- -
4.1.34	bakery ovens	nr	- -
4.1.35	refuse compactors	nr	- -
	Leisure furniture, fittings and equipment		
	General fittings		
4.1.36	internal planting	m2	- -
4.1.37	signs, notice-boards, shelving, fixed seating, curtains and blinds	m2	- -
4.1.38	electric hand-dryers, incinerators, mirrors	m2	- -
	Specific fittings		
4.1.39	lockers, coin return locks	nr	- -
4.1.40	kitchen units; excluding equipment	nr	- -
4.1.41	folding sun bed	nr	- -
4.1.42	security grille	nr	- -
4.1.43	entrance balustrading and control turnstile	nr	- -

INDUSTRIAL £ range £	RETAILING £ range £	LEISURE £ range £	OFFICES £ range £	HOTELS £ range £	Item nr.
- -	- -	- -	- -	- -	4.1.6
- -	- -	- -	- -	- -	4.1.7
- -	- -	- -	- -	- -	4.1.8
- -	- -	- -	- -	- -	4.1.9
- -	- -	- -	- -	- -	4.1.10
8.00 - 9.00	8.00 - 9.00	8.00 - 9.00	8.00 - 9.00	8.00 - 9.00	4.1.11
30.00 - 35.00	30.00 - 35.00	30.00 - 35.00	30.00 - 35.00	30.00 - 35.00	4.1.12
40.00 - 45.00	40.00 - 45.00	40.00 - 45.00	40.00 - 45.00	40.00 - 45.00	4.1.13
2.50 - 5.50	- -	- -	4.30 - 8.60	- -	4.1.14
4.50 - 9.00	- -	- -	6.50 - 11.50	- -	4.1.15
7.00 - 13.50	- -	- -	10.00 - 16.00	- -	4.1.16
- -	- -	- -	4.50 - 6.50	- -	4.1.17
6.50 - 8.00	- -	- -	7.00 - 9.00	- -	4.1.18
- -	- -	- -	5.50 - 7.00	- -	4.1.19
- -	- -	- -	7.00 - 11.00	- -	4.1.20
- -	- -	- -	12.50 - 18.00	- -	4.1.21
- -	6.50 - 12.50	- -	- -	- -	4.1.22
- -	27.00 - 32.50	- -	- -	- -	4.1.23
- -	14.50 - 18.00	- -	- -	- -	4.1.24
- -	475.00 -605.00	- -	- -	- -	4.1.25
- -	27000 -40500	- -	- -	- -	4.1.26
- -	170.00 -195.00	- -	- -	- -	4.1.27
- -	425.00 -475.00	- -	- -	- -	4.1.28
- -	630.00 -900.00	- -	- -	- -	4.1.29
- -	450.00 -600.00	- -	- -	- -	4.1.30
- -	630.00 -900.00	- -	- -	- -	4.1.31
- -	18.00 - 54.00	- -	- -	- -	4.1.32
- -	565.00 -610.00	- -	- -	- -	4.1.33
- -	7200 -10800	- -	- -	- -	4.1.34
- -	7700 -14000	- -	- -	- -	4.1.35
- -	- -	11.50 - 13.50	- -	- -	4.1.36
- -	- -	7.75 - 9.00	- -	- -	4.1.37
- -	- -	1.95 - 3.00	- -	- -	4.1.38
- -	- -	90.00 -125.00	- -	- -	4.1.39
- -	- -	1475 -2000	- -	- -	4.1.40
- -	- -	3475 -3950	- -	- -	4.1.41
- -	- -	4950 -7000	- -	- -	4.1.42
- -	- -	9000 -9900	- -	- -	4.1.43

Item nr.	SPECIFICATIONS	Unit	RESIDENTIAL £ range £	
4.1	**FITTINGS AND FURNISHINGS - cont'd** **gross internal area or individual units** (unless otherwise described)			
	Leisure furniture, fittings and equipment - cont'd			
4.1.44	sports nets, screens etc. in a medium sized sports hall	nr	-	-
	reception counter, fittings and reception counter			
4.1.45	screen	nr	-	-
4.1.46	bar and fittings	nr	-	-
4.1.47	telescopic seatings	nr	-	-
	Swimming pool fittings			
4.1.48	metal balustrades to pool areas	m	-	-
4.1.49	skimmer grilles to pool edge	nr	-	-
	stainless steel pool access ladder			
4.1.50	1700 mm high	nr	-	-
4.1.51	2500 mm high	nr	-	-
	Leisure pool fittings			
4.1.52	stainless steel lighting post	nr	-	-
4.1.53	water cannons	nr	-	-
4.1.54	fountain or water sculpture	nr	-	-
4.1.55	loudspeaker tower	nr	-	-
	grp water chute; 65 - 80 m long; steel supports,			
4.1.56	spiral stairs and balustrading	nr	-	-
	Hotel furniture, fittings and equipment			
	General fittings			
4.1.57	signs	m2	-	-
4.1.58	housekeeper and bolt	m2	-	-
	Fittings for specific areas			
4.1.59	bedrooms (including bathrooms)	bedroom	-	-
4.1.60	bedroom (operating supplies)	bedroom	-	-
4.1.61	bedroom corridors	m2	-	-
4.1.62	public areas	m2	-	-
4.1.63	restaurant (operating supplies)	m2	-	-
4.1.64	kitchen (including special equipment)	m2	-	-
4.1.65	laundry (including special equipment)	nr	-	-
5.1	**SANITARY AND DISPOSAL INSTALLATIONS** **gross internal area** (unless otherwise described)			
	comprising:-			
	Comparative sanitary fittings/sundries			
	Sanitary and disposal installations			
	Comparative sanitary fittings/sundries			
	Individual sanitary appliances (including fittings)			
	lavatory basins			
5.1.1	white	nr	135.00	-150.00
5.1.2	coloured	nr	160.00	-180.00
	low level WC's; on ground floor			
5.1.3	white	nr	120.00	-135.00
5.1.4	coloured	nr	145.00	-160.00
	low level WC's; one of a range; on upper floors			
5.1.5	white	nr	225.00	-250.00
5.1.6	coloured	nr	250.00	-275.00
	Extra for			
5.1.7	single storey PVC stack and connection to drain	nr	110.00	-135.00
5.1.8	bowl-type wall urinal; fireclay; white	nr	-	-
	shower tray; including fittings			
5.1.9	white	nr	315.00	-360.00
5.1.10	coloured	nr	350.00	-385.00

INDUSTRIAL £ range £		RETAILING £ range £		LEISURE £ range £		OFFICES £ range £		HOTELS £ range £		Item nr.
-	-	-	-	12500	-14500	-	-	-	-	4.1.44
-	-	-	-	14750	-24750	-	-	-	-	4.1.45
-	-	-	-	24750	-29750	-	-	-	-	4.1.46
-	-	-	-	34750	-39500	-	-	-	-	4.1.47
-	-	-	-	250.00	-400.00	-	-	-	-	4.1.48
-	-	-	-	200.00	-250.00	-	-	-	-	4.1.49
-	-	-	-	725.00	-850.00	-	-	-	-	4.1.50
-	-	-	-	1000	-1125	-	-	-	-	4.1.51
-	-	-	-	100.00	-125.00	-	-	-	-	4.1.52
-	-	-	-	675.00	-900.00	-	-	-	-	4.1.53
-	-	-	-	2750	-13500	-	-	-	-	4.1.54
-	-	-	-	3600	-4100	-	-	-	-	4.1.55
-	-	-	-	90000	-112500	-	-	-	-	4.1.56
-	-	-	-	-	-	-	-	1.60	- 2.70	4.1.57
-	-	-	-	-	-	-	-	5.50	- 9.00	4.1.58
-	-	-	-	-	-	-	-	3600	-7200	4.1.59
-	-	-	-	-	-	-	-	180.00	-270.00	4.1.60
-	-	-	-	-	-	-	-	54.00	-110.00	4.1.61
-	-	-	-	-	-	-	-	450.00	-900.00	4.1.62
-	-	-	-	-	-	-	-	63.00	- 80.00	4.1.63
-	-	-	-	-	-	-	-	650.00	-750.00	4.1.64
-	-	-	-	-	-	-	-	78750	-105750	4.1.65
120.00	-145.00	135.00	-180.00	135.00	-180.00	135.00	-180.00	135.00	-225.00	5.1.1
-	-	-	-	160.00	-220.00	160.00	-220.00	160.00	-220.00	5.1.2
110.00	-135.00	120.00	-135.00	120.00	-135.00	120.00	-135.00	120.00	-160.00	5.1.3
-	-	-	-	145.00	-180.00	145.00	-180.00	145.00	-205.00	5.1.4
220.00	-250.00	225.00	-250.00	225.00	-250.00	225.00	-250.00	225.00	-270.00	5.1.5
-	-	-	-	250.00	-280.00	250.00	-295.00	250.00	-315.00	5.1.6
100.00	-120.00	110.00	-135.00	110.00	-135.00	110.00	-135.00	110.00	-135.00	5.1.7
150.00	-180.00	170.00	-200.00	170.00	-200.00	170.00	-200.00	-	-	5.1.8
295.00	-330.00	320.00	-360.00	320.00	-360.00	320.00	-360.00	320.00	-450.00	5.1.9
-	-	-	-	320.00	-390.00	320.00	-400.00	320.00	-590.00	5.1.10

Item nr.	SPECIFICATIONS	Unit	RESIDENTIAL £ range £	
5.1	**SANITARY AND DISPOSAL INSTALLATIONS - cont'd**	**gross internal area** (unless otherwise described)		
	Comparative sanitary fittings/sundries - cont'd			
5.1.11	sink; white; fireclay	nr	-	-
	sink; stainless steel			
5.1.12	single drainer	nr	150.00	-180.00
5.1.13	double drainer	nr	180.00	-210.00
	bath			
5.1.14	acrylic; white	nr	225.00	-270.00
5.1.15	acrylic; coloured	nr	225.00	-270.00
5.1.16	enamelled steel; white	nr	250.00	-295.00
5.1.17	enamelled steel; coloured	nr	270.00	-315.00
	Soil waste stacks; 3.15 m storey height; branch and connection to drain			
5.1.18	100 mm PVC	nr	200.00	-225.00
	Extra for			
5.1.19	additional floors	nr	100.00	-115.00
5.1.20	100 mm cast iron; decorated	nr	-	-
	Extra for			
5.1.21	additional floors	nr	-	-
	Sanitary and disposal installations			
	Residential units			
5.1.22	range including WC, wash handbasin, bath and kitchen sink	nr	1000	-1700
5.1.23	range including WC, wash handbasin, bidet, bath and kitchen sink	nr	1250	-2000
5.1.24	range including two WC's, two wash handbasins, bath and kitchen sink	nr	1450	-2450
5.1.25	range including two WC's, two wash handbasins, bidet bath and kithen sink	nr	1700	-2600
	Extra for			
5.1.26	rainwater pipe per storey	nr	45.00	- 55.00
5.1.27	soil pipe per storey	nr	100.00	-110.00
5.1.28	shower over bath	nr	250.00	-350.00
	Industrial buildings			
	warehouse			
5.1.29	minimum provision	m2	-	-
5.1.30	high provision	m2	-	-
	production unit			
5.1.31	minimum provision	m2	-	-
5.1.32	minimum provision; area less than 1000 m2	m2	-	-
5.1.33	high provision	m2	-	-
	Retailing outlets			
5.1.34	to superstore	m2	-	-
5.1.35	to shopping centre malls; public conveniences; branch connections to shop shells	m2	-	-
5.1.36	fitting out public conveniences in shopping mall	block	-	-
5.1.37	Leisure buildings	m2	-	-
	Offices and industrial office buildings			
5.1.38	speculative; low rise; area less than 1000 m2	m2	-	-
5.1.39	speculative; low rise	m2	-	-
5.1.40	speculative; medium rise; area less than 1000 m2	m2	-	-
5.1.41	speculative; medium rise	m2	-	-
5.1.42	speculative; high rise	m2	-	-
5.1.43	owner-occupied; low rise; area less than 1000 m2	m2	-	-
5.1.44	owner-occupied; low rise	m2	-	-
5.1.45	owner-occupied; medium rise; area less than 1000 m2	m2	-	-
5.1.46	owner-occupied; medium rise	m2	-	-
5.1.47	owner-occupied; high rise	m2	-	-
	Hotels			
5.1.48	WC, bath, shower, basin to each bedroom; sanitary accommodation to public areas	m2	-	-

INDUSTRIAL £ range £	RETAILING £ range £	LEISURE £ range £	OFFICES £ range £	HOTELS £ range £	Item nr.
145.00 -180.00	145.00 -250.00	- -	145.00 -250.00	- -	5.1.11
- -	- -	- -	- -	- -	5.1.12
- -	- -	- -	- -	- -	5.1.13
- -	- -	- -	- -	225.00 -315.00	5.1.14
- -	- -	- -	- -	225.00 -315.00	5.1.15
- -	- -	- -	- -	250.00 -340.00	5.1.16
- -	- -	- -	- -	270.00 -360.00	5.1.17
200.00 -225.00	200.00 -225.00	200.00 -225.00	200.00 -225.00	200.00 -225.00	5.1.18
100.00 -115.00	100.00 -115.00	100.00 -115.00	100.00 -115.00	100.00 -115.00	5.1.19
400.00 -425.00	400.00 -425.00	400.00 -425.00	400.00 -425.00	400.00 -425.00	5.1.20
200.00 -225.00	200.00 -225.00	200.00 -225.00	200.00 -225.00	200.00 -225.00	5.1.21
- -	- -	- -	- -	- -	5.1.22
- -	- -	- -	- -	- -	5.1.23
- -	- -	- -	- -	- -	5.1.24
- -	- -	- -	- -	- -	5.1.25
- -	- -	- -	- -	- -	5.1.26
- -	- -	- -	- -	- -	5.1.27
- -	- -	- -	- -	- -	5.1.28
6.50 - 9.00	- -	- -	- -	- -	5.1.29
9.00 - 13.50	- -	- -	- -	- -	5.1.30
9.00 - 13.50	- -	- -	- -	- -	5.1.31
11.00 - 18.00	- -	- -	- -	- -	5.1.32
10.00 - 16.00	- -	- -	- -	- -	5.1.33
- -	2.25 - 5.50	- -	- -	- -	5.1.34
- -	4.50 - 7.25	- -	- -	- -	5.1.35
- -	3375 -5400	- -	- -	- -	5.1.36
- -	- -	7.75 - 10.00	- -	- -	5.1.37
3.25 - 8.50	- -	- -	3.25 - 8.50	- -	5.1.38
6.25 - 11.00	- -	- -	6.25 - 11.00	- -	5.1.39
- -	- -	- -	6.75 - 11.50	- -	5.1.40
- -	- -	- -	9.00 - 13.50	- -	5.1.41
- -	- -	- -	9.00 - 13.50	- -	5.1.42
- -	- -	- -	5.50 - 11.00	- -	5.1.43
5.50 - 11.00	- -	- -	9.00 - 13.50	- -	5.1.44
9.00 - 13.50	- -	- -	9.00 - 13.50	- -	5.1.45
- -	- -	- -	11.00 - 15.25	- -	5.1.46
- -	- -	- -	12.50 - 17.00	- -	5.1.47
- -	- -	- -	- -	15.75 - 38.25	5.1.48

Item nr.	SPECIFICATIONS	Unit	RESIDENTIAL £ range £
5.2	**WATER INSTALLATIONS**	**gross internal area**	
	Hot and cold water installations		
5.2.1	Complete installations	m2	11.00 - 18.00
5.2.2	To mall public conveniences; branch connections to shop shells	m2	- -
5.3	**HEATING, AIR-CONDITIONING AND VENTILATING INSTALLATIONS**	**gross internal area serviced** (unless otherwise described)	
	comprising:-		
	Solid fuel radiator heating		
	Gas or oil-fired radiator heating		
	Gas or oil-fired convector heating		
	Electric and under floor heating		
	Hot air systems		
	Ventilation systems		
	Heating and ventilation systems		
	Comfort cooling systems		
	Full air-conditioning systems		
	Solid fuel radiator heating		
5.3.1	Chimney stack, hearth and surround to independent residential unit	nr	1350 -1625
	Chimney, hot water service and central heating for		
5.3.2	two radiators	nr	2000 -2350
5.3.3	three radiators	nr	2350 -2600
5.3.4	four radiators	nr	2600 -2900
5.3.5	five radiators	nr	2900 -3150
5.3.6	six radiators	nr	3150 -3450
5.3.7	seven radiators	nr	3450 -4000
	Gas or oil-fired radiator heating		
	Gas-fired hot water service and central heating for		
5.3.8	three radiators	nr	1500 -2100
5.3.9	four radiators	nr	2100 -2300
5.3.10	five radiators	nr	2300 -2450
5.3.11	six radiators	nr	2450 -2600
5.3.12	seven radiators	nr	2600 -3100
	Oil-fired hot water service, tank and central heating for		
5.3.13	seven radiators	nr	2450 -3250
5.3.14	ten radiators	nr	2700 -3600
5.3.15	LPHW radiator system	m2	20.50 - 31.50
5.3.16	speculative; area less than 1000 m2	m2	- -
5.3.17	speculative	m2	- -
5.3.18	owner-occupied; area less than 1000 m2	m2	- -
5.3.19	owner-occupied	m2	- -
5.3.20	LPHW radiant panel system	m2	- -
5.3.21	speculative; less than 1000 m2	m2	- -
5.3.22	speculative	m2	- -
5.3.23	LPHW fin tube heating	m2	- -
	Gas or oil-fired convector heating		
5.3.24	LPHW convector system	m2	- -
5.3.25	speculative; area less than 1000 m2	m2	- -
5.3.26	speculative	m2	- -
5.3.27	owner-occupied; area less than 1000 m2	m2	- -
5.3.28	owner-occupied	m2	- -
	LPHW sill-line connector system		
5.3.29	owner-occupied	m2	- -

INDUSTRIAL £ range £	RETAILING £ range £	LEISURE £ range £	OFFICES £ range £	HOTELS £ range £	Item nr.
4.50 - 11.50	- -	14.50 - 20.75	8.50 - 11.75	22.50 - 27.00	5.2.1
- -	3.25 - 5.50	- -	- -	- -	5.2.2
- -	- -	- -	- -	- -	5.3.1
- -	- -	- -	- -	- -	5.3.2
- -	- -	- -	- -	- -	5.3.3
- -	- -	- -	- -	- -	5.3.4
- -	- -	- -	- -	- -	5.3.5
- -	- -	- -	- -	- -	5.3.6
- -	- -	- -	- -	- -	5.3.7
- -	- -	- -	- -	- -	5.3.8
- -	- -	- -	- -	- -	5.3.9
- -	- -	- -	- -	- -	5.3.10
- -	- -	- -	- -	- -	5.3.11
- -	- -	- -	- -	- -	5.3.12
- -	- -	- -	- -	- -	5.3.13
- -	- -	- -	- -	- -	5.3.14
30.00 - 41.50	30.00 - 54.00	38.00 - 48.50	- -	45.00 - 63.00	5.3.15
- -	- -	- -	37.00 - 46.00	- -	5.3.16
- -	- -	- -	40.00 - 54.00	- -	5.3.17
- -	- -	- -	40.00 - 53.00	- -	5.3.18
- -	- -	- -	42.50 - 59.50	- -	5.3.19
- -	36.00 - 54.00	45.00 - 54.00	- -	- -	5.3.20
39.50 - 47.00	- -	- -	- -	- -	5.3.21
41.50 - 48.50	- -	- -	- -	- -	5.3.22
52.00 - 76.50	- -	- -	- -	- -	5.3.23
31.50 - 41.50	30.00 - 56.50	40.50 - 50.50	- -	52.00 - 66.00	5.3.24
- -	- -	- -	40.50 - 48.50	- -	5.3.25
- -	- -	- -	42.50 - 56.50	- -	5.3.26
- -	- -	- -	45.00 - 54.00	- -	5.3.27
- -	- -	- -	49.50 - 63.00	- -	5.3.28
- -	- -	- -	65.00 - 90.00	- -	5.3.29

Item nr.	SPECIFICATIONS	Unit	RESIDENTIAL £ range £	
5.3	**HEATING, AIR-CONDITIONING AND VENTILATING INSTALLATIONS - cont'd** **gross internal area serviced** (unless otherwise described)			
	Electric and under floor heating			
5.3.30	Panel heaters	m2	-	-
5.3.31	Skirting heaters	m2	-	-
5.3.32	Storage heaters	m2	-	-
5.3.33	Underfloor heating in changing areas	m2	-	-
	Hot air systems			
	Hot water service and ducted hot air heating to			
5.3.34	three rooms	nr	1350	-1750
5.3.35	five rooms	nr	1850	-2100
5.3.36	Elvaco warm air heating	m2	-	
5.3.37	Gas-fired hot air space heating	m2	-	-
5.3.38	Warm air curtains; 1.6 m long electrically-operated	nr	-	-
5.3.39	Hot water-operated; including supply pipework	nr	-	-
	Ventilation systems			
	Local ventilation to			
5.3.40	WC's	nr	150.00	-200.00
5.3.41	toilet areas	m2	-	-
5.3.42	bathroom and toilet areas	m2	-	-
5.3.43	Air extract system	m2	-	-
5.3.44	Air supply and extract system	m2	-	-
5.3.45	Service yard vehicle extract system	m2	-	-
	Heating and ventilation systems			
5.3.46	Space heating and ventilation - economical	m2	-	-
5.3.47	Heating and ventilation	m2	-	-
5.3.48	Warm air heating and ventilation	m2	-	-
	Hot air heating and ventilation to shopping malls;			
5.3.49	including automatic remote vents in rooflights	m2	-	-
	Extra for			
5.3.50	comfort cooling	m2	-	-
5.3.51	full air-conditioning	m2	-	-
	Comfort cooling systems			
5.3.52	Fan coil/induction systems	m2	-	-
5.3.53	speculative; area less than 1000 m2	m2	-	-
5.3.54	speculative	m2	-	-
5.3.55	owner-occupied; area less than 1000 m2	m2	-	-
5.3.56	owner-occupied	m2	-	-
5.3.57	VAV system	m2	-	-
5.3.58	speculative; area less than 1000 m2	m2	-	-
5.3.59	speculative	m2	-	-
5.3.60	owner-occupied; area less than 1000 m2	m2	-	-
5.3.61	owner-occupied	m2	-	-
	Stand-alone air-conditioning unit systems			
5.3.62	air supply and extract	m2	-	-
5.3.63	air supply and extract; including heat re-claim	m2	-	-
	Extra for			
5.3.64	automatic control installation	m2	-	-
5.3.65	Full air-conditioning with dust and humidity control	m2	-	-
5.3.66	Fan coil/induction systems	m2	-	-
5.3.67	speculative; area less than 1000 m2	m2	-	-
5.3.68	speculative	m2	-	-
5.3.69	owner-occupied; area less than 1000 m2	m2	-	-
5.3.70	owner-occupied	m2	-	-

INDUSTRIAL £ range £	RETAILING £ range £	LEISURE £ range £	OFFICES £ range £	HOTELS £ range £	Item nr.
- -	- -	- -	9.00 - 11.50	- -	5.3.30
- -	- -	- -	13.50 - 18.00	- -	5.3.31
- -	- -	- -	16.00 - 21.00	- -	5.3.32
- -	- -	34.00 - 43.00	- -	- -	5.3.33
- -	- -	- -	- -	- -	5.3.34
- -	- -	- -	- -	- -	5.3.35
- -	- -	- -	48.50 - 61.00	- -	5.3.36
18.00 - 36.00	- -	- -	- -	- -	5.3.37
- -	1500 -2000	- -	- -	- -	5.3.38
- -	3000 -3400	- -	- -	- -	5.3.39
- -	- -	- -	- -	- -	5.3.40
- -	- -	- -	2.75 - 6.75	- -	5.3.41
- -	- -	- -	- -	15.75 - 20.25	5.3.42
- -	24.75 - 33.75	24.75 - 33.75	24.75 - 33.75	27.00 - 36.00	5.3.43
- -	- -	- -	36.00 - 54.00	- -	5.3.44
- -	24.75 - 33.75	- -	- -	- -	5.3.45
15.75 - 34.75	16.50 - 30.25	- -	- -	- -	5.3.46
40.50 - 49.50	- -	- -	- -	- -	5.3.47
- -	75.00 - 95.00	75.00 - 95.00	75.00 - 95.00	80.00 -110.00	5.3.48
- -	65.00 - 85.00	- -	- -	- -	5.3.49
- -	55.00 - 80.00	- -	- -	- -	5.3.50
- -	65.00 -120.00	- -	- -	- -	5.3.51
- -	- -	- -	- -	110.00 -150.00	5.3.52
- -	- -	- -	110.00 -130.00	110.00 -150.00	5.3.53
- -	- -	- -	120.00 -140.00	- -	5.3.54
- -	- -	- -	120.00 -150.00	- -	5.3.55
- -	- -	- -	125.00 -160.00	- -	5.3.56
- -	110.00 -160.00	- -	- -	135.00 -200.00	5.3.57
- -	- -	- -	125.00 -180.00	- -	5.3.58
- -	- -	- -	130.00 -185.00	- -	5.3.59
- -	- -	- -	135.00 -190.00	- -	5.3.60
- -	- -	- -	140.00 -200.00	- -	5.3.61
- -	- -	95.00 -125.00	- -	- -	5.3.62
- -	- -	105.00 -135.00	- -	- -	5.3.63
- -	- -	9.00 - 27.00	- -	- -	5.3.64
115.00 -180.00	- -	- -	- -	- -	5.3.65
- -	125.00 -185.00	- -	- -	130.00 -200.00	5.3.66
- -	- -	- -	120.00 -150.00	- -	5.3.67
- -	- -	- -	120.00 -165.00	- -	5.3.68
- -	- -	- -	125.00 -185.00	- -	5.3.69
- -	- -	- -	130.00 -200.00	- -	5.3.70

Item nr.	SPECIFICATIONS	Unit	RESIDENTIAL £ range £	
5.3	**HEATING, AIR-CONDITIONING AND VENTILATING INSTALLATIONS - cont'd** **gross internal area serviced** (unless otherwise described)			
	Full air-conditioning - cont'd			
5.3.71	VAV system	m2	-	-
5.3.72	speculative; area less than 1000 m2	m2	-	-
5.3.73	speculative	m2	-	-
5.3.74	owner-occupied; area less than 1000 m2	m2	-	-
5.3.75	owner-occupied	m2	-	-
5.4	**ELECTRICAL INSTALLATIONS** **gross internal area serviced** (unless otherwise described)			
	comprising:-			
	Mains and sub-mains switchgear and distribution			
	Lighting installation			
	Lighting and power installations			
	Comparative fittings/rates per point			
	Mains and sub-mains switchgear and distribution			
5.4.1	Mains intake only	m2	-	-
5.4.2	Mains switchgear only	m2	1.75 -	2.75
	Mains and sub-mains distribution			
5.4.3	to floors only	m2	-	-
5.4.4	to floors; including small power and supplies to equipment	m2	-	-
5.4.5	to floors; including lighting and power to landlords areas and supplies to equipment	m2	-	-
5.4.6	to floors; including power, communication and supplies to equipment	m2	-	-
5.4.7	to shop units; including fire alarms and telephone distribution	m2	-	-
	Lighting installation			
	Lighting to			
5.4.8	warehouse area	m2	-	-
5.4.9	production area	m2	-	-
5.4.10	General lighting; including luminaries	m2	-	-
5.4.11	Emergency lighting	m2	-	-
5.4.12	standby generators only	m2	-	-
5.4.13	Underwater lighting	m2	-	-
	Lighting and power installations			
	Lighting and power to residential units			
5.4.14	one person flat/bed-sit	nr	675	-1050
5.4.15	two person flat/house	nr	800	-1500
5.4.16	three person flat/house	nr	950	-1800
5.4.17	four person house	nr	1100	-2200
5.4.18	five/six person house	nr	1350	-2700
	Extra for			
5.4.19	intercom	nr	270.00 -	315.00
	Lighting and power to industrial buildings			
5.4.20	warehouse area	m2	-	-
5.4.21	production area	m2	-	-
5.4.22	production area; high provision	m2	-	-
5.4.23	office area	m2	-	-
5.4.24	office area; high provision	m2	-	-
	Lighting and power to retail outlets			
5.4.25	shopping mall and landlords' areas	m2	-	-

INDUSTRIAL £ range £	RETAILING £ range £	LEISURE £ range £	OFFICES £ range £	HOTELS £ range £	Item nr.
- -	- -	- -	- -	145.00 -225.00	5.3.71
- -	- -	- -	140.00 -185.00	- -	5.3.72
- -	- -	- -	145.00 -200.00	- -	5.3.73
- -	- -	- -	150.00 -215.00	- -	5.3.74
- -	- -	- -	150.00 -225.00	- -	5.3.75
1.20 - 2.25	- -	- -	- -	- -	5.4.1
2.90 - 6.00	2.75 - 6.25	- -	- -	- -	5.4.2
3.45 - 6.75	- -	13.50 - 24.75	10.00 - 18.00	- -	5.4.3
- -	- -	10.00 - 11.75	- -	- -	5.4.4
6.75 - 15.00	- -	- -	24.75 - 40.50	- -	5.4.5
- -	- -	- -	- -	40.50 - 58.50	5.4.6
- -	4.00 - 10.00	- -	- -	- -	5.4.7
14.50 - 27.00	- -	- -	- -	- -	5.4.8
18.00 - 30.00	- -	- -	- -	- -	5.4.9
- -	- -	18.00 - 29.50	- -	15.25 - 24.50	5.4.10
- -	- -	6.25 - 10.00	- -	2.75 - 5.00	5.4.11
1.75 - 7.25	1.75 - 7.25	1.75 - 7.25	1.75 - 7.25	1.75 - 7.25	5.4.12
- -	- -	4.50 - 9.00	- -	- -	5.4.13
- -	- -	- -	- -	- -	5.4.14
- -	- -	- -	- -	- -	5.4.15
- -	- -	- -	- -	- -	5.4.16
- -	- -	- -	- -	- -	5.4.17
- -	- -	- -	- -	- -	5.4.18
- -	- -	- -	- -	- -	5.4.19
27.00 - 45.00	- -	- -	- -	- -	5.4.20
31.50 - 49.50	- -	- -	- -	- -	5.4.21
45.00 - 58.50	- -	- -	- -	- -	5.4.22
67.50 - 80.00	- -	- -	- -	- -	5.4.23
90.00 -105.00	- -	- -	- -	- -	5.4.24
- -	45.00 - 75.00	- -	- -	- -	5.4.25

Item nr.	SPECIFICATIONS	Unit	RESIDENTIAL £ range £	
5.4	**ELECTRICAL INSTALLATIONS - cont'd** **gross internal area serviced** (unless otherwise described)			
	Lighting and power installations - cont'd			
	Lighting and power to offices			
5.4.26	speculative office areas; average standard	m2	-	-
5.4.27	speculative office areas; high standard	m2	-	-
5.4.28	owner-occupied office areas; average standard	m2	-	-
5.4.29	owner-occupied office areas; high standard	m2	-	-
	Comparative fittings/rates per point			
5.4.30	Consumer control unit	nr	115.00	-125.00
	Fittings; excluding lamps or light fittings			
5.4.31	lighting point; PVC cables	nr	35.00	- 39.00
5.4.32	lighting point; PVC cables in screwed conduits	nr	75.00	- 82.50
5.4.33	lighting point; MICC cables	nr	65.00	- 70.00
	switch socket outlet; PVC cables			
5.4.34	single	nr	38.00	- 42.00
5.4.35	double	nr	45.00	- 50.00
	switch socket outlet; PVC cables in screwed conduit			
5.4.36	single	nr	57.50	- 62.50
5.4.37	double	nr	61.00	- 68.00
	switch socket outlet; MICC cables			
5.4.38	single	nr	57.50	- 62.50
5.4.39	double	nr	63.00	- 70.00
5.4.40	Immersion heater point (excluding heater)	nr	54.00	- 63.00
5.4.41	Cooker point; including control unit	nr	76.50	-121.50
5.5	**GAS INSTALLATION**			
5.5.1	Connection charge	nr	400.00	-500.00
	Supply to heaters within shopping mall and capped off			
5.5.2	supply to shop shells	m2	-	-
5.6	**LIFT AND CONVEYOR INSTALLATIONS** **lift or escalator** (unless otherwise described)			
	comprising:-			
	Passenger lifts			
	Escalators			
	Goods lifts			
	Dock levellers			
	Electro-hydraulic passenger lifts			
	eight to twelve person lifts			
5.6.1	8 person; 0.3 m/sec ; 2 - 3 levels	nr	-	-
5.6.2	8 person; 0.63 m/sec ; 3 - 5 levels	nr	-	-
5.6.3	8 person; 1.5 m/sec ; 3 levels	nr	-	-
5.6.4	8 person; 1.6 m/sec ; 6 - 9 levels	nr	-	-
5.6.5	10 person; 0.63 m/sec ; 2 - 4 levels	nr	-	-
5.6.6	10 person; 1.0 m/sec ; 2 levels	nr	-	-
5.6.7	10 person; 1.0 m/sec ; 3 - 6 levels	nr	-	-
5.6.8	10 person; 1.0 m/sec ; 3 - 6 levels	nr	-	-
5.6.9	10 person; 1.6 m/sec ; 4 levels	nr	-	-
5.6.10	10 person; 1.6 m/sec ; 10 levels	nr	-	-
5.6.11	10 person; 1.6 m/sec ; 13 levels	nr	-	-
5.6.12	10 person; 2.5 m/sec ; 18 levels	nr	-	-
5.6.13	12 person; 1.0 m/sec ; 2 - 3 levels	nr	-	-
5.6.14	12 person; 1.0 m/sec ; 3 - 6 levels	nr	-	-

INDUSTRIAL £ range £	RETAILING £ range £	LEISURE £ range £	OFFICES £ range £	HOTELS £ range £	Item nr.
- -	- -	- -	60.00 - 85.00	- -	5.4.26
- -	- -	- -	75.00 - 90.00	- -	5.4.27
- -	- -	- -	80.00 -105.00	- -	5.4.28
- -	- -	- -	95.00 -115.00	- -	5.4.29
- -	- -	- -	- -	- -	5.4.30
- -	40.00 - 45.00	40.00 - 45.00	40.00 - 45.00	40.00 - 45.00	5.4.31
107.50 -115.00	92.50 -100.00	92.50 -100.00	92.50 -100.00	92.50 -100.00	5.4.32
92.50 -100.00	77.50 - 85.00	77.50 - 85.00	77.50 - 85.00	77.50 - 85.00	5.4.33
- -	46.00 - 50.00	46.00 - 50.00	46.00 - 50.00	46.00 - 50.00	5.4.34
90.00 - 97.50	55.00 - 60.00	55.00 - 60.00	55.00 - 60.00	55.00 - 60.00	5.4.35
70.00 - 78.50	65.00 - 67.50	65.00 - 67.50	65.00 - 67.50	65.00 - 67.50	5.4.36
80.00 - 87.00	70.00 - 76.50	70.00 - 76.50	70.00 - 76.50	70.00 - 76.50	5.4.37
75.00 - 79.00	65.00 - 68.50	65.00 - 68.50	65.00 - 68.50	65.00 - 68.50	5.4.38
80.00 -100.00	70.00 - 78.50	70.00 - 78.50	70.00 - 78.50	70.00 - 78.50	5.4.39
- -	- -	- -	- -	- -	5.4.40
- -	- -	- -	- -	- -	5.4.41
- -	- -	- -	- -	- -	5.5.1
- -	5.00 - 7.50	- -	- -	- -	5.5.2
- -	- -	20000 -28500	19000 -28000	- -	5.6.1
- -	- -	- -	35000 -50000	- -	5.6.2
47500 -60000	- -	- -	- -	- -	5.6.3
- -	- -	- -	62500 -75000	- -	5.6.4
- -	- -	- -	42000 -55000	- -	5.6.5
- -	- -	- -	35000 -38000	- -	5.6.6
- -	- -	- -	47500 -60000	- -	5.6.7
- -	- -	- -	50000 -63000	- -	5.6.8
- -	- -	- -	50000 -65000	- -	5.6.9
- -	- -	- -	70000 -80000	- -	5.6.10
- -	- -	- -	80000 -92000	- -	5.6.11
- -	- -	- -	140000 -155000	- -	5.6.12
25000 -47500	- -	- -	- -	- -	5.6.13
- -	- -	- -	- -	55000 -67500	5.6.14

Item nr.	SPECIFICATIONS	Unit	RESIDENTIAL £ range £	
5.6	**LIFT AND CONVEYOR INSTALLATIONS -cont'd**	lift or escalator (unless otherwise described)		
	Electro-hydraulic passenger lifts - cont'd			
	Thirteen person lifts			
5.6.15	13 person; 0.3 m/sec ; 2 - 3 levels	nr	-	-
5.6.16	13 person; 0.63 m/sec ; 4 levels	nr	-	-
5.6.17	13 person; 1.0 m/sec ; 2 levels	nr	-	-
5.6.18	13 person; 1.0 m/sec ; 2 - 3 levels	nr	-	-
5.6.19	13 person; 1.0 m/sec ; 3 - 6 levels	nr	-	-
5.6.20	13 person; 1.6 m/sec ; 7 - 11 levels	nr	-	-
5.6.21	13 person; 2.5 m/sec ; 7 - 11 levels	nr	-	-
5.6.22	13 person; 2.5 m/sec ; 12 - 15 levels	nr	-	-
	Sixteen person lifts			
5.6.23	16 person; 1.0 m/sec ; 2 levels	nr	-	-
5.6.24	16 person; 1.0 m/sec ; 3 levels	nr	-	-
5.6.25	16 person; 1.0 m/sec ; 3 - 6 levels	nr	-	-
5.6.26	16 person; 1.6 m/sec ; 3 - 6 levels	nr	-	-
5.6.27	16 person; 1.6 m/sec ; 7 - 11 levels	nr	-	-
5.6.28	16 person; 2.5 m/sec ; 7 - 11 levels	nr	-	-
5.6.29	16 person; 2.5 m/sec ; 12 - 15 levels	nr	-	-
5.6.30	16 person; 3.5 m/sec ; 12 - 15 levels	nr	-	-
	Twenty one person lifts			
5.6.31	21 person/bed; 0.4 m/sec ; 3 levels	nr	-	-
5.6.32	21 person; 0.6 m/sec ; 2 levels	nr	-	-
5.6.33	21 person; 1.0 m/sec ; 4 levels	nr	-	-
5.6.34	21 person; 1.6 m/sec ; 4 levels	nr	-	-
5.6.35	21 person; 1.6 m/sec ; 7 - 11 levels	nr	-	-
5.6.36	21 person; 1.6 m/sec ; 10 levels	nr	-	-
5.6.37	21 person; 1.6 m/sec ; 13 levels	nr	-	-
5.6.38	21 person; 2.5 m.sec ; 7 - 11 levels	nr	-	-
5.6.39	21 person; 2.5 m/sec ; 12 - 15 levels	nr	-	-
5.6.40	21 person; 2.5 m/sec ; 18 levels	nr	-	-
5.6.41	21 person; 3.5 m/sec ; 12 - 15 levels	nr	-	-
	Extra for			
5.6.42	enhanced finish to car	nr	-	-
5.6.43	glass backed observation car	nr	-	-
	Ten person wall climber lifts			
5.6.44	10 person; 0.5 m/sec ; 2 levels	nr	-	-
	Escalators			
	30° Escalator; 0.5 m/sec; enamelled steel glass balustrades			
5.6.45	3.5 m rise; 600 mm step width	nr	-	-
5.6.46	3.5 m rise; 800 mm step width	nr	-	-
5.6.47	3.5 m rise; 1000 mm step width	nr	-	-
	Extra for			
5.6.48	enhanced finish	nr	-	-
5.6.49	4.4 m rise; 800 mm step width	nr	-	-
5.6.50	4.4 m rise; 1000 mm step width	nr	-	-
5.6.51	5.2 m rise; 800 mm step width	nr	-	-
5.6.52	5.2 m rise; 1000 mm step width	nr	-	-
5.6.53	6.0 m rise; 800 mm step width	nr	-	-
5.6.54	6.0 m rise; 1000 mm step width	nr	-	-
	Extras (per escalator)			
5.6.55	under step lighting	nr	-	-
5.6.56	under handrail lighting	nr	-	-
5.6.57	stainless steel balustrades	nr	-	-
5.6.58	mirror glass cladding to sides and soffits	nr	-	-
5.6.59	heavy duty chairs	nr	-	-

INDUSTRIAL £ range £	RETAILING £ range £	LEISURE £ range £	OFFICES £ range £	HOTELS £ range £	Item nr.
- -	- -	- -	22500 -30000	- -	5.6.15
- -	- -	- -	47500 -60000	- -	5.6.16
- -	56000 -63000	- -	- -	- -	5.6.17
47500 -62000	- -	- -	- -	- -	5.6.18
- -	- -	- -	48000 -63000	55000 -70000	5.6.19
- -	- -	- -	62500 -77500	67500 -82500	5.6.20
- -	- -	- -	100000 -115000	- -	5.6.21
- -	- -	- -	115000 -133000	- -	5.6.22
- -	54000 -68000	- -	- -	- -	5.6.23
- -	63000 -73000	- -	- -	- -	5.6.24
- -	- -	- -	55000 -70000	- -	5.6.25
- -	- -	- -	57500 -73500	- -	5.6.26
- -	- -	- -	80000 -92500	- -	5.6.27
- -	- -	- -	105000 -120000	- -	5.6.28
- -	- -	- -	120000 -140000	- -	5.6.29
- -	- -	- -	125000 -147500	- -	5.6.30
- -	- -	- -	40000 -45000	- -	5.6.31
- -	- -	- -	57500 -70000	- -	5.6.32
- -	- -	- -	68000 -80000	- -	5.6.33
- -	- -	- -	60000 -70000	- -	5.6.34
- -	- -	- -	80000 -90000	- -	5.6.35
- -	- -	- -	90000 -100000	- -	5.6.36
- -	- -	- -	100000 -110000	- -	5.6.37
- -	- -	- -	115000 -135000	- -	5.6.38
- -	- -	- -	130000 -150000	- -	5.6.39
- -	- -	- -	155000 -170000	- -	5.6.40
- -	- -	- -	145000 -160000	- -	5.6.41
- -	- -	- -	- -	7500 -8000	5.6.42
- -	- -	- -	- -	8500 -15000	5.6.43
- -	125000 -160000	- -	125000 -160000	125000 -160000	5.6.44
- -	- -	- -	42000 -52000	- -	5.6.45
- -	- -	- -	45000 -55000	- -	5.6.46
- -	60000 -80000	48000 -58000	48000 -58000	50000 -70000	5.6.47
- -	18000 -25000	12500 -17500	12500 -17500	18000 -27000	5.6.48
- -	50000 -63000	- -	- -	- -	5.6.49
- -	57000 -69000	- -	- -	- -	5.6.50
- -	54000 -66000	- -	- -	- -	5.6.51
- -	60000 -72000	- -	- -	- -	5.6.52
- -	57000 -69000	- -	- -	- -	5.6.53
- -	63000 -75000	- -	- -	- -	5.6.54
- -	300 -350	- -	- -	- -	5.6.55
- -	1100 -2200	- -	- -	- -	5.6.56
- -	1200 -2400	- -	- -	- -	5.6.57
- -	8800 -11000	- -	- -	- -	5.6.58
- -	11000 -15000	- -	- -	- -	5.6.59

Item nr.	SPECIFICATIONS	Unit	RESIDENTIAL £ range £	
5.6	**LIFT AND CONVEYOR INSTALLATIONS -cont'd**	lift or escalator (unless otherwise described)		
	Goods lifts			
5.6.60	Hoist	nr	-	-
	Kitchen service hoist			
5.6.61	50 kg; 2 levels	nr	-	-
	Electric heavy duty goods lifts			
5.6.62	500 kg; 2 levels	nr	-	-
5.6.63	500 kg; 2 - 3 levels	nr	-	-
5.6.64	500 kg; 5 levels	nr	-	-
5.6.65	1000 kg; 2 levels	nr	-	-
5.6.66	1000 kg; 2 - 5 levels	nr	-	-
5.6.67	1000 kg; 5 levels	nr	-	-
5.6.68	1500 kg; 3 levels	nr	-	-
5.6.69	1500 kg; 4 levels	nr	-	-
5.6.70	1500 kg; 7 levels	nr	-	-
5.6.71	2000 kg; 2 levels	nr	-	-
5.6.72	2000 kg; 3 levels	nr	-	-
5.6.73	3000 kg; 2 levels	nr	-	-
5.6.74	3000 kg; 3 levels	nr	-	-
	Oil hydraulic heavy duty goods lifts			
5.6.75	500 kg; 3 levels	nr	-	-
5.6.76	1000 kg; 3 levels	nr	-	-
5.6.77	2000 kg; 3 levels	nr	-	-
5.6.78	3500 kg; 4 levels	nr	-	-
	Dock levellers			
5.6.79	Dock levellers	nr	-	-
5.6.80	Dock leveller and canopy	nr	-	-
5.7	**PROTECTIVE, COMMUNICATION AND SPECIAL INSTALLATIONS**	gross internal area served (unless otherwise described)		
	comprising:-			
	Fire fighting/protective installations			
	Security/communication installations			
	Special installations			
	Fire fighting/protective installations			
	Fire alarms/appliances			
5.7.1	loose fire-fighting equipment	m2	-	-
5.7.2	smoke detectors, alarms and controls	m2	-	-
5.7.3	hosereels, dry risers and extinguishers	m2	-	-
	Sprinkler installations			
5.7.4	landlords areas; supply to shop shells; including fire alarms, appliances etc.	m2	-	-
5.7.5	single level sprinkler systems, alarms and smoke detectors; low hazard	m2	-	-
	Extra for			
5.7.6	ordinary hazard	m2	-	-
5.7.7	single level sprinkler systems; alarms and smoke detectors; ordinary hazard	m2	-	-
5.7.8	double level sprinkler systems; alarms and smoke detectors; high hazard	m2	-	-
	Smoke vents			
5.7.9	automatic smoke vents over glazed shopping mall	m2	-	-
5.7.10	smoke control ventilation to atria	m2	-	-
5.7.11	Lightning protection	m2	-	-

INDUSTRIAL £ range £	RETAILING £ range £	LEISURE £ range £	OFFICES £ range £	HOTELS £ range £	Item nr.
4000 -15000	- -	- -	- -	- -	5.6.60
- -	- -	5000 -5600	- -	- -	5.6.61
- -	38000 -43000	- -	37500 -42500	37500 -42500	5.6.62
16000 -23000	- -	- -	- -	- -	5.6.63
- -	- -	- -	45000 -56000	- -	5.6.64
- -	42000 -50000	- -	40000 -48000	40000 -48000	5.6.65
21000 -30000	- -	- -	- -	- -	5.6.66
- -	- -	- -	- -	48000 -56000	5.6.67
45000 -54000	- -	- -	50000 -60000	- -	5.6.68
- -	- -	- -	55000 -65000	- -	5.6.69
- -	- -	- -	65000 -75000	- -	5.6.70
- -	- -	- -	50000 -60000	- -	5.6.71
- -	- -	- -	60000 -70000	- -	5.6.72
- -	- -	- -	55000 -65000	- -	5.6.73
- -	- -	- -	68000 -80000	- -	5.6.74
50000 -60000	- -	- -	- -	- -	5.6.75
54000 -63000	- -	- -	- -	- -	5.6.76
62000 -72000	- -	- -	- -	- -	5.6.77
55000 -66000	- -	- -	- -	- -	5.6.78
5000 -13000	5000 -13000	- -	- -	5000 -13000	5.6.79
7500 -18000	7500 -18000	- -	- -	- -	5.6.80
- -	- -	0.15 - 0.25	- -	- -	5.7.1
2.25 - 4.50	2.25 - 4.50	4.50 - 5.50	2.25 - 4.50	5.00 - 9.00	5.7.2
3.50 - 8.00	4.00 - 8.00	- -	4.00 - 8.00	3.75 - 7.75	5.7.3
- -	6.75 - 11.75	- -	- -	- -	5.7.4
8.00 - 12.50	9.00 - 12.50	- -	- -	- -	5.7.5
3.50 - 4.50	3.50 - 4.50	- -	- -	- -	5.7.6
11.75 - 16.25	12.50 - 16.25	- -	10.75 - 15.25	9.50 - 14.50	5.7.7
18.00 - 22.50	20.00 - 23.50	- -	18.00 - 22.50	- -	5.7.8
- -	12.50 - 20.00	- -	- -	- -	5.7.9
- -	- -	- -	38.50 - 45.00	38.50 - 47.50	5.7.10
0.45 - 0.55	0.90 - 1.80	0.45 - 0.55	0.65 - 1.15	0.65 - 1.15	5.7.11

Item nr.	SPECIFICATIONS	Unit	RESIDENTIAL £ range £	
5.7	**PROTECTIVE, COMMUNICATION AND SPECIAL INSTALLATIONS - cont'd** **gross internal area served (unless otherwise described)**			
	Security/communication installations			
5.7.12	Clock installation	m2	-	-
5.7.13	Security alarm system	m2	-	-
5.7.14	Telephone system	m2	-	-
5.7.15	Public address, television aerial and clocks	m2	-	-
5.7.16	Closed-circuit television	m2	-	-
5.7.17	Public address system	m2	-	-
5.7.18	Closed-circuit television and public address system	m2	-	-
	Special installations			
	Window cleaning equipment			
5.7.19	twin track	m	-	-
5.7.20	manual trolley/cradle	nr	-	-
5.7.21	automatic trolley/cradle	nr	-	-
5.7.22	Refrigeration installation for ice rinks	m2 ice	-	-
5.7.23	Pool water treatment installation	m2 pool	-	-
5.7.24	Laundry chute	nr	-	-
5.7.25	Sauna	nr	-	-
5.7.26	Jacuzzi installation	nr	-	-
5.7.27	Wave machine; four chamber wave generation equipment	nr	-	-
	Swimming pool; size 1300 x 6.00 m			
	Extra over cost including structure; finishings			
5.7.28	ventilation; heating and filtration	m2	-	-
5.8	**BUILDERS' WORK IN CONNECTION WITH SERVICES** **gross internal area**			
	General builders work to			
5.8.1	mains supplies; lighting and power to landlords areas	m2	-	-
5.8.2	central heating and electrical installation	m2	-	-
5.8.3	space heating and electrical installation	m2	-	-
5.8.4	central heating; electrical and lift installations	m2	-	-
	space heating; electrical and ventilation			
5.8.5	installations	m2	-	-
5.8.6	air-conditioning	m2	-	-
5.8.7	air-conditioning and electrical installation	m2	-	-
5.8.8	air-conditioning; electrical and lift installations	m2	-	-
	General builders work; including allowance for plant rooms; to			
5.8.9	central heating and electrical installations	m2	-	-
5.8.10	central heating, electrical and lift installations	m2	-	-
5.8.11	air-conditioning	m2	-	-
5.8.12	air-conditioning and electrical installation	m2	-	-
5.8.13	air-conditioning; electrical and lift installations	m2	-	-

INDUSTRIAL £ range £	RETAILING £ range £	LEISURE £ range £	OFFICES £ range £	HOTELS £ range £	Item nr.
- -	- -	0.10 - 0.90	- -	- -	5.7.12
- -	- -	1.25 - 1.75	1.25 - 1.75	- -	5.7.13
- -	- -	0.70 - 1.45	0.70 - 1.45	- -	5.7.14
1.75 - 2.75	- -	2.00 - 3.00	- -	1.75 - 3.75	5.7.15
- -	2.75 - 3.25	2.75 - 3.25	- -	- -	5.7.16
- -	7.25 - 8.00	7.25 - 8.00	- -	- -	5.7.17
- -	15.75 - 31.50	10.00 - 11.75	- -	- -	5.7.18
- -	100.00 -115.00	- -	90.00 -105.00	97.50 -110.00	5.7.19
- -	6500 -7500	- -	6500 -7500	6250 -7250	5.7.20
- -	14500 -18000	- -	15750 -18000	14500 -18000	5.7.21
- -	- -	350.00 -400.00	- -	- -	5.7.22
- -	- -	350.00 -375.00	- -	- -	5.7.23
- -	- -	9000 -10750	- -	9000 -10750	5.7.24
- -	- -	- -	- -	5500 -10750	5.7.25
- -	- -	10750 -13500	- -	10750 -13500	5.7.26
- -	- -	33750 -47250	- -	- -	5.7.27
- -	- -	- -	- -	675.00 -775.00	5.7.28
1.30 - 3.80	- -	- -	- -	- -	5.8.1
2.75 - 9.00	4.50 - 10.00	4.50 - 10.00	6.75 - 10.00	6.75 - 10.00	5.8.2
3.50 - 10.00	6.25 - 10.75	6.25 - 10.75	8.00 - 10.75	8.00 - 10.75	5.8.3
4.00 - 10.75	6.75 - 11.25	6.75 - 11.25	9.00 - 11.75	9.00 - 11.75	5.8.4
6.25 - 13.50	9.00 - 14.50	9.00 - 16.25	12.50 - 16.25	12.50 - 16.25	5.8.5
10.00 - 15.25	11.75 - 16.25	11.75 - 16.25	14.50 - 18.00	14.50 - 18.00	5.8.6
11.25 - 16.25	13.50 - 18.00	13.50 - 18.00	16.25 - 20.00	16.25 - 20.00	5.8.7
12.50 - 18.00	15.25 - 20.00	15.25 - 20.00	18.00 - 21.50	18.00 - 21.50	5.8.8
- -	18.00 - 22.50	18.00 - 22.50	22.50 - 27.00	22.50 - 27.00	5.8.9
- -	22.50 - 27.00	22.50 - 27.00	27.00 - 31.50	27.00 - 31.50	5.8.10
- -	31.50 - 36.00	31.50 - 36.00	40.50 - 45.00	40.50 - 45.00	5.8.11
- -	40.50 - 45.00	40.50 - 45.00	49.50 - 54.00	49.50 - 54.00	5.8.12
- -	49.50 - 54.00	49.50 - 54.00	58.50 - 63.00	58.50 - 63.00	5.8.13

Item nr.	SPECIFICATIONS	Unit	ALL AREAS £ range £
6.1	**SITE WORK**	surface area (unless otherwise described)	

comprising:-

Preparatory excavation and sub-bases
Seeded and planted areas
Sports Pitches
Parklands
Paved areas
Car parking alternatives
Roads and barriers
Road Crossing
Footbridges
Roadbridges
Underpasses
Roundabouts
Guard rails and parking bollards etc
Street furniture
Playground equipment
Fencing and screen walls, ancillary buildings etc.

Item nr.	SPECIFICATIONS	Unit	ALL AREAS £ range £
	Preparatory excavation and sub-bases		
	Excavating		
6.1.1	top soil; average 150 mm deep	m2	0.80 - 1.50
	top soil; average 225 mm deep; preserving in spoil heaps		
6.1.2	by machine	m2	1.50 - 2.00
6.1.3	by hand	m2	3.65 - 5.65
	to reduce levels; not exceeding 0.25 m deep		
6.1.4	by machine	m3	2.40 - 2.50
6.1.5	by hand	m3	10.00 - 11.00
6.1.6	to form new foundation levels and contours	m3	2.60 - 12.00
	Filling; imported top soil		
6.1.7	150 mm thick; spread and levelled for planting	m2	4.75 - 5.25
	Comparative sub-bases/beds		
6.1.8	50 mm sand	m2	1.60 - 1.80
6.1.9	75 mm ashes	m2	1.40 - 1.60
6.1.10	75 mm sand	m2	2.40 - 2.60
6.1.11	50 mm (1:8) blinding concrete	m2	2.75 - 3.00
6.1.12	100 mm granular fill	m2	2.75 - 3.00
6.1.13	150 mm granular fill	m2	4.25 - 4.50
6.1.14	75 mm (1:3:6) blinding concrete	m2	4.00 - 4.50
	Seeded and planted areas		
	Plant supply, planting, maintenance and 12 months guarantee		
6.1.15	seeded areas	m2	2.75 - 5.50
6.1.16	turfed areas	m2	3.50 - 7.25
	Planted areas (per m2 of surface area)		
6.1.17	herbaceous plants	m2	3.00 - 4.00
6.1.18	climbing plants	m2	4.00 - 6.75
6.1.19	general planting	m2	9.00 - 18.00
6.1.20	woodland	m2	13.50 - 27.00
6.1.21	shrubbed planting	m2	18.00 - 50.00
6.1.22	dense planting	m2	22.50 - 45.00
6.1.23	shrubbed area including allowance for small trees	m2	27.00 - 63.00
	Trees		
6.1.24	advanced nursery stock trees (12 - 20 cm girth)	tree	115.00 -135.00
	semi-mature trees; 5 - 8 m high		
6.1.25	coniferous	tree	360.00 -900.00
6.1.26	deciduous	tree	550 -1350

Item nr.	SPECIFICATIONS	Unit	ALL AREAS £ range £
6.1	**SITE WORK - cont'd** **surface area** (unless otherwise described)		
	Sports pitches		
	Costs include for cultivating ground, bringing to appropriate levels for the specified game, applying fertiliser, weedkiller, seeding and rolling and white line marking with nets, posts, etc as required		
6.1.27	Football pitch (114 x 72 m)	nr	13000
6.1.28	Cricket outfield (160 x 142 m)	nr	44000
6.1.29	Cricket square (20 x 20 m) incl imported marl or clay loam, bringing to accurate levels, seeding with cricket square type grass	nr	4000
6.1.30	Bowling green (38 x 38 m) rink, incl French drain and gravel path on four sides	nr	16000
6.1.31	Grass tennis courts 1 court (35 x 17 m) incl bringing to accurate levels, chain link perimeter fencing and gate, tennis posts and net	nr	16200
6.1.32	Two grass tennis courts (35 x 32 m) ditto	pair	27500
6.1.33	Artificial surface tennis courts (35 x 17 m) incl chain link fencing, gate, posts and net	nr	13000
6.1.34	Two courts (45 x 32 m) ditto	pair	24000
6.1.35	Artificial football pitch, incl sub-base, bitumen macadam open textured base and heavy duty Astroturf type carpet	nr	250000
6.1.36	Golf-putting green	hole	1300
6.1.37	Pitch and putt course	hole	4000-6000
6.1.38	Full length golf course, full specifications incl watering system	hole	16000-30000
6.1.39	Championship course	hole	up to 100000
	Parklands		
	Work on parklands will involve different techniques of earth shifting and cultivation. The following rates include for normal surface excavation, they exclude the provision of any land drainage.		
6.1.40	Parklands, incl cultivating ground, applying fertiliser, etc and seeding with parks type grass	ha	12300
6.1.41	General sportsfield	ha	14800
	Lakes incl excavation av 10 m deep, laying 1.5 mm thick butyl rubber sheet and spreading top soil evenly on top to depth of 300 mm		
6.1.42	Under 1 ha in area	ha	270000
6.1.43	Between 1 and 5 ha in area	ha	250000
6.1.44	Extra for planting aquatic plants in lake top soil	m2	40.00
	Land drainage		
	If land drainage is required on a project, the propensity of the land to flood will decide the spacing of the land drains. Costs include for excavation and backfilling of trenches and laying agricultural clay drain pipes with 75 mm diameter lateral runs average 60 mm deep and 100 mm diameter main runs average 750 mm deep.		
	Land drainage to parkland with laterals at		
6.1.45	30 m centres and main runs at 100 m centres	ha	2500
	Land drainage to sportsfields with laterals at		
6.1.46	10 cm centres and main runs at 33 m centres	ha	7500

Item nr.	SPECIFICATIONS	Unit	ALL AREAS £ range £
6.1	**SITE WORK - cont'd** **surface area** (unless otherwise described)		
	Paved areas		
	Gravel paving rolled to falls and cambers		
6.1.47	50 mm thick	m2	1.90 - 2.50
6.1.48	paving on sub-base; including excavation	m2	7.50 - 10.00
	Cold bitumen emulsion paving; in three layers		
6.1.49	25 mm thick	m2	3.50 - 4.50
	Tarmacadam paving; two layers; limestone or igneous chipping finish		
6.1.50	65 mm thick	m2	5.40 - 6.60
6.1.51	paving on sub-base; including excavation	m2	13.00 - 19.00
	Precast concrete paving slabs		
6.1.52	50 mm thick	m2	7.75 - 16.25
6.1.53	50 mm thick 'Texitone' slabs	m2	11.25 - 15.25
6.1.54	slabs on sub-base; including excavation	m2	18.00 - 25.00
	Precast concrete block paviours		
6.1.55	65 mm 'Keyblok' grey paving	m2	14.70 - 18.50
6.1.56	65 mm 'Mount Sorrel' grey paving	m2	14.85 - 17.40
6.1.57	65 mm 'Luttersett' paving	m2	15.20 - 18.50
6.1.58	60 mm 'Pedesta' paving	m2	15.80 - 18.50
6.1.59	paviours on sub-base; including excavation	m2	22.00 - 30.00
	Brick paviours		
	229 x 114 x 38 mm paving bricks		
6.1.60	laid flat	m2	25.00 - 30.00
6.1.61	laid to herring-bone pattern	m2	40.00 - 45.00
6.1.62	paviours on sub-base; including excavation	m2	47.50 - 55.00
	Granite setts		
6.1.63	200 x 100 x 100 mm setts	m2	39.00 - 43.00
6.1.64	setts on sub-base; including excavation	m2	50.00 - 55.00
	York stone slab paving		
6.1.65	paving on sub-base; including excavation	m2	50.00 - 55.00
	Cobblestone paving		
6.1.66	50 - 75 mm	m2	47.50 - 57.50
6.1.67	cobblestones on sub-base; including excavation	m2	55.00 - 70.00
	Car Parking alternatives		
	Surface level parking; including lighting and drainage		
6.1.68	tarmacadam on sub-base	car	775 -875
6.1.69	concrete interlocking blocks	car	850 -1000
6.1.70	Grasscrete precast concrete units filled with top soil and grass seed	car	900 -1100
6.1.71	at ground level with deck or building over	car	4300 -5000
	Garages etc		
6.1.72	single car park	nr	550 -900
6.1.73	single; traditional construction; in a block	nr	1800 -2500
6.1.74	single; traditional construction; pitched roof	nr	4100 -5200
6.1.75	double; traditional construction; pitched roof	nr	5600 -7200
	Multi-storey parking; including lighting and drainage on roof of two-storey shopping centre; including		
6.1.76	ramping and strengthening structure	car	4750 -5650
6.1.77	split level/parking ramp	car	4900 -5800
6.1.78	multi-storey flat slab	car	5900 -7200
6.1.79	multi-storey warped slab	car	6550 -7900
	Extra for		
6.1.80	feature cladding and pitched roof	car	800 -1350
	Underground parking; including lighting and drainage		
6.1.81	partially underground; natural ventilation; no sprinklers	car	8350 -9900
6.1.82	completely underground; mechanical ventilation and sprinklers	car	9900 -12400
6.1.83	completely underground; mechanical ventilation; sprinklers and landscaped roof	car	11700 -14000

Item nr.	SPECIFICATIONS	Unit	ALL AREAS £ range £
6.1	**SITE WORK - cont'd**	**surface area** (unless otherwise described)	
	Roads and barriers		
	Tarmacadam or reinforced concrete roads, including all earthworks, drainage, pavements, lighting, signs, fencing and safety barriers (where necessary); maximum cut 1.0 m		
6.1.84	Two lane road 7.3 m wide-rural location	m	850 -1100
6.1.85	Two lane road 7.3 m wide-urban location	m	1200 -1500
6.1.86	Two lane road 10.0 m wide-rural location	m	1000 -1300
6.1.87	Two lane road 10.0 m wide-urban location	m	1450 -1800
	Road crossing		
	Costs include road markings, beacons, lights, signs, advance danger signs etc.		
6.1.88	Zebra crossing	nr	3500 -4000
6.1.89	Pelican crossing	nr	14000 -15000
	Footbridges		
	Footbridge of either precast concrete or steel construction 4 m wide, 6 m high including deck, access stairs and ramp, parapets etc.		
6.1.90	15 - 20 m span to two lane road	nr	58000
6.1.91	30 m span to four lane dual carriageway	nr	94000
	Roadbridges		
	Roadbridges including all excavation, reinforcement, formwork, concrete, bearings, expansion joints, deck water proofing and finishings, parapets etc.		
	RC bridge with precast beams		
6.1.92	10 m span	m2 deck area	825.00
6.1.93	15 m span	m2 deck area	775.00
	RC bridge with prefabricated steel beams		
6.1.94	20 m span	m2 deck area	820.00
6.1.95	30 m span	m2 deck area	760.00
	Underpass		
	Provision of underpasses to new roads, constructed as part of a road building programme.		
	Precast concrete Pedestrian underpass		
6.1.96	3 m wide x 2.5 m high	m	2900 -3500
	Precast concrete Vehicle underpass		
6.1.97	7 m wide x 5 m high	m	8400 -10500
6.1.98	14 m wide x 5 m high	m	21000
	Bridge type structure vehicle underpass		
6.1.99	7 m wide x 5 m high	m	14700 -18000
6.1.100	14 m wide x 5 m high	m	36000
	Roundabouts		
6.1.101	Roundabout on existing dual carriageway; including perimeter road, drainage and lighting, signs and disruption while under construction	nr	270000 -400000
	Guard rails and parking bollards etc.		
6.1.102	Open metal post and rail fencing 1 m high	m	100.00 -120.00
6.1.103	Galvanised steel post and rail fencing 2 m high	m	110.00 -145.00
6.1.104	Steel guard rails and vehicle barriers	m	36.00 - 54.00
	Parking bollards		
6.1.105	precast concrete	nr	80.00 - 95.00
6.1.106	steel	nr	130.00 -170.00
6.1.107	cast iron	nr	155.00 -205.00
6.1.108	Vehicle control barrier; manual pole	nr	700.00 -750.00
6.1.109	Galvanised steel cycle stand	nr	30.00 - 38.00
6.1.110	Galvanised steel flag staff	nr	800 -1000

Item nr.	SPECIFICATIONS	Unit	ALL AREAS £ range £
6.1	**SITE WORK - cont'd**	**surface area** (unless otherwise described)	
	Street Furniture		
	Reflectorised traffic signs 0.25 m2 area		
6.1.111	on steel post	nr	70.00 -130.00
	Internally illuminated traffic signs		
6.1.112	dependent on area	nr	160.00 -210.00
	Externally illuminated traffic signs		
6.1.113	dependent on area	nr	375 -1000
	Lighting to pedestriam areas and estates		
6.1.114	roads on 4-6 m columns with up to 70 W lamps	nr	175.00 -250.00
	Lighting to main roads		
6.1.115	10-12 m columns with 250 W lamps	nr	400.00 -480.00
6.1.116	12-15 m columns with 400 W high pressure sodium lighting	nr	500.00 -600.00
6.1.117	Benches - hardwood and precast concrete	nr	150.00 -200.00
	Litter bins		
6.1.118	precast concrete	nr	150.00 -170.00
6.1.119	hardwood slatted	nr	60.00 - 80.00
6.1.120	cast iron	nr	250.00
6.1.121	large aluminium	nr	450.00
6.1.122	Bus stops	nr	275.00
6.1.123	Bus stops incl basic shelter	nr	625.00
6.1.124	Pillar box	nr	225.00
6.1.125	Telephone box	nr	2500.00
	Playground equipment		
	Modern swings with flat rubber safety seats:		
6.1.126	four seats, two bays	nr	1000.00
6.1.127	Stainless steel slide, 3.40 m long	nr	1100.00
6.1.128	Climbing frame - igloo type 3.20 x 3.75 m on plan x 2.00 m high	nr	1150.00
6.1.129	Seesaw comprising timber plank on sealed ball bearings 3960 x 230 x 70 mm thick	nr	800.00
6.1.130	Wicksteed Tumbleguard type safety surfacing around play equipment	m2	66.00
6.1.131	Bark particles type safety surfacing 150 mm thick on hardcore bed	m2	9.00
	Fencing and screen walls, ancillary buildings etc		
	Chain link fencing; plastic coated		
6.1.132	1.2 m high	m	12.50 - 14.50
6.1.133	1.8 m high	m	17.50 - 20.00
	Timber fencing		
6.1.134	1.2 m high chestnut pale fencing	m	14.00 - 16.00
6.1.135	1.8 m high close-boarded fencing	m	36.00 - 45.00
	Screen walls; one brick thick; including foundations etc		
6.1.136	1.8 m high facing brick screen wall	m	180.00 -225.00
6.1.137	1.8 m high coloured masonry block boundary wall	m	200.00 -250.00
6.1.138	Squash courts, independent building, including shower	nr	54000 -63000
	Demolish existing buildings		
6.1.139	of brick construction	m3	3.80 - 6.75

Item nr.	SPECIFICATIONS	Unit	ALL AREAS £ range £
6.2	**DRAINAGE** **gross internal area** (unless otherwise described)		
	comprising:-		
	Overall £/m2 allowances		
	Comparative pipework		
	Comparative manholes		
	Overall £/m2 allowances		
6.2.1	Site drainage (per m2 of paved areas)	m2	6.25 - 15.50
6.2.2	Building drainage (per m2 of gross floor area)	m2	6.25 - 13.50
6.2.3	Drainage work beyond the boundary of the site and final connection	nr	1800 -9000
	Comparative pipework		
	Vitrified clay pipes; flexible couplings; including 150 mm granular bed and benching; in trench 1 m deep		
6.2.4	100 mm 'Supersleve' pipe	m	14.50 - 17.50
6.2.5	150 mm 'Supersleve' pipe	m	18.50 - 21.00
6.2.6	100 mm 'Hepseal' pipe	m	17.50 - 20.50
6.2.7	150 mm 'Hepseal' pipe	m	20.00 - 23.50
	Vitrified clay pipes; cement and sand joints; including 150 mm concrete bed and benching in trench 1 m deep		
6.2.8	100 mm pipe	m	20.00 - 22.50
6.2.9	150 mm pipe	m	23.50 - 26.00
	UPVC pipes; including 150 mm granular bed and benching; in trench 1 m deep		
6.2.10	100 mm pipe	m	14.00 - 17.00
6.2.11	150 mm pipe	m	17.00 - 20.00
	Cast iron pipes; caulked lead joints; including 150 mm concrete bed; in trench 1 m deep		
6.2.12	100 mm pipe	m	37.50 - 45.00
6.2.13	150 mm pipe	m	54.00 - 60.00
	Extra for		
6.2.14	each additional 250 mm in trench not exceeding 2 m deep	m	2.60 - 3.00
6.2.15	each additional 250 mm in trench over 2 m and not exceeding 4 m deep	m	3.00 - 3.25
	Comparative manholes		
	Precast concrete ring manhole; including excavation; base, cover, channels and benching bottoms		
6.2.16	1.2 m dia x 1500 m deep	nr	1050 -1125
	Brick manhole; including excavation; concrete base, cover, channels and benching bottoms		
6.2.17	686 x 457 x 600 mm deep	nr	210.00 -250.00
	Extra for		
6.2.18	each additional 300 mm up to 2 m deep internally	nr	70.00 - 77.00
6.2.19	980 x 686 x 600 mm deep	nr	335.00 -380.00
	Extra for		
6.2.20	each additional 300 mm up to 2 m deep internally	nr	85.00 -100.00
6.2.21	1370 x 800 x 1500 mm deep	nr	765.00 -855.00
	Extra for		
6.2.22	685 x 685 mm shaft for each additional 1300 mm up to 4 m deep internally	nr	90.00 -105.00
6.2.23	100 mm three-quarter section branch bends	nr	7.40 - 9.00
6.2.24	150 mm three-quarter section branch bends	nr	12.00 - 13.50

Item nr.	SPECIFICATIONS	Unit	ALL AREAS £ range £
6.3	**EXTERNAL SERVICES** gross internal area		
	Service runs		
	All laid in trenches including excavation.		
	Water main		
6.3.1	75 mm uPVC main in 225 mm ductile iron pipe as duct	m	35.00
	Electric main		
6.3.2	600/1000 volt cables. Two core 25 mm cable incl 100 mm clayware duct	m	24.00
	Gas main		
6.3.3	150 mm ductile or cast iron gas pipe	m	35.00
	Telephone		
6.3.4	British Telecom installation in 100 mm uPVC duct	m	15.00
6.3.5	External lighting (per m2 of lighted area)	m2	1.75 - 2.75
	Connection charges		
	The privatisation of telephone, water and gas, and the future privatisation of electricity has complicated the assessment of service connection charges. Typically, service connection charges will include the actual cost of the direct connection plus an assessment of distribution costs from the main. The latter cost is difficult to estimate as it depends on the type of scheme and the distance from the mains. In addition, service charges are complicated by discounts that may be offered. For instance, the electricity boards will charge less for housing connections if the house is all electric.		
	However, typical charges for an estate of 200 houses might be as follows.		
6.3.6	Water	house	400.00 -800.00
	Electric		
6.3.7	all electric	house	200.00
6.3.8	gas/electric	house	400.00
	Extra cost of		
6.3.9	sub-station	nr	10000 -15000
6.3.10	Gas	house	400.00 -500.00
	Extra cost of		
6.3.11	governing station	nr	10000
6.3.12	Telephone	house	130.00
6.3.13	Sewerage	house	300.00 -400.00

Cost Limits and Allowances

Information given under this heading is based upon the cost targets currently in force for buildings financed out of public funds, i.e., hospitals, schools, universities and public authority housing. The information enables the cost limit for a scheme to be calculated and is not intended to be a substitiute for estimates prepared from drawings and specifications.

The cost limits are generally set as target costs based upon the user accommodation, i.e., in the case of universities they are given per metre of usable floor area. However ad-hoc additions can be agreed with the relevant Authority in exceptional circumstances.

The documents setting out cost targets are almost invariably complex and cover a range of differing circumstances. They should be studied carefully before being applied to any scheme; this study should preferably be undertaken in consultation with a Chartered Quantity Surveyor.

The cost limits for Public Authority housing generally known as the Housing Cost Yardstick have been replaced by a system of new procedures. The cost criteria prepared by the Department of the Enviroment that have superseded the Housing Cost Yardstick are intended as indicators and not cost limits.

HOSPITAL BUILDINGS

The information and tables which follow are contained in the CONCISE 4 database (issued to Health Authorities) version 6.0 and Capricode (obtainable from HMSO), reproduced here by kind permission of the Controller of Her Majesty's Stationery office. The tables, which should be read in conjunction with the Building notes (as issued to May 1990, including available provisional figures) concerned, provide the departmental cost limit related to the functional content of a project.

The cost allowances that follow relate to projects/schemes with a VOP (fluctuating) type main contract and are at an index level of 292 on the VOP series of the Median Index (unweighted) of the Public Sector Building Tender Prices Index (M). Firm price contract schemes attract an additional 9% on current cost allowances. These cost levels were promulgated on 4 April 1990. Health Authorities are notified by the Department of Health when revisions occur.

The reader should check the current levels of the following cost allowances when used for submitting new schemes.

The Works Cost comprises the Departmental cost for a scheme plus the On-costs.

ON COSTS

These are capital costs arising from interaction of the building and its site, (for example the cost of communications between departments, external works, and services, additional energy saving measures, auxiliary building and abnormals).

COMMUNICATIONS

Space and/or main engineering services between departments for movement of people supplies and services (including lifts, stairs and main ducts or shafts). EXTERNAL WORKS

Building and engineering works external to the outer wall of the building but forming an integral part of the scheme, and main engineering supply services internal to the building but outside the confines of individual departments.

ABNORMALS

Exceptional or ad-hoc factors, usually arising from site difficulties and constraints which increase capital costs, (for example, demolitions, adverse soil conditions, poor bearing capacity of ground works or alterations to existing buildings).

EXTENSION OF AN EXISTING HOSPITAL

Where it is intended to add to the departmental or service provision of an extant hospital by attaching to and/or extending the existing structure, the Works cost should be calculated as set out above for new separate hospital buildings.

UPGRADING OF EXISTING ACCOMMODATION

The Works cost for upgrading work should be estimated in each case. To establish that the cost is economic, it should be compared to a notional cost limit calculated on the cost of a functionally equivalent new building discounted for the life of the upgraded building and abated by the value of sound elements to be re-used in the existing building.

DEPARTMENTAL COSTS

The capital sum is calculated from the following tables of Allowances, based on (or derived from) the functional content, representing the total departmental cost. Where cost allowances are not appropriate, the costs for individual departments should be justified on the basis of similar accommodation in other schemes.

In interpolating for functional units which fall between two of the figures given in the tables the cost should be rounded to the nearest:-

- £1 000 for departments costing over £50 000
- £500 for departments costing £2 000 - £50 000
- £100 for departments costing £1 000 - £20 000
- £10 for departments costing less than £1 000

HOSPITAL BUILDINGS

REF. NO.	SERVICE	DEPARTMENT	ACCOMMODATION
01.01.01	In-Patients Services	General Acute Wards	Ward
02			Essential Complementary Accommodation: Orthopaedic Equipment Store
03			Seminar Room
04			Senior Nurse's Office
05			Clinical Teacher's Office
06			Relative's Room
07			Relative's WC
08			Relative's Shower/Wash
09			Shared Accommodation (for single ward)
10			Optional Accommodation: Cook Chill Trolley Holding Room
01.02.01		Children	Nursing Section: Shared Ancillary Accommodation
02			Independent Ancillary Accommodation
03			Day Unit
04			Essential Complementary Accommodation: Staff Locker Room (with central changing)
05			Staff Changing (with local changing)
06			Seminar Room
07			Nursing Officer's Office
08			Office/Interview Room
09			Parent's Bedroom
10			Classroom
11			Store
12			Teacher's Base
13			Adolescent Day Room
01.03.01		Elderly	Nursing Section Type A (acute)
02			Type B (rehabilitation)
03			Essential Complementary Accommodation: Seminar Room
04			Doctor's Office
05			Nursing Officer's Office
06			Staff Changing (with local changing)
07			Day Hospital
08			Administrative Centre
01.04.01		Maternity	Delivery Suite
02			Ward Accommodation
03			Neo-Natal Unit

HOSPITAL BUILDINGS

FUNCTIONAL SIZE		BUILDING NOTE NO.	INDEX	COST GUIDE £	REMARKS
1 Bed	)	4	M	17 342	
1 Store	)			1 454	
1 Room	)			10 708	
1 Office	)			7 114	
1 Office	)	4	M	7 114	
1 Room	)			5 315	
1 WC	)			3 501	
1 Shower	)			3 859	
1 Ward	)			13 660	
1 Room	)	4	M	17 641	
20 Beds	)			404 966	See HN(83)26.
20 Beds	)	23	M	416 040	(i) See HN(83)26.
8 Beds	)			129 675	(ii) For Children's OPD see 04.08.01
1 Room	)			5 266	)
1 Room	)			8 052	)
1 Room	)			7 795	)
1 Office	)			6 266	)
1 Office	)			6 748	)
1 Bedroom	)	23	M	7 543	) See HN(83)26.
10 Places	)			26 645	)
20 Places	)			31 977	)
1 Store	)			7 147	)
1 Base	)			5 468	)
1 Room	)			11 775	)
24 Beds	)	37	M	495 193	(i) See HN(80)21. (ii) Costs include, as additional accommodation, an allowance for a relatives'/hairdressing room.
24 Beds	)	37	M	514 223	(i) See HN(80)21. (ii) Costs include, as additional accommodation, an allowance for a relatives'/hairdressing room.
1 Room	)			11 753	)
1 Office	)	37	M	7 274	)See HN(80)21.
1 Office	)			6 972	)
1 Room	)			10 548	)
25 Places	)	37	M	493 993	)See HN(80)21.
40 Places	)			572 116	)
200 000 Pop.	)	37	M	69 987	See HN(80)21.
1 Suite	)	21	M	696 163	
1 Ward	)	21	M	398 729	
6 Cots	)			235 360	
8 Cots	)			245 224	
10 Cots	)	21	M	273 041	
12 Cots	)			300 542	
20 Cots	)			396 335	

HOSPITAL BUILDINGS

REF. NO.	SERVICE	DEPARTMENT	ACCOMMODATION
01.04.04	In-Patients' Service (continued)	Maternity (continued)	Essential Complementary Accommodation: Doctor's Office/Overnight Stay
05			Dayroom
06			Seminar/Staff Room
07			Consulting/Examination Room
08			Milk Kitchen Store
09			Cleaner's Room
10			Parents' Bedroom
11			Parents' Shower
12			Parents' WC
13			Office
14			Incubator Transport Bay
15			Optional Accommodation: Staff Cloaks
16			Interview/Examination Room
17			Emergency Obstetric Store
18			Milk Bank
01.05.01		Intensive Therapy	
02.01.01	Main Operating Facilities	Operating	Operating Suites
			Essential Complementary Accommodation:
02			Seminar Room
03			Clinical Nurse Teaching
			Optional Accommodation:
04			Ultra Clean Ventilation
03.01.01	Diagnostic and treatment Facilities	Radio-diagnostic	Department
02			Essential Complementary Accommodation: Seminar Room
03			workshop
04			Optional Accommodation: Darkroom
05			Lavage Room
06			Special Procedures Changing Room
07			Anaesthetic Room
08			Barium Preparation Room
09			Radiologist's Office
10			Typist's Room
11			Cooling
03.02.01		Radiosotopes (sub-regional)	

HOSPITAL BUILDINGS

FUNCTIONAL SIZE		BUILDING NOTE NO.	INDEX	COST GUIDE £	REMARKS
1 Office	)			6 221	
1 Room	)			7 171	
1 Room	)			7 033	
1 Room	)			9 142	
1 Store	)			7 229	
1 Room	)	21	M	4 640	
1 Bedroom	)			5 315	
1 Shower	)			3 859	
1 WC	)			3 501	
1 Office	)			5 266	
1 Bay	)			2 097	
1 Room	)			8 088	
1 Room	)			11 235	
1 Store	)	21	M	2 323	
1 Store	)			4 538	
8 Beds	)	27	M	228 334	Piped medical gases included.
4 Theatres	)			1 184 736	(i) BN 26 not applicable to 1,2 or
6	)	26	M	1 605 454	3 theatres.
8	)			2 159 894	
	)			11 371	
	)	26	M	6 058	
	)	26	M	24 973	Cooling and emergency sterilizer included
4 R/D Rooms	)			542 485	See HN(85)1.
5	)			623 546	
6	)			692 622	
7	)			781 860	
8	)	6	M	854 642	
9	)			977 433	
10	)			1 088 308	
11	)			1 199 629	
12	)			1 271 818	
1 Room	)	6	M	22 517	)See HN(85)1.
1 Room	)			9 892	)
1 Room	)			5 438	)
1 Room	)			10 523	)
1 Room	)			4 676	)
1 Room	)	6	M	15 703	)See HN(85)1.
1 Room	)			7 457	)
1 Office	)			8 205	)
1 Room	)			4 920	)
1 Unit	)			10 526	)
1 Department	)		M	135 686	See Design Guide issued October 1973.

HOSPITAL BUILDINGS

REF. NO.	SERVICE	DEPARTMENT	ACCOMMODATION
03.03.01	Diagnostic and treatment Facilities (continued)	Pathology	Area Laboratory
02			Public Health Services Laboratory (Supplementary)
03			Reference Laboratory (supplementary): Cytology
04			Trace Elements
05			Toxicology
06			Immunology
07			Chromosones
08			Automation
09			Neuropathology
03.04.01		Pharmacy	In-Patient Dispensing
02			Sterile Suite (small)
03			Sterile Suite (large)
04			Out-Patient Dispensing (attached)
05			Out-Patient Dispensing (remote)
06			Drug Information (hospital level)
07			Drug Information (district level)
08			Essential Complementary Accommodation: Seminar Room

HOSPITAL BUILDINGS

FUNCTIONAL SIZE		BUILDING NOTE NO.	INDEX	COST GUIDE £	REMARKS
39 LSUs	)	15	M	1 392 097	See Annex II to HN(80)21 for 'Howie' revisions to Building Note
6.5 LSUs	)	15	M	176 781	See Annex II to HN(80)21 for 'Howie' revisions to Building Note
1 LSUs 2	))	15	M	20 398 41 218	(i) See Annex II to HN(80)21 for 'Howie' revisions to Building Note (ii) 1 LSU serves up to 500 000 pop 2 LSUs serve up to 1 000 000 pop.
0.5 LSUs 1	))	15	M	18 213 36 858	(i) See Annex II to HN(80)21 for 'Howie' revisions to Building Note (ii) 0.5 LSU serves up to 500 000 pop 1 LSU serves up to 1 000 000 pop
0.5 LSUs 1 LSUs	))	15	M	18 213 36 858	(i) See Annex II to HN(80)21 for 'Howie' revisions to Building Note (ii) 0.5 LSU serves up to 500 000 pop 1 LSU serves up to 1 000 000 pop
1.5 LSUs 3 LSUs	))	15	M	35 647 70 857	(i) See Annex II to HN(80)21 for 'Howie' revisions to Building Note. (ii) 1.5 LSUs serve up to 500 000 pop 3 LSUs serve up to 1 000 000 pop.
1 LSUs	)	15	M	26 492	(i) See Annex II to HN(80)21 for 'Howie' revisions to Building Note. (ii) 1 LSU serves up to 1 000 000 pop.
2 LSUs 4	))	15	M	64 851 130 137	(i) See Annex II to HN(80)21 for 'Howie' revisions to Building Note (ii) 2 LSUs serve up to 500 000 pop 4 LSUs serve up to 1 000 000 pop.
2 LSUs 3	))	15	M	52 647 79 577	(i) See Annex II to HN(80)21 for 'Howie revisions to Building Note. (ii) 2 LSUs serve up to 500 000 pop 3 LSUs serve up to 1 000 000 pop.
450 Beds 600 601 1 200	))))	29	M	359 326 359 326 429 570 429 570	
1 Suite	)		M	74 664	
1 Suite	)	29	M	141 307	
1 Unit	)	29	M	24 045	
1 Suite	)	29	M	38 131	
1 Unit	)	29	M	14 442	
1 Unit	)	29	M	55 288	
8 Persons 20	))	29	M	10 537 18 955	

HOSPITAL BUILDINGS

REF. NO.	SERVICE	DEPARTMENT	ACCOMMODATION
03.04.09	**Diagnostic and**	Pharmacy	Flammable Store
10	**treatment**	(continued)	Medical Gas Cylinder Store
11	**Facilities**		Computer Facilities
12	**(continued)**		In-process Control Laboratory
13			Cytotoxic Preparation Room
14			Optional Accommodation: General Office/Reprographic Service
15			Sterile Manufacturing
16			Non-sterile Manufacturing
17			Assembly
18			Quality Control
19			Purchase and Distribution
20			Radio-Pharmaceuticals Preparation (closed procedure)
21			Radio-Pharmaceuticals Preparation (open procedure)
22			Essential Complementary Accommodation: Still Room
23			Porters' Base
24			Goods Reception/Loading Bay
25			Flammable/Hazardous Liquid Filling Area
26			Refuse Collection Store
27			Laboratory Workshop
28			Microbiological Testing Suite
29			Optional Accommodation: Tablet Packing
30			Engineer's Office
03.05.01		Mortuary and Post-Mortem	Body Viewing Facilities
02		(PROVISIONAL)	Body Store
03			Deep Freeze
04			Post Mortem Suite
05			Optional Accommodation: Post Mortem Table
06			Technician's Office
03.06.01		Medical Photography	Departmental Accommodation
03.07.01		Anaesthetic Services	Departmental Accommodation
03.08.01		Rehabilitation Physiotherapy, Occupational Therapy and Speech Therapy	Departmental Accommodation (1 = Small Unit 2 = Large Unit)

HOSPITAL BUILDINGS

FUNCTIONAL SIZE		BUILDING NOTE NO.	INDEX	COST GUIDE £	REMARKS
1 Store	)			5 092	
1 Store	)	29	M	12 190	
1 Room	)			12 781	
1 Laboratory	)			27 454	
1 Room	)			12 112	
1 Office	)	29	M	11 512	
1 Unit	)	29	M	734 848	
1 Unit	)	29	M	304 833	
1 Unit	)	29	M	160 807	
1 Unit	)	29	M	182 505	
1 Unit	)	29	M	301 914	
1 Area	)	29	M	75 146	
1 Area	)	29	M	110 290	
1 Room	)			83 416	
1 Base	)			4 922	
1 Bay	)			14 038	
1 Area	)	29	M	16 307	
1 Store	)			19 266	
1 Workshop	)			12 461	
1 Suite	)			71 535	
1 Line	)	29	M	58 192	
1 Office	)			4 858	
1 Suite	)	20	M	28 300	
6 Stores	)			56 603	
12	)	20	M	89 346	
18	)			108 480	
3 Stores	)	20	M	13 200	
1 Suite	)			172 386	
1 Table	)	20	M	27 556	
1 Office	)			8 562	
1 Department	)	19	M	50 816	
1 Department	)		M	71 563	See Design Guide issued Oct. 1971
1 Unit size	)	8	M	137 377	
2	)			184 655	

HOSPITAL BUILDINGS

REF. NO.	SERVICE	DEPARTMENT	ACCOMMODATION
02	**Diagnostic and Treatment Facilities (continued)**	Rehabilitation Physiotherapy, Occupational Therapy and Speech Therapy	Physiotherapy Section
03			Hydrotherapy Section
04			Occupational Therapy Section
05			Speech Therapy Section
			Essential Complementary Accommodation:
06			Splint Preparation Room
07			Assessment Wheelchair Store
08			Optional Accommodation:
			Consultant's Office
09			Consultant's C/E Room
10			District Therapist's Office
11			Secretarial Office
12			Therapy for Ward Areas
13			Staff Changing - Male
14			Staff Changing - Female
15			A.D.L. Bedroom
16			A.D.L. Utility/Laundry
17			Children's Therapy Area
18			Sub-waiting
19			Optional Service:
			Pool Counter-current Unit
20			Piped Oxygen and Medical Vacuum
04.01.01	**Out-Patient Services**	Out-Patient's Department-Genito Urinary Medicine	Clinic Suites
02			Plaster Facilities
03			Colposcopy C/E and Treatment Room
04			Primary Analysis Facility
05			Office/Interview
06			Health Records
04.02.01		Adult Acute Day Patients	Nursing Section
02			Treatment Suite
03			Treatment Room and Treatment Bathroom
04.03.01		Dental	Department

HOSPITAL BUILDINGS

FUNCTIONAL SIZE		BUILDING NOTE NO.	INDEX	COST GUIDE £	REMARKS	
40 Patients	)	8	M	294 389		
60	)			376 775		
1 Suite	)	8	M	290 068		
19 Patients	)	8	M	190 850		
33	)			270 375		
9 Patients	)	8	M	48 004		
18	)			78 951		
1 Room	)	8	M	15 400		
1 Store	)	8	M	3 500		
1 Office	)			7 325		
1 C/E Room	)			12 830		
1 Office	)			7 325		
1 Secrataries	)			5 840		
2	)			9 500		
1 Suite	)	8	M	19 382		
1 Room	)			6 800		
1 Room	)			8 650		
1 Bedroom	)			11 479		
1 Room	)			10 330		
1 Suite	)			36 082		
4 Spaces	)			3 500		
1 Unit	)	8	M	4 000		
2 Outlets	)			1 200		
6 C/E Rooms	)			250 417		
12	)	12	M	512 512		
18	)			665 455		
1 Room	)	12	M	36 461		
1 Room	)	12	M	17 383		
2 Workstation	)			11 848		
3	)			13 596		
1 Office	)			7 114		
7 000 Records	)			6 307		
10 000	)			8 343		
1 Section	)	38	M	281 764	See HN(81)33.	
1 Suite	)	38	M	123 218	See HN(81)33.	
1 Suite	)	38	M	30 387	(i) See (81)33. (ii) Alternative to Treatment Suite.	
					Sessions:	Surgeries:
1 Surgery	)			53 708	up to 9	1
2	)			74 037	10 to 18	2
3	)	28	M	166 621	19 to 27	3
4	)			190 964	28 to 36	4
5	)			214 692	37 to 45	5

HOSPITAL BUILDINGS

REF. NO.	SERVICE	DEPARTMENT	ACCOMMODATION
04.03.02	Out-Patient Services (continued	Dental (continued)	Orthodontic Supplementary
04.04.01		Opthalmic	Clinic
02			Essential Complementary Accommodation: Staff Locker Room (with central changing)
03			Optional Accommodation: Dispensing Optician
04.05.01		Dermatology (Supplementary to main OPD)	
04.06.01		Ear, Nose and Throat	
04.07.01		Special Treatment	Clinic
04.08.01		Children	Out-Patient
			Comprehensive Assessment:
02			Next to OPD
03			Independent
04.09.01		Accident and Emergency	Department
			Essential Complementary Accommodation:
04.09.02			Staff Changing (Male)
03			Staff Changing (Female)
04			Recovery (One Bed)
05			Nursing Officer's Office
06			Optional Accommodation: Major Treatment Room
04.10.01		Neuro-physiology	Clinic
04.11.01		Maternity	Clinic
			Essential Complementary Accommodation:
02			Ultra-Sound Room
03			Snack Bar and Store
			Optional Accommodation:
04			Treatment Room
05			Staff Cloaks

HOSPITAL BUILDINGS

FUNCTIONAL SIZE	BUILDING NOTE NO.	INDEX	COST GUIDE £	REMARKS	
				Sessions:	Surgeries:
1 Surgery)			65 788	up to 9	1
2)	28	M	84 274	10 to 18	2
3)			108 060	19 to 27	3
3 Consultants)			233 630	See HN(81)33.	
4)	39	M	258 798		
5)			297 648		
6)			335 353		
1 Room)	39	M	5 022	See HN(81)33.	
1 Optician)	39	M	11 756	See HN(81)33.	
1 Department)		M	29 573	(i) See HN(76)193 and Design Guide issued October 1976. (ii) Department serves a population of 250 000 to 30 000.	
150 000 Pop)			245 001	See Design Guide issued February 1974.	
200 000)		M	261 831		
250 000)			279 199		
300 000 Pop)		M	216 337	See Design Guide issued September 1974.	
1 Department)	23	M	130 763	(i) See HN(83)26. (ii) Department comprises of 3 C/E rooms.	
1 Unit)	23	M	111 991	)See HN(83)26.	
1 Unit)			166 688	)	
10 000 Attndncs)			364 042		
20 000)			414 358		
30 000)	22	M	545 518		
50 000)			641 281		
70 000)			736 825		
1 Room)			7 427		
1 Room)			12 978		
1 Room)	22	M	8 887		
1 Room)			7 113		
1 Room)	22	M	35 321		
9 Suite)		M	489 282		
1 Clinic)	21	M	314 534		
1 Room)			10 667		
1 Bar)	21	M	9 853		
1 Room)	21	M	10 548		
1 Room)			8 088		

HOSPITAL BUILDINGS

REF. NO.	SERVICE	DEPARTMENT	ACCOMMODATION
05.01.01	Psychiatric Patients' Services	Acute Mental Illness	Ward (annex to Day Hospital)
05.01.02			Ward (independent from Day Hospital)
03			Ward (elderly)
04			Essential Complementary Accommodation (Wards): Special Bedroom
05			En Suite WC/Wash
06			Day Hospital
07			Essential Complementary Accommodation (Day Hospital): Treatment/Clean Utility
08			Staff Dining/Rest Room (small)
09			Staff Dining/Rest Room (large)
10			Optional Accommodation (Day Hospital): Patients' Utility Room
11			Cleaner's Room
12			Disposal Room
13			Office/Interview Room
14			Day Hospital for the Elderly
15			Optional Accommodation (Day Hospital for the Elderly): Treatment Bathroom
16			Assisted Shower
17			Out-Patients' Suite
18			ECT Suite
20			Administrative Centre (small)
21			Administrative Centre (large)
05.02.01		Mental Handicap Hospital Units	Adult Residential Unit
02			Children's Residential Unit

HOSPITAL BUILDINGS

FUNCTIONAL SIZE		BUILDING NOTE NO.	INDEX	COST GUIDE £	REMARKS
10 Beds	)			185 438	
15	)	35	M	235 541	
20	)			302 178	
10 Beds	)			240 296	
15	)	35	M	322 277	
20	)			418 997	
25	)			491 680	
10 Beds	)			245 793	
15	)	35	M	336 512	
20	)			416 517	
25	)			487 494	
1 Bedroom	)	35	M	9 861	
1 Suite	)			7 657	
10 Places	)			139 831	Community Location
20	)	35	M	211 000	
30	)			298 307	
40	)			348 111	
1 Room	)			10 046	
1 Room	)	35	M	7 967	
1 Room	)			12 058	
1 Room	)			7 872	
1 Room	)	35	M	6 765	
1 Room	)			4 061	
1 Room	)			7 157	
10 Places	)			166 997	Community Location
20	)	35	M	235 607	
30	)			356 393	
40	)			441 110	
1 Bathroom	)	35	M	13 682	
1 Shower	)			8 205	
3 C/E Rooms	)	35	M	87 105	Only required when main OPD not
6	)			165 417	used.
1 Suite	)	35	M	78 841	
1 Centre	)	35	M	61 352	
1 Centre	)	35	M	109 614	
12 Places	)		M	147 782	See HN(77)58, Annex III to HN(80)21
24	)			251 796	and Design Bulletins Nos 1 and 2.
8 Places	)			121 002	See HN(77)58, Annex III to HN(80)21
16	)		M	205 623	and Design Bulletins Nos 1 and 2.
24	)			285 354	

HOSPITAL BUILDINGS

REF. NO.	SERVICE	DEPARTMENT	ACCOMMODATION
05.02.03	**Psychiatric Patients' Services (continued)**	Mental Handicap Hospital Units	Adult Day Care Unit
05.03.01		Mental Handicap Community Units	Normally Handicapped
05.03.02			Heavily Handicapped
06 01.01	**Teaching Facilities**	Education and Training	Teaching Accommodation (1 = Small Unit; 2 = Medium Unit; 3 = Large Unit;)
02			Lecture Hall
03			Library
04			Post Graduate Medical and Dental Education (1 = Small Unit; 2 = Large Unit)
05			Nurse and Midwife Education
06			Optional Accommodation: Projection room
07			Medical/Dental Tutor's Office
08			Centre Administrator's Office
06.02.01		Education Centre	
07.01.01	**Administration Services**	Administration	General
02			Medical Records
03			Group Accommodation
07.02.01	**Main Entrance**		Main Entrance Accommodation
02			Snack Bar Facilities

HOSPITAL BUILDINGS

FUNCTIONAL SIZE		BUILDING NOTE NO.	INDEX	COST GUIDE £	REMARKS
60 Places	)		M	313 329	See HN(77)58, Annex III to HN(80)21 and Design Bulletins Nos 1 and 2.
115	)			459 413	
8 Places	)			151 006	See HN(80)21 Annex III for on-cost details.
12	)			201 005	
16	)		M	250 999	
20	)			292 215	
24	)			333 429	
8 Places	)			158 606	See HN(80)21 Annex III for on-cost details.
12	)			211 308	
16	)		M	264 009	
20	)			307 249	
24	)			350 490	
1 Unit Size	)			240 324	
2	)	42	M	269 909	
3	)			347 058	
80 Places	)			121 231	
100	)	42	M	141 109	
120	)			157 681	
42 Places	)			139 249	
77	)	42	M	178 950	
160	)			273 768	
1 Unit size	)	42	M	110 373	
2	)			135 082	
185 Places	)			560 939	
400	)	42	M	879 328	
811	)			1 455 838	
1 Room	)			7 551	
1 Office	)	42	M	6 408	
1 Office	)			6 408	
1 Centre	)		M	65 652	(i) See DS 71/73. (ii) Serves 800 beds. (iii) Provides facilities for staff other than doctors or nurses.
9 Points	)			287 773	300 beds = 9 points
12	)	18	M	338 339	450 beds = 12 points
16	)			401 340	600 beds = 16 points
20	)			464 343	800 beds = 20 points
9 points	)			137 215	300 beds = 9 points
12	)	18	M	168 571	450 beds = 12 points
16	)			206 157	600 beds = 16 points
20	)			243 743	800 beds = 20 points
19 points	)			154 635	Points on same scale as Department 07.01.01 (General Administration).
30	)	18	M	211 578	
40	)			263 690	
1 Main Entrance	)	51	M	246 085	
1 Snack Bar	)	51	M	99 484	

HOSPITAL BUILDINGS

REF. NO.	SERVICE	DEPARTMENT	ACCOMMODATION
08.01.01	Staff Facilities	Catering (Dining)	Dining Rooms
02			Servery
03			Essential Complementary Accomodation: Furniture Store
08.02.01		Residential Accommodation for Staff	
08.03.01		Occupational Health Centre for Staff	
08.04.01		Staff Changing and Storage of Uniforms (excluding medical staff)	Manual System
02			Semi-Automatic System
03			Fully Automatic System
04			Essential Complementary Accommodation: Sewing Room
08.04.05			Optional Accommodation: Additional Storage for Uniforms (Manual system)
06			(Automatic system)
07			Addition for Storage on Hangers (Manual System)
09.01.01	Service Facilities	CSSD	
09.02.01		Telephone Services	Operator's Suite and Equipment Room

HOSPITAL BUILDINGS

FUNCTIONAL SIZE		BUILDING NOTE NO.	INDEX	COST GUIDE £	REMARKS
300 Meals	)	10	M	194 354	See HN(85)21.
600	)			300 908	
300 Meals	)	10	M	131 290	See HN(85)21.
600	)			170 928	
1 Store	)	10	M	7 861	See HN(85)21.
100 Points	)			117 164	(i) For each additional point after the first 500, allow £746.
150	)			164 830	(ii) Points scale per person:
200	)	24	M	211 381	Scale A: 6.5 points
250	)			254 397	Scale B: 8.5 points
500	)			489 963	Scale C: 12.5 points
					Scale D: 19.5 points
					Scale E: 24 points
					Scale F: 22 points
					Scale G: 28 points.
1 Centre	)		M	67 355	(i) See Design Guide issued April 1973. (ii)Centre serves 1000 beds.
300 Places	)			138 129	See HN(83)26.
600	)	41	M	204 592	
900	)			278 011	
1 200	)			356 266	
300 Places	)			145 226	See HN(83)26.
600	)	41	M	221 007	
900	)			302 778	
1 200	)			385 781	
300 Places	)			126 420	See HN(83)26.
600	)	41	M	173 543	
900	)			227 206	
1 200	)			286 274	
1 Seamstress	)	41	M	8 906	See HN(83)26.
2	)			13 706	
300 Places	)	41	M	1 884	)(i) For example, with central)issue of uniforms where staff use
300 Places	)			9 158	)local changing facilites.)(ii) See HN(83)26.
300 Places	)	41	M	4 094	See HN(83)26.
1 Department	)	13	M	240 223	
1 Cabinets	)			36 877	
2	)	48	M	45 542	
3	)			53 605	

HOSPITAL BUILDINGS

REF. NO.	SERVICE	DEPARTMENT	ACCOMMODATION
09.02.02	Service Facilities (continued)	Telephone Services (continued)	Telephone Equipment
09.03.01		Catering (Kithcens)	Central Kitchens
02			Optional Accommodation: Facilities for Future Expansion of 600-meal Kitchen
03			Bulk Store
04			Optional Accommodation: Storeman's Office
09.04.01		Laundries	Laundry
02			Infants' Napkin Section
03			Special Personal Clothing
04			Dry Cleaning Facility
09.04.05			Storage at Sending Hospital
10.01.01	Hospital Engineering and Works Services	Boiler Houses and Fuel Storage	

HOSPITAL BUILDINGS

FUNCTIONAL SIZE	BUILDING NOTE NO.	INDEX	COST GUIDE £	REMARKS
200 Extensions)				
300)				
400)				
500)				
600)				
700)				
800)				
900)				
1 000)				
1 100)				
1 200)				
1 300)				
1 400)				
1 500)				
1 600)				
600 Meals)	10	M	954 381	See HN(85)21.
1 200)			1 170 400	
1 Facility)	10	M	28 404	See HN(85)21.
600 Meals)	10	M	85 729	See HN(85)21.
1 200)			107 315	
1 Office)	10	M	4 537	See HN(85)21.
55 000 Art/wk)			1 115 703	See HN(77)65.
120 000)	25	M	2 104 158	
180 000)			2 988 838	
250 000)			3 899 099	
1 Section)	25	M	49 359	(i) See HN(77)65. (ii) Cost allowance applies to over 10 000 articles/week
3 500 Art/wk)			64 286	See HN(77)65
5 000)	25	M	85 568	
7 000)			118 401	
600 Kilos/wk)			62 300	See HN(77)65.
950)	25	M	77 430	
1 450)			105 520	
2 400)			151 547	
300 Bed)			23 060	See HN(77)65.
400)			32 262	
600)	25	M	43 977	
800)			49 627	
1 000)			64 231	
1 200)			72 773	
1 760 K/watts)			366 593	(i) See HN(81)33. (ii) Cost Allowances are for oil-fired installations: Coal-fired installations should be taken as approximately 50% increase on these figures.
4 400)			586 941	
6 150)	16	M	711 635	
8 800)			868 054	
11 700)			994 485	
20 500)			1 283 205	

HOSPITAL BUILDINGS

REF. NO.	SERVICE	DEPARTMENT	ACCOMMODATION
10.02.01	**Hospital Engineering and Works Services**	Works	Main Department
02			Lock-up Store and Workshop
11.01.01	**Community Health Services**	Health Centre	Primary Care Services: General Medical Practitioner Services
02			Community Health Services, Health Visiting, District Nursing and Midwifery and Welfare Foods
03			Shared Facilities: Related to General Medical Practitioners
04			Related to Community Health Services
05			Other Services: Chiropody
06			Speech Therapy and/or Child Health Assessment
07			Social Services
08			Dental Services: School and Priority Dental Services
09			General Dental Services

HOSPITAL BUILDINGS

FUNCTIONAL SIZE	BUILDING NOTE NO.	INDEX	COST GUIDE £	REMARKS
300 Bed)			154 386	(i) Compound, garages and a separate lock-up store are included. (ii) Beds are total in group served by the Department.
600)			154 386	
601)			202 052	
800)			202 052	
801)			249 716	
1 200)	34	M	249 716	
1 201)			299 811	
2 000)			299 811	
2 001)			349 397	
3 000)			349 397	
3 001)			389 689	
4 000)			389 689	
300 Bed)			4 775	(i) To serve hospital or group remote from the Works Department. (ii) Beds are total in group served by the Department.
600)			4 775	
601)			7 615	
800)			7 615	
801)			9 615	
1 200)	34	M	9 615	
1 201)			11 556	
2 000)			11 556	
2 001)			13 141	
3 000)			13 141	
3 001)			13 991	
4 000)			13 991	
3 GMPs)			58 818	)These allowances do not attract
6)	36	M	86 468	)the full on-costs of HNPN6. See
9)			135 132	)HN(76)54 for on-cost norms.
12)			168 194	)
7 500 Pop)			63 652	)These allowances do not attract
15 000)	36	M	87 590	)the full on-costs of HBNP6. See
22 500)			112 246	)HN(76)54 for on-cost norms.
30 000)			140 051	)
3 GMPs)			71 012	)
6)			103 792	)
9)			138 522	)These allowances do not attract
12)	36	M	171 657	)the full on-costs of HBPN6. See
7 500 Pop)			86 020	)HN(76)54 for on-cost norms.
15 000)			110 462	)
22 500)			130 514	)
30 000)			157 267	)
1 Chiropodist)			8 676	)
2)			17 404	)
1 Therapist)			13 670	)These allowances do not attract
2)	36	M	20 343	)the full on-costs of HBPN6. See
1 Soc. Worker)			5 130	)HN(76)54 for on-cost norms.
2)			5 662	)
3)			11 238	)
1 Dentist)			47 400	)
2)			60 954	)
3)			77 308	)These allowances do not attract
4)	36	M	91 427	)the full on-costs of HBPN6. See
1 Dentist)			31 254	)HN(76)54 for on-costs norms.
2)			45 195	)
3)			73 073	)
4)			87 585	)

HOSPITAL BUILDINGS

REF. NO.	SERVICE	DEPARTMENT	ACCOMMODATION
11.01.10	**Community Health Services (continued)**	Health Centre (continued)	Hospital Consultant Services
11			Pharmaceutical Services
11.02.01		Health Authority Clinic	
11.03.01		Ambulance Station	
20.01.01	**Teaching Hospital**	Administration	Medical Records: Reading Space for Students
02			Additional Storage Space for Records
03			Changing Accommodation
20.02.01	**Teaching Hospital (continued)**	Adult Nursing Units (general acute nursing sections, including geriatric assessment and rehabilitation)	Seminar Rooms
02			Clinical Investigation Area
03			Urine Testing Room
04			Student's Room
20.03.01		Psychiatric	Interview Room
02			Seminar/Demonstration Room
20.04.01		Geriatric Units (applicable to units not covered by Adult Nursing Units)	Seminar/Demonstration Room
20.05.01		Special Units: Intensive Therapy and Coronary Care	Seminar Room

HOSPITAL BUILDINGS

FUNCTIONAL SIZE	BUILDING NOTE NO.	INDEX	COST GUIDE £	REMARKS
9 Dr Sessions)	36	M	9 185	)These allowances do not attract
18)			18 501	)the full on-costs of HBPN6. See)HN(76)54 for on-cost norms.
15 000 Pop)	36	M	30 035	)(i) Cost allowances relate to
15 001)			39 874	)populations of up to 15 000.)(ii)These allowances do not)attract the full on-costs of)HBPN6. See HN(76)54 for on-cost)norms.
7 000 Pop)			121 636	See LABN3.
10 000)			121 636	
10 001)		M	136 470	
20 000)			136 470	
20 001)			154 270	
30 000)			154 270	
				Draft new guidance under way. Contact A. Maun at DHHS.
1 Space)		M		(i) See DS 65/74. (ii) UGC funds £1 866 (QSSD base 100) excluding VAT.
1 Space)		M		(i) See ds 65/74. (ii) UGC funds £8 049 (QSSD base 100) excluding VAT.
1 Space)		M		(i) See DS 65/74. (ii) UGC funds £79 (QSSD base 100) excluding VAT.
2 Rooms)		M	29 038	(i) See DS 65/74. (ii) Funded by the NHS.
1 Area)		M	26 210	(i) See DS 65/74. (ii) Funded by the NHS.
1 Room)		M	3 579	(i) See DS 65/74. (ii) Funded by the NHS.
1 Room)		M	6 623	(i) See DS 65/74. (ii) Funded by the NHS.
6 Rooms)		M	28 487	(i) See DS 65/74. (ii) Funded by the NHS.
1 Room)		M	21 889	(i) See DS 65/74. (ii) Funded by the NHS.
1 Room)		M	21 889	(i) See DS 65/74. (ii) Funded by the NHS.
1 Room)		M	16 415	(i) See DS 65/74. (ii) Funded by the NHS.

HOSPITAL BUILDINGS

REF. NO.	SERVICE	DEPARTMENT	ACCOMMODATION
20.06.01	Teaching Hospital (continued)	Maternity	Bed Areas: Seminar Room
02			Clinical Investigation Area
03			Urine Test Rooms
04			Ante-Natal Clinic: Seminar Room
05			Labour Suite: Students' Common Room
06			Operating Theatre Suite: CCTV to one Seminar Room
07			Student Changing Accommodation
08			Special Care Baby Units: Seminar Room
20.07.01		Children	Bed Areas: Seminar Room
02			Clinical Investigation Area
03			Urine Test Room
04			Students' Room
05			Out-Patients: Seminar Room
20.08.01		Operating Suites	Demonstration/Seminar Room
02			Student Changing Accommodation
20.09.01		Pathology	Additional Laboratory Space
20.10.01		Mortuary and Post Mortem Room	Observation Gallery
02			Student Changing Accommodation
20.11.01		Rehabilitation	Seminar Room

HOSPITAL BUILDINGS

FUNCTIONAL SIZE	BUILDING NOTE NO.	INDEX	COST GUIDE £	REMARKS
1 Room)			21 889	)
1 Area)		M	26 210	)(i) See DS 65/74. (ii) Funded by
2 Rooms)			7 158	)the NHS.
1 Room)		M	16 415	(i) See DS 65/74. (ii) Funded by the NHS.
1 Room)		M		(i) See DS 65/74. (ii) UGC funds £3 112 (QSSD base 100) excluding VAT.
1 Unit)		M		(i) See DS 65/74. (ii) UGC funds £101 (QSSD base 100) excluding VAT
1 Room)		M		(i) See DS 65/74. (ii) UGC funds £786 (QSSD base 100) excluding VAT
1 Room)		M	16 415	(i) See DS 65/74. (ii) Funded by the NHS.
1 Room)		M	16 415	(i) See DS 65/74. (ii) Funded by the NHS.
2 Areas)		M	52 417	(i) See DS 65/74. (ii) Funded by the NHS.
1 Room)		M	3 579	(i) See DS 65/74. (ii) Funded by the NHS.
1 Room)		M	6 623	(i) See DS 65/74. (ii) Funded by the NHS.
1 Room)		M	21 889	(i) See DS 65/74. (ii) Funded by the NHS.
1 Room)		M		(i) See DS 65/74. (ii) UGC funds £7 680 (M base 100) excluding VAT.
1 Room)		M		(i) See DS 65/74. (ii) UGC funds £786 (M base 100) excluding VAT
1 Space)		M	14 698	(i) See DS 65/74. (ii) Funded by the NHS.
1 Gallery)		M		(i) See DS 65/74. (ii) UGC funds £122 (M base 100) excluding VAT
1 Room)		M		(i) See DS 65.74. (ii) UGC funds £786 M base 100 excluding VAT
1 Room)		M	16 415	(i) See DS 65/74. (ii) Funded by the NHS.

HOSPITAL BUILDINGS

REF. NO.	SERVICE	DEPARTMENT	ACCOMMODATION
20.12.01	**Teaching Hospital (continued)**	Out-Patients	Consulting Suites: Increase in Doctor Sessions
02			Seminar Room
03			Adult Acute Day Patients Treatment Suite: Student Changing Accommodation
20.13.01		Accident and Emergency	Examination Rooms
02			Seminar Room
03			Students' Rest Room
04			Operating Theatre: Student Changing Accommodation

HOSPITAL BUILDINGS

FUNCTIONAL SIZE		BUILDING NOTE NO.	INDEX	COST GUIDE £	REMARKS
Dr Sessions	)	12	M		(i) See DS 65/74. (ii) Increase the number of doctor sessions by 10 percent. Cost allowances are on the same scale as those provided for in Department 04.01.01. (iii) Funded by the NHS.
1 Room	)		M	21 889	(i) See DS 65/74. (ii) Funded by the NHS.
1 Room	)		M		(i) See DS 65/74. (ii) UGC funds £786 (M base 100) excluding VAT
2 Rooms	)		M	13 049	(i) See DS 65/74. (ii) Funded by the NHS.
1 Room	)		M	16 415	(i) See DS 65/74. (ii) Funded by the NHS.
1 Room	)		M		(i) See DS 65/74. (ii) UGC funds £1 794 (M base 100) excluding VAT.
1 Room	)		M		(i) See DS 65/74. (ii) UGC funds £786 (M base 100) excluding VAT

UNIVERSITY BUILDINGS

The Universities Funding Council's document University Building Projects - Notes on Control and Guidance 1987, contains notes to assist universities in investment appraisal studies, preparing briefs and establishing expenditure limits for projects.

The notes contain areas derived from the Council's 'Planning Norms for University Buildings', and costs that are an expression of the standards and facilities considered appropriate by the Council.

The following notional unit areas taken from the notes are intended as a guide for planning purposes.

ACADEMIC AREAS		Academic Staff: Student ratio	Usable Area/FTE Ug m2	PgC m2	PgR m2	Inclusion for academic staff m2
Departmental areas including centrally timetabled seminar rooms for:						
1 + 2	Preclinical medicine and dentistry	1:8	14.1	14.1	20.4	3.8
3	Clinical medicine	1:6	6.5	6.5	22.2	5.2
4	Clinical dentistry	1:6	10.5	13.0	16.2	4.2
5	Studies allied to medicine	(as for Physical sciences)				
6	Biological sciences	1:9	9.2	12.2	19.1	3.4
	Experimental Psychology	1:11	8.2	10.5	19.5	3.0
7	Agriculture & Forestry	(as for Biological sciences with special additions)				
	Veterinary science	(as for Biological sciences with special additions)				
8	Physical sciences	1:8	9.8	12.3	18.4	3.8
9	Mathematics	1:11	3.6	3.6	5.0	1.3
10	Computer studies	1:11	7.3	10.2	11.2	2.2
11	Engineering and technology	1:9	9.8	16.6	17.9	3.2
12	Architecture, Building and Planning	1:8	9.8	9.8	9.4	2.3
13	Geography	1.14	5.4	6.8	7.1	1.0
	Economics	1.12	2.4	2.4	4.9	1.2
14	Politics, Law and other social studies	1:12	2.4	2.4	4.9	1.2
15	Business Management	1:11	3.3	3.3	5.0	1.3
	Accountancy	1.12	2.4	2.4	4.9	1.2
17	Languages	1.10	3.5	3.5	5.2	1.4
18	Humanities	1.11	2.6	2.6	5.2	1.3
	Archaeology	1.10	5.5	6.8	7.2	1.4
19	Art and Design	1.10	9.1	9.1	8.8	1.9
	Music, Drama	1:8	7.9	7.9	9.0	2.2
20	Education	1.11	5.0	5.0	4.8	1.3

Non departmental academic areas:	
Lecture Theatres, Provision for Ug and PgC only	0.5
Library	
Basic provision	1.25
Addition where necessary for law students	0.8
Expansion provision	0.2
Special collection	Ad hoc
Reserve book stores	50 m2 + 3.5 m2/1000 Vols

NON-ACADEMIC AREAS

Administration (including maintenance)	up to 3000 FTE	0.8
	over 3000 FTE	0.5
Social, Dining (excluding Kitchens), Health	up to 3000 FTE	1.4
	over 3000 FTE	1.3
Social, Health (clinical medical only)		1.3
Sports Facilities (buildings)		
Indoor	up to 3000 FTE	0.5
	300 to 6000 FTE	0.1
	over 6000 FTE	0.3
Outdoor	up to 3000 FTE	0.2
	over 3000 FTE	0.1

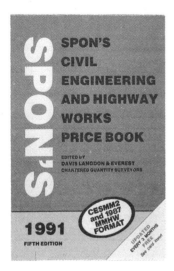

UNIVERSITY BUILDINGS

The following cost guidance taken from the notes is intended to assist in establishing expenditure limits for minor works projects (less than £1 m). The rates are expressed as the cost per metre of usable floor area at 1st quarter 1988 tender price levels and include for the normal balance area.

The Expenditure Limit for major projects (ie over £1 m) will be related to the facilities that are agreed to be necessary and may well differ from a limit based on the following rates.

	Subject Group	Basic Rate	Additions for specialist facilities	
		(1st quarter 1988 tender price levels)	Specialist area as % of total usable area	Rate (1st quarter 1988 tender price levels)
		£/usable m2	%	£/usable m2
1 + 2	Preclinical medicine and dentistry	745	80	290
3	Clinical medicine	745	55	180
4	Clinical dentistry	745	65	250
5	Studies allied to medicine - Nursing	700	50	60
	- Pharmacy	745	80	240
6	Biological sciences	745	80	240
	Experimental psychology	700	60	95
8	Physical sciences			
	Physics	710	75	85
	Geology	745	70	230
	Chemistry	745	75	260
9	Mathematics	700	45	5
10	Computer studies	700	60	35
11	Engineering and technology			
	Equipment dominated	605	To be agreed ad hoc	
	Person orientated	700	To be agreed ad hoc	
12	Architecture, Planning	700	60	20
13	Geography	700	50	25
14	Politics, Law and other social studies	700	--	--
15	Business management	700	15	5
17	Languages	700	15	30
18	Humanities	700	--	--
	Archaeology	700	45	25
19	Creative Arts			
	Music	700	50	105
	Drama	700	35	40
	Art and Design	700	60	20
20	Education	700	50	60

NOTE: Rates for specialist facilities to be applied to total usable area

Type of accommodation	Total rate (1st quarter 1988 tender price levels) £/usable m2
Lecture theatres	1000
Main libraries	850
Social, Dining, Unions	755
Maintenance Workshops	545
Internal Sports (exluding swimming pools)	650

Rates revised for 1st quarter 1988 tender price levels have been rebased using the DOE:PSA Public Sector Tender Price Index for firm price tenders (FP), apply to total area for specialist facilities, and also take into account the switch to fixed price tenders for Univeresities minor building works.

UNIVERSITY BUILDINGS

A separate document exists for the appraisal of and establishing expenditure limits for medical projects - the University Building Projects Notes on Control and Guidance for Medical Projects 1986. These notes deal with the provision of university accommodation for medecine and dentistry and should be read in conjunction with Notes on Control and Guidance of University Building Projects (NOCAG) 1986. These notes only apply to medical projects in which either the NHS has no share or for joint UFC/NHS projects which are managed by the university. When the Health Authority manages the work and controls expenditure, then these projects are subject to NHS building control procedures.

The following notional unit areas, taken from Annex C, are included in the Council's planning norms and notional limits on which the cost limits are based.

The figures, based on staff: student ratios of 1:8 for preclinical and 1:6 for clinical medicine and dentistry, are:-

Total academic usable area per FTE student

	UG m2	PGC m2	PGR m2	m2	
Preclinical M & D	14.1	14.1	20.4	3.8	) included in total
Clinical - M	6.5	6.5	22.2	5.2	) usable areas for
Clinical - D	10.5	13.0	16.1	4.2	) academic staff use

CATEGORY

General teaching:	Usable area m2
small groups (up to six)	6.5 for 6 places
larger groups	1.2 per place
class, seminar rooms (informal seating)	1.9 per place
class, seminar rooms (tables and chairs)	2.3 per place

Note: 0.35m2 per FTE UG and PGC student is included in notional unit area for general teaching.

Lecture theatres:	
lecture theatres (close seating)	1.0 per place
lecture theatres (clinical demonstration):	
additional area for patients in one clinical theatre per school:	
demonstration	18.5
waiting, preparation	18.5

	37.0

Note: Norm = 1 place per 2 FTE (UG and PGC) students.

Offices:	
Professors (including small group space)	18.5 per room
Tutorial staff (including small group space)	13.5 per room
Non tutorial staff (e.g. researchers, secretaries)	7.0 per person
Chief technician	9.3 per person

Laboratory places (teaching):	
(except Anatomy dissecting room)	
preparation + stores	5.0 per place
	2.5 per place

	7.5

UNIVERSITY BUILDINGS

	Usable area m2
Anatomy department:	
dissecting room	2.5 per place
preparation + stores	1.5 per place
museum	0.5 per place

	4.7
Clinical dental:	
preparation + stores	5.0 per place
	2.5 per place

	7.5
Laboratory places (individual research):	
Medical and all preclinical academic staff:	
research space	11.0 per place
ancillary	5.5 per place

	16.5
Medical and all preclinical research fellows, postgraduates and NHS staff:	
research space	10.0 per place
ancillary	6.0 per place

	16.0
Dental academic staff:	
research space	7.3 per place
ancillary	3.7 per place

	11.0
Dental research fellows, postgraduates, and NHS staff:	
research space	7.3 per place
ancillary	3.7 per place

	11.0
AVA facilities:	
Photographic and TV studios including ancillary areas (to be planned in association with hospital Medical Illustration Department if on hospital site).	0.3 per U/G student
Animal houses (other than in departmental ancillary areas):	Assessed ad hoc
Libraries, central administration including maintenance, social and indoor sports facilities.) for medical students	- See non-medical space standards and notional unit areas
Dining and kitchens (if not provided with NHS facilities)	- See non-medical notional unit areas

UNIVERSITY BUILDINGS

OTHER UNIVERSITY AREAS IN TEACHING HOSPITALS

Teaching and Research areas in Hospital Departments (Standard Teaching Additions)

The following basic areas were agreed in 1974 between the UFC and Health Departments (DHSS and SHHD) as being required for clinical teaching in a main teaching hospital in addition to:

a) teaching, research, and social facilities provided by the university for medical school staff and students, including postgraduate students taking a university degree or diploma, and

b) facilities provided by the NHS for the postgraduate education and training of doctors or dentists and the training of nurses, physiotherapists and laboratory technicians.

The basic areas listed below are subject to adjustment where specialised teaching is carried out in hospitals peripheral to the main teaching hospital(s).

Department	Net area m2 UFC-funded	NHS-funded
Administration:		
medical records	63.5	-
student changing	0.4 per FTE student	-
Acute nursing units (incl. Geriatric Assessment and Rehabilitation)	-	105.5 per 60.72 beds
Psychiatric Units	-	91.0
Geriatric Units	-	37.0
Intensive Care/Therapy Units	-	28.0
Maternity Department	18.5 + (4.0 per theatre)	142.0
Childrens Department	-	158.5
Main Operating Department	37.0 + (4.0 per theatre)	-
Pathology	-	20.5
Mortuary/Postmortem Room:		
gallery	$\frac{\text{1.0 (per annual student intake)}}{2}$	
student changing	4.0 per table	-
Rehabilitation Department (Physiotherapy, Occupational Therapy)	-	28.0
Outpatients Department	4.0 per theatre	37.0 per 100 doctor sessions
Accident and Emergency Department	11.0 + (4.0 per theatre)	50.0

For details see DHSS CIS notes and Health Departments/UFC document entitled 'Teaching Hospital Space Requirements'.

UNIVERSITY BUILDINGS

OTHER AREAS

Dining and Kitchens	- assessed on the basis of 1 meal per FTE student at NHS space and cost rates.
On-call residence	- where needed on hospital sites for teaching purposes in addition to the students' normal place of residence, these should normally be provided for numbers not exceeding one-third of the annual intake of clinical medical students at each medical school and in accordance with NHS nurses Scale A residences excluding common room facilities.
Car parking	- surface parking for medical school academic and technician staff on a hospital site need not exceed 1 place to every 3 FTE students.

UNIVERSITY ACCOMMODATION - DEPARTMENTAL COSTS

The following cost guidance, taken from Annex D, and since revised, are at 1st quarter 1988 tender price levels and should be applied to the departmental area or functional unit for university accommodation in each category, whether the space is to be used exclusively by the university or provided as part of a joint facility to be shared with NHS users, and aggregated to form the university's departmental cost limit.

Building rates have been rebased using the DOE:PSA Public Sector Tender Price Index for fluctuating (VOP) rather than firm price (FP) tenders, as non-medical rates.

The figures should be discussed with the Health Authority and included in their schedules of area and cost for the whole project at appropriate stages in the design of the project.

Departmental rates are not given for preclinical academic accommodation and if a case should arise where such accommodation is to be provided under Health Authority control using HBPN no. 6 procedures, reference should be made to the Universities Funding Council.

Rates for University Academic and Non-Academic Areas:

Academic areas		£/m2 @ Q1/1988 Departmental area
Basic rate for:		
Clinical medicine		450
Clinical dentistry		450
Additional rate for specialist facilities expressed as a percentage of the specialist usable area to be applied to the total departmental area:		
*40% (clinical medicine)		180
*50% (clinical dentistry)		225
75% (clinical medicine)	) whole usable area	290
	) with specialist area	
75% (clinical dentistry)	) facilities	310

*These percentages represent a full range of specialist facilities for a complete department included in the notional unit areas given in Annex C.

UNIVERSITY BUILDINGS

Non-academic areas

Libraries, central administration, social space and other non specialised areas	450
Lecture theatres	715

Other areas

Animal houses, AVA, photographic/medical illustration, etc.	ad hoc
Teaching and Research areas in Hospital departments (Standard Additions)	NHS rates
Dining Rooms and Kitchens	NHS rates
On-call residence places	NHS rates

For further information on Rates for other university areas, cost apportionment, on-costs and professional fees, refer to Annex D of Notes on Control and Guidance for Medical

The rates set out above from both the Notes on Control and Guidance and Notes on Control and Guidance for Medical Projects, both at 1st quarter 1988 price levels, are still current but the Universities Funding Council makes allowances based on normal tender indices. The latest DOE:PSA Public Sector Tender Price Index figures are 134 (FP) and 115 (VOP), both provisional figures for the 3rd quarter 1989. The indices for the 1st quarter 1988 were 122 (FP) and 110 (VOP).

EDUCATIONAL BUILDINGS (Procedures and Cost Guidance)

In January 1986 the Department of Education and Science introduced new procedures for the approval of educational building projects.

In its Administrative Memorandum 1/86 the Department currently requires detailed individual approval only for projects costing £2 millions and over. Minor works i.e., costing less than £200,000, require no approval except where the authority requires such or where statutory notices are involved. Projects of £200,000, up to £2 millions require approval which will be given automatically upon the submission of particulars of the scheme and performance data relating to it.

Capital works and repair projects supported by grant from the Department will continue to require individual approval.

Data relating to the performance of educational building (area and cost standards) will be published periodically and tender price data will continue to be issued quarterly. The current forecast cost of £515 per square metre (Q2 1990) is based upon a forecast public sector tender price index of 307 (1975 = 100) and relates to a mixture of types of accommodation normally found in complete new schools and further education establishments.

LOCAL AUTHORITY HOUSING AND HOUSING ASSSOCIATION SCHEMES

The cost limit and allowances arrangements for Local Authority and Housing Association schemes were previously very similar, with the Department of Enviroment admissable cost limits (ACL's) corresponding closely with the Housing Corporation's total indicative costs (TIC's) system.

However, 1989 has seen a divergence of procedures brought about by the Housing Corporation introducing a new system based on total cost indicator (TCI) tables.

Radically altering the previous arrangements, the new TCI's, which come into effect from 1st April 1989, represent not just development costs, but estimates of final costs at practical completion including professional fees, interest charges etc. There also now exist some divergence between groups and regions applicable to the two schemes.

LOCAL AUTHORITY HOUSING SCHEMES

In 1988 the Secretary of State for the Environment abolished the housing project control system as set out in DOE Circular 23/82, and substituted arrangements for assessing admissable costs for subsidy on local authority capital expenditure on housing. These arrangements, set out in DOE Circular 5/88 do not cover:-

(a) slum clearance (see DOE Circulars 129/74 and 42/75);
(b) improvement for sale and homesteading (see DOE Circular 20/80);
(c) local authority funded housing association schemes (see DOE Circular 14/83);
(d) grants and loans to the private sector. (For grants see DOE Circular 21/80 and the Department's letter of 5 February 1987 about revenue subsidies to private landlords);
(e) environmental works in General Improvement Areas and Housing Action Areas

The new system is based on a comparison of average actual resource costs with the appropriate figure from a table of admissible cost limits (ACLs). An ACL figure should represent a reasonable average cost for a dwelling of a specified size in a specified area; for 1989/90 the same figures will apply to both new build and renovation schemes, although this may not be the case in other years. As regards dwellings built to meet special housing needs, a further specified amount (only one of two options) will be added to the appropriate figure in the ACL table and it is the sum of these 2 amounts that will be the ACL for comparison with resource costs. The first step in calculating the amount of new admissible costs on housing capital expenditure is to determine the lower of the average actual resources costs per dwelling and the appropriate ACL figure. The intention is that an ACL should cover:-

for new build: the current market value of the land, construction costs and fees
for renovation: the pre-improvement value of the dwelling, works costs and fees

The latest ACLs and revised cost groups for 1989/90 have been published in DOE Circular 10/89, which replace those tables and annexes previously contained within DOE Circular 5/88.

For the benefit of readers, these revisions have been reproduced here in full.

DOE Circular 10/89 - Capital expenditures by Local Authorities under part II of the Housing act 1985: Admissible costs for subsidy 1989/90

1. **Since 1 April 1988 local authorities have been calculating new admissable costs for subsidy on housing capital expenditure by means of the admissable costs limits (ACLs) system. In consequence of paragraph 26 of DOE Circular 5/88 the distribution of local authorities into various cost groups has been reviewed and the local authority associations and local authorities have been consulted upon the outcome of the review. This circular now publishes the Secretary of State's decisions as regards the table of ACLs and the revised cost groups for the year 1989/90. Annexes A and B to this circular therefore replace Annexes A and B to DOE Circular 5/88 as from 1 April 1989. Expressions used in this circular which are used in the Housing Subsidies and Accounting Manual 1981 have, unless the contrary intention appears, the meaning they bear in the manual (e.g."year" means "financial year").**

LOCAL AUTHORITY HOUSING SCHEMES

Thresholds

2. The system of thresholds prescribing expenditure on renovation that is excluded from the averaging and comparison with ACLs is set out in paragraph 17 of DOE Circular 5/88. As local authorities will be aware, this has been reviewed. A proposal to dispense with the contract threshold was included in the list of amendments to the General Determination of Reckonable Expenditure (Appendix A to the Manual) on which the Department is currently consulting the local authority associations. Without any prejudice to those consultations, authorities are advised to proceed for the time being on the assumption that the proposed new rules on thresholds are to apply from 1 April 1989. Consultations close on 31 March. Responses will then be considered and a decision issued as soon as possible.

Building for Sale

3. The reference to the cost of dwellings built expressly for outright sale in Annex C to DOE Circular 5/88 is inconsistent with the provisions of the Manual. Even if the local authority puts up the working capital, the costs would be recouped if the dwelling was sold without being tenanted. Paragraph 1 of Annex C is therefore replaced with the following:

"The cost of dwellings built expressly for the outright sale even if met by the authority should not normally be included in the comparison between resource costs and ACLs. It may transpire however that after completion a local authority wishes or is obliged, for example under a buy-in guarantee provision, to transfer them into its rented stock, or to dispose of them on shared ownership terms. In such cases the Department's consent is required to an appropriate amount of admissible costs being added to the authority's total. In considering whether to give such consent the Department may have regard to the level of expenditure it considers reasonable in such cirumstances."

Provision for Review

4. The Department hopes that in future years it will be able to complete the annual review sooner than proved possible in this first year of the system, but it may not be able to do that by November. It will aim to provide authorities with more of the evidence on which the annual review is based; this has been a point of which the Department was pressed this year. Paragraph 26 of Circular 5/88 is therefore cancelled from 1 April 1989 and replaced with the following:

"The distribution of local authorities in ACL cost groups will be reviewed annually in consultation with the local authority associations. Details of the sources and data used by the Department in its review will be announced when proposals for the changes in cost groups for the ensuing year are issued for comment, together with the methodology for updating the ACLs. The table of ACLs and the cost groups will be published as early as possible before the year to which they relate."

Statistics

5. Local authorities are reminded of the need to complete forms LH 1-6 and the advantages of their doing so (see paragraph 25 of DOE Circular 5/88). It was proposed in the consultation letter of 22 December, on the Review of ACLs for 1989/90, that the revised General Determination of Reckonable Expenditure would render inadmissable any costs that have not been reported on these LH forms. The Department will give further thought to this proposal.

New Financial Regime

6. Authorities with a negative entitlement to subsidy for 1989/90 will know from paragraphs 15(i) and 25 of the consultation paper "New Financial Regime for Local Authority Housing in England and Wales" issued on 27 July 1988 and paragraph 4 of the Note issued on 20 October 1988 that they should assess costs of new tenders against ACLs because this will be taken into account in their base amount for HRA subsidy on 1 April 1990.

LOCAL AUTHORITY HOUSING SCHEMES

For further information on how to implement these new procedures the reader is referred to DOE Circulars 5/88 and 10/89 and also advised to check that the ACLs and other cost allowances provided here are valid at the time they are being used.

ANNEX A - NEW BUILD AND REHABILITATION

ADMISSIBLE COST LIMITS 1988/1989 - £ per dwelling

General needs accommodation

Dwelling floor area m2	Cost Groups A1 £	A £	B £	C £	D £	E £	F £	G £
25	62 400	54 700	45 700	38 200	32 000	26 600	22 300	18 600
35	73 200	64 200	53 700	44 800	37 400	31 300	26 100	21 800
45	83 600	73 300	61 300	51 200	42 700	35 700	29 800	24 900
55	93 900	82 400	68 900	57 500	48 100	40 100	33 500	28 000
65	104 400	91 600	76 500	63 900	53 400	44 600	37 200	31 100
75	114 300	100 300	83 900	70 000	58 400	48 800	40 800	34 100
85	124 400	109 100	91 100	76 100	63 600	53 100	44 400	37 100
95	133 500	117 100	97 900	81 800	68 300	57 000	47 600	39 800
105	143 100	125 500	104 800	87 600	73 200	61 100	51 100	42 700
115	152 300	133 600	111 500	93 200	77 800	65 000	54 400	45 400

Adjustments

Add for special needs (1 only of the following)

	A1	A	B	C	D	E	F	G
Old Person or Disabled Wheelchair	14 600	12 800	10 700	9 000	7 500	6 200	5 200	4 400
Old Person Warden Supervised or Frail Elderly	31 400	27 500	23 000	19 200	16 000	13 400	11 200	9 300

LOCAL AUTHORITY HOUSING SCHEMES

ANNEX B - ADMISSIBLE COST LIMITS GROUPS

Group A1 comprising:

Camden
Kensington & Chelsea
Westminster

Group A comprising:

The following **London Boroughs**

Barnet
Brent
Ealing
Enfield
Hackney
Hammersmith & Fulham
Haringey
Harrow
Hounslow
Islington
Kingston-upon-Thames
Lambeth
Richmond-upon-Thames
Southwark
Tower Hamlets
Wandsworth

City of London

The following Districts in the Counties of:

Essex
Epping Forest

Surrey
Elmbridge

Group B comprising:

The following **London Boroughs**

Bexley
Bromley
Croydon
Greenwich
Hillingdon
Lewisham
Merton
Newham
Redbridge
Sutton
Waltham Forest

The following Districts in the Counties of:

Berkshire
Bracknell Forest
Slough
Windsor & Maidenhead

Buckinghamshire
Chiltern
South Buckinghamshire

Hertfordshire
Broxbourne
East Hertfordshire
Hertsmere
St. Albans
Three Rivers
Watford
Welwyn Hatfield

Surrey
Epsom & Ewell
Mole Valley
Reigate & Banstead
Runneymede
Spelthorne
Tandridge
Woking

LOCAL AUTHORITY HOUSING SCHEMES

ANNEX B - ADMISSIBLE COST LIMITS GROUPS - cont'd

Group C comprising:

The following **London Boroughs**

Barking & Dagenham
Havering

The following Districts in the Counties of:

Bedfordshire
Luton

Berkshire
Reading
Wokingham

Buckinghamshire
Wycombe

Cambridgeshire
Cambridge

East Sussex
Brighton
Hove

Essex
Basildon
Brentwood
Castle Point
Chelmsford
Harlow
Rochford
Southend-on-Sea
Thurrock
Uttlesford

Hampshire
Basingstoke & Deane
Hart
Rushmore

Hertfordshire
all Districts not listed above

Isles of Scilly

Kent
Dartford
Gravesham
Sevenoaks
Tunbridge Wells

Oxfordshire
Oxford
South Oxfordshire

Surrey
all Districts not listed above

West Sussex
Chichester
Crawley
Horsham
Mid. Sussex

LOCAL AUTHORITY HOUSING SCHEMES

ANNEX B - ADMISSIBLE COST LIMITS GROUPS - cont'd

Group D comprising:

The following Districts in the Counties of:

Avon
Bath

Bedfordshire
all Districts not listed above

Berkshire
all Districts not listed above

Buckinghamshire
Aylesbury Vale

Cambridgeshire
Huntingdon
South Cambridgeshire

Dorset
Bournemouth
Christchurch
East Dorset
Poole

East Sussex
all Districts not listed above

Essex
all Districts not listed above

Hampshire
all Districts not listed above

Kent
Ashford
Gillingham
Maidstone
Rochester upon Medway
Shepway
Swale
Tonbridge & Malling

Norfolk
Broadland
Norwich

Oxfordshire
West Oxfordshire
Vale of White Horse

Suffolk
Babergh
Forest Heath
Ipswich
Mid-Suffolk
St. Edmundsbury
Suffolk Coastal

West Sussex
all Districts not listed above

Wiltshire
Kennet
North Wiltshire
Salisbury
Thamesdown

Group E comprising:

The following Districts in the Counties of:

Avon
all Districts not listed above

Buckinghamshire
all Districts not listed above

Cambridgeshire
East Cambridgeshire
Peterborough

Isle of Wight
Medina
South Wight

Kent
all Districts not listed above

Norfolk
all Districts not listed above

LOCAL AUTHORITY HOUSING SCHEMES

ANNEX B - ADMISSIBLE COST LIMITS GROUPS - cont'd

Group E (continued) comprising:

Devon
East Devon
Exeter
Mid-Devon
North Devon
Plymouth
South Hams
Teignbridge
Torbay

Dorset
all Districts not listed above

Gloucestershire
Cheltenham
Cotswold
Gloucester
Stroud
Tewkesbury

Hereford & Worcester
Bromsgrove

Northamptonshire
Northampton
South Northamptonshire

Oxfordshire
all Districts not listed above

Somerset
all Districts

Suffolk
all Districts not listed above

Warwickshire
Stratford-on-Avon
Warwick

Wiltshire
all Districts not listed above

The following District in the former Metropolitan County of:

West Midlands
Birmingham
Solihull

Group F comprising:

The following Districts in the former Metropolitan Counties of:

Greater Manchester
Bury
Manchester
Salford
Stockport
Tameside
Trafford

Merseyside
all Districts

South Yorkshire
Sheffield

Tyne & Wear
Newcastle upon Tyne

West Midlands
all Districts not listed above

West Yorkshire
Bradford
Leeds

LOCAL AUTHORITY HOUSING SCHEMES

ANNEX B - ADMISSIBLE COST LIMITS GROUPS - cont'd

Group F (continued) comprising:

The following Districts in the Counties of:

Cambridgeshire
all Districts not listed above

Cheshire
Chester
Congleton
Ellesmere Port & Neston
Halton
Macclesfield
Vale Royal

Cleveland
Hartlepool

Cornwall
all Districts

Cumbria
all Districts

Derbyshire
High Peak
Derbyshire Dales

Devon
all Districts not listed above

Durham
Darlington
Durham
Easington
Sedgefield
Teesdale

Gloucestershire
all Districts not listed above

Hereford & Worcester
all Districts not listed above

Humberside
Beverley
East Yorkshire
Hull

Lancashire
Blackpool
Burnley
Chorley
Fylde
Lancaster
Pendle
Preston
Ribble Valley
South Ribble
West Lancashire
Wyre

Leicestershire
all Districts

The following Districts in the Counties of:

Lincolnshire
Boston
Lincoln
South Holland
South Kesteven

Northamptonshire
all Districts not listed above

Nottinghamshire
Nottingham
Rushcliffe

Shropshire
Bridgnorth
Shrewsbury & Atcham
South Shropshire
The Wrekin

LOCAL AUTHORITY HOUSING SCHEMES

ANNEX B - ADMISSIBLE COST LIMITS GROUPS - cont'd

Group F (continued) comprising:

Northumberland
Alnwick
Berwick-upon-Tweed
Castle Morpeth
Tynedale

North Yorkshire
Craven
Harrowgate
Ryedale
Scarborough
Selby
York

Staffordshire
East Staffordshire
Lichfield
South Staffordshire
Tamworth

Warwickshire
all Districts not listed above

Group G comprising:

The following Districts in the former Metropolitan Counties of:

Greater Manchester
all Districts not listed above

South Yorkshire
all Districts not listed above

Tyne & Wear
all Districts not listed above

West Yorkshire
all Districts not listed above

The following Districts in the Counties of:

Cheshire
all Districts not listed above

Cleveland
all Districts not listed above

Derbyshire
all Districts not listed above

Durham
all Districts not listed above

Humberside
all Districts not listed above

Lancashire
all Districts not listed above

Lincolnshire
all Districts not listed above

Northumberland
all Districts not listed above

North Yorkshire
all Districts not listed above

Nottinghamshire
all Districts not listed above

Shropshire
all Districts not listed above

Staffordshire
all Districts not listed above

HOUSING ASSOCIATION SCHEMES

For the benefit of readers, the TCI system instigated by Housing Corporation Circular HC 14/90 has been reproduced here in full.

Total cost indicators for housing association schemes and related issues.

Summary

Advises associations of:

(i) the total cost indicators, tranche percentages and standard on-costs effective from 1 June 1990 for new build and rehabilitation for rent (tariff and non-tariff) schemes approved under the 1988 Act.

(ii) the total indicative costs effective from 1 June 1990 for schemes approved under the 1985 Act.

(iii) the standard on-costs for housing for sale schemes qualifying for grant.

This Circular supersedes Circulars HC 11/89, 50/89, 66/89 and 19/90. It amends Circular HC 4/89.

1. **Introduction**

1.1 The system of cost controls has the two-fold purpose of securing value for money and determining the maximum amount of grant appropriate to a particular scheme or programme of schemes. In particular, for tariff schemes, it is used as a basis for calculating eligible costs to which the appropriate grant rate for the scheme will be applied in the tariff calculation. For non-tariff schemes it is used to assess value for money in comparison with estimates of acquisition and works costs plus the appropriate standard on-cost.

1.2 The nucleus of the system is a series of Total Cost Indicator (TCI) tables which set out representative costs for different types of project in different parts of the country.

1.3 The TCI, in combination with the grant percentage applicable to a particular project or programme of projects, will ensure that public funding is confined to a proportion of reasonable costs for schemes of a particular type in a particular area. The basic TCI table is for new build, general needs dwellings, of specified floor areas and design occupancy. Standard multipliers are then applied to provide figures for rehabilitation, schemes not involving acquisition of a site or a property, and other variables. The TCI will apply equally to schemes funded by the Housing Corporation and by individual local authorities.

1.4 The TCI and grant rate framework has been modified for 1990/91 to include special need self contained and shared housing schemes. The introduction of the new Special Needs Management Allowance (SNMA) for shared and self-contained special needs housing has been postponed. Hostel Deficit Grant (HDG) will continue to be available for shared housing schemes funded under the residual Housing Association Grant (HAG) arrangements of the 1985 Act, which will also remain in operation for such schemes. For shared housing schemes which are viable without recourse to HDG and for self-contained special needs schemes which do not require the revenue support of SNMA, the TCI and on costs (and associated grant rates) have been developed to enable associations to use the new capital funding framework if they wish.

1.5 This Circular sets out the TCI tranche percentages and standard on-costs for schemes approved under the 1988 Act:

(i) the TCI are described in paragraph 2 and set out in Appendix 4;

(ii) the tranche percentages are defined and set out in paragraph 3;

(iii) the standard on-costs for housing for rent are described in paragraph 4.1 and set out in Appendix 5.

(iv) Standard on-costs for housing for sale schemes are described in paragraph 4.2 and set out in Appendix 6.

1.6 The adjustments to the TIC published in Circular HC 37/88 which will apply to schemes funded under the 1985 Act are set out in paragraph 5. The TCI and TIC apply to the same cost groups as defined in Appendix 3.

1.7 The main changes introduced in this circular are as follows:-

(i) allowance has been made for inflation, together with changes in Building Regulations, new charges for water and sewerage, higher standards of sound insulation in conversions etc.

(ii) cost groups have been restructured. There have been several steps in this process:-

(a) this years annual review indicated that there had been some changes in cost relativities between local authority areas which meant that some local authorities needed to be assigned to a higher or lower TCI cost group than shown in Circular HC 11/89.

(b) in order to minimise the number of year on year minor variations in costs groups, which have occurred in the past, the old cost groups F and G have been merged into a single cost group F.

(c) the cost differentials between TCI groups have been adjusted to take account of the restructuring described in (b) above.

The overall effect of restructering has been to produce clearer boundaries and definitions for the TCI groups.

(iii) Multipliers and on-costs have been revised, in particular:-

(a) the storey height multiplier has been replaced with one for lifts, and one for single storey units;

(b) the single family house multiplier has been replaced with a general increase in the rehabilitation multipliers;

(c) a set of new multipliers for special needs housing has been introduced. The Special Project Promotion Allowance (SPPA) will revert from being a multiplier to an allowance outside the TCI system (see Circular HC 6/90);

(d) a new package deal multiplier is included.

(iv) changes in the multipliers are reflected in corresponding changes in the standard on-costs.

(v) Standard on-costs for housing for sale schemes are now included in this circular. With the exception of DIYSO, they do not include capitalised interest for which a separate estimate is required.

(vi) the tranche percentages applying to the various cost groups have been modified in order to reflect the restructured cost groups.

(vii) standard percentage adjustments (SPAs) may apply to schemes within tariff agreements. Circular HC 4/89 explained in paragraph 6.4, that although the Procedure Guide for tariff funding refers to the application of SPAs to TCI and to grant rates, there had been no agreement to apply these adjustments and the references should be disregarded. This issue has been reconsidered and it has now been agreed that schemes developed under tariff agreements from 1 June 1990 may be subject to SPAs on the same basis as non-tariff schemes.

HOUSING ASSOCIATION SCHEMES

2. **Total cost indicators**

2.1 The TCI cover self-contained and shared accommodation designed to general needs, sheltered, frail elderly, special needs and wheelchair design criteria. The figures in Appendix 4 supersede the cost criteria published in Circular HC 11/89 (as updated by HC 66/89) and are for use from 1st June 1990 for schemes approved under the 1988 Act. They cover land/property purchase, works and on-costs, all as defined in Appendix 1.

2.2 TCI are set out in Appendix 4 as follows:

(i) A table of basic costs per dwelling (see Table 1) related to the floor area of the dwelling (in self-contained dwellings) or per person (in shared accommodation). Floor area for this purpose is defined in Appendix 2. The figures apply to both fixed price and variation of price building contracts.

(ii) Key multipliers in Table 2. Separate multipliers are shown for acquisition and works schemes, off-the-shelf new build schemes and for the purchase of existing satisfactory dwellings, works only, re-improvements etc. It should be noted that:

(a) only one key multiplier should be used for each dwelling type;

(b) the key multiplier for new build, acquisition and works is 1.00; i.e. the table of basic cost is for this dwelling type;

(c) separate multipliers are shown for each cost group.

(iii) Supplementary multipliers in Table 3. These apply to all cost groups. They apply both to new build and rehabilitation. They can be applied sequentially. Separate multipliers are shown for acquisition and works schemes and others, e.g. works only and re-improvements etc.

2.3 Further explanatory notes and definitions are set out in Appendix 2.

3. **Tranche percentages**

3.1 These are proportions of HAG payable at various stages of a scheme approved under the 1988 Act.

3.2 The following tranche percentages apply as appropriate from 1st June 1990. The tranches are payable at three key stages, at the time of:

(i) exchange of purchase contracts;

(ii) the main contract works start on site;

(iii) practical completion of the scheme or phase of the building contract.

3.3 The tranches, set out in order of payment, are as follows:

Scheme type

Cost Group	New build			Rehabilitation			Works only and Re-improvement	
	(i)	(ii)	(iii)	(i)	(ii)	(iii)	(i)	(ii)
A & B	30%	50%	20%	50%	30%	20%	80%	20%
C,D & E	20%	60%	20%	40%	40%	20%	80%	20%
F	10%	70%	20%	30%	50%	20%	80%	20%

HOUSING ASSOCIATION SCHEMES

4. **Standard on-costs**

4.1 TCI, as noted above, will cover all the costs of a scheme, a full list of which is set out at Appendix 1. It is therefore necessary to increase the estimate of all acquisition and works costs to allow for all the standard on-costs. The approach to be adopted in identifying the total eligible capital costs for a non-tariff scheme, either for rent or sale will be to add to the sum of acquisition and works costs, as defined in Appendix 1, the relevant standard percentage, as set out in Appendix 5 for housing for rent and Appendix 6 for housing for sale. (For housing for sale schemes apart from DIYSO capitalised interest will be added to the resulting total.) The appropriate on-cost percentage for a particular scheme is determined on the basis of predominant dwelling type, and will vary according to TCI cost group and the general purpose of the scheme.

4.2 Standard on-costs for housing for sale schemes, as for housing for rent, cover on-costs defined in Appendix 1. They are determined by scheme type and location according to groups defined in Appendix 3. They apply to acquisition and works costs as defined in Appendix 1.

5. **Total Indicative Costs**

5.1 TIC will continue to apply from 1st June 1990 to those shared housing schemes proceeding under the 1985 Act

5.2 Following the review the Secretary of State for the Environment has decided to increase the level of TIC in Circular HC 37/88 for new build and rehabilitation schemes by 16.5% in all TIC groups as from 1st June 1990. This supersedes the 11% increase announced in Circular HC 66/89

5.3 **The updating addition** to the acquisition costs will be calculated on form S6 using the updating percentages listed in the following table which supersedes the table given in paragraph 5.3 of Circular HC 11/89.

Year of acquisition (completion date of acquisition)	Cost criteria group		
	A & B %	C,D & E %	F & G %
1990 (from 1st Jun)	Nil	Nil	Nil
1990 (up to 31st May)	Nil	5	5
1989 (calendar year)	15	25	30
1988 "	35	45	50
1987 "	50	50	55
1986 "	60	65	60
1985 "	85	85	70
1984 "	105	100	75
1983 "	120	115	85
1982 "	120	115	85
1981 "	175	160	105
1980 "	265	235	140
1979 and earlier	330	290	155

6. **Further reviews of TCI and TIC**

TCI and TIC will be reviewed annually.

APPENDIX 1 - TCI DEFINITION

The TCI include the following:

1. **ACQUISITION**

(i) Purchase price of land/property

HOUSING ASSOCIATION SCHEMES

2. **WORKS**

(i) Main works contract costs (including where applicable, adjustments for additional claims and fluctuations, but excluding any costs defined as on-costs below)

(ii) Major Site Development Works (where applicable), including piling, soil stabilisation, road and sewer construction, major demolition, statutory agreements and associated bonds

(iii) Major Pre Works (Rehabilitation), where applicable

(iv) Home loss and associated costs, where applicable on new build schemes only

(v) VAT on the above, where applicable

3. **ON-COSTS**

(i) Legal fees, disbursements and expenses

(ii) Stamp duty

(iii) Interest charges on development period loans for housing for rent and interest charges on desposit for DIYSO

(iv) Building society or other valuation and administration fees

(v) Fees for building control and planning permission

(vi) In house or external consultants' fees, disbursements and expenses (where the development contract is a Design and Build package deal, the on-cost includes the builder's design fee element of the contract sum).

(vii) Insurance premiums (except contract insurance included in works cost)

(viii) Contract performance bond premiums

(ix) Borrowing administration charges

(x) Associations' development administration costs (formerly A & D allowances, excluding Co-operative promotional allowance and SPPA)

(xi) Furniture

(xii) Home loss and disturbance payments for rehabilitation

(xiii) Preliminary minor site development works (new build) and pre-works (Rehabilitation)

(xiv) VAT on the above, where applicable

APPENDIX 2 - EXPLANATORY NOTES AND DEFINITIONS

1. **Works costs**

Works costs are all-inclusive and thus include all dwelling and site works and car accommodation.

HOUSING ASSOCIATION SCHEMES

APPENDIX 2 - EXPLANATORY NOTES AND DEFINITIONS - cont'd

2. Unit Size and Number of Occupants

2.1 Generally

The unit size in square metres relates to the total floor area of the resultant unit. (For a definition of floor area see below). The probable occupancy figure is shown as a guide. The number of occupants is related to the number of single and double bedrooms provided. The TCI for a unit where the total floor area exceeds 120 m2 will be appropriate cost for a unit of 110 - 120 m2 plus, for each additional; 10 m2 or part thereof, the difference between the cost of a unit of 100 - 110 m2 and a unit of 110 - 120 m2.

2.2 Shared accommodation

The term shared housing is used to describe accommodation for two or more persons with shared facilities, e.g. bathroom, communal living room or dining room. The number of occupants/bedspaces in a shared unit of accommodation is exclusive of those in self-contained staff accommodation or other self-contained accommodation. There may be a number of shared units within the same scheme.

2.3 Total floor area

(i) Self-contained dwellings

Area is measured to the internal faces of the main containing walls on each floor of the dwelling and includes the space, on plan, taken up by private staircases, partitions, internal walls (but not 'party' or similar walls), chimney breasts, flues and heating appliances. It includes the area of tenant's storage space, except where a non-habitable basement or attic is used exclusively as storage space. It excludes:-

(a) any space where the height to the ceiling is less than 1.5 m (e.g. areas in rooms with sloping ceilings, external dustbin enclosures);

(b) any porch, covered way, etc. open to the air;

(c) any garage except that, where a garage is integral with or adjoining the dwelling, the excess over 12 m2 qualifies as storage space and as part of the area of the dwelling;

(d) all balconies (private, escape and access) and decks:

(e) all public access space (e.g. communal entrances, staircases, corridors);

(f) all space for communal facilities or services;

(g) all space for purposes other than housing (e.g. commercial premises).

(ii) Shared accommodation

Area is measured to the internal faces of the main containing walls on each floor of the unit and includes all space for shared facilities or services and the space, on the plan, taken up by staircases, partitions internal walls, chimney breasts, flues and heating appliances. It includes the area of the resident's storage space, except where a non-habitable basement or attic is used exclusively as storage space. It excludes:-

HOUSING ASSOCIATION SCHEMES

APPENDIX 2 - EXPLANATORY NOTES AND DEFINITIONS - cont'd

(a) any space where the height of the ceiling is less than 1.5 m (e.g. areas in rooms with sloping ceilings, external dustbin enclosures);
(b) any porch, covered way, etc. open to the air;
(c) all balconies (private, escape and access) and decks;
(d) any garage except that, where a garage is integral with or adjoining the dwelling, the excess over 12 m2 qualifies as storage space and as part of the area of the dwelling;
(e) all space for purposes other than housing (e.g. commercial premises).

The appropriate floor area should be calculated by taking the total floor area of the shared unit (see above) excluding self-contained staff accommodation or other self-contained accommodation, and dividing by the total number of bedspaces in the shared unit, excluding those in self-contained accommodation.

3. **Design and contracting requirments**

3.1 **Generally**

Supplementary multipliers and on-costs shall apply, as described elsewhere in the Circular, where the Design and Contracting Requirements are met.

3.2 **New lifts**

A dwelling shall be deemed to be served by lifts where it is in a block containing lifts and normal access to dwellings above ground level is by means of lifts.

3.3 **Extended Families**

Self-contained dwellings for extended families shall be for 8 or more persons and should incorporate particular design features relevant to the needs of extended families, for example, two living rooms, two kitchens or dining rooms, extra utility space, or special washing or sanitary facilities.

3.4 **Wheelchair accommodation**

The TCI multiplier allows for the standards for wheelchair accommodation as set out in the Design and Contracting Requirements, but not any fixed additional equipment to meet the particular needs of identified disabled persons being housed. These are met by 100% grants for Adaptations for the Physically Handicapped (see Procedure Guides).

3.5 **Common room etc**

The common room and associated communal facilities multipliers shall be applied as follows:-

(i) Common room. Where common room, chair store, tea kitchen and WC are provided.

(ii) Communal facilities. Where any three of the following are provided:-

Emergency alarm system
Wardens office
Laundry room
Guest room

HOUSING ASSOCIATION SCHEMES

3.6 Package deals

The term 'package deals' is deemed to include design and build, design and construct, and all other such building contract variants where the responsibility for design rests with the contractor.

APPENDIX 3 - GROUPINGS

The TCI shall apply to schemes in groups defined as follows:

Group A comprising:

The following **London Boroughs**:

Barnet
Brent
Camden
Ealing
Enfield
Hackney
Hammersmith & Fulham
Haringey
Harrow
Hillingdon
Hounslow
Islington

Kensington & Chelsea
Kingston-upon-Thames
Lambeth
Merton
Redbridge
Richmond-upon-Thames
Southwark
Tower Hamlets
Waltham Forest
Wandsworth
Westminster

City of London

The following Districts in the Counties of:

Buckinghamshire
Chiltern
South Bucks

Hertfordshire
Hertsmere
St. Albans
Three Rivers
Watford

Essex
Epping Forest

Surrey
Elmbridge
Epson and Ewell
Mole Valley
Runnymede
Spelthorne

Group B comprising:

The following **London Boroughs**:

Barking & Dagenham
Bexley
Bromley
Croydon
Greenwich

Havering
Lewisham
Newham
Sutton

The following Districts in the Counties of:

Berkshire
Slough
Windsor & Maidenhead
Wokingham

Kent
Dartford
Sevenoaks

HOUSING ASSOCIATION SCHEMES

Group B (continued) comprising:

Buckinghamshire
Wycombe

Essex
Brentwood
Harlow
Uttlesford

Hertfordshire
Broxbourne
Dacorum
East Hertfordshire
North Hertfordshire
Stevenage
Welwyn Hatfield

Surrey
Guildford
Reigate & Banstead
Tandridge
Waverly

West Sussex
Crawley
Horsham

Group C comprising:

The following Districts in the Counties of:

Avon
Bath

Berkshire
Bracknell
Newbury
Reading

Cambridgeshire
Cambridge
South Cambridgeshire

East Sussex
Brighton
Hove
Lewes
Rother

Essex
Basildon
Castle Point
Chelmsford
Maldon
Rochford
Southend-on-Sea
Tendring
Thurrock

Hampshire
Basingstoke & Dean
East Hampshire
Eastleigh
Fareham
Gosport
Hart
Havant
Portsmouth
Rushmoor
Southampton
Test Valley
Winchester

Isles of Scilly

Kent
Gravesham
Maidstone
Shepway
Tonbridge & Malling
Tunbridge Wells

Oxfordshire
Oxford
South Oxfordshire
Vale of White Horse

Surrey
Surrey Heath
Woking

West Sussex
Adiu
Arun
Chichester
Mid. Sussex
Worthing

HOUSING ASSOCIATION SCHEMES

Group D comprising:

The following Districts in the Counties of:

Avon
Bristol
Kingswood
Northavon
Wansdyke
Woodspring

Bedfordshire
Luton
Mid. Bedfordshire
North Bedfordshire
South Bedfordshire

Buckinghamshire
Aylesbury Vale
Milton Keynes

Cambridgeshire
East Cambridgeshire

Devon
East Devon
Exeter
North Devon
Teignbridge
Torbay

Dorset
Bournemouth
Christchurch
East Dorset
North Dorset
Poole
Purbeck
West Dorset
Weymouth & Portland

East Sussex
Eastbourne
Hastings
Wealden

Essex
Braintree
Colchester

Gloucestershire
Cheltenham
Tewkesbury

Hampshire
New Forest

Isle of Wight
Medina
South Wight

Kent
Ashford
Canterbury
Gillingham
Rochester upon Medway
Swale

Norfolk
Broadland
Norwich

Northamptonshire
South Northamptonshire

Oxfordshire
Cherwell
West Oxfordshire

Suffolk
Babergh
Forest Heath
Ipswich
Mid. Suffolk
St. Edmundsbury

Warwickshire
Warwick

West Midlands
Solihull

Wiltshire
Kennett
North Wiltshire
Salisbury
Thamesdown
West Wiltshire

Group E comprising:

The following Districts in the Counties of:

Cambridgeshire
Fenland
Huntingdon
Peterborough

Lincolnshire
South Holland

HOUSING ASSOCIATION SCHEMES

APPENDIX 3 - GROUPINGS - cont'd

Group E (continued) comprising:

Cornwall
Caradon
Carrick
Kerrier
North Cornwall
Penwith
Restormel

Devon
Mid. Devon
Plymouth
South Hams
Torridge
West Devon

Gloucestershire
Cotswold
Forest of Dean
Gloucester
Stroud

Hereford & Worcester
Bromsgrove
Malvern Hills
Redditch
Worcester
Wychavon

Kent
Dover
Thanet

Leicestershire
Blaby
Harborough
Hinckley & Bosworth
Leicester
Oadby & Wigston
Rutland

Norfolk
Breckland
Great Yarmouth
Kings Lynn & West Norfolk
North Norfolk
South Norfolk

Northamptonshire
Corby
Daventry
East Northamptonshire
Kettering
Northampton
Wellingborough

Somerset
Mendip
Sedgemoor
South Somerset
Taunton Deane
West Somerset

Suffolk
Suffolk Coastal
Waverley

Warwickshire
North Warwickshire
Nuneaton & Bedworth
Rugby
Stratford-on-Avon

West Midlands
Birmingham
Coventry

Group F comprising:

The following Districts in the Counties of:

Cheshire
Chester
Congleton
Crewe & Nantwich
Ellesmere Port & Neston
Halton
Macclesfield
Vale Royal
Warrington

Cleveland
Hartlepool
Langbaurgh
Middlesborough
Stockton-on-Tees

Greater Manchester
Bolton
Bury
Manchester
Oldham
Rochdale
Salford
Stockport
Tameside
Trafford
Wigan

HOUSING ASSOCIATION SCHEMES

APPENDIX 3 - GROUPINGS - cont'd

Group F (continued) comprising:

Cumbria
Allerdale
Barrow-in-Furness
Carlisle
Copeland
Eden|
South Lakeland

Derbyshire
Amber Valley
Bolsover
Chesterfield
Derby
Derbyshire Dales
Erewash
High Peak
North East Derbyshire
South Derbyshire

Durham
Chester-le-Street
Darlington
Derwentside
Durham
Easington
Sedgefield
Teesdale
Wear Valley

Hereford & Worcester
Hereford
Leominster
South Herefordshire
Wyre Forest

Humberside
Beverley
Boothferry
Cleethorpes
East Yorkshire
Glanford
Great Grimsby
Holderness
Kingston-upon-Hull
Scunthorpe

Lancashire
Blackburn
Blackpool
Burnley
Chorley
Fylde
Hyndburn
Lancaster
Pendle
Preston
Ribble Valley
Rossendale
South Ribble
West Lancashire
Wyre

Leicestershire
Charnwood
Melton
North West Leicestershire

Lincolnshire
Boston
East Lindsey
Lincoln
North Kesteven
South Kesteven
West Lindsey

Merseyside
Knowlsey
Liverpool
St. Helens
Sefton
Wirral

Northumberland
Alnwick
Berwick-upon-Tweed
Blyth Valley
Castle Morpeth
Tynedale
Wansbeck

North Yorkshire
Craven
Hambleton
Harrogate
Richmondshire
Ryedale
Scarborough
Selby
York

Nottinghamshire
Ashfield
Bassetlaw
Broxtowe
Gedling
Mansfield
Newark
Nottingham
Rushcliffe

Shropshire
Bridgnorth
North Shropshire
Oswestry
Shrewsbury & Atcham
South Shropshire
The Wrekin

HOUSING ASSOCIATION SCHEMES

Staffordshire
Cannock Chase
East Staffordshire
Lichfield
Newcastle-under-Lyme
South Staffordshire
Stafford
Staffordshire Moorlands
Stoke-on-Trent
Tamworth

South Yorkshire
Barnsley
Doncaster
Rotherham
Sheffield

Tyne & Wear
Gateshead
Newcastle upon Tyne
North Tyneside
South Tyneside
Sunderland

West Midlands
Dudley
Sandwell
Walsall
Wolverhampton

West Yorkshire
Bradford
Calderdale
Kirkless
Leeds
Wakefield

HOUSING ASSOCIATION SCHEMES

APPENDIX 4 - TABLE 1 - TOTAL COST INDICATORS

TOTAL DWELLING COSTS 1990/91 - £ PER DWELLING

Dwelling Dwelling floor area (m2)	Probable occupancy (persons)	Group A	B	C	D	E	F
Up to 30	1	65 400	56 100	48 200	41 400	35 500	30 500
Exceeding 30 and not exceeding 40	1 & 2	76 800	65 900	56 600	48 600	41 700	35 800
Exceeding 40 and not exceeding 50	2 & 3	88 000	75 500	64 800	55 600	47 700	40 900
Exceeding 50 and not exceeding 60	2 & 3	99 000	84 900	72 800	62 500	53 600	46 000
Exceeding 60 and not exceeding 70	3 & 4	109 600	94 000	80 600	69 200	59 300	50 900
Exceeding 70 and not exceeding 80	3,4 & 5	120 000	102 900	88 300	75 700	64 900	55 700
Exceeding 80 and not exceeding 90	4,5 & 6	130 200	111 600	95 700	82 100	70 400	60 300
Exceeding 90 and not exceeding 100	5 & 6	140 100	120 100	103 000	88 300	75 700	64 900
Exceeding 100 and not exceeding 110	6 & 7	149 700	128 300	110 000	94 300	80 900	69 300
Exceeding 110 and not exceeding 120	6 & over	159 100	136 400	116 900	100 200	85 900	73 600

APPENDIX 4 - TABLE 2 - KEY MULTIPLIERS By cost group

Note: Only one of the following to be used.

	Scheme Type	Form ref. TA1 & SFN1 Line No.	Group A	B	C	D	E	F
a)	**New build**							
	i) Acquisition & works	0010	1.00	1.00	1.00	1.00	1.00	1.00
	ii) Off-the-shelf	0020	0.80	0.80	0.79	0.79	0.78	0.78
	iii) Works only	0090	0.80	0.82	0.84	0.86	0.89	0.91
b)	**Rehabilitation**							
	i) Acquisition & works - Vacant	0100	1.16	1.11	1.07	1.03	0.99	0.95
	- Tenanted	0150	1.16	1.11	1.07	1.03	0.99	0.95
	ii) Existing Satisfactory	0120	0.80	0.80	0.79	0.79	0.78	0.78
	iii) Works Only - Vacant	0125	0.67	0.67	0.66	0.66	0.65	0.65
	- Tenanted	0130	0.90	0.88	0.87	0.85	0.84	0.82
	iv) Re-improvements	0135	0.90	0.88	0.87	0.85	0.84	0.82

HOUSING ASSOCIATION SCHEMES

APPENDIX 4 - TABLE 3 - SUPPLEMENTARY MULTIPLIERS All cost groups.

Note a) New build and Rehabilitation.

b) Can be applied sequentially, except that supplementary multipliers from sections (a), (b), (c) and (d) can not be used in conjunction with each other.

c) Only one supplementary multiplier from sections (a),(c), (e) and (h) may be used for each dwelling type.

Scheme type	Form Ref. TA1 & SFN1 Line No.	Acquisition and Works, Off-the-shelf & Existing Satisfactory	Works Only & Reimprovements
a) Sheltered:			
i) Category 1	0140	1.15	1.23
ii) Category 2	0170	1.31	1.48
b) Frail Elderly (includes wheelchair provision)	0180	1.56	1.88
c) Special Needs:			
i) Special Needs	0185	1.20	1.31
ii) Registered Homes	0190	1.31	1.48
d) Extended General Families	0195	1.10	1.16
e) Shared (General needs and Special needs only)			
Bedspaces per unit:			
i) 2 and 3	0200	1.00	1.00
ii) 4 to 6	0210	0.86	0.78
iii) 7 to 10	0220	0.82	0.72
iv) 11 to 50	0230	0.76	0.63
v) Over 50	0240	0.72	0.56
f) Served by new lifts	0245	1.10	1.16
g) One storey (Newbuid Only)	0095	1.06	1.09
h) Common room etc (Sheltered Category 1, Special Needs and Registered Homes only)			
i) Common room only	0250	1.06	1.09
ii) Associated communal facilities only	0260	1.05	1.08
iii) Common room and communal facilities	0270	1.11	1.17
j) Wheelchair (except where included as above)	0390	1.21	1.33
k) Package Deals (except Off-the-Shelf and Existing Satisfactory)	0410	0.98	0.97

HOUSING ASSOCIATION SCHEMES

APPENDIX 5 - STANDARD ON-COSTS FOR HOUSING FOR RENT

1. One standard on-cost shall be determined for each scheme. It is determined by:
   i) the TCI cost group;
   ii) the general purpose of the scheme;
   and is applied to the sum of the acquisition and works costs of a scheme.

2. The standard on-cost percentage will be the total of:
   i) the appropriate key on-cost (only one will be appropriate) plus
   ii) the total of one or more of the appropriate supplementary on-costs, except that supplementary on-costs from sections (a), (b), (c) and (d) can not be used in conjunction with each other. Where two are equally applicable, e.g. special needs and shared, the higher figures should be used.

Note The appropriate on-cost will be that which applies to:
i) for key on-costs - the predominant dwelling type determined by the largest number of persons in total;
ii) for supplementary on-costs - over 50% of the persons in the scheme.

TABLE 1 - KEY ON-COST (%)

Key On-Costs				Group		
	A	B	C	D	E	F
	%	%	%	%	%	%
a) New Build						
i) Acquisition and Works	21	21	22	22	23	24
ii) Off the Shelf	10	10	10	10	10	10
iii) Works Only	24	24	24	24	23	23
b) Rehabilitation						
i) Acquisition and Works						
- Vacant	19	20	21	21	22	23
- Tenanted	27	28	29	30	32	33
ii) Existing Satisfactory	10	10	10	10	10	10
iii) Works Only						
- Vacant	30	30	31	31	31	32
- Tenanted	38	38	39	40	40	41
iv) Reimprovements	38	38	39	40	40	41

HOUSING ASSOCIATION SCHEMES

TABLE 2 - SUPPLEMENTARY ON-COSTS (%)

Supplementary On-Costs	Acquisition and Works %	Works Only and Reimprovements %	Off-the-Shelf and Existing Satisfactory %
a) Sheltered with comman room or communal facilities	+1	+2	+1
b) Frail elderly	+4	+6	+2
c) Special Needs:			
i) Special Needs	+4	+6	+3
ii) Special Needs with common room or communal facilities	+5	+8	+4
iii) Registered Homes	+5	+8	+3
iv) Registered Homes with common room or communal facilities	+6	+9	+4
d) Shared	+6	+9	+5
e) Wheelchair	+1	+2	0
f) Package Deals	-2	-3	0

APPENDIX 6 - STANDARD ON-COSTS FOR SALE

Scheme type	Group A %	B %	C %	D %	E %	F %
a) Key on-costs						
i) New Build	19	19	19	20	20	21
ii) Rehabilitation	17	17	18	19	19	20
iii) Off-the-Shelf	10	10	10	10	10	10
iv) DIYSO	4	4	4	4	4	4

Scheme type	All Groups %
b) Supplementary on-costs:	
i) LSE schemes without communal facilities	+1
ii) LSE schemes with communal facilities	+2
iii) Package Deals	-2
iv) DIYSO mixed funded schemes	+1

Property Insurance

The problem of adequately covering by insurance the loss and damage caused to buildings by fire and other perils has been highlighted in recent years by the increasing rate of inflation.

There are a number of schemes available to the building owner wishing to insure his property against the usual risk. Traditionally the insured value must be sufficient to cover the actual cost of reinstating the building. This means that in addition to assessing the current value an estimate has also to be made of the increases likely to occur during the period of the policy and of rebuilding which, for a moderate size building, could amount to a total of three years. Obviously such an estimate is difficult to make with any degree of accuracy, if it is too low the insured may be penalized under the terms of the policy and if too high will result in the payment of unnecessary premiums. There are variations on the traditional method of insuring which aim to reduce the effects of over estimating and details of these are available from the appropriate offices. For the convenience of readers who may wish to make use of the information contained in this publication in calculating insurance cover required the following may be of interest.

1 PRESENT COST

The current rebuilding costs may be ascertained in a number of ways:

(a) Where the actual building cost is known this may be updated by reference to tender indices (page 770);
(b) By reference to average published prices per square metre of floor area (page 773). In this case it is important to understand clearly the method of measurement used to calculate the total floor area on which the rates have been based which are current at November 1990;
(c) By professional valuation;
(d) By comparison with the known cost of another similar building

Whichever of these methods is adopted regard must be paid to any special conditions that may apply, i.e., a confined site, complexity of design or any demolition and site clearance that may be required.

2 ALLOWANCE FOR INFLATION

The 'Present Cost' when established will usually, under the conditions of the policy, be the rebuilding cost on the first day of the policy period. To this must be added a sum to cover future increases. For this purpose, using the historical indices on pages 769 and 770, as a base and taking account of the likely change in building costs and tender climate the following annual average indices are predicted for the future.

	Cost index	Tender index
1987	276	259
1988	293	309
1989	314	340
1990	337	329
1991	362	337
1992	384	359
1993	407	383
1994	431	409

3 FEES

To the total of 1 and 2 above must be added an allowance for fees.

4 VALUE ADDED TAX

To the total of 1 to 3 above must be added Value Added Tax. Previously relief may have been given to total reconstructions following a fire damage etc. However revisions to previous VAT legislation contained in the 1989 Finance Bill require that such work now attracts VAT, and the limit of insurance cover should be raised to allow for this.

5 EXAMPLE

An assessment for insurance cover is required in mid 1990 for a property which cost £200 000.00 when completed in mid 1976.

Present cost		£
Known cost at mid 1976		200 000.00
Predicted tender index mid 1990 = 329		
Tender index mid 1976 = 100		
Increases in tender index = 229%		
applied to known cost =		458 000.00

Present cost (excluding any allowance for demolition)		658 000.00

Allowance for inflation		£
Present cost at day one of policy		658 000.00
Allow for changes in tender levels during 12 months currency of policy		
Predicted tender index at mid 1991 = 337		
Predicted tender index at mid 1990 = 329		
Increase in tender index = + 2.43%		
applied to present cost =	say	16 000.00
		----------
		674 000.00
Assuming that total damage is suffered on the last day of the currency of the policy and that planning and documentation would require a period of twelve months before re-building and could commence then a further similar allowance must be made		
Predicted tender index at mid 1992 = 359		
Predicted tender index at mid 1991 = 337		
Increase in tender index = + 6.53%		
applied to adjusted present cost =	say	44 000.00
		----------
		718 000.00
Assuming that total reinstatement would take two years allowance must be made for the increases in costs which would directly or indirectly be met under a building contract.		
Predicted cost index at mid 1994 = 431		
Predicted cost index at mid 1992 = 384		
Increases in cost index = + 12.24%. This is the total increase at the end of two years and the amount applicable to the contract cost incurred over this period might be about half, say + 6%	say	42 000.00

Estimated cost of reinstatement		760 000.00

SUMMARY

		£
Estimated cost of reinstatement		760 000.00
Add professional fees at say 16%	say	122 000.00
		882 000.00
Add for Value Added Tax (currently plus 15%)	say	133 000.00
Total insurance cover required, say		£ 1015 000.00

PART V

Tables and Memoranda

This part of the book contains the following sections:

Tables and Memoranda

CONVERSION TABLES

	Unit	Conversion factors	
Length			
Millimetre	mm	1 in = 25.4 mm	1 mm = 0.0394 in
Centimetre	cm	1 in = 2.54 cm	1 cm = 0.3937 in
Metre	m	1 ft = 0.3048 m	1 m = 3.2808 ft
Kilometre	km	1 yd = 0.0144 m	or 1.0936 yd
		1 mile= 1.6093 km	1 km = 0.6214 mile
Note: 1 cm = 10 mm		1 ft = 12 in	
1 m = 100 mm		1 yd = 3 ft	
1 km = 1 000 m		1 mile= 1 760 yd	
Area			
Square Millimetre	mm2	1 in2 = 645.2 mm2	1 mm2= 0.0016 in2
Square Centimetre	cm2	1 in2 = 6.4516 cm2	1 cm2= 1.1550 in2
Square Metre	m2	1 ft2 = 0.0929 m2	1 m = 10.764 ft2
		1 yd2 = 0.8361 m2	1 m2 = 1.1960 yd2
Square Kilometre	km2	1 mile2= 2.590 km2	1 km2= 0.3861 mile2
Note: 1 cm2 = 100 mm2		1 ft2 = 144 in2	
1 m2 = 10 000 cm2		1 yd2 = 9 ft2	
1 km2 = 100 hectares		1 acre= 4 840 yd2	
		1 mile2= 640 acres	
Volume			
Cubic Centimetre	cm3	1 cm3 = 0.0610 in3	1 in3 = 16.387 cm3
Cubic Decimetre	dm3	1 dm3 = 0.0353 ft3	1 ft3 = 28.329 dm3
Cubic metre	m3	1 m3 = 35.3147 ft3	1 ft3 = 0.0283 m3
		1 m3 = 1.3080 yd3	1 yd3 = 0.7646 m3
Litre	1	1 1 = 1.76 pint	1 pint= 0.5683 1
		= 2.113 US pt	= 0.4733 US 1
Note: 1 dm3 = 1 000 cm3		1 ft3 = 1 728 in3	
1 m3 = 1 000 dm3		1 yd3 = 27 ft3	
1 1 = 1 dm3		1 pint= 20 fl oz	
1 hl = 100 1		1 gal = 8 pints	

CONVERSION TABLES

	Unit	Conversion factors	
Mass			
Milligram	mg	1 mg = 0.0154 grain	1 grain= 64.935 mg
Gram	g	1 g = 0.0353 oz	1 oz = 28.35 g
Kilogram	kg	1 kg = 2.2046 lb	1 lb = 0.4536 kg
Tonne	t	1 t = 0.9842 ton	1 ton = 1.016 t
Note: 1 g = 1000 mg		1 oz = 437.5 grains	
1 kg= 1000 g		1 lb = 16 oz	
1 t = 1000 kg		1 stone= 14 lb	
		1 cwt = 112 lb	
		1 ton = 20 cwt	
Force			
Newton	N	1 lb f = 4.448 N	1 kg f = 9.807 N
Kilonewton	kN	1 lb f = 0.004448 kN	1 ton f= 9.964 kN
Meganewton	mN	100 ton f= 0.9964 mN	
Pressure and stress			
Kilonewton per square metre	kN/m2	1 lb f/in2 = 6.895 kN/m2	
		1 bar = 100 kN/m2	
Meganewton per square metre	mN/m2	1 ton f/ft2= 107.3 kN/m2 = 0.1073 mN/m2	
		1 kg f/cm2 = 98.07 kN/m2	
		1 lb f/ft2 = 0.04788 kN/m2	
Coefficient of consolidation (Cv) or swelling			
Square metre per year	m2/year	1 cm2/s = 3 154 m2/year	
		1 ft2/year = 0.0929 m2/year	
Coefficent of permeability			
Metre per second	m/s	1 cm/s = 0.01 m/s	
Metre per year	m/year	1 ft/year = 0.3048 m/year	
		= 0.9651 x (10)8 m/s	
Temperature			
Degree celcius ^C		$^\wedge C = \frac{5}{9} (^\wedge F - 32^\wedge)$	$^\wedge F = \frac{9}{5} (^\wedge C + 32^\wedge)$

FORMULAE

Two dimensional figures

Figure	**Area**
Square	(side) sq
Rectangle	Length x breadth
Triangle	0.5 x base x height or N(s(s - a)(s-b)(s-c)) where s = 0.5 x the sum of the three sides and a, b and c are the lengths of the three sides. or a2 = b2 + c2 - 2 x bc x COS A where A is the angle opposite side a
Hexagon	2.6 x (side) sq
Octagon	4.83 x (side) sq
Trapezoid	height x 0.5(base + top)
Circle	3.142 x radius sq or 0.7854 x diameter sq (circumference = 2 x 3.142 x radius or 3.142 x diameter)
Sector of a circle	0.5 x length of arc x radius
Segment of a circle	area of sector - area of triangle
Ellipse	3.142 x AB (where A = 0.5 x height and B = 0.5 x length)
Bellmouth	$\frac{3}{14}$ x radius sq

Three dimensional figures

Figure	**Volume**	**Surface Area**
Prism	Area of base x height	circumference of base x height
Cube	(side) cubed	6 x (side) sq
Cylinder	3.142 x radius sq x height	2 x 3.142 x radius x (height - radius)
Sphere	$\frac{4}{3}$ x 3.142 x radius cubed	4 x 3.142 x radius sq
Segment of a sphere	$\left(\left[\frac{3.142 \times h}{6}\right] \times [3 \times r2 + h2]\right)$	2 x 3.142 x r x h
Pyramid	$\frac{1}{3}$ of area of base x height	0.5 x circumference of base x slant height

FORMULAE

Three dimensional figures

Figure	**Volume**	**Surface area**
Cone	1 x 3.142 x radius sq x h - 3	3.142 x radius x slant height
Frustrum of a pyramid	0.33 x height [A + B + sq root (AB)] where A is the area of the large end and B is the area of the small end	0.5 mean circumference x slant height
Frustrum of a cone	(0.33 x 3.142 x height (R sq + r sq + R x r)) where R is the radius of the large end and r is the radius of the small end	(3.142 x slant height x (R + r))

Other formulae

Formulae	**Description**
Pythagoras theorem	A2 = B2 + C2 where A is the hypotenuse of a right-angled triangle and B and C are the two adjacent sides
Simpsons Rule	Volume = x(y^ + yn) + 2(y1 + y3 + y5) + 4(y2 + y4) - 3 trench must be split into even sections e.g. 1-7, the areas at intermediate even cross-sections (No 2, 4, 6 etc.) are each multiplied by 4 and the areas at intermediate uneven cross-sections (No 3,5 etc.) are each multiplied by 2 and the end cross-sections taken once only. The sum of these areas is multiplied by 0.33 of the distance between the cross-sections to give the total volume.
Trapezoidal Rule	(0.16 x [Total length of trench] x [area of first section x 4 times area of middle section + area of last section])

Note: Both Simpsons and Trapezoidal Rule are useful in calculating the volume of an irregular trench accurately.

DESIGN LOADINGS FOR BUILDINGS

Definitions

Dead load: The load due to the weight of all walls, permanent partitions, floors, roofs and finishes, including services and all other permanent construction.

Imposed load: The load assumed to be produced by the intended occupancy or use, including the weight of moveable partitions, distributed, concentrated, impact, inertia and snow loads, but excluding wind loads.

DESIGN LOADINGS FOR BUILDINGS

Distributed load: The uniformly distributed static loads per square metre of plan area which provide for the effects of normal use. Where no values are given for concentrated load it may be assumed that the tabulated distributed load is adequate for design purposes.

Note: The general recommendations are not applicable to certain a typical usages particularly where mechanical stacking, plant of machinery are to be installed and in these cases the designer should determine the loads from a knowledge of the equipment and processes likely to be employed.

The additional imposed load to provide for partitions, where their positions are not shown on the plans, on beams and floors, where these are capable of effective lateral distributional of the load, is a uniformly distributed load per square metre of not less than one-third of the weight per metre run by the partitions **but not less than 1kN/m2.**

Floor area usage	**Distributed load kN/m2**	**Concentrated load kN/300 mm2**
Industrial occupancy class (workshops, factories)		
Founderies	20.0	-
Cold storage	5.0 for each metre of storage height with a minimum of 15.0	9.0
Paper storage, for printing plants	4.0 for each metre of storage height	9.0
Storage, other than types listed seperately	2.4 for each metre of storage height	7.0
Type storage and other areas in printing plants	12.5	9.0
Boiler rooms, motor rooms, fan rooms and the like, including the weight of machinery	7.5	4.5
Factories, workshops and similar buildings	5.0	4.5
Corridors, hallways, foot-bridges, etc. subject to loads greater than for crowds, such as wheeled vehicles, trolleys and the like	5.0	4.5
Corridors, hallways, stairs, landings, footbridges, etc.	4.0	4.5

DESIGN LOADINGS FOR BUILDINGS

Floor area usage	Distributed load kN/m2	Concentrated load kN/300 mm2
Industrial occupancy class (workshops, factories) - cont'd		
Machinery halls, circulation spaces therein	4.0	4.5
Laboratories (including equipment), kitchens, laundries	3.0	4.5
Workrooms, light without storage	2.5	1.8
Toilet rooms	2.0	-
Cat walks	-	1.0 at 1m centres
Institutional and educational occupancy class (prisons, hospitals, schools, colleges)		
Dense mobile stacking (books) on mobile trolleys	4.8 for each metre of stack height but with a minimum of 9.6	7.0
Stack rooms (books)	2.4 for each metre of stack height but with a minimum of 6.5	7.0
Stationery stores	4.0 for each metre of storage height	9.0
Boiler rooms, motor rooms, fan rooms and the like, including the weight of machinery	7.5	4.5
Corridors, hallways, etc. subject to loads greater than from crowds, such as wheeled vehicles, trolleys and the like	5.0	4.5
Drill rooms and drill halls	5.0	9.0
Assembly areas without fixed seating, stages gymnasia	5.0	3.6
Bars	5.0	-
Projection rooms	5.0	-
Corridors, hallways, aisles, stairs, landings, foot-bridges, etc.	4.0	4.5

DESIGN LOADINGS FOR BUILDINGS

Floor area usage	Distributed load kN/m2	Concentrated load kN/300 mm2
Institutional and educational occupancy class (prisons, hospitals, schools, colleges - cont'd		
Reading rooms with book storage, e.g. libraries	4.0	4.5
Assembly areas with fixed seating	4.0	-
Laboratories (including equipment), kitchens, laundries	3.0	4.5
Classrooms, chapels	3.0	2.7
Reading rooms without book storage	2.5	4.5
Areas for equipment	2.0	1.8
X-ray rooms, operating rooms, utility rooms	2.0	4.5
Dining rooms, lounges, billiard rooms	2.0	2.7
Dressing rooms, hospital bedrooms and wards	2.0	1.8
Toilet rooms	2.0	-
Bedrooms, dormitories	1.5	1.8
Balconies	same as rooms to which they give access but with a minimum of 4.0	1.5 per metre run concentrated at the outer edge
Fly galleries	4.5 kN per metre run distributed uniformly over the width	-
Cat walks	-	1.0 at 1 m centres
Offices occupancy class (offices, banks)		
Stationery stores	4.0 for each metre of storage height	9.0
Boiler rooms, motor rooms, fan rooms and the like, including the weight of machinery	7.5	4.5

DESIGN LOADINGS FOR BUILDINGS

Floor area usage	Distributed load kN/m2	Concentrated load kN/300 mm2
Offices occupancy class (offices, banks) - cont'd		
Corridors, hallways, etc. subject to loads greater than from crowds, such as wheeled vehicles, trolleys and the like	5.0	4.5
File rooms, filing and storage space	5.0	4.5
Corridors, hallways, stairs, landings, footbridges, etc.	4.0	4.5
Offices with fixed computers or similar equipment	3.5	4.5
Laboratories (including equipment), kitchens, laundries	3.0	-
Banking halls	3.0	4.5
Offices for general use	2.5	2.7
Toilet rooms	2.0	-
Balconies	Same as rooms to which they give access but with a minimum of 4.0	1.5 per metre run concentrated at the outer edge
Cat walks	-	1.0 at 1m centre
Public assembly occupancy class (halls, auditoria, restaurants, museums, libraries, non-residential clubs, theatres, broadcasting studios, grandstands		
Dense mobile stacking (books) on mobile trucks	4.8 for each metre of stack height but with a minimum of 9.6	7.0
Stack rooms (books)	2.4 for each metre of stack height but with a minimum of 6.5	7.0
Boiler rooms, motor rooms, fan rooms and the like, including the weight of machinery	7.5	4.5
Stages	7.5	4.5

DESIGN LOADINGS FOR BUILDINGS

Floor area usage	Distributed load kN/m2	Concentrated load kN/300 mm2
Public assembly occupancy class (halls, auditoria, restaurants, museums, libraries, non-residential clubs, theatres, broadcasting studios, grandstands - cont'd		
Corridors, hallways, etc. subject to loads greater than from crowds, such as wheeled vehicles, trolleys and the like. Corridors, stairs, and passage-ways in grandstands	5.0	4.5
Drill rooms and drill halls	5.0	9.0
Assembly areas without fixed seating: dance halls, gymnasia, grandstands	5.0	3.6
Projection rooms, bars	5.0	-
Museum floors and art galleries for exhibition purposes	4.0	4.5
Corridors, hallways, stairs, landings, footbridges, etc.	4.0	4.5
Reading rooms with book storage, e.g. libraries	4.0	4.5
Assembly areas with fixed seating	4.0	-
Kitchens, laundries	3.0	4.5
Chapels, churches	3.0	2.7
Reading rooms without book storage	2.5	4.5
Grids	2.5	-
Areas for equipment	2.0	1.8
Dining rooms, lounges, billiard rooms	2.0	2.7
Dressing rooms	2.0	1.8
Toilet rooms	2.0	-
Balconies	Same as rooms to which they give access but with a minimum of 4.0	1.5 per metre run concentrated at the outer edge

DESIGN LOADINGS FOR BUILDINGS

Floor area usage	Distributed load kN/m2	Concentrated loads kN/300 mm2
Public assembly occupancy class (halls, auditoria, restaurants, museums, libraries, non-residential clubs, theatres, broadcasting studios, grandstands - cont'd		
Fly galleries	4.5 kN per metre run distributed uniformly over the width	
Cat walks	-	1.0 at 1 m centres
Residential occupancy class		
Self contained dwelling units		
All	1.5	1.4
Apartment houses, boarding houses, lodging houses, guest houses, hostels, residential clubs and communal areas in blocks of flats		
Boiler rooms, motor rooms, fan rooms and the like including the weight of machinery	7.5	4.5
Communal kitchens, laundries	3.0	4.5
Dining rooms, lounges, billiard rooms	2.0	2.7
Toilet rooms	2.0	-
Bedrooms, dormitories	1.5	1.8
Corridors, hallways, stairs, landings, footbridges, etc.	3.0	4.5
Balconies	Same as rooms to which they give access but with a minimum of 3.0	1.5 per metre run concentrated at the outer edge
Cat walks	-	1.0 at 1 m centres
Hotels and Motels		
Boiler rooms, motor rooms, fan rooms and the like, including the weight of machinery	7.5	4.5
Assembly areas without fixed seating, dance halls	5.0	3.6

DESIGN LOADINGS FOR BUILDINGS

Floor usage area	Distributed load kN/m2	Concentrated load kN/300 mm2
Hotels and Motels - cont'd		
Bars	5.0	-
Assembly areas with fixed seating	4.0	-
Corridors, hallways, stairs, landings, footbridges, etc.	4.0	4.5
Kitchens, laundries	3.0	4.5
Dining rooms, lounges, billiard rooms	2.0	2.7
Bedrooms	2.0	1.8
Toilet rooms	2.0	-
Balconies	Same as rooms to which they give access but with a minimum of 4.0	1.5 per metre run concentrated at the outer edge
Cat Walks	-	1.0 at 1 m centres
Retail occupancy class (shops, departmental stores, supermarkets)		
Cold storage	5.0 for each metre of storage height with a minimum of 15.0	9.0
Stationery stores	4.0 for each metre of storage height	9.0
Storage, other than types separately	2.4 for each metre of storage height	7.0
Boiler rooms, motor rooms, fan rooms and the like, including the weight of machinery	7.5	4.5
Corridors, hallways, etc. subject to loads greater than from crowds, such as wheeled vehicles, trolleys and the like	5.0	4.5

DESIGN LOADINGS FOR BUILDINGS

Floor area usage	Distribution load kN/m2	Concentrated load kN/300 mm2
Retail occupancy class (shops, departmental stores, supermarkets) - cont'd		
Corridors, hallways, stairs, landings, footbridges, etc.	4.0	4.5
Shop floors for the display and sale of merchandise	4.0	3.6
Kitchens, laundries	3.0	4.5
Toilet rooms	2.0	-
Balconies	Same as rooms to which they give access but with a minimum of 4.0	1.5 per metre run concentrated at the outer edge
Cat walks	-	1.0 at 1 m centres
Storage occupancy class (warehouses)		
Cold storage	5.0 for each metre of storage height with a mimimum of 15.0	9.0
Dense mobile stacking (books) on mobile trucks	4.8 for each metre of storage height with a minimum of 15.0	7.0
Paper storage, for printing plants	4.0 for each metre of storage height	9.0
Stationery stores	4.0 for each metre of storage height	9.0
Storage, other than types listed separately, warehouses	2.4 for each metre of storage height	7.0
Motor rooms, fan rooms and the like, including the weight of machinery	7.5	4.5
Corridors, hallways, foot-bridges, etc. subject to loads greater than for crowds, such as wheeled vehicles, trolleys and the like	5.0	4.5
Cat walks	-	1.0 at 1 m centres

DESIGN LOADINGS FOR BUILDINGS

Floor area usage	Distribution load kN/m2	Concentrated load kN/300 mm2
Vehicular occupancy class (garages, car parks, vehicle access ramps)		
Motor rooms, fan rooms and the like, including the weight of machinery	7.5	4.5
Driveways and vehicle ramps, other than in garages for the parking only of passenger vehicles and light vans not exceeding 2500 kg gross mass	5.0	9.0
Repair workshops for all types of vehicles, parking for vehicles exceeding 2500 kg gross mass including driveways and ramps	5.0	9.0
Footpaths, terraces and plazas leading from ground level with no obstruction to vehicular traffic, pavement lights	5.0	9.0
Corridors, hallways, stairs, landings, footbridges, etc. subject to crowd loading	4.0	4.5
Footpaths, terraces and plazas leading from ground level but restricted to pedestrian traffic only	4.0	4.5
Car parking only, for passenger vehicles and light vans not exceeding 2500 kg gross mass including garages, driveways and ramps.	2.5	9.0
Cat walks	-	1.0 at 1 m centres

PLANNING PARAMETERS

Functional unit areas

As a 'rule of thumb' guide to establish a cost per functional unit, or as a check on economy of design in terms of floor area, the following indicative functional unit areas have been derived from historical data:-

Car parking	- surface	19 m2/car
	- multi-storey	25 m2/car
	- basement	27 m2/car

PLANNING PARAMETERS

Concert Halls		8 m2/seat
Halls of residence	- college/polytechnic	25-35 m2/bedroom
	- university	30-50 m2/bedroom
Hospitals	- district general	65-85 m2/bed
	- teaching	120 + m2/bed
	- private	m2/bed
Hotels	- economy	40 m2/bedroom
	- five star	60 + m2/bedroom
Housing		
		Gross floor area
Private developer:	1 Bedroom Flat	45 - 40 m2
	2 Bedroom Flat	55 - 65 m2
	2 Bedroom House	55 - 65 m2
	3 Bedroom House	70 - 90 m2
	4 Bedroom House	90 - 100 m2
Offices	- high density open plan	10 m2/person to
	- low density cellular	20 m2/person
Schools	- nursery	3 - 5 m2/child
	- secondary	6 - 8 m2/child
	- boarding	10 -12 m2/child
Theatres	- small, local	3 m2/seat to
	- large, prestige	7 m2/seat

Recommended sizes of various sports facilities

Archery (Clout)	: 7.3 m firing area Range 109.728 (Women) 146.304 (Men) 182.88 (Normal range)
Baseball	: Overall 60 m x 70 m
Basketball	: 14.0 m x 26.0 m
Camogie	: 91 - 110 m x 54 - 68 m
Discus and Hammer	: Safety cage 2.74 m square Landing area 45 arc (65^ safety) 70 m radius
Football, American	: Pitch 109.8 m x 48.8 m Overall 118.94 m x 57.94 m

PLANNING PARAMETERS

Football, Association	:	NPFA rules Senior pitches 96 - 100 m x 60 - 64 m Junior pitches 90 m x 46 - 55 m International 100 - 110 m x 64 - 75 m
Football, Australian	:	Overall 135 - 185 m x 110 - 155 m
Football, Canadian	:	Overall 145.74 m x 59.47 m
Football, Gaelic	:	128 - 146.4 m x 76.8 - 91.50 m
Football, Rugby League	:	111 - 122 m x 68.0 m
Football, Rugby Union	:	144.00 m max x 69.0 m
Handball	:	91 - 110 m x 55 - 65 m
Hockey	:	91.5 m x 54.9 m
Hurling	:	137 m x 82 m
Javelin	:	Runway 36.5 m x 4.270 m Landing area 80 - 95 m long 48 m wide
Jump, High	:	Running area 38.8 m x 19 m Landing area 5 m x 4 m
Jump, Long	:	Runway 45 m x 1.220 m Landing area 9 m x 2.750 m
Jump, Triple	:	Runway 45 m x 1.22 m Landing area 7.3 m x 2.75 m
Korfball	:	90 m x 40 m
Lacrosse	:	(Mens) 100 m x 55 m (Womens) 110 m x 73 m
Netball		15.250 m x 30.480 m
Pole Vault	:	Runway 45 m x 1.220 m Landing area 5 m x 5 m
Polo	:	275 m x 183 m
Rounders	:	Overall 19 m x 17 m

PLANNING PARAMETERS

Recommended sizes of various sports facilities - cont'd

400m Running Track	:	115.61 m bend length x 2 84.39 m straight length x 2 Overall 176.91 m long x 92.52 m wide
Shot Putt	:	Base 2.135 m dia Landing area 65^ arc 25 m radius from base
Shinty	:	128 - 183 m x 64 - 91.5 m
Tennis	:	Court 23.77 m x 10.97 m Overall minimum 36.27 m x 18.29 m
Tug-of-war	:	46 m x 5 m

SOUND INSULATION

Sound reduction requirements as Building Regulations (E1/2/3)

The Building Regulations on airborne and impact sound (E1/2/3) state simply that both airborne and impact sound must be reasonably reduced in floors and walls. No minimum reduction is given but the following tables give example sound reductions for various types of constructions.

Sound reductions of typical walls

	Average sound reduction (dB)
13 mm Fibreboard	20
16 mm Plasterboard	25
6 mm Float glass	30
16 mm Plasterboard, plastered both sides	35
75 mm Plastered concrete blockwork (100 mm)	44
110 mm half brick wall, half brick thick, plastered both sides	43
SOUND INSULATION	
220 mm Brick wall one brick thick, plastered both sides	48
Timber stud partitioning with plastered metal lathing both sides	35
Cupboards used as partitions	30
Cavity block wall, plastered both sides	42

SOUND INSULATION	Average sound reductions (dB)
75 mm Breeze block cavity wall, plastered both sides	50
100 mm Breeze block cavity wall, plastered both sides including 50 mm air-gap and plasterboard suspended ceiling	55
As above with 150 mm Breeze blocks	65
19 mm T & G boarding on timber joists including plasterboard ceiling and plaster skim coat	32
As above including metal lash and plaster ceiling	37
As above with solid sound proofing material between joists approx 98 kg per sq metre	55
As above with floating floor of T & G boarding on batten and soundproofing quilt	75

Impact noise is particularly difficult to reduce satisfactorily. The following are the most efficient methods of reducing such sound.

1) Carpet on underlay of rubber or felt.
2) Pugging between joists (e.g. Slag Wool) and
3) A good suspended ceiling system.

THERMAL INSULATION

Thermal properties of various building elements

Thickness	Material	(m2k/W) R	(W/m2K) U - Value
-	Internal and external surface resistance	0.18	-
	Air-gap cavity	0.18	-
103 mm	Brick skin	0.12	-
	Dense concrete block		
100 mm	Arc conbloc	0.09	11.11
140 mm	Arc conbloc	0.13	7.69
190 mm	Arc conbloc	0.18	5.56
	Lightweight aggregate block		
100 mm	Celcon standard	0.59	1.69
125 mm	Celcon standard	0.74	1.35
150 mm	Celcon standard	0.88	1.14
200 mm	Celcon standard	1.18	0.85
	Lightweight aggregate thermal block		
125 mm	Celcon solar	1.14	0.88
150 mm	Celcon solar	1.36	0.74
200 mm	Celcon solar	1.82	0.55

THERMAL INSULATION

Thermal properties of various building elements - cont'd

Thickness	**Material**	**(m2K/W)** R	**(W/m2k)** U - Value
	Insulating board		
25 mm	Drithern	0.69	1.45
50 mm	Drithern	1.39	0.72
75 mm	Drithern	2.08	0.48
13 mm	Lightweight plaster 'Carlite'	0.07	14.29
13 mm	Dense plaster 'Thistle'	0.02	50.00
	Plasterboard		
9.5 mm	British gypsum	0.06	16.67
12.7 mm	British gypsum	0.08	12.50
40 mm	Screed	0.10	10.00
150 mm	Reinforced concrete	0.12	8.33
100 mm	Dow roofmate insultation	3.57	0.28

RESISTANCE TO THE PASSAGE OF HEAT

Provisions meeting the requirement set out in the Building Regulations (L2/3):-

	Minimum U - Value
a) **Dwellings**	
Roof	0.35
Exposed wall	0.60
Exposed floor	0.60
b) **Residential, Offices, Shops and Assembly Buildings**	
Roof	0.06
Exposed wall	0.60
Exposed floor	0.60
c) **Industrial Storage and Other Buildings**	
Roof	0.70
Exposed wall	0.70
Exposed floor	0.70

External wall, masonry construction:-

Concrete blockwork	**U - Value**
200 mm lightweight concrete block, 25 mm air-gap, 10 mm plasterboard	0.68
200 mm lightweight concrete block, 20 mm EPS slab, 10 mm plasterboard	0.54
200 mm lightweight concrete block, 25 mm air-gap, 25 mm EPS slab, 10 mm plasterboard	0.46

TYPICAL CONSTRUCTIONS MEETING THERMAL REQUIREMENTS

Brick/Cavity/Brick

Construction	
105 mm brickwork, 50 mm UF foam, 105 mm brickwork, 3 mm lightweight plaster	0.55
105 mm brickwork, 50 mm cavity, 125 mm Thermalite block, 3 mm lightweight plaster	0.59
105 mm brickwork, 50 mm cavity, 130 mm Thermalite block, 3 mm lightweight plaster	0.57
105 mm brickwork, 50 mm cavity, 130 mm Thermalite block, 3 mm dense plaster	0.59
105 mm brickwork, 50 mm cavity, 100 mm Thermalite block, foilbacked plasterboard	0.55
105 mm brickwork, 50 mm cavity, 115 mm Thermalite block, 9.5 mm plasterboard	0.58
105 mm brickwork, 50 mm cavity, 115 mm Thermalite block, foilbacked plasterboard	0.52
105 mm brickwork, 50 mm cavity, 125 mm Theramlite block, 9.5 mm plasterboard	0.55
105 mm brickwork, 50 mm cavity, 100 mm Thermalite block, 25 mm insulating plasterboard	0.53
105 mm brickwork, 50 mm cavity, 125 mm Thermalite block, 25 mm insulating plasterboard	0.47
105 mm brickwork, 25 mm cavity, 25 mm insulation, 100 mm Thermalite block, lightweight plaster	0.47
105 mm brickwork, 25 mm cavity, 25 mm insulation, 115 mm Thermalite block, lightweight plaster	0.44

Block/Cavity/Block

Construction	
Render, 100 mm 'SHIELD' block, 50 mm cavity, 100 mm Thermalite block, lightweight plaster	0.50
Render, 100 mm 'SHIELD' block, 50 mm cavity, 115 mm Thermalite block, lightweight plaster	0.47
Render, 100 mm 'SHIELD' block, 50 mm cavity, 125 mm Thermalite block, lightweight plaster	0.45

Tile hanging

Construction	
10 mm tile on battens and felt, 150 mm Thermalite block, lightweight plaster	0.57
25 mm insulating plasterboard	0.46

TYPICAL CONSTRUCTIONS MEETING THERMAL REQUIREMENTS

	U - Value
Tile hanging - cont'd	
10 mm tile on battens and felt, 190 mm Thermalite block, lightweight plaster	0.47
25 mm insulating plasterboard	0.40
10 mm tile on battens and felt, 200 mm Thermalite block, lightweight plaster	0.45
25 mm insulated plasterboard	0.38
10 mm tile on battens, breather paper, 25 mm air-gap, 50 mm glass fibre quilts, 10 mm plasterboard	0.56
10 mm tile on battens, breather paper, 25 mm air-gap, 75 mm glass fibre quilts, 10 mm plasterboard	0.41
10 mm tile on battens, breather paper, 25 mm air-gap, 100 mm glass fibre quilts, 10 mm plasterboard	0.33
Pitched roofs	
Slate or concrete tiles, felt, airspace, 'Rockwool' flexible slabs laid between rafters, plasterboard	
Slab 40 mm thick	0.62
50 mm thick	0.52
60 mm thick	0.45
75 mm thick	0.38
100 mm thick	0.29
Concrete tiles, sarking felt, rollbatts between joists, plasterboard	
Insulation 100 mm thick	0.31
120 mm thick	0.26
140 mm thick	0.23
160 mm thick	0.21
Steel frame 'Rockwool' insulation sandwiched between steel exterior profiled sheeting and interior sheet lining	
Insulation 60 mm thick	0.53
80 mm thick	0.41
100 mm thick	0.34

TYPICAL CONSTRUCTIONS MEETING THERMAL REQUIREMENTS

		U - Value
Steel frame, steel profiled sheeting, 'Rockwool' insulation over purlins and plasterboard lining		
Insulation	60 mm thick	0.51
	80 mm thick	0.38
	100 mm thick	0.32
	120 mm thick	0.27
	140 mm thick	0.24
	160 mm thick	0.21
Flat roofs		
Asphalt, 'Rockwool' roof slabs, 25 mm timber boarding, timber joists and 9.5 mm plasterboard		
Insulation	30 mm thick	0.68
	40 mm thick	0.57
	50 mm thick	0.49
	60 mm thick	0.44
	70 mm thick	0.39
	80 mm thick	0.35
	90 mm thick	0.32
	100 mm thick	0.29
Asphalt, 'Rockwool' roof slabs on 150 mm dense concrete deck and screed with 16 mm plaster finish		
Insulation	40 mm thick	0.68
	50 mm thick	0.57
	60 mm thick	0.49
	70 mm thick	0.43
	80 mm thick	0.39
	90 mm thick	0.35
	100 mm thick	0.32
Asphalt, 'Rockwool' roof slabs on 150 mm dense concrete deck and screed with suspended plasterboard ceiling		
Insulation	40 mm thick	0.60
	50 mm thick	0.52
	60 mm thick	0.45
	70 mm thick	0.40
	80 mm thick	0.36
	90 mm thick	0.33
	100 mm thick	0.30
Steel frame, asphalt on insulation slabs on troughed steel decking		
Insulation	50 mm thick	0.59
	60 mm thick	0.51
	70 mm thick	0.45
	80 mm thick	0.39
	90 mm thick	0.35
	100 mm thick	0.33

TYPICAL CONSTRUCTIONS MEETING THERMAL REQUIREMENTS

		U - Value
Steel frame, asphalt on insulation slabs on troughed steel decking including suspended plasterboard ceiling		
Insulation	40 mm thick	0.67
	50 mm thick	0.57
	60 mm thick	0.49
	70 mm thick	0.43
	80 mm thick	0.38
	90 mm thick	0.34
	100 mm thick	0.32

WEIGHTS OF VARIOUS MATERIALS

AGGREGATES	lbs/ft3	lbs/yd3	kg/m3
Ashes	50	1350	800
Cement (Portland)	90	2430	1441
Chalk	140	3780	2240
Chippings (stone)	110	2970	1762
Clinker (furnace)	50	1350	800
(concrete)	90	2430	1441
Ballast or stone	140	3780	2241
Pumice	70	1890	1121
Gravel	110	2970	1762
Lime:			
Chalk (lump)	44	1188	704
Ground	60	1620	961
Quick	55	1485	880
Sand:			
Dry	100	2700	1601
Wet	88	2376	1281
Water	62	1674	933
Shale	125	3371	2000
Whinstone	125	3371	2000
Broken stone	107	2881	1709
Pitch	72	1944	1152
METALS			
Aluminium	162	4374	2559
Brass	525	14175	8129
Bronze	524	14148	8113
Gunmetal	528	14256	8475
Iron:			
Cast	450	12150	7207
Wrought	480	12960	7687
Lead	708	19116	11260
Tin	465	12555	7448
Zinc	466	12582	7464

WEIGHTS OF VARIOUS MATERIALS

STONE AND BRICKWORK	lbs/ft3	lbs/yd3	kg/m3
Blockwork:			
Aerated	41	1095	650
Dense concrete	112	3034	1800
Lightweight concrete	75	2023	1200
Pumice concrete	67	1820	1080
Brickwork:			
Common Fletton	125	3375	1822
Glazed brick	130	3510	2080
Staffordshire Blue	135	3645	2162
Red Engineering	140	3780	2240
Concrete	115	3105	1841
Stone:			
Artificial	140	3780	2242
Bath	140	3780	2242
Blue Pennant	168	4536	2682
Cragleith	145	3915	2322
Darley Dale	148	3996	2370
Forest of Dean	149	4023	2386
Granite	166	4482	2642
Marble	170	4590	2742
Portland	135	3645	2170
Slate	180	4860	2882
York	150	4050	2402
Terra-cotta	132	3564	2116

WOOD			
Blockboard	31 - 44	843 - 1180	500 - 700
Cork Bark	5	135	80
Hardboard:			
Standard	59 - 62	1584 - 1686	940 - 1000
Tempered	59 - 66	1584 - 1787	940 - 1060
Wood chipboard:			
Type I	41 - 47	1096 - 1264	650 - 750
Type II	42 - 50	1146 - 1349	680 - 800
Type III	41 - 50	1096 - 1349	650 - 800
Type II/III	42 - 50	1146 - 1349	680 - 800
Laminboard	31 - 44	843 - 1180	500 - 700
Timber:			
Ash	50	1350	800
Baltic spruce	30	810	480
Beech	51	1377	816
Birch	45	1215	720
Box	60	1620	961
Cedar	30	810	480
Chestnut	40	1080	640
Ebony	76	2052	1217
Elm	39	1053	624
Greenheart	60	1620	961
Jarrah	51	1377	816

WEIGHTS OF VARIOUS MATERIALS

WOOD	lbs/ft3	lbs/yd3	kg/m3
Maple	47	1269	752
Mahogany:			
Honduras	36	972	576
Spanish	66	1782	1057
Oak:			
English	53	1431	848
American	45	1215	720
Austrian & Turkish	44	1188	704
Pine:			
Pitchpine	50	1350	800
Red Deal	36	972	576
Yellow Deal	33	891	528
Spruce	31	837	496
Sycamore	38	1026	530
Teak:			
African	60	1620	961
Indian	41	1107	656
Moulmein	46	1242	736
Walnut:			
English	41	1107	496
Black	45	1215	720

EXCAVATION AND EARTHWORK

Transport capacities

Type of vehicle	Capacity of vehicle	
	cu yards (solid)	cu metres (solid)
Standard wheelbarrow	0.10	0.08
2 ton truck (2.03 t)	1.50	1.15
3 ton truck (3.05 t)	2.25	1.72
4 ton truck (4.06 t)	2.90	2.22
5 ton truck (5.08 t)	3.50	2.68
6 ton truck (6.10 t)	4.50	3.44
2 cu yard dumper (1.53 m3)	1.50	1.15
3 cu yard dumper (2.29 m3)	2.25	1.72
6 cu yard dumper (4.59 m3)	4.50	3.44
10 cu yard dumper (7.65 m3)	7.50	5.73

EXCAVATION AND EARTHWORK

Planking and strutting

Maximum depth of excavation in various soils without the use of earthwork support

Ground conditions	Feet (ft)	Metres (m)
Compact soil	12	3.66
Drained loam	6	1.83
Dry sand	1	0.30
Gravelly earth	2	0.61
Ordinary earth	3	0.91
Stiff clay	10	3.05

It is important to note that the above table should only be used as a guide. Each case must be taken on its merits and, as the limited distances, given above, are approached, careful watch must be kept for the slightest signs of caving in.

Baulkage of soils after excavation

	Approximate bulk of 1 m3 after excavation
Vegetable soil and loam	25 - 30%
Soft clay	30 - 40%
Stiff clay	10 - 15%
Gravel	20 - 25%
Sand	40 - 50%
Chalk	40 - 50%
Rock, weathered	30 - 40%
Rock, unweathered	50 - 60%

CONCRETE WORK

Approximate average weights of materials

Materials	Percentage of voids (%)	Weight per m3 (kg)
Sand	39	1660
Gravel 10 - 20 mm	45	1440
Gravel 35 - 75 mm	42	1555
Crushed stone	50	1330
Crushed granite		
(over 15 mm)	50	1345
(n.e. 15 mm)	47	1440
'All-in' ballast	32	1800

CONCRETE WORK

Common mixes for various types of work per m3

Recommended mix	Class of work suitable for:-	Cement (kg)	Sand (kg)	Course Aggregate (kg)	No. of 50 kg bags of cement per m3 of combined aggregate
1:3:6	Roughest type of mass concrete such as footings, road haunching over 300 mm thick.	208	905	1509	4.00
1:2.5:5	Mass concrete of better class than 1:3:6 such as bases for machinery, walls below ground etc.	249	881	1474	5.00
1:2:4	Most ordinary uses of concrete, such as mass walls above ground, road slabs etc. and general reinforced concrete work.	304	889	1431	6.00
1:1.5:3	Watertight floors, pavements and walls, tanks, pits, steps, paths, surface of 2 course roads, reinforced concrete where extra strength is required.	371	801	1336	7.50
1:1:2	Work of thin section such as fence posts and small precast work.	511	720	1206	10.50

Bar reinforcement

Cross-sectional area and mass

Nominal sizes (m)	Cross-sectional area (mm2)	Mass per metre run (kg)
6*	28.3	0.222
8	50.3	0.395
10	78.5	0.616
12	113.1	0.888
16	201.1	1.579
20	314.2	2.466
25	490.9	3.854
32	804.2	6.313
40	1256.6	9.864
50*	1963.5	15.413

*Where a bar larger than 40 mm is to be used the recommended size is 50 mm. Where a bar smaller than 8 mm is to be used the recommended size is 6 mm.

CONCRETE WORK

Fabric reinforcement

Preferred range of designated fabric types and stock sheet sizes

Fabric reference	Longitudinal wires			Cross wires			
	Nominal wire size (mm)	Pitch (mm)	Area (mm2/m)	Nominal wire size (mm)	Pitch (mm)	Area (mm2/m)	Mass (kg/m2/m)
Square mesh							
A393	10	200	393	10 200	393	6.16	
A252	8	200	252	8 200	252	3.95	
A193	7	200	193	7 200	193	3.02	
A142	6	200	142	6 200	142	2.22	
A98	5	200	98	5 200	98	1.54	
Structural mesh							
B1131	12	100	1131	8 200	252	10.90	
B785	10	100	785	8 200	252	8.14	
B503	8	100	503	8 200	252	5.93	
B385	7	100	385	7 200	193	4.53	
B283	6	100	283	7 200	193	3.73	
B196	5	100	196	7 200	193	3.05	
Long mesh							
C785	10	100	785	6 400	70.8	6.72	
C636	9	100	636	6 400	70.8	5.55	
C503	8	100	503	5 400	49.0	4.34	
C385	7	100	385	5 400	49.0	3.41	
C283	6	100	283	5 400	49.0	2.61	
Wrapping mesh							
D98	5	200	98	5 200	98	1.54	
D49	2.5	100	49	2.5	100	49	0.77

	Longitudinal wires	Cross wires	Sheet area
Stock sheet size	Length 4.8 m	Width 2.4 m	11.52 m2

Average weight kg/m3 of steelwork reinforcement in concrete for various building elements

Substructure	kg/m3 concrete
Pile caps	110 - 150
Tie beams	130 - 170
Ground beams	230 - 330
Bases	125 - 180
Footings	100 - 150
Retaining walls	150 - 210
Raft	60 - 70

CONCRETE WORK

Superstructure

Slabs - one way	120 - 200
Slabs - two way	110 - 220
Plate slab	150 - 220
Cant slab	145 - 210
Ribbed floors	130 - 200
Topping to block floor	30 - 40
Columns	210 - 310
Beams	250 - 350
Stairs	130 - 170
Walls - normal	40 - 100
Walls - wind	70 - 125

Note: For exposed elements add the following % :

Walls 50%, Beams 100%, Columns 15%

BRICKWORK AND BLOCKWORK

Number of bricks required for various types of work

Description	**Brick size**	
	215 x 103.5 x 50 mm	**215 x 103.5 x 65 mm**
Half brick thick		
Stretcher bond	72	58
English bond	108	86
English garden wall bond	90	72
Flemish bond	96	79
Flemish garden wall bond	83	66
One brick thick and cavity wall of two half brick skins		
Stretcher bond	144	116

Standard bricks

Brick size 215 x 102.5 x 50 mm				
half brick wall (103 mm)	m2	72	0.022	0.0270.032
2 x half brick cavity wall (270 mm)	m2	144	0.044	0.0540.064
one brick wall (215 mm)	m2	144	0.052	0.0640.076
one and a half brick wall (322 mm)	m2	216	0.073	0.0910.108
Mass brickwork	m3	576	0.347	0.4130.480

BRICKWORK AND BLOCKWORK

Quantities of bricks and mortar required per m2 of walling

Description	Unit	No of bricks required	Mortar required (cubic metres)		
			No frogs	Single frogs	Double frogs
Brick size 215 x 102.5 x 65 mm					
half brick wall (103 mm)	m2	58	0.019	0.022	0.026
2 x half brick cavity wall (270 mm)	m2	116	0.038	0.045	0.055
one brick wall (215 mm)	m2	116	0.046	0.055	0.064
one and a half brick wall (322 mm)	m2	174	0.063	0.074	0.088
Mass brickwork	m3	464	0.307	0.360	0.413
Metric modular bricks			**Perforated**		
Brick size 200 x 100 x 75 mm					
90 mm thick	m2	67	0.016	0.019	
190 mm thick	m2	133	0.042	0.048	
290 mm thick	m2	200	0.068	0.078	
Brick size 200 x 100 x 100 mm					
90 mm thick	m2	50	0.013	0.016	
190 mm thick	m2	100	0.036	0.041	
290 mm thick	m2	150	0.059	0.067	
Brick size 300 x 100 x 75 mm					
90 mm thick	m2	33	-	0.015	
Brick size 300 x 100 x 100 mm					
90 mm thick	m2	44	0.015	0.018	

Note: Assuming 10 mm deep joints.

Standard available block sizes

Block	Length x height: Co-ordinating size	Work size	Thicknesses (work size)
A	400 x 100 400 x 200	390 x 90 440 x 190	(75, 90, 100, (140 & 190 mm
	450 x 225	440 x 215	(75, 90, 100 (140, 190, & 215 mm

BRICKWORK AND BLOCKWORK

Concrete blockwork

Block	Length x height Co-ordinating size	Work size	Thicknesses (work size)
B	400 x 100	390 x 90	(75, 90, 100
	400 x 200	390 x 190	(140 & 190 mm
	450 x 200	440 x 190	(
	450 x 225	440 x 215	(75, 90, 100
	450 x 300	440 x 290	(140, 190, & 215 mm
	600 x 200	590 x 190	(
	600 x 225	590 x 215	(
C	400 x 200	390 x 190	(
	450 x 200	440 x 190	(
	450 x 225	440 x 215	(60 & 75 mm
	450 x 300	440 x 290	(
	600 x 200	590 x 190	(
	600 x 225	590 x 215	(

Mortar required per m2 blockwork (9.88 blocks/m2)

Wall thickness	75	90	100	125	140	190	215
Mortar m3/m2	0.005	0.006	0.007	0.008	0.009	0.013	0.014

ROOFING

Total roof loadings for various types of tiles/slates

	Roof load (slope) kg/m2		
	Slate/Tile	Roofing underlay and battens	Total dead load kg/m2
Asbestos cement slate (600 x 300 mm)	21.50	3.14	24.64
Clay tile			
interlocking	67.00	5.50	72.50
plain	43.50	2.87	46.37
Concrete tile			
interlocking	47.20	2.69	49.89
plain	78.20	5.50	83.70
Natural slate (18" x 10")	35.40	3.40	38.80

ROOFING

Total roof loadings for various types of tiles/slates - cont'd

	Roof load (plan) kg/m2		
	Dead load	**Imposed load**	**Total roof load kg/m2**
Asbestos cement slate (600 x 300 mm)	28.45	76.50	104.95
Clay tile			
interlocking	53.54	76.50	130.04
plain	83.71	76.50	160.21
Concrete tile			
interlocking	57.60	76.50	134.10
plain	96.64	76.50	173.14
Natural slate (18" x 10")	44.80	76.50	121.30

Tiling data

Product	**Size of slates/tiles**	**Lap (mm)**	**Gauge of battens**	**No. slates per m2**	**Battens (m/m2)**	**Weight as laid (kg/m2)**
CEMENT SLATES						
Eternit T.A.C. slates (Duracem)						
	600 x 300 mm	100	250	13.4	4.00	19.50
		90	255	13.1	3.92	19.20
		80	260	12.9	3.85	19.00
		70	265	12.7	3.77	18.60
	600 x 350 mm	100	250	11.5	4.00	19.50
		90	255	11.2	3.92	19.20
	500 x 250 mm	100	200	20.0	5.00	20.00
		90	205	19.5	4.88	19.50
		80	210	19.1	4.76	19.00
		70	215	18.6	4.65	18.60
	400 x 200 mm	90	155	32.3	6.45	20.80
		80	160	31.3	6.25	20.20
		70	165	30.3	6.06	19.60
CONCRETE TILES/SLATES						
Redland Roofing						
Stonewold slate	430 x 380 mm	75	355	8.2	2.82	51.20
Double Roman tile	418 x 330 mm	75	355	8.2	2.91	45.50
Grovebury pantile	418 x 332 mm	75	343	9.7	2.91	47.90

ROOFING

Product	Size of slate	Lap (mm)	Gauge of battens	No. slates per m2	Battens (m/m2)	Weight as laid (kg/m2)
Redland Roofing						
Norfolk pantile	381 x 227 mm	75	306	16.3	3.26	44.01
		100	281	17.8	3.56	48.06
Renown interlocking tile	418 x 330 mm	75	343	9.7	2.91	46.40
'49' tile	381 x 227 mm	75	306	16.3	3.26	44.80
		100	281	17.8	3.56	48.95
Plain, vertical tiling	265 x 165 mm	35	115	52.7	8.70	62.20
Marley Roofing						
Bold roll tile	419 x 330 mm	75	344	9.7	2.90	47.00
		100	-	10.5	3.20	51.00
Modern roof tile	413 x 330 mm	75	338	10.2	3.00	54.00
		100	-	11.0	3.20	58.00
Ludlow major	413 x 330 mm	75	338	10.2	3.00	45.00
		100	-	11.0	3.20	49.00
Ludlow plus	380 x 230 mm	75	305	16.1	3.30	47.00
		100	-	17.5	3.60	51.00
Mendip tile	413 x 330 mm	75	338	10.2	3.00	47.00
		100	-	11.0	3.20	51.00
Wessex	413 x 330 mm	75	338	10.2	3.00	54.00
		100	-	11.0	3.20	58.00
Plain tile	265 x 165 mm	65	100	60.0	10.00	76.00
		75	95	64.0	10.50	81.00
		85	90	68.0	11.30	86.00
Plain vertical tiles (feature)	265 x 165 mm	35	110	53.0	8.70	67.00
		34	115	56.0	9.10	71.00
CLAY TILES						
Crossley Ltd						
Old English pantile	342 x 241 mm	75	267	18.5	3.76	47.00
French pattern pantiles	342 x 241 mm	75	267	18.8	3.76	48.00
Bold roll Roman tiles	342 x 266 mm	75	267	17.8	3.76	50.00

Slate nails, quantity per kilogram

	Type			
	Plain wire	Galvanised wire	Copper nail	Zinc nail
28.5 mm long	325	305	325	415
34.4 mm long	286	256	254	292
50.8 mm long	242	224	194	200

ROOFING

Metal sheet coverings

Thicknesses and weights of sheet metal coverings

Lead to BS 1178

BS Code No	3 4	5 6	7	8		
Colour Code	Green	Blue	Red	Black	White	Orange
Thickness (mm)	1.25	1.80	2.24	2.50	3.15	3.55
kg/m2	14.18	20.41	25.40	28.36	35.72	40.26

Copper to BS 2870

Thickness (mm)	0.60	0.70
Bay width		
Roll (mm)	500	650
Seam (mm)	525	600
Standard width to form bay	600	750
Normal length of sheet	1.80	1.80
kg/m2		

Zinc to BS 849

Zinc Gauge (Nr)	9	10	11	12	13	14	15	16
Thickness (mm)	0.43	0.48	0.56	0.64	0.71	0.79	0.91	1.04
kg/m2	3.1	3.2	3.8	4.3	4.8	5.3	6.2	7.0

Aluminium to BS 4868

Thickness (mm)	0.5	0.6	0.7	0.8	0.9	1.0	1.2
kg/m2	12.8	15.4	17.9	20.5	23.0	25.6	30.7

Type of felt	**Nominal mass per unit area (kg/10m)**	**Nominal mass per unit area of fibre base (g/m2)**	**Nominal length of roll (m)**
Class 1			
1B fine granule surfaced bitumen	14	220	10 or 20
	18	330	10 or 20
	25	470	10
1E mineral surfaced bitumen	38	470	10

ROOFING

Type of felt	Nominal mass per unit area (kg/10m)	Nominal mass per unit area of fibre base (g/m2)	Nominal length of roll (m)
1F reinforced bitumen	15	160 (fibre) 110 (hessian)	15
1F reinforced bitumen, aluminium faced	13	160 (fibre) 110 (hessian)	15
Class 2			
2B fine granule surfaced bitumen asbestos	18	500	10 or 20
2E mineral surfaced bitumen asbestos	38	600	10
Class 3			
3B fine granule surfaced bitumen glass fibre	18	60	20
3E mineral surfaced bitumen glass fibre	28	60	10
3E venting base layer bitumen glass fibre	32	60*	10
3H venting base layer bitumen glass fibre	17	60*	20

* Excluding effect of perforations

WOODWORK

Conversion tables (for timber only)

Inches	Millimetres	Feet	Metres
1	25	1	0.300
2	50	2	0.600
3	75	3	0.900
4	100	4	1.200
5	125	5	1.500
6	150	6	1.800
7	175	7	2.100
8	200	8	2.400
9	225	9	2.700
10	250	10	3.000
11	275	11	3.300
12	300	12	3.600
13	325	13	3.900
14	350	14	4.200
15	375	15	4.500
16	400	16	4.800
17	425	17	5.100
18	450	18	5.400
19	475	19	5.700
20	500	20	6.000
21	525	21	6.300
22	550	22	6.600
23	575	23	6.900
24	600	24	7.200

Planed softwood

The finished end section size of planed timber is usually 3/16" less than the original size from which it is produced. This however varies slightly dependant upon availability of material and origin of the species used.

Standards (timber) to cubic metres and cubic metres to standards (timber)

Cubic metres	Cubic metres standards	Standards
4.672	1	0.214
9.344	2	0.428
14.017	3	0.642
18.689	4	0.856
23.361	5	1.070
28.033	6	1.284
32.706	7	1.498
37.378	8	1.712
42.050	9	1.926
46.722	10	2.140
93.445	20	4.281
140.167	30	6.421
186.890	40	8.561
233.612	50	10.702
280.335	60	12.842
327.057	70	14.982
373.779	80	17.122
420.502	90	19.263
467.224	100	21.403

WOODWORK

Standards (timber) to cubic metres and cubic metres to standards (timber)

1 cu metre = 35.3148 cu ft = 0.21403 std

1 cu ft = 0.028317 cu metres

1 std = 4.67227 cu metres

Basic sizes of sawn softwood available (cross sectional areas)

Thickness (mm)	Width (mm)								
	75	100	125	150	175	200	225	250	300
16	X	X	X	X					
19	X	X	X	X					
22	X	X	X	X					
25	X	X	X	X	X	X	X	X	X
32	X	X	X	X	X	X	X	X	X
36	X	X	X	X					
38	X	X	X	X	X	X	X		
44	X	X	X	X	X	X	X	X	X
47*	X	X	X	X	X	X	X	X	X
50	X	X	X	X	X	X	X	X	X
63	X	X	X	X	X	X	X		
75		X	X	X	X	X	X	X	X
100		X		X		X		X	X
150				X		X			X
200						X			
250								X	
300									X

* This range of widths for 47 mm thickness will usually be found to be available in construction quality only.

Note: The smaller sizes below 100 mm thick and 250 mm width are normally but not exclusively of European origin. Sizes beyond this are usually of North and South American origin.

WOODWORK

Basic lengths of sawn softwood available (metres)

1.80	2.10	3.00	4.20	5.10	6.00	7.20
	2.40	3.30	4.50	5.40	6.30	
	2.70	3.60	4.80	5.70	6.60	
		3.90	6.90			

Note; Lengths of 6.00 m and over will generally only be available from North American species and may have to be recut from larger sizes.

Reductions from basic size to finished size by planing of two opposed faces

	Reductions from basic sizes for timber			
Purpose	**15 - 35 mm**	**36 - 100 mm**	**101 - 150 mm**	**over 150 mm**
a) constructional timber	3 mm	3 mm	5 mm	6 mm
b) Matching interlocking boards	4 mm	4 mm	6 mm	6 mm
c) Wood trim not specified in BS 584	5 mm	7 mm	7 mm	9 mm
d) Joinery and cabinet work	7 mm	9 mm	11 mm	13 mm

Note: The reduction of width or depth is overall the extreme size and is exclusive of any reduction of the face by the machining of a tongue or lap joints.

Maximum spans for various roof trusses

Maximum permissible spans for rafters for Fink trussed rafters

Basic size (mm)	**Actual size (mm)**	Pitch (degrees) **15 (m)**	**17.5 (m)**	**20 (m)**	**22.5 (m)**	**25 (m)**	**27.5 (m)**	**30 (m)**	**32.5 (m)**	**35 (m)**
38 x 75	35 x 72	6.03	6.16	6.29	6.41	6.51	6.60	6.70	6.80	6.90
38 x 100	35 x 97	7.48	7.67	7.83	7.97	8.10	8.22	8.34	8.47	8.61
38 x 125	35 x 120	8.80	9.00	9.20	9.37	9.54	9.68	9.82	9.98	10.16
44 x 75	41 x 72	6.45	6.59	6.71	6.83	6.93	7.03	7.14	7.24	7.35
44 x 100	41 x 97	8.05	8.23	8.40	8.55	8.68	8.81	8.93	9.09	9.22
44 x 125	41 x 120	9.38	9.60	9.81	9.99	10.15	10.31	10.45	10.64	10.81
50 x 75	47 x 72	6.87	7.01	7.13	7.25	7.35	7.45	7.53	7.67	7.78
50 x 100	47 x 97	8.62	8.80	8.97	9.12	9.25	9.38	9.50	9.66	9.80
50 x 125	47 x 120	10.01	10.24	10.44	10.62	10.77	10.94	11.00	11.00	11.00

WOODWORK

Sizes of internal and external doorsets

Description	Internal Size (mm)	Internal Permissible deviation	External Size (mm)	External Permissible deviation
Co-ordinating dimension: height of door leaf height sets	2100		2100	
Co-ordinating dimension: height of ceiling height set	2300		2300	
	2350		2350	
	2400		2400	
	2700		2700	
	3000		3000	
Co-ordinating dimension: width of all door sets	600 S		900 S	
	700 S		1000 S	
S = Single leaf set	800 S&D		1200 D	
D = Double leaf set	900 S&D		1500 D	
	1000 S&D		1800 D	
	1200 D		2100 D	
	1500 D			
	1800 D			
	2100 D			
Work size: height of door leaf height set	2090	± 2.0	2095	± 2.0
Work size: height of ceiling height set	2285)		2295)	
	2335)		2345)	
	2385)	± 2.0	2395)	± 2.0
	2685)		2695)	
	2985)		2995)	
Work size: width of all door sets	590 S)		895 S)	
	690 S)		995 S)	
	790 S&D)		1195 D)	± 2.0
S = Single leaf set	890 S&D)		1495 D)	
D = Double leaf set	990 S&D)	± 2.0	1795 D)	
	1190 D)		2095 D)	
	1490 D)			
	1790 D)			
	2090 D)			
Width of door leaf in single leaf sets	526 F)		806 F&P)	
	626 F)		906 F&P)	± 1.5
F = Flush leaf	726 F&P)	± 1.5		
P = Panel leaf	826 F&P)			
	926 F&P)			

WOODWORK

Description	Internal Size (mm)	Internal Permissible deviation	External Size (mm)	External Permissible deviation
Width of door leaf in double leaf sets F = Flush leaf P = Panel leaf	362 F 412 F 426 F 562 F&P 712 F&P 826 F&P 1012 F&P	± 1.5	552 F&P 702 F&P 852 F&P 1002 F&P	± 1.5
Door leaf height for all door sets	2040	± 1.5	1994	± 1.5

STRUCTURAL STEELWORK

Tables showing the mass and surface area per metre run for various steel members

Size (mm)	(kg/m)	Surface area per m2
Universal beams		
914 x 3419	388	3.404
	343	3.382
914 x 305	289	2.988
	253	2.967
	224	2.948
	201	2.932
838 x 292	226	2.791
	194	2.767
	176	2.754
762 x 267	197	2.530
	173	2.512
	147	2.493
686 x 254	170	2.333
	152	2.320
	140	2.310
	125	2.298
610 x 305	238	2.421
	179	2.381
	149	2.361
610 x 229	140	2.088
	125	2.075
	113	2.064
	101	2.053

STRUCTURAL STEELWORK

Size (mm)	(kg/m)	Surface area per m2
Universal beams		
533 x 210	122	1.872
	109	1.860
	101	1.853
	92	1.844
	82	1.833
457 x 191	98	1.650
	89	1.641
	82	1.633
	74	1.625
	67	1.617
457 x 152	82	1.493
	74	1.484
	67	1.474
	60	1.487
	52	1.476
406 x 178	74	1.493
	67	1.484
	60	1.476
	54	1.468
406 x 140	46	1.332
	39	1.320
356 x 171	67	1.371
	57	1.358
	51	1.351
	45	1.343
356 x 127	39	1.169
	33	1.160
305 x 165	54	1.245
	46	1.235
	40	1.227
305 x 127	48	1.079
	42	1.069
	37	1.062
305 x 102	33	1.006
	28	0.997
	25	0.988
254 x 146	43	1.069
	37	1.060
	31	1.050
254 x 102	28	0.900
	25	0.893
	22	0.887

STRUCTURAL STEELWORK

Size (mm)	(kg/m)	Surface area per m2
Universal beams		
203 x 133	30	0.912
	25	0.904
Universal columns		
356 x 406	634	2.525
	551	2.475
	467	2.425
	393	2.379
	340	2.346
	287	2.132
	235	2.279
356 x 368	202	2.187
	177	2.170
	153	2.154
	129	2.137
305 x 305	283	1.938
	240	1.905
	198	1.872
	158	1.839
	137	1.822
	118	1.806
	97	1.789
254 x 254	167	1.576
	132	1.543
	107	1.519
	89	1.502
	73	1.485
203 x 203	86	1.236
	71	1.218
	60	1.204
	52	1.194
	46	1.187
152 x 152	37	0.912
	30	0.900
	23	0.889

STRUCTURAL STEELWORK

Tables showing the mass and surface area per metre run for various steel members - cont'd

size (mm)	kg/m	Surface area per m2
Joists		
254 x 203	81.85	1.213
254 x 114	37.20	0.898
203 x 152	52.09	0.931
152 x 127	37.20	0.735
127 x 114	29.76	0.644
127 x 114	26.79	0.649
114 x 114	26.79	0.617
102 x 102	23.07	0.547
89 x 89	19.35	0.475
76 x 76	12.65	0.410
Circular hollow sections - outside dia (mm)		
21.3	1.43	0.067
26.9	1.43	0.085
33.7	1.87	0.106
	2.41	0.106
	2.93	0.106
42.4	2.55	0.133
	3.09	0.133
	3.79	0.133
48.3	3.56	0.152
	4.37	0.152
	5.34	0.152
60.3	4.51	0.189
	5.55	0.189
	6.82	0.189
76.1	5.75	0.239
	7.11	0.239
	8.77	0.239
88.9	6.76	0.279
	8.38	0.279
	10.30	0.279

STRUCTURAL STEELWORK

Outside diameter (mm)	**kg/m**	**Surface area per m2**
114.3	9.83	0.359
	13.50	0.359
	16.80	0.395
139.7	16.60	0.439
	20.70	0.439
	26.00	0.439
	32.00	0.439
168.3	20.10	0.529
	25.20	0.529
	31.60	0.529
	39.00	0.529
193.7	25.10	0.609
	29.10	0.609
	36.60	0.609
	45.30	0.609
	55.90	0.609
	70.10	0.609
219.1	33.10	0.688
	41.60	0.688
	51.60	0.688
	63.70	0.688
	80.10	0.688
	98.20	0.688
273.0	41.40	0.858
	52.30	0.858
	64.90	0.858
	80.30	0.858
	101.00	0.858
	125.00	0.858
	153.00	0.858
323.9	62.30	1.020
	77.40	1.020
	96.00	1.020
	121.00	1.020
	150.00	1.020
	184.00	1.020
406.4	97.80	1.280
	121.00	1.280
	154.00	1.280
	191.00	1.280
	235.00	1.280
	295.00	1.280
457.0	110.0	1.440
	137.0	1.440
	174.0	1.440
	216.0	1.440
	266.0	1.440
	335.0	1.440
	411.0	1.440

STRUCTURAL STEELWORK

Tables showing the mass and surface area per metre run for various steel members - cont'd

Outside diameter (mm)	kg/m	Surface area per m2
Square hollow sections - size (mm)		
20 x 20	1.12	0.076
	1.39	0.074
30 x 30	2.21	0.114
	2.65	0.113
40 x 40	3.03	0.154
	3.66	0.153
	4.46	0.151
50 x 50	4.66	0.193
	5.72	0.191
	6.97	0.189
60 x 60	5.67	0.233
	6.97	0.231
	8.54	0.229
70 x 70	7.46	0.272
	10.10	0.269
80 x 80	8.59	0.312
	11.70	0.309
	14.40	0.306
90 x 90	9.72	0.352
	13.30	0.349
	16.40	0.346
100 x 100	12.00	0.391
	14.80	0.389
	18.40	0.386
	22.90	0.383
	27.90	0.379
120 x 120	18.00	0.469
	22.30	0.466
	27.90	0.463
	34.20	0.459
150 x 150	22.70	0.589
	28.30	0.586
	35.40	0.583
	43.60	0.579
	53.40	0.573
	66.40	0.566

STRUCTURAL STEELWORK

Size (mm)	kg/m	Surface area per m2
180 x 180	34.20	0.706
	43.00	0.703
	53.00	0.699
	65.20	0.693
	81.40	0.686
200 x 200	38.20	0.786
	48.00	0.783
	59.30	0.779
	73.00	0.773
	91.50	0.766
250 x 250	48.10	0.986
	60.50	0.983
	75.00	0.979
	92.60	0.973
	117.00	0.966
300 x 300	90.70	1.180
	112.00	1.170
	142.00	1.170
350 x 350	106.00	1.380
	132.00	1.370
	167.00	1.370
400 x 400	122.00	1.580
	152.00	1.570
Rectangular hollow sections		
50 x 30	3.03	0.154
	3.66	0.153
60 x 40	4.66	0.193
	5.72	0.191
80 x 40	5.67	0.232
	6.97	0.231
90 x 50	7.46	0.272
	10.10	0.269
100 x 50	7.18	0.293
	8.86	0.291
	10.90	0.289
100 x 60	8.59	0.312
	11.70	0.309
	14.40	0.306

STRUCTURAL STEELWORK

Tables showing the mass and surface area per metre run for various steel members - cont'd

Size (mm)	kg/m	Surface area per m2
120 x 60	9.72	0.352
	13.30	0.349
	16.40	0.346
120 x 80	14.80	0.389
	18.40	0.386
	22.90	0.383
	27.90	0.379
150 x 100	18.70	0.489
	23.30	0.486
	29.10	0.483
	35.70	0.479
160 x 80	18.00	0.469
	22.30	0.466
	27.90	0.463
	34.20	0.459
200 x 100	22.70	0.589
	28.30	0.586
	35.40	0.583
	43.60	0.579
	53.40	0.573
250 x 150	38.20	0.785
	48.00	0.783
	59.30	0.779
	73.00	0.773
	91.50	0.766
300 x 200	48.10	0.986
	60.50	0.983
	75.00	0.979
	92.60	0.973
	117.00	0.966
400 x 200	90.70	1.180
	112.00	1.170
	142.00	1.170
450 x 250	106.00	1.380
	132.00	1.370
	167.00	1.370

STRUCTURAL STEELWORK

Size (mm)	kg/m	Surface area per m2
Channels		
432 x 102	65.54	1.217
381 x 102	55.10	1.118
305 x 102	46.18	0.966
305 x 89	41.69	0.920
254 x 89	35.74	0.820
254 x 76	28.29	0.774
229 x 89	32.76	0.770
229 x 76	26.06	0.725
203 x 89	29.78	0.720
203 x 76	23.82	0.675
178 x 89	26.81	0.671
178 x 76	20.84	0.625
152 x 89	23.84	0.621
152 x 76	17.88	0.575
127 x 64	14.90	0.476

Angles - sum of leg lengths	Thickness (mm)	Kg/m	Surface area per m2
50	3	1.11	0.10
	4	1.45	0.10
	5	1.77	0.10
80	4	2.42	0.16
	5	2.97	0.16
	6	3.52	0.16
90	4	2.74	0.18
	5	3.38	0.18
	6	4.00	0.18
100	5	3.77	0.20
	6	4.47	0.20
	8	5.82	0.20
115	5	4.35	0.23
	6	5.16	0.23
	8	6.75	0.23

STRUCTURAL STEELWORK

Tables showing the mass and surface area per metre run for various steel members - cont'd

Angles - sum of leg lengths	**Thickness (mm)**	**Kg/m**	**Surface area per m2**
120	5	4.57	0.24
	6	5.42	0.24
	8	7.09	0.24
	10	8.69	0.24
125	6	5.65	0.25
	8	7.39	0.25
200	8	12.20	0.40
	10	15.00	0.40
	12	17.80	0.40
	15	21.90	0.40
225	10	17.00	0.45
	12	20.20	0.45
	15	24.80	0.45
240	8	14.70	0.48
	10	18.20	0.48
	12	21.60	0.48
	15	26.60	0.48
300	10	23.00	0.60
	12	27.30	0.60
	15	33.80	0.60
	18	40.10	0.60
350	12	32.00	0.70
	15	39.60	0.70
	18	47.10	0.70
400	16	48.50	0.80
	18	54.20	0.80
	20	59.90	0.80
	24	71.10	0.80

PLUMBING AND MECHANICAL INSTALLATIONS

Dimensions and weights of tubes

Outside diameter (mm)	Internal dia (mm)	Weight per m (kg)	Internal dia (mm)	Weight per m (kg)	Internal dia (mm)	Weight per m (kg)
Copper to BS 2871 Part 1						
	Table X		**Table Y**		**Table Z**	
6	4.80	0.0911	4.40	0.1170	5.00	0.0774
8	6.80	0.1246	6.40	0.1617	7.00	0.1054
10	8.80	0.1580	8.40	0.2064	9.00	0.1334
12	10.80	0.1914	10.40	0.2511	11.00	0.1612
15	13.60	0.2796	13.00	0.3923	14.00	0.2031
18	16.40	0.3852	16.00	0.4760	16.80	0.2918
22	20.22	0.5308	19.62	0.6974	20.82	0.3589
28	26.22	0.6814	25.62	0.8985	26.82	0.4594
35	32.63	1.1334	32.03	1.4085	33.63	0.6701
42	39.63	1.3675	39.03	1.6996	40.43	0.9216
54	51.63	1.7691	50.03	2.9052	52.23	1.3343
76.1	73.22	3.1287	72.22	4.1437	73.82	2.5131
108	105.12	4.4666	103.12	7.3745	105.72	3.5834
133	130.38	5.5151	-	-	130.38	5.5151
159	155.38	8.7795	-	-	156.38	6.6056

PLUMBING AND MECHANICAL INSTALLATIONS

Dimensions and weights of tubes - cont'd

Nominal size	Outside diameter max	min	Wall thickness	Weight	Weight screwed and socketed
(mm)	(mm)	(mm)	(mm)	(kg/m)	kg/m
Steel pipes to BS 1387					
Light gauge					
6	10.1	9.7	1.80	0.361	0.364
8	13.6	13.2	1.80	0.517	0.521
10	17.1	16.7	1.80	0.674	0.680
15	21.4	21.0	2.00	0.952	0.961
20	26.9	26.4	2.35	1.410	1.420
25	33.8	33.2	2.65	2.010	2.030
32	42.5	41.9	2.65	2.580	2.610
40	48.4	47.8	2.90	3.250	3.290
50	60.2	59.6	2.90	4.110	4.180
65	76.0	75.2	3.25	5.800	5.920
80	88.7	87.9	3.25	6.810	6.980
100	113.9	113.0	3.65	9.890	10.200
Medium gauge					
6	10.4	9.8	2.00	0.407	0.410
8	13.9	13.3	2.35	0.650	0.654
10	17.4	16.8	2.35	0.852	0.858
15	21.7	21.1	2.65	1.220	1.230
20	27.2	26.6	2.65	1.580	1.590
25	34.2	33.4	3.25	2.440	2.460
32	42.9	42.1	3.25	3.140	3.170
40	48.8	48.0	3.25	3.610	3.650
50	60.8	59.8	3.65	5.100	5.170
65	76.6	75.4	3.65	6.510	6.630
80	89.5	88.1	4.05	8.470	8.640

PLUMBING AND MECHANICAL INSTALLATIONS

Nominal size (mm)	Outside diameter max (mm)	Outside diameter min (mm)	Wall thickness (mm)	Weight (kg/m)	Weight screwed and socketed (kg/m)
100	114.9	113.3	4.50	12.100	12.400
125	140.6	138.7	4.85	16.200	16.700
150	166.1	164.1	4.85	19.200	19.800
Heavy gauge					
6	10.4	9.8	2.65	0.493	0.496
8	13.9	13.3	2.90	0.769	0.773
10	17.4	16.8	2.90	1.020	1.030
15	21.7	21.1	3.25	1.450	1.460
20	27.2	26.6	3.25	1.900	1.910
25	34.2	33.4	4.05	2.970	2.990
32	42.9	42.1	4.05	3.840	3.870
40	48.8	48.0	4.05	4.430	4.470
50	60.8	59.8	4.50	6.170	6.240
65	76.6	75.4	4.50	7.900	8.020
80	89.5	88.1	4.85	10.100	10.300
100	114.9	113.3	5.40	14.400	14.700
125	140.6	138.7	5.40	17.800	18.300
150	166.1	164.1	5.40	21.200	21.800
Stainless steel pipes to BS 4127 Part 2					
8	8.045	7.940	0.60	0.1120	
10	10.045	9.940	0.60	0.1419	
12	12.045	11.940	0.60	0.1718	
15	15.045	14.940	0.60	0.2174	
18	18.045	17.940	0.70	0.3046	
22	22.055	21.950	0.70	0.3748	
28	28.055	27.950	0.80	0.5469	
35	35.070	34.965	1.00	0.8342	

PLUMBING AND MECHANICAL INSTALLATIONS

Maximum distances between pipe supports

Pipe material	BS nominal pipe size inch	BS nominal pipe size mm	Pipes fitted vertically support distances in metres	Pipes fitted horizontally onto low gradients support distances in metres
Copper	0.50	15.0	1.90	1.3
	0.75	22.0	2.50	1.9
	1.00	28.0	2.50	1.9
	1.25	35.0	2.80	2.5
	1.50	42.0	2.80	2.5
	2.00	54.0	3.90	2.5
	2.50	67.0	3.90	2.8
	3.00	76.1	3.90	2.8
	4.00	108.0	3.90	2.8
	5.00	133.0	3.90	2.8
	6.00	159.0	3.90	2.8
muPVC	1.25	32.0	1.20	0.5
	1.50	40.0	1.20	0.5
	2.00	50.0	1.20	0.6
Polypropylene	1.25	32.0	1.20	0.5
	1.50	40.0	1.20	0.5
uPVC	-	82.4	1.20	0.5
	-	110.0	1.80	0.9
	-	160.0	1.80	1.2

Litres of water storage required per person in various types of building

Type of building	Storage per person (litres)
Houses and flats	90
Hostels	90
Hotels	135
Nurse's home and medical quarters	115
Offices with canteens	45
Offices without canteens	35
Restaurants, per meal served	7
Boarding school	90
Day schools	30

PLUMBING AND MECHANICAL INSTALLATIONS

Cold water plumbing - thickness of insulation required against frost

Bore of tube		Pipework within buildings declared thermal conductivity (W/m degrees C)		
		Up to 0.040	0.041 to 0.055	0.056 to 0.070
(mm)	(in)	Minimum thickness of insulation (mm)		
15		32	50	75
20		32	50	75
25		32	50	75
32		32	50	75
40		32	50	75
50		25	32	50
65		25	32	50
80		25	32	50
100		19	25	38

Cisterns

Capacities and dimensions of galvanised mild steel cisterns from BS 417

Capacity (litres)	BS type	Dimensions (mm) length	width	depth
18	SCM 45	457	305	305
36	SCM 70	610	305	371
54	SCM 90	610	406	371
68	SCM 110	610	432	432
86	SCM 135	610	457	482
114	SCM 180	686	508	508
159	SCM 230	736	559	559
191	SCM 270	762	584	610
227	SCM 320	914	610	584
264	SCM 360	914	660	610
327	SCM 450/1	1220	610	610
336	SCM 450/2	965	686	686
423	SCM 570	965	762	787
491	SCM 680	1090	864	736
709	SCM 910	1170	889	889

Capacities of cold water polypropylene storage cisterns from BS 4213

Capacity (litres)	BS type	Maximum height (mm)
18	PC 4	310
36	PC 8	380
68	PC 15	430
91	PC 20	510
114	PC 25	530
182	PC 40	610
227	PC 50	660
273	PC 60	660
318	PC 70	660
455	PC 100	760

HEATING AND HOT WATER INSTALLATIONS

Storage capacity and recommended power of hot water storage boilers

Type of building	Storage at 65^C (litres per person)	Boiler power to 65^C (kW per person)
Flats and dwellings		
(a) Low rent properties	25	0.5
(b) Medium rent properties	30	0.7
(c) High rent properties	45	1.2
Nurses homes	45	0.9
Hostels	30	0.7
Hotels		
(a) Top quality - upmarket	45	1.2
(b) Average quality - low market	35	0.9
Colleges and schools		
(a) Live-in accommodation	25	0.7
(b) Public comprehensive	5	0.1
Factories	5	0.1
Hospitals		
(a) General	30	1.5
(b) Infectious	45	1.5
(c) Infirmaries	25	0.6
(d) Infirmaries (inc. laundry facilities)	30	0.9
(e) Maternity	30	2.1
(f) Mental	25	0.7
Offices	5	0.1
Sports pavilions	35	0.3

Thickness of thermal insulation for heating installations

Size of tube (mm)	Declared thermal conductivity: Up to 0.025	0.026 to 0.040	0.041 to 0.055	0.056 to 0.070
	Minimum thickness of insulation			
LTHW Systems				
15	25	25	38	38
20	25	32	38	38
25	25	38	38	38
32	32	38	38	50
40	32	38	38	50
50	38	38	50	50
65	38	50	50	50
80	38	50	50	50

HEATING AND HOT WATER INSTALLATIONS

Size of tube (mm)	Up to 0.025	0.026 to 0.040	0.041 to 0.055	0.056 to 0.070
LTHW Systems				
100		38	50	5063
125		38	50	5063
150	50	50	63	63
200	50	50	63	63
250	50	63	63	63
300	50	63	63	63
Flat surfaces	50	63	63	63
MTHW Systems and condensate				
		Declared thermal conductivity		
15	25	38	38	38
20	32	38	38	50
25	38	38	38	50
32	38	50	50	50
40	38	50	50	50
50	38	50	50	50
65	38	50	50	50
80	50	50	50	63
100	50	63	63	63
125	50	63	63	63
150	50	63	63	63
200	50	63	63	63
250	50	63	63	75
300	63	63	63	75
Flat surfaces	63	63	63	75
HTHW Systems and steam				
15	38	50	50	50
20	38	50	50	50
25	38	50	50	50
32	50	50	50	63
40	50	50	50	63
50	50	50	75	75
65	50	63	75	75
80	50	63	75	75
100	63	63	75	100
125	63	63	100	100
150	63	63	100	100
200	63	63	100	100
250	63	75	100	100
300	63	75	100	100
Flat surfaces	63	75	100	100

HEATING AND HOT WATER INSTALLATIONS

Capacities and dimensions of copper indirect cylinders (coil type) from BS 1566

Capacity (litres)	BS Type	External diameter (mm)		External height over dome (mm)
96	0	300	1600	
72	1	350		900
96	2	400		900
114	3	400	1050	
84	4	450		675
95	5	450		750
106	6	450		825
117	7	450		900
140	8	450	1050	
162	9	450	1200	
206	9 E	450	1500	
190	10	500	1200	
245	11	500	1500	
280	12	600	1200	
360	13	600	1500	
440	14	600	1800	

		Dimensions (mm) Internal diameter		Height
109	BSG 1M	457		762
136	BSG 2M	457		914
159	BSG 3M	457	1067	
227	BSG 4M	508	1270	
273	BSG 5M	508	1473	
364	BSG 6M	610	1372	
455	BSG 7M	610	1753	
123	BSG 8M	457		838

Comparison of energy costs (January 1987)

Energy form	Calorific value per unit of supply			Average price	
	Average price	Gross (MJ)	Nett (MJ)	Gross (£/GJ)	Nett (£/GJ)
Electricity - direct	3.66p/kwh	3.6	3.6	10.17	10.17
- off peak	1.70p/kwh	3.6	3.6	4.72	4.72
Natural gas	36.2p/therm	105.5	95.2	3.43	3.80
Fuel oil - 35 seconds	15.63p/l	38.1	35.7	4.10	4.38
- 200 seconds	14.11p/l	40.4	38.2	3.49	3.69
- 950 seconds	12.04p/l	40.6	38.4	2.97	3.14
- 3500 seconds	11.03p/l	41.1	38.8	2.68	2.84
Propane	£226.00/tonne	50000	46300	4.52	4.88
Butane	£190.00/tonne	49300	45800	3.85	4.04
Coal - singles	£66.60/tonne	28400	27200	2.35	2.45
- smalls	£61.60/tonne	28100	26900	2.19	2.29
Industrial coke	£95.80/tonne	27900	27500	3.43	3.48

VENTILATION AND AIR-CONDITIONING

Typical fresh air supply factors in typical situations

Building type	**Litres of fresh air per second per person**	**Litres of fresh air per second per m2 floor area**
General offices	5 - 8	1.3
Board rooms	18 - 25	6.0
Private offices	5 - 12	1.2 - 2.0
Dept. stores	5 - 8	3.0
Factories	20 - 30	0.8
Garages	-	8.0
Bars	12 - 18	-
Dance halls	8 - 12	-
Hotel rooms	8 - 12	1.7
Schools	14	-
Assembly halls	14	-
Drawing offices	16	-

Note: As a global figure for fresh air allow per 1000 m2 1.2 m3/second.

Typical air-changes per hour in typical situations

Building type	**Air changes per hour**
Residences	1 - 2
Churches	1 - 2
Storage buildings	1 - 2
Libraries	3 - 4
Book stacks	1 - 2
Banks	5 - 6
Offices	4 - 6
Assembly halls	5 - 10
Laboratories	4 - 6
Internal bathrooms	5 - 6
Laboratories - internal	6 - 8
Restaurants/cafes	10 - 15
Canteens	8 - 12
Small kitchens	20 - 40
Large kitchens	10 - 20
Boiler houses	15 - 30

GLAZING

Float and polished plate glass

Nominal thickness (mm)	**Tolerance on thickness (mm) (kg/m2)**	**Approximate weight (mm)**	**Normal maximum size**
3	± 0.2	7.5	2140 x 1220
4	± 0.2	10.0	2760 x 1220
5	± 0.2	12.5	3180 x 2100
6	± 0.2	15.0	4600 x 3180
10	± 0.3	25.0)	
12	± 0.3	30.0)	6000 x 3300
15	± 0.5	37.5	3050 x 3000
19	± 1.0	47.5)	
25	± 1.0	63.5)	3000 x 2900
Clear sheet glass			
2 *	± 0.2	5.0	1920 x 1220
3	± 0.3	7.5	2130 x 1320
4	± 0.3	10.0	2760 x 1220
5 *	± 0.3	12.5)	
6 *	± 0.3	15.0)	2130 x 2400
Cast glass			
3	+ 0.4		
	- 0.2	6.0)	
4	± 0.5	7.5)	2140 x 1280
5	± 0.5	9.5	2140 x 1320
6	± 0.5	11.5)	
10	± 0.8	21.5)	3700 x 1280
Wired glass			
(Cast wired glass)			
6	+ 0.3	-)	
	- 0.7	)	3700 x 1840
7	± 0.7	-)	
(Polished wire glass)			
6	± 1.0	-	330 x 1830

* The 5 mm and 6 mm thickness are known as 'thick drawn sheet'. Although 2 mm sheet glass is available it is not recommended for general glazing purposes.

DRAINAGE

Width required for trenches for various diameters of pipes

Pipe diameter (mm)	Trench n.e. 1.5 m deep	Trench over 1.5 m deep
n.e. 100 mm	450 mm	600 mm
100 - 150 mm	500 mm	650 mm
150 - 225 mm	600 mm	750 mm
225 - 300 mm	650 mm	800 mm
300 - 400 mm	750 mm	900 mm
400 - 450 mm	900 mm	1050 mm
450 - 600 mm	1100 mm	1300 mm

Weights and dimensions of typically sized uPVC pipes

Nominal size	Mean outside diameter (mm)		Wall thickness	Weight kg per metre
Standard pipes	min	max		
82.4	82.4	82.7	3.2	1.2
110.0	110.0	110.4	3.2	1.6
160.0	160.0	160.6	4.1	3.0
200.0	200.0	200.6	4.9	4.6
250.0	250.0	250.7	6.1	7.2

Perforated pipes

Heavy grade as above

Thin wall

Nominal size	Mean outside diameter min (mm)	Mean outside diameter max (mm)	Wall thickness	Weight kg per metre
82.4	82.4	82.7	1.7	-
110.0	110.0	110.4	2.2	-
160.0	160.0	160.6	3.2	-

Vitrified clay pipes

Product	Nominal diameter (mm)	Effective pipe length (mm)	Limits of bore load per metre length min	max	Crushing strength (kN/m)	Weight kg/pipe (/m)
Supersleve	100	1600	96	105	35.00	15.63 (9.77)
Hepsleve	150	1600	146	158	22.00 (normal)	36.50 (22.81)
Hepseal	150	1500	146	158	22.00	37.04 (24.69)
	225	1750	221	235	28.00	95.24 (54.42)
	300	2500	295	313	34.00	196.08 (78.43)
	400	2500	394	414	44.00	357.14 (142.86)

DRAINAGE

Vitrified clay pipes - cont'd

Product	Nominal diameter (mm)	Effective pipe length (mm)	Limits of bore load per metre length min	max	Crushing strength (kN/m)	Weight kg/pipe (/m)
Hepseal cont-d	450	2500	444	464	44.00	500.00 (200.00)
	500	2500	494	514	48.00	555.56 (222.22)
	600	3000	591	615	70.00	847.46 (282.47)
	700	3000	689	719	81.00	1111.11 (370.37)
	800	3000	788	822	86.00	1351.35 (450.35)
	1000	3000	985	1027	120.00	2000.00 (666.67)
Hepline	100	1250	95	107	22.00	15.15 (12.12)
	150	1500	145	160	22.00	32.79 (21.86)
	225	1850	219	239	28.00	74.07 (40.04)
	300	1850	292	317	34.00	105.26 (56.90)
Hepduct (Conduit)	90	1500	-	-	28.00	12.05 (8.03)
	100	1600	-	-	28.00	14.29 (8.93)
	125	1250	-	-	22.00	21.28 (17.02)
	150	1250	-	-	22.00	28.57 (22.86)
	225	1850	-	-	28.00	64.52 (34.88)
	300	1850	-	-	34.00	111.11 (60.06)

USEFUL ADDRESSES FOR FURTHER INFORMATION

ACOUSTICAL INVESTIGATION & RESEARCH ORGANISATION LTD
Duxon's Turn, Maylands Avenue,
Hemel Hempstead, Herts HP2 4SB
Hemel Hempstead (0442) 47146/7

AGGREGATE CONCRETE BLOCK ASSOCIATION
60 Charles Street, Leicester LE1 1FB
Leicester (0533) 536161

ALUMINIUM EXTRUDERS ASSOCIATION
Broadway House, Calthorpe Road,
Five ways, Birmingham B15 1TN
021-456 1103
Telex BIRCOM G 338024 ALFED

ALUMINIUM FEDERATION LTD
Broadway House, Calthorpe Road,
Five Ways, Birmingham B15 1TN
021-456 1103
Telex BIRCOM G 338024 ALFED

ALUMINIUM WINDOW ASSOCIATION
Suites 323 & 324 Golden House
29 Great Pultney Street
London 1R 3DD
071-494 4650

AMERICAN PLYWOOD ASSOCIATION
101 Wigmore Street, London W1H 9AB
071-629 3437/8
Telex 296009 USAGOF G

ARBORICULTURAL ADVISORY & INFORMATION SERVICE
Forst Research Station,
Alice Holt Lodge, Farnham,
Surrey GU10 4LH
Bentley (0420) 22255

ARBORICULTURAL ASSOCIATION
Ampfield House, Ampfield,
Romsey, Hants S05 9PA
Braishfield (0794) 68717

ARCHITECTURAL ADVISORY SERVICE (APPLIED & ANODIC METAL FINISHES)
Unit 5, Royal London Estate,
29, North Acton Road, Willesden,
London NW10 6PD
081-965 4677/0833

ARCHITECTURAL ALUMINIUM ASSOCIATION
193 Forest Road, Tunbridge Wells
Kent TN2 5TA
Tunbridge Wells (0892) 30630

ARCHITECTURAL ASSOCIATION
34/36 Bedford Square, London WC1B 3ES
071-636 0974

ARMS (ASSOCIATION OF ROOFING MATERIALS SUPPLIERS
c/o Allan Harris & Sons Ltd, Station Road,
St. Georges, Weston-Super-Mare, Avon
BS22 0XN
Weston-Super-Mare (0934) 511166
Alloa (0259) 721010
Darwen (0254) 771722

ASBESTOS INFORMATION CENTRE LTD
St. Andrews House, 22 - 28 High Street
Epsom, Surrey KT19 8AH
Epsom (037 27) 42055
Telex 21120 Ref 2526

ASBESTOS REMOVAL CONTRACTOR ASSOCIATION
1 High Street, Chelmsford,
Essex CM1 1BE
Chelmsford (0245) 259744
Telex 927298
Fax (0245) 490722

BRITISH AGGREGATE CONSTRUCTION MATERIALS INDUSTRIES LTD
156 Buckingham Palace Road,
London SW1W 9TR
071-730 8194
Fax 071-730 4355

BRITISH AIR CONDITIONING APPROVALS BOARD
30 Millbank, London SW1P 4RD
071-834 8827

BRITISH AIRPORTS AUTHORITY
Corporate Office, 130 Wilton Road
London SW1 1LQ
071-834 9449
Telex 919268 BAA PLC
Fax 071-932 6699

BRITISH ANODISING ASSOCIATION
Broadway House, Calthorpe Road,
Five Ways, Birmingham B15 1TN
021-456 1103
Telex BIRCOM-G-338024 ALFED

USEFUL ADDRESSES FOR FURTHER INFORMATION

BRITISH ARCHITECTURAL LIBRARY
RIBA, 66 Portland Place, London
W1N 4AD
071-580 5533

BRITISH ASSOCIATION OF LANDSCAPE INDUSTRIES
9 Henry Street, Keighley, West Yorkshire
BD21 3DR
Keighley (0535) 606139

BRITISH BATH MANUFACTURERS ASSOCIATION
Fleming House, 134 Renfrew Street,
Glasgow G3 6TG
041-332 0826
Telex 779433

BRITISH BLIND AND SHUTTER ASSOCIATION
5 Greenfield Crescent, Edgbaston,
Birmingham B15 3BE
021-454 2177
Telex 336006

BRITISH BOARD OF AGREEMENT
PO Box 195, Bucknall Lane, Garston,
Watford Herts WD2 7NG
Garston (0923) 670844
Telex 94016145 BBAG G
Fax (0923) 662133

BRITISH CARPET MANUFACTURERS ASSOCIATION
Fourth Floor, Royalty House,
72 Dean Street,
London W1V 5HB
071-734 9853

BRITISH CERAMIC RESEARCH ASSOCIATION LTD
Queens Road, Penkhull, Stoke-on-Trent
Staffs ST4 2SA
Stoke-on-Trent (0782) 45431
Telex 36228 DCRA G

BRITISH CERAMIC TILE COUNCIL
Federation House, Station Road,
Stoke-on-Trent, Staffs ST4 2RV
Stoke-on-Trent (0782) 45147

BRITISH CLAYWARE LAND DRAIN INDUSTRY
Federation House, Station Road,
Stoke-on-Trent, Staffs ST4 2SA
Stoke-on-Trent (0782) 747256
Telex 367446

BRITISH COMBUSTION EQUIPMENT MANUFACTURERS ASSOCIATION
The Fernery, Market Place, Midhurst,
West Sussex GU29 9DP
Midhurst (073081) 2782

BRITISH CONCRETE MASONRY ASSOCIATION
St. John's Works, Bedford MK42 0DL
Bedford (0234) 63171/9

BRITISH CONCRETE PUMPING ASSOCIATION LTD
Mossland Road, Hillington Ind. Estate,
Glasgow G52 4XW
041-882 9027

BRITISH CONSTRUCTIONAL STEELWORK ASSOCIATION LTD
35 Old Queen Street, London SW1H 9HZ
071-222 2254
Telex 27523

BRITISH CONTRACT FURNISHING ASSOCIATION
PO Box 384, London N12 8HF
081-445 8694

BRITISH CUBICLE MANUFACTURERS ASSOCIATION
17 Bridge Street, Evesham, Worcs WR11 4SQ
Evesham (0386) 6560

BRITISH DECORATORS ASSOCIATION
6 Haywra Street, Harrogate,
North Yorkshire HG1 5BL
Harrogate (0423) 67292/3

BRITISH EFFLUENT & WATER ASSOCIATION
5 Castle Street, High Wycombe, Bucks
HP13 6RZ
High Wycombe (0494) 444544

BRITISH ELECTRICAL & ALLIED MANUFACTURERS ASSOCIATION
Leicester House, 8 Leicester St, London
WC2H 7BN
071-437 0678

BRITISH ELECTRICAL SYSTEMS ASSOCIATION (BESA)
Granville Chambers, 2 Radford St, Stone,
Staffs ST15 8DA
Stone (0785) 812426

BRITISH FIRE PROTECTION SYSTEMS ASSOCIATION LTD
48a Eden Street, Kingston-upon-Thames,
Surrey KT1 1EE
081-549 5855

USEFUL ADDRESSES FOR FURTHER INFORMATION

BRITISH FIRE SERVICE ASSOCIATION
86 London Road, Leicester LE2 0QR
Leicester (0533) 542879

BRITISH FLAT ROOFING COUNCIL
PO Box 125, Haywards Heath, West Susses
RH16 3TJ
Haywards Heath (0444) 416681/2

BRITISH FLAT ROOFING COUNCIL
PO Box 125, Haywards Heath,
West Sussex RH16 3TJ
Haywards Heath (0444) 416681/2

BRITISH FLOOR COVERING MANUFACTURERS ASSOCIATION
125 Queens Road, Brighton BN1 3YW
Brighton (0273) 29271
Telex 87595

BRITISH FLUE & CHIMNEY MANUFACTURERS ASSOCIATION
Nicholson House, High St, Maidenhead,
Berks SL6 1LF
Maidenhead (0628) 34667

BRITISH FURNITURE MANUFACTURERS FEDERATED ASSOCIATIONS
30 Harcourt Street, London W1H 2AA
071-724 0854 Telex 269592 EXTURN G

BRITISH GLASS INDUSTRY RESEARCH ASSOCIATION
Northumberland Road, Sheffield S10 2UA
Sheffield (0742) 686201

BRITISH GYPSUM ARCHITECTS ADVISORY SERVICE
Westfield, 360 Singlewell Rd, Gravesend,
Kent DA11 7RZ
Gravesend (0474) 534251
Telex 96439

BRITISH INDEPENDENT STEEL PRODUCERS ASSOCIATION
5 Cromwell Road, London SW7 2HX
071-581 0231

BRITISH INDUSTRIAL FASTENERS FEDERATION
Queens House, Queens Rd, Coventry CV1 3EG
Coventry (0203) 22325
Telex 311650

BRITISH INSTITUTE OF INTERIOR DESIGN
1c Devonshire Avenue, Beeston, Notts
NG9 1BS
Nottingham (0602) 221255

BRITISH KITCHEN FURNITURE MANUFACTURERS
c/o Building Employers Confederation, 82
New Cavendish St, London W1M 8AD
071-580 5588
Telex 265763

BRITISH LAMINATED PLASTICS FABRICATORS ASSOCIATION
5 Belgrave Square, London SW1X 8PH
071-235 9483
Telex 8951528 PLAFED G

BRITISH LEAD MANUFACTURERS ASSOCIATION
22 High Street, Epsom, Surrey, KT13 8BL
Epsom (03727) 43976

BRITISH LIBRARY LENDING DIVISION
Boston Spa, Wetherby, West Yorks LS23 7BQ
Wetherby (0937) 843434
Telex 557381

BRITISH LIBRARY, SCIENCE REFERENCE LIBRARY
25 Southampton Buildings, Chancery Lane,
London WC2A 1AW
071-323 7494/7496
Telex 266959 SCIREF G

BRITISH LOCK MANUFACTURERS ASSOCIATION
5 Greenfield Crescent, Edgbaston
Birmingham B15 3BE
021-454 2177
Telex 336006

BRITISH MALLEABLE TUBE FITTINGS ASSOCIATION
105 Meadow View Road, Catford, London
SE6 3NJ
081-698 8856

BRITISH NON-FERROUS METALS FEDERATION
10 Greenfield Crescent, Edgbaston,
Birmingham B15 3AU
021-456 3322
Telex 339161

BRITISH NON-FERROUS METALS TECHNICAL CENTRE
Grove Laboratories, Detchworth Rd,
Wantage, Oxon OX12 9BJ
Wantage (023 57) 2992
Telex 837166

BRITISH PLASTICS FEDERATION
5 Belgrave Square, London SW1X 8PH
071-235 9483
Information 071-235 9888
Telex 8951528

USEFUL ADDRESSES FOR FURTHER INFORMATION

BRITISH PRECAST CONCRETE FEDERATION
60 Charles St, Leicester LE1 1FB
Leicester (0533) 536161

BRITISH PUMP MANUFACTURERS ASSOCIATION
Artillery House, Artillery Row, London SW1P 1RT
071-222 0830

BRITISH READY MIXED CONCRETE ASSOCIATION
1 Bramber Court/2 Bramber Road, London W14 9PB
071-381 6582

BRITISH REFRIGERATION ASSOCIATION
Nicholson House, High St, Maidenhead, Berks SL6 1LF
Maidenhead (0628) 3466745

BRITISH REINFORCEMENT MANUFACTURERS ASSOCIATION (BRMA)
20/21 Tooks Court, London EC4A 1LA
071-831 7581
Telex 23485 ARMADA G

BRITISH ROAD FEDERATION
Cowdray House, 6 Portugal St, London WC2 2HG
071-242 1285

BRITISH RUBBER MANUFACTURERS ASSOCIATION LTD
90/91 Tottenham Court Road, London W1P 0BR
071-580 2794
Telex 267059 BRMA G

BRITISH STANDARDS INSTITUTION
2 Park Street, London W1A 2BS
071-629 9000
Enquiries: Milton Keynes (0908) 220908

BRITISH STEEL CORPORATION
Swinden Laboratories, Moorgate, Rotherham S60 3AR
Rotherham (0709) 820166
Telex 547279

BRITISH VALVE MANUFACTURERS ASSOCIATION LTD
3 Pannells Court, Chertsey Street, Guildford, Surrey GU1 4EU
Guildford (0483) 37379

BRITISH WELDED STEEL TUBE MANUFACTURERS ASSOCIATION
38 Pamela Road, Northfield, Birmingham B31 2QG
021-475 3583

BRITISH WOOD PRESERVING ASSOCIATION
Premier House, 150 Southampton Row, London WC1B 5AL
071-837 8217

BRITISH WOODWORKING FEDERATION
82 New Cavendish St, London W1M 8AD
071-580 5588
Telex 265763

BUILDING CENTRE: BRISTOL
The Building Centre,
35 King Street, Bristol BS1 4DZ
Management Bristol : (0272) 262953
Information Bristol: (0272) 260264

BUILDING CENTRE: COVENTRY
Coventry Building Information Centre, Dept, of Architecture & Planning, Tower Block, Council Offices, Much Park St, Coventry CV1 5RT
Coventry (0203) 25555, ext 2512
Telex 31469

BUILDING CENTRE: LONDON
The Building Centre, 26 Store Street, London WC1E 7BT
Administration: 071-637 1022
Telex 261446
Bookshop: 071-637 3151
Information Service: (0344) 884999

BUILDING CENTRE: MANCHESTER
113-115 Portland Street, Manchester MI 6FB
061-236 6933/9802

BUILDING CENTRE: PETERBOROUGH
Building Materials Information Service, 22 Broadway, Peterborough PE1 1RU
Peterborough (0733) 314239

BUILDING CENTRE: SCOTLAND
The Building Centre Scotland 1971 Ltd, Macdata Unit, 47 High Street, Paisley PA1 2AL
041-840 1199

USEFUL ADDRESSES FOR FURTHER INFORMATION

BUILDING CENTRE: STOKE-ON-TRENT
The Building Information Centre,
Stoke-on-Trent Cauldon College,
The Concourse, Stoke Rd, Shelton,
Stoke-on-Trent ST4 2DG
Stoke-on-Trent (0782) 29561

BUILDING EMPLOYERS CONFEDERATION
82 New Cavendish St, London W1M 8AD
071-580 5588

BUILDING MAINTENANCE INFORMATION
85/87 Clarence St, Kingston-upon-Thames,
Surrey KT1 1RB
081-546 7554

BUILDING RESEARCH ESTABLISHMENT: SCOTLAND
Kelvin Rd, East Kilbride, Glasgow G75 0RZ
East Kilbride (035 52) 33001

BUILDING SERVICES RESEARCH AND INFORMATION ASSOCIATION
Old Bracknell Lane West, Bracknell,
Berks RG12 4AH
Bracknell (0344) 426511
Telex 848288 BSRIAC G

CALCIUM SILICATE BRICK ASSOCIATION
24 Fearnley Rd, Welwyn Garden City,
Herts AL8 6HW
Welwyn Garden (07073) 24538

CATERING EQUIPMENT MANUFACTURERS ASSOCIATION (CEMA)
14 Pall Mall, London SW1Y 5LZ
071-930 0461
Telex 24282

CAVITY FOAM BUREAU
9-11 The Hayes, Cardiff CF1 1NU
Cardiff (0222) 388621
Telex 497629

CEMENT ADMIXTURES ASSOCIATION
2a High St, Hythe, Southampton SO4 6YW
Southampton (0703) 842765

CEMENT AND CONCRETE ASSOCIATION
Wexham Springs, Slough, Berks SL3 6PL
Fulmer (028 16) 2727
Telex 848352

CEMENT MAKERS' FEDERATION
Terminal House, 52 Grovsenor Gardens,
London SW1W 0AH
071-730 2148
Telex 261700 CEMFED

CHARTERED INSTITUTE OF ARBITRATORS
75 Cannon Street, London EC4N 5BH
071-236 8761
Telex 893466 CIARB G

CHARTERED INSTITUTE OF BUILDING
Englemere, Kings Ride, Ascot,
Berks SL5 8BJ
Ascot (0990) 23355

CHARTERED INSTITUTION OF BUILDING SERVICE ENGINEERS
Delta House, 222 Balham High Road,
London SW12 9BS
081-675 5211

CHIPBOARD PROMOTION ASSOCIATION LTD
50 Station Road, Marlow, Bucks SL7 1NN
Marlow (06284) 3022

CLAY PIPE DEVELOPMENT ASSOCIATION
Drayton House, 30 Gordon Street,
London WC1H 0AN
071-388 0025

CLAY ROOFING TILE COUNCIL
Federation House, Station Road,
Stoke-on-Trent ST4 2SA
Stoke-on-Trent (0782) 747256

COLD ROLLED SECTIONS ASSOCIATIONS
Centre City Tower, 7 Hill Street,
Birmingham, B5 4UU
021-643 5494
Telex 339420

COMMITEE OF ASSOCIATIONS OF SPECIALIST ENGINEERING CONTRACTORS
ESCA House, 34 Palace Court, Bayswater,
London W2 4JG
071-229 2488
Telex 27929

CONCRETE BRICK MANUFACTURERS ASSOCIATION
c/o British Precast Concrete Federation,
60 Charles St, Leicester LE1 1FP
Leicester (0533) 536161

USEFUL ADDRESSES FOR FURTHER INFORMATION

CONCRETE PIPE ASSOCIATION OF GREAT BRITAIN
60 Charles St, Leicester LE1 1FB
Leicester (0533) 536161

CONCRETE SOCIETY
Devon House, 12-15 Dartmouth St,
London SW1H 9BL
071-222 1822

CONFEDERATION OF BRITISH INDUSTRY
Centre Point, 103 New Oxford St,
London WC1A 1DU
071-379 7400
Telex 21332 DOCFAC

CONSTRADO (CONSTRUCTIONAL STEEL RESEARCH AND DEVELOPMENT ORGANISATION)
NLA Tower, 12 Addiscombe Rd, Croydon
CR9 3JH
Croydon 081-688 2688, 081-686 0366

CONTRACT FLOORING ASSOCIATION
Long Furlow House, Holt
Norfolk NR25 7DD
Holt (0263) 740223/740370

CONTRACTORS MECHANICAL PLANT ENGINEERS
20 Knave Wood Rd, Kemsing, Sevenoaks,
Kent TN15 6RH
Otford (09592) 2628

COPPER CYLINDER AND BOILER MANUFACTURERS ASSOCIATION
35/37 Whitworth Street West,
Manchester M1 5NG
061-236 0384

COPPER DEVELOPMENT ASSOCIATION
Orchard House, Mutton Lane
Potters Bar, Herts EN6 3AP
Potters Bar (0707) 50711
Telex 265451 MONREF (72:MAG 30836)

COPPER TUBE FITTINGS MANUFACTURERS ASSOCIATION
7 Highfield Rd, Birmingham B15 3ED
021-454 7766
Telex 339161

COUNCIL OF BRITISH CERAMIC SANITARY WARE MANUFACTURERS
Federation House, Station Rd,
Stoke-on-Trent ST4 2RT
Stoke-on-Trent (0782) 48675

CP/M USERS GROUP (UK)
72 Mill Rd, Hawley, Dartford,
Kent DA2 7RZ
Dartford (0322) 22669

DECORATIVE LIGHTING ASSOCIATION LTD
Bishops Castle, Shropshire SY9 5LE
Clun (05884) 658

DOOR & SHUTTER MANUFACTURERS ASSOCIATION
5 Greenfield Crescent, Edgbaston,
Birmingham B15 3BE
021-454 2177
Telex 336006

DRAUGHT PROOFING ADVISORY ASSOCIATION
PO Box 12, Haslemere, Surrey GU27 3AN
Haslemere (0428) 54011

DRY LINING AND PARTITION ASSOCIATION
82 New Cavendish Street, London W1M 8AD
071-580 5588
Telex 265763

DRY STONE WALLING ASSOCIATION
YFC Centre, National Agricultural Centre
Kenilworth, Warwickshire CV8 2LG
021-378 0493

DUCTILE IRON PIPE ASSOCIATION
8th Floor, Bridge House, 121 Smallbrook,
Queensway, Birmingham B5 4JP
021-643 3377

ELECTRIC CABLE MAKERS CONFEDERATION
56 Palace Road, East Molesey,
Surrey KT8 9DW
081-941 4079
Telex 24893

ELECTRICAL CONTRACTORS ASSOCIATION (ECA)
ESCA House, 34 Palace Court, Bayswater,
London W2 4HY
071-229 1266
Telex 27929

USEFUL ADDRESSES FOR FURTHER INFORMATION

ELECTRICAL CONTRACTORS ASSOCIATION OF SCOTLAND
23 Heriot Row, Edinburgh EH3 6EW
031-225 7221

ELECTRICAL INSTALLATION EQUIPMENT MANUFACTURERS ASSOCIATION
Leicester House, 8 Leicester St,
London WC2H 7BN
071-437 0678
Telex 263536

ENERGY EFFICIENCY OFFICE
c/o Building Centre,
21 Store Street,
London WC1E 7BT
071-637 1022

ENERGY SYSTEMS TRADE ASSOCIATION LTD (ESTA)
PO Box 16, Stroud, Glos GL5 5EB
Amberley (0453) 873568

EUROBUILD
26 Rue la Perouse, 75116 Paris
(010 33) 7201020
Telex 611975 F

EXTERNAL WALL INSULATION ASSOCIATION
PO Box 12, Haslemere, Surrey GU27 3AN
Haslemere (0428) 54011
Fax (0428) 51401

FABRIC CARE RESEARCH ASSOCIATION
Forest House Laboratories,
Knaresborough Rd, Harrogate,
North Yorks HG2 7LZ
Harrogate (0423) 882301

FARM BUILDINGS ASSOCIATION
National Agricultural Centre, Stoneleigh,
Kenilworth, Warwickshire CV8 2LG
Coventry (0203) 26444

FEDERATION OF BUILDING AND CIVIL ENGINEERING CONTRACTORS (NI) LTD
143 malone Rd, Belfast BT9 6SU
Belfast (0232) 661711

FEDERATION OF CIVIL ENGINEERING CONTRACTORS
Cowdray House, 6 Portugal Street,
London WC2A 2HH
071-404 4020

FEDERATION OF RESIN FORMULATORS AND APPLICATORS LTD (FERFA)
16 Courtmoor Avenue, Fleet,
Aldershot, Hampshire
GU13 9UF
Fleet (02514) 6936

FEDERATION OF MANUFACTURERS OF CONSTRUCTION EQUIPMENT & CRANES
22-26 Dingwell Rd, Croydon, Surrey CR0 9XF
081-688 2727
Telex 9419625 FMCEC G

FEDERATION OF MASTER BUILDERS
33 John Street, London WC1N 2BB
071-242 7583

FEDERATION OF PILING SPECIALIST
Dickens House, 15 Tooks Court, London
EC4A 1LB
071-831 7581
Telex 897051

FEDERATION OF WIRE ROPE MANUFACTURERS OF GREAT BRITIAN
PO Box 121, The Fountain Precinct,
1 Balm Green, Sheffield S1 3AF
Sheffield (0742) 766789
Telex 54170

FLAT ROOFING CONTRACTORS ADVISORY BOARD
Maxwelton House, 41/43 Boltro Road,
Haywards Heath, West Sussex RH16 1BJ
Haywards Heath (0444) 440027

FENCING CONTRACTORS ASSOCIATION
St Johns House, 23 St Johns Rd,
Watford WD1 1PY
Watford (0923) 248895

FIBRE BONDED CARPET MANUFACTURERS ASSOCIATION
3 Manchester Rd, Bury, Lancs BL9 0DR
Bury 061-764 1114

FIBRE CEMENT MANUFACTURERS ASSOCIATION LTD (FCMA)
PO Box 92, Elmswell, Bury St Edmunds,
Suffolk IP30 9HS
Elmswell (0359) 40963

FINNISH PLYWOOD INTERNATIONAL
PO Box 99, Welywn Garden City,
Herts AL6 0HS
Bulls Green (043879) 746

USEFUL ADDRESSES FOR FURTHER INFORMATION

FLAT GLASS MANUFACTURERS ASSOCIATION
Prescot Road, St Helens, Merseyside
WA10 3TT
St Helens (0744) 28882

FLEXIBLE ROOFING ASSOCIATION
125 Queen's Road, Brighton,
East Sussex BN1 3YW
Brighton (0273) 33322

GLASS AND GLAZING FEDERATION
44/48 Borough High Street
London SE1
071-403 7177
Fax 071-457 7458

GLASS MANUFACTURERS FEDERATION
19 Portland Place, London W1N 4BH
071-580 6952
Telex 27470

GLASSFIBRE REINFORCEMENT CEMENT ASSOCIATION
5 Upper Bar, Newport, Shropshire TF10 7EH
Newport (0952) 811397

GLAZED AND FLOOR TILE HOME TRADE ASSOCIATION
Federation House, Station Road,
Stoke-on-Trent ST4 2RT
Stoke-on-Trent (0782) 747147

GYPSUM PRODUCTS DEVELOPMENT ASSOCIATION
Hillside, Kingston-on Soar, Nottingham
NG11 0DF
Nottingham (0602) 830781

HEAT PUMP AND AIR CONDITIONING BUREAU
30 Millbank, London SW1P 4RD
071-834 8827

HEAT PUMP MANUFACTURERS ASSOCIATION
Nicholson House, High St, Maidenhead,
Berks SL6 1LF
Maidenhead (0628) 34667

HEATING AND VENTILATING CONTRACTORS ASSOCIATION
ESCA House, 34 Palace Court, Bayswater,
London W2 4JG
071-229 2488
Telex 27929 ESCA G

HEATING, VENTILATING AND AIR CONDITIONING MANUFACTURERS ASSOCIATION
Nicholson House, High St, Maidenhead,
Berks SL6 1LF
Maidenhead (0628) 34667

HOUSING CORPORATION
149 Tottenham Court Road, London W1P 0BN
071-387 9466

INDUSTRIAL BUILDING BUREAU
33 Upper St, London N1 0PN
071-359 9877

INSTITUTE OF ACOUSTICS
25 Chambers St, Edinburgh EH1 1HU
031-225 2143

INSTITUTE OF ASPHALT TECHNOLOGY
Unit 18, Central Trading Estate, Staines,
Middlesex TW18 4XF
Staines (0784) 65387

INSTITUTE OF CONCRETE TECHNOLOGY
PO Box 255, Beaconsfield, Herts HP9 1JE
Beaconsfield (04946) 4572

INSTITUTE OF DOMESTIC HEATING AND ENVIRONMENTAL ENGINEERS
37a High Road, Benfleet, Essex SS7 5LH
South Benfleet (0268) 754266

INSTITUTE OF PLUMBING
64 Station Lane, Hornchurch, Essex
RM12 6NB Hornchurch
(04024) 72791

INSTITUTE OF SHEET METAL ENGINEERING
174 Dry Sandford, Abingdon,
Oxon OX13 6JW
Oxford (0865) 390868

INSTITUTE OF WASTES MANAGEMENT
3 Albion Place, Northampton NN1 1UD
Northampton (0604) 20426

INSTITUTE OF WATER AND ENVIRONMENTAL MANAGEMENT
15 John Street,
London WC1N 2EB
071-430 0899

USEFUL ADDRESSES FOR FURTHER INFORMATION

INSTITUTION OF BRITISH ENGINEERS
Regency House, 6 Hampton Place,
Brighton, East Sussex
Brighton (0273) 734274

INSTITUTION OF CIVIL ENGINEERS
1-7 Great George St, London SW1P 3AA
071-222 7722

INSTITUTION OF ELECTRICAL AND ELECTRONICS INCORPORATED ENGINEERS
Savoy Hill House, Savoy Hill,
London W2 4JG
071-836 3357

INSTITUTION OF ELECTRICAL ENGINEERS
Savoy Place, London WC2R 0BL
071-240 1871
Telex 261176

INSTITUTION OF STRUCTURAL ENGINEERS
11 Upper Belgrave Street, London
071-235 4535
Fax 071-235 4224

JOINT CONTRACTS TRIBUNAL
66 Portland Place, London W1N 4AD
071-580 5533,

LEAD DEVELOPMENT ASSOCIATION
34 Berkeley Square, London W1X 6AJ
081-499 8422
Telex 261286

NATIONAL HOUSE BUILDING COUNCIL
Clitters Avenue, Amersham,
Bucks HP6 5AP
Amersham (0494) 434477
071-637 1248/9

NATIONAL PAVING AND KERB ASSOCIATION
60 Charles St, Leicester LE1 1FB
Leicester (0533) 536161
Fax (0533) 514568

PARTITIONING INDUSTRY ASSOCIATION
692 Warwick Road, Solihull,
West Midlands B91 3DX
021-705 9270

PATENT GLAZING CONFERENCE
13 Upper High Street, Epsom,
Surrey KT17 4QY
Epsom, (03727) 29191

PIPELINE INDUSTRIES GUILD
17 Grosvenor Crescent, London SW1X 7ES
071-235 7938

PITCH FIBRE PIPE ASSOCIATION OF GREAT BRITAIN
c/o Croda Hydrocarbons Ltd, Pipes
Division, PO Box 16, Weeland Rd,
Knottingley, West Yorkshire WF11 8DZ
Knottingley (0977) 87161
Telex 8814171

PLASTERERS CRAFT GUILD
56 Burton Rd, Kingston-upon-Thames,
Surrey KT2 5TF
Kingston-upon-Thames 01-546 1470

PLASTIC PIPE MANUFACTURERS SOCIETY
89 Cornwall Street, Birmingham B3 3BY
021-236 1866

PLASTICS AND RUBBER INSTITUTE
11 Hobart Place, London SW1W 0HL
071-245 9555
Telex 915719 PRIUK G

PLASTICS BATH MANUFACTURERS ASSOCIATION
12th Floor, Fleming House, 134 Renfrew St,
Glasgow G3 6TG
041-332 0826

PLASTICS TANKS AND CISTERNS MANUFACTURERS ASSOCIATION
8 Belmain Close, Grange Road, Ealing,
London W5 5BY
081-579 6081

POST-TENSIONING ASSOCIATION
CCL Systems Ltd, Cabca House, 296 Elwell
Road, Surbiton, Surrey, KT6 7AH
081-390 1122

PRECAST CONCRETE FRAME ASSOCIATION
60 Charles St, Leicester LE1 1FB
Leicester (0533) 536161

USEFUL ADDRESSES FOR FURTHER INFORMATION

PREFABRICATED BUILDING MANUFACTURERS ASSOCIATION OF GREAT BRITAIN
Westgate House, Chalk Lane, Epsom, Surrey
KT18 7AJ
Epsom (037 27) 40044

PRESTRESSED CONCRETE ASSOCIATION
60 Charles St, Leicester LE1 1FB
Leicester (0533) 536161

PROPERTY SERVICES AGENCY DEPARTMENT OF THE ENVIRONMENT
Whitgift Centre, Wellesley Road,
Croydon CR9 3LY
Croydon 081-686 8710
Architectural Services: 081-686 8710
Civil Accommodation/Estate Surveying
Services: 071-928 7999
Defence Services 1 (Army, Overseas):
081-397 5266
Defence Services 2 (Air/Navy):
081-686 8710
Diplomatic/Post Offices Services:
081-686 5622
Engineering/QS Services:
081 686 3499

ROYAL INSTITUTE OF BRITISH ARCHITECTS
66 Portland Place, London W1N 4AD
071-580 5533

ROYAL INSTITUTION OF CHARTERED SURVEYORS
12 Great George Street, London SW1P 3AD
071-222 7000
Telex 915443 RICS G

SAND AND GRAVEL ASSOCIATION
1 Bramber Court, 2 Bramber Road
London W14 9PB
071-381 1443

SCOTTISH BUILDING EMPLOYERS FEDERATION
13 Woodside Crescent, Glasgow G3 7UP
041-332 7144
Telex 779657
Aberdeen (0224) 643838
Edinburgh 031-226 4907
Inverness (0463) 237626

SCOTTISH DEVELOPMENT AGENCY
120 Bothwell Street, Glasgow G2 7JP
041-248 2700
Road, Surbiton, Surrey, KT6 7AH

SCOTTISH PRECAST CONCRETE MANUFACTURERS ASSOCIATION
9 Princes Street, Falkirk FK1 1LS
Falkirk (0324) 22088

SCOTTISH SPECIAL HOUSING ASSOCIATION
15/21 Palmerston Place, Edinburgh
EH12 5AJ
031-225 1281

SCOTTISH WIREWORK MANUFACTURERS ASSOCIATION
36 Renfield Street, Glasgow, G2 1BD
041-248 6161

SOCIETY OF CHAIN LINK FENCING MANUFACTURERS
PO Box 121, The Fountain Precinct,
1 Balm Green, Sheffield S1 3AF
Sheffield (0742) 751234
Telex 54170

SOCIETY OF GLASS TECHNOLOGY
20 Hallamgate Road, Broom Hill,
Sheffield S10 5BT
Sheffield (0742) 663168

SPONS A & B EDITORS
Davis Langdon Computer Systems
11 Portland Square
Bristol BS2 8ST
Bristol (0272) 232283

STAINLESS STEEL ADVISORY CENTRE
Shepcote Lane, PO Box 161,
Sheffield S9 1TR
Sheffield (0742) 440060
Telex 547025

STAINLESS STEEL FABRICATORS ASSOCIATION OF GREAT BRITAIN
14 Knoll Road, Dorking, Surrey RH4 3EW
Dorking (0306) 884079

STEEL CASTINGS RESEARCH & TRADE ASSOCIATION
East Bank Road, Sheffield S2 3PT
Sheffield (0742) 728647

USEFUL ADDRESSES FOR FURTHER INFORMATION

STEEL LINTEL MANUFACTURERS ASSOCIATION
c/o PO Box 10, British Steel Corporation,
Newport, Gwent NP9 0XN
Newport (0633) 272281

STEEL WINDOW ASSOCIATION
102 Great Russell Street, London WC1
071-637 3571

STONE FEDERATION
82 New Cavendish Street, London W1M 8AD
071-580 5588

STRUCTURAL INSULATION ASSOCIATION
45 Sheen Lane, London SW14 8AB
081-876 4415/6
Telex 927298 ALLEN G

SUSPENDED ACCESS EQUIPMENT MANUFACTURERS ASSOCIATION
82 New Cavendish Street, London W1M 8AD
071-580 5588
Telex 265763

SUSPENDED CEILINGS ASSOCIATION
29 High Street, Hemel Hempstead,
Herts HP1 3AA
Hemel Hempstead (0442) 40313

SWEDISH FINNISH TIMBER COUNCIL
21 Carolgate, Retford, Notts DN22 6BZ
Retford (0777) 706616

SWIMMING POOL AND ALLIED TRADES ASSOCIATION (SPATA)
SPATA House, Junction Road,
Andover, Hants SP10 3QT
Andover (0264) 56210
Fax (0264) 332628

TAR INDUSTRIES SERVICES
Mill Lane, Wingerworth, Chesterfield,
Derbyshire S42 6NG
Chesterfield (0246) 276823
Telex 547061

THERMAL INSULATION CONTRACTORS ASSOCIATION
Kensway House, 388 High Rd, Ilford,
Essex IG1 1TL
081-514 2120

THERMAL INSULATION MANUFACTURERS AND SUPPLIERS ASSOCIATION
45 Sheen Lane, London, SW14 8AB
081-876 4415
Telex 927298 ALLEN

TIMBER TRADE FEDERATION OF THE UNITED KINGDOM
Clareville House, 26-27 Oxenden Street,
London SW1Y 4EL
071-839 1891
Telex 8954628

TOWN AND COUNTRY PLANNING ASSOCIATION
17 Carlton House Terrace, London SW1Y 5AS
071-930 8903

TRADA
Stocking Lane, Hughenden Valley, High
Wycombe, Bucks HP14 4ND
Naphill (024024) 3091
Telex 83292

UNITED KINGDOM WOOD WOOL ASSOCIATION
Gordon House, Oakleigh Road South,
New Southgate, London N11 1HL
081-368 1266
Telex 21252

VERMICULITE INFORMATION SERVICES COUNCIL LTD
Mark House, The Square, Lightwater,
Surrey GU18 5SS
Bagshot (0276) 71617

VITREOUS ENAMEL DEVELOPMENT
New House, High Street, Ticehurst,
Wadhurst, East Sussex TN5 7AL
Ticehurst (0580) 200152

WALLCOVERING MANUFACTURERS ASSOCIATION
Alembic House, 93 Albert Embankment,
London SE1 7TY
071-582 1185

USEFUL ADDRESSES FOR FURTHER INFORMATION

WELDING INSTITUTE
Abingdon Hall, Abingdon, Cambridge
CB1 6AL
Cambridge (0223) 891162
Fax (0223) 892588
Telex 81183

WESTERN WOOD PRODUCTS ASSOCIATION (USA)
69 Wigmore St, London W1H 9LG
071-486 7488/9

WIRE GOODS MANUFACTURERS ASSOCIATION
Kensington House, 136 Suffolk St,
Queensway, Birmingham B1 1LL
021-643 4488
Telex 338876 GRBHAMG

WOOD WOOL SLAB MANUFACTURERS ASSOCIATION
10 Great George St, London SW1P
071-222 5315

ZINC DEVELOPMENT ASSOCIATION
34 Berkeley Square, London W1X 6AJ
071-499 6636
Telex 261286
Fax 071-493 1555

Index to Advertisers

Advertising agent:
Edwin Bray
Building Materials Market Research
7 St. Peter's Place
Brighton BN1 6TB

Index

References in brackets after page numbers, refer to SMM7 and the New Common Arrangement sections.